Y0-BWB-697

THE FACTS ON FILE ENCYCLOPEDIA OF

SCIENCE

Volume I A–K

Contributors

Anthony Burton
Dougal Dixon
Nigel Dudley
J. M. Dunwell
David Evans
Barry Fox
Tim Furniss
David Gould
Wendy Grossman
Tony Jones
Peter Lafferty
James Le Fanu
Graham Littler
Paul Meehan
Pamela Morley
Douglas Palmer
Roy Porter
Paulette Pratt
Ian Ridpath
Peter Rodgers
Jack Schofield
A. D. Smith
Chris Stringer
Catherine Thompson
Colin Tudge
Edward Young
Keith Wallbanks
Cathy Walsh
Martin Walters
Stephen Webster
Gordon Woods

Editors

Project Editors
Sharon Brimblecombe
Diana Gallannaugh
Catherine Thompson

Database Manager
Nick Andrews

Production
Tony Ballsdon

Art and Design Manager
Terence Caven

THE FACTS ON FILE ENCYCLOPEDIA OF SCIENCE

Volume I A–K

Facts On File, Inc.

The Facts On File Encyclopedia of Science

Facts On File, Inc.
11 Penn Plaza
New York, NY 10001

Library of Congress Cataloging-in-Publication Data

The Facts on File encyclopedia of science.
p. cm.
ISBN 0-8160-4008-7 (set) (alk. paper)
ISBN 0-8160-4006-0 (Vol. 1)
ISBN 0-8160-4007-9 (Vol. 2)
1. Science—Dictionaries. I. Facts on File, Inc. II. Title:
Encyclopedia of science.
Q123.F334 1999
503—dc21 98-53201

Typesetting by TechType, Abingdon, Oxon
Cover design by Cathy Rincon

Printed in the United States of America

VB Helicon 10 9 8 7 6 5 4 3 2 1

This book is printed on acid-free paper

CONTENTS

INTRODUCTION

The *Facts On File Encyclopedia of Science,* is a science reference book for everyone. The material is accurate and appropriate for academic purposes, but it is also clearly written and far from dull.

As science concerns us all, and not just scientists and students of science, this encyclopedia contains informative features examining contemporary issues, innovations, and landmark discoveries. Alongside the alphabetical entries, designed to be of use for students and anyone else seeking definitions for scientific terms, there are also links to web sites to facilitate further research; capsule biographies and quotes to give an insight into the scientists themselves; and amazing facts, just for fun! With chronologies, tables, and a wealth of diagrams and photographs, it is both a browser's delight and an invaluable reference source.

Subject coverage

One of the aims of this encyclopedia has been to offer an extremely broad coverage of science-related topics, so as well as biology, chemistry, physics, and earth sciences, there are agriculture, archaeology, astronomy, computing, environmental science, mathematics, medicine, natural history, technology, and transportation.

Arrangement of entries

Entries are ordered alphabetically as if there were no spaces between words. Thus entries for words beginning "sulphur" follow the order:

sulphur
sulphur dioxide
sulphuric acid
sulphur trioxide

Common or technical names

Terms are usually placed under the name by which they are most widely known, rather than their technical name. For example, under **pertussis** there is cross-reference to **whooping cough**, where the main entry is given. Similarly, if the abbreviated form of a term is so widely used that the expanded form is quite unlikely to be looked up, the main definition is given under the abbreviation: for example, **DNA** not **deoxyribonucleic acid**.

Cross-references

These are shown by a ◊ immediately preceding the reference to alert the reader to related articles.

A symbol for ◊mass number, the number of neutrons and protons in an atomic nucleus.

A in physics, symbol for ◊ampere, a unit of electrical current.

aardvark *Afrikaans 'earth pig'* nocturnal mammal *Orycteropus afer,* the only species in the order Tubulidentata, found in central and southern Africa. A timid, defenceless animal about the size of a pig, it has a long head, a piglike snout, large ears, sparse body hair, a thick tail, and short legs.

It can burrow rapidly with its clawed front feet. It spends the day in its burrow, and at night digs open termite and ant nests, licking up the insects with its long sticky tongue. Its teeth are unique, without enamel, and are the main reason for the aardvark being placed in its own order. When fully grown, it is about 1.5 m/5 ft long and its tongue is 30 cm/12 in long.

AARDVARK

http://www.oit.itd.umich.edu/bio108/Chordata/Mammalia/Tubulidentata.shtml

Detailed description of this strange-looking, insectivorous mammal. There is information about the aardvark's skeletal and dental structure, and links to explain unfamiliar terms.

aardwolf nocturnal mammal *Proteles cristatus* of the ◊hyena family, Hyaenidae. It is yellowish grey with dark stripes and, excluding its bushy tail, is around 70 cm/30 in long. It is found in eastern and southern Africa, usually in the burrows of the ◊aardvark. It feeds almost exclusively on termites, eating up to 300,000 per day, but may also eat other insects and small mammals.

Males mate monogamously but females, which have a much larger range, will mate with other males.

abacus ancient calculating device made up of a frame of parallel wires on which beads are strung. The method of calculating with a handful of stones on a 'flat surface' (Latin *abacus*) was familiar to the Greeks and Romans, and used by earlier peoples, possibly even in ancient Babylon; it survives in the more sophisticated bead-frame form of the Russian *schoty* and the Japanese *soroban.* The abacus has been superseded by the electronic calculator.

ABACUS: THE ART OF CALCULATING WITH BEADS

http://members.aol.com/stubbs3/abacus.htm

Clear explanation of how the abacus works – with demonstrations of how to add and subtract. This site requires a Java-enabled Web browser or Netscape 2.0+.

The wires of a bead-frame abacus define place value (for example, in the decimal number system each successive wire, counting from right to left, would stand for ones, tens, hundreds, thousands, and so on) and beads are slid to the top of each wire in order to represent the digits of a particular number. On a simple decimal abacus, for example, the number 8,493 would be entered by sliding three beads on the first wire (three ones), nine beads on the second wire (nine tens), four beads on the third wire (four hundreds), and eight beads on the fourth wire (eight thousands).

abalone edible marine snail of the worldwide genus *Haliotis,* family Haliotidae. Abalones have flattened, oval, spiralled shells, which have holes around the outer edge and a bluish mother-of-pearl lining. This lining is used in ornamental work.

abdomen in vertebrates, the part of the body below the ◊thorax, containing the digestive organs; in insects and other arthropods, it is the hind part of the body. In mammals, the abdomen is separated from the thorax by the ◊diaphragm, a sheet of muscular tissue; in arthropods, commonly by a narrow constriction. In mammals, the female reproductive organs are in the abdomen. In insects and spiders, it is characterized by the absence of limbs.

aberration of starlight apparent displacement of a star from its true position, due to the combined effects of the speed of light and the speed of the Earth in orbit around the Sun (about 30 km per second/18.5 mi per second).

Aberration, discovered in 1728 by English astronomer James Bradley (1693–1762), was the first observational proof that the Earth orbits the Sun.

aberration, optical any of a number of defects that impair the image in an optical instrument. Aberration occurs because of minute variations in lenses and mirrors, and because different

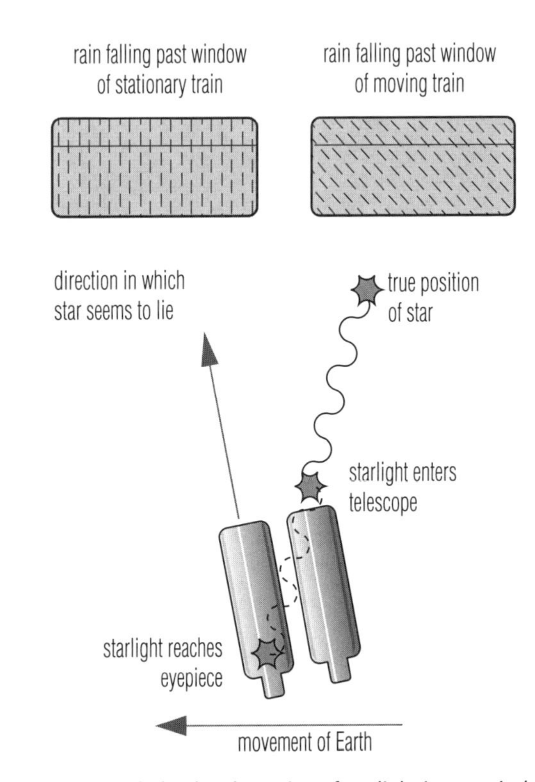

aberration of starlight *The aberration of starlight is an optical illusion caused by the motion of the Earth. Rain falling appears vertical when seen from the window of a stationary train; when seen from the window of a moving train, the rain appears to follow a sloping path. In the same way, light from a star 'falling' down a telescope seems to follow a sloping path because the Earth is moving. This causes an apparent displacement, or aberration, in the position of the star.*

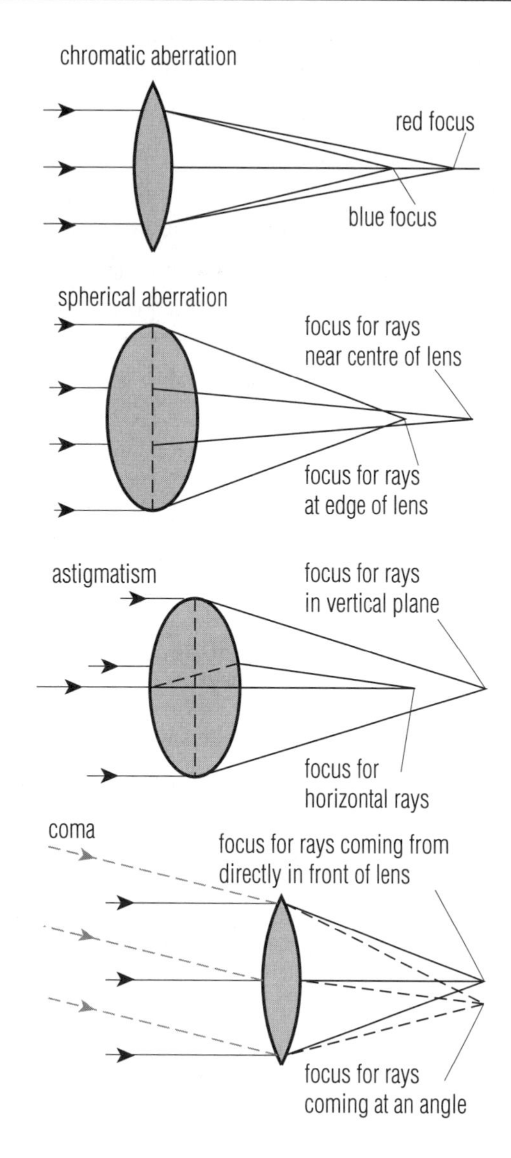

***aberration, optical** The main defects, or aberrations, of optical systems. Chromatic aberration, or coloured fringes around images, arises because light of different colours is focused at different points by a lens, causing a blurred image. Spherical aberration arises because light that passes through the centre of the lens is focused at a different point from light passing through the edge of the lens. Astigmatism arises if a lens has different curvatures in the vertical and horizontal directions. Coma arises because light passing directly through a lens is focused at a different point to light entering the lens from an angle.*

parts of the light ◊spectrum are reflected or refracted by varying amounts.

In **chromatic aberration** the image is surrounded by coloured fringes, because light of different colours is brought to different focal points by a lens. In **spherical aberration** the image is blurred because different parts of a spherical lens or mirror have different focal lengths. In **astigmatism** the image appears elliptical or cross-shaped because of an irregularity in the curvature of the lens. In **coma** the images appear progressively elongated towards the edge of the field of view. Elaborate computer programs are now used to design lenses in which the aberrations are minimized.

abiotic factor a nonorganic variable within the ecosystem, affecting the life of organisms. Examples include temperature, light, and soil structure. Abiotic factors can be harmful to the environment, as when sulphur dioxide emissions from power stations produce acid rain.

abortion ***Latin** aborire **'to miscarry'*** ending of a pregnancy before the fetus is developed sufficiently to survive outside the uterus. Loss of a fetus at a later gestational age is termed premature stillbirth. Abortion may be accidental (◊miscarriage) or deliberate (termination of pregnancy).

deliberate termination In the first nine weeks of pregnancy, medical termination may be carried out using the 'abortion pill' (◊mifepristone) in conjunction with a ◊prostaglandin. There are also various procedures for surgical termination, such as ◊dilatation and curettage, depending on the length of the pregnancy.

Worldwide, an estimated 150,000 unwanted pregnancies are terminated each day by induced abortion. One-third of these abortions are performed illegally and unsafely, and cause one in eight of all maternal deaths.

abortion as birth control Abortion as a means of birth control has long been controversial. The argument centres largely upon whether a woman should legally be permitted to have an abortion and, if so, under what circumstances. Another aspect is whether, and to what extent, the law should protect the fetus.

Those who oppose abortion generally believe that human life begins at the moment of conception, when a sperm fertilizes an egg. This is the view held, for example, by the Roman Catholic Church. Those who support unrestricted legal abortion may believe in a woman's right to choose whether she wants a child, and may take into account the large numbers of deaths and injuries from unprofessional back-street abortions.

Others approve abortion for specific reasons. For example, if a woman's life or health is jeopardized, abortion may be recommended; and if there is a strong likelihood that the child will be born with severe mental or physical disability. Other grounds for abortion include pregnancy resulting from sexual assault such as rape or incest.

abrasive ***Latin 'to scratch away'*** substance used for cutting and polishing or for removing small amounts of the surface of hard materials. There are two types: natural and artificial abrasives, and their hardness is measured using the ◊Mohs' scale. Natural abrasives include quartz, sandstone, pumice, diamond, emery, and corundum; artificial abrasives include rouge, whiting, and carborundum.

abscess collection of ◊pus in solid tissue forming in response to infection. Its presence is signalled by pain and inflammation.

abscissa in ◊coordinate geometry, the x-coordinate of a point – that is, the horizontal distance of that point from the vertical or y-axis. For example, a point with the coordinates (4, 3) has an abscissa of 4. The y-coordinate of a point is known as the ◊ordinate.

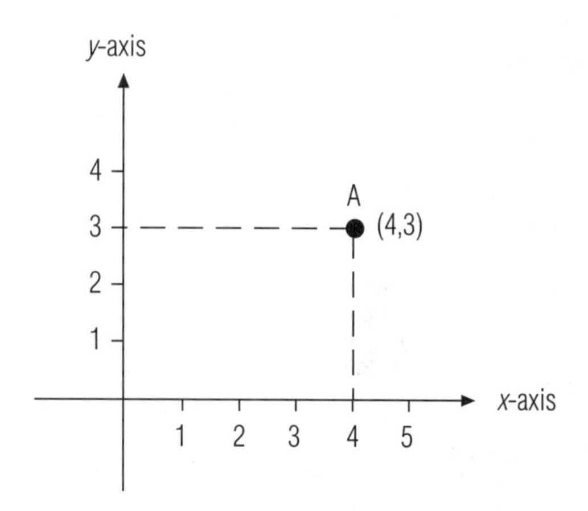

abscissa

abscissin or ***abscissic acid*** plant hormone found in all higher plants. It is involved in the process of ◊abscission and also inhibits stem elongation, germination of seeds, and the sprouting of buds.

abscission in botany, the controlled separation of part of a plant from the main plant body – most commonly, the falling of leaves or the dropping of fruit controlled by ◊abscissin. In ◊deciduous plants the leaves are shed before the winter or dry season, whereas ◊evergreen plants drop their leaves continually throughout the year. Fruitdrop, the abscission of fruit while still immature, is a naturally occurring process.

Abscission occurs after the formation of an abscission zone at the point of separation. Within this, a thin layer of cells, the abscission layer, becomes weakened and breaks down through the conversion of pectic acid to pectin. Consequently the leaf, fruit, or other part can easily be dislodged by wind or rain. The process is thought to be controlled by the amount of ◊auxin present. Fruitdrop is particularly common in fruit trees such as apples, and orchards are often sprayed with artificial auxin as a preventive measure.

absolute (of a value) in computing, real and unchanging. For example, an **absolute address** is a location in memory and an **absolute cell reference** is a single fixed cell in a spreadsheet display. The opposite of absolute is ◊relative.

absolute value or ***modulus*** in mathematics, the value, or magnitude, of a number irrespective of its sign. The absolute value of a number n is written $|n|$ (or sometimes as mod n), and is defined as the positive square root of n^2. For example, the numbers –5 and 5 have the same absolute value:

$$|5| = |-5| = 5$$

For a ◊complex number, the absolute value is its distance to the origin when it is plotted on an Argand diagram, and can be calculated (without plotting) by applying ◊Pythagoras' theorem. By definition, the absolute value of any complex number $a + ib$ (where a and b are real numbers and i is $\sqrt{-1}$) is given by the expression:

$$|a + ib| = \sqrt{(a^2 + b^2)}$$

absolute zero lowest temperature theoretically possible according to kinetic theory, zero kelvin (0 K), equivalent to –273.15°C/–459.67°F, at which molecules are in their lowest energy state. Although the third law of ◊thermodynamics indicates the impossibility of reaching absolute zero in practice, a temperature of 2.8×10^{-10} K (0.28 billionths of a degree above absolute zero) has been produced in 1993 at the Low Temperature Laboratory in Helsinki, Finland, using a technique called nuclear demagnetization. Near absolute zero, the physical properties of some materials change substantially; for example, some metals lose their electrical resistance and become superconducting.

absorption the taking up of one substance by another, such as a liquid by a solid (ink by blotting paper) or a gas by a liquid (ammonia by water). In physics, absorption is the phenomenon by which a substance retains radiation of particular wavelengths; for example, a piece of blue glass absorbs all visible light except the wavelengths in the blue part of the spectrum; it also refers to the partial loss of energy resulting from light and other electromagnetic waves passing through a medium. In nuclear physics, absorption is the capture by elements, such as boron, of neutrons produced by fission in a reactor.

absorption lines in astronomy, dark line in the spectrum of a hot object due to the presence of absorbing material along the line of sight. Absorption lines are caused by atoms absorbing light from the source at sharply defined wavelengths. Numerous absorption lines in the spectrum of the Sun (Fraunhofer lines) allow astronomers to study the composition of the Sun's outer layers. Absorption lines in the spectra of stars give clues to the composition of interstellar gas.

absorption spectroscopy or ***absorptiometry*** in analytical chemistry, a technique for determining the identity or amount present of a chemical substance by measuring the amount of electromagnetic radiation the substance absorbs at specific wavelengths; see ◊spectroscopy.

abutilon one of a group of 90 related species of tropical or semitropical ornamental plants. The Indian mallow or velvet leaf (*Abutilon theophrastus*) is one of the more common; it has bell-shaped yellow flowers and is the source of a jutelike fibre. Many of the species are pollinated by hummingbirds. (Genus *Abutilon*, family Malvaceae.)

abyssal plain broad expanse of sea floor lying 3–6 km/2–4 mi below sea level. Abyssal plains are found in all the major oceans, and they extend from bordering continental rises to mid-oceanic ridges.

abyssal zone dark ocean region 2,000–6,000 m/6,500–19,500 ft deep; temperature 4°C/39°F. Three-quarters of the area of the deep-ocean floor lies in the abyssal zone, which is too far from the surface for photosynthesis to take place. Some fish and crustaceans living there are blind or have their own light sources. The region above is the bathyal zone; the region below, the hadal zone.

Abyssinian cat or ***rabbit cat*** breed of domestic shorthaired cat, possibly descended from antiquity. In modern times, it was imported from Abyssinia (now Ethiopia) to Britain in the 1860s. The coat of the usual variety is ruddy brown (similar to a rabbit or hare) with each hair ringed with two or three darker coloured bands. It has a medium-length body, long, slender legs, large wideset ears, and deep gold or green eyes. It resembles cats that appear in ancient Egyptian wall paintings.

The breed was recognized in Britain 1882 and is now most widely bred in the USA. There are many varieties.

abzyme in biotechnology, an artificially created antibody that can be used like an enzyme to accelerate reactions.

AC in physics, abbreviation for ◊alternating current.

> To remember the order of items in an AC circuit:
>
> THE **VOLTAGE** (V) ACROSS A **CAPACITOR** (C) LAGS THE **CURRENT** (I) BY 90 DEGREES. THE VOLTAGE ACROSS THE **INDUCTOR** (L) LEADS THE CURRENT BY 90 DEGREES. THIS IS GIVEN BY THE WORD CIVIL, WHEN SPLIT INTO CIV AND VIL.

acacia any of a large group of shrubs and trees that includes the thorn trees of the African savanna and the gum arabic tree (*Acacia senegal*) of N Africa, and several North American species of the southwestern USA and Mexico. The hardy tree commonly known as acacia is the false acacia (*Robinia pseudacacia*, of the subfamily Papilionoideae). True acacias are found in warm regions of the world, particularly Australia. (Genus *Acacia*, family Leguminosae.)

A. dealbata is grown in the open air in some parts of France and the warmer European countries, and is remarkable for its clusters of fluffy, scented yellow flowers, sold by florists as mimosa. The leaves of the genus are normally bipinnate (leaflets on both sides of each stem and stems growing on both sides of a larger stem), and the flowers grow in a head.

acanthus herbaceous plant with handsome lobed leaves. Twenty species are found in the Mediterrranean region and Old World tropics, including bear's-breech (*Acanthus mollis*) whose leaves were used as a motif in classical architecture, especially on Corinthian columns. (Genus *Acanthus*, family Acanthaceae.)

accelerated freeze drying common method of food preservation. See ◊food technology.

acceleration rate of change of the velocity of a moving body. It is usually measured in metres per second per second (m s^{-2}) or feet per second per second (ft s^{-2}). Because velocity is a ◊vector quantity (possessing both magnitude and direction) a body travelling at constant speed may be said to be accelerating if its direction of motion changes. According to Newton's second law of motion, a body will accelerate only if it is acted upon by an unbalanced, or resultant, ◊force.

Acceleration due to gravity is the acceleration of a body falling freely under the influence of the Earth's gravitational field; it varies slightly at different latitudes and altitudes. The value adopted internationally for gravitational acceleration is 9.806 m s^{-2}/32.174 ft s^{-2}.

acceleration, secular in astronomy, the continuous and nonperiodic change in orbital velocity of one body around another, or the axial rotation period of a body.

An example is the axial rotation of the Earth. This is gradually slowing down owing to the gravitational effects of the Moon and the resulting production of tides, which have a frictional effect on the Earth. However, the angular ◊momentum of the Earth–Moon system is maintained, because the momentum lost by the Earth is passed to the Moon. This results in an increase in the Moon's orbital period and a consequential moving away from the Earth. The overall effect is that the Earth's axial rotation period is increasing by about 15 millionths of a second a year, and the Moon is receding from the Earth at about 4 cm/1.5 in a year.

acceleration, uniform in physics, acceleration in which the velocity of a body changes by equal amounts in successive time intervals. Uniform acceleration is represented by a straight line on a ◊speed–time graph.

accelerator in physics, a device to bring charged particles (such as protons and electrons) up to high speeds and energies, at which they can be of use in industry, medicine, and pure physics. At low energies, accelerated particles can be used to produce the image on a television screen and generate X-rays (by means of a ◊cathode-ray tube), destroy tumour cells, or kill bacteria. When high-energy particles collide with other particles, the fragments formed reveal the nature of the fundamental forces.

The first accelerators used high voltages (produced by ◊van de Graaff generators) to generate a strong, unvarying electric field. Charged particles were accelerated as they passed through the electric field. However, because the voltage produced by a generator is limited, these accelerators were replaced by machines where the particles passed through regions of alternating electric fields, receiving a succession of small pushes to accelerate them.

The first of these accelerators was the **linear accelerator** or **linac**. The linac consists of a line of metal tubes, called drift tubes, through which the particles travel. The particles are accelerated by electric fields in the gaps between the drift tubes.

Another way of making repeated use of an electric field is to bend the path of a particle into a circle so that it passes repeatedly through the same electric field. The first accelerator to use this idea was the **cyclotron** pioneered in the early 1930s by US physicist Ernest Lawrence. A cyclotron consists of an electromagnet with two hollow metal semicircular structures, called dees, supported between the poles of an electromagnet. Particles such as protons are introduced at the centre of the machine and travel outwards in a spiral path, being accelerated by an oscillating electric field each time they pass through the gap between the dees. Cyclotrons can accelerate particles up to energies of 25 MeV (25 million electron volts); to produce higher energies, new techniques are needed.

In the ◊synchrotron, particles travel in a circular path of constant radius, guided by electromagnets. The strengths of the electromagnets are varied to keep the particles on an accurate path. Electric fields at points around the path accelerate the particles.

ACCELERATOR PHYSICS PAGE

http://www-laacg.atdiv.lanl.gov/accphys.html

Virtual library dedicated to accelerator physics, with pages on design and components, as well as direct links to laboratories throughout the world.

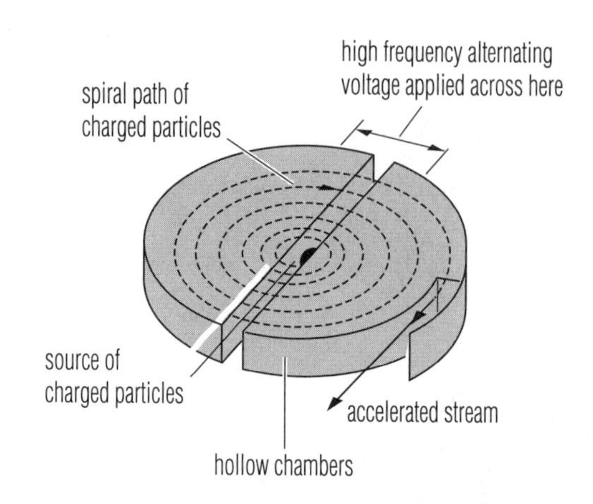

***accelerator** The cyclotron, an early accelerator, consisted of two D-shaped hollow chambers enclosed in a vacuum. An alternating voltage was applied across the gap between the hollows. Charged particles spiralled outward from the centre, picking up energy and accelerating each time they passed through the gap.*

Early accelerators directed the particle beam onto a stationary target; large modern accelerators usually collide beams of particles that are travelling in opposite directions. This arrangement doubles the effective energy of the collision.

The world's most powerful accelerator is the 2 km/1.25 mi diameter machine at ◊Fermilab near Batavia, Illinois, USA. This machine, the Tevatron, accelerates protons and antiprotons and then collides them at energies up to a thousand billion electron volts (or 1 TeV, hence the name of the machine). The largest accelerator is the ◊Large Electron Positron Collider at ◊CERN near Geneva, which has a circumference of 27 km/16.8 mi around which electrons and positrons are accelerated before being allowed to collide. The world's longest linac is also a colliding beam machine: the Stanford Linear Collider, in California, in which electrons and positrons are accelerated along a straight track, 3.2 km/2 mi long, and then steered to a head-on collision with other particles, such as protons and neutrons. Such experiments have been instrumental in revealing that protons and neutrons are made up of smaller elementary particles called ◊quarks.

accelerator board type of ◊expansion board that makes a computer run faster. It usually contains an additional processor (see ◊central processing unit).

accelerometer apparatus, either mechanical or electromechanical, for measuring ◊acceleration or deceleration – that is, the rate of increase or decrease in the ◊velocity of a moving object.

The mechanical types have a spring-supported mass with a damper system, with indication of acceleration on a scale on which a light beam is reflected from a mirror on the mass. The electromechanical types use (1) a slide wire, (2) a strain gauge, (3) variable inductance, or (4) a piezoelectric or similar device that produces electrically measurable effects of acceleration.

Accelerometers are used to measure the efficiency of the braking systems on road and rail vehicles; those used in aircraft and spacecraft can determine accelerations in several directions simultaneously. There are also accelerometers for detecting vibrations in machinery.

acceptable use in computing, set of rules enforced by a service provider or backbone network restricting the use to which their facilities may be put. Every organization on the Internet has its own **acceptable use policy** (AUP); schools, for example, may ban the use of their facilities to find or download pornography from the Internet.

Originally, when the Internet was publicly funded, acceptable use banned advertising, and although funding is moving to private enterprise and advertising has now become commonplace, some service providers still do not allow commercial exploitation. The US National Science Foundation's NSFnet, for example, imposes a strict AUP to prohibit commercial organizations from using the network.

access in computing, the way in which ◊file access is provided so that the data can be stored, retrieved, or updated by the computer.

access privilege in computer networking, authorized access to files. The ability to authorize or restrict access selectively to files or directories, including separate privileges such as reading, writing, or changing data, is a key element in computer security systems. This kind of system ensures that, for example, a company's employees cannot read its personnel files or alter payroll data unless they work for the appropriate departments, or that freelance or temporary staff can be given access to some areas of the computer system but not others.

Certain types of restrictions may be applied by users themselves to files on their own desktop machines, such as private e-mail; others may be granted only by the system administrator.

On all client–server systems, all data, even private e-mail and personal letters, can be accessed by the system administrator, who needs system-wide privileges in order to manage the network properly. Data which is encrypted, however, will not be readable unless the system administrator knows the user's individual password.

The privacy of employee e-mail is a contentious issue, as many employees assume their e-mail is private, while many companies presume ownership of all data stored on company systems.

access provider in computing, another term for ◊Internet Service Provider.

access time or ***reaction time*** in computing, the time taken by a computer, after an instruction has been given, to read from or write to ◊memory.

acclimation or ***acclimatization*** the physiological changes induced in an organism by exposure to new environmental conditions. When humans move to higher altitudes, for example, the number of red blood cells rises to increase the oxygen-carrying capacity of the blood in order to compensate for the lower levels of oxygen in the air.

In evolutionary terms, the ability to acclimate is an important adaptation as it allows the organism to cope with the environmental changes occurring during its lifetime.

accommodation in biology, the ability of the ◊eye to focus on near or far objects by changing the shape of the lens.

For an object to be viewed clearly its image must be precisely focused on the retina, the light-sensitive layer of cells at the rear of the eye. Close objects can be seen when the lens takes up a more spherical shape, distant objects when the lens is flattened. These changes in shape are caused by the movement of ligaments attached to a ring of ciliary muscles lying beneath the iris.

From about the age of 40, the lens in the human eye becomes less flexible, causing the defect of vision known as **presbyopia** or lack of accommodation. People with this defect need different spectacles for reading and distance vision.

account on a computer network, a ◊user-ID issued to a specific individual to enable access to the system for purposes of billing, administration, or private messaging. The existence of accounts allows system administrators to assign ◊access privileges to specific individuals (which in turn enables those individuals to receive private messages such as e-mail) and also to track the use of the computer system and its resources.

On commercial systems such as CompuServe or America Online, users are given an account when they dial up the system and give the number of a credit card to which usage may be billed. On other types of systems, accounts are typically issued by the system administrator. In all cases, accounts are protected by a password, which should be carefully chosen.

accretion in astrophysics, a process by which an object gathers up surrounding material by gravitational attraction, so simultaneously increasing in mass and releasing gravitational energy. Accretion on to compact objects such as ◊white dwarfs, ◊neutron stars and ◊black holes can release large amounts of gravitational energy, and is believed to be the power source for active galaxies. Accreted material falling towards a star may form a swirling disc of material known as an ◊accretion disc that can be a source of X-rays.

accretion disc in astronomy, a flattened ring of gas and dust orbiting an object in space, such as a star or ◊black hole. The orbiting material is accreted (gathered in) from a neighbouring object such as another star. Giant accretion discs are thought to exist at the centres of some galaxies and ◊quasars.

If the central object of the accretion disc has a strong gravitational field, as with a neutron star or a black hole, gas falling onto the accretion disc releases energy, which heats the gas to extreme temperatures and emits short-wavelength radiation, such as X-rays.

accumulator in computing, a special register, or memory location, in the ◊arithmetic and logic unit of the computer processor. It is used to hold the result of a calculation temporarily or to store data that is being transferred.

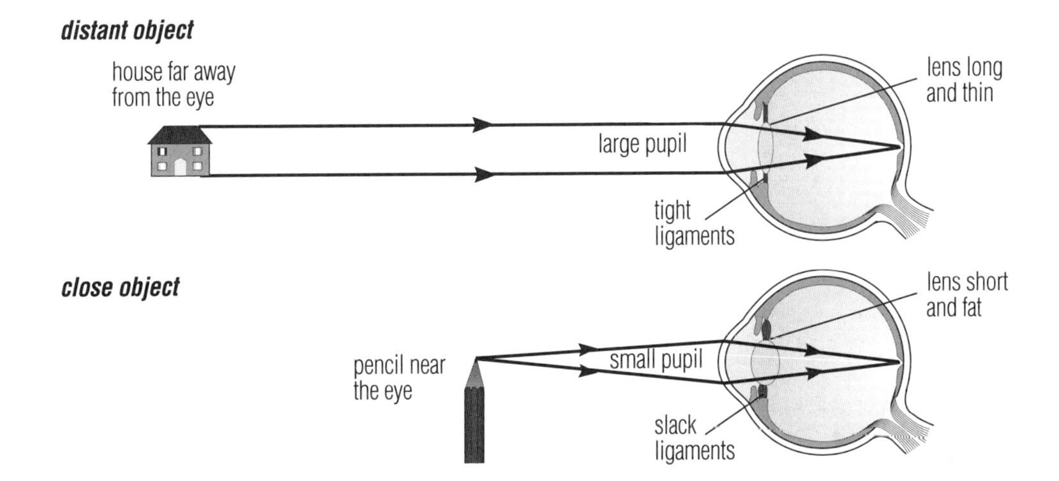

accommodation *The mechanism by which the shape of the lens in the eye is changed so that clear images of objects, whether distant or near, can be focused on the retina.*

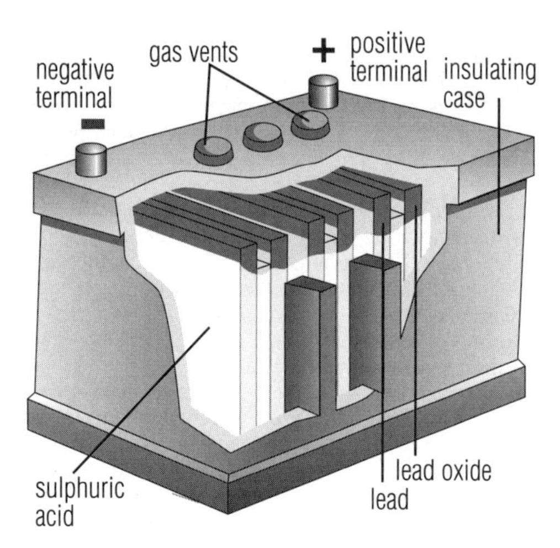

accumulator The lead–acid car battery is a typical example of an accumulator. The battery has a set of grids immersed in a sulphuric acid electrolyte. One set of grids is made of lead (Pb) and acts as the anode and the other set made of lead oxide (PbO2) acts as the cathode.

accumulator in electricity, a storage ◊battery – that is, a group of rechargeable secondary cells. A familiar example is the lead–acid car battery.

An ordinary 12-volt car battery consists of six lead–acid cells which are continually recharged when the motor is running by the car's alternator or dynamo. It has electrodes of lead and lead oxide in an electrolyte of sulphuric acid. Another common type of accumulator is the 'nife' or Ni Fe cell, which has electrodes of nickel and iron in a potassium hydroxide electrolyte.

accuracy in mathematics, a measure of the precision of a number. The degree of accuracy depends on how many figures or decimal places are used in rounding off the number. For example, the result of a calculation or measurement (such as 13.429314) might be rounded off to three decimal places (13.429), to two decimal places (13.43), to one decimal place (13.4), or to the nearest whole number (13). The first answer is more accurate than the second, the second more accurate than the third, and so on.

Accuracy also refers to a range of errors. For example, an accuracy of ± 5% means that a value may lie between 95% and 105% of a given answer.

acer group of over 115 related species of trees and shrubs of the temperate regions of the northern hemisphere, many of them popular garden specimens. They include ◊sycamore and ◊maple. Some species have pinnate leaves (leaflets either side of a stem), including the box elder (*Acer negundo*) of North America. (Genus *Acer.*)

That is the essence of science: ask an impertinent question, and you are on the way to a pertinent answer.

JACOB BRONOWSKI Polish-born scientist, broadcaster, and writer. *Ascent of Man*

acesulfame-K noncarbohydrate sweetener that is up to 300 times as sweet as sugar. It is used in soft drinks and desserts.

acetaldehyde common name for ◊ethanal.

acetate common name for ◊ethanoate.

acetic acid common name for ◊ethanoic acid.

acetone common name for ◊propanone.

acetylcholine (ACh) chemical that serves as a ◊neuro-transmitter, communicating nerve impulses between the cells of the nervous system. It is largely associated with the transmission of impulses across the ◊synapse (junction) between nerve and muscle cells, causing the muscles to contract.

ACh is produced in the synaptic knob (a swelling at the end of a nerve cell) and stored in vesicles until a nerve impulse triggers its discharge across the synapse. When the ACh reaches the membrane of the receiving cell it binds with a specific site and brings about depolarization – a reversal of the electric charge on either side of the membrane – causing a fresh impulse (in nerve cells) or a contraction (in muscle cells). Its action is shortlived because it is quickly destroyed by the enzyme cholinesterase.

Anticholinergic drugs have a number of uses in medicine to block the action of ACh, thereby disrupting the passage of nerve impulses and relaxing certain muscles, for example in premedication before surgery.

acetylene common name for ◊ethyne.

acetylsalicylic acid chemical name for the painkilling drug ◊aspirin.

achene dry, one-seeded ◊fruit that develops from a single ◊ovary and does not split open to disperse the seed. Achenes commonly occur in groups – for example, the fruiting heads of buttercup *Ranunculus* and clematis. The outer surface may be smooth, spiny, ribbed, or tuberculate, depending on the species.

Achernar or ***Alpha Eridani*** brightest star in the constellation Eridanus, and the ninth brightest star in the sky. It is a hot, luminous, blue star with a true luminosity 250 times that of the Sun. It is 125 light years away.

Achilles tendon tendon at the back of the ankle attaching the calf muscles to the heel bone. It is one of the largest tendons in the human body, and can resist great tensional strain, but is sometimes ruptured by contraction of the muscles in sudden extension of the foot.

Ancient surgeons regarded wounds in this tendon as fatal, probably because of the Greek legend of Achilles, which relates how the mother of the hero Achilles dipped him when an infant into the river Styx, so that he became invulnerable except for the heel by which she held him.

achondrite type of ◊meteorite. They comprise about 15% of all meteorites and lack the **chondrules** (silicate spheres) found in ◊chondrites.

achromatic lens combination of lenses made from materials of different refractive indexes, constructed in such a way as to minimize chromatic aberration (which in a single lens causes coloured fringes around images because the lens diffracts the different wavelengths in white light to slightly different extents).

acid compound that, in solution in an ionizing solvent (usually water), gives rise to hydrogen ions (H^+ or protons). In modern chemistry, acids are defined as substances that are proton donors and accept electrons to form ◊ionic bonds. Acids react with ◊bases to form salts, and they act as solvents. Strong acids are corrosive; dilute acids have a sour or sharp taste, although in some organic acids this may be partially masked by other flavour characteristics.

Acids can be detected by using coloured indicators such as ◊litmus and methyl orange. The strength of an acid is measured by its hydrogen-ion concentration, indicated by the ◊pH value. Acids are classified as monobasic, dibasic, tribasic, and so forth, according to the

To remember how to mix acid and water safely:

ADD ACID TO WATER, JUST AS YOU OUGHTA!

number of hydrogen atoms, replaceable by bases, in a molecule. The first known acid was vinegar (ethanoic or acetic acid). Inorganic acids include boric, carbonic, hydrochloric, hydrofluoric, nitric, phosphoric, and sulphuric. Organic acids include acetic, benzoic, citric, formic, lactic, oxalic, and salicylic, as well as complex substances such as ◊nucleic acids and ◊amino acids.

acidic oxide oxide of a ◊nonmetal. Acidic oxides are covalent compounds. Those that dissolve in water, such as sulphur dioxide, give acidic solutions.

$$SO_2 + H_2O \leftrightarrow H_2SO_{3(aq)} \leftrightarrow H^+_{(aq)} + HSO_3^-{}_{(aq)}$$

All acidic oxides react with alkalis to form salts.

$$CO_2 + NaOH \rightarrow NaHCO_3$$

acid rain acidic precipitation thought to be caused principally by the release into the atmosphere of sulphur dioxide (SO_2) and oxides of nitrogen. Sulphur dioxide is formed by the burning of fossil fuels, such as coal, that contain high quantities of sulphur; nitrogen oxides are contributed from various industrial activities and from car exhaust fumes.

Acid deposition occurs not only as wet precipitation (mist, snow, or rain), but also comes out of the atmosphere as dry particles or is absorbed directly by lakes, plants, and masonry as gases. Acidic gases can travel over 500 km/310 mi a day so acid rain can be considered an example of transboundary pollution.

Acid rain is linked with damage to and the death of forests and lake organisms in Scandinavia, Europe, and eastern North America. It also results in damage to buildings and statues. US and European power stations that burn fossil fuels release about 8 g/0.3 oz of sulphur dioxide and 3 g/0.1 oz of nitrogen oxides per kilowatt-hour.

ACID RAIN

http://www.environment-agency.gov.uk/s-enviro/states/3-2.html

Britain's Environmental Agency report on the environmental consequences of acid rain. This well-written article, supported by graphs and statistics, gives a good understanding of the scale of the problems caused by acid rain.

acid salt chemical compound formed by the partial neutralization of a dibasic or tribasic ◊acid (one that contains two or three hydrogen atoms). Although a salt, it contains replaceable hydrogen, so it may undergo the typical reactions of an acid. Examples are sodium hydrogen sulphate ($NaHSO_4$) and acid phosphates.

ack radio-derived term for 'acknowledge'. It is used on the Internet as a brief way of indicating agreement with or receipt of a message or instruction.

aclinic line the magnetic equator, an imaginary line near the Equator, where a compass needle balances horizontally, the attraction of the north and south magnetic poles being equal.

ACM abbreviation for the US ◊Association for Computing Machinery.

acne skin eruption, mainly occurring among adolescents and young adults, caused by inflammation of the sebaceous glands which secrete an oily substance (sebum), the natural lubricant of the skin. Sometimes the openings of the glands become blocked, causing the formation of pus-filled swellings. Teenage acne is seen mainly on the face, back, and chest.

There are other, less common types of acne, sometimes caused by contact with irritant chemicals (chloracne).

aconite or ***monkshood*** or ***wolfsbane*** herbaceous plant belonging to the buttercup family, with hooded blue–mauve flowers, native to

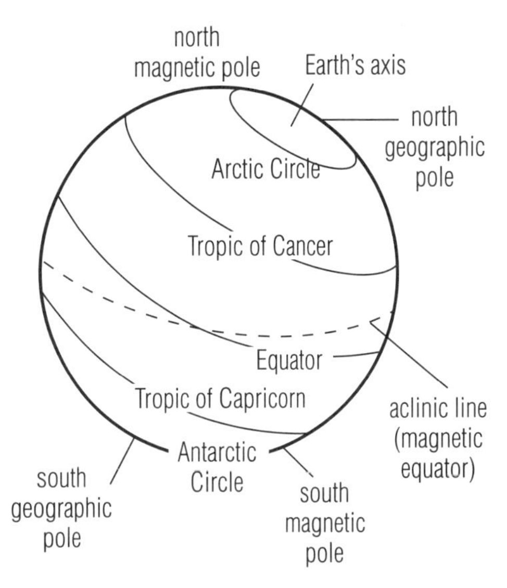

aclinic line The magnetic equator, or the line at which the attraction of both magnetic poles is equal. Along the aclinic line, a compass needle swinging vertically will settle in a horizontal position.

Europe and Asia. It produces aconitine, a poison with pain-killing and sleep-inducing properties. (*Aconitum napellus,* family Ranunculaceae.)

There are about 100 species throughout the northern temperate regions, all hardy ◊herbaceous plants containing poison. Summer aconite (*Aconitum uncinatum*) is a common North American flower. Winter aconite (*Eranthis hyemalis*) belongs to another genus of the buttercup family; it has yellow buttercuplike flowers with six petals which appear in February and March; the leaves follow later.

acorn fruit of the ◊oak tree, a ◊nut growing in a shallow cup.

Acorn UK computer manufacturer. In the early 1980s, Acorn produced a series of home microcomputers, including the Electron and the Atom. Its most successful computer, the BBC Microcomputer, was produced in conjunction with the BBC. Subsequent computers (the Master and the Archimedes) were less successful. Acorn was rescued by the Italian company Olivetti in 1985 but it has since sold off its majority shareholding.

acouchi any of several small South American rodents, genus *Myoprocta*. They have white-tipped tails, and are smaller relatives of the ◊agouti.

acoustic coupler device that enables computer data to be transmitted and received through a normal telephone handset; the handset rests on the coupler to make the connection. A small speaker within the device is used to convert the computer's digital output data into sound signals, which are then picked up by the handset and transmitted through the telephone system. At the receiving telephone, a second acoustic coupler or modem converts the sound signals back into digital data for input into a computer.

Unlike a ◊modem, an acoustic coupler does not require direct connection to the telephone system. However, interference from background noise means that the quality of transmission is poorer than with a modem, and more errors are likely to arise.

acoustic ohm c.g.s. unit of acoustic impedance (the ratio of the sound pressure on a surface to the sound flux through the surface). It is analogous to the ohm as the unit of electrical ◊impedance.

acoustics in general, the experimental and theoretical science of sound and its transmission; in particular, that branch of the science that has to do with the phenomena of sound in a particular space such as a room or theatre. In architecture, the sound-reflecting character of an internal space.

Acoustical engineering is concerned with the technical control of sound, and involves architecture and construction, studying control

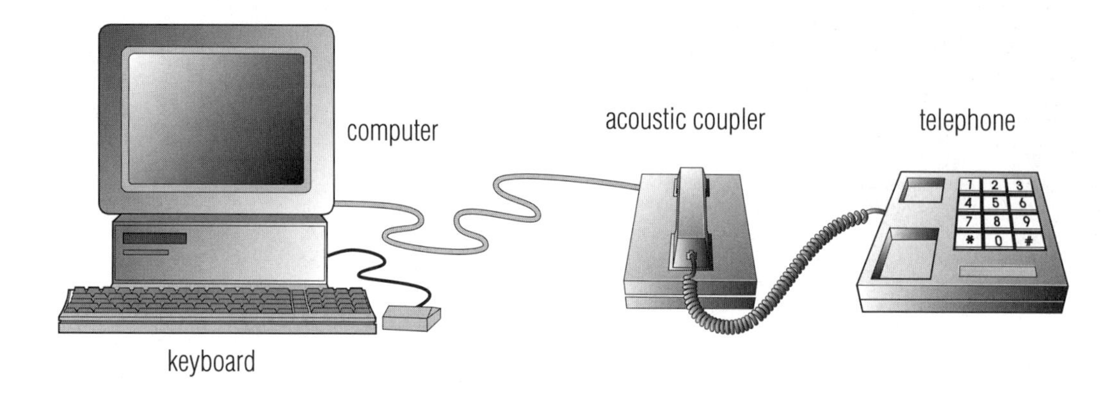

acoustic coupler *The acoustic coupler converts digital output from a computer to sound signals that can be sent via a telephone line.*

of vibration, soundproofing, and the elimination of noise. It also includes all forms of sound recording and reinforcement, the hearing and perception of sounds, and hearing aids.

ACOUSTIC ILLUSIONS

http://www.uni-bonn.de/~uzs083/akustik.html

Dedicated to the Shepard effect: a scale that gives the listener the impression of an endlessly rising melody, when in fact the pitch of the tones does not rise. Sound files illustrate this, and accompany an account of an experiment in which the Shepard effect was applied to J S Bach's *Das Musikalische Opfer/Musical Offering*.

acquired character feature of the body that develops during the lifetime of an individual, usually as a result of repeated use or disuse, such as the enlarged muscles of a weightlifter.

French naturalist Jean Baptiste ◊Lamarck's theory of evolution assumed that acquired characters were passed from parent to offspring. Modern evolutionary theory does not recognize the inheritance of acquired characters because there is no reliable scientific evidence that it occurs, and because no mechanism is known whereby bodily changes can influence the genetic material. The belief that this does not occur is known as ◊central dogma.

acquired immune deficiency syndrome full name for the disease ◊AIDS.

acre traditional English land measure equal to 4,840 square yards (4,047 sq m/0.405 ha). Originally meaning a field, it was the size that a yoke of oxen could plough in a day.

As early as Edward I's reign, (1272–1307) the acre was standardized by statute for official use, although local variations in Ireland, Scotland, and some English counties continued. It may be subdivided into 160 square rods (one square rod equalling 25.29 sq m/30.25 sq yd).

acre-foot unit sometimes used to measure large volumes of water, such as the capacity of a reservoir (equal to its area in acres multiplied by its average depth in feet). One acre-foot equals 1,233.5 cu m/43,560 cu ft or the amount of water covering one acre to a depth of one foot.

acridine $C_{13}H_9N$, a heterocyclic organic compound that occurs in coal tar. It is crystalline, melting at 108°C/226.4°F. Acridine is extracted by dilute acids but can also be obtained synthetically. It is used to make dyes and drugs.

Acrobat program developed by Adobe to allow users of different types of computers to view the same documents complete with graphics and layout. Launched in 1993, Acrobat was designed to get around the limitations of existing systems when transferring data between different types of computers, which typically required all formatting to be stripped from the documents. The program to generate the code that makes the documents transferable with formatting intact must be bought, but the program for reading the documents is available free of charge.

By 1996 Acrobat was in common use on the World Wide Web for distributing certain types of company documents, and the program had been enhanced to integrate with Web ◊browsers.

acrolein or ***acraldehyde*** CH_2:CHCHO, a colourless liquid formed during the partial combustion of fats. It is an unsaturated aldehyde, boils at 52°C/125.6°F, has an irritating action on the skin, and its vapours cause a copious flow of tears. It was used in gas form as a chemical weapon in World War I by the French under the codename *Papite*.

acronym abbreviation that can be pronounced as a word, for example **RISC** (Reduced Instruction Set Computer) and **MUD** (multi-user dungeon). People in the computer industry often incorrectly refer to all abbreviations as acronyms. Both are frequently used as industry jargon and as shorthand to save typing on the Net. See also ◊TLA (three letter acronym).

acrylic acid common name for ◊propenoic acid.

ACTH (abbreviation for ***adrenocorticotrophic hormone***) ◊hormone secreted by the anterior lobe of the ◊pituitary gland. It controls the production of corticosteroid hormones by the ◊adrenal gland and is commonly produced as a response to stress.

actinide any of a series of 15 radioactive metallic chemical elements with atomic numbers 89 (actinium) to 103 (lawrencium). Elements 89 to 95 occur in nature; the rest of the series are synthesized elements only. Actinides are grouped together because of their chemical similarities (for example, they are all bivalent), the properties differing only slightly with atomic number. The series is set out in a band in the ◊periodic table of the elements, as are the ◊lanthanides.

actinium ***Greek aktis 'ray'*** white, radioactive, metallic element, the first of the actinide series, symbol Ac, atomic number 89, relative atomic mass 227; it is a weak emitter of high-energy alpha particles.

Actinium occurs with uranium and radium in ◊pitchblende and other ores, and can be synthesized by bombarding radium with neutrons. The longest-lived isotope, Ac-227, has a half-life of 21.8 years (all the other isotopes have very short half-lives). Chemically, it is exclusively trivalent, resembling in its reactions the lanthanides and the other actinides. Actinium was discovered 1899 by the French chemist André Debierne (1874–1949).

actinium K original name of the radioactive element ◊francium, given 1939 by its discoverer, the French scientist Marguerite Perey (1909–1975).

action and reaction in physical mechanics, equal and opposite forces which act together. For example, the pressure of expanding gases from the burning of fuel in a rocket engine produces an equal and opposite reaction, which causes the rocket to move. This is Newton's third law of motion (see ◊Newton's laws of motion.

action potential in biology, a change in the ◊potential difference (voltage) across the membrane of a nerve cell when an impulse passes along it. A change in potential (from about –60 to +45 millivolts) accompanies the passage of sodium and potassium ions across the membrane.

activation analysis in analytical chemistry, a technique used to reveal the presence and amount of minute impurities in a substance or element. A sample of a material that may contain traces of a certain element is irradiated with ◊neutrons, as in a reactor. Gamma rays emitted by the material's radioisotope have unique energies and relative intensities, similar to the spectral lines from a luminous gas. Measurements and interpretation of the gamma-ray spectrum, using data from standard samples for comparison, provide information on the amount of impurities present.

activation energy in chemistry, the energy required in order to start a chemical reaction. Some elements and compounds will react together merely by bringing them into contact (spontaneous reaction). For others it is necessary to supply energy in order to start the reaction, even if there is ultimately a net output of energy. This initial energy is the activation energy.

active galaxy in astronomy, a type of galaxy that emits vast quantities of energy from a small region at its centre called the active galactic nucleus (AGN). Active galaxies are subdivided into ◊radio galaxies, ◊Seyfert galaxies, ◊BL Lacertae objects, and ◊quasars.

Active galaxies are thought to contain black holes with a mass some 108 times that of the Sun, drawing stars and interstellar gas towards it in a process of accretion. The gravitational energy released by the in-falling material is the power source for the AGN. Some of the energy may appear as a pair of opposed jets emerging from the nucleus. The orientation of the jets to the line of sight and their interaction with surrounding material determines the type of active galaxy that is seen by observers. See also ◊starburst galaxy.

active matrix LCD (or ***TFT* (*thin film transistor*) *display***) type of colour ◊liquid crystal display (LCD) commonly used in laptop computers. Active matrix displays are made by sandwiching a film containing tiny transistors between two plates of glass. They achieve high contrast and brightness by applying voltage across the horizontal and vertical wires between the two glass plates, balanced by using a small transistor inside each ◊pixel to amplify the voltage when so instructed.

To create ◊VGA colour, each pixel must also integrate colour filters; essentially, each logical pixel is made up of three physical pixels, one for each of red, blue, and green, the primary colours of light. The consequence of this – and the reason active matrix screens are such expensive options – is that a VGA display requires approximately a billion transistors, and even minute imperfections render the screens useless for computing purposes. A high refresh rate means that the screens are extremely responsive, so the cursor does not disappear as a mouse is moved quickly across the screen.

Active matrix displays began to appear on laptops in 1992, and are expected by many to eventually replace the older cathode-ray technology for television sets as well as display monitors.

active transport in cells, the use of energy to move substances, usually molecules or ions, across a membrane.

Energy is needed because movement occurs against a concentration gradient, with substances being passed into a region where they are already present in significant quantities. Active transport thus differs from ◊diffusion, the process by which substances move towards a region where they are in lower concentration, as when oxygen passes into the blood vessels of the lungs. Diffusion requires no input of energy.

active window on graphical operating systems, the ◊window containing the program actually in use at any given time. Usually active windows are easily identified by the use of colour schemes which assign a different colour to the window's title bar (a thin strip along the top of each window bearing the name of the window's specific program or function) from that of the title bars of inactive windows.

On a true ◊multitasking system, each window may represent an active program, but the active window is the one into which the user may enter data. A user might, for example, be typing a document into a word processor in the active window while in the background other programs back up files or sort data in a database.

ActiveX in computing, Microsoft's umbrella name for a collection of technologies used to create applications that run on the World Wide Web or on ◊intranets.

ActiveX is based on DCOM (Microsoft's Distributed Component Object Model) and uses ActiveX Controls, which are a lightweight version of OLE (◊object linking and embedding) Custom Controls or OCXs. It also includes scripting languages such as JavaScript and VB Script (Visual Basic Script), and a ◊Java Virtual Machine (JVM).

ActiveX was announced in 1996, and later that year was handed to the ◊Open Group, to manage its development and turn it into a cross-platform industry standard.

activity in physics, the number of particles emitted in one second by a radioactive source. The term is used to describe the radioactivity or the potential danger of that source. The unit of activity is the becquerel (Bq), named after the French physicist Henri Becquerel.

activity series in chemistry, alternative name for ◊reactivity series.

acupuncture in alternative medicine, a system of inserting long, thin metal needles into the body at predetermined points to relieve pain, as an anaesthetic in surgery, and to assist healing. The needles are rotated manually or electrically. The method, developed in ancient China and increasingly popular in the West, is thought to work by stimulating the brain's own painkillers, the ◊endorphins.

Acupuncture is based on a theory of physiology that posits a network of life-energy pathways, or 'meridians', in the human body and some 800 'acupuncture points' where metal needles may be inserted to affect the energy flow for purposes of preventative or remedial therapy or to produce a local anaesthetic effect. Numerous studies and surveys have attested the efficacy of the method, which is widely conceded by orthodox practitioners despite the lack of an acceptable scientific explanation.

ACUPUNCTURE.COM

http://acupuncture.com/

Thorough introduction to the alternative world of acupuncture with descriptions of the main notions and powers of herbology, yoga, Qi Gong, and Chinese nutrition. Consumers are given access to lists of practitioners and an extensive section on acupuncture research. Practitioners can browse through journal listings and the latest industry news and announcements.

acute in medicine, term used to describe a disease of sudden and severe onset which resolves quickly; for example, pneumonia and meningitis. In contrast, a **chronic** condition develops and remains over a long period.

acute angle an angle between 0° and 90°; that is, an amount of turn that is less than a quarter of a circle.

Ada high-level computer-programming language, developed and owned by the US Department of Defense, designed for use in situations in which a computer directly controls a process or machine, such as a military aircraft. The language took more than five years to specify, and became commercially available only in the late 1980s. It is named after English mathematician Ada Augusta Byron.

adaptation ***Latin adaptare 'to fit to'*** in biology, any change in the structure or function of an organism that allows it to survive and reproduce more effectively in its environment. In ◊evolution, adaptation is thought to occur as a result of random variation in the genetic make-up of organisms coupled with ◊natural selection. Species become extinct when they are no longer adapted to their environment – for instance, if the climate suddenly becomes colder.

adaptive radiation in evolution, the formation of several species, with ◊adaptations to different ways of life, from a single ancestral type. Adaptive radiation is likely to occur whenever members of a species migrate to a new habitat with unoccupied ecological niches. It is thought that the lack of competition in such niches allows sections of the migrant population to develop new adaptations, and eventually to become new species.

The colonization of newly formed volcanic islands has led to the development of many unique species. The 13 species of Darwin's finch on the Galápagos Islands, for example, are probably descended from a single species from the South American mainland. The parent stock evolved into different species that now occupy a range of diverse niches.

ADC in electronics, abbreviation for ◊analogue-to-digital converter.

addax light-coloured ◊antelope *Addax nasomaculatus* of the family Bovidae. It lives in N Africa around the Sahara Desert where it exists on scanty vegetation without drinking. It is about 1.1 m/3.5 ft at the shoulder, and both sexes have spirally twisted horns. Its hooves are broad, enabling it to move easily on soft sand.

adder electronic circuit in a computer or calculator that carries out the process of adding two binary numbers. A separate adder is needed for each pair of binary ◊bits to be added. Such circuits are essential components of a computer's ◊arithmetic and logic unit (ALU).

adder ***Anglo-Saxon naedre 'serpent'*** European venomous snake, the common ◊viper *Vipera berus*. Growing on average to about 60 cm/24 in in length, it has a thick body, triangular head, a characteristic V-shaped mark on its head and, often, zigzag markings along the back. It feeds on small mammals and lizards. The puff adder *Bitis arietans* is a large, yellowish, thick-bodied viper up to 1.6 m/5 ft long, living in Africa and Arabia.

addiction state of dependence caused by habitual use of drugs, alcohol, or other substances. It is characterized by uncontrolled craving, tolerance, and symptoms of withdrawal when access is denied. Habitual use produces changes in body chemistry and treatment must be geared to a gradual reduction in dosage.

Initially, only opium and its derivatives (morphine, heroin, codeine) were recognized as addictive, but many other drugs, whether therapeutic (for example, tranquillizers) or recreational (such as cocaine and alcohol), are now known to be addictive.

Research points to a genetic predisposition to addiction; environment and psychological make-up are other factors. Although physical addiction always has a psychological element, not all psychological dependence is accompanied by physical dependence. A carefully controlled withdrawal programme can reverse the chemical changes of habituation. Cure is difficult because of the many other factors contributing to addiction.

adding machine device for adding (and usually subtracting, multiplying, and dividing) numbers, operated mechanically or electromechanically; now largely superseded by electronic ◊calculators.

addition in arithmetic, the operation of combining two numbers to form a sum; thus, 7 + 4 = 11. It is one of the four basic operations of arithmetic (the others are subtraction, multiplication, and division).

addition polymerization ◊polymerization reaction in which a single monomer gives rise to a single polymer, with no other reaction products.

addition reaction chemical reaction in which the atoms of an element or compound react with a double bond or triple bond in an organic compound by opening up one of the bonds and becoming attached to it, for example

$$CH_2{=}CH_2 + HCl \rightarrow CH_3CH_2Cl$$

An example is the addition of hydrogen atoms to ◊unsaturated compounds in vegetable oils to produce margarine. Addition reactions are used to make useful polymers from ◊alkenes.

add-on small program written to extend the features of a larger one. The earliest successful add-on for personal computer users in the UK was a small routine which allowed the original version of the spreadsheet Lotus 1-2-3 to print out a pound sign (£), something the program's US developers had thought unnecessary.

address in a computer memory, a number indicating a specific location.

At each address, a single piece of data can be stored. For microcomputers, this normally amounts to one ◊byte (enough to represent a single character, such as a letter or digit).

The maximum capacity of a computer memory depends on how many memory addresses it can have. This is normally measured in units of 1,024 bytes (known as kilobytes, or K).

address means of specifying either a computer or a person for the purpose of directing messages or other data across a network. Addressing e-mail to a person across the Internet involves typing in a string of characters such as 'userID@machine.system.type.country'. To send mail to Jane Doe, for example, whose user ID is 'janed' and who works at a company called Anyco in the UK, a user would type in 'janed@anyco.co.uk'.

Computers do not, however, use these named addresses in routing data. The portion after the address is known as a ◊domain, and the domain name is an easy-to-remember alias for a numbered address that is understandable by a computer. This numbered address, which takes a form similar to 127.000.000.001, is known as an IP (Internet protocol) address. Both the numbered IP addresses and domain names are assigned by the ◊InterNIC.

address book facility in most e-mail software that allows the storage and retrieval of e-mail addresses. Address books remove the problem of trying to remember a particular user's exact e-mail address – and it must be exact, as computers are unable to correct human errors. The best address-book software allows a user to type in just the correspondent's name and fills in the rest automatically.

address bus in computing, the electrical pathway or ◊bus used to select the route for any particular data item as it is moved from one part of a computer to another.

adenoids masses of lymphoid tissue, similar to ◊tonsils, located in the upper part of the throat, behind the nose. They are part of a child's natural defences against the entry of germs but usually shrink and disappear by the age of ten.

Adenoids may swell and grow, particularly if infected, and block the breathing passages. If they become repeatedly infected, they may be removed surgically (**adenoidectomy**), usually along with the tonsils.

ADH (abbreviation for ***antidiuretic hormone***) in biology, part of the system maintaining a correct salt/water balance in vertebrates.

Its release is stimulated by the ◊hypothalamus in the brain, which constantly receives information about salt concentration from receptors situated in the neck. In conditions of water shortage increased ADH secretion from the brain will cause more efficient conservation of water in the kidney, so that fluid is retained by the body. When an animal is able to take in plenty of water, decreased ADH secretion will cause the urine to become dilute and plentiful. The system allows the body to compensate for a varying fluid intake and maintain a correct balance.

Adhara star 600 light years from Earth. It plays a greater part in ionizing hydrogen in our region of the Galaxy than all the other 3 million stars lying closer to our local cloud, according to US astronomers in 1995.

Most of space contains unionized hydrogen in tiny amounts but Adhara is connected to our local cloud by a tunnel of almost hydrogen-free space. This means its ionizing radiation reaches our local cloud without obstacle.

The surface temperature of Adhara is 21,000 K (almost four times that of the Sun) and it is the brightest source of extreme UV radiation apart from the Sun.

ADHESIVE

The Florida leaf beetle *Hemisphaerota cyanen* has 60,000 adhesive pads on its feet. These enable it to resist the pull of a 2 g/0.07 oz weight (the equivalent of a human hanging on to 200 grand pianos). Ant predators are unable to move the beetle.

adhesive substance that sticks two surfaces together. Natural adhesives (glues) include gelatin in its crude industrial form (made from bones, hide fragments, and fish offal) and vegetable gums. Synthetic adhesives include thermoplastic and thermosetting resins, which are often stronger than the substances they join; mixtures of ◊epoxy resin and hardener that set by chemical reaction; and elastomeric (stretching) adhesives for flexible joints. Superglues are fast-setting adhesives used in very small quantities.

adiabatic in physics, a process that occurs without loss or gain of heat, especially the expansion or contraction of a gas in which a change takes place in the pressure or volume, although no heat is allowed to enter or leave.

adipose tissue type of ◊connective tissue of vertebrates that serves as an energy reserve, and also pads some organs. It is commonly called fat tissue, and consists of large spherical cells filled with fat. In mammals, major layers are in the inner layer of skin and around the kidneys and heart.

Fatty acids are transported to and from it via the blood system. An excessive amount of adipose tissue is developed in the course of some diseases, especially obesity.

adit in mining, a horizontal shaft from the surface to reach a mineral seam. It was a common method of mining in hilly districts, and was also used to drain water. The mineral-bearing rock is excavated by digging horizontally into the side of a valley. It is used, for example, in ◊coal mining.

adjacent angles pair of angles meeting at a common vertex (corner) and sharing a common arm. Two adjacent angles lying on the same side of a straight line add up to 180° and are said to be supplementary.

adjacent side in a ◊right-angled triangle, the side that is next to a given angle but is not the hypotenuse (the side opposite the right angle). The third side is the **opposite side** to the given angle.

admiral any of several species of butterfly in the same family (Nymphalidae) as the tortoiseshells. The red admiral *Vanessa atalanta,* wingspan 6 cm/2.5 in, is found worldwide in the northern hemisphere. It either hibernates, or migrates south each year from northern areas to subtropical zones.

Adobe US company specializing in graphics and desktop publishing software. Founded in 1982 by former Xerox PARC researchers John Warnock and Chuck Geschke, Adobe was the inventor of ◊PostScript and is the publisher of ◊Acrobat and Pagemaker. Adobe's enduring contribution to the computer industry is that it facilitated the use of computers to produce the fancy fonts without which desktop publishing would not have been possible.

Adobe Type Manager in computing, program from Adobe that manages fonts under Windows 3.1 and allows the printing and display of ◊PostScript fonts.

adolescence in the human life cycle, the period between the beginning of puberty and adulthood.

The four stages of man are infancy, childhood, adolescence and obsolescence.

ART LINKLETTER US writer and broadcaster. *Child's Garden of Misinformation*

ADP abbreviation for ***adenosine diphosphate***, the chemical product formed in cells when ◊ATP breaks down to release energy.

adrenal gland or ***suprarenal gland*** triangular gland situated on top of the ◊kidney. The adrenals are soft and yellow, and consist of two parts: the cortex and medulla. The **cortex** (outer part) secretes various steroid hormones and other hormones that control salt and water metabolism and regulate the use of carbohydrates, proteins, and fats. The **medulla** (inner part) secretes the hormones adrenaline and noradrenaline which, during times of stress, cause the heart to beat faster and harder, increase blood flow to the heart and muscle cells, and dilate airways in the lungs, thereby delivering more oxygen to cells throughout the body and in general preparing the body for 'fight or flight'.

adrenaline or ***epinephrine*** hormone secreted by the medulla of the ◊adrenal glands. Adrenaline is synthesized from a closely related substance, noradrenaline, and the two hormones are released into the bloodstream in situations of fear or stress.

Adrenaline's action on the ◊liver raises blood-sugar levels by stimulating glucose production and its action on adipose tissue raises blood fatty-acid levels; it also increases the heart rate, increases blood flow to muscles, reduces blood flow to the skin with the production of sweat, widens the smaller breathing tubes (bronchioles) in the lungs, and dilates the pupils of the eyes.

adrenocorticotrophic hormone hormone secreted by the anterior lobe of the ◊pituitary gland; see ◊ACTH.

ADSL (abbreviation for ***asymmetric digital subscriber loop***) standard for transmitting video data through existing copper telephone wires. ADSL was developed by US telephone companies as a way of competing with cable television companies in delivering both TV and phone services. By 1996 it was developing into a possible alternative means for high-speed Internet access. ADSL is one of several types of digital subscriber loops (DSLs) in progress.

adsorption taking up of a gas or liquid at the surface of another substance, most commonly a solid (for example, activated charcoal adsorbs gases). It involves molecular attraction at the surface, and should be distinguished from ◊absorption (in which a uniform solution results from a gas or liquid being incorporated into the bulk structure of a liquid or solid).

Advanced Earth Observing Satellite (ADEOS) Japanese ◊remote sensing satellite launched August 1996. It gathers data on climate change, the environment, and Earth and ocean processes.

ADEOS is carrying eight instruments: Ocean Color and Temperature Scanner (OCTS); Advanced Visible and Near-Infrared Radiometer (AVNIR), Interferometric Monitor for Greenhouse Gases, Improved Limb Atmospheric Spectrometer (ILAS), Retroreflector in Space (RIS), Scatterometer (NSCAT), Total Ozone Mapping Spectrometer (TOMS), and Polarization and Directionality of the Earth's Reflectances (POLDER). NSCAT and TOMS are NASA instruments and POLDER is French, the rest are Japanese.

advanced gas-cooled reactor (AGR) type of ◊nuclear reactor used in W Europe. The AGR uses a fuel of enriched uranium dioxide in stainless-steel cladding and a moderator of graphite. Carbon dioxide gas is pumped through the reactor core to extract the heat produced by the ◊fission of the uranium. The heat is transferred to water in a steam generator, and the steam drives a turbogenerator to produce electricity.

Advanced Technology Attachment Packet Interface in computing, enhancement to integrated drive electronics (IDE), usually abbreviated to ◊ATAPI.

adventitious in botany, arising in an abnormal position, as in roots developing on the stem of a cutting or buds developing on roots.

advertising in computing, practice of paying to place information about a company's services or products in front of consumers. The earliest advertisers on the Net used to distribute their information as widely as possible in a practice quickly dubbed 'spamming'. By 1996 the practice of advertising on the World Wide Web was becoming commonplace.

Much Web advertising is sold in the same way as advertising in traditional media such as the print and broadcasting industries. Advertisers pay to place a small graphic known as an 'advertising banner' on a particular Web page in a spot (usually the top) where users are expected to see it clearly and click on it to follow the link to the advertiser's own site for more information. More sophisticated systems are under development which allow an advertising agency to track users' interests by watching which Web sites they visit and using that information to choose banners to insert which match those users' interests.

The US-based research company Jupiter Communications puts the value of online advertising at $940 million for 1997, up from $55 million in 1995. It is expected to reach $4.4 billion in 2000.

Aepyornis genus of large, extinct, flightless birds living in Madagascar until a few thousand years ago. Some stood 3 m/10 ft high and laid eggs with a volume of 43 l/9.5 gal. They had long, thick legs, four-toed feet, and rudimentary wings.

aerated water water that has had air (oxygen) blown through it. Such water supports aquatic life and prevents the growth of putrefying bacteria. Polluted waterways may be restored by artificial aeration.

aerenchyma plant tissue with numerous air-filled spaces between the cells. It occurs in the stems and roots of many aquatic plants where it aids buoyancy and facilitates transport of oxygen around the plant.

aerial or ***antenna*** in radio and television broadcasting, a conducting device that radiates or receives electromagnetic waves. The design of an aerial depends principally on the wavelength of the signal. Long waves (hundreds of metres in wavelength) may employ long wire aerials; short waves (several centimetres in wavelength) may employ rods and dipoles; microwaves may also use dipoles – often with reflectors arranged like a toast rack – or highly directional parabolic dish aerials. Because microwaves travel in straight lines, requiring line-of-sight communication, microwave aerials are usually located at the tops of tall masts or towers.

aerial oxidation in chemistry, a reaction in which air is used to oxidize another substance, as in the contact process for the manufacture of sulphuric acid:

$$2SO_2 + O_2 \leftrightarrow 2SO_3$$

and in the ◊souring of wine.

aerobic in biology, term used to describe those organisms that require oxygen (usually dissolved in water) for the efficient release of energy contained in food molecules, such as glucose. They include almost all organisms (plants as well as animals) with the exception of certain bacteria, yeasts, and internal parasites.

Aerobic reactions occur inside every cell and lead to the formation of energy-rich ◊ATP, subsequently used by the cell for driving its metabolic processes. Oxygen is used to convert glucose to carbon dioxide and water, thereby releasing energy.

Most aerobic organisms die in the absence of oxygen, but certain organisms and cells, such as those found in muscle tissue, can function for short periods anaerobically (without oxygen). ◊Anaerobic organisms can survive without oxygen.

aerodynamics branch of fluid physics that studies the forces exerted by air or other gases in motion. Examples include the airflow around bodies moving at speed through the atmosphere (such as land vehicles, bullets, rockets, and aircraft), the behaviour of gas in engines and furnaces, air conditioning of buildings, the deposition of snow, the operation of air-cushion vehicles (hovercraft), wind loads on buildings and bridges, bird and insect flight, musical wind instruments, and meteorology. For maximum efficiency, the aim is usually to design the shape of an object to produce a streamlined flow, with a minimum of turbulence in the moving air. The behaviour of aerosols or the pollution of the atmosphere by foreign particles are other aspects of aerodynamics.

aerogel light, transparent, highly porous material composed of more than 90% air. Such materials are formed from silica, metal oxides, and organic chemicals, and are produced by drying gels – networks of linked molecules suspended in a liquid – so that air fills the spaces previously occupied by the liquid. They are excellent heat insulators and have unusual optical, electrical, and acoustic properties.

Aerogels were first produced by US scientist Samuel Kristler in the early 1930s by drying silica gels at high temperatures and pressures.

aeronautics science of travel through the Earth's atmosphere, including aerodynamics, aircraft structures, jet and rocket propulsion, and aerial navigation.

In **subsonic aeronautics** (below the speed of sound), aerodynamic forces increase at the rate of the square of the speed.

Transsonic aeronautics covers the speed range from just below to just above the speed of sound and is crucial to aircraft design. Ordinary sound waves move at about 1,225 kph/760 mph at sea level, and air in front of an aircraft moving slower than this is 'warned' by the waves so that it can move aside. However, as the flying speed approaches that of the sound waves, the warning is too late for the air to escape, and the aircraft pushes the air aside, creating shock waves, which absorb much power and create design problems. On the ground the shock waves give rise to a ◊sonic boom. It was once thought that the speed of sound was a speed limit to aircraft, and the term ◊sound barrier came into use.

Supersonic aeronautics concerns speeds above that of sound and in one sense may be considered a much older study than aeronautics itself, since the study of the flight of bullets, known as ◊ballistics, was undertaken soon after the introduction of firearms. **Hypersonics** is the study of airflows and forces at speeds above five times that of sound (Mach 5); for example, for guided missiles, space rockets, and advanced concepts such as ◊HOTOL (horizontal takeoff and landing). For all flight speeds streamlining is necessary to reduce the effects of air resistance.

Aeronautics is distinguished from astronautics, which is the science of travel through space. Astronavigation (navigation by reference to the stars) is used in aircraft as well as in ships and is a part of aeronautics.

aeroplane (US ***airplane***) powered heavier-than-air craft supported in flight by fixed wings. Aeroplanes are propelled by the thrust of a jet engine or airscrew (propeller). They must be designed aerodynamically, since streamlining ensures maximum flight efficiency. The Wright brothers flew the first powered plane (a biplane) in Kitty Hawk, North Carolina, USA in 1903. For the history of aircraft and aviation, see ◊flight.

design Efficient streamlining prevents the formation of shock waves over the body surface and wings, which would cause instability and power loss. The wing of an aeroplane has the cross-sectional shape of an aerofoil, being broad and curved at the front, flat underneath (sometimes slightly curved), curved on top, and tapered to a sharp point at the rear. It is so shaped that air passing above it is speeded up, reducing pressure below atmospheric pressure and air passing below it is slower thus increasing pressure and providing a double effect. This follows from ◊Bernoulli's principle and results in a force acting vertically upwards, called lift, which counters the plane's weight. In level flight lift equals weight. The wings develop sufficient lift to support the plane when they move quickly through the air. The thrust that causes propulsion comes from the reaction to the air stream accelerated backwards by the propeller or the gases shooting backwards from the jet exhaust. In flight the engine thrust must overcome the air resistance, or ◊drag. Drag depends on frontal area (for example, large, airliner; small, fighter plane) and shape (drag coefficient); in level

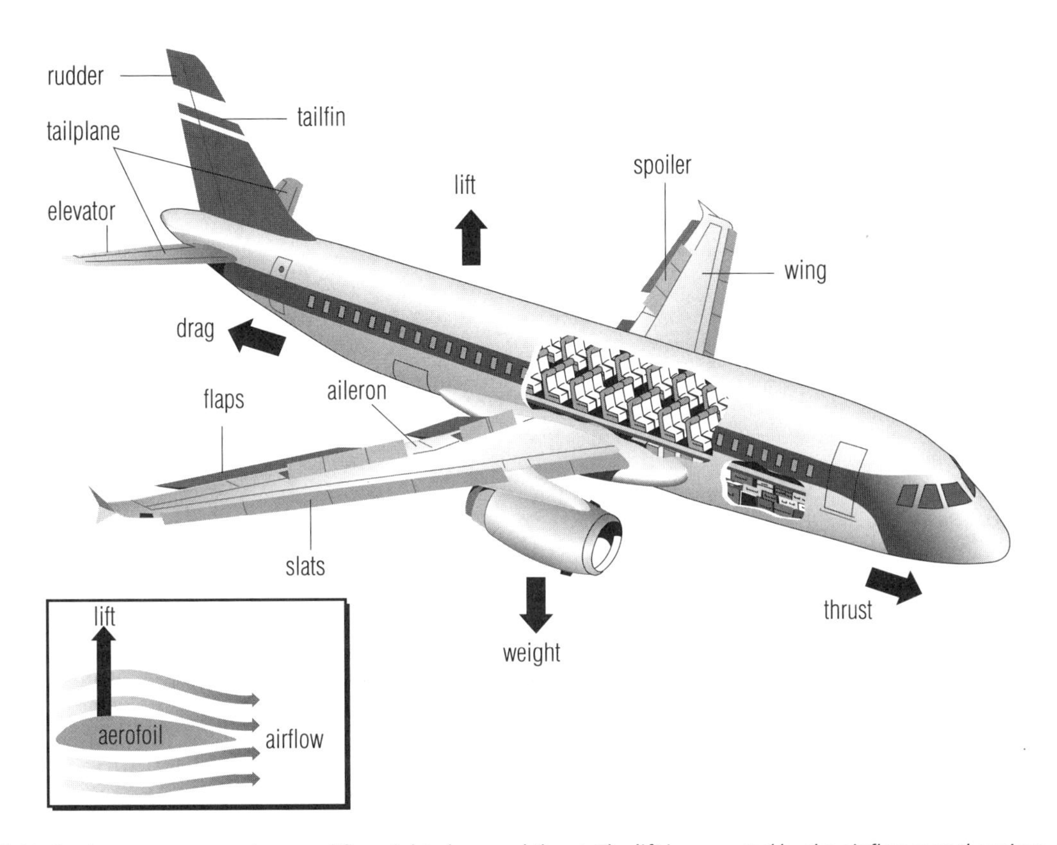

aeroplane *In flight, the forces on an aeroplane are lift, weight, drag, and thrust. The lift is generated by the air flow over the wings, which have the shape of an aerofoil. The engine provides the thrust. The drag results from the resistance of the air to the aeroplane's passage through it. Various moveable flaps on the wings and tail allow the aeroplane to be controlled. The rudder is moved to turn the aeroplane. The elevators allow the craft to climb or dive. The ailerons are used to bank the aeroplane while turning. The flaps, slats, and spoilers are used to reduce lift and speed during landing.*

flight, drag equals thrust. The drag is reduced by streamlining the plane, resulting in higher speed and reduced fuel consumption for a given power. Less fuel need be carried for a given distance of travel, so a larger payload (cargo or passengers) can be carried.

aerosol particles of liquid or solid suspended in a gas. Fog is a common natural example. Aerosol cans contain a substance such as scent or cleaner packed under pressure with a device for releasing it as a fine spray. Most aerosols used chlorofluorocarbons (CFCs) as propellants until these were found to cause destruction of the ◊ozone layer in the stratosphere.

The international community has agreed to phase out the use of CFCs, but most so-called 'ozone-friendly' aerosols also use ozone-depleting chemicals, although they are not as destructive as CFCs. Some of the products sprayed, such as pesticides, can be directly toxic to humans.

aestivation in zoology, a state of inactivity and reduced metabolic activity, similar to ◊hibernation, that occurs during the dry season in species such as lungfish and snails. In botany, the term is used to describe the way in which flower petals and sepals are folded in the buds. It is an important feature in ◊plant classification.

aether alternative form of ◊ether, the hypothetical medium once believed to permeate all of space.

AFD abbreviation for ***accelerated freeze drying***, a common method of food preservation. See ◊food technology.

affine geometry geometry that preserves parallelism and the ratios between intervals on any line segment.

affinity in chemistry, the force of attraction (see ◊bond) between atoms that helps to keep them in combination in a molecule. The

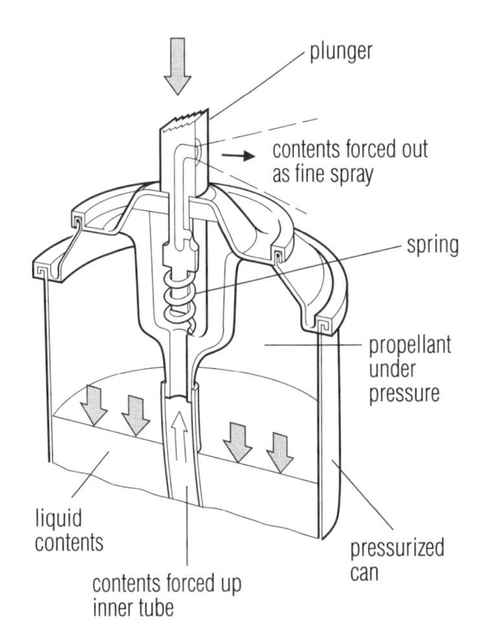

aerosol *The aerosol can produces a fine spray of liquid particles, called an aerosol. When the top button is pressed, a valve is opened, allowing the pressurized propellant in the can to force out a spray of the liquid contents. As the liquid sprays from the can, the small amount of propellant dissolved in the liquid vaporizes, producing a fine spray of small droplets.*

term is also applied to attraction between molecules, such as those of biochemical significance (for example, between ◊enzymes and substrate molecules). This is the basis for affinity ◊chromat-ography, by which biologically important compounds are separated.

The atoms of a given element may have a greater affinity for the atoms of one element than for another (for example, hydrogen has a great affinity for chlorine, with which it easily and rapidly combines to form hydrochloric acid, but has little or no affinity for argon).

afforestation planting of trees in areas that have not previously held forests. (**Reafforestation** is the planting of trees in deforested areas.) Trees may be planted (1) to provide timber and wood pulp; (2) to provide firewood in countries where this is an energy source; (3) to bind soil together and prevent soil erosion; and (4) to act as windbreaks.

Afforestation is a controversial issue because while many ancient woodlands of mixed trees are being lost, the new plantations consist almost exclusively of conifers. It is claimed that such plantations acidify the soil and conflict with the interests of ◊biodiversity (they replace more ancient and ecologically valuable species and do not sustain wildlife).

Afghan hound breed of fast hunting dog resembling the ◊saluki in build, though slightly smaller.

African palm squirrel smallish rodent *Epixerus ebii* living in dense tropical rainforest in western Africa and little known, having seldom been seen alive. The species has probably always been quite scarce, but is now threatened by rainforest destruction over a large part of its range.

African violet herbaceous plant from tropical central and E Africa, with velvety green leaves and scentless purple flowers. Different colours and double-flowered varieties have been bred. (*Saintpaulia ionantha,* family Gesneriaceae.)

afterbirth in mammals, the placenta, umbilical cord, and ruptured membranes, which become detached from the uterus and expelled soon after birth.

afterburning method of increasing the thrust of a gas turbine (jet) aeroplane engine by spraying additional fuel into the hot exhaust duct between the turbojet and the tailpipe where it ignites. Used for short-term increase of power during takeoff, or during combat in military aircraft.

afterimage persistence of an image on the retina of the eye after the object producing it has been removed. This leads to persistence of vision, a necessary phenomenon for the illusion of continuous movement in films and television. The term is also used for the persistence of sensations other than vision.

after-ripening process undergone by the seeds of some plants before germination can occur. The length of the after-ripening period in different species may vary from a few weeks to many months.

It helps seeds to germinate at a time when conditions are most favourable for growth. In some cases the embryo is not fully mature at the time of dispersal and must develop further before germination can take place. Other seeds do not germinate even when the embryo is mature, probably owing to growth inhibitors within the seed that must be leached out or broken down before germination can begin.

agama small central African lizard. It lives in groups of up to 25 in tropical forest and feeds mainly on insects.
classification The agama *Agama agama* is in the Old World family Agamidae, order Squamata.

agamid lizard in the family Agamidae, containing about 300 species.

Agamids include the common ◊agama; the Australian frilled lizard *Chlamydosaurus,* which runs on its hind legs and has a frill on each side of its neck; the thorny devil *Moloch horridus,* whose body is covered with large spikes; and the Malaysia flying dragon *Draco volans.*
classification Agamids are in family Agamidae, suborder Sauria, order Squamata, class Reptilia.

agar jellylike carbohydrate, obtained from seaweeds. It is used mainly in microbiological experiments as a culture medium for growing bacteria and other microorganisms. The agar is resistant to breakdown by microorganisms, remaining a solid jelly throughout the course of the experiment.

agaric any of a group of fungi (see ◊fungus) of typical mushroom shape. Agarics include the field mushroom *Agaricus campestris* and the cultivated edible mushroom *A. brunnesiens.* Closely related is the often poisonous *Amanita,* ◊which includes the fly agaric *A. muscaria.* (Genus *Agaricus,* family Agaricaceae.)

agate cryptocrystalline (with crystals too small to be seen with an optical microscope) silica, SiO_2, composed of cloudy and banded ◊chalcedony, sometimes mixed with ◊opal, that forms in rock cavities.

Agate stones, being hard, are also used to burnish and polish gold applied to glass and ceramics and as clean vessels for grinding rock samples in laboratories.

agave any of several related plants with stiff, sword-shaped, spiny leaves arranged in a rosette. All species come from the warmer parts of the New World. They include *Agave sisalina,* whose fibres are used for rope making, and the Mexican century plant *A. americana,* which may take many years to mature (hence its common name). Alcoholic drinks such as tequila and pulque are made from the sap of agave plants. (Genus *Agave,* family Agavaceae.)

ageing in common usage, the period of deterioration of the physical condition of a living organism that leads to death; in biological terms, the entire life process.

Three current theories attempt to account for ageing. The first suggests that the process is genetically determined, to remove individuals that can no longer reproduce. The second suggests that it is due to the accumulation of mistakes during the replication of ◊DNA at cell division. The third suggests that it is actively induced by fragments of DNA that move between cells, or by cancer-causing viruses; these may become abundant in old cells and induce them to produce unwanted ◊proteins or interfere with the control functions of their DNA.

agent software that mimics intelligence by automating tasks according to user-defined rules. The most visible agent on the Internet in 1995 was Firefly, which recommends music that users might like based on information they have already given about their favourite artists. Agents might also select news stories of

agama *During the breeding season male agamids usually develop bright colours, especially on the head. The male of this* Agama atricollis, *a widespread species in the savannas of S and E Africa, is able to fade out his conspicuous blue hues and change to a camouflage brown coloration within a minute or two of sensing danger.* *Premaphotos Wildlife*

agave Agaves are desert plants from the New World. They normally flower once when they reach maturity – which in some species may take as long as 60 years – and then die. Fibres such as sisal can be produced from some species, and the alcoholic drink tequila is made from others. Premaphotos Wildlife

interest, arrange scheduling with other agents, and filter out unwanted junk e-mail.

Much research on agents is proceeding at the ◊MIT Media Lab, where Professor Pattie Maes directs the group studying the capability and potential of autonomous agents. See also ◊crawler and ◊bot.

Agent Orange selective ◊weedkiller, notorious for its use in the 1960s during the Vietnam War by US forces to eliminate ground cover which could protect enemy forces. It was subsequently discovered to contain highly poisonous ◊dioxin. Thousands of US troops who had handled it, along with many Vietnamese people who came into contact with it, later developed cancer or produced deformed babies.

Agent Orange, named after the distinctive orange stripe on its packaging, combines equal parts of 2,4-D (2,4-dichlorophenoxyacetic acid) and 2,4,5-T (2,4,5-trichlorophenoxyacetic acid), both now banned in the USA. Companies that had manufactured the chemicals faced an increasing number of lawsuits in the 1970s. All the suits were settled out of court in a single class action, resulting in the largest ever payment of its kind ($180 million) to claimants.

age–sex graph graph of the population of an area showing age and sex distribution.

aggression in biology, behaviour used to intimidate or injure another organism (of the same or of a different species), usually for the purposes of gaining territory, a mate, or food. Aggression often involves an escalating series of threats aimed at intimidating an opponent without having to engage in potentially dangerous physical contact. Aggressive signals include roaring by red deer, snarling by dogs, the fluffing-up of feathers by birds, and the raising of fins by some species of fish.

agonist in biology, a ◊muscle that contracts and causes a movement. Contraction of an agonist is complemented by relaxation of its ◊antagonist. For example, the biceps (in the front of the upper arm) bends the elbow whilst the triceps (lying behind the biceps) straightens the arm.

agoraphobia ◊phobia involving fear of open spaces and public places. The anxiety produced can be so severe that some sufferers are unable to leave their homes for many years.

Agoraphobia affects 1 person in 20 at some stage in their lives. The most common time of onset is between the ages of 18 and 28.

agouti small rodent of the genus *Dasyprocta*, family Dasyproctidae. It is found in the forests of Central and South America. The agouti is herbivorous, swift-running, and about the size of a rabbit.

AGR abbreviation for ◊advanced gas-cooled reactor, **a type of nuclear reactor**.

agribusiness commercial farming on an industrial scale, often financed by companies whose main interests lie outside agriculture; for example, multinational corporations. Agribusiness farms are mechanized, large in size, highly structured, and reliant on chemicals.

agriculture ***Latin ager 'field', colere 'to cultivate'*** the practice of farming, including the cultivation of the soil (for raising crops) and the raising of domesticated animals. The units for managing agricultural production vary from smallholdings and individually owned farms to corporate-run farms and collective farms run by entire communities.

Crops are for human or animal food, or commodities such as cotton and sisal. For successful production, the land must be prepared (ploughed, cultivated, harrowed, and rolled). Seed must be planted and the growing plants nurtured. This may involve ◊fertilizers, ◊irrigation, pest control by chemicals, and monitoring of acidity or nutrients. When the crop has grown, it must be harvested and, depending on the crop, processed in a variety of ways before it is stored or sold.

Greenhouses allow cultivation of plants that would otherwise find the climate too harsh. ◊Hydroponics allows commercial cultivation of crops using nutrient-enriched solutions instead of soil. Special methods, such as terracing, may be adopted to allow cultivation in hostile terrain and to retain topsoil in mountainous areas with heavy rainfall.

Animals are raised for wool, milk, leather, dung (as fuel), or meat. They may be semidomesticated, such as reindeer, or fully domesticated but nomadic (where naturally growing or cultivated food supplies are sparse), or kept in one location. Animal farming involves accommodation (buildings, fencing, or pasture), feeding, breeding, gathering the produce (eggs, milk, or wool), slaughtering, and further processing such as tanning.

agrimony herbaceous plant belonging to the rose family, with small yellow flowers on a slender spike. It grows along hedges and in fields. (*Agrimonia eupatoria*, family Rosaceae.)

agronomy study of crops and soils, a branch of agricultural science. Agronomy includes such topics as selective breeding (of plants and animals), irrigation, pest control, and soil analysis and modification.

AI abbreviation for ◊artificial intelligence.

Aichi Japanese aircraft of World War II, principally used by the navy.

The **B7A**, known to the Allies as 'Grace', was a torpedo-bomber produced in small numbers. The **D3A**, known as 'Val', was a carrier dive bomber of great strength and efficiency; it was the principal dive bomber used at Pearl Harbor Dec 1941, and sank numerous Allied warships throughout the war. The **E13A**, 'Jake', was a floatplane used for reconnaissance; it was used in this role to prepare the way for the Pearl Harbor raid, and was carried by almost all Japanese warships.

AIDS acronym for ***acquired immune deficiency syndrome***, the gravest of the sexually transmitted diseases, or ◊STDs. It is caused by the human immunodeficiency virus (◊HIV), now known to be a ◊retrovirus, an organism first identified in 1983. HIV is transmitted in body fluids, mainly blood and genital secretions.

diagnosis of AIDS The effect of the virus in those who become ill is the devastation of the immune system, leaving the victim sus-

AIDS AND HIV INFORMATION

http://www.thebody.com/

AIDS/HIV site offering safe sex and AIDS prevention advice, information about treatments and testing, and health/nutritional guidance for those who have contracted the disease.

Agriculture: chronology

10000–8000 BC	Holocene (post-glacial) period of hunters and gatherers. Harvesting and storage of wild grains in southwest Asia. Herding of reindeer in northern Eurasia. Domestic sheep in northern Iraq.
8000	Neolithic revolution with cultivation of domesticated wheats and barleys, sheep, and goats in southwest Asia. Domestication of pigs in New Guinea.
7000–6000	Domestic goats, sheep, and cattle in Anatolia, Greece, Persia, and the Caspian basin. Planting and harvesting techniques transferred from Asia Minor to Europe.
5000	Beginning of Nile valley civilization. Millet cultivated in China.
3400	Flax used for textiles in Egypt. Widespread corn production in the Americas.
3200	Records of ploughing, raking, and manuring by Egyptians.
C. 3100	River Nile dammed during the rule of King Menes.
3000	First record of asses used as beasts of burden in Egypt. Sumerian civilization used barley as main crop with wheat, dates, flax, apples, plums, and grapes.
2900	Domestication of pigs in eastern Asia.
2640	Reputed start of Chinese silk industry.
2500	Domestic elephants in the Indus valley. Potatoes a staple crop in Peru.
2350	Wine-making in Egypt.
2250	First known irrigation dam.
1600	Important advances in the cultivation of vines and olives in Crete.
1500	*Shadoof* (mechanism for raising water) used for irrigation in Egypt.
1400	Iron ploughshares in use in India.
1300	Aqueducts and reservoirs used for irrigation in Egypt.
1200	Domestic camels in Arabia.
1000–500	Evidence of crop rotation, manuring, and irrigation in India.
600	First windmills used for corn grinding in Persia.
350	Rice cultivation well established in parts of western Africa. Hunting and gathering in the east, central, and south parts of the continent.
C. 200	Use of gears to create ox-driven water wheel for irrigation. Archimedes screw used for irrigation.
100	Cattle-drawn iron ploughs in use in China.
AD 65	*De Re Rustica/On Rural Things*, Latin treatise on agriculture and irrigation.
500	'Three fields in two years' rotation used in China.
630	Cotton introduced into Arabia.
800	Origins of the 'open field' system in northern Europe.
900	Wheeled ploughs in use in western Europe. Horse collar, originating in China, allowed horses to be used for ploughing as well as carrying.
1000	Frisians (NW Netherlanders) began to build dykes and reclaim land. Chinese began to introduce Champa rice which cropped much more quickly than other varieties.
11TH CENTURY	Three-field system replaced the two-field system in western Europe. Concentration on crop growing.
1126	First artesian wells, at Artois, France.
12TH CENTURY	Increasing use of water mills and windmills. Horses replaced oxen for pulling work in many areas.
12TH–14TH CENTURIES	Expansion of European population brought more land into cultivation. Crop rotations, manuring, and new crops such as beans and peas helped increase productivity. Feudal system at its height.
13TH–14TH CENTURIES	Agricultural recession in western Europe with a series of bad harvests, famines, and pestilence.
1347	Black Death killed about a third of the European population.
16TH CENTURY	Decline of the feudal system in western Europe. More specialist forms of production were now possible with urban markets. Manorial estates and serfdom remained in eastern Europe. Chinese began cultivation of non-indigenous crops such as corn, sweet potatoes, potatoes, and peanuts.
17TH CENTURY	Potato introduced into Europe. Norfolk crop rotation became widespread in England, involving wheat, turnips, barley and then ryegrass/clover.
1700–1845	Agricultural revolution began in England. Two million hectares of farmland in England enclosed. Removal of open fields in other parts of Europe followed.
C. 1701	Jethro Tull developed the seed drill and the horse-drawn hoe.
1747	First sugar extracted from sugar beet in Prussia.
1762	Veterinary school founded in Lyon, France.
1783	First plough factory in England.
1785	Cast-iron ploughshare patented.
1793	Invention of the cotton gin.
1800	Early threshing machines developed in England.
1820S	First nitrates for fertilizer imported from South America.
1830	Reaping machines developed in Scotland and the US. Steel plough made by John Deere in Illinois, US.
1840S	Extensive potato blight in Europe.
1850S	Use of clay pipes for drainage well established throughout Europe.
1862	First steam plough used in the Netherlands.
1850–1890S	Major developments in transport and refrigeration technology altered the nature of agricultural markets with crops, dairy products, and wheat being shipped internationally.
1890S	Development of stationary engines for ploughing.
1892	First petrol-driven tractor in the USA.
1921	First attempt at crop dusting with pesticides from an aeroplane near Dayton, Ohio, US.
1938	First self-propelled grain combine harvester used in the USA.
1942–62	Huge increase in the use of pesticides, later curbed by disquiet about their effects and increasing resistance of pests to standard controls such as DDT.
1945 ONWARDS	Increasing use of scientific techniques, crop specialization and larger scale of farm enterprises.
1985	First cases of bovine spongiform encephalopathy (BSE) recorded by UK vets.
1992	Number of cases of BSE in cattle was at its peak (700 cases per week).
1995	Increase in the use of genetic engineering with nearly 3,000 transgenic crops being field-tested.
1996	Organic farming was on the increase in EU countries. The rise was 11% per year in Britain, 50% in Germany, and 40% in Italy.

ceptible to diseases that would not otherwise develop. Diagnosis of AIDS is based on the appearance of rare tumours or opportunistic infections in unexpected candidates. *Pneumocystis carinii* pneumonia, for instance, normally seen only in the malnourished or those whose immune systems have been deliberately suppressed, is common among AIDS victims and, for them, a leading cause of death.

treatment In the West the time-lag between infection with HIV and the development of AIDS seems to be about ten years, but progression is far more rapid in developing countries. Some AIDS victims die within a few months of the outbreak of symptoms, some survive for several years; roughly 50% are dead within three years. There is no cure for the disease and the four antivirals currently in use against AIDS have not lived up to expectations. Trials began in 1994 using a new AIDS drug called 3TC in conjunction with ◊zidovudine (formerly ◊AZT). Although individually the drugs produce little effect, when the drugs were used together in 1995, the levels of virus in the blood were ten times lower than at the beginning of the trial. Treatment of opportunistic infections extended the average length of survival with AIDS (in Western countries) from about 11 months in 1985 to 23 months in 1994.

HIV/AIDS – worldwide statistics Allowing for under-diagnosis, incomplete reporting, and reporting delay, and based on the available data on HIV infections around the world, it is estimated (1997) that approximately 8.4 million AIDS cases in adults and children have occurred worldwide since the pandemic began. WHO estimate that of these cases, which include active AIDS cases and people

AIDS – Recent Developments

BY PAUL MOSS

The global count of people infected with the human immunodeficiency virus (HIV) continues to increase at an alarming rate. Despite some notable successes with health education campaigns, millions are infected, with the worst infection rates being in the developing world. Nevertheless, there are signs that the enormous scientific effort that has been mobilized against the disease is now paying dividends. Current drug regimes are able to prolong life and are even allowing scientists to debate the prospect of cure.

When HIV enters the body, the main cell type infected is the CD4+ T lymphocyte, a circulating white cell that plays an important role in controlling immune responses. In addition, the virus can enter a variety of other cell types that have the CD4 molecule on their surface. The net result of infection is a relentless fall in the number of CD4+ T cells and a gradual dismantling of the immune system's ability to fight off infectious agents such as bacteria, fungi, and other viruses. This process leaves patients very susceptible to a wide variety of infections, many of which are virtually never seen in people with a normally functioning immune system. One important advance in the last few years has been the use of molecular assays to measure the amount of virus in the blood of infected patients. This procedure is valuable in predicting how rapidly they are likely to progress to an advanced state of the disease.

Early treatment

The initial successes in the drug treatment of HIV infection were with agents that could treat or prevent the infectious complications of the disease. These drugs remain very valuable but do not have any activity against HIV itself. The first drug with proven activity against HIV was zidovudine (AZT). Zidovudine resembles one of the building blocks of DNA, and when HIV undergoes replication, zidovudine can bind to an essential HIV enzyme and prevent the virus from completing its life cycle. Zidovudine can improve the symptoms of HIV infection, is valuable in asymptomatic patients with low CD4+ lymphocyte counts, and is effective at reducing the rate of transmission of HIV from pregnant women to their babies. However, when zidovudine is used alone, the virus is usually able to escape from the effects of zidovudine by mutating its DNA sequence. There is now an increasing appreciation of the need to use zidovudine in combination with some of the new antiviral drugs.

New developments

At the moment, probably the most exciting class of drugs that inhibit HIV replication is the protease inhibitors. When HIV replicates inside a cell it has to make a copy of its DNA, and then this genetic message is decoded into a protein. Some HIV proteins need to be broken down into smaller pieces in order to function, and this is done by a protease molecule. Normal function of the HIV protease appears to be vital for efficient replication of the virus, and over the last few years researchers have spent a great deal of effort in developing drugs that can block its function.

At least four protease inhibitors have been tried in clinical practice: saquinavir, ritonavir, nelfinavir, and indinavir. All have slightly different properties and different side effects. In clinical trials, these drugs have demonstrated a spectacular ability to reduce the amount of virus in the body. Sensitive molecular assays such as the polymerase chain reaction (PCR) have shown that the amount of HIV in blood can be reduced by over a thousandfold and sometimes may reach undetectable levels. Although effective on their own, most of the current drive in HIV therapeutics is to use these agents in combination with other anti-HIV agents. Typically this would include zidovudine, a protease inhibitor, and another agent such as didanosine. In a recent trial this combination led to the virus being undetectable in 60 % of patients after 24 weeks of treatment. A very encouraging observation with protease inhibitors is that they can be used at a very advanced state of the disease. It seems that the drugs should be used at quite large doses, to avoid the development of a resistant virus, and unfortunately they do have several side effects. Although most of these effects are not serious, many patients are unable to tolerate a particular drug combination; in these cases a change to another combination is indicated.

The role of combination therapy

The exact role of combination therapy in the overall management of HIV infection is a subject of considerable debate at present. In an attempt to achieve consensus, a panel of the International AIDS Society-USA met in 1996. After results from many clinical trials, the group suggested that combination therapy was now the treatment of choice. It remains unclear, however, whether or not protease inhibitors should be used with all patients or just those at particular risk of rapid progression to full-blown AIDS based on measurement of the amount of HIV in their blood. Patients with symptoms should be offered treatment, but for those who are asymptomatic the situation is less clear and the decision will be based on the CD4+ count and the viral load in the blood. There are relatively little data to recommend how to treat patients who have been infected in the last month or two and are suffering from the typical symptoms of acute infection: fever, swollen lymph nodes, and headaches. As this is a time of intense viral replication, there is a theoretical advantage in using the strongest available treatment to limit the initial multiplication of the virus. This may also reduce the chance of the virus making mutations that would allow it to resist drug treatment.

The last few years have seen valuable advances in the treatment of HIV infection. Several powerful drugs are now available and are being tested in trials around the world. The human immunodeficiency virus has an astounding ability to mutate itself in order to evade drugs, and there are likely to be setbacks ahead. It is too soon to say whether or not some patients may be offered the prospect of complete cure. Nevertheless, many AIDS researchers are hoping that they can now maintain the upper hand in the battle against this formidable virus.

who have died of AIDS, not HIV infections, more than 70% were in Africa, with about 9% in the USA, 9% in the rest of the Americas, 6% in Asia, and 4% in Europe. Of the total number of AIDS cases reported, 39% were in the USA, 34% in Africa, 20% in Asia, 12.5% in Europe, and 12% in the Americas.

Estimates released by WHO in Nov 1996 and Jan 1997 revealed that 22.6 million men, women, and children had been infected by HIV. Of these, 21.8 million were adults and 830,000 children. Approximately 42% of adult sufferers were female, with the proportion of women infected by HIV/AIDS steadily increasing. The majority of newly infected adults were under 25 years of age. WHO estimated that by the year 2000 30–40 million people would have been infected by the virus. The United Nations AIDS programme (UNAIDS) concluded in 1997 that the world total number of HIV infections was just under 30 million.

sub-Saharan Africa The worst affected area is sub-Saharan Africa, where the United Nations AIDS programme (UNAIDS) estimated that 20 million people had been infected. About two-thirds of this total were in east and central Africa, an area that accounts for only about one-sixth of the total population of the sub-Saharan region. In Kenya, Malawi, Rwanda, Uganda, Tanzania, Zambia, and Zimbabwe, surveys showed that over 10% of women attending antenatal clinics in urban areas were HIV-infected, with rates exceeding 40% in some surveillance sites. The number of AIDS cases in Africa has also continued to increase. An estimated 5 million people in sub-Saharan Africa have developed AIDS.

S and SE Asia The most alarming trends of HIV infection are in south and southeast Asia, where the epidemic is spreading as fast as it was a decade ago in sub-Saharan Africa. The majority of the 6 million HIV infections estimated to have occurred in adults in these regions (1997) appeared in India, Thailand, and Myanmar (Burma), but high rates of HIV spread have been seen elsewhere too.

Latin America and the Caribbean WHO estimated (1996) that about 1.6 million HIV infections had occurred in Latin America and the Caribbean since the epidemic began. In these regions epidemics are increasingly occurring among women and adolescents. The future course of the epidemic in the region depends greatly on the rate at which the virus spreads in Brazil, which has more AIDS cases than any other country outside Africa and the USA.

Middle East and north Africa About 200,000 HIV infections were estimated to have occurred so far in the Middle East and north Africa (1996). These figures are of particular concern because other factors – the presence of other sexually transmitted diseases and intravenous drug use, for example – suggest that there is an increased exposure to the risk of HIV infection.

E Europe and central Asia It was estimated (1996) that over 50,000 adults in eastern Europe and central Asia had been infected with HIV. HIV is spreading in these regions – sometimes quite rapidly – to communities and countries that were hardly affected by the epidemic only a few years ago. Ukraine recently reported a substantial increase in newly infected drug users in cities bordering the Black Sea. The Russian Federation may experience a similar progression.

North America In 1993 AIDS for the first time became the USA's leading cause of death among all people aged between 25 and 44. By Nov 1996, 565,097 cumulative AIDS cases had been reported in the USA. However, in 1996 the AIDS death rate in the USA fell from 15.6 per 100,000 people to 11.6 (26%, which was the first decline in the 15 years since the pandemic had began. In 1996 AIDS was no longer the main killer of adults between the ages of 25 to 44 but it remained so for African Americans in that age group. Although the overall number of new HIV infections in the USA has decreased, results from several studies suggest that the HIV epidemic has now spread to a new generation of homosexual and bisexual men. In Jan 1997, AIDS experts reported a 30% drop in AIDS deaths in New York. The actual count fell from 7,046 to 4,944. The researchers said this was as a result of improved treatments and better access to care. An estimated 860,000 adults in North America were living with HIV infection at the end of 1997.

W Europe The United Nations AIDS programme (UNAIDS) estimated in 1997 that 150,000 HIV infections had occurred in western Europe. As of mid-1996, 167,021 AIDS cases had been reported throughout the European Union (EU). The highest rate of reported AIDS cases within the EU was in Spain. Information from the UK indicates that the declining trend in male-to-male transmission observed in the late 1980s may have begun to reverse as early as 1990.

E Asia and the Pacific WHO estimated that by the end of 1996 over 113,000 HIV infections had occurred in east Asia and the Pacific. An estimated 15,100 AIDS cases had occurred in these regions, 11,700 of these in Australia and New Zealand. There was evidence that HIV infection rates had reached a plateau in Australia and were declining in New Zealand. However, an HIV epidemic recently developed in Papua New Guinea. By the end of 1994, this island of about 4 million people had an estimated 4,000 adults living with HIV, overtaking Australia as the country with the highest number of HIV cases per head of population in the Pacific region.

The cumulative direct and indirect costs of HIV and AIDS in the 1980s have been conservatively estimated at $240 billion. The global cost – direct and indirect – of HIV and AIDS by the year 2000 could be as high as $500 billion a year – equivalent to more than 2% of global GDP.

ailanthus any of several trees or shrubs with compound leaves made up of pointed leaflets and clusters of small greenish flowers with an unpleasant smell. The tree of heaven (*Ailanthus altissima*), native to E Asia, is grown worldwide as an ornamental; it can grow to 30 m/100 ft in height and the trunk can reach 1 m/3 ft in diameter. (Genus *Ailanthus*, family Simaroubaceae.)

air the mixture of gases making up the Earth's ◊atmosphere.

airbrush small fine spray-gun used by artists, graphic designers, and photographic retouchers. Driven by air pressure from a compressor or pressurized can, it can apply a thin, very even layer of ink or paint, allowing for subtle gradations of tone.

air conditioning system that controls the state of the air inside a building or vehicle. A complete air-conditioning unit controls the temperature and humidity of the air, removes dust and odours from it, and circulates it by means of a fan. US inventor Willis Haviland Carrier (1876–1950) developed the first effective air-conditioning unit in 1902 for a New York printing plant.

The air in an air conditioner is cooled by a type of ◊refrigeration unit comprising a compressor and a condenser. The air is cleaned by means of filters and activated charcoal. Moisture is extracted by condensation on cool metal plates. The air can also be heated by electrical wires or, in large systems, pipes carrying hot water or steam; and cool, dry air may be humidified by circulating it over pans of water or through a water spray.

The first air conditioners were used in 19th century textile mills, where a fine water spray was used to cool and humidify the atmosphere.

A specialized air-conditioning system is installed in spacecraft as part of the life-support system. This includes the provision of oxygen to breathe and the removal of exhaled carbon dioxide.

Outdoor spaces may also be cooled using overhead cool air jets.

aircraft any aeronautical vehicle capable of flying through the air. It may be lighter than air (supported by buoyancy) or heavier than air (supported by the dynamic action of air on its surfaces). ◊Balloons and ◊airships are lighter-than-air craft. Heavier-than-air craft include the ◊aeroplane, glider, autogiro, and helicopter.

air-cushion vehicle (ACV) craft that is supported by a layer, or cushion, of high-pressure air. The ◊hovercraft is one form of ACV.

Airedale terrier breed of large terrier, about 60 cm/24 in tall, with a wiry red-brown coat and black saddle patch. It originated about 1850 in England, as a cross between the otterhound and Irish and Welsh terriers.

airlock airtight chamber that allows people to pass between areas of different pressure; also an air bubble in a pipe that impedes fluid flow. An airlock may connect an environment at ordinary pressure and an environment that has high air pressure (such as a submerged caisson used for tunnelling or building dams or bridge foundations).

An airlock may also permit someone wearing breathing apparatus to pass into an airless environment (into water from a submerged submarine or into the vacuum of space from a spacecraft).

air passages in biology, the nose, pharynx, larynx, trachea, and bronchi. When a breath is taken, air passes through high narrow passages on each side of the nose where it is warmed and moistened and particles of dust are removed. Food and air passages meet and cross in the pharynx. The larynx lies in front of the lower part of the pharynx and it is the organ where the voice is produced using the vocal cords. The air passes the glottis (the opening between the vocal cords) and enters the trachea. The trachea leads into the chest and divides above the heart into two bronchi. The bronchi carry the air to the lungs and they subdivide to form a succession of fine tubes and, eventually, a network of capillaries that allow the exchange of gases between the inspired air and the blood.

air pollution contamination of the atmosphere caused by the discharge, accidental or deliberate, of a wide range of toxic airborne substances. Often the amount of the released substance is relatively high in a certain locality, so the harmful effects become more noticeable. The cost of preventing any discharge of pollutants into the air is prohibitive, so attempts are more usually made to reduce the amount of discharge gradually and to disperse it as quickly as possible by using a very tall chimney, or by intermittent release.

Possibly the world's worst ever human-made air pollution disas-

Air pollution: major pollutants

pollutant	*sources*	*effects*
sulphur dioxide SO_2	oil, coal combustion in power stations	acid rain formed, which damages plants, trees, buildings, and lakes
oxides of nitrogen NO, NO_2	high-temperature combustion in cars, and to some extent power stations	acid rain formed
lead compounds	from leaded petrol used by cars	nerve poison
carbon dioxide CO_2	oil, coal, petrol, diesel combustion	greenhouse effect
carbon monoxide CO	limited combustion of oil, coal, petrol, diesel fuels	poisonous, leads to photochemical smog in some areas
nuclear waste	nuclear power plants, nuclear weapon testing, war	radioactivity, contamination of locality, cancers, mutations, death

ter occurred in Indonesia in September 1997. It was caused by forest clearance fires. Smoke pollution in the city of Palangkaraya reached 7.5 mg per cu m (nearly 3 mg more than in the London smog of 1952 in which 4,000 people died). The pollutants spread to Malaysia and other countries of the region.

The 1997 Kyoto protocol commits the industrialized nations of the world to cutting their levels of harmful gas emissions to 5.2% by 2012. Europe is expected to take the biggest cut of 8%, the USA 7%, and Japan 6%. The agreement covers Russia and Eastern Europe as well.

AIR POLLUTION – COMMITTEE ON THE MEDICAL ASPECTS OF AIR POLLUTANTS

http://www.open.gov.uk/doh/hef/airpol/airpolh.htm

Comprehensive report on the state of Britain's air from the Ministry of Health. There is a large amount of textual and statistical information on all aspects of air pollution, description of improved warning and detection measures, and details of how the general public may access advice and information.

air sac in birds, a thin-walled extension of the lungs. There are nine of these and they extend into the abdomen and bones, effectively increasing lung capacity. In mammals, it is another name for the alveoli in the lungs, and in some insects, for widenings of the trachea.

The sacs subdivide into further air spaces which partially replace the marrow in many of the bird's bones. The air space in these bones assists flight by making them lighter.

airship or ***dirigible*** any aircraft that is lighter than air and power-driven, consisting of an ellipsoidal balloon that forms the streamlined envelope or hull and has below it the propulsion system (propellers), steering mechanism, and space for crew, passengers, and/or cargo. The balloon section is filled with lighter-than-air gas, either the nonflammable helium or, before helium was industrially available in large enough quantities, the easily ignited and flammable hydrogen. The envelope's form is maintained by internal pressure in the nonrigid (blimp) and semirigid (in which the nose and tail sections have a metal framework connected by a rigid keel) types. The rigid type (zeppelin) maintains its form using an internal metal framework. Airships have been used for luxury travel, polar exploration, warfare, and advertising.

AIRSHIP AND BLIMP RESOURCES

http://www.hotairship.com/index.html

Information about airships – with the main focus on contemporary development. This site includes sections such as 'hot news', 'manufacturer database', 'homebuilding', and 'museums'.

Rigid airships predominated from about 1900 until 1940. As the technology developed, the size of the envelope was increased from about 45 m/150 ft to more than 245 m/800 ft for the last two zeppelins built. In 1852 the first successful airship was designed and flown by Henri Giffard of France. In 1900 the first rigid type was designed by Count (*Graf*) Ferdinand von ◊Zeppelin of Germany (though he did not produce a successful model till his L-24 in1908). Airships were used by both sides during World War I, but they were not seriously used for military purposes after that as they were largely replaced by aeroplanes. The British mainly used small machines for naval reconnaissance and patrolling the North Sea; Germany used Schutte-Lanz and Zeppelin machines for similar patrol work and also for long-range bombing attacks against English and French cities, mainly Paris and London.

air transport means of conveying goods or passengers by air from one place to another. See ◊flight.

ajolote Mexican reptile of the genus *Bipes*. It and several other tropical burrowing species are placed in the Amphisbaenia, a group separate from lizards and snakes among the Squamata. Unlike the others, however, which have no legs, it has a pair of short but well-developed front legs. In line with its burrowing habits, the skull is very solid, the eyes small, and external ears absent.

The scales are arranged in rings, giving the body a wormlike appearance.

AKITA

http://club.infocom.net/~akita/index.html

Profile of the Akita from the American Kennel Club. There is a history of the breed, description of the ideal breed standard, a note on the dog's temperament, and some quirky facts about the breed. There is also a high resolution photo of this large and powerful dog.

Akita breed of guard dog from the prefecture of Akita in NW Japan. It is strongly built and stands about 69 cm/27 in tall. It has small, pointed ears, a thick coat, which may be beige, grey, brindled or black, and carries its tail curled over its back.

alabaster naturally occurring fine-grained white or light-coloured translucent form of gypsum, often streaked or mottled. A soft material, it is easily carved, but seldom used for outdoor sculpture.

Alamogordo town in New Mexico, USA, associated with nuclear testing. The first atom bomb was exploded nearby at Trinity Site 16 July 1945.

It is now a test site for guided missiles.

Alaskan malamute breed of dog. It is a type of ◊husky.

albacore name loosely applied to several species of fish found in warm regions of the Atlantic and Pacific oceans, in particular to a large tuna, *Thunnus alalunga*, and to several other species of the mackerel family.

albatross large seabird, genus *Diomedea*, with long narrow wings adapted for gliding and a wingspan of up to 3 m/10 ft, mainly found in the southern hemisphere. It belongs to the family Diomedeidae, order Procellariiformes, the same group as petrels and shearwaters. The external nostrils of birds in this order are more or less tubular, and the bills are hooked.

Albatrosses feed mainly on squid and fish, and nest on remote oceanic islands. Albatrosses can cover enormous distances, flying as far as 16,100 km/10,000 mi in 33 days, or up to 640 km/600 mi in one day. They continue flying even after dark, at speeds of up to 53.5 kph/50 mph, though they may stop for an hour's rest and to feed during the night. They are sometimes called 'gooney birds', probably because of their clumsy way of landing. Albatrosses are becoming increasingly rare, and are in danger of extinction. In the southern hemisphere, more than 40,000 albatrosses drown each year as a result of catching squid attached to bait lines.

The Diomedeidae family contains 14 species of albatross found in the Southern Atlantic and the Pacific oceans. The **wandering albatross** *D. exulans*, which has a wingspan of up to 3.4 m/11 ft, is the largest oceanic bird and can live for up to 80 years. Its huge wingspan means that it has difficulty in taking off unless there are strong winds. For this reason it nests on cliffs on islands. A single white egg is laid. The chick's full weight is 12 kg/26 lb, heavier than the parents, which typically weigh around 9 kg/20 lb. The chick needs this extra body weight to survive the Antarctic winter; the parents only return to the chick if and when they can find food for it.

albedo the fraction of the incoming light reflected by a body such as a planet. A body with a high albedo, near 1, is very bright, while a body with a low albedo, near 0, is dark. The Moon has an average albedo of 0.12, Venus 0.76, Earth 0.37.

albinism rare hereditary condition in which the body has no tyrosinase, one of the enzymes that form the pigment ◊melanin, normally found in the skin, hair, and eyes. As a result, the hair is white and the skin and eyes are pink. The skin and eyes are abnormally sensitive to light, and vision is often impaired. The condition occurs among all human and animal groups.

albumin or ***albumen*** any of a group of sulphur-containing ◊proteins. The best known is in the form of egg white; others occur in milk, and as a major component of serum. They are soluble in water and dilute salt solutions, and are coagulated by heat.

The presence of serum albumin in the urine, termed albuminuria or proteinuria, may be indicative of kidney or heart disease.

Alcan Canadian aluminium-producing company. Alcan Aluminium was founded in Montréal 1928 as the Canadian subsidiary of **Alcoa**, the US aluminium company founded in Pittsburgh, Pennsylvania, 1888. Alcan became independent from Alcoa 1945 and is the fifth largest company in Canada today, second in the world in aluminium production only to Alcoa.

ALCHEMICAL SUBSTANCES

http://www.levity.com/alchemy/substanc.html

Modern explanations of the lyrical names alchemists gave to their substances. Part of the huge Virtual Alchemy Library Web site this section is slightly disorganized, not being listed in alphabetical or any other particular order. Some of the exotic substances described include the bizarrely named 'Thion hudor' and the rather poetic 'Purple of Cassius'.

alchemy ***Arabic*** *al-Kimya* supposed technique of transmuting base metals, such as lead and mercury, into silver and gold by the philosopher's stone, a hypothetical substance, to which was also attributed the power to give eternal life.

This aspect of alchemy constituted much of the chemistry of the Middle Ages. More broadly, however, alchemy was a system of philosophy that dealt both with the mystery of life and the formation of inanimate substances. Alchemy was a complex and indefinite conglomeration of chemistry, astrology, occultism, and magic, blended with obscure and abstruse ideas derived from various religious systems and other sources. It was practised in Europe from ancient times to the Middle Ages but later fell into disrepute when ◊chemistry and ◊physics developed.

What is accomplished with fire is alchemy, whether in the furnace or the kitchen stove.

PARACELSUS Swiss physician, alchemist, and scientist.
In J Bronowski *The Ascent of Man* 1975

alcohol any member of a group of organic compounds characterized by the presence of one or more aliphatic OH (hydroxyl) groups

Alkane	Alcohol	Aldehyde	Ketone	Carboxylic acid	Alkene
CH_4 methane	CH_3OH methanol	HCHO methanal	—	HCO_2H methanoic acid	—
CH_3CH_3 ethane	CH_3CH_2OH ethanol	CH_3CHO ethanal	—	CH_3CO_2H ethanoic acid	CH_2CH_2 ethene
$CH_3CH_2CH_3$ propane	$CH_3CH_2CH_2OH$ propanol	CH_3CH_2CHO propanal	CH_3COCH_3 propanone	$CH_3CH_2CO_2H$ propanoic acid	CH_2CHCH_3 propene
methane	methanol	methanal	propanone	methanoic acid	ethene

alcohol The systematic naming of simple straight-chain organic molecules.

in the molecule, and which form ◊esters with acids. The main uses of alcohols are as solvents for gums, resins, lacquers, and varnishes; in the making of dyes; for essential oils in perfumery; and for medical substances in pharmacy. The alcohol produced naturally in the ◊fermentation process and consumed as part of alcoholic beverages is called ◊ethanol.

Alcohols may be liquids or solids, according to the size and complexity of the molecule. The five simplest alcohols form a series in which the number of carbon and hydrogen atoms increases progressively, each one having an extra CH_2 (methylene) group in the molecule: methanol or wood spirit (methyl alcohol, CH_3OH); ethanol (ethyl alcohol, C_2H_5OH); propanol (propyl alcohol, C_3H_7OH); butanol (butyl alcohol, C_4H_9OH); and pentanol (amyl alcohol, $C_5H_{11}OH$). The lower alcohols are liquids that mix with water; the higher alcohols, such as pentanol, are oily liquids immiscible with water; and the highest are waxy solids – for example, hexadecanol (cetyl alcohol, $C_{16}H_{33}OH$) and melissyl alcohol ($C_{30}H_{61}OH$), which occur in sperm-whale oil and beeswax respectively. Alcohols containing the CH_2OH group are primary; those containing CHOH are secondary; while those containing COH are tertiary.

alcoholic solution solution produced when a solute is dissolved in ethanol.

alcoholism dependence on alcohol. It is characterized as an illness when consumption of alcohol interferes with normal physical or emotional health. Excessive alcohol consumption, whether through sustained ingestion or irregular drinking bouts or binges, may produce physical and psychological addiction and lead to nutritional and emotional disorders. Long-term heavy consumption of alcohol leads to diseases of the heart, liver, and peripheral nerves. Support groups such as Alcoholics Anonymous are helpful.

Aldebaran or ***Alpha Tauri*** brightest star in the constellation Taurus and the 14th brightest star in the sky; it marks the eye of the 'bull'. Aldebaran is a red giant 60 light years away, shining with a true luminosity of about 100 times that of the Sun.

aldehyde any of a group of organic chemical compounds prepared by oxidation of primary alcohols, so that the OH (hydroxyl) group loses its hydrogen to give an oxygen joined by a double bond to a carbon atom (the aldehyde group, with the formula CHO).

alder any of a group of trees or shrubs belonging to the birch family, found mainly in cooler parts of the northern hemisphere and characterized by toothed leaves and catkins. (Genus *Alnus*, family Betulaceae.)

alewife fish *Alosa pseudoharengus* of the ◊herring group, up to 30 cm/1 ft long, found in the NW Atlantic and in the Great Lakes of North America.

alexanders strong-smelling tall ◊herbaceous plant belonging to the carrot family. It is found along hedgerows and on cliffs throughout S Europe. Its yellow flowers appear in spring and early summer. (*Smyrnium olusatrum*, family Umbelliferae.)

Alexander technique in alternative medicine, a method of correcting bad habits of posture, breathing, and muscular tension, which Australian therapist F M Alexander maintained cause many ailments. The technique is also used to promote general health and relaxation and enhance vitality.

Back troubles, migraine, asthma, hypertension, and some gastric and gynaecological disorders are among the conditions said to be alleviated by the technique, which is also said to be effective in the prevention of disorders, particularly those of later life.

alfalfa or ***lucerne*** perennial tall ◊herbaceous plant belonging to the pea family. It is native to Europe and Asia and has spikes of small purple flowers in late summer. It is now a major fodder crop, commonly processed into hay, meal, or silage. Alfalfa sprouts, the sprouted seeds, have become a popular salad ingredient. (*Medicago sativa*, family Leguminosae.)

algae (singular ***alga***) highly varied group of plants, ranging from single-celled forms to large and complex seaweeds. They live in both fresh and salt water, and in damp soil. Algae do not have true roots, stems, or leaves.

Marine algae help combat ◊global warming by removing carbon dioxide from the atmosphere during ◊photosynthesis.

Alembert, Jean le Rond d' (1717–1783)

French mathematician, encyclopedist, and theoretical physicist. In association with Denis Diderot, he helped plan the great Encyclopédie, for which he also wrote the 'Discours préliminaire' 1751. He framed several theorems and principles – notably d'Alembert's principle – in dynamics and celestial mechanics, and devised the theory of partial differential equations.

The principle that now bears his name was first published in his *Traité de dynamique* 1743, and was an extension of the third of Isaac ◊Newton's laws of motion. D'Alembert maintained that the law was valid not merely for a static body, but also for mobile bodies. Within a year he had found a means of applying the principle to the theory of equilibrium and the motion of fluids. Using also the theory of partial differential equations, he studied the properties of sound, and air compression, and also managed to relate his principle to an investigation of the motion of any body in a given figure.

Mary Evans Picture Library

algebra branch of mathematics in which the general properties of numbers are studied by using symbols, usually letters, to represent variables and unknown quantities. For example, the algebraic statement $(x + y)^2 = x^2 + 2xy + y^2$ is true for all values of x and y. If $x = 7$ and $y = 3$, for instance:

$$(7 + 3)^2 = 7^2 + 2(7 \times 3) + 3^2 = 100$$

An algebraic expression that has one or more variables (denoted by letters) is a ◊polynomial equation. Algebra is used in many areas of mathematics – for example, matrix algebra and Boolean algebra (the latter is used in working out the logic for computers).

In ordinary algebra the same operations are carried on as in arithmetic, but, as the symbols are capable of a more generalized and extended meaning than the figures used in arithmetic, it facilitates calculation where the numerical values are not known, or are inconveniently large or small, or where it is desirable to keep them in an analysed form.

Within an algebraic equation the separate calculations involved must be completed in a set order. Any elements in brackets should always be calculated first, followed by multiplication, division, addition, and subtraction.

quadratic equation This is a polynomial equation of second degree (that is, an equation containing as its highest power the square of a variable, such as x^2). The general formula of such equations is $ax^2 + bx + c = 0$, in which a, b, and c are real numbers, and only the coefficient a cannot equal 0.

Some quadratic equations can be solved by factorization, or the values of x can be found by using the formula for the general solution

$$x = [-b \pm \sqrt{(b^2 - 4ac)}]/2a$$

Depending on the value of the discriminant $b^2 - 4ac$, a quadratic equation has two real, two equal, or two complex roots (solutions). When $b^2 - 4ac > 0$ there are two distinct real roots. When $b^2 - 4ac = 0$ there are two equal real roots. When $b^2 - 4ac < 0$ there are two distinct complex roots.

simultaneous equations If there are two or more algebraic equations that contain two or more unknown quantities that may have

> To remember the order of operations in complex algebraic or numerical expression:
>
> BLESS MY DEAR AUNT SALLY!
>
> BRACKETS, MULTIPLY, DIVIDE, ADD, SUBTRACT

a unique solution they can be solved simultaneously. For example, in the case of two linear equations with two unknown variables, such as:

$$\text{(i) } x + 3y = 6 \text{ and (ii) } 3y - 2x = 4$$

the solution will be those unique values of x and y that are valid for both equations. Linear simultaneous equations can be solved by using algebraic manipulation to eliminate one of the variables. For example, both sides of equation (i) could be multiplied by 2, which gives $2x + 6y = 12$. This can be added to equation (ii) to get $9y = 16$, which is easily solved: $y = \frac{16}{9}$. The variable x can now be found by inserting the known y value into either original equation and solving for x.

'Algebra' was originally the name given to the study of equations. In the 9th century, the Arab mathematician Muhammad ibn-Mūsā al-Khwārizmī used the term *al-jabr* for the process of adding equal quantities to both sides of an equation. When his treatise was later translated into Latin, *al-jabr* became 'algebra' and the word was adopted as the name for the whole subject.

algebraic fraction fraction in which letters are used to represent numbers – for example, $\frac{a}{b}$, $\frac{xy}{z}2$, and $\frac{1}{x\ 1\ y}$. Like numerical fractions, algebraic f\ractions may be simplified or factorized. Two equivalent algebraic fractions can be cross-multiplied; for example, if $\frac{a}{b} = \frac{b}{c}$ then $ad = bc$

(In the same way, the two equivalent numerical fractions $\frac{2}{3}$ and $\frac{4}{6}$ can be cross-multiplied to give cross-products that are both 12.)

alginate salt of alginic acid, $(C_6H_8O_6)_n$, obtained from brown seaweeds and used in textiles, paper, food products, and pharmaceuticals.

ALGOL (acronym for **algo*rithmic* language**) in computing, an early high-level programming language, developed in the 1950s and 1960s for scientific applications. A general-purpose language, ALGOL is best suited to mathematical work and has an algebraic style. Although no longer in common use, it has greatly influenced more recent languages, such as Ada and PASCAL.

Algol or **Beta Persei** ◊eclipsing binary, a pair of orbiting stars in the constellation Perseus, one of which eclipses the other every 69 hours, causing its brightness to drop by two-thirds.

The brightness changes were first explained in 1782 by English amateur astronomer John Goodricke (1764–1786). He pointed out that the changes between magnitudes 2.2 and 3.5 repeated themselves exactly after an interval of 2.867 days and supposed this to be due to two stars orbiting round and eclipsing each other.

Algonquin Radio Observatory site in Ontario, Canada, of the radio telescope, 46 m/150 ft in diameter, of the National Research Council of Canada, opened 1966.

algorithm procedure or series of steps that can be used to solve a problem.

In computer science, it describes the logical sequence of operations to be performed by a program. A ◊flow chart is a visual representation of an algorithm.

The word derives from the name of 9th-century Arab mathematician Muhammad ibn-Mūsā al-Khwārizmī.

alias name representing a particular user or group of users in e-mail systems. This feature, which is not available on all systems, is a matter of convenience as it allows a user to substitute shorter or easier-to-remember real names for e-mail addresses. In 1995 CompuServe announced a system of named aliases for its long, numbered addresses.

aliasing or ***jaggies*** effect seen on computer screen or printer output, when smooth curves appear to be made up of steps because the resolution is not high enough. The steps are caused by clumps of pixels that become visible when the monitor's definition is lower than that of the image that it is trying to show. ◊Anti-aliasing is a software technique that reduces this effect by using intermediate shades of colour to create an apparently smoother curve.

ALife in computing, contraction of ◊artificial life.

alimentary canal in animals, the tube through which food passes; it extends from the mouth to the anus. It is a complex organ, adapted for ◊digestion. In human adults, it is about 9 m/30 ft long, consisting of the mouth cavity, pharynx, oesophagus, stomach, and the small and large intestines.

A constant stream of enzymes from the canal wall and from the pancreas assists the breakdown of food molecules into smaller, soluble nutrient molecules, which are absorbed through the canal wall into the bloodstream and carried to individual cells. The muscles of the alimentary canal keep the incoming food moving, mix it with the enzymes and other juices, and slowly push it in the direction of the anus, a process known as ◊peristalsis. The wall of the canal receives an excellent supply of blood and is folded so as to increase its surface area. These two adaptations ensure efficient absorption of nutrient molecules.

aliphatic compound any organic chemical compound in which the carbon atoms are joined in straight chains, as in hexane (C_6H_{14}), or in branched chains, as in 2-methylpentane ($CH_3CH(CH_3)CH_2CH_2CH_3$).

Aliphatic compounds have bonding electrons localized within the vicinity of the bonded atoms. ◊Cyclic compounds that do not have delocalized electrons are also aliphatic, as in the alicyclic compound cyclohexane (C_6H_{12}) or the heterocyclic piperidine ($C_5H_{11}N$). Compare ◊aromatic compound.

alkali in chemistry, a compound classed as a ◊base that is soluble in water. Alkalis neutralize acids and are soapy to the touch.

The hydroxides of metals are alkalis; those of sodium and potassium being chemically powerful; both were historically derived from the ashes of plants.

The four main alkalis are sodium hydroxide (caustic soda, NaOH); potassium hydroxide (caustic potash, KOH); calcium hydroxide (slaked lime or limewater, $Ca(OH)_2$); and aqueous ammonia ($NH_{3\,(aq)}$). Their solutions all contain the hydroxide ion OH^-, which gives them a characteristic set of properties.

Alkalis react with acids to form a salt and water (neutralization).

$$KOH + HNO_3 \rightarrow KNO_3 + H_2OOH^- + H^+ \rightarrow H_2O$$

They give a specific colour reaction with indicators; for example, litmus turns blue.

alkali metal any of a group of six metallic elements with similar chemical properties: lithium, sodium, potassium, rubidium, caesium, and francium. They form a linked group (Group One) in the ◊periodic table of the elements. They are univalent (have a valency of one) and of very low density (lithium, sodium, and potassium float on water); in general they are reactive, soft, low-melting-point metals. Because of their reactivity they are only found as compounds in nature.

alkaline-earth metal any of a group of six metallic elements with similar bonding properties: beryllium, magnesium, calcium, strontium, barium, and radium. They form a linked group in the ◊periodic table of the elements. They are strongly basic, bivalent (have a valency of two), and occur in nature only in compounds.

alkaloid any of a number of physiologically active and frequently poisonous substances contained in some plants. They are usually organic bases and contain nitrogen. They form salts with acids and, when soluble, give alkaline solutions.

Substances in this group are included by custom rather than by scientific rules. Examples include morphine, cocaine, quinine, caffeine, strychnine, nicotine, and atropine.

In 1992, epibatidine, a chemical extracted from the skin of an Ecuadorian frog, was identified as a member of an entirely new

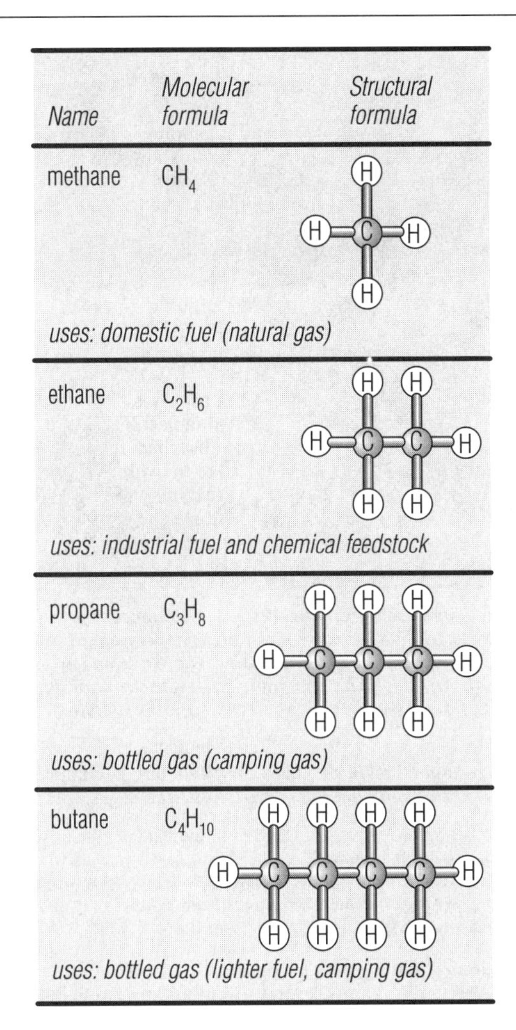

alkane The lighter alkanes methane, ethane, propane, and butane, showing the aliphatic chains. A hydrogen atom bonds to a carbon atom at all available sites.

class of alkaloid. It is an organochlorine compound, which is rarely found in animals, and a powerful painkiller, about 200 times as effective as morphine.

alkane member of a group of ◊hydrocarbons having the general formula C_nH_{2n+2}, commonly known as **paraffins**. As they contain only single ◊covalent bonds, alkanes are said to be saturated. Lighter alkanes, such as methane, ethane, propane, and butane, are colourless gases; heavier ones are liquids or solids. In nature they are found in natural gas and petroleum.

alkene member of the group of ◊hydrocarbons having the general formula C_nH_{2n}, formerly known as **olefins**. Alkenes are unsaturated compounds, characterized by one or more double bonds between adjacent carbon atoms. Lighter alkenes, such as ethene and propene, are gases, obtained from the ◊cracking of oil fractions. Alkenes react by addition, and many useful compounds, such as poly(ethene) and bromoethane, are made from them.

alkyne member of the group of ◊hydrocarbons with the general formula C_nH_{2n-2}, formerly known as the **acetylenes**. They are unsaturated compounds, characterized by one or more triple bonds between adjacent carbon atoms. Lighter alkynes, such as ethyne, are gases; heavier ones are liquids or solids.

allele one of two or more alternative forms of a ◊gene at a given position (locus) on a chromosome, caused by a difference in the ◊DNA. Blue and brown eyes in humans are determined by different alleles of the gene for eye colour.

Organisms with two sets of chromosomes (diploids) will have two copies of each gene. If the two alleles are identical the individual is said to be ◊homozygous at that locus; if different, the individual is ◊heterozygous at that locus. Some alleles show ◊dominance over others.

ALLERGY FACTS

http://www.sig.net/~allergy/facts.html

Extended fact sheet about allergies. It provides basic information about the causes, symptoms, and treatment of allergy attacks, and helps clarify typical misunderstandings about allergy sufferers. It also provides tips for the everyday life of an allergy sufferer and an extended list of allergy symptoms.

allergy special sensitivity of the body that makes it react with an exaggerated response of the natural immune defence mechanism to the introduction of an otherwise harmless foreign substance (**allergen**).

alligator ***Spanish** el lagarto* **'the lizard'** reptile of the genus *Alligator*, related to the crocodile. There are only two living species: *A. mississipiensis*, the Mississippi alligator of the southern states of the USA, and *A. sinensis* from the swamps of the lower Chang Jiang River in China. The former grows to about 4 m/12 ft, but the latter only to 1.5 m/5 ft. Alligators lay their eggs in waterside nests of mud and vegetation and are good mothers. They swim well with lashing movements of the tail and feed on fish and mammals but seldom attack people.

The skin is of value for fancy leather, and alligator farms have been established in the USA. Closely related are the caymans of South America; these belong to the genus *Caiman*. Alligators ranged across N Europe from the Upper Cretaceous to the Pliocene period.

AMERICAN ALLIGATOR

http://www.seaworld.org/animal_bytes/alligatorab.html

Illustrated guide to the American alligator including information about genus, size, life span, habitat, gestation, diet, and a series of fun facts.

alligator clip small metal clip wired to other similar clips to allow temporary connections. Today's modular phone jacks generally make it easy to hook up modems and telephones. However, in some situations, such as a hotel room where a telephone is hard-wired to the wall or in a foreign country where a visitor's modem plug is incompatible with the local telephone network, the only answer is to take the phone apart and hook the modem directly to the phone line using these small clips.

allium any of a group of plants of the lily family, usually strong-smelling with a sharp taste; they form bulbs in which sugar is stored. Cultivated species include onion, garlic, chive, and leek. Some species are grown in gardens for their decorative globular heads of white, pink, or purple flowers. (Genus *Allium*, family Liliaceae.)

allometry in biology, a regular relationship between a given feature (for example, the size of an organ) and the size of the body as a whole, when this relationship is not a simple proportion of body size. Thus, an organ may increase in size proportionately faster, or slower, than body size does. For example, a human baby's head is much larger in relation to its body than is an adult's.

alloparental care in animal behaviour, the care of another animal's offspring. 'Fostering' is common in some birds, such as pigeons, and social mammals, such as meerkats. Usually both the adoptive parent and the young benefit.

allopathy *Greek allos 'other', pathos 'suffering'* in ◊homoeopathy a term used for orthodox medicine, using therapies designed to counteract the manifestations of the disease. In strict usage, allopathy is the opposite of homoeopathy.

allotropy property whereby an element can exist in two or more forms (allotropes), each possessing different physical properties but the same state of matter (gas, liquid, or solid). The allotropes of carbon are diamond and graphite. Sulphur has several allotropes (flowers of sulphur, plastic, rhombic, and monoclinic). These solids have different crystal structures, as do the white and grey forms of tin and the black, red, and white forms of phosphorus.

Oxygen exists as two gaseous allotropes: one used by organisms for respiration (O_2), and the other a poisonous pollutant, ozone (O_3).

alloy metal blended with some other metallic or nonmetallic substance to give it special qualities, such as resistance to corrosion, greater hardness, or tensile strength. Useful alloys include bronze, brass, cupronickel, duralumin, German silver, gunmetal, pewter, solder, steel, and stainless steel.

Among the oldest alloys is bronze (mainly an alloy of copper and tin), the widespread use of which ushered in the Bronze Age. Complex alloys are now common; for example, in dentistry, where a cheaper alternative to gold is made of chromium, cobalt, molybdenum, and titanium. Among the most recent alloys are superplastics: alloys that can stretch to double their length at specific temperatures, permitting, for example, their injection into moulds as easily as plastic.

allspice spice prepared from the dried berries of the evergreen pimento tree, also known as the West Indian pepper tree, (*Pimenta dioica*) of the myrtle family, cultivated chiefly in Jamaica. It has an aroma similar to that of a mixture of cinnamon, cloves, and nutmeg.

alluvial deposit layer of broken rocky matter, or sediment, formed from material that has been carried in suspension by a river or stream and dropped as the velocity of the current decreases. River plains and deltas are made entirely of alluvial deposits, but smaller pockets can be found in the beds of upland torrents.

Alluvial deposits can consist of a whole range of particle sizes, from boulders down through cobbles, pebbles, gravel, sand, silt, and clay. The raw materials are the rocks and soils of upland areas that are loosened by erosion and washed away by mountain streams. Much of the world's richest farmland lies on alluvial deposits. These deposits can also provide an economic source of minerals. River currents produce a sorting action, with particles of heavy material deposited first while lighter materials are washed downstream.

Hence heavy minerals such as gold and tin, present in the original rocks in small amounts, can be concentrated and deposited on stream beds in commercial quantities. Such deposits are called 'placer ores'.

Almagest *Arabic al 'the' and a corruption of the Greek megiste 'greatest'* book compiled by the Greek astronomer ◊Ptolemy during the 2nd century AD, which included the idea of an Earth-centred universe; it was translated into Arabic in the 9th century. Some medieval books on astronomy, astrology, and alchemy were given the same title.

Each of the 13 sections of the book deals with a different branch of astronomy. The introduction describes the universe as spherical and contains arguments for the Earth being stationary at the centre. From this mistaken assumption, it goes on to describe the motions of the Sun, Moon, and planets; eclipses; and the positions, brightness, and precession of the 'fixed stars'. The book drew on the work of earlier astronomers such as Hipparchus.

almond tree related to the peach and apricot. Dessert almonds are the kernels of the fruit of the sweet variety *Prunus amygdalus dulcis,* which is also used to produce a low-cholesterol cooking oil. Oil of bitter almonds, from the variety *P. amygdalus amara,* is used in flavouring. Almond oil is also used for cosmetics, perfumes, and fine lubricants. (*Prunus amygdalus,* family Rosaceae.)

aloe *Aloes vary in size from dwarf species, no more than a few centimetres in diameter (such as the popular houseplant* Aloe variegata*), to species as large as a small tree. This* Aloe arborescens *from South Africa is of intermediate size. In their natural habitat most aloes flower during winter.* *Premaphotos Wildlife*

aloe one of a group of plants native to southern Africa, with long, fleshy, spiny-edged leaves. The drug usually referred to as 'bitter aloes' is a powerful purgative (agent that causes the body to expel impurities) prepared from the juice of the leaves of several of the species. (Genus *Aloe,* family Liliaceae.)

alpaca domesticated South American hoofed mammal *Lama pacos* of the camel family, found in Chile, Peru, and Bolivia, and herded at high elevations in the Andes. It is bred mainly for its long, fine, silky wool, and stands about 1 m/3 ft tall at the shoulder with neck and head another 60 cm/2 ft.

The alpaca is also used for food at the end of its fleece-producing years. Like the ◊llama, it was probably bred from the wild ◊guanaco and is a close relative of the ◊vicuna.

alpha in computing, the first version of a new software program. Developing modern software requires much testing and many versions before the definitive product is achieved. The first versions of any new product are typically full of ◊bugs, and are tested by the developers and their assistants. Later versions, known as ◊beta versions, are given to outside users to test.

Alpha Centauri or **Rigil Kent** brightest star in the constellation Centaurus and the third-brightest star in the sky. It is actually a triple star (see ◊binary star); the two brighter stars orbit each other every 80 years, and the third, Proxima Centauri, is the closest star to the Sun, 4.2 light years away, 0.1 light years closer than the other two.

alpha channel in ◊24-bit colour, a channel for controlling colour information. Describing colour for a computer display requires three channels of information per ◊pixel, one for each of the primary colours of light: red, blue, and green. A 24-bit graphics adapter with a 32-bit ◊bus can use the remaining 8 bits to send control information for the remaining 24 bits.

alpha decay disintegration of the nucleus of an atom to produce an ◊alpha particle. See also ◊radioactivity.

alphanumeric data data made up of any of the letters of the alphabet and any digit from 0 to 9. The classification of data according to the type or types of character contained enables computer ◊validation systems to check the accuracy of data: a computer can be programmed to reject entries that contain the wrong type of character. For example, a person's name would be rejected if it contained any numeric data, and a bank-account number would be rejected if it contained any alphabetic data. A car's registration number, by comparison, would be expected to contain alphanumeric data but no punctuation marks.

alpha particle positively charged, high-energy particle emitted from the nucleus of a radioactive atom. It is one of the products of the spontaneous disintegration of radioactive elements (see ◊radioactivity) such as radium and thorium, and is identical with the nucleus of a helium atom – that is, it consists of two protons and two neutrons. The process of emission, **alpha decay**, transforms one element into another, decreasing the atomic (or proton) number by two and the atomic mass (or nucleon number) by four.

Because of their large mass alpha particles have a short range of only a few centimetres in air, and can be stopped by a sheet of paper. They have a strongly ionizing effect (see ◊ionizing radiation) on the molecules that they strike, and are therefore capable of damaging living cells. Alpha particles travelling in a vacuum are deflected slightly by magnetic and electric fields.

Alps, Lunar conspicuous mountain range on the Moon, NE of the Sea of Showers (Mare Imbrium), cut by a valley 150 km/93 mi long. The highest peak is Mont Blanc, about 3,660 m/12,000 ft.

Alsatian another name for the ◊German shepherd dog.

Altair or ***Alpha Aquilae*** brightest star in the constellation Aquila and the 12th brightest star in the sky. It is a white star 16 light years away and forms the so-called Summer Triangle with the stars Deneb (in the constellation Cygnus) and Vega (in Lyra).

AltaVista search engine on the World Wide Web run by DEC (Digital Equipment Corporation). AltaVista runs an automated program to index all the pages it can find on the Web, enabling visitors to enter search terms such as a name or subject and quickly retrieve a list of pages to visit to look for specific information. It has a similar indexing program for UseNet.

In mid-1996 DEC claimed the service indexed 30 million pages on 275,600 servers, as well as 3 million articles from 14,000 USENET newsgroups, and was accessed over 16 million times per weekday.

alternate angles a pair of angles that lie on opposite sides and at opposite ends of a transversal (a line that cuts two or more lines in the same plane). The alternate angles formed by a transversal of two parallel lines are equal.

alternating current (AC) electric current that flows for an interval of time in one direction and then in the opposite direction, that is, a current that flows in alternately reversed directions through or around a circuit. Electric energy is usually generated as alternating current in a power station, and alternating currents may be used for both power and lighting.

The advantage of alternating current over direct current (DC), as from a battery, is that its voltage can be raised or lowered economically by a transformer: high voltage for generation and transmission, and low voltage for safe utilization. Railways, factories, and domestic appliances, for example, use alternating current.

My personal desire would be to prohibit entirely the use of alternating currents. They are unnecessary as they are dangerous ... I can therefore see no justification for the introduction of a system which has no element of permanency and every element of danger to life and property.

THOMAS ALVA EDISON US scientist and inventor.
Quoted in R L Weber, *A Random Walk in Science*

alternation of generations typical life cycle of terrestrial plants and some seaweeds, in which there are two distinct forms occurring alternately: **diploid** (having two sets of chromosomes) and **haploid** (one set of chromosomes). The diploid generation produces haploid spores by ◊meiosis, and is called the sporophyte, while the haploid generation produces gametes (sex cells), and is called the gametophyte. The gametes fuse to form a diploid ◊zygote which develops into a new sporophyte; thus the sporophyte and gametophyte alternate.

alternative energy see ◊energy, alternative.

alternative medicine see ◊medicine, alternative.

alternator electricity ◊generator that produces an alternating current.

alt hierarchy on USENET, the 'alternative' set of ◊newsgroups, set up so that anyone can start a newsgroup on any topic. Most areas of USENET, such as the ◊Big Seven hierarchies, allow the creation of newsgroups only after structured discussion and a vote to demonstrate that demand for the newsgroup exists. The alt hierarchy was created to allow users to bypass this process.

altimeter instrument used in aircraft that measures altitude, or height above sea level. The common type is a form of aneroid ◊barometer, which works by sensing the differences in air pressure at different altitudes. This must continually be recalibrated because of the change in air pressure with changing weather conditions. The ◊radar altimeter measures the height of the aircraft above the ground, measuring the time it takes for radio pulses emitted by the aircraft to be reflected. Radar altimeters are essential features of automatic and blind-landing systems.

altimetry method of measuring changes in sea level using an ◊altimeter attached to a satellite. The altimeter measures the distance between the satellite and water surface.

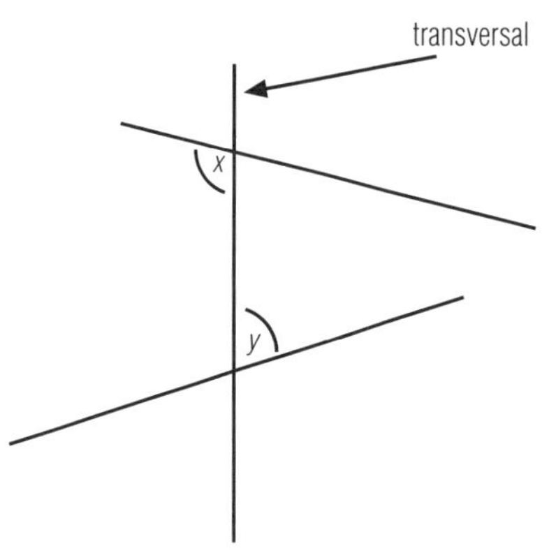

x and *y* are alternate angles

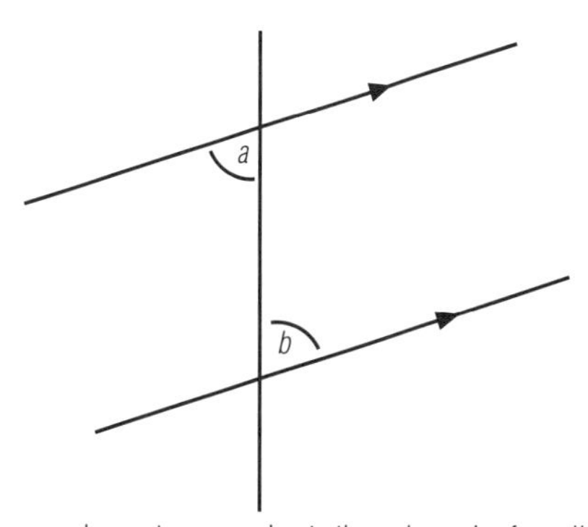

where a transversal cuts through a pair of parallel lines the alternate angles *a* and *b* are equal

alternate angles

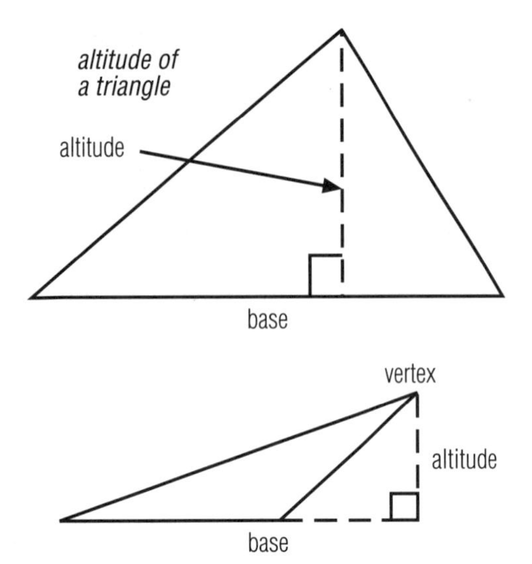

two altitudes of a quadrilateral

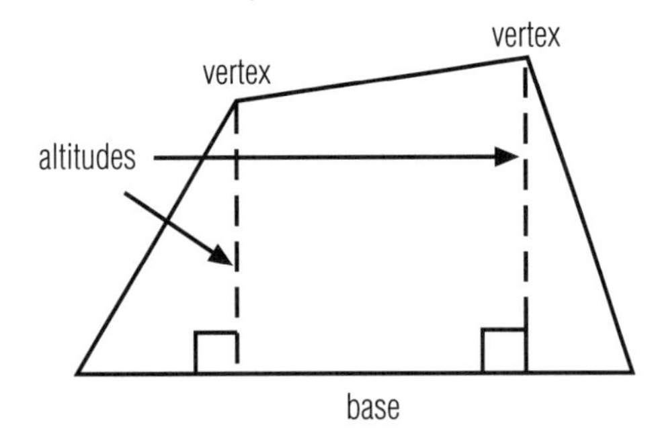

altitude *The altitude of a figure is the perpendicular distance from a vertex (corner) to the base (the side opposite the vertex).*

altitude or ***elevation*** in astronomy, the angular distance of an object above the horizon, ranging from 0° on the horizon to 90° at the zenith. Together with ◊azimuth, it forms the system of horizontal coordinates for specifying the positions of celestial bodies.

altitude in geometry, the perpendicular distance from a ◊vertex (corner) of a figure, such as a triangle, to the base (the side opposite the vertex).

altitude measurement of height, usually given in metres above sea level.

altricial animals born in a very dependent state and in need of a high degree of parental care. Examples include those mammals that are born naked and blind, such as mice and most other rodents, and most baby birds.

altruism in biology, helping another individual of the same species to reproduce more effectively, as a direct result of which the altruist may leave fewer offspring itself. Female honey bees (workers) behave altruistically by rearing sisters in order to help their mother, the queen bee, reproduce, and forgo any possibility of reproducing themselves.

ALU abbreviation for ◊arithmetic and logic unit.

alum any double sulphate of a monovalent metal or radical (such as sodium, potassium, or ammonium) and a trivalent metal (such as aluminium, chromium, or iron). The commonest alum is the double sulphate of potassium and aluminium, $K_2Al_2(SO_4)_4.24H_2O$, a white crystalline powder that is readily soluble in water. It is used in curing animal skins. Other alums are used in papermaking and to fix dye in the textile industry.

alumina or ***corundum*** Al_2O_3 oxide of aluminium, widely distributed in clays, slates, and shales. It is formed by the decomposition of the feldspars in granite and used as an abrasive. Typically it is a white powder, soluble in most strong acids or caustic alkalis but not in water. Impure alumina is called 'emery'. Rubies, sapphires, and topaz are corundum gemstones.

aluminium lightweight, silver-white, ductile and malleable, metallic element, symbol Al, atomic number 13, relative atomic mass 26.9815, melting point 658°C. It is the third most abundant element (and the most abundant metal) in the Earth's crust, of which it makes up about 8.1% by mass. It is non-magnetic, an excellent conductor of electricity, and oxidizes easily, the layer of oxide on its surface making it highly resistant to tarnish. In the USA the original name suggested by the scientist Humphry Davy, 'aluminum', is retained.

pure aluminium Aluminium is a reactive element with stable compounds, so a great deal of energy is needed in order to separate aluminium from its ores, and the pure metal was not readily obtainable until the middle of the 19th century. Commercially, it is prepared by the electrolysis of alumina (aluminium oxide), which is obtained from the ore ◊bauxite. In its pure state aluminium is a weak metal, but when combined with elements such as copper, silicon, or magnesium it forms alloys of great strength.

uses Aluminium is widely used in the shipbuilding and aircraft industries because of its light weight (relative density 2.70). It is also used in making cooking utensils, cans for beer and soft drinks, and foil. It is much used in steel-cored overhead cables and for canning uranium slugs for nuclear reactors. Aluminium is an essential constituent in some magnetic materials; and, as a good conductor of electricity, is used as foil in electrical capacitors. A plastic form of aluminium, developed 1976, which moulds to any shape and extends to several times its original length, has uses in electronics, cars, and building construction.

aluminium chloride $AlCl_3$ white solid made by direct combination of aluminium and chlorine.

$$2Al + 3Cl_2 \rightarrow 2AlCl_3$$

The anhydrous form is a typical covalent compound.

aluminium hydroxide or ***alumina cream*** $Al(OH)_3$ gelatinous precipitate formed when a small amount of alkali solution is added to a solution of an aluminium salt.

$$Al_{(aq)} + 3OH_{(aq)} \rightarrow Al(OH)_{3(s)}$$

It is an ◊amphoteric compound as it readily reacts with both acids and alkalis.

aluminium oxide or ***alumina*** Al_2O_3 white solid formed by heating aluminium hydroxide. It is an ◊amphoteric oxide, since it reacts readily with both acids and alkalis, and it is used as a refractory (furnace lining) and in column ◊chromatography.

alveolus (plural ***alveoli***) one of the many thousands of tiny air sacs in the ◊lungs in which exchange of oxygen and carbon dioxide takes place between air and the bloodstream.

Alzheimer's disease common manifestation of ◊dementia, thought to afflict one in 20 people over 65. After heart disease, cancer, and strokes it is the most common cause of death in the Western world. Attacking the brain's 'grey matter', it is a disease of mental processes rather than physical function, characterized by memory loss and progressive intellectual impairment. It was first

ALZHEIMER'S DISEASE WEB PAGE

http://med-www.bu.edu/Alzheimer/home.html

Although some of the pages on this site are aimed at the local community, there is much to interest those further afield – with information for families, caregivers, and investigators.

The Ageing Brain and Alzheimer's Disease

BY A D SMITH

'Omnia fert aetas, animum quoque; saepe ego longas cantando puerum memini me condere soles: nunc oblita mihi tot carmina ...'

'Time wastes all things, the mind too: often I remember how in boyhood I outwore long sunlit days in singing: now I have forgotten so many a song ...'

(Virgil, 'Eclogue IX', translation by Mackail)

Was Virgil right? Is it really true that dementia, a decline in our mental and cognitive functions illustrated here by a declining memory, is an inevitable part of ageing? If so, then we will be facing a very serious problem in the next 30 years because the world's population is ageing rapidly as advances in medicine, surgery and nutrition help to extend the human life span. The future demographic predictions are alarming. Between 1990 and 2030 the proportion of people over 60 will grow by about 75% in OECD countries, with a dramatic rise of almost 100% in Japan. It has been known for a long time that the incidence of dementia increases with age, so we can expect the proportion with dementia to increase by similar amount. Dementia, of which Alzheimer's disease is by far the most common cause, is a dreadful devastation of the human mind and spirit; it involves a progressive loss of those very features that make us human: the mind, normal behaviour and even our very personality disintegrates. Modern medicine has followed Virgil in the view that dementia is an inevitable consequence of the ageing of the brain. However, recent discoveries have cast serious doubt on this view and have led to a different idea, that dementia is a true disease, which, if its causes can be identified, should be possible to treat or even to prevent. In other words, to understand dementia we do not have to unravel the secrets of ageing but should consider it just like any other common disease such as cancer, heart disease or diabetes.

Alzheimer's disease (AD) was first described in a woman in her 50s by Alois Alzheimer in 1907. He examined the brain after she died and found in the microscope two kinds of structures that were stained by silver salts. One is called the amyloid plaque and lies outside the cells, while the other is called the neurofibrillary tangle and is found mainly within nerve cells in the cerebral cortex. Both these silver-staining structures are composed of highly insoluble proteins in a fibrillar form. The amyloid plaque is mainly composed of a protein called beta-amyloid peptide which has 40–43 amino acid residues per molecule but which is normally completely insoluble because the individual molecules aggregate together to form fibrils. The protein that makes up the neurofibrillary tangle is different and is derived from a protein called tau whose normal function is to stabilize the microtubules in the nerve's axon, so maintaining the function of this long process of the nerve cell.

Most research into dementia has concentrated on these two proteins and on the question: 'What causes them to accumulate in the brain in people with AD?' However, it has been realized that the deposition of these proteins might not be the primary cause of the symptoms of dementia; perhaps they are the end products of a process that is more closely related to the loss of normal functioning of the brain. This process is in fact the loss of nerve cells and, in particular, the loss of the connections between nerve cells that occur at synapses. The long nerve fibres in the cerebral cortex that pass from one part of the cortex to another die back in AD and so the different functional parts of the cortex are no longer connected to each other. The disease first strikes in the part of the brain that is known to be crucial for memory, the medial temporal lobe which lies deep in the middle of the brain. Nerve cells in some parts of the medial temporal lobe (the hippocampus) die earliest, so that more than 70% of them have disappeared by the time the patient dies. This process of continual death of nerve cells (neurodegeneration) can be followed in life by a simple brain scan, using computed tomography or magnetic resonance imaging. The loss of tissue in the medial temporal lobe occurs at 15% per year and correlates very well with the decline in cognitive function as measured by standard neuropsychological tests. No such rapid loss of tissue occurs with normal ageing and healthy individuals followed each year also show no obvious loss in cognitive function. It is likely that the finding that old people on average have a lower performance in some cognitive tests is mainly explained by the fact that any group of elderly people will include some in whom AD has already begun even though the symptoms are not readily detected.

The nerve cells of the medial temporal lobe are connected to almost all other parts of the cerebral cortex and must serve some kind of integrative role, such as the consolidation of memory. The loss of these connections as the neurons die may be one of the causes of the different symptoms of AD, which arise from abnormal functioning of different parts of the cortex. The death of nerve cells in the medial temporal lobe is accompanied by the death of another small group of nerve cells that lie in the basal nucleus of the forebrain. These basal forebrain neurons make the transmitter called acetylcholine and their main function is to regulate all parts of the cerebral cortex. The only available current treatments for AD act by inhibiting the enzyme that normally destroys acetylcholine, but they are not very effective and they do not slow down the progression of the disease. What is this disease process that strikes at nerve cells in the middle of the cerebral cortex and in the base of the forebrain? Why are these nerve cells particularly vulnerable? These are crucial questions for current research. If we can find the answers to these questions we should be able to devise drugs that slow down or stop the death of cells from occurring and so slow down the decline in cognitive function. There are several current views about what causes the nerve cells to die. One is that the cells are killed by the accumulation of beta-amyloid, and it has indeed been found that beta-amyloid peptide is toxic to nerve cells under certain circumstances. However, the paradox is that amyloid plaques occur most densely in parts of the brain where relatively few cells die. Nevertheless, several of the next generation of drugs being developed for AD are based on the view that if we can stop beta-amyloid being generated from its precursor protein (amyloid precursor protein) then we may be able to slow down progression of the disease. Another view is that the nerve cells die because they are 'strangled' from within by the accumulation of abnormal tau protein. Again, if ways can be found to prevent this from occurring then the disease might be treatable. A third view is that the nerve cells die because there is an excessive production of free radicals in the brain. Free radicals are molecules that have an extra electron and so they are very reactive and combine readily with many of the cell's important proteins. The consequence is that the protein's normal functioning is disturbed. Because free radicals are so reactive, they cannot readily be detected in brain tissue but the damaged proteins have been found in larger amounts than normal in the brains of people with AD. Even if we cannot identify the origin of these free radicals it might be possible to 'capture' them by administering drugs to which they combine in preference to the proteins of cells. Indeed, one such 'drug' is vitamin E and clinical trials are in progress to see whether feeding large amounts of vitamin E can slow down the progression of AD.

A major breakthrough in our understanding of AD was made at St Mary's Hospital Medical School, London, in 1991 where scientists discovered that one of the rare forms of familial AD (i.e. AD that is inherited in a classical Mendelian manner) is due to a mutation in the gene for amyloid precursor protein on chromosome 21. People with this mutation develop the symptoms of AD before the age of 60, so-called early-onset AD. Other forms of early-onset AD have also been found to be due to mutations in specific genes on chromosomes 1 and 14, but these familial forms of AD are quite rare, comprising about 5 % of all cases. It seems likely that these familial forms of AD are all due to abnormalities in the way the body handles the amyloid precursor protein, leading to excessive production of the toxic beta-amyloid. The common sporadic form of AD is usually later in onset, over the age of 65. Sporadic AD and familial AD show the same pathological changes in the brain and yet they must have very different causes. The causes of sporadic AD are now known to be multi-factorial, which means that several factors have to come together in one person before the disease develops. A discovery by scientists at Duke University, North Carolina in 1993 has transformed our understanding of AD; they found that people who carry a common variant (*E4*) of the gene for apolipoprotein E had a five-fold greater risk of developing AD than people who do not carry this variant. This is the first of many 'susceptibility genes' discovered for AD. A susceptibility gene only increases the risk of the disease developing; having this gene does not mean that the person will inevitably get AD. Clearly, other risk factors have to be present. As well as genetic risk factors, a variety of nongenetic factors may influence our risk of getting AD. A lot of current research is directed to identifying these risk factors because some of them may turn out to be modifiable by drugs, life style or diet. Some of these have obvious social and medical implications. Thus, if low education really is a risk factor, we must clearly do our best to give everyone the same high level of schooling. If dietary deficiencies are a risk then we must make sure that everyone has an adequate diet. The possibility of preventing AD from developing in a proportion of the elderly must be pursued with as much vigour as the search for drug treatments for those who are unfortunate enough to develop the disease. These are two of medicine's major challenges for the 21st century.

described by Alois Alzheimer 1906. It affects up to 4 million people in the USA and around 600,000 in Britain.

causes Various factors have been implicated in causing Alzheimer's disease including high levels of aluminium in drinking water and the presence in the brain of an abnormal protein, known as beta-amyloid.

In 1993 the gene coding for apolipoprotein (APOE) was implicated. US researchers established that people who carry a particular version of this gene (APOE-E4) are at greatly increased risk of developing the disease. It is estimated that one person in thirty carries this protein mutation; in the USA the figure is as high as 15%. The suspect gene can be detected with a test, so it is technically possible to identify those most at risk. As no cure is available such testing is unlikely to be widespread.

A second Alzheimer's gene was identified in 1997. The gene, HLA-A2, increases the speed at which the disease begins. The *Neurology* report stated that if a person had both HLA-A2 and APOE-E4, he or she may reach dementia a decade earlier than those with only one of the genes.

diagnosis US researchers began trialling a simple eye test in 1994 that could be used to diagnose sufferers. The drug tropicamide causes marked pupil dilation in those with the disease, and only slight dilation in healthy individuals.

treatment Some researchers are convinced that, whatever its cause, Alzheimer's disease is essentially an inflammatory condition, similar to rheumatoid arthritis. Although there is no cure, trials of anti-inflammatory drugs have shown promising results. Also under development are drugs which block the toxic effects of beta-amyloid. A 1996 study by US neuroscientists found that oestrogen skin patches were also beneficial in the treatment of female Alzheimer's patients, improving concentration and memory.

AM abbreviation for ◊amplitude modulation.

amalgam any alloy of mercury with other metals. Most metals will form amalgams, except iron and platinum. Amalgam is used in dentistry for filling teeth, and usually contains copper, silver, and zinc as the main alloying ingredients. This amalgam is pliable when first mixed and then sets hard, but the mercury leaches out and may cause a type of heavy-metal poisoning.

Amalgamation, the process of forming an amalgam, is a technique sometimes used to extract gold and silver from their ores. The ores are ground to a fine sand and brought into contact with mercury, which dissolves the gold and silver particles. The amalgam is then heated to distil the mercury, leaving a residue of silver and gold. The mercury is recovered and reused.

Almagamation to extract gold from its ore has been in use since Roman times.

amanita any of a group of fungi (see ◊fungus) distinguished by a ring (or volva) around the base of the stalk, warty patches on the cap, and the clear white colour of the gills. Many of the species are brightly coloured and highly poisonous. (Genus *Amanita*, family Agaricaceae.)

amatol explosive consisting of ammonium nitrate and TNT (trinitrotoluene) in almost any proportions.

amber fossilized ◊resin from coniferous trees of the Middle ◊Tertiary period. It is often washed ashore on the Baltic coast with plant and animal specimens preserved in it; many extinct species have been found preserved in this way. It ranges in colour from red to yellow, and is used to make jewellery.

When amber is rubbed with cloth, it attracts light objects, such as feathers. The effect, first noticed by the ancient Greeks, is due to acquisition of negative electric charge, hence the adaptation of the Greek word for amber, *elektron*, for electricity (see ◊static electricity).

Amber has been coveted for its supposed special properties since prehistoric times. Archaeologists have found amulets made of amber dating back as far as 35000 BC.

Amber's preservative properties were demonstrated 1992 when DNA was extracted from insects estimated to be around 30 million years old which were found fossilized in amber, and in 1995 US scientists succeeded in extracting bacterial spores from a bee in amber that was 40 million years old. Despite their lengthy dormancy, the bacterial spores were successfully germinated.

ambergris fatty substance, resembling wax, found in the stomach and intestines of the sperm ◊whale. It is found floating in warm seas, and is used in perfumery as a fixative.

Basically intestinal matter, ambergris is not the result of disease, but the product of an otherwise normal intestine. The name derives from the French *ambre gris* (grey amber).

American Association for the Advancement of Science (AAAS) US scientific society founded in 1848 with the aim of informing the public of the progress of science, and furthering scientific responsibility; its headquarters are in Washington, DC. It holds a large annual meeting, with many speakers describing the latest scientific developments, publishes journals, and sponsors awards.

It was modelled on the British Association for the Advancement of Science.

American National Standards Institute (ANSI) US national standards body. It sets official procedures in (among other areas) computing and electronics. The ANSI ◊character set is the standard set of characters used by Windows-based computers.

America Online (AOL) US market-leading commercial information service. America Online was launched 1986 with a bright, colourful graphical interface and a marketing campaign that issued free discs on almost every US magazine cover. In 1995 it overtook the then market leader, CompuServe, and by 1996 had more than 5 million users worldwide. America Online combined with the German publishing conglomerate Bertelsmann to launch a UK version of the service, known as AOL, in early 1996.

Because a number of America Online users, many of them using temporary accounts set up from the many free discs, acted in breach of ◊netiquette when America Online opened its Internet ◊gateway 1994, America Online users are held in contempt by many parts of the Net.

americium radioactive metallic element of the ◊actinide series, symbol Am, atomic number 95, relative atomic mass 243.13; it was first synthesized 1944. It occurs in nature in minute quantities in ◊pitchblende and other uranium ores, where it is produced from the decay of neutron-bombarded plutonium, and is the element with the highest atomic number that occurs in nature. It is synthesized in quantity only in nuclear reactors by the bombardment of plutonium with neutrons. Its longest-lived isotope is Am-243, with a half-life of 7,650 years.

The element was named by Glenn Seaborg, one of the team who first synthesized it in 1944, after the United States of America. Ten isotopes are known.

Ames Research Center US space-research (NASA) installation at Mountain View, California, USA, for the study of aeronautics and life sciences. It has managed the Pioneer series of planetary probes and is involved in the search for extraterrestrial life.

amethyst variety of ◊quartz, SiO_2, coloured violet by the presence of small quantities of impurities such as manganese or iron; used as a semiprecious stone. Amethysts are found chiefly in the Ural Mountains, India, the USA, Uruguay, and Brazil.

amide any organic chemical derived from a fatty acid by the replacement of the hydroxyl group (–OH) by an amino group ($-NH_2$).

One of the simplest amides is acetamide (CH_3CONH_2), which has a strong mousy odour.

Amiga microcomputer produced by US company Commodore 1985 to succeed the Commodore C64 home computer. The original Amiga was based on the Motorola 68000 microprocessor and achieved significant success in the domestic market.

Despite a failure to sell to the general business market, the latest versions of the Amiga are widely used in the film and video industries, where the Amiga's specialized graphics capabilities are used to create a variety of visual effects.

amine any of a class of organic chemical compounds in which one or more of the hydrogen atoms of ammonia (NH_3) have been replaced by other groups of atoms.

alanine $CH_3CH{\cdot}(NH_2){\cdot}COOH$

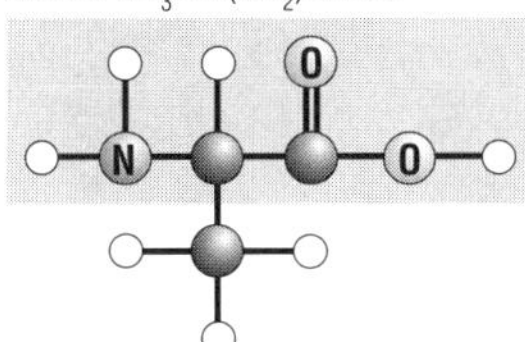

cysteine $SH{\cdot}CH_2CH{\cdot}(NH_2){\cdot}COOH$

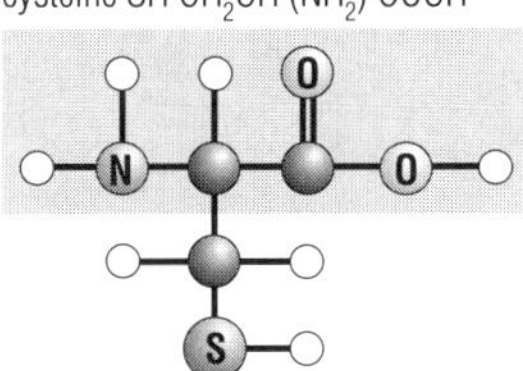

glycine NH_2CH_2COOH

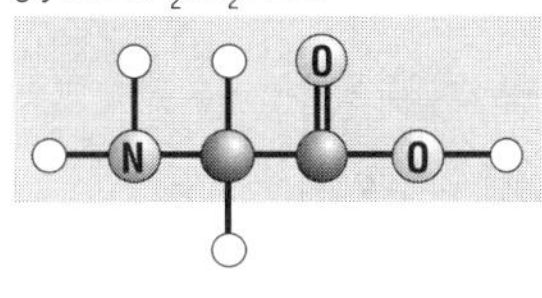

tyrosine $C_6H_4OH{\cdot}CH_2CH{\cdot}(NH_2){\cdot}COOH$

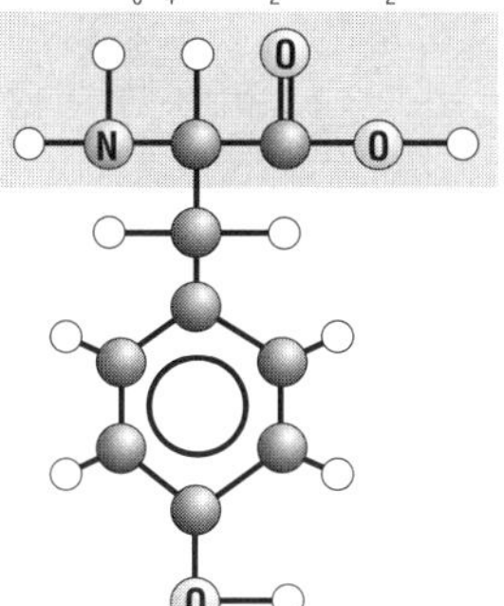

— covalent bond
○ hydrogen atom
carbon atom
(O) oxygen atom
(N) nitrogen atom
(S) sulphur atom

***amino acid** Amino acids are natural organic compounds that make up proteins and can thus be considered the basic molecules of life. There are 20 different common amino acids. They consist mainly of carbon, oxygen, hydrogen, and nitrogen. Each amino acid has a common core structure (consisting of two carbon atoms, two oxygen atoms, a nitrogen atom, and four hydrogen atoms) to which is attached a variable group, known as the R group. In glycine, the R group is a single hydrogen atom; in alanine, the R group consists of a carbon and three hydrogen atoms.*

Methyl amines have unpleasant ammonia odours and occur in decomposing fish. They are all gases at ordinary temperature.

Aromatic amine compounds include aniline, which is used in dyeing.

AMINO ACIDS

http://www.chemie.fu-berlin.de/chemistry/bio/amino-acids.html

Small but interesting site giving the names and chemical structures of all the amino acids.

amino acid water-soluble organic ◊molecule, mainly composed of carbon, oxygen, hydrogen, and nitrogen, containing both a basic amino group (NH_2) and an acidic carboxyl (COOH) group. They are small molecules able to pass through membranes. When two or more amino acids are joined together, they are known as ◊peptides; ◊proteins are made up of peptide chains folded or twisted in characteristic shapes.

Many different proteins are found in the cells of living organisms, but they are all made up of the same 20 amino acids, joined together in varying combinations (although other types of amino acid do occur infrequently in nature). Eight of these, the **essential amino acids**, cannot be synthesized by humans and must be obtained from the diet. Children need a further two amino acids that are not essential for adults. Other animals also need some preformed amino acids in their diet, but green plants can manufacture all the amino acids they need from simpler molecules, relying on energy from the Sun and minerals (including nitrates) from the soil.

To remember the ten essential amino acids:

THESE TEN VALUABLE AMINO ACIDS HAVE LONG PRESERVED LIFE IN MAN.

THREONINE / TRYPTOPHAN / VALINE / ARGENINE / HISTIDINE / LYSINE / PHENYLALANINE / LEUCINE / ISOLEUCINE / METHIONINE

ammeter instrument that measures electric current, usually in ◊amperes. The ammeter is placed in series with the component through which current is to be measured, and is constructed with a low internal resistance in order to prevent the reduction of that current as it flows through the instrument itself. A common type is the ◊moving-coil meter, which measures direct current (DC), but can, in the presence of a rectifier, measure alternating current (AC) also. Hot-wire, moving-iron and dynamometer ammeters can be used for both DC and AC.

ammonia NH_3 colourless, pungent-smelling gas, lighter than air and very soluble in water. It is made on an industrial scale by the ◊Haber (or Haber–Bosch) process, and used mainly to produce nitrogenous fertilizers, nitric acid, and some explosives.

In aquatic organisms and some insects, nitrogenous waste (from the breakdown of amino acids and so on) is excreted in the form of ammonia, rather than as urea in mammals.

ammoniacal solution in chemistry, a solution produced by dissolving a solute in aqueous ammonia.

ammonite extinct marine ◊cephalopod mollusc of the order Ammonoidea, related to the modern nautilus. The shell was curled in a plane spiral and made up of numerous gas-filled chambers, the outermost containing the body of the animal. Many species flourished between 200 million and 65 million years ago, ranging in size from that of a small coin to 2 m/6 ft across.

ammonium carbonate $(NH_4)_2CO_3$ white, crystalline solid that readily sublimes at room temperature into its constituent gases: ammonia, carbon dioxide, and water. It was formerly used in ◊smelling salts.

ammonium chloride or ***sal ammoniac*** NH_4Cl a volatile salt that forms white crystals around volcanic craters. It is prepared synthetically for use in 'dry-cell' batteries, fertilizers, and dyes.

ammonium nitrate NH_4NO_3 colourless, crystalline solid, prepared by ◊neutralization of nitric acid with ammonia; the salt is crystallized from the solution. It sublimes on heating.

amnesia loss or impairment of memory. As a clinical condition it may be caused by disease or injury to the brain, by some drugs, or by shock; in some cases it may be a symptom of an emotional disorder.

amniocentesis sampling the amniotic fluid surrounding a fetus in the womb for diagnostic purposes. It is used to detect Down's syn-

AMNIOCENTESIS

http://www.aomc.org/amnio.html#_wmh4_822004218

Comprehensive plain English guide to amniocentesis. For a pregnant woman considering the procedure, this is an invaluable source of information. The advantages and the risks of sampling the amniotic fluid are clearly presented.

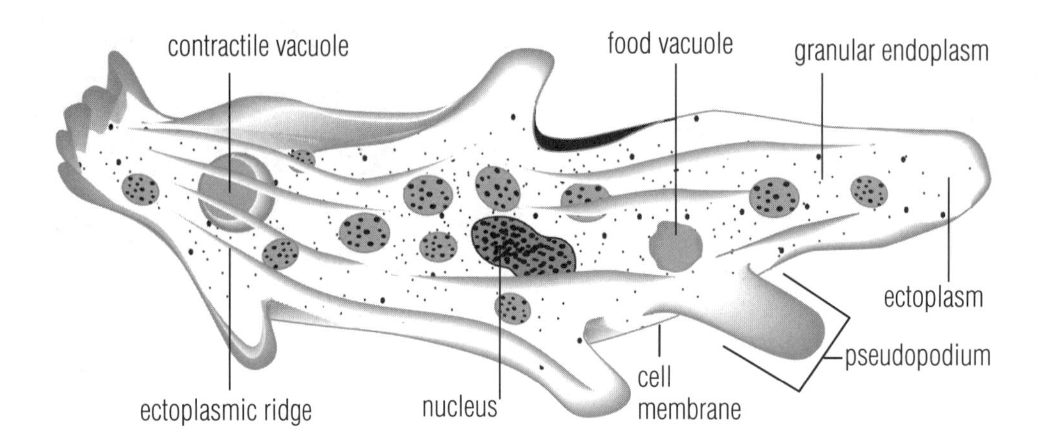

***amoeba** The amoebae are among the simplest living organisms, consisting of a single cell. Within the cell, there is a nucleus, which controls cell activity, and many other microscopic bodies and vacuoles (fluid-filled spaces surrounded by a membrane) with specialized functions. Amoebae eat by flowing around food particles, engulfing the particle, a process called phagocytosis.*

drome and other genetic abnormalities. The procedure carries a 1 in 200 risk of miscarriage.

amnion innermost of three membranes that enclose the embryo within the egg (reptiles and birds) or within the uterus (mammals). It contains the amniotic fluid that helps to cushion the embryo.

amoeba (plural ***amoebae***) one of the simplest living animals, consisting of a single cell and belonging to the ◊protozoa group. The body consists of colourless protoplasm. Its activities are controlled by the nucleus, and it feeds by flowing round and engulfing organic debris. It reproduces by ◊binary fission. Some species of amoeba are harmful parasites.

amp in physics, abbreviation for **ampere**, a unit of electrical current. Its use is deprecated.

ampere SI unit (symbol A) of electrical current. Electrical current is measured in a similar way to water current, in terms of an amount per unit time; one ampere represents a flow of about 6.28 $\times 10^{18}$ ◊electrons per second, or a rate of flow of charge of one coulomb per second.

The ampere is defined as the current that produces a specific magnetic force between two long, straight, parallel conductors placed 1m/3.3 ft apart in a vacuum. It is named after the French scientist André Ampère.

Ampère, André Marie
(1775–1836)

French physicist and mathematician who made many discoveries in electromagnetism and electrodynamics. He followed up the work of Hans Oersted on the interaction between magnets and electric currents, developing a rule for determining the direction of the magnetic field associated with an electric current. The unit of electric current, the **ampere**, is named after him.

Ampère's law is an equation that relates the magnetic force produced by two parallel current-carrying conductors to the product of their currents and the distance between the conductors. Today Ampère's law is usually stated in the form of calculus: the line integral of the magnetic field around an arbitrarily chosen path is proportional to the net electric current enclosed by the path.

Mary Evans Picture Library

Ampère's rule rule developed by French physicist André Ampère connecting the direction of an electric current and its associated magnetic currents. It states that if a person were travelling along a current-carrying wire in the direction of conventional current flow (from the positive to the negative terminal), and carrying a magnetic compass, then the north pole of the compass needle would be deflected to the left-hand side.

amphetamine or ***speed*** powerful synthetic ◊stimulant. Benzedrine was the earliest amphetamine marketed, used as a 'pep pill' in World War II to help soldiers overcome fatigue, and until the 1970s amphetamines were prescribed by doctors as an appetite suppressant for weight loss; as an antidepressant, to induce euphoria; and as a stimulant, to increase alertness.

Indications for its use today are very restricted because of severe side effects, including addiction. Amphetamine is a sulphate or phosphate form of $C_9H_{13}N$.

amphibian ***Greek 'double life'*** member of the vertebrate class Amphibia, which generally spend their larval (tadpole) stage in

***amphibian** Toads are among the commonest amphibians. This striking green toad* Bufo viridis *is distributed in a number of habitats in the E Mediterranean area, such as these coastal sand flats on the island of Corfu. Premaphotos Wildlife*

AMPHIBIAN

Frogs and toads never eat with their eyes open. When eating, they have to push down with the back of the eyeball in order to force food into the stomach.

fresh water, transferring to land at maturity (after ◊metamorphosis) and generally returning to water to breed. Like fish and reptiles, they continue to grow throughout life, and cannot maintain a temperature greatly differing from that of their environment. The class contains 4,553 known species, 4,000 of which are frogs and toads, 390 salamanders, and 163 caecilians (worm-like in appearance).

Amphioxus genus name of the ◊lancelet.

amphoteric term used to describe the ability of some chemical compounds to behave either as an ◊acid or as a ◊base depending on their environment. For example, the metals aluminium and zinc, and their oxides and hydroxides, act as bases in acidic solutions and as acids in alkaline solutions.

Amino acids and proteins are also amphoteric, as they contain both a basic (amino, $-NH_2$) and an acidic (carboxyl, $-COOH$) group.

amplifier electronic device that magnifies the strength of a signal, such as a radio signal. The ratio of output signal strength to input signal strength is called the **gain** of the amplifier. As well as achieving high gain, an amplifier should be free from distortion and able to operate over a range of frequencies. Practical amplifiers are usually complex circuits, although simple amplifiers can be built from single transistors or valves.

amplitude or ***argument*** in mathematics, the angle in an ◊Argand diagram between the line that represents the complex number and the real (positive horizontal) axis. If the complex number is written in the form $r(\cos\theta + i \sin\theta)$, where r is radius and $i = \sqrt{-1}$, the amplitude is the angle θ (theta). The amplitude is also the peak value of an oscillation.

amplitude maximum displacement of an oscillation from the equilibrium position. For a wave motion, it is the height of a crest (or the depth of a trough). With a sound wave, for example, amplitude corresponds to the intensity (loudness) of the sound. In AM (amplitude modulation) radio broadcasting, the required audio-frequency signal is made to modulate (vary slightly) the amplitude of a continuously transmitted radio carrier wave.

amplitude modulation (AM) method by which radio waves are altered for the transmission of broadcasting signals. AM waves are constant in frequency, but the amplitude of the transmitting wave varies in accordance with the signal being broadcast.

ampulla in the inner ◊ear, a slight swelling at the end of each semicircular canal, able to sense the motion of the head.

The sense of balance depends largely on sensitive hairs within the ampullae responding to movements of fluid within the canal.

amygdala almond-shaped region of the ◊brain adjacent to the hippocampus, that links the cortex, responsible for conscious thought, with the regions controlling emotions. A 1994 US study showed that it was involved in interpreting fear-provoking information and linking it to fear responses. For example, where the amygdala is damaged patients are unable to recognize fearful expressions.

Emotionally charged events are more easily recalled than neutral events, and in 1996 US researchers demonstrated a link between the amygdala and the memorizing of emotionally loaded images.

amyl alcohol former name for ◊pentanol.

amylase one of a group of ◊enzymes that break down starches into their component molecules (sugars) for use in the body. It occurs widely in both plants and animals. In humans, it is found in saliva and in pancreatic juices.

Human amylase has an optimum pH of 7.2–7.4. Like most enzymes amylase is denatured by temperatures above 60°C.

anabolic steroid any ◊hormone of the ◊steroid group that stimulates tissue growth. Its use in medicine is limited to the treatment of some anaemias and breast cancers; it may help to break up blood clots. Side effects include aggressive behaviour, masculinization in women, and, in children, reduced height.

anabolism process of building up body tissue, promoted by the influence of certain hormones. It is the constructive side of ◊metabolism, as opposed to ◊catabolism.

anaconda South American snake *Eunectes murinus*, a member of the python and boa family, the Boidae. One of the largest snakes, growing to 9 m/30 ft or more, it is found in and near water, where it lies in wait for the birds and animals on which it feeds. The anaconda is not venomous, but kills its prey by coiling round it and squeezing until the creature suffocates.

Females are up to 5 times larger than males. They have litters of up to 80 babies, born live, and each weighing only 250–300 g. The gestation period last six to eight months, during which time the female will not eat at all.

anaemia condition caused by a shortage of haemoglobin, the oxygen-carrying component of red blood cells. The main symptoms are fatigue, pallor, breathlessness, palpitations, and poor resistance to infection. Treatment depends on the cause.

Anaemia arises either from abnormal loss or defective production of haemoglobin. Excessive loss occurs, for instance, with chronic slow bleeding or with accelerated destruction (◊haemolysis) of red blood cells. Defective production may be due to iron deficiency, vitamin B_{12} deficiency (pernicious anaemia), certain blood diseases (sickle-cell disease and thalassaemia), chronic infection, kidney disease, or certain kinds of poisoning. Untreated anaemia taxes the heart and may prove fatal.

anaerobic (of living organisms) not requiring oxygen for the release of energy from food molecules such as glucose. Anaerobic organisms include many bacteria, yeasts, and internal parasites.

Obligate anaerobes, such as certain primitive bacteria, cannot function in the presence of oxygen; but **facultative anaerobes**, like the fermenting yeasts and most bacteria, can function with or without oxygen. Anaerobic organisms release much less of the available energy from their food than do ◊aerobic organisms.

anaesthetic drug that produces loss of sensation or consciousness; the resulting state is **anaesthesia**, in which the patient is insensitive to stimuli. Anaesthesia may also happen as a result of nerve disorder.

Ever since the first successful operation in 1846 on a patient rendered unconscious by ether, advances have been aimed at increasing safety and control. Sedatives may be given before the anaesthetic to make the process easier. The level and duration of unconsciousness are managed precisely. Where general anaesthesia may be inappropriate (for example, in childbirth, for a small procedure, or in the elderly), many other techniques are available. A topical substance may be applied to the skin or tissue surface; a local agent may be injected into the tissues under the skin in the area to be treated; or a regional block of sensation may be achieved by injection into a nerve. Spinal anaesthetic, such as epidural, is injected into the tissues surrounding the spinal cord, producing loss of feeling in the lower part of the body.

ANAESTHETIC

Although the science of anaesthetics is 150 years old, it is only comparatively recently that safe doses have begun to be calculated correctly. It is said that when the Japanese bombed Pearl Harbor in 1941, more US servicemen were killed by anaesthetists than by bombs.

History of Anaesthesia

BY PAULETTE PRATT

While the need for some form of anaesthesia had been recognized since earliest times, for centuries all that was available to deaden the pain of surgery was a copious draft of alcohol. As late as 1839 a leading French surgeon predicted: 'To escape pain in surgical operations is a chimera which we cannot expect in our time.' Mercifully, he was mistaken for the first use of anaesthesia in surgery was recorded only seven years later.

By this time there had been some experimentation with volatile agents such as ether, chloroform and nitrous oxide. Ether, for example, enjoyed a certain notoriety arising from medical students' 'ether frolics'. But this was the substance used in what is credited as the first operation undertaken in an anaesthetized patient – in the famous 'ether dome' at Massachusetts General Hospital in Boston on October 16, 1846. Its use was demonstrated by the dentist William Thomas Green Morton (1819–68).

The following day, not content with the apparatus he had used to administer the ether, Morton commissioned an improved version. Later he abandoned this, too, in favour of a simple sponge. 'This should be about the size of the open hand, or a little larger, and concave, to suit over the nose and mouth,' he wrote in the *Lancet* (June 30, 1847).

'The sponge is then thoroughly saturated with ether, applied to the nose and mouth, and, with the latter open, the patient directed to inhale as fully and freely as possible. In this way, I have found the result more sure and satisfactory, and the difficulty of inhalation very much reduced, or entirely removed. The most delicate or nervous females, or aged persons, as well as young children, are thus rapidly and almost imperceptibly narcotized...'

Only a month after its Boston debut, ether anaesthesia was first used in Britain, at London's University College Hospital. In January 1847, James Young Simpson (1811–70), Professor of Obstetrics at Edinburgh, was the first to introduce its use for women in childbirth, initially for forceps or operative deliveries but soon also for uncomplicated cases.

However, disliking ether's 'disagreeable and very persistent smell', Simpson soon began casting around for an alternative and decided to try chloroform. In November 1847, to the consternation of his household, he and two junior colleagues were accidentally exposed to chloroform when a bottle was knocked over at his house in Queen Street. He later wrote: 'The first night we took chloroform, Dr Duncan, Dr Keith and I all tried it simultaneously and were under the table in a minute or two.'

The first obstetric delivery under chloroform took place the day after this mishap and, ten days later, Simpson published his first pamphlet on its use. A forceful figure, he insisted that liberal administration of chloroform on the corner of a towel or handkerchief was harmless.

Not surpisingly, however, since it had gone from first testing to clinical use within the space of little more than a week, the safety of chloroform as an anaesthetic agent was never established. In fact the first UK death from chloroform anaesthesia took place only three months after its introduction.

Elsewhere, once its toxicity became apparent, chloroform fell out of favour. Predictably its use was banned at Massachusetts General, the birthplace of ether anaesthesia. But in Britain Simpson had done such a good job of publicizing chloroform that its popularity grew regardless. Here controversy centred not on safety but on the morality of using anaesthesia at all to relieve the pain of childbirth. There were those who pointed to its apparent biblical proscription – 'In pain shalt thou bear offspring.'

However, the role of anaesthesia in childbirth was confirmed when, on April 7 1853, the well-known London physician John Snow held out a chloroform-soaked handkerchief to assist the birth of Queen Victoria's ninth child, Prince Leopold. The Queen subsequently noted in her diary: 'Dr Snow administered that blessed chloroform and the effect was soothing, quieting and delightful beyond measure.'

Despite a rising death toll (and the appointment of various commissions to investigate the action of chloroform), the drug remained in use until well into the new century.

Meanwhile a third volatile agent, nitrous oxide, had surfaced at around the same time as ether and chloroform. Its pain-killing properties were discovered in 1800 by the British chemist Humphry Davy (1778–1829), who foresaw that 'it may probably be used with advantage in surgical operations'. In 1845, nitrous oxide was first used for pain relief, also at Massachusetts General Hospital, in what proved to be a disastrous demonstration by another dentist, Horace Wells (1815–48). Then it was lost to view for more than two decades.

Nitrous oxide – popularly known as 'laughing gas' because of the excitement it induces – was finally adopted as an anaesthetic agent by way of its use by travelling salesmen. One of these, Gardner Quincy Colton (1814–98), founder of the New York Dental Institute, demonstrated use of the gas at the 1867 International Exhibition in Paris. It was taken up by an influential US dental surgeon, Thomas Wiltberger (1823–97), who in due course introduced it to British dentists. Nitrous oxide is still used as an inhalation agent in surgery today.

There were to be further colourful chapters in the development of safe, effective anaesthesia, including the loss of a few surgical patients to exploding gases. Also memorable was the era of 'twilight sleep' in obstetrics, when a light state of scopolamine–morphine narcosis was used to ease pain and eradicate unpleasant recollections. The main effect of this, in addition to prolonging labour and expunging any memory on the part of the mother of her child's birth, was to produce babies almost too dopey to breathe.

While some specialist training was initiated in the 1930s, it was only after World War 2 – a century after the first use of anaesthesia in surgery – that the speciality really came into its own. Today it is the biggest UK medical speciality, practised with a precision unimaginable in the early days under the 'ether dome'.

analgesic agent for relieving ◊pain. Opiates alter the perception or appreciation of pain and are effective in controlling 'deep' visceral (internal) pain. Non-opiates, such as ◊aspirin, ◊paracetamol, and NSAIDs (nonsteroidal anti-inflammatory drugs), relieve musculoskeletal pain and reduce inflammation in soft tissues.

Pain is felt when electrical stimuli travel along a nerve pathway, from peripheral nerve fibres to the brain via the spinal cord.

An anaesthetic agent acts either by preventing stimuli from being sent (local), or by removing awareness of them (general). Analgesic drugs act on both.

Temporary or permanent analgesia may be achieved by injection of an anaesthetic agent into, or the severing of, a nerve. Implanted devices enable patients to deliver controlled electrical stimulation to block pain impulses. Production of the body's natural opiates, ◊endorphins, can be manipulated by techniques such as relaxation and biofeedback. However, for the severe pain of, for example, terminal cancer, opiate analgesics are required.

US researchers found 1996 that some painkillers were more effective and provided longer-lasting relief for women than men.

analogous in biology, term describing a structure that has a similar function to a structure in another organism, but not a similar evolutionary path. For example, the wings of bees and of birds have the same purpose – to give powered flight – but have different origins.

Compare ◊homologous.

analogue (of a quantity or device) changing continuously; by contrast a ◊digital quantity or device varies in series of distinct steps. For example, an analogue clock measures time by means of a continuous movement of hands around a dial, whereas a digital clock measures time with a numerical display that changes in a series of discrete steps.

Most computers are digital devices. Therefore, any signals and data from an analogue device must be passed through a suitable ◊analogue-to-digital converter before they can be received and processed by computer. Similarly, output signals from digital computers must be passed through a digital-to-analogue converter before they can be received by an analogue device.

analogue computer computing device that performs calculations through the interaction of continuously varying physical quantities, such as voltages (as distinct from the more common ◊digital computer, which works with discrete quantities). An analogue computer is said to operate in real time (corresponding to time in the real world), and can therefore be used to monitor and control other events as they happen.

Although common in engineering since the 1920s, analogue computers are not general-purpose computers, but specialize in solving ◊differential calculus and similar mathematical problems. The earliest analogue computing device is thought to be the flat, or planispheric, astrolabe, which originated in about the 8th century.

analogue signal in electronics, current or voltage that conveys or stores information, and varies continuously in the same way as the information it represents (compare ◊digital signal). Analogue signals are prone to interference and distortion.

The bumps in the grooves of a vinyl record form a mechanical analogue of the sound information stored, which is then is converted into an electrical analogue signal by the record player's pick-up device.

analogue-to-digital converter (ADC) electronic circuit that converts an analogue signal into a digital one. Such a circuit is needed to convert the signal from an analogue device into a digital signal for input into a computer. For example, many ◊sensors designed to measure physical quantities, such as temperature and pressure, produce an analogue signal in the form of voltage and this must be passed through an ADC before computer input and processing. A ◊digital-to-analogue converter performs the opposite process.

analogy in mathematics and logic, a form of argument or process of reasoning from one case to another parallel case. Arguments from analogy generally have the following form: if some event or thing has the properties *a* and *b*, and if another event or thing has the properties *b* and *c*, then the former event or thing has the property *c*, too. Arguments from analogy are not always sound and can mislead. False analogies arise when the cases are insufficiently similar to support the reasoning. For example, a whale lives in water and resembles a fish, but we cannot conclude from this that it is a fish. When arguments from analogy are compressed, they are called metaphors.

analysis in chemistry, the determination of the composition of substances; see ◊analytical chemistry.

analysis branch of mathematics concerned with limiting processes on axiomatic number systems; ◊calculus of variations and infinitesimal calculus is now called analysis.

analyst job classification for ◊computer personnel. An analyst prepares a report on an existing data processing system and makes proposals for changes and improvements.

ANALYTICAL CHEMISTRY BASICS

http://www.scimedia.com/chem-ed/analytic/ac-basic.htm

Detailed online course, designed for those at undergraduate level, that provides the user with an introduction to some of the fundamental concepts and methods of analytical chemistry. Some of the sections included are gravimetric analysis, titration and spectroscopy.

analytical chemistry branch of chemistry that deals with the determination of the chemical composition of substances. **Qualitative analysis** determines the identities of the substances in a given sample; **quantitative analysis** determines how much of a particular substance is present.

Simple qualitative techniques exploit the specific, easily observable properties of elements or compounds – for example, the flame test makes use of the different flame colours produced by metal cations when their compounds are held in a hot flame. More sophisticated methods, such as those of ◊spectroscopy, are required where substances are present in very low concentrations or where several substances have similar properties.

Most quantitative analyses involve initial stages in which the substance to be measured is extracted from the test sample, and purified. The final analytical stages (or 'finishes') may involve measurement of the substance's mass (gravimetry) or volume (volumetry, titrimetry), or a number of techniques initially developed for qualitative analysis, such as fluorescence and absorption spectroscopy, chromatography, electrophoresis, and polarography. Many modern methods enable quantification by means of a detecting device that is integrated into the extraction procedure (as in gas–liquid chromatography).

For him [the scientist], truth is so seldom the sudden light that shows new order and beauty; more often, truth is the uncharted rock that sinks his ship in the dark.

JOHN CORNFORTH Australian chemist. Nobel prize address 1975

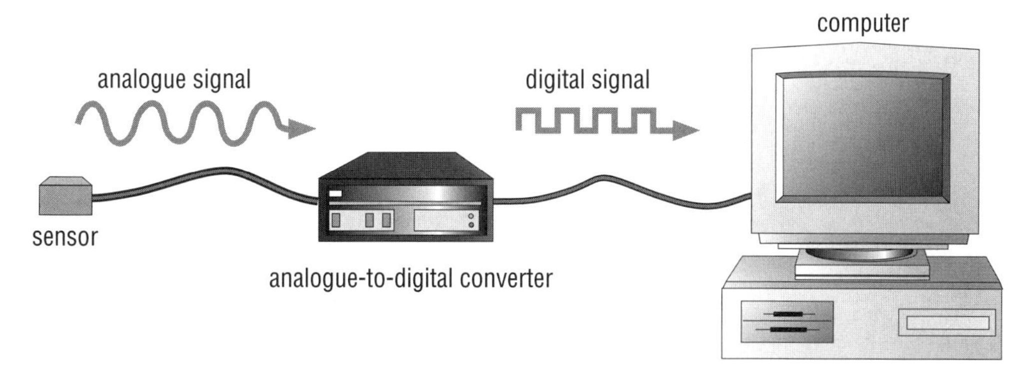

analogue-to-digital converter *An analogue-to-digital converter, or ADC, converts a continuous analogue signal produced by a sensor to a digital ('off and on') signal for computer processing.*

analytical engine programmable computing device designed by English mathematician Charles ◊Babbage in 1833.

It was based on the ◊difference engine but was intended to automate the whole process of calculation. It introduced many of the concepts of the digital computer but, because of limitations in manufacturing processes, was never built.

Among the concepts introduced were input and output, an arithmetic unit, memory, sequential operation, and the ability to make decisions based on data. It would have required at least 50,000 moving parts. The design was largely forgotten until some of Babbage's writings were rediscovered in 1937.

The whole of the developments and operations of analysis are now capable of being executed by machinery. ... As soon as an Analytical Engine exists, it will necessarily guide the future course of science.

CHARLES BABBAGE English mathematician. *Passages from the Life of a Philosopher* 1864

analytical geometry another name for ◊coordinate geometry.

anamorphic projection technique used in film and in ◊virtual reality to squeeze wide-frame images so that they fit into the dimensions of a 35-mm frame of film. In film projection, the projector has a complementary lens which reverses the process. In virtual reality, the computer must calculate the amount of deformation and reverse it.

anatomy study of the structure of the body and its component parts, especially the ◊human body, as distinguished from physiology, which is the study of bodily functions.

Herophilus of Chalcedon (c. 330–c. 260 BC) is regarded as the founder of anatomy. In the 2nd century AD, the Graeco-Roman physician Galen produced an account of anatomy that was the only source of anatomical knowledge until *On the Working of the Human Body* 1543 by Belgian physician Andreas Vesalius. In 1628, English physician William Harvey published his demonstration of the circulation of the blood. With the invention of the microscope, Italian physiologist Marcello Malpighi and Dutch microscopist Anton van Leeuwenhoek were able to found the study of ◊histology. In 1747, Albinus (1697–1770), with the help of the artist Wandelaar (1691–1759), produced the most exact account of the bones and muscles, and in 1757–65 Swiss biologist Albrecht von Haller gave the most complete and exact description of the organs that had yet appeared. Among the anatomical writers of the early 19th century are the surgeon Charles Bell (1774–1842), Jonas Quain (1796–1865), and Henry Gray (1825–1861). Radiographic anatomy (using X-rays; see ◊radiography) has been one of the triumphs of the 20th century, which has also been marked by immense activity in embryological investigation.

anchor in computing, an HTML (hypertext markup language) tag that turns ordinary text into a ◊hyperlink. Anchors are used to enable easy navigation within a single large document or to link to remote documents on distant computers. On the World Wide Web, anchor text is underlined, coloured differently, or surrounded by a dotted line in order to mark it out from normal text.

anchovy small fish *Engraulis encrasicholus* of the ◊herring family. It is fished extensively, being abundant in the Mediterranean, and is also found on the Atlantic coast of Europe and in the Black Sea. It grows to 20 cm/8 in.

Pungently flavoured, it is processed into fish pastes and essences, and used as a garnish, rather than eaten fresh.

andesite volcanic igneous rock, intermediate in silica content between rhyolite and basalt. It is characterized by a large quantity of feldspar ◊minerals, giving it a light colour. Andesite erupts from volcanoes at destructive plate margins (where one plate of the Earth's surface moves beneath another; see ◊plate tectonics), including the Andes, from which it gets its name.

AND gate in electronics, a type of ◊logic gate.

Anderson, Elizabeth Garrett (1836–1917)

English physician, the first English woman to qualify in medicine. Unable to attend medical school, Anderson studied privately and was licensed by the Society of Apothecaries in London in 1865. She was physician to the Marylebone Dispensary for Women and Children (later renamed the Elizabeth Garrett Anderson Hospital), a London hospital now staffed by women and serving female patients.

She became the first woman member of the British Medical Association in 1873 and the first woman mayor in Britain. She lectured at the London School of Medicine for Women (1875–97), and was its dean (1883–1903).

Mary Evans Picture Library

androecium male part of a flower, comprising a number of ◊stamens.

androgen general name for any male sex hormone, of which ◊testosterone is the most important.

They are all ◊steroids and are principally involved in the production of male ◊secondary sexual characteristics (such as beard growth).

Andromeda major constellation of the northern hemisphere, visible in autumn. Its main feature is the Andromeda galaxy. The star Alpha Andromedae forms one corner of the Square of Pegasus. It is named after the princess of Greek mythology.

Andromeda galaxy galaxy 2.2 million light years away from Earth in the constellation Andromeda, and the most distant object visible to the naked eye. It is the largest member of the ◊Local Group of galaxies.

Like the Milky Way, it is a spiral orbited by several companion galaxies but contains about twice as many stars. It is about 200,000 light years across.

AND rule rule used for finding the combined probability of two or more independent events both occurring. If two events E_1 and E_2 are independent (have no effect on each other) and the probabilities of their taking place are p_1 and p_2, respectively, then the combined probability p that both E_1 and E_2 will happen is given by:

$$p = p_1 \times p_2$$

For example, if a blue die and a red die are thrown together, the probability of a blue six is $\frac{1}{6}$, and the probability of a red six is $\frac{1}{6}$. Therefore, the probability of both a red six and a blue six being thrown is $\frac{1}{6} \times \frac{1}{6} = \frac{1}{36}$.

By contrast, the **OR rule** is used for finding the probability of either one event or another taking place.

anechoic chamber room designed to be of high sound absorbency. All surfaces inside the chamber are covered by sound-absorbent materials such as rubber. The walls are often covered with inward-facing pyramids of rubber, to minimize reflections. It is used for experiments in ◊acoustics and for testing audio equipment.

anemometer device for measuring wind speed and liquid flow. The most basic form, the **cup-type anemometer**, consists of cups at the ends of arms, which rotate when the wind blows. The speed of rotation indicates the wind speed.

Vane-type anemometers have vanes, like a small windmill or propeller, that rotate when the wind blows. **Pressure-tube anemometers** use the pressure generated by the wind to indicate speed. The wind blowing into or across a tube develops a pressure, proportional to the wind speed, that is measured by a manometer or pressure gauge. **Hot-wire anemometers** work on the principle that the rate at which heat is transferred from a hot wire to the surrounding air is a measure of the air speed. Wind speed is determined by measuring either the electric current required to maintain a hot wire at a constant temperature, or the variation of resistance while a constant current is maintained.

ANEMONE

http://www.actwin.com/fish/species/anemone.html

Article on keeping anemones in an aquarium environment. The site offers information on species, feeding, and tank maintenance practices to help these creatures thrive in captivity.

anemone flowering plant belonging to the buttercup family, found in northern temperate regions, mainly in woodland. It has ◊sepals which are coloured to attract insects. (Genus *Anemone*, family Ranunculaceae.)

The garden anemone (*Anemone coronaria*) is blue, purple, red, or white. The European and Asian white wood anemone (*A. nemorosa*), or windflower, grows in shady woods, flowering in spring. *Hepatica nobilis*, once included within the genus *Anemone*, is common in the Alps. The ◊pasqueflower is now placed in a separate genus.

anemophily *A familiar example of anemophily (wind pollination) is provided by the silver birch* Betula pendula, *whose pendant male catkins produce vast amounts of pollen. The much slimmer and shorter female catkins point upwards.* *Premaphotos Wildlife*

anemophily type of ◊pollination in which the pollen is carried on the wind. Anemophilous flowers are usually unscented, have either very reduced petals and sepals or lack them altogether, and do not produce nectar. In some species they are borne in ◊catkins. Male and female reproductive structures are commonly found in separate flowers. The male flowers have numerous exposed stamens, often on long filaments; the female flowers have long, often branched, feathery stigmas.

aneroid barometer kind of ◊barometer.

aneurysm weakening in the wall of an artery, causing it to balloon outwards with the risk of rupture and serious, often fatal, blood loss. If detected in time, some accessible aneurysms can be repaired by bypass surgery, but such major surgery carries a high risk for patients in poor health.

angel dust popular name for the anaesthetic **phencyclidine**, a depressant drug.

angelfish any of a number of unrelated fish. The freshwater **angelfish**, genus *Pterophyllum*, of South America, is a tall, side-to-side flattened fish with a striped body, up to 26 cm/10 in long, but usually smaller in captivity. The **angelfish** or **monkfish** of the genus *Squatina* is a bottom-living shark up to 1.8 m/6 ft long with a body flattened from top to bottom. The **marine angelfishes**, *Pomacanthus* and others, are long narrow-bodied fish with spiny fins, often brilliantly coloured, up to 60 cm/2 ft long, living around coral reefs in the tropics.

ANGELFISH

http://www.actwin.com/fish/species/angelfish.html

Angelfish page by Cindy Hawley. This site includes details of angelfish breeds, feeding, keeping in an aquarium, and possible diseases. Images and multimedia clips are also downloadable from this site.

angelica any of a group of tall, perennial herbs with divided leaves and clusters of white or greenish flowers, belonging to the carrot family. Most are found in Europe and Asia. The roots and fruits have long been used in cooking and in medicine. (Genus *Angelica*, family Umbelliferae.)

angina or ***angina pectoris*** severe pain in the chest due to impaired blood supply to the heart muscle because a coronary artery is narrowed. Faintness and difficulty in breathing accompany the pain. Treatment is by drugs or bypass surgery.

angiosperm flowering plant in which the seeds are enclosed within an ovary, which ripens into a fruit. Angiosperms are divided into ◊monocotyledons (single seed leaf in the embryo) and ◊dicotyledons (two seed leaves in the embryo). They include the majority of flowers, herbs, grasses, and trees except conifers.

There are over 250,000 different species of angiosperm, found in a wide range of habitats. Like ◊gymnosperms, they are seed plants, but differ in that ovules and seeds are protected within the

ANGIOSPERM ANATOMY

http://www.botany.uwc.ac.za:80/sci_ed/std8/anatomy/

Good general guide to angiosperms. The differences between monocotyledons and dicotyledons are set out here. The functions of roots, stems, leaves, and flowers are explained by readily understandable text and good accompanying diagrams.

carpel. Fertilization occurs by male gametes passing into the ovary from a pollen tube. After fertilization the ovule develops into the seed while the ovary wall develops into the fruit.

Angkor site of the ancient capital of the Khmer Empire in NW Cambodia, north of Tonle Sap. The remains date mainly from the 10th–12th centuries AD, and comprise temples originally dedicated to the Hindu gods, shrines associated with Theravāda Buddhism, and royal palaces. Many are grouped within the enclosure called **Angkor Thom**, but the great temple of **Angkor Wat** (early 12th century) lies outside.

Angkor was abandoned in the 15th century, and the ruins were overgrown by jungle and not adequately described until 1863. Buildings on the site suffered damage during the civil war 1970–75; restoration work is in progress.

angle in mathematics, the amount of turn or rotation; it may be defined by a pair of rays (half-lines) that share a common endpoint but do not lie on the same line. Angles are measured in ◊degrees (°) or ◊radians (rads) – a complete turn or circle being 360° or 2π rads.

Angles are classified generally by their degree measures: **acute angles** are less than 90°; **right angles** are exactly 90° (a quarter turn); **obtuse angles** are greater than 90° but less than 180°; **reflex angles** are greater than 180° but less than 360°.

<> in documentation, brackets that indicate places where the user should input information of the type described between the brackets. Angled brackets are also used in online services and on the Internet to indicate that the name used is a user-ID rather than a real name, and on CompuServe as part of certain ◊emoticons.

<g> on CompuServe, an indicator that the message writer is smiling. The ‘g’, which stands for ‘grin’, is similar to an ◊emoticon in that it helps to identify an online writer's state of mind in the absence of facial expressions, tone of voice, and other real-world clues. Variants include <vbg> for ‘very big grin’.

angle of declination angle at a particular point on the Earth's surface between the direction of the true or geographic North Pole and the magnetic north pole. The angle of declination has varied over time because of the slow drift in the position of the magnetic north pole.

angle of dip or ***angle of inclination*** angle at a particular point on the Earth's surface between the direction of the Earth's magnetic field and the horizontal; see ◊magnetic dip.

angle of incidence angle between a ray of light striking a mirror (incident ray) and the normal to that mirror. It is equal to the ◊angle of reflection.

angle of reflection angle between a ray of light reflected from a mirror and the normal to that mirror. It is equal to the ◊angle of incidence.

angle of refraction angle between a refracted ray of light and the normal to the surface at which ◊refraction occurred. When a ray passes from air into a denser medium such as glass, it is bent towards the normal so that the angle of refraction is less than the ◊angle of incidence.

angler any of an order of fish Lophiiformes, with flattened body and broad head and jaws. Many species have small, plantlike tufts on their skin. These act as camouflage for the fish as it waits, either

ANGLER

The parasitic male anglerfish stays with his partner for life – literally. Once the tiny male is attached to the much larger female, his tissues fuse with hers, he loses his senses of sight and smell, and receives his nourishment from her bloodstream.

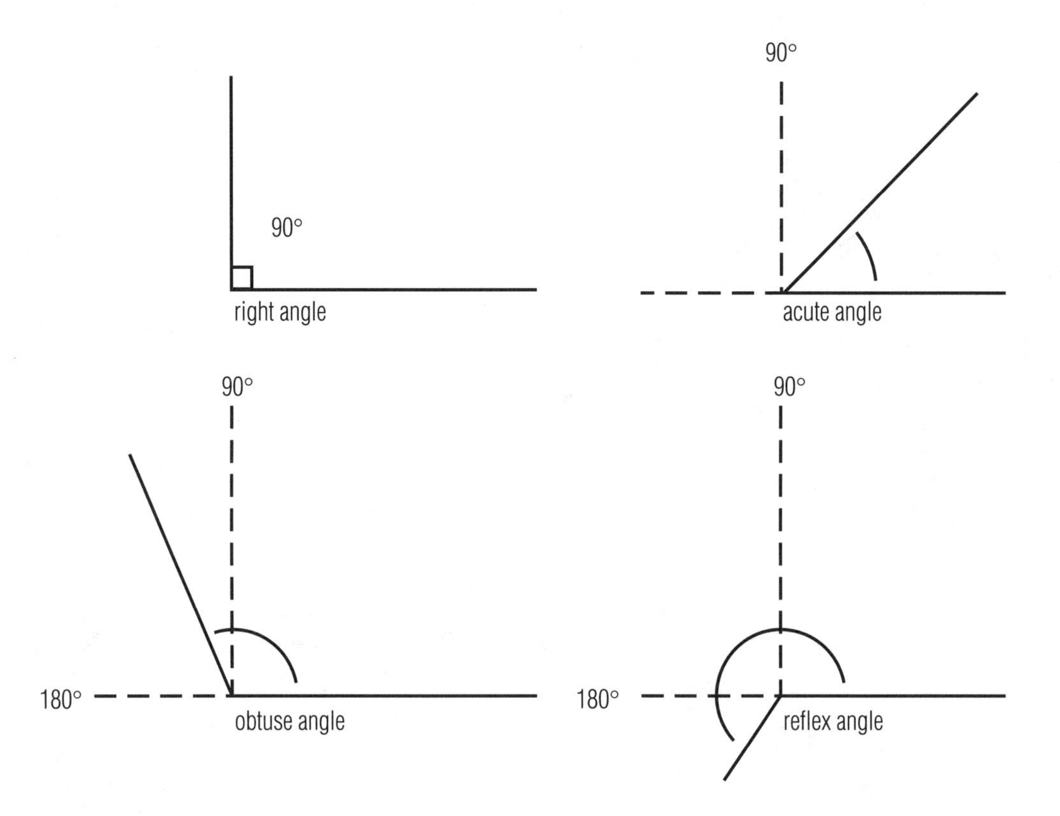

angle The four types of angle, as classified by their degree measures. No angle is classified as having a measure of 180°, as by definition such an ‘angle’ is actually a straight line.

floating among seaweed or lying on the sea bottom, twitching the enlarged tip of the threadlike first ray of its dorsal fin to entice prey.

There are over 200 species of angler fish, living in both deep and shallow water in temperate and tropical seas. The males of some species have become so small that they live as parasites on the females.

Anglo-Australian Telescope large telescope on ◊Siding Spring Mountain, New South Wales.

angstrom unit (symbol Å) of length equal to 10^{-10} metres or one-ten-millionth of a millimetre, used for atomic measurements and the wavelengths of electromagnetic radiation.

It is named after the Swedish scientist A J Ångström.

Anguilliformes order of bony fish comprising the ◊eels. All Anguilliformes have snakelike bodies and no pelvic fins. The order includes freshwater eels (Anguillidae), moray eels (Muraenidae), and conger eels (Congridae).

Anguis genus of legless lizards containing the ◊slow-worm.
classification *Anguis* is in family Anguidae, suborder Sauria,order Squamata, class Reptilia.

angular momentum in physics, a type of ◊momentum.

anhydride chemical compound obtained by the removal of water from another compound; usually a dehydrated acid. For example, sulphur(VI) oxide (sulphur trioxide, SO_3) is the anhydride of sulphuric acid (H_2SO_4).

anhydrite naturally occurring anhydrous calcium sulphate ($CaSO_4$).

It is used commercially for the manufacture of plaster of Paris and builders' plaster.

anhydrous of a chemical compound, containing no water. If the water of crystallization is removed from blue crystals of copper(II) sulphate, a white powder (anhydrous copper sulphate) results. Liquids from which all traces of water have been removed are also described as being anhydrous.

aniline ***Portuguese*** *anil* ***'indigo'*** $C_6H_5NH_2$ or ***phenylamine*** one of the simplest aromatic chemicals (a substance related to benzene, with its carbon atoms joined in a ring). When pure, it is a colourless oily liquid; it has a characteristic odour, and turns brown on contact with air. It occurs in coal tar, and is used in the rubber industry and to make drugs and dyes.

It is highly poisonous.

Aniline was discovered 1826, and was originally prepared by the dry distillation of ◊indigo, hence its name.

What is man without the beasts? If all the beasts were gone, man would die from a great loneliness of spirit.

CHIEF SEATTLE Native American chief.
Reputed letter to US President Franklin Pierce 1854, shown 1992 to have been largely a forgery created 1971 by TV scriptwriter Ted Perry

animal or ***metazoan Latin*** *anima* ***'breath', 'life'*** member of the ◊kingdom Animalia, one of the major categories of living things, the science of which is **zoology**. Animals are all ◊heterotrophs (they obtain their energy from organic substances produced by other organisms); they have eukaryotic cells (the genetic material is contained within a distinct nucleus) bounded by a thin cell membrane rather than the thick cell wall of plants. Most animals are capable of moving around for at least part of their life cycle.

In the past, it was common to include the single-celled ◊protozoa with the animals, but these are now classified as protists, together with single-celled plants. Thus all animals are multicellular. The oldest land animals known date back 440 million years. Their remains were found 1990 in a sandstone deposit near Ludlow, Shropshire, UK, and included fragments of two centipedes a few centimetres long and a primitive spider measuring about 1 mm.

Speeds of Animals

Animal	*Speed*	
	kph	*mph*
Cheetah	103	64
Wildebeest	98	61
Lion	81	50
Quarterhorse	76	47.5
Elk	72	45
Cape hunting dog	72	45
Coyote	69	43
Grey fox	68	42
Hyena	64	40
Zebra	64	40
Greyhound	63	39
Whippet	57	35.5
Rabbit (domestic)	56	35
Jackal	56	35
Reindeer	51	32
Giraffe	51	32
White-tailed deer	48	30
Wart hog	48	30
Grizzly bear	48	30
Cat (domestic)	48	30
Human	45	28
Elephant	40	25
Black mamba snake	32	20
Squirrel	19	12
Pig (domestic)	18	11
Chicken	14	9
Giant tortoise	0.27	0.17
Three-toed sloth	0.24	0.15
Garden snail	0.05	0.03

animal behaviour scientific study of the behaviour of animals, either by comparative psychologists (with an interest mainly in the psychological processes involved in the control of behaviour) or by ethologists (with an interest in the biological context and relevance of behaviour; see ◊ethology).

animal, domestic in general, a tame animal. In agriculture, it is an animal brought under human control for exploitation of its labour; use of its feathers, hide, or skin; or consumption of its eggs, milk, or meat. Common domestic animals include poultry, cattle (including buffalo), sheep, goats, and pigs. Starting about 10,000 years ago, the domestication of animals has only since World War II led to intensive ◊factory farming.

Increasing numbers of formerly wild species have been domesticated, with stress on scientific breeding for desired characteristics. At least 60% of the world's livestock is in developing countries, but the Third World consumes only 20% of all meat and milk produced. Most domestic animals graze plants that are not edible to humans, and 40% of the world's cereal production becomes animal feed; in the USA it is 90%.

animation, computer computer-generated graphics that appear to move across the screen. Traditional animation involves a great deal of drudgery in creating the 24 frames per second needed to deceive the human eye into seeing a moving picture on film. In computer-generated animation, while humans still create the key frames that specify the starting and ending points of a particular sequence – a character running through a landscape, for example – computers are faster and more accurate at calculating the in-between positions and generating the frames.
first achievements The first completely computer-generated character to appear in a major motion picture was the sea-water creature in James Cameron's film *The Abyss* 1990, developed at the leading special effects shop ◊Industrial Light & Magic. It was quickly followed by the liquid-metal man in Cameron's *Terminator 2* 1991. The first entirely computer-animated full-length feature film was ◊Pixar's *Toy Story* 1995, which was the first film ever to

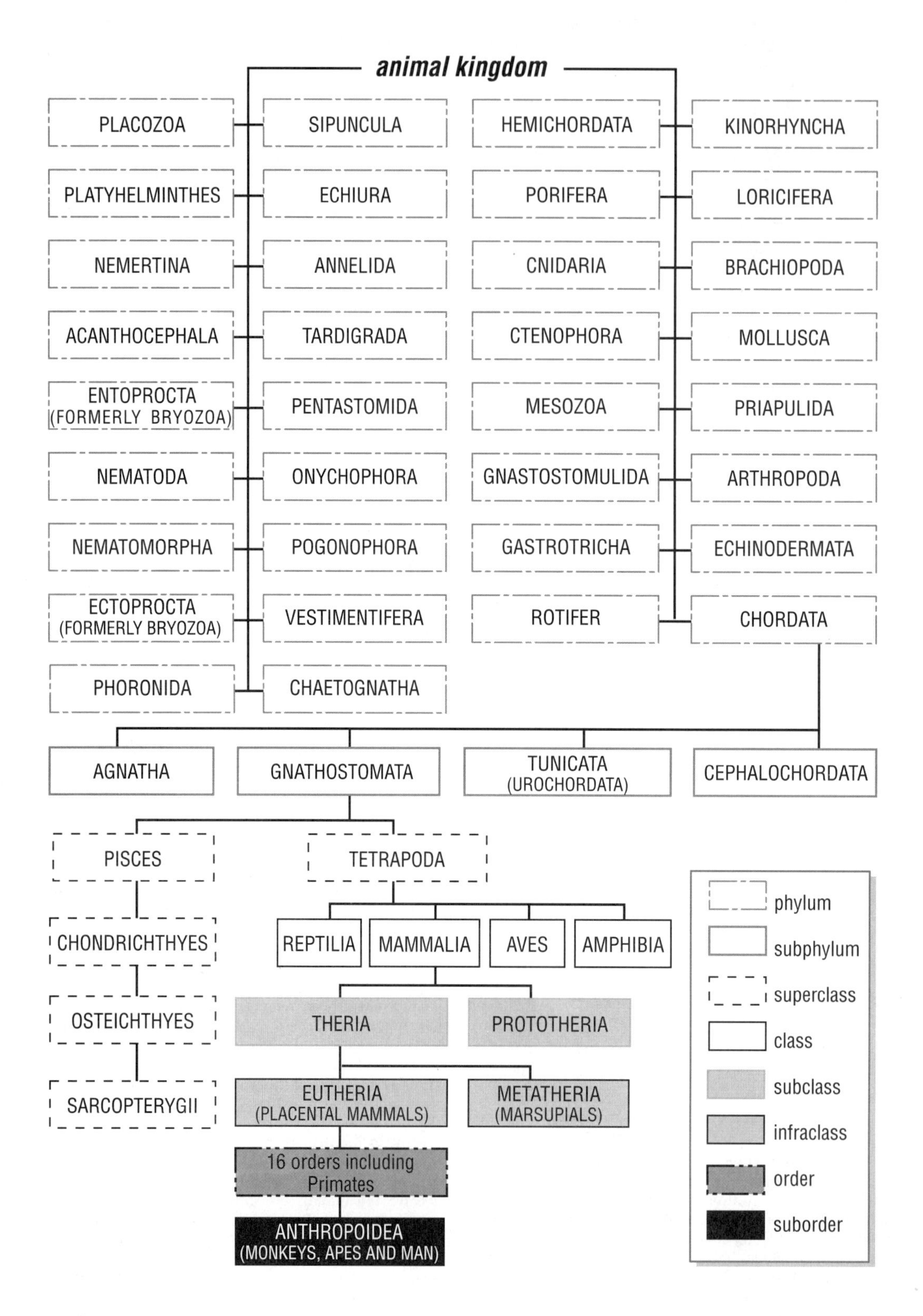

animal *The animal kingdom is divided into 34 major groups or phyla. The large phylum Chordata (animals that have, at some time in their life, a notochord, or stiff rod of cells running along the length of their body) is subdivided into four subphyla, of which two, the Agnatha and the Gnathostomata, are vertebrates (animals with backbones). The subphyla are divided into classes and subclasses.*

achieve independent motion of characters and backgrounds in the same sequence.

algorithms and initial image creation The basis of computer animation is ◊algorithms developed by academic researchers. These are used to develop software routines that handle the complex calculations needed to work out the precise colour of each ◊pixel in each of the finished frames; the process demands exceptionally powerful hardware with massive storage capacity. For the animator, an image begins as an on-screen collection of lines that look much like a wire frame. There are a variety of techniques the animator can use to develop 3-D objects – they can be extruded from a cross section, or 'swept', which is the on-screen equivalent of turning a cross section on a lathe to produce an evenly curved surface. Less symmetrical objects may be defined by a series of ◊Bezier curves.

adding solidity and colour The object then has to be **rendered**, which essentially means making it into an image of a solid object. To do this, the computer needs four types of information. First, the object has to be located in space. Second, it has to be assigned a colour, specified either by levels of red, blue, and green or by levels of hue, saturation, and brightness. Third, the location and focal point of the camera photographing the object have to be specified – these determine how the object appears on screen, in perspective. Fourth, the location and type of light sources must be specified: colour, brightness and, in the case of spotlighting, the size of the cone-shaped pool of illumination. ◊Ray-tracing, meanwhile, calculates how the light directed at the object reaches it, with what intensity, and in what areas. From all this information, the computer can calculate the colour intensity of each pixel making up the object.

light reflection There is another element, too: how the object itself reflects light. Two techniques model this, each named after its creator. If the object's surface can be described as a mosaic of polygons, **Gouraud shading** works by measuring the colour and brightness at the vertices of the polygons and mixing these to get values for the areas inside them. **Phong shading** extends this by taking into account the angle of reflection; it is therefore a more accurate technique for creating spectacular highlights. Gouraud, because it is simpler, is faster, and there is specialized hardware available for it; Phong has to be implemented in software. Both methods produce an object that looks as though it is made of soft, smooth plastic – the smoothness comes from **anti-aliasing**, a process that removes the jagged edges or stepped effect which mars the edges of diagonal lines on a computer display screen.

mapping for realism **Mappings** are what make the objects look as though they are made of real-world materials. There are four main types of mapping: texture, environment, bump, and transparency. Texture is the actual texture of the material the object is made of: brick, water, wood, and so on. The system essentially wraps the object in the texture the animator chooses. Environment mapping adds the reflections on the object's surface of its surroundings; a shiny, round, metal object rolling down a hill, for example, must show accurate reflections of the trees and other objects it rolls past. Bump mapping takes into account the shape of the object itself and the way this affects reflections and shadings in its surface colour. Transparency mapping defines what can be seen through the object, and the distortion caused by the substance of the object; this was a key element in animating the monster in *The Abyss,* which was made of sea water.

fog and haze Finally, fog and haze are important elements of computer animation, particularly for backgrounds, as computerized images tend to look too flat and sharp. The introduction of a little fog hides the sharp edges and makes the scene look more realistic. This is vital for one of the largest growth areas in computer animation for film and video: simulated flyovers, which are impossible in live action and expensive and difficult in model work.

animism in anthropology, the belief that everything, whether animate or inanimate, possesses a soul or spirit. It is a fundamental system of belief in certain religions, particularly those of some preindustrial societies. Linked with this is the worship of natural objects such as stones and trees, thought to harbour spirits (naturism); fetishism; and ancestor worship.

In psychology and physiology, animism is the view of human personality that attributes human life and behaviour to a force distinct from matter. In developmental psychology, an animistic stage in the early thought and speech of the child has been described, notably by Swiss psychologist Jean Piaget.

In philosophy, the view that in all things consciousness or something mindlike exists.

In religious theory, the conception of a spiritual reality behind the material one: for example, beliefs in the soul as a shadowy duplicate of the body capable of independent activity, both in life and death.

anion ion carrying a negative charge. During electrolysis, anions in the electrolyte move towards the anode (positive electrode).

An electrolyte, such as the salt zinc chloride ($ZnCl_2$), is dissociated in aqueous solution or in the molten state into doubly charged Zn^{2+} zinc ◊cations and singly-charged Cl^- anions. During electrolysis, the zinc cations flow to the cathode (to become discharged and liberate zinc metal) and the chloride anions flow to the anode.

anise Mediterranean plant belonging to the carrot family, with small creamy-white flowers in clusters; its fragrant seeds, similar to liquorice in taste, are used to flavour foods. Aniseed oil is used in cough medicines. (*Pimpinella anisum,* family Umbelliferae.)

annealing controlled cooling of a material to increase ductility and strength. The process involves first heating a material (usually glass or metal) for a given time at a given temperature, followed by slow cooling. It is a common form of ◊heat treatment.

annelid any segmented worm of the phylum Annelida. Annelids include earthworms, leeches, and marine worms such as lugworms.

They have a distinct head and soft body, which is divided into a number of similar segments shut off from one another internally by membranous partitions, but there are no jointed appendages. Annelids are noted for their ability to regenerate missing parts of their bodies.

annihilation in nuclear physics, a process in which a particle and its 'mirror image' particle called an **antiparticle** collide and disappear, with the creation of a burst of energy.

The energy created is equivalent to the mass of the colliding particles in accordance with the ◊mass–energy equation. For example, an electron and a positron annihilate to produce a burst of high-energy X-rays.

Not all particle–antiparticle interactions result in annihilation; the exception concerns the group called ◊mesons, which are composed of ◊quarks and their antiquarks. See ◊antimatter.

annotate to add one's own comments to Web pages or graphical computerized documents such as stored faxes.

annual percentage rate (APR) the true annual rate of interest charged for a loan. Lenders usually increase the return on their money by compounding the interest payable on a loan to that loan on a monthly or even daily basis. This means that each time that interest is payable on a loan it is charged not only on the initial sum (principal) but also on the interest previously added to that principal. As a result, APR is usually approximately double the flat rate of interest, or simple interest.

annual plant plant that completes its life cycle within one year, during which time it germinates, grows to maturity, bears flowers, produces seed, and then dies.

annual rings or ***growth rings*** concentric rings visible on the wood of a cut tree trunk or other woody stem. Each ring represents a period of growth when new ◊xylem is laid down to replace tissue being converted into wood (secondary xylem). The wood formed from xylem produced in the spring and early summer has larger and more numerous vessels than the wood formed from xylem produced in autumn when growth is slowing down. The result is a clear boundary between the pale spring wood and the denser, darker autumn wood. Annual rings may be used to estimate the age of the plant (see ◊dendrochronology), although occasionally more than one growth ring is produced in a given year.

annular eclipse solar ◊eclipse in which the Moon does not completely obscure the Sun and a thin ring of sunlight remains visible. Annular eclipses occur when the Moon is at its furthest point from the Earth.

annulus ***Latin 'ring'*** in geometry, the plane area between two concentric circles, making a flat ring.

anode in chemistry, the positive electrode of an electrolytic ◊cell, towards which negative particles (anions), usually in solution, are attracted. See ◊electrolysis.

anode in electronics, the positive electrode of a thermionic valve, cathode ray tube, or similar device, towards which electrons are drawn after being emitted from the ◊cathode.

anodizing process that increases the resistance to ◊corrosion of a metal, such as aluminium, by building up a protective oxide layer on the surface. The natural corrosion resistance of aluminium is provided by a thin film of aluminium oxide; anodizing increases the thickness of this film and thus the corrosion protection.

It is so called because the metal becomes the ◊anode in an electrolytic bath containing a solution of, for example, sulphuric or chromic acid as the ◊electrolyte. During ◊electrolysis oxygen is produced at the anode, where it combines with the metal to form an oxide film.

anomalocaris prehistoric marine predator resembling a crustacean but with a softer exoskeleton. Up to 2 m in length, it was the largest animal on Earth when it thrived around 525 million years ago. See ◊prehistoric life.

anomalous expansion of water expansion of water as it is cooled from 4°C to 0°C. This behaviour is unusual because most substances contract when they are cooled. It means that water has a greater density at 4°C than at 0°C. Hence ice floats on water, and the water at the bottom of a pool in winter is warmer than at the surface. As a result lakes and ponds freeze slowly from the surface downwards, usually remaining liquid near the bottom, where aquatic life is more likely to survive.

anonymous FTP (***file transfer protocol***) method of retrieving a file from a remote computer without having an account on that computer. Many organizations, such as universities and software companies, maintain publicly accessible archives of files that may be retrieved across the Internet via ◊FTP. An ordinary user who is not affiliated to the organization may retrieve files by entering the FTP address and then typing in either 'anonymous' or 'ftp' when asked for a user-ID or log-in name, followed by the user's e-mail address in place of a password. These users are typically offered ◊access privileges to only a small part of the company's stored files, and the rest may be cordoned off from access by a ◊firewall.

anonymous remailer service that allows Internet users to post to USENET and send e-mail without revealing their true identity or e-mail address. To send an anonymous message, a user first sends the message to the remailer, which strips all identifying information from the message before sending it on to its specified destination, identified only as coming from the anonymous server.

The ability to post anonymously also removes user accountability, and so these servers are controversial. However, they provide a useful function on the Net in support groups and other areas where the ability to post anonymously allows people to speak freely about confidential matters without the risk of being identified by friends, family, or anyone else.

The best-known anonymous server, the Finnish **anon.penet.fi** was closed down in Aug 1996 as it could no longer guarantee anonymity following a court case ordering the operator to reveal a user's name. The more elaborate servers use encryption to make the message even more difficult to trace.

anorak term used interchangeably with **geek**, **techie**, or **nerd**. It derives from the stereotype that all technical people resemble the stereotypical anorak-wearing trainspotter; in other words, that they are obsessive, slightly antisocial, and overly knowledgeable about matters that interest very few other people.

anorexia lack of desire to eat, or refusal to eat, especially the pathological condition of **anorexia nervosa**, most often found in adolescent girls and young women. Compulsive eating, or ◊bulimia, distortions of body image, and depression often accompany anorexia.

Anorexia nervosa is characterized by severe self-imposed restriction of food intake. The consequent weight loss may lead, in women, to absence of menstruation. Anorexic patients sometimes commit suicide. Anorexia nervosa is often associated with increased physical activity and symptoms of mental disorders. Psychotherapy is an important part of the treatment.

anoxia or ***hypoxia*** in biology, deprivation of oxygen, a condition that rapidly leads to collapse or death, unless immediately reversed.

ANSI abbreviation for **American National Standards Institute**, ◊a US national standards body.

ant insect belonging to the family Formicidae, and to the same order (Hymenoptera) as bees and wasps. Ants are characterized by a conspicuous waist and elbowed antennae. About 10,000 different species are known; all are social in habit, and all construct nests of various kinds. Ants are found in all parts of the world, except the polar regions. It is estimated that there are about 10 million billion ants.

Ant behaviour is complex, and serves the colony rather than the individual. Ants find their way by light patterns, gravity (special sense organs are located in the joints of their legs), and chemical trails between food areas and the nest.

specialized roles Communities include **workers**, sterile, wingless females, often all alike, although in some species large-headed 'soldiers' are differentiated; **fertile females**, fewer in number and usually winged; and **males**, also winged and smaller than their consorts, with whom they leave the nest on a nuptial flight at certain times of the year. After aerial mating, the males die, and the fertilized queens lose their wings when they settle, laying eggs to found their own new colonies. The eggs hatch into wormlike larvae, which then pupate in silk cocoons before emerging as adults.

remarkable species Some species conduct warfare. Others are pastoralists, tending herds of ◊aphids and collecting a sweet secretion ('honeydew') from them. Army (South American) and driver (African) ants march nomadically in huge columns, devouring even tethered animals in their path. Leaf-cutter ants, genus *Atta*, use pieces of leaf to grow edible fungus in underground 'gardens' which can be up to 5 m/16 ft deep and cover hundreds of square metres. Weaver ants, genus *Oecophylla*, use their silk-producing larvae as living shuttles to bind the edges of leaves together to form the nest. Eurasian robber ants *Formica sanguinea* raid the nests of

ant *With a length of over 30 mm/1.2 in, this* Diponera *species ant from the rainforests of South America is among the giants of its kind. Like most members of its primitive subfamily Ponerinae, it forages singly rather than in bands and the nests generally contain only a few dozen individuals. Premaphotos Wildlife* another ant species, *Formica fusca*, for pupae, then use the adults as 'slaves' when they hatch. Among honey ants, some workers serve as distended honey stores.

antacid any substance that neutralizes stomach acid, such as sodium bicarbonate or magnesium hydroxide ('milk of magnesia').

Antacids are weak ◊bases, swallowed as solids or emulsions. They may be taken between meals to relieve symptoms of hyperacidity, such as pain, bloating, nausea, and 'heartburn'. Excessive or prolonged need for antacids should be investigated medically.

antagonist in biology, a ◊muscle that relaxes in response to the contraction of its agonist muscle. The biceps, in the front of the upper arm, bends the elbow whilst the triceps, lying behind the biceps, straightens the arm.

antagonistic muscles in the body, a pair of muscles allowing coordinated movement of the skeletal joints. The extension of the arm, for example, requires one set of muscles to relax, while another set contracts. The individual components of antagonistic pairs can be classified into extensors (muscles that straighten a limb) and flexors (muscles that bend a limb).

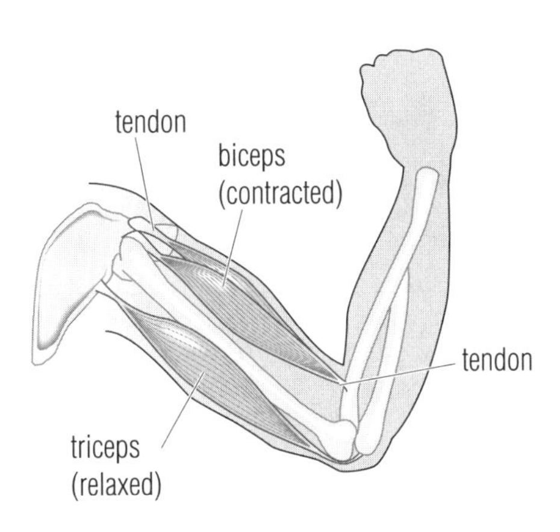

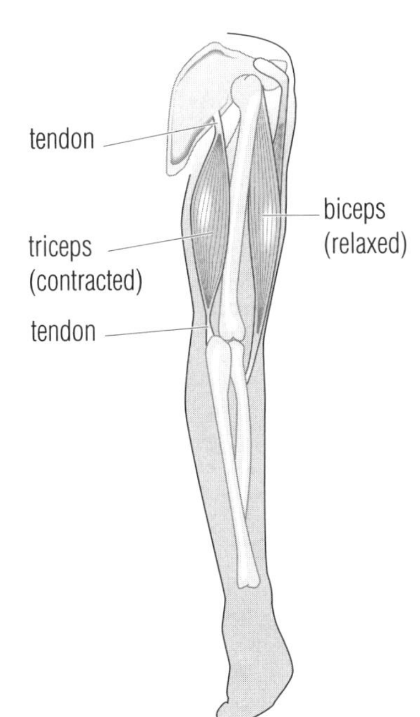

***antagonistic muscle** Even simple movements such as bending and straightening the arm require muscle pairs to contract and relax synchronously.*

Antarctic Circle imaginary line that encircles the South Pole at latitude 66° 32' S. The line encompasses the continent of Antarctica and the Antarctic Ocean.

The region south of this line experiences at least one night in the southern summer during which the Sun never sets, and at least one day in the southern winter during which the Sun never rises.

Antares or ***Alpha Scorpii*** brightest star in the constellation Scorpius and the 15th brightest star in the sky. It is a red supergiant several hundred times larger than the Sun and perhaps 10,000 times as luminous, lies about 300 light years away, and fluctuates slightly in brightness.

anteater mammal of the family Myrmecophagidae, order Edentata, native to Mexico, Central America, and tropical South America. The anteater lives almost entirely on ants and termites. It has toothless jaws, an extensile tongue, and claws for breaking into the nests of its prey.

Species include the giant anteater *Myrmecophaga tridactyla*, about 1.8 m/6 ft long including the tail, the tamandua or collared anteater *Tamandua tetradactyla*, about 90 cm/3.5 ft long, and the silky anteater *Cyclopes didactyla*, about 35 cm/14 in long. The name is also incorrectly applied to the aardvark, the echidna, and the pangolin.

antelope any of numerous kinds of even-toed, hoofed mammals belonging to the cow family, Bovidae. Most antelopes are lightly built and good runners. They are grazers or browsers, and chew the cud. They range in size from the dik-diks and duikers, only 30 cm/1 ft high, to the eland, which can be 1.8 m/6 ft at the shoulder.

The majority of antelopes are African, including the eland, wildebeest, kudu, springbok, and waterbuck, although other species live in parts of Asia, including the deserts of Arabia and the Middle East. The pronghorn antelope *Antilocapra americana* of North America belongs to a different family, the Antilocapridae.

antenatal in medicine, before birth. Antenatal care refers to health services provided to ensure the health of pregnant women and their babies.

antenna in radio and television, another name for ◊aerial.

antenna in zoology, an appendage ('feeler') on the head. Insects, centipedes, and millipedes each have one pair of antennae but there are two pairs in crustaceans, such as shrimps. In insects, the antennae are involved with the senses of smell and touch; they are frequently complex structures with large surface areas that increase the ability to detect scents.

anterior in biology, the front of an organism, usually the part that goes forward first when the animal is moving. The anterior end of the nervous system, over the course of evolution, has developed into a brain with associated receptor organs able to detect stimuli including light and chemicals.

anther in a flower, the terminal part of a stamen in which the ◊pollen grains are produced. It is usually borne on a slender stalk or filament, and has two lobes, each containing two chambers, or pollen sacs, within which the pollen is formed.

antheridium organ producing the male gametes, ◊antherozoids, in algae, bryophytes (mosses and liverworts), and pteridophytes (ferns, club mosses, and horsetails). It may be either single-celled, as in most algae, or multicellular, as in bryophytes and pteridophytes.

antherozoid motile (or independently moving) male gamete produced by algae, bryophytes (mosses and liverworts), pteridophytes (ferns, club mosses, and horsetails), and some gymnosperms (notably the cycads). Antherozoids are formed in an antheridium and, after being released, swim by means of one or more ◊flagella, to the female gametes. Higher plants have nonmotile male gametes contained within ◊pollen grains.

anthracene white, glistening, crystalline, tricyclic, aromatic hydrocarbon with a faint blue fluorescence when pure. Its melting point is about 216°C/421°F and its boiling point 351°C/664°F. It occurs in the high-boiling-point fractions of coal tar, where it was discovered in 1832 by the French chemists Auguste Laurent (1808–1853) and Jean Dumas (1800–1884).

anthracite ***from Greek anthrax, 'coal'*** hard, dense, shiny variety of ◊coal, containing over 90% carbon and a low percentage of ash and impurities, which causes it to burn without flame, smoke, or smell. Because of its purity, anthracite gives off relatively little sulphur dioxide when burnt.

Anthracite gives intense heat, but is slow-burning and slow to light; it is therefore unsuitable for use in open fires. Its characteristic composition is thought to be due to the action of bacteria in disintegrating the coal-forming material when it was laid down during the ◊Carboniferous period.

Among the chief sources of anthracite coal are Pennsylvania in the USA; S Wales, UK; the Donbas, Ukraine and Russia; and Shanxi province, China.

anthrax disease of livestock, occasionally transmitted to humans, usually via infected hides and fleeces. It may develop as black skin pustules or severe pneumonia. Treatment is with antibiotics. Vaccination is effective.

Anthrax is caused by a bacillus (*Bacillus anthracis*). In the 17th century, some 60,000 cattle died in a European pandemic known as the Black Bane, thought to have been anthrax. The disease is described by the Roman poet Virgil and may have been the cause of the biblical fifth plague of Egypt.

anthropic principle in science, the idea that 'the universe is the way it is because if it were different we would not be here to observe it'. The principle arises from the observation that if the laws of science were even slightly different, it would have been impossible for intelligent life to evolve. For example, if the electric charge on the electron were only slightly different, stars would have been unable to burn hydrogen and produce the chemical elements that make up our bodies. Scientists are undecided whether the principle is an insight into the nature of the universe or a piece of circular reasoning.

antenna *The antennae of some male moths are capable of picking up incredibly low concentrations of airborne female sex pheromones at distances of several kilometres. This picture shows the enormous antennae of the American moon moth* Actias luna (Saturniidae) *Premaphotos Wildlife*

ANTHRAX

http://www.outbreak.org/cgi-unreg/dynaserve.exe/cb/anthrax.html

Basic medical details of the livestock disease and biological weapon anthrax. The Web site contains brief sections covering characteristics of the disease, the symptoms it causes, cautions and precautions, first aid therapy for victims of the disease, and a list of neutralization and decontamination methods.

Man appears to be the missing link between anthropoid apes and human beings.

KONRAD LORENZ Austrian zoologist. *The New York Times Magazine* 11 April 1965

anthropoid ***Greek anthropos 'man', eidos 'resemblance'*** any primate belonging to the suborder Anthropoidea, including monkeys, apes, and humans.

You will die but the carbon will not; its career does not end with you ... it will return to the soil, and there a plant may take it up again in time, sending it once more on a cycle of plant and animal life.

JACOB BRONOWSKI Polish-born British scientist, broadcaster, and writer. 'Biography of an Atom – and the Universe *New York Times* 13 Oct 1968

anthropology ***Greek anthropos 'man', logos 'discourse'*** the study of humankind. It investigates the cultural, social, and physical diversity of the human species, both past and present. It is divided into two broad categories: biological or physical anthropology, which attempts to explain human biological variation from an evolutionary perspective; and the larger field of social or cultural anthropology, which attempts to explain the variety of human cultures. This differs from sociology in that anthropologists are concerned with cultures and societies other than their own.

biological anthropology Biological anthropology is concerned with human ◊palaeontology, primatology, human adaptation, demography, ◊population genetics, and human growth and development.

social anthropology Social or cultural anthropology is divided into three subfields: social or cultural anthropology proper, ◊prehistory or prehistoric archaelogy, and anthropological linguistics. The term 'anthropology' is frequently used to refer solely to social anthropology. With a wide range of theoretical perspectives and topical interests, it overlaps with many other disciplines. It is a uniquely Western social science.

participant observation Anthropology's primary method involves the researcher living for a year or more in another culture, speaking the local language and participating in all aspects of everyday life; and writing about it afterwards. By comparing these accounts, anthropologists hope to understand who we are.

anthropometry science dealing with the measurement of the human body, particularly stature, body weight, cranial capacity, and length of limbs, in samples of living populations, as well as the remains of buried and fossilized humans.

anti-aliasing in computer graphics, a software technique for diminishing ◊aliasing ('jaggies') – steplike lines that should be smooth. Jaggies occur because the output device, the monitor or printer, does not have a high enough resolution to represent a smooth line. Anti-aliasing reduces the prominence of jaggies by surrounding the steps with intermediate shades of grey (for grey-scaling devices) or colour (for colour devices). Although this reduces the jagged appearance of the lines, it also makes them fuzzier.

ANTIBIOTICS: HOW DO ANTIBIOTICS WORK

http://ericir.syr.edu/Projects/Newton/12/Lessons/antibiot.html

Introduction to the use and importance of antibiotics in easy to understand language. The site also includes a glossary of scientific and difficult terms, a further reading list, and an activities sheet.

antibiotic drug that kills or inhibits the growth of bacteria and fungi. It is derived from living organisms such as fungi or bacteria, which distinguishes it from synthetic antimicrobials.

The earliest antibiotics, the ◊penicillins, came into use from 1941 and were quickly joined by chloramphenicol, the ◊cephalosporins, erythromycins, tetracyclines, and aminoglycosides. A range of broad-spectrum antibiotics, the 4-quinolones, was developed 1989, of which ciprofloxacin was the first. Each class and individual antibiotic acts in a different way and may be effective against either a broad spectrum or a specific type of disease-causing agent. Use of antibiotics has become more selective as side effects, such as toxicity, allergy, and resistance, have become better understood. Bacteria have the ability to develop resistance following repeated or subclinical (insufficient) doses, so more advanced and synthetic antibiotics are continually required to overcome them.

ANTIBODY RESOURCE PAGE

http://www.antibodyresource.com/

Fascinating annotated access site for all you could ever want to know about antibodies and some you'd rather not! This site contains several educational resources (aimed at university level), but also contains images, animations, and descriptions of research into many different types of antibodies.

antibody protein molecule produced in the blood by ◊lymphocytes in response to the presence of foreign or invading substances (◊antigens); such substances include the proteins carried on the surface of infecting microorganisms. Antibody production is only one aspect of ◊immunity in vertebrates.

Each antibody acts against only one kind of antigen, and combines with it to form a 'complex'. This action may render antigens harmless, or it may destroy microorganisms by setting off chemical changes that cause them to self-destruct.

In other cases, the formation of a complex will cause antigens to form clumps that can then be detected and engulfed by white blood cells, such as ◊macrophages and ◊phagocytes.

Each bacterial or viral infection will bring about the manufacture of a specific antibody, which will then fight the disease. Many diseases can only be contracted once because antibodies remain in the blood after the infection has passed, preventing any further invasion. Vaccination boosts a person's resistance by causing the production of antibodies specific to particular infections.

Antibodies were discovered 1890 by German physician Emil von Behring and Japanese bacteriologist Shibasaburo Kitasato.

Large quantities of specific antibodies can now be obtained by the monoclonal technique (see ◊monoclonal antibody).

anticline in geology, rock layers or beds folded to form a convex arch (seldom preserved intact) in which older rocks comprise the core. Where relative ages of the rock layers, or stratigraphic ages, are not known, convex upward folded rocks are referred to as **antiforms**.

The fold of an anticline may be undulating or steeply curved. A steplike bend in otherwise gently dipping or horizontal beds is a **monocline**. The opposite of an anticline is a ◊syncline.

anticlockwise direction of rotation, opposite to the way the hands of a clock turn.

anticoagulant substance that inhibits the formation of blood clots. Common anticoagulants are heparin, produced by the liver and some white blood cells, and derivatives of coumarin. Anticoagulants are used medically in the prevention and treatment of thrombosis and heart attacks. Anticoagulant substances are also produced by blood-feeding animals, such as mosquitoes, leeches, and vampire bats, to keep the victim's blood flowing.

Most anticoagulants prevent the production of thrombin, an enzyme that induces the formation from blood plasma of fibrinogen, to which blood platelets adhere and form clots.

anticyclone area of high atmospheric pressure caused by descending air, which becomes warm and dry. Winds radiate from a calm centre, taking a clockwise direction in the northern hemisphere and an anticlockwise direction in the southern hemisphere. Anticyclones are characterized by clear weather and the absence of rain and violent winds. In summer they bring hot, sunny days and in winter they bring fine, frosty spells, although fog and low cloud are not uncommon in the UK. **Blocking anticyclones**, which prevent the normal air circulation of an area, can cause summer droughts and severe winters.

antidepressant any drug used to relieve symptoms in depressive illness. The two main groups are the tricyclic antidepressants (TCADs) and the monoamine oxidase inhibitors (MAOIs), which act by altering chemicals available to the central nervous system. Both may produce serious side effects and are restricted.

anti-emetic any substance that counteracts nausea or vomiting.

antiferromagnetic material material with a very low magnetic ◊susceptibility that increases with temperature up to a certain temperature, called the ◊Néel temperature. Above the Néel temperature, the material is only weakly attracted to a strong magnet.

antifreeze substance added to a water-cooling system (for example, that of a car) to prevent it freezing in cold weather.

antigen any substance that causes the production of ◊antibodies by the body's immune system. Common antigens include the proteins carried on the surface of bacteria, viruses, and pollen grains. The proteins of incompatible blood groups or tissues also act as antigens, which has to be taken into account in medical procedures such as blood transfusions and organ transplants.

antihistamine any substance that counteracts the effects of ◊histamine. Antihistamines may occur naturally or they may be synthesized.

H_1 antihistamines are used to relieve allergies, alleviating symptoms such as runny nose, itching, swelling, or asthma. H_2 antihistamines suppress acid production by the stomach, and are used in the treatment of peptic ulcers, often making surgery unnecessary.

antiknock substance added to ◊petrol to reduce knocking in car engines. It is a mixture of dibromoethane and tetraethyl lead. Its use in leaded petrol has resulted in atmospheric pollution by lead compounds.

antilogarithm or ***antilog*** the inverse of ◊logarithm, or the number whose logarithm to a given base is a given number. If $y = \log_a x$, then $x = \text{antilog}_a y$.

antimatter in physics, a form of matter in which most of the attributes (such as electrical charge, magnetic moment, and spin) of ◊elementary particles are reversed. Such particles (◊antiparticles) can be created in particle accelerators, such as those at ◊CERN in Geneva, Switzerland, and at ◊Fermilab in the USA. In 1996 physicists at CERN created the first atoms of antimatter: nine atoms of antihydrogen survived for 40 nanoseconds.

Antibiotic Resistance: A Rising Toll

BY PAULETTE PRATT

When, in 1969, the US Surgeon-General announced that we could soon 'close the book' on infectious diseases, he was speaking prematurely. For already, two years earlier, first reports had surfaced of penicillin resistance developing in *Pneumococcus*, a bacterium which causes a number of potentially fatal diseases, including pneumonia and meningitis. Within little more than a decade, epidemics of pneumococcal disease were breaking out in many countries.

In 1995, when workers from the Centers for Disease Control in Atlanta looked at samples taken from patients with severe pneumonia, they found that a quarter had been hit by pneumococcal (*Staphylococcus pneumoniae*) strains resistant to penicillin; some 15 % were also resistant to erythromycin. Today, growing resistance of this and other organisms to antibiotics is a global public health problem.

Rise of the superbugs

Bacteria vary greatly in their sensitivity to antibiotics, and there is no such thing as a 'magic bullet' with blanket activity against all pathogens. The trouble is that, after more than half a century of antibiotic use, bacteria are mutating faster than new drugs can be found. The growth of super-resistance is being hastened by the indiscriminate use of antibiotics, including over-prescribing.

In the developed world, the danger is greatest in hospitals, where people are already very sick. Currently some 14,000 Americans are dying each year from hospital-acquired infections caused by resistant bacteria. Most at risk are people whose immune systems are in some way impaired, including the very young or the frail elderly, AIDS patients, organ transplant recipients, and patients receiving chemotherapy for cancer.

Most notorious of the so-called 'superbugs' stalking hospital floors is methicillin-resistant *Staphylococcus aureus* (MRSA), a pathogen with an awesome talent for acquiring resistance traits from its microscopic neighbours. So far outbreaks of MRSA, which can cause temporary closure of operating rooms and intensive care units, have been met with vancomycin, a 'last resort' antibiotic normally reserved for life-threatening infections. But in spring of 1997, the Japanese reported the most convincing evidence yet of the appearance of vancomycin-resistant strains.

The Japanese report brings one step closer the spectre of the unstoppable 'superbug' overrunning hospitals. In fact, microbiologists have been predicting just this scenario, not least because strains of the intestinal bacterium *Enterococcus* have for some time been defying all existing antibiotics, including vancomycin and the closely related teicoplanin. In the laboratory, it has been shown that genes for resistance can pass from *Enterococcus* to the more deadly *S. aureus* by plasmid transfer.

It was fear of an epidemic of untreatable infections that prompted the recent European ban on avoparcin, a drug administered to farm animals to promote growth. The rationale for antibiotic use in this context is that it improves feeding efficiency. However, avoparcin is close in chemical structure to vancomycin, and many scientists argue that its use as a growth promoter in livestock creates a potential reservoir of vancomycin-resistant bacteria that would be transmissible to human beings.

Return of old-time diseases

The phenomenon of super-resistance means also that many one-time killer diseases, including tuberculosis (TB), typhoid, cholera, and diphtheria, are returning in force. TB, always a major problem in the Third World, is the biggest threat, since now it is making a comeback in countries where previously it had been brought under control. Moreover, some strains of the bacterium have become resistant. Parts of the US worst affected are deprived inner-city areas such as Newark, New Jersey, and Brooklyn, the Bronx, and Harlem in New York, where resistance is widespread.

A big factor in the spread of multi-drug-resistant tuberculosis (MDRTB) has been the failure of control programs in the industrialized countries, including the US. This has meant that many patients starting out on medication lapse before the six-month course is completed. If a patient carrying a resistant strain takes only one drug instead of the prescribed combination, or fails to complete the course, the effect is to promote resistant strains.

Fresh strategies

Bacteria developing the ability to foil an antibiotic can become permanently resistant, according to researchers at Emory University in Atlanta. They demonstrated this in another rising superbug, *Escherichia coli*, which causes gastrointestinal infections. This finding implies that, contrary to what many doctors previously believed, reducing antibiotic use will not eliminate resistant strains.

While the quest for new drugs to fight infection is now paramount, many researchers are developing fresh strategies. These include: tinkering with the structure of antibiotics to add in helper molecules; seeking to disable genes for resistance; and developing laser-activated chemical compounds to blitz the superbugs. Some teams, too, are reviving the old idea of turning bacteriophages (bacteria-eating viruses) loose on resistant bacteria.

All these strategies and more may be needed to overcome the rising toll exacted by antibiotic resistance. Certainly, with infectious diseases claiming more than 17 million lives a year worldwide, we are still no nearer to 'closing the book'.

Mechanisms of resistance

Bacteria may be naturally resistant to some antibiotics, or resistance may be acquired, mostly by the phenomenon of plasmid transfer. A plasmid is a free-floating fragment of DNA adrift in the cell cytoplasm. It carries some genetic material, including data governing the cell's resistance to antibiotics. Plasmids can be transmitted from one bacterium to another.

Occasionally resistance may be due to spontaneous mutation, which is the result of an error in replication of the cell's nuclear material during reproduction. Further reproduction causes the development of a resistant strain.

Bacteria demonstrate resistance in two ways. One way is by producing enzymes that disable drugs. A drug-defying enzyme is not always expressed by the organism targeted by therapy. The normally innocuous *Staphylococcus epidermidis* produces an enzyme that disables penicillin before it can act against harmful staphylococcal species; or some bacteria can contrive metabolic changes that foil the action of drugs. Sulphonamides – antibacterials introduced before the discovery of antibiotics – are often defeated by these metabolic readjustments on the part of bacteria.

antimony silver-white, brittle, semimetallic element (a metalloid), symbol Sb (from Latin *stibium*), atomic number 51, relative atomic mass 121.75. It occurs chiefly as the ore stibnite, and is used to make alloys harder; it is also used in photosensitive substances in colour photography, optical electronics, fireproofing, pigment, and medicine. It was employed by the ancient Egyptians in a mixture to protect the eyes from flies.

antinode in physics, the position in a ◊standing wave pattern at which the amplitude of vibration is greatest (compare ◊node). The standing wave of a stretched string vibrating in the fundamental mode has one antinode at its midpoint. A vibrating air column in a pipe has an antinode at the pipe's open end and at the place where the vibration is produced.

anti-oxidant any substance that prevents deterioration of fats, oils, paints, plastics, and rubbers by oxidation. When used as food additives, anti-oxidants prevent fats and oils from becoming rancid when exposed to air, and thus extend their shelf life.

Vegetable oils contain natural anti-oxidants, such as vitamin E, which prevent spoilage, but anti-oxidants are nevertheless added to most oils. They are not always listed on food labels because if a food manufacturer buys an oil to make a food product, and the oil has anti-oxidant already added, it does not have to be listed on the label of the product.

antiparticle in nuclear physics, a particle corresponding in mass and properties to a given ◊elementary particle but with the opposite electrical charge, magnetic properties, or coupling to other fundamental forces. For example, an electron carries a negative charge whereas its antiparticle, the positron, carries a positive one. When a particle and its antiparticle collide, they destroy each other, in the process called 'annihilation', their total energy being converted to lighter particles and/or photons. A substance consisting entirely of antiparticles is known as ◊antimatter.

antipodes ***Greek 'opposite feet'*** places at opposite points on the globe.

antipyretic any drug, such as aspirin, used to reduce fever.

antirrhinum any of several plants in the figwort family, including the snapdragon (*Antirrhinum majus*). Foxgloves and toadflax are relatives. Antirrhinums are native to the Mediterranean region and W North America. (Genus *Antirrhinum,* family Scrophulariaceae.)

antiseptic any substance that kills or inhibits the growth of microorganisms. The use of antiseptics was pioneered by Joseph Lister. He used carbolic acid (◊phenol), which is a weak antiseptic; antiseptics such as TCP are derived from this.

antitussive any substance administered to suppress a cough. Coughing, however, is an important reflex in clearing secretions from the airways; its suppression is usually unnecessary and possibly harmful, unless damage is being done to tissue during excessive coughing spasms.

antivirus software computer program that detects ◊viruses and/or cleans viruses from an infected computer system. There are many types of antivirus software. Scanners check a computer system and detect viruses; these must be updated regularly, as new viruses are written and released. Other utilities allow a user to edit the data on hard and floppy discs directly or repair system damage. Still other types, which may come with specialized hardware, function by detecting and blocking changes to files or system activities which are typical of how viruses behave.

antler 'horn' of a deer, often branched, and made of bone rather than horn. Antlers, unlike true horns, are shed and regrown each year. Reindeer of both sexes grow them, but in all other types of deer, only the males have antlers.

During growth the antler is covered by a sensitive, hairy skin, known as 'velvet', which dries up and is rubbed off when maturity is attained. The number and complexity of the branches increase year by year.

ant lion larva of one of the insects of the family Myrmeleontidae, order Neuroptera, which traps ants by waiting at the bottom of a pit dug in loose, sandy soil. Ant lions are mainly tropical, but also occur in parts of Europe, where there are more than 40 species, and in the USA, where they are called doodlebugs.

anus or **anal canal** the opening at the end of the alimentary canal that allows undigested food and other waste materials to pass out of the body, in the form of faeces. In humans, the term is also used to describe the last 4 cm/1.5 in of the alimentary canal. The anus is found in all types of multicellular animal except the coelenterates (sponges) and the platyhelminths (flatworms), which have a mouth only.

It is normally kept closed by rings of muscle called sphincters. The commonest medical condition associated with the anus is haemorrhoids (piles).

anxiety unpleasant, distressing emotion usually to be distinguished from fear. Fear is aroused by the perception of actual or threatened danger; anxiety arises when the danger is imagined or cannot be identified or clearly perceived. It is a normal response in stressful situations, but is frequently experienced in many mental disorders.

Anxiety is experienced as a feeling of suspense, helplessness, or alternating hope and despair together with excessive alertness and characteristic bodily changes such as tightness in the throat, disturbances in breathing and heartbeat, sweating, and diarrhoea.

In psychiatry, an anxiety state is a type of neurosis in which the anxiety either seems to arise for no reason or else is out of proportion to what may have caused it. 'Phobic anxiety' refers to the irrational fear that characterizes ◊phobia.

AOL abbreviation for ◊America Online; the UK version of the US service.

aorta the body's main ◊artery, arising from the left ventricle of the heart in birds and mammals. Carrying freshly oxygenated blood, it arches over the top of the heart and descends through the trunk, finally splitting in the lower abdomen to form the two iliac arteries. Loss of elasticity in the aorta provides evidence of ◊atherosclerosis, which may lead to heart disease.

In fish a ventral aorta carries deoxygenated blood from the heart to the ◊gills, and the dorsal aorta carries oxygenated blood from the gills to other parts of the body.

a.p. in physics, abbreviation for **atmospheric pressure**.

Apache Point Observatory US observatory in the Sacramento Mountains of New Mexico containing a 3.5-m/138-in reflector, opened 1994, and operated by the Astrophysical Research Consortium (the universities of Washington, Chicago, Princeton, New Mexico, and Washington State).

apastron the point at which an object travelling in an elliptical orbit around a star is at its furthest from the star. The term is usually applied to the position of the minor component of a ◊binary star in relation to the primary. Its opposite is ◊periastron.

apatite common calcium phosphate mineral, $Ca_5(PO_4)_3(F,OH,Cl)$. Apatite has a hexagonal structure and occurs widely in igneous rocks, such as pegmatite, and in contact metamorphic rocks, such as marbles. It is used in the manufacture of fertilizer and as a source of phosphorus. Carbonate hydroxylapatite, $Ca_5(PO_4,CO_3)_3(OH)_2$, is the chief constituent of tooth enamel and, together with other related phosphate minerals, is the inorganic constituent of bone. Apatite ranks 5 on the ◊Mohs' scale of hardness.

apatosaurus large plant-eating dinosaur, formerly called **brontosaurus**, which flourished about 145 million years ago. Up to 21 m/69 ft long and 30 tonnes in weight, it stood on four elephantlike legs and had a long tail, long neck, and small head. It probably snipped off low-growing vegetation with peglike front teeth, and swallowed it whole to be ground by pebbles in the stomach.

ape ◊primate of the family Pongidae, closely related to humans, including gibbon, orang-utan, chimpanzee, and gorilla.

> *If it is true that we have sprung from the ape, there are occasions when my own spring appears not to have been very far.*
>
> Cornelia Otis Skinner US writer and actress.
> *The Ape in Me*, title essay

Ape City Yerkes Regional Primate Center, Atlanta, Georgia, where large numbers of primates are kept for physiological and psychological experiment. A major area of research at Ape City is language.

aperture in photography, an opening in the camera that allows light to pass through the lens to strike the film. Controlled by the iris diaphragm, it can be set mechanically or electronically at various diameters.

The **aperture ratio** or **relative aperture**, more commonly known as the ◊f-number, is a number defined as the focal length of the lens divided by the effective diameter of the aperture. A smaller f-number implies a larger diameter lens and therefore more light available for high-speed photography, or for work in poorly illuminated areas. However, small f-numbers involve small depths of focus.

aperture synthesis in astronomy, a technique used in ◊radio astronomy in which several small radio dishes are linked together to simulate the performance of one very large radio telescope, which can be many kilometres in diameter. See ◊radio telescope.

apex the highest point of a triangle, cone, or pyramid – that is, the vertex (corner) opposite a given base.

NATIONAL APHASIA ASSOCIATION

http://www.aphasia.org/

US-based site with a lot of straightforward information about aphasia; for example, there is a page of important facts and a true/false quiz. The site also contains information about current research and links to support groups for sufferers, their friends, and families.

aphasia general term for the many types of disturbance in language that are due to brain damage, especially in the speech areas of the dominant hemisphere.

aphelion the point at which an object, travelling in an elliptical orbit around the Sun, is at its furthest from the Sun. The Earth is at its aphelion on 5 July.

aphid any of the family of small insects, Aphididae, in the order Hemiptera, suborder Homoptera, that live by sucking sap from plants. There are many species, often adapted to particular plants; some are agricultural pests.

aphid *A colony of aphids* Cavariella konoi, *commonly known as greenfly, feeding on green stems of the almond willow. Ants often attend such colonies to feed on the honeydew (a sweet, sticky substance) that aphids excrete.* *Premaphotos Wildlife*

In some stages of their life cycle, wingless females rapidly produce large numbers of live young by ◊parthenogenesis, leading to enormous infestations, and numbers can approach 2 billion per hectare/1 billion per acre. They can also cause damage by transmitting viral diseases. An aphid that damages cypress and cedar trees appeared in Malawi in 1985 and by 1991 was attacking millions of trees in central and E Africa. Some research suggests, however, that aphids may help promote fertility in the soil through the waste they secrete, termed 'honeydew'. Aphids are also known as plant lice, greenflies, or blackflies.

aphrodisiac ***from Aphrodite, the Greek goddess of love*** any substance that arouses or increases sexual desire.

API abbreviation for ◊Applications Program Interface, a standard environment in which computer programs are written.

apogee the point at which an object, travelling in an elliptical orbit around the Earth, is at its furthest from the Earth.

Apollo asteroid member of a group of ◊asteroids whose orbits cross that of the Earth. They are named after the first of their kind, Apollo, discovered in 1932 and then lost until 1973. Apollo asteroids are so small and faint that they are difficult to see except when close to Earth (Apollo is about 2 km/1.2 mi across).

Apollo project US space project to land a person on the Moon, achieved 20 July 1969, when Neil Armstrong was the first to set foot there. He was accompanied on the Moon's surface by 'Buzz' Aldrin; Michael Collins remained in the orbiting command module.

The programme was announced in 1961 by President Kennedy. The world's most powerful rocket, *Saturn V*, was built to launch the Apollo spacecraft, which carried three astronauts. When the spacecraft was in orbit around the Moon, two astronauts would descend to the surface in a lunar module to take samples of rock and set up experiments that would send data back to Earth. After three other preparatory flights, *Apollo 11* made the first lunar landing. Five more crewed landings followed, the last in 1972. The total cost of the programme was over $24 billion.

APOLLO 11

http://www.nasa.gov/hqpao/apollo_11.html

This NASA page relives the excitement of the *Apollo 11* mission, with recollections from the participating astronauts, images, audio clips, access to key White House documents, and a bibliography.

Apollo–Soyuz test project joint US–Soviet space mission in which an Apollo and a Soyuz craft docked while in orbit around the Earth on 17 July 1975. The craft remained attached for two days and crew members were able to move from one craft to the other through an airlock attached to the nose of the Apollo. The mission was designed to test rescue procedures as well as having political significance.

apoptosis or ***cell suicide*** a cell's destruction of itself. All cells contain genes that cause them to self-destruct if damaged, diseased, or as part of the regulation of cell numbers during the organism's normal development. Many cancer cells have mutations in genes controlling apoptosis, so understanding apoptosis may lead to new cancer treatments where cells can be instructed to destroy themselves.

During apoptosis, a cell first produces the enzymes needed for self-destruction before shrinking to a characteristic spherical shape with balloon-like bumps on its outer surface. The enzymes break down its contents into small fragments which are easily digestible by surrounding cells.

aposematic coloration in biology, the technical name for **warning coloration** markings that make a dangerous, poisonous, or foul-tasting animal particularly conspicuous and recognizable to a predator. Examples include the yellow and black stripes of bees and wasps, and the bright red or yellow colours of many poisonous frogs. See also ◊mimicry.

apothecaries' weights obsolete units of mass, formerly used in pharmacy: 20 grains equal one scruple; three scruples equal one dram; eight drams equal an apothecary's ounce (oz apoth.), and 12 such ounces equal an apothecary's pound (lb apoth.). There are 7,000 grains in one pound avoirdupois (0.454 kg).

apparent depth depth that a transparent material such as water or glass appears to have when viewed from above. This is less than its real depth because of the ◊refraction that takes place when light passes into a less dense medium. The ratio of the real depth to the apparent depth of a transparent material is equal to its ◊refractive index.

appendicitis inflammation of the appendix, a small, blind extension of the bowel in the lower right abdomen. In an acute attack, the pus-filled appendix may burst, causing a potentially lethal spread of infection. Treatment is by removal (appendicectomy).

appendix a short, blind-ended tube attached to the ◊caecum. It has no known function in humans, but in herbivores it may be large, containing millions of bacteria that secrete enzymes to digest grass (as no vertebrate can secrete enzymes that will digest cellulose, the main constituent of plant cell walls).

apple fruit of several species of apple tree. There are several hundred varieties of cultivated apples, grown all over the world, which may be divided into eating, cooking, and cider apples. All are derived from the wild ◊crab apple. (Genus *Malus,* family Rosaceae.)

Apple trees grow best in temperate countries with a cool climate and plenty of rain during the winter. The desired variety is grafted onto rootstocks, and the tree must grow for six to eight years before it produces a good crop of fruit. The tree requires a winter period, in which it is dormant, in order to fruit in the spring, but must be protected from frost while the flowers and fruit are young. Pruning is necessary to produce strong branches, and sprays are used to protect the fruit from pests and to influence its development. The apple has been an important food plant in Europe and Asia for thousands of years.

Apple US computer company, manufacturer of the ◊Macintosh range of computers.

applet mini-software application. Examples of applets include Microsoft WordPad, the simple word processor in Windows 95, or the single-purpose applications that in 1996 were beginning to appear on the World Wide Web, written in Java. These include small animations, such as a moving ticker tape of stock prices.

Appleton layer or ***F layer*** band containing ionized gases in the Earth's upper atmosphere, at a height of 150–1,000 km/94–625 mi, above the ◊E layer (formerly the Kennelly–Heaviside layer). It acts as a dependable reflector of radio signals as it is not affected by atmospheric conditions, although its ionic composition varies with the sunspot cycle.

The Appleton layer has the highest concentration of free electrons and ions of the atmospheric layers. It is named after the English physicist Edward Appleton.

application in computing, a program or job designed for the benefit of the end user, such as a payroll system or a ◊word processor. The term is used to distinguish such programs from those that control the computer (◊systems programs) or assist the programmer, such as a ◊compiler.

application in mathematics, a curved line that connects a series of points (or 'nodes') in the smoothest possible way. The shape of the curve is governed by a series of complex mathematical formulae. Applications are used in ◊computer graphics and ◊CAD.

applications package in computing, the set of programs and related documentation (such as instruction manuals) used in a particular application. For example, a typical payroll applications package would consist of separate programs for the entry of data, updating the master files, and printing the pay slips, plus documentation in the form of program details and instructions for use.

Applications Program Interface (API) in computing, standard environment, including tools, protocols, and other routines, in which programs can be written. An API ensures that all applications are consistent with the operating system and have a similar ◊user interface.

appropriate technology simple or small-scale machinery and tools that, because they are cheap and easy to produce and maintain, may be of most use in the developing world; for example, hand ploughs and simple looms. This equipment may be used to supplement local crafts and traditional skills to encourage small-scale industrialization.

Many countries suffer from poor infrastructure and lack of capital but have the large supplies of labour needed for this level of technology. The use of appropriate technology was one of the recommendations of the Brandt Commission.

approximation rough estimate of a given value. For example, for ◊pi (which has a value of 3.1415926 correct to seven decimal places), 3 is an approximation to the nearest whole number.

APR abbreviation for ◊annual percentage rate.

apricot yellow-fleshed fruit of the apricot tree, which is closely related to the almond, peach, plum, and cherry. Although native to the Far East, it has long been cultivated in Armenia, from where it was introduced into Europe and the USA. (Genus *Prunus armeniaca,* family Rosaceae.)

aquaculture the cultivation of fish and shellfish for human consumption; see ◊fish farming.

aqualung or ***scuba*** underwater breathing apparatus worn by divers, developed in the early 1940s by French diver Jacques Cousteau. Compressed-air cylinders strapped to the diver's back are regulated by a valve system and by a mouth tube to provide air to the diver at the same pressure as that of the surrounding water (which increases with the depth).

aquamarine blue variety of the mineral ◊beryl. A semiprecious gemstone, it is used in jewellery.

aquaplaning phenomenon in which the tyres of a road vehicle cease to make direct contact with the road surface, owing to the presence of a thin film of water. As a result, the vehicle can go out of control (particularly if the steered wheels are involved).

aqua regia ***Latin 'royal water'*** mixture of three parts concentrated hydrochloric acid and one part concentrated nitric acid, which dissolves all metals except silver.

aquarium tank or similar container used for the study and display of living aquatic plants and animals. The same name is used for institutions that exhibit aquatic life. These have been common since Roman times, but the first modern public aquarium was opened in Regent's Park, London in 1853. A recent development is the oceanarium or seaquarium, a large display of marine life forms.

Aquarius zodiacal constellation a little south of the celestial equator near Pegasus. Aquarius is represented as a man pouring water from a jar. The Sun passes through Aquarius from late February to early March. In astrology, the dates for Aquarius, the 11th sign of the zodiac, are between about 20 January and 18 February (see ◊precession).

aquatic living in water. All life on Earth originated in the early oceans, because the aquatic environment has several advantages for organisms. Dehydration is almost impossible, temperatures usually remain stable, and the density of water provides physical support.

Life forms that cannot exist out of water, amphibians that take to the water on occasions, animals that are also perfectly at home on land, and insects that spend a stage of their life cycle in water can all be described as aquatic. Aquatic plants are known as ◊hydrophytes.

aquatic insect insect that spends all or part of its life in water. Of the 29 insect orders, 11 members have some aquatic stages. Most of these have aquatic, immature stages, which usually take place in fresh water, sometimes in brackish water (very few species are truly marine); the adults are terrestrial, but in some orders there are species where all stages (egg, larva, and adult) live in the water.
partially aquatic Three orders, Ephemeroptera (mayflies), Odonata (dragonflies), and Plecoptera (stone-flies) have aquatic larvae, but the adults are terrestrial. In the orders Neuroptera (alder flies), Tricoptera (caddis flies), Lepidoptera (butterflies and moths), and Diptera (true flies), some species have aquatic larvae, but the adults of all species are terrestrial. Hymenoptera, the social insect order which includes the ants and bees, has some aquatic species of ichneumon fly: immature stages of *Agriotypus* are aquatic, and adults of *Caraphractus* and *Prestwitchia*.
totally aquatic The order Collembola (springtails) has two species in which all stages are aquatic: *Hydropodura aquatica* and *Isotoma palustris*. In Hemiptera (bugs), and Coleoptera (beetles), some members, for example the water bugs, spend all stages of the life-cycle in the water.

aqueduct any artificial channel or conduit for water, originally applied to water supply tunnels, but later used to refer to elevated structures of stone, wood, or ironcarrying navigable canals across valleys.One of the first great aqueducts was built in 691 BC, carrying water for 80 km/50 mi to Ninevah, capital of the ancient Assyrian Empire. Many Roman aqueducts are still standing, for example the one carried by the Pont du Gard at Nîmes in S France, built about 8 BC (48 m/160 ft high).

The largest Roman aqueduct, at Carthage in Tunisia, is 141 km/87 mi long and was built during the reign of Publius Aelius Hadrianus between AD 117 and 138. A recent aqueduct is the California State Water Project taking water from Lake Oroville in the north, through two power plants and across the Tehachapi Mountains, more than 177 km/110 mi to S California.

aqueous humour watery fluid found in the chamber between the cornea and lens of the vertebrate eye. Similar to blood serum in composition, it is constantly renewed.

aqueous solution solution in which the solvent is water.

aquifer a body of rock through which appreciable amounts of water can flow. The rock of an aquifer must be porous and permeable (full of interconnected holes) so that it can conduct water. Aquifers are an important source of fresh water, for example, for drinking and irrigation, in many arid areas of the world, and are exploited by the use of ◊artesian wells.

An aquifer may be underlain, overlain, or sandwiched between less permeable layers, called aquicludes or **aquitards**, which impede water movement. Sandstones and porous limestones make the best aquifers.

Aquila constellation on the celestial equator (see ◊celestial sphere). Its brightest star is first-magnitude ◊Altair, flanked by the stars Beta and Gamma Aquilae. It is represented by an eagle.

Nova Aquilae, which appeared June 1918, shone for a few days nearly as brightly as Sirius.

arable farming cultivation of crops, as opposed to the keeping of animals. Crops may be ◊cereals, vegetables, or plants for producing oils or cloth. Arable farming generally requires less attention than livestock farming. In a ◊mixed farming system, crops may therefore be found farther from the farm centre than animals.

arachnid or ***arachnoid*** type of arthropod of the class Arachnida, including spiders, scorpions, ticks, and mites. They differ from insects in possessing only two main body regions, the cephalothorax and the abdomen, and in having eight legs.

araucaria coniferous tree related to the firs, with flat, scalelike needles. Once widespread, it is now native only to the southern hemisphere. Some grow to gigantic size. Araucarias include the monkey-puzzle tree (*Araucaria araucana*), the Australian bunya bunya pine (*A. bidwillii*), and the Norfolk Island pine (*A. heterophylla*). (Genus *Araucaria*, family Araucariaceae.)

ARACHNOLOGY

http://dns.ufsia.ac.be/Arachnology/Arachnology.html

Largely a collection of annotated links, organized under headings such as taxonomy and classification, palaeontology, poison, bites, diseases, pests, and phobia. The site also contains spider-related myths, stories, poems, songs, and art.

arboretum collection of trees. An arboretum may contain a wide variety of species or just closely related species or varieties – for example, different types of pine tree.

arbor vitae any of several coniferous trees or shrubs belonging to the cypress family, with flattened branchlets covered in overlapping aromatic green scales. The northern white cedar (*Thuja occidentalis*) and the western red cedar (*T. plicata*) are found in North America. The Chinese or Oriental species *T. orientalis*, reaching 18 m/60 ft in height, is widely grown as an ornamental. (Genus *Thuja*, family Cupressaceae.)

arbutus any of a group of evergreen shrubs belonging to the heath family, found in temperate regions. The strawberry tree (*Arbutus unedo*) is grown for its ornamental, strawberrylike fruit. (Genus *Arbutus*, family Ericaceae.)

arc in geometry, a section of a curved line or circle. A circle has three types of arc: a **semicircle**, which is exactly half of the circle; **minor arcs**, which are less than the semicircle; and **major arcs**, which are greater than the semicircle.

An arc of a circle is measured in degrees, according to the angle formed by joining its two ends to the centre of that circle. A semicircle is therefore 180°, whereas a minor arc will always be less than 180° (acute or obtuse) and a major arc will always be greater than 180° but less than 360° (reflex).

Archaea group of microorganisms that are without a nucleus and have a single chromosome. All are strict anaerobes, that is, they are killed by oxygen. This is thought to be a primitive condition and to indicate that Archaea are related to the earliest life forms, which appeared about 4 billion years ago, when there was little oxygen in the Earth's atmosphere. They are found in undersea vents, hot springs, the Dead Sea, and salt pans, and have even adapted to refuse tips.

Archaea was originally classified as bacterial, but in 1996 when the genome of *Methanococcus jannaschii* (an archaeaon that lives in undersea vents at temperatures around 100°C/212°F) was sequenced, US geneticists found that 56% of its genes were unlike those of any other organism, making Archaea unique.

In 1994 US biologists detected archaeans in the Antarctic (where they make up 30% of the single-celled marine biomass), Arctic, Mediterranean, and Baltic Sea.

Archaean or ***Archaeozoic*** widely used term for the earliest era of geological time; the first part of the Precambrian **Eon**, spanning the interval from the formation of Earth to about 2,500 million years ago.

archaeology ***Greek** archaia **'ancient things'**, logos **'study'*** study of prehistory and history, based on the examination of physical remains. Principal activities include preliminary field (or site) surveys, ◊excavation (where necessary), and the classification, ◊dating, and interpretation of finds.
history A museum found at the ancient Sumerian city of Ur indicates that interest in the physical remains of the past stretches back into prehistory. In the Renaissance this interest gained momentum among dealers in and collectors of ancient art and was further stimulated by discoveries made in Africa, the Americas, and Asia by Europeans during the period of imperialist colonization in the 16th–19th centuries, such as the antiquities discovered during Napoleon's Egyptian campaign in the 1790s. Romanticism in Europe stimulated an enthusiasm for the mouldering skull, the

ARCHAEOLOGY

http://www.archaeology.org/

Online edition of the prestigious journal of the Archaeological Institute of America. The well-written articles give a glimpse of the range of archaeological research being done around the globe. The full texts of short articles are available online, together with abstracts (and often the full text) of a large number of longer articles. Two years of back issues can be accessed.

ancient potsherds, ruins, and dolmens; relating archaeology to a wider context of art and literature.

Towards the end of the 19th century archaeology became an academic study, making increasing use of scientific techniques and systematic methodologies such as aerial photography. Since World War II new developments within the discipline include medieval, postmedieval, landscape, and industrial archaeology; underwater reconnaissance enabling the excavation of underwater sites; and rescue archaeology (excavation of sites risking destruction).

related disciplines Useful in archaeological studies are ◊dendrochronology (tree-ring dating), ◊geochronology (science of measuring geological time), ◊stratigraphy (study of geological strata), palaeobotany (study of ancient pollens, seeds, and grains), archaeozoology (analysis of animal remains), epigraphy (study of inscriptions), and numismatics (study of coins).

archaeopteryx ***Greek** archaios **'ancient'**, pterux **'wing'*** extinct primitive bird, known from fossilized remains, about 160 million years old, found in limestone deposits in Bavaria, Germany. It is popularly known as 'the first bird', although some earlier bird ancestors are now known. It was about the size of a crow and had feathers and wings, with three clawlike digits at the end of each wing, but in many respects its skeleton is reptilian (teeth and a long, bony tail) and very like some small meat-eating dinosaurs of the time.

archegonium ***Greek** arche **'origin'**, gonos **'offspring'*** female sex organ found in bryophytes (mosses and liverworts), pteridophytes (ferns, club mosses, and horsetails), and some gymnosperms. It is a multicellular, flask-shaped structure consisting of two parts: the swollen base or venter containing the egg cell, and the long, narrow neck. When the egg cell is mature, the cells of the neck dissolve, allowing the passage of the male gametes, or ◊antherozoids.

archerfish surface-living fish of the family Toxotidae, such as the genus *Toxotes*, native to SE Asia and Australia. The archerfish grows to about 25 cm/10 in and is able to shoot down insects up to 1.5 m/5 ft above the water by spitting a jet of water from its mouth.

Archie software tool for locating information on the ◊Internet. It can be difficult to locate a particular file because of the relatively unstructured nature of the Internet. Archie uses indexes of files and their locations on the Internet to find them quickly.

Archimedes
(c. 287–212 BC)

Greek mathematician who made major discoveries in geometry, hydrostatics, and mechanics, and established the sciences of statics and hydrostatics. He formulated a law of fluid displacement (Archimedes' principle), and is credited with the invention of the Archimedes screw, a cylindrical device for raising water. His method of finding mathematical proof to substantiate experiment and observation became the method of modern science in the High Renaissance.

Hydrostatics and Archimedes' principle The best-known result of Archimedes' work on hydrostatics is Archimedes' principle, which states that a body immersed in water will displace a volume of fluid that weighs as much as the body would weigh in air. It is alleged that Archimedes' principle was discovered when he stepped into the public bath and saw the water overflow. He was so delighted that he rushed home naked, crying 'Eureka! Eureka!' ('I have found it! I have found it!').

He used his discovery to prove that the goldsmith of Hieron II, King of Syracuse, had adulterated a gold crown with silver. Archimedes realized that if the gold had been mixed with silver (which is less dense than gold), the crown would have a greater volume and therefore displace more water than an equal weight of pure gold. The story goes that the crown was found to be impure, and that the unfortunate goldsmith was executed.

Statics and the lever In the field of statics, he is credited with working out the rigorous mathematical proofs behind the law of the lever. The lever had been used by other scientists, but it was Archimedes who demonstrated mathematically that the ratio of the effort applied to the load raised is equal to the inverse ratio of the distances of the effort and load from the pivot or fulcrum of the lever. Archimedes is credited with having claimed that if he had a sufficiently distant place to stand, he could use a lever to move the world.

This claim is said to have given rise to a challenge from King Hieron to Archimedes to show how he could move a truly heavy object with ease, even if he could not move the world. In answer to this, Archimedes developed a system of compound pulleys. According to Plutarch's Life of Marcellus (who sacked Syracuse), Archimedes used this to move with ease a ship that had been lifted with great effort by many men out of the harbour on to dry land. The ship was laden with passengers, crew and freight, but Archimedes – sitting at a distance from the ship – was reportedly able to pull it over the land as though it were gliding through water.

Mathematics Archimedes wrote many mathematical treatises, some of which still exist in altered forms in Arabic. Archimedes' approximation for the value for π was more accurate than any previous estimate – the value lying between $\frac{223}{71}$ and $\frac{220}{70}$. The average of these two numbers is less than 0.0003 different from the modern approximation for π. He also examined the expression of very large numbers, using a special notation to estimate the number of grains of sand in the Universe. Although the result, 10^{63}, was far from accurate, Archimedes demonstrated that large numbers could be considered and handled effectively.

Archimedes also evolved methods to solve cubic equations and to determine square roots by approximation. His formulae for the determination of the surface areas and volumes of curved surfaces and solids anticipated the development of integral calculus, which did not come for another 2,000 years. Archimedes had decreed that his gravestone be inscribed with a cylinder enclosing a sphere together with the formula for the ratio of their volumes – a discovery that he regarded as his greatest achievement.

Mary Evans Picture Library

Eureka! I have found it!

ARCHIMEDES Greek mathematician.
Remark, quoted in Vitruvius Pollio *De Architectura* IX

Archimedes' principle in physics, the principle that the weight of the liquid displaced by a floating body is equal to the weight of the body. The principle is often stated in the form: 'an object totally or partially submerged in a fluid displaces a volume of fluid that weighs the same as the apparent loss in weight of the object (which, in turn, equals the upwards force, or upthrust, experienced by that object).' It was discovered by the Greek mathematician Archimedes.

Archimedes screw one of the earliest kinds of pump, associated with the Greek mathematician Archimedes. It consists of an enormous spiral screw revolving inside a close-fitting cylinder. It is used, for example, to raise water for irrigation.

The lowest portion of the screw just dips into the water, and as the cylinder is turned a small quantity of water is scooped up. The inclination of the cylinder is such that at the next revolution the water is raised above the next thread, whilst the lowest thread scoops up another quantity. The successive revolutions, therefore, raise the water thread by thread until it emerges at the top of the cylinder.

archipelago group of islands, or an area of sea containing a group of islands. The islands of an archipelago are usually volcanic in origin, and they sometimes represent the tops of peaks in areas around continental margins flooded by the sea.

Volcanic islands are formed either when a hot spot within the Earth's mantle produces a chain of volcanoes on the surface, such as the Hawaiian Archipelago or at a destructive plate margin (see ◊plate tectonics) where the subduction of one plate beneath another produces an arc-shaped island group called an 'island arc', such as the Aleutian Archipelago. Novaya Zemlya in the Arctic Ocean, the northern extension of the Ural Mountains, resulted from continental flooding.

architecture in computing, the overall design of a computer system, encompassing both hardware and software. The architecture of a particular system includes the specifications of individual components and the ways they interact. Because the operating system defines how these elements interact with each other and with application software, it is also included in the term.

archive collection of computer files. The term is commonly used to refer to the files created by ◊data compression programs, such as the popular PKZIP, which contain one or more files. On the Internet it is also used to refer to a large store of files from which visitors can select the ones they want.

Archimedes screw The Archimedes screw, a spiral screw turned inside a cylinder, was once commonly used to lift water from canals. The screw is still used to lift water in the Nile delta in Egypt, and is often used to shift grain in mills and powders in factories.

arc lamp or ***arc light*** electric light that uses the illumination of an electric arc maintained between two electrodes. The English chemist Humphry Davy demonstrated the electric n arc in 1802 and electric arc lighting was first introduced by English electrical engineer W E Staite (1809–1854) in 1846. The lamp consists of two carbon electrodes, between which a very high voltage is maintained. Electric current arcs (jumps) between the two electrolytes, creating a brilliant light. Its main use in recent years has been in cinema projectors.

arc minute, arc second units for measuring small angles, used in geometry, surveying, map-making, and astronomy. An arc minute (symbol ') is one-sixtieth of a degree, and an arc second (symbol ') is one-sixtieth of an arc minute. Small distances in the sky, as between two close stars or the apparent width of a planet's disc, are expressed in minutes and seconds of arc.

arctic animals animals inhabiting the Arctic. The birds are chiefly sea birds, such as petrels, eider ducks, cormorants, auks, gulls, puffins, and guillemots; all are migratory. The mammals include the walrus, seals, and several varieties of whale; the polar bear, reindeer, elk, fox, wolf, ermine, and musk-ox are the principal terrestrial mammals.

Insectivorous and herbivorous habits are almost absent in arctic animals, which are either fish- or meat-eating. Many of them become snowy white in winter; among these are birds, such as the ptarmigan, and mammals, such as the hare and lemming, which are brown in summer, and the arctic fox, which is salty-blue in summer; the polar bear is white all year round.

Molluscs, annelids, and jellyfish are common to all the northern seas, while such fish as salmon, cod, and halibut are plentiful. Insects found in the far north include bees, flies, and butterflies, but as the flora is scanty they do not occur in great abundance.

ARCTIC CIRCLE

http://www.lib.uconn.edu/ArcticCircle/

Well-written site with excellent information about all aspects of life in the Arctic. There are sections on history, natural resources, the rights of indigenous peoples, and issues of environmental concern.

Arctic Circle imaginary line that encircles the North Pole at latitude 66° 33' north. Within this line there is at least one day in the summer during which the Sun never sets, and at least one day in the winter during which the Sun never rises.

Arcturus or ***Alpha Boötis*** brightest star in the constellation Boötes and the fourth-brightest star in the sky. Arcturus is a red giant about 28 times larger than the Sun and 70 times more luminous, 36 light years away from Earth.

are metric unit of area, equal to 100 square metres (119.6 sq yd); 100 ares make one ◊hectare.

area the size of a surface. It is measured in square units, usually square centimetres (cm^2), square metres (m^2), or square kilome-

Areas: Common Areas

figure	*rule for calculating area*
rectangle	length× breadth
triangle	half base length × vertical height
parallelogram	base length × vertical height
trapezium	average length of parallel sides × perpendicular distance between them
circle	πr^2, where r is the radiussector $\frac{x\pi r^2}{360}$ where x is the angle of the sector

tres (km^2). Surface area is the area of the outer surface of a solid.

The areas of geometrical plane shapes with straight edges are determined using the area of a rectangle. Integration may be used to determine the area of shapes enclosed by curves.

areca any of a group of palm trees native to Asia and Australia. The ◊betel nut comes from the species *Areca catechu.* (Genus *Areca.*)

Arecibo site in Puerto Rico of the world's largest single-dish ◊radio telescope, 305 m/1,000 ft in diameter. It is built in a natural hollow and uses the rotation of the Earth to scan the sky. It has been used both for radar work on the planets and for conventional radio astronomy, and is operated by Cornell University, USA.

In 1996 it received a $25 million upgrade, increasing the sensitivity of the disc tenfold. Two new mirrors were also added and the observation frequency increased from 3,000 megahertz to up to 10,000 megahertz. Another upgrade took place in 1997, when a new receiver capable of monitoring 168 million radio channels, SERENDIP IV, was added to the facility.

Water, water, everywhere, / Nor any drop to drink.

Samuel Coleridge English poet.
The Ancient Mariner pt 2

arête (***German*** **grat**; ***North American*** **combe-ridge**) sharp narrow ridge separating two ◊glacial troughs (valleys), or ◊corries. The typical U-shaped cross sections of glacial troughs give arêtes very steep sides. Arêtes are common in glaciated mountain regions such as the Rockies, the Himalayas, and the Alps.

argali wild sheep from the mountains of central Asia. It is the largest species of sheep with a shoulder height of up to 1.2 m/4 ft.
classification Argali *Ovis ammon* is in family Bovidae, order Artiodactyla.

Argand diagram in mathematics, a method for representing complex numbers by Cartesian coordinates (x, y). Along the x-axis (horizontal axis) are plotted the real numbers, and along the y-axis (vertical axis) the nonreal, or ◊imaginary, numbers.

argon ***Greek*** ***argos*** ***'idle'*** colourless, odourless, nonmetallic, gaseous element, symbol Ar, atomic number 18, relative atomic mass 39.948. It is grouped with the ◊inert gases, since it was long believed not to react with other substances, but observations now indicate that it can be made to combine with boron fluoride to form compounds. It constitutes almost 1% of the Earth's atmosphere, and was discovered in 1894 by British chemists John Rayleigh (1842–1919) and William Ramsay after all oxygen and nitrogen had been removed chemically from a sample of air. It is used in electric discharge tubes and argon lasers.

argonaut or ***paper nautilus*** octopus living in the open sea, genus *Argonauta*. The female of the common paper nautilus, *A. argo,* is 20 cm/8 in across, and secretes a spiralled papery shell for her eggs from the web of the first pair of arms. The male is a shell-less dwarf, 1 cm/0.4 in across.

argument in computing, the value on which a ◊function operates. For example, if the argument 16 is operated on by the function 'square root', the answer 4 is produced.

argument in mathematics, a specific value of the independent variable of a ◊function of x. It is also another name for ◊amplitude.

Ariane launch vehicle built in a series by the European Space Agency (first flight 1979). The launch site is at Kourou in French Guiana. Ariane is a three-stage rocket using liquid fuels. Small solid-fuel and liquid-fuel boosters can be attached to its first stage to increase carrying power.

Since 1984 it has been operated commercially by Arianespace, a private company financed by European banks and aerospace industries. A more powerful version, *Ariane 5,* was launched 4 June 1996, and was intended to carry astronauts aboard the Hermes spaceplane. However, it went off course immediately after takeoff , turned on its side, broke into two and disintegrated. A fault in the software controlling the takeoff trajectory was to blame.

A mostly successful test flight for *Ariane 5* was completed in November 1997.

arid region in earth science, a region that is very dry and has little vegetation. Aridity depends on temperature, rainfall, and evaporation, and so is difficult to quantify, but an arid area is usually defined as one that receives less than 250 mm/10 in of rainfall each year. (By comparison, New York City receives 1,120 mm/44 in per year.) There are arid regions in North Africa, Pakistan, Australia, the USA, and elsewhere. Very arid regions are ◊deserts.

Ariel series of six UK satellites launched by the USA 1962–79, the most significant of which was *Ariel 5* in 1974, which made a pioneering survey of the sky at X-ray wavelengths.

Aries zodiacal constellation in the northern hemisphere between Pisces and Taurus, near Auriga, represented as the legendary ram whose golden fleece was sought by Jason and the Argonauts.

Its most distinctive feature is a curve of three stars of decreasing brightness. The brightest of these is Hamal or Alpha Arietis, 65 light years from Earth.

The Sun passes through Aries from late April to mid-May. In astrology, the dates for Aries, the first sign of the zodiac, are between about 21 March and 19 April (see ◊precession). The spring ◊equinox once lay in Aries, but has now moved into Pisces through the effect of the Earth's precession (wobble).

aril accessory seed cover other than a ◊fruit; it may be fleshy and sometimes brightly coloured, woody, or hairy. In flowering plants (◊angiosperms) it is often derived from the stalk that originally attached the ovule to the ovary wall. Examples of arils include the bright-red, fleshy layer surrounding the yew seed (yews are ◊gymnosperms so they lack true fruits), and the network of hard filaments that partially covers the nutmeg seed and yields the spice known as mace.

Another aril, the horny outgrowth found towards one end of the seed of the castor-oil plant *Ricinus communis,* is called a caruncle. It is formed from the integuments (protective layers enclosing the ovule) and develops after fertilization.

arithmetic branch of mathematics concerned with the study of numbers and their properties. The fundamental operations of arithmetic are addition, subtraction, multiplication, and division. Raising to powers (for example, squaring or cubing a number), the extraction of roots (for example, square roots), percentages, fractions, and ratios are developed from these operations.

Forms of simple arithmetic existed in prehistoric times. In China, Egypt, Babylon, and early civilizations generally, arithmetic was used for commercial purposes, records of taxation, and astronomy. During the Dark Ages in Europe, knowledge of arithmetic was preserved in India and later among the Arabs. European mathematics revived with the development of trade and overseas exploration. Hindu-Arabic numerals replaced Roman numerals, allowing calculations to be made on paper, instead of by the ◊abacus.

The essential feature of this number system was the introduction of zero, which allows us to have a **place-value** system. The decimal numeral system employs ten numerals (0,1,2,3,4,5,6,7,8,9) and is said to operate in 'base ten'. In a base-ten number, each position has a value ten times that of the position to its immediate right; for example, in the number 23 the numeral 3 represents three units (ones), and the numeral 2 represents two tens. The Babylonians, however, used a complex base-sixty system, residues of which are found today in the number of minutes in each hour and in angular measurement (6×60 degrees). The Mayas used a base-twenty system.

There have been many inventions and developments to make the manipulation of the arithmetic processes easier, such as the invention of ◊logarithms by Scottish mathematician John ◊Napier in 1614 and of the slide rule in the period 1620–30. Since then, many forms of ready reckoners, mechanical and electronic calculators, and computers have been invented.

Modern computers fundamentally operate in base two, using only two numerals (0,1), known as a binary system. In binary, each position has a value twice as great as the position to its immediate right, so that for example binary 111 (or 111_2) is equal to 7 in the decimal system, and binary 1111 (or 1111_2) is equal to 15. Because the main operations of subtraction, multiplication, and division can be reduced mathematically to addition, digital computers carry out calculations by adding, usually in binary numbers in which the numerals 0 and 1 can be represented by off and on pulses of electric current.

Modular or modulo arithmetic, sometimes known as residue arithmetic or clock arithmetic, can take only a specific number of digits, whatever the value. For example, in modulo 4 (mod 4) the only values any number can take are 0, 1, 2, or 3. In this system, 7 is written as 3 mod 4, and 35 is also 3 mod 4. Notice 3 is the residue, or remainder, when 7 or 35 is divided by 4. This form of arithmetic is often illustrated on a circle. It deals with events recurring in regular cycles, and is used in describing the functioning of petrol engines, electrical generators, and so on. For example, in the mod 12, the answer to a question as to what time it will be in five hours if it is now ten o'clock can be expressed 10 + 5 = 3.

arithmetic and logic unit (ALU) in a computer, the part of the ◊central processing unit (CPU) that performs the basic arithmetic and logic operations on data.

arithmetic mean the average of a set of n numbers, obtained by adding the numbers and dividing by n. For example, the arithmetic mean of the set of 5 numbers 1, 3, 6, 8, and 12 is (1 + 3 + 6 + 8 + 12)/5 = 30/5 = 6.

The term 'average' is often used to refer only to the arithmetic mean, even though the mean is in fact only one form of average (the others include ◊median and ◊mode).

arithmetic progression or ***arithmetic sequence*** sequence of numbers or terms that have a common difference between any one term and the next in the sequence. For example, 2, 7, 12, 17, 22, 27, ... is an arithmetic sequence with a common difference of 5.

The nth term in any arithmetic progression can be found using the formula:

$$n\text{th term} = a + (n - 1)d$$

where a is the first term and d is the common difference.

An **arithmetic series** is the sum of the terms in an arithmetic sequence. The sum S of n terms is given by:

$$S = \frac{n}{2}[2a + (n - 1)d]$$

ARM (abbreviation for ***Advanced RISC Machine***) microprocessor developed by Acorn in 1985 for use in the Archimedes microcomputer. In 1990 the company Advanced RISC Machines was formed to develop the ARM microprocessor. The ARM is the microprocessor in Apple's ◊Newton.

armadillo mammal of the family Dasypodidae, with an armour of bony plates along its back or, in some species, almost covering the entire body. Around 20 species live between Texas and Patagonia and range in size from the fairy armadillo, or pichiciego, *Chlamyphorus truncatus,* at 13 cm/5 in, to the giant armadillo *Priodontes giganteus,* 1.5 m/4.5 ft long. Armadillos feed on insects, snakes, fruit, and carrion. Some can roll into an armoured ball if attacked; others defend themselves with their claws or rely on rapid burrowing for protection.

ARMADILLO

Nine-banded armadillos almost always produce litters of quadruplets. The one fertilized egg divides into four to produce a single-sex litter of identical babies. If times are hard, the female armadillo can delay the implantation of her egg for up to three years to maximize the chances of her offspring's survival.

They belong to the order Edentata ('without teeth') which also includes sloths and anteaters. However, only the latter are toothless. Some species of armadillos can have up to 90 peglike teeth.

armature in a motor or generator, the wire-wound coil that carries the current and rotates in a magnetic field. (In alternating-current machines, the armature is sometimes stationary.) The pole piece of a permanent magnet or electromagnet and the moving, iron part of a ◊solenoid, especially if the latter acts as a switch, may also be referred to as armatures.

armillary sphere earliest known astronomical device, in use from the 3rd century BC. It showed the Earth at the centre of the universe, surrounded by a number of movable metal rings representing the Sun, Moon, and planets. The armillary sphere was originally used to observe the heavens and later for teaching navigators about the arrangements and movements of the heavenly bodies.

aromatherapy in alternative medicine, use of oils and essences derived from plants, flowers, and wood resins. Bactericidal properties and beneficial effects upon physiological functions are attributed to the oils, which are sometimes ingested but generally massaged into the skin.

Aromatherapy was first used in ancient Greece and Egypt, but became a forgotten art until the 1930s, when a French chemist accidentally spilt lavender over a cut and found that the wound healed without a scar. However, it was not until the 1970s that it began to achieve widespread popularity.

aromatic compound organic chemical compound in which some of the bonding electrons are delocalized (shared among several atoms within the molecule and not localized in the vicinity of the atoms involved in bonding). The commonest aromatic compounds have ring structures, the atoms comprising the ring being either all carbon or containing one or more different atoms (usually nitrogen, sulphur, or oxygen). Typical examples are benzene (C_6H_6) and pyridine (C_6H_5N).

ARPANET (acronym for ***Advanced Research Projects Agency Network***) early US network that forms the basis of the ◊Internet. It was set up in 1969 by ARPA to provide services to US academic institutions and commercial organizations conducting computer science research.

ARPANET pioneered many of today's networking techniques.

It was renamed DARPANET when ARPA changed its name to Defense Advanced Research Projects Agency. In 1975 responsibility for DARPANET was passed on to the Defense Communication Agency.

array in computer programming, a list of values that can all be referred to by a single ◊variable name. Separate values are distinguished by using a **subscript** with each variable name.

Arrays are useful because they allow programmers to write general routines that can process long lists of data. For example, if every price stored in an accounting program used a different variable name, separate program instructions would be needed to process each price. However, if all the prices were stored in an array, a general routine could be written to process, say, 'price (J)', and, by allowing J to take different values, could then process any individual price.

For example, consider this list of highest daily temperatures: day 1 – 22°C; day 2 – 23°C; day 3 – 19°C; day 4 – 21°C. This array might be stored with the single variable name 'temp'. Separate elements of the array would then be identified with subscripts. So, for example, the array element 'temp($_1$)' would store the value '22', and the array element 'temp($_3$)' would store the value '19'.

array collection of numbers (or letters representing numbers) arranged in rows and columns. A ◊matrix is an array shown inside a pair of brackets; it indicates that the array should be treated as a single entity.

arrhythmia disturbance of the normal rhythm of the heart. There are various kinds of arrhythmia, some benign, some indicative of heart disease. In extreme cases, the heart may beat so fast as to be potentially lethal and surgery may be used to correct the condition.

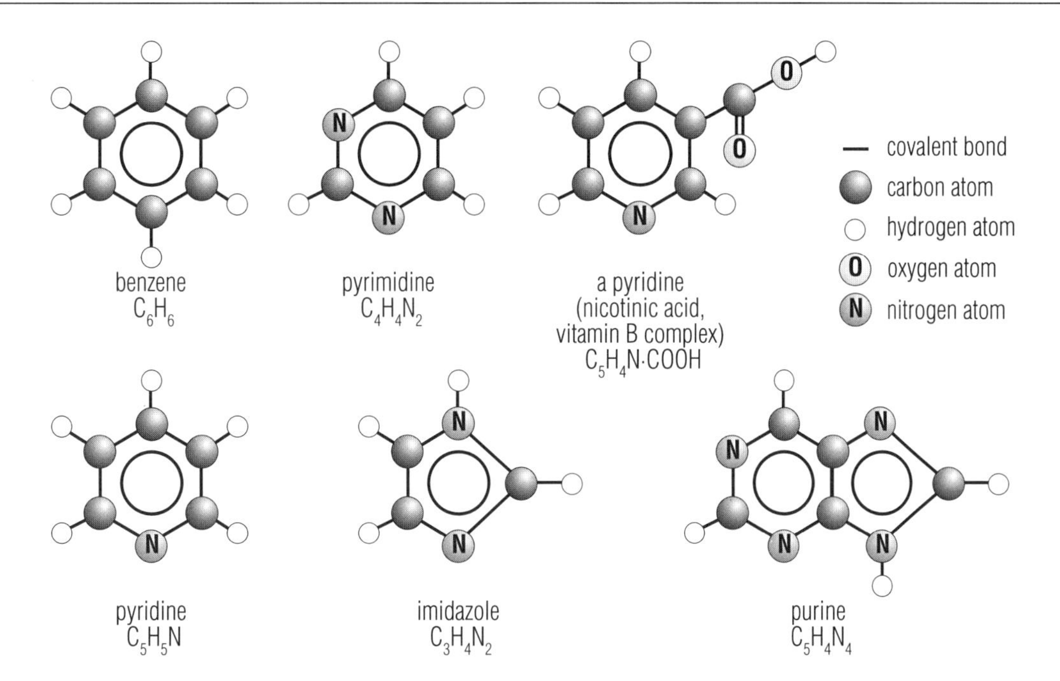

***aromatic compound** Compounds whose molecules contain the benzene ring, or variations of it, are called aromatic. The term was originally used to distinguish sweet-smelling compounds from others.*

Extra beats between the normal ones are called **extrasystoles**; abnormal slowing is known as **bradycardia** and speeding up is known as **tachycardia**.

arrowroot starchy substance used as a thickener in cooking, produced from the clumpy roots of various tropical plants. The true arrowroot (*Maranta arundinacea*) was used by native South Americans as an antidote against the effects of poisoned arrows.

The West Indian island of St Vincent is the main source of supply today. The plant roots and tubers are dried, finely powdered, and filtered. Because of the small size of the starch particles, the powder becomes translucent when cooked.

arrowwood any of various North American trees and shrubs, especially of the genus *Viburnum*, named for their long, straight branches, which were used by American Indians to make arrows.

arsenic brittle, greyish-white, semimetallic element (a metalloid), symbol As, atomic number 33, relative atomic mass 74.92. It occurs in many ores and occasionally in its elemental state, and is widely distributed, being present in minute quantities in the soil, the sea, and the human body. In larger quantities, it is poisonous. The chief source of arsenic compounds is as a by-product from metallurgical processes. It is used in making semiconductors, alloys, and solders.

As it is a cumulative poison, its presence in food and drugs is very dangerous. The symptoms of arsenic poisoning are vomiting, diarrhoea, tingling and possibly numbness in the limbs, and collapse. It featured in some drugs, including Salvarsan, the first specific treatment for syphilis. Its name derives from the Latin *arsenicum*.

arteriosclerosis hardening of the arteries, with thickening and loss of elasticity. It is associated with smoking, ageing, and a diet high in saturated fats. The term is used loosely as a synonym for ◊atherosclerosis.

artery vessel that carries blood from the heart to the rest of the body. It is built to withstand considerable pressure, having thick walls which contain smooth muscle fibres. During contraction of the heart muscle, arteries expand in diameter to allow for the sudden increase in pressure that occurs; the resulting ◊pulse or pressure wave can be felt at the wrist. Not all arteries carry oxygenated (oxygen-rich) blood; the pulmonary arteries convey deoxygenated (oxygen-poor) blood from the heart to the lungs.

Arteries are flexible, elastic tubes, consisting of three layers, the middle of which is muscular; its rhythmic contraction aids the pumping of blood around the body. In middle and old age, the walls degenerate and are vulnerable to damage by the build-up of fatty deposits. These reduce elasticity, hardening the arteries and decreasing the internal bore. This condition, known as ◊atherosclerosis, can lead to high blood pressure, loss of circulation, heart disease, and death.

Research indicates that a typical Western diet, high in saturated fat, increases the chances of arterial disease developing.

artesian well well that is supplied with water rising naturally from an underground water-saturated rock layer (◊aquifer). The water rises from the aquifer under its own pressure. Such a well may be drilled into an aquifer that is confined by impermeable rocks both above and below. If the water table (the top of the region of water saturation) in that aquifer is above the level of the well head, hydrostatic pressure will force the water to the surface.

Artesian wells are often overexploited because their water is fresh and easily available, and they eventually become unreliable. There is also some concern that pollutants such as pesticides or nitrates can seep into the aquifers.

arthritis inflammation of the joints, with pain, swelling, and restricted motion. Many conditions may cause arthritis, including gout, infection, and trauma to the joint. There are three main forms of arthritis: ◊rheumatoid arthritis; osteoarthritis; and septic arthritis.

arthropod member of the phylum Arthropoda; an invertebrate animal with jointed legs and a segmented body with a horny or chitinous casing (exoskeleton), which is shed periodically and replaced as the animal grows. Included are arachnids such as spiders and mites, as well as crustaceans, millipedes, centipedes, and insects.

artichoke either of two plants belonging to the sunflower family, parts of which are eaten as vegetables. The **common** or **globe artichoke** (*Cynara scolymus*) is a form of thistle native to the Mediterranean. It is tall, with purplish-blue flowers; the leaflike structures (bracts) around the unopened flower are eaten. The **Jerusalem artichoke** (*Helianthus tuberosus*), which has edible tubers, is a native of North America (its common name is a corruption of the Italian for sunflower, *girasole*). (Family Compositae.)

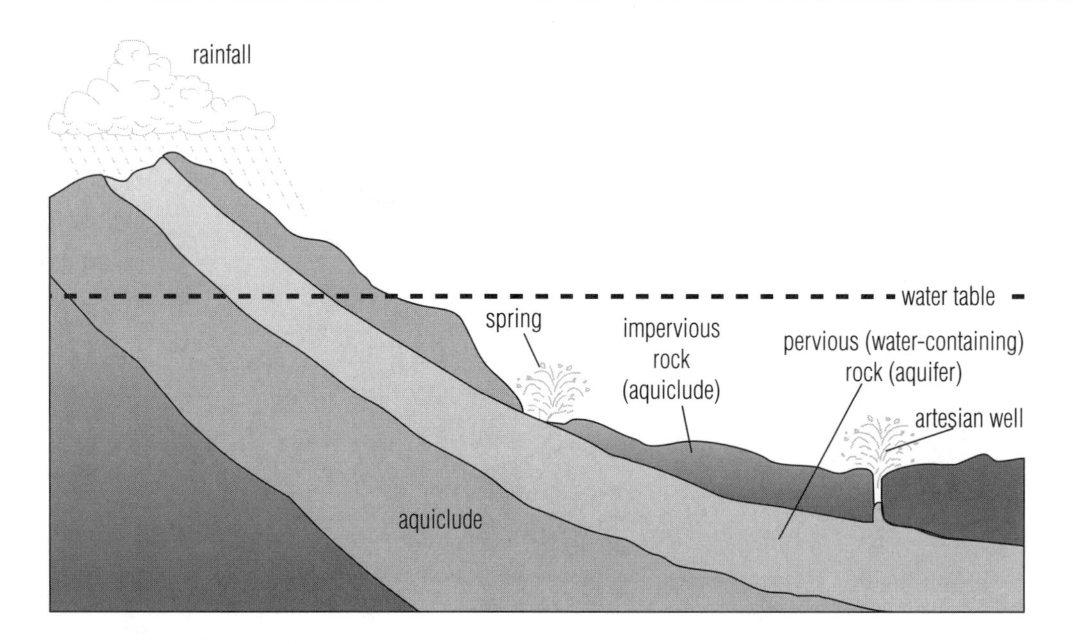

artesian well *In an artesian well, water rises from an underground water-containing rock layer under its own pressure. Rain falls at one end of the water-bearing layer, or aquifer, and percolates through the layer. The layer fills with water up to the level of the water table. Water will flow from a well under its own pressure if the well head is below the level of the water table.*

article or ***posting*** on USENET, an individual public message.

artificial insemination (AI) introduction by instrument of semen from a sperm bank or donor into the female reproductive tract to bring about fertilization. Originally used by animal breeders to improve stock with sperm from high-quality males, in the 20th century it has been developed for use in humans, to help the infertile. See ◊in vitro fertilization.

The whole thinking process is rather mysterious to us, but I believe that the attempt to make a thinking machine will help us greatly in finding out how we think ourselves.

ALAN MATHISON TURING English mathematician.
Quoted in A Hodges *Alan Turing: The Enigma of Intelligence* 1985

artificial intelligence (AI) branch of science concerned with creating computer programs that can perform actions comparable with those of an intelligent human. Current AI research covers such areas as planning (for robot behaviour), language understanding, pattern recognition, and knowledge representation.

The possibility of artificial intelligence was first proposed by the English mathematician Alan ◊Turing in 1950. Early AI programs, developed in the 1960s, attempted simulations of human intelligence or were aimed at general problem-solving techniques. By the mid-1990s, scientists were concluding that AI was more difficult to create than they had imagined. It is now thought that intelligent behaviour depends as much on the knowledge a system possesses as on its reasoning power. Present emphasis is on ◊knowledge-based systems, such as ◊expert systems, while research projects focus on ◊neural networks, which attempt to mimic the structure of the human brain.

On the ◊Internet, small bits of software that automate common routines or attempt to predict human likes or behaviour based on past experience are called intelligent agents or bots.

artificial life (**contracted to *ALife***) in computing, area of scientific research that attempts to simulate biological phenomena via computer programs. The first ALife workshop was held at Los Alamos, USA, in 1987. Research in this area is being conducted all around the world; one of the most significant centres is the ◊MIT Media Lab.

artificial limb device to replace a limb that has been removed by surgery or lost through injury, or one that is malformed because of genetic defects. It is one form of ◊prosthesis.

artificial radioactivity natural and spontaneous radioactivity arising from radioactive isotopes or elements that are formed when elements are bombarded with subatomic particles – protons, neutrons, or electrons – or small nuclei.

artificial respiration emergency procedure to restart breathing once it has stopped; in cases of electric shock or apparent drowning, for example, the first choice is the expired-air method, the **kiss of life** by mouth-to-mouth breathing until natural breathing is restored.

artificial selection in biology, selective breeding of individuals that exhibit the particular characteristics that a plant or animal breeder wishes to develop. In plants, desirable features might include resistance to disease, high yield (in crop plants), or attractive appearance. In animal breeding, selection has led to the development of particular breeds of cattle for improved meat production (such as the Aberdeen Angus) or milk production (such as Jerseys).

Artificial selection was practised by the Sumerians at least 5,500 years ago and carried on through the succeeding ages, with the result that all common vegetables, fruit, and livestock are long modified by selective breeding. Artificial selection, particularly of pigeons, was studied by the English evolutionist Charles Darwin who saw a similarity between this phenomenon and the processes of natural selection.

Artiodactyla order of even-toed mammals containing pigs, camels, hippos, and ruminant animals, such as antelope, deer, and sheep.

The order is divided into nine living families (the other 20 families are all extinct). Suidae includes the eight species of pig; Tayassuidae, the two species of peccary; and Hippopotamidae, the two species of hippopotamuses. Camelidae has three to five species of camels and the guanaco. The five ruminant (cud-chewing) families are Tragulidae, with four species of chevrotain; Cervidae, 41 species of deer; Bovidae, 128 species of cattle, sheep, and

arum Arum dioscoridis *comes from the Mediterranean region, where arum lilies grow in abundance. What appears to be a single flower is in fact a flower head, the spadix, bearing separate rings of male and female flowers enclosed in a leaflike spathe. Premaphotos Wildlife*

antelopes; Antilocapridae, one species, the pronghorn, often misnamed antelope, of North America; and Giraffidae, two species of giraffe.

arum any of a group of mainly European plants with narrow leaves and a single, usually white, special leaf (spathe) surrounding the spike of tiny flowers. The ornamental arum called the trumpet lily (*Zantedeschia aethiopica*) is a native of South Africa. (Genus *Arum*, family Araceae.)

asbestos any of several related minerals of fibrous structure that offer great heat resistance because of their nonflammability and poor conductivity. Commercial asbestos is generally either made from serpentine ('white' asbestos) or from sodium iron silicate ('blue' asbestos). The fibres are woven together or bound by an inert material. Over time the fibres can work loose and, because they are small enough to float freely in the air or be inhaled, asbestos usage is now strictly controlled; exposure to its dust can cause cancer.

ASCII (acronym for ***American standard code for information interchange***) in computing, a coding system in which numbers are assigned to letters, digits, and punctuation symbols. Although computers work in code based on the ◊binary number system, ASCII numbers are usually quoted as decimal or ◊hexadecimal numbers. For example, the decimal number 45 (binary 0101101) represents a hyphen, and 65 (binary 1000001) a capital A. The first 32 codes are used for control functions, such as carriage return and backspace.

Strictly speaking, ASCII is a 7-bit binary code, allowing 128 different characters to be represented, but an eighth bit is often used to provide ◊parity or to allow for extra characters. The system is widely used for the storage of text and for the transmission of data between computers.

ASCII art pictures or fancy graphics created entirely out of ◊ASCII characters such as letters of the alphabet or punctuation marks. ASCII art has existed since the invention of computers. Today it is found in USENET ◊signatures (.sigs), special ◊newsgroups such as alt.art.ascii, and occasionally in messages, both public and private.

ascorbic acid $C_6H_8O_6$ or ***vitamin C*** a relatively simple organic acid found in citrus fruits and vegetables. It is soluble in water and destroyed by prolonged boiling, so soaking or overcooking of vegetables reduces their vitamin C content. Lack of ascorbic acid results in scurvy.

In the human body, ascorbic acid is necessary for the correct synthesis of ◊collagen. Lack of vitamin C causes skin sores or ulcers, tooth and gum problems, and burst capillaries (scurvy symptoms) owing to an abnormal type of collagen replacing the normal type in these tissues.

The Australian billygoat plum, *Terminalia ferdiandiana*, is the richest natural source of vitamin C, containing 100 times the concentration found in oranges.

asepsis practice of ensuring that bacteria are excluded from open sites during surgery, wound dressing, blood sampling, and other medical procedures. Aseptic technique is a first line of defence against infection.

asexual reproduction in biology, reproduction that does not involve the manufacture and fusion of sex cells, nor the necessity for two parents. The process carries a clear advantage in that there is no need to search for a mate nor to develop complex pollinating mechanisms; every asexual organism can reproduce on its own. Asexual reproduction can therefore lead to a rapid population build-up.

In evolutionary terms, the disadvantage of asexual reproduction arises from the fact that only identical individuals, or clones, are produced – there is no variation.

In the field of horticulture, where standardized production is needed, this is useful, but in the wild, an asexual population that cannot adapt to a changing environment or evolve defences against a new disease is at risk of extinction. Many asexually reproducing organisms are therefore capable of reproducing sexually as well.

Asexual processes include ◊binary fission, in which the parent organism splits into two or more 'daughter' organisms, and ◊budding, in which a new organism is formed initially as an out-

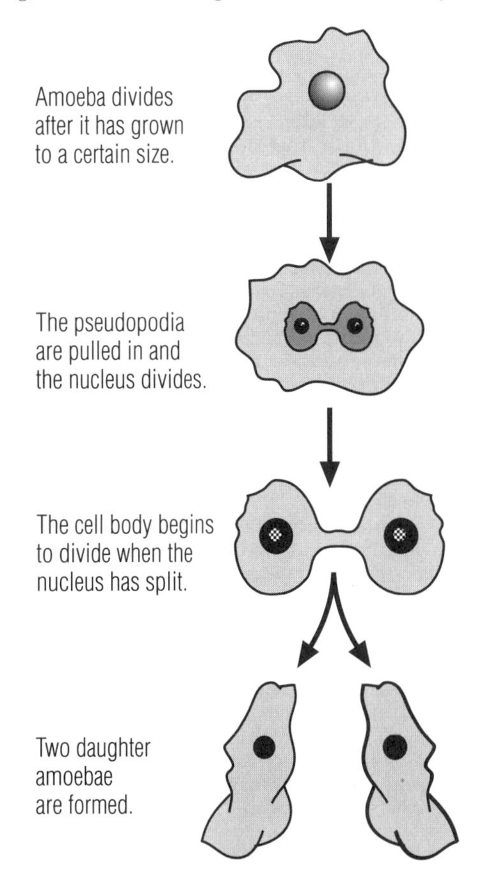

asexual reproduction Asexual reproduction is the simplest form of reproduction, occurring in many simple plants and animals. Binary fission, shown here occurring in an amoeba, is one of a number of asexual reproduction processes.

growth of the parent organism. The asexual reproduction of spores, as in ferns and mosses, is also common and many plants reproduce asexually by means of runners, rhizomes, bulbs, and corms; see also ◊vegetative reproduction.

ash any tree of a worldwide group belonging to the olive family, with winged fruits. The ◊mountain ash **or rowan**, which resembles the ash, belongs to the family Rosaceae. (Genus *Fraxinus*, family Oleaceae.)

ashen light in astronomy, a faint glow occasionally reported in the dark hemisphere of ◊Venus when the planet is in a crescent phase. Its origin is unknown, but it may be related to the terrestrial airglow caused by interaction of high-energy solar radiation with the upper atmosphere of the Earth.

Asiatic wild ass alternative name for both the ◊kiang and the ◊onager.

ASIC (abbreviation for ***application-specific integrated circuit***) integrated circuit built for a specific application.

asp any of several venomous snakes, including *Vipera aspis* of S Europe, allied to the adder, and the Egyptian cobra *Naja haje*, reputed to have been used by the Egyptian queen Cleopatra for her suicide.

asparagus any of a group of plants with small scalelike leaves and many fine, feathery branches. Native to Europe and Asia, *Asparagus officinalis* is cultivated and the tender young shoots (spears) are greatly prized as a vegetable. (Genus *Asparagus*, family Liliaceae.)

aspartame noncarbohydrate sweetener used in foods under the tradename Nutrasweet. It is about 200 times as sweet as sugar and, unlike saccharine, has no aftertaste.

The aspartame molecule consists of two amino acids (aspartic acid and phenylalanine) linked by a methylene ($-CH_2-$) group. It breaks down slowly at room temperature and rapidly at higher temperatures. It is not suitable for people who suffer from phenylketonuria.

aspen any of several species of ◊poplar tree. The European quaking aspen (*Populus tremula*) has flattened leafstalks that cause the leaves to flutter in the slightest breeze. The soft, light-coloured wood is used for matches and paper pulp. (Genus *Populus*.)

asphalt mineral mixture containing semisolid brown or black ◊bitumen, used in the construction industry. Asphalt is mixed with rock chips to form paving material, and the purer varieties are used for insulating material and for waterproofing masonry. It can be produced artificially by the distillation of ◊petroleum.

The availability of recycled coloured glass led in 1988 to the invention of **glassphalt**, asphalt that is 15% crushed glass. It is used to pave roads in New York.

Considerable natural deposits of asphalt occur around the Dead Sea and in the Philippines, Cuba, Venezuela, and Trinidad. Bituminous limestone occurs at Neufchâtel, France.

asphodel either of two related Old World plants of the lily family. The white asphodel or king's spear (*Asphodelus albus*) is found in Italy and Greece, sometimes covering large areas, and providing grazing for sheep. The other asphodel is the yellow asphodel (*Asphodeline lutea*). (Genera *Asphodelus* and *Asphodeline*, family Liliaceae.)

The ancient Greeks connected the beautiful plants of *A. lutea* with the dead, and they were supposed to grow in the mythological Elysian fields where heroes enjoyed new life after death.

asphyxia suffocation; a lack of oxygen that produces a potentially lethal build-up of carbon dioxide waste in the tissues.

Asphyxia may arise from any one of a number of causes, including inhalation of smoke or poisonous gases, obstruction of the windpipe (by water, food, vomit, or a foreign object), strangulation, or smothering. If it is not quickly relieved, brain damage or death ensues.

aspidistra any of several Asiatic plants of the lily family. The Chinese *Aspidistra elatior* has broad leaves which taper to a point and, like all aspidistras, grows well in warm indoor conditions. (Genus *Aspidistra*, family Liliaceae.)

aspirin acetylsalicylic acid, a popular pain-relieving drug (◊analgesic) developed in the late 19th century as a household remedy for aches and pains. It relieves pain and reduces inflammation and fever. It is derived from the white willow tree *Salix alba*, and is the world's most widely used drug.

Aspirin was first refined from salicylic acid by German chemist Felix Hoffman, and marketed in 1899. Although salicylic acid occurs naturally in willow bark (and has been used for pain relief since 1763) the acetyl derivative is less bitter and less likely to cause vomiting.

ass any of several horselike, odd-toed, hoofed mammals of the genus *Equus*, family Equidae. Species include the African wild ass *E. asinus*, and the Asian wild ass *E. hemionus*. They differ from horses in their smaller size, larger ears, tufted tail, and characteristic bray. Donkeys and burros are domesticated asses.

assassin bug member of a family of blood-sucking bugs that contains about 4,000 species. Assassin bugs are mainly predators, feeding on other insects, but some species feed on birds and mammals, including humans. They are found, mainly in tropical regions, although some have established themselves in Europe and North America.

classification Assassin bugs are in family Reduviidae, suborder Heteroptera, order Hemiptera (true bugs), class Insecta, phylum Arthropoda.

The general characteristics of the family include bright coloration, a long four-segmented antenna, and a cone-shaped proboscis which, when the insect is not feeding, is folded under the head. Because of this they are sometimes called **cone-nosed** bugs.

species *Reduvius personatus* is about 15 mm/0.5 in long and dark brown. In the wild it inhabits hollow trees where it feeds on the blood of other insects. It can invade houses, where it hides in holes and crevices in the wall. Like other assassin bugs, it is nocturnal, emerging at night to feed on ◊bedbugs and other insects. It is widely distributed in Europe and North America.

Kissing bugs comprise several Central and South American genera that have earned their name from their habit of biting sleeping people on the face. *Rhodnius*, *Triatoma*, and *Panstrongylus*, are vectors for the parasite that causes ◊Chagas's disease.

assay in chemistry, the determination of the quantity of a given substance present in a sample. Usually it refers to determining the purity of precious metals.

The assay may be carried out by 'wet' methods, when the sample is wholly or partially dissolved in some reagent (often an acid), or by 'dry' or 'fire' methods, in which the compounds present in the sample are combined with other substances.

assembler in computing, a program that translates a program written in an assembly language into a complete ◊machine code program that can be executed by a computer. Each instruction in the assembly language is translated into only one machine-code instruction.

assembly industry manufacture that involves putting together many prefabricated components to make a complete product; for example, a car or television set. The inputs for this type of industry are therefore outputs from others. Some assembly industries are surrounded by their suppliers; others use components from far afield.

assembly language low-level computer-programming language closely related to a computer's internal codes. It consists chiefly of a set of short sequences of letters (mnemonics), which are translated, by a program called an assembler, into ◊machine code for the computer's ◊central processing unit (CPU) to follow directly. In assembly language, for example, 'JMP' means 'jump' and 'LDA' means 'load accumulator'. Assembly code is used by programmers who need to write very fast or efficient programs.

Because they are much easier to use, high-level languages are normally used in preference to assembly languages. An assembly language may still be used in some cases, however, particularly when no suitable high-level language exists or where a very efficient machine-code program is required.

assembly line method of mass production in which a product is built up step-by-step by successive workers adding one part at a time. It is commonly used in industries such as the car industry.

US inventor Eli Whitney pioneered the concept of industrial assembly in the 1790s, when he employed unskilled labour to assemble muskets from sets of identical precision-made parts produced by machine tools. In 1901 Ransome Olds in the USA began mass-producing motor cars on an assembly-line principle, a method further refined by the introduction of the moving conveyor belt by Henry Ford in 1913 and the time-and-motion studies of Fredriech Winslow Taylor. On the assembly line human workers now stand side by side with ◊robots.

assimilation in animals, the process by which absorbed food molecules, circulating in the blood, pass into the cells and are used for growth, tissue repair, and other metabolic activities. The actual destiny of each food molecule depends not only on its type, but also on the body requirements at that time.

Association for Computing Machinery (ACM) US organization made up of computer professionals of all types. Its monthly journal, the *Communications of the Association for Computing Machinery*, is peer-reviewed. Its subsidiary special interest groups, or **SIGs**, focus on areas such as graphics and human–computer interaction. Several of these run major conferences for their areas such as SIGGRAPH (graphics) and SIGCHI (human–computer interaction).

The equivalent UK organization is the **British Computer Society** (BCS).

associative operation in mathematics, an operation in which the outcome is independent of the grouping of the numbers or symbols concerned. For example, multiplication is associative, as $4 \times (3 \times 2) = (4 \times 3) \times 2 = 24$; however, division is not, as $12 \div (4 \div 2) = 6$, but $(12 \div 4) \div 2 = 1.5$. Compare ◊commutative operation and ◊distributive operation.

assortative mating in ◊population genetics, selective mating in a population between individuals that are genetically related or have similar characteristics. If sufficiently consistent, assortative mating can theoretically result in the evolution of new species without geographical isolation (see ◊speciation).

astatine ***Greek astatos 'unstable'*** nonmetallic, radioactive element, symbol At, atomic number 85, relative atomic mass 210. It is a member of the ◊halogen group, and is very rare in nature. Astatine is highly unstable, with at least 19 isotopes; the longest lived has a half-life of about eight hours.

aster any plant of a large group belonging to the same subfamily as the daisy. All asters have starlike flowers with yellow centres and outer rays (not petals) varying from blue and purple to white. Asters come in many sizes. Many are cultivated as garden flowers, including the Michaelmas daisy (*Aster nova-belgii*). (Genus *Aster*, family Compositae.)

The China aster (*Callistephus chinensis*) belongs to a closely related genus; it was introduced to Europe and the USA from China in the early 18th century.

asterisk (*) or ***star*** wild card character standing for multiple characters in most operating systems. It allows a user to specify a group of files for mass handling. Typing 'dir *.bat' in DOS, for example, will return a list of all files with the extension .BAT in the current directory. On USENET, * is used to denote a group of ◊newsgroups; the phrase 'alt.music.*' means all the newsgroups in the alt.music hierarchy, such as alt.music.pop, alt.music.jazz, and so on. On the Internet, an asterisk before and after a word is a way of indicating emphasis.

asteroid or ***minor planet*** any of many thousands of small bodies, composed of rock and iron, that orbit the Sun. Most lie in a belt between the orbits of Mars and Jupiter, and are thought to be fragments left over from the formation of the ◊Solar System. About 100,000 may exist, but their total mass is only a few hundredths the mass of the Moon.

They include ◊Ceres (the largest asteroid, 940 km/584 mi in diameter), Vesta (which has a light-coloured surface, and is the brightest as seen from Earth), ◊Eros, and ◊Icarus. Some asteroids are in orbits that bring them close to Earth, and some, such as the ◊Apollo asteroids, even cross Earth's orbit; at least some of these may be remnants of former comets. One group, the Trojans, moves along the same orbit as Jupiter, 60° ahead and behind the planet. One unusual asteroid, ◊Chiron, orbits beyond Saturn.

NASA's Near Earth Asteroid Rendezvous (NEAR) was launched in February 1996 to study Eros to ascertain what asteroids are made of and whether they are similar in structure to meteorites. In 1997 it flew past asteroid Mathilde, revealing a 25 km crater covering the 53-km asteroid. The Near Earth Asteroid Tracking (NEAT) system had detected more than 10,000 asteroids by August 1997.

The first asteroid was discovered by the Italian astronomer Giuseppe Piazzi at the Palermo Observatory, Sicily, 1 January 1801. The first asteroid moon was observed by the space probe *Galileo* ◊in 1993 orbiting asteroid Ida.

Bifurcated asteroids, first discovered 1990, are in fact two chunks of rock that touch each other. It may be that at least 10% of asteroids approaching the Earth are bifurcated.

asthenosphere a layer within Earth's ◊mantle lying beneath the ◊lithosphere, typically beginning at a depth of approximately 100 km/63 mi and extending to depths of approximately 260 km/160 mi. Sometimes referred to as the 'weak sphere', it is characterized by being weaker and more elastic than the surrounding mantle.

The asthenosphere's elastic behaviour and low viscosity allow the overlying, more rigid plates of lithosphere to move laterally in a process known as ◊plate tectonics. Its elasticity and viscosity also allow overlying crust and mantle to move vertically in response to gravity to achieve **isostatic equilibrium** (see ◊isostasy).

asthma chronic condition characterized by difficulty in breathing due to spasm of the bronchi (air passages) in the lungs. Attacks may be provoked by allergy, infection, and stress. The incidence of asthma may be increasing as a result of air pollution and occupational hazard. Treatment is with ◊bronchodilators to relax the bronchial muscles and thereby ease the breathing, and in severe cases by inhaled ◊steroids that reduce inflammation of the bronchi.

Extrinsic asthma, which is triggered by exposure to irritants such as pollen and dust, is more common in children and young adults. In February 1997 Brazilian researchers reported two species of dust mite actually living on children's scalps. This explains why vacuuming of bedding sometimes fails to prevent asthma attacks. The use of antidandruff shampoo should keep numbers of mites down by reducing their food supply. Less common, intrinsic asthma tends to start in the middle years.

Approximately 5–10% of children suffer from asthma, but about a third of these will show no symptoms after adolescence, while another 5–10% of people develop the condition as adults. Growing evidence that the immune system is involved in both

ASTHMA – TUTORIAL FOR CHILDREN AND PARENTS

http://sln.fi.edu/inquirer/warming.html

Online tutorial for parents and children on asthma. It provides an explanation of what happens during an asthma attack, a description of the symptoms normally registered during an attack, and a discussion of the available medications. Movies and sound clips of asthmatic breathing are also included.

ASTHMA

Dust mites live in large numbers on the human scalp, feeding on skin flakes – an activity which causes allergic reactions in asthmatic children. This discovery was announced by Brazilian researchers in 1997, and may explain why repeated vacuuming of bedding does not get rid of dust mites.

forms of asthma has raised the possibility of a new approach to treatment.

Although the symptoms are similar to those of bronchial asthma, **cardiac asthma** is an unrelated condition and is a symptom of heart deterioration.

astigmatism aberration occurring in the lens of the eye. It results when the curvature of the lens differs in two perpendicular planes, so that rays in one plane may be in focus while rays in the other are not. With astigmatic eyesight, the vertical and horizontal cannot be in focus at the same time; correction is by the use of a cylindrical lens that reduces the overall focal length of one plane so that both planes are seen in sharp focus.

astrolabe ancient navigational instrument, forerunner of the sextant. Astrolabes usually consisted of a flat disc with a sighting rod that could be pivoted to point at the Sun or bright stars.

From the altitude of the Sun or star above the horizon, the local time could be estimated.

ASTROLABE – AN INSTRUMENT WITH A PAST AND A FUTURE

http://myhouse.com/mc/planet/astrodir/astrolab.htm

Good guide to this ancient astronomical computer. The long history of the instrument, and its importance in the history of astrology and astronomy, is explained with the help of good photographs. A section on the parts of the astrolabe explains how the instrument works and its uses.

astrometry measurement of the precise positions of stars, planets, and other bodies in space. Such information is needed for practical purposes including accurate timekeeping, surveying and navigation, and calculating orbits and measuring distances in space. Astrometry is not concerned with the surface features or the physical nature of the body under study.

Before telescopes, astronomical observations were simple astrometry. Precise astrometry has shown that stars are not fixed in position, but have a ◊proper motion caused as they and the Sun orbit the Milky Way Galaxy. The nearest stars also show ◊parallax (apparent change in position), from which their distances can be calculated. Above the distorting effects of the atmosphere, satellites such as ◊Hipparcos *can make even more precise measurements than ground telescopes, so refining the distance scale of space.*

astronaut person making flights into space; the term **cosmonaut** is used in the West for any astronaut from the former Soviet Union.

ASTRONAUT

Astronauts in space cannot belch. It is gravity that causes bubbles to rise to the top of a liquid, so space shuttle crews were forced to request less gas in their fizzy drinks to avoid discomfort.

ASK AN ASTRONAUT

http://www.nss.org/askastro/

Apollo 17 multimedia archive with video and audio files, plus mission details and astronaut biographies. Sponsored by the National Space Society, this site includes numerous images, a list of questions and answers, and links to related sites.

astronautics science of space travel. See ◊rocket; ◊satellite; ◊space probe.

Astronomer Royal honorary post in British astronomy. Originally it was held by the director of the Royal Greenwich Observatory; since 1972 the title of Astronomer Royal has been awarded separately as an honorary title to an outstanding British astronomer. The Astronomer Royal from 1995 is Martin Rees. There is a separate post of Astronomer Royal for Scotland.

Astronomical Almanac in astronomy, an international work of reference published jointly every year by the ◊Royal Greenwich Observatory and the ◊US Naval Observatory containing detailed tables of planetary motions, ◊eclipses, and other astronomical phenomena.

astronomical unit unit (symbol AU) equal to the mean distance of the Earth from the Sun: 149,597,870 km/92,955,800 mi. It is used to describe planetary distances. Light travels this distance in approximately 8.3 minutes.

astronomy science of the celestial bodies: the Sun, the Moon, and the planets; the stars and galaxies; and all other objects in the universe. It is concerned with their positions, motions, distances, and physical conditions and with their origins and evolution. Astronomy thus divides into fields such as astrophysics, celestial mechanics, and ◊cosmology. See also ◊gamma-ray astronomy, ◊infrared astronomy, ◊radio astronomy, ◊ultraviolet astronomy, and ◊X-ray astronomy.

Greek astronomers Astronomy is perhaps the oldest recorded science; there are observational records from ancient Babylonia, China, Egypt, and Mexico. The first true astronomers, however, were the Greeks, who deduced the Earth to be a sphere and attempted to measure its size. Ancient Greek astronomers included Thales and ◊Pythagoras. Eratosthenes of Cyrene measured the size of the Earth with considerable accuracy. Star catalogues were drawn up, the most celebrated being that of Hipparchus. The *Almagest,* by ◊Ptolemy of Alexandria, summarized Greek astronomy and survived in its Arabic translation. The Greeks still regarded the Earth as the centre of the universe, although this was doubted by some philosophers, notably Aristarchus of Samos, who maintained that the Earth moves around the Sun.

Ptolemy, the last famous astronomer of the Greek school, died about AD 180, and little progress was made for some centuries.

Arab revival The Arabs revived the science, developing the astrolabe and producing good star catalogues. Unfortunately, a general belief in the pseudoscience of astrology continued until the end of the Middle Ages (and has been revived from time to time).

the Sun at the centre The dawn of a new era came in 1543, when a Polish canon, ◊Copernicus, published a work entitled *De revolutionibus orbium coelestium/On the Revolutions of the Heavenly Spheres,* in which he demonstrated that the Sun, not the Earth, is the centre of our planetary system. (Copernicus was wrong in many respects – for instance, he still believed that all celestial orbits must be perfectly circular.) Tycho ◊Brahe, a Dane, increased the accuracy of observations by means of improved instruments allied to his own personal skill, and his observations were used by German mathematician Johannes ◊Kepler to prove the validity of the Copernican system. Considerable opposition existed, however, for removing the Earth from its central position in the universe; the Catholic church was openly hostile to the idea, and, ironically, Brahe never accepted the idea that the Earth could move around the Sun. Yet before the end of the 17th century, the theoretical work of Isaac ◊Newton had established celestial mechanics.

Astronomy: chronology

2300 BC	Chinese astronomers made their earliest observations.
2000	Babylonian priests made their first observational records.
1900	Stonehenge was constructed: first phase.
434	Anaxagoras claims the Sun is made up of hot rock.
365	The Chinese observed the satellites of Jupiter with the naked eye.
3RD CENTURY	Aristarchus argued that the Sun is the centre of the Solar System.
2ND CENTURY AD	Ptolemy's complicated Earth-centred system was promulgated, which dominated the astronomy of the Middle Ages.
1543	Copernicus revived the ideas of Aristarchus in *De Revolutionibus*.
1608	Hans Lippershey invented the telescope, which was first used by Galileo in 1609.
1609	Johannes Kepler's first two laws of planetary motion were published (the third appeared in 1619).
1632	The world's first official observatory was established in Leiden in the Netherlands.
1633	Galileo's theories were condemned by the Inquisition.
1675	The Royal Greenwich Observatory was founded in England.
1687	Isaac Newton's *Principia* was published, including his 'law of universal gravitation'.
1705	Edmond Halley correctly predicted that the comet that had passed the Earth in 1682 would return in 1758; the comet was later to be known by his name.
1781	William Herschel discovered Uranus and recognized stellar systems beyond our Galaxy.
1796	Pierre Laplace elaborated his theory of the origin of the solar system.
1801	Giuseppe Piazzi discovered the first asteroid, Ceres.
1814	Joseph von Fraunhofer first studied absorption lines in the solar spectrum.
1846	Neptune was identified by Johann Galle, following predictions by John Adams and Urbain Leverrier.
1859	Gustav Kirchhoff explained dark lines in the Sun's spectrum.
1887	The earliest photographic star charts were produced.
1889	Edward Barnard took the first photographs of the Milky Way.
1908	Fragment of comet fell at Tunguska, Siberia.
1920	Arthur Eddington began the study of interstellar matter.
1923	Edwin Hubble proved that the galaxies are systems independent of the Milky Way, and by 1930 had confirmed the concept of an expanding universe.
1930	The planet Pluto was discovered by Clyde Tombaugh at the Lowell Observatory, Arizona, USA.
1931	Karl Jansky founded radio astronomy.
1945	Radar contact with the Moon was established by Z Bay of Hungary and the US Army Signal Corps Laboratory.
1948	The 5-m/200-in Hale reflector telescope was installed at Mount Palomar, California, USA.
1957	The Jodrell Bank telescope dish in England was completed.
1957	The first Sputnik satellite (USSR) opened the age of space observation.
1962	The first X-ray source was discovered in Scorpius.
1963	The first quasar was discovered.
1967	The first pulsar was discovered by Jocelyn Bell and Antony Hewish.
1969	The first crewed Moon landing was made by US astronauts.
1976	A 6 m/240 in reflector telescope was installed at Mount Semirodniki, USSR.
1977	Uranus was discovered to have rings.
1977	The spacecraft *Voyager* 1 and 2 were launched, passing Jupiter and Saturn 1979–1981.
1978	The spacecraft *Pioneer Venus* 1 and 2 reached Venus.
1978	A satellite of Pluto, Charon, was discovered by James Christy of the US Naval Observatory.
1986	Halley's comet returned. *Voyager 2* flew past Uranus and discovered six new moons.
1987	Supernova SN1987A flared up, becoming the first supernova to be visible to the naked eye since 1604. The 4.2-m/165-in William Herschel Telescope on La Palma, Canary Islands, and the James Clerk Maxwell Telescope on Mauna Kea, Hawaii, began operation.
1988	The most distant individual star was recorded – a supernova, 5 billion light years away, in the AC118 cluster of galaxies.
1989	*Voyager 2* flew by Neptune and discovered eight moons and three rings.
1990	Hubble Space Telescope was launched into orbit by the US space shuttle.
1991	The space probe *Galileo* flew past the asteroid Gaspra, approaching it to within 26,000 km/16,200 mi.
1992	COBE satellite detected ripples from the Big Bang that mark the first stage in the formation of galaxies.
1994	Fragments of comet Shoemaker–Levy struck Jupiter.
1996	US astronomers discovered the most distant galaxy so far detected. It is in the constellation Virgo and is 14 billion light years from Earth.
1997	Data from the satellite *Hipparicos* improved estimates of the age of the universe, and the distances to many nearby stars.

Galileo and the telescope The refracting telescope was invented about 1608, by Hans Lippershey in Holland, and was first applied to astronomy by Italian scientist ◊Galileo in the winter of 1609–10. Immediately, Galileo made a series of spectacular discoveries. He found the four largest satellites of Jupiter, which gave strong support to the Copernican theory; he saw the craters of the Moon, the phases of Venus, and the myriad faint stars of our ◊Galaxy, the Milky Way.

Galileo's most powerful telescope magnified only 30 times, but it was not long before larger telescopes were built and official observatories were established.

Galileo's telescope was a refractor; that is to say, it collected its light by means of a glass lens or object glass. Difficulties with his design led Newton, in 1671, to construct a reflector, in which the light is collected by means of a curved mirror.

further discoveries In the 17th and 18th centuries astronomy was mostly concerned with positional measurements. Uranus was discovered 1781 by William ◊Herschel, and this was soon followed by the discovery of the first four asteroids, Ceres in 1801, Pallas in 1802, Juno in 1804, and Vesta in 1807. In 1846 Neptune was located by Johann Galle, following calculations by British astronomer John Couch Adams and French astronomer Urbain Jean Joseph Leverrier. Also significant was the first measurement of the distance of a star, when in 1838 the German astronomer Friedrich Bessel measured the ◊parallax of the star 61 Cygni, and calculated that it lies at a distance of about 6 light years (about half the correct value).

Astronomical spectroscopy was developed, first by Fraunhofer in Germany and then by people such as Pietro Angelo Secchi and William Huggins, while Gustav ◊Kirchhoff successfully interpreted the spectra of the Sun and stars. By the 1860s good photographs of the Moon had been obtained, and by the end of the century photographic methods had started to play a leading role in research.

galaxies William Herschel, probably the greatest observer in the history of astronomy, investigated the shape of our Galaxy during the latter part of the 18th century and concluded that its stars are arranged roughly in the form of a double-convex lens. Basically Herschel was correct, although he placed our Sun near the centre of the system; in fact, it is well out towards the edge, and lies 25,000 light years from the galactic nucleus. Herschel also studied the luminous 'clouds' or nebulae, and made the tentative suggestion that those nebulae capable of resolution into stars might be separate galaxies, far outside our own Galaxy.

It was not until 1923 that US astronomer Edwin Hubble, using the 2.5 m/100 in reflector at the Mount Wilson Observatory, was able to verify this suggestion. It is now known that the 'spiral nebulae' are galaxies in their own right, and that they lie at immense distances. The most distant galaxy visible to the naked eye, the Great Spiral in ◊Andromeda, is 2.2 million light years away; the most remote galaxy so far measured lies over 10 billion light years away. It was also found that galaxies tended to form groups, and that the groups were apparently receding from each other at speeds proportional to their distances.

a growing universe This concept of an expanding and evolving universe at first rested largely on Hubble's law, relating the distance of objects to the amount their spectra shift towards red – the ◊red shift. Subsequent evidence derived from objects studied in other parts of the ◊electromagnetic spectrum, at radio and X-ray wavelengths, has provided confirmation. ◊Radio astronomy established its place in probing the structure of the universe by demonstrating in 1954 that an optically visible distant galaxy was identical with a powerful radio source known as Cygnus A. Later analysis of the comparative number, strength, and distance of radio sources suggested that in the distant past these, including the ◊quasars discovered in 1963, had been much more powerful and numerous than today. This fact suggested that the universe has been evolving from an origin, and is not of infinite age as expected under a ◊steady-state theory.

The discovery in 1965 of microwave background radiation was evidence for the enormous temperature of the giant explosion, or Big Bang, that brought the universe into existence.

further exploration Although the practical limit in size and efficiency of optical telescopes has apparently been reached, the siting of these and other types of telescope at new observatories in the previously neglected southern hemisphere has opened fresh areas of the sky to search. Australia has been in the forefront of these developments. The most remarkable recent extension of the powers of astronomy to explore the universe is in the use of rockets, satellites, space stations, and space probes. Even the range and accuracy of the conventional telescope may be greatly improved free from the Earth's atmosphere. When the USA launched the Hubble Space Telescope into permanent orbit in 1990, it was the most powerful optical telescope yet constructed, with a 2.4 m/94.5 in mirror. It detects celestial phenomena seven times more distant (up to 14 billion light years) than any Earth-based telescope.

See also ◊black hole and ◊infrared radiation.

We now have direct evidence of the birth of the Universe and its evolution ... ripples in space-time laid down earlier than the first billionth of a second. If you're religious it's like seeing God.

GEORGE SMOOT US astrophysicist.
Attributed remark 1992

---

astrophotography use of photography in astronomical research. The first successful photograph of a celestial object was the daguerreotype plate of the Moon taken by John W Draper (1811–1882) of the USA in March 1840. The first photograph of a star, Vega, was taken by US astronomer William C Bond (1789–1859) in 1850. Modern-day astrophotography uses techniques such as ◊charge-coupled devices (CCDs).

astrophysics study of the physical nature of stars, galaxies, and the universe. It began with the development of spectroscopy in the 19th century, which allowed astronomers to analyse the composition of stars from their light. Astrophysicists view the universe as a vast natural laboratory in which they can study matter under conditions of temperature, pressure, and density that are unattainable on Earth.

asymmetric digital subscriber loop in computing, standard for transmitting video data; see ◊ADSL.

asymptote in ◊coordinate geometry, a straight line that a curve approaches progressively more closely but never reaches. The x and y axes are asymptotes to the graph of xy = constant (a rectangular ◊hyperbola).

If a point on a curve approaches a straight line such that its distance from the straight line is d, then the line is an asymptote to the curve if limit d tends to zero as the point moves towards infinity. Among ◊conic sections (curves obtained by the intersection of a plane and a double cone), a hyperbola has two asymptotes, which in the case of a rectangular hyperbola are at right angles to each other.

asynchronous irregular or not synchronized. In computer communications, the term is usually applied to data transmitted irregularly rather than as a steady stream. Asynchronous communication uses ◊start bits and ◊stop bits to indicate the beginning and end of each piece of data. Most personal computer communications are asynchronous, including connections across the Internet.

asynchronous transfer mode (ATM) in computing, high-speed computer ◊networking standard suitable for all types of data, including voice and video, that can be used on both private and public networks. ATM is used mainly on the core 'backbones' of large communications networks and in wide-area networks.

The basic technology was developed as part of the Cambridge Ring in the late 1970s, and is now being adopted by companies such as IBM and AT&T.

ATAPI (abbreviation for ***Advanced Technology Attachment Packet Interface***) in computing, enhancement to integrated drive electronics (IDE) that allows easier installation and support of CD-ROM drives and other devices. Part of the Enhanced IDE standard introduced by hard disc manufacturer Western Digital in 1994, ATAPI uses a standard software device driver and does away with the need for older, proprietary interfaces.

atavism (***Latin** atavus **'ancestor'***) in genetics, the reappearance of a characteristic not apparent in the immediately preceding generations; in psychology, the manifestation of primitive forms of behaviour.

ataxia loss of muscular coordination due to neurological damage or disease.

AT command set (abbreviation for ***attention command set***) set of standard commands allowing a ◊modem to be controlled via software. These commands are used via special communications software to control a modem's actions from the computer console. The most common are ATZ to reset the modem and ATH to hang the modem up at the end of a call. The set was invented by Hayes Computer Products for its earliest modems.

Ateles genus of ◊spider monkey.

atheroma furring-up of the interior of an artery by deposits, mainly of cholesterol, within its walls.

Associated with atherosclerosis, atheroma has the effect of narrowing the lumen (channel) of the artery, thus restricting blood flow. This predisposes to a number of conditions, including thrombosis, angina, and stroke.

atherosclerosis thickening and hardening of the walls of the arteries, associated with ◊atheroma.

Atlas rocket US rocket, originally designed and built as an intercontinental missile, but subsequently adapted for space use. Atlas rockets launched astronauts in the Mercury series into orbit, as well as numerous other satellites and space probes.

ATM in computing, abbreviation for ◊asynchronous transfer mode, **automated teller machine**, or ◊Adobe Type Manager, depending on context.

atmosphere mixture of gases surrounding a planet. Planetary atmospheres are prevented from escaping by the pull of gravity. On Earth, atmospheric pressure decreases with altitude. In its lowest layer, the atmosphere consists of 78% nitrogen and 21% oxygen, both in molecular form (two atoms bonded together), and 1% argon. Small quantities of other gases, including water and carbon dioxide, are important in the chemistry and physics of Earth's atmosphere. The atmosphere plays a major part in the various cycles of nature (the ◊water cycle, the ◊carbon cycle, and the ◊nitrogen cycle). It is the principal industrial source of nitrogen, oxygen, and argon, which are obtained by the fractional distillation of liquid air.

The combination of gases, moisture, and dust particles in the Earth's atmosphere filter, reflect, refract, and scatter the rays of

◊light energy travelling from the Sun. Visible light varies in colour, the shorter wavelengths being blue and the longer wavelengths being red. Blue light waves are readily scattered by tiny particles of matter in the atmosphere, while the remaining light waves travel on uninterrupted unless they meet up with larger particles. Thus, the clearer and less polluted the atmosphere, the bluer the sky will appear; when there are smoke particles or other pollutants in the atmosphere, more of the colour rays are scattered and the sky will appear greyer and darker. Water droplets reflect the light rays rather than scatter them, and so in a moist atmosphere the sky will appear a paler blue. Hence, the sky will appear deepest blue where the atmosphere is driest and least polluted, and when the sun is directly overhead. At sunrise or sunset, when sunlight has further to travel through the atmosphere, more of the light waves are scattered, and any undisturbed red light waves will give the sun and sky near the horizon a red or orange appearance. At higher altitudes, as the atmosphere thins, there are fewer particles to scatter the light rays and so more are absorbed, and the sky

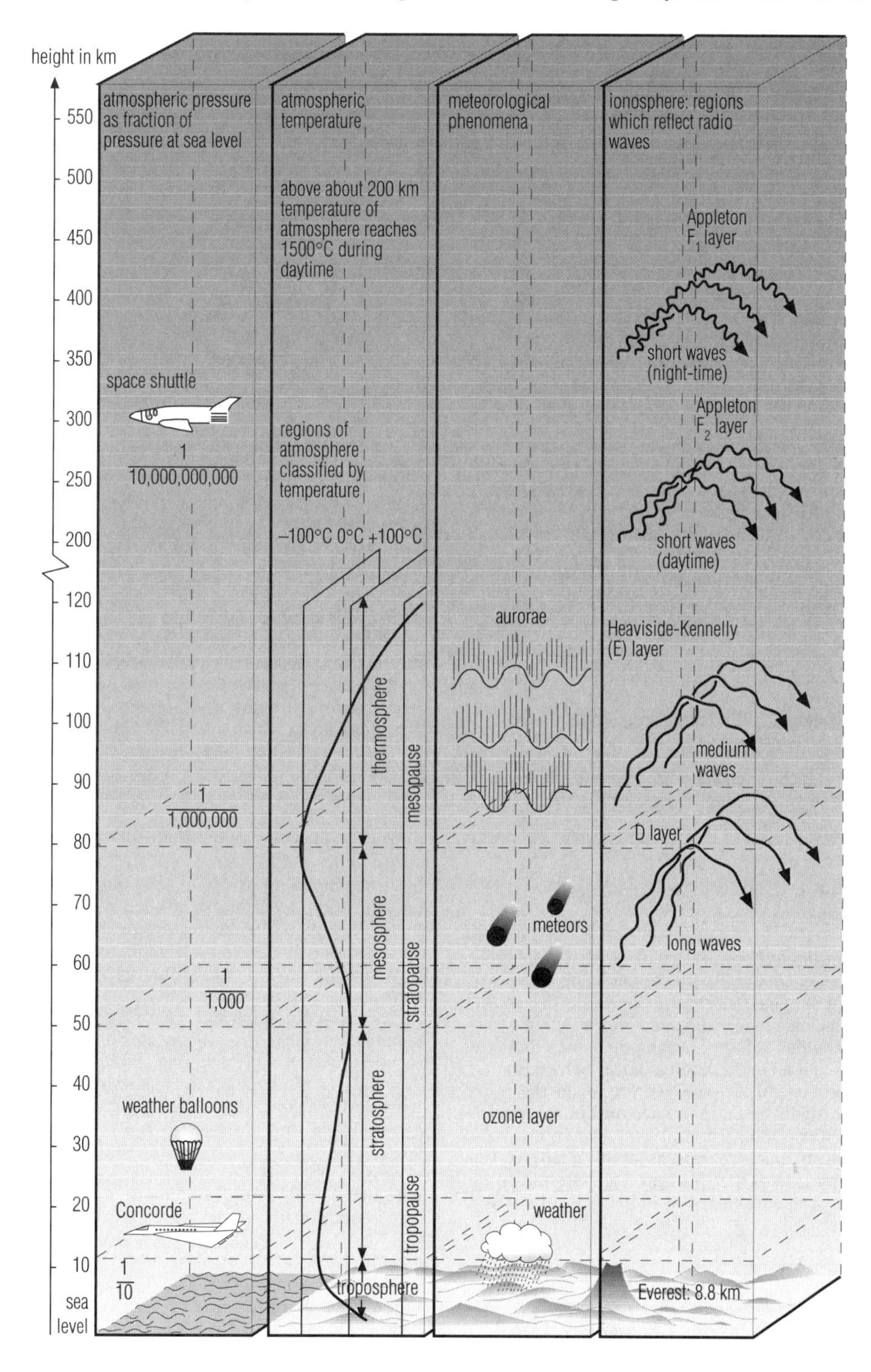

***atmosphere** All but 1% of the Earth's atmosphere lies in a layer 30 km/19 mi above the ground. At a height of 5,500 m/18,000 ft, air pressure is half that at sea level. The temperature of the atmosphere varies greatly with height; this produces a series of layers, called the troposphere, stratosphere, mesosphere, and thermosphere.*

Atmosphere: composition

gas	*symbol*	*volume (%)*	*role*
nitrogen	N_2	78.08	cycled through human activities and through the action of microorganisms on animal and plant waste
oxygen	O_2	20.94	cycled mainly through the respiration of animals and plants and through the action of photosynthesis
carbon dioxide	CO_2	0.03	cycled through respiration and photosynthesis in exchange reactions with oxygen. It is also a product of burning fossil fuels
argon	Ar	0.093	chemically inert and with only a few industrial uses
neon	Ne	0.0018	as argon
helium	He	0.0005	as argon
krypton	Kr	trace	as argon
xenon	Xe	trace	as argon
ozone	O_3	0.00006	a product of oxygen molecules split into single atoms by the Sun's radiation and unaltered oxygen molecules
hydrogen	H_2	0.00005	unimportant; it is so light it escapes into space

tends to lose its colour and appear darker. The phenomenon of why the sky appears blue was first demonstrated by British physicist John Tyndall in the mid-19th century, and later explained theoretically by Lord Rayleigh.

atmosphere or ***standard atmosphere*** in physics, a unit (symbol atm) of pressure equal to 760 torr, 1013.25 millibars, or 1.01325×10^5 newtons per square metre. The actual pressure exerted by the atmosphere fluctuates around this value, which is assumed to be standard at sea level and 0°C/32°F, and is used when dealing with very high pressures.

atmospheric pressure the pressure at any point on the Earth's surface that is due to the weight of the column of air above it; it therefore decreases as altitude increases. At sea level the average pressure is 101 kilopascals (1,013 millibars, 760 mmHg, or 14.7 lb per sq in).

Changes in atmospheric pressure, measured with a barometer, are used in weather forecasting. Areas of relatively high pressure are called ◊anticyclones; areas of low pressure are called ◊depressions.

To remember the effects of falling and rising atmospheric pressure:

WHEN PRESSURE IS FALLING, STORMS MAY COME A'CALLING.
WHEN PRESSURE IS HIGH, EXPECT CLEAR BLUE SKY.

atoll continuous or broken circle of ◊coral reef and low coral islands surrounding a lagoon.

atom *Greek atomos 'undivided'* smallest unit of matter that can take part in a chemical reaction, and which cannot be broken down chemically into anything simpler. An atom is made up of protons and neutrons in a central nucleus surrounded by electrons (see ◊atomic structure). The atoms of the various elements differ in atomic number, relative atomic mass, and chemical behaviour.

Atoms are much too small to be seen by even the most powerful optical microscope (the largest, caesium, has a diameter of 0.0000005 mm/0.00000002 in), and they are in constant motion. However, modern electron microscopes, such as the ◊scanning tunnelling microscope (STM) and the ◊atomic force microscope (AFM), can produce images of individual atoms and molecules.

The unleashed power of the atom has changed everything save our modes of thinking and we thus drift toward unparalleled catastrophe.

ALBERT EINSTEIN German-born US physicist. Telegram sent to prominent Americans 24 May 1946

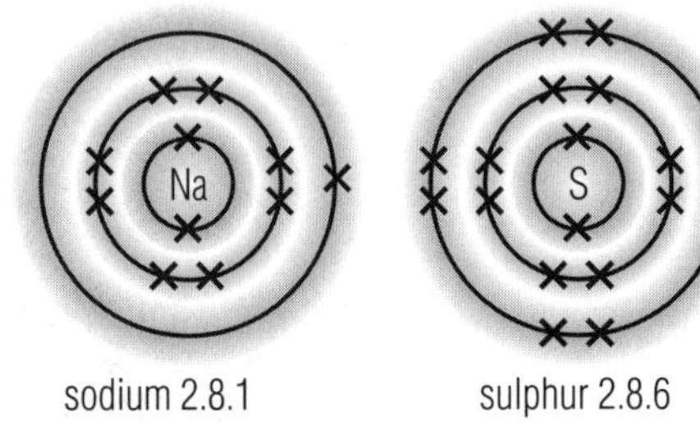

atom, electronic structure *The arrangement of electrons in a sodium atom and a sulphur atom. The number of electrons in a neutral atom gives that atom its atomic number: sodium has an atomic number of 11 and sulphur has an atomic number of 16.*

atom, electronic structure of the arrangement of electrons around the nucleus of an atom, in distinct energy levels, also called orbitals or shells (see ◊orbital, ◊atomic). These shells can be regarded as a series of concentric spheres, each of which can contain a certain maximum number of electrons; the noble gases have an arrangement in which every shell contains this number (see ◊noble gas structure). The energy levels are usually numbered beginning with the shell nearest to the nucleus. The outermost shell is known as the ◊valency shell as it contains the valence electrons.

The lowest energy level, or innermost shell, can contain no more than two electrons. Outer shells are considered to be stable when they contain eight electrons but additional electrons can sometimes be accommodated provided that the outermost shell has a stable configuration. Electrons in unfilled shells are available to take part in chemical bonding, giving rise to the concept of valency. In ions, the electron shells contain more or fewer electrons than are required for a neutral atom, generating negative or positive charges.

The atomic number of an element indicates the number of electrons in a neutral atom. From this it is possible to deduce its electronic structure. For example, sodium has atomic number 11 ($Z = 11$) and its electronic arrangement (configuration) is two electrons in the first energy level, eight electrons in the second energy level and one electron in the third energy level – generally written as 2.8.1. Similarly for sulphur ($Z = 16$), the electron arrangement will be 2.8.6. The electronic structure dictates whether two elements will combine by ionic or covalent bonding (see ◊bond) or not at all.

atomic absorption spectrometry technique used in archaeology to determine quantitatively the chemical composition of artefactual metals, minerals, and rocks, in order to identify raw material sources, to relate artefacts of the same material, or to trace trade routes. A sample of the material is atomized in a flame, and its light intensity measured. The method is slow and destroys the sample.

atomic clock timekeeping device regulated by various periodic processes occurring in atoms and molecules, such as atomic vibration or the frequency of absorbed or emitted radiation.

The first atomic clock was the **ammonia clock**, invented at the US National Bureau of Standards in 1948. It was regulated by measuring the speed at which the nitrogen atom in an ammonia molecule vibrated back and forth. The rate of molecular vibration is not affected by temperature, pressure, or other external influences, and can be used to regulate an electronic clock.

A more accurate atomic clock is the **caesium clock**. Because of its internal structure, a caesium atom produces or absorbs radiation of a very precise frequency (9,192,631,770 Hz) that varies by less than one part in 10 billion. This frequency has been used to define the second, and is the basis of atomic clocks used in international timekeeping.

Hydrogen maser clocks, based on the radiation from hydrogen atoms, are the most accurate. The hydrogen maser clock at the US Naval Research Laboratory, Washington DC, is estimated to lose one second in 1,700,000 years. Cooled hydrogen maser clocks could theoretically be accurate to within one second in 300 million years.

Atomic clocks are so accurate that minute adjustments must be made periodically to the length of the year to keep the calendar exactly synchronized with the Earth's rotation, which has a tendency to slow down. There have been 17 adjustments made since 1972 addding a total of 20 seconds to the calendar. In 1997 the northern hemisphere's summer was longer than usual – by one second. An extra second was added to the world's time at precisely 23 hours, 59 minutes, and 60 seconds on 30 June 1997. The adjustment was called for by the International Earth Rotation Service in Paris, which monitors the difference between Earth time and atomic time.

atomic energy another name for ◊nuclear energy.

atomic force microscope (AFM) microscope developed in the late 1980s that produces a magnified image using a diamond probe, with a tip so fine that it may consist of a single atom, dragged over the surface of a specimen to 'feel' the contours of the surface. In effect, the tip acts like the stylus of a record player, reading the surface. The tiny up-and-down movements of the probe are converted to an image of the surface by computer and displayed on a screen. The AFM is useful for examination of biological specimens since, unlike the ◊scanning tunnelling microscope, the specimen does not have to be electrically conducting.

atomicity number of atoms of an ◊element that combine together to form a molecule. A molecule of oxygen (O_2) has atomicity 2; sulphur (S_8) has atomicity 8.

atomic mass see ◊relative atomic mass.

atomic mass unit or ***dalton unit*** (symbol amu or u) unit of mass that is used to measure the relative mass of atoms and molecules. It is equal to one-twelfth of the mass of a carbon-12 atom, which is equivalent to the mass of a proton or 1.66×10^{-27} kg. The ◊relative atomic mass of an atom has no units; thus oxygen-16 has an atomic mass of 16 daltons but a relative atomic mass of 16.

atomic number or ***proton number*** the number (symbol *Z*) of protons in the nucleus of an atom. It is equal to the positive charge on the nucleus.

In a neutral atom, it is also equal to the number of electrons surrounding the nucleus. The chemical elements are arranged in the ◊periodic table of the elements according to their atomic number. See also ◊nuclear notation.

atomic radiation energy given out by disintegrating atoms during ◊radioactive decay, whether natural or synthesized. The energy may be in the form of fast-moving particles, known as ◊alpha particles and ◊beta particles, or in the form of high-energy electromagnetic waves known as ◊gamma radiation. Overlong exposure to atomic radiation can lead to ◊radiation sickness.

Radiation biology studies the effect of radiation on living organisms. Exposure to atomic radiation is linked to chromosomal damage, cancer, and, in laboratory animals at least, hereditary disease.

atomic size or ***atomic radius*** size of an atom expressed as the radius in ◊angstroms or other units of length.

The sodium atom has an atomic radius of 1.57 angstroms (1.57×10^{-8} cm). For metals, the size of the atom is always greater than the size of its ion. For non-metals the reverse is true.

atomic structure internal structure of an ◊atom.

the nucleus The core of the atom is the **nucleus**, a dense body only one ten-thousandth the diameter of the atom itself. The simplest nucleus, that of hydrogen, comprises a single stable positively charged particle, the **proton**. Nuclei of other elements contain more protons and additional particles, called **neutrons**, of about the same mass as the proton but with no electrical charge. Each element has its own characteristic nucleus with a unique number of protons, the atomic number. The number of neutrons may vary. Where atoms of a single element have different numbers of neutrons, they are called ◊isotopes. Although some isotopes tend to be unstable and exhibit ◊radioactivity, they all have identical chemical properties.

electrons The nucleus is surrounded by a number of moving **electrons**, each of which has a negative charge equal to the positive charge on a proton, but which weighs only $\frac{1}{1839}$ times as much. In a neutral atom, the nucleus is surrounded by the same number of electrons as it contains protons. According to ◊quantum theory, the position of an electron is uncertain; it may be found at any point. However, it is more likely to be found in some places than others. The region of space in which an electron is most likely to be found is called an orbital (see ◊orbital, atomic). The chemical properties of an element are determined by the ease with which its atoms can gain or lose electrons from its outer orbitals.

attraction and repulsion Atoms are held together by the electrical forces of attraction between each negative electron and the positive protons within the nucleus. The latter repel one another with enormous forces; a nucleus holds together only because an even stronger force, called the **strong nuclear force**, attracts the protons and neutrons to one another. The strong force acts over a very short range – the protons and neutrons must be in virtual contact with one another (see ◊forces, fundamental). If, therefore, a fragment of a complex nucleus, containing some protons, becomes only slightly loosened from the main group of neutrons and protons, the natural repulsion between the protons will cause this fragment to fly apart from the rest of the nucleus at high speed. It is by such fragmentation of atomic nuclei (nuclear ◊fission) that nuclear energy is released.

atomic time time as given by ◊atomic clocks, which are regulated by natural resonance frequencies of particular atoms, and display a continuous count of seconds.

In 1967 a new definition of the second was adopted in the SI system of units: the duration of 9,192,631,770 periods of the radiation corresponding to the transition between two hyperfine levels of the ground state of the caesium-133 atom. The International Atomic Time Scale is based on clock data from a number of countries; it is a continuous scale in days, hours, minutes, and seconds from the origin on 1 January 1958, when the Atomic Time Scale was made 0 h 0 min 0 sec when Greenwich Mean Time was at 0 h 0 min 0 sec.

atomic weight another name for ◊relative atomic mass.

atomizer device that produces a spray of fine droplets of liquid. A vertical tube connected with a horizontal tube dips into a bottle of liquid, and at one end of the horizontal tube is a nozzle, at the other a rubber bulb. When the bulb is squeezed, air rushes over the top of the vertical tube and out through the nozzle. Following ◊Bernoulli's principle, the pressure at the top of the vertical tube is reduced, allowing the liquid to rise. The air stream picks up the liquid, breaks it up into tiny drops, and carries it out of the nozzle as a spray. Scent spray, paint spray guns, and carburettors all use the principle of the atomizer.

ATP (abbreviation for ***adenosine triphosphate***), a nucleotide molecule found in all cells. It can yield large amounts of energy, and is used to drive the thousands of biological processes needed to sustain life, growth, movement, and reproduction. Green plants use light energy to manufacture ATP as part of the process of ◊photosynthesis. In animals, ATP is formed by the breakdown of glucose molecules, usually obtained from the carbohydrate component of a diet, in a series of reactions termed ◊respiration. It is the driving force behind muscle contraction and the synthesis of complex molecules needed by individual cells.

atrium either of the two upper chambers of the heart. The left atrium receives freshly oxygenated blood from the lungs via the pulmonary vein; the right atrium receives deoxygenated blood from the ◊vena cava. Atrium walls are thin and stretch easily to allow blood into the heart. On contraction, the atria force blood into the thick-walled ventricles, which then give a second, more powerful beat.

atrophy in medicine, a diminution in size and function, or output, of a body tissue or organ. It is usually due to nutritional impairment, disease, or disuse (muscle).

atropine alkaloid derived from ◊belladonna, a plant with toxic properties. It acts as an anticholinergic, inhibiting the passage of certain nerve impulses. It is used in premedication, to reduce bronchial and gastric secretions. It is also administered as a mild antispasmodic drug, and to dilate the pupil of the eye.

attachment way of incorporating a file into an e-mail message for transmission. Within a single system, such as a corporate local area network (LAN) or a commercial online service, ◊binary files can be sent intact. Over the Internet, attached files must be encoded into ◊ASCII characters and then decoded by the receiver. See ◊MIME.

attar of roses perfume derived from the essential oil of roses (usually damask roses), obtained by crushing and distilling the petals of the flowers.

attention-deficit hyperactivity disorder (ADHD) psychiatric condition occurring in young children characterized by impaired attention and hyperactivity. The disorder, associated with disruptive behaviour, learning difficulties, and under-achievement, is more common in boys. It is treated with methylphenidate (Ritalin). There was a 50% increase in the use of the drug in the USA 1994–96, with an estimated 5% of school-age boys diagnosed as suffering from ADHD.

In 1996, US researchers found that 50% of children diagnosed as ADHD sufferers carry a gene that affects brain cell response to the neurotransmitter dopamine. The same gene has also been linked to impulsiveness in adults. Diagnosis requires the presence, for at least six months, of eight behavioural problems, first developing before the age of seven. In addition to their hyperactivity, such children are found to be reckless, impulsive, and accident prone; they are often aggressive and tend to be unpopular with other children. The outlook for ADHD sufferers varies, with up to a quarter being diagnosed with antisocial personality disorder as adults.

aubergine or ***eggplant*** plant belonging to the nightshade family, native to tropical Asia. Its purple-skinned, sometimes white, fruits are eaten as a vegetable. (*Solanum melongena,* family Solanaceae.)

aubrieta any of a group of spring-flowering dwarf perennial plants native to the Middle East. All are trailing plants with showy, purple flowers. They are widely cultivated in rock gardens. (Genus *Aubrieta,* family Cruciferae.)

audio file computer file that encodes sounds which can be played back using the appropriate software and hardware. On the World Wide Web, the latest types of audio files can be played on the user's computer system in real time while they are being downloaded. Apple Macintosh computers have sound capabilities built in, as do multimedia personal computers (MPCs). Older PCs need to have a ◊sound card installed in order to achieve good playback quality.

audio–video interleave in computing, ◊file format for video clips.

auditory canal tube leading from the outer ◊ear opening to the eardrum. It is found only in animals whose eardrums are located inside the skull, principally mammals and birds.

audit trail record of computer operations, showing what has been done and, if available, who has done it. The term is taken from accountancy, but audit trails are now widely used to check many aspects of computer security, in addition to use in accounts programs.

auger tool for boring holes. Originally, a carpenter's tool, with a cutting edge and a screw point, manipulated by means of a handle at right angles to the shank. In archaeology a large auger is used to collect sediment and soil samples below ground without hand excavation, or to determine the depth and type of archaeological deposits. The auger may be hand- or machine-powered.

auk *The Atlantic, or common, puffin* Fratercula arctica *lives in the open seas and breeds on the rocky coasts of the N Atlantic. With its short tail and narrow wings, which beat rapidly in flight, the puffin is a typical auk. Like all of the auks, the puffin spends most of its life at sea, coming ashore only during the breeding season. Its colourful striped beak has serrated edges to allow it to catch and grip many small fish before flying back to the nest. Premaphotos Wildlife*

augmented reality use of computer systems and data to overlay video or other real-life representations. For example, a video of a car engine with the mechanical drawings overlaid.

auk oceanic bird belonging to the family Alcidae, order Charadriiformes, consisting of 22 species of marine diving birds including razorbills, puffins, murres, and guillemots. Confined to the northern hemisphere, their range extends from well inside the Arctic Circle to the lower temperate regions. They feed on fish, and use their wings to 'fly' underwater in pursuit.

Most auks are colonial, breeding on stack tops or cliff edges, although some nest in crevices or holes. With the exception of one species they all lay a single large, very pointed egg.

AUP abbreviation for **acceptable use policy**; see ◊acceptable use.

auricula species of ◊primrose, *a* plant whose leaves are said to resemble a bear's ears. It grows wild in the Alps but is popular in cool-climate areas and often cultivated in gardens. *(Primula auricula.)*

Auriga constellation of the northern hemisphere, represented as a charioteer. Its brightest star is the first-magnitude ◊Capella, about 45 light years from Earth; Epsilon Aurigae is an ◊eclipsing binary star with a period of 27 years, the longest of its kind (last eclipse 1983).

aurochs (plural ***aurochs***) extinct species of long-horned wild cattle *Bos primigenius* that formerly roamed Europe, SW Asia, and N Africa. It survived in Poland until 1627. Black to reddish or grey, it was up to 1.8 m/6 ft at the shoulder. It is depicted in many cave paintings, and is considered the ancestor of domestic cattle.

aurora coloured light in the night sky near the Earth's magnetic poles, called **aurora borealis** ('northern lights') in the northern hemisphere and **aurora australis** in the southern hemisphere. Although aurorae are usually restricted to the polar skies, fluctuations in the ◊solar wind occasionally cause them to be visible at lower latitudes. An aurora is usually in the form of a luminous arch with its apex towards the magnetic pole followed by arcs, bands, rays, curtains, and coronas, usually green but often showing shades of blue and red, and sometimes yellow or white. Aurorae are caused at heights of over 100 km/60 mi by a fast stream of charged particles from solar flares and low-density 'holes' in the Sun's corona.

These are guided by the Earth's magnetic field towards the north and south magnetic poles, where they enter the upper atmosphere and bombard the gases in the atmosphere, causing them to emit visible light.

auscultation evaluation of internal organs by listening, usually with the aid of a stethoscope.

AUSSAT organization formed in 1981 by the federal government of Australia and Telecom Australia to own and operate Australia's domestic satellite system. The first stage, *Aussat 1,* was launched by the US space shuttle *Discovery* in 1985 and the third and final stage was launched from French Guiana, South America, in 1987. The AUSSAT satellite system enables people in remote outback areas of Australia to receive television broadcasts.

Australian cattle dog breed of herding dog known also as a 'heeler' from its technique of controlling cattle by nipping their heels. It has a short coat flecked with red or blue, pricked ears, and a long tail, and stands about 51 cm/20 in tall.

Bred from imported collies, Australian cattle dogs are also claimed to have ◊dingo blood.

Australian Museum the original museum of Australia, which dates from 1827, when it was known as the Colonial Museum. Housed in Sydney, New South Wales, its collection covers anthropology, geology, palaeontology, and the natural sciences (with the exception of botany). It is rated as one of the major natural history museums in the world.

Australian sheepdog breed of dog. See ◊kelpie.

Australian terrier small low-set dog with a long body and straight back. Its straight, rough coat is about 5–6.5 cm/2–2.5 in long, and blue or silver-grey, and tan or clear red or sandy in colour. It has a long head with a topknot of soft hair and ears either pricked or dropped forwards towards the front. Australian terriers are about 25 cm/10 in high and weigh 4.5–5 kg/10–11 lb

The Australian terrier traces its ancestry to the Cairn, Norwich, and Yorkshire terriers.

Australia Prize annual award for achievement internationally in science and technology, established 1990 and worth £115,000.

The first winners were Allan Kerr of Adelaide University, Australia; Eugene Nester of Washington University, USA; and Jeff Schell of the Max Planck Institute in Cologne, Germany. Their studies of the genetic systems of the crown-gall bacterium *Agrobacterium tumefaciens* led to the creation of genetically engineered plants resistant to herbicides, pests, and viruses.

Australia Telescope giant radio telescope in New South Wales, Australia, operated by the Commonwealth Scientific and Industrial Research Organization (CSIRO). It consists of six 22-m/72-ft antennae at Culgoora, a similar antenna at Siding Spring Mountain, and the 64-m/210-ft ◊Parkes radio telescope – the whole simulating a dish 300 m/186 mi across.

Australopithecus the first hominid, living 3.5–4 million years ago; see ◊human species, ◊origins of.

authentication system for certifying the origin of an electronic communication. In the real world, a handwritten signature authenticates a document, for example a contract, as coming from a particular person. In the electronic world, encryption systems provide the same function via ◊digital signatures and other techniques.

In ◊public-key cryptography, for example, the ability to decrypt a message with a particular user's public key authenticates the message as coming from that user and no one else. Authentication is an essential requirement for electronic commerce.

authoring development of multimedia presentations. Authoring includes pulling together the necessary audio, video, graphics, and text files and formatting them for display.

authoring tool software that allows developers to create multimedia presentations or World Wide Web pages. Typically, these tools automate some of the more difficult parts of generating program source codes so that developers can work on a higher, more abstract level. Popular authoring tools for the World Wide Web include Hot Metal and HTML Assistant, both available in ◊shareware versions.

authorization permission to access a particular system. Unauthorized access to private computer systems was made illegal in many countries during the late 1980s.

autism, infantile rare disorder, generally present from birth, characterized by a withdrawn state and a failure to develop normally in language or social behaviour. Although the autistic child may, rarely, show signs of high intelligence (in music or with numbers, for example), many have impaired intellect. The cause is unknown, but is thought to involve a number of factors, possibly including an inherent abnormality of the child's brain. Special education may bring about some improvement.

Autism was initially defined by four common traits – preference for aloneness; insistence on sameness; need for elaborate routines; and the possession of some abilities that seem exceptional compared with other deficits – but current clinical diagnosis involves wider criteria.

AutoCAD the leading computer-aided design (CAD) software package. It is published by the specialist US company AutoDesk (founded 1982). Users include engineers, architects, and designers.

autochrome in photography, a single-plate additive colour process devised by the ◊Lumière brothers in 1903. It was the first commercially available process, in use 1907–35.

autoclave pressurized vessel that uses superheated steam to sterilize materials and equipment such as surgical instruments. It is similar in principle to a pressure cooker.

autoexec.bat in computing, a file in the ◊MS-DOS operating system that is automatically run when the computer is ◊booted.

autogiro or ***autogyro*** heavier-than-air craft that supports itself in the air with a rotary wing, or rotor. The Spanish aviator Juan de la Cierva designed the first successful autogiro in 1923. The autogiro's rotor provides only lift and not propulsion; it has been superseded by the helicopter, in which the rotor provides both. The autogiro is propelled by an orthodox propeller.

The three- or four-bladed rotor on an autogiro spins in a horizontal plane on top of the craft, and is not driven by the engine. The blades have an aerofoil cross section, as a plane's wings. When the autogiro moves forward, the rotor starts to rotate by itself, a state known as autorotation. When travelling fast enough, the rotor develops enough lift from its aerofoil blades to support the craft.

autoimmunity in medicine, condition where the body's immune responses are mobilized not against 'foreign' matter, such as invading germs, but against the body itself. Diseases considered to be of autoimmune origin include ◊myasthenia gravis, ◊rheumatoid arthritis, and ◊lupus erythematous.

In autoimmune diseases T-lymphocytes reproduce to excess to home in on a target (properly a foreign disease-causing molecule); however, molecules of the body's own tissue that resemble the target may also be attacked, for example insulin-producing cells, resulting in insulin-dependent diabetes; if certain joint membrane cells are attacked, then rheumatoid arthritis may result; and if myelin, the basic protein of the nervous system, then multiple sclerosis results. In 1990 in Israel a T-cell vaccine was produced that arrests the excessive reproduction of T-lymphocytes attacking healthy target tissues.

autolysis in biology, the destruction of a ◊cell after its death by the action of its own ◊enzymes, which break down its structural molecules.

automatic fallback in computing, feature allowing ◊modems to drop to a slower speed if conditions such as line noise make it necessary. Modem speeds are typically rated according to one or another ◊CCITT standard (known as a **V number**). All modems rated for a specific standard are ◊backwards compatible.

automatic pilot control device that keeps an aeroplane flying automatically on a given course at a given height and speed.

The automatic pilot contains a set of gyroscopes that provide references for the plane's course. Sensors detect when the plane deviates from this course and send signals to the control surfaces – the ailerons, elevators, and rudder – to take the appropriate action. Autopilot is also used in missiles. Most airliners cruise on automatic pilot, also called autopilot and gyropilot, for much of the time.

US business executive Lawrence Sperry first used a ◊gyroscope in 1912 to create an artificial horizon. This entered production in 1924 and was soon linked to aircraft controls to increase stability. More gyroscopes were added later to control altitude and course. The first automatic pilot was introduced in the 1930s using pneumatic power.

automation widespread use of self-regulating machines in industry. Automation involves the addition of control devices, using electronic sensing and computing techniques, which often follow the pattern of human nervous and brain functions, to already mechanized physical processes of production and distribution; for example, steel processing, mining, chemical production, and road, rail, and air control.

automation *Industrial automation has led to greater productivity and improved quality control. On this production line at a Rover car plant in England car bodies are welded by computer-controlled robots, a task that is complex yet boring and potentially hazardous for human workers. Rover Group*

Civilization advances by extending the number of important operations which we can perform without thinking about them.

ALFRED NORTH WHITEHEAD English philosopher and mathematician. *An Introduction to Mathematics*

automatism performance of actions without awareness or conscious intent. It is seen in sleepwalking and in some (relatively rare) psychotic states.

automaton mechanical figure imitating human or animal performance. Automatons are usually designed for aesthetic appeal as opposed to purely functional robots. The earliest recorded automaton is an Egyptian wooden pigeon of 400 BC.

autonomic nervous system in mammals, the part of the nervous system that controls those functions not controlled voluntarily, including the heart rate, activity of the intestines, and the production of sweat.

There are two divisions of the autonomic nervous system. The **sympathetic** system responds to stress, when it speeds the heart rate, increases blood pressure, and generally prepares the body for action. The **parasympathetic** system is more important when the body is at rest, since it slows the heart rate, decreases blood pressure, and stimulates the digestive system.

At all times, both types of autonomic nerves carry signals that bring about adjustments in visceral organs. The actual rate of heartbeat is the net outcome of opposing signals. Today, it is known that the word 'autonomic' is misleading – the reflexes managed by this system are actually integrated by commands from the brain and spinal cord (the central nervous system).

autopsy or ***postmortem*** examination of the internal organs and tissues of a dead body, performed to try to establish the cause of death.

autoradiography in biology, a technique for following the movement of molecules within an organism, especially a plant, by labelling with a radioactive isotope that can be traced on photographs. It is used to study ◊photosynthesis, where the pathway of radioactive carbon dioxide can be traced as it moves through the various chemical stages.

autoresponder on the Internet, a ◊server that responds automatically to specific messages or input. A common use for autoresponders is to automate the dispatch of sales information via e-mail. A user requesting such information typically sends a message with specified words such as 'send info' in the subject line or the body of the message. The words trigger the autoresponder to send the prepared information file.

Autoresponders are also used in e-mail systems which can be configured to notify correspondents that the user is on holiday.

autosome any ◊chromosome in the cell other than a sex chromosome. Autosomes are of the same number and kind in both males and females of a given species.

autosuggestion conscious or unconscious acceptance of an idea as true, without demanding rational proof, but with potential subsequent effect for good or ill. Pioneered by French psychotherapist Emile Coué (1857–1926) in healing, it is sometimes used in modern psychotherapy to conquer nervous habits and dependence on addictive substances such as tobacco and alcohol.

autotroph any living organism that synthesizes organic substances from inorganic molecules by using light or chemical energy. Autotrophs are the **primary producers** in all food chains since the materials they synthesize and store are the energy sources of all other organisms. All green plants and many planktonic organisms are autotrophs, using sunlight to convert carbon dioxide and water into sugars by ◊photosynthesis.

The total ◊biomass of autotrophs is far greater than that of animals, reflecting the dependence of animals on plants, and the ultimate dependence of all life on energy from the Sun – green plants convert light energy into a form of chemical energy (food) that animals can exploit. Some bacteria use the chemical energy of sulphur compounds to synthesize organic substances. It is estimated that 10% of the energy in autotrophs can pass into the next stage of the ◊food chain, the rest being lost as heat or indigestible matter. See also ◊heterotroph.

autumnal equinox see ◊equinox.

autumn crocus any of a group of late-flowering plants belonging to the lily family. The mauve **meadow saffron** (*Colchicum autumnale*) yields **colchicine**, which is used in treating gout and in plant breeding. (Genus *Colchicum*, family Liliaceae.)

Colchicine causes plants to double the numbers of their chromosomes, forming ◊polyploids)

auxin plant ◊hormone that promotes stem and root growth in plants. Auxins influence many aspects of plant growth and development, including cell enlargement, inhibition of development of axillary buds, ◊tropisms, and the initiation of roots. **Synthetic auxins** are used in rooting powders for cuttings, and in some weedkillers, where high auxin concentrations cause such rapid growth that the plants die. They are also used to prevent premature fruitdrop in orchards. The most common naturally occurring auxin is known as indoleacetic acid, or IAA. It is produced in the shoot apex and transported to other parts of the plant.

avalanche ***from French avaler 'to swallow'*** fall or flow of a mass of snow and ice down a steep slope under the force of gravity. Avalanches occur because of the unstable nature of snow masses in mountain areas.

Changes of temperature, sudden sound, or earth-borne vibrations may trigger an avalanche, particularly on slopes of more than 35°. The snow compacts into ice as it moves, and rocks may be carried along, adding to the damage caused.

Avalanches leave slide tracks, long gouges down the mountainside that can be up to 1 km/0.6 mi long and 100 m/330 ft wide. These slides have a similar beneficial effect on biodiversity as do forest fires, clearing the land of snow and mature mountain forest enabling plants and shrubs that cannot grow in shade, to recolonize and creating wildlife corridors.

AVALANCHE AWARENESS

http://www-nsidc.colorado.edu/NSIDC/EDUCATION/AVALANCHE/

Excellent description of avalanches, what causes them, and how to minimize dangers if caught in one. There is advice on how to determine the stability of a snowpack, what to do if caught out, and how to locate people trapped under snow. Nobody skiing off piste should set off without reading this.

avatar computer-generated character that represents a human in on-screen interaction. In the mid-1990s, avatars were primarily used in computer games, but because they take up much less memory or bandwidth than full video, companies such as British Telecom were researching the possibility of building multiparty videoconferencing systems using this technology.

Avebury Europe's largest stone circle (diameter 412 m/1,350 ft), in Wiltshire, England. This megalithic henge monument, probably a ritual complex, contains 650 massive blocks of stone, arranged in circles and avenues. It was probably constructed around 3,500 years ago, and is linked with nearby ◊Silbury Hill.

The henge, an earthen bank and interior ditch with opposed entrances, originally rose 15 m/49 ft above the bottom of the ditch. This earthwork and an outer ring of stones surround the inner circles. The stones vary in size from 1.5 m/5 ft to 5.5 m/18 ft high and 1 m/3 ft to 3.65 m/12 ft broad. They were erected by a late Neolithic or early Bronze Age culture. Visible remains seen today may cover an earlier existing site – a theory applicable to a number of prehistoric sites.

avens any of several low-growing plants found throughout Europe, Asia, and N Africa. (Genus *Geum*, family Rosaceae.)

Mountain avens (*Dryas octopetala*) belongs to a different genus and grows in mountain and arctic areas of Europe, Asia, and North America. A creeping perennial, it has white flowers with yellow stamens.

average in statistics, a term used inexactly to indicate the typical member of a set of data. It usually refers to the ◊arithmetic mean. The term is also used to refer to the middle member of the set when it is sorted in ascending or descending order (the Gmedian), and the most commonly occurring item of data (the ◊mode), as in 'the average family'.

AVI (abbreviation for ***Audio-Visual Interleave***) file format capable of storing moving images (such as video) with accompanying sound. AVI files can be replayed by any multimedia PC with Microsoft ◊Windows and a ◊sound card. AVI files are frequently very large (around 50 Mbyte for a five-minute rock video, for example), so they are usually stored on ◊CD-ROM.

aviation term used to describe the science of powered ◊flight.

AVIATION ENTHUSIASTS' CORNER

http://www.brooklyn.cuny.edu/rec/air/air.html

Forum dedicated to furthering interest in aviation-related hobbies. It includes links to museums and displays, features on key events in aviation history, and indexes of aircraft by type and manufacturer.

avocado tree belonging to the laurel family, native to Central America. Its dark-green, thick-skinned, pear-shaped fruit has buttery-textured flesh and is used in salads. (*Persea americana*, family Lauraceae.)

avocet wading bird, with a characteristic long, narrow, upturned bill, which it uses to sift water as it feeds in the shallows. It is about 45 cm/18 in long, has long legs, partly webbed feet, and black and white plumage. There are four species of avocet, genus *Recurvirostra*, family Recurvirostridae, order Charadriiformes. They are found in Europe, Africa, and central and southern Asia. Stilts belong to the same family.

Avogadro, Amedeo, Conte di Quaregna (1776–1856) Italian physicist, one of the founders of physical chemistry, who proposed ◊Avogadro's hypothesis on gases in 1811. His work enabled scientists to calculate ◊Avogadro's number, and still has relevance for atomic studies.

Avogadro made it clear that the gas particles need not be individual atoms but might consist of molecules, the term he introduced to describe combinations of atoms. No previous scientist had made this fundamental distinction between the atoms of a substance and its molecules.

Avogadro's hypothesis in chemistry, the law stating that equal volumes of all gases, when at the same temperature and pressure,

AVOGADRO'S HYPOTHESIS OF 1811

http://dbhs.wvusd.k12.ca.us/Chem-History/Avogadro.html

Avogadro's hypothesis was contained in the *Essay on a Manner of Determining the Relative Masses of the Elementary Molecules of Bodies, and the Proportions in Which They Enter Into These Compounds from the Journal de physique*, 73: 58-76 (1811). This Web site is a transcript of a translation of that essay taken from Alembic Club Reprints, No. 4, *Foundations of the Molecular Theory: Comprising Papers and Extracts by John Dalton, Joseph Louis Gay-Lussac, and Amadeo Avogadro (1808–11)*.

have the same numbers of molecules. It was first propounded by Amedeo Avogadro.

Avogadro's number or ***Avogadro's constant*** the number of carbon atoms in 12 g of the carbon-12 isotope (6.022045×10^{23}). The relative atomic mass of any element, expressed in grams, contains this number of atoms. It is named after Amedeo Avogadro.

avoirdupois system of units of mass based on the pound (0.45 kg), which consists of 16 ounces (each of 16 drams) or 7,000 grains (each equal to 65 mg).

axil upper angle between a leaf (or bract) and the stem from which it grows. Organs developing in the axil, such as shoots and buds, are termed axillary, or lateral.

axiom in mathematics, a statement that is assumed to be true and upon which theorems are proved by using logical deduction; for example, two straight lines cannot enclose a space. The Greek mathematician Euclid used a series of axioms that he considered could not be demonstrated in terms of simpler concepts to prove his geometrical theorems.

axis (plural ***axes***) in geometry, one of the reference lines by which a point on a graph may be located. The horizontal axis is usually referred to as the x-axis, and the vertical axis as the y-axis. The term is also used to refer to the imaginary line about which an object may be said to be symmetrical **(axis of symmetry)** – for example, the diagonal of a square – or the line about which an object may revolve **(axis of rotation)**.

axis deer or ***chital*** species of deer found in India and the East Indies. It is profusely spotted with white on a fawn background, shading from almost black on the back to white on the underparts.

classification The axis deer *Axis axis* is in family Cervidae, order Artiodactyla.

axolotl ***Aztec 'water monster'*** aquatic larval form ('tadpole') of the Mexican salamander *Ambystoma mexicanum,* belonging to the family Ambystomatidae. Axolotls may be up to 30 cm/12 in long. They are remarkable because they can breed without changing to the adult form, and will metamorphose into adults only in response to the drying-up of their ponds. The adults then migrate to another pond.

Axolotls resemble a newt in shape, having a powerful tail, two pairs of weak limbs, and three pairs of simple external gills. They lay eggs like a frog's in strings attached to water plants by a viscous substance, and the young, hatched in two to three weeks, resemble the parents. See also ◊neoteny.

axon long threadlike extension of a ◊nerve cell that conducts electrochemical impulses away from the cell body towards other nerve cells, or towards an effector organ such as a muscle. Axons terminate in ◊synapses, junctions with other nerve cells, muscles, or glands.

aye-aye nocturnal tree-climbing prosimian *Daubentonia madagascariensis* of Madagascar, related to the lemurs. It is just over 1 m/3 ft long, including a tail 50 cm/20 in long.

It has an exceptionally long middle finger with which it probes for insects and their larvae under the bark of trees, and gnawing, rodentlike front teeth, with which it tears off the bark to get at its prey. The aye-aye has become rare through loss of its forest habitat, and is now classified as an endangered species.

Ayurveda basically naturopathic system of medicine widely practised in India and based on principles derived from the ancient Hindu scriptures, the Vedas. Hospital treatments and remedial prescriptions tend to be nonspecific and to coordinate holistic therapies for body, mind, and spirit.

azalea any of a group of deciduous flowering shrubs belonging to the heath family. Several species are native to Asia and North America, and many cultivated varieties have been derived from these. Azaleas are closely related to the mostly evergreen ◊rhododendrons. (Genus *Rhododendron,* family Ericaceae.)

Azaleas, particularly the Japanese varieties, make fine ornamental shrubs. Several species are highly poisonous.

Azilian archaeological period following the close of the Old Stone (Palaeolithic) Age and regarded as the earliest culture of the Mesolithic Age in W Europe. It was first recognized at Le Mas d'Azil, a cave in Ariège, France.

azimuth in astronomy, the angular distance of an object eastwards along the horizon, measured from due north, between the astronomical ◊meridian (the vertical circle passing through the centre of the sky and the north and south points on the horizon) and the vertical circle containing the celestial body whose position is to be measured.

azo dye synthetic dye containing the azo group of two nitrogen atoms (N=N) connecting aromatic ring compounds. Azo dyes are usually red, brown, or yellow, and make up about half the dyes produced. They are manufactured from aromatic ◊amines.

AZT drug used in the treatment of AIDS; see ◊zidovudine.

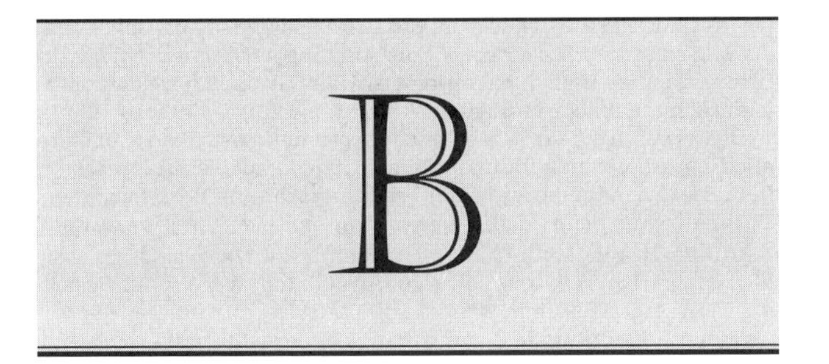

Babbage, Charles (1792–1871)

English mathematician who devised a precursor of the computer. He designed an analytical engine, a general-purpose mechanical computing device for performing different calculations according to a program input on punched cards (an idea borrowed from the Jacquard loom). This device was never built, but it embodied many of the principles on which digital computers are based.

Mary Evans Picture Library

BABBAGE, CHARLES

Charles Babbage, who was perhaps the first person to conceive the idea of a mechanical computer, was a stickler for accuracy. He once wrote to the poet Tennyson to object to the poet's lines 'Every moment dies a man, every moment one is born'. According to Babbage, Tennyson's recipe gave zero population growth, and should be corrected to:'Every moment one and one-sixteenth is born'.

Babbit metal soft, white metal, an ◊alloy of tin, lead, copper, and antimony, used to reduce friction in bearings, developed by the US inventor Isaac Babbit in 1839.

babbler bird of the thrush family Muscicapidae with a loud babbling cry. Babblers, subfamily Timaliinae, are found in the Old World, and there are some 250 species in the group.

babirusa wild pig *Babirousa babyrussa,* becoming increasingly rare, found in the moist forests and by the water of Sulawesi, Buru, and nearby Indonesian islands. The male has large upper tusks which grow upwards through the skin of the snout and curve back towards the forehead. The babirusa is up to 80 cm/2.5 ft at the shoulder. It is nocturnal, and swims well.

baboon large monkey of the genus *Papio,* with a long doglike muzzle and large canine teeth, spending much of its time on the ground in open country. Males, with head and body up to 1.1 m/3.5 ft long, are larger than females, and dominant males rule the 'troops' in which baboons live. They inhabit Africa and SW Arabia.

Species include the **olive baboon** *P. anubis* from W Africa to Kenya, the **chacma** *P. ursinus* from S Africa, and the **sacred baboon** *P. hamadryas* from NE Africa and SW Arabia. The male sacred baboon has a 'cape' of long hair.

bacille Calmette-Guérin tuberculosis vaccine ◊BCG.

bacillus member of a group of rodlike ◊bacteria that occur everywhere in the soil and air. Some are responsible for diseases such as ◊anthrax or for causing food spoilage.

backbone in networking, a high-◊bandwidth trunk to which smaller networks connect. The original backbone of the Internet was NSFnet, funded by the US National Science Foundation, which linked together the five regional supercomputing centres.

backcross breeding technique used to determine the genetic makeup of an individual organism.

background radiation radiation that is always present in the environment. By far the greater proportion (87%) of it is emitted from natural sources. Alpha and beta particles, and gamma radiation are radiated by the traces of radioactive minerals that occur naturally in the environment and even in the human body, and by radioactive gases such as radon and thoron, which are found in soil and may seep upwards into buildings. Radiation from space (◊cosmic radiation) also contributes to the background level.

backing storage in computing, memory outside the ◊central processing unit used to store programs and data that are not in current use. Backing storage must be nonvolatile – that is, its contents must not be lost when the power supply to the computer system is disconnected.

back pain aches in the region of the spine. Low back pain can be caused by a very wide range of medical conditions. About half of all episodes of back pain will resolve within a week, but severe back pain can be chronic and disabling. The causes include muscle sprain, a prolapsed intervertebral disc, and vertebral collapse due to ◊osteoporosis or cancer. Treatment methods include rest, analgesics, physiotherapy, osteopathy, and exercises.

backswimmer or *water boatman* aquatic predatory bug living mostly in fresh water. The adults are about 15 mm/0.5 in long and rest upside down at the water surface to breathe. When disturbed they dive, carrying with them a supply of air trapped under the wings. They have piercing beaks, used in feeding on tadpoles and small fish.

classification Backswimmers belong to the genus *Notonecta,* family Notonectidae in suborder Heteroptera, order Hemiptera (true bugs), class Insecta, phylum Arthropoda.

Females have a sharp ovipositor to pierce the stems of aquatic plants. In each notch one egg is laid; each female lays a total of approximately 60 eggs over a period of a few weeks. Backswimmers fly readily from pond to pond.

backup in computing, a copy file that is transferred to another medium, usually a ◊floppy disc or tape. The purpose of this is to have available a copy of a file that can be restored in case of a fault in the system or the file itself. Backup files are also created by many applications (with the extension .BAC or .BAK); a version is therefore available of the original file before it was modified by the current application.

backup system in computing, a duplicate computer system that can take over the operation of a main computer system in the event of equipment failure. A large interactive system, such as an airline's ticket-booking system, cannot be out of action for even a few hours without causing considerable disruption. In such cases a complete duplicate computer system may be provided to take over and run the system should the main computer develop a fault or need maintenance.

Backup systems include **incremental backup** and **full backup**.

backwards compatible in computing, term describing a product that is designed to be compatible with its predecessors. In software, a word processor is backwards compatible if it can read and write the files of earlier versions of the same software, and an operating system is backwards compatible if it can run programs designed for earlier versions of the operating system. Similarly, all modems are compatible with all the standards (V numbers) which precede the fastest one they can handle.

bacon beetle destructive species of beetle that attacks bacon, dried foods, and hides. It is related to the ◊carpet beetles.

classification The bacon beetle *Dermestes lardarius* is in the family Dermestidae, order Coleoptera, class Insecta, phylum Arthropoda.

bacteria (singular ***bacterium***) microscopic single-celled organisms lacking a nucleus. Bacteria are widespread, present in soil, air, and water, and as parasites on and in other living things. Some parasitic bacteria cause disease by producing toxins, but others are harmless and may even benefit their hosts. Bacteria usually reproduce by ◊binary fission (dividing into two equal parts), and this may occur approximately every 20 minutes. It is thought that 1–10% of the world's bacteria have been identified.

classification Bacteria are now classified biochemically, but their varying shapes provide a rough classification; for example, **cocci** are round or oval, **bacilli** are rodlike, **spirilla** are spiral, and **vibrios** are shaped like commas. Exceptionally, one bacterium has been found, *Gemmata obscuriglobus,* that does have a nucleus. Unlike ◊viruses, bacteria do not necessarily need contact with a live cell to become active.

Bacteria can be classified into two broad classes (called Gram positive and negative) according to their reactions to certain stains, or dyes, used in microscopy. The staining technique, called the Gram test after Danish bacteriologist Hans Gram, allows doctors to identify many bacteria quickly.

Bacteria have a large loop of ◊DNA, sometimes called a bacterial chromosome. In addition there are often small, circular pieces of DNA known as ◊plasmids that carry spare genetic information. These plasmids can readily move from one bacterium to another, even though the bacteria may be of different species. In a sense, they are parasites within the bacterial cell, but they survive by coding characteristics that promote the survival of their hosts. For example, some plasmids confer antibiotic resistance on the bacteria they inhabit. The rapid and problematic spread of antibiotic resistance among bacteria is due to plasmids, but they are also useful to humans in ◊genetic engineering. There are ten times more bacterial cells than human cells in the human body.

functions Certain types of bacteria are vital in many food and industrial processes, while others play an essential role in the ◊nitrogen cycle, which maintains soil fertility. For example, bacteria are used to break down waste products, such as sewage; make butter, cheese, and yoghurt; cure tobacco; tan leather; and (by virtue of the ability of certain bacteria to attack metal) clean ships' hulls and derust their tanks, and even extract minerals from mines. Several species of bacteria in the stomach of a bowhead whale are

Phages

BY DAVID EVANS

Introduction

Phages, or bacteriophages (literally 'bacteria eaters'), are viruses that infect bacteria. They were first identified by F W Twort in 1905 and, independently, by F d'Herelle in 1917 as filterable infectious agents that caused the lysis (disintegration) of *Shigella dynsenteriae*, the bacterium that causes dysentery. They were initially of interest for their specificity, they would not infect other bacteria, and potential therapeutic use, and have become some of the best characterized viruses.

Life cycle of phages

Phages, like other viruses, are obligate intracellular parasites. They cannot reproduce independently and must instead infect a host bacterium within which they replicate, usually resulting in the lysis of the bacteria and release of progeny phage particles. The phage particle is a metabolically inert entity consisting of a protein coat (capsid) surrounding the genome which consists of single- or double-stranded DNA or RNA. Two distinct replication patterns exist; lytic, in which genome replication, transcription and translation of phage genes and the subsequent assembly of progeny phage particles occurs immediately (so called virulent phages) and lysogenic, in which the phage genome becomes integrated into the host chromosome (the temperate phages). In the latter case the phage genome is replicated as part of the bacterial chromosome, remaining latent as a prophage until induction at which point the phage enters a normal lytic cycle.

Phage structure and classification

The diverse range of phages are primarily classified according to their capsid structure (filamentous – e.g. M13, tailless icosahedral – e.g. [phgr]X174, phages with noncontractile tails – e.g. [lambda] and phages with contractile tails – e.g. T2) and the type of genetic material. The T-even phages of *Escherichia coli* (T2, T4, T6) possess large double-stranded DNA genomes within an icosahedral capsid bearing a contractile tail, and are some of the most complex of all viruses.

Phages in research

The short replication cycle (as little as 15 min) and ease of manipulation in the laboratory have meant that phages have been used for some of the most significant biological, and in particular genetic, experiments of the last 50 years. The identification of DNA as the genetic material was based on a study of phage T2 infection of *E. coli* by Alfred Hershey and Martha Chase in 1952. Phage [phgr]X174, a single-stranded DNA phage was the first complete genome to be sequenced (1977), leading to the discovery of overlapping genes, and the switch between lytic and lysogenic life cycles by phage lambda ([lambda]) is the paradigm for a 'genetic switch', which has broad similarities with similar controlling events in many other organisms (including humans). The natural ability of phages to exchange genetic material between bacteria has led to them being exploited for experiments in molecular genetics –both phage M13 and [lambda] are widely used DNA cloning vectors.

Phages and human disease

The specificity of bacterial infection by phages means that they can be used to distinguish between different strains of bacteria (particularly *Salmonella* sp.), a technique called 'phage typing' widely used in reference laboratories. Furthermore, phages may have a role in causing human disease as well as diagnosing it. *Vibrio cholerae*, the bacterium that causes cholera, requires two factors for full virulence; the cholera toxin (CT) and a toxin-coregulated pili (TCP) required for intestinal colonization. Recent research has shown that the genes encoding both CT and TCP are carried by the filamentous phage CTXθ, which can transfer them to apathogenic strains of *V. cholerae* and so lead to the formation of novel virulent strains capable of causing cholera.

d'Herelle suggested that the specific lysis of bacteria by phages could be exploited in the development of therapeutic treatments for bacterial infections. The intervening 80 years have seen the discovery and widespread use of antibiotics for treating bacterial infection. Although very effective, there are an increasing number of multiply drug-resistant bacteria, e.g. *Mycobacterium tuberculosis* (TB) and *Staphylococcus aureus* (septicaemia) as a direct consequence of the widespread use of antibiotics, and there has been a resurgence in the interest in the use of phages for treating bacterial infection.

capable of digesting pollutants (naphthalene and anthracene, two carcinogenic fractions of oil difficult to break down, and PCBs, also carcinogenic).

Bacteria cannot normally survive temperatures above 100°C/212°F, such as those produced in pasteurization, but those in deep-sea hot vents in the eastern Pacific are believed to withstand temperatures of 350°C/662°F. *Thermus aquaticus*, or taq, grows freely in the boiling waters of hot springs, and an enzyme derived from it is used in genetic engineering to speed up the production of millions of copies of any DNA sequence, a reaction requiring very high temperatures.

interaction Certain bacteria can influence the growth of others; for example, lactic acid bacteria will make conditions unfavourable for salmonella bacteria. Other strains produce nisin, which inhibits growth of listeria and botulism organisms. Plans in the food industry are underway to produce super strains of lactic acid bacteria to avoid food poisoning.

An estimated 99% of bacteria live in **biofilms** rather than in single-species colonies. These are complex colonies made up of a number of different species of bacteria structured on a layer of slime produced by the bacteria. Fungi, algae, and protozoa may also inhabit the biofilms.

prehistoric bacteria Bacterial spores 40 million years old were extracted from a fossilized bee and successfully germinated by US scientists in 1995. It is hoped that prehistoric bacteria can be tapped as a source of new chemicals for use in the drugs industry. Any bacteria resembling extant harmful pathogens will be destroyed, and all efforts are being to made to ensure no bacteria escape the laboratory.

bacteriology the study of ◊bacteria.

bacteriophage virus that attacks ◊bacteria. Such viruses are now of use in genetic engineering.

Bactrian species of ◊camel *Camelus bactrianus* found in the Gobi Desert in Central Asia. Body fat is stored in two humps on the back. It has very long winter fur which is shed in ragged lumps. The head and body length is about 3 m/10 ft, and the camel is up to 2.1 m/6.8 ft tall at the shoulder. Most Bactrian camels are domesticated and are used as beasts of burden in W Asia.

badger large mammal of the weasel family with molar teeth of a crushing type adapted to a partly vegetable diet, and short strong legs with long claws suitable for digging. The Eurasian **common badger** *Meles meles* is about 1 m/3 ft long, with long, coarse, greyish hair on the back, and a white face with a broad black stripe along each side. Mainly a woodland animal, it is harmless and nocturnal, and spends the day in a system of burrows called a 'sett'. It feeds on roots, a variety of fruits and nuts, insects, worms, mice, and young rabbits.

The Eurasian badger lives for up to 15 years. It mates February to March, and again July to Septemebr if the earlier mating has not resulted in fertilization. Implantation of the ◊blastocyst (early embryo) is however delayed until December. Cubs are born January to March, and remain below ground for eight weeks. They remain with the sow at least until autumn.

The **American badger** *Taxidea taxus* is slightly smaller than the Eurasian badger, and lives in open country in North America. Various species of hog badger, ferret badger, and stink badger occur in S and E Asia, the last having the well-developed anal scent glands characteristic of the weasel family.

badlands barren landscape cut by erosion into a maze of ravines, pinnacles, gullies and sharp-edged ridges. Areas in South Dakota and Nebraska, USA, are examples.

Baikonur launch site for spacecraft, located at Tyuratam, Kazakhstan, near the Aral Sea: the first satellites and all Soviet space probes and crewed Soyuz missions were launched from here. It covers an area of 12,200 sq km/4,675 sq mi, much larger than its US equivalent, the ◊Kennedy Space Center in Florida.

Bailey bridge prefabricated bridge developed by the British Army in World War II; made from a set of standardized components so that bridges of varying lengths and load-carrying ability could be assembled to order.

Baird, John Logie (1888–1946)

Scottish electrical engineer who pioneered television. In 1925 he gave the first public demonstration of television, transmitting an image of a recognizable human face. The following year, he gave the world's first demonstration of true television before an audience of about 50 scientists at the Royal Institution, London. By 1928 Baird had succeeded in demonstrating colour television.

Baird used a mechanical scanner which temporarily changed an image into a sequence of electronic signals that could then be reconstructed on a screen as a pattern of half-tones. The neon discharge lamp Baird used offered a simple means for the electrical modulation of light at the receiver. His first pictures were formed of only 30 lines repeated approximately 10 times a second. The results were crude but it was the start of television as a practical technology.

By 1927, Baird had transmitted television over 700 km/435 mi of telephone line between London and Glasgow and soon after made the first television broadcast using radio, between London and the *SS Berengaria*, halfway across the Atlantic Ocean. He also made the first transatlantic television broadcast between Britain and the United States when signals transmitted from the Baird station in Coulson, Kent, were picked up by a receiver in Hartsdale, New York.

Baird's black-and-white system was used by the BBC in an experimental television service in 1929. In 1936, when the public television service was started, his system was threatened by one promoted by Marconi-EMI. The following year the Baird system was dropped in favour of the Marconi electronic system, which gave a better definition.

Mary Evans Picture Library

They were used in every theatre of the war and many remained in place for several years after the war until the civil authorities could replace them with more permanent structures.

Baily's beads bright spots of sunlight seen around the edge of the Moon for a few seconds immediately before and after a total ◊eclipse of the Sun, caused by sunlight shining between mountains at the Moon's edge. Sometimes one bead is much brighter than the others, producing the so-called **diamond ring** effect. The effect was described in 1836 by the English astronomer Francis Baily (1774–1844), a wealthy stockbroker who retired in 1825 to devote himself to astronomy.

Bakelite first synthetic ◊plastic, created by Leo Baekeland in 1909. Bakelite is hard, tough, and heatproof, and is used as an

BADLANDS NATIONAL PARK

http://www.nps.gov/badl/htmlfiles/expanded.htm

Impressive US National Park Service guide to the Badlands National Park. There is comprehensive information on the geology, flora and fauna, and history of the Badlands. Hikers are provided with detailed guidance on routes and there is a listing of the park's educational activities.

electrical insulator. It is made by the reaction of phenol with formaldehyde, producing a powdery resin that sets solid when heated. Objects are made by subjecting the resin to compression moulding (simultaneous heat and pressure in a mould).

It is one of the thermosetting plastics, which do not remelt when heated, and is often used for electrical fittings.

baking powder mixture of ◊bicarbonate of soda, an acidic compound, and a nonreactive filler (usually starch or calcium sulphate), used in baking as a raising agent. It gives a light open texture to cakes and scones, and is used as a substitute for yeast in making soda bread.

Several different acidic compounds (for example, tartaric acid, cream of tartar, sodium or calcium acid phosphates, and glucono-delta-lactone) may be used, any of which will react with the sodium hydrogencarbonate, in the presence of water and heat, to release the carbon dioxide that causes the cake mix or dough to rise.

balance apparatus for weighing or measuring mass. The various types include the **beam balance**, consisting of a centrally pivoted lever with pans hanging from each end, and the **spring balance**, in which the object to be weighed stretches (or compresses) a vertical coil spring fitted with a pointer that indicates the weight on a scale. Kitchen and bathroom scales are balances.

balanced diet diet that includes carbohydrate, protein, fat, vitamins, water, minerals, and fibre (roughage). Although all these substances are needed if a person is to be healthy, there is disagreement over how much of each type a person needs.

balance of nature in ecology, the idea that there is an inherent equilibrium in most ◊ecosystems, with plants and animals interacting so as to produce a stable, continuing system of life on Earth. The activities of human beings can, and frequently do, disrupt the balance of nature.

Organisms in the ecosystem are adapted to each other – for example, waste products produced by one species are used by another and resources used by some are replenished by others; the oxygen needed by animals is produced by plants while the waste product of animal respiration, carbon dioxide, is used by plants as a raw material in photosynthesis. The nitrogen cycle, the water cycle, and the control of animal populations by natural predators are other examples.

BALDNESS

The best way for a man to know whether he will go bald is to look at his mother's father. Baldness is hereditary, but the gene controlling it is on the sex-linked chromosome and skips one generation.

baldness loss of hair from the scalp, common in older men. Its onset and extent are influenced by genetic make-up and the level of male sex ◊hormones. There is no cure, and expedients such as hair implants may have no lasting effect. Hair loss in both sexes may also occur as a result of ill health or radiation treatment, such as for cancer. **Alopecia**, a condition in which the hair falls out, is different from the 'male-pattern baldness' described above.

Experience is a comb which nature gives to men when they are bald.

ANONYMOUS Eastern proverb

Balistidae family of fish containing the ◊triggerfish.

ball-and-socket joint joint allowing considerable movement in three dimensions, for instance the joint between the pelvis and the femur. To facilitate movement, such joints are rimmed with cartilage and lubricated by synovial fluid. The bones are kept in place by ligaments and moved by muscles.

ballistics study of the motion and impact of projectiles such as bullets, bombs, and missiles. For projectiles from a gun, relevant exterior factors include temperature, barometric pressure, and wind strength; and for nuclear missiles these extend to such factors as the speed at which the Earth turns.

balloon lighter-than-air craft that consists of a gasbag filled with gas lighter than the surrounding air and an attached basket, or gondola, for carrying passengers and/or instruments. In 1783, the first successful human ascent was in Paris, in a hot-air balloon designed by the Montgolfier brothers Joseph Michel and Jacques Etienne. In 1785, a hydrogen-filled balloon designed by French physicist Jacques Charles travelled across the English Channel.

balloon help small cartoon-style bubble which pops up in a graphical computer system to convey ◊online help. In many new products, balloon help is activated by holding the mouse over an icon or other type of control for a few seconds. Such help is context-sensitive.

ball valve valve that works by the action of external pressure raising a ball and thereby opening a hole.

balm, lemon garden herb, see ◊lemon balm.

balsam any of various garden plants belonging to the balsam family. They are usually annuals with spurred red or white flowers and pods that burst and scatter their seeds when ripe. (Genus *Impatiens*, family Balsaminaceae.)

In medicine and perfumery, balsam refers to various oily or gummy aromatic plant ◊resins, such as balsam of Peru from the Central American tree *Myroxylon pereirae*.

bamboo any of a large group of giant grass plants, found mainly in tropical and subtropical regions. Some species grow as tall as 36 m/

bamboo *The tall, closely packed stems of many bamboos, such as this example from the tropical rainforest of Brazil, frequently form dense clumps in the forest understorey. Bamboo grows rapidly and has a wide variety of uses.* *Premaphotos Wildlife*

120 ft. The stems are hollow and jointed and can be used in furniture, house, and boat construction. The young shoots are edible; paper is made from the stems. (Genus *Bambusa,* family Gramineae.)

Bamboos flower and seed only once before the plant dies, sometimes after growing for as long as 120 years.

banana any of several treelike tropical plants which grow up to 8 m/25 ft high. The edible banana is the fruit of a sterile hybrid form. (Genus *Musa,* family Musaceae.)

The curved yellow fruits of the commercial banana, arranged in clusters known as 'hands', form cylindrical masses of a hundred or more fruits.

They are picked and exported green and ripened aboard refrigerated ships. The plant is destroyed after cropping. The **plantain**, a larger, coarser hybrid variety that is used green as a cooked vegetable, is a dietary staple in many countries. In the wild, bananas depend on bats for pollination.

banded brush beetle type of ◊rose chafer.
classification The banded brush beetle *Trichius fasciatus,* in the subfamily Trichiinae, family Scarabaeidae, order Coleoptera, class Insecta, phylum Arthropoda.

bandfish marine bony fish. It is elongated with very long dorsal fins that run the length of its body. Bandfish spend most of their time on the sea bed, occasionally swimming to midwater to feed on small planktonic crustacea.

bandicoot small marsupial mammal inhabiting Australia and New Guinea. There are about 11 species, family Peramelidae, rat- or rabbit-sized, and living in burrows. They have long snouts, eat insects, and are nocturnal. A related group, the rabbit bandicoots or bilbies, is reduced to a single species that is now endangered and protected by law.

bandwidth in computing and communications, the rate of data transmission, measured in ◊bits per second (bps).

bandy-bandy venomous Australian snake *Vermicella annulata* of the cobra family, which grows to about 75 cm/2.5 ft. It is banded in black and white. It is not aggressive toward humans.

bang in UNIX, an exclamation mark (!). It appears in some older types of Internet addresses and is used in dictating the commands necessary to run UNIX systems.

bang path list of routing that appears in the header of a message sent across the Internet, showing how it travelled from the sender to its destination. It is named after the ◊bangs separating the sites in the list.

banksia any shrub or tree of a group native to Australia, including the honeysuckle tree. They are named after the British naturalist and explorer Joseph Banks. (Genus *Banksia,* family Proteaceae.)

Banksias have spiny evergreen leaves and large flower spikes, made up of about 1,000 individual flowers formed around a central axis. The colours of the flower spikes can be gold, red, brown, purple, greenish-yellow, and grey.

banner in computing, advertisement on a World Wide Web page, usually but not always in the form of a horizontal rectangle. Clicking on a banner usually takes the user to the advertised Web site. Noncommercial sites display one another's banners via organizations like LinkExchange and BannerExchange.

bantam small ornamental variety of domestic chicken weighing about 0.5–1 kg/1–2 lb. Bantams can either be a small version of one of the larger breeds, or a separate type. Some are prolific egg layers. Bantam cocks have a reputation as spirited fighters.

banteng wild species of cattle *Bos banteng,* now scarce, but formerly ranging from Myanmar (Burma) through SE Asia to Malaysia and Java, inhabiting hilly forests. Its colour varies from pale brown to blue-black, usually with white stockings and rump patch, and it is up to 1.5 m/5 ft at the shoulder.

banyan tropical Asian fig tree. It produces aerial roots that grow down from its spreading branches, forming supporting pillars that look like separate trunks. (*Ficus benghalensis,* family Moraceae.)

baobab tree with rootlike branches, hence the nickname 'upside-down tree', and a disproportionately thick girth, up to 9 m/30 ft in diameter. The pulp of its fruit is edible and is known as monkey bread. (Genus *Adansonia,* family Bombacaceae.)

bar unit of pressure equal to 10^5 pascals or 10^6 dynes/cm^2, approximately 750 mmHg or 0.987 atm. Its diminutive, the **millibar** (one-thousandth of a bar), is commonly used by meteorologists.

barb general name for fish of the genus *Barbus* and some related genera of the family Cyprinidae. As well as the ◊barbel, barbs include many small tropical Old World species, some of which are familiar aquarium species. They are active egg-laying species, usually of 'typical' fish shape and with barbels at the corner of the mouth.

banyan *This banyan tree in the Ranthambhore National Park in N India is reputed to be the second largest specimen in the country. Its shade provides a midday resting place for many people and animals. As a result, banyans are planted along roadsides over much of India. Premaphotos Wildlife*

Barbary ape tailless, yellowish-brown macaque monkey *Macaca sylvanus,* 55–75 cm/20–30 in long. Barbary apes are found in the mountains and wilds of Algeria and Morocco, especially in the forests of the Atlas Mountains. They were introduced to Gibraltar, where legend has it that the British will leave if the ape colony dies out.

The macaque is threatened by illegal logging, which is devastating some of the ancient forests in the area. Although it is breeding well in captivity, forest loss may confound attempts to reintroduce this species into the wild.

Barbary sheep or ***aoudad*** or ***udad*** species of bovid related to the goat and sheep. It has powerful horns and a goatlike odour, but is distinguished from goats by its longer tail and the mane of long hair on the throat and upper parts of the forelegs. It is found in North Africa and parts of the Sudan.
classification The Barbary sheep *Ammotragus lervia* is in family Bovidae, order Artiodactyla

barbastelle insect-eating bat *Barbastella barbastellus* with hairy cheeks and lips, 'frosted' black fur, and a wingspan of about 25 cm/10 in. It lives in hollow trees and under roofs, and is occasionally found in the UK but more commonly in Europe.

barbed wire cheap fencing material made of strands of galvanized wire (see ◊galvanizing), twisted together with sharp barbs at close intervals. In 1873 an American, Joseph Glidden, devised a machine to mass-produce barbed wire. Its use on the open grasslands of 19th-century America led to range warfare between farmers and cattle ranchers; the latter used to drive their herds cross-country.

barbel freshwater fish *Barbus barbus* found in fast-flowing rivers with sand or gravel bottoms in Britain and Europe. Long-bodied, and up to 1 m/3 ft long in total, the barbel has four **barbels** ('little beards' – sensory fleshy filaments) near the mouth.

barberry any spiny shrub belonging to the barberry family, with sour red berries and yellow flowers. These shrubs are often used as hedges. (Genus *Berberis*, family Berberidaceae.)

barbet ***Latin barbatus, 'bearded'*** small, tropical bird, often brightly coloured. There are about 78 species of barbet in the family Capitonidae, order Piciformes, common to tropical Africa, Asia, and America. Barbets eat insects and fruit and, being distant relations of woodpeckers, drill nest holes with their beaks. The name comes from the 'little beard' of bristles about the mouth that assists them in catching insects.

barbiturate hypnosedative drug, commonly known as a 'sleeping pill', consisting of any salt or ester of barbituric acid $C_4H_4O_3N_2$. It works by depressing brain activity. Most barbiturates, being highly addictive, are no longer prescribed and are listed as controlled substances.

Tolerance develops quickly in the user so that increasingly large doses are required to induce sleep. A barbiturate's action persists for hours or days, causing confused, aggressive behaviour or disorientation. Overdosage causes death by inhibiting the breathing centre in the brain. Short-acting barbiturates are used as ◊anaesthetics to induce general anaesthesia; slow-acting ones may be prescribed for epilepsy.

bar chart in statistics, a way of displaying data, using horizontal or vertical bars. The heights or lengths of the bars are proportional to the quantities they represent.

bar code pattern of bars and spaces that can be read by a computer. Bar codes are widely used in retailing, industrial distribution, and public libraries. The code is read by a scanning device; the computer determines the code from the widths of the bars and spaces.

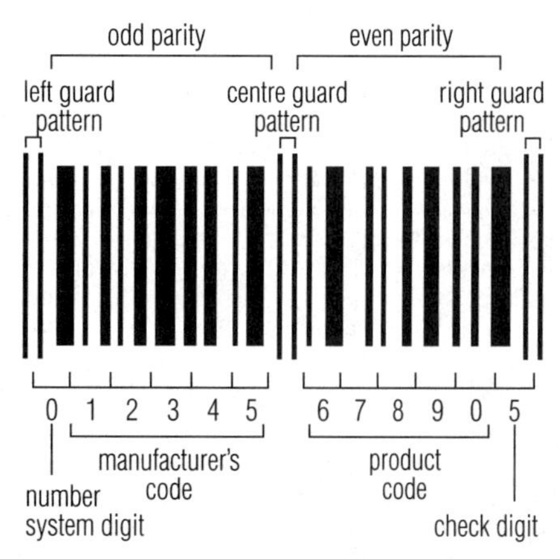

***bar code** The bars of varying thicknesses and spacings represent two series of numbers, identifying the manufacturer and the product. Two longer, thinner bars mark the beginning and end of the manufacturer and product codes. The bar code is used on most articles for sale in shops.*

The technique was patented in 1949 but became popular only in 1973, when the food industry in North America adopted the Universal Product Code system.

barium ***Greek barytes 'heavy'*** soft, silver-white, metallic element, symbol Ba, atomic number 56, relative atomic mass 137.33. It is one of the alkaline-earth metals, found in nature as barium carbonate and barium sulphate. As the sulphate it is used in medicine: taken as a suspension (a 'barium meal'), its movement along the gut is followed using X-rays. The barium sulphate, which is opaque to X-rays, shows the shape of the gut, revealing any abnormalities of the alimentary canal. Barium is also used in alloys, pigments, and safety matches and, with strontium, forms the emissive surface in cathode-ray tubes. It was first discovered in barytes or heavy spar.

bark protective outer layer on the stems and roots of woody plants, composed mainly of dead cells. To allow for expansion of the stem, the bark is continually added to from within, and the outer surface often becomes cracked or is shed as scales. Trees deposit a variety of chemicals in their bark, including poisons. Many of these chemical substances have economic value because they can be used in the manufacture of drugs. Quinine, derived from the bark of the *Cinchona* tree, is used to fight malarial infections; curare, an anaesthetic used in medicine, comes from the *Strychnus toxifera* tree in the Amazonian rainforest.

Bark technically includes all the tissues external to the vascular ◊cambium (the ◊phloem, cortex, and periderm), and its thickness may vary from 2.5 mm/0.1 in to 30 cm/12 in or more, as in the giant redwood *Sequoia* where it forms a thick, spongy layer.

bark beetle any one of a number of species of mainly wood-boring beetles. Bark beetles are cylindrical, brown or black, and 1–9 mm/0.04–0.4 in long. Some live just under the bark and others bore deeper into the hardwood. The detailed tunnelling pattern that they make within the trunk varies with the species concerned, and is used for identification.

Most bark beetles live in forest trees; some, however, attack fruit trees. Generally, but not always, dead or dying timber is attacked. Some species transmit pathogens, for example, the fungus that causes ◊Dutch elm disease.

Examples include the **birch bark beetle** *Scolytus ratzeburgi* and the **greater fruit-tree bark beetle** *S. meli.*

classification Bark beetles are in the families Curculionidae or Scolytidae, order Coleoptera, class Insecta, phylum Arthropoda.

barking deer another name for the ◊muntjac.

barley cereal belonging to a family of grasses. It resembles wheat but is more tolerant of cold and draughts. Cultivated barley (*Hordeum vulgare*) comes in three main varieties – six-rowed, four-rowed, and two-rowed. (Family Gramineae.)

Barley was one of the earliest cereals to be cultivated, about 5000 BC in Egypt, and no other cereal can thrive in so wide a range of climatic conditions; polar barley is sown and reaped well within the Arctic Circle in Europe. Barley is no longer much used in bread making, but it is used in soups and stews and as a starch. Its high-protein form is widely used as animal feed, and its low-protein form is used in brewing and distilling alcoholic drinks.

barley midge another name for the ◊hessian fly.

barn farm building traditionally used for the storage and processing of cereal crops and hay. On older farmsteads, the barn is usually the largest building. It is often characterized by ventilation openings rather than windows and has at least one set of big double doors for access. Before mechanization, wheat was threshed by hand on a specially prepared floor inside these doors.

barnacle marine crustacean of the subclass Cirripedia. The larval form is free-swimming, but when mature, it fixes itself by the head to rock or floating wood. The animal then remains attached, enclosed in a shell through which the cirri (modified legs) protrude to sweep food into the mouth. Barnacles include the stalked **goose barnacle** *Lepas anatifera* found on ships' bottoms, and the **acorn barnacles**, such as *Balanus balanoides*, common on rocks.

barnacle *Acorn barnacles, such as* Chthalamus stellatus, *are a familiar sight on rocky seashores between the tidemarks, where they are the most abundant life form. Identification of the various species is based on the arrangement of the calcareous plates and requires the use of a powerful hand lens or microscope.*
Premaphotos Wildlife

Barnard's star second-closest star to the Sun, six light years away in the constellation Ophiuchus. It is a faint red dwarf of 10th magnitude, visible only through a telescope. It is named after the US astronomer Edward E Barnard (1857–1923), who discovered in 1916 that it has the fastest proper motion of any star, crossing 1 degree of sky every 350 years.

Some observations suggest that Barnard's star may be accompanied by planets.

BARNACLE

The parasitic barnacle *Loxothylacus panopaei* uses a crab as its host. The barnacle larva attaches itself to a crab and injects it with a squirming worm-shaped bag of cells. One of these cells develops into the adult barnacle that spends its life as a lump on the crab's carapace.

Barnard, Christiaan Neethling (1922–)

South African surgeon who performed the first human heart transplant 1967 at Groote Schuur Hospital in Cape Town. The 54-year-old patient lived for 18 days.

Barnard also discovered that intestinal artresia – a congenital deformity in the form of a hole in the small intestine – is the result of an insufficient supply of blood to the fetus during pregnancy. It was a fatal defect before he developed the corrective surgery.

barograph device for recording variations in atmospheric pressure. A pen, governed by the movements of an aneroid ◊barometer, makes a continuous line on a paper strip on a cylinder that rotates over a day or week to create a **barogram**, or permanent record of variations in atmospheric pressure.

barometer instrument that measures atmospheric pressure as an indication of weather. Most often used are the **mercury barometer** and the **aneroid barometer**.

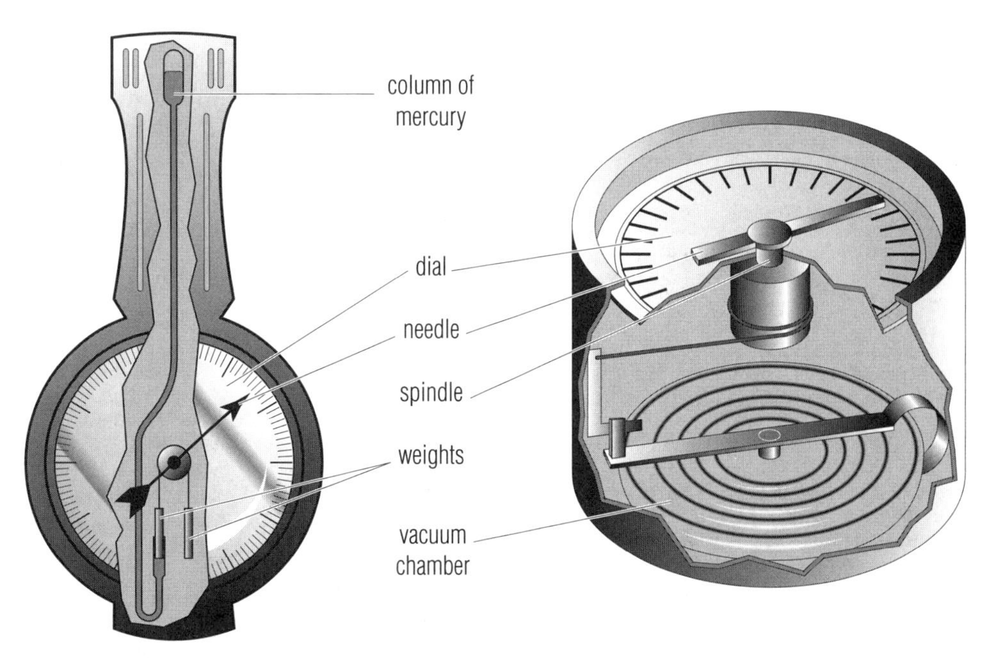

barometer *(left) The mercury barometer and (right) the aneroid barometer. In the mercury barometer, the weight of the column of mercury is balanced by the pressure of the atmosphere on the lower end. A change in height of the column indicates a change in atmospheric pressure. In the aneroid barometer, any change of atmospheric pressure causes the metal box which contains the vacuum to be squeezed or to expand slightly. The movements of the box sides are transferred to a pointer and scale via a chain of levers.*

In a mercury barometer a column of mercury in a glass tube, roughly 0.75 m/2.5 ft high (closed at one end, curved upwards at the other), is balanced by the pressure of the atmosphere on the open end; any change in the height of the column reflects a change in pressure. In an aneroid barometer, a shallow cylindrical metal box containing a partial vacuum expands or contracts in response to changes in pressure.

baroreceptor in biology, a specialized nerve ending that is sensitive to pressure. There are baroreceptors in various regions of the heart and circulatory system (carotid sinus, aortic arch, atria, pulmonary veins, and left ventricle). Increased pressure in these structures stimulates the baroreceptors, which relay information to the medulla providing an important mechanism in the control of blood pressure.

barracuda large predatory fish *Sphyraena barracuda* found in the warmer seas of the world. It can grow over 2 m/6 ft long and has a superficial resemblance to a pike. Young fish shoal, but the older ones are solitary. The barracuda has very sharp shearing teeth and may attack people.

BARRACUDA

Barracudas are mysteriously attracted to all things yellow. This means that they may be caught by the use of yellow-feathered lines.

barrel cylindrical container, tapering at each end, made of thick strips of wood bound together by metal hoops. Barrels are used for the bulk storage of fine wines and spirits.

Barrels were made by craftsmen known as coopers, whose main skill was the shaping and bending of the wooden strips (staves) so that they fitted together without gaps when secured by the hoops. Barrels were widely used for storing liquids and dry goods until the development of plastic containers.

barrel unit of liquid capacity, the value of which depends on the liquid being measured. It is used for petroleum, a barrel of which contains 159 litres/35 imperial gallons; a barrel of alcohol contains 189 litres/41.5 imperial gallons.

barrier island long island of sand, lying offshore and parallel to the coast.

Some are over 100 km/60 mi in length. Most barrier islands are derived from marine sands piled up by shallow longshore currents that sweep sand parallel to the seashore. Others are derived from former spits, connected to land and built up by drifted sand, that were later severed from the mainland.

Often several islands lie in a continuous row offshore. Coney Island and Jones Beach near New York City are well-known examples, as is Padre Island, Texas. The Frisian Islands are barrier islands along the coast of the Netherlands.

barrier reef ◊coral reef that lies offshore, separated from the mainland by a shallow lagoon.

Barringer Crater or *Arizona Meteor Crater* or *Coon Butte* impact crater near Winslow in Arizona caused by the impact of a 50 m/165 ft iron ◊meteorite some 25,000 years ago. It is 1.2 km/0.7 mi in diameter, 200 m/660 ft deep and the walls are raised 50–60 m/165–198 ft above the surrounding desert.

It is named after the US mining engineer Daniel Barringer who proposed in 1902 that it was an impact crater rather than a volcanic feature, an idea confirmed in the 1960s by US geologist Eugene Shoemaker.

barrow *Old English beorgh 'hill or mound'* burial mound, usually composed of earth but sometimes of stones. Examples are found in many parts of the world. The two main types are **long**, dating from the Neolithic (New Stone Age), and **round**, dating from the early Bronze Age. Barrows made entirely of stones are known as cairns.

long barrow Long barrows may be mere mounds, typically higher and wider at one end. They usually contain a chamber of wood or stone slabs, or a turf-lined cavity, in which the body or bodies of the deceased were placed. Secondary chambers may be added in the sides of the mound. They are common in South England from Sussex to Dorset. Earthen (or unchambered) long barrows belong to the early and middle Neolithic, whereas others were constructed over megalithic (great stone) tombs which generally served as collective burial chambers. The stones are arranged to form one, often large, chamber with a single entrance, and are buried under a mound of earth. The remains of these stone chambers, once their earth covering has disappeared, are known as **dolmens**, and in Wales as **cromlechs**.

round barrow Round barrows belong mainly to the Bronze Age, although in historic times there are examples from the Roman period, and some of the Saxon and most of the Danish invaders were barrow-builders. In northern Europe, round barrows were sometimes built above a tree-trunk coffin in which waterlogged conditions have preserved nonskeletal material, such as those found in Denmark dating from around 1000 BC.

In Britain the most common type is the bell barrow, consisting of a circular mound enclosed by a ditch and an outside bank of earth. Other types include the bowl barrow, pond barrow, saucer barrow, ring barrow, and disc barrow, all of which are associated with the Wessex culture (early Bronze Age culture of S England dating from approximately 2000–1500 BC).

Barrows from the Roman era have a distinctive steep and conical outline, and in SE Britain usually cover the graves of wealthy merchant traders. They are also found in Belgic Gaul, where the traders had commercial links. Not all burials in the Roman era were in barrows; cemeteries were also used.

In eastern European and Asiatic areas where mobility was afforded by the horse and wagon, a new culture developed of pit graves marked by a *kurgan*, or round mound, in which a single body lay, often accompanied by grave goods which might include a wagon. These date from around 3000 BC.

boat burial The placing of a great person's body in a ship is seen in Viking burials, such as the Oseberg ship in Norway, which was buried and sealed around AD 800. Barrows were erected over boat burials during the Saxon period, and the Sutton Hoo boat burial excavated in Suffolk, England during 1938–39 was that of an East Anglian king of Saxon times.

baryon in nuclear physics, a heavy subatomic particle made up of three indivisible elementary particles called quarks. The baryons form a subclass of the ◊hadrons and comprise the nucleons (protons and neutrons) and hyperons.

basal metabolic rate (BMR) minimum amount of energy needed by the body to maintain life. It is measured when the subject is awake but resting, and includes the energy required to keep the heart beating, sustain breathing, repair tissues, and keep the brain and nerves functioning. Measuring the subject's consumption of oxygen gives an accurate value for BMR, because oxygen is needed to release energy from food.

A cruder measure of BMR estimates the amount of heat given off, some heat being released when food is used up. BMR varies from one species to another, and from males to females. In humans, it is highest in children and declines with age. Disease, including mental illness, can make it rise or fall. Hormones from the ◊thyroid gland control the BMR.

basalt commonest volcanic ◊igneous rock in the solar system. Much of the surfaces of the terrestrial planets Mercury, Venus, Earth, and Mars, as well as the Moon, are composed of basalt. Earth's ocean floor is virtually entirely made of basalt. Basalt is mafic, that is, it contains relatively little ◊silica: about 50% by weight. It is usually dark grey but can also be green, brown, or black. Its essential constituent minerals are calcium-rich ◊feldspar and calcium and magnesium-rich ◊pyroxene.

The groundmass may be glassy or finely crystalline, sometimes with large ◊crystals embedded. Basaltic lava tends to be runny and flows for great distances before solidifying. Successive eruptions of basalt have formed the great plateaus of Colorado and the Deccan plateau region of southwest India. In some places, such as Fingal's

Cave in the Inner Hebrides of Scotland and the Giant's Causeway in Antrim, Northern Ireland, shrinkage during the solidification of the molten lava caused the formation of hexagonal columns.

The dark-coloured lowland maria regions (see ◊mare) of the Moon are underlain by basalt. Lunar mare basalts have higher concentrations of titanium and zirconium and lower concentrations of volatile elements like potassium and sodium relative to terrestrial basalts. Martian basalts are characterized by low ratios of iron to manganese relative to terrestrial basalts, as judged from some martian meteorites (shergottites, a class of the **SNC** meteorites) and spacecraft analyses of rocks and soils on the Martian surface.

bascule bridge type of drawbridge in which one or two counterweighted deck members pivot upwards to allow shipping to pass underneath. One example is the double bascule Tower Bridge, London.

base in chemistry, a substance that accepts protons, such as the hydroxide ion (OH^-) and ammonia (NH_3). Bases react with acids to give a salt. Those that dissolve in water are called alkalis.

Inorganic bases are usually oxides or hydroxides of metals, which react with dilute acids to form a salt and water. A number of carbonates also react with dilute acids, additionally giving off carbon dioxide. Many organic compounds that contain nitrogen are bases.

Binary (Base 2)	*Octal (Base 8)*	*Decimal (Base 10)*	*Hexadecimal (Base 16)*
0	0	0	0
1	1	1	1
10	2	2	2
11	3	3	3
100	4	4	4
101	5	5	5
110	6	6	6
111	7	7	7
1000	10	8	8
1001	11	9	9
1010	12	10	A
1011	13	11	B
1100	14	12	C
1101	15	13	D
1110	16	14	E
1111	17	15	F
10000	20	16	10
11111111	377	255	FF
11111010001	3721	2001	7D1

base in mathematics, the number of different single-digit symbols used in a particular number system. In our usual (decimal) counting system of numbers (with symbols 0, 1, 2, 3, 4, 5, 6, 7, 8, 9) the base is 10. In the ◊binary number system, which has only the symbols 1 and 0, the base is two. A base is also a number that, when raised to a particular power (that is, when multiplied by itself a particular number of times as in $10^2 = 10 \times 10 = 100$), has a ◊logarithm equal to the power. For example, the logarithm of 100 to the base ten is 2.

In geometry, the term is used to denote the line or area on which a polygon or solid stands.

Science is an essentially anarchistic enterprise: theoretical anarchism is more humanitarian and more likely to encourage progress than its law-and-order alternatives.

PAUL K FEYERABEND Austrian-born US philosopher of science. *Against Method* 1975

Binary (base 2)	*Octal (base 8)*	*Decimal (base 10)*	*Hexadecimal (base 16)*
0	0	0	0
1	1	1	1
10	2	2	2
11	3	3	3
100	4	4	4
101	5	5	5
110	6	6	6
111	7	7	7
1000	10	8	8
1001	11	9	9
1010	12	10	A
1011	13	11	B
1100	14	12	C
1101	15	13	D
1110	16	14	E
1111	17	15	F
10000	20	16	10
11111111	377	255	FF
11111010001	3721	2001	7D1

base *Four different numerical systems showing the numbers 1–16, with some examples of greater numbers. In the hexadecimal (base 16) system, all numbers up to 15 must be represented by a single character. To achieve this the decimal values 10–15 are represented by the letters A–F.*

baseband in computing, type of ◊network that transmits a computer signal without modulation (conversion of ◊digital signals to ◊analogue). To be able to send a computer's signal over the analogue telephone network, a ◊modem is required to convert – or modulate – the signal. On baseband networks, which include the most popular standards such as ◊Ethernet, the signal can be sent directly, without such processing.

basenji breed of dog originating in Central Africa, where it is used for hunting. About 41 cm/16 in tall, it has pointed ears, curled tail, and short glossy coat of black or red, often with white markings. It is remarkable because it has no true bark.

base pair in biochemistry, the linkage of two base (purine or pyrimidine) molecules in ◊DNA. They are found in nucleotides, and form the basis of the genetic code.

One base lies on one strand of the DNA double helix, and one on the other, so that the base pairs link the two strands like the rungs of a ladder. In DNA, there are four bases: adenine and guanine (purines) and cytosine and thymine (pyrimidines). Adenine always pairs with thymine, and cytosine with guanine.

BASIC (acronym for ***beginner's all-purpose symbolic instruction code***) high-level computer-programming language, developed 1964, originally designed to take advantage of ◊multiuser systems (which can be used by many people at the same time). The language is relatively easy to learn and is popular among microcomputer users.

Most versions make use of an ◊interpreter, which translates BASIC into ◊machine code and allows programs to be entered and run with no intermediate translation. Some more recent versions of BASIC allow a ◊compiler to be used for this process.

basicity number of replaceable hydrogen atoms in an acid. Nitric acid (HNO_3) is monobasic, sulphuric acid (H_2SO_4) is dibasic, and phosphoric acid (H_3PO_4) is tribasic.

basic–oxygen process most widely used method of steelmaking, involving the blasting of oxygen at supersonic speed into molten pig iron.

Pig iron from a blast furnace, together with steel scrap, is poured into a converter, and a jet of oxygen is then projected into the mixture. The excess carbon in the mix and other impurities

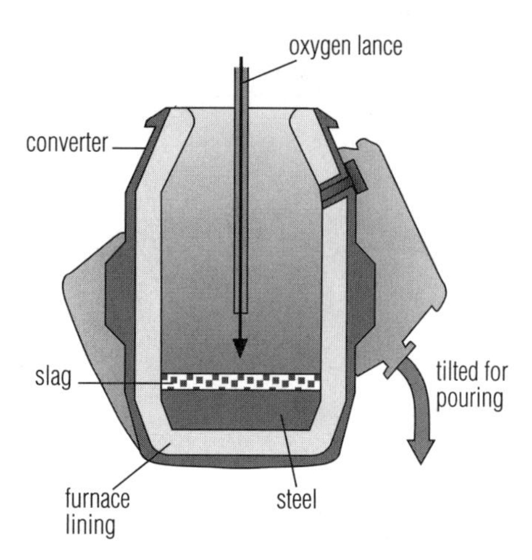

basic–oxygen process The basic–oxygen process is the primary method used to produce steel. Oxygen is blown at high pressure through molten pig iron and scrap steel in a converter lined with basic refractory materials. The impurities, principally carbon, quickly burn out, producing steel.

quickly burn out or form a slag, and the converter is emptied by tilting. It takes only about 45 minutes to refine 350 tonnes/400 tons of steel. The basic–oxygen process was developed 1948 at a steelworks near the Austrian towns of Linz and Donawitz. It is a version of the ◊Bessemer process.

basidiocarp spore-bearing body, or 'fruiting body', of all basidiomycete fungi (see ◊fungus), except the rusts and smuts. A well known example is the edible mushroom *Agaricus brunnescens.* Other types include globular basidiocarps (puffballs) or flat ones that project from tree trunks (brackets). They are made up of a mass of tightly packed, intermeshed ◊hyphae.

The tips of these hyphae develop into the reproductive cells, or **basidia**, that form a fertile layer known as the hymenium, or **gills**, of the basidiocarp. Four spores are budded off from the surface of each basidium.

basil or ***sweet basil*** plant with aromatic leaves, belonging to the mint family. A native of the tropics, it is cultivated in Europe as a herb and used to flavour food. Its small white flowers appear on spikes. (Genus *Ocimum basilicum,* family Labiatae.)

basilisk Central and South American lizard, genus *Basiliscus.* It is about 50 cm/20 in long and weighs about 90 g/0.2 lb. Its rapid speed (more than 2 m/6.6 ft per second) and the formation of air pockets around the feet enable it to run short distances across the surface of water. The male has a well-developed crest on the head, body, and tail.

basket-star any ◊brittle-star of the order Phrynaphiurida, whose spiny arms branch repeatedly to form a coiled mass. Unlike other brittle-stars, which tend to be carnivorous, the hundred or so species of basket-star are suspension feeders, trapping large particles in their extended arms.

bass long-bodied scaly sea fish *Morone labrax* found in the N Atlantic and Mediterranean. They grow to 1 m/3 ft, and are often seen in shoals.

Other fish of the same family (Serranidae) are also called bass, as are North American freshwater fishes of the family Centrarchidae, such as black bass and small-mouthed bass.

basset any of several breeds of hound with a long low body and long pendulous ears, of a type originally bred in France for hunting hares by scent.

BAT CONSERVATION INTERNATIONAL

`http://www.batcon.org/`

Articles, photographs, and miscellaneous bat trivia, plus sound files of bat echolocation signals. As well as promoting membership of their bat society, this page contains plenty of bat-related information, including US and European species' lists and tips on photographing these elusive animals.

bat any mammal of the order Chiroptera, related to the Insectivora (hedgehogs and shrews), but differing from them in being able to fly. Bats are the only true flying mammals. Their forelimbs are developed as wings capable of rapid and sustained flight. There are two main groups of bats: **megabats**, which eat fruit, and **microbats**, which mainly eat insects. Although by no means blind, many microbats rely largely on ◊echolocation for navigation and finding prey, sending out pulses of high-pitched sound and listening for the echo. Bats are nocturnal, and those native to temperate countries hibernate in winter. There are about 977 species forming the order Chiroptera, making this the second-largest mammalian order; bats make up nearly one-quarter of the world's mammals. Although bats are widely distributed, populations have declined alarmingly and many species are now endangered.

megabats The Megachiroptera live in the tropical regions of the Old World, Australia, and the Pacific, and feed on fruit, nectar, and pollen. The hind feet have five toes with sharp hooked claws which suspend the animal head downwards when resting. There are 162 species of Megachiroptera. Relatively large, weighing up to 900 g/2 lb and with a wingspan as great as 1.5 m/5 ft, they have large eyes and a long face, earning them the name 'flying fox'. Most orient by sight.

Many rainforest trees depend on bats for pollination and seed dispersal, and around 300 bat-dependent plant species yield more than 450 economically valuable products. Some bats are keystone species on whose survival whole ecosystems may depend. Bat-pollinated flowers tend to smell of garlic, rotting vegetation, or fungus.

microbats Most bats are Microchiroptera, mainly small and insect-eating. Some eat fish as well as insects; others consume small rodents, frogs, lizards, or birds; a few feed on the blood of mammals (◊vampire bats). There are about 750 species. They roost in caves, crevices, and hollow trees. A single bat may eat 3,000

bat The Gambian epauletted bat Epomophorus gambianus *is a common species in W Africa. It often roosts quite low in trees during the day, undisturbed by the presence of people. Like many tropical bats, it feeds on fruit. A young bat is visible tucked under the wing of one of the females pictured. Premaphotos Wildlife* insects in one night. The bumblebee bat, inhabiting SE Asian rainforests, is the smallest mammal in the world.

Many microbats have poor sight and orientation and hunt their prey principally by echolocation. They have relatively large ears and many have nose-leaves, fleshy appendages around the nose and mouth, that probably help in sending or receiving the signals, which are squeaks pitched so high as to be inaudible to the human ear.

ancestors The difference in the two bat groups is so marked that many biologists believe that they must have had different ancestors: microbats descending from insectivores and megabats descending from primates. Analysis of the proteins in blood serum from megabats and primates by German biologists in 1994 showed enough similarities to suggest a close taxonomic relationship between the two groups.

biology A bat's wings consist of a thin hairless skin expansion, stretched between the four fingers of the hand, from the last finger down to the hindlimb, and from the hindlimb to the tail. The thumb is free and has a sharp claw to help in climbing. The shoulder girdle and breastbone are large, the latter being keeled, and the pelvic girdle is small. The bones of the limbs are hollow, other bones are slight, and the ribs are flattened.

An adult female bat usually rears only one pup a year, which she carries with her during flight. In species that hibernate, mating may take place before hibernation, the female storing the sperm in the genital tract throughout the winter and using it to fertilize her egg on awakening in spring.

batch file file that runs a group (batch) of commands. The most commonly used batch file is the ◊DOS start-up file ◊AUTOEXEC.BAT.

batch processing in computing, a system for processing data with little or no operator intervention. Batches of data are prepared in advance to be processed during regular 'runs' (for example, each night). This allows efficient use of the computer and is well suited to applications of a repetitive nature, such as a company payroll.

In ◊interactive computing, by contrast, data and instructions are entered while the processing program is running.

bat-eared fox small African fox *Otocyon megalotis,* with huge ears, sandy or greyish coat, black legs, and black-tipped bushy tail. They measure about 80 cm/31.5 in in length, including tail, and are 30 cm/12 in at the shoulder; weight 3–5 kg/6.5–11 lb. Bat-eared foxes feed on insects, particularly termites. There are East African and South African subspecies.

Litters of two to five cubs are born after a gestation period of 60 days. Both parents help to raise the young, sometimes aided by a cub from the previous year, but cub mortality is high as they are vulnerable to predators and disease. Cubs are fully grown at four months.

Bates eyesight training method developed by US ophthalmologist William Bates (1860–1931) to enable people to correct problems of vision without wearing glasses. The method is of proven effectiveness in relieving all refractive conditions, correcting squints, lazy eyes, and similar problems, but does not claim to treat eye disease.

bat fly wingless parasitic fly. Bat flies are tiny, bloodsucking, external parasites of bats and look rather spiderlike.

Penicillidia dufouri measures about 5 mm/0.2 in in length, is rust-brown in colour and parasitizes mouse-eared bats.

classification Bat flies are in the families Nycteribiidae and Streblidae, order Diptera, class Insecta, phylum Arthropoda.

batholith large, irregular, deep-seated mass of intrusive ◊igneous rock, usually granite, with an exposed surface of more than 100 sq km/40 sq mi. The mass forms by the intrusion or upswelling of magma (molten rock) through the surrounding rock. Batholiths form the core of some large mountain ranges like the Sierra Nevada of western North America.

According to ◊plate tectonic theory, magma rises in subduction zones along continental margins where one plate sinks beneath another. The solidified magma becomes the central axis of a rising mountain range, resulting in the deformation (folding and overthrusting) of rocks on either side. Gravity measurements indicate that the downward extent or thickness of many batholiths is some 6–9 mi/10–15 km.

Bates, H(enry) W(alter) (1825–1892)

English naturalist and explorer. He spent 11 years collecting animals and plants in South America and identified 8,000 new species of insects. He made a special study of camouflage in animals, and his observation of insect imitation of species that are unpleasant to predators is known as 'Batesian mimicry'.

Mary Evans Picture Library

bathyal zone upper part of the ocean, which lies on the continental shelf at a depth of between 200 m/650 ft and 2,000 m/6,500 ft.

Bathyal zones (both temperate and tropical) have greater biodiversity than coral reefs, according to a 1995 study by the Natural History Museum in London. Maximum biodiversity occurs between 1,000 m/3,280 ft and 3,000 m/9,800 ft.

bathyscaph or ***bathyscaphe*** or ***bathyscape*** deep-sea diving apparatus used for exploration at great depths in the ocean. In 1960, Jacques Piccard and Don Walsh took the bathyscaph *Trieste* to a depth of 10,917 m/35,820 ft in the Challenger Deep in the ◊Mariana Trench off the island of Guam in the Pacific Ocean.

battery any energy-storage device allowing release of electricity on demand. It is made up of one or more electrical ◊cells. Primary-cell batteries are disposable; secondary-cell batteries, or ◊accumulators, are rechargeable. Primary-cell batteries are an extremely uneconomical form of energy, since they produce only 2% of the power used in their manufacture. It is dangerous to try to recharge a primary-cell battery.

The common **dry cell** is a primary-cell battery based on the Leclanché cell and consists of a central carbon electrode immersed

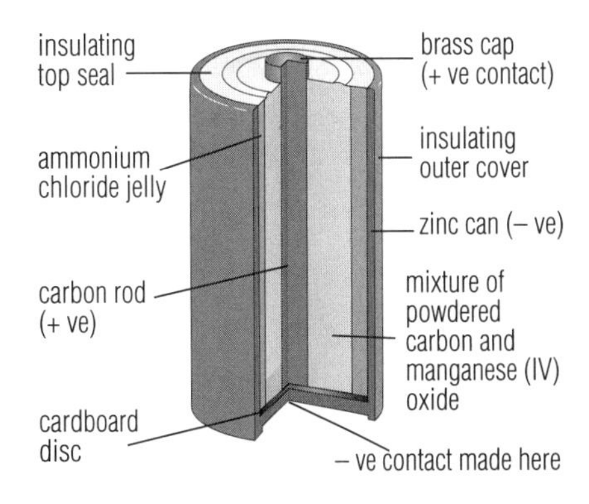

battery *The common dry cell relies on chemical changes occurring between the electrodes – the central carbon rod and the outer zinc casing – and the ammonium chloride electrolyte to produce electricity. The mixture of carbon and manganese is used to increase the life of the cell.*

in a paste of manganese dioxide and ammonium chloride as the electrolyte. The zinc casing forms the other electrode.

The lead–acid **car battery** is a secondary-cell battery. The car's generator continually recharges the battery when the engine is running. It consists of sets of lead (positive) and lead peroxide (negative) plates in an electrolyte of sulphuric acid (◊battery acid). Hydrogen cells and sodium–sulphur batteries were developed in 1996 to allow cars to run entirely on battery power for up to 60 km/100 mi.

The introduction of rechargeable nickel–cadmium batteries has revolutionized portable electronic news gathering (sound recording, video) and information processing (computing). These batteries offer a stable, short-term source of power free of noise and other electrical hazards.

battery acid ◊sulphuric acid of approximately 70% concentration used in lead–acid cells (as found in car batteries).

The chemical reaction within the battery that is responsible for generating electricity also causes a change in the acid's composition. This can be detected as a change in its specific gravity: in a fully charged battery the acid's specific gravity is 1.270–1.290; in a half-charged battery it is 1.190–1.210; in a flat battery it is 1.110–1.130.

baud in engineering, a unit of electrical signalling speed equal to one pulse per second, measuring the rate at which signals are sent between electronic devices such as telegraphs and computers; 300 baud is about 300 words a minute.

Bauds were used as a measure to identify the speed of ◊modems until the early 1990s because at the lower modem speeds available then the baud rate generally equalled the rate of transmission measured in ◊bps (bits per second). At higher speeds, this is not the case, and modem speeds now are generally quoted in bps.

Baudot code five-bit code developed in France by engineer Emil Baudot (1845–1903) in the 1870s. It is still in use for telex.

bauxite principal ore of ◊aluminium, consisting of a mixture of hydrated aluminium oxides and hydroxides, generally contaminated with compounds of iron, which give it a red colour. It is formed by the ◊chemical weathering of rocks in tropical climates. Chief producers of bauxite are Australia, Guinea, Jamaica, Russia, Kazakhstan, Surinam, and Brazil.

To extract aluminium from bauxite, high temperatures (about 800°C/1,470°F) are needed to make the ore molten. Strong electric currents are then passed through the molten ore. The process is only economical if cheap electricity is readily available, usually from a hydroelectric plant.

bay any of various species of ◊laurel tree. The aromatic evergreen leaves are used for flavouring in cookery. There is also a golden-leaved variety. (Genus *Laurus*, family Lauraceae.)

Bayesian statistics form of statistics that uses the knowledge of prior probability together with the probability of actual data to determine posterior probabilities, using Bayes' theorem.

Bayes' theorem in statistics, a theorem relating the ◊probability of particular events taking place to the probability that events conditional upon them have occurred.

For example, the probability of picking an ace at random out of a pack of cards is $\frac{4}{52}$. If two cards are picked out, the probability of the second card being an ace is conditional on the first card: if the first card is an ace the probability of drawing a second ace will be $\frac{3}{51}$; if not it will be $\frac{4}{51}$. Bayes' theorem gives the probability that given that the second card is an ace, the first card is also.

BCE abbreviation for **before the Common Era**, used with dates instead of BC.

B cell or ***B lymphocyte*** immune cell that produces antibodies. Each B cell produces just one type of ◊antibody, specific to a single ◊antigen. Lymphocytes are related to ◊T cells.

BCG (abbreviation for ***bacille Calmette-Guérin***), bacillus injected as a vaccine to confer active immunity to ◊tuberculosis (TB).

BCG was developed by Albert Calmette and Camille Guérin in France in 1921 from live bovine TB bacilli. These bacteria were bred in the laboratory over many generations until they became attenuated (weakened). Each inoculation contains just enough live, attenuated bacilli to provoke an immune response: the formation of specific antibodies. The vaccine provides protection for 50–80% of infants vaccinated.

beach strip of land bordering the sea, normally consisting of boulders and pebbles on exposed coasts or sand on sheltered coasts. It is usually defined by the high- and low-water marks. A berm, a ridge of sand and pebbles, may be found at the farthest point that the water reaches.

The material of the beach consists of a rocky debris eroded from exposed rocks and headlands, or carried in by rivers. The material is transported to the beach, and along the beach, by waves that hit the coastline at an angle, resulting in a net movement of the material in one particular direction. This movement is known as **longshore drift**. Attempts are often made to halt longshore drift by erecting barriers (groynes), at right angles to the movement. Pebbles are worn into round shapes by being battered against one another by wave action and the result is called **shingle**. The finer material, the **sand**, may be subsequently moved about by the wind, forming sand dunes.

Apart from the natural process of longshore drift, a beach may be threatened by the commercial use of sand and aggregate, by the mineral industry – since particles of metal ore are often concentrated into workable deposits by the wave action – and by pollution (for example, by oil spilled or dumped at sea).

beagle short-haired hound with pendant ears, sickle tail, and a bell-like voice, bred for hunting hares on foot ('beagling').

beak horn-covered projecting jaws of a bird (see ◊bill), or other horny jaws such as those of the octopus, platypus, or tortoise.

Beaker people prehistoric people thought to have been of Iberian origin, who spread out over Europe from the 3rd millennium BC. They were skilled in metalworking, and are identified by their use of distinctive earthenware drinking vessels with various designs.

A type of beaker with an inverted bell-shaped profile was widely distributed throughout Europe. These bell beakers are associated with the spread of alcohol consumption, probably mead. The Beaker people favoured individual inhumation (burial of the intact body), often in round ◊barrows, with an associated set of small stone and metal artefacts, or secondary burials in some form of chamber tomb. A beaker typically accompanied male burials, possibly to hold a drink for the deceased on their final journey. The inclusion of flint, later metal, daggers in grave goods may signify a warrior, and suggests that the incursion of Bell Beaker culture may have come as an intrusion into traditional pre-existing cultures.

beam balance instrument for measuring mass (or weight). A simple form consists of a beam pivoted at its midpoint with a pan hanging at each end. The mass to be measured, in one pan, is compared with a variety of standard masses placed in the other. When the beam is balanced, the masses' turning effects or moments under gravity, and hence the masses themselves, are equal.

beam engine engine that works by providing an up and down motion to one end of a beam, which is translated into working machinery at the other end. Beam machines may be powered by a number of sources, including steam and water.

beam weapon weapon capable of destroying a target by means of a high-energy beam. Beam weapons similar to the 'death ray' of science fiction have been explored, most notably during Ronald Reagan's presidential term in the 1980s in the USA.

The **high-energy laser** (HEL) produces a beam of high accuracy that burns through the surface of its target. The **charged particle beam** uses either electrons or protons, which have been accelerated almost to the speed of light, to slice through its target.

bean seed of a large number of leguminous plants (see ◊legume). Beans are rich in nitrogen compounds and proteins and are grown both for human consumption and as food for cattle and horses. Varieties of bean are grown throughout Europe, the USA, South America, China, Japan, SE Asia, and Australia.

The broad bean (*Vicia faba*) has been cultivated in Europe since prehistoric times. The French bean, kidney bean, or haricot (*Phaseolus vulgaris*) is probably of South American origin; the runner bean (*P. coccineus*) is closely related to it, but differs in its climbing habit. Among beans of warmer countries are the lima or butter bean (*P. lunatus*) of South America; the soya bean (*Glycine max*), extensively used in China and Japan; and the winged bean (*Psophocarpus tetragonolobus*) of SE Asia. The tuberous root of the winged bean has potential as a main crop in tropical areas where protein deficiency is common. The Asian mung bean (*Phaseolus mungo*) yields the bean sprouts used in Chinese cookery. Canned baked beans are usually a variety of (*P. vulgaris*), which grows well in the USA.

BEAR DEN

http://www.nature-net.com/bears/index.html

Invaluable resource for information on all types of bears. As well as general information on the evolution and history of bears in general, there are more specific details on each of the eight species of bear, including habitat, reproduction, food, and much more. The site also has photographs and sound effects.

bear large mammal with a heavily built body, short powerful limbs, and a very short tail. Bears breed once a year, producing one to four cubs. In northern regions they hibernate, and the young are born in the winter den. They are found mainly in North America and N Asia. The skin of the polar bear is black to conserve 80–90% of the solar energy trapped and channelled down the hollow hairs of its fur.

Bears walk on the soles of the feet and have long, nonretractable claws. The bear family, Ursidae, is related to carnivores such as dogs and weasels, and all are capable of killing prey. (The panda is probably related to both bears and raccoons.)

species There are seven species of bear. The **brown bear** *Ursus arctos* formerly ranged across most of Europe, N Asia, and North America, but is now reduced in number. It varies in size from under 2 m/7 ft long in parts of the Old World to 2.8 m/9 ft long and 780 kg/1,700 lb in Alaska. The **grizzly bear** is a North American variety of this species, and another subspecies, the **Kodiak bear** of Alaska, is the largest living land carnivore. The white **polar bear** *Thalarctos maritimus* is up to 2.5 m/8 ft long, has furry undersides to the feet, and feeds mainly on seals. It is found in the north polar region. The North American **black bear** *Euarctos americanus* and the **Asian black bear** *Selenarctos thibetanus* are smaller, only about 1.6 m/5 ft long. The latter has a white V-mark on its chest. The **spectacled bear** *Tremarctos ornatus* of the Andes is similarly sized, as is the **sloth bear** *Melursus ursinus* of India and Sri Lanka, which has a shaggy coat and uses its claws and protrusile lips to obtain termites, one of its preferred foods. The smallest bear is the Malaysian **sun bear** *Helarctos malayanus*, rarely more than 1.2 m/4 ft long, a good climber, whose favourite food is honey.

threat of extinction Of the seven species of bear, five are currently reckoned to be endangered and all apart from the polar bear and the American black bear are in decline. The population of brown bears in the Pyrenees was estimated at eight in 1994, and it is feared they will be extinct in 20 years unless new bears are introduced. In May 1996 two female Slovenian brown bears were released into the central Pyrenees; the Slovenian brown bear is closest genetically to the Pyrenean one.

In 1992, American black bears were upgraded to Appendix 2 of CITES (Convention on International Trade in Endangered Species) to stem the trade in their gall bladders, which are used in Asian traditional medicine to treat liver disease. The gall bladders contain an active substance, ursodiol, which is tapped through surgically-implanted tubes. Although an inexpensive synthetic version of ursodiol is available, in 1995 there were at least 10,000 bears being kept in farms in China for their gall bladders, for which many people still prefer to pay thousands of dollars. Trade in Asian black bears and their parts is illegal.

bearberry any of a group of evergreen trailing shrubs belonging to the heath family, found in high and rocky places. Most bearberries are North American but *Arctostaphylos uva-ursi* is also found in Asia and Europe in northern mountainous regions. It has small pink flowers in spring, followed by red berries that are dry but edible. (Genus *Arctostaphylos*, family Ericaceae.)

bearded collie breed of British ◊sheepdog with shaggy hair on its muzzle. Standing about 53 cm/21 in tall, it has a long coat, which is often grey, or sometimes sandy, with white on the head, chest, and feet.

Bear, Great and Little common names (and translations of the Latin) for the constellations ◊Ursa Major and ◊Ursa Minor respectively.

bearing device used in a machine to allow free movement between two parts, typically the rotation of a shaft in a housing.

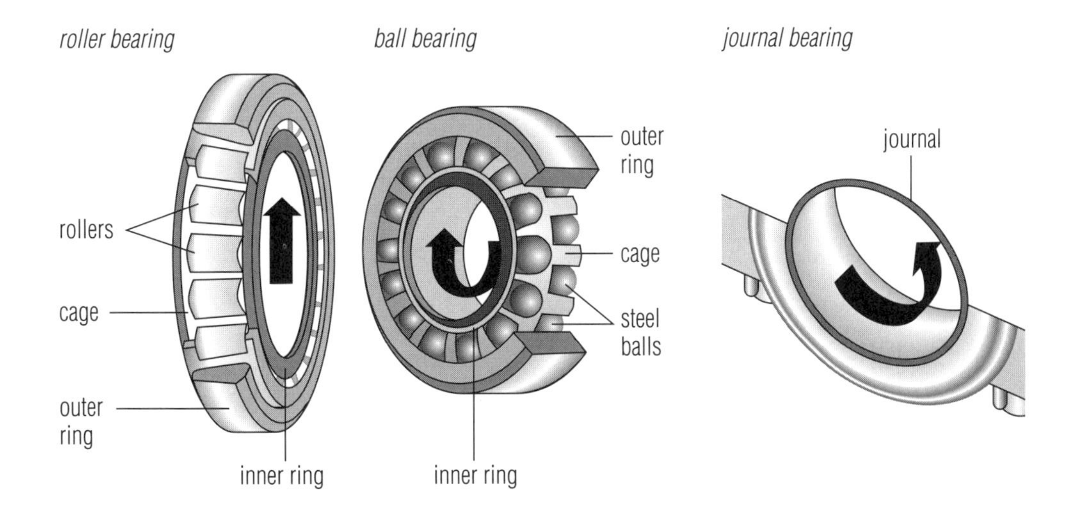

bearing *Three types of bearing. The roller and the ball bearing are similar, differing only in the shape of the parts that roll when the middle shaft turns. The simpler journal bearing consists of a sleeve, or journal, lining the surface of the rotating shaft. The bearing is lubricated to reduce friction and wear.*

Ball bearings consist of two rings, one fixed to a housing, one to the rotating shaft. Between them is a set, or race, of steel balls. They are widely used to support shafts, as in the spindle in the hub of a bicycle wheel.

The **sleeve**, or **journal bearing**, is the simplest bearing. It is a hollow cylinder, split into two halves. It is used for the big-end and main bearings on a car ◊crankshaft.

In some machinery the balls of ball bearings are replaced by cylindrical rollers or thinner **needle bearings**.

In precision equipment such as watches and aircraft instruments, bearings may be made from material such as ruby and are known as **jewel bearings**.

For some applications bearings made from nylon and other plastics are used. They need no lubrication because their surfaces are naturally waxy.

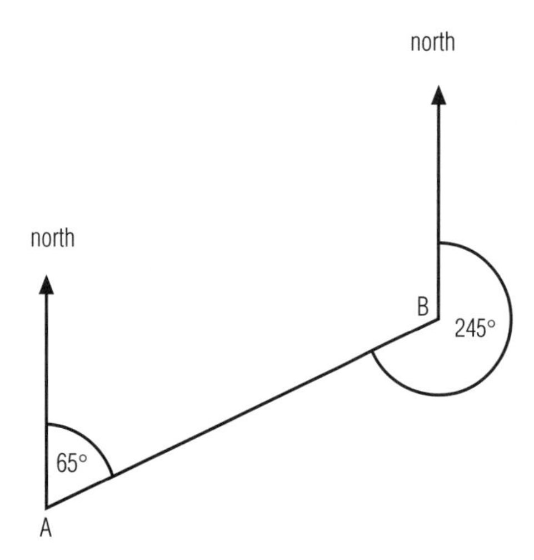

***bearing** A bearing is the direction of a fixed point, or the path of a moving object, from a point of observation on the Earth's surface, expressed as an angle from the north. In the diagram, the bearing of a point A from an observer at B is the angle between the line BA and the north line through B, measured in a clockwise direction from the north line.*

bearing the direction of a fixed point, or the path of a moving object, from a point of observation on the Earth's surface, expressed as an angle from the north. Bearings are taken by ◊compass and are measured in degrees (°), given as three-digit numbers increasing clockwise. For instance, north is 000°, northeast is 045°, south is 180°, and southwest is 225°.

True north differs slightly from magnetic north (the direction in which a compass needle points), hence NE may be denoted as 045M or 045T, depending on whether the reference line is magnetic (M) or true (T) north. True north also differs slightly from grid north since it is impossible to show a spherical Earth on a flat map.

beat frequency in musical acoustics, fluctuation produced when two notes of nearly equal pitch or ◊frequency are heard together. Beats result from the ◊interference between the sound waves of the notes. The frequency of the beats equals the difference in frequency of the notes.

Beaufort scale system of recording wind velocity (speed), devised by Francis Beaufort in 1806. It is a numerical scale ranging from 0 to 17, calm being indicated by 0 and a hurricane by 12; 13–17 indicate degrees of hurricane force.

In 1874 the scale received international recognition; it was modified in 1926. Measurements are made at 10 m/33 ft above ground level.

BEAVER

By building dams, beavers create extremely fertile living conditions for other animals. The ponds formed by the dams trap a wealth of nourishing minerals. They are also full of zooplankton, the combined mass of which is a thousand times greater than that found elsewhere.

beaver aquatic rodent with webbed hind feet, a broad flat scaly tail, and thick waterproof fur. It has very large incisor teeth and fells trees to feed on the bark and to use the logs to construct the 'lodge', in which the young are reared, food is stored, and much of the winter is spent. There are two species, the Canadian *Castor canadensis* and the European *C. fiber.* They grow up to 1.4 m/4.6 ft in length and weigh about 20 kg/44 lb.

Beavers are monogamous and a pair will produce a litter of twins each year. Their territory consists of about 3 km/2 mi of river. Beavers can construct dams on streams, and thus modify the environment considerably; beaver ponds act as traps for minerals and provide fertile living conditions for other species – zooplankton biomass may be 1,000 times greater within a beaver pond than elsewhere. Beavers once ranged across Europe, N Asia, and North America, but in Europe now only survive where they are protected, and are reduced elsewhere, partly through trapping for their fur.

becquerel SI unit (symbol Bq) of ◊radioactivity, equal to one radioactive disintegration (change in the nucleus of an atom when a particle or ray is given off) per second.

Beaufort Scale

Number and description	*Features*	*Air speed*	
		kph	*mph*
0 calm	smoke rises vertically; water smooth	0–2	0–1
1 light air	smoke shows wind direction; water ruffled	2–5	1–3
2 light breeze	leaves rustle; wind felt on face	6–11	4–7
3 gentle breeze	loose paper blows around	12–19	8–12
4 moderate breeze	branches sway	20–29	13–18
5 fresh breeze	small trees sway, leaves blown off	30–39	19–24
6 strong breeze	whistling in telephone wires; sea spray from waves	40–50	25–31
7 near gale	large trees sway	51–61	32–38
8 gale	twigs break from trees	62–74	39–46
9 strong gale	branches break from trees	75–87	47–54
10 storm	trees uprooted; weak buildings collapse	88–101	55–63
11 violent storm	widespread damage	102–117	64–73
12 hurricane	widespread structural damage	above 118	above 74

Becquerel, (Antoine) Henri (1852–1908)

French physicist. He discovered penetrating radiation coming from uranium salts, the first indication of radioactivity, and shared a Nobel prize with Marie and Pierre Curie in 1903.

Mary Evans Picture Library

The becquerel is much smaller than the previous standard unit, the ◊curie (3.7×10^{10} Bq). It is named after French physicist Henri Becquerel.

bed in geology, a single ◊sedimentary rock unit with a distinct set of physical characteristics or contained fossils, readily distinguishable from those of beds above and below. Well-defined partings called **bedding planes** separate successive beds or strata.

The depth of a bed can vary from a fraction of a centimetre to several metres or yards, and can extend over any area. The term is also used to indicate the floor beneath a body of water (lake bed) and a layer formed by a fall of particles (ash bed).

bedbug flattened wingless red-brown insect *Cimex lectularius* with piercing mouthparts. It hides by day in crevices or bedclothes and emerges at night to suck human blood.

Bedlington breed of ◊terrier with a short body, long legs, and curly hair, usually grey, named after a district of Northumberland, England.

bee four-winged insect of the superfamily Apoidea in the order Hymenoptera, usually with a sting. There are over 12,000 species, of which fewer than 1 in 20 are social in habit. The **hive bee** or **honeybee** *Apis mellifera* establishes perennial colonies of about 80,000, the majority being infertile females (workers), with a few larger fertile males (drones), and a single very large fertile female (the queen). Worker bees live for no more than a few weeks, while a drone may live a few months, and a queen several years. Queen honeybees lay two kinds of eggs: fertilized, female eggs, which have two sets of chromosomes and develop into workers or queens, and unfertilized, male eggs, which have only one set of chromosomes and develop into drones.

Bees transmit information to each other about food sources by 'dances', each movement giving rise to sound impulses which are picked up by tiny hairs on the back of the bee's head, the orientation of the dance also having significance. They use the Sun in navigation (see ◊migration). Besides their use in crop pollination and production of honey and wax, bees (by a measure of contaminants brought back to their hives) can provide an inexpensive and effective monitor of industrial and other pollution of the atmosphere and soil.

The most familiar species is the ◊bumblebee, genus *Bombus*, which is larger and stronger than the hive bee and so is adapted to fertilize plants in which the pollen and nectar lie deep, as in red clover; they can work in colder weather than the hive bee.

Social bees, apart from the bumblebee and the hive bee, include the stingless South American **vulture bee** *Trigona hypogea*, **discovered in 1982, which is solely carnivorous.**

Solitary bees include species useful in pollinating orchards in spring, and may make their nests in tunnels under the ground or in hollow plant stems; 'cuckoo' bees lay their eggs in the nests of bumblebees, which they closely resemble.

The killer bees of South America are a hybrid type, created when an African subspecies of honeybee escaped from a research establishment in Brazil in 1957. They mated with, and supplanted, the honeybees of European origin in most of South and Central America, and by 1990 had spread as far north as Texas, USA. As well as being more productive and resistant to disease than European honeybees, they also defend their hives more aggressively, in larger numbers, and for a greater length of time than other honeybees. However, their stings are no more venomous, and although they have killed hundreds of thousands of animals and probably more than 1,000 people, most individuals survive an attack, and almost all deaths have occurred where the victim has somehow been prevented from fleeing.

Most bees are passive unless disturbed, but some species are aggressive. One bee sting may be fatal to a person who is allergic to them, but this is comparatively rare (about 1.5% of the population), and most adults can survive 300–500 stings without treatment. A vaccine treatment against bee stings, which uses concentrated venom, has been developed.

BEEKEEPING

http://weber.u.washington.edu/~jlks/bee.html

Everything you could possibly wish to know about beekeeping is here. There are, for example, sections on the diseases afflicting bees, as well as advice on honey and mead-making. The site also includes an archived list of articles and some multimedia items.

beech one of several European hardwood trees or related trees growing in Australasia and South America. The common beech (*Fagus sylvaticus*), found in European forests, has a smooth grey trunk and edible nuts, or 'mast', which are used as animal feed or processed for oil. The timber is used in furniture. (Genera *Fagus* and *Nothofagus*, family Fagaceae.)

bee-eater brightly-coloured bird *Merops apiaster*, family Meropidae, order Coraciiformes, found in Africa, S Europe, and Asia. Bee-eaters are slender, with chestnut, yellow, and blue-green plumage, a long bill and pointed wings, and a flight like that of the swallow, which they resemble in shape. They feed on bees, wasps, and other insects, and nest in colonies in holes dug out with their long bills in sandy river banks.

bee louse any of a number of species of wingless flies parasitic on bees. They look more like lice than flies. Most have claws at the ends of their legs.

classification Bee lice are in the family Braulidae, order Diptera (suborder Cyclorrhapha), class Insecta, phylum Arthropoda.

Bee lice lay their eggs on the walls or under the cappings of honey cells. The larvae work their way from one cell to the next, feeding on the honey. In small numbers the bee louse is insignificant; if many are present however, they can be serious pests to bees and beekeepers.

Braula caeca is rust-brown and about 1.5 mm/0.06 in long. It is a parasite of honeybees.

beet any of several plants belonging to the goosefoot family, used as food crops. One variety of the common beet (*Beta vulgaris*) is used in to produce sugar and another, the mangelwurzel, is grown as a cattle feed. The beetroot, or red beet (*B. rubra*), is a salad plant. (Genus *Beta*, family Chenopodiaceae.)

beetfly small grey fly with black hairs. Its maggots feed on beet leaves. As soon as the maggots hatch they begin to feed, continuing for one month, when they turn into chestnut-brown pupae. The flies emerge a fortnight later.

classification The beet fly *Anthomyia betae* is in the family Anthomyiidae, order Diptera, class Insecta, phylum Arthropoda.

beetle common name of insects in the order Coleoptera (Greek 'sheath-winged') with leathery forewings folding down in a protective sheath over the membranous hindwings, which are those used for flight. They pass through a complete metamorphosis. They include some of the largest and smallest of all insects: the largest is the **Hercules beetle** *Dynastes hercules* of the South American rainforests, 15 cm/6 in long; the smallest is only 0.05 cm/0.02 in long. Comprising more than 50% of the animal kingdom, beetles number some 370,000 named species, with many not yet described.

BEETLE

When it jumps, the flea beetle accelerates at 260 times more than gravity while spinning head over heels about 70 times per second. Flea beetles always land the right way up.

Beetles are found in almost every land and freshwater habitat, and feed on almost anything edible. Examples include **click beetle** or **skipjack** species of the family Elateridae, so called because if they fall on their backs they right themselves with a jump and a loud click; the larvae, known as **wireworms**, feed on the roots of crops. In some tropical species of Elateridae the beetles have luminous organs between the head and abdomen and are known as **fireflies**. The potato pest **Colorado beetle** *Leptinotarsa decemlineata* is striped in black and yellow. The **blister beetle** *Lytta vesicatoriaf,* a shiny green species from S Europe, was once sold pulverized as an aphrodisiac and contains the toxin cantharidin. The larvae of the **furniture beetle** *Anobium punctatum* and the **deathwatch beetle** *Xestobium rufovillosum* and their relatives are serious pests of structural timbers and furniture (see ◊woodworm).

begonia any of a group of tropical and subtropical plants. They have fleshy and succulent leaves, and some have large, brilliant flowers. There are numerous species in the tropics, especially in South America and India. (Genus *Begonia,* family Begoniaceae.)

behaviourism school of psychology originating in the USA, of which the leading exponent was John B Watson.

Behaviourists maintain that all human activity can ultimately be explained in terms of conditioned reactions or reflexes and habits formed in consequence. Leading behaviourists include Ivan ◊Pavlov and B F Skinner.

It is a good morning exercise for a research scientist to discard a pet hypothesis every day before breakfast. It keeps him young.

Konrad Lorenz Austrian zoologist. *The So-Called Evil*

behaviour therapy in psychology, the application of behavioural principles, derived from learning theories, to the treatment of clinical conditions such as ◊phobias, ◊obsessions, and sexual and interpersonal problems.

The symptoms of these disorders are regarded as learned patterns of behaviour that therapy can enable the patient to unlearn. For example, in treating a phobia, the patient is taken gradually into the feared situation in about 20 sessions until the fear noticeably reduces.

bel unit of sound measurement equal to ten ◊decibels. It is named after Scottish scientist Alexander Graham Bell.

belemnite extinct relative of the squid, with rows of little hooks rather than suckers on the arms. The parts of belemnites most frequently found as fossils are the bullet-shaped shells that were within the body. Like squid, these animals had an ink sac which could be used to produce a smokescreen when attacked.

Belgian sheepdog any of four varieties of herding and guarding dog developed in Belgium: the Groenedael, Turvuren, Malinois, and Lakenois. Similar in build and size, they stand about 62 cm/24 in tall, the main difference between them being the variations in colour and type of coat.

The Groenedael has a long all-black coat; the Tervuren has a long tawny coat with black markings on the face; the Malinois is similar to the Tervuren in colour, but with a short coat; the Lakenois also shares the Tervuren's colouring, but with a short, coarse, wavy coat.

Bell, Alexander Graham (1847–1922)

Scottish-born US scientist and inventor. He was the first person ever to transmit speech from one point to another by electrical means. This invention – the telephone – was made in 1876. Later Bell experimented with a type of phonograph and, in aeronautics, invented the tricycle undercarriage.

Bell also invented a photophone, which used selenium crystals to apply the telephone principle to transmitting words in a beam of light. He thus achieved the first wireless transmission of speech.

Mary Evans Picture Library

BELL, ALEXANDER GRAHAM

Alexander Graham Bell's invention of the telephone stemmed from his interest in hearing defects. His wife was profoundly deaf. On the day that Bell was buried in 1922, the entire telephone network in the USA was closed down for one minute in tribute.

belladonna or ***deadly nightshade*** poisonous plant belonging to the nightshade family, found in Europe and Asia. It grows to 1.5 m/5 ft in height, with dull green leaves growing in unequal pairs, up to 20 cm/8 in long, and single purplish flowers that produce deadly black berries. Drugs are made from the leaves. (*Atropa belladonna,* family Solanaceae.)

The dried powdered leaves are used to produce the drugs atropine and hyoscine. Belladonna extract acts medicinally as an anticholinergic (blocking the passage of certain nerve impulses), and is highly poisonous in large doses.

Bell Burnell, (Susan) Jocelyn (1943–) British astronomer. In 1967 she discovered the first ◊pulsar (rapidly flashing star) with Antony Hewish and colleagues at Cambridge University, England.

bellflower general name for many plants with bell-shaped flowers. The ◊harebell (*Campanula rotundifolia*) is a wild bellflower. The Canterbury bell (*C. medium*) is the garden variety, originally from S Europe. (Genus *Campanula,* family Campanulaceae.)

bells nautical term applied to half-hours of watch. A day is divided into seven watches, five of four hours each and two, called dog-watches, of two hours. Each half-hour of each watch is indicated by the striking of a bell, eight bells signalling the end of the watch.

benchmark in computing, a measure of the performance of a piece of equipment or software, usually consisting of a standard program or suite of programs. Benchmarks can indicate whether a computer is powerful enough to perform a particular task, and so enable machines to be compared. However, they provide only a very rough guide to practical performance, and may lead manufacturers to design systems that get high scores with the artificial benchmark programs but do not necessarily perform well with day-to-day programs or data.

bends or ***compressed-air sickness*** or ***caisson disease*** popular name for a syndrome seen in deep-sea divers, arising from too rapid a release of nitrogen from solution in their blood. If a diver surfaces too quickly, nitrogen that had dissolved in the blood under increasing water pressure is suddenly released, forming bubbles in the bloodstream and causing pain (the 'bends') and paralysis. Immediate treatment is gradual decompression in a decompression chamber, whilst breathing pure oxygen.

Benioff zone seismically active zone inclined from a deep sea trench beneath a continent or continental margin. Earthquakes along Benioff zones define the top surfaces of plates of ◊lithosphere that descend in to the mantle beneath another, overlying plate. The zone is named after Hugo Benioff, a US seismologist who first described this feature.

bent or **bent grass** any of a group of grasses. Creeping bent grass (*Agrostis stolonifera*), also known as fiorin, is common in N North America, Europe, and Asia, including lowland Britain. It spreads by ◊runners and has large attractive clusters (panicles) of yellow or purple flowers on thin stalks, like oats. It is often used on lawns and golf courses. (Genus *Agrostris,* family Gramineae.)

benzaldehyde C_6H_5CHO colourless liquid with the characteristic odour of almonds. It is used as a solvent and in the making of perfumes and dyes. It occurs in certain leaves, such as the cherry, laurel, and peach, and in a combined form in certain nuts and kernels. It can be extracted from such natural sources, but is usually made from ◊toluene.

Benzedrine trade name for ◊amphetamine, a stimulant drug.

benzene C_6H_6 clear liquid hydrocarbon of characteristic odour, occurring in coal tar. It is used as a solvent and in the synthesis of many chemicals.

The benzene molecule consists of a ring of six carbon atoms, all of which are in a single plane, and it is one of the simplest ◊cyclic compounds. Benzene is the simplest of a class of compounds collectively known as **aromatic compounds**. Some are considered carcinogenic (cancer-inducing).

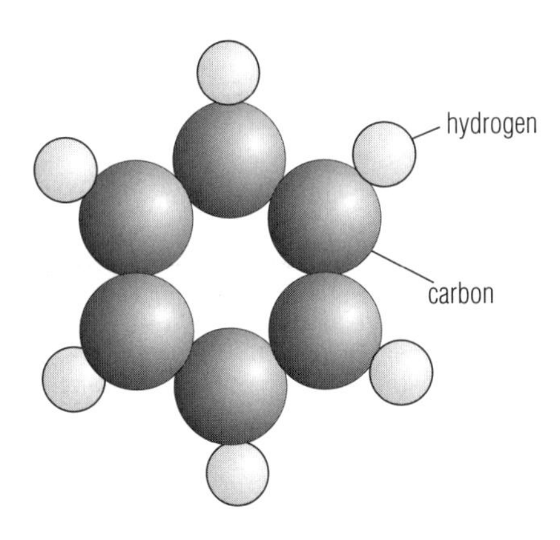

***benzene** The molecule of benzene consists of six carbon atoms arranged in a ring, with six hydrogen atoms attached. The benzene ring structure is found in many naturally occurring organic compounds.*

benzodiazepine any of a group of mood-altering drugs (tranquillizers), for example Librium and Valium. They are addictive and interfere with the process by which information is transmitted between brain cells, and various side effects arise from continued use. They were originally developed as muscle relaxants, and then excessively prescribed in the West as anxiety-relieving drugs.

Today the benzodiazepines are recommended only for short-term use in alleviating severe anxiety or insomnia.

benzoic acid C_6H_5COOH white crystalline solid, sparingly soluble in water, that is used as a preservative for certain foods and as an antiseptic. It is obtained chemically by the direct oxidation of benzaldehyde and occurs in certain natural resins, some essential oils, and as hippuric acid.

benzoin resin (thick liquid that hardens in the air) obtained by making cuts in the bark of the tree *Styrax benzoin,* which grows in the East Indies. Benzoin is used in cosmetics, perfumes, and incense.

benzpyrene one of a number of organic compounds associated with a particular polycyclic ring structure. Benzpyrenes are present in coal tar at low levels and are considered carcinogenic (cancer-inducing). Traces of benzpyrenes are present in wood smoke, and this has given rise to some concern about the safety of naturally smoked foods.

bergamot small evergreen tree belonging to the rue family. A fragrant citrus-scented essence is obtained from the rind of its fruit and used as a perfume and food flavouring, for example in Earl Grey tea. The sole source of supply is S Calabria, Italy, but the name comes from the town of Bergamo, in Lombardy. (*Citrus bergamia,* family Rutaceae.)

Bergmann musquete ***Machine Pistol 18*** German automatic weapon of World War I, the forerunner of the modern submachine gun. A simple automatic weapon with a short barrel and wooden stock, it fired the standard 9 mm pistol cartridge at 400 rounds per minute.

It was first issued in 1916 for trench defence, but the adoption of the infiltration tactic by General von Hutier led to the weapon being issued to 'Storm Troops' since it provided the ideal combination of firepower and portability demanded in this role.

beriberi nutritional disorder occurring mostly in the tropics and resulting from a deficiency of vitamin B_1 (◊thiamine). The disease takes two forms: in one ◊oedema (waterlogging of the tissues) occurs; in the other there is severe emaciation. There is nerve degeneration in both forms and many victims succumb to heart failure.

Beringia or **Bering Land Bridge** former land bridge 1,600 km/1,000 mi wide between Asia and North America; it existed during the ice ages that occurred before 35,000 BC and during the period 24,000–9,000 BC. It is now covered by the Bering Strait and Chukchi Sea.

berkelium synthesized, radioactive, metallic element of the actinide series, symbol Bk, atomic number 97, relative atomic mass 247.

It was first produced in 1949 by Glenn Seaborg and his team, at the University of California at Berkeley, USA, after which it is named.

Bernoulli's principle law stating that the pressure of a fluid varies inversely with speed, an increase in speed producing a decrease in pressure (such as a drop in hydraulic pressure as the fluid speeds up flowing through a constriction in a pipe) and vice versa. The principle also explains the pressure differences on each surface of an aerofoil, which gives lift to the wing of an aircraft. The principle was named after Swiss mathematician and physicist Daniel Bernoulli.

berry fleshy, many-seeded ◊fruit that does not split open to release the seeds. The outer layer of tissue, the exocarp, forms an outer skin that is often brightly coloured to attract birds to eat the fruit and thus disperse the seeds. Examples of berries are the tomato and the grape.

A **pepo** is a type of berry that has developed a hard exterior, such as the cucumber fruit. Another is the **hesperidium**, which has a thick, leathery outer layer, such as that found in citrus fruits, and fluid-containing vesicles within, which form the segments.

beryl mineral, beryllium aluminium silicate, $Be_3Al_2Si_6O_{18}$, which forms crystals chiefly in granite. It is the chief ore of beryllium. Two of its gem forms are aquamarine (light-blue crystals) and emerald (dark-green crystals).

beryllium hard, light-weight, silver-white, metallic element, symbol Be, atomic number 4, relative atomic mass 9.012. It is one of the ◊alkaline-earth metals, with chemical properties similar to those of magnesium. In nature it is found only in combination with other elements and occurs mainly as beryl ($3BeO.Al_2O_3.6SiO_2$). It is used to make sturdy, light alloys and to control the speed of neutrons in nuclear reactors. Beryllium oxide was discovered in 1798 by French chemist Louis-Nicolas Vauquelin (1763–1829), but the element was not isolated until 1828, by Friedrich Wöhler and Antoine-Alexandre-Brutus Bussy independently.

In 1992 large amounts of beryllium were unexpectedly discovered in six old stars in the Milky Way.

Berzelius, Jöns Jakob (1779–1848)

Swedish chemist. He accurately determined more than 2,000 relative atomic and molecular masses. In 1813–14, he devised the system of chemical symbols and formulae now in use and proposed oxygen as a reference standard for atomic masses. His discoveries include the elements cerium 1804, selenium 1817, and thorium 1828; he was the first to prepare silicon in its amorphous form and to isolate zirconium. The words 'isomerism', 'allotropy', and 'protein' were coined by him.

Berzelius noted that some reactions appeared to work faster in the presence of another substance which itself did not appear to change, and postulated that such a substance contained a catalytic force. Platinum, for example, was capable of speeding up reactions between gases. Although he appreciated the nature of catalysis, he was unable to give any real explanation of the mechanism.

Mary Evans Picture Library

Bessemer process the first cheap method of making ◊steel, invented by Henry Bessemer in England 1856. It has since been superseded by more efficient steel-making processes, such as the ◊basic–oxygen process. In the Bessemer process compressed air is

Bessemer, Henry (1813–1898)

British engineer and inventor who developed a method of converting molten pig iron into steel (the Bessemer process) in 1856. Knighted in 1879.

Mary Evans Picture Library

blown into the bottom of a converter, a furnace shaped like a cement mixer, containing molten pig iron. The excess carbon in the iron burns out, other impurities form a slag, and the furnace is emptied by tilting.

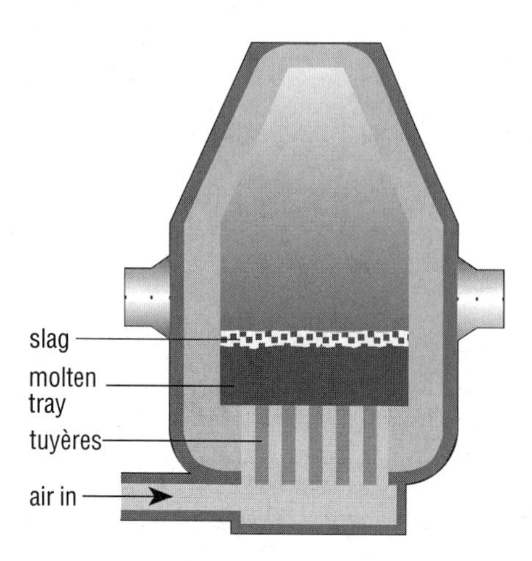

Bessemer process *In a Bessemer converter, a blast of high-pressure air oxidizes impurities in molten iron and converts it to steel.*

beta pre-release version of a new software program still in development, which is handed out to users for testing. The worst ◊bugs are usually eliminated at the earlier alpha stage of development. Beta testers use the software to do real work and report any bugs or badly implemented features they find to the developers, who incorporate this information in refining the product for release. Companies that assist with such testing are known as beta sites.

beta-blocker any of a class of drugs that block impulses that stimulate certain nerve endings (beta receptors) serving the heart muscle. This reduces the heart rate and the force of contraction, which in turn reduces the amount of oxygen (and therefore the blood supply) required by the heart. Beta-blockers may be useful in the treatment of angina, arrhythmia (abnormal heart rhythms), and raised blood pressure, and following heart attacks. They must be withdrawn from use gradually.

beta decay the disintegration of the nucleus of an atom to produce a beta particle, or high-speed electron, and an electron-antineutrino. During beta decay, a neutron in the nucleus changes into a proton, thereby increasing the atomic number by one while the mass number stays the same. The mass lost in the change is converted into kinetic (movement) energy of the beta particle.

Beta decay is caused by the weak nuclear force, one of the fundamental ◊forces of nature operating inside the nucleus.

beta index mathematical measurement of the connectivity of a transport network. If the network is represented as a simplified topological map, made up of nodes (junctions or places) and edges (links), the beta index may be calculated by dividing the number of nodes by the number of edges. If the number of nodes is n and the number of edges is e, then the beta index ß is given by the formula:

$$ß = n/e$$

The higher the index number, the better connected the network is. If ß is greater than 1, then a complete circuit exists.

beta particle electron ejected with great velocity from a radioactive atom that is undergoing spontaneous disintegration. Beta particles do not exist in the nucleus but are created on disintegration, beta decay, when a neutron converts to a proton to emit an electron.

Beta particles are more penetrating than ◊alpha particles, but less so than ◊gamma radiation; they can travel several metres in air, but are stopped by 2–3 mm of aluminium. They are less strongly ionizing than alpha particles and, like cathode rays, are easily deflected by magnetic and electric fields.

beta version in computing, a pre-release version of ◊software or an ◊application program, usually distributed to a limited number of expert users (and often reviewers). Distribution of beta versions allows user testing and feedback to the developer, so that any necessary modifications can be made before release.

Betelgeuse or ***Alpha Orionis*** red supergiant star in the constellation of ◊Orion. It is the tenth brightest star in the night sky, although its brightness varies. It is 1,100 million km/700 million mi across, about 800 times larger than the Sun, roughly the same size as the orbit of Mars. It is over 10,000 times as luminous as the Sun, and lies 650 light years from Earth. Light takes 60 minutes to travel across the giant star.

Its magnitude varies irregularly between 0.4 and 1.3 in a period of 5.8 years. It was the first star whose angular diameter was measured with the Mount Wilson ◊interferometer in 1920. The name is a corruption of the Arabic, describing its position in the shoulder of Orion.

betel nut fruit of the areca palm (*Areca catechu*), which is chewed together with lime and betel pepper as a stimulant by peoples of the East and Papua New Guinea. Chewing it blackens the teeth and stains the mouth deep red.

betony plant belonging to the mint family, formerly used in medicine and dyeing. It has a hairy stem and leaves, and dense heads of reddish-purple flowers. (*Stachys* (formerly *Betonica*) *officinalis*, family Labiatae.)

Bezier curve curved line invented by Pierre Bézier that connects a series of points (or 'nodes') in the smoothest possible way. The shape of the curve is governed by a series of complex mathematical formulae. They are used in ◊computer graphics and ◊CAD.

bezoar or **bezoar stone** hardened mass occasionally found in the stomach or intestines of ruminating animals, such as goats, llamas, antelopes, and cows. They appear to be formed through the presence of some irritating substance in the alimentary tract.

bhang name for a weak form of the drug ◊cannabis used in India.

bhp abbreviation for **brake horsepower**.

bicarbonate common name for ◊hydrogencarbonate

bicarbonate indicator pH indicator sensitive enough to show a colour change as the concentration of the gas carbon dioxide increases. The indicator is used in photosynthesis and respiration experiments to find out whether carbon dioxide is being liberated. The initial red colour changes to yellow as the pH becomes more acidic.

Carbon dioxide, even in the concentrations found in exhaled air, will dissolve in the indicator to form a weak solution of carbonic acid, which will lower the pH and therefore give the characteristic colour change.

bicarbonate of soda or ***baking soda*** (technical name **sodium hydrogencarbonate**) $NaHCO_3$ white crystalline solid that neutralizes acids and is used in medicine to treat acid indigestion. It is also used in baking powders and effervescent drinks.

bichir African fish, genus *Polypterus*, found in tropical swamps and rivers. Cylindrical in shape, some species grow to 70 cm/2.3 ft or more. They show many 'primitive' features, such as breathing air by using the swimbladder, having a spiral valve in the intestine, having heavy bony scales, and having larvae with external gills. These, and the fleshy fins, lead some scientists to think they are related to lungfish and coelacanths.

bichon frise breed of small dog probably originating in France or Spain and characterized by its pure white, softly curling coat. Compactly built, it carries its tail curved over its back and stands 23–8 cm/9–11 in at the shoulder.

bicuspid valve or ***mitral valve*** in the left side of the ◊heart, a flap of tissue that prevents blood flowing back into the atrium when the ventricle contracts.

bicycle pedal-driven two-wheeled vehicle used in cycling. It consists of a metal frame mounted on two large wire-spoked wheels, with handlebars in front and a seat between the front and back wheels. The bicycle is an energy-efficient, nonpolluting form of transport, and it is estimated that 800 million bicycles are in use throughout the world – outnumbering cars three to one. China, India, Denmark, and the Netherlands are countries with a high use of bicycles. More than 10% of road spending in the Netherlands is on cycleways and bicycle parking.

history The first bicycle was seen in Paris in 1791 and was a form of hobby-horse (though there are versions of the hobby-horse that date back even earlier) that had to be propelled by pushing the feet against the ground. The first treadle-propelled cycle was designed by the Scottish blacksmith Kirkpatrick Macmillan in 1839. The Rover 'safety bike' of 1885 may be considered the forerunner of the modern bicycle, with a chain and sprocket drive on the rear wheel. By the end of the 19th century wire wheels, metal frames (replacing wood), and pneumatic tyres (invented by the Scottish veterinary surgeon John B Dunlop 1888) had been added. Among the bicycles of that time was the front-wheel-driven penny farthing with a large front wheel.

technology Recent technological developments have been related to reducing wind resistance caused by the frontal area and the turbulent drag of the bicycle. Most of an Olympic cyclist's energy is taken up in fighting wind resistance in a sprint. The first major innovation was the solid wheel, first used in competitive cycling 1984, but originally patented as long ago as 1878. Further developments include handlebars that allow the cyclist to crouch and use the shape of the hands and forearms to divert air away from the chest. Modern racing bicycles now have a monocoque structure produced by laying carbon fibre around an internal mould and then baking them in an oven. Using all these developments, Chris Boardman set a speed record of 54.4 kph (34 mph) on his way to winning a gold medal at the 1992 Barcelona Olympics. To manufacture a bicycle requires only 1% of the energy and materials used to build a car.

When I see an adult on a bicycle, I have hope for the human race.

H G Wells English writer.
Attributed remark

---

biennial plant plant that completes its life cycle in two years. During the first year it grows vegetatively and the surplus food produced is stored in its ◊perennating organ, usually the root. In the following year these food reserves are used for the production of leaves, flowers, and seeds, after which the plant dies. Many root vegetables are biennials, including the carrot *Daucus carota* and parsnip *Pastinaca sativa*. Some garden plants that are grown as biennials are actually perennials, for example, the wallflower *Cheiranthus cheiri*.

Big Bang in astronomy, the hypothetical 'explosive' event that marked the origin of the universe as we know it. At the time of the Big Bang, the entire universe was squeezed into a hot, superdense state. The Big Bang explosion threw this compact material outwards, producing the expanding universe (see ◊red shift). The cause of the Big Bang is unknown; observations of the current rate of expansion of the universe suggest that it took place about 10–20 billion years ago. The Big Bang theory began modern ◊cosmology.

According to a modified version of the Big Bang, called the **inflationary theory**, the universe underwent a rapid period of expansion shortly after the Big Bang, which accounts for its current large size and uniform nature. The inflationary theory is supported by the most recent observations of the ◊cosmic background radiation.

Scientists have calculated that one 10^{-36} second (equivalent to one million-million-million-million-million-millionth of a second) after the Big Bang, the universe was the size of a pea, and the temperature was 10 billion million million million°C (18 billion million million million°F). One second after the Big Bang, the temperature was about 10 billion°C (18 billion°F).

Big Blue popular name for ◊IBM, derived from the company's size and its blue logo.

Big Crunch in cosmology, a possible fate of the universe in which it ultimately collapses to a point following the halting and reversal of the present expansion. See also ◊Big Bang and ◊critical density.

Big Dipper North American name for the Plough, the seven brightest and most prominent stars in the constellation ◊Ursa Major.

big horn sheep or ***Rocky Mountain sheep*** species of large North American sheep with a brown coat, which turns to bluish-grey in winter. It is so named from the size of the horns of the ram, which often measure over 1 m/3.3 ft round the curve.
classification The big horn sheep *Ovis canadensis* is in family Bovidae, order Artiodactyla

bight coastal indentation, crescent-shaped or gently curving, such as the Bight of Biafra in W Africa and the Great Australian Bight.

Big Seven hierarchies on UseNet, the original seven ◊hierarchies of ◊newsgroups. They are: comp (computing), misc (miscellaneous), news, rec (recreation), sci (science), soc (social issues), and talk (debate). These categories of newsgroups are managed according to specific rules which govern the creation of new groups, in contrast to the ◊alt hierarchy.

bilberry any of several shrubs belonging to the heath family, closely related to North American blueberries. They have blue or black edible berries. (Genus *Vaccinium*, family Ericaceae.)

bilby rabbit-eared bandicoot *Macrotis lagotis*, a lightly built marsupial with big ears and long nose. This burrowing animal is mainly carnivorous, and its pouch opens backwards.

bile brownish alkaline fluid produced by the liver. Bile is stored in the gall bladder and is intermittently released into the duodenum (small intestine) to aid digestion. Bile consists of bile salts, bile pigments, cholesterol, and lecithin. **Bile salts** assist in the breakdown and absorption of fats; **bile pigments** are the breakdown products of old red blood cells that are passed into the gut to be eliminated with the faeces.

To remember the properties of bile:

BILE FROM THE LIVER EMULSIFIES GREASES
TINGES THE URINE AND COLOURS THE FAECES
AIDS PERISTALSIS, PREVENTS PUTREFACTION

IF YOU REMEMBER ALL THIS YOU'LL GIVE SATISFACTION.

bilharzia or ***schistosomiasis*** disease that causes anaemia, inflammation, formation of scar tissue, dysentery, enlargement of the spleen and liver, cancer of the bladder, and cirrhosis of the liver. It is contracted by bathing in water contaminated with human sewage. Some 200 million people are thought to suffer from this disease in the tropics, and 750,000 people a year die.

Freshwater snails act as host to the first larval stage of blood flukes of the genus *Schistosoma*; when these larvae leave the snail in their second stage of development, they are able to pass through human skin, become sexually mature, and produce quantities of eggs, which pass to the intestine or bladder. Numerous eggs are excreted from the body in urine or faeces to continue the cycle. Treatment is by means of drugs, usually containing antimony, to kill the parasites.

bill in birds, the projection of the skull bones covered with a horny sheath. It is not normally sensitive, except in some aquatic birds, rooks, and woodpeckers, where the bill is used to locate food that is not visible. The bills of birds are adapted by shape and size to specific diets, for example, shovellers use their bills to sieve mud in order to extract food; birds of prey have hooked bills adapted to tearing flesh; the bills of the avocet, and the curlew are long and narrow for picking tiny invertebrates out of the mud; and those of woodpeckers are sharp for pecking holes in trees and plucking out insects. The bill is also used by birds for preening, fighting, display, and nest-building.

billion the cardinal number represented by a 1 followed by nine zeros (1,000,000,000 or 10^9), equivalent to a thousand million.

bimetallic strip strip made from two metals each having a different coefficient of ◊thermal expansion; it therefore bends when subjected to a change in temperature. Such strips are used widely for temperature measurement and control, for instance in the domestic thermostat.

bimodal in statistics, having two distinct peaks of ◊frequency distribution.

binary file any file that is not plain text. Program (.EXE or .COM), sound, video, and graphics files are all types of binary files. Such files require special treatment for inclusion in e-mail sent across the Internet, which can transmit only ◊ASCII text and imposes a size limit of 64Kb per message. Several programs have been developed to code binary files into ASCII for transmission, splitting them into smaller parts as necessary. The most commonly used such program is ◊UUencode, but there are others including base64 and BinHex. See also ◊MIME.

binary fission in biology, a form of ◊asexual reproduction, whereby a single-celled organism, such as the amoeba, divides into two smaller 'daughter' cells. It can also occur in a few simple multicellular organisms, such as sea anemones, producing two smaller sea anemones of equal size.

binary large object (contracted to ***BLOB***) in computing, any large single block of data stored in a database, such as a picture or sound file. A BLOB does not include record fields, and so cannot be directly searched by the database's search engine.

binary newsgroup any UseNet ◊newsgroup set up for the transmission of picture and other nontext files. The binary newsgroups have their own sub-hierarchy, alt.binaries, and include groups such as alt.binaries.pictures.fine-art.digitized and alt.binaries.pictures.erotica.

Because newsgroups are subject to the same restrictions as Internet e-mail for the transmission of ◊binary files, pictures, programs, and other files posted to these newsgroups are ◊UUencoded and split into sections. To view the pictures, all the parts must be downloaded and then UUdecoded and stitched back together to form the original file, which can then be viewed using the appropriate graphics program.

Other binary newsgroups distribute sound files (alt.binaries. sound.*) or user-contributed new levels for games such as *Doom* (alt.binaries.doom). These newsgroups take up a lot of ◊bandwidth and therefore not all sites elect to carry them; blocking software typically bars access to many of these groups. It is considered a breach of ◊netiquette to post binary files to nonbinary newsgroups.

binary number system system of numbers to ◊base two, using combinations of the digits 1 and 0. Codes based on binary numbers are used to represent instructions and data in all modern digital computers, the values of the binary digits (contracted to 'bits') being stored or transmitted as, for example, open/closed switches, magnetized/unmagnetized discs and tapes, and high/low voltages in circuits.

The value of any position in a binary number increases by powers of 2 (doubles) with each move from right to left (1, 2, 4, 8, 16, and so on). For example, 1011 in the binary number system means $(1 \times 8) + (0 \times 4) + (1 \times 2) + (1 \times 1)$, which adds up to 11 in the decimal system.

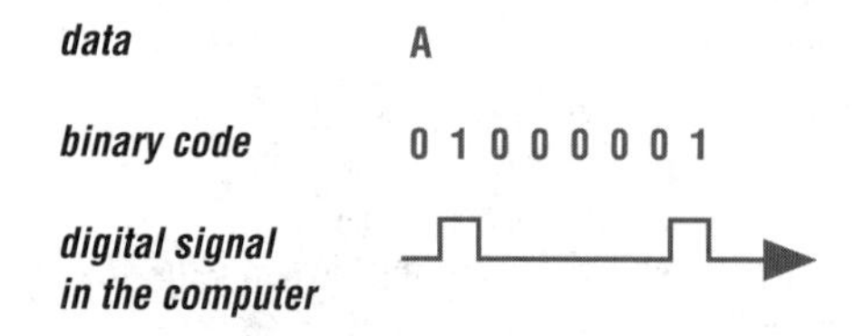

binary number system The capital letter A represented in binary form.

binary search in computing, a rapid technique used to find any particular record in a list of records held in sequential order. The computer is programmed to compare the record sought with the record in the middle of the ordered list. This being done, the computer discards the half of the list in which the record does not appear, thereby reducing the number of records left to search by half. This process of selecting the middle record and discarding the unwanted half of the list is repeated until the required record is found.

binary star pair of stars moving in orbit around their common centre of mass. Observations show that most stars are binary, or even multiple – for example, the nearest star system to the Sun, ◊Alpha Centauri.

One of the stars in the binary system Epsilon Aurigae may be the largest star known. Its diameter is 2,800 times that of the Sun. If it were in the position of the Sun, it would engulf Mercury, Venus, Earth, Mars, Jupiter, and Saturn. A spectroscopic binary is a binary in which two stars are so close together that they cannot be seen separately, but their separate light spectra can be distinguished by a spectroscope.

Another type is the ◊eclipsing binary.

binding energy in physics, the amount of energy needed to break the nucleus of an atom into the neutrons and protons of which it is made.

BinHex program for coding ◊binary files into ◊ASCII for transmission over the Internet via e-mail.

binoculars optical instrument for viewing an object in magnification with both eyes; for example, field glasses and opera glasses. Binoculars consist of two telescopes containing lenses and prisms, which produce a stereoscopic effect as well as magnifying the image.

Use of prisms has the effect of 'folding' the light path, allowing for a compact design.

The first binocular telescope was constructed by the Dutch inventor Hans Lippershey in 1608. Later development was largely due to the German Ernst Abbe of Jena, who at the end of the 19th century designed prism binoculars that foreshadowed the instruments of today, in which not only magnification but also stereoscopic effect is obtained.

binomial in mathematics, an expression consisting of two terms, such as $a + b$ or $a - b$.

binomial system of nomenclature in biology, the system in which all organisms are identified by a two-part Latinized name. Devised by the biologist ◊Linnaeus, it is also known as the Linnaean system. The first name is capitalized and identifies the ◊genus; the second identifies the ◊species within that genus.

binomial theorem formula whereby any power of a binomial quantity may be found without performing the progressive multiplications.

It was discovered by Isaac ◊Newton and first published in 1676.

binturong shaggy-coated mammal *Arctitis binturong,* the largest member of the mongoose family, nearly 1 m/3 ft long excluding a long muscular tail with a prehensile tip. Mainly nocturnal and tree-dwelling, the binturong is found in the forests of SE Asia, feeding on fruit, eggs, and small animals.

biochemistry science concerned with the chemistry of living organisms: the structure and reactions of proteins (such as enzymes), nucleic acids, carbohydrates, and lipids.

Its study has led to an increased understanding of life processes, such as those by which organisms synthesize essential chemicals from food materials, store and generate energy, and pass on their characteristics through their genetic material. A great deal of medical research is concerned with the ways in which these processes are disrupted. Biochemistry also has applications in agriculture and in the food industry (for instance, in the use of enzymes).

biodegradable capable of being broken down by living organisms, principally bacteria and fungi. In biodegradable substances, such as food and sewage, the natural processes of decay lead to compaction and liquefaction, and to the release of nutrients that are then recycled by the ecosystem.

This process can have some disadvantageous side effects, such as the release of methane, an explosive greenhouse gas. However, the technology now exists for waste tips to collect methane in

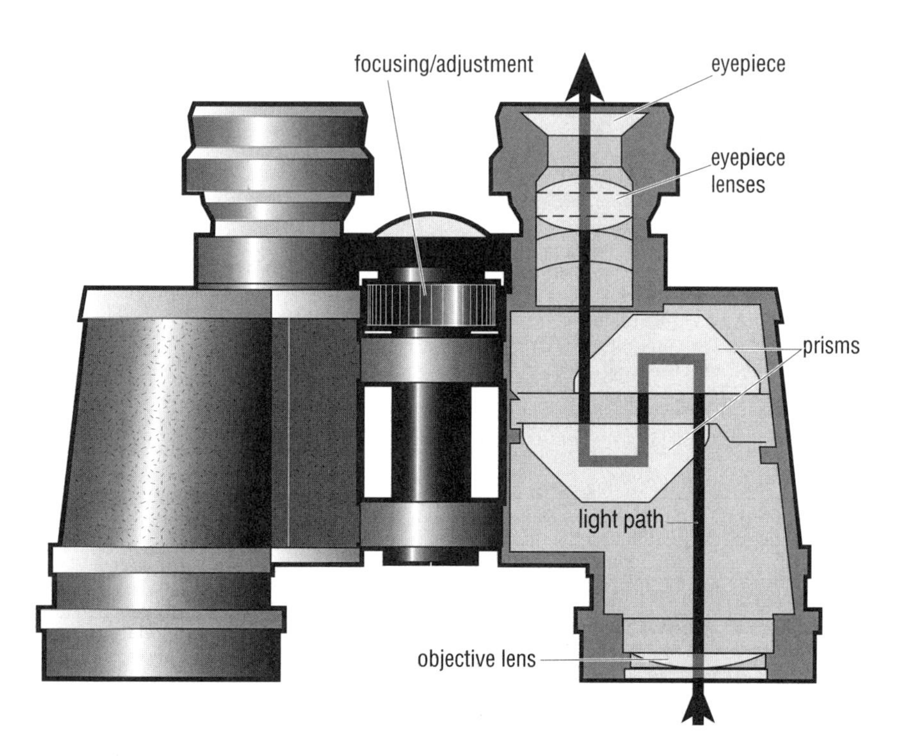

***binoculars** An optical instrument that allows the user to focus both eyes on the magnified image at the same time. The essential components of binoculars are objective lenses, eyepieces, and a system of prisms to invert and reverse the image. A focusing system provides a sharp image by adjusting the relative positions of these components.*

underground pipes, drawing it off and using it as a cheap source of energy. Nonbiodegradable substances, such as glass, heavy metals, and most types of plastic, present serious problems of disposal.

biodiversity (contraction of ***biological diversity***) measure of the variety of the Earth's animal, plant, and microbial species; of genetic differences within species; and of the ecosystems that support those species. Its maintenance is important for ecological stability and as a resource for research into, for example, new drugs and crops. In the 20th century, the destruction of habitats is believed to have resulted in the most severe and rapid loss of biodiversity in the history of the planet.

Estimates of the number of species vary widely because many species-rich ecosystems, such as tropical forests, contain unexplored and unstudied habitats. Especially among small organisms, many are unknown; for instance, it is thought that only 1–10% of the world's bacterial species have been identified.

The most significant threat to biodiversity comes from the destruction of rainforests and other habitats in the southern hemisphere. It is estimated that 7% of the Earth's surface hosts 50–75% of the world's biological diversity. Costa Rica, for example, has an area less than 10% of the size of France but possesses three times as many vertebrate species.

biodynamic farming agricultural practice based on the principle of ◊homeopathy: tiny quantities of a substance are applied to transmit vital qualities to the soil. It is a form of ◊organic farming, and was developed by the Austrian holistic mystic Rudolf Steiner and Ehrenfried Pfiffer.

bioengineering the application of engineering to biology and medicine. Common applications include the design and use of artificial limbs, joints, and organs, including hip joints and heart valves.

biofeedback in medicine, the use of electrophysiological monitoring devices to 'feed back' information about internal processes and thus facilitate conscious control. Developed in the USA in the 1960s, independently by neurophysiologist Barbara Brown and neuropsychiatrist Joseph Kamiya, the technique is effective in alleviating hypertension and preventing associated organic and physiological dysfunctions.

biofuel any solid, liquid, or gaseous fuel produced from organic (once living) matter, either directly from plants or indirectly from industrial, commercial, domestic, or agricultural wastes. There are three main methods for the development of biofuels: the burning of dry organic wastes (such as household refuse, industrial and agricultural wastes, straw, wood, and peat); the fermentation of wet wastes (such as animal dung) in the absence of oxygen to produce biogas (containing up to 60% methane), or the fermentation of sugar cane or corn to produce alcohol and esters; and energy forestry (producing fast-growing wood for fuel).

Fermentation produces two main types of biofuels: alcohols and esters. These could theoretically be used in place of fossil fuels but, because major alterations to engines would be required, biofuels are usually mixed with fossil fuels. The EU allows 5% ethanol, derived from wheat, beet, potatoes, or maize, to be added to fossil fuels. A quarter of Brazil's transportation fuel in 1994 was ethanol.

biogenesis biological term coined in 1870 by English scientist Thomas Henry Huxley to express the hypothesis that living matter always arises out of other similar forms of living matter. It superseded the opposite idea of ◊spontaneous generation or abiogenesis (that is, that living things may arise out of nonliving matter).

biogeography study of how and why plants and animals are distributed around the world, in the past as well as in the present; more specifically, a theory describing the geographical distribution of ◊species developed by Robert MacArthur and US zoologist Edward O Wilson. The theory argues that for many species, ecological specializations mean that suitable habitats are patchy in their occurrence. Thus for a dragonfly, ponds in which to breed are separated by large tracts of land, and for edelweiss adapted to alpine peaks the deep valleys between cannot be colonized.

biological clock regular internal rhythm of activity, produced by unknown mechanisms, and not dependent on external time signals. Such clocks are known to exist in almost all animals, and also in many plants, fungi, and unicellular organisms; the first biological clock gene in plants was isolated in 1995 by a US team of researchers. In higher organisms, there appears to be a series of clocks of graded importance. For example, although body temperature and activity cycles in human beings are normally 'set' to 24 hours, the two cycles may vary independently, showing that two clock mechanisms are involved.

biological control *The beetle* Cryptolaemus montrouzieri *being released on to a mealybug-infested passion flower. Because mealybugs, which are a serious pest, have a waxy coating and so are resistant to insecticides, biological control agents are used against them.* *Premaphotos Wildlife*

biological control control of pests such as insects and fungi through biological means, rather than the use of chemicals. This can include breeding resistant crop strains; inducing sterility in the pest; infecting the pest species with disease organisms; or introducing the pest's natural predator. Biological control tends to be naturally self-regulating, but as ecosystems are so complex, it is difficult to predict all the consequences of introducing a biological controlling agent.

The introduction of the cane toad to Australia 50 years ago to eradicate a beetle that was destroying sugar beet provides an example of the unpredictability of biological control. Since the cane toad is poisonous it has few Australian predators and it is now a pest, spreading throughout eastern and northern Australia at a rate of 35 km/22 mi a year.

BIOLOGICAL CONTROL

http://www.nysaes.cornell.edu:80/ent/biocontrol/

University-based site on the various methods of biological control used by farmers in the USA. This includes sections on parasitoids, predators, pathogens, and weed feeders. Each sections contains images and sections on 'relative effectiveness' and 'pesticide susceptibility'.

biological oxygen demand (BOD) the amount of dissolved oxygen taken up by microorganisms in a sample of water. Since these microorganisms live by decomposing organic matter, and the amount of oxygen used is proportional to their number and metabolic rate, BOD can be used as a measure of the extent to which the water is polluted with organic compounds.

biological shield shield around a nuclear reactor that is intended to protect personnel from the effects of ◊radiation. It usually consists of a thick wall of steel and concrete.

biology *Greek bios 'life', logos 'discourse'* science of life. Biology includes all the life sciences – for example, anatomy and physiology (the study of the structure of living things), cytology (the study of cells), zoology (the study of animals) and botany (the study of plants), ecology (the study of habitats and the interaction of living species), animal behaviour, embryology, and taxonomy, and plant breeding. Increasingly in the 20th century biologists have concentrated on molecular structures: biochemistry, biophysics, and genetics (the study of inheritance and variation).

Biological research has come a long way towards understanding the nature of life, and during the 1990s our knowledge will be further extended as the international ◊Human Genome Project attempts to map the entire genetic code contained in the 23 pairs of human chromosomes.

> *The greatest stride in biology, in our century, was its shift to the molecular dimension. The next will be its shift toward the submolecular, electronic dimension.*
>
> ALBERT SZENT-GYÖRGYI Hungarian-born US biochemist. *Bioelectronics* 1968

bioluminescence production of light by living organisms. It is a feature of many deep-sea fishes, crustaceans, and other marine animals. On land, bioluminescence is seen in some nocturnal insects such as glow-worms and fireflies, and in certain bacteria and fungi. Light is usually produced by the oxidation of luciferin, a reaction catalysed by the ◊enzyme luciferase. This reaction is unique, being the only known biological oxidation that does not produce heat. Animal luminescence is involved in communication, camouflage, or the luring of prey, but its function in other organisms is unclear.

biomass the total mass of living organisms present in a given area. It may be specified for a particular species (such as earthworm biomass) or for a general category (such as herbivore biomass). Estimates also exist for the entire global plant biomass. Measurements of biomass can be used to study interactions between organisms, the stability of those interactions, and variations in population numbers. Where dry biomass is measured, the material is dried to remove all water before weighing.

Some two-thirds of the world's population cooks and heats water by burning biomass, usually wood. Plant biomass can be a renewable source of energy as replacement supplies can be grown relatively quickly. Fossil fuels however, originally formed from biomass, accumulate so slowly that they cannot be considered renewable. The burning of biomass (defined either as natural areas of the ecosystem or as forest, grasslands, and fuelwoods) produces 3.5 million tonnes of carbon in the form of carbon dioxide each year, accounting for up to 40% of the world's annual carbon dioxide production.

BIOMASS

http://www.nrel.gov/research/industrial_tech/biomass.html

Well-presented information on biomass from the US Department of Energy. A graph supports the textual explanation of the fact that the world is only using 7 % of annual biomass production. There is a clear explanation of the chemical composition of biomass and development of technologies to transform it into usable fuel sources.

biome broad natural assemblage of plants and animals shaped by common patterns of vegetation and climate. Examples include the tundra biome and the desert biome.

biomechanics application of mechanical engineering principles and techniques in the field of medicine and surgery, studying natural structures to improve those produced by humans. For example, mother-of-pearl is structurally superior to glass fibre, and deer antlers have outstanding durability because they are composed of microscopic fibres. Such natural structures may form the basis of high-tech composites. Biomechanics has been responsible for many recent advances in ◊orthopaedics, anaesthesia, and intensive care. Biomechanical assessment of the requirements for replacement of joints, including evaluation of the stresses and strains between parts, and their reliability, has allowed development of implants with very low friction and long life.

biometrics in computing, biometrics is applied loosely to the measurement of biological (human) data, usually for security purposes, rather than the statistical analysis of biological data. For example, when someone wants to enter a building or cash a cheque, their finger or eyeball may be scanned and compared with a fingerprint or eyeball scan stored earlier. Biometrics saves people from having to remember PINs (personal identification numbers) and passwords.

biometry literally, the measurement of living things, but generally used to mean the application of mathematics to biology. The term is now largely obsolete, since mathematical or statistical work is an integral part of most biological disciplines.

bionics *from 'biological electronics'* design and development of electronic or mechanical artificial systems that imitate those of living things. The bionic arm, for example, is an artificial limb (◊prosthesis) that uses electronics to amplify minute electrical signals generated in body muscles to work electric motors, which operate the joints of the fingers and wrist.

The first person to receive two bionic ears was Peter Stewart, an Australian journalist, 1989. His left ear was fitted with an array of 22 electrodes, replacing the hairs that naturally convert sounds into electrical impulses. Five years previously he had been fitted with a similar device in his right ear.

biophysics application of physical laws to the properties of living organisms. Examples include using the principles of ◊mechanics to calculate the strength of bones and muscles, and ◊thermodynamics to study plant and animal energetics.

biopsy removal of a living tissue sample from the body for diagnostic examination.

biorhythm rhythmic change, mediated by ◊hormones, in the physical state and activity patterns of certain plants and animals that have seasonal activities. Examples include winter hibernation, spring flowering or breeding, and periodic migration. The hormonal changes themselves are often a response to changes in day length (◊photoperiodism); they signal the time of year to the animal or plant. Other biorhythms are innate and continue even if external stimuli such as day length are removed. These include a 24-hour or ◊circadian rhythm, a 28-day or circalunar rhythm (corresponding to the phases of the Moon), and even a year-long rhythm in some organisms.

Such innate biorhythms are linked to an internal or ◊biological clock, whose mechanism is still poorly understood.

Often both types of rhythm operate; thus many birds have a circalunar rhythm that prepares them for the breeding season, and a photoperiodic response. There is also a nonscientific and unproven theory that human activity is governed by three biorhythms: the **intellectual** (33 days), the **emotional** (28 days), and the **physical** (23 days). Certain days in each cycle are regarded as 'critical', even more so if one such day coincides with that of another cycle.

BIOS (acronym for ***basic input/output system***) in computing, the part of the ◊operating system that handles input and output. The term is also used to describe the programs stored in ◊ROM (and called ROM BIOS), which are automatically run when a computer is switched on allowing it to ◊boot. BIOS is unaffected by upgrades to the operating system stored on disc.

biosensor device based on microelectronic circuits that can directly measure medically significant variables for the purpose of diagnosis or monitoring treatment. One such device measures the blood sugar level of diabetics using a single drop of blood, and shows the result on a liquid crystal display within a few minutes.

biosphere the narrow zone that supports life on our planet. It is limited to the waters of the Earth, a fraction of its crust, and the lower regions of the atmosphere.

BioSphere 2 (BS2) ecological test project, a 'planet in a bottle', in Arizona, USA. Under a sealed glass and metal dome, different habitats are recreated, with representatives of nearly 4,000 species, to test the effects that various environmental factors have on ecosystems. Simulated ecosystems, or 'mesocosms', include savanna, desert, rainforest, marsh and Caribbean reef. The response of such systems to elevated atmospheric concentrations of carbon dioxide gas (CO_2) are among the priorities of Biosphere 2 researchers.
BioSphere 1 Experiments with biospheres that contain relatively simple life forms have been carried out for decades, and a 21-day trial period in 1989 that included humans preceded the construction of BS2. However, BS2 is not in fact the second in a series: the Earth is considered to be Biosphere 1.
human inhabitants Originally, people, called 'Biospherians', were sealed in the dome. The people within were self-sufficient, except for electricity, which was supplied by a 3.7 megawatt power station on the outside (solar panels were considered too expensive). The original team of eight in residence 1991–1993 was replaced March 1994 with a new team of seven people sealed in for six-and-a-half months. In 1995 it was decided that further research would not involve sealing people within the biosphere. Researchers, students, and visitors routinely go in and out of the Biosphere 2 facility.
organization Biosphere 2 was originally run by a private company partly funded by ecology-minded oil millionaire Edward P Bass (1945–) and was called Space Biosphere Ventures. Space Biosphere Ventures investors expected to find commercial applications for the techniques developed in the course of the project. As of 1 January 1996 Columbia University, USA, joined Space Bisphere Ventures to form Biosphere 2 Center, Inc.. The purpose of the joint venture is to use the facility for conferences and classes as well as further short-term experiments with the artificial ecosystems that do not involve isolating humans inside.

biosynthesis synthesis of organic chemicals from simple inorganic ones by living cells – for example, the conversion of carbon dioxide and water to glucose by plants during ◊photosynthesis.

Other biosynthetic reactions produce cell constituents including proteins and fats.

biotechnology industrial use of living organisms to manufacture food, drugs, or other products. The brewing and baking industries have long relied on the yeast microorganism for ◊fermentation purposes, while the dairy industry employs a range of bacteria and fungi to convert milk into cheeses and yoghurts. ◊Enzymes, whether extracted from cells or produced artificially, are central to most biotechnological applications.

Recent advances include ◊genetic engineering, in which single-celled organisms with modified ◊DNA are used to produce insulin and other drugs.

In 1993 two-thirds of biotechnology companies were concentrating on human health developments, whilst only 1 in 10 were concerned with applications for food and agriculture.

biotic factor organic variable affecting an ecosystem – for example, the changing population of elephants and its effect on the African savanna.

biotin or ***vitamin H*** vitamin of the B complex, found in many different kinds of food; egg yolk, liver, legumes, and yeast contain large amounts. Biotin is essential to the metabolism of fats. Its absence from the diet may lead to dermatitis.

birch any of a group of slender trees with small leaves and fine, peeling bark. About 40 species are found in cool temperate parts of the northern hemisphere. Birches grow rapidly, and their hard, beautiful wood is used for veneers and cabinet work. (Genus *Betula*, family Betulaceae.)

bird backboned animal of the class Aves, the biggest group of land vertebrates, characterized by warm blood, feathers, wings, breathing through lungs, and egg-laying by the female. Birds are bipedal; feet are usually adapted for perching and never have more than four toes. Hearing and eyesight are well developed, but the sense of smell is usually poor. No existing species of bird possesses teeth.

Most birds fly, but some groups (such as ostriches) are flightless, and others include flightless members. Many communicate by sounds (nearly half of all known species are songbirds) or by visual displays, in connection with which many species are brightly coloured, usually the males. Birds have highly developed patterns of instinctive behaviour. There are nearly 8,500 species of birds.

According to the Red List of endangered species published by the World Conservation Union (IUCN) for 1996, 11% of bird species are threatened with extinction.
wing structure The wing consists of the typical bones of a forelimb, the humerus, radius and ulna, carpus, metacarpus, and digits. The first digit is the pollex, or thumb, to which some feathers, known as ala spuria, or bastard wing, are attached; the second digit is the index, which bears the large feathers known as the primaries or manuals, usually ten in number. The primary feathers, with the secondaries or cubitals, which are attached to the ulna, form the large wing-quills, called remiges, which are used in flight.
anatomy The sternum, or breastbone, of birds is affected by their powers of flight: those birds which are able to fly have a keel projecting from the sternum and serving as the basis of attachment of the great pectoral muscles which move the wings. In birds that do not fly the keel is absent or greatly reduced. The vertebral column is completed in the tail region by a flat plate known as the pygostyle, which forms a support for the rectrices, or steering tailfeathers.

The legs are composed of the femur, tibia and fibula, and the bones of the foot; the feet usually have four toes, but in many cases there are only three. In swimming birds the legs are placed well back.

BIRD

Rifleman, short-tailed pygmy tyrant, frilled coquette, bobwhite, tawny frogmouth, trembler, wattle-eye, fuscous honeyeater, dickcissel, common grackle, and forktailed drongo are all common names for species of bird.

The uropygial gland on the pygostyle (bone in the tail) is an oil gland used by birds in preening their feathers, as their skin contains no sebaceous glands. The eyes have an upper and a lower eyelid and a semi-transparent nictitating membrane with which the bird can cover its eyes at will.

The **vascular system** contains warm blood, which is kept usually at a higher temperature (about 41°C/106°F) than that of mammals; death from cold is rare unless the bird is starving or ill. The aortic arch (main blood vessel leaving the heart) is on the right side of a bird, whereas it is on the left in a mammal. The heart of a bird consists of a right and a left half with four chambers.

The **lungs** are small and prolonged into air-sacs connected to a number of air-spaces in the bones. These air-spaces are largest in powerful fliers, but they are not so highly developed in young, small, aquatic, and terrestrial birds. These air-spaces increase the efficiency of the respiratory system and reduce the weight of the bones. The lungs themselves are more efficient than those of mammals; the air is circulated through a system of fine capillary tubes, allowing continuous gas exchange to take place, whereas in mammals the air comes to rest in blind air sacs.

The organ of voice is not the larynx, but usually the syrinx, a peculiarity of this class formed at the bifurcation of the trachea (windpipe) and the modulations are effected by movements of the adjoining muscles.

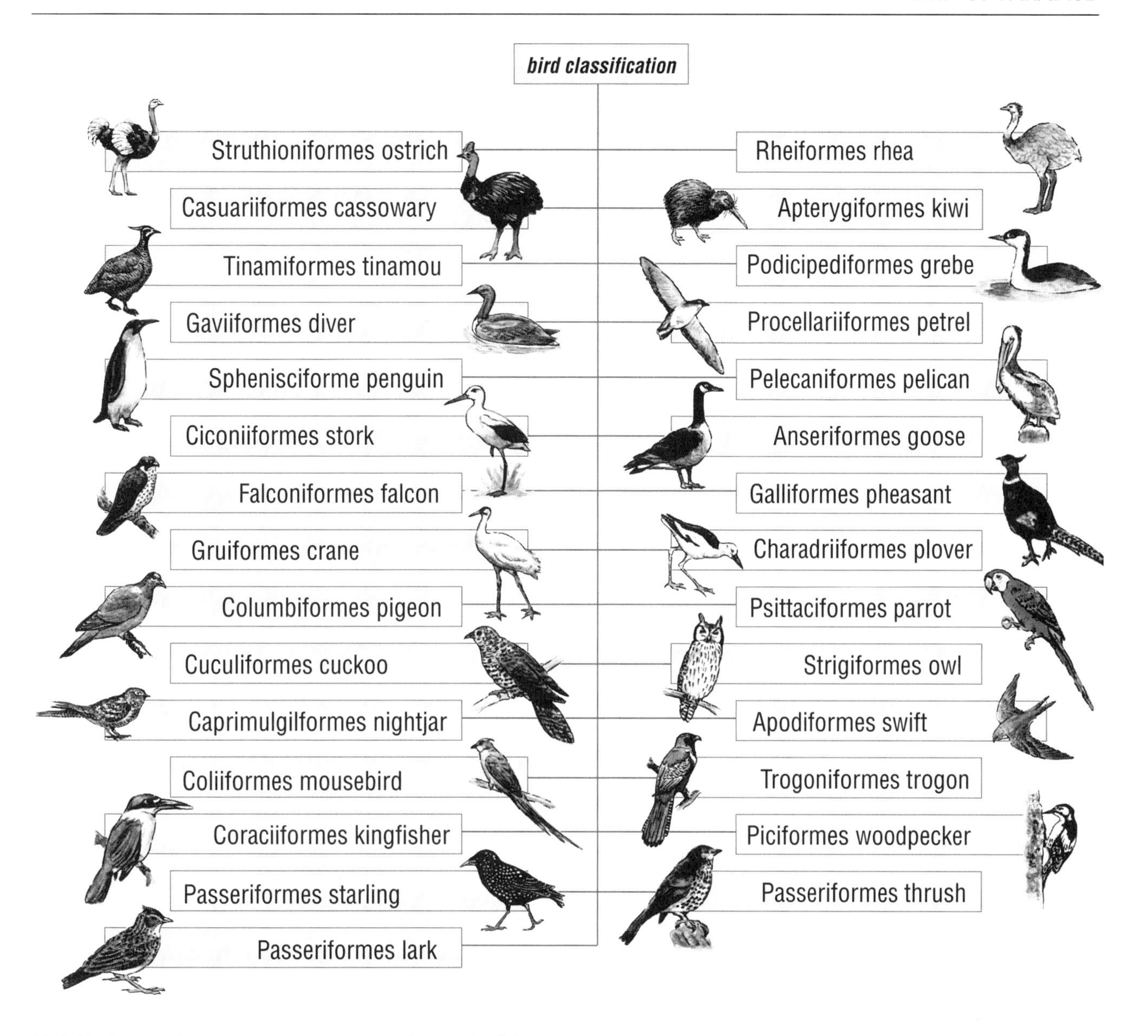

bird *This diagram shows a representative species from each of the 29 orders. There are nearly 8,500 species of birds, of which the largest is the N African ostrich, reaching a height of 2.74 m/9 ft and weighing 156 kg/345 lb. The smallest bird, the bee hummingbird of Cuba and the Isle of Pines, measures 57 mm/2.24 in in length and weighs a mere 1.6 g/0.056 oz.*

digestion Digestion takes place in the oesophagus, stomach, and intestines in a manner basically similar to mammals. The tongue aids in feeding, and there is frequently a **crop**, a dilation of the oesophagus, where food is stored and softened. The stomach is small with little storage capacity and usually consists of the proventriculus, which secretes digestive juices, and the gizzard, which is tough and muscular and grinds the food, sometimes with the aid of grit and stones retained within it. Digestion is completed, and absorption occurs, in the intestine and the digestive caeca. The intestine ends in a cloaca through which both urine and faeces are excreted.

nesting and eggs Typically eggs are brooded in a nest and, on hatching, the young receive a period of parental care. The collection of nest material, nest building, and incubation may be carried out by the male, female, or both. The cuckoo neither builds a nest nor rears its own young, but places the eggs in the nest of another bird and leaves the foster parents to care for them.

The study of birds is called ◊ornithology.

bird louse parasitic biting louse, found mainly on birds, less frequently on mammals. Bird lice are wingless ectoparasites (living on the skin of their hosts), have biting mouthparts (as opposed to true lice which have sucking mouthparts), and reduced eyes.

classification Bird lice are in the order Mallophaga, class Insecta, phylum Arthropoda.

The **chicken body louse** *Menacanthus stramineus* is small, yellowish, and about 2.8–3.3 mm/0.1–0.13 in long. The parasite feeds on the skin debris and feathers of the chicken. A heavily infested bird can carry over 8,000 lice on its body. The host appears to withstand the usual degree of infestation without apparent ill-effects, but heavy infestation can result in loss of plumage and a decline in the bird's health.

bird of paradise one of 40 species of crowlike birds in the family Paradiseidae, native to New Guinea and neighbouring islands. Females are generally drably coloured, but the males have bright and elaborate plumage used in courtship display. Hunted almost to extinction for their plumage, they are now subject to conservation.

They are smallish birds, extremely active, and have compressed beaks, large toes, and strong feet. Their food consists chiefly of fruits, seeds, and nectar, but it may also include insects and small animals, such as worms. The Australian ◊bowerbirds are closely related.

Birman or ***'sacred cat of Burma'*** breed of domestic cat with medium-length fur, possibly originating in the temples of Burma. Similar to a ◊Colourpoint Longhair or Himalayan cat, the Birman has shorter hair, legs with a longer body. In Britain, the Blue-point variety's fur is beige-gold with blue-grey points (dark face, tail, and legs), with white paws; the US standard calls for bluish-white fur and deep blue points. It has brilliant blue eyes. There are several other varieties.

Birmans were recognized as a breed in France in 1925 and in the USA in 1926.

birth act of producing live young from within the body of female animals. Both viviparous and ovoviviparous animals give birth to young. In viviparous animals, embryos obtain nourishment from the mother via a ◊placenta or other means.

In ovoviviparous animals, fertilized eggs develop and hatch in the oviduct of the mother and gain little or no nourishment from maternal tissues. See also ◊pregnancy.

> *There's a time when you have to explain to your children why they're born, and it's a marvelous thing if you know the reason by then.*
>
> HAZEL SCOTT US entertainer. Quoted in *Ms.* Nov 1974

birth control another name for ◊family planning; see also ◊contraceptive.

bisect to divide a line or angle into two equal parts.

bisector a line that bisects an angle or another line (known as a **perpendicular bisector** when it bisects at right angles).

bismuth hard, brittle, pinkish-white, metallic element, symbol Bi, atomic number 83, relative atomic mass 208.98. It has the highest atomic number of all the stable elements (the elements from atomic number 84 up are radioactive). Bismuth occurs in ores and occasionally as a free metal (◊native metal). It is a poor conductor of heat and electricity, and is used in alloys of low melting point and in medical compounds to soothe gastric ulcers. The name comes from the Latin *besemutum,* from the earlier German *Wismut.*

bison large, hoofed mammal of the bovine family. There are two species, both brown. The **European bison** or **wisent** *Bison bonasus,* of which only a few protected herds survive, is about 2 m/7 ft high and weighs up to 1,100 kg/2,500 lb. The **North American bison** (often known as 'buffalo') *Bison bison* is slightly smaller, with a heavier mane and more sloping hindquarters. Formerly roaming the prairies in vast numbers, it was almost exterminated in the 19th century, but survives in protected areas. There were about 14,000 bison in North American reserves in 1994.

Crossed with domestic cattle, the latter has produced a hardy hybrid, the 'beefalo', producing a lean carcass on an economical grass diet.

bistable circuit or ***flip-flop*** simple electronic circuit that remains in one of two stable states until it receives a pulse (logic 1 signal) through one of its inputs, upon which it switches, or 'flips', over to the other state. Because it is a two-state device, it can be used to store binary digits and is widely used in the ◊integrated circuits used to build computers.

bisulphate another term for ◊hydrogen sulphate.

bit (contraction of ***binary digit***) in computing, a single binary digit, either 0 or 1. A bit is the smallest unit of data stored in a computer; all other data must be coded into a pattern of individual bits. A ◊byte represents sufficient computer memory to store a single ◊character of data, and usually contains eight bits. For example, in the ◊ASCII code system used by most microcomputers the capital letter A would be stored in a single byte of memory as the bit pattern 01000001.

The maximum number of bits that a computer can normally process at once is called a **word**. Microcomputers are often described according to how many bits of information they can handle at once. For instance, the first microprocessor, the Intel 4004 (launched 1971), was a 4-bit device. In the 1970s several different 8-bit computers, many based on the Zilog Z80 or Rockwell 6502 processors, came into common use. In 1981, the IBM Personal Computer (PC) was introduced, using the Intel 8088 processor, which combined a 16-bit processor with an 8-bit ◊data bus. Business micros of the later 1980s began to use 32-bit processors such as the Intel 80386 and Motorola 68030. Machines based on the first 64-bit microprocessor appeared in 1993.

The higher the number of bits a computer can process simultaneously, the more powerful the computer is said to be. However, other factors influence the overall speed of a computer system, such as the ◊clock rate of the processor and the amount of ◊RAM available. Tasks that require a high processing speed include sorting a data-

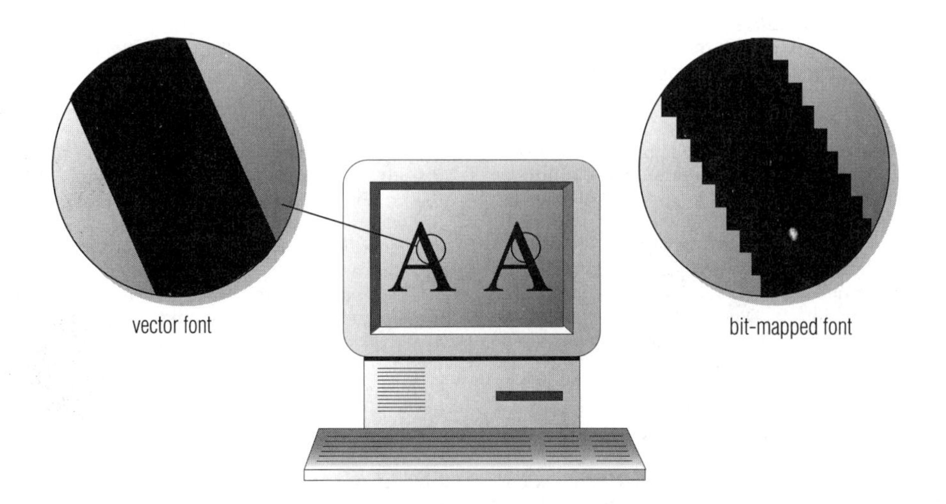

bit map *The difference in close-up between a bit-mapped and vector font. As separate sets of bit maps are required for each different type size, scaleable vector graphics (outline) is the preferred medium for fonts.*

base or doing long, complex calculations in spreadsheets. A system running slowly with a ◊graphical user interface may benefit more from the addition of extra RAM than from a faster processor.

In the PC industry, new hardware is most readily adopted when it is compatible with old software, which slows the adoption of new software. For example, most people were still using Microsoft's 16-bit Windows 3 program with 16-bit applications in 1995–96, a decade after 32-bit processors like Intel's 80386 became widely available. This was true even though 32-bit operating systems – Unix, IBM's OS/2 and Microsoft's Windows NT – had been available for some years.

bit map in computing, a pattern of ◊bits used to describe the organization of data. Bit maps are used to store typefaces or graphic images (bit-mapped or ◊raster graphics), with 1 representing black (or a colour) and 0 white.

Bit maps may be used to store a typeface or ◊font, but a separate set of bit maps is required for each typesize. A vector font, by contrast, can be held as one set of data and scaled as required. Bit-mapped graphics are not recommended for images that require scaling (compare ◊vector graphics – those stored in the form of geometric formulas).

bit-mapped font ◊font held in computer memory as sets of bit maps.

Bitnet (acronym for ***Because It's Time Network***) news ◊network developed in 1983 at the City University of New York, USA. Bitnet operates as a collection of mailing lists using ◊Listserv, which was picked up by the rest of the Internet and is widely used, although Bitnet itself is falling into disuse.

bit pad computer input device; see ◊graphics tablet.

bits per second (bps) commonly used measure of the speed of transmission of a ◊modem. In 1997 the fastest modems readily available were rated at 33,600 bps, with two incompatible 56K systems also competing for sales. Modem speeds should conform to standards, known as ◊V numbers, laid down by the ◊Comité Consultatif International Téléphonique et Télégraphique (CCITT) so that modems from different manufacturers can connect to each other. Many modems transfer data much faster than their nominal speeds via techniques such as ◊data compression.

bitterling freshwater fish *Rhodeus sericeus* of Northern Europe, introduced to North America. It grows to a length of 90 mm/3.5 in, and has an attractive bluish stripe along each side towards the tail. It is found in lakes and slow-moving rivers.

Having selected a mate, the female bitterling deposits her eggs inside a freshwater mussel via a long tube protruding from her body, and the male then deposits his sperm into the mussel. The eggs are nurtured safe from predators (and unharmed by the mussel) within the mussel's gills for two to three weeks before hatching and swimming away.

bittern any of several species of small herons, in particular the common bittern *Botaurus stellaris* of Europe and Asia. It is shy, stoutly built, buff-coloured, speckled with black and tawny brown, with a long bill and a loud, booming call. Its habit of holding its neck and bill in a vertical position conceals it among the reeds, where it rests by day, hunting for frogs, reptiles, and fish towards nightfall. An inhabitant of marshy country, it is now quite rare in Britain.

bittersweet alternative name for the woody ◊nightshade plant.

bitumen impure mixture of hydrocarbons, including such deposits as petroleum, asphalt, and natural gas, although sometimes the term is restricted to a soft kind of pitch resembling asphalt.

Solid bitumen may have arisen as a residue from the evaporation of petroleum. If evaporation took place from a pool or lake of petroleum, the residue might form a pitch or asphalt lake, such as Pitch Lake in Trinidad. Bitumen was used in ancient times as a mortar, and by the Egyptians for embalming.

bivalent in biology, a name given to the pair of homologous chromosomes during reduction division (◊meiosis). In chemistry, the term is sometimes used to describe an element or group with a ◊valency of two, although the term 'divalent' is more common.

bivalve marine or freshwater mollusc whose body is enclosed between two shells hinged together by a ligament on the dorsal side of the body.

The shell is closed by strong 'adductor' muscles. Ventrally, a retractile 'foot' can be put out to assist movement in mud or sand. Two large platelike gills are used for breathing and also, with the ◊cilia present on them, make a mechanism for collecting the small particles of food on which bivalves depend. The bivalves form one of the five classes of molluscs, the Lamellibranchiata, otherwise known as Bivalvia or Pelycypoda, containing about 8,000 species.

black beetle another name for ◊cockroach, although cockroaches belong to an entirely different order of insects (Dictyoptera) from the beetles (Coleoptera).

blackberry prickly shrub, closely related to raspberries and dewberries. Native to northern parts of Europe, it produces pink or white blossom and edible black compound fruits. (*Rubus fruticosus,* family Rosaceae.)

blackbird bird *Turdus merula* of the thrush family, Muscicapidae, order Passeriformes, about 25 cm/10 in long. The male is black with a yellow bill and eyelids, the female dark brown with a dark beak. It lays three to five blue-green eggs with brown spots in a nest of grass and moss, plastered with mud, built in thickets or creeper-clad trees. The blackbird feeds on fruit, seeds, worms, grubs, and snails. Its song is rich and flutelike.

Found across Europe, Asia, and North Africa, the blackbird adapts well to human presence and gardens, and is one of the most common British birds. North American 'blackbirds' belong to a different family of birds, the Icteridae.

black body in physics, a hypothetical object that completely absorbs all electromagnetic radiation striking it. It is also a perfect emitter of thermal radiation.

Although a black body is hypothetical, a practical approximation can be made by using a small hole in the wall of a constant-temperature enclosure. The radiation emitted by a black body is of all wavelengths, but with maximum radiation at a particular wavelength that depends on the body's temperature. As the temperature increases, the wavelength of maximum intensity becomes shorter (see ◊Wien's displacement law). The total energy emitted at all wavelengths is proportional to the fourth power of the temperature (see ◊Stefan–Boltzmann law). Attempts to explain these facts failed until the development of ◊quantum theory in 1900.

black box popular name for the unit containing an aeroplane's flight and voice recorders. These monitor the plane's behaviour and the crew's conversation, thus providing valuable clues to the cause of a disaster. The box is nearly indestructible and usually painted orange for easy recovery. The name also refers to any compact electronic device that can be quickly connected or disconnected as a unit.

The maritime equivalent is the **voyage recorder**, installed in ships from 1989. It has 350 sensors to record the performance of engines, pumps, navigation lights, alarms, radar, and hull stress.

blackbuck antelope *Antilope cervicapra* found in central and NW India. It is related to the gazelle, from which it differs in having spirally-twisted horns. The male is black above and white beneath, whereas the female and young are fawn-coloured above.

It is about 76 cm/2.5 ft in height.

blackcap ◊warbler *Sylvia atricapilla,* family Muscicapidae, order Passeriformes. The male has a black cap, the female a reddish-brown one. The general colour of the bird is an ashen-grey, turning to an olive-brown above and pale or whitish-grey below. About 14 cm/5.5 in long, the blackcap likes wooded areas, and is a summer visitor to N Europe, wintering in Africa.

blackcock or ***heathcock*** large grouse *Lyrurus tetrix* found on moors and in open woods in N Europe and Asia. The male is mainly black with a lyre-shaped tail, and grows up to 54 cm/1.7 ft in height. The female is speckled brown and only 40 cm/1.3 ft tall.

Their food consists of buds, young shoots, berries, and insects.

Blackcocks are polygamous, and in the spring males attract females by curious crowings. In males a piece of bright red skin above the eyes also becomes more intense during the pairing season.

They are related to the quail, partridge, and capercaillie, in the order Galliformes.

blackcurrant variety of ◊currant.

Black Death great epidemic of bubonic ◊plague that ravaged Europe in the mid-14th century, killing between one-third and half of the population (about 75 million people). The cause of the plague was the bacterium *Yersinia pestis,* transmitted by fleas borne by migrating Asian black rats. The name Black Death was first used in England in the early 19th century.

black earth exceedingly fertile soil that covers a belt of land in NE North America, Europe, and Asia.

In Europe and Asia it extends from Bohemia through Hungary, Romania, S Russia, and Siberia, as far as Manchuria, having been deposited when the great inland ice sheets melted at the close of the last ◊ice age. In North America, it extends from the Great Lakes east through New York State, having been deposited when the last glaciers melted and retreated from the terminal moraine.

blackfly plant-sucking insect, a type of ◊aphid.

blackfly small but stoutly built blood-sucking flies with short antennae. Blackflies have broad wings with all the obvious veins in the anterior part of the wing. The family is widely distributed, the adults often occurring in such large numbers as to make them a nuisance. They are most abundant in north temperate and subarctic regions.
classification Blackflies are in family Simuliidae, order Diptera, class Insecta, phylum Arthropoda.

There are six larval stages that are found in running water, including cascades and waterfalls; they have a well capsulated head, a solitary thoracic proleg and a posterior sucker composed of small hooks by which they anchor themselves against the current. They are found on stones, reeds, mayfly larvae, and other aquatic forms. The pupae usually rest in a tent of silk in similar situations to the larvae.

Simulium species are the vectors of ◊onchocerciasis in Central and South America, Africa, and the Yemen. They also transmit other filarial worms to cattle and to ducks. Blackflies are vectors of a large number of avian malarias to many birds including domestic stock, turkeys, ducks, and geese in North America and Canada. In addition, number of blackflies attacking livestock can be so great, and the attacks so fierce, as to kill the livestock, and human deaths have also occurred.

black hole object in space whose gravity is so great that nothing can escape from it, not even light. Thought to form when massive stars shrink at the end of their lives, a black hole sucks in more matter, including other stars, from the space around it. Matter that falls into a black hole is squeezed to infinite density at the centre of the hole. Black holes can be detected because gas falling towards them becomes so hot that it emits X-rays.

Black holes containing the mass of millions of stars are thought to lie at the centres of ◊quasars. Satellites have detected X-rays from a number of objects that may be black holes, but only four likely black holes in our Galaxy had been identified by 1994.

blacksnake any of several species of snake. The blacksnake *Pseudechis porphyriacus* is a venomous snake of the cobra family found in damp forests and swamps in E Australia. The blacksnake, *Coluber constrictor* from the eastern USA, is a relative of the European grass snake, growing up to 1.2 m/4 ft long, and without venom.

blackthorn densely branched spiny European bush. It produces white blossom on bare black branches in early spring. Its sour plumlike blue-black fruit, the sloe, is used to make sloe gin. (*Prunus spinosa,* family Rosaceae.)

Blackwell, Elizabeth (1821–1910) English-born US physician, the first woman to qualify in medicine in the USA in 1849, and the first woman to be recognized as a qualified physician in the UK in 1869.

black widow *The term 'black widow' covers a number of different species of* Latrodectus *spiders including* L. mactans, *the southern black widow from the New World. The name derives from the generally held belief that the female spider invariably eats the male after mating. Recent observations indicate that this may be the case in only a few species. Premaphotos Wildlife*

black widow North American spider *Latrodectus mactans.* The male is small and harmless, but the female is 1.3 cm/0.5 in long with a red patch below the abdomen and a powerful venomous bite. The bite causes pain and fever in human victims, but they usually recover.

bladder hollow elastic-walled organ which stores the urine produced in the kidneys. It is present in the ◊urinary systems of some fishes, most amphibians, some reptiles, and all mammals. Urine enters the bladder through two ureters, one leading from each kidney, and leaves it through the urethra.

bladderwort any of a large group of carnivorous aquatic plants. They have leaves with bladders (hollow sacs) that trap small animals living in the water. (Genus *Utricularia,* family Lentibulariaceae.)

blast freezing industrial method of freezing substances such as foods by blowing very cold air over them.

blast furnace smelting furnace used to extract metals from their ores, chiefly pig iron from iron ore. The temperature is raised by the injection of an air blast.

In the extraction of iron the ingredients of the furnace are iron ore, coke (carbon), and limestone. The coke is the fuel and provides the carbon monoxide for the reduction of the iron ore; the limestone acts as a flux, removing impurities.

blastocyst in mammals, the hollow ball of cells which is an early stage in the development of the ◊embryo, roughly equivalent to the ◊blastula of other animal groups.

blastomere in biology, a cell formed in the first stages of embryonic development, after the splitting of the fertilized ovum, but before the formation of the ◊blastula or blastocyst.

blastula early stage in the development of a fertilized egg, when the egg changes from a solid mass of cells (the morula) to a hollow ball of cells (the blastula), containing a fluid-filled cavity (the blastocoel). See also ◊embryology.

bleaching decolorization of coloured materials. The two main types of bleaching agent are the **oxidizing bleaches**, which bring about the ◊oxidation of pigments and include the ultraviolet rays in sunshine, hydrogen peroxide, and chlorine in household bleaches, and the **reducing bleaches**, which bring about ◊reduction and include sulphur dioxide.

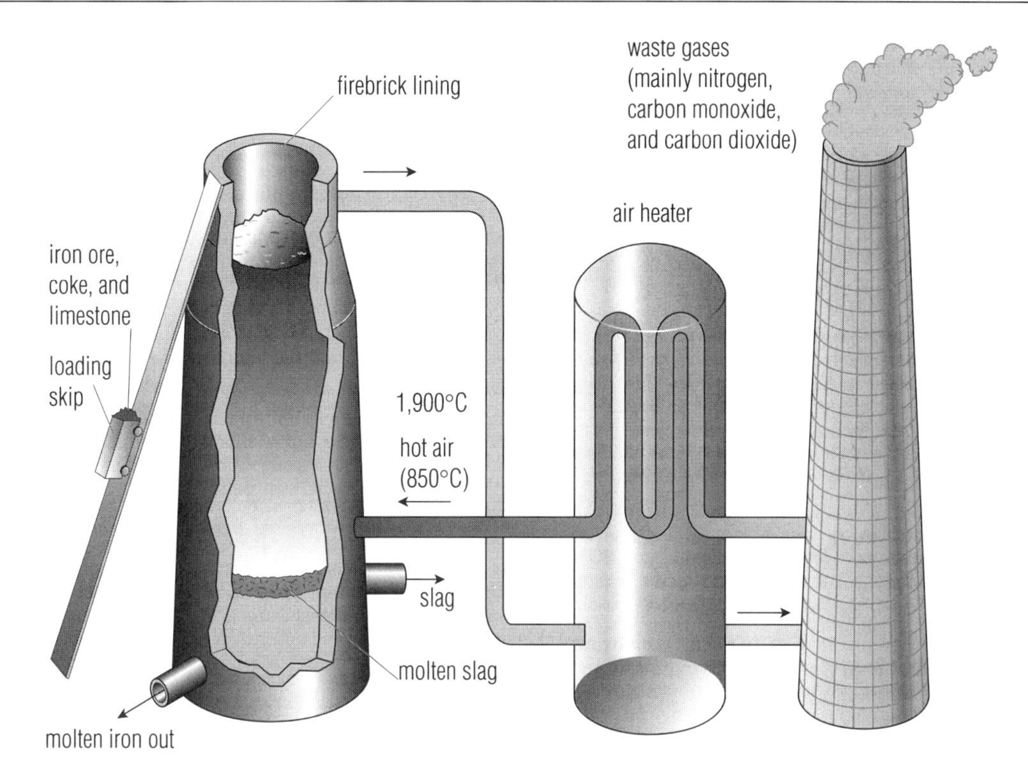

blast furnace *The blast furnace is used to extract iron from a mixture of iron ore, coke, and limestone. The less dense impurities float above the molten iron and are tapped off as slag. The molten iron sinks to the bottom of the furnace and is tapped off into moulds referred to as pigs. The iron extracted this way is also known as pig iron.*

blenny *Blennies are normally found in coastal rockpools. However, the E African species* Omobranchus striatus*, seen here, spends much of its time out of water, in groups on rocks beside the pool, ready to flip quickly into the water when danger threatens. Premaphotos Wildlife*

Bleaching processes have been known from antiquity, mainly those acting through sunlight. Both natural and synthetic pigments usually possess highly complex molecules, the colour property often being due only to a part of the molecule. Bleaches usually attack only that small part, yielding another substance similar in chemical structure but colourless.

bleak freshwater fish *Alburnus alburnus* of the carp family. It is up to to 20 cm/8 in long, and lives in still or slow-running clear water in Britain and Europe.

In Eastern Europe its scales are used in the preparation of artificial pearls.

bleeding loss of blood from the circulation; see ◊haemorrhage.

blenny any fish of the family Blenniidae, mostly small fish found near rocky shores, with elongated slimy bodies tapering from head to tail, no scales, and long pelvic fins set far forward.

blesbok African antelope *Damaliscus albifrons*, about 1 m/3 ft high, with curved horns, brownish body, and a white blaze on the face. It was seriously depleted in the wild at the end of the 19th century. A few protected herds survive in South Africa. It is farmed for meat.

blesbok *Blesbok antelopes, which live on the African savanna, are now scarce. Blesbok males are highly territorial and spend a great deal of time standing near a central dung-heap, attracting females to their harem and driving off rival males. Premaphotos Wildlife*

blight any of a number of plant diseases caused mainly by parasitic species of ◊fungus, which produce a whitish appearance on leaf and stem surfaces; for example, **potato blight** *Phytophthora infestans*. General damage caused by aphids or pollution is sometimes known as blight.

In 1998 a new virulent strain of *P. infestans*, US-8, was decimating potato and tomato crops throughout the US and eastern Canada, proving to be resistant to previously effective fungicides.

blind carbon copy e-mail message sent to multiple recipients who do not know each other's identities. The facility for blind carbon copies is built into some e-mail software, and is useful in eliminating long lists of recipients which clutter up a mass-distribution message; it also protects the confidentiality of a particular user's contact list.

blindness complete absence or impairment of sight. It may be caused by heredity, accident, disease, or deterioration with age.

Age-related macular degeneration (AMD), the commonest form of blindness, occurs as the retina gradually deteriorates with age. It affects 1% of people over the age of 70, with many more experiencing marked reduction in sight.

Retinitis pigmentosa, a common cause of blindness, is a hereditary disease affecting 1.2 million people worldwide.

Education of the blind was begun by Valentin Haüy, who published a book with raised lettering in 1784, and founded a school. Aids to the blind include the use of the Braille and Moon alphabets in reading and writing. Guide dogs for the blind were first trained in Germany for soldiers blinded in World War I.

> *Science without religion is lame. Religion without science is blind.*
>
> ALBERT EINSTEIN German-born US scientist. Quoted in A Pais *'Subtle is the Lord…': The Science and the Life of Albert Einstein* 1982

blind signature encryption technique that authenticates a message without revealing any information about the sender. Blind signatures are one element in the attempt to develop technology that protects individual privacy as more and more transactions take place over public networks where users' activities can be tracked.

blind spot area where the optic nerve and blood vessels pass through the retina of the ◊eye. No visual image can be formed as there are no light-sensitive cells in this part of the retina.

Thus the organism is blind to objects that fall in this part of the visual field.

blindworm another name for a ◊slow-worm.

blink in communications, to ◊log on using an offline reader or other software that uses automated ◊scripts. Blinking saves on communications and telephone charges, but it changes the nature of online interaction because users cannot use chat facilities.

Blinking also encourages repetition, since users replying off-line are unlikely to realize they are echoing each others' comments.

On Americal OnLine, a blink is called a **flashsession**.

blister beetle or ***oil beetle*** any of a small group of medium sized (3–20 mm/0.1–0.8 in) often brightly coloured beetles. Most give off an evil-smelling liquid, containing the irritant cantharidin, from the joints of their legs as a defence mechanism. When in contact with human skin, the liquid causes inflammation and blisters.

classification Blister beetles are members of the family Meloidae, order Coleoptera, class Insecta, phylum Arthropoda.

The general characteristics of the group include: head strongly bent downwards, narrow neck, and cylindrical and fairly soft body.

The **Spanish fly** *Lytta vesicatoria* was used to produce cantharidin for medicinal purposes, when blistering was a common medical treatment.

BL Lacertae object starlike object that forms the centre of a distant galaxy, with a prodigious energy output. BL Lac objects, as they are called, seem to be related to ◊quasars and are thought to be the brilliant nuclei of elliptical galaxies. They are so named because the first to be discovered lies in the small constellation Lacerta.

BLOB in computing, contraction of ◊binary large object.

block in computing, a group of records treated as a complete unit for transfer to or from ◊backing storage. For example, many disc drives transfer data in 512-byte blocks.

block and tackle type of ◊pulley.

blocking software any of various software programs that work on the World Wide Web to block access to categories of information considered offensive or dangerous. Typically used by parents or teachers to ensure that children do not see pornographic or other adult material, some blocking products additionally allow the blocking of personal information such as home addresses and telephone numbers; some people regard this as censorship.

Blocking software became even more controversial in mid-1996 when Washington DC-based reporters Brock Meeks and Declan McCullagh revealed that the list of banned sites in some popular products included political material and that some sites were blocked indiscriminately.

Popular blocking software products include Net Nanny, SurfWatch, CyberPatrol, and CyberSitter.

blood fluid circulating in the arteries, veins, and capillaries of vertebrate animals; the term also refers to the corresponding fluid in those invertebrates that possess a closed ◊circulatory system. Blood carries nutrients and oxygen to each body cell and removes waste products, such as carbon dioxide. It is also important in the immune response and, in many animals, in the distribution of heat throughout the body.

In humans blood makes up 5% of the body weight, occupying a volume of 5.5 l/10 pt in the average adult. It is composed of a colourless, transparent liquid called **plasma**, in which are suspended microscopic cells of three main varieties:

Red cells (erythrocytes) form nearly half the volume of the blood, with about 6 million red cells in every millilitre of an adult's blood. Their red colour is caused by ◊haemoglobin.

White cells (leucocytes) are of various kinds. Some (phagocytes) ingest invading bacteria and so protect the body from disease; these also help to repair injured tissues. Others (lymphocytes) produce antibodies, which help provide immunity.

Blood **platelets** (thrombocytes) assist in the clotting of blood.

Blood cells constantly wear out and die and are replaced from the bone marrow. Red blood cells die at the rate of 200 billion per day but the body produces new cells at an average rate of 9,000 million per hour.

> To remember the functions of the blood:
>
> OLD CHARLIE FOSTER HATES WOMEN HAVING DULL CLOTHES.
>
> OXYGEN (TRANSPORT) / CARBON DIOXIDE (TRANSPORT) / FOOD / HEAT / WASTE / HORMONES / DISEASE / CLOTTING

blood clotting complex series of events (known as the blood clotting cascade) that prevents excessive bleeding after injury. It is triggered by ◊vitamin K. The result is the formation of a meshwork of protein fibres (fibrin) and trapped blood cells over the cut blood vessels.

When platelets (cell fragments) in the bloodstream come into contact with a damaged blood vessel, they and the vessel wall itself release the enzyme **thrombokinase**, which brings about the conversion of the inactive enzyme **prothrombin** into the active **thrombin**. Thrombin in turn catalyses the conversion of the soluble protein **fibrinogen**, present in blood plasma, to the insoluble **fibrin**. This fibrous protein forms a net over the wound that traps red blood cells and seals the wound; the resulting jellylike clot hardens on exposure to air to form a scab. Calcium, vitamin K, and a variety of enzymes called factors are also necessary for efficient blood clotting. ◊Haemophilia is one of several diseases in which the clotting mechanism is impaired.

blood group any of the types into which blood is classified according to the presence or otherwise of certain ◊antigens on the surface of its red cells. Red blood cells of one individual may carry molecules on their surface that act as antigens in another individual whose red blood cells lack these molecules. The two main antigens are designated A and B. These give rise to four blood groups: having A only (A), having B only (B), having both (AB), and having neither (O). Each of these groups may or may not contain the ◊rhesus factor. Correct typing of blood groups is vital in transfusion, since incompatible types of donor and recipient blood will result in coagulation, with possible death of the recipient.

The ABO system was first described by Austrian scientist Karl Landsteiner in 1902. Subsequent research revealed at least 14 main types of blood group systems, 11 of which are involved with induced ◊antibody production. Blood typing is also of importance in forensic medicine, cases of disputed paternity, and in anthropological studies.

bloodhound breed of dog that originated as a hunting dog in Belgium in the Middle Ages. Black and tan in colour, it has long, pendulous ears and distinctive wrinkled head and face. It grows to a height of about 65 cm/26 in. Its excellent powers of scent have been employed in tracking and criminal detection from very early times.

blood poisoning presence in the bloodstream of quantities of bacteria or bacterial toxins sufficient to cause serious illness.

blood pressure pressure, or tension, of the blood against the inner walls of blood vessels, especially the arteries, due to the muscular pumping activity of the heart. Abnormally high blood pressure (◊hypertension) may be associated with various conditions or arise with no obvious cause; abnormally low blood pressure (hypotension) occurs in ◊shock and after excessive fluid or blood loss from any cause.

In mammals, the left ventricle of the ◊heart pumps blood into the arterial system. This pumping is assisted by waves of muscular contraction by the arteries themselves, but resisted by the elasticity of the inner and outer walls of the same arteries. Pressure is greatest when the heart ventricle contracts (**systole**) and lowest when the ventricle relaxes (**diastole**), and pressure is solely maintained by the elasticity of the arteries. Blood pressure is measured in millimetres of mercury (the height of a column on the measuring instrument, a sphygmomanometer). Normal human blood pressure varies with age, but in a young healthy adult it is around 120/80 mm Hg; the first number represents the systolic pressure and the second the diastolic. Large deviations from this reading usually indicate ill health.

blood test laboratory evaluation of a blood sample. There are numerous blood tests, from simple typing to establish the ◊blood group to sophisticated biochemical assays of substances, such as hormones, present in the blood only in minute quantities.

The majority of tests fall into one of three categories: **haematology** (testing the state of the blood itself), **microbiology** (identifying infection), and **blood chemistry** (reflecting chemical events elsewhere in the body). Before operations, a common test is haemoglobin estimation to determine how well a patient might tolerate blood loss during surgery.

blood transfusion see ◊transfusion.

blood vessel tube that conducts blood either away from or towards the heart in multicellular animals. Freshly oxygenated blood is carried in the arteries – major vessels which give way to the arterioles (small arteries) and finally capillaries; deoxygenated blood is returned to the heart by way of capillaries, then venules (small veins) and veins.

bloodworm larvae of the ◊midge. They are red because their blood plasma contains haemoglobin like human blood, which increases its ability to take up oxygen. This is of value to the larvae, which commonly burrow in the oxygen-poor mud bottom of pools and rivers. They feed on algae and detritus.

Bloodworms are long, with a distinct head, and segmentation of the abdomen. Prolegs (leglike projections) are found on the first thoracic and last abdominal segments. Gills are present on the last abdominal segment, and often on the segment preceding it. On average they measure 6 mm/0.2 in in length.

Bloodworms frequently build tubes of mud around themselves, which may be attached to stones. They constitute a major part of the diet of fish, hence they are often used as bait by anglers.

Not all midge larvae are red. Those that do not live in mud tubes, but frequent the surface waters, are green, and some species have blue bands.

bloom whitish powdery or waxlike coating over the surface of certain fruits that easily rubs off when handled. It often contains ◊yeasts that live on the sugars in the fruit. The term bloom is also used to describe a rapid increase in number of certain species of algae found in lakes, ponds, and oceans.

Such blooms may be natural but are often the result of nitrate pollution, in which artificial fertilizers, applied to surrounding fields, leach out into the waterways. This type of bloom can lead to the death of almost every other organism in the water; because light cannot penetrate the algal growth, the plants beneath can no longer photosynthesize and therefore do not release oxygen into the water. Only those organisms that are adapted to very low levels of oxygen survive.

BLOWFLY

Blowfly larvae develop in less than two weeks. During this time they can gain 5% of their final larval weight each hour.

blowfly any fly of the genus *Calliphora*, also known as bluebottle, or of the related genus *Lucilia*, when it is greenbottle. It lays its eggs in dead flesh, on which the maggots feed.

blowfly *Bluebottles* Calliphora vicina, *commonly known as blowfly, lay eggs on rotting flesh.* *Premaphotos Wildlife*

blubber thick layer of ◊fat under the skin of marine mammals, which provides an energy store and an effective insulating layer, preventing the loss of body heat to the surrounding water. Blubber has been used (when boiled down) in engineering, food processing, cosmetics, and printing, but all of these products can now be produced synthetically.

bluebell name given in Scotland to the ◊harebell (*Campanula rotundifolia*), and in England to the wild hyacinth (*Endymion non-scriptus*), belonging to the lily family (Liliaceae).

blueberry any of various North American shrubs belonging to the heath family, growing in acid soil. The genus also includes huckleberries, bilberries, deerberries, and cranberries, many of which resemble each other and are difficult to tell apart from blueberries. All have small oval short-stalked leaves, slender green or reddish twigs, and whitish bell-like blossoms. Only true blueberries, however, have tiny granular speckles on their twigs. Blueberries have black or blue edible fruits, often covered with a white bloom. (Genus *Vaccinium,* family Ericaceae.)

bluebird or ***blue robin*** or ***blue warbler*** three species of a North American bird, genus *Sialia,* belonging to the thrush subfamily, Turdinae, order Passeriformes. The eastern bluebird *Sialia sialis* is regarded as the herald of spring as it returns from migration. About 18 cm/7 in long, it has a reddish breast, the upper plumage being sky-blue, and a distinctive song. It lays about six pale-blue eggs.

bluebottle another name for ◊blowfly.

bluebuck any of several species of antelope, including the blue ◊duiker *Cephalophus monticola* of South Africa, about 33 cm/13 in high. The male of the Indian ◊nilgai antelope is also known as the bluebuck.

The bluebuck or blaubok, *Hippotragus leucophaeus,* was a large blue-grey South African antelope. Once abundant, it was hunted to extinction, the last being shot in 1800.

bluegrass dense spreading grass, which is blue-tinted and grows in clumps. Various species are known from the northern hemisphere. Kentucky bluegrass (*Poa pratensis*), introduced to the USA from Europe, provides pasture for horses. (Genus *Poa,* family Gramineae.)

blue-green algae or ***cyanobacteria*** single-celled, primitive organisms that resemble bacteria in their internal cell organization, sometimes joined together in colonies or filaments. Blue-green algae are among the oldest known living organisms and, with bacteria, belong to the kingdom Monera; remains have been found in rocks up to 3.5 billion years old. They are widely distributed in aquatic habitats, on the damp surfaces of rocks and trees, and in the soil.

Blue-green algae and bacteria are prokaryotic organisms. Some can fix nitrogen and thus are necessary to the nitrogen cycle, while others follow a symbiotic existence – for example, living in association with fungi to form lichens. Fresh water can become polluted by nitrates and phosphates from fertilizers and detergents. This eutrophication, or overenrichment, of the water causes multiplication of the algae in the form of algae blooms. The algae multiply and cover the water's surface, remaining harmless until they give off toxins as they decay. These toxins kill fish and other wildlife and can be harmful to domestic animals, cattle, and people.

blue gum either of two Australian trees: Tasmanian blue gum (*Eucalyptus globulus*) of the myrtle family, with bluish bark, a chief source of eucalyptus oil; or the tall, straight Sydney blue gum (*E. saligna*). The former is widely cultivated in California and has also been planted in South America, India, parts of Africa, and S Europe.

blueprint photographic process used for copying engineering drawings and architectural plans, so called because it produces a white copy of the original against a blue background.

The plan to be copied is made on transparent tracing paper, which is placed in contact with paper sensitized with a mixture of iron ammonium citrate and potassium hexacyanoferrate. The paper is exposed to ◊ultraviolet radiation and then washed in water. Where the light reaches the paper, it turns blue (Prussian blue). The paper underneath the lines of the drawing is unaffected, so remains white.

blue-ribbon campaign campaign for free speech on the Internet. It was launched to protest against various international moves towards censorship on the Internet, especially the ◊Communications Decency Act in 1996. Participation in the campaign is indicated by the small graphic of a looped blue ribbon displayed on many sites on the World Wide Web and available from the campaign's Web site http://www.eff.org/blueribbon.html.

blue shark species of ◊shark with a blue back and white underside. It grows to a length of at least 7 m/23 ft and is found in all oceanic waters except the polar seas.
classification The blue shark *Odontaspis glaucas* belongs in the family Odontaspididae, order Lamniformes (typical sharks), subclass Elasmobranchii, class Chondrichthyes.

blue shift in astronomy, a manifestation of the ◊Doppler effect in which an object appears bluer when it is moving towards the observer or the observer is moving towards it (blue light is of a higher frequency than other colours in the spectrum). The blue shift is the opposite of the ◊red shift.

blue whale the world's largest animal; see ◊whale.

BMP in Windows, a file extension indicating a graphics file in ◊bitmap format. Bit-mapped files are commonly used for icons and wallpaper.

BMR abbreviation for ◊basal metabolic rate.

boa any of various nonvenomous snakes of the family Boidae, found mainly in tropical and subtropical parts of the New World. Boas feed mainly on small mammals and birds. They catch these in their teeth or kill them by constriction (crushing the creature within their coils until it suffocates). The boa constrictor *Constrictor constrictor* can grow up to 5.5 m/18.5 ft long, but rarely reaches more than 4 m/12 ft. Other boas include the anaconda and the emerald tree boa *Boa canina,* about 2 m/6 ft long and bright green.

Some small burrowing boas live in N Africa and W Asia, while other species live on Madagascar and some Pacific islands, but the majority of boas live in South and Central America. The name boa is sometimes used loosely to include the pythons of the Old World, which also belong to the Boidae family, and which share with boas vestiges of hind limbs and constricting habits.

boar wild member of the pig family, such as the Eurasian wild boar *Sus scrofa,* from which domestic pig breeds derive. The wild boar is sturdily built, being 1.5 m/4.5 ft long and 1 m/3 ft high, and possesses formidable tusks. Of gregarious nature and mainly woodland-dwelling, it feeds on roots, nuts, insects, and some carrion.

The dark coat of the adult boar is made up of coarse bristles with varying amounts of underfur, but the young are striped. The male domestic pig is also known as a boar, the female as a sow.

boarfish marine bony fish found chiefly in the Mediterranean and northeast Atlantic. It has a flat oval body that is reddish coloured with seven transverse orange bands on the back. Boarfish are related to the dory.
classification Boarfish *Capros aper* belongs to the order Zeiformes, class Osteichthyes.

The name boarfish is derived from its projecting hoglike snout.

bobcat wild cat *Lynx rufus* living in a variety of habitats from S Canada through to S Mexico. It is similar to the lynx, but only 75 cm/2.5 ft long, with reddish fur and less well-developed ear tufts.

bobolink North American songbird *Dolichonyx oryzivorus,* family Icteridae, order Passeriformes, that takes its common name from the distinctive call of the male. It has a long middle toe and pointed tailfeathers. Breeding males are mostly black, with a white rump; females are buff-coloured with dark streaks. Bobolinks are about 18 cm/7 in long and build their nests on the ground in hayfields and weedy meadows.

Bode's law numerical sequence that gives the approximate distances, in astronomical units (distance between Earth and Sun = one astronomical unit), of the planets from the Sun by adding 4 to each term of the series 0, 3, 6, 12, 24, ... and then dividing by 10. Bode's law predicted the existence of a planet between ◊Mars and ◊Jupiter, which led to the discovery of the asteroids.

The 'law' breaks down for ◊Neptune and ◊Pluto. The relationship was first noted in 1772 by the German mathematician Johann Titius (1729–1796) (it is also known as the Titius–Bode law).

Bohr, Niels Henrik David (1885–1962)

Danish physicist whose theoretical work established the structure of the atom and the validity of quantum theory by showing that the nuclei of atoms are surrounded by shells of electrons, each assigned particular sets of quantum numbers according to their orbits. For this work he was awarded the Nobel Prize for Physics in 1922. He explained the structure and behaviour of the nucleus, as well as the process of nuclear fission. Bohr also proposed the doctrine of complementarity, the theory that a fundamental particle is neither a wave nor a particle, because these are complementary modes of description.

Mary Evans Picture Library

quantum theory and atomic structure Bohr's first model of the atom was developed working in Manchester, UK with Ernest Rutherford, who had proposed a nuclear theory of atomic structure from his work on the scattering of alpha rays in 1911. It was not, however, understood how electrons could continually orbit the nucleus without radiating energy, as classical physics demanded.

In 1913, Bohr developed his theory of atomic structure by applying quantum theory to the observations of radiation emitted by atoms. Ten years earlier, Max Planck had proposed that radiation is emitted or absorbed by atoms in discrete units, or quanta, of energy. Bohr postulated that an atom may exist in only a certain number of stable states, each with a certain amount of energy in which electrons orbit the nucleus without emitting or absorbing energy. He proposed that emission or absorption of energy occurs only with a transition from one stable state to another. When a transition occurs, an electron moving to a higher orbit absorbs energy and an electron moving to a lower orbit emits energy. In so doing, a set number of quanta of energy are emitted or absorbed at a particular frequency. Bohr's atomic theory was validated in 1922 by the discovery of an element he had predicted, hafnium.

the liquid-droplet model In 1939, Bohr proposed his liquid-droplet model for the nucleus, in which nuclear particles are pulled together by short-range forces, similar to the way in which molecules in a drop of liquid are attracted to one another. The extra energy produced by the absorption of a neutron causes the nuclear particles to separate into two groups of approximately the same size, thus breaking the nucleus into two smaller nuclei – as happens in nuclear fission. The model was vindicated when Bohr correctly predicted the differing behaviour of nuclei of uranium-235 and uranium-238 from the fact that the number of neutrons in each nucleus is odd and even respectively.

BODMAS (mnemonic for **brackets, of, division, multiplication, addition, subtraction**) – the order in which an arithmetical expression should be calculated.

Boeing US military and commercial aircraft manufacturer. Among the models Boeing has produced are the B-17 Flying Fortress, 1935; the B-52 Stratofortress, 1952; the Chinook helicopter, 1961; the first jetliner, the Boeing 707, 1957; the ◊jumbo jet or Boeing 747, 1969; the ◊jetfoil, 1975; and the 777-300 jetliner, 1997.

The company was founded in 1916 near Seattle, Washington, by William E Boeing (1881–1956) as the Pacific Aero Products Company. Renamed the following year, the company built its first seaplane and in 1919 set up an airmail service between Seattle and Victoria, British Columbia. The company announced in December 1996 that they would merge with US aircraft manufacturers McDonnell Douglas to create the world's largest aerospace company, with sales of $48 billion/£29 billion, some 200,000 employees, and an order book of civil and military aircraft worth $100 billion/£60 billion. The new US group would manufacture about three-quarters of the world's commercial airliners. It would operate under the Boeing name and with its principal headquarters in Seattle, WA. The single aerospace giant would transform the whole industry and threaten all of its rivals, including the European consortium Airbus Industrie, which had a 20% share of the commercial airline market. The merger of Boeing and McDonnell Douglas was approved by the European Union in 1997. Also in 1997 Boeing unveiled its 777-300 jetliner, the world's longest and largest twin-engine aircraft of its kind at the time, which would replace the four-engine Boeing 747.

bog type of wetland where decomposition is slowed down and dead plant matter accumulates as ◊peat. Bogs develop under conditions of low temperature, high acidity, low nutrient supply, stagnant water, and oxygen deficiency. Typical bog plants are sphagnum moss, rushes, and cotton grass; insectivorous plants such as sundews and bladderworts are common in bogs (insect prey make up for the lack of nutrients).

bogbean or ***buckbean*** aquatic or bog plant belonging to the gentian family, with a creeping rhizome (underground stem) and leaves and pink flower spikes held above water. It is found over much of the northern hemisphere. (*Menyanthes trifoliata*, family Gentianaceae.)

bohrium synthesized, radioactive element of the ◊transactinide series, symbol Bh, atomic number 107, relative atomic mass 262. It was first synthesized by the Joint Institute for Nuclear Research in Dubna, Russia in 1976; in 1981 the Laboratory for Heavy Ion Research in Darmstadt, Germany, confirmed its existence. It was named in 1997 after Danish physicist Niels ◊Bohr. Its temporary name was unnilseptium.

Bohr model model of the atom conceived by Danish physicist Neils Bohr in 1913. It assumes that the following rules govern the behaviour of electrons: (1) electrons revolve in orbits of specific radius around the nucleus without emitting radiation; (2) within each orbit, each electron has a fixed amount of energy; electrons in orbits farther away from the nucleus have greater energies; (3) an electron may 'jump' from one orbit of high energy to another of lower energy causing the energy difference to be emitted as a ◊photon of electromagnetic radiation such as light. The Bohr model has been superseded by wave mechanics (see ◊quantum theory).

boiler any vessel that converts water into steam. Boilers are used in conventional power stations to generate steam to feed steam ◊turbines, which drive the electricity generators. They are also used in steamships, which are propelled by steam turbines, and in steam locomotives. Every boiler has a furnace in which fuel (coal, oil, or gas) is burned to produce hot gases, and a system of tubes in which heat is transferred from the gases to the water.

The common kind of boiler used in ships and power stations is the **water-tube** type, in which the water circulates in tubes surrounded by the hot furnace gases. The water-tube boilers at power stations produce steam at a pressure of up to 300 atmospheres and at a temperature of up to 600°C/1,100°F to feed to the steam turbines. It is more efficient than the **fire-tube** type that is used in steam locomotives. In this boiler the hot furnace gases are drawn through tubes surrounded by water.

boiling process of changing a liquid into its vapour, by heating it at the maximum possible temperature for that liquid (see ◊boiling point) at atmospheric pressure.

boiling point for any given liquid, the temperature at which the application of heat raises the temperature of the liquid no further, but converts it into vapour.

The boiling point of water under normal pressure is 100°C/ 212°F. The lower the pressure, the lower the boiling point and vice versa.

boletus any of several fleshy fungi (see ◊fungus) with thick stems and caps of various colours. The European *Boletus edulis* is edible, but some species are poisonous. (Genus *Boletus,* class Basidiomycetes.)

boletus The red crack boletus Boletus chrysenteron. *Like other members of the genus, it releases its spores through a mass of pores beneath the cap, rather than from the more familiar gills seen beneath the cap of an edible mushroom. Premaphotos Wildlife*

boll weevil small American beetle *Anthonomus grandis* of the weevil group. The female lays her eggs in the unripe pods or 'bolls' of the cotton plant, and on these the larvae feed, causing great destruction.

bolometer sensitive ◊thermometer that measures the energy of radiation by registering the change in electrical resistance of a fine wire when it is exposed to heat or light.

The US astronomer Samuel Langley devised it in 1880 for measuring radiation from stars.

bolometric magnitude in astronomy, a measure of the brightness of a star over all wavelengths. Bolometric magnitude is related the total radiation output of the star. See ◊magnitude.

Boltzmann, Ludwig Eduard (1844–1906) Austrian physicist who studied the kinetic theory of gases, which explains the properties of gases by reference to the motion of their constituent atoms and molecules. He established the branch of physics now known as statistical mechanics.

He derived a formula, the **Boltzmann distribution**, which gives the number of atoms or molecules with a given energy at a specific temperature. The constant in the formula is called the **Boltzmann constant**.

BOLTZMANN, LUDWIG

http://www-history.mcs.st-and.ac.uk/~history/Mathematicians/Boltzmann.html

Biography of the physicist and mathematician. The site contains a description of the work of Boltzmann, and also describes his working relationships with his contemporaries, which due to the radical nature of his work were often strained. Also included are several literature references for further reading, and a photograph of Boltzmann.

Boltzmann constant in physics, the constant (symbol *k*) that relates the kinetic energy (energy of motion) of a gas atom or molecule to temperature. Its value is 1.38066×10^{-23} joules per kelvin. It is equal to the gas constant *R*, divided by ◊Avogadro's number.

bolus mouthful of chewed food mixed with saliva, ready for swallowing. Most vertebrates swallow food immediately, but grazing mammals chew their food a great deal, allowing a mechanical and chemical breakdown to begin.

bombardier beetle beetle that emits an evil-smelling fluid from its abdomen, as a defence mechanism. This fluid rapidly evaporates into a gas, which appears like a minute jet of smoke when in contact with air, and blinds the predator about to attack.
classification Bombardier beetles in genus *Brachinus,* family Carabidae, order Coleoptera, class Insecta, phylum Arthropoda.

Bombay duck or ***bummalow*** small fish *Harpodon nehereus* found in the Indian Ocean. It has a thin body, up to 40 cm/16 in long, and sharp, pointed teeth. It feeds on shellfish and other small fish. It is valuable as a food fish, and is eaten, salted and dried, with dishes such as curry.

bond in chemistry, the result of the forces of attraction that hold together atoms of an element or elements to form a molecule. The principal types of bonding are ◊ionic, ◊covalent, ◊metallic, and ◊intermolecular (such as hydrogen bonding).

bone hard connective tissue comprising the ◊skeleton of most vertebrate animals. Bone is composed of a network of collagen fibres impregnated with mineral salts (largely calcium phosphate and calcium carbonate), a combination that gives it great density and strength, comparable in some cases with that of reinforced concrete. Enclosed within this solid matrix are bone cells, blood vessels, and nerves. The interior of the long bones of the limbs consists of a spongy matrix filled with a soft marrow that produces blood cells.

There are two types of bone: those that develop by replacing ◊cartilage and those that form directly from connective tissue. The latter, which includes the bones of the cranium, are usually plate-like in shape and form in the skin of the developing embryo. Humans have about 206 distinct bones in the skeleton, of which the smallest are the three ossicles in the middle ear.

To remember the bones of the upper limb:

SOME CRIMINALS HAVE UNDERESTIMATED ROYAL CANADIAN MOUNTED POLICE.

SCAPULA / CLAVICLE / HUMERUS / ULNA / RADIUS / CARPALS / METACARPALS / PHALANGES

bone marrow substance found inside the cavity of bones. In early life it produces red blood cells but later on lipids (fat) accumulate and its colour changes from red to yellow.

Bone marrow may be transplanted in the treatment of some diseases, such as leukaemia, using immunosuppressive drugs in the recipient to prevent rejection. Transplants to adult monkeys from early aborted monkey fetuses have successfully bypassed rejection.

section through a long bone (the femur)

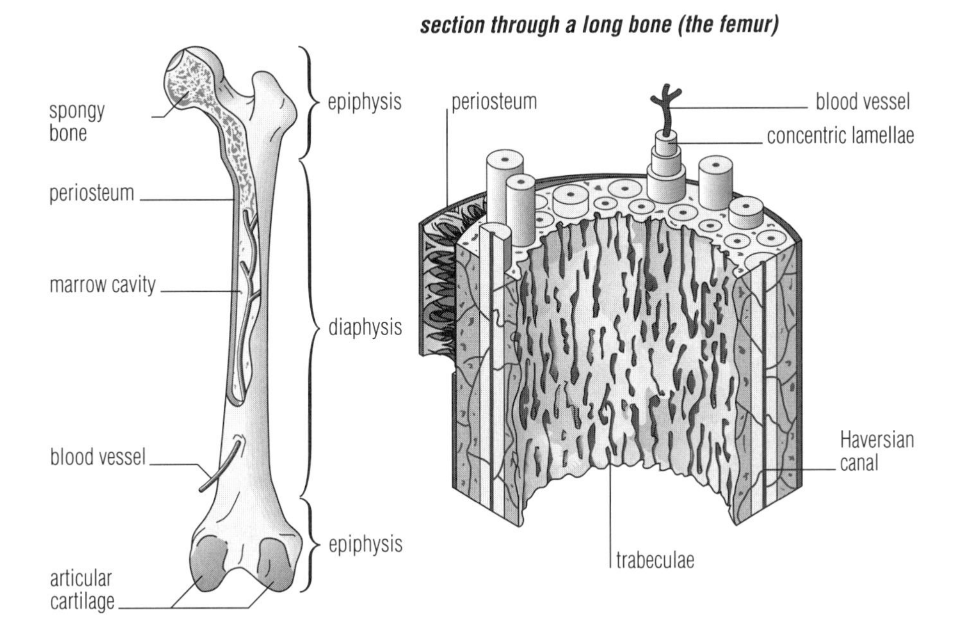

bone *Bone is a network of fibrous material impregnated with mineral salts and as strong as reinforced concrete. The upper end of the thighbone or femur is made up of spongy bone, which has a fine lacework structure designed to transmit the weight of the body. The shaft of the femur consists of hard compact bone designed to resist bending. Fine channels carrying blood vessels, nerves, and lymphatics interweave even the densest bone.*

bongo Central African antelope *Boocercus eurycerus,* living in dense humid forests. Up to 1.4 m/4.5 ft at the shoulder, it has spiral horns which may be 80 cm/2.6 ft or more in length. The body is rich chestnut, with narrow white stripes running vertically down the sides, and a black belly.

bonito any of various species of medium-sized tuna, predatory fish of the genus *Sarda,* in the mackerel family. The ocean bonito *Katsuwonus pelamis* grows to 1 m/3 ft and is common in tropical seas. The Atlantic bonito *Sarda sarda* is found in the Mediterranean and tropical Atlantic and grows to the same length but has a narrower body.

bonobo species of ◊chimpanzee.

boobook owl *Ninox novaeseelandiae* found in Australia, so named because of its call.

booby tropical seabird of the genus *Sula,* in the same family, Sulidae, as the northern ◊gannet, order Pelicaniformes. There are six species, including the circumtropical brown booby *S. leucogaster.* Plumage is white and black or brown, with no feathers on the throat and lower jaw. They inhabit coastal waters, and dive to catch fish. The name was given by sailors who saw the bird's tameness as stupidity.

One species, **Abbott's booby**, breeds only on Christmas Island, in the western Indian Ocean. Unlike most boobies and gannets, it nests high up in trees. Large parts of its breeding ground have been destroyed by phosphate mining, but conservation measures now protect the site.

booklouse any of numerous species of tiny wingless insects of the order Psocoptera, especially *Atropus pulsatoria,* which lives in books and papers, feeding on starches and moulds.

Most of the other species live in bark, leaves, and lichens. They thrive in dark, damp conditions.

bookmark facility for marking a specific place in electronic documentation to enable easy return to it. It is used in several types of software, including electronic help files and tutorials. Bookmarks are especially important on the World Wide Web, where it can be difficult to remember a uniform resource locator (◊URL) in order to

Boole, George
(1815–1864)

English mathematician. His work *The Mathematical Analysis of Logic 1847* established the basis of modern mathematical logic, and his Boolean algebra can be used in designing computers.

Boole's system is essentially two-valued. By subdividing objects into separate classes, each with a given property, his algebra makes it possible to treat different classes according to the presence or absence of the same property. Hence it involves just two numbers, 0 and 1 – the binary system used in the computer.

Mary Evans Picture Library

BOOLE, GEORGE

http://www-history.mcs.st-and.ac.uk/~history/Mathematicians/Boole.html

Extensive biography of the mathematician. The site contains a clear description of his working relationship with his contemporaries, and also includes the title page of his famous book *Investigation of the Laws of Thought*. Several literature references for further reading on the mathematician are also listed, and the Web site also features a portrait of Boole.

return to it. Most Web browsers therefore have built-in bookmark facilities, whereby the browser stores the URL with the page name attached. To return directly to the site, the user picks the page name from the list of saved bookmarks.

Boolean algebra set of algebraic rules, named after mathematician George Boole, in which TRUE and FALSE are equated to 0 and 1. Boolean algebra includes a series of operators (AND, OR, NOT, NAND (NOT AND), NOR, and XOR (exclusive OR)), which can be used to manipulate TRUE and FALSE values (see ◊truth table). It is the basis of computer logic because the truth values can be directly associated with ◊bits.

These rules are used in searching databases either locally or across the ◊Internet via services like AltaVista to limit the number of hits to those which most closely match a user's requirements. A search instruction such as 'tennis NOT table' would retrieve articles about tennis and reject those about ping-pong.

boomslang rear-fanged venomous African snake *Dispholidus typus,* often green but sometimes brown or blackish, and growing to a length of 2 m/6 ft. It lives in trees, and feeds on tree-dwelling lizards such as chameleons. Its venom can be fatal to humans; however, boomslangs rarely attack people.

boot or ***bootstrap*** in computing, the process of starting up a computer. Most computers have a small, built-in boot program that starts automatically when the computer is switched on – its only task is to load a slightly larger program, usually from a hard disc, which in turn loads the main ◊operating system.

In microcomputers the operating system is often held in the permanent ◊ROM memory and the boot program simply triggers its operation.

Some boot programs can be customized so that, for example, the computer, when switched on, always loads and runs a program from a particular backing store or always adopts a particular mode of screen display.

boot disc (also known as an ***emergency disc***) floppy disc containing the necessary files to ◊boot a computer without needing to access its hard disc. Boot discs are vital in recovering from virus attacks, when it is not known which files on a computer's hard disc may be infected; in recovering from a system crash which has corrupted existing files; or in correcting mistakes introduced into files necessary for starting up the computer by newly installed software programs.

Boötes constellation of the northern hemisphere represented by a herdsman driving a bear (◊Ursa Major) around the pole. Its brightest star is ◊Arcturus (or Alpha Boötis), which is about 37 light years from Earth. The herdsman is assisted by the neighbouring ◊Canes Venatici, 'the Hunting Dogs'.

borage plant native to S Europe, used in salads and in medicine. It has small blue flowers and hairy leaves. (*Borago officinalis,* family Boraginaceae.)

borax hydrous sodium borate, $Na_2B_4O_7.10H_2O$, found as soft, whitish crystals or encrustations on the shores of hot springs and in the dry beds of salt lakes in arid regions, where it occurs with other borates, halite, and gypsum. It is used in bleaches and washing powders.

A large industrial source is Borax Lake, California. Borax is also used in glazing pottery, in soldering, as a mild antiseptic, and as a metallurgical flux.

Bordeaux mixture a solution made up of equal quantities of copper(II) sulphate and lime in water, used in horticulture and in the wine industry as a ◊fungicide.

border collie breed of ◊sheepdog originating in the Borders region of Scotland and still much prized as a versatile working dog with a powerful herding instinct. It has a smooth or fairly long, dense, black and white coat and stands about 53 cm/21 in tall.

border terrier small, hardy, short-tailed dog with an otterlike head, moderately broad skull and short, strong muzzle. Its small, V-shaped ears drop forward. The coat is hard and dense with a close undercoat and is red, beige, and tan, or blue and tan. Dogs weigh 6–7 kg/13–15.5 lb; bitches 5–6.5 kg/ 11–14.5 lb.

bore surge of tidal water up an estuary or a river, caused by the funnelling of the rising tide by a narrowing river mouth. A very high tide, possibly fanned by wind, may build up when it is held back by a river current in the river mouth. The result is a broken wave, a metre or a few feet high, that rushes upstream.

Famous bores are found in the rivers Severn (England), Seine (France), Hooghly (India), and Chiang Jiang (China), where bores of over 4 m/13 ft have been reported.

boric acid or ***boracic acid*** $B(OH)_3$ acid formed by the combination of hydrogen and oxygen with nonmetallic boron. It is a weak antiseptic and is used in the manufacture of glass and enamels. It is also an efficient insecticide against ants and cockroaches.

boron nonmetallic element, symbol B, atomic number 5, relative atomic mass 10.811. In nature it is found only in compounds, as with sodium and oxygen in borax. It exists in two allotropic forms (see ◊allotropy): brown amorphous powder and very hard, brilliant crystals. Its compounds are used in the preparation of boric acid, water softeners, soaps, enamels, glass, and pottery glazes. In alloys it is used to harden steel. Because it absorbs slow neutrons, it is used to make boron carbide control rods for nuclear reactors. It is a necessary trace element in the human diet. The element was named by Humphry Davy, who isolated it in 1808, from *bor*ax + -on, as in carb*on*.

borzoi ***Russian 'swift'*** breed of large dog originating in Russia. It is of the greyhound type, white with darker markings, with a thick, silky coat, and stands 75 cm/30 in or more at the shoulder.

The borzoi's original quarry was hares and foxes, but it was selectively bred in the 19th century to produce a larger, stronger dog suitable for hunting wolves.

Bosch, Carl (1874–1940)

German metallurgist and chemist. He developed the GHaber process from a small-scale technique for the production of ammonia into an industrial high-pressure process that made use of water gas as a source of hydrogen. He shared the Nobel Prize for Chemistry in 1931 with Friedrich Bergius.

Mary Evans Picture Library

Bose–Einstein condensate hypothesis put forward 1925 by Albert Einstein and Indian physicist Satyendra Bose, suggesting that when a dense gas is cooled to a little over absolute zero it will condense and its atoms will lose their individuality and act as an organized whole. The first Bose–Einstein condensate was produced in June 1995 by US physicists cooling rubidinum atoms to 10 billionths of a degree above zero. The condensate existed for about a minute before becoming rubidinum ice.

boson in physics, an elementary particle whose spin can only take values that are whole numbers or zero. Bosons may be classified as ◊gauge bosons (carriers of the four fundamental forces) or ◊mesons. All elementary particles are either bosons or ◊fermions.

Unlike fermions, more than one boson in a system (such as an atom) can possess the same energy state. When developed mathematically, this statement is known as the Bose–Einstein law, after its discoverers Indian physicist Satyendra Bose and Albert Einstein.

Boston terrier breed of dog developed in the USA for dog fighting during the second half of the 19th century from crosses of English and French ◊bulldogs. It is bred in three sizes, ranging from 7 kg/15 lb to 11 kg/24 lb in weight, has a brindled or black coat with symmetrical white markings on face, chest, and legs, and carries its ears upright.

The name 'Boston terrier' is misleading as it is related to the bulldog breeds rather than terriers.

'bot (short for ***robot***) on the Internet, automated piece of software that performs specific tasks. 'Bots are commonly found on multi-user dungeons (◊MUDs) and other multi-user role-playing game sites, where they maintain a constant level of activity even when few human users are logged on. On the World Wide Web, 'bots automate maintenance tasks such as indexing Web pages and tracing broken links.

botanical garden place where a wide range of plants is grown, providing the opportunity to see a botanical diversity not likely to be encountered naturally. Among the earliest forms of botanical garden was the **physic garden**, devoted to the study and growth of medicinal plants; an example is the Chelsea Physic Garden in London, established in 1673 and still in existence. Following increased botanical exploration, botanical gardens were used to test the commercial potential of new plants being sent back from all parts of the world.

Today a botanical garden serves many purposes: education, science, and conservation. Many are associated with universities and also maintain large collections of preserved specimens (see ◊herbarium), libraries, research laboratories, and gene banks. There are 1,600 botanical gardens worldwide.

botany ***Greek botane 'herb'*** the study of living and fossil ◊plants, including form, function, interaction with the environment, and classification.

Botany is subdivided into a number of specialized studies, such as the identification and classification of plants (taxonomy), their external formation (plant morphology), their internal arrangement (plant anatomy), their microscopic examination (plant histology), their functioning and life history (plant physiology), and their distribution over the Earth's surface in relation to their surroundings (plant ecology). Palaeobotany concerns the study of fossil plants, while economic botany deals with the utility of plants. ◊Horticulture, ◊agriculture, and ◊forestry are branches of botany.

botfly any fly of the family Oestridae. The larvae are parasites that feed on the skin (warblefly of cattle) or in the nasal cavity (nostrilflies of sheep and deer). The horse botfly belongs to another family, the Gasterophilidae. It has a parasitic larva that feeds in the horse's stomach.

bo tree or ***peepul*** Indian ◊fig tree, said to be the tree under which the Buddha became enlightened. (*Ficus religiosa*, family Moraceae.)

bottlebrush any of several trees or shrubs common in Australia, belonging to the myrtle family. They have cylindrical, composite flower heads in green, yellow, white, various shades of red, and violet. (Genus *Callistemon*, family Myrtaceae.)

bottlenose species of ◊dolphin.

botulism rare, often fatal type of ◊food poisoning. Symptoms include vomiting, diarrhoea, muscular paralysis, breathing difficulties and disturbed vision.

It is caused by a toxin produced by the bacterium *Clostridium botulinum*, found in soil and sometimes in improperly canned foods.

Thorough cooking destroys the toxin, which otherwise suppresses the cardiac and respiratory centres of the brain. In neurology, botulinum toxin is sometimes used to treat rare movement disorders.

bougainvillea any plant of a group of South American tropical vines of the four o'clock family, now cultivated in warm countries around the world for the colourful red and purple bracts (leaflike structures) that cover the flowers. They are named after the French navigator Louis de Bougainville. (Genus *Bougainvillea*, family Nyctaginaceae.)

bounce in computing, system by which an electronic mail message that cannot be delivered to its addressee is returned ('bounced back') to the sender, with a note advising of its failure to reach its destination. Failed delivery is usually due to an incorrect e-mail address or a network problem.

boundary a line around the edge of an area; a perimeter. The boundary of a circle is known as the **circumference**. The boundary which marks the limit of land may be indicated by a post, ditch, hedge, march of stones, road, or river, or it may be indicated by reference to a plan, or to possession of tenants, or by actual measurement.

Bourbaki, Nicolas pseudonym adopted by a group of mathematicians, most of them French, who, collectively and anonymously, published a definitive survey of mathematics 1939–67. The group, which at any one time contained about 20 members, was centred at the Ecole Normale Supérieure in Paris. The group's founder was André Weil.

Bourdon gauge instrument for measuring pressure, patented by French watchmaker Eugène Bourdon in 1849. The gauge contains a C-shaped tube, closed at one end. When the pressure inside the tube increases, the tube uncurls slightly causing a small movement at its closed end. A system of levers and gears magnifies this movement and turns a pointer, which indicates the pressure on a circular scale. Bourdon gauges are often fitted to cylinders of compressed gas used in industry and hospitals.

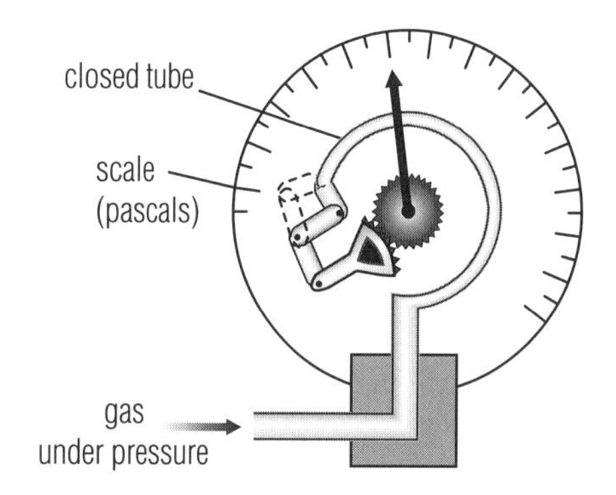

Bourdon gauge *The most common form of Bourdon gauge is the C-shaped tube. However, in high-pressure gauges spiral tubes are used; the spiral rotates as pressure increases and the tip screws forwards.*

Bovidae mammal family that consists of 128 species of antelopes, sheep, goats, and cattle. They are ruminants (chew the cud) and also artiodactylate (even-toed); all the males and some of the females have horns consisting of solid bony extensions of the skull encased in a sheath of true horn.

Bovids occur in all parts of the Old World and in North America, but are not native to Australia and South America.

classification Bovidae is a family in order Artiodactyla.

bovine somatotropin (BST) hormone that increases an injected cow's milk yield by 10–40%. It is a protein naturally occurring in milk and breaks down within the human digestive tract into harmless amino acids. However, doubts have arisen recently as to whether such a degree of protein addition could in the long term be guaranteed harmless either to cattle or to humans.

Although no evidence of adverse side effects to consumers have been found, BST was banned in Europe 1993 until the year 2000. In the USA genetically engineered BST has been in use since February 1994; in Vermont a law requiring milk containing BST to be labelled as such was passed September 1995.

The incidence of mastitis in herds injected with BST is 15–45% higher.

Facts do not cease to exist because they are ignored.

ALDOUS HUXLEY English novelist.
Proper Studies, 'Note on Dogma'

bovine spongiform encephalopathy (BSE) or ***mad cow disease*** disease of cattle, related to ◊scrapie in sheep, which attacks the nervous system, causing aggression, lack of coordination, and collapse. First identified in 1986, it is almost entirely confined to the UK. By 1996 it had claimed 158,000 British cattle.

BSE is one of a group of diseases known as the transmissible spongiform encephalopathies, since they are characterized by the appearance of spongy changes in brain tissue. Some scientists believe that all these conditions, including Creutzfeldt–Jakob disease (CJD) in humans, are in effect the same disease, and in 1996 a link was established between the deaths of 10 young people from CJD and the consumption of beef products.

The cause of these universally fatal diseases is not fully understood, but they may be the result of a rogue protein called a prion. A prion may be inborn or it may be transmitted in contaminated tissue.

According to an official European Commission Report released in Mar 1997, consumers throughout Europe were being exposed to BSE-infected meat. The report also highlighted lax health controls, and supported the view that the extent of BSE throughout the EU was much wider then governments were prepared to admit.

It was also revealed in 1997 that the British government had allowed more than 6,000 carcasses suspected of having BSE to be buried in landfill sites across Britain – in direction contravention of its own regulations. Because of fears that BSE could get into drinking water, or the food chain, both the British government and the EU have insisted the carcasses should be incinerated.

bower bird New Guinean and N Australian bird of the family Ptilonorhynchidae, order Passeriformes, related to the ◊bird of paradise. The males are dull-coloured, and build elaborate bowers of sticks and grass, decorated with shells, feathers, or flowers, and even painted with the juice of berries, to attract the females. There are 17 species.

bowfin North American fish *Amia calva* with a swim bladder highly developed as an air sac, enabling it to breathe air. It is the only surviving member of a primitive group of bony fishes.

bowhead Arctic whale *Balaena mysticetus* with strongly curving upper jawbones supporting the plates of baleen with which it sifts planktonic crustaceans from the water. Averaging 15 m/50 ft long and 90 tonnes/100 tons in weight, these slow-moving, placid whales were once extremely common, but by the 17th century were already becoming scarce through hunting. Only an estimated 3,000 remain, and continued hunting by the Inuit may result in extinction.

Bowman's capsule in the vertebrate kidney, a cup-shaped structure enclosing the glomerulus, which is the initial site of filtration of the blood leading to urine formation.

There are approximately a million of these capsules in a human kidney, each containing a tight knot of capillaries and leading into a kidney tubule or nephron where unwanted fluid and waste molecules are filtered from the blood to be excreted in the form of urine.

box any of several small evergreen trees and shrubs, with small, leathery leaves. Some species are used as hedging plants and for shaping into garden ornaments. (Genus *Buxus,* family Buxaceae.)

boxer breed of dog, about 60 cm/24 in tall, with a smooth coat and a set-back nose. The tail is usually docked. A boxer is usually brown, often with white markings, but may be fawn or brindled.

boxfish or ***trunkfish***, any fish of the family Ostraciodontidae, with scales that are hexagonal bony plates fused to form a box covering the body, only the mouth and fins being free of the armour. Boxfish swim slowly. The cowfish, genus *Lactophrys,* with two 'horns' above the eyes, is a member of this group.

Boyle, Robert (1627–1691)

Irish chemist and physicist who published the seminal The Sceptical Chymist 1661. He formulated Boyle's law in 1662. He was a pioneer in the use of experiment and scientific method.

Boyle questioned the alchemical basis of the chemical theory of his day and taught that the proper object of chemistry was to determine the compositions of substances. The term 'analysis' was coined by Boyle and many of the reactions still used in qualitative work were known to him. He introduced certain plant extracts, notably litmus, for the indication of acids and bases. He was also the first chemist to collect a sample of gas.

Mary Evans Picture Library

Father of Chemistry and Uncle of the Earl of Cork.

ROBERT BOYLE Irish chemist.
On his tombstone in Dublin, quoted in
R L Weber *More Random Walks in Science*

Boyle's law law stating that the volume of a given mass of gas at a constant temperature is inversely proportional to its pressure. For example, if the pressure of a gas doubles, its volume will be reduced by a half, and vice versa. The law was discovered in 1662 by Irish physicist and chemist Robert Boyle. See also ◊gas laws.

bozo filter facility to eliminate messages from irritating users. It is also known as a ◊killfile.

BP abbreviation for ***British Pharmacopoeia***; **British Petroleum**.

bps abbreviation for **bits per second**, measure used in specifying data transmission rates.

brachiopod or ***lamp shell*** any member of the phylum Brachiopoda, marine invertebrates with two shells, resembling but totally unrelated to bivalves.

There are about 300 living species; they were much more numerous in past geological ages. They are suspension feeders, ingesting minute food particles from water. A single internal organ, the lophophore, handles feeding, aspiration, and excretion.

Brachyteles primate genus consisting of the single species commonly called the woolly ◊spider monkey.

bracken any of several large ferns (especially *Pteridium aquilinum*) which grow abundantly in the northern hemisphere. The rootstock produces coarse fronds each year, which die down in autumn.

bracket fungus any of a group of fungi (see ◊fungus) with fruiting bodies that grow like shelves from the trunks and branches of trees. (Class Basidiomycetes.)

braconid small parasitic wasp closely related to the ◊ichneumon flies, but differing from them mainly by having fewer wing veins. Braconids chiefly parasitize caterpillars, but also some beetle and fly larvae, so they are selectively used in ◊biological control programmes.

bracket fungus This varicoloured Coriolus versicolor *is one of a large number of often unrelated fungi referred to collectively as bracket fungi because they form bracketlike growths on trees and timber. The fruiting body is the only part that is immediately visible. The mass of hyphae, which form its mycelium, penetrates and feeds upon the dead wood on which the fungus is living. The spore-producing fruiting body is derived from the mycelium. Premaphotos Wildlife*

classification Braconids are in the family Braconidae, order Hymenoptera, class Insecta, phylum Arthropoda.

Apanteles glomeratus is a tiny black braconid that parasitizes the cabbage white butterfly. The female lays her eggs in the caterpillar. About autumn, masses of sulphur-yellow cocoons of the parasite can be seen attached to the skin casts of the dead (pupated) caterpillar.

Brahe, Tycho
(1546–1601)

Danish astronomer. His accurate observations of the planets enabled German astronomer and mathematician Johannes Kepler to prove that planets orbit the Sun in ellipses. Brahe's discovery and report of the 1572 supernova brought him recognition, and his observations of the comet of 1577 proved that it moved in an orbit among the planets, thus disproving Aristotle's view that comets were in the Earth's atmosphere.

Brahe was a colourful figure who wore a silver nose after his own was cut off in a duel, and who took an interest in alchemy. In 1576 Frederick II of Denmark gave him the island of Hven, where he set up an observatory. Brahe was the greatest observer in the days before telescopes, making the most accurate measurements of the positions of stars and planets. He moved to Prague as imperial mathematician in 1599, where he was joined by Kepler, who inherited his observations when he died.

Mary Evans Picture Library

BRAHE, TYCHO

Tycho Brahe, the 16th-century Danish astronomer, wore a false nose made of silver. His own had been cut off in a duel while he was a student in Rostock.

bract leaflike structure in whose ◊axil a flower or inflorescence develops. Bracts are generally green and smaller than the true leaves. However, in some plants they may be brightly coloured and conspicuous, taking over the role of attracting pollinating insects to the flowers, whose own petals are small; examples include poinsettia *Euphorbia pulcherrima* and bougainvillea.

A whorl of bracts surrounding an ◊inflorescence is termed an **involucre**. A **bracteole** is a leaf-like organ that arises on an individual flower stalk, between the true bract and the ◊calyx.

brain in higher animals, a mass of interconnected ◊nerve cells forming the anterior part of the ◊central nervous system, whose activities it coordinates and controls. In ◊vertebrates, the brain is contained by the skull. At the base of the ◊brainstem, the **medulla oblongata** contains centres for the control of respiration, heartbeat rate and strength, and blood pressure. Overlying this is the **cerebellum**, which is concerned with coordinating complex muscular processes such as maintaining posture and moving limbs.

The cerebral hemispheres (**cerebrum**) are paired outgrowths of the front end of the forebrain, in early vertebrates mainly concerned with the senses, but in higher vertebrates greatly developed and involved in the integration of all sensory input and motor output, and in thought, emotions, memory, and behaviour.

In vertebrates, many of the nerve fibres from the two sides of the body cross over as they enter the brain, so that the left cerebral hemisphere is associated with the right side of the body and vice versa. In humans, a certain asymmetry develops in the two halves of the cerebrum. In right-handed people, the left hemisphere seems to play a greater role in controlling verbal and some mathematical skills, whereas the right hemisphere is more involved in spatial perception. In general, however, skills and abilities are not closely localized. In the brain, nerve impulses are passed across ◊synapses by neurotransmitters, in the same way as in other parts of the nervous system.

In mammals the cerebrum is the largest part of the brain, carrying the **cerebral cortex**. This consists of a thick surface layer of cell bodies (grey matter), below which fibre tracts (white matter) connect various parts of the cortex to each other and to other points in the central nervous system. As cerebral complexity grows, the surface of the brain becomes convoluted into deep folds. In higher mammals, there are large unassigned areas of the brain that seem to be connected with intelligence, personality, and higher mental faculties. Language is controlled in two special regions usually in the left side of the brain: **Broca's area** governs the ability to talk, and **Wernicke's area** is responsible for the comprehension of spoken and written words. In 1990, scientists at Johns Hopkins University, Baltimore, succeeded in culturing human brain cells.

If the cells and fibre of the human brain were stretched out end to end, they would certainly reach to the moon and back. Yet the fact that they are not arranged end to end enabled man to go there himself. The astonishing tangle within our heads makes us what we are.

COLIN BLAKEMORE English physiologist.
BBC Reith Lecture 1976

brain damage impairment which can be caused by trauma (for example, accidents) or disease (such as encephalitis), or which may be present at birth. Depending on the area of the brain that is affected, language, movement, sensation, judgement, or other abilities may be impaired.

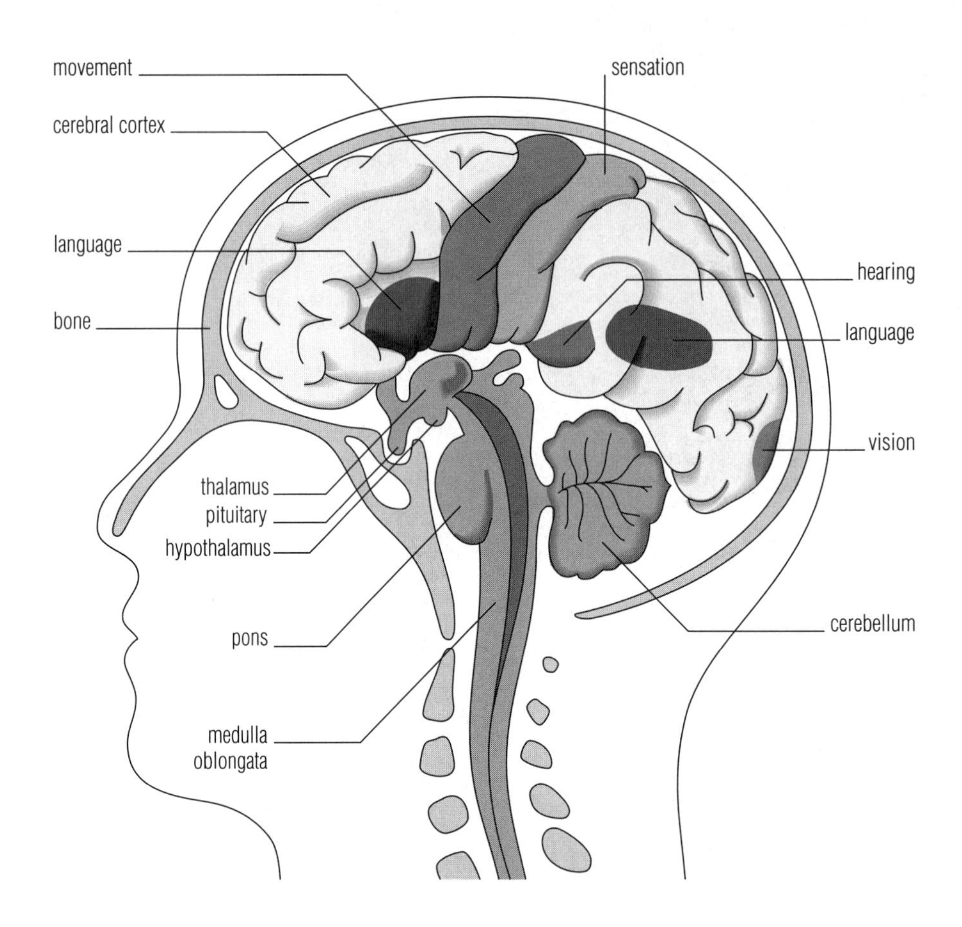

brain *The structure of the human brain. At the back of the skull lies the cerebellum, which coordinates reflex actions that control muscular activity. The medulla controls respiration, heartbeat, and blood pressure. The hypothalamus is concerned with instinctive drives and emotions. The thalamus relays signals to and from various parts of the brain. The pituitary gland controls the body's hormones. Distinct areas of the large convoluted cerebral hemispheres that fill most of the skull are linked to sensations, such as hearing and sight, and voluntary activities, such as movement.*

brainstem region where the top of the spinal cord merges with the undersurface of the brain, consisting largely of the medulla oblongata and midbrain.

The oldest part of the brain in evolutionary terms, the brainstem is the body's life-support centre, containing regulatory mechanisms for vital functions such as breathing, heart rate, and blood pressure. It is also involved in controlling the level of consciousness by acting as a relay station for nerve connections to and from the higher centres of the brain.

In many countries, death of the brainstem is now formally recognized as death of the person as a whole. Such cases are the principal donors of organs for transplantation. So-called 'beating-heart donors' can be maintained for a limited period by life-support equipment.

brake device used to slow down or stop the movement of a moving body or vehicle. The mechanically applied calliper brake used on bicycles uses a scissor action to press hard rubber blocks against the wheel rim. The main braking system of a car works hydraulically: when the driver depresses the brake pedal, liquid pressure forces pistons to apply brakes on each wheel.

Two types of car brakes are used. **Disc brakes** are used on the front wheels of some cars and on all wheels of sports and performance cars, since they are the more efficient and less prone to fading (losing their braking power) when they get hot. Braking pressure forces brake pads against both sides of a steel disc that rotates with the wheel. **Drum brakes** are fitted on the rear wheels of some cars and on all wheels of some passenger cars. Braking pressure forces brake shoes to expand outwards into contact with a drum rotating with the wheels. The brake pads and shoes have a tough ◊friction lining that grips well and withstands wear.

Many trucks and trains have **air brakes**, which work by compressed air. On landing, jet planes reverse the thrust of their engines to reduce their speed quickly. Space vehicles use retro-rockets for braking in space and use the air resistance, or drag of the atmosphere, to slow down when they return to Earth.

bramble any of a group of prickly bushes belonging to the rose family. Examples are ◊blackberry, raspberry, and dewberry. (Genus *Rubus,* family Rosaceae.)

brambling or ***bramble finch*** bird *Fringilla montifringilla* belonging to the finch family Fringillidae, order Passeriformes. It is about 15 cm/6 in long, and breeds in N Europe and Asia.

brass metal ◊alloy of copper and zinc, with not more than 5% or 6% of other metals. The zinc content ranges from 20% to 45%, and the colour of brass varies accordingly from coppery to whitish yellow. Brasses are characterized by the ease with which they may be shaped and machined; they are strong and ductile, resist many forms of corrosion, and are used for electrical fittings, ammunition cases, screws, household fittings, and ornaments.

Brasses are usually classed into those that can be worked cold (up to 25% zinc) and those that are better worked hot (about 40% zinc).

brassica any of a group of plants, many of which are cultivated as vegetables. The most familiar is the common cabbage (*Brassica*

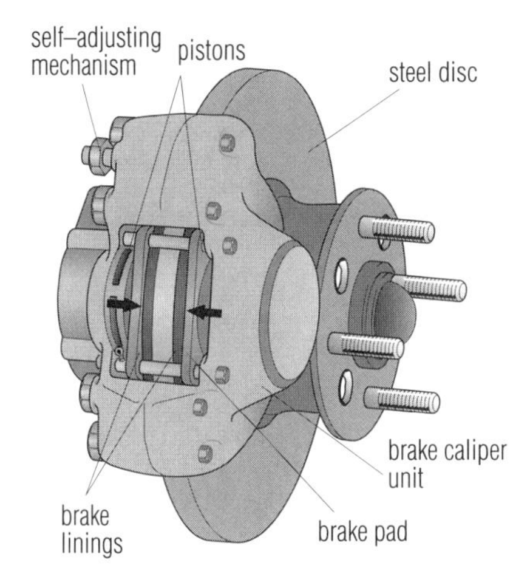

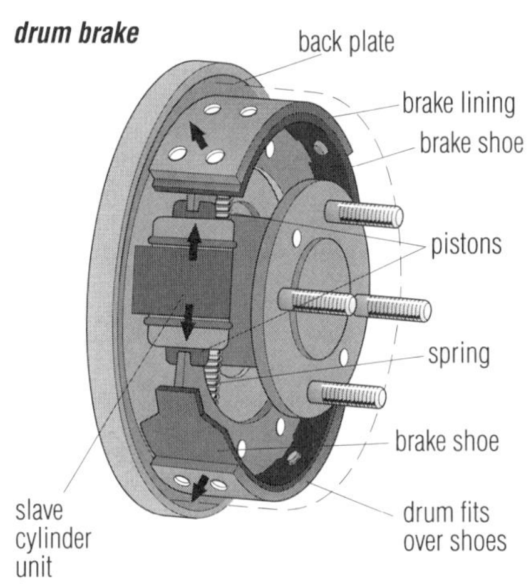

brake Two common braking systems: the disc brake (left) and the drum brake (right). In the disc brake, increased hydraulic pressure of the brake fluid in the pistons forces the brake pads against the steel disc attached to the wheel. A self-adjusting mechanism balances the force on each pad. In the drum brake, increased pressure of the brake fluid within the slave cylinder forces the brake pad against the brake drum attached to the wheel.

oleracea), with its varieties broccoli, cauliflower, kale, and Brussels sprouts. (Genus *Brassica*, family Cruciferae.)

In 1990 US experiments in cross-pollinating the wild cabbage (*B. campestris*) with related varieties of cultivated cabbage, turnip, and swede produced a new plant with a life cycle of only five weeks. This is now being used in US schools to enable pupils to carry out plant-breeding experiments that can produce ten generations in one year.

Braun AEG German company, manufacturer of sound equipment and domestic appliances, founded 1921 by Max Braun and based in Frankfurt.

The factory was rebuilt 1945 and in 1951 when Max Braun died his son Artur (1925–) began commissioning a number of young and highly innovative German industrial designers associated with the new design school (Hochschüle für Gestaltung) at Ulm – among them Dieter Rams and Hans Gugelot (1920–1965) – to update the company's product range. Their radically new designs quickly became hallmarks of the stark, geometric design style that developed in Germany at that time.

Brazil nut gigantic South American tree; also its seed, which is rich in oil and highly nutritious. The seeds (nuts) are enclosed in a hard outer casing, each fruit containing 10–20 seeds arranged like the segments of an orange. The timber of the tree is also valuable. (*Bertholletia excelsa*, family Lecythidaceae.)

brazing method of joining two metals by melting an ◊alloy or metal into the joint. It is similar to soldering (see ◊solder) but takes place at a much higher temperature. Copper and silver alloys are widely used for brazing, at temperatures up to about 900°C/1,650°F.

Where high precision is needed, as in space technology, nickel based filters are used in the temperature range 1,000–1,200°C/ 1,832–2,192°F.

breadfruit fruit of two tropical trees belonging to the mulberry family. It is highly nutritious and when baked is said to taste like bread. It is native to many South Pacific islands. (*Artocarpus communis* and *A. altilis*, family Moraceae.)

breadfruit Breadfruit trees, native to Polynesia, are now a familiar sight in many tropical countries around the world, where they have been planted for their large nutritious fruits. When cooked, the fruit has a breadlike texture, hence the name. Premaphotos Wildlife

breadth thickness, another name for width. The area of a rectangle is given by the formula: area = length times breadth.

bream deep-bodied, flattened fish *Abramis brama* of the carp family, growing to about 50 cm/1.6 ft, typically found in lowland rivers across Europe.

breast one of a pair of organs on the chest of the human female, also known as a ◊mammary gland. Each of the two breasts contains milk-producing cells and a network of tubes or ducts that lead to openings in the nipple.

Milk-producing cells in the breast do not become active until a woman has given birth to a baby. Breast milk is made from sub-

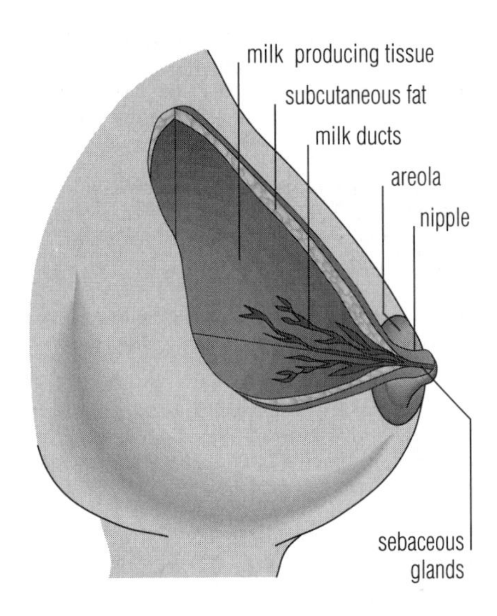

breast *The human breast or mammary gland. Milk produced in the tissue of the breast to feed a baby after a woman has given birth passes along ducts which lead to openings in the nipple.*

stances extracted from the mother's blood as it passes through the breasts, and contains all the nourishment a baby needs. Breast-fed newborns develop fewer infections than bottle-fed babies because of the antibodies and white blood cells contained in breast milk. These are particularly abundant in the colostrum produced in the first few days of breast-feeding.

breast cancer in medicine, ◊cancer of the ◊breast. It is usually diagnosed following the detection of a painless lump in the breast (either through self-examination or ◊mammography). Other, less common symptoms, include changes in the shape or texture of the breast and discharge from the nipple. It is the commonest cancer amongst women: there are 28,000 new cases of breast cancer in Britain each year and 185,700 in the USA.
treatment If the tumour is caught early, only it and the immediate surrounding tissue needs removing, in a process called lumpectomy, usually accompanied by radiotherapy. In more advanced cases a mastectomy is performed. Chemotherapy or hormone-blocking drugs (such as tamoxifen) may also accompany either procedure. The average survival rate after 5 years was 83.2% in 1996.
risk factors Possible risk factors include a family history of breast cancer (mutations in the genes *BRCA1* and *BRCA2* were found to cause more than 50% of inherited breast cancer cases in the 1990s); childlessness or late childbearing; early onset of menstruation and late menopause.

breast screening in medicine, examination of the breast to detect the presence of breast cancer at an early stage. Screening methods include self-screening by monthly examination of the breasts and formal programmes of screening by palpation (physical examination) and mammography in special clinics. Screening may be offered to older women on a routine basis and it is important in women with a family history of breast cancer.

BREAST CANCER AWARENESS

http://avon.avon.com/showpage.asp?thepage=crusade

Promotes awareness of this disease through a library of frequently asked questions about breast cancer and mammograms, as well as a glossary of common terms and access to support groups. There is also information here about their fundraising activities and recipients of awards for people and groups seen to be contributing the most to the fight against this disease. It is sponsored by the make-up company Avon.

Breathalyzer trademark for an instrument for on-the-spot checking by police of the amount of alcohol consumed by a suspect driver. The driver breathes into a plastic bag connected to a tube containing a chemical (such as a diluted solution of potassium dichromate in 50% sulphuric acid) that changes colour in the presence of alcohol. Another method is to use a gas chromatograph, again from a breath sample.

breathing in terrestrial animals, the muscular movements whereby air is taken into the lungs and then expelled, a form of ◊gas exchange. Breathing is sometimes referred to as external respiration, for true respiration is a cellular (internal) process.

Lungs are specialized for gas exchange but are not themselves muscular, consisting of spongy material. In order for oxygen to be passed to the blood and carbon dioxide removed, air is drawn into the lungs (inhaled) by the contraction of the diaphragm and intercostal muscles; relaxation of these muscles enables air to be breathed out (exhaled). The rate of breathing is controlled by the brain. High levels of activity lead to a greater demand for oxygen and an increased rate of breathing.

breathing rate the number of times a minute the lungs inhale and exhale. The rate increases during exercise because the muscles require an increased supply of oxygen and nutrients. At the same time very active muscles produce a greater volume of carbon dioxide, a waste gas that must be removed by the lungs via the blood.

The regulation of the breathing rate is under both voluntary and involuntary control, although a person can only forcibly stop breathing for a limited time. The regulatory system includes the use of chemoreceptors, which can detect levels of carbon dioxide in the blood. High concentrations of carbon dioxide, occurring for example during exercise, stimulate a fast breathing rate.

breccia coarse-grained clastic ◊sedimentary rock, made up of broken fragments (clasts) of pre-existing rocks held together in a fine-grained matrix. It is similar to ◊conglomerate but the fragments in breccia are jagged in shape.

breed recognizable group of domestic animals, within a species, with distinctive characteristics that have been produced by ◊artificial selection.

breeder reactor or ***fast breeder*** alternative names for ◊fast reactor, a type of nuclear reactor.

breeding in biology, the crossing and selection of animals and plants to change the characteristics of an existing ◊breed or ◊cultivar (variety), or to produce a new one.

Cattle may be bred for increased meat or milk yield, sheep for thicker or finer wool, and horses for speed or stamina. Plants, such as wheat or maize, may be bred for disease resistance, heavier and more rapid cropping, and hardiness to adverse weather.

breeding in nuclear physics, a process in a reactor in which more fissionable material is produced than is consumed in running the reactor.

For example, plutonium-239 can be made from the relatively plentiful (but nonfissile) uranium-238, or uranium-233 can be produced from thorium. The Pu-239 or U-233 can then be used to fuel other reactors.

brewing making of beer, ale, or other alcoholic beverage, from ◊malt and ◊barley by steeping (mashing), boiling, and fermenting.

Mashing the barley releases its sugars. Yeast is then added, which contains the enzymes needed to convert the sugars into ethanol (alcohol) and carbon dioxide. Hops are added to give a bitter taste.

brewster unit (symbol B) for measuring the reaction of optical materials to stress, defined in terms of the slowing down of light passing through the material when it is stretched or compressed.

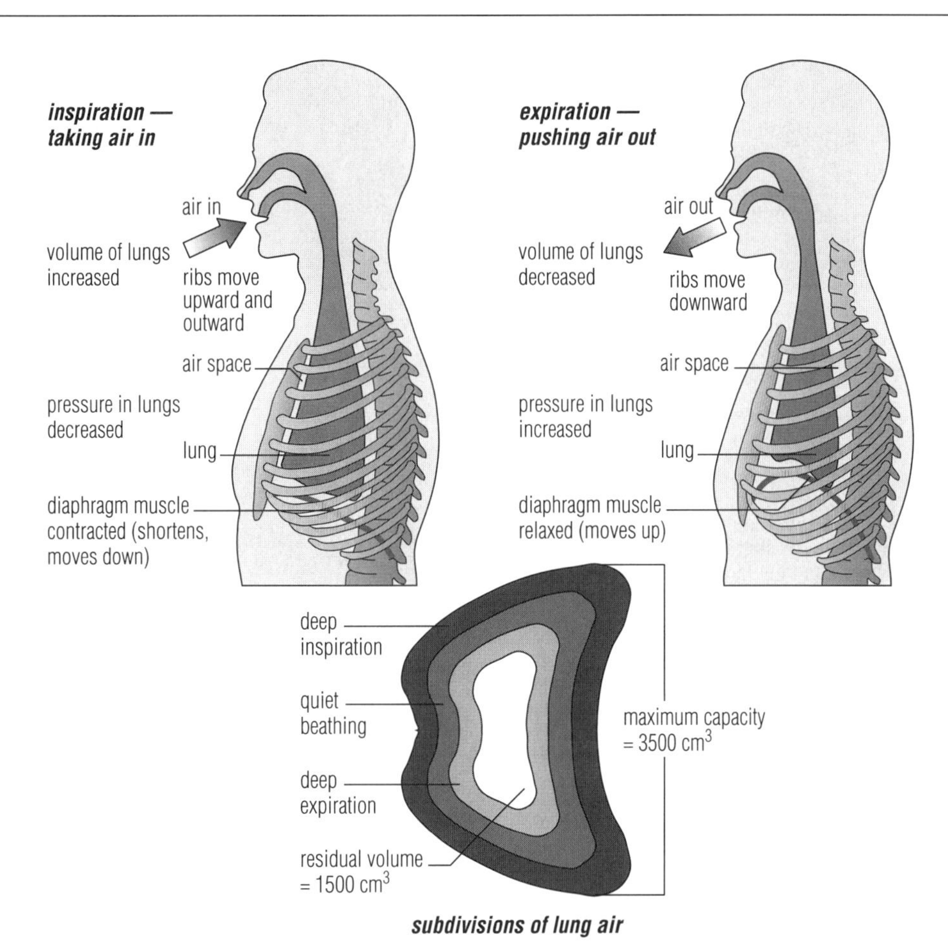

breathing *The two phases of the process of breathing, or respiration. Gas exchange occurs in the alveoli, tiny air tubes in the lungs.*

brick common block-shaped building material, with all opposite sides parallel. It is made of clay that has been fired in a kiln. Bricks are made by kneading a mixture of crushed clay and other materials into a stiff mud and extruding it into a ribbon. The ribbon is cut into individual bricks, which are fired at a temperature of up to about 1,000°C/1,800°F. Bricks may alternatively be pressed into shape in moulds.

Refractory bricks used to line furnaces are made from heat-resistant materials such as silica and dolomite. They must withstand operating temperatures of 1,500°C/2,700°F or more.

Facing bricks are designed to be visually more attractive than than common bricks, and include specially moulded bricks.

Sun-dried bricks of mud reinforced with straw were first used in Mesopotamia some 8,000 years ago. Similar mud bricks, called adobe, are still used today in Mexico and other areas where the climate is warm and dry.

brickwork method of construction using bricks made of fired clay or sun-dried earth. In wall building, bricks are either laid out as stretchers (long side facing out) or as headers (short side facing out). The two principal patterns of brickwork are **English bond** in which alternate courses, or layers, are made up of stretchers or headers only, and **Flemish bond** in which stretchers and headers alternate within courses.

Some evidence exists of the use of fired bricks in ancient Mesopotamia and Egypt, although the Romans were the first to make extensive use of this technology. Today's mass production of fired bricks tends to be concentrated in temperate regions where there are plentiful supplies of fuel available.

bridge in computing, a device that connects two similar local area networks (LANs). Bridges transfer data in packets between the two networks, without making any changes or interpreting the data in any way. See also ◊router and ◊brouter.

bridge structure that provides a continuous path or road over water, valleys, ravines, or above other roads. The basic designs and composites of these are based on the way they bear the weight of the structure and its load. **Beam**, or **girder**, bridges are supported at each end by the ground with the weight thrusting downwards. **Cantilever** bridges are a complex form of girder in which only one end is supported. **Arch** bridges thrust outwards and downwards at their ends. **Suspension** bridges use cables under tension to pull inwards against anchorages on either side of the span, so that the roadway hangs from the main cables by the network of vertical cables. The **cable-stayed** bridge relies on diagonal cables connected directly between the bridge deck and supporting towers at each end. Some bridges are too low to allow traffic to pass beneath easily, so they are designed with movable parts, like swing and draw bridges.

history In prehistory, people used logs or wove vines into ropes that were thrown across the obstacle. Clapper bridges, made from flat stones simply laid across or supported by piles of stones, were some of the earliest bridges. By 4000 BC arched structures of stone and/or brick were used in the Middle East, and the Romans built long arched spans, many of which are still standing. Cast iron bridges were introduced in 1779. The ◊Bessemer process produced steel that made it possible to build long-lived framed structures that support great weight over long spans.

brill flatfish *Scophthalmus laevis,* living in shallow water over sandy bottoms in the NE Atlantic and Mediterranean. It is a freckled sandy brown, and grows to 60 cm/2 ft.

brine common name for a solution of sodium chloride (NaCl) in water.

Brines are used extensively in the food-manufacturing industry for canning vegetables, pickling vegetables (sauerkraut manufacture), and curing meat. Industrially, brine is the source from which chlorine, caustic soda (sodium hydroxide), and sodium carbonate are made.

Brinell hardness test test of the hardness of a substance according to the area of indentation made by a 10 mm/0.4 in hardened steel or sintered tungsten carbide ball under standard loading conditions in a test machine. The resulting Brinell number is equal to the load (kg) divided by the surface area (mm^2) and is named after its inventor, Swedish metallurgist Johann Brinell.

brisling processed form of sprat *Sprattus sprattus* a small herring, fished in Norwegian fjords, then seasoned and canned.

bristlecone pine the oldest living species of ◊pine tree.

bristletail primitive wingless insect of the order Thysanura. Up to 2 cm/0.8 in long, bristletails have a body tapering from front to back, two long antennae, and three 'tails' at the rear end. They include the **silverfish** *Lepisma saccharina* and the **firebrat** *Thermobia domestica*. Two-tailed bristletails constitute another insect order, the Diplura. They live under stones and fallen branches, feeding on decaying material.

bristle-worm or ***polychaete*** segmented worm of the class Polychaeta, characterized by having a pair of fleshy paddles (parapodia) on each segment, together with prominent bristles (setae), and a well-developed head with a pair each of eyes, antennae, and palps. Most bristle-worms are marine, and live in burrows in sand and mud or in rock crevices and under stones. More than 5,300 species are recognized.

The bristle-worms can be conveniently divided into two loosely defined groups: the **Errantia**, which are free-living (that is swimming, crawling, or actively burrowing), and the **Sedentaria**, which live within a tube or a permanent burrow.

British Blue breed of domestic shorthaired cat. It has a fuller face than the ◊Russian Blue, a more compact body, and shorter legs. The coat is a solid blue-grey colour and the eyes should be either orange or copper.

British Standards Institute (BSI) UK national standards body. Although government funded, the institute is independent. The BSI interprets international technical standards for the UK, and also sets its own.

For consumer goods, it sets standards which products should reach (the BS standard), as well as testing products to see that they conform to that standard (as a result of which the product may be given the BSI 'kite' mark).

British Telecom (BT) British ◊telecommunications company. Its principal activity is the supply of local, long-distance, and international telecommunications services and equipment in the UK, serving 27 million exchange lines. BT also offers an international direct-dialled telephone service to more than 200 countries and other overseas territories – covering 99% of the world's 800 million telephones. One of the world's leading providers of telecommunications services and one of the largest private-sector companies in Europe, in 1997 BT had a market capitalization in excess of £28 billion and had established operations in more than 30 countries worldwide, with joint ventures in Spain, Germany, Italy, the Netherlands, Sweden, South Africa, New Zealand, Japan, and India.

BT formed part of the Post Office until 1980, and was privatized in 1984. Previously a monopoly, it now faces commercial competition for some of its services. BT is not allowed to offer other cable services apart from telephones.

British thermal unit imperial unit (symbol Btu) of heat, now replaced in the SI system by the ◊joule (one British thermal unit is approximately 1,055 joules). Burning one cubic foot of natural gas releases about 1,000 Btu of heat.

One British thermal unit is defined as the amount of heat required to raise the temperature of 0.45 kg/1 lb of water by 1°F. The exact value depends on the original temperature of the water.

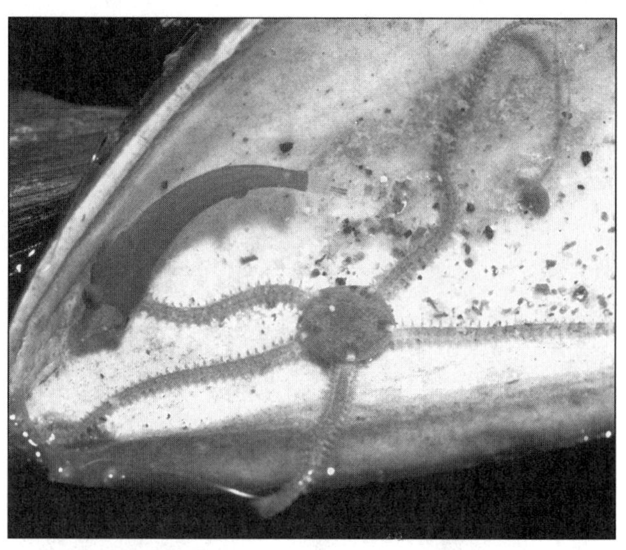

brittle-star *Living mostly on soft seabeds, the brittle-star* Ophiothrix fragilis *feeds on small crustaceans and molluscs. Brittle-stars are distinguished from starfish by the more or less clear demarcation between the five arms and the body. Both starfish and brittle-stars have the ability to regrow their arms if they are broken off. Premaphotos Wildlife*

brittle material material that breaks suddenly under stress at a point just beyond its elastic limit (see ◊elasticity). Brittle materials may also break suddenly when given a sharp knock. Pottery, glass, and cast iron are examples of brittle materials. Compare ◊ductile material.

brittle-star any member of the echinoderm class Ophiuroidea. A brittle-star resembles a starfish, and has a small, central, rounded body and long, flexible, spiny arms used for walking. The small brittle-star *Amphipholis squamata* is greyish, about 4.5 cm/2 in across, and found on sea bottoms worldwide. It broods its young, and its arms can be luminous.

About 2,000 species of brittle-stars and basket-stars, whose arms are tangled and rootlike, are included in this group.

broadband in computing, term indicating a high ◊bandwidth.

broadbill primitive perching bird of the family Eurylaimidae, found in Africa and S Asia. Broadbills are forest birds and are often found near water. They are gregarious and noisy, have brilliant coloration and wide bills, and feed largely on insects.

broadcasting the transmission of sound and vision programmes by ◊radio and ◊television. Broadcasting may be organized under private enterprise, as in the USA, or may operate under a compromise system, as in Britain, where a television and radio service controlled by the state-regulated British Broadcasting Corporation (BBC) operates alongside the commercial ◊Independent Television Commission (known as the Independent Broadcasting Authority before 1991).

In the USA, broadcasting is limited only by the issue of licences from the Federal Communications Commission to competing commercial companies; in Britain, the BBC is a centralized body appointed by the state and responsible to Parliament, but with policy and programme content not controlled by the state; in Japan, which ranks next to the USA in the number of television sets owned, there is a semigovernmental radio and television broadcasting corporation (NHK) and numerous private television companies.

Television broadcasting entered a new era with the introduction of high-powered communications satellites in the 1980s. The signals broadcast by these satellites are sufficiently strong to be picked up by a small dish aerial located, for example, on the roof of a house. Direct broadcast by satellite thus became a feasible alternative to land-based television services. See also ◊cable television. A similar revolution will take place when digital television becomes widely available.

broad-leaved tree another name for a tree belonging to the ◊angiosperms, such as ash, beech, oak, maple, or birch. The leaves are generally broad and flat, in contrast to the needlelike leaves of most ◊conifers. See also ◊deciduous tree.

broccoli variety of ◊cabbage. It contains high levels of the glucosinolate compound glucoraphanin. A breakdown product of this was found to neutralize damage to cells and so help to prevent cancer.

brocket name for a male European red deer in its second year, when it has short, straight, pointed antlers. Brocket deer, genus *Mazama,* include a number of species of small, shy, solitary deer found in Central and South America. They are up to 1.3 m/4 ft in body length and 65 cm/2 ft at the shoulder, and have similar small, straight antlers even when adult.

brolga or ***native companion***, Australian crane *Grus rubicunda,* about 1.5 m/5 ft tall, mainly grey with a red patch on the head.

brome grass any of several annual grasses found in temperate regions; some are used as food for horses and cattle, but many are weeds. (Genus *Bromus,* family Gramineae.)

bromeliad any tropical or subtropical plant belonging to the pineapple family, usually with stiff leathery leaves, which are often coloured and patterned, and bright, attractive flower spikes. There are about 1,400 species in tropical America; several are cultivated as greenhouse plants. (Family Bromeliaceae.)

Some grow in habitats ranging from scrub desert to tropical forest floor. Many, however, grow on rainforest trees. These are epiphytes: they are supported by the tree but do not take nourishment from it, using rain and decayed plant and animal remains for independent sustenance.

bromeliad *The large* Bromelia balansae, *shown here, is one of numerous terrestrial bromeliads present on the sand dunes of the Brazilian coast near Rio; it is also widespread in dry areas inland. The flower spikes are regularly visited by hummingbirds, which are the main pollinators of many members of the family. Premaphotos Wildlife*

bromide salt of the halide series containing the Br^- ion, which is formed when a bromine atom gains an electron.

The term 'bromide' is sometimes used to describe an organic compound containing a bromine atom, even though it is not ionic. Modern naming uses the term 'bromo-' in such cases. For example, the compound C_2H_5Br is now called bromoethane; its traditional name, still used sometimes, is ethyl bromide.

bromine ***Greek bromos 'stench'*** dark, reddish-brown, nonmetallic element, a volatile liquid at room temperature, symbol Br, atomic number 35, relative atomic mass 79.904. It is a member of the ◊halogen group, has an unpleasant odour, and is very irritating to mucous membranes. Its salts are known as bromides.

Bromine was formerly extracted from salt beds but is now mostly obtained from sea water, where it occurs in small quantities. Its compounds are used in photography and in the chemical and pharmaceutical industries.

bronchiole small-bore air tube found in the vertebrate lung responsible for delivering air to the main respiratory surfaces. Bronchioles lead off from the larger bronchus and branch extensively before terminating in the many thousand alveoli that form the bulk of lung tissue.

bronchitis inflammation of the bronchi (air passages) of the lungs, usually caused initially by a viral infection, such as a cold or flu. It is aggravated by environmental pollutants, especially smoking, and results in a persistent cough, irritated mucus-secreting glands, and large amounts of sputum.

bronchodilator drug that relieves obstruction of the airways by causing the bronchi and bronchioles to relax and widen. It is most useful in the treatment of ◊asthma.

bronchus one of a pair of large tubes (bronchi) branching off from the windpipe and passing into the vertebrate lung. Apart from their size, bronchi differ from the bronchioles in possessing cartilaginous rings, which give rigidity and prevent collapse during breathing movements.

Numerous glands in the wall of the bronchus secrete a slimy mucus, which traps dust and other particles; the mucus is constantly being propelled upwards to the mouth by thousands of tiny hairs or cilia. The bronchus is adversely effected by several respiratory diseases and by smoking, which damages the cilia and therefore the lung-cleansing mechanism.

brontosaurus former name of a type of large, plant-eating dinosaur, now better known as ◊apatosaurus.

bronze alloy of copper and tin, yellow or brown in colour. It is harder than pure copper, more suitable for ◊casting, and also resists ◊corrosion. Bronze may contain as much as 25% tin, together with small amounts of other metals, mainly lead.

Bronze is one of the first metallic alloys known and used widely by early peoples during the period of history known as the ◊Bronze Age. The first bronze objects date from 3000 BC.

Bell metal, the bronze used for casting bells, contains 15% or more tin. **Phosphor bronze** is hardened by the addition of a small percentage of phosphorus. **Silicon bronze** (for telegraph wires) and **aluminium bronze** are similar alloys of copper with silicon or aluminium and small amounts of iron, nickel, or manganese, but usually no tin.

Bronze Age stage of prehistory and early history when copper and bronze (an alloy of tin and copper) became the first metals worked extensively and used for tools and weapons. One of the classifications of the Danish archaeologist Christian Thomsen's Three Age System, it developed out of the Stone Age and generally preceded the Iron Age. It first began in the Far East and may be dated 5000–1200 BC in the Middle East and about 2000–500 BC in Europe.

Mining and metalworking were the first specialized industries, and the invention of the wheel during this time revolutionized transport.

Agricultural productivity (which began during the New Stone Age, or Neolithic period, about 6000 BC) was transformed by the ox-drawn plough, increasing the size of the population that could be supported by farming.

In some areas, including most of Africa, there was no Bronze Age, and ironworking was introduced directly into the Stone Age economy.

broom any of a group of shrubs (especially species of *Cytisus* and *Spartium*), often cultivated for their bright yellow flowers. (Family Leguminosae.)

brouter device for connecting computer networks that incorporates the facilities of both a ◊bridge and a ◊router. Brouters usually offer routing over a limited number of ◊protocols, operating by routing where possible and bridging the remaining protocols.

Brown, Robert (1773–1858)

Scottish botanist who in 1827 discovered Brownian motion. As a botanist, his more lasting work was in the field of plant morphology. He was the first to establish the real basis for the distinction between gymnosperms (pines) and angiosperms (flowering plants).

On an expedition to Australia in 1801–05 Brown collected 4,000 plant species and later classified them using the 'natural' system of Bernard de Jussieu (1699–1777) rather than relying upon the system of Carolus Linnaeus.

Mary Evans Picture Library

brown bear species of ◊bear found in N Europe.

brown dwarf in astronomy, an object less massive than a star, but heavier than a planet. Brown dwarfs do not have enough mass to ignite nuclear reactions at their centres, but shine by heat released during their contraction from a gas cloud. Some astronomers believe that vast numbers of brown dwarfs exist throughout the Galaxy. Because of the difficulty of detection, none were spotted until 1995, when US astronomers discovered a brown dwarf, GI229B, in the constellation Lepus. It is about 20–40 times as massive as Jupiter but emits only 1% of the radiation of the smallest known star. In 1996 UK astronomers discovered four possible brown dwarfs within 150 light years of the Sun.

Brownian movement the continuous random motion of particles in a fluid medium (gas or liquid) as they are subjected to impact from the molecules of the medium. The phenomenon was explained by German physicist Albert Einstein in 1905 but was probably observed as long ago as 1827 by the Scottish botanist Robert Brown. It provides evidence for the ◊kinetic theory of matter.

BROWNIAN MOTION

`http://dbhs.wvusd.k12.ca.us/Chem-History/Brown-1829.html`

Transcript of 'Remarks on Active Molecules' by Robert Brown from *Additional Remarks on Active Molecules* (1829). The text describes Robert Brown's observations of the random motion of particles, which in turn led to the placement of the subject on millions of school exam papers.

brown ring test in analytical chemistry, a test for the presence of ◊nitrates.

To an aqueous solution containing the test substance is added iron(II) sulphate. Concentrated sulphuric acid is then carefully poured down the inside wall of the test tube so that it forms a distinct layer at the bottom. The formation of a brown colour at the boundary between the two layers indicates the presence of nitrate.

browse to explore a computer system or network for particular files or information. To browse in Windows is to search for a particular file to open or run. On the World Wide Web, browsing is the activity of moving from site to site to view information. This is sometimes also called 'surfing'.

browser in computing, any program that allows the user to search for and view data. Browsers are usually limited to a particular type of data, so, for example, a graphics browser will display graphics files stored in many different file formats. Browsers usually do not permit the user to edit data, but are sometimes able to convert data from one file format to another.

Web browsers allow access to the ◊World Wide Web. Netscape Navigator and Microsoft's Internet Explorer were the leading Web browsers in 1996–97. They act as a graphical interface to information available on the Internet – they read ◊HTML (hypertext markup language) documents and display them as graphical documents which may include images, video, sound, and ◊hypertext links to other documents.

The first widespread browser for personal computers (PCs) was the text-based program Lynx, which is still used via ◊gateways from text-based online systems such as Delphi and CIX. Browsers using ◊graphical user interfaces became widely available from 1993 with the release of ◊Mosaic, written by Marc Andreessen and Eric Bina. For some specialist applications such as viewing the virtual reality sites beginning to appear on the Web, a special virtual reality modelling language (◊VRML) browser is needed.

brucellosis disease of cattle, goats, and pigs, also known when transmitted to humans as **undulant fever** since it remains in the body and recurs. It was named after Australian doctor David Bruce (1855–1931), and is caused by bacteria (genus *Brucella*). It is transmitted by contact with an infected animal or by drinking contaminated milk.

Brunel, Isambard Kingdom (1806–1859)

British engineer and inventor. In 1833 he became engineer to the Great Western Railway, which adopted the 2.1-m/7-ft gauge on his advice. He built the Clifton Suspension Bridge over the river Avon at Bristol and the Saltash Bridge over the river Tamar near Plymouth. His shipbuilding designs include the *Great Western* 1837, the first steamship to cross the Atlantic regularly; the *Great Britain* 1843, the first large iron ship to have a screw propeller; and the *Great Eastern* 1858, which laid the first transatlantic telegraph cable.

Mary Evans Picture Library

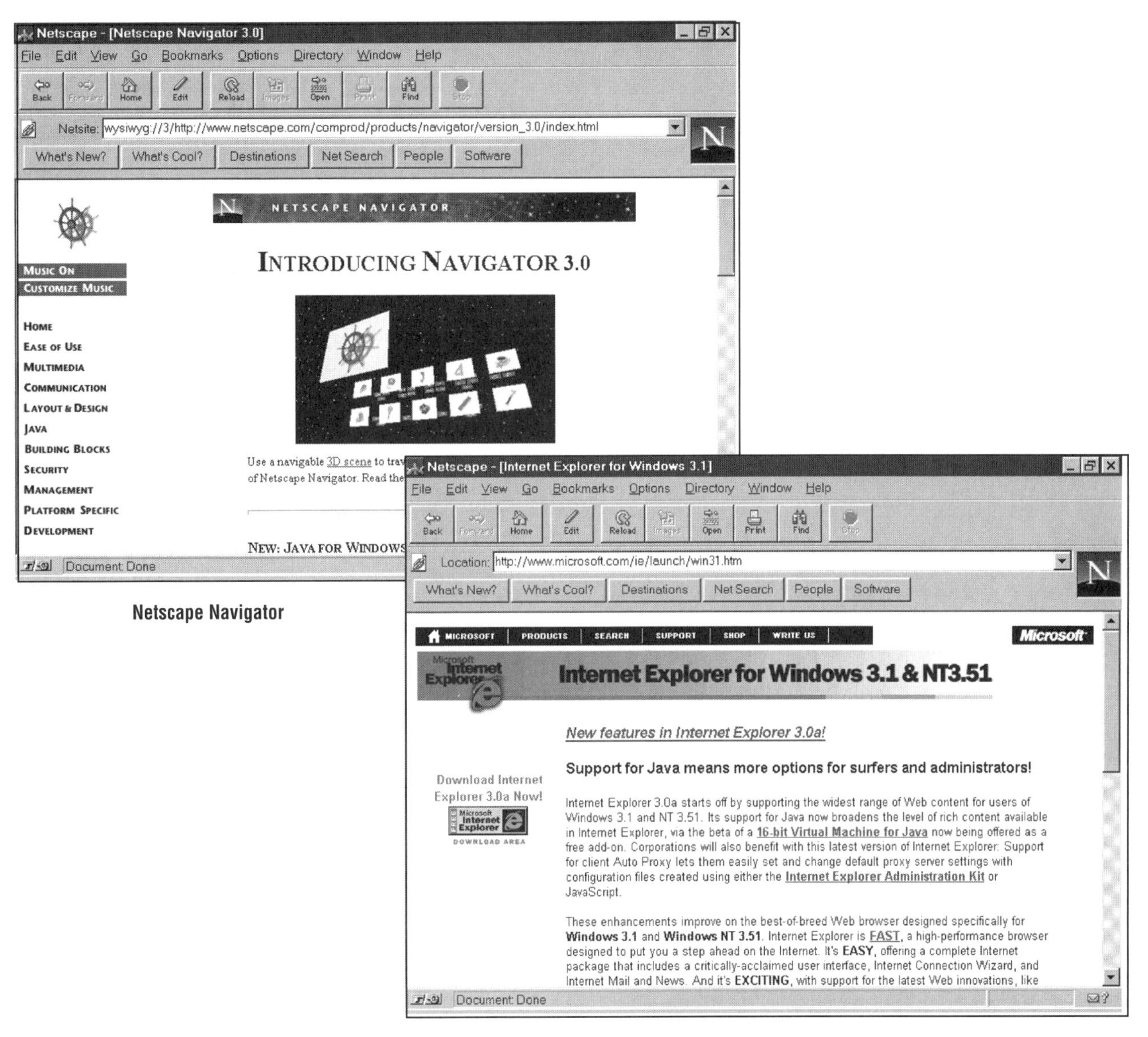

Netscape Navigator

Microsoft Internet Explorer

browser *Two popular World Wide Web browsers, Netscape Navigator and Microsoft Internet Explorer, which provide the user with a straightforward method of accessing information available online.*

It has largely been eradicated in the West through vaccination of livestock and pasteurization of milk. Brucellosis is especially prevalent in the Mediterranean and in Central and South America. It can be treated with antibacterial drugs.

brush in certain electric motors, one of a pair of contacts that pass electric current into and out of the rotating coils by means of a device known as a ◊commutator. The brushes, which are often replaceable, are usually made of a soft, carbon material to reduce wear of the copper contacts on the rotating commutator.

Brussels sprout one of the small edible buds along the stem of a variety of ◊cabbage. (*Brassica oleracea* var. *gemmifera.*) They are high in the glucosinolate compound sinigrin. Sinigrin was found to destroy precancerous cells in laboratory rats in 1996.

bryony either of two climbing hedgerow plants found in Britain: **white bryony** (*Bryonia dioca*) belonging to the gourd family (Cucurbitaceae), and **black bryony** (*Tamus communis*) of the yam family (Dioscoreaceae).

bryophyte member of the Bryophyta, a division of the plant kingdom containing three classes: the Hepaticae (◊liverwort), Musci (◊moss), and Anthocerotae (◊hornwort). Bryophytes are generally small, low-growing, terrestrial plants with no vascular (water-conducting) system as in higher plants. Their life cycle shows a marked ◊alternation of generations. Bryophytes chiefly occur in damp habitats and require water for the dispersal of the male gametes (◊antherozoids).

In bryophytes, the ◊sporophyte, consisting only of a spore-bearing capsule on a slender stalk, is wholly or partially dependent on the ◊gametophyte for water and nutrients. In some liverworts the plant body is a simple ◊thallus, but in the majority of bryophytes it is differentiated into stem, leaves, and ◊rhizoids.

BSI abbreviation for ◊British Standards Institute.

BST abbreviation for ***British Summer Time*** and ◊bovine somatotropin.

To remember when we lose an hour, and when we gain an hour:

SPRING FORWARD, FALL BACK

OR

REMEMBER THAT CLOCKS GO FORWARD AN HOUR IN SPRING, SINCE THEY ARE EAGER FOR THE SUMMER; CLOCKS GO BACK AN HOUR IN AUTUMN, SINCE THEY ARE SHYING AWAY FROM THE BITTER CHILL OF WINTER

BT abbreviation for ◊British Telecom.

Btu symbol for ◊British thermal unit.

bubble chamber in physics, a device for observing the nature and movement of atomic particles, and their interaction with radiation. It is a vessel filled with a superheated liquid through which ionizing particles move and collide. The paths of these particles are shown by strings of bubbles, which can be photographed and studied. By using a pressurized liquid medium instead of a gas, it overcomes drawbacks inherent in the earlier ◊cloud chamber. It was invented by US physicist Donald Glaser in 1952. See ◊particle detector.

bubble-jet printer in computing, an ◊ink-jet printer in which the ink is heated to boiling point so that it forms a bubble at the end of a nozzle. When the bubble bursts, the ink is transferred to the paper.

bubble memory in computing, a memory device based on the creation of small 'bubbles' on a magnetic surface. Bubble memories typically store up to 4 megabits (4 million ◊bits) of information. They are not sensitive to shock and vibration, unlike other memory devices such as disc drives, yet, like magnetic discs, they are nonvolatile and do not lose their information when the computer is switched off.

bubbleshell marine mollusc with a tiny shell. It burrows in the sand and feeds on animal matter.

bubble sort in computing, a technique for ◊sorting data. Adjacent items are continually exchanged until the data are in sequence.

bubonic plague epidemic disease of the Middle Ages; see ◊plague and ◊Black Death.

buckminsterfullerene form of carbon, made up of molecules (buckyballs) consisting of 60 carbon atoms arranged in 12 pentagons and 20 hexagons to form a perfect sphere. It was named after the US architect and engineer Richard Buckminster Fuller because of its structural similarity to the geodesic dome that he designed. See ◊fullerene.

buckthorn any of several thorny shrubs. The buckthorn (*Rhamnus catharticus*) is native to Britain, but is also found throughout Europe, W Asia, and N Africa. Its berries were formerly used in medicine as a purgative, to clean out the bowels. (Genus *Rhamnus*, family Rhamnaceae.)

buckwheat any of a group of cereal plants. The name usually refers to *Fagopyrum esculentum*, which reaches about 1 m/3 ft in height and can grow on poor soil in a short summer. The highly nutritious black triangular seeds (groats) are eaten by both animals and humans. They can be cooked and eaten whole or as a cracked meal (kasha), or ground into flour, often made into pancakes. (Genus *Fagopyrum*, family Polygonaceae.)

buckyballs popular name for molecules of ◊buckminsterfullerene.

bud undeveloped shoot usually enclosed by protective scales; inside is a very short stem and numerous undeveloped leaves, or flower parts, or both. Terminal buds are found at the tips of shoots, while axillary buds develop in the ◊axils of the leaves, often remaining dormant unless the terminal bud is removed or damaged. Adventitious buds may be produced anywhere on the plant, their formation sometimes stimulated by an injury, such as that caused by pruning.

budding type of ◊asexual reproduction in which an outgrowth develops from a cell to form a new individual. Most yeasts reproduce in this way.

In a suitable environment, yeasts grow rapidly, forming long chains of cells as the buds themselves produce further buds before being separated from the parent. Simple invertebrates, such as ◊hydra, can also reproduce by budding.

In horticulture, the term is used for a technique of plant propagation whereby a bud (or scion) and a sliver of bark from one plant are transferred to an incision made in the bark of another plant (the stock). This method of ◊grafting is often used for roses.

buddleia any of a group of ornamental shrubs or trees with spikes of fragrant flowers. The purple or white flower heads of the butterfly bush (*Buddleia davidii*) attract large numbers of butterflies. (Genus *Buddleia*, family Buddleiaceae.)

budgerigar small Australian parakeet *Melopsittacus undulatus* of the parrot family, Psittacidae, order Psittaciformes, that feeds mainly on grass seeds. Normally it is bright green, but yellow, white, blue, and mauve varieties have been bred for the pet market. It breeds freely in captivity.

buffalo either of two species of wild cattle. The Asiatic water buffalo *Bubalis bubalis* is found domesticated throughout S Asia and wild in parts of India and Nepal. It likes moist conditions. Usually grey or black, up to 1.8 m/6 ft high, both sexes carry large horns. The African buffalo *Syncerus caffer* is found in Africa, south of the Sahara, where there is grass, water, and cover in which to retreat. There are a number of subspecies, the biggest up to 1.6 m/5 ft high, and black, with massive horns set close together over the head. The name is also commonly applied to the American ◊bison.

buffer in chemistry, mixture of compounds chosen to maintain a steady ◊pH. The commonest buffers consist of a mixture of a weak organic acid and one of its salts or a mixture of acid salts of phosphoric acid. The addition of either an acid or a base causes a shift in the ◊chemical equilibrium, thus keeping the pH constant.

buffer in computing, a part of the ◊memory used to store data temporarily while it is waiting to be used. For example, a program might store data in a printer buffer until the printer is ready to print it.

bug in computing, an ◊error in a program. It can be an error in the logical structure of a program or a syntax error, such as a spelling mistake. Some bugs cause a program to fail immediately; others remain dormant, causing problems only when a particular combination of events occurs. The process of finding and removing errors from a program is called **debugging**.

bug in entomology, an insect belonging to the order Hemiptera. All these have two pairs of wings with forewings partly thickened.

They also have piercing mouthparts adapted for sucking the juices of plants or animals, the 'beak' being tucked under the body when not in use.

They include: the bedbug, which sucks human blood; the shieldbug, or stinkbug, which has a strong odour and feeds on plants; and the water boatman and other water bugs.

bugle any of a group of low-growing plants belonging to the mint family, with spikes of white, pink, or blue flowers. The leaves may be smooth-edged or slightly toothed, the lower ones with a long stalk. They are often grown as ground cover. (Genus *Ajuga*, family Labiatae.)

bugloss any of several plants native to Europe and Asia, distinguished by their rough, bristly leaves and small blue flowers. (Genera *Anchusa*, *Lycopsis*, and *Echium*, family Boraginaceae.)

bulb underground bud with fleshy leaves containing a reserve food supply and with roots growing from its base. Bulbs function in vegetative reproduction and are characteristic of many monocotyledonous plants such as the daffodil, snowdrop, and onion. Bulbs are grown on a commercial scale in temperate countries, such as England and the Netherlands.

Millennium Bug: Preparing Computers for the Year 2000

BY SCOTT KIRSNER

Digital Disarray

It sounds like a bad riddle: how will two missing digits create a $600 billion industry when the calendar flips from 1999 to 2000?

Unfortunately, it's not a riddle; it's reality. As the 20th century draws to a close, an expensive computer problem dubbed 'the millennium bug' has emerged. In brief, most computers handle dates using a two-digit shorthand: 98 instead of the four-digit 1998. When presented with a date like 00, they become hopelessly confused. Since they're missing the two digits that indicate what millennium and century the date is in (19 or 20), computers tend to assume that 00 is actually 1900. So they'll either begin making errors of calculation or they'll stop working altogether. Reprogramming them to be capable of comprehending dates in the new millennium is expected to cost as much as $600 billion worldwide.

The origins of the problem are simple. First, the programmers who wrote software in the 1960s for the first generation of commercial mainframe computers were shortsighted. They didn't imagine that the programs they were creating – or the machines they were creating them for – would still be in service in the far-off year of 2000. So they conserved the computers' memory by using a two-digit shorthand for the year. Every byte of memory was precious in those days, and lopping off 19 from dates was an obvious way to save a few bytes here and there.

Banks were among the first institutions to notice the downside to that approach. When they began writing long-term mortgages and approving loans that lasted past 1999, they were forced to confront the millennium bug. But the problem received little widespread attention until the mid-1990s. Technology consultants began writing articles and giving speeches about 'the year 2000 problem'. Business executives began to take notice, and even consumers couldn't ignore the problem when credit card issuers and driver's licence organizations began to renew cards and licences for shorter periods of time, because their systems couldn't handle an expiration date past 1999.

How might businesses and consumers be affected by the millennium bug? Computers, both new and old, in every industry are vulnerable. They might shut down as a result of being asked to process the date 00. Some say that is the best-case scenario, because at least businesses will know something is wrong. Worse would be if computers continued to operate, making numerous date-related errors that would be difficult to identify and fix.

The areas of greatest concern are defence, health care, transportation, telecommunications, financial services, and national and local governments. Technology experts warn of the hazards of air travel if the Federal Aviation Administration's computers can't manage data properly, the danger of hospital stays if the computers that monitor patients go awry, and the possibility of social unrest if the federal government can't provide services in 2000.

Date Expansion or Windowing?

Fixing the millennium bug is a labour-intensive endeavour. An organization can opt to replace its systems entirely with new ones that can function in the 21st century, or it may pursue one of two basic repair strategies – 'date expansion' or 'windowing'.

Date expansion involves changing the two-digit dates to four. That entails converting all of the data an organization has stored from one format to the other, and reprogramming systems to handle four-digit dates like 2001.

Windowing is considered a simpler, less expensive solution, but it's only a temporary patch. Rather than converting all of a company's data, the windowing approach merely adds logic to a program to help it determine whether a two-digit date belongs in the 20th century or the 21st. Programmers might create a 'window' of time – from 00 to 30, for example – and then instruct the computer to assume that those dates should all be preceded by 20, whereas dates between 31 and 99 should be preceded by 19. But when 2031 rolls around, that hypothetical company would have a new problem on its hands. Windowing assumes that an organization will either replace its older systems before the window of time closes, or reprogram them yet again.

Eradicating the millennium bug is a multi-stage process. An organization must first assess which of its systems will be unable to handle dates in the 21st century. Then, it must convert those systems, either through expansion or windowing. Finally, it has to test the systems to ensure that they will work after the clock ticks past midnight on 31 December, 1999.

Ripple Effects

But even if companies successfully repair their own systems, they're still vulnerable to what has been dubbed 'the ripple effect'. One of their suppliers or customers, or a government regulator, could send them unconverted data and contaminate their systems. Or even worse, a key supplier might be unable to provide services or raw materials as a result of the bug, hamstringing its customers. For those reasons, organizations must make sure that everyone else with whom they do business is solving their own year 2000 problems. Certain sectors of the economy, like the financial services arena, are even coordinating massive, interorganizational tests to make sure that stock exchanges, banks, regulators, and clearing houses will be able to work together in the new millennium.

And waiting in the wings are the lawyers. If software or hardware fails, they'll be scrutinizing contracts to see who is liable. If a conversion project turns out to have been defective, they may bring litigation against the service provider that was contracted to perform the fix. And if a company's stock takes a dive as a result of year 2000-related failures, lawyers may file negligence lawsuits against the Board of Directors. Once litigation and damages are figured into the cost of the millennium bug, some analysts believe the total worldwide cost could skyrocket to as much as $3.6 trillion.

The sudden emergence of the year 2000 problem has created an entire mini-economy. Programmers and technology managers are finding that they can demand and receive higher salaries, computer consultants have more work than they can handle, and software companies have begun to market tools aimed at making assessment, conversion, and testing more efficient. There are dozens of Web sites and books devoted to the problem. The American Stock Exchange has even created an options index that enables investors to speculate on the fortunes of 18 companies selling software or services intended to solve the year 2000 problem.

Few participants in this mini-economy are willing to speculate about the extent to which the world will be affected by the millennium bug. Will 1 January, 2000, arrive without a hitch, or will, as some technology experts predict, the front pages of every major newspaper be filled with stories about date-related computer crises? All that's certain is that programmers and their technology managers won't be among the celebrants on New Year's Eve, 1999; they'll be huddled over their mainframes, fingers crossed.

bulbil small bulb that develops above ground from a bud. Bulbils may be formed on the stem from axillary buds, as in members of the saxifrage family, or in the place of flowers, as seen in many species of onion *Allium*. They drop off the parent plant and develop into new individuals, providing a means of ◊vegetative reproduction and dispersal.

bulbul fruit-eating bird of the family Pycnonotidae, order Passeriformes, that ranges in size from that of a sparrow to a blackbird. They are mostly rather dull coloured and very secretive, living in dense forests. They are widely distributed throughout Africa and Asia; there are about 120 species.

bulimia ***Greek 'ox hunger'*** eating disorder in which large amounts of food are consumed in a short time ('binge'), usually followed by depression and self-criticism. The term is often used for **bulimia nervosa**, an emotional disorder in which eating is followed by deliberate vomiting and purging. This may be a chronic stage in ◊anorexia nervosa.

Bulinus large genus of tiny freshwater and land snails, comprising over 1,000 species. They have external shells, and are related to the hedge- and grass-snails.

Certain species of this snail act as the intermediate host for the microscopic worm that causes ◊bilharzia, one of the most common parasitic diseases.
classification *Bulinus* is in class Gastropoda of phylum Mollusca.

bulldog British breed of dog of ancient but uncertain origin, formerly bred for bull-baiting. The head is broad and square, with deeply wrinkled cheeks, small folded ears, very short muzzle, and massive jaws, the peculiar set of the lower jaw making it difficult for the dog to release its grip. Thickset in build, the bulldog grows to about 45 cm/18 in and has a smooth beige, tawny, or brindle coat. The French bulldog is much lighter in build and has large upright ears.

bulldozer earth-moving machine widely used in construction work for clearing rocks and tree stumps and levelling a site. The bulldozer is a kind of tractor with a powerful engine and a curved, shovel-like blade at the front, which can be lifted and forced down by hydraulic rams. It usually has ◊caterpillar tracks so that it can move easily over rough ground.

bulletin board in computing, a centre for the electronic storage of messages, usually accessed over the telephone network via a ◊modem but also sometimes accessed via ◊Telnet across the Internet. Bulletin board systems (often abbreviated to BBSs) are usually dedicated to specific interest groups, and may carry public and private messages, notices, and programs.

bullfinch Eurasian finch with a thick head and neck, and short heavy bill, genus *Pyrrhula pyrrhula,* family Fringillidae, order Passeriformes. It is small and blue-grey or black in colour, the males being reddish and the females brown on the breast. Bullfinches are 15 cm/6 in long, and usually seen in pairs. They feed on tree buds as well as seeds and berries, and are usually seen in woodland. They also live in the Aleutians and on the Alaska mainland.

bullfrog North American species of ◊frog.

bullhead or ***miller's thumb*** small fish *Cottus gobio* found in fresh water in the northern hemisphere, often under stones. It has a large head, a spine on the gill cover, and grows to 10 cm/4 in.

Related bullheads, such as the **father lasher** *Myxocephalus scorpius,* live in coastal waters. They are up to 30 cm/1 ft long. The male guards the eggs and fans them with his tail.

bull terrier breed of dog, originating as a cross between a terrier and a bulldog. Very powerfully built, it grows to about 40 cm/16 in tall, and has a short, usually white, coat, narrow eyes, and distinctive egg-shaped head. It was formerly used in bull-baiting. Pit bull terriers are used in illegal dog fights. The ◊Staffordshire bull terrier is a distinct breed.

bulrush either of two plants: the great reed mace or cat's tail (*Typha latifolia*) with velvety chocolate-brown spikes of tightly packed flowers reaching up to 15 cm/6 in long; and a type of sedge (*Scirpus lacustris*) with tufts of reddish-brown flowers at the top of a rounded, rushlike stem.

bumblebee any large ◊bee, 2–5 cm/1–2 in, usually dark-coloured but banded with yellow, orange, or white, belonging to the genus *Bombus.*

Most species live in small colonies, usually underground, often in an old mousehole. The queen lays her eggs in a hollow nest of moss or grass at the beginning of the season. The larvae are fed on pollen and honey, and develop into workers. All the bees die at the end of the season except fertilized females, which hibernate and produce fresh colonies in the spring. Bumblebees are found naturally all over the world, with the exception of Australia, where they have been introduced to facilitate the pollination of some cultivated varieties of clover.

bundling computer industry practice of selling different, often unrelated, products in a single package. Bundles may consist of hardware or software or both; for example, a modem or a selection of software may be bundled with a personal computer to make the purchase of the computer seem more attractive.

Bunsen burner gas burner used in laboratories, consisting of a vertical metal tube through which a fine jet of fuel gas is directed. Air is drawn in through airholes near the base of the tube and the mixture is ignited and burns at the tube's upper opening.

The invention of the burner is attributed to German chemist Robert von Bunsen in 1855 but English chemist and physicist Michael Faraday is known to have produced a similar device at an earlier date. A later refinement was the metal collar that can be turned to close or partially close the airholes, thereby regulating the amount of air sucked in and hence the heat of the burner's flame.

bunting any of a number of sturdy, finchlike birds with short, thick bills, of the family Emberizidae, order Passeriformes, especially the genera *Passerim* and *Emberiza.* Most of these brightly coloured birds are native to the New World.

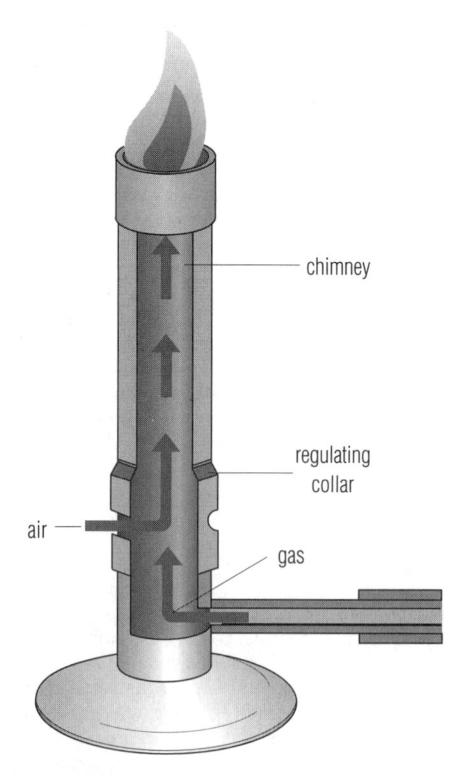

***Bunsen burner** The Bunsen burner, used for heating laboratory equipment and chemicals. The flame can reach temperatures of 1,500°C/2,732°F and is at its hottest when the collar is open.*

Bunsen, Robert Wilhelm (1811–1899)

German chemist credited with the invention of the Bunsen burner. His name is also given to the carbon–zinc electric cell, which he invented in 1841 for use in arc lamps. In 1860 he discovered two new elements, caesium and rubidium.

Mary Evans Picture Library

buoy floating object used to mark channels for shipping or warn of hazards to navigation. Buoys come in different shapes, such as a pole (spar buoy), cylinder (car buoy), and cone (nun buoy). Light buoys carry a small tower surmounted by a flashing lantern, and bell buoys house a bell, which rings as the buoy moves up and down with the waves. Mooring buoys are heavy and have a ring on top to which a ship can be tied.

buoyancy lifting effect of a fluid on a body wholly or partly immersed in it. This was studied by ◊Archimedes in the 3rd century BC.

bur or ***burr*** in botany, a type of 'false fruit' or ◊pseudocarp, surrounded by numerous hooks; for instance, that of burdock *Arctium*, where the hooks are formed from bracts surrounding the flowerhead. Burs catch in the feathers or fur of passing animals, and thus may be dispersed over considerable distances.

burbot or **eelpout** long, rounded fish *Lota lota* of the cod family, the only one living entirely in fresh water. Up to 1 m/3 ft long, it lives on the bottom of clear lakes and rivers, often in holes or under rocks, throughout Europe, Asia, and North America.

burdock any of several bushy herbs characterized by hairy leaves and ripe fruit enclosed in ◊burs with strong hooks. (Genus *Arctium*, family Compositae.)

burette in chemistry, a piece of apparatus, used in ◊titration, for the controlled delivery of measured variable quantities of a liquid.

It consists of a long, narrow, calibrated glass tube, with a tap at the bottom, leading to a narrow-bore exit.

Burmese cat breed of domestic shorthaired cat of ancient origin. The modern breed is descended for a cat introduced into the USA in 1930 from Burma and crossed with a Siamese. The original Burmese has a sable brown coat with lighter shading on the underside; the medium-length body is muscular and more rounded than a Siamese.

It was first recognized as a true breed in the USA in 1936 and in the UK in 1952. The standards vary for the Burmese in the USA and Britain; for instance, the British eye shape is specified as oval, while the American standard demands a rounder look. There are many varieties.

burn in medicine, destruction of body tissue by extremes of temperature, corrosive chemicals, electricity, or radiation. **First-degree burns** may cause reddening; **second-degree burns** cause blistering and irritation but usually heal spontaneously; **third-degree burns** are disfiguring and may be life-threatening.

Burns cause plasma, the fluid component of the blood, to leak from the blood vessels, and it is this loss of circulating fluid that engenders ◊shock. Emergency treatment is needed for third-degree burns in order to replace the fluid volume, prevent infection (a serious threat to the severely burned), and reduce the pain. Plastic, or reconstructive, surgery, including skin grafting, may be required to compensate for damaged tissue and minimize disfigurement. If a skin graft is necessary, dead tissue must be removed from a burn (a process known as debridement) so that the patient's blood supply can nourish the graft.

burnet herb belonging to the rose family, also known as **salad burnet**. It smells of cucumber and can be used in salads. The name is also used for other members of the genus. (*Sanguisorba minor*, family Rosaceae.)

burning common name for ◊combustion.

burying beetle another name for the ◊sexton beetle.

bus in computing, the electrical pathway through which a computer processor communicates with some of its parts and/or peripherals. Physically, a bus is a set of parallel tracks that can carry digital signals; it may take the form of copper tracks laid down on the computer's ◊printed circuit boards (PCBs), or of an external cable or connection.

A computer typically has three internal buses laid down on its main circuit board: a **data bus**, which carries data between the components of the computer; an **address bus**, which selects the route to be followed by any particular data item travelling along the data bus; and a **control bus**, which is used to decide whether data is written to or read from the data bus. An external **expansion bus** is used for linking the computer processor to peripheral devices, such as modems and printers.

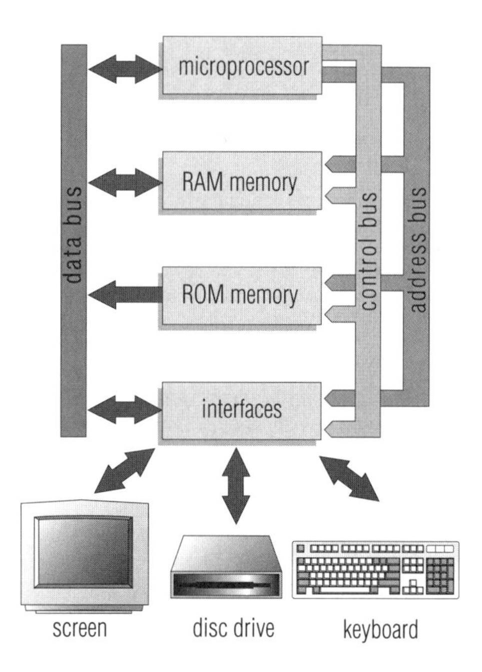

bus *The communication path used between the component parts of a computer.*

Bush, Vannevar (1890–1974)

US electrical engineer and scientist. During the 1920s and 1930s he developed several mechanical and mechanical–electrical analogue computers which were highly effective in the solution of differential equations. The standard electricity meter is based on one of his designs.

bushbaby small nocturnal African prosimian with long feet, long, bushy tail, and large ears. Bushbabies are active tree dwellers and feed on fruit, insects, eggs, and small birds.
classification Bushbabies are members of the loris family Lorisidae, order Primates.

bushbuck antelope *Tragelaphus scriptus* found over most of Africa south of the Sahara. Up to 1 m/3 ft high, the males have keeled horns twisted into spirals, and are brown to blackish. The females are generally hornless, lighter, and redder. All have white markings, including stripes or vertical rows of dots down the sides. Rarely far from water, bushbuck live in woods and thick brush.

bushel dry or liquid measure equal to eight gallons or four pecks (2,219.36 cu in/36.37 litres) in the UK; some US states have different standards according to the goods measured.

bushman's rabbit or ***riverine rabbit*** a wild rodent *Bunolagus monticularis* found in dense riverine bush in South Africa. It lives in small populations, and individuals are only seen very occasionally; it is now at extreme risk of extinction owing to loss of habitat to agriculture. Very little is known about its life or habits.

bushmaster large snake *Lachesis muta*. It is a type of pit viper, and is related to the rattlesnakes. Up to 4 m/12 ft long, it is found in wooded areas of South and Central America, and is the largest venomous snake in the New World. When alarmed, it produces a noise by vibrating its tail among dry leaves.

bush pig wild pig found in forested regions of Africa. Typically it is dark brown in colour, about 60 cm/24 in high and has short tusks no more than 15 cm/6 in long.

The **red river hog** is a West African variety of the same species but is rusty red in colour with prominent black and white markings on the face.
classification The bush pig *Potamochoerus porcus* is in family Suidae, order Artiodactyla.

bustard bird of the family Otididae, order Gruiformes, related to ◊cranes but with a rounder body, thicker neck, and a relatively short beak. Bustards are found on the ground on open plains and fields.

The great bustard *Otis tarda* is one of the heaviest flying birds at 18 kg/40 lb, and the larger males may have a length of 1 m/3 ft and wingspan of 2.3 m/7.5 ft. It is found in N Asia and Europe, although there are fewer than 30,000 great bustards left in Europe; two-thirds of these live on the Spanish steppes.

butadiene or ***buta-1,3-diene*** CH_2:CHCH:CH_2 inflammable gas derived from petroleum, used in making synthetic rubber and resins.

butane C_4H_{10} one of two gaseous alkanes (paraffin hydrocarbons) having the same formula but differing in structure. Normal butane is derived from natural gas; isobutane is a by-product of petroleum manufacture. Liquefied under pressure, it is used as a fuel for industrial and domestic purposes (for example, in portable cookers).

butcherbird another name for a ◊shrike.

butene C_4H_8 fourth member of the ◊alkene series of hydrocarbons. It is an unsaturated compound, containing one double bond.

buttercup any plant of the buttercup family with divided leaves and yellow flowers. (Genus *Ranunculus*, family Ranunculaceae.)

butterfly insect belonging, like moths, to the order Lepidoptera, in which the wings are covered with tiny scales, often brightly coloured. There are some 15,000 species of butterfly, many of which are under threat throughout the world because of the destruction of habitat.

Butterflies have a tubular proboscis through which they suck up nectar, or, in some species, carrion, dung, or urine. ◊Metamorphosis is complete; the pupa, or chrysalis, is usually without the protection of a cocoon. Adult lifespan may be only a few weeks, but some species hibernate and lay eggs in the spring.

The largest family, Nymphalidae, has some 6,000 species; it includes the peacock, tortoiseshells, and fritillaries. The family Pieridae includes the **cabbage white**, one of the few butterflies injurious to crops. The Lycaenidae are chiefly small, often with metallic coloration, for example the blues, coppers, and hairstreaks. The **large blue** *Lycaena arion* has a complex life history: it lays its eggs on wild thyme, and the caterpillars are then taken by Myrmica ants to their nests. The ants milk their honey glands, while the caterpillars feed on the ant larvae. In the spring, the caterpillars finally pupate and emerge as butterflies. The mainly tropical Papilionidae, or swallowtails, are large and very beautiful, especially the South American species. The world's largest butterfly is Queen Alexandra's birdwing *Ornithoptera alexandrae* of Papua New Guinea, with a body 7.5 cm/3 in long and a wingspan of 25 cm/10 in. The most spectacular migrant is the orange and black monarch butterfly *Danaus plexippus*, which may fly from N Canada to Mexico in the autumn.

Butterflies usually differ from moths in having the antennae club-shaped rather than plumed or feathery, no 'lock' between the fore- and hindwing, and resting with the wings in the vertical position rather than flat or sloping.

BUTTERFLY

The sensors on the feet of a red admiral butterfly are 200 times more sensitive to sugar than the human tongue.

butterfly fish any of several fishes, not all related. They include the freshwater butterfly fish *Pantodon buchholzi* of western Africa and the tropical marine butterfly fishes in family Chaetodontidae.

P. buchholzi can leap from the water and glide for a short distance on its large winglike pectoral fins. Up to 10 cm/4 in long, it lives in stagnant water, at the water surface during the day, lying beneath floating leaves. At night it hunts for insects, jumping out of the water to catch them.

The members of Chaetodontidae are brightly coloured with laterally flattened bodies, often with long snouts which they poke into crevices in rocks and coral when feeding. They have a flattened narrow body with one dorsal fin and a fairly long snout. Their bristlelike teeth are closely set in rows for feeding on small animals and green algae. The commonest colours are black, yellow, and brilliant metallic blues and greens.
classification Chaetodontidae is in order Perciformes; *Pantodon buchholzi* is in order Osteoglossiformes; both are in class Osteichthyes.

butterwort insectivorous plant belonging to the bladderwort family, with purplish flowers and a rosette of flat leaves covered with a sticky substance that traps insects. (Genus *Pinguicula*, family Lentibulariaceae.)

buzzard species of medium-sized hawk with broad wings, often seen soaring. Buzzards are in the falcon family, Falconidae, order Falconiformes. The **common buzzard** *Buteo buteo* of Europe and Asia is about 55 cm/1.8 ft long with a wingspan of over 1.2 m/4 ft. It preys on a variety of small animals up to the size of a rabbit.

The **rough-legged buzzard** *B. lagopus* lives in the northern tundra and eats lemmings. The **honey buzzard** *Pernis apivora* feeds largely, as its name suggests, on honey and insect larvae. It spends the summer in Europe and W Asia and winters in Africa. The **red-shouldered hawk** *B. lineatus* and **red-tailed hawk** *B. jamaicensis* occur in North America.

by-product substance formed incidentally during the manufacture of some other substance; for example, slag is a by-product of the production of iron in a ◊blast furnace. For industrial processes to be economical, by-products must be recycled or used in other ways as far as possible; in this example, slag is used for making roads.

Often, a poisonous by-product is removed by transforming it into another substance, which although less harmful is often still inconvenient. For example, the sulphur dioxide produced as a by-product of electricity generation can be removed from the smoke stack using ◊flue-gas desulphurization. This process produces large amounts of gypsum, some of which can be used in the building industry.

byssus tough protein fibres secreted by the foot of sessile (fixed) bivalves, such as ◊mussels, as a means of attachment to rocks.

The byssus of some rock creatures can be woven into fabrics. A delicate silk called byssus is made from the byssus of mussels found in the Mediterranean.

byte sufficient computer memory to store a single ◊character of data. The character is stored in the byte of memory as a pattern of ◊bits (binary digits), using a code such as ◊ASCII. A byte usually contains eight bits – for example, the capital letter F can be stored as the bit pattern 01000110.

A single byte can specify 256 values, such as the decimal numbers from 0 to 255; in the case of a single-byte ◊pixel (picture element), it can specify 256 different colours. Three bytes (24 bits) can specify 16,777,216 values. Computer memory size is measured in **kilobytes** (1,024 bytes) or **megabytes** (1,024 kilobytes).

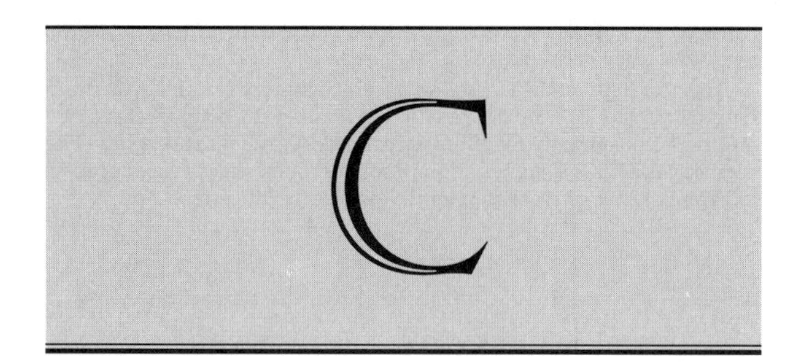

°C symbol for degrees ◊Celsius, sometimes called centigrade.

C++ in computing, a high-level programming language used in ◊object-oriented applications. It is derived from the language C.

C in computing, a high-level, general-purpose programming language popular on minicomputers and microcomputers. Developed in the early 1970s from an earlier language called BCPL, C was first used as the language of the operating system ◊UNIX, though it has since become widespread beyond UNIX. It is useful for writing fast and efficient systems programs, such as operating systems (which control the operations of the computer).

cabbage vegetable plant related to the turnip and wild mustard, or charlock. It was cultivated as early as 2000 BC, and the many commercial varieties include kale, Brussels sprouts, common cabbage, savoy, cauliflower, sprouting broccoli, and kohlrabi. (*Brassica oleracea,* family Cruciferae.)

cabbage butterfly one of several butterfly species, the caterpillars of which feed on the leaves of members of the cabbage family, particularly as pests on cabbages. ◊Ichneumon flies parasitize the caterpillars, thereby controlling their numbers.

classification Cabbage butterflies are in genus *Pieris* in order Lepidoptera, class Insecta, phylum Arthropoda.

species The **large white** *Pieris brassicae* is found in Europe and North Africa. The expanded wings measure 7.5 cm/3 in across, and are white with black edgings. The female, which has black spots on the upper surfaces of the wings, lays her yellow eggs in clusters on cabbage leaves. The fully grown caterpillar sometimes measures 4 cm/1.5 in, and will eat twice its own weight of leaf in 24 hours. After it has hung for some time by its tail from a ledge, it changes into a shining pale green chrysalis. The butterfly, which in the case of the autumn brood waits till winter is past before emerging, lives on nectar.

The **small white butterfly** *Pieris rapae,* has a wing expansion of about 5 cm/2 in, lays its eggs singly on the underside of vegetable leaves, and produces a velvety caterpillar which devours the hearts, instead of merely the leaves, of cabbages. The chrysalis is brownish-yellow with black spots.

A third species, the **green-veined white butterfly** *Pieris napi,* which is similar to the small white, cannot multiply so fast, because both the butterfly and its caterpillar are a favourite food of small birds.

cabbage-tree palm tall, fan-leaf palm *Livistona australis* of the coastal areas of E Australia. Aborigines eat the cabbagelike hearts of the young leaves.

cable unit of length, used on ships, originally the length of a ship's anchor cable or 120 fathoms (219 m/720 ft), but now taken as one-tenth of a ◊nautical mile (185.3 m/608 ft).

cable car method of transporting passengers up steep slopes by cable. In the **cable railway**, passenger cars are hauled along rails by a cable wound by a powerful winch. A pair of cars usually operates together on the funicular principle, one going up as the other goes down. The other main type is the **aerial cable car**, where the passenger car is suspended from a trolley that runs along an aerial cableway.

A cable-car system has operated in San Francisco since 1873. The streetcars travel along rails and are hauled by moving cables under the ground.

cable modem box supplied by cable companies to provide television and telephone services, including Internet. The advantages of cable modems over traditional ◊modems, which operate over standard telephone lines, are greatly increased speed of communications as well as the ability to transmit video and two-way audio, and lower costs.

cable television distribution of broadcast signals through cable relay systems.

Narrow-band systems were originally used to deliver services to areas with poor regular reception; systems with wider bands, using coaxial and fibreoptic cable, are increasingly used for distribution and development of home-based interactive services, typically telephones.

In 1997, the USA had 65 million cable television subscribers using more than 11,000 cable systems. The systems were to spend more than $3.5 billion on basic programming in 1997, up 14% from 1996. In 1998, the 65 million US subscribers paid an average of more than $31 a month. The cost for 1998 represented the third straight year the price has increased by 8%, four times the inflation rate.

cacao tropical American evergreen tree, now also cultivated in W Africa and Sri Lanka. Its seeds are cocoa beans, from which cocoa and chocolate are prepared. (*Theobroma cacao,* family Sterculiaceae.)

The trees mature at five to eight years and produce two crops a year. The fruit is 17–25 cm/6.5–9.5 in long, hard, and ridged, with the beans inside. The seeds are called cocoa nibs; when left to ferment, then roasted and separated from the husks, they contain about 50% fat, part of which is removed to make chocolate and cocoa.

The Aztecs revered cacao and made a drink exclusively for the nobility from cocoa beans and chillies, which they called chocolatl. In the 16th century Spanish traders brought cacao to Europe. It was used to make a drink, which came to rival coffee and tea in popularity.

cachalot alternative name for the sperm whale; see ◊whale.

cache memory in computing, a reserved area of the ◊immediate access memory used to increase the running speed of a computer program.

The cache memory may be constructed from ◊SRAM, which is faster but more expensive than the normal ◊DRAM. Most programs access the same instructions or data repeatedly. If these frequently used instructions and data are stored in a fast-access SRAM memory cache, the program will run more quickly. In other cases, the memory cache is normal DRAM, but is used to store frequently used instructions and data that would normally be accessed from ◊backing storage. Access to DRAM is faster than access to backing storage so, again, the program runs more quickly. This type of cache memory is often called a **disc cache**.

cactus (plural ***cacti***) strictly, any plant of the family Cactaceae, although the word is commonly used to describe many different succulent and prickly plants. True cacti have a woody axis (central core) surrounded by a large fleshy stem, which takes various forms and is usually covered with spines (actually reduced leaves). They are all specially adapted to growing in dry areas.

Cactus flowers are often large and brightly coloured; the fruit is fleshy and often edible, as in the case of the prickly pear. The Cactaceae are a New World family and include the treelike saguaro and the night-blooming cereus with blossoms 30 cm/12 in across.

CACTUS AND SUCCULENT PLANT MALL

http://www.cactus-mall.com/index.html

Huge source of information on cacti and how to grow them. There are also reports on conservation of cacti and succulents. There are links to a large number of international cacti associations.

CAD (acronym for ***computer-aided design***) the use of computers in creating and editing design drawings. CAD also allows such things as automatic testing of designs and multiple or animated three-dimensional views of designs. CAD systems are widely used in architecture, electronics, and engineering, for example in the motor-vehicle industry, where cars designed with the assistance of computers are now commonplace.

A related development is ◊CAM (computer-assisted manufacturing).

Cadarache French nuclear research site, NE of Aix-en-Provence.

caddis fly insect of the order Trichoptera. Adults are generally dull brown, mothlike, with wings covered in tiny hairs. Mouthparts are poorly developed, and many caddis flies do not feed as adults. They are usually found near water.

The larvae are aquatic, and many live in cases, open at both ends, which they make out of sand or plant remains. Some species make silk nets among aquatic vegetation to help trap food.

caddis fly Caddis fly adults are generally rather drab insects most likely to be seen by ponds, lakes, and rivers where they may form dense mating swarms. The larvae fashion intricate cases from materials such as sand grains, twigs, empty snail shells, or other debris. Premaphotos Wildlife

cadmium soft, silver-white, ductile, and malleable metallic element, symbol Cd, atomic number 48, relative atomic mass 112.40. Cadmium occurs in nature as a sulphide or carbonate in zinc ores. It is a toxic metal that, because of industrial dumping, has become an environmental pollutant. It is used in batteries, electroplating, and as a constituent of alloys used for bearings with low coefficients of friction; it is also a constituent of an alloy with a very low melting point.

Cadmium is also used in the control rods of nuclear reactors, because of its high absorption of neutrons. It was named in 1817 by the German chemist Friedrich Strohmeyer (1776–1835) after the Greek mythological character Cadmus.

caecilian tropical amphibian of wormlike appearance. There are about 170 species known in the family Caeciliidae, forming the amphibian order Apoda (also known as Caecilia or Gymnophiona). Caecilians have a grooved skin that gives a 'segmented' appearance; they have no trace of limbs or pelvis. The body is 20–130 cm/8–50 in long, beige to black in colour. The eyes are very small and weak or blind. They eat insects and small worms. Some species bear live young, others lay eggs.

Caecilians live in burrows in damp ground in the tropical Americas, Africa, Asia, and the Seychelles Islands.

caecum in the ◊digestive system of animals, a blind-ending tube branching off from the first part of the large intestine, terminating in the appendix. It has no function in humans but is used for the digestion of cellulose by some grass-eating mammals.

The rabbit caecum and appendix contains millions of bacteria that produce cellulase, the enzyme necessary for the breakdown of cellulose to glucose. In order to be able to absorb nutrients released by the breakdown of cellulose, rabbits pass food twice down the intestine. They egest soft pellets which are then re-eaten. This is known as coprophagy.

CAECILIANS WEB SITE

http://www.sfo.com/~morriss/eels.html

Many facts about these legless amphibians, especially *Typhlonectes natans*, an aquatic caecilian often found in aquariums in the USA. Tips on care, health, and breeding can also be found at this site.

Caesarean section surgical operation to deliver a baby by way of an incision in the mother's abdominal and uterine walls. It may be recommended for almost any obstetric complication implying a threat to mother or baby.

Caesarean section was named after the Roman emperor Julius Caesar, who was born this way. In medieval Europe, it was performed mostly in attempts to save the life of a child whose mother had died in labour. The Christian Church forbade cutting open the mother before she was dead.

caesium ***Latin caesius 'bluish-grey'*** soft, silvery-white, ductile metallic element, symbol Cs, atomic number 55, relative atomic mass 132.905. It is one of the ◊alkali metals, and is the most electropositive of all the elements. In air it ignites spontaneously, and it reacts vigorously with water. It is used in the manufacture of photocells. The name comes from the blueness of its spectral line.

The rate of vibration of caesium atoms is used as the standard of measuring time. Its radioactive isotope Cs-137 (half-life 30.17 years) is a product of fission in nuclear explosions and in nuclear reactors; it is one of the most dangerous waste products of the nuclear industry, being a highly radioactive biological analogue for potassium.

caffeine ◊alkaloid organic substance found in tea, coffee, and kola nuts; it stimulates the heart and central nervous system. When isolated, it is a bitter crystalline compound, $C_8H_{10}N_4O_2$. Too much caffeine (more than six average cups of tea or coffee a day) can be detrimental to health.

caiman or **cayman** large reptile, related to the ◊alligator.

All caimans are found only in Central and South America.

types of caiman The black caiman *Melanosuchus niger* is the largest South American predator, reaching 6 m in length. Cuvier's dwarf caiman *Paleosuchus palpebrosus* may be as little as 1.2 m in length when fully grown (females). Schneider's dwarf caiman *Paleosuchus trigonatus* reaches 1.7 m and is reasonably common in the Amazonian rainforests. Broad-snouted caiman *Caiman latirostris* is found is found mainly in freshwater swamps and grows up to 3.5 m. Common caiman *Caiman crocodilus* is found from southern Mexico to northern Argentina and can grow up to 3 m.

cairn Scottish breed of ◊terrier. Shaggy, short-legged, and compact, it can be sandy, greyish brindle, or red. It was formerly used for flushing out foxes and badgers.

caisson hollow cylindrical or boxlike structure, usually of reinforced ◊concrete, sunk into a riverbed to form the foundations of a bridge.

An **open caisson** is open at the top and at the bottom, where there is a wedge-shaped cutting edge. Material is excavated from inside, allowing the caisson to sink. A **pneumatic caisson** has a pressurized chamber at the bottom, in which workers carry out the excavation. The air pressure prevents the surrounding water entering; the workers enter and leave the chamber through an airlock, allowing for a suitable decompression period to prevent ◊decompression sickness (the so-called bends).

cal symbol for ◊calorie.

CAL (acronym for ***computer-assisted learning***) the use of computers in education and training: the computer displays instructional material to a student and asks questions about the information given; the student's answers determine the sequence of the lessons.

calabash tropical South American evergreen tree with gourds (fruits) 50 cm/20 in across, whose dried skins are used as water containers. The Old World tropical-vine bottle gourd (*Lagenaria siceraria*, of the gourd family Cucurbitaceae) is sometimes also called a calabash, and it produces equally large gourds. (*Crescentia cujete*, family Bignoniaceae.)

calamine $ZnCO_3$ zinc carbonate, an ore of zinc. The term also refers to a pink powder made of a mixture of zinc oxide and iron(II) oxide used in lotions and ointments as an astringent for treating, for example, sunburn, eczema, measles rash, and insect bites and stings.

In the USA the term refers to zinc silicate $Zn_4Si_2O_7(OH)_2.H_2O$.

calceolaria plant with brilliantly coloured slipper-shaped flowers. Native to South America, calceolarias were introduced to Europe and the USA in the 1830s. (Genus *Calceolaria*, family Scrophulariaceae.)

calcination ◊oxidation of metals by burning in air.

calcite colourless, white, or light-coloured common rock-forming mineral, calcium carbonate, $CaCO_3$. It is the main constituent of ◊limestone and marble and forms many types of invertebrate shell.

Calcite often forms ◊stalactites and stalagmites in caves and is also found deposited in veins through many rocks because of the ease with which it is dissolved and transported by groundwater; ◊oolite is a rock consisting of spheroidal calcite grains. It rates 3 on the ◊Mohs' scale of hardness. Large crystals up to 1 m/3 ft have been found in Oklahoma and Missouri, USA. Iceland spar is a transparent form of calcite used in the optical industry; as limestone it is used in the building industry.

calcium *Latin calcis 'lime'* soft, silvery-white metallic element, symbol Ca, atomic number 20, relative atomic mass 40.08. It is one of the ◊alkaline-earth metals. It is the fifth most abundant element (the third most abundant metal) in the Earth's crust. It is found mainly as its carbonate $CaCO_3$, which occurs in a fairly pure condition as chalk and limestone (see ◊calcite). Calcium is an essential component of bones, teeth, shells, milk, and leaves, and it forms 1.5% of the human body by mass.

Calcium ions in animal cells are involved in regulating muscle contraction, blood clotting, hormone secretion, digestion, and glycogen metabolism in the liver. It is acquired mainly from milk and cheese, and its uptake is facilitated by vitamin D. Calcium deficiency leads to chronic muscle spasms (tetany); an excess of calcium may lead to the formation of stones in the kidney or gall bladder.

The element was discovered and named by the English chemist Humphry Davy in 1808. Its compounds include slaked lime (calcium hydroxide, $Ca(OH)_2$); plaster of Paris (calcium sulphate, $CaSO_4.2H_2O$); calcium phosphate ($Ca_3(PO_4)_2$), the main constituent of animal bones; calcium hypochlorite ($CaOCl_2$), a bleaching agent; calcium nitrate ($Ca(NO_3)_2.4H_2O$), a nitrogenous fertilizer; calcium carbide (CaC_2), which reacts with water to give ethyne (acetylene); calcium cyanamide ($CaCN_2$), the basis of many pharmaceuticals, fertilizers, and plastics, including melamine; calcium cyanide ($Ca(CN)_2$), used in the extraction of gold and silver and in electroplating; and others used in baking powders and fillers for paints.

calcium carbonate $CaCO_3$ white solid, found in nature as limestone, marble, and chalk. It is a valuable resource, used in the making of iron, steel, cement, glass, slaked lime, bleaching powder, sodium carbonate and bicarbonate, and many other industrially useful substances.

calcium hydrogencarbonate $Ca(HCO_3)_2$ substance found in ◊hard water, formed when rainwater passes over limestone rock.

$$CaCO_{3\,(s)} + CO_{2\,(g)} + H_2O_{(l)} \rightarrow Ca(HCO_3)_{2\,(aq)}$$

calabash *Like many tropical trees, the flowers of the calabash tree grow directly from the wood of the trunk and branches. At night the flowers give off an unpleasant odour, attracting numerous pollinating bats.* *Premaphotos Wildlife*

When this water is boiled it reforms calcium carbonate, removing the hardness; this type of hardness is therefore known as temporary hardness.

calcium hydrogenphosphate $Ca(H_2PO_4)_2$ substance made by heating calcium phosphate with 70% sulphuric acid. It is more soluble in water than calcium phosphate, and is used as a fertilizer.

calcium hydroxide $Ca(OH)_2$ or ***slaked lime*** white solid, slightly soluble in water. A solution of calcium hydroxide is called ◊limewater and is used in the laboratory to test for the presence of carbon dioxide. It is manufactured industrially by adding water to calcium oxide (quicklime) in a strongly exothermic reaction.

$$CaO + H_2O \rightarrow Ca(OH)_2$$

It is used to reduce soil acidity and as a cheap alkali in many industrial processes.

calcium nitrate $Ca(NO_3)_2$ white, crystalline compound that is very soluble in water. The solid decomposes on heating to form oxygen, brown nitrogen(IV) oxide gas, and the white solid calcium oxide.

$$2Ca(NO_3)_2 \rightarrow 2CaO + 4NO_2 + O_2$$

calcium oxide or ***quicklime*** CaO white solid compound, formed by heating ◊calcium carbonate.

$$CaCO_3 \rightarrow CaO + CO_2$$

When water is added it forms calcium hydroxide (slaked lime) in an ◊exothermic reaction.

$$CaO + H_2O \rightarrow Ca(OH)_2$$

It is a typical basic oxide, turning litmus blue.

calcium phosphate $Ca_3(PO_4)_2$ or ***calcium orthophosphate*** white solid, the main constituent of animal bones. It occurs naturally as the mineral apatite and in rock phosphate, and is used in the preparation of phosphate fertilizers.

calcium sulphate $CaSO_4$ white, solid compound, found in nature as gypsum and anhydrite. It dissolves slightly in water to form ◊hard water; this hardness is not removed by boiling, and hence is sometimes called permanent hardness.

calcium superphosphate common name for ◊calcium hydrogen-phosphate.

calculator pocket-sized electronic computing device for performing numerical calculations. It can add, subtract, multiply, and divide; many calculators also compute squares and roots and have advanced trigonometric and statistical functions. Input is by a small keyboard and results are shown on a one-line computer screen, typically a ◊liquid-crystal display (LCD) or a light-emitting diode (LED). The first electronic calculator was manufactured by the Bell Punch Company in the USA in 1963.

calculus ***Latin 'pebble'*** branch of mathematics which uses the concept of a derivative (see ◊differentiation) to analyse the way in which the values of a ◊function vary. Calculus is probably the most widely used part of mathematics. Many real-life problems are analysed by expressing one quantity as a function of another – position of a moving object as a function of time, temperature of an object as a function of distance from a heat source, force on an object as a function of distance from the source of the force, and so on – and calculus is concerned with such functions.

There are several branches of calculus. Differential and integral calculus, both dealing with small quantities which during manipulation are made smaller and smaller, compose the **infinitesimal calculus**. **Differential equations** relate to the derivatives of a set of variables and may include the variables. Many give the mathematical models for physical phenomena such as ◊simple harmonic ◊motion. Differential equations are solved generally by ◊integration, depending on their degree. If no analytical processes are available, integration can be performed numerically. Other branches of calculus include calculus of variations and calculus of errors.

caldera in geology, a very large basin-shaped ◊crater. Calderas are found at the tops of volcanoes, where the original peak has collapsed into an empty chamber beneath. The basin, many times larger than the original volcanic vent, may be flooded, producing a crater lake, or the flat floor may contain a number of small volcanic cones, produced by volcanic activity after the collapse.

Typical calderas are Kilauea, Hawaii; Crater Lake, Oregon, USA; and the summit of Olympus Mons, on Mars. Some calderas are wrongly referred to as craters, such as Ngorongoro, Tanzania.

calendar division of the ◊year into months, weeks, and days and the method of ordering the years. From year one, an assumed date of the birth of Jesus, dates are calculated backwards (BC 'before Christ' or BCE 'before common era') and forwards (AD, Latin *anno Domini* 'in the year of the Lord', or CE 'common era'). The **lunar month** (period between one new moon and the next) naturally averages 29.5 days, but the Western calendar uses for convenience a **calendar month** with a complete number of days, 30 or 31 (February has 28). For adjustments, since there are slightly fewer than six extra hours a year left over, they are added to Feb as a 29th day every fourth year (**leap year**), century years being excepted unless they are divisible by 400. For example, 1896 was a leap year; 1900 was not.

The **month names** in most European languages were probably derived as follows: January from Janus, Roman god; February from *Februar,* Roman festival of purification; March from Mars, Roman god; April from Latin *aperire,* 'to open'; May from Maia, Roman goddess; June from Juno, Roman goddess; July from Julius Caesar, Roman general; August from Augustus, Roman emperor; September, October, November, December (originally the seventh to tenth months) from the Latin words meaning seventh, eighth, ninth, and tenth, respectively.

The **days of the week** are Monday named after the Moon; Tuesday from Tiu or Tyr, Anglo-Saxon and Norse god; Wednesday from Woden or Odin, Norse god; Thursday from Thor, Norse god; Friday from Freya, Norse goddess; Saturday from Saturn, Roman god; and Sunday named after the Sun.

All early calendars except the ancient Egyptian were lunar. The word calendar comes from the Latin *Kalendae* or *calendae,* the first day of each month on which, in ancient Rome, solemn proclamation was made of the appearance of the new moon.

The **Western** or **Gregorian calendar** derives from the **Julian calendar** instituted by Julius Caesar 46 BC. It was adjusted by Pope Gregory XIII 1582, who eliminated the accumulated error caused by a faulty calculation of the length of a year and avoided its recurrence by restricting century leap years to those divisible by 400. Other states only gradually changed from Old Style to New Style; Britain and its colonies adopted the Gregorian calendar 1752, when the error amounted to 11 days, and 3 September 1752 became 14 September (at the same time the beginning of the year was put back from 25 March to 1 January). Russia did not adopt it until the October Revolution of 1917, so that the event (then 25 October) is currently celebrated 7 November.

The **Jewish calendar** is a complex combination of lunar and solar cycles, varied by considerations of religious observance. A year may have 12 or 13 months, each of which normally alternates between 29 and 30 days; the New Year (Rosh Hashanah) falls between 5 September and 5 October. The calendar dates from the hypothetical creation of the world (taken as 7 October 3761 BC).

The **Chinese calendar** is lunar, with a cycle of 60 years. Both the traditional and, from 1911, the Western calendar are in use in China.

The **Muslim calendar**, also lunar, has 12 months of alternately 30 and 29 days, and a year of 354 days. This results in the calendar rotating around the seasons in a 30-year cycle. The era is counted as beginning on the day Muhammad fled from Mecca AD 622.

calibration the preparation of a usable scale on a measuring instrument. A mercury ◊thermometer, for example, can be calibrated with a Celsius scale by noting the heights of the mercury column at two standard temperatures – the freezing point (0°C) and boiling point (100°C) of water – and dividing the distance between them into 100 equal parts and continuing these divisions above and below.

Science is all those things which are confirmed to such a degree that it would be unreasonable to withhold one's provisional consent.

STEPHEN JAY GOULD US palaeontologist and writer.
Lecture on Evolution

California current cold ocean ◊current in the E Pacific Ocean flowing southwards down the West coast of North America. It is part of the North Pacific ◊gyre (a vast, circular movement of ocean water).

californium synthesized, radioactive, metallic element of the actinide series, symbol Cf, atomic number 98, relative atomic mass 251. It is produced in very small quantities and used in nuclear reactors as a neutron source. The longest-lived isotope, Cf-251, has a half-life of 800 years.

It is named after the state of California, where it was first synthesized 1950 by US nuclear chemist Glenn Seaborg and his team at the University of California at Berkeley.

calla alternative name for ◊arum lily.

call for votes in computing, on ◊USENET, process by which the nature and scope of a new ◊newsgroup is determined. Calls for votes are posted to **news.announce.newgroups**. In the case of the ◊Big Seven hierarchies, the call for votes is a requirement; it is recommended but not compulsory for ◊alt hierarchy groups. The point is to ensure that newsgroup names follow a consistent pattern and that new newsgroups are formed in response to genuine interest.

Callichthys genus of South American ◊catfish. Its body is protected by large, hard, scaly plates.
classification *Callichthys* belongs to the family Callichthydae (armoured catfish), order Siluriformes, class Osteichthyes.

callipers measuring instrument used, for example, to measure the internal and external diameters of pipes. Some callipers are made like a pair of compasses, having two legs, often curved, pivoting about a screw at one end. The ends of the legs are placed in contact with the object to be measured, and the gap between the ends is then measured against a rule. The slide calliper looks like an adjustable spanner, and carries a scale for direct measuring, usually with a ◊vernier scale for accuracy.

callistemon the genus of the Australian ◊bottlebrush.

Callisto second-largest moon of Jupiter, 4,800 km/3,000 mi in diameter, orbiting every 16.7 days at a distance of 1.9 million km/1.2 million mi from the planet. Its surface is covered with large craters.

The space probe *Galileo* detected molecules containing both carbon and nitrogen atoms on the surface of Callisto, US astronomers announced in March 1997. Their presence may indicate that Callisto harboured life at some time.

callus in botany, a tissue that forms at a damaged plant surface. Composed of large, thin-walled ◊parenchyma cells, it grows over and around the wound, eventually covering the exposed area.

In animals, a callus is a thickened pad of skin, formed where there is repeated rubbing against a hard surface. In humans, calluses often develop on the hands and feet of those involved in heavy manual work.

calomel Hg_2Cl_2 (technical name ***mercury(I) chloride***) white, heavy powder formerly used as a laxative, now used as a pesticide and fungicide.

calorie c.g.s. unit of heat, now replaced by the ◊joule (one calorie is approximately 4.2 joules). It is the heat required to raise the temperature of one gram of water by 1°C. In dietetics, the Calorie or kilocalorie is equal to 1,000 calories.

The kilocalorie measures the energy value of food in terms of its heat output: 28 g/1 oz of protein yields 120 kilocalories, of carbohydrate 110, of fat 270, and of alcohol 200.

calorific value the amount of heat generated by a given mass of fuel when it is completely burned. It is measured in joules per kilogram. Calorific values are measured experimentally with a bomb calorimeter.

calorimeter instrument used in physics to measure various thermal properties, such as heat capacity or the heat produced by fuel. A simple calorimeter consists of a heavy copper vessel that is polished (to reduce heat losses by radiation) and covered with insulating material (to reduce losses by convection and conduction).

In a typical experiment, such as to measure the heat capacity of a piece of metal, the calorimeter is filled with water, whose temperature rise is measured using a thermometer when a known mass of the heated metal is immersed in it. Chemists use a bomb calorimeter to measure the heat produced by burning a fuel completely in oxygen.

calotype paper-based photograph using a wax paper negative, the first example of the ◊negative/positive process invented by the English photographer Fox Talbot around 1834.

calyptra in mosses and liverworts, a layer of cells that encloses and protects the young ◊sporophyte (spore capsule), forming a sheathlike hood around the capsule. The term is also used to describe the root cap, a layer of ◊parenchyma cells covering the end of a root that gives protection to the root tip as it grows through the soil. This is constantly being worn away and replaced by new cells from a special ◊meristem, the calyptrogen.

calyx collective term for the ◊sepals of a flower, forming the outermost whorl of the ◊perianth. It surrounds the other flower parts and protects them while in bud. In some flowers, for example, the campions *Silene,* the sepals are fused along their sides, forming a tubular calyx.

CAM (acronym for ***computer-aided manufacturing***) the use of computers to control production processes; in particular, the control of machine tools and ◊robots in factories. In some factories, the whole design and production system has been automated by linking ◊CAD (computer-aided design) to CAM.

Linking flexible CAD/CAM manufacturing to computer-based sales and distribution methods makes it possible to produce semi-customized goods cheaply and in large numbers.

Calvin, Melvin (1911–1997)

Mary Evans Picture Library

US chemist who, using radioactive carbon-14 as a tracer, determined the biochemical processes of photosynthesis, in which green plants use chlorophyll to convert carbon dioxide and water into sugar and oxygen. Nobel prize 1961.

cam part of a machine that converts circular motion to linear motion or vice versa. The **edge cam** in a car engine is in the form of a rounded projection on a shaft, the camshaft. When the camshaft turns, the cams press against linkages (plungers or followers) that open the valves in the cylinders.

A **face cam** is a disc with a groove in its face, in which the follower travels. A **cylindrical cam** carries angled parallel grooves, which impart a to-and-fro motion to the follower when it rotates.

Camberwell beauty species of ◊butterfly.

cambium in botany, a layer of actively dividing cells (lateral ◊meristem), found within stems and roots, that gives rise to ◊secondary growth in perennial plants, causing an increase in girth. There are two main types of cambium: **vascular cambium**, which gives rise to secondary ◊xylem and ◊phloem tissues, and **cork cambium** (or phellogen), which gives rise to secondary cortex and cork tissues (see ◊bark).

Cambrian period of geological time 570–510 million years ago; the first period of the Palaeozoic era. All invertebrate animal life appeared, and marine algae were widespread. The **Cambrian Explosion** 530–520 million years ago saw the first appearance in the fossil record of all modern animal phyla; the earliest fossils with hard shells, such as trilobites, date from this period.

The name comes from Cambria, the medieval Latin name for Wales, where Cambrian rocks are typically exposed and were first described.

camcorder another name for a ◊video camera.

camel large cud-chewing mammal of the even-toed hoofed order Artiodactyla. Unlike typical ruminants, it has a three-chambered stomach. It has two toes which have broad soft soles for walking on sand, and hooves resembling nails. There are two species, the single-humped **Arabian camel** *Camelus dromedarius* and the twin-humped **Bactrian camel** *C. bactrianus* from Asia. They carry a food reserve of fatty tissue in the hump, can go without drinking for long periods, can feed on salty vegetation, and withstand extremes of heat and cold, thus being well adapted to desert conditions.

The Arabian camel has long been domesticated, so that its original range is not known. It is used throughout Arabia and N Africa, and has been taken to other places such as North America and Australia, in the latter country playing a crucial part in the development of the interior. The **dromedary** is, strictly speaking, a lightly built, fast, riding variety of the Arabian camel, but often the name is applied to all one-humped camels. Arabian camels can be used as pack animals, for riding, racing, milk production, and for meat. The Bactrian camel is native to the central Asian deserts, where a small number still live wild, but most are domestic animals. With a head and body length of 3 m/10 ft and shoulder height of about 2 m/6 ft, the Bactrian camel is a large animal, but not so long in the leg as the Arabian. It has a shaggy winter coat. In 1995 there were only about 730–880 Bactrian camels remaining in the wild. Smaller, flat-backed members of the camel family include the ◊alpaca, the ◊guanaco, the ◊llama, and the ◊vicuna.

camellia any oriental evergreen shrub with roselike flowers belonging to the tea family. Many species, including *Camellia japonica* and *C. reticulata*, have been introduced into Europe, the USA, and Australia; they are widely cultivated as ornamental shrubs. (Genus *Camellia*, family Theaceae.)

CAMERAS: THE TECHNOLOGY OF PHOTOGRAPHIC IMAGING

http://www.mhs.ox.ac.uk/cameras/index.htm

General presentation of the camera collection at the Museum of the History of Science, Oxford, UK. Of the many highlights perhaps the most distinguished are some very early photographs, the photographic works of Sarah A Acland, and the cameras of T E Lawrence.

camera apparatus used in ◊photography, consisting of a lens system set in a light-proof box inside of which a sensitized film or plate can be placed. The lens collects rays of light reflected from the subject and brings them together as a sharp image on the film. The opening or hole at the front of the camera, through which light enters, is called an ◊aperture. The aperture size controls the amount of light that can enter. A shutter controls the amount of time light has to affect the film. There are small-, medium-, and large-format cameras; the format refers to the size of recorded image and the dimensions of the image obtained.

A simple camera has a fixed shutter speed and aperture, chosen so that on a sunny day the correct amount of light is admitted. More complex cameras allow the shutter speed and aperture to be adjusted; most have a built-in exposure meter to help choose the correct combination of shutter speed and aperture for the ambient conditions and subject matter. The most versatile camera is the single lens reflex (◊SLR) which allows the lens to be removed and special lenses attached. A pin-hole camera has a small (pin-sized) hole instead of a lens. It must be left on a firm support during exposures, which are up to ten seconds with slow film, two seconds with fast film and five minutes for paper negatives in daylight. The pinhole camera gives sharp images from close-up to infinity.

camera obscura darkened box with a tiny hole for projecting the inverted image of the scene outside on to a screen inside. For its development as a device for producing photographs, see ◊photography.

camouflage colours or structures that allow an animal to blend with its surroundings to avoid detection by other animals. Camouflage can take the form of matching the background colour, of countershading (darker on top, lighter below, to counteract natural shadows), or of irregular patterns that break up the outline of the animal's body. More elaborate camouflage involves closely resembling a feature of the natural environment, as with the stick insect; this is closely akin to ◊mimicry. Camouflage is also important as a military technique, disguising either equipment, troops, or a position in order to conceal them from an enemy.

camphor $C_{10}H_{16}O$ volatile, aromatic ◊ketone substance obtained from the camphor tree *Cinnamomum camphora*. It is distilled from chips of the wood, and is used in insect repellents and medicinal inhalants and liniments, and in the manufacture of celluloid.

The camphor tree, a member of the family Lauraceae, is native to China, Taiwan, and Japan.

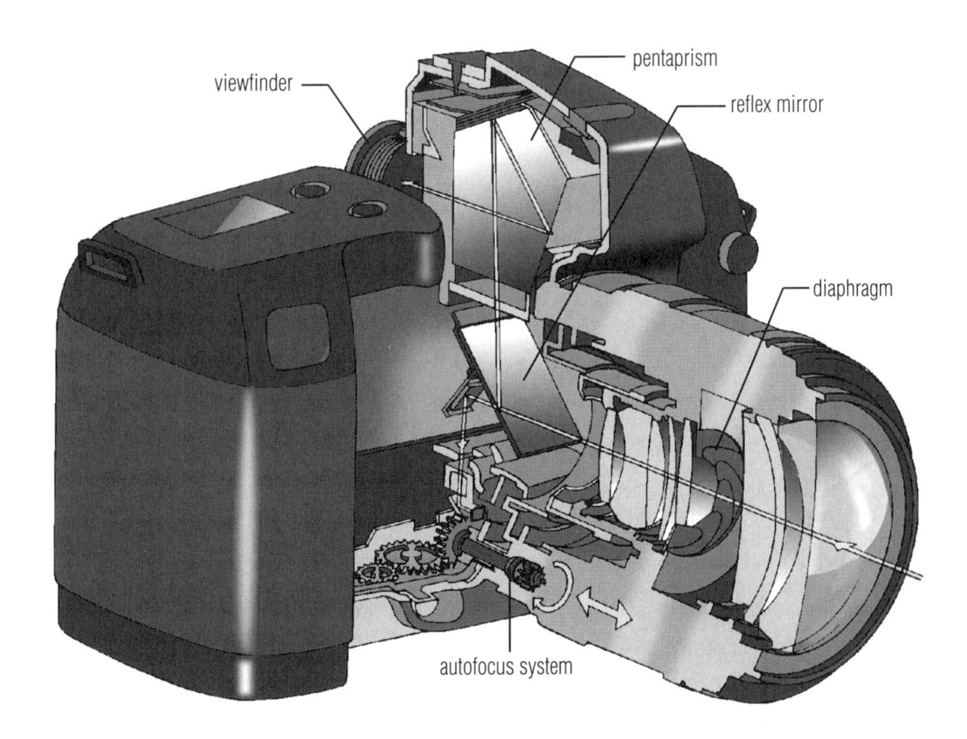

camera The single-lens reflex (SLR) camera in which an image can be seen through the lens before a picture is taken. The reflex mirror directs light entering the lens to the viewfinder. The SLR allows different lenses, such as close-up or zoom, to be used because the photographer can see exactly what is being focused on.

campion any of several plants belonging to the pink family. They include the garden campion (*Lychnis coronaria*), the wild white and red campions (*Silene alba* and *S. dioica*), and the bladder campion (*S. vulgaris*). (Genera *Lychnis* and *Silene*, family Caryophyllaceae.)

campus-wide information system (CWIS) computerized information service used on US university campuses, often hooked to the Internet. These systems typically include local events listings, general campus information, access to the library catalogue, weather reports, directories, and even ◊bulletin-board and messaging services.

One of the first such systems was Cornell University's CUINFO, developed by a team led by technical administrator Steve Worona 1982. The development of ◊Gopher servers 1991 made these systems much easier to navigate, and many systems were redesigned to take advantage of the new technology. In the mid-1990s these systems began moving to the World Wide Web.

Campylobacter genus of bacteria that cause serious outbreaks of gastroenteritis. They grow best at 43°C, and so are well suited to the digestive tract of birds. Poultry is therefore the most likely source of a *Campylobacter* outbreak, although the bacteria can also be transmitted via beef or milk. *Campylobacter* can survive in water for up to 15 days, so may be present in drinking water if supplies are contaminated by sewage or reservoirs are polluted by seagulls.

canal artificial waterway constructed for drainage, irrigation, or navigation. **Irrigation canals** carry water for irrigation from rivers, reservoirs, or wells, and are designed to maintain an even flow of water over the whole length. **Navigation and ship canals** are constructed at one level between ◊locks, and frequently link with rivers or sea inlets to form a waterway system. The Suez Canal in 1869 and the Panama Canal in 1914 eliminated long trips around continents and dramatically shortened shipping routes.

irrigation canals The river Nile has fed canals to maintain life in Egypt since the earliest times. The division of the waters of the Upper Indus and its tributaries, which form an extensive system in Pakistan and Punjab, India, was, for more than ten years, a major cause of dispute between India and Pakistan, settled by a treaty 1960. The Murray basin, Victoria, Australia, and the Imperial and Central Valley projects in California, USA, are examples of 19th- and 20th-century irrigation-canal development. Excessive extraction of water for irrigation from rivers and lakes can cause environmental damage.

ship canals Probably the oldest ship canal to be still in use, as well as the longest, is the Grand Canal in China, which links Tianjin and Hangzhou and connects the Huang He (Yellow River) and Chang Jiang. It was originally built in three stages 485 BC–AD 283, reaching a total length of 1,780 km/1,110 mi. Large sections silted up in later years, but the entire system was dredged, widened, and rebuilt between 1958 and 1972 in conjunction with work on flood protection, irrigation, and hydroelectric schemes. It carries millions of tonnes of freight every year.

Where speed is not a prime factor, the cost-effectiveness of transporting goods by canal has encouraged a revival; Belgium, France, Germany, and the states of the former USSR are among countries that have extended and streamlined their canals. The Baltic–Volga waterway links the Lithuanian port of Klaipeda with Kahovka, at the mouth of the Dnieper on the Black Sea, a distance of 2,430 km/1,510 mi. A further canal cuts across the north Crimea, thus shortening the voyage of ships from the Dnieper through the Black Sea to the Sea of Azov. In Central America, the Panama Canal 1904–14 links the Atlantic and Pacific oceans (64 km/40 mi). In North America, the Erie Canal 1825 links the Great Lakes with the Hudson River and opened up the northeast and Midwest to commerce; the St Lawrence Seaway 1954–59 extends from Montréal to Lake Ontario (290 km/180 mi) and, with the deepening of the Welland Ship Canal and some of the river channels, provides a waterway that enables ocean going vessels to travel (during the ice-free months) between the Atlantic and Duluth, Minnesota, USA, at the western end of Lake Superior, some 3,770 km/2,342 mi.

Canaries current cold ocean current in the North Atlantic Ocean flowing SW from Spain along the NW coast of Africa. It meets the northern equatorial current at a latitude of 20° N.

canary bird *Serinus canaria* of the finch family Fringillidae, found wild in the Canary Islands and Madeira. In its wild state the plumage is green, sometimes streaked with brown. The wild canary builds its nest of moss, feathers, and hair in thick high shrubs or trees, and produces two to four broods in a season.

Canaries have been bred as cage birds in Europe since the 15th century, and many domestic varieties are yellow or orange as a result of artificial selection.

Some canaries were used in mines as detectors of traces of poison gas in the air.

CANARY FAQ

http://www2.upatsix.com/faq/canary.htm

Answers to frequently asked questions about canaries provide a wealth of information on all breeds of these cage birds. Every conceivable aspect of caring for canaries is covered and there are useful links to other canary sites.

cancel in mathematics, to simplify a fraction or ratio by dividing both numerator and denominator by the same number (which must be a ◊common factor of both of them). For example, $\frac{5x}{25}$ cancels to $\frac{x}{5}$ when divided top and bottom by 5.

cancelbot automated software program (see ◊bot) that cancels messages on UseNet. The arrival of ◊spamming (advertising) on the Net prompted the development of technology to use features built into UseNet to cancel messages. While single messages are easily cancelled manually, an automated routine is needed to handle mass postings, which may go out to more than 14,000 newsgroups. Cancelbot is activated by the ◊CancelMoose.

CancelMoose anonymous individual who fires off the ◊cancelbot. The CancelMoose (usually written as 'CancelMoose™' on the Net) monitors newsgroups such as alt.current-events.net-abuse and news.admin.net-abuse for complaints about ◊spamming (advertising), usually defined as messages posted to more than 25 newsgroups of widely varying content. The CancelMoose's identity is kept secret for reasons of personal safety.

cancer group of diseases characterized by abnormal proliferation of cells. Cancer (malignant) cells are usually degenerate, capable only of reproducing themselves (tumour formation). Malignant cells tend to spread from their site of origin by travelling through the bloodstream or lymphatic system. Cancer kills about 6 million people a year worldwide.

causes There are more than 100 types of cancer. Some, like lung or bowel cancer, are common; others are rare. The likely causes remain unexplained. Triggering agents (◊carcinogens) include chemicals such as those found in cigarette smoke, other forms of smoke, asbestos dust, exhaust fumes, and many industrial chemicals. Some viruses can also trigger the cancerous growth of cells (see ◊oncogenes), as can X-rays and radioactivity. Dietary factors are important in some cancers; for example, lack of fibre in the diet may predispose people to bowel cancer and a diet high in animal fats and low in fresh vegetables and fruit increases the risk of breast cancer. Psychological ◊stress may increase the risk of cancer, more so if the person concerned is not able to control the source of the stress.

cancer genes In some families there is a genetic tendency towards a particular type of cancer. In 1993 researchers isolated the first gene that predisposes individuals to cancer. About 1 in 200 people in the West carry the gene. If the gene mutates, those with the altered gene have a 70% chance of developing colon cancer, and female carriers have a 50% chance of developing cancer of the uterus. This accounts for an estimated 10% of all colon cancer.

In 1994 a gene that triggers breast cancer was identified. **BRCA1** was found to be responsible for almost half the cases of inherited breast cancer, and most cases of ovarian cancer. In 1995 a link between BRCA1 and non-inherited breast cancer was discovered. Women with the gene have an 85% chance of developing breast or ovarian cancer during their lifetime. A second breast cancer gene **BRCA2** was identified later in 1995.

The commonest cancer in young men is testicular cancer, the incidence of which has been rising by 3% a year since 1974 (1998).

Cancer faintest of the zodiacal constellations (its brightest stars are fourth magnitude). It lies in the northern hemisphere between ◊Leo and ◊Gemini, and is represented as a crab. The Sun passes through the constellation during late July and early August. In astrology, the dates for Cancer are between about 22 June and 22 July (see ◊precession).

Cancer's most distinctive feature is the open star cluster Praesepe, popularly known as the Beehive, visible to the naked eye as a nebulous patch.

candela SI unit (symbol cd) of luminous intensity, which replaced the old units of candle and standard candle. It measures the brightness of a light itself rather than the amount of light falling on an object, which is called **illuminance** and measured in ◊lux.

One candela is defined as the luminous intensity in a given direction of a source that emits monochromatic radiation of frequency 540×10^{-12} Hz and whose radiant energy in that direction is 1/683 watt per steradian.

Candida albicans yeastlike fungus present in the human digestive tract and in the vagina, which causes no harm in most healthy people. However, it can cause problems if it multiplies excessively, as in vaginal candidiasis or ◊thrush, the main symptom of which is intense itching.

The most common form of thrush is oral, which often occurs in those taking steroids or prolonged courses of antibiotics.

Newborn babies may pick up the yeast during birth and suffer an infection of the mouth and throat. There is also some evidence that overgrowth of *Candida* may occur in the intestines, causing diarrhoea, bloating, and other symptoms such as headache and fatigue, but this is not yet proven. Occasionally, *Candida* can infect immunocompromised patients, such as those with AIDS. Treatment for candidiasis is based on antifungal drugs.

candle cylinder of wax (such as tallow or paraffin wax) with a central wick of string. A flame applied to the end of the wick melts the wax, thereby producing a luminous flame. The wick is treated with a substance such as alum so that it carbonizes but does not rapidly burn out.

Candles and oil lamps were an early form of artificial lighting. Accurately made candles – which burned at a steady rate – were calibrated along their lengths and used as a type of clock. The candle was also the name of a unit of luminous intensity, replaced in 1940 by the ◊candela (cd), equal to $\frac{1}{60}$ of the luminance of 1 sq cm of a black body radiator at a temperature of 2,042K (the temperature of solidification of platinum).

CANDLE MAKING

http://www.southwest.com.au/~snorth/crafts.htm#contents

Mainly text guide to making and designing your own candles. There are a few general pictures of the finished product, but the text is very instructive.

cane reedlike stem of various plants such as the sugar cane, bamboo, and, in particular, the group of palms called rattans, consisting of the genus *Calamus* and its allies. Their slender stems are dried and used for making walking sticks, baskets, and furniture.

Canes Venatici constellation of the northern hemisphere near ◊Ursa Major, identified with the hunting dogs of ◊Boötes, the herder. Its stars are faint, and it contains the Whirlpool galaxy (M51), the first spiral galaxy to be recognized.

It contains many objects of telescopic interest, including the relatively bright ◊globular cluster M3. The brightest star, a third magnitude double, is called Cor Caroli or Alpha Canum Venaticorum.

cane toad toad of the genus *Bufo marinus*, family Bufonidae. Also known as the giant or marine toad, the cane toad is the largest in the world. It acquired its name after being introduced to Australia during the 1930s to eradicate the cane beetle, which had become a serious pest there. However, having few natural enemies, the cane toad itself has now become a pest in Australia.

The cane toad's defence system consists of highly developed glands on each side of its neck which can squirt a poisonous fluid to a distance of around 1 m/3.3 ft.

CANE TOAD

http://share.jcu.edu.au/dept/PHTM/staff/rsbufo.htm

Profile of the world's largest toad, its unwise introduction to Australia, and the damage it has wrought. There's a photo of a cane toad, another of the massive number of eggs it can produce, and a complete bibliography of books and articles written about *Bufo marinus*.

canine in mammalian carnivores, any of the long, often pointed teeth found at the front of the mouth between the incisors and premolars. Canine teeth are used for catching prey, for killing, and for tearing flesh. They are absent in herbivores such as rabbits and sheep, and are much reduced in humans.

Canis Major brilliant constellation of the southern hemisphere, represented (with Canis Minor) as one of the two dogs following at the heel of ◊Orion. Its main star, ◊Sirius, is the brightest star in the night sky.

Epsilon Canis Majoris is also of the first magnitude, and there are three second magnitude stars.

Canis Minor small constellation along the celestial equator (see ◊celestial sphere), represented as the smaller of the two dogs of ◊Orion (the other dog being ◊Canis Major). Its brightest star is the first magnitude ◊Procyon.

Procyon and Beta Canis Minoris form what the Arabs called 'the Short Cubit', in contrast to 'the Long Cubit' formed by ◊Castor and ◊Pollux (Alpha and Beta Geminorum).

cannabis dried leaves and female flowers (marijuana) and ◊resin (hashish) of certain varieties of ◊hemp, which are smoked or swallowed to produce a range of effects, including feelings of great happiness and altered perception. (*Cannabis sativa*, family Cannabaceae.)

Cannabis is a soft drug in that any dependence is psychological rather than physical. It is illegal in many countries and has not been much used in medicine since the 1930s. However, recent research has led to the discovery of cannabis receptors (sensory nerve endings) in the brain, and the discovery of a naturally occurring brain chemical which produces the same effects as smoking cannabis. Researchers believe this work could lead to the use of cannabislike compounds to treat physical illness without affecting the mind. Cannabis is claimed to have beneficial effects in treating chronic diseases such as AIDS and ◊multiple sclerosis. The main psychoactive ingredient in cannabis is delta-9-tetrahydrocannabinol (THC) which is available legally as a prescribed drug in capsule form.

canning food preservation in hermetically sealed containers by the application of heat. Originated by Nicolas Appert in France 1809 with glass containers, it was developed by Peter Durand in

England in 1810 with cans made of sheet steel thinly coated with tin to delay corrosion. Cans for beer and soft drinks are now generally made of aluminium.

Canneries were established in the USA before 1820, but the US canning industry expanded considerably in the 1870s when the manufacture of cans was mechanized and factory methods of processing were used. The quality and taste of early canned food was frequently inferior but by the end of the 19th century, scientific research made greater understanding possible of the food-preserving process, and standards improved. More than half the aluminium cans used in the USA are now recycled.

Canopus or *Alpha Carinae* second brightest star in the night sky (after Sirius), lying in the southern constellation Carina. It is a first-magnitude yellow-white supergiant about 120 light years from Earth, and thousands of times more luminous than the Sun.

cantaloupe any of several small varieties of muskmelon *Cucumis melo,* distinguished by their round, ribbed fruits with orange-coloured flesh.

cantilever beam or structure that is fixed at one end only, though it may be supported at some point along its length; for example, a diving board. The cantilever principle, widely used in construction engineering, eliminates the need for a second main support at the free end of the beam, allowing for more elegant structures and reducing the amount of materials required. Many large-span bridges have been built on the cantilever principle.

A typical cantilever bridge consists of two beams cantilevered out from either bank, each supported part way along, with their free ends meeting in the middle. The multiple-cantilever Forth Rail Bridge (completed 1890) across the Firth of Forth in Scotland has twin main spans of 521 m/1,710 ft.

canyon *Spanish cañon 'tube'* deep, narrow valley or gorge running through mountains. Canyons are formed by stream down-cutting, usually in arid areas, where the rate of down-cutting is greater than the rate of weathering, and where the stream or river receives water from outside the area.

There are many canyons in the western USA and in Mexico, for example the Grand Canyon of the Colorado River in Arizona, the canyon in Yellowstone National Park, and the Black Canyon in Colorado.

cap another name for a ◊diaphragm contraceptive.

capacitance, electrical property of a capacitor that determines how much charge can be stored in it for a given potential difference between its terminals. It is equal to the ratio of the electrical charge stored to the potential difference. It is measured in ◊farads.

capacitor or *condenser* device for storing electric charge, used in electronic circuits; it consists of two or more metal plates separated by an insulating layer called a dielectric.

Its **capacitance** is the ratio of the charge stored on either plate to the potential difference between the plates. The SI unit of capacitance is the farad, but most capacitors have much smaller capacitances, and the microfarad (a millionth of a farad) is the commonly used practical unit.

capacity alternative term for ◊volume, generally used to refer to the amount of liquid or gas that may be held in a container. Units of capacity include litre and millilitre (metric); pint and gallon (imperial).

Cape Canaveral promontory on the Atlantic coast of Florida, USA, 367 km/228 mi N of Miami, used as a rocket launch site by ◊NASA.

Cape gooseberry plant *Physalis peruviana* of the potato family. Originating in South America, it is grown in South Africa, from where it takes its name. It is cultivated for its fruit, a yellow berry surrounded by a papery ◊calyx.

Capella or *Alpha Aurigae* brightest star in the constellation ◊Auriga and the sixth brightest star in the night sky. It is a visual and spectroscopic binary that consists of a pair of yellow-giant stars 45 light years from Earth, orbiting each other every 104 days.

It is a first-magnitude star, whose Latin name means the 'the Little Nanny Goat': its kids are the three adjacent stars Epsilon, Eta, and Zeta Aurigae.

Cape mountain zebra ◊zebra subspecies *Equus zebra zebra,* confined to South Africa. It almost became extinct in the 1940s, and in 1993 had a population of only 450, despite attempts at conservation. The main population is in Mountain Zebra Park in the east of the country, although some zebras have been moved to other parks in an attempt to build up other viable breeding herds.

caper trailing shrub native to the Mediterranean. Its flower buds are preserved in vinegar as a condiment. (*Capparis spinosa,* family Capparidaceae.)

capercaillie or *wood-grouse* or *cock of the wood* (*Gaelic capull coille, 'cock of the wood'*) large bird *Tetrao uroqallus* of the ◊grouse type, family Tetraonidae, order Galliformes. Found in coniferous woodland in Europe and N Asia, it is about the size of the turkey and resembles the ◊blackcock in appearance and polygamous habit. The general colour of the male is blackish-grey above, black below, with a dark green chest, and rounded tail which is fanned out in courtship. The female is smaller, mottled, and has a reddish breast barred with black. The feathers on the legs and feet are longest in winter time, and the toes are naked. The capercaillie feeds on insects, worms, berries, and young pine-shoots. At nearly 1 m/3 ft long, the male is the biggest gamebird in Europe. The female is about 60 cm/2 ft long.

Hunted to extinction in Britain in the 18th century, the capercaillie was reintroduced from Sweden in the 1830s and has re-established itself in Scotland.

capillarity spontaneous movement of liquids up or down narrow tubes, or capillaries. The movement is due to unbalanced molecular attraction at the boundary between the liquid and the tube. If liquid molecules near the boundary are more strongly attracted to molecules in the material of the tube than to other nearby liquid molecules, the liquid will rise in the tube. If liquid molecules are less attracted to the material of the tube than to other liquid molecules, the liquid will fall.

capillary in biology, narrowest blood vessel in vertebrates, 0.008–0.02 mm in diameter, barely wider than a red blood cell. Capillaries are distributed as **beds**, complex networks connecting arteries and veins. Capillary walls are extremely thin, consisting of a single layer of cells, and so nutrients, dissolved gases, and waste products can easily pass through them. This makes the capillaries the main area of exchange between the fluid (◊lymph) bathing body tissues and the blood.

To remember the principles of capillary action:

WATER RISES TO THE TOPS OF PLANTS BY CAPILLARY ACTION (CA), WHICH DEPENDS UPON COHESIVE AND ADHESIVE (C+A) PROPERTIES OF WATER

capillary in physics, a very narrow, thick-walled tube, usually made of glass, such as in a thermometer. Properties of fluids, such as surface tension and viscosity, can be studied using capillary tubes.

capitulum in botany, a flattened or rounded head (inflorescence) of numerous, small, stalkless flowers. The capitulum is surrounded by a circlet of petal-like bracts and has the appearance of a large, single flower.

Capricorn alternative term for Capricornus.

Capricornus zodiacal constellation in the southern hemisphere next to ◊Sagittarius. It is represented as a fish-tailed goat, and its brightest stars are third magnitude. The Sun passes through it late January to mid-February. In astrology, the dates for Capricornus (popularly known as Capricorn) are between about 22 December and 19 January(see ◊precession).

capsicum any of a group of pepper plants belonging to the nightshade family, native to Central and South America. The different species produce green to red fruits that vary in size. The small ones are used whole to give the hot flavour of chilli, or ground to produce cayenne or red pepper; the large pointed or squarish pods, known as sweet peppers or pimientos (green, red, or yellow peppers), are mild-flavoured and used as a vegetable. (Genus *Capsicum*, family Solanaceae.)

Capstone long-term US government project to develop a set of standards for publicly available ◊cryptography as authorized by the Computer Security Act 1987. The initiative has four elements: a data encryption ◊algorithm (Skipjack), a ◊hash function, a key exchange protocol, and a ◊digital signature algorithm.

The project is managed primarily by the National Security Agency (NSA) and the National Institute of Standards and Technology (NIST).

capsule in botany, a dry, usually many-seeded fruit formed from an ovary composed of two or more fused ◊carpels, which splits open to release the seeds. The same term is used for the spore-containing structure of mosses and liverworts; this is borne at the top of a long stalk or seta.

Capsules burst open (dehisce) in various ways, including lengthwise, by a transverse lid – for example, scarlet pimpernel *Anagallis arvensis* – or by a number of pores, either towards the top of the capsule, as in the poppy *Papaver*, or near the base, as in certain species of bellflower *Campanula*.

capture saving of user actions as digital data that can be read by a computer. In real-time data communications, it refers to using software to log a session so that the session can be saved to a file. The term is also used with reference to screens, where the graphical material displayed on a computer screen may be saved as a picture file. In the study of ◊human–computer interaction, the data captured are user keystrokes, mouse movements, and even facial expressions and muttered complaints so that developers can replay the session to help them design better ◊user interfaces.

captured rotation or ***synchronous rotation*** in astronomy, the circumstance in which one body in orbit around another, such as the moon of a planet, rotates on its axis in the same time as it takes to complete one orbit. As a result, the orbiting body keeps one face permanently turned towards the body about which it is orbiting. An example is the rotation of our own ◊Moon, which arises because of the tidal effects of the Earth over a long period of time.

capuchin monkey of the genus *Cebus* found in Central and South America, so called because the hairs on the head resemble the cowl of a Capuchin monk. Capuchins live in small groups, feed on fruit and insects, and have a long tail that is semiprehensile and can give support when climbing through the trees.

There are now thought to be only 800 yellow-breasted capuchins left in the wild, found only in the Atlantic forest in Bahia state, Brazil.

capybara world's largest rodent *Hydrochoerus hydrochaeris*, up to 1.3 m/4 ft long and 50 kg/110 lb in weight. It is found in South America, and belongs to the guinea-pig family. The capybara inhabits marshes and dense vegetation around water. It has thin, yellowish hair, swims well, and can rest underwater with just eyes, ears, and nose above the surface.

car small, driver-guided, passenger-carrying motor vehicle; originally the automated version of the horse-drawn carriage, meant to convey people and their goods over streets and roads.

Over 50 million motor cars are produced each year worldwide. The number of cars in the world in 1997 exceeded 500 million. Most are four-wheeled and have water-cooled, piston-type internal-combustion engines fuelled by petrol or diesel. Variations have existed for decades that use ingenious and often nonpolluting power plants, but the motor industry long ago settled on this general formula for the consumer market. Experimental and sports models are streamlined, energy-efficient, and hand-built.

origins Although it is recorded that in 1479 Gilles de Dom was paid 25 livres (the equivalent of 25 pounds of silver) by the treasurer of Antwerp in the Low Countries for supplying a self-propelled vehicle, the ancestor of the automobile is generally agreed to be the cumbersome steam carriage made by Nicolas-Joseph Cugnot in 1769, still preserved in Paris. Steam was an attractive form of power to the English pioneers, and in 1803 Richard Trevithick built a working steam carriage. Later in the 19th century, practical steam coaches were used for public transport until stifled out of existence by punitive road tolls and legislation.

the first motorcars Although a Frenchman, Jean Etienne Lenoir, patented the first internal-combustion engine (gas-driven but immobile) in 1860, and an Austrian, Siegfried Marcus, built a vehicle which was shown at the Vienna Exhibition (1873), two Germans, Gottleib Daimler and Karl Benz are generally regarded as the creators of the motorcar. In 1885 Daimler and Benz built and ran the first petrol-driven motorcar (they worked independently with Daimler building a very efficient engine and Benz designing a car but with a poor engine). The pattern for the modern motorcar was set by Panhard in 1891 (front radiator, Daimler engine under bonnet, sliding-pinion gearbox, wooden ladder-chassis) and Mercedes in 1901 (honeycomb radiator, in-line four-cylinder engine, gate-change gearbox, pressed-steel chassis) set the pattern for the modern car. Emerging with Haynes and Duryea in the early 1890s, US demand was so fervent that 300 makers existed by 1895; only 109 were left by 1900.

In Britain, cars were still considered to be light locomotives in the eyes of the law and, since the Red Flag Act 1865, had theoretically required someone to walk in front with a red flag (by night, a lantern). Despite these obstacles, which put UK development another ten years behind all others, in 1896 (after the Red Flag Act had been repealed) Frederick Lanchester produced an advanced and reliable vehicle, later much copied.

motorcars as an industry The period 1905–06 inaugurated a world motorcar boom continuing to the present day. Among the legendary cars of the early 20th century are: De Dion Bouton, with the first practical high-speed engines; Mors, notable first for racing and later as a silent tourer; Napier, the 24-hour record-holder at Brooklands 1907, unbeaten for 17 years; the incomparable Rolls-Royce Silver Ghost; the enduring Model T Ford; and the many types of Bugatti and Delage, from record-breakers to luxury tourers. After World War I popular motoring began with the era of cheap, light (baby) cars made by Citroên, Peugeot, and Renault (France); Austin, Morris, Clyno, and Swift (England); Fiat (Italy); Volkswagen (Germany); and the cheap though bigger Ford, Chevrolet, and Dodge in the USA. During the interwar years a great deal of racing took place, and the experience gained benefited the everyday motorist in improved efficiency, reliability, and safety. There was a divergence between the lighter, economical European car, with its good handling, and the heavier US car, cheap, rugged, and well adapted to long distances on straight roads at speed. By this time motoring had become a universal pursuit.

After World War II small European cars tended to fall into three categories, in about equal numbers: front engine and rear drive, the classic arrangement; front engine and front-wheel drive; rear engine and rear-wheel drive. Racing cars have the engine situated in the middle for balance. From the 1950s a creative resurgence produced in practical form automatic transmission for small cars, rubber suspension, transverse engine mounting, self-levelling ride, disc brakes, and safer wet-weather tyres.

By the mid-1980s, Japan was building 8 million cars a year, on par with the US. The largest Japanese manufacturer, Toyota, was producing 2.5 million cars per year.

The car has become the carapace, the protective and aggressive shell, of urban and suburban man.

MARSHALL MCLUHAN Canadian communications theorist. *Understanding Media* ch 22

caracal cat *Felis caracal* related to the ◊lynx. It has long black ear tufts, a short tail, and short reddish-fawn fur. It lives in bush and desert country in Africa, Arabia, and India, hunting birds and small mammals at night. Head and body length is about 75 cm/2.5 ft.

carambola small evergreen tree of SE Asia. The fruits, called **star fruit**, are yellowish, about 12 cm/4 in long, with a five-pointed star-shaped cross-section. They can be eaten raw, cooked, or pickled, and are juicily acidic. The juice is also used to remove stains from hands and clothes. (*Averrhoa carambola,* family Averrhoaceae.)

caramel complex mixture of substances produced by heating sugars, without charring, until they turn brown. Caramel is used as colouring and flavouring in foods. Its production in the manufacture of sugar confection gives rise to a toffeelike sweet of the same name.

The intricate chemical reactions involved in the production of caramel (caramelization) are not fully understood, but are known to result in the formation of a number of compounds. Two compounds in particular (acetylformoin and 4-hydroxy-2,5-dimethyl-3-furanone) are thought to contribute to caramel's characteristic flavour.

Commercially, the caramelization process is speeded up by the addition of selected ◊amino acids.

carapace protective covering of many animals, particularly the arched bony plate characteristic of the order Chelonia (tortoises, terrapins, and turtles), and the shield that protects the fore parts of crustaceans, such as crabs.

carat ***Arabic** quirrat* ***'seed'*** unit for measuring the mass of precious stones; it is equal to 0.2 g/0.00705 oz, and is part of the troy system of weights. It is also the unit of purity in gold (US karat). Pure gold is 24-carat; 22-carat (the purest used in jewellery) is 22 parts gold and two parts alloy (to give greater strength).

Originally, one carat was the weight of a carob seed.

caraway herb belonging to the carrot family. Native to northern temperate regions of Europe and Asia, it is grown for its spicy, aromatic seeds, which are used in cookery, medicine, and perfumery. (*Carum carvi,* family Umbelliferae.)

carbide compound of carbon and one other chemical element, usually a metal, silicon, or boron.

Calcium carbide (CaC_2) can be used as the starting material for many basic organic chemical syntheses, by the addition of water and generation of ethyne (acetylene). Some metallic carbides are used in engineering because of their extreme hardness and strength. Tungsten carbide is an essential ingredient of carbide tools and high-speed tools. The 'carbide process' was used during World War II to make organic chemicals from coal rather than from oil.

carbohydrate chemical compound composed of carbon, hydrogen, and oxygen, with the basic formula $C_m(H_2O)_n$, and related compounds with the same basic structure but modified ◊functional groups. As sugar and starch, carbohydrates form a major energy-providing part of the human diet.

The simplest carbohydrates are sugars (**monosaccharides**, such as glucose and fructose, and **disaccharides**, such as sucrose), which are soluble compounds, some with a sweet taste. When these basic sugar units are joined together in long chains or branching structures they form **polysaccharides**, such as starch and glycogen, which often serve as food stores in living organisms. Even more complex carbohydrates are known, including ◊chitin, which is found in the cell walls of fungi and the hard outer skeletons of insects, and ◊cellulose, which makes up the cell walls of plants. Carbohydrates form the chief foodstuffs of herbivorous animals.

carbolic acid common name for the aromatic compound ◊phenol.

CHEMISTRY OF CARBON

http://cwis.nyu.edu/pages/mathmol/modules/carbon/carbon1.html

Excellent introduction to carbon, the element at the heart of life as we know it. This site is illustrated throughout and explains the main basic forms of carbon and the importance of the way scientists choose to represent these various structures.

carbon ***Latin** carbo, carbonaris* ***'coal'*** nonmetallic element, symbol C, atomic number 6, relative atomic mass 12.011. It occurs on its own as diamond, graphite, and as fullerenes (the allotropes), as compounds in carbonaceous rocks such as chalk and limestone, as carbon dioxide in the atmosphere, as hydrocarbons in petroleum, coal, and natural gas, and as a constituent of all organic substances.

In its amorphous form, it is familiar as coal, charcoal, and soot. The atoms of carbon can link with one another in rings or chains, giving rise to innumerable complex compounds. Of the inorganic carbon compounds, the chief ones are **carbon dioxide**, a colourless gas formed when carbon is burned in an adequate supply of air; and **carbon monoxide** (CO), formed when carbon is oxidized in a limited supply of air. **Carbon disulphide** (CS_2) is a dense liquid with a sweetish odour. Another group of compounds is the **carbon halides**, including ◊carbon tetrachloride (tetrachloromethane, CCl_4).

When added to steel, carbon forms a wide range of alloys with useful properties. In pure form, it is used as a moderator in nuclear reactors; as colloidal graphite it is a good lubricant and, when deposited on a surface in a vacuum, reduces photoelectric and secondary emission of electrons. Carbon is used as a fuel in the form of coal or coke. The radioactive isotope carbon-14 (half-life 5,730 years) is used as a tracer in biological research and in radiocarbon dating. Analysis of interstellar dust has led to the discovery of discrete carbon molecules, each containing 60 carbon atoms. The C_{60} molecules have been named ◊buckminsterfullerenes because of their structural similarity to the geodesic domes designed by US architect and engineer Buckminster Fuller.

Life exists in the universe only because the carbon atom possesses certain exceptional properties.

JAMES HOPWOOD JEANS English mathematician and scientist. *Mysterious Universe*

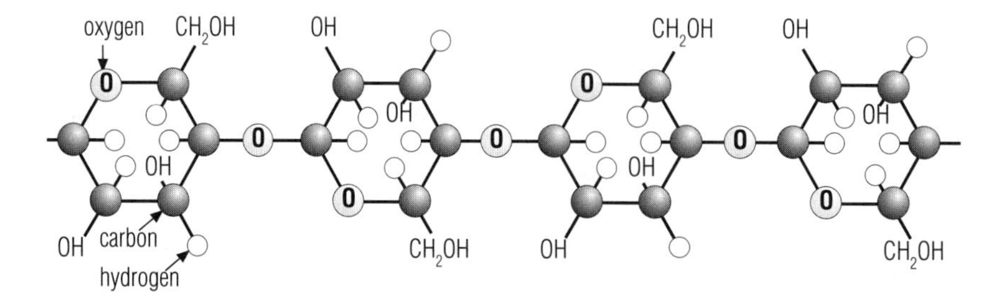

carbohydrate *A molecule of the polysaccharide glycogen (animal starch) is formed from linked glucose (C6H12O6) molecules. A typical glycogen molecule has 100–1,000 glucose units.*

carbonate CO_3^{2-} ion formed when carbon dioxide dissolves in water; any salt formed by this ion and another chemical element, usually a metal.

Carbon dioxide (CO_2) dissolves sparingly in water (for example, when rain falls through the air) to form carbonic acid (H_2CO_3), which unites with various basic substances to form carbonates. Calcium carbonate ($CaCO_3$) (chalk, limestone, and marble) is one of the most abundant carbonates known, being a constituent of mollusc shells and the hard outer skeletons of crustaceans.

carbonated water water in which carbon dioxide is dissolved under pressure. It forms the basis of many fizzy soft drinks such as soda water and lemonade.

carbon copy in e-mail, a duplicate copy of a message sent to multiple recipients; a nod to traditional office systems. It is often abbreviated in software and on line to 'cc'.

carbon cycle sequence by which ◊carbon circulates and is recycled through the natural world. The carbon element from carbon dioxide, released into the atmosphere by living things as a result of ◊respiration, is taken up by plants during ◊photosynthesis and converted into carbohydrates; the oxygen component is released back into the atmosphere. Some of this carbon becomes locked up in coal and petroleum and other sediments. The simplest link in the carbon cycle occurs when an animal eats a plant and carbon is transferred from, say, a leaf cell to the animal body. The oceans absorb 25–40% of all carbon dioxide released into the atmosphere.

Today, the carbon cycle is in danger of being disrupted by the increased consumption and burning of fossil fuels, and the burning of large tracts of tropical forests, as a result of which levels of carbon dioxide are building up in the atmosphere and probably contributing to the ◊greenhouse effect.

carbon cycle in astrophysics, a sequence of nuclear fusion reactions in which carbon atoms act as a catalyst to convert four hydrogen atoms into one helium atom with the release of energy. The carbon cycle is the dominant energy source for ordinary stars of mass greater than about 1.5 times the mass of the Sun.

Nitrogen and oxygen are also involved in the sequence so it is sometimes known as the **carbon-nitrogen-oxygen cycle** or **CNO cycle**.

carbon dating alternative name for ◊radiocarbon dating.

carbon dioxide CO_2 colourless, odourless gas, slightly soluble in water and denser than air. It is formed by the complete oxidation of carbon.

It is produced by living things during the processes of respiration and the decay of organic matter, and plays a vital role in the carbon cycle. It is used as a coolant in its solid form (known as 'dry ice'), and in the chemical industry. Its increasing density contributes to the ◊greenhouse effect and ◊global warming. Britain has 1% of the world's population, yet it produces 3% of CO_2 emissions; the USA has 5% of the world's population and produces 25% of CO_2 emissions. Annual releases of carbon dioxide reached 23 billion tones in 1997. According to a 1997 estimate by the World Energy council, carbon dioxide emissions rose by 7.8% between 1986 and 1996.

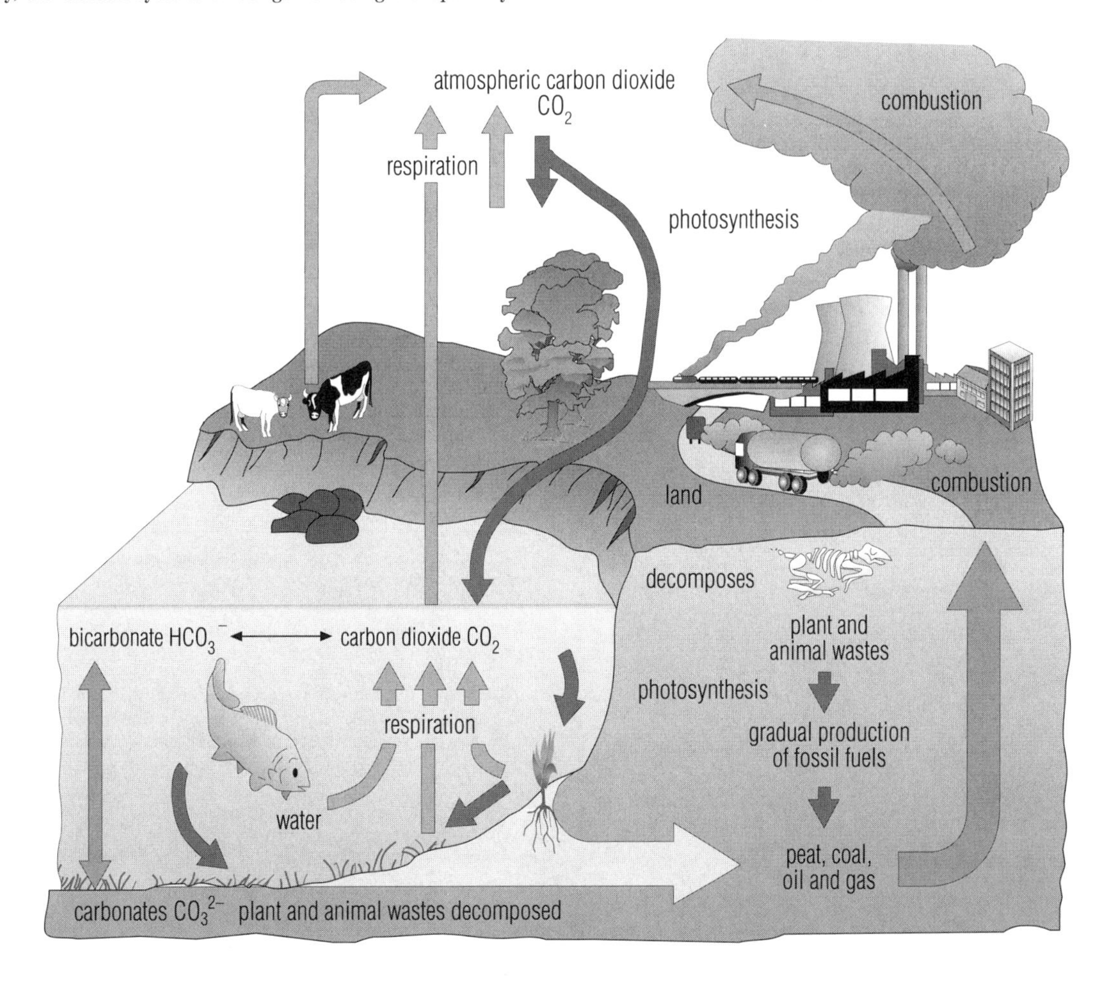

***carbon cycle** The carbon cycle is necessary for the continuation of life. Since there is only a limited amount of carbon in the Earth and its atmosphere, carbon must be continuously recycled if life is to continue. Other chemicals necessary for life – nitrogen, sulphur, and phosphorus, for example – also circulate in natural cycles.*

carbon fibre fine, black, silky filament of pure carbon produced by heat treatment from a special grade of Courtelle acrylic fibre and, used for reinforcing plastics. The resulting composite is very stiff and, weight for weight, has four times the strength of high-tensile steel. It is used in the aerospace industry, cars, and electrical and sports equipment.

carbonic acid H_2CO_3 weak, dibasic acid formed by dissolving carbon dioxide in water.

$$H_2O + CO_2 \leftrightarrow H_2CO_3$$

It forms two series of salts: ◊carbonates and ◊hydrogencarbonates. Fizzy drinks are made by dissolving carbon dioxide in water under pressure; soda water is a solution of carbonic acid.

Carboniferous period of geological time 362.5–290 million years ago, the fifth period of the Palaeozoic era. In the USA it is divided into two periods: the Mississippian (lower) and the Pennsylvanian (upper).

Typical of the lower-Carboniferous rocks are shallow-water ◊limestones, while upper-Carboniferous rocks have ◊delta deposits with ◊coal (hence the name). Amphibians were abundant, and reptiles evolved during this period.

carbon monoxide CO colourless, odourless gas formed when carbon is oxidized in a limited supply of air. It is a poisonous constituent of car exhaust fumes, forming a stable compound with haemoglobin in the blood, thus preventing the haemoglobin from transporting oxygen to the body tissues.

In industry, carbon monoxide is used as a reducing agent in metallurgical processes – for example, in the extraction of iron in ◊blast furnaces – and is a constituent of cheap fuels such as water gas. It burns in air with a luminous blue flame to form carbon dioxide.

carbon sequestration disposal of carbon dioxide waste in solid or liquid form. From 1993 energy conglomerates such as Shell, Exxon, and British coal have been researching ways to reduce their carbon dioxide emissions by developing efficient technologies to trap the gas and store it securely – for example, by burying it or dumping it in the oceans. See also ◊greenhouse effect.

carbon tetrachloride former name for the chlorinated organic compound ◊tetrachloromethane.

Carborundum trademark for a very hard, black abrasive, consisting of silicon carbide (SiC), an artificial compound of carbon and silicon. It is harder than ◊corundum but not as hard as ◊diamond.

It was first produced 1891 by US chemist Edward Acheson (1856–1931).

carboxyl group –COOH in organic chemistry, the acidic functional group that determines the properties of fatty acids (carboxylic acids) and amino acids.

carboxylic acid organic acid containing the carboxyl group (–COOH) attached to another group (R), which can be hydrogen (giving methanoic acid, HCOOH) or a larger molecule (up to 24 carbon atoms). When R is a straight-chain alkyl group (such as CH_3 or CH_3CH_2), the acid is known as a ◊fatty acid.

Examples of carboxylic acids include acetic acid, found in vinegar, malic acid, found in rhubarb, and citric acid, contained in oranges and lemons.

> To remember the (alpha, omega) dicarboxylic acids from $C_2 - C_{10}$:
>
> OH MY, SUCH GOOD APPLE PIE, SWEET AS SUGAR
>
> OXALIC, MALONIC, SUCCINIC, GLUTARIC, ADIPIC, PIMELIC, SUBERIC, AZELAIC, SEBACIC

carburation mixing of a gas, such as air, with a volatile hydrocarbon fuel, such as petrol, kerosene, or fuel oil, in order to form an explosive mixture. The process, which ensures that the maximum amount of heat energy is released during combustion, is used in internal-combustion engines. In most petrol engines the liquid fuel is atomized and mixed with air by means of a device called a **carburettor**.

carcinogen any agent that increases the chance of a cell becoming cancerous (see ◊cancer), including various chemical compounds, some viruses, X-rays, and other forms of ionizing radiation. The term is often used more narrowly to mean chemical carcinogens only.

carcinoma malignant ◊tumour arising from the skin, the glandular tissues, or the mucous membranes that line the gut and lungs.

cardinal number in mathematics, one of the series of numbers 0, 1, 2, 3, 4, ... Cardinal numbers relate to quantity, whereas ordinal numbers (first, second, third, fourth, ...) relate to order.

cardioid heart-shaped curve traced out by a point on the circumference of a circle, resulting from the circle rolling around the edge of another circle of the same diameter.

The polar equation of the cardioid is of the form:

$$r = a(1 + \cos \theta)$$

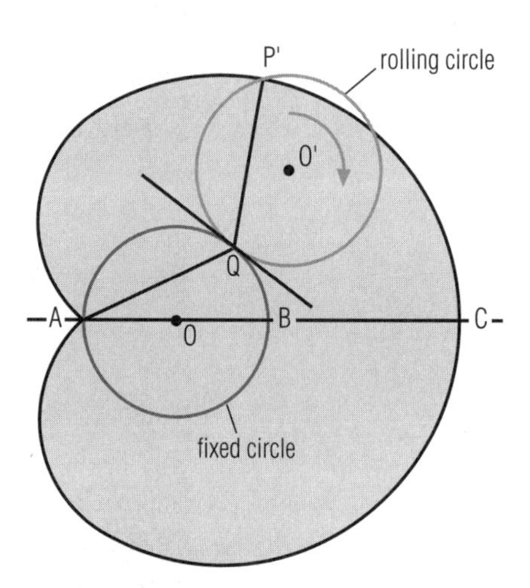

cardioid The cardioid is the curve formed when one circle rolls around the edge of another circle of the same size. It is named after its heart shape.

caribou the ◊reindeer of North America.

caries decay and disintegration, usually of the substance of teeth (cavity) or bone, caused by acids produced when the bacteria that live in the mouth break down sugars in the food. Fluoride, a low sugar intake, and regular brushing are all protective. Caries form mainly in the 45 minutes following consumption of sugary food.

Carina constellation of the southern hemisphere, represented as a ship's keel. Its brightest star is ◊Canopus, the second brightest in the night sky; it also contains Eta Carinae, a massive and highly luminous star embedded in a gas cloud, perhaps 8,000 light years away.

Carina was formerly regarded as part of Argo, and is situated in one of the brightest parts of the ◊Milky Way.

Carnac site of prehistoric ◊megaliths in Brittany, France, where remains of tombs and stone alignments of the period 2000–1500 BC (Neolithic and early Bronze Age) are found. Stones removed for local building have left some gaps in the alignments.

There are various groups of menhirs round the village of Carnac in the *département* of Morbihan, situated at Kermario (place of the dead), Kerlescan (place of burning), Erdeven, and St-Barbe. The largest of the stone alignments has 1,000 blocks of grey granite up to 4 m/13 ft high, extending over 2 km/1.2 mi. These ◊menhirs (standing stones) are arranged in 11 parallel rows, with a circle at the western end.

Stone circles and alignments are thought to be associated with astronomical and religious ritual, and those at Carnac may possibly have been used for calculating the phases of the moon.

carnassial tooth one of a powerful scissorlike pair of molars, found in all mammalian carnivores except seals. Carnassials are formed from an upper premolar and lower molar, and are shaped to produce a sharp cutting surface. Carnivores such as dogs transfer meat to the back of the mouth, where the carnassials slice up the food ready for swallowing.

carnation any of a large number of double-flowered cultivated varieties of a plant belonging to the ◊pink family. The flowers smell like cloves; they are divided into flake, bizarre, and picotees, according to whether the petals have one or more colours on their white base, have the colour appearing in strips, or have a coloured border to the petals. (*Dianthus caryophyllus,* family Carophyllaceae.)

carnauba palm tree native to South America, especially Brazil. It produces fine timber and a hard wax, used for polishes and lipsticks. (*Copernicia cerifera.*)

carnivore in zoology, mammal of the order Carnivora. Although its name describes the flesh-eating ancestry of the order, it includes pandas, which are herbivorous, and civet cats, which eat fruit.

Carnivores have the greatest range of body size of any mammalian order, from the 100 g/3.5 oz weasel to the 800 kg/1,764 lb polar bear.

The characteristics of the Carnivora are sharp teeth, small incisors, a well-developed brain, a simple stomach, a reduced or absent caecum, and incomplete or absent clavicles (collarbones); there are never less than four toes on each foot; the scaphoid and lunar bones are fused in the hand; and the claws are generally sharp and powerful.

Carnot, (Nicolas Leonard) Sadi (1796–1832) French scientist and military engineer who founded the science of thermodynamics. His pioneering work was *Reflexions sur la puissance motrice du feu/On the Motive Power of Fire,* which considered the changes that would take place in an idealized, frictionless steam engine.

Carnot's theorem showed that the amount of work that an engine can produce depends only on the temperature difference that occurs in the engine. He described the maximum amount of heat convertible into work by the formula $(T_1 - T_2)/T_2$, where T_1 is the temperature of the hottest part of the machine and T_2 is the coldest part.

Carnot cycle series of changes in the physical condition of a gas in a reversible heat engine, necessarily in the following order: (1) isothermal expansion (without change of temperature), (2) adiabatic expansion (without change of heat content), (3) isothermal compression, and (4) adiabatic compression.

The principles derived from a study of this cycle are important in the fundamentals of heat and ◊thermodynamics.

carob small Mediterranean tree belonging to the ◊legume family. Its pods, 20 cm/8 in long, are used as an animal feed; they are also the source of a chocolate substitute. (*Ceratonia siliqua,* family Leguminosae.)

carotene naturally occurring pigment of the ◊carotenoid group. Carotenes produce the orange, yellow, and red colours of carrots, tomatoes, oranges, and crustaceans.

carotenoid any of a group of yellow, orange, red, or brown pigments found in many living organisms, particularly in the ◊chloroplasts of plants. There are two main types, the **carotenes** and the **xanthophylls**. Both types are long-chain lipids (◊fats).

Some carotenoids act as accessory pigments in ◊photosynthesis, and in certain algae they are the principal light-absorbing pigments functioning more efficiently than ◊chlorophyll in low-intensity light. Carotenoids can also occur in organs such as petals, roots, and fruits, giving them their characteristic colour, as in the yellow and orange petals of wallflowers *Cheiranthus.* They are also responsible for the autumn colours of leaves, persisting longer than the green chlorophyll, which masks them during the summer.

carotid artery one of a pair of major blood vessels, one on each side of the neck, supplying blood to the head.

carp fish *Cyprinus carpio* found all over the world. It commonly grows to 50 cm/1.8 ft and 3 kg/7 lb, but may be even larger. It lives in lakes, ponds, and slow rivers. The wild form is drab, but cultivated forms may be golden, or may have few large scales (mirror carp) or be scaleless (leather carp). **Koi** carp are highly prized and can grow up to 1 m/3 ft long with a distinctive pink, red, white, or black colouring.

A large proportion of European freshwater fish belong to the carp family, Cyprinidae, and related fishes are found in Asia, Africa, and North America. The carp's fast growth, large size, and ability to live in still water with little oxygen have made it a good fish to farm, and it has been cultivated for hundreds of years and spread by human agency. Members of this family have a single non-spiny dorsal fin, pelvic fins well back on the body, and toothless jaws, although teeth in the throat form an efficient grinding apparatus. Minnows, roach, rudd, and many others, including goldfish, belong to this family. Chinese **grass carp** *Ctenopharyngodon idella* have been introduced (one sex only) to European rivers for weed control.

carpel female reproductive unit in flowering plants (◊angiosperms). It usually comprises an ◊ovary containing one or more ovules, the stalk or style, and a ◊stigma at its top which receives the pollen. A flower may have one or more carpels, and they may be separate or fused together. Collectively the carpels of a flower are known as the ◊gynoecium.

carpet beetle small black or brown beetle. The larvae are covered with hairs and often known as **woolly bears**; they feed on carpets, fabrics, and hides causing damage.

The **common carpet beetle** *Anthrenus scrophularia* measures 3–4 mm/0.1–0.2 in long. It is oval, black, and has yellow scales on the shield (pronotum) and ginger red on the wing cases (elytra). This species prefers fabrics to carpets, and the larvae, which are brownish, hairy grubs, eat holes into fabrics or make slits in the carpet.

Larvae of *Anthrenus museorum* can do extensive damage to museum collections. Specimens of stuffed animals or entire insect collections, which have been inadequately treated prior to storage, can be destroyed.

classification Carpet beetles are in the genus *Anthrenus* and *Attagenus* in the family Dermestidae of order Coleoptera, class Insecta, phylum Arthropoda.

carragheen species of deep-reddish branched seaweed. Named after Carragheen, near Waterford, in the Republic of Ireland, it is found on rocky shores on both sides of the Atlantic. It is exploited commercially in food and medicines and as cattle feed. (*Chondrus crispus.*)

carriage return (CR) in computing, a special code (◊ASCII value 13) that moves the screen cursor or a print head to the beginning of the current line. Most word processors and the ◊MS-DOS operating system use a combination of CR and line feed (LF – ASCII value 10) to represent a hard return. The ◊UNIX system, however, uses only LF and therefore files transferred between MS-DOS and UNIX require a conversion program.

carrier in medicine, anyone who harbours an infectious organism without ill effects but can pass the infection to others. The term is also applied to those who carry a recessive gene for a disease or defect without manifesting the condition.

carrion crow British bird *Corvus corone* of family Corvidae, order Passeriformes, closely related to *C. cornix,* the hooded crow. In the USA the name refers to the black vulture.

carrot hardy European biennial plant with feathery leaves and an orange tapering root that is eaten as a vegetable. It has been cultivated since the 16th century. The root has a high sugar content and also contains ◊carotene, which is converted into vitamin A by the human liver. (*Daucus carota,* family Umbelliferae.)

carrying capacity in ecology, the maximum number of animals of a given species that a particular area can support. When the carrying capacity is exceeded, there is insufficient food (or other resources) for the members of the population. The population may then be reduced by emigration, reproductive failure, or death through starvation.

Cartesian coordinates in ◊coordinate geometry, components used to define the position of a point by its perpendicular distance from a set of two or more axes, or reference lines. For a two-dimensional area defined by two axes at right angles (a horizontal x-axis and a vertical y-axis), the coordinates of a point are given by its perpendicular distances from the y-axis and x-axis, written in the form (x,y). For example, a point P that lies three units from the y-axis and four units from the x-axis has Cartesian coordinates (3,4) (see ◊abscissa and ◊ordinate). In three-dimensional coordinate geometry, points are located with reference to a third, z-axis, mutually at right angles to the x and y axes.

The Cartesian coordinate system can be extended to any finite number of dimensions (axes), and is used thus in theoretical mathematics. It is named after the French mathematician, René Descartes. The system is useful in creating technical drawings of machines or buildings, and in computer-aided design (◊CAD).

cartilage flexible bluish-white ◊connective tissue made up of the protein collagen. In cartilaginous fish it forms the skeleton; in other vertebrates it forms the greater part of the embryonic skeleton, and is replaced by ◊bone in the course of development, except in areas of wear such as bone endings, and the discs between the backbones. It also forms structural tissue in the larynx, nose, and external ear of mammals.

Cartilage does not heal itself, so where injury is severe the joint may need to be replaced surgically. In a 1994 trial, Swedish doctors repaired damaged knee joints by implanting cells cultured from the patient's own cartilage.

cartography art and practice of drawing ◊maps.

caryopsis dry, one-seeded ◊fruit in which the wall of the seed becomes fused to the carpel wall during its development. It is a type of ◊achene, and therefore develops from one ovary and does not split open to release the seed. Caryopses are typical of members of the grass family (Gramineae), including the cereals.

casein main protein of milk, from which it can be separated by the action of acid, the enzyme rennin, or bacteria (souring); it is also the main protein in cheese. Casein is used as a protein supplement in the treatment of malnutrition. It is used commercially in cosmetics, glues, and as a sizing for coating paper.

case-sensitive term describing a system that distinguishes between capitals and lower-case letters. Domain names and Internet addresses are typically not case-sensitive; however, a particular system may be case-sensitive for user IDs. On most systems, the software controlling passwords is case-sensitive, making it harder for an unauthorized user to guess a password.

cash crop crop grown solely for sale rather than for the farmer's own use, for example, coffee, cotton, or sugar beet. Many Third World countries grow cash crops to meet their debt repayments rather than grow food for their own people. The price for these crops depends on financial interests, such as those of the multinational companies and the International Monetary Fund.

In 1990 Uganda, Rwanda, Nicaragua, and Somalia were the countries most dependent on cash crops for income.

cashew tropical American tree. Widely cultivated in India and Africa, it produces poisonous kidney-shaped nuts that become edible after being roasted. (*Anacardium occidentale,* family Anacardiaceae.)

cassava or **manioc** plant belonging to the spurge family. Native to South America, it is now widely grown throughout the tropics for its starch-containing roots, from which tapioca and bread are made. (*Manihot utilissima,* family Euphorbiaceae.)

Cassava is grown as a staple crop in rural Africa, Asia, and South America. Altogether, it provides a staple crop for approximately 200 million people. The root cells contain the poison cyanoglucoside (converted to hydrogen cyanide in the body) but the plant's latex (milky fluid) contains enzymes that break down the poison. During the processing of cassava the two must mix; the commonest method is by fermentation, although some poison may remain. In Congo (formerly Zaire) and Mozambique at least 10,000 women and children suffered from chronic poisoning between 1985 and 1995.

Cassegrain telescope or ***Cassegrain reflector*** type of reflecting ◊telescope in which light collected by a concave primary mirror is reflected on to a convex secondary mirror, which in turn directs it back through a hole in the primary mirror to a focus behind it. As a result, the telescope tube can be kept short, allowing equipment for analyzing and recording starlight to be mounted behind the main mirror. All modern large astronomical telescopes are of the Cassegrain type.

It is named after the 17th century French astronomer, Cassegrain who first devised it as an improvement to the simpler ◊Newtonian telescope.

cassia bark of an aromatic SE Asian plant (*Cinnamomum cassia*) belonging to the laurel family (Lauraceae). It is very similar to cinnamon, and is often used as a substitute for it. *Cassia* is also a genus of pod-bearing tropical plants of the family Caesalpiniaceae, many of which have strong purgative (cleansing) properties; *C. senna* is the source of the laxative drug senna (which causes the bowels to empty).

Cassini joint space probe of the US agency NASA and the European Space Agency to the planet Saturn. *Cassini* was launched in October1997, to go into orbit around Saturn 2004, dropping off a sub-probe, *Huygens,* to land on Saturn's largest moon, Titan.

It was launched on a Titan 4 rocket, with its electricity supplied by 32 kg/70 lb of plutonium. This is the largest amount of plutonium ever to be sent into space, and provoked fears of contamination should Cassini, or its rocket, malfunction. A Titan 4 exploded in flight 1993.

CASSINI MISSION TO SATURN

http://miranda.colorado.edu/cassini/

Well-presented information on the purposes of the Cassini space probe. The complexities of designing the probe and its instruments and navigating it to the distant planet are explained in easy to understand language with the aid of good graphics. There are also interesting details of what is already known about Saturn's moons.

Cassiopeia prominent constellation of the northern hemisphere, named after the mother of Andromeda. It has a distinctive W-shape, and contains one of the most powerful radio sources in the sky, Cassiopeia A. This is the remains of a ◊supernova (star explosion) that occurred c. AD 1702, too far away to be seen from Earth.

It was in Cassiopeia that Tycho ◊Brahe observed a new star 1572, probably a supernova, since it was visible in daylight and outshone ◊Venus for ten days.

cassiterite or ***tinstone*** mineral consisting of reddish-brown to black stannic oxide (SnO_2), usually found in granite rocks. It is the chief ore of tin. When fresh it has a bright ('adamantine') lustre. It was formerly extensively mined in Cornwall, England; today Malaysia is the world's main supplier. Other sources of cassiterite are Africa, Indonesia, and South America.

cassowary large flightless bird, genus *Casuarius,* of the family Casuariidae, order Casuariiformes, found in New Guinea and N Australia, usually in forests. Related to the emu, the cassowary has a bare head with a horny casque, or helmet, on top, and brightly-coloured skin on the neck. Its loose plumage is black and its wings tiny, but it can run and leap well and defends itself by kicking.

Cassowaries stand up to 1.5 m/5 ft tall. They live in pairs and the male usually incubates the eggs, about six in number, which the female lays in a nest of leaves and grass.

casting process of producing solid objects by pouring molten material into a shaped mould and allowing it to cool. Casting is used to shape such materials as glass and plastics, as well as metals and alloys.

The casting of metals has been practised for more than 6,000 years, using first copper and bronze, then iron, and now alloys of zinc and other metals. The traditional method of casting metal is **sand casting**. Using a model of the object to be produced, a hollow mould is made in a damp sand and clay mix. Molten metal is then poured into the mould, taking its shape when it cools and solidifies. The sand mould is broken to release the casting. Permanent metal moulds called **dies** are also used for casting, in particular, small items in mass-production processes where molten metal is injected under pressure into cooled dies. **Continuous casting** is a method of shaping bars and slabs that involves pouring molten metal into a hollow, water-cooled mould of the desired cross section.

cast iron cheap but invaluable constructional material, most commonly used for car engine blocks. Cast iron is partly refined pig (crude) ◊iron, which is very fluid when molten and highly suitable for shaping by casting; it contains too many impurities (for example, carbon) to be readily shaped in any other way. Solid cast iron is heavy and can absorb great shock but is very brittle.

Castor or *Alpha Geminorum* second brightest star in the constellation ◊Gemini and the 23rd brightest star in the night sky. Along with the brighter ◊Pollux, it forms a prominent pair at the eastern end of Gemini, representing the head of the twins.

Second-magnitude Castor is 45 light years from Earth, and is one of the finest ◊binary stars in the sky for small telescopes. The two main components orbit each other over a period of 467 years. A third, much fainter, star orbits the main pair over a period probably exceeding 10,000 years. Each of the three visible components is a spectroscopic binary, making Castor a sextuple star system.

castoreum preputial follicles of the beaver, abbreviated as 'castor', and used in perfumery.

castor-oil plant tall tropical and subtropical shrub belonging to the spurge family. The seeds, called 'castor beans' in North America, yield the purgative castor oil (which cleans out the bowels) and also ricin, one of the most powerful poisons known. Ricin can be used to destroy cancer cells, leaving normal cells untouched. (*Ricinus communis,* family Euphorbiaceae.)

castration removal of the sex glands (either ovaries or testes). Male domestic animals may be castrated to prevent reproduction, to make them larger or more docile, or to eradicate disease.

Castration of humans was used in ancient and medieval times and occasionally later to preserve the treble voice of boy singers or, by Muslims, to provide eunuchs, trustworthy harem guards. If done in childhood, it inhibits sexual development: for instance, the voice remains high, and growth of hair on the face and body is reduced, owing to the absence of the hormones normally secreted by the testes.

CASTRATION

Exceptional boy singers used to be castrated to preserve their very high voices. Many of these **castrati** earned very high fees as opera singers.

casuarina any of a group of trees or shrubs with many species in Australia and New Guinea, also found in Africa and Asia. Commonly known as she-oaks, casuarinas have taken their Latin name from the similarity of their long, drooping branchlets to the feathers of the cassowary bird (whose genus is *Casuarius*). (Genus *Casuarina,* family Casuarinaceae.)

cat small, domesticated, carnivorous mammal *Felis catus,* often kept as a pet or for catching small pests such as rodents. Found in many colour variants, it may have short, long, or no hair, but the general shape and size is constant. Cats have short muzzles, strong limbs, and flexible spines which enable them to jump and climb. All walk on the pads of their toes (**digitigrade**) and have retractile claws, so are able to stalk their prey silently. They have large eyes and an acute sense of hearing. The canine teeth are long and well-developed, as are the shearing teeth in the side of the mouth.

origins Domestic cats have a common ancestor, the **African wild cat** *Felis libyca,* found across Africa and Arabia. This is similar to the **European wild cat** *F. silvestris.* Domestic cats can interbreed with either of these wild relatives. Various other species of small wild cat live in all continents except Antarctica and Australia. Large cats such as the lion, tiger, leopard, puma, and jaguar also belong to the cat family Felidae.

CAT

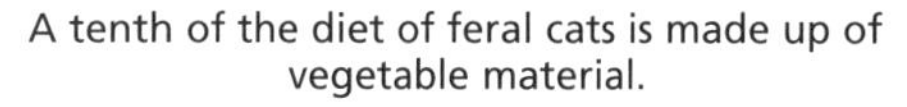

A tenth of the diet of feral cats is made up of vegetable material.

catabolism in biology, the destructive part of ◊metabolism where living tissue is changed into energy and waste products.

It is the opposite of ◊anabolism. It occurs continuously in the body, but is accelerated during many disease processes, such as fever, and in starvation.

catalpa any of a group of trees belonging to the trumpet creeper family, found in North America, China, and the West Indies. The northern catalpa (*Catalpa speciosa*) of North America grows to 30 m/100 ft and has heart-shaped deciduous leaves and tubular white flowers with purple borders. (Genus *Catalpa,* family Bignoniaceae.)

catalyst substance that alters the speed of, or makes possible, a chemical or biochemical reaction but remains unchanged at the end of the reaction. ◊Enzymes are natural biochemical catalysts. In practice most catalysts are used to speed up reactions.

catalytic converter device fitted to the exhaust system of a motor vehicle in order to reduce toxic emissions from the engine. It converts harmful exhaust products to relatively harmless ones by passing the exhaust gases over a mixture of catalysts coated on a metal or ceramic honeycomb (a structure that increases the surface area and therefore the amount of active catalyst with which the exhaust gases will come into contact). **Oxidation catalysts** (small amounts of precious palladium and platinum metals) convert hydrocarbons (unburnt fuel) and carbon monoxide into carbon dioxide and water, but do not affect nitrogen oxide emissions. **Three-way catalysts** (platinum and rhodium metals) convert nitrogen oxide gases into nitrogen and oxygen.

Over the lifetime of a vehicle, a catalytic converter can reduce hydrocarbon emissions by 87%, carbon monoxide emissions by 85%, and nitrogen oxide emissions by 62%, but will cause a slight increase in the amount of carbon dioxide emitted. Catalytic converters are standard in the USA, where a 90% reduction in pollution from cars was achieved without loss of engine performance or fuel economy. Only 10% of cars in Britain had catalytic converters in 1993.

Catalytic converters are destroyed by emissions from leaded petrol and work best at a temperature of 300°C. The benefits of catalytic converters are offset by any increase in the number of cars in use.

catamaran *Tamil 'tied log'* twin-hulled sailing vessel, based on the native craft of South America and the Indies, made of logs lashed together, with an outrigger. A similar vessel with three hulls is known as a trimaran. Car ferries with a wave-piercing catamaran design are also in use in parts of Europe and North America. They have a pointed main hull and two outriggers and travel at a speed of 35 knots (84.5 kph/52.5 mph).

cataract eye disease in which the crystalline lens or its capsule becomes cloudy, causing blindness. Fluid accumulates between the fibres of the lens and gives place to deposits of ◊albumin. These coalesce into rounded bodies, the lens fibres break down, and areas of the lens or the lens capsule become filled with opaque products of degeneration. The condition is estimated to have blinded more than 25 million people worldwide, and 150,000 in the UK.

The condition nearly always affects both eyes, usually one more than the other. In most cases, the treatment is replacement of the opaque lens with an artificial implant.

catarrh inflammation of any mucous membrane, especially of the nose and throat, with increased production of mucus.

Catarrhini Old World monkeys and apes, including macaques and baboons (Cercopithecidae) and chimpanzees and orang-utans (Pongidae). The term is now rarely used.

Catarrhines are characterized by having their nostrils close to each other and facing downwards. Many species have cheek pouches and brightly coloured buttocks. They have 32 teeth and do not have prehensile tails. Compare with the New World ◊Platyrrhini.

catastrophe theory mathematical theory developed by René Thom 1972, in which he showed that the growth of an organism proceeds by a series of gradual changes that are triggered by, and in turn trigger, large-scale changes or 'catastrophic' jumps. It also has applications in engineering – for example, the gradual strain on the structure of a bridge that can eventually result in a sudden collapse – and has been extended to economic and psychological events.

It is characteristic of science that the full explanations are often seized in their essence by the percipient scientist long in advance of any possible proof.

John Desmond Bernal
The Origin of Life 1967

catecholamine chemical that functions as a ◊neurotransmitter or a ◊hormone. Dopamine, adrenaline (epinephrine), and noradrenaline (norepinephrine) are catecholamines.

catenary curve taken up by a flexible cable suspended between two points, under gravity; for example, the curve of overhead suspension cables that hold the conductor wire of an electric railway or tramway.

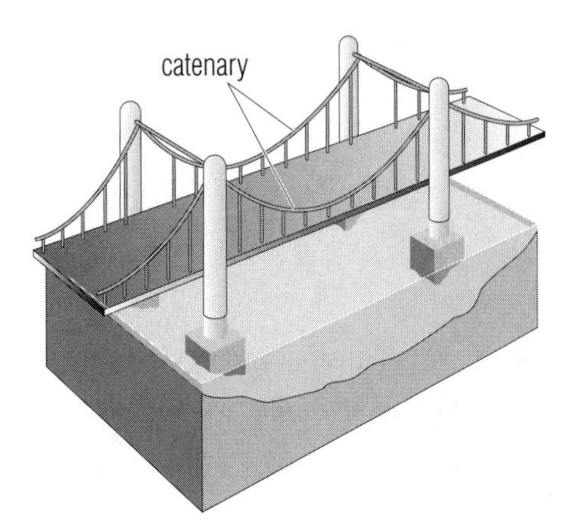

catenary Construction utilizing the form of a catenary can be seen in this suspension bridge.

caterpillar larval stage of a ◊butterfly or ◊moth. Wormlike in form, the body is segmented, may be hairy, and often has scent glands. The head has strong biting mandibles, silk glands, and a spinneret.

caterpillar A caterpillar of a notodontid moth from the rainforest of New Guinea. Its aposematic (warning) coloration is a signal to potential predators that it has an unpleasant taste. *Premaphotos Wildlife*

Many caterpillars resemble the plant on which they feed, dry twigs, or rolled leaves. Others are highly coloured and rely for their protection on their irritant hairs, disagreeable smell, or on their power to eject a corrosive fluid. Yet others take up a 'threat attitude' when attacked. Caterpillars emerge from eggs that have been laid by the female insect on the food plant and feed greedily, increasing greatly in size and casting their skins several times, until the pupal stage is reached. The abdominal segments bear a varying number of 'prolegs' as well as the six true legs on the thoracic segments.

CATERPILLAR

The caterpillars of some swallowtail butterflies deter predators by their uncanny resemblance to fresh bird droppings.

caterpillar track trade name for an endless flexible belt of metal plates on which certain vehicles such as tanks and bulldozers run, which takes the place of ordinary tyred wheels and improves performance on wet or uneven surfaces.

A track-laying vehicle has a track on each side, and its engine drives small cogwheels that run along the top of the track in contact with the ground. The advantage of such tracks over wheels is that they distribute the vehicle's weight over a wider area and are thus ideal for use on soft and waterlogged as well as rough and rocky ground.

catfish fish belonging to the order Siluriformes, in which barbels (feelers) on the head are well-developed, so giving a resemblance to the whiskers of a cat. Catfishes are found worldwide, mainly but not exclusively in fresh water, and are plentiful in South America.

The E European **giant catfish** or **wels** *Silurus glanis* grows to 1.5 m/5 ft long or more. It has been introduced to several places in Britain. The unrelated marine **wolffish** *Anarhicas lupus,* a deep-sea relative of the blenny, growing 1.2 m/4 ft long, is sometimes called a catfish.

catheter fine tube inserted into the body to introduce or remove fluids. The urinary catheter, passed by way of the urethra (the duct that leads urine away from the bladder) was the first to be used. In today's practice, catheters can be inserted into blood vessels, either in the limbs or trunk, to provide blood samples and local pressure measurements, and to deliver drugs and/or nutrients directly into the bloodstream.

cathode in chemistry, the negative electrode of an electrolytic ◊cell, towards which positive particles (cations), usually in solution, are attracted. See ◊electrolysis.

A cathode is given its negative charge by connecting it to the negative side of an external electrical supply. This is in contrast to the negative electrode of an electrical (battery) cell, which acquires its charge in the course of a spontaneous chemical reaction taking place within the cell.

cathode in electronics, the part of an electronic device in which electrons are generated. In a thermionic valve, electrons are produced by the heating effect of an applied current; in a photocell, they are produced by the interaction of light and a semiconducting material. The cathode is kept at a negative potential relative to the device's other electrodes (anodes) in order to ensure that the liberated electrons stream away from the cathode and towards the anodes.

cathode-ray oscilloscope (CRO) instrument used to measure electrical potentials or voltages that vary over time and to display the waveforms of electrical oscillations or signals. Readings are displayed graphically on the screen of a ◊cathode-ray tube.

cathode-ray tube vacuum tube in which a beam of electrons is produced and focused onto a fluorescent screen. It is an essential component of television receivers, computer visual display units, and oscilloscopes.

The electrons' kinetic energy is converted into light energy as they collide with the screen.

cation ◊ion carrying a positive charge. During electrolysis, cations in the electrolyte move to the cathode (negative electrode).

To remember that cations are atoms that have lost an electron:

CAT LOST AN EYE

catkin in flowering plants (◊angiosperms), a pendulous inflorescence, bearing numerous small, usually unisexual flowers. The tiny flowers are stalkless and the petals and sepals are usually absent or much reduced in size. Many types of trees bear catkins, including willows, poplars, and birches. Most plants with catkins are wind-pollinated, so the male catkins produce large quantities of pollen. Some ◊gymnosperms also have catkin-like structures that produce pollen, for example, the swamp cypress *Taxodium*.

CAT scan or *CT scan* (acronym for ***computerized axial tomography scan***) sophisticated method of X-ray imaging. Quick and non-invasive, CAT scanning is used in medicine as an aid to diagnosis, helping to pinpoint problem areas without the need for exploratory surgery. It is also used in archaeology to investigate mummies.

The CAT scanner passes a narrow fan of X-rays through successive slices of the suspect body part. These slices are picked up by crystal detectors in a scintillator and converted electronically into cross-sectional images displayed on a viewing screen. Gradually, using views taken from various angles, a three-dimensional picture of the organ or tissue can be built up and irregularities analysed.

cattle any large, ruminant, even-toed, hoofed mammal of the genus *Bos*, family Bovidae, including wild species such as the yak, gaur, gayal, banteng, and kouprey, as well as domestic breeds. Asiatic water buffaloes *Bubalus*, African buffaloes *Syncerus*, and American bison *Bison* are not considered true cattle. Cattle are bred for meat (beef cattle) or milk (dairy cattle).

Cattle were first domesticated in the Middle East during the Neolithic period, about 8000 BC. They were brought north into Europe by migrating Neolithic farmers. Fermentation in the four-chambered stomach allows cattle to make good use of the grass that is normally the main part of the diet. There are two main types of domesticated cattle: the European breeds, variants of *Bos taurus* descended from the ◊aurochs, **and the various breeds of zebu** ***Bos indicus***, the humped cattle of India, which are useful in the tropics for their ability to withstand the heat and diseases to which European breeds succumb. The old-established beef breeds are mostly British in origin. The Hereford, for example, is the premier English breed, ideally suited to rich lowland pastures but it will also thrive on poorer land such as that found in the US Midwest and the Argentine pampas.

Of the Scottish beef breeds, the Aberdeen Angus, a black and hornless variety, produces high-quality meat through intensive feeding methods. Other breeds include the Devon, a hardy early-maturing type, and the Beef Shorthorn, now less important than formerly, but still valued for an ability to produce good calves when crossed with less promising cattle. In recent years, more interest has been shown in other European breeds, their tendency to have less fat being more suited to modern tastes. Examples include the Charolais and the Limousin from central France, and the Simmental, originally from Switzerland. In the USA, four varieties of zebus, called Brahmans, have been introduced. They interbreed with *B. taurus* varieties and produce valuable hybrids that resist heat, ticks, and insects. For dairying purposes, a breed raised in many countries is variously known as the Friesian, Holstein, or Black and White. It can give enormous milk yields, up to 13,000 l/3,450 gal in a single lactation, and will produce calves ideally suited for intensive beef production. Other dairying types include the Jersey and Guernsey, whose milk has a high butterfat content, and the Ayrshire, a smaller breed capable of staying outside all year.

cauda tail, or taillike appendage; part of the *cauda equina*, a bundle of nerves at the bottom of the spinal cord in vertebrates.

cauliflower variety of ◊cabbage, with a large edible head of fleshy, cream-coloured flowers which do not fully mature. It is similar to broccoli but less hardy. (*Brassica oleracea botrytis*, family Cruciferae.)

caustic soda former name for ◊sodium hydroxide (NaOH).

cauterization in medicine, the use of special instruments to burn or fuse small areas of body tissue to destroy dead cells, prevent the spread of infection, or seal tiny blood vessels to minimize blood loss during surgery.

Cavalier King Charles spaniel breed of toy dog very similar to the related ◊King Charles spaniel. It is slightly larger (up to 8 kg/18 lb) and with a longer muzzle, and may be black and tan, tricoloured, ruby (a rich red) or Blenheim (chestnut and white).

cave roofed-over cavity in the Earth's crust usually produced by the action of underground water or by waves on a seacoast. Caves of the former type commonly occur in areas underlain by limestone, such as Kentucky and many Balkan regions, where the rocks are soluble in water. A **pothole** is a vertical hole in rock caused by water descending a crack; it is thus open to the sky.

Cave animals often show loss of pigmentation or sight, and under isolation, specialized species may develop. The scientific study of caves is called **speleology**. During the ◊ice age, humans began living in caves leaving many layers of debris that archaeologists have unearthed and dated in the Old World and the New. They also left cave art, paintings of extinct animals often with hunters on their trail. Celebrated caves include the Mammoth Cave in Kentucky, USA, 6.4 km/4 mi long and 38 m/125 ft high; the Caverns of Adelsberg (Postumia) near Trieste, Italy, which extend for many miles; Carlsbad Cave, New Mexico, the largest in the USA; the Cheddar Caves, England; Fingal's Cave, Scotland, which has a range of basalt columns; and Peak Cavern, England.

VIRTUAL CAVE

http://www.vol.it/MIRROR2/EN/CAVE/virtcave.html

Browse the mineral wonders unique to the cave environment – from bell canopies and bottlebrushes to splattermites and stalactites.

cave animal animal that has adapted to life within caves. The chief characteristics of cave animals are reduced or absent eyes and consequently other well-developed sense organs, such as antennae and feelers, and their lack of colour. Most are predators, owing to the lack of vegetable matter in this dark habitat.

Cave-dwellers include several species of snails found in Austrian caves that have developed blindness as a result of their mode of life and a genus of small cockroaches in caves of the Philippine Islands where the females are blind and flightless. There are numerous species of ◊cavefish including *Amblyopsis,* which occurs in the Mammoth Cave of Kentucky, USA. Cave-dwelling amphibians include several blind salamanders, such as the ◊olm.

cavefish cave-dwelling fish, which may belong to one of several quite unrelated groups, independently adapted to life in underground waters. Cavefish have in common a tendency to blindness and atrophy of the eye, enhanced touch-sensitive organs in the skin, and loss of pigment.

The **Kentucky blindfish** *Amblyopsis spelaea,* which lives underground in limestone caves, has eyes that are vestigial and beneath the skin, and a colourless body. The Mexican **cave characin** is a blind, colourless form of *Astyanax fasciatus* found in surface rivers of Mexico.

Cavendish experiment measurement of the gravitational attraction between lead spheres, which enabled English physicist and chemist Henry Cavendish to calculate 1798 a mean value for the mass and density of Earth, using Isaac Newton's law of universal gravitation.

cave spider small spider *Meta menardi* found in caves, cellars, and other dark places. It spins a small open web to catch insects. Its presence can be detected by its spherical egg sacs suspended by thick silk.

cavitation formation of partial vacuums (or cavities) in fluids at high velocities, produced by propellers or other machine parts in hydraulic engines, in accordance with Bernoulli's principle. When these vacuums collapse, pitting, vibration, and noise can occur in the metal parts in contact with the fluids.

cavy short-tailed South American rodent, family Caviidae, of which the guinea-pig *Cavia porcellus* is an example. Wild cavies are greyish or brownish with rather coarse hair. They live in small groups in burrows, and have been kept for food since ancient times.

cayenne pepper or ***red pepper*** spice produced from the dried fruits of several species of ◊capsicum (especially *Capsicum frutescens*), a tropical American group of plants. Its origins are completely different from black or white pepper, which comes from an East Indian plant (*Piper nigrum*).

cc symbol for ***cubic centimetre***; abbreviation for **carbon copy/copies**.

CCITT abbreviation for ◊Comité Consultatif International Téléphonique et Télégraphique, an organization that sets international communications standards.

CD abbreviation for ◊compact disc; **Corps Diplomatique** (French 'Diplomatic Corps'); **certificate of deposit**.

CD-I or **CD-i** (abbreviation for ***compact disc-interactive***) ◊compact disc developed by Philips for storing a combination of video, audio, text, and pictures. It was intended principally for the consumer market to be used in systems using a combination of computer and television. It flopped as a consumer system but is still used in education and training.

CD-quality sound digitized sound at 44.1 KHz and 16 bits, the standard defined in ISO 10149, known as the Red Book. CD-quality sound was designed to be the minimum standard required to reproduce every sound the human ear can hear. Most audio CDs are recorded to this level.

CD-R (abbreviation for ***compact disc-recordable***) compact disc on which data can be overwritten (compare ◊CD-ROM, compact disc

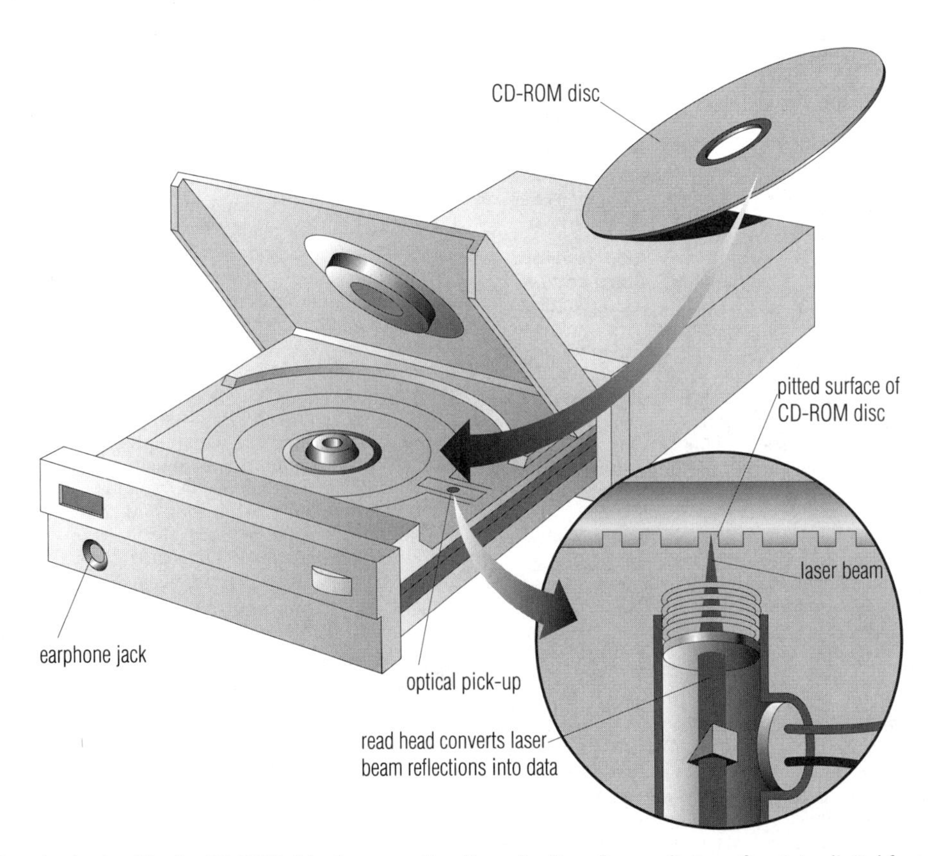

CD-ROM drive *Data is obtained by the CD-ROM drive by converting the reflections from a disc's surface into digital form.*

read-only memory). The disc combines magnetic and optical technology: during the writing process, a laser melts the surface of the disc, thereby allowing the magnetic elements of the surface layer to be realigned.

CD-ROM (abbreviation for ***compact-disc read-only memory***) computer storage device developed from the technology of the audio ◊compact disc. It consists of a plastic-coated metal disc, on which binary digital information is etched in the form of microscopic pits. This can then be read optically by passing a laser beam over the disc. CD-ROMs typically hold about 650 ◊megabytes of data, and are used in distributing large amounts of text, graphics, audio, and video, such as encyclopedias, catalogues, technical manuals, and games.

Standard CD-ROMs cannot have information written onto them by computer, but must be manufactured from a master, although recordable CDs, called CD-R discs, have been developed for use as computer discs. A compact disc, CD-RW, that can be overwritten repeatedly by a computer has also been developed. The compact disc, with its enormous storage capability, may eventually replace the magnetic disc as the most common form of backing store for computers.

The technology is being developed rapidly: a standard CD-ROM disc spins at between 240–1170 rpm, but faster discs have been introduced which speed up data retrieval to many times the standard speed. Research is being conducted into high-density CDs capable of storing many ◊gigabytes of data, made possible by using multiple layers on the surface of the disc, and by using double-sided discs. The first commercial examples of this research include DVD players and DVD-ROM computer discs launched in 1997.

PhotoCD, developed by Kodak and released in 1992, transfers ordinary still photographs onto CD-ROM discs.

CD-ROM drive in computing, a disc drive for reading CD-ROM discs. The vast majority of CD-ROM drives conform to the Yellow Book standard, defined by Philips and Sony. Because of this, all drives are essentially interchangeable. CD-ROM drives are available either as stand-alone or built-in units with a variety of interfaces (connections) and access times. (*See illustration on page 140.*)

CD-ROM XA (***CD-ROM extended architecture***) in computing, a set of standards for storing multimedia information on CD-ROM. Developed by Philips, Sony, and Microsoft, it is a partial development of the ◊CD-I standard. It interleaves data (as in CD-I) so that blocks of audio data are sandwiched between blocks of text, graphics, or video. This allows parallel streams of data to be handled, so that information can be seen and heard simultaneously.

Cebidae largest family of South American monkeys. The Cebidae is divided into five subfamilies represented by titis, squirrel monkeys, sakis, howlers, capuchins, and spider monkeys. They are all tree-living and are found in the rainforest where they feed on vegetable matter and small animals.

Cebus genus of South American monkey. The species have a well-developed big toe, a hairy prehensile tail, and 36 teeth. They include the ◊capuchin monkeys.

classification *Cebus* monkeys are typical of the family Cebidae in the order Primates.

cedar any of an Old World group of coniferous trees belonging to the pine family. The cedar of Lebanon (*Cedrus libani*) grows to great height and age in the mountains of Syria and Asia Minor. Of the historic forests on Mount Lebanon itself, only a few groups of trees remain. (Genus *Cedrus*, family Pinaceae.)

Ceefax ***'see facts'*** one of Britain's two ◊teletext systems (the other is Teletext), or 'magazines of the air', developed by the BBC and first broadcast in 1973.

In 1995 the BBC began testing a scheme to allow Ceefax (repackaged in HTML, hypertext markup language, to enable it to behave like Web pages) to be viewed on a PC by connecting a DAB (digital audio broadcasting) radio to the PC like a modem.

CEGB abbreviation for the former (until 1990) UK ***Central Electricity Generating Board***. For current industry structure see ◊electricity.

celandine either of two plants belonging to different families, the only similarity being their bright yellow flowers. The **greater celandine** (*Chelidonium majus*) belongs to the poppy family and is common in hedgerows. The **lesser celandine** (*Ranunculus ficaria*) is a member of the buttercup family and is a common wayside and meadow plant in Europe.

celeriac variety of garden celery belonging to the carrot family, with an edible turniplike root and small bitter stems. (*Apium graveolens* var. *rapaceum*, family Umbelliferae.)

celery Old World plant belonging to the carrot family. It grows wild in ditches and salt marshes and has a coarse texture and sharp taste. Cultivated varieties of celery are grown under cover to make the edible stalks less bitter. (*Apium graveolens*, family Umbelliferae.)

celestial mechanics the branch of astronomy that deals with the calculation of the orbits of celestial bodies, their gravitational attractions (such as those that produce the Earth's tides), and also the orbits of artificial satellites and space probes. It is based on the laws of motion and gravity laid down by Isaac ◊Newton.

celestial sphere imaginary sphere surrounding the Earth, on which the celestial bodies seem to lie. The positions of bodies such as stars, planets, and galaxies are specified by their coordinates on the celestial sphere. The equivalents of latitude and longitude on the celestial sphere are called ◊declination and ◊right ascension (which is measured in hours from 0 to 24). The **celestial poles** lie directly above the Earth's poles, and the **celestial equator** lies over the Earth's Equator. The celestial sphere appears to rotate once around the Earth each day, actually a result of the rotation of the Earth on its axis.

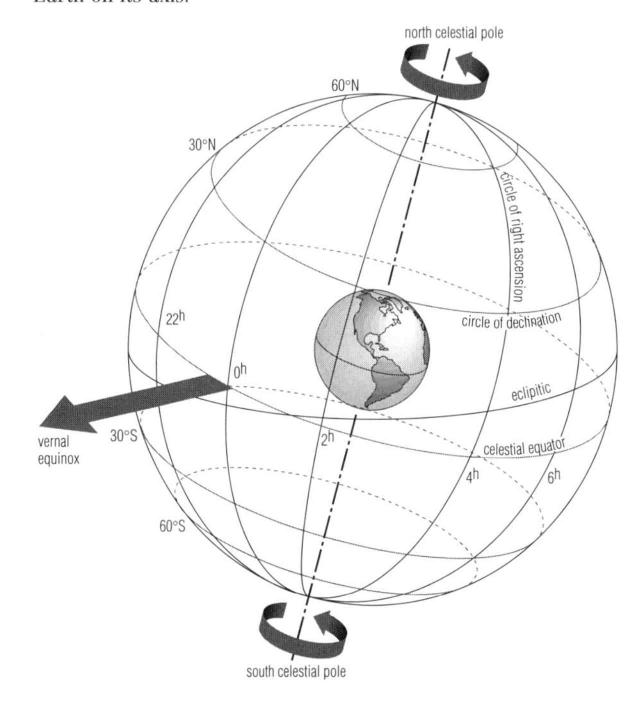

celestial sphere *The main features of the celestial sphere. The equivalents of latitude and longitude on the celestial sphere are declination and right ascension. Declination runs from 0° at the celestial equator to 90° at the celestial poles. Right ascension is measured in hours eastwards from the vernal equinox, one hour corresponding to 15° of longitude.*

cell in biology, a discrete, membrane-bound portion of living matter, the smallest unit capable of an independent existence. All living organisms consist of one or more cells, with the exception of

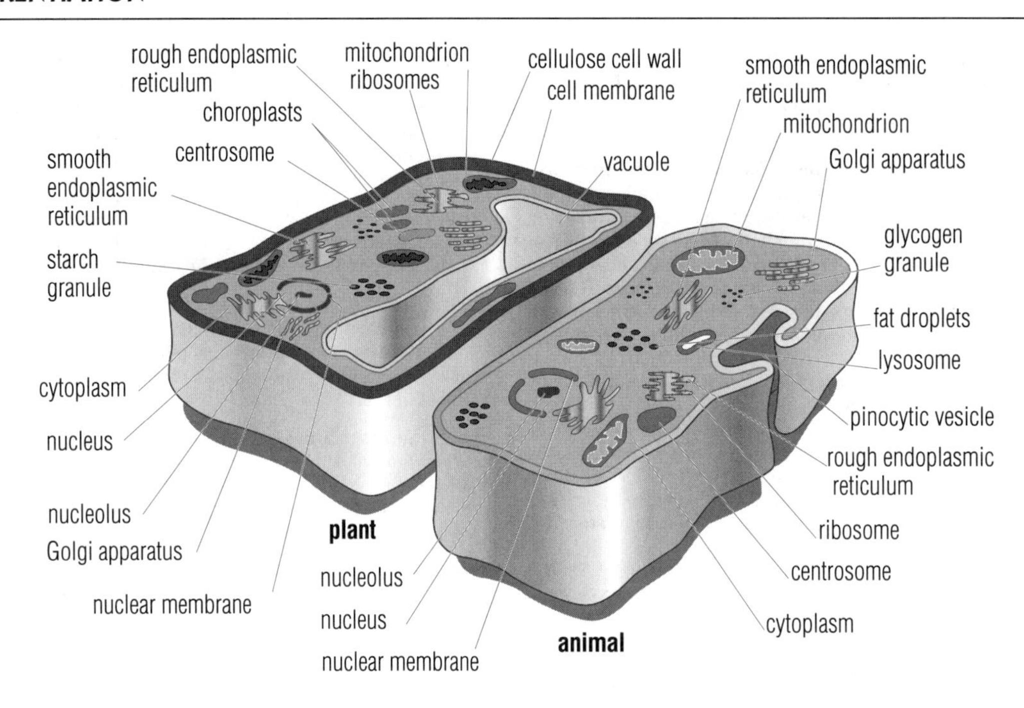

cell *Typical plant and animal cell. Plant and animal cells share many structures, such as ribosomes, mitochondria, and chromosomes, but they also have notable differences: plant cells have chloroplasts, a large vacuole, and a cellulose cell wall. Animal cells do not have a rigid cell wall but have an outside cell membrane only.*

◊viruses. Bacteria, protozoa, and many other microorganisms consist of single cells, whereas a human is made up of billions of cells. Essential features of a cell are the membrane, which encloses it and restricts the flow of substances in and out; the jellylike material within, the ◊cytoplasm; the ◊ribosomes, which carry out protein synthesis; and the ◊DNA, which forms the hereditary material.

Truth is ever to be found in simplicity, and not in the multiplicity and confusion of things ... He is the God of order and not of confusion.

ISAAC NEWTON English physicist and mathematician. Quoted in R L Weber, *More Random Walks in Science*

cell differentiation in developing embryos, the process by which cells acquire their specialization, such as heart cells, muscle cells, skin cells, and brain cells. The seven-day-old human pre-embryo consists of thousands of individual cells, each of which is destined to assist in the formation of individual organs in the body.

Research has shown that the eventual function of a cell, in for example, a chicken embryo, is determined by the cell's position. The embryo can be mapped into areas corresponding with the spinal cord, the wings, the legs, and many other tissues. If the embryo is relatively young, a cell transplanted from one area to another will develop according to its new position. As the embryo develops the cells lose their flexibility and become unable to change their destiny.

CELLS ALIVE

http://www.cellsalive.com/

Lively and attractive collection of microscopic and computer-generated images of living cells and microorganisms. It includes sections on HIV infection, penicillin, and how antibodies are made.

To remember the phases of cell division:

PRIME MINISTER – A TOAD!

PROPHASE, METAPHASE, ANAPHASE, TELOPHASE

cell division the process by which a cell divides, either ◊meiosis, associated with sexual reproduction, or ◊mitosis, associated with growth, cell replacement, or repair. Both forms involve the duplication of DNA and the splitting of the nucleus.

cell, electrical or ***voltaic cell*** or ***galvanic cell*** device in which chemical energy is converted into electrical energy; the popular name is ◊'battery', but this actually refers to a collection of cells in one unit. The reactive chemicals of a **primary cell** cannot be replenished, whereas **secondary cells** – such as storage batteries – are rechargeable: their chemical reactions can be reversed and the original condition restored by applying an electric current. It is dangerous to attempt to recharge a primary cell.

cell, electrolytic device to which electrical energy is applied in order to bring about a chemical reaction; see ◊electrolysis.

cell membrane or ***plasma membrane*** thin layer of protein and fat surrounding cells that controls substances passing between the cytoplasm and the intercellular space. The cell membrane is semipermeable, allowing some substances to pass through and some not.

Generally, small molecules such as water, glucose, and amino acids can penetrate the membrane, while large molecules, such as starch, cannot. Membranes also play a part in ◊active transport, hormonal response, and cell metabolism.

cellophane transparent wrapping film made from wood ◊cellulose, widely used for packaging, first produced by Swiss chemist Jacques Edwin Brandenberger in 1908.

Cellophane is made from wood pulp, in much the same way that the artificial fibre ◊rayon is made: the pulp is dissolved in chemicals to form a viscose solution, which is then pumped through a long narrow slit into an acid bath where the emergent viscose stream turns into a film of pure cellulose.

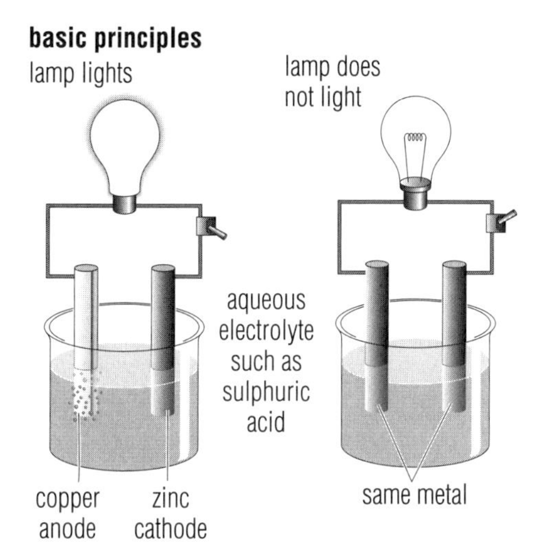

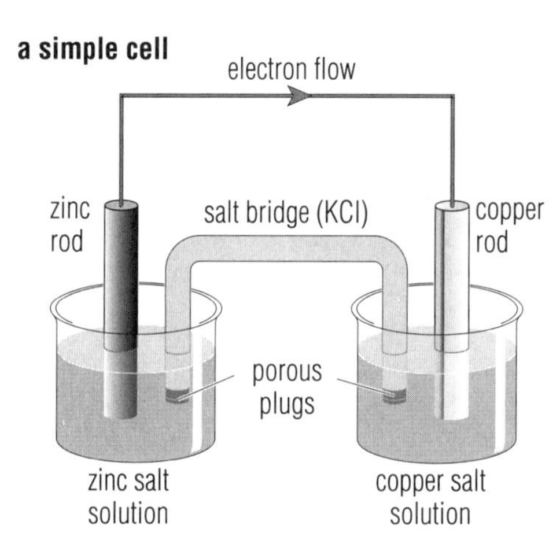

cell, electrical *When electrical energy is produced from chemical energy using two metals acting as electrodes in a aqueous solution, it is sometimes known as a galvanic cell or voltaic cell. Here the two metals copper (+) and zinc (–) are immersed in dilute sulphuric acid, which acts as an electrolyte. If a light bulb is connected between the two, an electric current will flow with bubbles of gas being deposited on the electrodes in a process known as polarization.*

cellphone short for ◊cellular phone.

cell sap dilute fluid found in the large central vacuole of many plant cells. It is made up of water, amino acids, glucose, and salts. The sap has many functions, including storage of useful materials, and provides mechanical support for non-woody plants.

cellular modem type of ◊modem that connects to a ◊cellular phone for the wireless transmission of data.

cellular phone or ***cellphone*** mobile radio telephone, one of a network connected to the telephone system by a computer-controlled communication system. Service areas are divided into small 'cells', about 5 km/3 mi across, each with a separate low-power transmitter.

The cellular system allows the use of the same set of frequencies with the minimum risk of interference. Nevertheless, in crowded city areas, cells can become overloaded. This has led to a move away from analogue transmissions to digital methods that allow more calls to be made within a limited frequency range.

cellular phone wireless phone that operates over radio frequencies and links calls to the public telephone system via a base station; the area covered by each base station is called a cell. Unlike phones connected up by telephone lines, cellular phones allow mobility, as calls can be made while moving from one radio cell to another. A network of connected base stations and exchanges connects the cellular calls to the public telephone system.

cellulite fatty compound alleged by some dietitians to be produced in the body by liver disorder and to cause lumpy deposits on the hips and thighs. Medical opinion generally denies its existence, attributing the lumpy appearance to a type of subcutaneous fat deposit.

celluloid transparent or translucent, highly flammable, plastic material (a ◊thermoplastic) made from cellulose nitrate and camphor. It was once used for toilet articles, novelties, and photographic film, but has now been replaced by the nonflammable substance ◊cellulose acetate.

cellulose complex ◊carbohydrate composed of long chains of glucose units, joined by chemical bonds called glycosidic links. It is the principal constituent of the cell wall of higher plants, and a vital ingredient in the diet of many ◊herbivores. Molecules of cellulose are organized into long, unbranched microfibrils that give support to the cell wall. No mammal produces the enzyme cellulase, necessary for digesting cellulose; mammals such as rabbits and cows are only able to digest grass because the bacteria present in their gut can manufacture it.

Cellulose is the most abundant substance found in the plant kingdom. It has numerous uses in industry: in rope-making; as a source of textiles (linen, cotton, viscose, and acetate) and plastics (cellophane and celluloid); in the manufacture of nondrip paint; and in such foods as whipped dessert toppings.

Japanese chemists produced the first synthetic cellulose in 1996 and the gene for the plant enzyme that makes cellulose was identified by Australian biologists in 1998.

cellulose acetate or **cellulose ethanoate** chemical (an ◊ester) made by the action of acetic acid (ethanoic acid) on cellulose. It is used in making transparent film, especially photographic film; unlike its predecessor, celluloid, it is not flammable.

cellulose nitrate or ***nitrocellulose*** series of esters of cellulose with up to three nitrate (NO_3) groups per monosaccharide unit. It is made by the action of concentrated nitric acid on cellulose (for example, cotton waste) in the presence of concentrated sulphuric acid. Fully nitrated cellulose (gun cotton) is explosive, but esters with fewer nitrate groups were once used in making lacquers, rayon, and plastics, such as coloured and photographic film, until replaced by the nonflammable cellulose acetate. ◊Celluloid is a form of cellulose nitrate.

cell wall in plants, the tough outer surface of the cell. It is constructed from a mesh of ◊cellulose and is very strong and relatively inelastic. Most living cells are turgid (swollen with water; see ◊turgor) and develop an internal hydrostatic pressure (wall pressure) that acts against the cellulose wall. The result of this turgor pressure is to give the cell, and therefore the plant, rigidity. Plants that are not woody are particularly reliant on this form of support.

The cellulose in cell walls plays a vital role in global nutrition. No vertebrate is able to produce cellulase, the enzyme necessary for the breakdown of cellulose into sugar. Yet most mammalian herbivores rely on cellulose, using secretions from microorganisms living in the gut to break it down. Humans cannot digest the cellulose of the cell walls; they possess neither the correct gut microorganisms nor the necessary grinding teeth. However, cellulose still forms a necessary part of the human diet as ◊fibre (roughage).

Celsius scale of temperature, previously called centigrade, in which the range from freezing to boiling of water is divided into 100 degrees, freezing point being 0 degrees and boiling point 100 degrees.

The degree centigrade (°C) was officially renamed Celsius in 1948 to avoid confusion with the angular measure known as the centigrade (one hundredth of a grade). The Celsius scale is named after the Swedish astronomer Anders Celsius (1701–1744), who devised it in 1742 but in reverse (freezing point was 100°; boiling point 0°).

Celsius, Anders (1701–1744)

Swedish astronomer, physicist, and mathematician who introduced the Celsius scale of temperature. His other scientifc works include a paper on accurately determining the shape and size of the Earth, some of the first attempts to gauge the magnitude of the stars in the constellation Aries, and a study of the falling water level of the Baltic Sea.

Mary Evans Picture Library

CELSIUS, ANDERS (1701–1744)

http://www.astro.uu.se/history/Celsius_eng.html

Biography of Swedish astronomer Anders Celsius. Famous for his creation of the Celsius temperature scale, the astronomer was also the first to realise that the aurora phenomenon was magnetic in nature. A portrait of Celsius is also available from this page.

cement any bonding agent used to unite particles in a single mass or to cause one surface to adhere to another. **Portland cement** is a powder obtained from burning together a mixture of lime (or chalk) and clay, which when mixed with water and sand or gravel turns into mortar or concrete.

In geology, cement refers to a chemically precipitated material such as carbonate that occupies the interstices of clastic rocks.

The term 'cement' covers a variety of materials, such as fluxes and pastes, and also bituminous products obtained from tar. In 1824 English bricklayer Joseph Aspdin (1779–1855) created and patented the first Portland cement, so named because its colour in the hardened state resembled that of Portland stone, a limestone used in building.

Cenozoic or ***Caenozoic*** era of geological time that began 65 million years ago and continues to the present day. It is divided into the Tertiary and Quaternary periods. The Cenozoic marks the emergence of mammals as a dominant group, including humans, and the formation of the mountain chains of the Himalayas and the Alps.

censorship banning of certain types of information from public access. Concerns over the ready availability of material such as bomb recipes and pornography have led a number of countries to pass laws attempting to censor the Internet. The best known of these is the US ◊Communications Decency Act 1996, but initiatives have been taken in other countries, for example Singapore, which announced 1996 new regulations bringing the Internet under the Singapore Broadcasting Authority and requiring all access providers and users to be registered and licensed. Less formal pressures have been applied against ◊Internet Service Providers in Germany and the UK to block specific types of material.

centaur in astronomy, cometlike object with an unstable orbit of less than 200 years. They are 100–400 km in diameter and are redder than other asteroids. The six known centaurs originated in the ◊Kuiper belt. ◊Chiron and ◊Pholus are centaurs.

Centaurus large, bright constellation of the southern hemisphere, represented as a centaur. Its brightest star, ◊Alpha Centauri, is a triple star, and contains the closest star to the Sun, Proxima Centauri, which is only 4.3 light years away. Omega Centauri, which is just visible to the naked eye as a hazy patch, is the largest and brightest ◊globular cluster of stars in the sky, 16,000 light years away.

Alpha and Beta Centauri are both of the first magnitude and, like Alpha and Beta Ursae Majoris, are known as 'the Pointers', as a line joining them leads to ◊Crux. Centaurus A, a galaxy 15 million light years away, is a strong source of radio waves and X-rays.

centigrade former name for the ◊Celsius temperature scale.

centipede jointed-legged animal of the group Chilopoda, members of which have a distinct head and a single pair of long antennae. Their bodies are composed of segments (which may number nearly 200), each of similar form and bearing a single pair of legs. Most are small, but the tropical *Scolopendra gigantea* may reach 30 cm/1 ft in length. **Millipedes**, class Diplopoda, have fewer segments (up to 100), but have two pairs of legs on each.

Nocturnal, frequently blind, and all carnivorous, centipedes live in moist, dark places, and protect themselves by a poisonous secretion. They have a pair of poison claws, and strong jaws with poison fangs. The bite of some tropical species is dangerous to humans. Several species live in Britain, *Lithobius forficatus* being the most common.

centipede *A giant centipede of the species* Scolopendra, *active at night in the forests of Madagascar. Unlike millipedes, which have two pairs of legs to each body segment, centipedes have one pair of legs to a segment. They also have venomous fangs with which to immobilize their prey, mostly worms and insects but occasionally small vertebrates.* *Premaphotos Wildlife*

central dogma in genetics and evolution, the fundamental belief that ◊genes can affect the nature of the physical body, but that changes in the body (◊acquired character, for example, through use or accident) cannot be translated into changes in the genes.

central heating system of heating from a central source, typically of a house, larger building, or group of buildings, as opposed to heating each room individually. Steam heat and hot-water heat are the most common systems in use. Water is heated in a furnace burning oil, gas, or solid fuel, and, as steam or hot water, is then pumped through radiators in each room. The level of temperature can be selected by adjusting a ◊thermostat on the burner or in a room.

Central heating has its origins in the hypocaust heating system introduced by the Romans nearly 2,000 years ago. From the 18th century, steam central heating, usually by pipe, was available in the West and installed in individual houses on an ad hoc basis. The Scottish engineer James ◊Watt heated his study with a steam pipe connected to a boiler, and Matthew Boulton installed steam heating in a friend's Birmingham house. Not until the latter half of the

20th century was central heating in general use. Central heating systems are usually switched on and off by a time switch. Another kind of central heating system uses hot air, which is pumped through ducts (called risers) to grills in the rooms. Underfloor heating (called radiant heat) is used in some houses, the heat coming from electric elements buried in the floor. New energy-efficient houses use heat from the Sun and good insulation to replace some central heating.

central nervous system (CNS) the brain and spinal cord, as distinct from other components of the ◊nervous system. The CNS integrates all nervous function.

In invertebrates it consists of a paired ventral nerve cord with concentrations of nerve-cell bodies, known as ◊ganglia in each segment, and a small brain in the head. Some simple invertebrates, such as sponges and jellyfishes, have no CNS but a simple network of nerve cells called a nerve net.

central processing unit (CPU) main component of a computer, the part that executes individual program instructions and controls the operation of other parts. It is sometimes called the central processor or, when contained on a single integrated circuit, a microprocessor.

The CPU has three main components: the **arithmetic and logic unit** (ALU), where all calculations and logical operations are carried out; a **control unit**, which decodes, synchronizes, and executes program instructions; and the **immediate access memory**, which stores the data and programs on which the computer is currently working. All these components contain ◊registers, which are memory locations reserved for specific purposes.

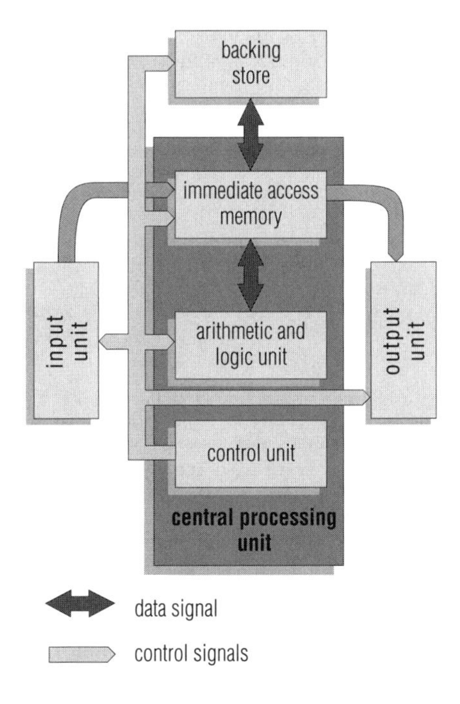

__central processing unit__ The relationship between the three main areas of a computer's central processing unit. The arithmetic and logic unit (ALU) does the arithmetic, using the registers to store intermediate results, supervised by the control unit. Input and output circuits connect the ALU to external memory, input, and output devices.

centre of gravity the point in an object about which its weight is evenly balanced. In a uniform gravitational field, this is the same as the centre of mass.

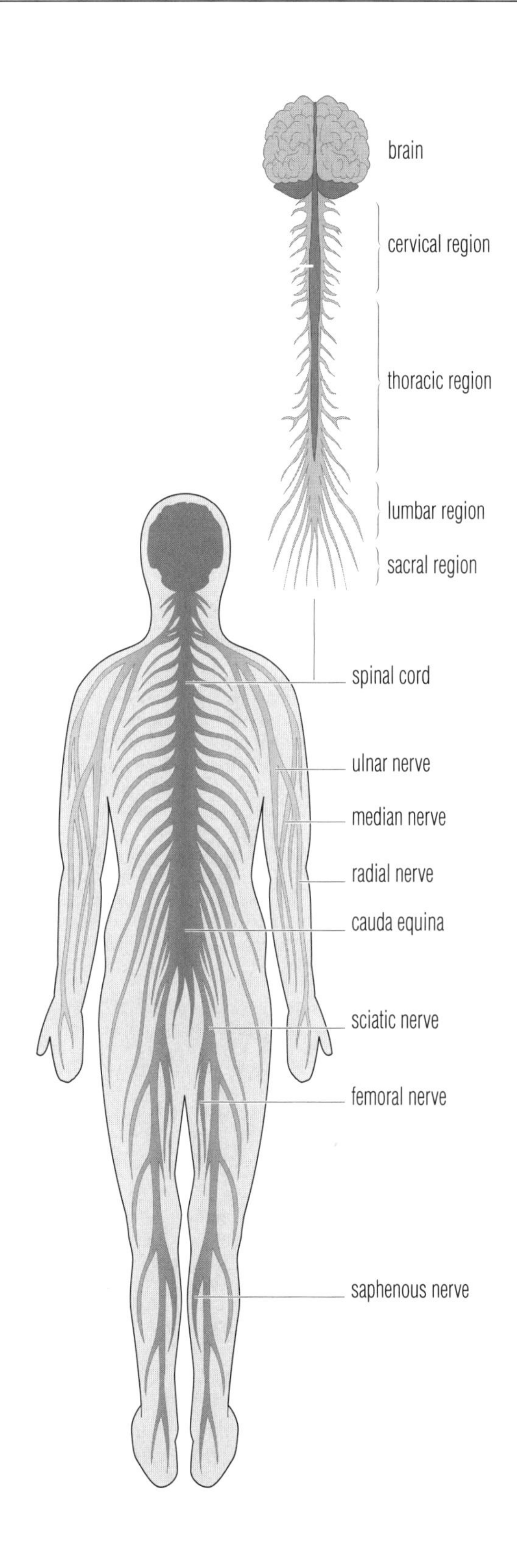

__central nervous system__ The central nervous system (CNS) with its associated nerves. The CNS controls and integrates body functions. In humans and other vertebrates it consists of a brain and a spinal cord, which are linked to the body's muscles and organs by means of the peripheral nervous system.

> *Were it not for gravity one man might hurl another by a puff of his breath into the depths of space, beyond recall for all eternity.*
>
> RUGGERIO GIUSEPPE BOSCOVICH Croatian-born Italian scientist. *Theoria*

centre of mass point in or near an object at which the whole mass of the object may be considered to be concentrated. A symmetrical homogeneous object such as a sphere or cube has its centre of mass at its geometrical centre; a hollow object (such as a cup) may have its centre of mass in space inside the hollow.

For an object to be in stable equilibrium, a vertical line down through its centre of mass must run within the boundaries of its base; if tilted until this line falls outside the base, the object becomes unstable and topples over.

centrifugal force useful concept in physics, based on an apparent (but not real) force. It may be regarded as a force that acts radially outward from a spinning or orbiting object, thus balancing the ◊centripetal force (which is real). For an object of mass m moving with a velocity v in a circle of radius r, the centrifugal force F equals mv^2/r (outward).

centrifuge apparatus that rotates containers at high speeds, creating centrifugal forces. One use is for separating mixtures of substances of different densities.

The mixtures are usually spun horizontally in balanced containers ('buckets'), and the rotation sets up centrifugal forces, causing their components to separate according to their densities. A common example is the separation of the lighter plasma from the heavier blood corpuscles in certain blood tests. The **ultracentrifuge** is a very high-speed centrifuge, used in biochemistry for separating ◊colloids and organic substances; it may operate at several million revolutions per minute. The centrifuges used in the industrial separation of cream from milk, and yeast from fermented wort (infused malt), operate by having mixtures pumped through a continually rotating chamber, the components being tapped off at different points. Large centrifuges are used for physiological research – for example, in astronaut training where bodily responses to gravitational forces many times the normal level are tested.

centriole structure found in the ◊cells of animals that plays a role in the processes of ◊meiosis and ◊mitosis (cell division).

centripetal force force that acts radially inward on an object moving in a curved path. For example, with a weight whirled in a circle at the end of a length of string, the centripetal force is the tension in the string. For an object of mass m moving with a velocity v in a circle of radius r, the centripetal force F equals mv^2/r (inward). The reaction to this force is the ◊centrifugal force.

centromere part of the ◊chromosome where there are no ◊genes. Under the microscope, it usually appears as a constriction in the strand of the chromosome, and is the point at which the spindle fibres are attached during ◊meiosis and ◊mitosis (cell division).

Centronics interface standard type of computer ◊interface, used to connect computers to ◊parallel devices, usually printers. (Centronics was an important printer manufacturer in the early days of microcomputing.)

centrosome cell body that contains the ◊centrioles. During cell division the centrosomes organize the microtubules to form the spindle that divides the chromosomes into daughter cells. Centrosomes were first described in 1887, independently by German biologist Theodor Boveri (1862–1915) and Edouard van Beneden.

cephalopod any predatory marine mollusc of the class Cephalopoda, with the mouth and head surrounded by tentacles. Cephalopods are the most intelligent, the fastest-moving, and the largest of all animals without backbones, and there are remarkable luminescent forms which swim or drift at great depths. They have the most highly developed nervous and sensory systems of all invertebrates, the eye in some closely paralleling that found in vertebrates. Examples include squid, ◊octopus, and ◊cuttlefish. Shells are rudimentary or absent in most cephalopods.

Typically, they move by swimming with the mantle (fold of outer skin) aided by the arms, but can squirt water out of the siphon (funnel) to propel themselves backwards by jet propulsion. Squid, for example, can escape predators at speeds of 11 kph/7mph. Cephalopods grow very rapidly and may be mature in a year. The female common octopus lays 150,000 eggs after copulation, and stays to brood them for as long as six weeks. After they hatch the female dies, and, although reproductive habits of many cephalopods are not known, it is thought that dying after spawning may be typical.

CEPHALOPOD PAGE

http://is.dal.ca/~ceph/wood.html

Introduction to the world of cephalopods, the class which includes the squids, cuttlefish, and octopuses, maintained by a graduate student at Dalhousie University, Canada. As well as some excellent images of marine life, this site also contains biological information about each subgroup.

cephalosporin any of a class of broad-spectrum antibiotics derived from a fungus (genus *Cephalosporium*). They are similar to penicillins and are used on penicillin-resistant infections.

The first cephalosporin was extracted from sewage-contaminated water, and other naturally occurring ones have been isolated from moulds taken from soil samples. Side effects include allergic reactions and digestive upsets. Synthetic cephalosporins can be designed to be effective against a particular ◊pathogen.

Cepheid variable yellow supergiant star that varies regularly in brightness every few days or weeks as a result of pulsations. The time that a Cepheid variable takes to pulsate is directly related to its average brightness; the longer the pulsation period, the brighter the star.

This relationship, the **period luminosity law** (discovered by US astronomer Henrietta Leavitt), allows astronomers to use Cepheid variables as 'standard candles' to measure distances in our Galaxy and to nearby galaxies. They are named after their prototype, Delta Cephei, whose light variations were observed 1784 by English astronomer John Goodricke (1764–1786).

Cepheus constellation of the north polar region, named after King Cepheus of Greek mythology, husband of Cassiopeia and father of Andromeda. It contains the Garnet Star (Mu Cephei), a red supergiant of variable brightness that is one of the reddest-coloured stars known, and Delta Cephei, prototype of the Cepheid ◊variables, which are important both as distance indicators and for the information they give about stellar evolution.

Cerberus genus of viviparous (bearing their young live) and aquatic snakes. They are common to the rivers and estuaries of the East Indies from Bengal to Australia.

C. rhynchops has large ventral scales. None of the species is fatally poisonous to humans.

classification *Cerberus* is in family Colubridae, suborder Serpentes, order Squamata, class Reptilia.

Cercopithecus genus of monkeys consisting of the guenons.

cereal grass grown for its edible, nutrient-rich, starchy seeds. The term refers primarily to wheat, oats, rye, and barley, but may also refer to maize (corn), millet, and rice. Cereals contain about 75% complex carbohydrates and 10% protein, plus fats and fibre (roughage). They store well. If all the world's cereal crop were consumed as whole-grain products directly by humans, everyone could obtain adequate protein and carbohydrate; however, a large proportion of cereal production in affluent nations is used as animal feed to boost the production of meat, dairy products, and eggs.

The term also refers to breakfast foods prepared from the seeds of cereal crops. Some cereals require cooking (porridge oats), but most are ready to eat. Mass-marketed cereals include refined and sweetened varieties as well as whole cereals such as muesli. Whole cereals are more nutritious and provide more fibre than the refined cereals, which often have vitamins and flavourings added to replace those lost in the refining process.

cerebellum part of the brain of ◊vertebrate animals which controls muscle tone, movement, balance, and coordination. It is relatively small in lower animals such as newts and lizards, but large in birds since flight demands precise coordination. The human cerebellum is also well developed, because of the need for balance when walking or running, and for finely coordinated hand movements.

cerebral haemorrhage or ***apoplectic fit*** in medicine, a form of ◊stroke in which there is bleeding from a cerebral blood vessel into the surrounding brain tissue. It is generally caused by degenerative disease of the arteries and high blood pressure. Depending on the site and extent of bleeding, the symptoms vary from transient weakness and numbness to deep coma and death. Damage to the brain is permanent, though some recovery can be made. Strokes are likely to recur.

cerebral hemisphere one of the two halves of the ◊cerebrum.

cerebral palsy any nonprogressive abnormality of the brain occurring during or shortly after birth. It is caused by oxygen deprivation, injury during birth, haemorrhage, meningitis, viral infection, or faulty development. Premature babies are at greater risk of being born with cerebral palsy, and in 1996 US researchers linked this to low levels of the thyroid hormone thyroxine. The condition is characterized by muscle spasm, weakness, lack of coordination, and impaired movement; or there may be spastic paralysis, with fixed deformities of the limbs. Intelligence is not always affected.

CEREBRAL PALSY TUTORIAL

http://galen.med.virginia.edu/~smb4v/tutorials/cp/cp.htm

Multimedia tutorial on cerebral palsy for children and parents. It discusses the causes and different kinds of the disorder, describes a series of therapeutic interventions, and presents equipment of different kinds that can prove useful for children with this debilitating disease.

cerebrum part of the vertebrate ◊brain, formed from the two paired cerebral hemispheres, separated by a central fissure. In birds and mammals it is the largest and most developed part of the brain. It is covered with an infolded layer of grey matter, the cerebral cortex, which integrates brain functions. The cerebrum coordinates all voluntary activity.

Ceres the largest asteroid, 940 km/584 mi in diameter, and the first to be discovered (by Italian astronomer Giuseppe Piazzi 1801). Ceres orbits the Sun every 4.6 years at an average distance of 414 million km/257 million mi. Its mass is about one-seventieth of that of the Moon.

cerium malleable and ductile, grey, metallic element, symbol Ce, atomic number 58, relative atomic mass 140.12. It is the most abundant member of the lanthanide series, and is used in alloys, electronic components, nuclear fuels, and lighter flints. It was discovered 1804 by the Swedish chemists Jöns Berzelius and Wilhelm Hisinger (1766–1852), and, independently, by Martin Klaproth. The element was named after the then recently discovered asteroid Ceres.

cermet (contraction of ***ceramics and metal***) bonded material containing ceramics and metal, widely used in jet engines and nuclear reactors. Cermets behave much like metals but have the great heat resistance of ceramics. Tungsten carbide, molybdenum boride, and aluminium oxide are among the ceramics used; iron, cobalt, nickel, and chromium are among the metals.

CERN particle physics research organization founded 1954 as a cooperative enterprise among European governments. It has laboratories at Meyrin, near Geneva, Switzerland. It was originally known as the **Conseil Européen pour la Recherche Nucléaire** but subsequently renamed **Organisation Européenne pour la Recherche Nucléaire**, although still familiarly known as CERN. It houses the world's largest particle ◊accelerator, the ◊Large Electron Positron Collider (LEP), with which notable advances have been made in ◊particle physics.

In 1965 the original laboratory was doubled in size by extension across the border from Switzerland into France. In 1994 the 19 member nations of CERN approved the construction of the Large Hadron Collider. It is expected to cost £1.25 million and to be fully functional in 2005.

Cerro Tololo Inter-American Observatory observatory on Cerro Tololo mountain in the Chilean Andes operated by AURA (the Association of Universities for Research into Astronomy). Its main instrument is a 4-m/158-in reflector, opened in 1974, a twin of that at Kitt Peak, Arizona, USA.

CERT abbreviation for ◊Computer Emergency Response Team.

cervical cancer in medicine, ◊cancer of the cervix (neck of the womb).

cervical smear in medicine, removal of a small sample of tissue from the cervix (neck of the womb) to screen for changes implying a likelihood of cancer. The procedure is also known as the **Pap test** after its originator, George Papanicolau.

CET abbreviation for ***Central European Time***.

Cetus ***Latin 'whale'*** large constellation on the celestial equator (see ◊celestial sphere), represented as a sea monster or a whale. Cetus contains the long-period variable star ◊Mira, and Tau Ceti, one of the nearest stars, which is visible with the naked eye.

It is named after the sea monster sent to devour Andromeda. Mira is sometimes the most conspicuous object in the constellation, but it is more usually invisible to the naked eye.

CFC abbreviation for ◊chlorofluorocarbon.

CGA (abbreviation for ***colour graphics adapter***) in computing, first colour display system for IBM PCs and compatible machines. It has been superseded by ◊EGA, ◊VGA, ◊SVGA, and◊ XGA.

CGI abbreviation for ◊common gateway interface.

c.g.s. system system of units based on the centimetre, gram, and second, as units of length, mass, and time, respectively. It has been replaced for scientific work by the ◊SI units to avoid inconsistencies in definition of the thermal calorie and electrical quantities.

chacma species of ◊baboon.

Chaetodon genus of spiny-rayed fishes known as ◊butterfly fishes.

Chaetopterus large sedentary bristle-worm (polychaete) that inhabits U-shaped tubes in sand or mud. The tube can be 23 cm/9 in long and the two ends protrude from the surface and can be seen at a very low spring tide.

The worm's body, which is phosphorescent, is divided into three regions with various sizes of fans and bristles. These aid in spinning a mucous net, which filters out particles suspended in the sea water and is then eaten by the worm, another being spun in its place about once an hour.

classification *Chaetopterus* is a member of class Polychaeta in phylum Annelida.

chafer beetle of the family Scarabeidae. The adults eat foliage or flowers, and the underground larvae feed on roots, chiefly those of grasses and cereals, and can be very destructive. Examples are the ◊cockchafer **and the rose chafer** *Cetonia aurata*, about 2 cm/0.8 in long and bright green.

chaffinch bird *Fringilla coelebs* of the finch family, common throughout much of Europe and W Asia. About 15 cm/6 in long, the male is olive-brown above, with a bright chestnut breast, a bluish-grey cap, and two white bands on the upper part of the wing; the female is duller. During winter they form single-sex flocks.

Chagas's disease disease common in Central and South America, infecting approximately 18 million people worldwide. It is caused by a trypanosome parasite, *Trypanosoma cruzi*, transmitted by several species of blood-sucking insect; it results in incurable damage to the heart, intestines, and brain. It is named after Brazilian doctor Carlos Chagas (1879–1934).

It is the world's third most prevalent parasitic disease, after malaria and schistosomiasis. In its first stage symptoms resemble flu but 20–30% of sufferers develop inflammation of the heart muscles up to 20 years later.

chain reaction in chemistry, a succession of reactions, usually involving ◊free radicals, where the products of one stage are the reactants of the next. A chain reaction is characterized by the continual generation of reactive substances.

A chain reaction comprises three separate stages: **initiation** – the initial generation of reactive species; **propagation** – reactions that involve reactive species and generate similar or different reactive species; and **termination** – reactions that involve the reactive species but produce only stable, nonreactive substances. Chain reactions may occur slowly (for example, the oxidation of edible oils) or accelerate as the number of reactive species increases, ultimately resulting in explosion.

chain reaction in nuclear physics, a fission reaction that is maintained because neutrons released by the splitting of some atomic nuclei themselves go on to split others, releasing even more neutrons. Such a reaction can be controlled (as in a nuclear reactor) by using moderators to absorb excess neutrons. Uncontrolled, a chain reaction produces a nuclear explosion (as in an atom bomb).

chalaza glutinous mass of transparent albumen supporting the yolk inside birds' eggs. The chalaza is formed as the egg slowly passes down the oviduct, when it also acquires its coiled structure.

chalcedony form of the mineral quartz, SiO_2, in which the crystals are so fine-grained that they are impossible to distinguish with a microscope (cryptocrystalline). Agate, onyx, and carnelian are ◊gem varieties of chalcedony.

chalcopyrite copper iron sulphide mineral, $CuFeS_2$, the most common ore of copper. It is brassy yellow in colour and may have an iridescent surface tarnish. It occurs in many different types of mineral vein, in rocks ranging from basalt to limestone.

chalk soft, fine-grained, whitish sedimentary rock composed of calcium carbonate, $CaCO_3$, extensively quarried for use in cement, lime, and mortar, and in the manufacture of cosmetics and toothpaste. **Blackboard chalk** in fact consists of gypsum (calcium sulphate, $CaSO_4.2H_2O$).

Chalk was once thought to derive from the remains of microscopic animals or foraminifera. In 1953, however, it was seen under the electron microscope to be composed chiefly of ◊coccolithophores, unicellular lime-secreting algae, and hence primarily of plant origin. It is formed from deposits of deep-sea sediments called oozes.

Challenger orbiter used in the US ◊space shuttle programme which on 28 January 1986 exploded on takeoff, killing all seven crew members.

chameleon any of 80 or so species of lizard of the family Chameleontidae. Some species have highly developed colour-changing abilities, caused by stress and changes in the intensity of light and temperature, which alter the dispersal of pigment granules in the layers of cells beneath the outer skin.

The tail is long and highly prehensile, assisting the animal when climbing. Most chameleons live in trees and move very slowly.

The tongue is very long, protrusile, and covered with a viscous secretion; it can be shot out with great rapidity to 20 cm/8 in for the capture of insects. The eyes are on 'turrets', move independently, and can swivel forward to give stereoscopic vision for 'shooting'. Most live in Africa and Madagascar, but the **common chameleon** *Chameleo chameleon* is found in Mediterranean countries; two species live in SW Arabia, and one species in India and Sri Lanka.

Some species of chameleon, such as the African species *C. bitaeniatus* give birth to live young; the female 'gives birth' to a fully-formed young enclosed in a membrane, which is immediately shed.

chamois goatlike mammal *Rupicapra rupicapra* found in mountain ranges of S Europe and Asia Minor. It is brown, with dark patches running through the eyes, and can be up to 80 cm/2.6 ft high. Chamois are very sure-footed, and live in herds of up to 30 members.

Both sexes have horns which may be 20 cm/8 in long. These are set close together and go up vertically, forming a hook at the top. Chamois skin is very soft, and excellent for cleaning glass, but the chamois is now comparatively rare and 'chamois leather' is often made from the skin of sheep and goats.

champignon any of a number of edible fungi (see ◊fungus). The fairy ring champignon (*Marasmius oreades*) has this name because its fruiting bodies (mushrooms) grow in rings around the outer edge of the underground mycelium (threadlike body) of the fungus. (Family Agaricaceae.)

chance likelihood, or ◊probability, of an event taking place, expressed as a fraction or percentage. For example, the chance that a tossed coin will land heads up is 50%.

As a science, it originated when the Chevalier de Méré consulted Blaise ◊Pascal about how to reduce his gambling losses. In 1664, in correspondence with another mathematician, Pierre de ◊Fermat, Pascal worked out the foundations of the theory of chance. This underlies the science of statistics.

I have been trying to point out that in our lives chance may have an astonishing influence and, if I may offer advice to the young laboratory worker, it would be this – never to neglect an extraordinary appearance or happening. It may be – usually is, in fact – a false alarm that leads to nothing, but it may on the other hand be the clue provided by fate to lead you to some important advance.

ALEXANDER FLEMING Scottish bacteriologist.
Lecture at Harvard

---

chancroid or ***soft sore*** acute localized, sexually transmitted ulcer on or about the genitals caused by the bacterium *Haemophilus ducreyi*.

It causes painful enlargement and suppuration of lymph nodes in the groin area.

Chandrasekhar limit or ***Chandrasekhar mass*** in astrophysics, the maximum possible mass of a ◊white dwarf star. The limit depends slightly on the composition of the star but is equivalent to 1.4 times the mass of the Sun. A white dwarf heavier than the Chandrasekhar limit would collapse under its own weight to form a ◊neutron star or a ◊black hole. The limit is named after the

CHAMELEONS

http://www.skypoint.com/members/mikefry/chams.html

Excellent guide to the expensive and time-consuming business of looking after a pet chameleon. Would-be owners are warned not to support the trade in endangered species. Dietary and housing requirements are well-explained, together with advice on what to do if your lizard is stressed, goes on hunger strike, has parasites, or does not get enough unfiltered sunlight. There are links to a number of specialist sites related to chameleons in the wild or in captivity.

Indian-US astrophysicist Subrahmanyan Chandrasekhar who developed the theory of white dwarfs in the 1930s.

change of state in science, a change in the physical state (solid, liquid, or gas) of a material. For instance, melting, boiling, evaporation, and their opposites, solidification and condensation, are changes of state. The former set of changes are brought about by heating or decreased pressure; the latter by cooling or increased pressure.

These changes involve the absorption or release of heat energy, called ◊latent heat, even though the temperature of the material does not change during the transition between states.

See also ◊states of matter. In the unusual change of state called **sublimation**, a solid changes directly to a gas without passing through the liquid state. For example, solid carbon dioxide (dry ice) sublimes to carbon dioxide gas.

channel in computing, path connecting a computer to peripheral devices along which data can be transferred.

channel efficiency measure of the ability of a river channel to discharge water. Channel efficiency can be assessed by calculating the channel's ◊hydraulic radius. The most efficient channels are generally semicircular in cross-section, and it is this shape that water engineers try to create when altering a river channel to reduce the risk of flooding.

Surely John Bull will not endanger his birthright, his liberty, his property, simply in order that men and women may cross between England and France without running the risk of sea-sickness.

Garnet Wolseley English soldier.
On the Channel Tunnel proposals of 1882

Channel Tunnel tunnel built beneath the English Channel, linking Britain with mainland Europe. It comprises twin rail tunnels, 50 km/31 mi long and 7.3 m/24 ft in diameter, located 40 m/130 ft beneath the seabed. Construction began in 1987, and the French and English sections were linked in December 1990. It was officially opened on 6 May 1994. The shuttle train service, Le Shuttle, opened to lorries in May 1994 and to cars in December 1994. The tunnel's high-speed train service, Eurostar, linking London to Paris and Brussels, opened in November 1994.

The estimated cost of the tunnel has continually been revised upwards to a figure of £8 billion (1995). In 1995 Eurotunnel plc, the Anglo-French company that built the tunnel, made a loss of £925 million.

chanterelle edible ◊fungus that is bright yellow and funnel-shaped. It grows in deciduous woodland. (*Cantharellus cibarius*, family Cantharellaceae.)

Chaos Theory: The Mathematics of Chaos

BY IAN STEWART

The mathematics of chaos

Why are tides predictable years ahead, whereas weather forecasts often go wrong within a few days?

Both tides and weather are governed by natural laws. Tides are caused by the gravitational attraction of the Sun and Moon; the weather by the motion of the atmosphere under the influence of heat from the Sun. The law of gravitation is not noticeably simpler than the laws of fluid dynamics; yet for weather the resulting behaviour seems to be far more complicated.

The reason for this is 'chaos', which lies at the heart of one of the most exciting and most rapidly expanding areas of mathematical research, the theory of nonlinear dynamic systems.

Random behaviour in dynamic systems

It has been known for a long time that dynamic systems – systems that change with time according to fixed laws – can exhibit regular patterns, such as repetitive cycles. Thanks to new mathematical techniques, emphasizing shape rather than number, and to fast and sophisticated computer graphics, we now know that dynamic systems can also behave randomly. The difference lies not in the complexity of the formulae that define their mathematics, but in the geometrical features of the dynamics. This is a remarkable discovery: random behaviour in a system whose mathematical description contains no hint whatsoever of randomness.

Simple geometric structure produces simple dynamics. For example, if the geometry shrinks everything towards a fixed point, then the motion tends towards a steady state. But if the dynamics keep stretching things apart and then folding them together again, the motion tends to be chaotic – like food being mixed in a bowl. The motion of the Sun and Moon, on the kind of timescale that matters when we want to predict the tides, is a series of regular cycles, so prediction is easy. The changing patterns of the weather involve a great deal of stretching and folding, so here chaos reigns.

Fractals

The geometry of chaos can be explored using theoretical mathematical techniques such as topology – 'rubber-sheet geometry' – but the most vivid pictures are obtained using computer graphics. The geometric structures of chaos are 'fractals': they have detailed form on all scales of magnification. Order and chaos, traditionally seen as opposites, are now viewed as two aspects of the same basic process, the evolution of a system in time. Indeed, there are now examples where both order and chaos occur naturally within a single geometrical form.

Predicting the unpredictable

Does chaos make randomness predictable? Sometimes. If what looks like random behaviour is actually governed by a dynamic system, then short-term prediction becomes possible. Long-term prediction is not as easy, however. In chaotic systems any initial error of measurement, however small, will grow rapidly and eventually ruin the prediction. This is known as the butterfly effect: if a butterfly flaps its wings, a month later the air disturbance created may cause a hurricane.

Chaos can be applied to many areas of science, such as chemistry, engineering, computer science, biology, electronics, and astronomy. For example, although the short-term motions of the Sun and Moon are not chaotic, the long-term motion of the Solar System **is** chaotic. It is impossible to predict on which side of the Sun Pluto will lie in 200 million years' time. Saturn's satellite Hyperion tumbles chaotically. Chaos caused by Jupiter's gravitational field can fling asteroids out of orbit, towards the Earth. Disease epidemics, locust plagues, and irregular heartbeats are more down-to-earth examples of chaos, on a more human timescale.

Making sense of chaos

Chaos places limits on science: it implies that even when we know the equations that govern a system's behaviour, we may not in practice be able to make effective predictions. On the other hand, it opens up new avenues for discovery, because it implies that apparently random phenomena may have simple, nonrandom explanations. So chaos is changing the way scientists think about what they do: the relation between determinism and chance, the role of experiment, the computability of the world, the prospects for prediction, and the interaction between mathematics, science, and nature. Chaos cuts right across traditional subject boundaries, and distinctions between pure and applied mathematicians, between mathematicians and physicists, between physicists and biologists, become meaningless when compared to the unity revealed by their joint efforts.

chaos theory or ***chaology*** or ***complexity theory*** branch of mathematics that attempts to describe irregular, unpredictable systems – that is, systems whose behaviour is difficult to predict because there are so many variable or unknown factors. Weather is an example of a chaotic system.

Chaos theory, which attempts to predict the *probable* behaviour of such systems, based on a rapid calculation of the impact of as wide a range of elements as possible, emerged in the 1970s with the development of sophisticated computers. First developed for use in meteorology, it has also been used in such fields as economics.

char or **charr** fish *Salvelinus alpinus* related to the trout, living in the Arctic coastal waters, and also in Europe and North America in some upland lakes. It is one of Britain's rarest fish, and is at risk from growing acidification.

Numerous variants have been described, but they probably all belong to the same species.

characin freshwater fish belonging to the family Characidae. There are over 1,300 species, mostly in South and Central America, but also in Africa. Most are carnivores. In typical characins, unlike the somewhat similar carp family, the mouth is toothed, and there is a small dorsal adipose fin just in front of the tail.

Characins are small fishes, often colourful, and they include ◊tetras and ◊piranhas.

character one of the symbols that can be represented in a computer. Characters include letters, numbers, spaces, punctuation marks, and special symbols.

characteristic in mathematics, the integral (whole-number) part of a ◊logarithm. The fractional part is the ◊mantissa.

For example, in base ten, $10^0 = 1$, $10^1 = 10$, $10^2 = 100$, and so on; the powers to which 10 is raised are the characteristics. To determine the power to which 10 must be raised to obtain a number between 10 and 100, say 20 (2×10, or log 2 + log 10), the logarithm for 2 is found (0.3010), and the characteristic 1 added to make 1.3010.

character printer computer ◊printer that prints one character at a time.

character set in computing, the complete set of symbols that can be used in a program or recognized by a computer. It may include letters, digits, spaces, punctuation marks, and special symbols.

extended character set in PC-based computing, the set of 254 characters stored in ◊ROM. Besides the 128 ◊ASCII characters, the set includes block graphics and foreign language characters.

character type check in computing, a ◊validation check to ensure that an input data item does not contain invalid characters. For example, an input name may be checked to ensure that it contains only letters of the alphabet or an input six-figure date may be checked to ensure it contains only numbers.

charcoal black, porous form of ◊carbon, produced by heating wood or other organic materials in the absence of air. It is used as a fuel in the smelting of metals such as copper and zinc, and by artists for making black line drawings. **Activated charcoal** has been powdered and dried so that it presents a much increased surface area for adsorption; it is used for filtering and purifying liquids and gases – for example, in drinking-water filters and gas masks.

Charcoal was traditionally produced by burning dried wood in a kiln, a process lasting several days. The kiln was either a simple hole in the ground, or an earth-covered mound. Today kilns are of brick or iron, both of which allow the waste gases to be collected and used. Charcoal had many uses in earlier centuries. Because of the high temperature at which it burns (2,012°F/1,100°C), it was used in furnaces and blast furnaces before the development of ◊coke. It was also used in an industrial process for obtaining ethanoic acid (acetic acid), in producing wood tar and ◊wood pitch, and (when produced from alder or willow trees) as a component of gunpowder.

charge see ◊electric charge.

charge-coupled device (CCD) device for forming images electronically, using a layer of silicon that releases electrons when struck by incoming light. The electrons are stored in ◊pixels and read off into a computer at the end of the exposure. CCDs have now almost entirely replaced photographic film for applications such as astrophotography where extreme sensitivity to light is paramount.

charged particle beam high-energy beam of electrons or protons. Such beams are being developed as weapons.

Charles, Jacques Alexandre César (1746–1823)

French physicist who studied gases and made the first ascent in a hydrogen-filled balloon in 1783. His work on the expansion of gases led to the formulation of Charles's law.

Hearing of the hot-air balloons of the Montgolfier brothers, Charles and his brothers began experimenting with hydrogen balloons and made their ascent only ten days after the Montgolfiers' first flight. In later flights Charles ascended to an altitude of 3,000 m/10,000 ft.

Mary Evans Picture Library

Charles's law law stating that the volume of a given mass of gas at constant pressure is directly proportional to its absolute temperature (temperature in kelvin). It was discovered by French physicist Jacques Charles 1787, and independently by French chemist Joseph Gay-Lussac in 1802.

The gas increases by 1/273 of its volume at 0°C for each °C rise of temperature. This means that the coefficient of expansion of all gases is the same. The law is only approximately true and the coefficient of expansion is generally taken as 0.003663 per °C.

charlock or ***wild mustard*** annual plant belonging to the cress family, found in Europe and Asia. It has hairy stems and leaves and yellow flowers. (*Sinapis arvensis,* family Cruciferae.)

charm in physics, a property possessed by one type of ◊quark (very small particles found inside protons and neutrons), called the charm quark. The effects of charm are only seen in experiments with particle ◊accelerators. See ◊elementary particles.

chat real-time exchange of messages between users of a particular system. Chat allows people who are geographically far apart to type messages to each other which are sent and received instantly. On a system like ◊America Online, users may chat while playing competitive games or while reading messages, as well as joining public or private 'rooms' to talk with a variety of other users. The biggest chat system is Internet Relay Chat (IRC), which is used for the exchange of information and software as well as for social interaction.

check box small, square box used as a control in ◊dialog boxes. Check boxes ◊toggle functions and are operated by moving the cursor over the box and clicking the mouse button to check or clear the box.

check digit in computing, a digit attached to an important code number as a ◊validation check.

checksum in computing, a ◊control total of specific items of data. A checksum is used as a check that data have been input or transmitted correctly. It is used in communications and in, for example, accounts programs. See also ◊validation.

cheetah large wild cat *Acinonyx jubatus* native to Africa, Arabia, and SW Asia, but now rare in some areas. Yellowish with black spots, it has a slim lithe build. It is up to 1 m/3 ft tall at the shoulder, and up to 1.5 m/5 ft long. It can reach 103 kph/64 mph, but tires after about 400 yards. Cheetahs live in open country where they hunt small antelopes, hares, and birds.

A cheetah's claws do not retract as fully as in most cats. It is the world's fastest mammal. Cheetahs face threats both from ranchers who shoot them as vermin and from general habitat destruction that is reducing the prey on which they feed, especially gazelles. As a result the wild population is thought to have fallen by over half since the 1970s; there are now thought to be no more than 5,000–12,000 left.

CHEETAH

The cheetah can accelerate from zero to 72 kmph/45 mph in two seconds.

chelate chemical compound whose molecules consist of one or more metal atoms or charged ions joined to chains of organic residues by coordinate (or dative covalent) chemical ◊bonds.

The parent organic compound is known as a **chelating agent** – for example, EDTA (ethylene-diaminetetraacetic acid), used in chemical analysis. Chelates are used in analytical chemistry, in agriculture and horticulture as carriers of essential trace metals, in water softening, and in the treatment of thalassaemia by removing excess iron, which may build up to toxic levels in the body. Metalloproteins (natural chelates) may influence the performance of enzymes or provide a mechanism for the storage of iron in the spleen and plasma of the human body.

Chelonia order of reptiles containing some 250 species of ◊tortoises, ◊terrapins, and ◊turtles. Members possess a shell made up of bony plates. The upper part of the shell, the carapace, may be fused to the vertebrae and ribs. Chelonians are toothless, with a hardened horny beak.

chemical change change that occurs when two or more substances (reactants) interact with each other, resulting in the production of different substances (products) with different chemical compositions. A simple example of chemical change is the burning of carbon in oxygen to produce carbon dioxide.

chemical element alternative name for ◊element.

chemical equation method of indicating the reactants and products of a chemical reaction by using chemical symbols and formulae. A chemical equation gives two basic pieces of information: (1) the reactants (on the left-hand side) and products (right-hand side); and (2) the reacting proportions (stoichiometry) – that is, how many units of each reactant and product are involved. The equation must balance; that is, the total number of atoms of a particular element on the left-hand side must be the same as the number of atoms of that element on the right-hand side.

$$Na_2CO_3 + 2HCl \rightarrow 2NaCl + CO_2 + H_2O$$

$$\text{reactants} \rightarrow \text{products}$$

This equation states that one molecule of sodium carbonate combines with two molecules of hydrochloric acid to form two molecules of sodium chloride, one of carbon dioxide, and one of water. Double arrows indicate that the reaction is reversible – in the formation of ammonia from hydrogen and nitrogen, the direction depends on the temperature and pressure of the reactants.

$$3H_2 + N_2 \leftrightarrow 2NH_3$$

chemical equilibrium condition in which the products of a reversible chemical reaction are formed at the same rate at which they decompose back into the reactants, so that the concentration of each reactant and product remains constant.

The amounts of reactant and product present at equilibrium are defined by the **equilibrium constant** for that reaction and specific temperature.

chemical family collection of elements that have very similar chemical and physical properties. In the ◊periodic table of the elements such collections are to be found in the vertical columns (groups). The groups that contain the most markedly similar elements are group I, the ◊alkali metals; group II, the ◊alkaline-earth metals; group VII, the ◊halogens; and group 0, the noble or ◊inert gases.

chemical oxygen demand (COD) measure of water and effluent quality, expressed as the amount of oxygen (in parts per million) required to oxidize the reducing substances present.

Under controlled conditions of time and temperature, a chemical oxidizing agent (potassium permanganate or dichromate) is added to the sample of water or effluent under consideration, and the amount needed to oxidize the reducing materials present is measured. From this the chemically equivalent amount of oxygen can be calculated. Since the reducing substances typically include remains of living organisms, COD may be regarded as reflecting the extent to which the sample is polluted. Compare ◊biological oxygen demand.

chemical weathering form of ◊weathering brought about by chemical attack of rocks, usually in the presence of water. Chemical weathering involves the 'rotting', or breakdown, of the original minerals within a rock to produce new minerals (such as ◊clay minerals, ◊bauxite, and ◊calcite). Some chemicals are dissolved and carried away from the weathering source.

A number of processes bring about chemical weathering, such as carbonation (breakdown by weakly acidic rainwater), hydrolysis (breakdown by water), hydration (breakdown by the absorption of water), and oxidation (breakdown by the oxygen in water). The reaction of carbon dioxide gas in the atmosphere with ◊silicate minerals in rocks to produce carbonate minerals (see ◊calcite) is called the 'Urey reaction' after the chemist who proposed it, Harold Urey. The Urey reaction is an important link between Earth's climate and the geology of the planet. It has been proposed that chemical weathering of large mountain ranges like the Himalayas of Nepal can remove carbon dioxide from the atmosphere by the Urey reaction (or other more complicated reactions like it) , leading to a cooler climate as the ◊greenhouse effects of the lost carbon dioxide are diminished.

chemiluminescence the emission of light from a substance as a result of a chemical reaction (rather than raising its temperature). See ◊luminescence.

chemisorption the attachment, by chemical means, of a single layer of molecules, atoms, or ions of gas to the surface of a solid or, less frequently, a liquid. It is the basis of catalysis (see ◊catalyst) and is of great industrial importance.

chemistry branch of science concerned with the study of the structure and composition of the different kinds of matter, the changes which matter may undergo and the phenomena which occur in the course of these changes.

Organic chemistry is the branch of chemistry that deals with carbon compounds. **Inorganic chemistry** deals with the description, properties, reactions, and preparation of all the elements and their compounds, with the exception of carbon compounds. **Physical chemistry** is concerned with the quantitative explanation of chemical phenomena and reactions, and the measurement of data required for such explanations. This branch studies in particular the movement of molecules and the effects of temperature and pressure, often with regard to gases and liquids.

molecules, atoms, and elements All matter can exist in three states: gas, liquid, or solid. It is composed of minute particles termed **molecules**, which are constantly moving, and may be further divided into ◊atoms.

Molecules that contain atoms of one kind only are known as **elements**; those that contain atoms of different kinds are called **compounds**.

compounds and mixtures Chemical compounds are produced by a chemical action that alters the arrangement of the atoms in the

Chemistry: chronology

c. 3000 BC Egyptians were producing bronze – an alloy of copper and tin.

c. 450 BC Greek philosopher Empedocles proposed that all substances are made up of a combination of four elements – Earth, air, fire, and water – an idea that was developed by Plato and Aristotle and persisted for over 2,000 years.

c. 400 BC Greek philosopher Democritus theorized that matter consists ultimately of tiny, indivisible particles, *atomos*.

AD 1 Gold, silver, copper, lead, iron, tin, and mercury were known.

200 The techniques of solution, filtration, and distillation were known.

7th–17th centuries Chemistry was dominated by alchemy, the attempt to transform nonprecious metals such as lead and copper into gold. Though misguided, it led to the discovery of many new chemicals and techniques, such as sublimation and distillation.

12th century Alcohol was first distilled in Europe.

1242 Gunpowder introduced to Europe from the Far East.

1620 Scientific method of reasoning expounded by Francis Bacon in his *Novum Organum*.

1650 Leyden University in the Netherlands set up the first chemistry laboratory.

1661 Robert Boyle defined an element as any substance that cannot be broken down into still simpler substances and asserted that matter is composed of 'corpuscles' (atoms) of various sorts and sizes, capable of arranging themselves into groups, each of which constitutes a chemical substance.

1662 Boyle described the inverse relationship between the volume and pressure of a fixed mass of gas (Boyle's law).

1697 Georg Stahl proposed the erroneous theory that substances burn because they are rich in a certain substance, called phlogiston.

1755 Joseph Black discovered carbon dioxide.

1774 Joseph Priestley discovered oxygen, which he called 'dephlogisticated air'. Antoine Lavoisier demonstrated his law of conservation of mass.

1777 Lavoisier showed air to be made up of a mixture of gases, and showed that one of these – oxygen – is the substance necessary for combustion (burning) and rusting to take place.

1781 Henry Cavendish showed water to be a compound.

1792 Alessandra Volta demonstrated the electrochemical series.

1807 Humphry Davy passed electric current through molten compounds (the process of electrolysis) in order to isolate elements, such as potassium, that had never been separated by chemical means. Jöns Berzelius proposed that chemicals produced by living creatures should be termed 'organic'.

1808 John Dalton published his atomic theory, which states that every element consists of similar indivisible particles – called atoms – which differ from the atoms of other elements in their mass; he also drew up a list of relative atomic masses. Joseph Gay-Lussac announced that the volumes of gases that combine chemically with one another are in simple ratios.

1811 Publication of Amedeo Avogadro's hypothesis on the relation between the volume and number of molecules of a gas, and its temperature and pressure.

1813–14 Berzelius devised the chemical symbols and formulae still used to represent elements and compounds.

1828 Franz Wöhler converted ammonium cyanate into urea – the first synthesis of an organic compound from an inorganic substance.

1832–33 Michael Faraday expounded the laws of electrolysis, and adopted the term 'ion' for the particles believed to be responsible for carrying current.

1846 Thomas Graham expounded his law of diffusion.

1853 Robert Bunsen invented the Bunsen burner.

1858 Stanislao Cannizzaro differentiated between atomic and molecular weights (masses).

1861 Organic chemistry was defined by German chemist Friedrich Kekulé as the chemistry of carbon compounds.

1864 John Newlands devised the first periodic table of the elements.

1869 Dmitri Mendeleyev expounded his periodic table of the elements (based on atomic mass), leaving gaps for elements that had not yet been discovered.

1874 Jacobus van't Hoff suggested that the four bonds of carbon are arranged tetrahedrally, and that carbon compounds can therefore be three-dimensional and asymmetric.

1884 Swedish chemist Svante Arrhenius suggested that electrolytes (solutions or molten compounds that conduct electricity) dissociate into ions, atoms or groups of atoms that carry a positive or negative charge.

1894 William Ramsey and Lord Rayleigh discovered the first inert gases, argon.

1897 The electron was discovered by J J Thomson.

1901 Mikhail Tsvet invented paper chromatography as a means of separating pigments.

1909 Sören Sörensen devised the pH scale of acidity.

1912 Max von Laue showed crystals to be composed of regular, repeating arrays of atoms by studying the patterns in which they diffract X-rays.

1913–14 Henry Moseley equated the atomic number of an element with the positive charge on its nuclei, and drew up the periodic table, based on atomic number, that is used today.

1916 Gilbert Newton Lewis explained covalent bonding between atoms as a sharing of electrons.

1927 Nevil Sidgwick published his theory of valency, based on the numbers of electrons in the outer shells of the reacting atoms.

1930 Electrophoresis, which separates particles in suspension in an electric field, was invented by Arne Tiselius.

1932 Deuterium (heavy hydrogen), an isotope of hydrogen, was discovered by Harold Urey.

1940 Edwin McMillan and Philip Abelson showed that new elements with a higher atomic number than uranium can be formed by bombarding uranium with neutrons, and synthesized the first transuranic element, neptunium.

1942 Plutonium was first synthesized by Glenn T Seaborg and Edwin McMillan.

1950 Derek Barton deduced that some properties of organic compounds are affected by the orientation of their functional groups (the study of which became known as conformational analysis).

1954 Einsteinium and fermium were synthesized.

1955 Ilya Prigogine described the thermodynamics of irreversible processes (the transformations of energy that take place in, for example, many reactions within living cells).

1962 Neil Bartlett prepared the first compound of an inert gas, xenon hexafluoroplatinate; it was previously believed that inert gases could not take part in a chemical reaction.

1965 Robert B Woodward synthesized complex organic compounds.

1981 Quantum mechanics applied to predict course of chemical reactions by US chemist Roald Hoffmann and Kenichi Fukui of Japan.

1982 Element 109, unnilennium, synthesized.

1985 Fullerenes, a new class of carbon solids made up of closed cages of carbon atoms, were discovered by Harold Kroto and David Walton at the University of Sussex, England.

1987 US chemists Donald Cram and Charles Pederson, and Jean-Marie Lehn of France created artificial molecules that mimic the vital chemical reactions of life processes.

1990 Jean-Marie Lehn, Ulrich Koert, and Margaret Harding reported the synthesis of a new class of compounds, called nucleohelicates, that mimic the double helical structure of DNA, turned inside out.

1993 US chemists at the University of California and the Scripps Institute synthesized rapamycin, one of a group of complex, naturally occurring antibiotics and immunosuppressants that are being tested as anticancer agents.

1994 Elements 110 (ununnilium) and 111 (unununium) discovered at the GSI heavy-ion cyclotron, Darmstadt, Germany.

1995 German chemists built the largest ever wheel molecule made up of 154 molybdenum atoms surrounded by oxygen atoms. It has a relative molecular mass of 24,000 and is soluble in water.

1996 Element 112 discovered at the GSI heavy-ion cyclotron, Darmstadt, Germany.

1997 The International Union of Pure and Applied Chemistry (IUPAC) stated that elements 104–109 should be named rutherfordium (104), dubnium (105), seaborgium (106), bohrium (107), hassium (108), and meitnerium (109).

reacting molecules. Heat, light, vibration, catalytic action, radiation, or pressure, as well as moisture (for ionization), may be necessary to produce a chemical change. Examination and possible breakdown of compounds to determine their components is **analysis**, and the building up of compounds from their components is **synthesis**. When substances are brought together without changing their molecular structures they are said to be **mixtures**.

formulas and equations Symbols are used to denote the elements. The symbol is usually the first letter or letters of the English or Latin name of the element – for example, C for carbon; Ca for calcium; Fe for iron (*ferrum*). These symbols represent one atom of the element; molecules containing more than one atom of an element are denoted by a subscript figure – for example, water is H_2O. In some substances a group of atoms acts as a single entity, and these are enclosed in parentheses in the symbol – for example $(NH_4)_2SO_4$ denotes ammonium sulphate. The symbolic representation of a molecule is known as a **formula**. A figure placed before a formula represents the number of molecules of a substance taking part in, or being produced by, a chemical reaction – for example, $2H_2O$ indicates two molecules of water. Chemical reactions are expressed by means of **equations** as in:

$$NaCl + H_2SO_4 \rightarrow NaHSO_4 + HCl$$

This equation states the fact that sodium chloride (NaCl) on being treated with sulphuric acid (H_2SO_4) is converted into sodium bisulphate (sodium hydrogensulphate, $NaHSO_4$) and hydrogen chloride (HCl). See also ◊chemical equation.

metals, nonmetals, and the periodic system Elements are divided into **metals**, which have lustre and conduct heat and electricity, and **nonmetals**, which usually lack these properties. The **periodic system**, developed by John Newlands in 1863 and established by Dmitri ◊Mendeleyev in 1869, classified elements according to their relative atomic masses. Those elements that resemble each other in general properties were found to bear a relation to one another by weight, and these were placed in groups or families. Certain anomalies in this system were later removed by classifying the elements according to their atomic numbers. The latter is equivalent to the positive charge on the nucleus of the atom.

The true use of chemistry is not to make gold but to prepare medicines.

PARACELSUS Swiss physician, alchemist, and scientist.
Attributed remark

chemosynthesis method of making ◊protoplasm (contents of a cell) using the energy from chemical reactions, in contrast to the use of light energy employed for the same purpose in ◊photosynthesis. The process is used by certain bacteria, which can synthesize organic compounds from carbon dioxide and water using the energy from special methods of ◊respiration.

Nitrifying bacteria are a group of chemosynthetic organisms which change free nitrogen into a form that can be taken up by plants; nitrobacteria, for example, oxidize nitrites to nitrates. This is a vital part of the ◊nitrogen cycle. As chemosynthetic bacteria can survive without light energy, they can live in dark and inhospitable regions, including the hydrothermal vents of the Pacific ocean. Around these vents, where temperatures reach up to 350°C/662°F, the chemosynthetic bacteria are the basis of a food web supporting fishes and other marine life.

chemotaxis in biology, the property that certain cells have of attracting or repelling other cells. For example, white blood cells are attracted to the site of infection by the release of substances during certain types of immune response.

chemotherapy any medical treatment with chemicals. It usually refers to treatment of cancer with cytotoxic and other drugs. The term was coined by the German bacteriologist Paul Ehrlich for the use of synthetic chemicals against infectious diseases.

chemotropism movement by part of a plant in response to a chemical stimulus. The response by the plant is termed 'positive' if the growth is towards the stimulus or 'negative' if the growth is away from the stimulus.

Fertilization of flowers by pollen is achieved because the ovary releases chemicals that produce a positive chemotropic response from the developing pollen tube.

Cherenkov radiation a type of electromagnetic radiation emitted by charged particles entering a transparent medium at a speed greater than the speed of light in the medium. It appears as a bluish light. Cherenkov radiation can be detected from high-energy cosmic rays entering the Earth's atmosphere. It is named after Pavel Alexseevich Cherenkov, the Russian physicist who first observed it.

Chernobyl town in northern Ukraine, 100 km/62 mi north of Kiev; site of a nuclear power station. On 26 April 1986, two huge explosions occurred at the plant, destroying a central reactor and breaching its 1,000-tonne roof. In the immediate vicinity of Chernobyl, 31 people died (all firemen or workers at the plant) and 135,000 were permanently evacuated. It has been estimated that there will be an additional 20–40,000 deaths from cancer in the following 60 years; 600,000 are officially classified as at risk. According to WHO figures of 1995, the incidence of thyroid cancer in children has increased 200-fold in Belarus as a result of fallout from the disaster.

The Chernobyl disaster occurred as the result of an unauthorized test being conducted, in which the recactor was run while its cooling system was inoperative. The resulting clouds of radioactive isotopes spread all over Europe, from Ireland to Greece. A total of 9 tonnes of radioactive material were released into the atmosphere, 90 times the amount produced by the Hiroshima A-bomb. In all, 5 million people are thought to have been exposed to radioactivity following the blast. In Ukraine, Belarus, and Russia more than 500,000 people were displaced from affected towns and villages and thousands of square miles of land were contaminated.

cherry any of a group of fruit-bearing trees distinguished from plums and apricots by their fruits, which are round and smooth and not covered with a bloom. They are cultivated in temperate regions with warm summers and grow best in deep fertile soil. (Genus *Prunus*, family Rosaceae.)

chervil any of several plants belonging to the carrot family. The garden chervil (*Anthriscus cerefolium*) has leaves with a sweetish smell, similar to parsley. It is used as a garnish and in soups. Chervil originated on the borders of Europe and Asia and was introduced to W Europe by the Romans. (Genus *Anthriscus*, family Umbelliferae.)

chestnut any of a group of trees belonging to the beech family. The Spanish or sweet chestnut (*Castanea sativa*) produces edible nuts inside husks; its timber is also valuable. ◊Horse chestnuts are quite distinct, belonging to the genus *Aesculus*, family Hippocastanaceae. (True chestnut genus *Castanea*, family Fagaceae.)

chevrotain or ***mouse deer*** small forest-dwelling mammals resembling deer. Horns are absent and they reach a maximum height at the shoulder of about 35 cm/14 in. They are active at night and feed mainly on plants and fruit.

There are four species in two genera, *Tragulus* and *Hyemoschus*. *Tragulus* contains three species of small animals, which have more or less the characteristics and habits of some rodents. They inhabit Asia, Malaysia, Sri Lanka, and India. *Hyemoschus* contains only one species, the African **water chevrotain**.

classification Chevrotains are members of the family Tragulidae in the mammalian order Artiodactyla.

chicken domestic fowl; see under ◊poultry.

chickenpox or ***varicella*** common, usually mild disease, caused by a virus of the ◊herpes group and transmitted by airborne droplets. Chickenpox chiefly attacks children under the age of ten. The incubation period is two to three weeks. One attack normally gives immunity for life.

The temperature rises and spots (later inflamed blisters) develop on the torso, then on the face and limbs. The sufferer recovers within a week, but remains infectious until the last scab disappears.

The US Food and Drug Administration approved a chickenpox vaccine in March 1995. Based on a weakened form of the live virus, the vaccine is expected to be 70–90% effective. A vaccine is available in Europe, but is only used in children with an impaired immune system.

chickpea annual leguminous plant (see ◊legume), grown for food in India and the Middle East. Its short hairy pods contain edible seeds similar to peas. (*Cicer arietinum*, family Leguminosae.)

chickweed any of several low-growing plants belonging to the pink family, with small white starlike flowers. (Genera *Stellaria* and *Cerastium*, family Caryophyllaceae.)

chicle milky juice from the sapodilla tree *Achras zapota* of Central America; it forms the basis of chewing gum.

chicory plant native to Europe and W Asia, with large, usually blue, flowers. Its long taproot is used dried and roasted as a coffee substitute. As a garden vegetable, grown under cover, its blanched leaves are used in salads. It is related to ◊endive. (*Cichorium intybus*, family Compositae.)

chiffchaff small songbird *Phylloscopus collybita* of the warbler family, Muscicapidae, order Passeriformes. It is found in woodlands and thickets in Europe and N Asia during the summer, migrating south for winter. About 11 cm/4.3 in long, olive above, greyish below, with yellow-white nether parts, an eyestripe, and usually dark legs, it looks similar to a willow warbler but has a distinctive song.

chigger or ***harvest mite*** scarlet or rusty brown ◊mite genus *Trombicula*, family Trombiculidae, in the order Acarina, common in summer and autumn. Chiggers are parasitic, and their tiny red larvae cause intensely irritating bites in places where the skin is thin, such as behind the knees or between the toes. After a time they leave their host and drop to the ground where they feed upon minute insects.

Chiggers are medically important and can be harmful to humans in two ways: either by the feeding activities of the larval mite, on folds of skin or the edges of hair follicles resulting in painful lesions 0.4–2 cm/0.16–0.8 in long, or by acting as carriers of disease, for example scrub ◊typhus or haemorrhagic fever.

chihuahua smallest breed of dog, 15 cm/10 in high, developed in the USA from Mexican origins. It may weigh only 1 kg/2.2 lb. The domed head and wide-set ears are characteristic, and the skull is large compared to the body. It can be almost any colour, and occurs in both smooth (or even hairless) and long-coated varieties.

chilblain painful inflammation of the skin of the feet, hands, or ears, due to cold. The parts turn red, swell, itch violently, and are very tender. In bad cases, the skin cracks, blisters, or ulcerates.

childbirth the expulsion of a baby from its mother's body following ◊pregnancy. In a broader sense, it is the period of time involving labour and delivery of the baby.

chilli (***North American* chili**) pod, or powder made from the pod, of a variety of ◊capsicum (*Capsicum frutescens*), a small, hot, red pepper. It is widely used in cooking. The hot ingredient of chilli is capsaicin. It causes a burning sensation in the mouth by triggering nerve branches in the eyes, nose, tongue, and mouth.

Capsaicin does not activate the taste buds and therefore has no flavour. It is claimed that people can become physically addicted to it.

CHILLI!

http://www.tpoint.net/~wallen/chili.html

Online adulation of the hottest food in the world with personal recommendations on the many varieties and recipes, as well as how to prepare the chillies themselves.

chimaera fish of the group Holocephali. Chimaeras have thick bodies that taper to a long thin tail, large fins, smooth skin, and a cartilaginous skeleton. They can grow to 1.5 m/4.5 ft. Most chimaeras are deep-water fish, and even *Chimaera monstrosa*, a relatively shallow-living form caught around European coasts, lives at a depth of 300–500 m/1,000–1,600 ft.

chimera or ***chimaera*** in biology, an organism composed of tissues that are genetically different. Chimeras can develop naturally if a ◊mutation occurs in a cell of a developing embryo, but are more commonly produced artificially by implanting cells from one organism into the embryo of another.

CHIMPANZEE

http://www.seaworld.org/animal_bytes/chimpanzeeab.html

Illustrated guide to the chimpanzee including information about genus, size, life span, habitat, gestation, diet, and a series of fun facts.

chimpanzee highly intelligent African ape *Pan troglodytes* that lives mainly in rain forests but sometimes in wooded savanna. Chimpanzees are covered in thin but long black body hair, except for the face, hands, and feet, which may have pink or black skin. They normally walk on all fours, supporting the front of the body on the knuckles of the fingers, but can stand or walk upright for a short distance. They can grow to 1.4 m/4.5 ft tall, and weigh up to 50 kg/110 lb. They are strong and climb well, but spend time on the ground, living in loose social groups. The bulk of the diet is fruit, with some leaves, insects, and occasional meat. Chimpanzees can use 'tools', fashioning twigs to extract termites from their nests.

The **bonobo** or pygmy chimpanzee, *Pan paniscus* is found only in a small area of rainforest in Congo (formerly Zaire). Bonobos are about the same height as 'common' chimpanzees, but they are of a slighter build, with less hair, and stand upright more frequently. They are a distinct species, numbering approximately 10,000.

Chimpanzees are found in an area from W Africa to W Uganda and Tanzania in the east. Studies of chromosomes suggest that chimpanzees are the closest apes to humans, perhaps sharing 99% of the same genes. Trained chimpanzees can learn to communicate with humans with the aid of machines or sign language, but are probably precluded from human speech by the position of the voicebox.

CHIMPANZEE

Sick chimpanzees use natural remedies to heal themselves. Internal parasites are purged by eating certain leaves, and stomach upsets are treated with mouthfuls of earth rich in clay minerals, sodium, iron, and aluminium.

china clay commercial name for ◊kaolin.

chincherinchee poisonous plant *Ornithogalum thyrsoides* of the lily family Liliaceae. It is native to South Africa, and has spikes of long-lasting, white or yellow, waxlike flowers.

chinchilla South American rodent *Chinchilla laniger* found in high, rather barren areas of the Andes in Bolivia and Chile. About the size of a small rabbit, it has long ears and a long bushy tail, and shelters in rock crevices. These gregarious animals have thick, soft, silver-grey fur, and were hunted almost to extinction for it. They are now farmed and protected in the wild.

Chinese crested dog breed of toy dog developed in China, notable for being hairless except for large plumes on its head, feet and tail. Its skin has a blue or pink tinge, sometimes with black spots. Lightly built, it weighs up to 5.5 kg/12 lb.

The 'Powder Puff' variety has long, plume-like, very fine hair all over its body.

chip or ***silicon chip*** another name for an ◊integrated circuit, a complete electronic circuit on a slice of silicon (or other semiconductor) crystal only a few millimetres square.

chipmunk any of several species of small ground squirrel with characteristic stripes along its side. Chipmunks live in North America and E Asia, in a variety of habitats, usually wooded, and take shelter in burrows. They have pouches in their cheeks for carrying food. They climb well but spend most of their time on or near the ground.

The **Siberian chipmunk** *Eutamias sibiricus,* about 13 cm/5 in long, is found in N Russia, N China, and Japan.

chip-set in computing, group of ◊chips that work together to perform a particular set of functions. Standard chip-sets, for example, manage graphics or form the working parts of a modem.

Chiron unusual Solar-System object orbiting between Saturn and Uranus, discovered 1977 by US astronomer Charles T Kowal (1940–).

Initially classified as an asteroid, it is now believed to be a giant cometary nucleus about 200 km/120 mi across, composed of ice with a dark crust of carbon dust. It has a 51-year orbit and a coma (cloud of gas and dust) caused by evaporation from its surface, resembling that of a comet. It is classified as a ◊centaur.

chiropractic in alternative medicine, technique of manipulation of the spine and other parts of the body, based on the principle that physical disorders are attributable to aberrations in the functioning of the nervous system, which manipulation can correct.

Developed in the 1890s by US practitioner Daniel David Palmer, chiropractic is widely practised today by accredited therapists, although orthodox medicine remains sceptical of its efficacy except for the treatment of back problems.

chiru Tibetan species of antelope. It is pale fawn in colour with coarse hair; the male alone has horns, and these are long, straight, ringed and gazellelike. It is nearly 1 m/3.3 ft in height.
classification The chiru *Pantholops hodgsoni* is in the family Bovidae (cattle and antelopes) of order Artiodactyla.

chital another name for the ◊axis deer.

chitin complex long-chain compound, or ◊polymer; a nitrogenous derivative of glucose. Chitin is widely found in invertebrates. It forms the ◊exoskeleton of insects and other arthropods. It combines with protein to form a covering that can be hard and tough, as in beetles, or soft and flexible, as in caterpillars and other insect larvae. It is insoluble in water and resistant to acids, alkalis, and many organic solvents. In crustaceans such as crabs, it is impregnated with calcium carbonate for extra strength.

Chitin also occurs in some ◊protozoans and coelenterates (such as certain jellyfishes), in the jaws of annelid worms, and as the cell-wall polymer of fungi. Its uses include coating apples (still fresh after six months), coating seeds, and dressing wounds. In 1993 chemists at North Carolina State University found that it can be used to filter pollutants from industrial waste water.

chiton group of marine molluscs, ranging in size from 1 cm/0.4 in to 15 cm/6 in; some are littoral (intertidal) whilst others live at greater depths. They live on vegetable matter and are like limpets in habit; they usually attach themselves to rocks, but can crawl by means of their long foot, and are capable of rolling themselves up.

All the species are bilaterally symmetrical, have eight shell-plates embedded partially or entirely in the mantle, and are covered with spicules.
classification Chitons are in class Polyplacophora, phylum Mollusca.

chive or ***chives*** perennial European plant belonging to the lily family, related to onions and leeks. It has an underground bulb, long hollow tubular leaves, and globe-shaped purple flower heads. The leaves are used as a garnish for salads. (*Allium schoenoprasum,* family Liliaceae.)

chlamydia viruslike bacteria which live parasitically in animal cells, and cause disease in humans and birds. Chlamydiae are thought to be descendants of bacteria that have lost certain metabolic processes. In humans, a strain of chlamydia causes ◊trachoma, a disease found mainly in the tropics (a leading cause of blindness); venereally transmitted chlamydiae cause genital and urinary infections.

Protein from *C. Pneumoniae* (which accounts for 10% of pneumonia cases) has been found in 79% of cases of atheroma (furring up of the arteries) in a US study, and it has also been cultured from a diseased coronary artery, providing a possible link between chlamydia infection and heart disease. A link has also been established between *C. Pneumoniae* infection and chronic high blood pressure.

chloral or ***trichloroethanal*** CCl_3CHO oily, colourless liquid with a characteristic pungent smell, produced by the action of chlorine on ethanol. It is soluble in water and its compound chloral hydrate is a powerful sleep-inducing agent.

chlorate any salt derived from an acid containing both chlorine and oxygen and possessing the negative ion ClO^-, ClO_2^-, ClO_3^-, or ClO_4^-. Common chlorates are those of sodium, potassium, and barium. Certain chlorates are used in weedkillers.

chlorella any single-celled, green, freshwater alga of the genus *Chlorella,* 3–10 micrometres in diameter, which can increase its weight by four times in 12 hours. Nutritive content: 50% protein, 20% fat, 20% carbohydrate, 10% phosphate, calcium, and ◊trace elements.

chloride Cl^- negative ion formed when hydrogen chloride dissolves in water, and any salt containing this ion, commonly formed by the action of hydrochloric acid (HCl) on various metals or by direct combination of a metal and chlorine. Sodium chloride (NaCl) is common table salt.

chlorinated solvent any liquid organic compound that contains chlorine atoms, often two or more. These compounds are very effective solvents for fats and greases, but many have toxic properties.

They include trichloromethane (chloroform, $CHCl_3$), tetrachloromethane (carbon tetrachloride, CCl_4), and trichloroethene ($CH_2ClCHCl_2$).

chlorination the treatment of water with chlorine in order to disinfect it; also, any chemical reaction in which a chlorine atom is introduced into a chemical compound.

chlorine ***Greek*** *chloros* ***'green'*** greenish-yellow, gaseous, non-metallic element with a pungent odour, symbol Cl, atomic number 17, relative atomic mass 35.453. It is a member of the ◊halogen group and is widely distributed, in combination with the ◊alkali metals, as chlorates or chlorides.

Chlorine was discovered in 1774 by the German chemist Karl Scheele, but English chemist Humphry Davy first proved it to be an element in 1810 and named it after its colour. In nature it is always found in the combined form, as in hydrochloric acid, produced in the mammalian stomach for digestion. Chlorine is obtained commercially by the electrolysis of concentrated brine and is an important bleaching agent and ermicide, used for sterilizing both drinking water and swimming-pools. As an oxidizing agent it finds many

CHLORINE

http://c3.org/

Designed to promote better understanding of the science of chlorine chemistry, examining how it contributes to an enhancement of our standard of living and quality of life.

applications in organic chemistry. The pure gas (Cl_2) is a poison and was used in gas warfare in World War I, where its release seared the membranes of the nose, throat, and lungs, producing pneumonia. Chlorine is a component of chlorofluorocarbons (CFCs) and is partially responsible for the depletion of the ◊ozone layer; it is released from the CFC molecule by the action of ultraviolet radiation in the upper atmosphere, making it available to react with and destroy the ozone. The concentration of chlorine in the atmosphere in 1997 reached just over 3 parts per billion. It is expected to reach its peak in 1999 and then start falling rapidly due to international action to curb ozone-destroying chemicals.

PROBLEM OF CHLOROFLUOROCARBONS

http://pooh.chem.wm.edu/chemWWW/courses/chem105/projects/group2/page5.html

Excellent graphical presentation of the effect of CFCs on the ozone layer. The information can be readily understood by a general reader wishing to learn more about the chemistry of ozone depletion and why more ultraviolet radiation is reaching the surface of the earth.

chlorofluorocarbon (CFC) a class of synthetic chemicals that are odourless, nontoxic, nonflammable, and chemically inert. The first CFC was synthesized in 1892, but no use was found for it until the 1920s. Since then their stability and apparently harmless properties have made CFCs popular as propellants in ◊aerosol cans, as refrigerants in refrigerators and air conditioners, as degreasing agents, and in the manufacture of foam packaging. They are partly responsible for the destruction of the ◊ozone layer. In June 1990 representatives of 93 nations, including the UK and the USA, agreed to phase out production of CFCs and various other ozone-depleting chemicals by the end of the 20th century.

When CFCs are released into the atmosphere, they drift up slowly into the stratosphere, where, under the influence of ultraviolet radiation from the Sun, they react with ozone (O_3) to form free chlorine (Cl) atoms and molecular oxygen (O_2), thereby destroying the ozone layer that protects Earth's surface from the Sun's harmful ultraviolet rays. The chlorine liberated during ozone breakdown can react with still more ozone, making the CFCs particularly dangerous to the environment. CFCs can remain in the atmosphere for more than 100 years. Replacements for CFCs are being developed, and research into safe methods for destroying existing CFCs is being carried out. In 1996 it was reported that US chemists at Yale University had developed a process for breaking down freons and other gases containing CFCs into nonhazardous compounds.

Since their initial introduction, different CFC compounds have been introduced into the environment as new applications have evolved. Comparisons of the concentrations of different CFC compounds can therefore be used to date ground water flows and even ocean circulation patterns. Ironically, CFC dating provides the environmental industry and research scientists with a powerful new tool for understanding the factors that affect Earth's environment.

The European Union agreed to ban by the end of 1995 the five 'full hydrogenated' CFCs that are restricted under the ◊Montréal Protocol and a range of CFCs used as industrial solvents, refrigerants, and in fire extinguishers.

The tragedy of a scientific man is that he has found no way to guide his own discoveries to a constructive end.

CHARLES A LINDBERGH US aviator.
Attributed remark

chloroform (technical name ***trichloromethane***) $CHCl_3$ clear, colourless, toxic, carcinogenic liquid with a characteristic pungent, sickly sweet smell and taste, formerly used as an anaesthetic (now superseded by less harmful substances).

It is used as a solvent and in the synthesis of organic chemical compounds.

chlorophyll green pigment present in most plants; it is responsible for the absorption of light energy during ◊photosynthesis.

The pigment absorbs the red and blue-violet parts of sunlight but reflects the green, thus giving plants their characteristic colour.

Chlorophyll is found within chloroplasts, present in large numbers in leaves. Cyanobacteria (blue-green algae) and other photosynthetic bacteria also have chlorophyll, though of a slightly different type. Chlorophyll is similar in structure to ◊haemoglobin, but with magnesium instead of iron as the reactive part of the molecule.

chloroplast structure (◊organelle) within a plant cell containing the green pigment chlorophyll. Chloroplasts occur in most cells of the green plant that are exposed to light, often in large numbers. Typically, they are flattened and disclike, with a double membrane enclosing the stroma, a gel-like matrix. Within the stroma are stacks of fluid-containing cavities, or vesicles, where ◊photosynthesis occurs.

It is thought that the chloroplasts were originally free-living cyanobacteria (blue-green algae) which invaded larger, non-photosynthetic cells and developed a symbiotic relationship with them. Like ◊mitochondria, they contain a small amount of DNA and divide by fission. Chloroplasts are a type of ◊plastid.

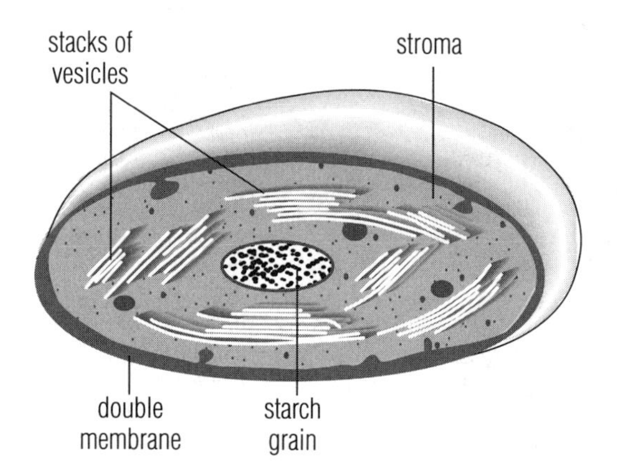

chloroplast *Green chlorophyll molecules on the membranes of the vesicle stacks capture light energy to produce food by photosynthesis.*

chlorosis abnormal condition of green plants in which the stems and leaves turn pale green or yellow. The yellowing is due to a reduction in the levels of the green chlorophyll pigments. It may be caused by a deficiency in essential elements (such as magnesium, iron, or manganese), a lack of light, genetic factors, or viral infection.

choke coil in physics, a coil employed to limit or suppress alternating current without stopping direct current, particularly the type used as a 'starter' in the circuit of fluorescent lighting.

cholecalciferol or ***vitamin D*** fat-soluble chemical important in the uptake of calcium and phosphorous for bones. It is found in liver, fish oils and margarine. It can be produced in the skin, provided that the skin is adequately exposed to sunlight. Lack of vitamin D leads to rickets and other bone diseases.

cholecystectomy surgical removal of the ◊gall bladder. It is carried out when gallstones or infection lead to inflammation of the gall bladder, which may then be removed either by conventional surgery or by a 'keyhole' procedure (see ◊endoscopy).

cholera disease caused by infection with various strains of the bacillus *Vibrio cholerae,* transmitted in contaminated water and characterized by violent diarrhoea and vomiting. It is prevalent in many tropical areas.

The formerly high death rate during epidemics has been much reduced by treatments to prevent dehydration and loss of body salts, together with the use of antibiotics. There is an effective vaccine that must be repeated at frequent intervals for people exposed to continuous risk of infection. The worst epidemic in the Western hemisphere for 70 years occurred in Peru in 1991, with 55,000 confirmed cases and 258 deaths. It was believed to have been spread by the consumption of seafood contaminated by untreated sewage. 1991 was also the worst year on record for cholera in Africa with 13,000 deaths.

cholesterol white, crystalline ◊sterol found throughout the body, especially in fats, blood, nerve tissue, and bile; it is also provided in the diet by foods such as eggs, meat, and butter. A high level of cholesterol in the blood is thought to contribute to atherosclerosis (hardening of the arteries).

Cholesterol is an integral part of all cell membranes and the starting point for steroid hormones, including the sex hormones. It is broken down by the liver into bile salts, which are involved in fat absorption in the digestive system, and it is an essential component of **lipoproteins**, which transport fats and fatty acids in the blood. **Low-density lipoprotein cholesterol** (LDL-cholesterol), when present in excess, can enter the tissues and become deposited on the surface of the arteries, causing ◊atherosclerosis. **High-density lipoprotein cholesterol** (HDL-cholesterol) acts as a scavenger, transporting fat and cholesterol from the tissues to the liver to be broken down. The composition of HDL-cholesterol can vary and some forms may not be as effective as others. Blood cholesterol levels can be altered by reducing the amount of alcohol and fat in the diet and by substituting some of the saturated fat for polyunsaturated fat, which gives a reduction in LDL-cholesterol. HDL-cholesterol can be increased by exercise.

CHOLESTEROL

The Inuit people rarely suffer from heart disease. One explanation for this may be the high amount of fish, notably salmon, in their diet, which reduces the level of blood cholesterol.

chondrite type of ◊meteorite characterized by **chondrules**, small spheres, about 1 mm in diameter, made up of the silicate minerals olivine and orthopyroxene.

chondrule in astronomy a small, round mass of ◊silicate material found. Chondrites (stony ◊meteorites) are characterized by the presence of chondrules.

Chondrules are thought to be mineral grains that condensed from hot gas in the early Solar System, most of which were later incorporated into larger bodies from which the ◊planets formed.

chord in geometry, a straight line joining any two points on a curve. The chord that passes through the centre of a circle (its longest chord) is the diameter. The longest and shortest chords of an ellipse (a regular oval) are called the major and minor axes, respectively.

Chordata phylum of animals, members of which are called ◊chordates.

chordate animal belonging to the phylum Chordata, which includes vertebrates, sea squirts, amphioxi, and others. All these animals, at some stage of their lives, have a supporting rod of tissue (notochord or backbone) running down their bodies.

Chordates are divided into three major groups: ◊tunicates, cephalochordates (see ◊lancelet), and craniates (including all vertebrates).

chorea condition featuring involuntary movements of the face muscles and limbs. It is seen in a number of neurological diseases, including ◊Huntington's chorea.

chorion outermost of the three membranes enclosing the embryo of reptiles, birds, and mammals; the ◊amnion is the innermost membrane.

chorionic villus sampling (CVS) ◊biopsy of a small sample of placental tissue, carried out in early pregnancy at 10–12 weeks' gestation. Since the placenta forms from embryonic cells, the tissue obtained can be tested to reveal genetic abnormality in the fetus. The advantage of CVS over ◊amniocentesis is that it provides an earlier diagnosis, so that if any abnormality is discovered, and the parents opt for an abortion, it can be carried out more safely.

choroid layer found at the rear of the ◊eye beyond the retina. By absorbing light that has already passed through the retina, it stops back-reflection and so prevents blurred vision.

chough bird *Pyrrhocorax pyrrhocorax* of the crow family, Corvidae, order Passeriformes, about 38 cm/15 in long, black-feathered, with red bill and legs, and long hooked claws. Choughs are frugivorous and insectivorous. They make mud-walled nests and live on sea cliffs and mountains from Europe to E Asia, but are now rare.

The **alpine chough** *Pyrrhocorax graculus* is similar, but has a shorter yellow bill and is found up to the snowline in mountains from the Pyrenees to Central Asia.

chow chow breed of dog originating in China in ancient times. About 45 cm/1.5 ft tall, it has a broad neck and head, round catlike feet, a soft woolly undercoat with a coarse outer coat, and a mane. Its coat should be of one colour, and it has an unusual blue-black tongue.

Christmas rose decorative species of ◊hellebore.

Christmas tree any evergreen tree brought indoors and decorated for Christmas. The custom was a medieval German tradition and is now practised in many Western countries.

chroma key in television, technique for substituting backgrounds. For example, the empty studio behind a newscaster may be replaced with an outdoor scene or a frame of video footage. This technique is commonly used on news programmes and other shows that feature 'talking heads'.

A computer analyses the image of the newscaster, who is placed in front of a plain background, usually blue, to identify the exact ◊pixels where the talking figure begins and ends. It can then substitute a new image for just the area specified. The technique allows broadcasters to add visual interest while keeping costs down.

chromatography ***Greek chromos 'colour'*** technique for separating or analysing a mixture of gases, liquids, or dissolved substances. This is brought about by means of two immiscible substances, one of which (**the mobile phase**) transports the sample mixture through the other (**the stationary phase**). The mobile phase may be a gas or a liquid; the stationary phase may be a liquid or a solid, and may be in a column, on paper, or in a thin layer on a glass or plastic support. The components of the mixture are absorbed or impeded by the stationary phase to different extents

CHROMATOGRAPHY

http://www.eng.rpi.edu/dept/chem-eng/Biotech-Environ/CHROMO/chromintro.html

Explanation of the theory and practice of chromatography. Designed for school students (and introduced by a Biotech Bunny), the contents include equipment, analysing a chromatogram, and details of the various kinds of chromatography.

and therefore become separated. The technique is used for both qualitative and quantitive analyses in biology and chemistry.

In **paper chromatography**, the mixture separates because the components have differing solubilities in the solvent flowing through the paper and in the chemically bound water of the paper.

In **thin-layer chromatography**, a wafer-thin layer of adsorbent medium on a glass plate replaces the filter paper. The mixture separates because of the differing solubilities of the components in the solvent flowing up the solid layer, and their differing tendencies to stick to the solid (adsorption). The same principles apply in **column chromatography**.

In **gas–liquid chromatography**, a gaseous mixture is passed into a long, coiled tube (enclosed in an oven) filled with an inert powder coated in a liquid. A carrier gas flows through the tube. As the mixture proceeds along the tube it separates as the components dissolve in the liquid to differing extents or stay as a gas. A detector locates the different components as they emerge from the tube. The technique is very powerful, allowing tiny quantities of substances (fractions of parts per million) to be separated and analysed.

Preparative chromatography is carried out on a large scale for the purification and collection of one or more of a mixture's constituents; for example, in the recovery of protein from abattoir wastes.

Analytical chromatography is carried out on far smaller quantities, often as little as one microgram (one-millionth of a gram), in order to identify and quantify the component parts of a mixture. It is used to determine the identities and amounts of amino acids in a protein, and the alcohol content of blood and urine samples. The technique was first used in the separation of coloured mixtures into their component pigments.

chromite $FeCr_2O_4$, iron chromium oxide, the main chromium ore. It is one of the ◊spinel group of minerals, and crystallizes in dark-coloured octahedra of the cubic system. Chromite is usually found in association with ultrabasic and basic rocks; in Cyprus, for example, it occurs with ◊serpentine, and in South Africa it forms continuous layers in a layered ◊intrusion.

chromium ***Greek chromos 'colour'*** hard, brittle, grey-white, metallic element, symbol Cr, atomic number 24, relative atomic mass 51.996. It takes a high polish, has a high melting point, and is very resistant to corrosion. It is used in chromium electroplating, in the manufacture of stainless steel and other alloys, and as a catalyst. Its compounds are used for tanning leather and for ◊alums. In human nutrition it is a vital trace element. In nature, it occurs chiefly as chrome iron ore or chromite ($FeCr_2O_4$). Kazakhstan, Zimbabwe, and Brazil are sources.

The element was named 1797 by the French chemist Louis Vauquelin (1763–1829) after its brightly coloured compounds.

chromium ore essentially the mineral chromite, $FeCr_2O_4$, from which chromium is extracted. South Africa and Zimbabwe are major producers.

chromogranin protein released by the ◊adrenal gland, along with the hormone adrenaline, during times of stress. There are three types, chromagranin A, B, and C. The function of chromagranins is poorly understood, but they do have an antibacterial affect that could boost the immune system when it is suppressed during times of stress. Chromagranin A also stimulates ◊insulin release and relaxes blood vessels – functions similarly reduced by adrenaline during stress.

chromosome structure in a cell nucleus that carries the ◊genes. Each chromosome consists of one very long strand of DNA, coiled and folded to produce a compact body. The point on a chromosome where a particular gene occurs is known as its locus. Most higher organisms have two copies of each chromosome (they are ◊diploid) but some have only one (they are ◊haploid). There are 46 chromosomes in a normal human cell. See also ◊mitosis and ◊meiosis.

Chromosomes are only visible during cell division; at other times they exist in a less dense form called chromatin.

The first artificial human chromosome was built by US geneticists in 1997. They constructed telomeres, centromeres, and DNA containing genetic information, which they removed from white blood cells, and inserted them into human cancer cells. The cells assembled the material into chromosomes. The artificial chromosome was successfully passed onto all daughter cells.

XY

chromosome *The 23 pairs of chromosomes of a normal human male.*

chromosphere ***Greek 'colour' and 'sphere'*** layer of mostly hydrogen gas about 10,000 km/6,000 mi deep above the visible surface of the Sun (the photosphere). It appears pinkish red during ◊eclipses of the Sun.

chronic in medicine, term used to describe a condition that is of slow onset and then runs a prolonged course, such as rheumatoid arthritis or chronic bronchitis. In contrast, an **acute** condition develops quickly and may be of relatively short duration.

chronic fatigue syndrome a common debilitating condition also known as myalgic encephalomyelitis (ME), postviral fatigue syndrome, or 'yuppie flu'. It is characterized by a diffuse range of symptoms present for at least six months including extreme fatigue, muscular pain, weakness, depression, poor balance and coordination, joint pains, and gastric upset. It is usually diagnosed after exclusion of other diseases and frequently follows a flulike illness.

CHRONIC FATIGUE SYNDROME FAQ

http://www.cais.com/cfs-news/faq.htm

Extensive answer sheet shedding light on the hottest questions regarding chronic fatigue syndrome. The site deals with issues such as the relation to stress and depression, the possible causes and duration of the illness, and the onset and clinical symptoms of the disease. It also discuss a series of common misunderstandings related to the syndrome.

The cause of CFS remains unknown, but it is believed to have its origin in a combination of viral, social, and psychological factors. Theories based on one specific cause (such as Epstein-Barr virus) have been largely discredited. There is no definitive treatment, but with time the symptoms become less severe. Depression is treated if present, and recent research has demonstrated the effectiveness of ◊cognitive therapy.

chronometer instrument for measuring time precisely, originally used at sea. It is designed to remain accurate through all conditions of temperature and pressure. The first accurate marine chronometer, capable of an accuracy of half a minute a year, was made in 1761 by John Harrison in England.

chrysalis pupa of an insect, but especially that of a ◊butterfly or ◊moth. It is essentially a static stage of the creature's life, when the adult insect, benefiting from the large amounts of food laid down by the actively feeding larva, is built up from the disintegrating larval tissues. The chrysalis may be exposed or within a cocoon.

chrysanthemum any of a large group of plants with colourful, showy flowers, containing about 200 species. There are hundreds of cultivated varieties, whose exact wild ancestry is uncertain. In the Far East the common chrysanthemum has been cultivated for more than 2,000 years and is the imperial emblem of Japan. Chrysanthemums can be grown from seed, but new plants are more commonly produced from cuttings or by dividing up established plants. (Genus *Chrysanthemum*, family Compositae.)

chrysotile mineral in the ◊serpentine group, $Mg_3Si_2O_5(OH)_4$. A soft, fibrous, silky mineral, the primary source of asbestos.

chub freshwater fish *Leuciscus cephalus* of the carp family. Thickset and cylindrical, it grows up to 60 cm/2 ft, is dark greenish or grey on the back, silvery yellow below, with metallic flashes on the flanks. It lives generally in clean rivers throughout Europe.

chyme general term for the stomach contents. Chyme resembles a thick creamy fluid and is made up of partly digested food, hydrochloric acid, and a range of enzymes.

The muscular activity of the stomach churns this fluid constantly, continuing the mechanical processes initiated by the mouth. By the time the chyme leaves the stomach for the duodenum, it is a smooth liquid ready for further digestion and absorption by the small intestine.

Cibachrome in photography, a process of printing directly from transparencies. It can be home-processed and the rich, saturated colours are highly resistant to fading. It was introduced 1963.

cicada any of several insects of the family Cicadidae. Most species are tropical, but a few occur in Europe and North America. Young cicadas live underground, for up to 17 years in some species. The adults live on trees, whose juices they suck. The males produce a loud, almost continuous, chirping by vibrating membranes in resonating cavities in the abdomen.

CICHLID HOME PAGE

http://trans4.neep.wisc.edu/~gracy/fish/opener.html

Comprehensive source of information on cichlidae, their habitats, and how to keep them in an aquarium. There are a large number of photographs. Fish can be searched for by scientific or common names.

cichlid any freshwater fish of the family Cichlidae. Cichlids are somewhat perchlike, but have a single nostril on each side instead of two. They are mostly predatory, and have deep, colourful bodies, flattened from side to side so that some are almost disc-shaped. Many are territorial in the breeding season and may show care of the young. There are more than 1,000 species found in South and Central America, Africa, and India.

The **discus fish** *Symphysodon* produces a skin secretion on which the young feed. Other cichlids, such as those of the genus *Tilapia*, brood their young in the mouth.

cigarette beetle small beetle that feeds preferentially on tobacco products, such as cigarettes and cigars. It may, however, feed on a wide range of other products for example, raisins, ginger, cocoa, drugs, and even straw.

classification The cigarette beetle *Lasioderma serricorne* is a member of the family Anobiidae (furniture beetles) in order Coleoptera, class Insecta, phylum Arthropoda.

cilia (singular *cilium*) small hairlike organs on the surface of some cells, particularly the cells lining the upper respiratory tract. Their wavelike movements waft particles of dust and debris towards the exterior. Some single-celled organisms move by means of cilia. In multicellular animals, they keep lubricated surfaces clear of debris. They also move food in the digestive tracts of some invertebrates.

ciliary muscle ring of muscle surrounding and controlling the lens inside the vertebrate eye, used in ◊accommodation (focusing). Suspensory ligaments, resembling spokes of a wheel, connect the lens to the ciliary muscle and pull the lens into a flatter shape when the muscle relaxes. When the muscle is relaxed the lens has its longest ◊focal length and focuses rays from distant objects. On contraction, the lens returns to its normal spherical state and therefore has a shorter focal length and focuses images of near objects.

cinchona any of a group of tropical American shrubs or trees belonging to the madder family. The drug ◊quinine is produced from the bark of some species, and these are now cultivated in India, Sri Lanka, the Philippines, and Indonesia. (Genus *Chinchona*, family Rubiaceae.)

cine camera camera that takes a rapid sequence of still photographs called frames. When the frames are projected one after the other on to a screen, they appear to show movement, because our eyes hold on to the image of one picture until the next one appears.

The cine camera differs from an ordinary still camera in having a motor that winds the film on. The film is held still by a claw mechanism while each frame is exposed. When the film is moved between frames, a semicircular disc slides between the lens and the film and prevents exposure.

CinePak in computing, software method of compressing and decompressing ◊QuickTime 'movies', also called a software codec. CinePak takes a recorded QuickTime file and reduces it in size, frame by frame. This is a slow process, but the result is a file that can be played back efficiently by computers with QuickTime installed.

cinnabar mercuric sulphide mineral, HgS, the only commercially useful ore of mercury. It is deposited in veins and impregnations near recent volcanic rocks and hot springs. The mineral itself is used as a red pigment, commonly known as **vermilion**. Cinnabar is found in the USA (California), Spain (Almadén), Peru, Italy, and Slovenia.

cinnamon dried inner bark of a tree belonging to the laurel family, grown in India and Sri Lanka. The bark is ground to make the spice used in curries and confectionery. Oil of cinnamon is obtained from waste bark and is used as flavouring in food and medicine. (*Cinnamomum zeylanicum*, family Lauraceae.)

cinquefoil any of a group of plants that usually have five-lobed leaves and brightly coloured flowers. They is widespread in northern temperate regions. (Genus *Potentilla*, family Rosaceae.)

circadian rhythm metabolic rhythm found in most organisms, which generally coincides with the 24-hour day. Its most obvious manifestation is the regular cycle of sleeping and waking, but body temperature and the concentration of ◊hormones that influence mood and behaviour also vary over the day. In humans, alteration of habits (such as rapid air travel round the world) may result in the circadian rhythm being out of phase with actual activity patterns, causing malaise until it has had time to adjust.

In mammals the circadian rhythm is controlled by the suprachiasmatic nucleus in the ◊hypothalamus. US researchers discovered a second circadian control mechanism in 1996; they found that cells within the retina also produced the hormone melatonin. In 1997, US geneticists identified a gene, *clock*, in chromosome 5 in mice, that regulated the circadian rhythm.

circle perfectly round shape, the path of a point that moves so as to keep a constant distance from a fixed point (the centre). Each circle has a **radius** (the distance from any point on the circle to the centre), a **circumference** (the boundary of the circle), **diameters** (straight lines crossing the circle through the centre), **chords** (lines joining two points on the circumference), **tangents** (lines that touch the circumference at one point only), **sectors** (regions inside the circle between two radii), and **segments** (regions between a chord and the circumference).

The ratio of the distance all around the circle (the circumference) to the diameter is an ◊irrational number called π (**pi**), roughly equal to 3.1416. A circle of radius r and diameter d has a circumference $C = \pi d$, or $C = 2\pi r$, and an area $A = \pi r^2$. The area of a circle can be shown by dividing it into very thin sectors and reassembling them to make an approximate rectangle. The proof of $A = \pi r^2$ can be done only by using ◊integral calculus.

> To remember the circumference and area of a circle:
>
> FIDDLEDEDUM, FIDDLEDEDEE, A RING ROUND THE MOON IS π TIMES D. IF A HOLE IN YOUR SOCK YOU WANT REPAIRED, YOU USE THE FORMULA πR SQUARED.

circuit in physics or electrical engineering, an arrangement of electrical components through which a current can flow. There are two basic circuits, series and parallel. In a ◊series circuit, the components are connected end to end so that the current flows through all components one after the other. In a ◊parallel circuit, components are connected side by side so that part of the current passes through each component. A circuit diagram shows in graphical form how components are connected together, using standard symbols for the components.

circuit breaker switching device designed to protect an electric circuit by breaking the circuit if excessive current flows. It has the same action as a ◊fuse, and many houses now have a circuit breaker between the incoming mains supply and the domestic circuits. Circuit breakers usually work by means of ◊solenoids. Those at electricity-generating stations have to be specially designed to prevent dangerous arcing (the release of luminous discharge) when the high-voltage supply is switched off. They may use an air blast or oil immersion to quench the arc.

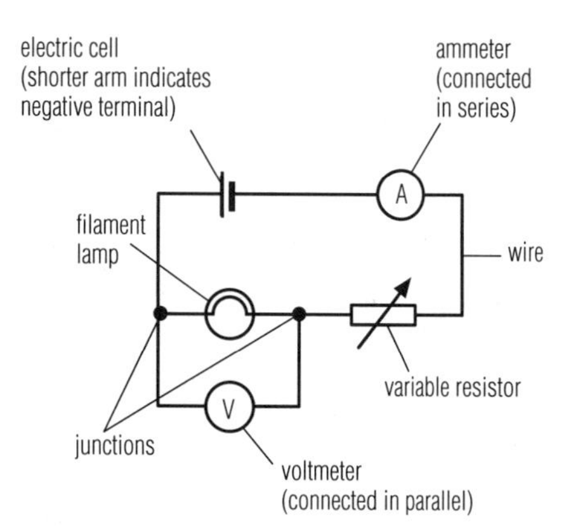

circuit diagram *A circuit diagram shows in graphical form how the components of an electric circuit are connected together. Each component is represented by an internationally recognized symbol, and the connecting wires are shown by straight lines. A dot indicates where wires join.*

circuit diagram simplified drawing of an electric circuit. The circuit's components are represented by internationally recognized symbols, and the connecting wires by straight lines. A dot indicates where wires join.

circulatory system system of vessels in an animal's body that transports essential substances (blood or other circulatory fluid) to and from the different parts of the body. Except for simple animals such as sponges and coelenterates (jellyfishes, sea anemones, corals), all animals have a circulatory system.

In fishes, blood passes once around the body before returning to a two-chambered heart (single circulation). In birds and mammals, blood passes to the lungs and back to the heart before circulating around the remainder of the body (double circulation). In all vertebrates, blood flows in one direction. Valves in the heart, large arteries, and veins prevent backflow, and the muscular walls of the arteries assist in pushing the blood around the body.

Although most animals have a heart or hearts to pump the blood, normal body movements circulate the fluid in some small invertebrates. In the **open system**, found in snails and other mol-

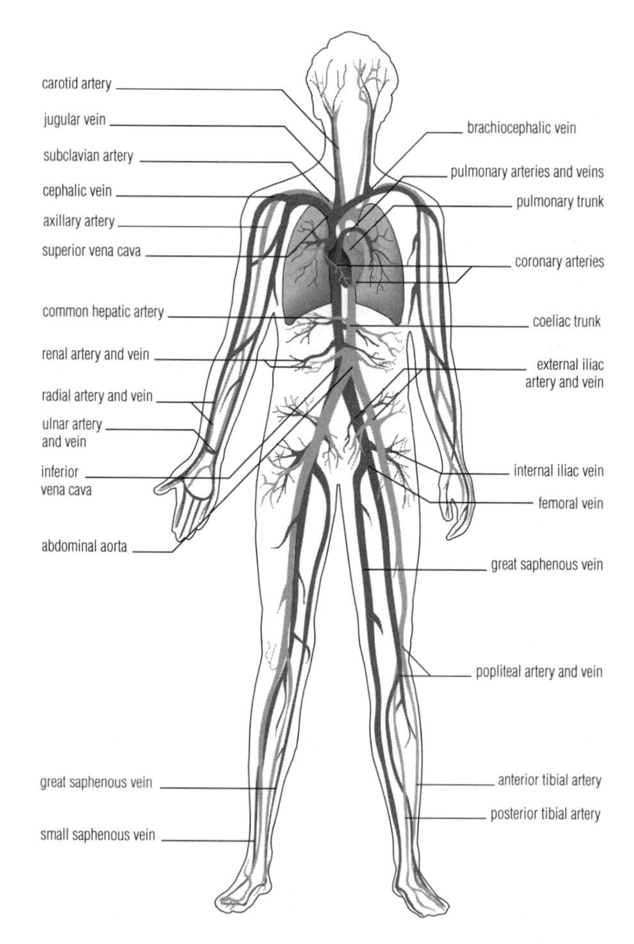

circulatory system *Blood flows through 96,500 km/60,000 mi of arteries and veins, supplying oxygen and nutrients to organs and limbs. Oxygen-poor blood (blue) circulates from the heart to the lungs where oxygen is absorbed. Oxygen-rich blood (red) flows back to the heart and is then pumped round the body through the aorta, the largest artery, to smaller arteries and capillaries. Here oxygen and nutrients are exchanged with carbon dioxide and waste products and the blood returns to the heart via the veins. Waste products are filtered by the liver, spleen, and kidneys, and nutrients are absorbed from the stomach and small intestine.*

Blood: The Discovery of Circulation

BY JULIAN ROWE

Background
After completing a preliminary medical course at the University of Cambridge, where would an ambitious young man in the 17th century go to get a really good medical training? To the University of Padua in Italy, where the great Italian anatomist Hieronymous Fabricius (1537–1619) taught. So this is where English physician William Harvey (1578–1657) naturally went.

William Harvey had a consuming interest in the movement of the blood in the body. In 1579, Fabricius had publicly demonstrated the valves, which he termed 'sluice gates', in the veins: his principal anatomical work was an accurate and detailed description of them.

Galen's theory
Galen, a Greek physician (c. 130– c. 200), had 1,500 years previously written a monumental treatise covering every aspect of medicine. In this work, he asserted that food turned to blood in the liver, ebbed and flowed in vessels and, on reaching the heart, flowed through pores in the septum (the dividing wall) from the right to left side, and was sent on its way by heart spasms. The blood did not circulate. This doctrine was still accepted and taught well into the 16th century.

One-way flow
Harvey was unconvinced. He had done a simple calculation. He worked out that for each human heart beat, about 60 cm^3 of blood left the heart, which meant that the heart pumped out 259 litres every hour. This is more than three times the weight of the average man.

Harvey examined the heart and blood vessels of 128 mammals and found that the valve which separated the left side of the heart from the right ventricle is a one-way structure, as were the valves in the veins discovered by his tutor Fabricius. For this reason he decided that the blood in the veins must flow only towards the heart.

Harvey's experiment
Harvey was now in a position to do his famous experiment. He tied a tourniquet round the upper part of his arm. It was just tight enough to prevent the blood from flowing through the veins back into his heart – but not so tight that arterial blood could not enter the arms. Below the tourniquet, the veins swelled up; above it, they remained empty. This showed that the blood could be entering the arm only through the arteries. Further, by carefully stroking the blood out of a short length of vein, Harvey showed that it could fill up only when blood was allowed to enter it from the end that was furthest away from the heart. He had proved that blood in the veins must flow only towards the heart.

A new theory of circulation
Galen's pores in the septum of the heart had never been found. Belgian physician Andreas Versalius (1514–1564) was another alumnus of Padua University. Although brought up in the Galen tradition, he had carried out secret dissections to discover the pores, and had failed. He did, however, show that men and women had the same number of ribs!

Harvey clinched his researches into the movement of the blood when he demonstrated that no blood seeps through the septum of the heart. He reasoned that blood must pass from the right side of the heart to the left through the lungs. He had discovered the circulation of the blood, and thus, some 20 years after he left Padua, became the father of modern physiology. In 1628 Harvey published his proof of the circulation of the blood in his classic book *On the Motion of the Heart and Blood in Animals*. A new age in medicine and biology had begun.

luscs, the blood (more correctly called ◊haemolymph) passes from the arteries into a body cavity (haemocoel), and from here is gradually returned to the heart, via the gills, by other blood vessels. Insects and other arthropods have an open system with a heart. In the **closed system** of earthworms, blood flows directly from the main artery to the main vein, via smaller lateral vessels in each body segment. Vertebrates, too, have a closed system with a network of tiny ◊capillaries carrying the blood from arteries to veins.

circumcision surgical removal of all or part of the foreskin (prepuce) of the penis, usually performed on the newborn; it is practised among Jews and Muslims. In some societies in Africa and the Middle East, female circumcision or clitoridectomy (removal of the labia minora and/or clitoris) is practised on adolescents as well as babies; it is illegal in the West.

Female circumcision has no medical benefit and often causes disease and complications in childbirth; in 1994 there were at least 90 million women and girls worldwide who had undergone circumcision. Male circumcision too is usually carried out for cultural reasons and not as a medical necessity, apart from cases where the opening of the prepuce is so small as to obstruct the flow of urine. Some evidence indicates that it protects against the development of cancer of the penis later in life and that women with circumcised partners are at less risk from cancer of the cervix.

circumference in geometry, the curved line that encloses a curved plane figure, for example a ◊circle or an ellipse. Its length varies according to the nature of the curve, and may be ascertained by the appropriate formula. The circumference of a circle is πd or $2\pi r$, where d is the diameter of the circle, r is its radius, and π is the constant pi, approximately equal to 3.1416.

circumpolar in astronomy, a description applied to celestial objects that remain above the horizon at all times and do not set as seen from a given location. The amount of sky that is circumpolar depends on the observer's latitude on Earth. At the Earth's poles, all the visible sky is circumpolar, but at the Earth's equator none of it is circumpolar.

circumscribe in geometry, to surround a figure with a circle which passes through all the vertices of the figure. Any triangle may be circumscribed and so may any regular polygon. Only certain quadrilaterals may be circumscribed (their opposite angles must add up to 180°).

cirque French name for a ◊corrie, a steep-sided hollow in a mountainside.

cirrhosis any degenerative disease in an organ of the body, especially the liver, characterized by excessive development of connective tissue, causing scarring and painful swelling. Cirrhosis of the liver may be caused by an infection such as viral hepatitis, chronic obstruction of the common bile duct, chronic alcoholism or drug use, blood disorder, heart failure, or malnutrition. However, often no cause is apparent. If cirrhosis is diagnosed early, it can be arrested by treating the cause; otherwise it will progress to coma and death.

CIS, CI$ abbreviations for ***CompuServe Information Service***; see ◊CompuServe.

CISC (acronym for ***complex instruction-set computer***) in computing, a microprocessor (processor on a single chip) that can carry out a large number of ◊machine code instructions – for example, the Intel 80386. The term was introduced to distinguish them from the more rapid ◊RISC (reduced instruction-set computer) processors, which handle only a smaller set of instructions.

cistron in genetics, the segment of ◊DNA that is required to synthesize a complete polypeptide chain. It is the molecular equivalent of a ◊gene.

CITES (abbreviation for ***Convention on International Trade in Endangered Species***) international agreement under the auspices of the IUCN with the aim of regulating trade in ◊endangered species of animals and plants. The agreement came into force 1975 and by 1997 had been signed by 138 states. It prohibits any trade in a category of 8,000 highly endangered species and controls trade in a further 30,000 species.

Animals and plants listed in Appendix 1 of CITES are classified endangered; those listed in Appendix 2 are classified vulnerable.

citizens' band (CB) short-range radio communication facility (around 27 MHz) used by members of the public in the USA and many European countries to talk to one another or call for emergency assistance.

citric acid $HOOCCH_2C(OH)(COOH)CH_2COOH$ organic acid widely distributed in the plant kingdom; it is found in high concentrations in citrus fruits and has a sharp, sour taste. At one time it was commercially prepared from concentrated lemon juice, but now the main source is the fermentation of sugar with certain moulds.

citronella lemon-scented oil used in cosmetics and insect repellents, obtained from a S Asian grass (*Cymbopogon nardus*).

citrus any of a group of evergreen and aromatic trees or shrubs, found in warm parts of the world. Several species – the orange, lemon, lime, citron, and grapefruit – are cultivated for their fruit. (Genus *Citrus,* family Rutaceae.)

civet small to medium-sized carnivorous mammal found in Africa and Asia, belonging to the family Viverridae, which also includes ◊mongooses **and** ◊**genets**. Distant relations of cats, they generally have longer jaws and more teeth. All have a scent gland in the inguinal (groin) region. Extracts from this gland are taken from the ◊**African civet** *Civettictis civetta* and used in perfumery.

As well as eating animal matter, many species, for example, the SE Asian **palm civet** *Arctogalidia trivirgata,* are fond of fruit.

civil aviation operation of passenger and freight transport by air. With increasing traffic, control of air space is a major problem, and in 1963 Eurocontrol was established by Belgium, France, West Germany, Luxembourg, the Netherlands, and the UK to supervise both military and civil movement in the air space over member countries. There is also a tendency to coordinate services and other facilities between national airlines; for example, the establishment of Air Union in 1963 by France (Air France), West Germany (Lufthansa), Italy (Alitalia), and Belgium (Sabena).

In the UK there are about 170 airports. Heathrow, City, Gatwick, and Stansted (all serving London), Prestwick, and Edinburgh are managed by the British Airports Authority (founded 1965). Close cooperation is maintained with authorities in other countries, including the Federal Aviation Agency, which is responsible for regulating development of aircraft, air navigation, traffic control, and communications in the USA. The Civil Aeronautics Board is the US authority prescribing safety regulations and investigating accidents. There are no state airlines in the USA, although many of the private airlines are large. The world's largest airline was the USSR's government-owned Aeroflot (split among republics in 1992), which operated 1,300 aircraft over 1 million km/620,000 mi of routes. It once carried over 110 million passengers a year, falling to 62 million by 1992.

civil engineering branch of engineering that is concerned with the construction of roads, bridges, airports, aqueducts, waterworks, tunnels, canals, irrigation works, and harbours.

The term is thought to have been used for the first time by British engineer John Smeaton in about 1750 to distinguish civilian from military engineering projects.

Civil Engineers, Institution of the first national body concerned with the engineering profession in England, founded in 1818. The celebrated builder of roads, bridges and canals, Thomas Telford, became its first president.

CIX abbreviation for ◊Compulink Information eXchange.

cladistics method of biological ◊classification (taxonomy) that uses a formal step-by-step procedure for objectively assessing the extent to which organisms share particular characters, and for assigning them to taxonomic groups. Taxonomic groups (for example, ◊species, ◊genus, family) are termed **clades**.

cladode in botany, a flattened stem that is leaflike in appearance and function. It is an adaptation to dry conditions because a stem contains fewer ◊stomata than a leaf, and water loss is thus minimized. The true leaves in such plants are usually reduced to spines or small scales. Examples of plants with cladodes are butcher's-broom *Ruscus aculeatus,* asparagus, and certain cacti. Cladodes may bear flowers or fruit on their surface, and this distinguishes them from leaves.

clam common name for a ◊bivalve mollusc. The giant clam *Tridacna gigas* of the Indopacific can grow to 1 m/3 ft across in 50 years and weigh, with the shell, 500 kg/1,000 lb.

A giant clam produces a billion eggs in a single spawning.

The term is usually applied to edible species, such as the North American hard clam *Venus mercenaria,* used in clam chowder, and whose shells were formerly used as money by North American Indians. A giant clam may produce a billion eggs in a single spawning.

ClariNet commercial news service distributed via USENET. It is not available on all sites since companies must pay to receive ClariNet, which is owned by Clarinet Communications Corp. Under the service's terms and conditions, professional media personnel are banned from using ClariNet news as a source in their work.

Clarke orbit alternative name for ◊geostationary orbit, an orbit 35,900 km/22,300 mi high, in which satellites circle at the same speed as the Earth turns. This orbit was first suggested by space writer Arthur C Clarke in 1945.

class in biological classification, a group of related ◊orders. For example, all mammals belong to the class Mammalia and all birds to the class Aves. Among plants, all class names end in 'idae' (such as Asteridae) and among fungi in 'mycetes'; there are no equivalent conventions among animals. Related classes are grouped together in a ◊phylum.

class in mathematics another name for a ◊set.

classification in biology, the arrangement of organisms into a hierarchy of groups on the basis of their similarities in biochemical, anatomical, or physiological characters. The basic grouping is a ◊species, several of which may constitute a ◊genus, which in turn are grouped into families, and so on up through orders, classes, phyla (in plants, sometimes called divisions), to kingdoms.

To remember the order of taxonomic classification:

KRAKATOA POSITIVELY CASTS OFF FUMES GENERATING SULPHUROUS VAPOURS.

OR

KINDLY PLACE COVER ON FRESH GREEN SPRING VEGETABLES.

KINGDOM / PHYLUM / CLASS / ORDER / FAMILY / GENUS / SPECIES / VARIETY

Classification of Living Things

Classification is the grouping of organisms based on similar traits and evolutionary histories. Taxonomy and systematics are the two sciences that attempt to classify living things. In taxonomy, organisms are generally assigned to groups based on their characteristics. In modern systematics, the placement of organisms into groups is based on evolutionary relationships among organisms. Thus, the groupings are based on evolutionary relatedness or family histories called phylogenies.

The groups into which organisms are classified are called taxa (singular, taxon). The taxon that includes the fewest members is the species, which consists of a single organism. Closely related species are placed into a genus (plural, genera). Related genera are placed into families, families into orders, orders into classes, classes into phyla (singular, phylum) or, in the case of plants and fungi, into divisions, and phyla into divisions or kingdoms. The kingdom level, of which five are generally recognized, is the broadest taxonomic group and includes the greatest number of species. The table below provides an example of the classification of an organism representative of the animal kingdom and the plant kingdom.

Taxonomic Groups[1]	*Common name Genus*[3]	*Kingdom*	*Phylum/division*[2]	*Class*	*Order*	*Family Species*[3]
human	Animalia	Chordata	Mammalia	Primates	Hominoidea	*Homo sapiens*
Douglas fir	Plantae	Tracheophyta	Gymnospermae	Coniferales	Pinaceae	*Pseudotsuga douglasii*

[1] Intermediate taxonomic levels can be created by adding the prefixes 'super-' or 'sub-' to the name of any taxonomic level.

[2] The term division is generally used in place of phylum/phyla for the classification of plants and fungi.

[3] An individual organism is given a two-part name made up of its genus and species names. For example, Douglas fir is correctly known as *Pseudotsuga douglasii*.

It takes a very unusual mind to undertake the analysis of the obvious.

ALFRED NORTH WHITEHEAD English philosopher and mathematician. *Science and the Modern World*

classify in mathematics, to put into separate classes, or ◊sets, which may be uniquely defined.

class interval in statistics, the range of each class of data, used when dealing with large amounts of data. To obtain an idea of the distribution, the data are broken down into convenient classes, which must be mutually exclusive and are usually equal. The class interval defines the range of each class; for example, if the class interval is five and the data begin at zero, the classes are 0–4, 5–9, 10–14, and so on.

clathrate compound formed when the small molecules of one substance fill in the holes in the structural lattice of another, solid, substance – for example, sulphur dioxide molecules in ice crystals. Clathrates are therefore intermediate between mixtures and true compounds (which are held together by ◊ionic or covalent chemical bonds).

clathration in chemistry, a method of removing water from an aqueous solution, and therefore increasing the solution's concentration, by trapping it in a matrix with inert gases such as freons.

clausius in engineering, a unit of ◊entropy (the loss of energy as heat in any physical process). It is defined as the ratio of energy to temperature above absolute zero.

claustrophobia ◊phobia involving fear of enclosed spaces.

clavicle ***Latin clavis 'key'*** the collar bone of many vertebrates. In humans it is vulnerable to fracture, since falls involving a sudden force on the arm may result in very high stresses passing into the chest region by way of the clavicle and other bones. It is connected at one end with the sternum (breastbone), and at the other end with the shoulder-blade, together with which it forms the arm socket. The wishbone of a chicken is composed of its two fused clavicles.

claw hard, hooked, pointed outgrowth of the digits of mammals, birds, and most reptiles. Claws are composed of the protein keratin, and grow continuously from a bundle of cells in the lower skin layer. Hooves and nails are modified structures with the same origin as claws.

clay very fine-grained ◊sedimentary deposit that has undergone a greater or lesser degree of consolidation. When moistened it is plastic, and it hardens on heating, which renders it impermeable. It may be white, grey, red, yellow, blue, or black, depending on its composition. Clay minerals consist largely of hydrous silicates of aluminium and magnesium together with iron, potassium, sodium, and organic substances. The crystals of clay minerals have a layered structure, capable of holding water, and are responsible for its plastic properties. According to international classification, in mechanical analysis of soil, clay has a grain size of less than 0.002 mm/0.00008 in.

clay mineral one of a group of hydrous silicate minerals that form most of the fine-grained particles in clays. Clay minerals are normally formed by weathering or alteration of other silicate minerals. Virtually all have sheet silicate structures similar to the micas. They exhibit the following useful properties: loss of water on heating; swelling and shrinking in different conditions;, cation exchange with other media; and plasticity when wet. Examples are kaolinite, illite, and montmorillonite.

Kaolinite $Al_2Si_2O_5(OH)_4$ is a common white clay mineral derived from alteration of aluminium silicates, especially feldspars. Illite contains the same constituents as kaolinite, plus potassium, and is the main mineral of clay sediments, mudstones, and shales; it is a weathering product of feldspars and other silicates. Montmorillonite contains the constituents of kaolinite plus sodium and magnesium; along with related magnesium- and iron-bearing clay minerals, it is derived from alteration and weathering of mafic igneous rocks. Kaolinite (the mineral name for kaolin or china clay) is economically important in the ceramic and paper industries. Illite, along with other clay minerals, may also be used in ceramics. Montmorillonite is the chief constituent of fuller's earth, and is also used in drilling muds (muds used to cool and lubricate drilling equipment). Vermiculite (similar to montmorillonite) will expand on heating to produce a material used in insulation.

cleanliness unit unit for measuring air pollution: the number of particles greater than 0.5 micrometres in diameter per cubic foot of air. A more usual measure is the weight of contaminants per cubic metre of air.

cleartext or ***plaintext*** in encryption, the original, unencrypted message.

cleavage in mineralogy, the tendency of a mineral to split along defined, parallel planes related to its internal structure. It is a useful distinguishing feature in mineral identification. Cleavage occurs where bonding between atoms is weakest, and cleavages may be perfect, good, or poor, depending on the bond strengths; a given mineral may possess one, two, three, or more orientations along which it will cleave.

Some minerals have no cleavage, for example, quartz will fracture to give curved surfaces similar to those of broken glass. Some other minerals, such as apatite, have very poor cleavage that is sometimes known as a parting. Micas have one perfect cleavage and therefore split easily into very thin flakes. Pyroxenes have two good cleavages and break (less perfectly) into long prisms. Galena has three perfect cleavages parallel to the cube edges, and readily breaks into smaller and smaller cubes. Baryte has one perfect cleavage plus good cleavages in other orientations.

cleg another name for ◊horsefly.

cleistogamy production of flowers that never fully open and that are automatically self-fertilized. Cleistogamous flowers are often formed late in the year, after the production of normal flowers, or during a period of cold weather, as seen in several species of violet *Viola*.

It frequently happens that in the ordinary affairs ... of life opportunities present themselves of contemplating the most curious operations of nature.

Benjamin Thompson, Count Rumford Rumford US-born British physicist. Addressing the Royal Society 1798

clematis any of a group of temperate woody climbing plants with colourful showy flowers. They belong to the buttercup family. (Genus *Clematis*, family Ranunculaceae.)

click in computing, to press down and then immediately release a button on a ◊mouse. The phrase 'to click on' means to select an ◊icon on a computer screen by moving the mouse cursor to the icon's position and clicking a mouse button. See also ◊double click.

click beetle ◊beetle that can regain its feet from lying on its back by jumping into the air and turning over, clicking as it does so.

clickstream unedited log of mouse-clicks that records visitor actions on a site on the World Wide Web. This data is analysed to create feedback for advertisers, enabling them to check whether their strategies are successful in attracting user attention.

client in ◊client–server architecture, software that enables a user to access a store of data or programs on a ◊server. On the Internet, client software is the software that users need to run on home computers in order to be able to use services such as the World Wide Web.

client–server architecture in computing, a system in which the mechanics of looking after data are separated from the programs that use the data. For example, the 'server' might be a central database, typically located on a large computer that is reserved for this purpose. The 'client' would be an ordinary program that requests data from the server as needed.

Most Internet services are examples of client–server applications, including the World Wide Web, FTP, Telnet, and Gopher.

climactichnite flat, soft-bodied animal of the Cambrian period, about 25 cm/10 in in length. Climactichnites pulled themselves along the sand, possibly using flaps on either side of the body, feeding on microorganisms. Although an evolutionary dead end, they may have been some of the first animals to move on land.

climate combination of weather conditions at a particular place over a period of time – usually a minimum of 30 years. A classification of climate encompasses the averages, extremes, and frequencies of all meteorological elements such as temperature, atmospheric pressure, precipitation, wind, humidity, and sunshine, together with the factors that influence them. The primary factors involved are: the Earth's rotation and latitudinal effects; ocean currents; large-scale movements of wind belts and air masses over the Earth's surface; temperature differences between land and sea surfaces; and topography. Climatology , the scientific study of climate, includes the construction of computer-generated models, and considers not only present-day climates, their effects and their classification, but also long-term climate changes, covering both past climates (paleoclimates) and future predictions. Climatologists are especially concerned with the influence of human activity on climate change, among the most important of which, at both a local and global level, are those currently linked with ◊ozone depleters and the ◊greenhouse effect.

climate classification The word climate comes from the Greek *klima*, meaning an inclination or slope (referring to the angle of the Sun's rays, and thus latitude) and the earliest known classification of climate was that of the ancient Greeks, who based their system on latitudes. In recent times, many different systems of classifying climate have been devised, most of which follow that formulated by the German climatologist Wladimir Köppen (1846–1940) in 1900.

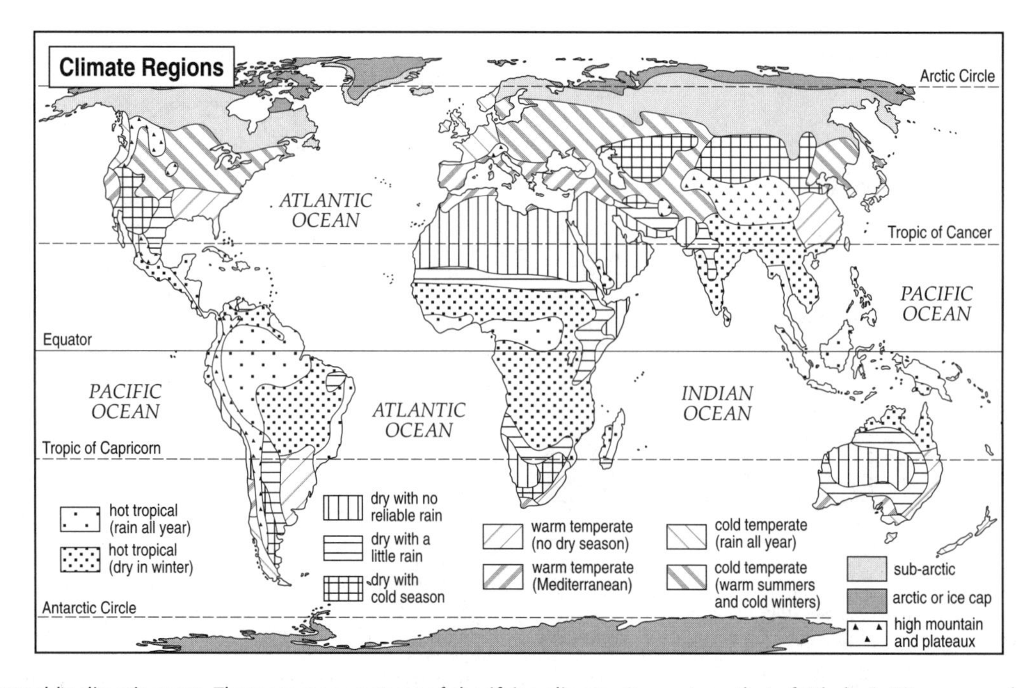

climate *The world's climatic zones. There are many systems of classifying climate. One system, that of Wladimir Köppen, was based on temperature and plant type. Other systems take into account the distribution of global winds.*

These systems use vegetation-based classifications such as desert, tundra, and rainforest. Classification by air mass is used in conjunction with this method. This idea was first introduced in 1928 by the Norwegian meteorologist Tor Bergeron, and links the climate of an area with the movement of the air masses it experiences.

In the 18th century, the British scientist George Hadley developed a model of the general circulation of atmosphere based on convection. He proposed a simple pattern of cells of warm air rising at the Equator and descending at the poles. In fact, due to the rotation of the Earth, there are three such cells in each hemisphere. The first two of these consist of air that rises at the Equator and sinks at latitudes north and south of the tropics; the second two exist at the mid-latitudes where the rising air from the sub-tropics flows towards the cold air masses of the third pair of cells circulating from the two polar regions. Thus, in this model, there are six main circulating cells of air above ground producing seven terrestrial zones – three rainy regions (at the Equator and the temperate latitudes) resulting from the moisture-laden rising air, interspersed and bounded by four dry or desert regions (at the poles and subtropics) resulting from the dry descending air.

climax community assemblage of plants and animals that is relatively stable in its environment. It is brought about by ecological ◊succession, and represents the point at which succession ceases to occur.

In temperate or tropical conditions, a typical climax community comprises woodland or forest and its associated fauna (for example, an oak wood in the UK). In essence, most land management is a series of interferences with the process of succession.

The theory, created by Frederic Clement in 1916, has been criticized for not explaining 'retrogressive' succession, when some areas revert naturally to pre-climax vegetation.

climax vegetation the plants in a ◊climax community.

clinical psychology branch of psychology dealing with the understanding and treatment of health problems, particularly mental disorders. The main problems dealt with include anxiety, phobias, depression, obsessions, sexual and marital problems, drug and alcohol dependence, childhood behavioural problems, psychoses (such as schizophrenia), mental disability, and brain disease (such as dementia) and damage. Other areas of work include forensic psychology (concerned with criminal behaviour) and health psychology.

Assessment procedures assess intelligence and cognition (for example, in detecting the effects of brain damage) by using psychometric tests. **Behavioural approaches** are methods of treatment that apply learning theories to clinical problems. **Behaviour therapy** helps clients change unwanted behaviours (such as phobias, obsessions, sexual problems) and to develop new skills (such as improving social interactions). **Behaviour modification** relies on operant conditioning, making selective use of rewards (such as praise) to change behaviour. This is helpful for children, the mentally disabled, and for patients in institutions, such as mental hospitals. **Cognitive therapy** is a new approach to treating emotional problems, such as anxiety and depression, by teaching clients how to deal with negative thoughts and attitudes. **Counselling**, developed by Carl Rogers, is widely used to help clients solve their own problems. **Psychoanalysis**, as developed by Sigmund Freud and Carl Jung, is little used by clinical psychologists today. It emphasizes childhood conflicts as a source of adult problems.

clinometer hand-held surveying instrument for measuring angles of slope.

clip art small graphics used to liven up documents and presentations. Many software packages such as word processors and presentation graphics packages come with a selection of clip art.

clipboard in computing, a temporary file or memory area where data can be stored before being copied into an application file. It is used, for example, in cut-and-paste operations.

Clipper chip controversial encryption hardware system that contains built-in facilities to allow authorized third parties access to the encrypted data. Adopted as a US government standard in 1994, the Clipper chip was a chip that used ◊public-key cryptography and a proprietary ◊algorithm called Skipjack, and could be built into any communications device, such as a telephone or modem. It was developed by the US National Security Agency as part of its ◊Capstone project.

Clipper was instantly unpopular on the Net because of privacy concerns: it contained a system for depositing a copy of the user's private key in escrow (see ◊key escrow), from where it could be obtained by law enforcement officials equipped with an appropriate court order.

Clipper suffered further defeat when Matt Blaze, a researcher at AT&T Bell Labs cracked the technology in 1995. In 1996, the US government proposed the development of a network of trusted third parties to hold keys in escrow; the initiative was dubbed 'Clipper III'.

clo unit of thermal insulation of clothing. Standard clothes have an insulation of about 1 clo; the warmest clothing is about 4 clo per 2.5 cm/1 in of thickness. See also ◊tog.

cloaca the common posterior chamber of most vertebrates into which the digestive, urinary, and reproductive tracts all enter; a cloaca is found in most reptiles, birds, and amphibians; many fishes; and, to a reduced degree, marsupial mammals. Placental mammals, however, have a separate digestive opening (the anus) and urinogenital opening. The cloaca forms a chamber in which products can be stored before being voided from the body via a muscular opening, the cloacal aperture.

clock any device that measures the passage of time, usually shown by means of pointers moving over a dial or by a digital display. Traditionally a timepiece consists of a train of wheels driven by a spring or weight controlled by a balance wheel or pendulum. Many clocks now run by batteries rather than clockwork. The watch is a portable clock.

history In ancient Egypt the time during the day was measured by a shadow clock, a primitive form of ◊sundial, and at night the water clock was used. Up to the late 16th century the only clock available for use at sea was the sand clock, of which the most familiar form is the hourglass. During the Middle Ages various types of sundial were widely used, and portable sundials were in use from the 16th to the 18th century. Watches were invented in the 16th century – the first were made in Nürnberg, Germany, shortly after 1500 – but it was not until the 19th century that they became cheap enough to be widely available. The first known public clock was set up in Milan, Italy, in 1353. The timekeeping of both clocks and watches was revolutionized in the 17th century by the application of pendulums to clocks and of balance springs to watches.

types of clock The **marine chronometer** is a precision timepiece of special design, used at sea for giving Greenwich mean time (GMT). **Electric timepieces** were made possible by the discovery early in the 19th century of the magnetic effects of electric currents. One of the earliest and most satisfactory methods of electrical control of a clock was invented by Matthaeus Hipp in 1842. In one kind of electric clock, the place of the pendulum or spring-controlled balance wheel is taken by a small synchronous electric motor, which counts up the alternations (frequency) of the incoming electric supply and, by a suitable train of wheels, records the time by means of hands on a dial. The **quartz crystal clock** (made possible by the ◊piezoelectric effect of certain crystals) has great precision, with a short-term variation in accuracy of about one-thousandth of a second per day. More accurate still is the ◊atomic clock. This utilizes the natural resonance of certain atoms (for example, caesium) as a regulator controlling the frequency of a quartz crystal ◊oscillator. Atomic clocks can be accurate to within one second in 300 million years.

clock interrupt in computing, an ◊interrupt signal generated by the computer's internal electronic clock.

clock rate the frequency of a computer's internal electronic clock. Every computer contains an electronic clock, which produces a sequence of regular electrical pulses used by the control unit to synchronize the components of the computer and regulate the ◊fetch–execute cycle by which program instructions are processed.

A fixed number of time pulses is required in order to execute each particular instruction. The speed at which a computer can process instructions therefore depends on the clock rate: increasing the clock rate will decrease the time required to complete each particular instruction.

Clock rates are measured in **megahertz** (MHz), or millions of pulses a second. Microcomputers commonly have a clock rate of 8–50 MHz.

clockwise the direction in which the hands of a traditional clock turn.

clone in genetics, any one of a group of genetically identical cells or organisms. An identical ◊twin is a clone; so, too, are bacteria living in the same colony. The term 'clone' has also been adopted by computer technology to describe a (nonexistent) device that mimics an actual one to enable certain software programs to run correctly.

In August 1996, scientists in Oregon, USA, cloned two rhesus monkeys from embryo cells. President Clinton announced in March 1997 a ban on using federal funds to support human cloning research, and called for a moratorium on this type of scientific research. He also asked the National Bioethics Advisory Commission to review and issue a report on the ramifications that cloning would have on humans.

British scientists confirmed in February 1997 that they had cloned an adult sheep from a single cell to produce a lamb with the same genes as its mother. A cell was taken from the udder of the mother sheep, and its DNA combined with an unfertilized egg that had had its DNA removed. The fused cells were grown in the laboratory and then implanted into the uterus of a surrogate mother sheep. The resulting lamb, Dolly, came from an animal that was six years old, whose genes have already been damaged by environmental toxins and cosmic rays; the sheep could therefore develop cancers abnormally early.

It was the first time cloning had been achieved using cells other then reproductive cells. The cloning breakthrough has ethical implications, as the same principle could be used with human cells and eggs. The news was met with international calls to prevent the cloning of humans. The UK, Spain, Germany, Canada, and Denmark already have laws against cloning humans, as do some individual states in the USA (legislators introduced bills to ban human cloning and associated research nationally March 1997). France and Portugal also have very restrictive laws on cloning.

In June 1997, in response to the recommendations of the National Bioethics Advisory Commission, President Clinton proposed a five-year ban on cloning a human being. He said this would not stop the cloning of animals or of human DNA. The first binding international ban on human cloning was signed in January 1998 by 19 European countries. The text, which was an addition to the European Convention on Human Rights and Biomedicine, placed a total ban on human cloning although it allowed the cloning of cells for research purposes. The 40-member Council of Europe called the protocol 'Europe's response to the threat' of human cloning.

Britain did not sign the protocol because it was not yet a signatory to the convention of which it was a part. The cloning protocol, agreed to by European leaders at a summit October 1997, would also not include Germany, which claimed the measure was weaker than the current German law that forbids all research on human embryos.

A calf cloned from fetal muscle cells by French geneticists was born near Paris in 1998. She only survived for about a month.

People must understand that science is inherently neither a potential for good nor for evil. It is a potential to be harnessed by man to do his bidding.

GLENN T SEABORG US physicist.
Associated Press interview with Alton Blakeslee, 29 Sept 1964

clone in computing, copy of hardware or software that may not be identical to the original design but provides the same functions. All personal computers (PCs) are to some extent clones of the original IBM PC and PC AT launched by IBM in 1981 and 1984, respectively – including IBM's current machines. Clones typically compete by being cheaper and are sometimes less well made than the branded product but this is not always the case. Compaq, for example, competed with IBM by producing the first portable PC and by building better desktop machines, while Dell competed by building PCs to individual orders and supplying customers direct.

Cloning a disc drive or workstation, however, means making an exact copy of all the files or software so that the new drive or machine functions identically to the original one.

closed in mathematics, descriptive of a set of data for which an operation (such as addition or multiplication) done on any members of the set gives a result that is also a member of the set.

For example, the set of even numbers is closed with respect to addition, because two even numbers added to each other always give another even number.

closed-circuit television (CCTV) localized television system in which programmes are sent over relatively short distances, the camera, receiver, and controls being linked by cable. Closed-circuit TV systems are used in department stores and large offices as a means of internal security, monitoring people's movements.

clothes moth moth whose larvae feed on clothes, upholstery, and carpets. The adults are small golden or silvery moths. The natural

Dolly – The Cloning Debate

BY STEPHEN WEBSTER

On Sunday, 23 February 1997, the *New York Times* announced the existence of a new kind of animal. Dolly, a 7-month-old Finn Dorset sheep, was alive and well and living in a guarded pen in the Roslin Institute, a research station just outside Edinburgh, Scotland. What made Dolly uniquely interesting to the newspapers, and soon after to almost everyone, was that she arose not by the union of egg and sperm but by cloning. For the first time a mammal had been cloned from an adult cell: a tiny fleck of skin, scraped from the udder of another sheep and stored for months in a refrigerator, had been treated in such a way that it started to grow. As reported in the newspapers, the embryo, implanted in a surrogate mother, became a fetus and was born in the ordinary way. These were the facts of Dolly's life; yet why Dolly had come and whether any good could come from her were issues largely ignored in the ensuing media frenzy. Along with many others, President Clinton was quick to see the popular significance: if Dolly proves that clones can be made from the cells of adult mammals, then presumably the technique can be used on that other well-known mammal, *Homo sapiens*. On the day following the announcement, Clinton declared that Dolly 'raises serious ethical questions, particularly with respect to the possible use of this technology to clone human embryos'. Meanwhile, across the Atlantic in the UK, a House of Commons committee summoned Ian Wilmut, the team leader of Roslin's Dolly project, to explain the meaning of his work. Wilmut confirmed that his technology could be applied to humans and that, given the resources, it might lead to cloned humans 'within a couple of years'. He said, however, that such a use of the technique would be 'pointless', and declared himself glad that there were laws in the UK banning the cloning of humans.

Advantages of cloning

It is important, therefore, in assessing the significance of Dolly, to understand the background to cloning, and to appreciate why Wilmut's small agricultural research institute persisted for so long in its attempts to clone an adult mammal. The motive is simple, and relates to the commercial breeding of animals. If a farmer has a successful animal, for example a cow that produces a great quantity of excellent milk, similar cows would also be welcome. Normally, breeders obtain the animals they want by mating one favoured individual with another. Yet sexual reproduction produces variation among animals, so the offspring are always a little different from the parents. If it were possible to reproduce an animal without using sex, then the offspring would be identical to its single parent: a clone. Any useful characteristics in the parent would then be found in its genetically identical offspring.

With plants the application was obvious: tomatoes, strawberries, and carrots can all be cloned from individuals judged successful by farmer and consumer – and have been. It is harder to clone animals, yet soon another scientific development made cloning even more attractive: genetic engineering. Much time and money has been invested in making transgenic animals: creatures that contain one or more genes from another species, particularly humans. The Roslin Institute, with its commercial links to the pharmaceutical company PPL Therapeutics, was interested in making sheep with genes that altered the composition of the milk, so that it contained valuable medicines. Clones of such sheep would be guaranteed to have the same ability, and the investment would be secure.

The basis of cloning is that all the cells of an organism, with the exception of the sex cells, contain a full set of genes. A liver cell, for example, contains all the genes for making a brain, the skin, the skeleton, and indeed every other part of the body, yet when a liver cell reproduces it only ever makes other liver cells. It is as if all the other genes it contains were permanently switched off. Therefore, in order to grow an animal from a single cell, a way had to be found to switch back on every gene.

Early cloning attempts

Early cloning experiments, unsurprisingly, used cells taken from an embryo. The method followed was always this: take the nucleus (where the genes are found) out of a cell and then inject it into a fertilized egg – one that has been prepared by having its own nucleus removed. Success came in 1952 in Philadelphia, Pennsylvania, when Robert Briggs and Tom King took a frog embryo, separated out all the cells, and inserted each nucleus into a prepared egg. Twenty-seven tadpoles developed, each genetically identical. It was a world first: an animal had been experimentally cloned.

Frogs are not economically important; mammals are. If a sheep or cow could be cloned after its excellence had been established, in other words when it was fully developed, the technique would be highly lucrative. However, efforts to clone mammals were at first unsuccessful. Then, in the mid-1980s scientists at the University of Wisconsin, funded by the US beef giant W R Grace and Company, made a breakthrough – they managed to clone cow embryo cells, with the cloned embryos growing into adults. The same procedure was followed: take the nucleus from an embryo cell, inject it into a prepared egg, and look for signs of development. If all went well the growing egg, now itself an embryo, would be implanted into a surrogate mother and after the normal gestation period the cow would be born. Venture capitalists saw an opportunity here, related to the beef industry's requirements for productive, disease-resistant cows. Yet while the technique worked, the cloned cows were expensive. The quality of American beef was good enough using ordinary breeding techniques. Cloning cattle was an expensive luxury and within just a few years, research money for cloning was once more in short supply.

Recent experiments

Researchers at the Roslin Institute had meanwhile been developing their interest in transgenic sheep. Sheep embryos were being injected with human genes, and some of these genes were finding their way into the sheep genetic apparatus. One such gene caused the sheep to produce in their milk the drug alpha-1 antitrypsin, used in treating some lung diseases. However, the technique is hit-and-miss: the sheep embryos only incorporate the human genes occasionally. Yet if it were possible to clone transgenic sheep, especially from those individuals with a proven drug-producing history, then the offspring would be guaranteed to contain the gene, and there would be no need for those gene injections with their low success rates. Ian Wilmut had already had success with cloning from embryonic cells, but believed that it should be possible to use adult cells instead. He argued that all the genes contained within an adult nucleus could be reactivated; it was just a matter of finding the right method. Oddly enough, a period of starvation was found to produce the desired effect. Cells taken from a sheep's udder were starved for a short period and this produced in the nucleus a change: the genes became active again. Egg cells were prepared by having their own nuclei removed and replaced by the udder nuclei. Out of 277 eggs that received the nuclei, just 29 developed into embryos, all of which were implanted in surrogate mothers. Fourteen pregnancies began, but most miscarried; only one pregnancy went through to term – this was Dolly.

Ethical issues

It is no exaggeration to say that the scientific world was astonished by the achievement. Scientists' widespread feeling had been that cloning from adult mammals was impossible; indeed, throughout the long-running but sporadic debate about the ethics of cloning, running since the 1970s, scientific commentators tended to downplay the possibility of cloning from adults. In any event, with the attention of the media focused on the concept of human cloning, scientists have had little opportunity to explain that Dolly is not simply an example of scientists in white coats 'playing God' but might constitute instead a serious medical advance.

Following the birth of Dolly, the public debate focused entirely on this question: will humans be cloned? The prospect of dictators cloning themselves and of women giving birth to their father were all discussed as serious possibilities. One reader, writing to the UK newspaper *The Times*, suggested that if his son, who had died in a car accident, were cloned 'he would be able to resume his relationship with my wife and myself and our younger son in a meaningful way and our family would be complete again'. Yet a clone of a person would have their own personality, wrought by the environment. They would have their own identity. In reality no-one could clone themselves and predict the outcome, any more than one can predict the future. Controlling the outcome of human cloning, in any civilized society, would most likely be illegal, as human clones would have the same rights as any other human. The scientific reasons commonly given for cloning humans all fail when considered alonside the social and ethical problems. However, if research into human cloning gets under way, there are useful applications that involve the cloning of tissues, not individuals. Cloned bone marrow, genetically identical to the patient in need, would save lives. More lives would be saved if whole organs could be grown, genetically matched to someone with heart or kidney disease. For the moment research in such areas is banned in the UK. In the USA federal funds cannot be used for human embryo research, but private laboratories have greater freedom. Research into mammal cloning, if not human cloning, is bound to continue. The debate too will continue, and will raise the most profound of questions about human and animal rights, about the question of personal identity, and about the purposes and methods of science. Meanwhile, Dolly the sheep thrives in her pen in Scotland, the centre of attention, but a reminder that the end of the century has seen the arrival of another troubling scientific development.

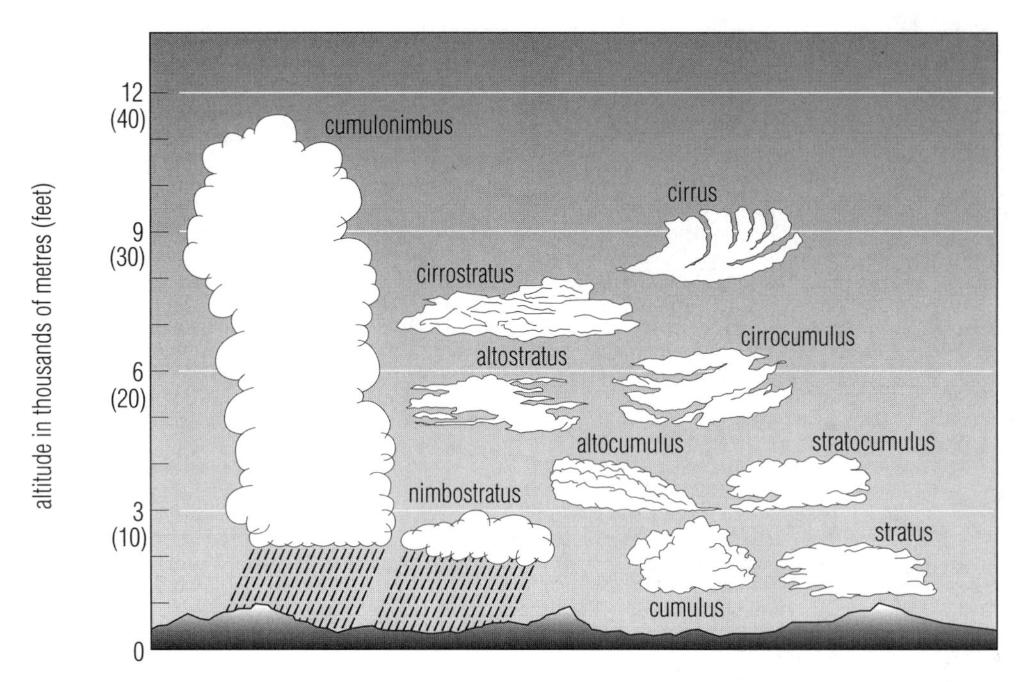

cloud *Standard types of cloud. The height and nature of a cloud can be deduced from its name. Cirrus clouds are at high levels and have a wispy appearance. Stratus clouds form at low level and are layered. Middle-level clouds have names beginning with 'alto'. Cumulus clouds, ball or cottonwool clouds, occur over a range of height.*

habitat of the larvae is in the nests of animals, feeding on remains of hair and feathers, but they have adapted to human households and can cause considerable damage, for example, the common clothes moth *Tineola bisselliella*.

cloud water vapour condensed into minute water particles that float in masses in the atmosphere. Clouds, like fogs or mists, which occur at lower levels, are formed by the cooling of air containing water vapour, which generally condenses around tiny dust particles.

Clouds are classified according to the height at which they occur and their shape. **Cirrus** and **cirrostratus** clouds occur at around 10 km/33,000 ft. The former, sometimes called mares'-tails, consist of minute specks of ice and appear as feathery white wisps, while cirrostratus clouds stretch across the sky as a thin white sheet. Three types of cloud are found at 3–7 km/10,000–23,000 ft: cirrocumulus, altocumulus, and altostratus. **Cirrocumulus** clouds occur in small or large rounded tufts, sometimes arranged in the pattern called mackerel sky. **Altocumulus** clouds are similar, but larger, white clouds, also arranged in lines. **Altostratus** clouds are like heavy cirrostratus clouds and may stretch across the sky as a grey sheet. **Stratocumulus** clouds are generally lower, occurring at 2–6 km/6,500–20,000 ft. They are dull grey clouds that give rise to a leaden sky that may not yield rain. Two types of clouds, **cumulus** and **cumulonimbus**, are placed in a special category because they are produced by daily ascending air currents, which take moisture into the cooler regions of the atmosphere. Cumulus clouds have a flat base generally at 1.4 km/4,500 ft where condensation begins, while the upper part is dome-shaped and extends to about 1.8 km/6,000ft. Cumulonimbus clouds have their base at much the same level, but extend much higher, often up to over 6 km/20,000 ft. Short heavy showers and sometimes thunder may accompany them. **Stratus** clouds, occurring below 1–2.5 km/3,000–8,000 ft, have the appearance of sheets parallel to the horizon and are like high fogs.

In addition to their essential role in the water cycle, clouds are important in the regulation of radiation in the Earth's atmosphere. They reflect short-wave radiation from the Sun, and absorb and re-emit long-wave radiation from the Earth's surface.

CLOUD CATALOGUE

http://covis.atmos.uiuc.edu/guide/clouds/html/cloud.home.html

Illustrated guide to how clouds form and to the various different types. The site contains plenty of images and a glossary of key terms.

cloud chamber apparatus for tracking ionized particles. It consists of a vessel fitted with a piston and filled with air or other gas, saturated with water vapour. When the volume of the vessel is suddenly expanded by moving the piston outwards, the vapour cools and a cloud of tiny droplets forms on any nuclei, dust, or ions present. As fast-moving ionizing particles collide with the air or gas molecules, they show as visible tracks.

Much information about interactions between such particles and radiations has been obtained from photographs of these tracks.

The system has been improved upon in recent years by the use of liquid hydrogen or helium instead of air or gas (see ◊particle detector). The cloud chamber was devised in 1897 by Charles Thomson Rees Wilson (1869–1959) at Cambridge University.

clove dried, unopened flower bud of the clove tree. A member of the myrtle family, the tree is a native of the Maluku Islands, Indonesia. Cloves are used for flavouring in cookery and confectionery. Oil of cloves, which has tonic qualities and relieves wind, is used in medicine. The aroma of cloves is also shared by the leaves, bark, and fruit of the tree. (*Eugenia caryophyllus*, family Myrtaceae.)

clover any of an Old World group of low-growing leguminous plants (see ◊legume), usually with leaves consisting of three leaflets and small flowers in dense heads. Sweet clover refers to various species belonging to the related genus *Melilotus*. (True clover genus *Trifolium*, family Leguminosae.)

club moss or **lycopod** any of a group of mosslike plants that do not produce seeds but reproduce by ◊spores. They are related to the ferns and horsetails. (Order Lycopodiales, family Pteridophyta.)

These plants have a wide distribution, but were far more numerous in Palaeozoic times, especially the Carboniferous period

(363–290 million years ago), when members of the group were large trees. The species that exist now are all small in size.

clubroot disease affecting cabbages, turnips, and allied plants of the Cruciferae family. It is caused by a ◊slime mould, *Plasmodiophora brassicae*. This attacks the roots of the plant, which send out knotty outgrowths. Eventually the whole plant decays.

clubshell or ***watering-pot shell*** bivalve shelled mollusc. It is hermaphrodite and usually lives on corals and rocks.
classification Clubshells are in genus *Clavagella*, order Pholadomyoida, class Bivalvia, phylum Mollusca.

Clumber spaniel breed of medium-sized gundog (a dog trained to aid in hunting) that takes its name from Clumber Park, Nottinghamshire, the estate of the dukes of Newcastle who imported the dogs into England from France probably in the late 18th century. Its thick, soft coat is white with lemon-coloured flecks. One of the largest spaniels, it weighs up to 36 kg/80 lb.

clusec unit for measuring the power of a vacuum pump.

Cluster a ◊European Space Agency project to study the interaction of the solar wind with the Earth's ◊magnetosphere from an array of four identical satellites. Cluster works in conjunction with *SOHO* (Solar and Heliospheric Observatory).

clutch any device for disconnecting rotating shafts, used especially in a car's transmission system. In a car with a manual gearbox, the driver depresses the clutch when changing gear, thus disconnecting the engine from the gearbox.

The clutch consists of two main plates, a pressure plate and a driven plate, which is mounted on a shaft leading to the gearbox. When the clutch is engaged, the pressure plate presses the driven plate against the engine ◊flywheel, and drive goes to the gearbox. Depressing the clutch springs the pressure plate away, freeing the driven plate. Cars with **automatic transmission** have no clutch. Drive is transmitted from the flywheel to the automatic gearbox by a liquid coupling or ◊torque converter.

cm symbol for **centimetre**.

CMOS (abbreviation for ***complementary metal-oxide semiconductor***) family of integrated circuits (chips) widely used in building electronic systems.

CMYK (abbreviation for ***cyan–magenta–yellow–black***) four-colour separation used in most (subtractive) colour printing processes. Representation on computer screens normally uses the additive ◊RGB method and so conversion is usually necessary on output for printing either on colour printers or as separations.

CNC abbreviation for ◊computer numerical control.

CNO cycle in astrophysics, alternative name for ◊carbon cycle.

coachwood tree *Ceratopetalum apetalum* with light, easily worked timber, found in gullies in E Australia.

coal black or blackish mineral substance formed from the compaction of ancient plant matter in tropical swamp conditions. It is used as a fuel and in the chemical industry. Coal is classified according to the proportion of carbon it contains. The main types are ◊anthracite **(shiny, with about 90% carbon), bituminous coal** (shiny and dull patches, about 75% carbon), and **lignite** (woody, grading into peat, about 50% carbon). Coal burning is one of the main causes of ◊acid rain.

coal gas gas produced when coal is destructively distilled or heated out of contact with the air. Its main constituents are methane, hydrogen, and carbon monoxide. Coal gas has been superseded by ◊natural gas for domestic purposes.

coal mining extraction of coal from the Earth's crust. Coal mines may be opencast ◊adit, or deepcast. The least expensive is opencast but this may result in scars on the landscape.

coal tar black oily material resulting from the destructive distillation of bituminous coal.

Further distillation of coal tar yields a number of fractions: light oil, middle oil, heavy oil, and anthracene oil; the residue is called pitch. On further fractionation a large number of substances are obtained, about 200 of which have been isolated. They are used as dyes and in medicines.

coastal erosion the erosion of the land by the constant battering of the sea's waves. This produces two effects. The first is a hydraulic effect, in which the force of the wave compresses air pockets in coastal rocks and cliffs, and the air then expands explosively. The second is the effect of ◊corrasion, in which rocks and pebbles are flung against the cliffs, wearing them away. Frost shattering (or freeze-thaw), caused by the expansion of frozen seawa-

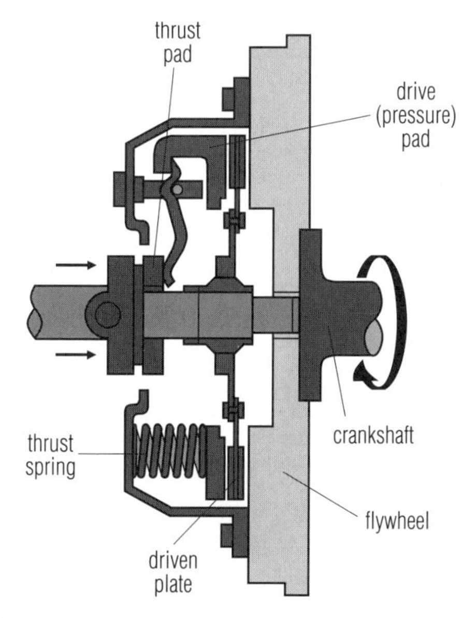

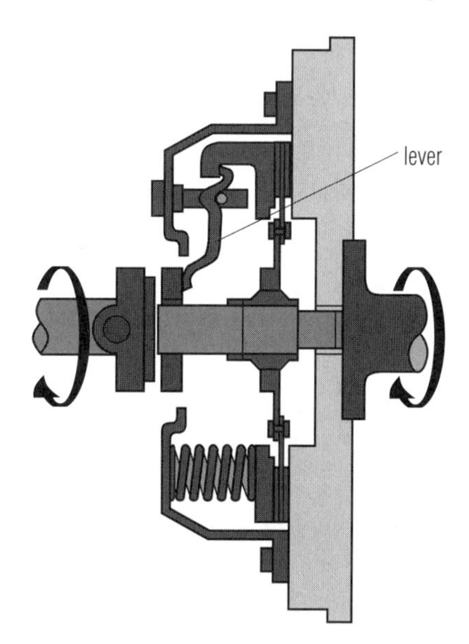

***clutch** The clutch consists of two main plates: a drive plate connected to the engine crankshaft and a driven plate connected to the wheels. When the clutch is disengaged, the drive plate does not press against the driven plate. When the clutch is engaged, the two plates are pressed into contact and the rotation of the crankshaft is transmitted to the wheels.*

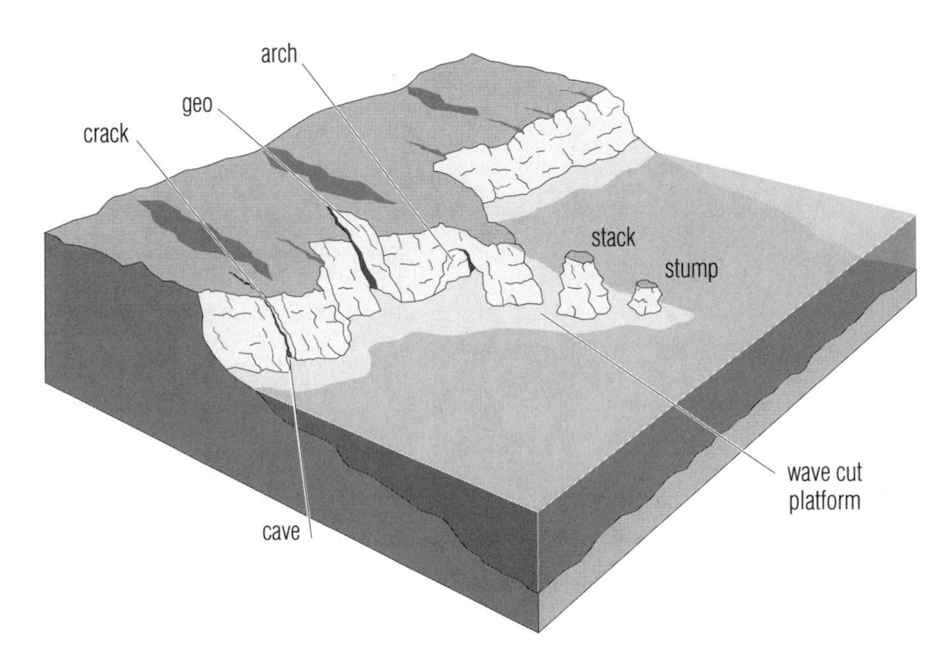

coastal erosion Typical features of coastal erosion: from the initial cracks in less resistant rock through to arches, stacks, and stumps that can occur as erosion progresses.

ter in cavities, and biological weathering, caused by the burrowing of rock-boring molluscs, also result in the breakdown of the rock.

In areas where there are beaches, the waves cause longshore drift, in which sand and stone fragments are carried parallel to the shore, causing buildups (sandspits) in some areas and beach erosion in others.

coastal protection measures taken to prevent ◊coastal erosion. Many stretches of coastline are so severely affected by erosion that beaches are swept away, threatening the livelihood of seaside resorts, and buildings become unsafe.

To reduce erosion, several different forms of coastal protection may be employed. Structures such as sea walls attempt to prevent waves reaching the cliffs by deflecting them back to sea. Such structures are expensive and of limited success. Adding sediment (beach nourishment) to make a beach wider causes waves to break early so that they have less power when they reach the cliffs. Wooden or concrete barriers called groynes may also be constructed at right angles to the beach in order to block the movement of sand along the beach (longshore drift).

coati or ***coatimundi*** any of several species of carnivores of the genus *Nasua,* in the same family, Procyonidae, as the raccoons. A coati is a good climber and has long claws, a long tail, a good sense of smell, and a long, flexible piglike snout used for digging. Coatis live in packs in the forests of South and Central America.

The common coati *Nasua nasua* of South America is about 60 cm/2 ft long, with a tail about the same length.

coaxial cable electric cable that consists of a solid or stranded central conductor insulated from and surrounded by a solid or braided conducting tube or sheath. It can transmit the high-frequency signals used in television, telephone, and other telecommunications transmissions.

cobalt ***German Kobalt 'evil spirit'*** hard, lustrous, grey, metallic element, symbol Co, atomic number 27, relative atomic mass 58.933. It is found in various ores and occasionally as a free metal, sometimes in metallic meteorite fragments. It is used in the preparation of magnetic, wear-resistant, and high-strength alloys; its compounds are used in inks, paints, and varnishes.

The isotope Co-60 is radioactive (half-life 5.3 years) and is produced in large amounts for use as a source of gamma rays in industrial radiography, research, and cancer therapy. Cobalt was named in 1730 by Swedish chemist Georg Brandt (1694–1768); the name derives from the fact that miners considered its ore malevolent because it interfered with copper production.

cobalt-60 radioactive (half-life 5.3 years) isotope produced by neutron radiation of cobalt in heavy-water reactors, used in large amounts for gamma rays in cancer therapy, industrial radiography, and research, substituting for the much more costly radium.

cobalt chloride $CoCl_2$ compound that exists in two forms: the hydrated salt ($CoCl_2.6H_2O$), which is pink, and the anhydrous salt, which is blue. The anhydrous form is used as an indicator because it turns pink if water is present. When the hydrated salt is gently heated the blue anhydrous salt is reformed.

cobalt ore cobalt is extracted from a number of minerals, the main ones being **smaltite**, $(CoNi)As_3$; **linnaeite**, Co_3S_4; **cobaltite**, CoAsS; and **glaucodot**, (CoFe)AsS.

All commercial cobalt is obtained as a by-product of other metals, usually associated with other ores, such as copper. Congo (formerly Zaire) is the largest producer of cobalt, and it is obtained there as a byproduct of the copper industry. Other producers include Canada and Morocco. Cobalt is also found in the manganese nodules that occur on the ocean floor, and was successfully refined in 1988 from the Pacific Ocean nodules, although this process has yet to prove economic.

COBOL (acronym for ***common business-oriented language***) high-level computer-programming language, designed in the late 1950s for commercial data-processing problems; it has become the major language in this field. COBOL features powerful facilities for file handling and business arithmetic. Program instructions written in this language make extensive use of words and look very much like English sentences. This makes COBOL one of the easiest languages to learn and understand.

cobra any of several poisonous snakes, especially the genus *Naja,* of the family Elapidae, found in Africa and southern Asia, species of which can grow from 1 m/3 ft to over 4.3 m/14 ft. The neck stretches into a hood when the snake is alarmed. Cobra venom contains nerve toxins powerful enough to kill humans.

The Indian cobra *Naja naja* is about 1.5 m/5 ft long, and found over most of southern Asia. Some individuals have 'spectacle' markings on the hood. The hamadryad *N. hannah* of southern and southeast Asia can be 4.3 m/14 ft or more, and eats snakes. The ringhals *Hemachatus hemachatus* of S Africa and the black-necked cobra

N. nigricollis, of the African savanna are both about 1 m/3 ft long. Both are able to spray venom towards the eyes of an attacker.

coca South American shrub belonging to the coca family, whose dried leaves are the source of the drug cocaine. It was used as a holy drug by the Andean Indians. (*Erythroxylon coca,* family Erythroxylaceae.)

cocaine alkaloid $C_{17}H_{21}NO_4$ extracted from the leaves of the coca tree. It has limited medical application, mainly as a local anaesthetic agent that is readily absorbed by mucous membranes (lining tissues) of the nose and throat. It is both toxic and addictive. Its use as a stimulant is illegal. ◊Crack is a derivative of cocaine.

Cocaine was first extracted from the coca plant in Germany in the 19th century. Most of the world's cocaine is produced from coca grown in Peru, Bolivia, Colombia, and Ecuador. Estimated annual production totals 215,000 tonnes, with most of the processing done in Colombia. Long-term use may cause mental and physical deterioration.

coccolithophorid microscopic, planktonic marine alga, which secretes a calcite shell. The shells (coccoliths) of coccolithophores are a major component of deep sea ooze. Coccolithophores were particularly abundant during the late ◊Cretaceous period and their remains form the northern European chalk deposits, such as the white cliffs of Dover.

coccus (plural *cocci*) member of a group of globular bacteria, some of which are harmful to humans. The cocci contain the subgroups **streptococci**, where the bacteria associate in straight chains, and **staphylococci**, where the bacteria associate in branched chains.

cochineal red dye obtained from the cactus-eating Mexican ◊scale insect *Dactylopius coccus,* used in colouring food and fabrics.

cochlea part of the inner ◊ear. It is equipped with approximately 10,000 hair cells, which move in response to sound waves and thus stimulate nerve cells to send messages to the brain. In this way they turn vibrations of the air into electrical signals.

cockatiel Australian parrot *Nymphicus hollandicus,* about 20 cm/8 in long, with greyish or yellow plumage, yellow cheeks, a long tail, and a crest like a cockatoo. Cockatiels are popular as pets and aviary birds.

cockatoo any of several crested parrots, especially of the genus *Cacatua,* family Psittacidae, of the order Psittaciformes. They usually have light-coloured plumage with tinges of red, yellow, or orange on the face, and an erectile crest on the head. They are native to Australia, New Guinea, and nearby islands.

There are about 17 species, one of the most familiar being the sulphur-crested cockatoo *C. galerita* of Australia and New Guinea, about 50 cm/20 in long, white with a yellow crest and dark beak.

cockchafer or ***maybug*** European beetle *Melolontha melolontha,* of the scarab family, up to 3 cm/1.2 in long, with clumsy, buzzing flight, seen on early summer evenings. Cockchafers damage trees by feeding on the foliage and flowers.

cocker spaniel breed of small gundog developed in Britain for hunting woodcock (hence its name). It stands about 40 cm/15 in tall and weighs about 14 kg/30 lb. The American cocker is recognized as distinct from the English breed but both have a long, dense coat, which may be a solid colour (red, gold, black) or bi-coloured (black and white).

cockle any of over 200 species of bivalve mollusc with ribbed, heart-shaped shells. Some are edible and are sold in W European markets.

cock-of-the-rock South American bird of the genus *Rupicola.* It belongs to the family Cotingidae, which also includes the cotingas and umbrella birds. There are two species: *R. peruviana,* the Andean cock-of-the-rock, and *R. rupicola,* the Guyanan cock-of-the-rock. The male has brilliant orange plumage including the head crest, the female is a duller brown. Males display at a communal breeding area.

cockroach *The so-called Australian cockroach* Periplaneta australasiae. *Mostly of African origin, cockroaches have become a cosmopolitan pest, infesting in particular kitchens and food stores, and often being linked to outbreaks of food poisoning.* *Premaphotos Wildlife*

cockroach any of numerous insects of the family Blattidae, distantly related to mantises and grasshoppers. There are 3,500 species, mainly in the tropics. They have long antennae and biting mouthparts. They can fly, but rarely do so.

The common cockroach, or black-beetle *Blatta orientalis,* is found in human dwellings, is nocturnal, omnivorous, and contaminates food. The German cockroach *Blattella germanica* and American cockroach *Periplaneta americana* are pests in kitchens, bakeries, and warehouses. In Britain only two innocuous species are native, but several have been introduced with imported food and have become severe pests. They are very difficult to eradicate. Cockroaches have a very high resistance to radiation, making them the only creatures likely to survive a nuclear holocaust.

COCKROACHES

http://www.ex.ac.uk/~gjlramel/blatodea.html

General information about the cockroach from its original arrival on Earth to the present day. Includes life history, its relationship with humans, and even how to keep a cockroach as a pet!

cocktail effect the effect of two toxic, or potentially toxic, chemicals when taken together rather than separately. Such effects are known to occur with some mixtures of drugs, with the active ingredient of one making the body more sensitive to the other.

This sometimes occurs because both drugs require the same ◊enzyme to break them down. Chemicals such as pesticides and food additives are only ever tested singly, not in combination with other chemicals that may be consumed at the same time, so no allowance is made for cocktail effects.

'Gulf War syndrome' may have resulted from the cocktail effect of an anti-nerve gas drug and two different insecticides.

coconut fruit of the coconut palm, which grows throughout the lowland tropics. The fruit has a large outer husk of fibres, which is removed and used to make coconut matting and ropes. Inside this is the nut which is exported to temperate countries. Its hard shell contains white flesh and clear coconut milk, both of which are tasty and nourishing. (*Cocos nucifera,* family Arecaceae.)

The white flesh of the coconut can be eaten fresh, or it can be dried before extracting the oil which makes up nearly two-thirds of it. The oil is used to make soap and margarine and in cooking; the remains are used in cattle feed.

cocoon pupa-case of many insects, especially of ◊moths and ◊silkworms. This outer web or ball is spun from the mouth by caterpillars before they pass into the ◊chrysalis state.

cod any fish of the family Gadidae, especially the Atlantic cod, *Gadus morhua* found in the N Atlantic and Baltic. It is brown to grey with spots, white below, and can grow to 1.5 m/5 ft.

The main cod fisheries are in the North Sea, and off the coasts of Iceland and Newfoundland, Canada. Much of the catch is salted and dried. Formerly one of the cheapest fish, decline in numbers from overfishing has made it one of the most expensive.

COD abbreviation for ◊chemical oxygen demand, a measure of water and effluent quality.

code the expression of an ◊algorithm in a ◊programming language. The term is also used as a verb, to describe the act of programming.

codec device that codes and decodes an ◊analogue stream to or from ◊digital data. It is used in applications such as remote broadcast-quality voiceovers recorded in a remote studio and transmitted via codecs and ◊Integrated Services Digital Network (ISDN) lines to a central studio for final mixing.

codeine opium derivative that provides ◊analgesia in mild to moderate pain. It also suppresses the cough centre of the brain. It is an alkaloid, derived from morphine but less toxic and addictive.

codominance in genetics, the failure of a pair of alleles, controlling a particular characteristic, to show the classic recessive-dominant relationship. Instead, aspects of both alleles may show in the phenotype.

The snapdragon shows codominance in respect to colour. Two alleles, one for red petals and the other for white, will produce a pink colour if the alleles occur together as a heterozygous form.

codon in genetics, a triplet of bases (see ◊base pair) in a molecule of DNA or RNA that directs the placement of a particular amino acid during the process of protein (polypeptide) synthesis. There are 64 codons in the ◊genetic code.

coefficient the number part in front of an algebraic term, signifying multiplication. For example, in the expression $4x^2 + 2xy - x$, the coefficient of x^2 is 4 (because $4x^2$ means $4 \times x^2$), that of xy is 2, and that of x is -1 (because $-1 \times x = -x$).

In general algebraic expressions, coefficients are represented by letters that may stand for numbers; for example, in the equation $ax^2 + bx + c = 0$, a, b, and c are coefficients, which can take any number.

coefficient of relationship the probability that any two individuals share a given gene by virtue of being descended from a common ancestor. In sexual reproduction of diploid species, an individual shares half its genes with each parent, with its offspring, and (on average) with each sibling; but only a quarter (on average) with its grandchildren or its siblings' offspring; an eighth with its great-grandchildren, and so on.

In certain species of insects (for example honey bees), females have only one set of chromosomes (inherited from the mother), so that sisters are identical in genetic make-up; this produces a different set of coefficients. These coefficients are important in calculations of ◊inclusive fitness.

coelacanth large dark brown to blue-grey fish that lives in the deep waters (200 m/650 ft) of the western Indian Ocean around the Comoros Islands. They can grow to about 2 m/6 ft in length, and weigh up to 73 kg/160 lb. They have bony, overlapping scales, and muscular lobe (limblike) fins sometimes used like oars when swimming and for balance while resting on the sea floor. They feed on other fish, and give birth to live young rather than shedding eggs as most fish do. Coelacanth fossils exist dating back over 400 million years and coelacanth were believed to be extinct until one was caught in 1938 off the coast of South Africa. For this reason they are sometimes referred to as 'living fossils'.

classification Coelacanths belong to the animal phylum Chordata, superclass Pisces (fish), class Sarcopterygii, subclass Crossopterygii, order Actinistia or coelacanthiformes, represented by a single family Lateimeriidae. There is only one known surviving species of coelacanth, *Latimeria chalumnae*. Coelacanths are now threatened, and have been listed as endangered by ◊CITES since 1991.

Nature's oddities are more than good theories. They are material for probing the limits of interesting theories about life's history and meaning.

Stephen Jay Gould US palaeontologist and writer. *The Panda's Thumb* 1980

---

coelenterate any freshwater or marine organism of the phylum Coelenterata, having a body wall composed of two layers of cells. They also possess stinging cells. Examples are jellyfish, hydra, and coral.

coeliac disease disease in which the small intestine fails to digest and absorb food. The disease can appear at any age but has a peak incidence in the 30–50 age group; it is more common in women. It is caused by an intolerance to gluten (a constituent of wheat, rye and barley) and characterized by diarrhoea and malnutrition. Treatment is by a gluten-free diet.

coelom in all but the simplest animals, the fluid-filled cavity that separates the body wall from the gut and associated organs, and allows the gut muscles to contract independently of the rest of the body.

coevolution evolution of those structures and behaviours within a species that can best be understood in relation to another species. For example, insects and flowering plants have evolved together: insects have produced mouthparts suitable for collecting pollen or drinking nectar, and plants have developed chemicals and flowers that will attract insects to them.

Coevolution occurs because both groups of organisms, over millions of years, benefit from a continuing association, and will evolve structures and behaviours that maintain this association.

coffee drink made from the roasted and ground beanlike seeds found inside the red berries of any of several species of shrubs, originally native to Ethiopia and now cultivated throughout the tropics. It contains a stimulant, ◊caffeine. (Genus *Coffea*, family Rubiaceae.)

cultivation The shrub, naturally about 5 m/17 ft high, is pruned to about 2 m/7 ft; it is fully fruit-bearing in 5 or 6 years, and lasts for 30 years. Coffee grows best on frost-free hillsides with moderate rainfall. The world's largest producers are Brazil, Colombia, and Côte d'Ivoire; others include Indonesia (Java), Ethiopia, India, Hawaii, and Jamaica. In recent years the world coffee market has suffered from over-supply, and in the early 1990s the price of coffee was well below the cost of production.

history Coffee drinking began in Arab regions in the 14th century but did not become common in Europe until three hundred years later, when the first coffee houses were opened in Vienna, and soon after in Paris and London. In the American colonies, coffee became the substitute for tea when tea was taxed by the British.

coffee machine on the Internet, the coffee machine at Cambridge University, England, whose supplies may be monitored via the World Wide Web. It derives from an idea originally developed at a US university, where the Coke machine was some distance from the programming lab. A system of switches was installed so that a programmer could check the machine's supply of Cokes and their temperature before going to collect a drink.

cognition in psychology, a general term covering the functions involved in synthesizing information – for example, perception (seeing, hearing, and so on), attention, memory, and reasoning.

cognitive therapy or ***cognitive behaviour therapy*** treatment for emotional disorders such as ◊depression and ◊anxiety states. It encourages the patient to challenge the distorted and unhelpful thinking that is characteristic of depression, for example. The treatment may include ◊behaviour therapy.

coherence in physics, property of two or more waves of a beam of light or other electromagnetic radiation having the same frequency and the same ◊phase, or a constant phase difference.

cohesion in physics, a phenomenon in which interaction between two surfaces of the same material in contact makes them cling together (with two different materials the similar phenomenon is called adhesion). According to kinetic theory, cohesion is caused by attraction between particles at the atomic or molecular level. ◊Surface tension, which causes liquids to form spherical droplets, is caused by cohesion.

coil in medicine, another name for an ◊intrauterine device.

coke clean, light fuel produced, along with town gas, when coal is strongly heated in an airtight oven. Coke contains 90% carbon and makes a useful domestic and industrial fuel (used, for example in the iron and steel industries).

The process was patented in England 1622, but it was only in 1709 that Abraham Darby devised a commercial method of production.

cola or ***kola*** any of several tropical trees, especially *Cola acuminata*. In W Africa the nuts are chewed for their high ◊caffeine content, and in the West they are used to flavour soft drinks. (Genus *Cola*, family Sterculiaceae.)

cold, common minor disease of the upper respiratory tract, caused by a variety of viruses. Symptoms are headache, chill, nasal discharge, sore throat, and occasionally cough. Research indicates that the virulence of a cold depends on psychological factors and either a reduction or an increase of social or work activity, as a result of stress, in the previous six months.

There is little immediate hope of an effective cure since the viruses transform themselves so rapidly.

cold-blooded of animals, dependent on the surrounding temperature; see ◊poikilothermy.

cold dark matter theory in cosmology, a theory in which the bulk of the matter in the universe is in the form of dark, unseen material consisting of slow-moving particles. The gravitational clumping of this dark matter in the early universe is may have lead to the formation of clusters and superclusters of ◊galaxies.

cold fusion in nuclear physics, the fusion of atomic nuclei at room temperature. If cold fusion were possible it would provide a limitless, cheap, and pollution-free source of energy, and it has therefore been the subject of research around the world.

In 1989, Martin Fleischmann (1927–) and Stanley Pons (1943–) of the University of Utah, USA, claimed that they had achieved cold fusion in the laboratory, but their results could not be substantiated. The University of Utah announced in 1998 that they would allow the cold fusion patent to elapse, given that the work of Pons and Fleischmann has never been reproduced.

An important scientific innovation rarely makes its way by gradually winning over and converting its opponents: it rarely happens that Saul becomes Paul. What does happen is that its opponents gradually die out, and that the growing generation is familiarized with the ideas from the beginning.

Max Planck German physicist.
In G Holton *Thematic Origins of Scientific Thought* 1973, *Scientific Autobiography* 1949

---

cold-working method of shaping metal at or near atmospheric temperature.

coleoptile the protective sheath that surrounds the young shoot tip of a grass during its passage through the soil to the surface. Although of relatively simple structure, most coleoptiles are very sensitive to light, ensuring that seedlings grow upwards.

colic spasmodic attack of pain in the abdomen, usually coming in waves. Colicky pains are caused by the painful muscular contraction and subsequent distension of a hollow organ; for example, the bowels, gall bladder (biliary colic), or ureter (renal colic).

Intestinal colic is due to partial or complete blockage of the intestine, or constipation; **infantile colic** is usually due to wind in the intestine.

colitis inflammation of the colon (large intestine) with diarrhoea (often bloody). It is usually due to infection or some types of bacterial dysentery.

collagen protein that is the main constituent of ◊connective tissue. Collagen is present in skin, cartilage, tendons, and ligaments. Bones are made up of collagen, with the mineral calcium phosphate providing increased rigidity.

It was identified in a yeastlike fungus in 1996, the first time it has been found in a nonanimal source.

collective farm ***Russian*** **kolkhoz** farm in which a group of farmers pool their land, domestic animals, and agricultural implements, retaining as private property enough only for the members' own requirements. The profits of the farm are divided among its members. In cooperative farming, farmers retain private ownership of the land.

Collective farming was first developed in the USSR in 1917, where it became general after 1930. Stalin's collectivization drive 1929–33 wrecked a flourishing agricultural system and alienated the Soviet peasants from the land: 15 million people were left homeless, 1 million of whom were sent to labour camps and some 12 million deported to Siberia. In subsequent years, millions of those peasants forced into collectives died. Collective farming is practised in other countries; it was adopted from 1953 in China, and Israel has a large number of collective farms.

collective unconscious in psychology, a shared pool of memories, ideas, modes of thought, and so on, which, according to the Swiss psychiatrist Carl Jung, comes from the life experience of one's ancestors, indeed from the entire human race. It coexists with the personal ◊unconscious, which contains the material of individual experience, and may be regarded as an immense depository of ancient wisdom.

Primal experiences are represented in the collective unconscious by archetypes, symbolic pictures, or personifications that appear in dreams and are the common element in myths, fairy tales, and the literature of the world's religions. Examples include the serpent, the sphinx, the Great Mother, the anima (representing the nature of woman), and the mandala (representing balanced wholeness, human or divine).

collenchyma plant tissue composed of relatively elongated cells with thickened cell walls, in particular at the corners where adjacent cells meet.

It is a supporting and strengthening tissue found in nonwoody plants, mainly in the stems and leaves.

collie any of several breeds of sheepdog originally bred in Britain. They include the ◊border collie, the ◊bearded collie, and the ◊rough collie and its smooth-haired counterpart.

colligative property property that depends on the concentration of particles in a solution. Such properties include osmotic pressure (see ◊osmosis), elevation of ◊boiling point, depression of ◊freezing point, and lowering of ◊vapour pressure.

collimator (1) small telescope attached to a larger optical instrument to fix its line of sight; (2) optical device for producing a nondivergent beam of light; (3) any device for limiting the size and angle of spread of a beam of radiation or particles.

collinear in mathematics, lying on the same straight line.

collision detection in ◊virtual reality, the ability of software to detect when two on-screen objects make contact.

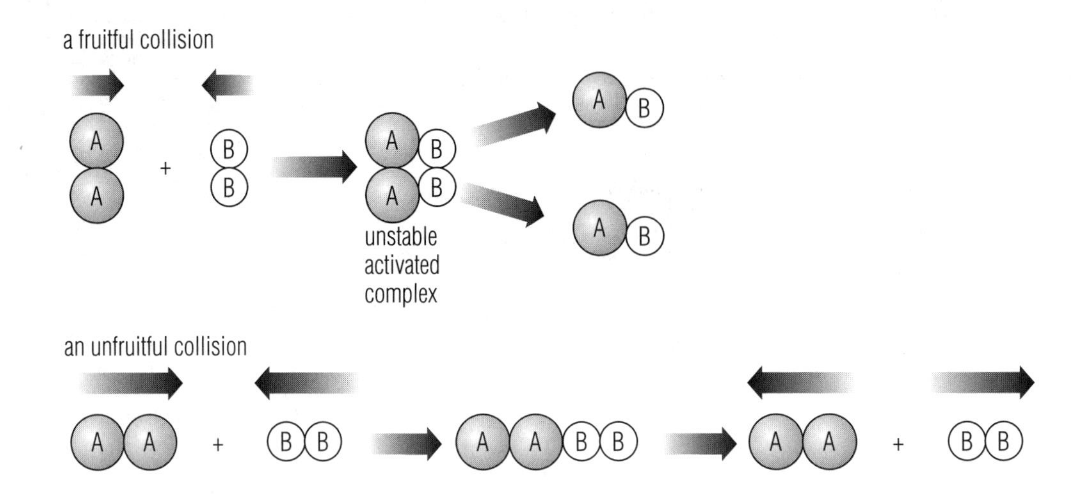

collision theory *Collision theory explains how chemical reactions occur and why rates of reaction differ. For a reaction to occur, particles must collide. If the collision causes a chemical change it is referred to as a fruitful collision.*

collision theory theory that explains how chemical reactions take place and why rates of reaction alter. For a reaction to occur the reactant particles must collide. Only a certain fraction of the total collisions cause chemical change; these are called **fruitful collisions**. The fruitful collisions have sufficient energy (activation energy) at the moment of impact to break the existing bonds and form new bonds, resulting in the products of the reaction. Increasing the concentration of the reactants and raising the temperature bring about more collisions and therefore more fruitful collisions, increasing the rate of reaction.

When a ◊catalyst undergoes collision with the reactant molecules, less energy is required for the chemical change to take place, and hence more collisions have sufficient energy for reaction to occur. The reaction rate therefore increases.

colloid substance composed of extremely small particles of one material (the dispersed phase) evenly and stably distributed in another material (the continuous phase). The size of the dispersed particles (1–1,000 nanometres across) is less than that of particles in suspension but greater than that of molecules in true solution. Colloids involving gases include **aerosols** (dispersions of liquid or solid particles in a gas, as in fog or smoke) and **foams** (dispersions of gases in liquids).

Those involving liquids include **emulsions** (in which both the dispersed and the continuous phases are liquids) and **sols** (solid particles dispersed in a liquid). Sols in which both phases contribute to a molecular three-dimensional network have a jellylike form and are known as **gels**; gelatin, starch 'solution', and silica gel are common examples.

Milk is a natural emulsion of liquid fat in a watery liquid; synthetic emulsions such as some paints and cosmetic lotions have chemical emulsifying agents to stabilize the colloid and stop the two phases from separating out. Colloids were first studied thoroughly by the British chemist Thomas Graham, who defined them as substances that will not diffuse through a semipermeable membrane (as opposed to what he termed crystalloids, solutions of inorganic salts, which will diffuse through).

colobus or ***guereza*** large tree-dwelling African monkey characterized by the almost complete suppression of the thumb. There six species divided into two groups: the black-and-white colobus and the red colobus. They live in groups and feed on fruit, leaves, flowers, and twigs.

Black-and-white colobus monkeys are slightly larger, with a head and body length of 62.5 cm/24.5 in and a long tail (80 cm/31.5 in).

classification Colobuses are in the genus *Colobus,* family Cercopithecidae, order Primates.

colon in anatomy, the main part of the large intestine, between the caecum and rectum. Water and mineral salts are absorbed from undigested food in the colon, and the residue passes as faeces towards the rectum.

colonization in ecology, the spread of species into a new habitat, such as a freshly cleared field, a new motorway verge, or a recently flooded valley. The first species to move in are called **pioneers**, and may establish conditions that allow other animals and plants to move in (for example, by improving the condition of the soil or by providing shade). Over time a range of species arrives and the habitat matures; early colonizers will probably be replaced, so that the variety of animal and plant life present changes. This is known as ◊succession.

Colorado beetle or ***potato beetle*** North American black and yellow striped beetle that is a pest on potato crops. Although it was once a serious pest, it can now usually be controlled by using insecticides. It has also colonized many European countries.
classification Colarado beetles *Leptinotarsa decemlineata* are in the family Chrysomelidae, order Coleoptera, class Insecta, phylum Arthropoda.

colour quality or wavelength of light emitted or reflected from an object. Visible white light consists of electromagnetic radiation of various wavelengths, and if a beam is refracted through a prism, it can be spread out into a spectrum, in which the various colours correspond to different wavelengths. From long to short wavelengths (from about 700 to 400 nanometres) the colours are red, orange, yellow, green, blue, indigo, and violet.

The light entering our eyes is either reflected from the objects we see, or emitted by hot or luminous objects.

TO REMEMBER THE ADDITIVE AND SUBTRACTIVE MIXTURES OF COLOURS:

BETTER GET READY WHILE YOUR MISTRESS COMES BACK.

BLUE + GREEN + RED = WHITE (ADDITIVE) YELLOW + MAGENTA + CYAN = BLACK (SUBTRACTIVE)

emitted light Sources of light have a characteristic ◊spectrum or range of wavelengths. Hot solid objects emit light with a broad range of wavelengths, the maximum intensity being at a wavelength which depends on the temperature. The hotter the object, the shorter the wavelengths emitted, as described by ◊Wien's displacement law. Hot gases, such as the vapour of sodium street lights, emit light at discrete wavelengths. The pattern of wavelengths emitted is unique to each gas and can be used to identify the gas (see ◊spectroscopy).
reflected light When an object is illuminated by white light, some of the wavelengths are absorbed and some are reflected to the eye of an observer. The object appears coloured because of the mixture of wavelengths in the reflected light. For instance, a red object absorbs all wavelengths falling on it except those in the red end of the spectrum. This process of subtraction also explains why certain mixtures of paints produce different colours. Blue and yellow paints when mixed together produce green because between them the yellow and blue pigments absorb all wavelengths except those around green. A suitable combination of three pigments – cyan (blue-green), magenta (blue-red), and yellow – can produce any colour when mixed. This fact is used in colour printing, although additional black pigment is also added.
primary colours In the light-sensitive lining of our eyeball (the ◊retina), cells called cones are responsible for colour vision. There are three kinds of cones. Each type is sensitive to one colour only, either red, green, or blue. The brain combines the signals sent from the set of cones to produce a sensation of colour. When all cones are stimulated equally the sensation is of white light. The three colours to which the cones respond are called the **primary colours**. By mixing lights of these three colours, it is possible to produce any colour. This process is called colour mixing by addition, and is used to produce the colour on a television screen, where glowing phosphor dots of red, green, and blue combine.
complementary colours Pairs of colours that produce white light, such as yellow and blue, are called complementary colours.
classifying colours Many schemes have been proposed for classifying colours. The most widely used is the Munsell scheme, which classifies colours according to their hue (dominant wavelength), saturation (the degree of whiteness), and brightness (intensity).

colour in particle physics, a property of ◊quarks analogous to electric charge but having three states, denoted by red, green, and blue. The colour of a quark is changed when it emits or absorbs a ◊gluon. The term has nothing to do with colour in its usual sense. See ◊quantum chromodynamics.

colour blindness hereditary defect of vision that reduces the ability to discriminate certain colours, usually red and green. The condition is sex-linked, affecting men more than women.

colour depth in computing, the maximum number of colours that can be displayed simultaneously in an image by a particular computer system.

The most common modes are 16, 256, 32K, 64K and 16.7 million (true colour). The greater the colour depth, the larger the size of the picture file but the more detailed and realistic the quality of the picture.

colour index in astronomy, a measure of the colour of a star made by comparing its brightness through different coloured filters. It is defined as the difference between the ◊magnitude of the star measured through two standard photometric filters. Colour index is directly related to the surface temperature of a star and its spectral classification.

colouring food additive used to alter or improve the colour of processed foods. Colourings include artificial colours, such as tartrazine and amaranth, which are made from petrochemicals, and the 'natural' colours such as chlorophyll, caramel, and carotene. Some of the natural colours are actually synthetic copies of the naturally occurring substances, and some of these, notably the synthetically produced caramels, may be injurious to health.

Colourpoint Longhair or ***Himalayan*** breed of domestic longhaired cat originally developed in the 1920s in America and Sweden from a cross between Persian and Siamese cats. The Sealpoint was one of the first varieties to be developed and has a warm cream coat with contrasting deep brown markings on the face, ears, feet, paws, and tail. It has small ears, blue eyes, and short legs.

The breed was recognized in 1955 in Europe and in the USA in 1957 where it is most widely bred. It has many varieties.

colour vision the ability of the eye to recognize different frequencies in the visible spectrum as colours. In most vertebrates, including humans, colour vision is due to the presence on the ◊retina of three types of light-sensitive cone cell, each of which responds to a different primary colour (red, green, or blue).

Colour vision is one of the ways in which the brain can acquire knowledge of the unchanging characteristics of objects. Colours are constructs of the brain, rather than physical features of objects or their surface. They remain more or less stable, and objects remain recognizable, in spite of the continuously changing illumination in which they are seen, a phenomenon known as **colour constancy**.

Colt, Samuel
(1814–1862)

US gunsmith who invented the revolver in 1835 that bears his name. With its rotating cylinder which turned, locked, and unlocked by cocking the hammer, the Colt was superior to other revolving pistols, and it revolutionized military tactics.

mass production for war Colt built a large factory in Hartford, Connecticut in 1854. He introduced mass-production techniques, and his weapons had interchangeable parts, making them easy to maintain and repair. During the Crimean War 1853–56 he also manufactured arms in Pimlico, London. By 1855 he had the largest private armoury in the world. When the American Civil War broke out in 1861, he supplied thousands of guns to the US government.

Mary Evans Picture Library

coltsfoot perennial plant belonging to the daisy family. The single yellow flower heads have many narrow rays (not petals), and the stems have large purplish scales. The large leaf, up to 22 cm/9 in across, is shaped like a horse's foot and gives the plant its name. Coltsfoot grows in Europe, N Asia, and N Africa, often on bare ground and in waste places, and has been introduced to North America. It was formerly used in medicine. (*Tussilago farfara*, family Compositae.)

colugo or ***flying lemur*** SE Asian climbing mammal of the genus *Cynocephalus*, order Dermoptera, about 60 cm/2 ft long including the tail. It glides between forest trees using a flap of skin that extends from head to forelimb to hindlimb to tail. It may glide 130 m/425 ft or more, losing little height. It feeds largely on buds and leaves, and rests hanging upside down under branches.

There are two species, *C. variegatus* of Indochina and Indonesia, and *C. volans* of the Philippines.

columbine any of a group of plants belonging to the buttercup family. All are perennial herbs with divided leaves and hanging flower heads with spurred petals. (Genus *Aquilegia*, family Ranunculaceae.)

The wild columbine (*Aquilegia vulgaris*), with blue flowers, has been developed by repeated crossing to produce modern garden species (*A. x hybrida*), with larger flowers and a wider range of colours. The eastern columbine (*A. canadensis*), with red flowers, is native to E North America.

columbium (Cb) former name for the chemical element ◊niobium. The name is still used occasionally in metallurgy.

column a vertical list of numbers or terms, especially in matrices.

COM acronym for ◊computer output on microfilm/microfiche.

coma in astronomy, the hazy cloud of gas and dust that surrounds the nucleus of a ◊comet.

coma in medicine, a state of deep unconsciousness from which the subject cannot be roused. Possible causes include head injury, brain disease, liver failure, cerebral haemorrhage, and drug overdose.

coma in optics, one of the geometrical aberrations of a lens, whereby skew rays from an object make a comet-shaped spot on the image plane instead of meeting at a point.

combe or ***coombe*** steep-sided valley found on the scarp slope of a chalk ◊escarpment. The inclusion of 'combe' in a placename usually indicates that the underlying rock is chalk.

combination in mathematics, a selection of a number of objects from some larger number of objects when no account is taken of order within any one arrangement. For example, 123, 213, and 312 are regarded as the same combination of three digits from 1234. Combinatorial analysis is used in the study of ◊probability.

The number of ways of selecting r objects from a group of n is given by the formula:

$$n!/[r!(n-r)!]$$

(see ◊factorial). This is usually denoted by ^{n}Cr.

combine in ◊probability theory, to work out the chances of two or more events occurring at the same time.

combined cycle generation system of electricity generation that makes use of both a gas turbine and a steam turbine. Combined cycle plants are more efficient than conventional generating plants.

In combined cycle generation, the gas turbine is powered by burning gas fuel, and turns an electric generator. The exhaust gases are then used to heat water to produce steam. The steam powers a steam turbine attached to an electric generator, producing additional electricity.

combined heat and power generation (CHP generation) simultaneous production of electricity and useful heat in a power station. The heat is often in the form of hot water or steam, which can be used for local district heating or in industry. The electricity output from a CHP plant is lower than from a conventional station, but the overall efficiency of energy conversion is higher. A typical CHP plant may convert 80% of the original fuel energy into a mix of electricity and useful heat, whereas a conventional power station rarely even manages a 40% conversion rate.

combine harvester or ***combine*** machine used for harvesting cereals and other crops, so called because it combines the actions of reaping (cutting the crop) and threshing (beating the ears so that the grain separates).

A combine, drawn by horses, was first built in Michigan in 1836. Today's mechanical combine harvesters are capable of cutting a swath of up to 9 m/30 ft or more.

combustion burning, defined in chemical terms as the rapid combination of a substance with oxygen, accompanied by the evolution of heat and usually light. A slow-burning candle flame and the explosion of a mixture of petrol vapour and air are extreme examples of combustion.

comet small, icy body orbiting the Sun, usually on a highly elliptical path. A comet consists of a central nucleus a few kilometres across, and has been likened to a dirty snowball because it consists mostly of ice mixed with dust. As a comet approaches the Sun its nucleus heats up, releasing gas and dust which form a tenuous coma, up to 100,000 km/60,000 mi wide, around the nucleus. Gas and dust stream away from the coma to form one or more tails, which may extend for millions of kilometres. US astronomers concluded in 1996 that there are two distinct types of comet: one rich in methanol and one low in methanol. Evidence for this comes in part from observations of the spectrum of Comet Hyakutake.

Comets are believed to have been formed at the birth of the Solar System. Billions of them may reside in a halo (the ◊Oort cloud) beyond Pluto. The gravitational effect of passing stars pushes some towards the Sun, when they eventually become visible from Earth. Most comets swing around the Sun and return to distant space, never to be seen again for thousands or millions of years, although some, called periodic comets, have their orbits altered by the gravitational pull of the planets so that they reappear every 200 years or less. Periodic comets are thought to come from the ◊Kuiper belt, a zone just beyond Neptune. Of the 800 or so comets whose orbits have been calculated, about 160 are periodic. The one with the shortest known period is ◊Encke's comet, which orbits the Sun every 3.3 years. A dozen or more comets are discovered every year, some by amateur astronomers.

Comet Hale-Bopp (C/1995–01) large and exceptionally active comet, which in March 1997 made its closest flyby to Earth since 2000 BC, coming within 190 million km/118 million mi. It has a diameter of approximately 40 km/25 mi and an extensive gas coma (when close to the Sun Hale-Bopp released 10 tonnes of gas every second). Unusually, Hale-Bopp has three tails: one consisting of dust particles, one of charged particles, and a third of sodium particles.

Comet Hale-Bopp was discovered independently in July 1995 by two amateur US astronomers, Alan Hale and Thomas Bopp.

Comet Shoemaker-Levy 9 a ◊comet that crashed into ◊Jupiter in July 1994. The fragments crashed into Jupiter at 60 kps/37 mps over the period 16–22 July 1994. The impacts occurred on the far side of Jupiter, but the impact sites came into view of Earth about 25 minutes later. Analysis of the impacts shows that most of the pieces were solid bodies about 1 km/0.6 mi in diameter, but that at least three of them were clusters of smaller objects.

When first sighted in 24 March 1993 by US astronomers Carolyn and Eugene Shoemaker and David Levy, it was found to consist of at least 21 fragments in an unstable orbit around Jupiter. It is believed to have been captured by Jupiter in about 1930 and fragmented by tidal forces on passing within 21,000 km/ 13,050 mi of the planet in July 1992.

COMET SHOEMAKER-LEVY HOME PAGE

`http://www.jpl.nasa.gov/sl9/`

Description of the comet's collision with Jupiter 1994, the first collision of two Solar System bodies ever to be observed. There is background information, latest theories about the effects of the collision, and even some animations of Jupiter and the impact.

comfort index estimate of how tolerable conditions are for humans in hot climates. It is calculated as the temperature in degrees Fahrenheit plus a quarter of the relative ◊humidity, expressed as a percentage. If the sum is less than 95, conditions are tolerable for those unacclimatized to the tropics.

comfrey any of a group of plants belonging to the borage family, with rough, hairy leaves and small bell-shaped flowers (blue, purple-pink, or white). They are found in Europe and W Asia. (Genus *Symphytum*, family Boraginaceae.)

The European species (*Symphytum officinale*) was once used to make ointment for treating wounds and various ailments, and is still sometimes used as a poultice which is applied to the skin to treat inflammation. It grows up to 1.2 m/4 ft tall and has hairy,

Comet Hale-Bopp

BY CHARLES ARTHUR

The appearance of the Hale-Bopp comet in the skies of the northern hemisphere in February 1997 excited scientists and the public alike; but nobody could have foreseen that by the time it had disappeared from our skies, 41 people would have taken their lives because of its presence.

Comets throughout history

Comets have been viewed as harbingers of doom for thousands of years: their inexplicable appearance in an otherwise orderly sky perplexed ancient astronomers. Now they are known to be chunks of rock, dust, ice, and other chemicals, following enormously elliptical orbits around the Sun, which take them within the Earth's orbit before they swing past into the darkness of space beyond the most distant planet, Pluto. As a comet nears the Sun, its body is heated by solar radiation causing some of its constituent chemicals to boil off into space, producing the characteristic 'tail'. This tail is pushed away from the Sun by the pressure of the 'solar wind': thus the tail's direction is not related to the direction of travel, only to the position of the Sun.

This elliptical orbit means that a comet travels through a regular path; each comet differs only in how long it takes to appear. Halley's Comet, for example, turns up every 76 years; its last visit was in 1986. In 1996, many people were disappointed when Comet Hyakutake, discovered by an amateur Japanese astronomer, turned out to be a drab show.

Hale-Bopp, also discovered by amateur astronomers, takes roughly 4,000 years to complete its orbit. It first became visible in the northern hemisphere in February 1997, because its orbit is tilted compared to ours – so that it approached from 'above' the plane of the solar system.

Ablaze in the night sky

Astronomers and the general public were delighted. 'The comet of the century', commented the scientists. It quickly became so bright that outside the light pollution of cities it was easily visible at night without telescopes or binoculars, and most notably early in the morning.

Beautiful time-lapse photographs – taken in deserts, over monuments such as Stonehenge (which was seeing the comet for at least the second time), and open country – filled newspapers, showing off its dual blue (dust) and yellow (gas) tails stretching for up to 20 million miles. It was, astronomers agreed, one of the brightest comets in the past 500 years. On the Internet, the discussion about the comet drew outlandish comments from some people – including some who said that the comet was being shadowed by a UFO that was hiding behind it in order to approach the Earth. Scientists laughed it off as the sort of misconception that can be commonplace on the global network. 'If I was in a UFO and wanted to sneak up on earth, I wouldn't do it behind the most-watched celestial object for years', said one.

Comet Hale-Bopp grew brighter by the day. Its closest approach to the Earth was on 22 March, when it flew by just 123 million miles away from us; its closest approach to the Sun was on 1 April, when it was just 85 million miles away from it. (The Earth is 93 million miles from the Sun.)

Heaven's gate

However, on Wednesday, March 26, the idea that Hale-Bopp was simply a cosmic phenomenon all changed. It was late in the afternoon when San Diego police drove to a five-bedroom Spanish mansion at 18241 Colina Norte, after receiving two anonymous phone calls telling them to check on the residents.

They had no reason to be suspicious: anyone who knew the house had heard that it was the centre for a group of people who made their living programming Web pages for corporate customers: not unusual for San Diego.

Arriving there, the police found the front doors locked, the windows shut and blinds drawn. A side door was unlocked – and through there they found ten dead men on the floor. The bodies were already decomposing. In all, the corpses of 21 women and 18 men, passports and IDs tucked in the top pockets of their shirts, were discovered. It had been a mass suicide, timed to coincide with Hale-Bopp's closest approach to Earth.

Videotapes made by the suicides, who had dressed in matching black with crewcut hair, showed that despite the name of their business – WWW Higher Source – they were quite simply a cult. They believed their leader, who told them that the comet was really the cloak for a UFO; they believed the time had come to 'shed their earthly containers' (bodies) and regain their true identities as angels from another planet. They died in a mass suicide carried out in three waves over a period of days, using a mixture of alcohol and phenobarbitol (swallowed with pudding or apple sauce) and plastic bags over their heads; the last survivors took the bags off the earlier dead and draped them with purple scarves. Each had videotaped a statement saying they were 'going to a better place'. The cult called themselves 'Heaven's Gate', and had its own Web site, which had been crammed with invisible text to give it a high profile on the Internet for anyone searching for information about UFOs. The site invited people to join them in their mission. Sadly, two more people did, committing suicide in April and May, in the belief that they would join the group in leaving on the comet.

Yet away from the obsessed, the comet brought good news to makers of binoculars and telescopes, and of camera tripods: people wanted to get pictures of the comet of the century. Undoubtedly, it fulfilled all expectations.

---

winged stems, lanceolate (tapering) leaves, and white, yellowish, purple, or pink flowers in drooping clusters.

Comité Consultatif International Téléphonique et Télégraphique (CCITT) international organization that determines international communications standards and protocols for data communications, including ◊fax. It was subsumed into the International Telecommunications Union (ITU) in 1993.

Natura non nisi parendo vincitur.
Nature, to be commanded, must be obeyed.

FRANCIS BACON English politician, philosopher, and essayist.
Novum Organum 1620 Aphorism 43

command language in computing, a set of commands and the rules governing their use, by which users control a program. For example, an ◊operating system may have commands such as SAVE and DELETE, or a payroll program may have commands for adding and amending staff records.

command line interface (CLI) in computing, a character-based interface in which a prompt is displayed on the screen at which the user types a command, followed by ◊carriage return, at which point the command, if valid, is executed.

commensalism in biology, a relationship between two ◊species whereby one (the commensal) benefits from the association, whereas the other neither benefits nor suffers. For example, certain species of millipede and silverfish inhabit the nests of army ants and live by scavenging on the refuse of their hosts, but without affecting the ants.

Commercial Internet eXchange (CIX) US-based international nonprofit-making organization of ◊Internet Service Providers and other data network suppliers. It is part of the Internet's US ◊backbone funded by commercial service providers.

Common Agricultural Policy (CAP) system of financial support for farmers in European Union (EU) countries. The most important way in which EU farmers are supported is through guaranteeing them minimum prices for part of what they produce. The CAP has been criticized for its role in creating overproduction, and consequent environmental damage, and for the high price of food subsidies. ***aims*** The CAP permits the member countries of the EU jointly to organize and control agricultural production within their boundaries. The objectives of the CAP were outlined in the Treaty of Rome: to increase agricultural productivity, to provide a fair standard of living for farmers and their employees, to stabilize markets, and to assure the availability of supply at a price that was reasonable to the consumer.
history The policy, applied to most types of agricultural product, was evolved and introduced between 1962 and 1967, and has since been amended to take account of changing conditions and the entry of additional member states. At the heart of the CAP is a price support system based on setting a target price for a commodity, imposing a levy on cheaper imports, and intervening to buy produce at a predetermined level to maintain the stability of the internal market. When the CAP was devised, the six member states were net importers of most essential agricultural products, and the intervention mechanism was aimed at smoothing out occasional surpluses caused by an unusually productive season.
overproduction The CAP became extremely expensive in the 1970s and 1980s due to overproduction of those agricultural products that were subsidized. In many years, far more was produced than could be sold and it had to be stored, creating 'mountains' and 'lakes' of produce. This put the CAP under intense financial and political strain, and led to mounting pressure for reform.

common denominator denominator that is a common multiple of, and hence exactly divisible by, all the denominators of a set of fractions, and which therefore enables their sums or differences to be found.

For example, $\frac{2}{3}$ and $\frac{3}{4}$ can both be converted to equivalent fractions of denominator 12, $\frac{2}{3}$ being equal to $\frac{8}{12}$ and $\frac{3}{4}$ to $\frac{9}{12}$. Hence their sum is $\frac{17}{12}$ and their difference is $\frac{1}{12}$. The **lowest common denominator** (lcd) is the smallest common multiple of the denominators of a given set of fractions.

common difference the difference between any number and the next in an ◊arithmetic progression. For example, in the set 1, 4, 7, 10, ... , the common difference is 3.

common factor number that will divide two or more others without leaving a remainder. For example, 3 is a common factor of 15, 21, and 24.

common gateway interface (CGI) on the World Wide Web, a facility for adding scripts to handle user input. It allows a Web ◊server to communicate with other programs running on the same server in order to process data input by visitors to the Web site. CGI scripts 'parse' the input data, identifying each element and feeding it to the correct program for action, normally a ◊search engine or e-mail program. The results are then fed back to the user in the form of search results or sent e-mail.

common logarithm another name for a ◊logarithm to the base ten.

comms program contraction of ◊communications program.

communication in biology, the signalling of information by one organism to another, usually with the intention of altering the recipient's behaviour. Signals used in communication may be **visual** (such as the human smile or the display of colourful plumage in birds), **auditory** (for example, the whines or barks of a dog), **olfactory** (such as the odours released by the scent glands of a deer), **electrical** (as in the pulses emitted by electric fish), or **tactile** (for example, the nuzzling of male and female elephants).

What hath God wrought?

SAMUEL FINLEY BREESE MORSE US inventor.
First message sent on his electric telegraph 24 May 1844

communications in computing, see ◊data communications.

Communications Decency Act 1996 rider (supplement) to the US Telecommunications Bill seeking to prohibit the transmission of indecent material to minors via the Internet.

Within hours of the bill's passage into law on 8 February 1996, suits were filed by 46 plaintiffs including the American Civil Liberties Union, Voter Telecom Watch, the Electronic Frontier Foundation, and the Center for Democracy and Technology to block the law's enforcement. On 12 June the Philadelphia federal court struck the law down with a judgement that read in part: 'Just as the strength of the Internet is chaos, so the strength of our liberty depends upon the chaos and cacophony of the unfettered speech the First Amendment (to the US Constitution) protects'. A second judgement from a New York court agreed. The government was expected to appeal both rulings to the Supreme Court.

communications program or ***comms program*** general-purpose program for accessing older ◊online systems and ◊bulletin board systems which use a ◊command-line interface; also known as a terminal emulator.

Most operating systems include a trimmed-down comms program, but full-featured programs include facilities to store phone numbers and settings for frequently called services, address books, and the ability to write scripts to automate logging on. Popular comms programs include ProComm, Smartcom, Qmodem, and Odyssey.

communications satellite relay station in space for sending telephone, television, telex, and other messages around the world. Messages are sent to and from the satellites via ground stations. Most communications satellites are in ◊geostationary orbit, appearing to hang fixed over one point on the Earth's surface.

The first satellite to carry TV signals across the Atlantic Ocean was *Telstar* in July 1962. The world is now linked by a system of communications satellites called Intelsat. Other satellites are used by individual countries for internal communications, or for business or military use. A new generation of satellites, called **direct broadcast satellites**, are powerful enough to transmit direct to small domestic aerials. The power for such satellites is produced by solar cells (see ◊solar energy). The total energy requirement of a satellite is small; a typical communications satellite needs about 2 kW of power, the same as an electric heater.

community in ecology, an assemblage of plants, animals, and other organisms living within a circumscribed area. Communities are usually named by reference to a dominant feature such as characteristic plant species (for example, a beech-wood community), or a prominent physical feature (for example, a freshwater-pond community).

commutative operation in mathematics, an operation that is independent of the order of the numbers or symbols concerned. For example, addition is commutative: the result of adding 4 + 2 is the same as that of adding 2 + 4; subtraction is not as 4 – 2 = 2, but 2 – 4 = –2. Compare ◊associative operation and ◊distributive operation.

commutator device in a DC (direct-current) electric motor that reverses the current flowing in the armature coils as the armature rotates.

A DC generator, or ◊dynamo, uses a commutator to convert the AC (alternating current) generated in the armature coils into DC. A commutator consists of opposite pairs of conductors insulated from one another, and contact to an external circuit is provided by carbon or metal brushes.

compact disc (CD) disc for storing digital information, about 12 cm/4.5 in across, mainly used for music, when it can have over an hour's playing time. The compact disc is etched by a ◊laser beam

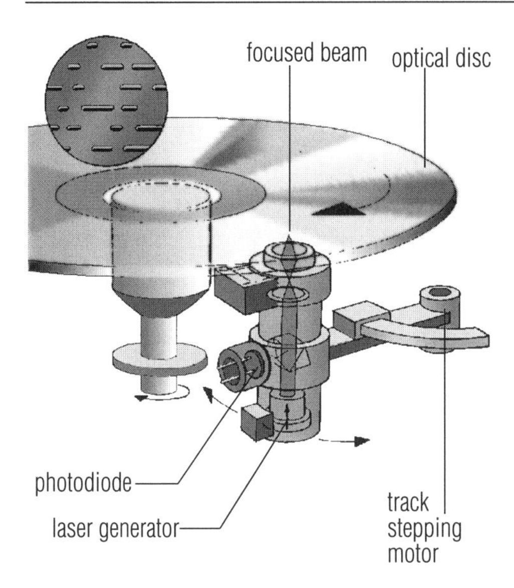

compact disc The compact disc is a digital storage device; music is recorded as a series of etched pits representing numbers in digital code. During playing, a laser scans the pits and the pattern of reflected light reveals the numbers representing the sound recorded. The optical signal is converted to electrical form by a photocell and sent to the amplifiers and loudspeakers.

with microscopic pits that carry a digital code representing the sounds; the pitted surface is then coated with aluminium. During playback, a laser beam reads the code and produces signals that are changed into near-exact replicas of the original sounds.

CD-ROM, or **Compact-Disc Read-Only Memory**, is used to store written text, pictures, and video clips in addition to music. The discs are ideal for large works, such as catalogues and encyclopedias. CD-I, or **Compact-Disc Interactive**, is a form of CD-ROM used with a computerized reader, which responds intelligently to the user's instructions. Recordable CDs, called **WORMs** ('write once, read many times'), are used as computer discs, but are as yet too expensive for home use. **Video CDs**, on sale since 1994, store an hour of video. High-density video discs, first publicly demonstrated in 1995, can hold full-length features. Erasable CDs, which can be erased and recorded many times, are also used by the computer industry. These are coated with a compound of cobalt and the rare earth metal gadolinium, which alters the polarization of light falling on it. In the reader, the light reflected from the disc is passed through polarizing filters and the changes in polarization are converted into electrical signals.

Multi-layer CDs with increased storage capacity were developed in 1996. Two layers are enough to store a film 2 hours long, and up to 16 layers have been reliably read.

comparative anatomy study of the similarity and differences in the anatomy of different groups of animals. It helps to reveal how animals are related to each other and how they have changed through evolution.

Structures are **homologous** if they have arisen from the same ancestral structure through evolution, but perform either similar or different functions in modern animals. Examples of homologous structures are the wings of birds, the human arm, and the forelimb of whales.

Analogous structures have developed from different ancestral structures, but perform similar functions, such as the wings of insects and those of birds.

compass any instrument for finding direction. The most commonly used is a magnetic compass, consisting of a thin piece of magnetic material with the north-seeking pole indicated, free to rotate on a pivot and mounted on a compass card on which the points of the compass are marked. When the compass is properly adjusted and used, the north-seeking pole will point to the magnetic north, from which true north can be found from tables of magnetic corrections.

Compasses not dependent on the magnet are gyrocompasses, dependent on the ◊gyroscope, and radiocompasses, dependent on the use of radio. These are unaffected by the presence of iron and by magnetic anomalies of the Earth's magnetic field, and are widely used in ships and aircraft. See ◊navigation.

To remember the points of the compass, in the correct order:

NOBODY EVER SWALLOWS WHALES

OR

NEVER EAT SHREDDED WHEAT

OR

NEVER EAT SLIMY WORMS

OR

NEVER EAT SOGGY WAFFLES

OR

NEVER EAT SOUR WATERMELON

PLACE THE FIRST LETTER OF EACH WORD IN A CLOCKWISE CIRCLE STARTING AT THE 12 O'CLOCK (NORTH) POSITION

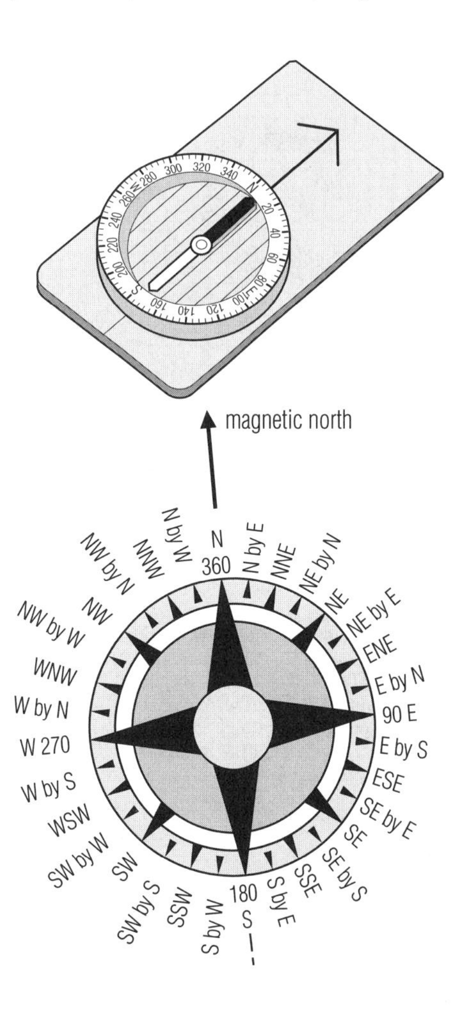

compass As early as 2500 BC, the Chinese were using pieces of magnetic rock, magnetite, as simple compasses. By the 12th century, European navigators were using compasses consisting of a needle-shaped magnet floating in a bowl of water.

compensation point in biology, the point at which there is just enough light for a plant to survive. At this point all the food produced by ◊photosynthesis is used up by ◊respiration. For aquatic plants, the compensation point is the depth of water at which there is just enough light to sustain life (deeper water = less light = less photosynthesis).

competition in ecology, the interaction between two or more organisms, or groups of organisms (for example, species), that use a common resource which is in short supply. Competition invariably results in a reduction in the numbers of one or both competitors, and in ◊evolution contributes both to the decline of certain species and to the evolution of ◊adaptations.

Thus plants may compete with each other for sunlight, or nutrients from the soil, while animals may compete amongst themselves for food, water, or refuge.

compiler computer program that translates programs written in a ◊high-level language into machine code (the form in which they can be run by the computer). The compiler translates each high-level instruction into several machine-code instructions – in a process called **compilation** – and produces a complete independent program that can be run by the computer as often as required, without the original source program being present.

Different compilers are needed for different high-level languages and for different computers. In contrast to using an ◊interpreter, using a compiler adds slightly to the time needed to develop a new program because the machine-code program must be recompiled after each change or correction. Once compiled, however, the machine-code program will run much faster than an interpreted program.

complement in mathematics, the set of the elements within the universal set that are not contained in the designated set. For example, if the universal set is the set of all positive whole numbers and the designated set S is the set of all even numbers, then the complement of S (denoted S') is the set of all odd numbers.

complementary angles two angles that add up to 90°.

complementary medicine in medicine, systems of care based on methods of treatment or theories of disease that differ from those taught in most western medical schools. See ◊medicine, ◊alternative.

complementary metal-oxide semiconductor (CMOS) in electronics, a particular way of manufacturing integrated circuits (chips). The main advantage of CMOS chips is their low power requirement and heat dissipation, which enables them to be used in electronic watches and portable microcomputers. However, CMOS circuits are expensive to manufacture and have lower operating speeds than have circuits of the ◊transistor–transistor logic (TTL) family.

complementary number in number theory, the number obtained by subtracting a number from its base. For example, the complement of 7 in numbers to base 10 is 3. Complementary numbers are necessary in computing, as the only mathematical operation of which digital computers (including pocket calculators) are directly capable is addition. Two numbers can be subtracted by adding one number to the complement of the other; two numbers can be divided by using successive subtraction (which, using complements, becomes successive addition); and multiplication can be performed by using successive addition.

complementation in genetics, the interaction that can occur between two different mutant alleles of a gene in a ◊diploid organism, to make up for each other's deficiencies and allow the organism to function normally.

completing the square method of converting a quadratic expression such as $x^2 + px + q$ into a perfect square by adding $\frac{p^2}{4}$ and subtracting q in order to solve the quadratic equation $x^2 + px + q = 0$. The steps are:

$$x^2 + px = -q$$

$$x^2 + px + \frac{p^2}{4} = \frac{p^2}{4} - q$$

$$(x^2 + \frac{p}{2})^2 = \frac{p^2}{4} - q$$

$$x + \frac{p}{2} = \pm \frac{p^2}{4} - q$$

$$x = -\frac{p}{2} \pm \frac{p^2}{4} - q$$

complex in psychology, a group of ideas and feelings that have become repressed because they are distasteful to the person in whose mind they arose, but are still active in the depths of the person's unconscious mind, continuing to affect his or her life and actions, even though he or she is no longer fully aware of their existence. Typical examples include the ◊Oedipus complex and the ◊inferiority complex.

complex number in mathematics, a number written in the form $a + ib$, where a and b are ◊real numbers and i is the square root of -1 (that is, $i^2 = -1$); i used to be known as the 'imaginary' part of the complex number. Some equations in algebra, such as those of the form $x^2 + 5 = 0$, cannot be solved without recourse to complex numbers, because the real numbers do not include square roots of negative numbers.

The sum of two or more complex numbers is obtained by adding separately their real and imaginary parts, for example:

$$(a + bi) + (c + di) = (a + c) + (b + d)i$$

Complex numbers can be represented graphically on an Argand diagram, which uses rectangular ◊Cartesian coordinates in which the x-axis represents the real part of the number and the y-axis the imaginary part. Thus the number $z = a + bi$ is plotted as the point (a, b). Complex numbers have applications in various areas of science, such as the theory of alternating currents in electricity.

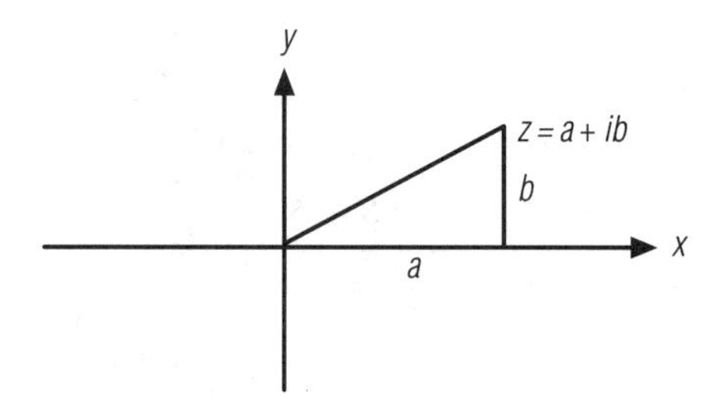

complex number *A complex number can be represented graphically as a line whose end-point coordinates equal the real and imaginary parts of the complex number. This type of diagram is called an Argand diagram after the French mathematician Jean Argand (1768–1822) who devised it.*

component in physics, one of two or more vectors, normally at right angles to each other, that add together to produce the same effect as a single resultant vector. Any ◊vector quantity, such as force, velocity, or electric field, can be resolved into components chosen for ease of calculation. For example, the weight of a body resting on a slope can be resolved into two force components (see ◊resolution of forces); one normal to the slope and the other parallel to the slope.

COM port (contraction of ***communication port*** on a personal computer (PC), one of the serial ◊ports through which ◊data communications take place. PCs may have up to four COM ports. However, these cannot all be used simultaneously as COM1 and COM3 share an ◊interrupt, as do COM2 and COM4. A modem added to a machine with a mouse on COM1 must be attached to COM2 or COM4.

Compositae daisy family, comprising dicotyledonous flowering plants characterized by flowers borne in composite heads (see ◊capitulum). It is the largest family of flowering plants, the majority being herbaceous. Birds seem to favour the family for use in nest 'decoration', possibly because many species either repel or kill insects (see ◊pyrethrum). Species include the daisy and dandelion; food plants such as the artichoke, lettuce, and safflower; and the garden varieties of chrysanthemum, dahlia, and zinnia.

composite in industry, any purpose-designed engineering material created by combining materials with complementary properties into a composite form. Most composites have a structure in which one component consists of discrete elements, such as fibres, dispersed in a continuous matrix. For example, lengths of asbestos, glass, or carbon steel, or 'whiskers' (specially grown crystals a few millimetres long) of substances such as silicon carbide may be dispersed in plastics, concrete, or steel.

composite function in mathematics, a function made up of two or more other functions carried out in sequence, usually denoted by * or ^, as in the relation $(f * g)\ x = f\,[g(x)]$.

Usually, composition is not commutative: $(f * g)$ is not necessarily the same as $(g * f)$.

compost organic material decomposed by bacteria under controlled conditions to make a nutrient-rich natural fertilizer for use in gardening or farming. A well-made compost heap reaches a high temperature during the composting process, killing most weed seeds that might be present.

compound chemical substance made up of two or more ◊elements bonded together, so that they cannot be separated by physical means. Compounds are held together by ionic or covalent bonds.

compound document in Windows, a document containing elements that have been created using other programs. Usually managed through a word processor, such a document might include a table created in a spreadsheet and pictures created in a drawing program. These items may be linked using ◊object linking and embedding (OLE), so that any changes made to the table in the word processor will also be made to the original table developed in the spreadsheet.

Compressed Serial Line Internet Protocol in computing, protocol usually abbreviated to ◊CSLIP.

compression in computing, see ◊data compression.

compressor machine that compresses a gas, usually air, commonly used to power pneumatic tools, such as road drills, paint sprayers, and dentist's drills.

Reciprocating compressors use pistons moving in cylinders to compress the air. Rotary compressors use a varied rotor moving eccentrically inside a casing. The air compressor in jet and ◊gas turbine engines consists of a many-varied rotor rotating at high speed within a fixed casing, where the rotor blades slot between fixed, or stator, blades on the casing.

Compton effect in physics, the increase in wavelength (loss of energy) of a photon by its collision with a free electron (**Compton scattering**). The Compton effect was first demonstrated with X-rays and provided early evidence that electromagnetic waves consisted of particles – photons – which carried both energy and momentum. It is named after US physicist Arthur Compton.

Compulink Information eXchange (CIX) London-based electronic conferencing system founded in 1987. Owned by Frank and Sylvia Thornley, CIX is the oldest and largest native British conferencing system. In 1996 it had approximately 16,000 users, including most of the country's technology journalists.

CompuServe large (US-based) public online information service. It is widely used for ◊electronic mail and ◊bulletin boards, as well as ◊gateway access to large periodical databases.

CompuServe was established in 1979. It is easier to use than the Internet and most computer hardware and software suppliers provide support for their products on CompuServe. Worldwide subscribers to CompuServe have risen from half a million in 1988 to about 5.2 million in 1997 CompuServe started moving some of its services to the ◊World Wide Web in 1996 and was taken over by rival ◊America Online in 1997.

COMPUTER MUSEUM

http://www.net.org/

Well-designed interactive museum, examining the history, development, and future of computer technology. As well as plenty of illustrations and detailed explanations, it is possible to change your route through the museum by indicating whether you are a kid, student, adult, or educator.

computer programmable electronic device that processes data and performs calculations and other symbol-manipulation tasks. There are three types: the ◊digital computer, which manipulates information coded as binary numbers (see ◊binary number

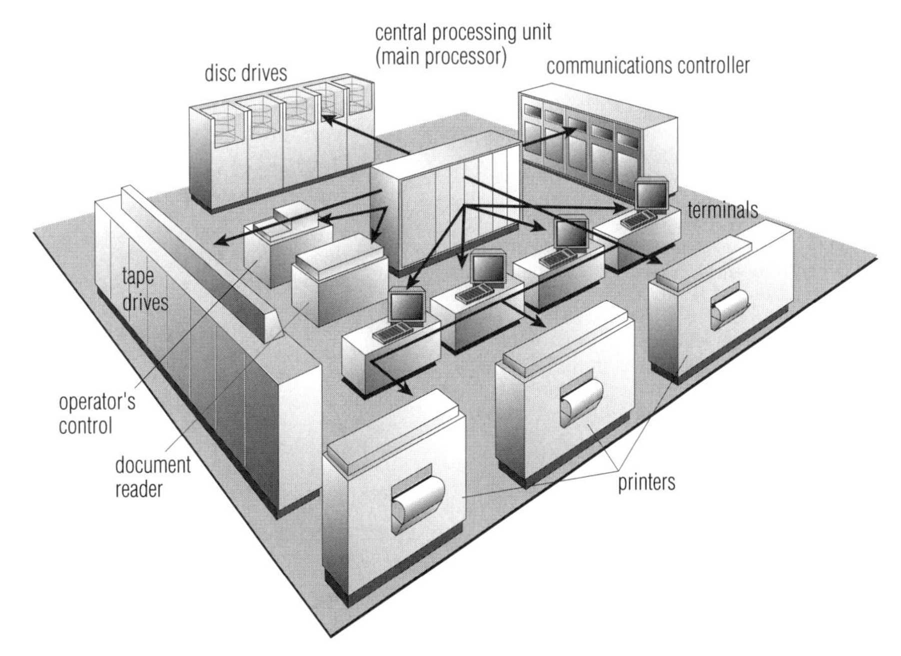

computer *A mainframe computer. Functionally, it has the same component parts as a microcomputer, but on a much larger scale. The central processing unit is at the hub, and controls all the attached devices.*

Computer: chronology

1614	John Napier invented logarithms.
1615	William Oughtred invented the slide rule.
1623	Wilhelm Schickard (1592–1635) invented the mechanical calculating machine.
1645	Blaise Pascal produced a calculator.
1672–74	Gottfried Leibniz built his first calculator, the Stepped Reckoner.
1801	Joseph-Marie Jacquard developed an automatic loom controlled by punch cards.
1820	The first mass-produced calculator, the Arithometer, was developed by Charles Thomas de Colmar (1785–1870).
1822	Charles Babbage completed his first model for the difference engine.
1830s	Babbage created the first design for the analytical engine.
1890	Herman Hollerith developed the punched-card ruler for the US census.
1936	Alan Turing published the mathematical theory of computing.
1938	Konrad Zuse constructed the first binary calculator, using Boolean algebra.
1939	US mathematician and physicist J V Atanasoff (1903–1995) became the first to use electronic means for mechanizing arithmetical operations.
1943	The Colossus electronic code-breaker was developed at Bletchley Park, England. The Harvard University Mark I or Automatic Sequence Controlled Calculator (partly financed by IBM) became the first program-controlled calculator.
1946	ENIAC (acronym for electronic numerator, integrator, analyser, and computer), the first general purpose, fully electronic digital computer, was completed at the University of Pennsylvania, USA.
1948	Manchester University (England) Mark I, the first stored-program computer, was completed. William Shockley of Bell Laboratories invented the transistor.
1951	Launch of Ferranti Mark I, the first commercially produced computer. Whirlwind, the first real-time computer, was built for the US air-defence system. Grace Murray Hopper of Remington Rand invented the compiler computer program.
1952	EDVAC (acronym for electronic discrete variable computer) was completed at the Institute for Advanced Study, Princeton, USA (by John Von Neumann and others).
1953	Magnetic core memory was developed.
1958	The first integrated circuit was constructed.
1963	The first minicomputer was built by Digital Equipment (DEC). The first electronic calculator was built by Bell Punch Company.
1964	Launch of IBM System/360, the first compatible family of computers. John Kemeny and Thomas Kurtz of Dartmouth College invented BASIC (Beginner's All-purpose Symbolic Instruction Code), a computer language similar to FORTRAN.
1965	The first supercomputer, the Control Data CD6600, was developed.
1971	The first microprocessor, the Intel 4004, was announced.
1974	CLIP–4, the first computer with a parallel architecture, was developed by John Backus at IBM.
1975	Altair 8800, the first personal computer (PC), or microcomputer, was launched.
1981	The Xerox Star system, the first WIMP system (acronym for windows, icons, menus, and pointing devices), was developed. IBM launched the IBM PC.
1984	Apple launched the Macintosh computer.
1985	The Inmos T414 transputer, the first 'off-the-shelf' microprocessor for building parallel computers, was announced.
1988	The first optical microprocessor, which uses light instead of electricity, was developed.
1989	Wafer-scale silicon memory chips, able to store 200 million characters, were launched.
1990	Microsoft released Windows 3, a popular windowing environment for PCs.
1992	Philips launched the CD-I (Compact-Disc Interactive) player, based on CD audio technology, to provide interactive multimedia programs for the home user.
1993	Intel launched the Pentium chip containing 3.1 million transistors and capable of 100 MIPs (millions of instructions per second). The Personal Digital Assistant (PDA), which recognizes user's handwriting, went on sale.
1995	Intel launched the Pentium Pro microprocessor (formerly codenamed P6).
1996	IBM's computer Deep Blue beat grand master Gary Kaspanov at chess, the first time a computer has beaten a human grand master.
1997	In the USA, an attempt to bring legislation to control the Internet, intended to prevent access to sexual material, is rejected as unconstitutional.

system); the ◊analogue computer, which works with continuously varying quantities; and the hybrid computer, which has characteristics of both analogue and digital computers.

There are four types of digital computer, corresponding roughly to their size and intended use. **Microcomputers** are the smallest and most common, used in small businesses, at home, and in schools. They are usually single-user machines. **Minicomputers** are found in medium-sized businesses and university departments. They may support from around 10 to 200 users at once. **Mainframes**, which can often service several hundred users simultaneously, are found in large organizations, such as national companies and government departments. **Supercomputers** are mostly used for highly complex scientific tasks, such as analysing the results of nuclear physics experiments and weather forecasting.

We do not need to have an infinity of different machines doing different jobs. A single one will suffice. The engineering problem of producing various machines for various jobs is replaced by the office work of 'programming' the universal machine to do these jobs.

ALAN MATHISON TURING English mathematician.
Quoted in A Hodges *Alan Turing: The Enigma of Intelligence* 1985

computer-aided design use of computers to create and modify design drawings; see ◊CAD.

computer-aided manufacturing use of computers to regulate production processes in industry; see ◊CAM.

computer art art produced with the help of a computer.

Since the 1950s the aesthetic use of computers has been increasingly evident in most artistic disciplines, including film animation, architecture, and music. ◊Computer graphics has been the most developed area, with the 'paint-box' computer liberating artists from the confines of the canvas. It is now also possible to programme computers in advance to generate graphics, music, and sculpture, according to 'instructions' which may include a preprogrammed element of unpredictability. In this last function, computer technology has been seen as a way of challenging the elitist nature of art by putting artistic creativity within relatively easy reach of anyone owning a computer.

computer-assisted learning use of computers in education and training; see ◊CAL.

computer-assisted reporting use of computers to do journalistic research. At its simplest, computer-assisted reporting involves searching an online database for basic information such as addresses and phone numbers. At their most sophisticated, computer systems allow journalists to sift through large quantities of data to find patterns of behaviour or connections that would not be visible by other means.

computer crime broad term applying to any type of crime committed via a computer, including unauthorized access to files. Most computer crime is committed by disgruntled former employees or subcontractors. Examples include the releasing of ◊viruses, ◊hacking, and computer fraud. Many countries, including the USA and the UK, have specialized law enforcement units to supply the technical knowledge needed to investigate computer crime.

Computer Emergency Response Team (CERT) team of engineers based at Carnegie-Mellon University in Pittsburgh, Pennsylvania, USA, that issues security advice and helps resolve emergencies on the Internet by providing technical expertise.

In 1996 the US government announced the formation of a national emergency response team.

computer engineer job classification for ◊computer personnel. A computer engineer repairs and maintains computer hardware.

computer game or ***video game*** any computer-controlled game in which the computer (sometimes) opposes the human player. Computer games typically employ fast, animated graphics on a ◊VDU (visual display unit) and synthesized sound.

Commercial computer games became possible with the advent of the ◊microprocessor in the mid-1970s and rapidly became popular as amusement-arcade games, using dedicated chips. Available games range from chess to fighter-plane simulations.

Some of the most popular computer games in the early 1990s were id Software's *Wolfenstein 3D* and *Doom,* which were designed to be played across networks including the Internet. A whole subculture built up around those particular games, as users took advantage of id's help to create their own additions to the game.

The computer games industry has been criticized for releasing many violent games with little intellectual content.

computer generation any of the five broad groups into which computers may be classified: **first generation** the earliest computers, developed in the 1940s and 1950s, made from valves and wire circuits; **second generation** from the early 1960s, based on transistors and printed circuits; **third generation** from the late 1960s, using integrated circuits and often sold as families of computers, such as the IBM 360 series; **fourth generation** using ◊microprocessors, large-scale integration (LSI), and sophisticated programming languages, still in use in the 1990s; and **fifth generation** based on parallel processing and very large-scale integration, currently under development.

computer graphics use of computers to display and manipulate information in pictorial form. Input may be achieved by scanning an image, by drawing with a mouse or stylus on a graphics tablet, or by drawing directly on the screen with a light pen.

The output may be as simple as a pie chart, or as complex as an animated sequence in a science-fiction film, or a seemingly three-dimensional engineering blueprint. The drawing is stored in the computer as raster graphics or vector graphics.

Vector graphics are stored in the computer memory by using geometric formulas. They can be transformed (enlarged, rotated, stretched, and so on) without loss of picture resolution. It is also possible to select and transform any of the components of a vector-graphics display because each is separately defined in the computer memory. In these respects vector graphics are superior to raster graphics. They are typically used for drawing applications, allowing the user to create and modify technical diagrams such as designs for houses or cars.

Raster graphics are stored in the computer memory by using a map to record data (such as colour and intensity) for every ◊pixel that makes up the image. When transformed (enlarged, rotated, stretched, and so on), raster graphics become ragged and suffer loss of picture resolution, unlike vector graphics. They are typically used for painting applications, which allow the user to create artwork on a computer screen much as if they were painting on paper or canvas.

Computer graphics are increasingly used in computer-aided design (◊CAD), and to generate models and simulations in engineering, meteorology, medicine and surgery, and other fields of science.

computerized axial tomography medical technique, usually known as ◊CAT scan, for noninvasive investigation of disease or injury.

computer-mediated communication umbrella term for all types of communication via computers, such as ◊electronic conferencing and chat.

Computer Misuse Act British law passed in 1990 which makes it illegal to hack into computers (see ◊hacking). The first prosecution under the Act was that of British hacker Paul Bedworth, who in 1993 was acquitted on the grounds that he was addicted to computing.

computer numerical control control of machine tools, most often milling machines, by a computer. The pattern of work for the machine to follow, which often involves performing repeated sequences of actions, is described using a special-purpose programming language.

computer operator job classification for ◊computer personnel. Computer operators work directly with the computer, running the programs, changing discs and tapes, loading paper into printers, and ensuring all ◊data security procedures are followed.

computer output on microfilm/microfiche (COM) technique for producing computer output in very compact, photographically reduced form (◊microform).

computer personnel people who work with or are associated with computers. In a large computer department the staff may work under the direction of a **data processing manager**, who supervises and coordinates the work performed. Computer personnel can be broadly divided into two categories: those who run and maintain existing ◊applications programs (programs that perform a task for the benefit of the user) and those who develop new applications.

Personnel who run existing applications programs: **data control staff** receive information from computer users (for instance, from the company's wages clerks), ensure that it is processed as required, and return it to them in processed form; **data preparation staff**, or **keyboard operators**, prepare the information received by the data control staff so that it is ready for processing by computer. Once the information has been typed at the keyboard of a VDU (or at a ◊key-to-disc or key-to-tape station), it is placed directly onto a medium such as disc or tape; **computer operators** work directly with computers, running the programs, changing discs and tapes, loading paper into printers, and ensuring that all ◊data security procedures are followed; **computer engineers** repair and maintain computer hardware; **file librarians**, or **media librarians**, store and issue the data files used by the department; an **operations manager** coordinates all the day-to-day activities of these staff. Personnel who develop new applications: **systems analysts** carry out the analysis of an existing system (see ◊systems analysis), whether already computerized or not, and prepare proposals for a new system; **programmers** write the software needed for new systems.

Computer Professionals for Social Responsibility (CPSR) US organization advocating the responsible use of computers. Based in Washington DC, it was one of the first organizations to oppose President Reagan's Strategic Defense Initiative on the grounds that the many billions of lines of code it would take to program it could never be debugged successfully.

computer program coded instructions for a computer; see ◊program.

computer simulation representation of a real-life situation in a computer program. For example, the program might simulate the flow of customers arriving at a bank. The user can alter variables, such as the number of cashiers on duty, and see the effect.

More complex simulations can model the behaviour of chemical reactions or even nuclear explosions. The behaviour of solids and liquids at high temperatures can be simulated using ◊quantum simulation. Computers also control the actions of machines – for example, a ◊flight simulator models the behaviour of real aircraft and allows training to take place in safety. Computer simulations are very useful when it is too dangerous, time consuming, or simply impossible to carry out a real experiment or test.

computer-supported collaborative work (CSCW) work undertaken by individuals who, using computers, are able to function together as a group on a project despite being geographically separated. The technology to facilitate CSCW is still under development. Early initiatives include video and data conferencing so that two users can talk on the telephone while simultaneously viewing a document in progress. Changes made by either participant affect both participants' displays.

computer terminal the device whereby the operator communicates with the computer; see ◊terminal.

Computer Underground Digest widely distributed ◊e-zine covering such issues as ◊hacking, freedom of speech, and security risks.

computing device any device built to perform or help perform computations, such as the ◊abacus, ◊slide rule, or ◊computer.

The earliest known example is the abacus. Mechanical devices with sliding scales (similar to the slide rule) date from ancient Greece. In 1642, French mathematician Blaise Pascal built a mechanical adding machine and in 1671 German mathematician Gottfried Leibniz produced a machine to carry out multiplication. The first mechanical computer, the ◊analytical engine, was designed by British mathematician Charles Babbage in 1835. For the subsequent history of computing, see ◊computer.

concave of a surface, curving inwards, or away from the eye. For example, a bowl appears concave when viewed from above. In geometry, a concave polygon is one that has an interior angle greater than 180°. Concave is the opposite of ◊convex.

concave lens lens that possesses at least one surface that curves inwards. It is a diverging lens, spreading out those light rays that have been refracted through it. A concave lens is thinner at its centre than at its edges, and is used to correct short-sightedness.

Common forms include the **biconcave** lens (with both surfaces curved inwards) and the **plano-concave** (with one flat surface and one concave). The whole lens may be further curved overall, making a **convexo-concave** or diverging meniscus lens, as in some lenses used for corrective purposes.

concave mirror curved mirror that reflects light from its inner surface. It may be either circular or parabolic in section. A concave mirror converges light rays to form a reduced, inverted, real image in front, or an enlarged, upright, virtual image seemingly behind it, depending on how close the object is to the mirror.

Only a parabolic concave mirror has a true, single-point ◊principal focus for parallel rays. For this reason, parabolic mirrors are used as reflectors to focus light in telescopes, or to focus microwaves in satellite communication systems. The reflector behind a spot lamp or car headlamp is parabolic.

concentration in chemistry, the amount of a substance (◊solute) present in a specified amount of a solution. Either amount may be specified as a mass or a volume (liquids only). Common units used are ◊moles per cubic decimetre, grams per cubic decimetre, grams per 100 cubic centimetres, and grams per 100 grams.

The term also refers to the process of increasing the concentration of a solution by removing some of the substance (◊solvent) in which the solute is dissolved. In a **concentrated solution**, the solute is present in large quantities. Concentrated brine is around 30% sodium chloride in water; concentrated caustic soda (caustic liquor) is around 40% sodium hydroxide; and concentrated sulphuric acid is 98% acid.

concentration gradient change in the concentration of a substance from one area to another. Particles, such as sugar molecules, in a fluid move over time so that they become evenly distributed throughout the fluid. In particular, they move from an area of high concentration to an area of low concentration; that is, they diffuse along the concentration gradient (see ◊diffusion).

This explains why oxygen in the lungs will diffuse into the blood supply. The oxygen molecules are more concentrated in the lungs than they are in the capillaries surrounding the ◊alveoli (air sacs). As it diffuses along the concentration gradient, oxygen will tend to pass into the blood. Gas exchange therefore depends on the maintenance of a concentration gradient, so that oxygen will continue to diffuse across the respiratory surface.

concentric circles two or more circles that share the same centre.

conceptacle flask-shaped cavities found in the swollen tips of certain brown seaweeds, notably the wracks, *Fucus*. The gametes are formed within them and released into the water via a small pore in the conceptacle, known as an ostiole.

conch name applied to various shells, but especially to the fountain shell, a species of gastropod mollusc in the order Mesogastropoda.

conchology branch of zoology that studies molluscs with reference to their shells.

Concorde the only supersonic airliner, which cruises at Mach 2, or twice the speed of sound, about 2,170 kph/1,350 mph. Concorde, the result of Anglo-French cooperation, made its first flight 1969 and entered commercial service seven years later. It is 62 m/202 ft long and has a wing span of nearly 26 m/84 ft. Developing Concorde cost French and British taxpayers £2 billion.

concrete building material composed of cement, stone, sand, and water. It has been used since Egyptian and Roman times. Since the late 19th century, it has been increasingly employed as an economical alternative to materials such as brick and wood, and has been combined with steel to increase its tension capacity.

Reinforced concrete and prestressed concrete are strengthened by combining concrete with another material, such as steel rods or glass fibres. The addition of carbon fibres to concrete increases its conductivity. The electrical resistance of the concrete changes with increased stress or fracture, so this 'smart concrete' can be used as an early indicator of structural damage.

concurrent lines two or more lines passing through a single point; for example, the diameters of a circle are all concurrent at the centre of the circle.

concussion temporary unconsciousness resulting from a blow to the head. It is often followed by amnesia for events immediately preceding the blow.

condensation conversion of a vapour to a liquid. This is frequently achieved by letting the vapour come into contact with a cold surface. It is the process by which water vapour turns into fine water droplets to form ◊cloud.

Condensation in the atmosphere occurs when the air becomes completely saturated and is unable to hold any more water vapour. As air rises it cools and contracts – the cooler it becomes the less water it can hold. Rain is frequently associated with warm weather fronts because the air rises and cools, allowing the water vapour to condense as rain. The temperature at which the air becomes saturated is known as the ◊dew point. Water vapour will not condense in air if there are not enough condensation nuclei (particles of dust, smoke or salt) for the droplets to form on. It is then said to be supersaturated. Condensation is an important part of the ◊water cycle.

condensation in organic chemistry, a reaction in which two organic compounds combine to form a larger molecule, accompanied by the removal of a smaller molecule (usually water). This is also known as an addition–elimination reaction. Polyamides (such as nylon) and polyesters (such as Terylene) are made by condensation ◊polymerization.

condensation number in physics, the ratio of the number of molecules condensing on a surface to the total number of molecules touching that surface.

condensation polymerization ◊polymerization reaction in which one or more monomers, with more than one reactive functional group, combine to form a polymer with the elimination of water or another small molecule.

condenser laboratory apparatus used to condense vapours back to liquid so that the liquid can be recovered. It is used in ◊distillation and in reactions where the liquid mixture can be kept boiling without the loss of solvent.

condenser in electronic circuits, a former name for a ◊capacitor.

condenser in optics, a ◊lens or combination of lenses with a short focal length used for concentrating a light source onto a small area, as used in a slide projector or microscope substage lighting unit. A condenser can also be made using a concave mirror.

conditioning in psychology, two major principles of behaviour modification.

In **classical conditioning**, described by Russian psychologist Ivan Pavlov, a new stimulus can evoke an automatic response by being repeatedly associated with a stimulus that naturally provokes that response. For example, the sound of a bell repeatedly associated with food will eventually trigger salivation, even if sounded without food being presented. In **operant conditioning**, described by US psychologists Edward Lee Thorndike (1874–1949) and B F Skinner, the frequency of a voluntary response can be increased by following it with a reinforcer or reward.

School yourself to demureness and patience. Learn to innure yourself to drudgery in science. Learn, compare, collect the facts.

IVAN PAVLOV Russian physiologist.
Bequest to the Academic Youth of Soviet Russia 27 Feb 1936

condom or ***sheath*** or ***prophylactic*** barrier contraceptive, made of rubber, which fits over an erect penis and holds in the sperm produced by ejaculation. It is an effective means of preventing pregnancy if used carefully, preferably with a ◊spermicide. A condom with spermicide is 97% effective; one without spermicide is 85% effective as a contraceptive. Condoms can also give some protection against sexually transmitted diseases, including AIDS.

In 1996 the European Union agreed a standard for condoms, which is 17 cm/6.7 in long; although the width can be variable, a regular width was agreed as 5.2 cm/2 in.

condor name given to two species of birds in separate genera. The **Andean condor** *Vultur gryphus,* has a wingspan up to 3 m/10 ft, weighs up to 13 kg/28 lb, and can reach up to 1.2 m/3.8 ft in length. It is black, with some white on the wings and a white frill at the base of the neck. It lives in the Andes at heights of up to 4,500 m/14,760 ft, and along the South American coast, and feeds mainly on carrion. The **Californian condor** *Gymnogyps californianus* is a similar bird, with a wingspan of about 3 m/10 ft. It feeds entirely on carrion, and is on the verge of extinction.

The Californian condor lays only one egg at a time and may not breed every year. In 1994, only 89 Californian condors remained, of which only four were in the wild. It became the subject of a special conservation effort, and by July 1995 the number had increased to 104.

conductance ability of a material to carry an electrical current, usually given the symbol G. For a direct current, it is the reciprocal of ◊resistance: a conductor of resistance R has a conductance of $1/R$. For an alternating current, conductance is the resistance R divided by the ◊impedance Z: $G = R/Z$. Conductance was formerly expressed in reciprocal ohms (or mhos); the SI unit is the ◊siemens (S).

conduction, electrical flow of charged particles through a material giving rise to electric current. Conduction in metals involves the flow of negatively charged free ◊electrons. Conduction in gases and some liquids involves the flow of ◊ions that carry positive charges in one direction and negative charges in the other. Conduction in a ◊semiconductor such as silicon involves the flow of electrons and positive holes.

TO REMEMBER THE PRINCIPLES OF HEAT TRANSFER:

CONDUCTION – IMAGINE A LINE OF PASSENGERS ON A BUS BEING ASKED TO MOVE DOWN BY THE **CONDUCTOR**, EACH PASSENGER CAUSING THE NEXT TO BUSTLE ALONG (ANALOGY FOR THE MOVEMENT/VIBRATION OF ATOMS THAT IS PASSED ALONG, CAUSING HEAT TO BE TRANSFERRED) **CONVECTION** – CONSIDER **VECTOR**, A DISEASE-CARRYING INSECT, FOR EXAMPLE A MOSQUITO, WHICH TRAVELS IN SWARMS (VERY MUCH LIKE THE MOVEMENT OF CONVECTION CURRENTS) **RADIATION** – HEAT RADIATION IS A FORM OF **RADIATION** –(THINK OF NUCLEAR FALLOUT OR THE SUN'S RADIATION) AND THUS TRAVELS IN WAVES UNDETECTED UNTIL THEY FALL UPON ANOTHER BODY

conduction, heat flow of heat energy through a material without the movement of any part of the material itself (compare ◊conduction, ◊electrical). Heat energy is present in all materials in the form of the kinetic energy of their vibrating molecules, and may be conducted from one molecule to the next in the form of this mechanical vibration. In the case of metals, which are particularly good conductors of heat, the free electrons within the material carry heat around very quickly.

conductivity, thermal (unit $W\ m^{-1}\ K^{-1}$) measure of how well a material conducts heat. A good conductor, such as a metal, has a high conductivity; a poor conductor, called an insulator, has a low conductivity. See also ◊U-value.

conductor any material that conducts heat or electricity (as opposed to an insulator, or nonconductor). A good conductor has a high electrical or heat conductivity, and is generally a substance rich in free electrons such as a metal. A poor conductor (such as the nonmetals, glass and porcelain) has few free electrons. Carbon is exceptional in being nonmetallic and yet (in some of its forms) a relatively good conductor of heat and electricity. Substances such as silicon and germanium, with intermediate conductivities that are improved by heat, light, or impurities, are known as ◊semiconductors.

cone in botany, the reproductive structure of the conifers and cycads; also known as a ◊strobilus. It consists of a central axis surrounded by numerous, overlapping, scalelike, modified leaves

cone *The western yellow or ponderosa pine* Pinus ponderosa *var.* arizonica *exhibits the needlelike leaves and cones typical of gymnosperms. These are female cones which have shed their seeds; male cones are smaller and shed vast quantities of pollen into the air. Premaphotos Wildlife* (sporophylls) that bear the reproductive organs. Usually there are separate male and female cones, the former bearing pollen sacs containing pollen grains, and the larger female cones bearing the ovules that contain the ova or egg cells. The pollen is carried from male to female cones by the wind (◊anemophily). The seeds develop within the female cone and are released as the scales open in dry atmospheric conditions, which favour seed dispersal.

In some groups (for example, the pines) the cones take two or even three years to reach maturity. The cones of ◊junipers have fleshy cone scales that fuse to form a berrylike structure. One group of ◊angiosperms, the alders, also bear conelike structures; these are the woody remains of the short female catkins, and they contain the alder ◊fruits.

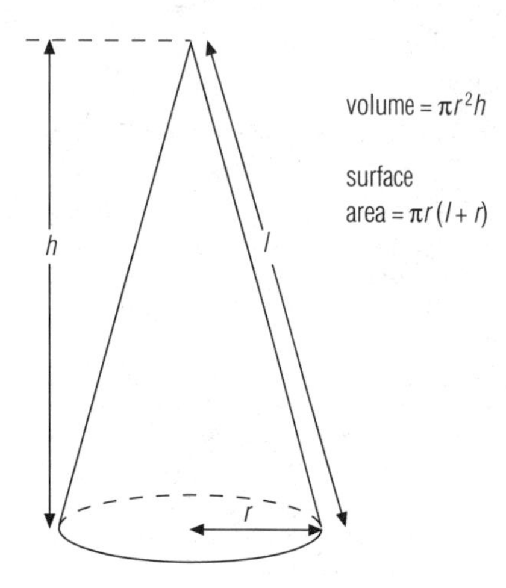

cone The volume and surface area of a cone are given by formulae involving a few simple dimensions.

cone in geometry, a solid or surface consisting of the set of all straight lines passing through a fixed point (the vertex) and the points of a circle or ellipse whose plane does not contain the vertex.

A circular cone of perpendicular height, with its apex above the centre of the circle, is known as a **right circular cone**; it is generated by rotating an isosceles triangle or framework about its line of symmetry. A right circular cone of perpendicular height h and base of radius r has a volume $V = \frac{1}{3}\pi r^2 h$.

The distance from the edge of the base of a cone to the vertex is called the slant height. In a right circular cone of slant height l, the curved surface area is πrl, and the area of the base is πr^2. Therefore the total surface area $A = \pi rl + \pi r^2 = \pi r(l + r)$.

cone in zoology, type of light-sensitive cell found in the retina of the ◊eye.

config.sys in computing, the ◊configuration file used by the MS-DOS and OS/2 ◊operating systems. It is read when the system is ◊booted.

configuration in chemistry, the arrangement of atoms in a molecule or of electrons in atomic orbitals.

configuration in computing, the way in which a system, whether it be ◊hardware and/or ◊software, is set up. A minimum configuration is often referred to for a particular application, and this will usually include a specification of processor, disc and memory size, and peripherals required.

congenital disease in medicine, a disease that is present at birth. It is not necessarily genetic in origin; for example, congenital herpes may be acquired by the baby as it passes through the mother's birth canal.

conger any large marine eel of the family Congridae, especially the genus *Conger*. Conger eels live in shallow water, hiding in crevices during the day and active by night, feeding on fish and crabs. They are valued for food and angling.

conglomerate in geology, coarse-grained clastic ◊sedimentary rock, composed of rounded fragments (clasts) of pre-existing rocks cemented in a finer matrix, usually sand.

The fragments in conglomerates are pebble- to boulder-sized, and the rock can be regarded as the lithified equivalent of gravel. A ◊bed of conglomerate is often associated with a break in a sequence of rock beds (an unconformity), where it marks the advance of the sea over an old eroded landscape. An **oligomict conglomerate** contains one type of pebble; a **polymict conglomerate** has a mixture of pebble types. If the rock fragments are angular, it is called a ◊breccia.

congruent in geometry, having the same shape and size, as applied to two-dimensional or solid figures. With plane congruent figures, one figure will fit on top of the other exactly, though this may first require rotation and/or rotation of one of the figures.

conic section curve obtained when a conical surface is intersected by a plane. If the intersecting plane cuts both extensions of the cone, it yields a ◊hyperbola; if it is parallel to the side of the cone, it produces a ◊parabola. Other intersecting planes produce ◊circles or ◊ellipses.

The Greek mathematician Apollonius wrote eight books with the title *Conic Sections*, which superseded previous work on the subject by Aristarchus and Euclid.

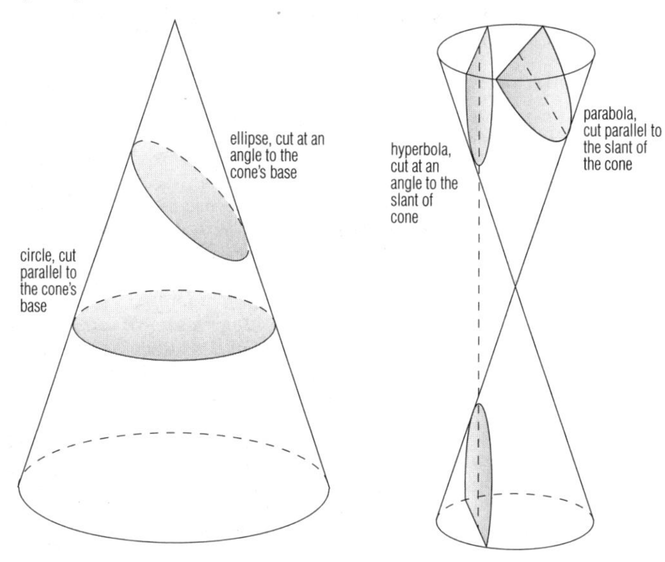

conic section The four types of curve that may be obtained by cutting a single or double right-circular cone with a plane (two-dimensional surface).

conidium (plural ***conidia***) asexual spore formed by some fungi at the tip of a specialized ◊hypha or conidiophore. The conidiophores grow erect, and cells from their ends round off and separate into conidia, often forming long chains. Conidia easily become detached and are dispersed by air movements.

conifer any of a large number of cone-bearing trees or shrubs. They are often pyramid-shaped, with leaves that are either scaled or needle-shaped; most are evergreen. Conifers include pines, spruces, firs, yews, junipers, monkey puzzles, and larches. (Order Coniferales.)

Conifers belong to the ◊gymnosperm or naked-seed-bearing group of plants. The reproductive organs are the male and female cones, and pollen is scattered in the wind. The seeds develop in the female cones. The processes of reaching maturity, fertilization, and seed ripening may take several years. Most conifers grow quickly and can survive in poor soil, on steep slopes, and in short growing seasons. Coniferous forests are widespread in Scandinavia and upland areas of the UK such as the Scottish Highlands, and are often planted in ◊afforestation schemes. Conifers also grow in ◊woodland.

conjugate in mathematics, a term indicating that two elements are connected in some way; for example, $(a + ib)$ and $(a - ib)$ are conjugate complex numbers.

conjugate angles two angles that add up to 360°.

conjugation in biology, the bacterial equivalent of sexual reproduction. A fragment of the ◊DNA from one bacterium is passed along a thin tube, the pilus, into another bacterium.

conjugation in organic chemistry, the alternation of double (or triple) and single carbon–carbon bonds in a molecule – for example, in penta-1,3-diene, $H_2C{=}CH{-}CH{=}CH{-}CH_3$. Conjugation imparts additional stability as the double bonds are less reactive than isolated double bonds.

conjunction in astronomy, the alignment of two celestial bodies as seen from Earth. A ◊superior planet (or other object) is in conjunction when it lies behind the Sun. An ◊inferior planet (or other object) comes to **inferior conjunction** when it passes between the Earth and the Sun; it is at **superior conjunction** when it passes behind the Sun.

Planetary conjunction takes place when a planet is closely aligned with another celestial object, such as the Moon, a star, or another planet.

Because the orbital planes of the inferior planets are tilted with respect to that of the Earth, they usually pass either above or below the Sun at inferior conjunction. If they line up exactly, a ◊transit will occur.

conjunctiva membrane covering the front of the vertebrate ◊eye. It is continuous with the epidermis of the eyelids, and lies on the surface of the cornea.

conjunctivitis inflammation of the conjunctiva, the delicate membrane that lines the inside of the eyelids and covers the front of the eye. Symptoms include redness, swelling, and a watery or pus-filled discharge. It may be caused by infection, allergy, or other irritant.

connective tissue in animals, tissue made up of a noncellular substance, the ◊extracellular matrix, in which some cells are embedded. Skin, bones, tendons, cartilage, and adipose tissue (fat) are the main connective tissues. There are also small amounts of connective tissue in organs such as the brain and liver, where they maintain shape and structure.

conodont extinct eel-like organism 520–240 million years old. Several thousand species have been described, ranging from 1–40 cm/0.4–15.8 in in length. They were predators, and, following the discovery of a large fossil in South Africa in 1995, are believed to be the first ◊vertebrates. This fossil of a 40-cm/15.8-in conodont shows fossilized soft tissue including muscle and eyes. It has eye muscles, an exclusively vertebrate feature, and 'real teeth'.

Conodont fossils consist mainly of teeth, and are often abundant, especially in Upper ◊Cambrian and ◊Triassic limestone. They were first discovered by Russian naturalist Christian Pander (1794–1865).

Conservation must come before recreation.

PRINCE OF WALES CHARLES Heir to the throne of Great Britain and Northern Ireland.
The Times 5 July 1989

conservation in the life sciences, action taken to protect and preserve the natural world, usually from pollution, overexploitation, and other harmful features of human activity. The late 1980s saw a great increase in public concern for the environment, with membership of conservation groups, such as ◊Friends of the Earth, ◊Greenpeace, and the US Sierra Club, rising sharply. Globally the most important issues include the depletion of atmospheric ozone by the action of ◊chlorofluorocarbons (CFCs), the build-up of carbon dioxide in the atmosphere (thought to contribute to an intensification of the ◊greenhouse effect), and ◊deforestation.

Conservation: chronology

1627	Last surviving aurochs, long-horned wild cattle that previously roamed Europe, southwest Asia, and North Africa, became extinct in Poland.
1664	A Dutch mandate drawn up to protect forest in Cape Colony, South Africa.
1681	The last dodo died on the island of Mauritius.
1764	The British established forest reserves on Tobago, after deforestation in Barbados and Jamaica resulted in widespread soil erosion.
1769	The French passed conservation laws in Mauritius.
1868	First laws passed in the UK to protect birds.
1948	The International Union for Conservation of Nature and Natural Resources (IUCN) was founded, with its sister organization, the World Wildlife Fund (WWF).
1970	The Man and the Biosphere Programme was initiated by UNESCO, providing for an international network of biosphere reserves.
1971	The Convention on Wetlands of International Importance (especially concerned with wildfowl habitat) signed in Ramsar, Iran, and started a List of Wetlands of International Importance.
1972	The Convention Concerning the Protection of the World Cultural and Natural Heritage adopted in Paris, France, providing for the designation of World Heritage Sites.
1972	The UN Conference on the Human Environment held in Stockholm, Sweden, lead to the creation of the UN Environment Programme (UNEP).
1973	The Convention on International Trade in Endangered Species of Wild Fauna and Flora (CITES) signed in Washington, DC.
1974	The world's largest protected area, the Greenland National Park covering 97 million hectares, created.
1980	The World Conservation Strategy launched by the IUCN, with the WWF and UNEP, showed how conservation contributes to development.
1982	The first herd of ten Arabian oryx bred from a 'captive breeding' programme released into the wild in Oman. The last wild oryx had been killed 1972.
1986	The first 'Red List' of endangered animal species compiled by IUCN.
1989	International trade in ivory banned under CITES legislation in an effort to protect the African elephant from poachers.
1992	The UN convened the 'Earth Summit' in Rio de Janeiro, Brazil, to discuss global planning for a sustainable future. The Convention on Biological Diversity and the Convention on Climate Change were opened for signing.
1993	The Convention on Biological Diversity came into force.
1995	The Arabian oryx conservation programme (began 1962), the most successful attempt at reintroducing zoo-bred animals to the wild, came to an end as the last seven animals were flown from the USA to join the 228-strong herd in Oman.
1996	The World Wide Fund for Nature had 5 million members in 28 countries.
1997	The world ban on the ivory trade was lifted in June 1997 at the tenth CITES convention. Trade is scheduled to resume in 1999 with Zimbabwe, Botswana, and Namibia the only countries allowed to export.

conservation of energy in chemistry, the principle that states that in a chemical reaction, the total amount of energy in the system remains unchanged.

For each component there may be changes in energy due to change of physical state, changes in the nature of chemical bonds, and either an input or output of energy. However, there is no net gain or loss of energy.

conservation of mass in chemistry, the principle that states that in a chemical reaction the sum of all the masses of the substances involved in the reaction (reactants) is equal to the sum of all of the masses of the substances produced by the reaction (products) – that is, no matter is gained or lost.

conservation of momentum in mechanics, a law that states that total ◊momentum is conserved (remains constant) in all collisions, providing no external resultant force acts on the colliding bodies. The principle may be expressed as an equation used to solve numerical problems: total momentum before collision = total momentum after collision.

console in computing, a combination of keyboard and screen (also described as a terminal). For a multiuser system, such as ◊UNIX, there is only one system console from which the system can be administered, while there may be many user terminals. See also ◊games console.

constant in mathematics, a fixed quantity or one that does not change its value in relation to ◊variables. For example, in the algebraic expression $y^2 = 5x - 3$, the numbers 3 and 5 are constants. In physics, certain quantities are regarded as universal constants, such as the speed of light in a vacuum.

constantan or ***eureka*** high-resistance alloy of approximately 40% nickel and 60% copper with a very low coefficient of ◊thermal expansion (measure of expansion on heating). It is used in electrical resistors.

constant composition, law of in chemistry, the law that states that the proportions of the amounts of the elements in a pure compound are always the same and are independent of the method by which the compound was produced.

constellation one of the 88 areas into which the sky is divided for the purposes of identifying and naming celestial objects. The first

CONSTELLATIONS AND THEIR STARS

http://www.vol.it/mirror/constellations/

Notes on the constellations (listed alphabetically, by month, and by popularity), plus lists of the 25 brightest stars and the 32 nearest stars, and photographs of the Milky Way.

Constellations

Constellation	*Abbreviation*	*Popular name(s)*
Andromeda	And	–
Antlia Ant	Airpump	
Apus Aps	Bird of Paradise	
Aquarius	Aqr	Water-Bearer
Aquila Aqi	Eagle	
Ara Ara	Altar	
Aries Ari	Ram	
Auriga Aur	Charioteer	
BoötesBoo	Herdsman	
Caelum	Cae	Chisel
Camelopardalis	Cam	Giraffe
CancerCnc	Crab	
Canes Venatici	CVn	Hunting Dogs
Canis Major	CMa	Great Dog
Canis Minor	CMi	Little Dog
Capricornus	Cap	Goat, Sea-Goat
Carina	Car	Keel
Cassiopeia	Cas	–
Centaurus	Cen	Centaur
Cepheus	Cep	–
Cetus	Cet	Whale
Chamaeleon	Cha	Chameleon
Circinus	Cir	Compasses
Columba	Col	Dove
Coma Berenices	Com	Berenice's Hair
Corona Australis	CrA	Southern Crown
Corona Borealis	CrB	Northern Crown
Corvus	Crv	Crow, Raven
Crater	Crt	Cup
Crux	Cru	Southern Cross
Cygnus	Cyn	Swan
Delphinus	Del	Dolphin
Dorado	Dor	Goldfish, Swordfish
Draco	Dra	Dragon
Equuleus	Equ	Filly, Foal
Eridanus	Eri	River
Fornax	For	Furnace
Gemini	Gem	Twins
Grus	Gru	Crane
Hercules	Her	–
Horologium	Hor	Clock
Hydra	Hya	Sea Serpent
Hydrus	Hyi	Watersnake
Indus	Ind	Indian
Lacerta	Lac	Lizard
Leo	Leo	Lion
Leo Minor	LMi	Little Lion
Lepus	Lep	Hare
Libra	Lib	Scales
Lupus	Lup	Wolf
Lynx	Lyn	–
Lyra	Lyr	Lyre, Harp
Mensa	Men	Table, Mountain
Microscopium	Mic	Microscope
Monoceros	Mon	Unicorn
Musca	Mus	Southern Fly
Norma	Nor	Rule, Straightedge
Octans	Oct	Octant
Ophiuchus	Oph	Serpent-Bearer
Orion	Ori	–
Pavo	Pav	Peacock
Pegasus	Peg	Flying Horse
Perseus	Per	–
Phoenix	Phe	–
Pictor	Pic	Painter, Easel
Pisces	Psc	Fishes
Piscis Austrinus	PsA	Southern Fish
Puppis	Pup	Poop
Pyxis	Pyx	Compass
Reticulum	Ret	Net
Sagitta	Sge	Arrow
Sagittarius	Sgr	Archer
Scorpius	Sco	Scorpion
Sculptor	Scl	–
Scutum	Sct	Shield
Serpens	Ser	Serpent
Sextans	Sex	Sextant
Taurus	Tau	Bull
Telescopium	Tel	Telescope
Triangulum	Tri	Triangle
Triangulum Australe	TrA	Southern Triangle
Tucana	Tuc	Toucan
Ursa Major	UMa	Big Dipper
Ursa Minor	UMi	Little Dipper
Vela	Vel	Sail
Virgo	Vir	Virgin
Volans	Vol	Flying Fish
Vulpecula	Vul	Fox

constellations were simple, arbitrary patterns of stars in which early civilizations visualized gods, sacred beasts, and mythical heroes.

The constellations in use today are derived from a list of 48 known to the ancient Greeks, who inherited some from the Babylonians. The current list of 88 constellations was adopted by the International Astronomical Union, astronomy's governing body, in 1930.

constipation in medicine, the infrequent emptying of the bowel. The intestinal contents are propelled by peristaltic contractions of the intestine in the digestive process. The faecal residue collects in the rectum, distending it and promoting defecation. Constipation may be due to illness, alterations in food consumption, stress, or as an adverse effect of certain drugs. An increased intake of dietary fibre (see ◊fibre, ◊dietary) can alleviate constipation. Laxatives may be used to relieve temporary constipation but they should not be used routinely.

contact force force or push produced when two objects are pressed together and their surface atoms try to keep them apart. Contact forces always come in pairs – for example, the downwards force exerted on a floor by the sole of a person's foot is matched by an equal upwards force exerted by the floor on that sole.

contact lens lens, made of soft or hard plastic, that is worn in contact with the cornea and conjunctiva of the eye, beneath the eyelid, to correct defective vision. In special circumstances, contact lenses may be used as protective shells or for cosmetic purposes, such as changing eye colour.

The earliest use of contact lenses in the late 19th century was protective, or in the correction of corneal malformation. It was not until the 1930s that simplification of fitting technique by taking eye impressions made general use possible. Recent developments are a type of soft lens that can be worn for lengthy periods without removal, and a disposable soft lens that needs no cleaning but should be discarded after a week of constant wear.

contact process the main industrial method of manufacturing the chemical ◊sulphuric acid. Sulphur dioxide (produced by burning sulphur) and air are passed over a hot (450°C) ◊catalyst of vanadium (V) oxide. The sulphur trioxide produced is absorbed in concentrated sulphuric acid to make fuming sulphuric acid (oleum), which is then diluted with water to give concentrated sulphuric acid (98%). Unreacted gases are recycled.

content provider organization or individual who creates intellectual property, such as information databases, which may be distributed via traditional media or via the World Wide Web.

context-sensitive help type of help built into software that displays information related to the particular function in use.

continent any one of the seven large land masses of the Earth, as distinct from the oceans. They are Asia, Africa, North America, South America, Europe, Australia, and Antarctica. Continents are constantly moving and evolving (see ◊plate tectonics). A continent does not end at the coastline; its boundary is the edge of the shallow continental shelf, which may extend several hundred kilometres out to sea.

At the centre of each continental mass lies a shield or ◊craton, a deformed mass of old ◊metamorphic rocks dating from Precambrian times. The shield is thick, compact, and solid (the Canadian Shield is an example), having undergone all the mountain-building activity it is ever likely to, and is usually worn flat. Around the shield is a concentric pattern of fold mountains, with older ranges, such as the Rockies, closest to the shield, and younger ranges, such as the coastal ranges of North America, farther away. This general concentric pattern is modified when two continental masses have drifted together and they become welded with a great mountain range along the join, the way Europe and N Asia are joined along the Urals. If a continent is torn apart, the new continental edges have no fold mountains; for instance, South America has fold mountains (the Andes) along its western flank, but none along the east where it tore away from Africa 200 million years ago.

To remember the seven continents:

EAT AN ASPIRIN AFTER A NIGHT-TIME SNACK

EUROPE, ANTARCTICA, ASIA, AFRICA, AUSTRALIA, NORTH AMERICA, SOUTH AMERICA (THE SECOND LETTER IN THE FIRST THREE 'A' WORDS HELPS TO REMEMBER THE 'A' CONTINENTS)

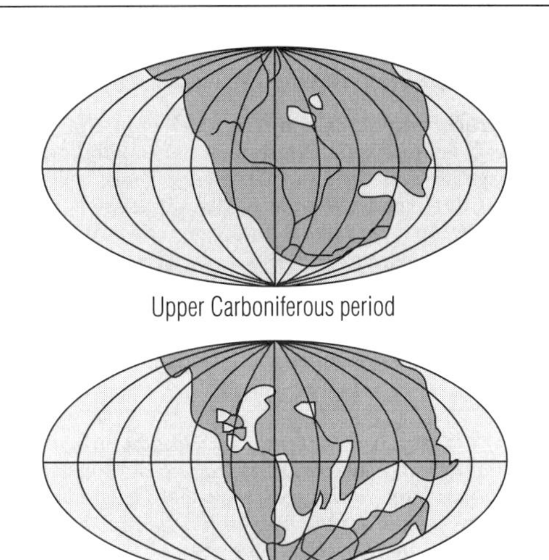

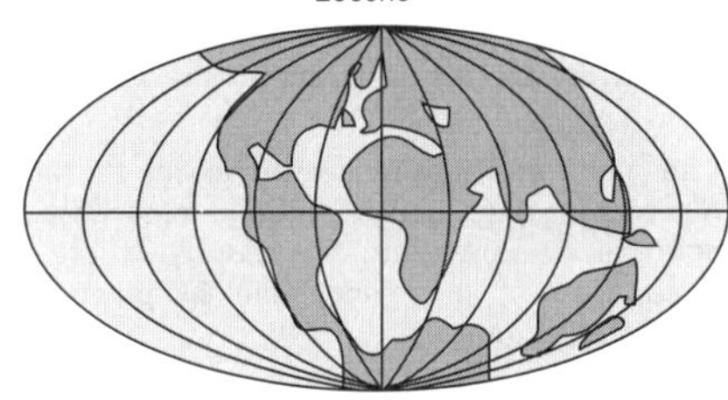

***continental drift** The continents are slowly shifting their positions, driven by fluid motion beneath the Earth's crust. Over 200 million years ago, there was a single large continent called Pangaea. By 200 million years ago, the continents had started to move apart. By 50 million years ago, the continents were approaching their present positions.*

continental drift in geology, the theory that, about 250–200 million years ago, the Earth consisted of a single large continent (◊Pangaea), which subsequently broke apart to form the continents known today. The theory was proposed 1912 by German meteorologist Alfred Wegener, but such vast continental movements could not be satisfactorily explained until the study of ◊plate tectonics in the 1960s.

The term 'continental drift' is not strictly correct, since land masses do not drift through the oceans. The continents form part of a plate, and the amount of crust created at divergent plate margins must equal the amount of crust destroyed at subduction zones.

continental shelf the submerged edge of a continent, a gently sloping plain that extends into the ocean. It typically has a gradient of less than 1°. When the angle of the sea bed increases to 1°–5° (usually several hundred kilometres away from land), it becomes known as the **continental slope**.

continuity in mathematics, property of functions of a real variable that have an absence of 'breaks'. A function f is said to be continuous at a point a if $\lim f(x) = f(a)$.

continuous data data that can take any of an infinite number of values between whole numbers and so may not be measured completely accurately. This type of data contrasts with ◊discrete data, in which the variable can only take one of a finite set of values. For example, the sizes of apples on a tree form continuous data, whereas the numbers of apples form discrete data.

continuous variation the slight difference of an individual character, such as height, across a sample of the population. Although there are very tall and very short humans, there are also many people with an intermediate height. The same applies to weight. Continuous variation can result from the genetic make-up of a population, or from environmental influences, or from a combination of the two.

continuum in mathematics, a ◊set that is infinite and everywhere continuous, such as the set of points on a line.

> *I have continued my work on the continuum problem last summer and I finally succeeded in proving the consistency of the continuum hypothesis (even the generalized form) with respect to generalized set theory. But for the time being please do not tell anyone of this.*
>
> KURT GÖDEL Austrian-born US mathematician. Letter to his teacher Karl Menger 1937

contouring in ◊computer graphics, a technique for enhancing the outline of a particular shape. This technique is used in applications such as mapping (see ◊animation, ◊computer), where a computer following the contours of an object ◊pixel by pixel can be much more precise than a human.

contraceptive any drug, device, or technique that prevents pregnancy. The contraceptive pill (the ◊Pill) contains female hormones that interfere with egg production or the first stage of pregnancy. The 'morning-after' pill can be taken up to 72 hours after unprotected intercourse. Barrier contraceptives include ◊condoms (sheaths) and ◊diaphragms, also called caps or Dutch caps; they prevent the sperm entering the cervix (neck of the womb).

◊Intrauterine devices, also known as IUDs or coils, cause a slight inflammation of the lining of the womb; this prevents the fertilized egg from becoming implanted. See also ◊family planning.

Other contraceptive methods include ◊sterilization (women) and ◊vasectomy (men); these are usually nonreversible. 'Natural' methods include withdrawal of the penis before ejaculation (coitus interruptus), and avoidance of intercourse at the time of ovulation (◊rhythm method). These methods are unreliable and normally only used on religious grounds. A new development is a sponge impregnated with spermicide that is inserted into the vagina. The use of any contraceptive (birth control) is part of family planning. The effectiveness of a contraceptive method is often given as a percentage. To say that a method has 95% effectiveness means that, on average, out of 100 healthy couples using that method for a year, 95 will not conceive.

contractile root in botany, a thickened root at the base of a corm, bulb, or other organ that helps position it at an appropriate level in the ground. Contractile roots are found, for example, on the corms of plants of the genus *Crocus*. After they have become anchored in the soil, the upper portion contracts, pulling the plant deeper into the ground.

contractile vacuole tiny organelle found in many single-celled fresh-water organisms. It slowly fills with water, and then contracts, expelling the water from the cell.

Fresh-water protozoa such as *Amoeba* absorb water by the process of ◊osmosis, and this excess must be eliminated. The rate of vacuole contraction slows as the external salinity is increased, because the osmotic effect weakens; marine protozoa do not have a contractile vacuole.

control in biology, the process by which a tissue, an organism, a population, or an ecosystem maintains itself in a balanced, stable state. Blood sugar must be kept at a stable level if the brain is to function properly, and this steady-state is maintained by an interaction between the liver, the hormone insulin, and a detector system in the pancreas.

In the ecosystem, the activities of the human race are endangering the balancing mechanisms associated with the atmosphere in general and the ◊greenhouse effect in particular.

control bus in computing, the electrical pathway, or ◊bus, used to communicate control signals.

control character in computing, any character produced by depressing the control key (Ctrl) on a keyboard at the same time as another (usually alphabetical) key. The control characters form the first 32 ◊ASCII characters and most have specific meanings according to the operating system used. They are also used in combination to provide formatting control in many word processors, although the user may not enter them explicitly.

control experiment essential part of a scientifically valid experiment, designed to show that the factor being tested is actually responsible for the effect observed. In the control experiment all factors, apart from the one under test, are exactly the same as in the test experiments, and all the same measurements are carried out. In drug trials, a placebo (a harmless substance) is given alongside the substance being tested in order to compare effects.

control total in computing, a ◊validation check in which an arithmetic total of a specific field from a group of records is calculated. This total is input together with the data to which it refers. The program recalculates the control total and compares it with the one entered to ensure that no entry errors have been made.

control unit the component of the ◊central processing unit that decodes, synchronizes, and executes program instructions.

convection heat energy transfer that involves the movement of a fluid (gas or liquid). Fluid in contact with the source of heat expands and tends to rise within the bulk of the fluid. Cooler fluid sinks to take its place, setting up a convection current. This is the principle of natural convection in many domestic hot-water systems and space heaters.

convection current current caused by the expansion of a liquid or gas as its temperature rises. The expanded material, being less dense, rises above colder and therefore denser material. Convection currents arise in the atmosphere above warm land masses or seas, giving rise to sea breezes and land breezes, respectively. In some heating systems, convection currents are used to carry hot water upwards in pipes.

Convection currents in the viscous rock of the Earth's mantle help to drive the movement of the rigid plates making up the Earth's surface (see ◊plate tectonics).

conventional current direction in which an electric current is considered to flow in a circuit. By convention, the direction is that in which positive-charge carriers would flow – from the positive terminal of a cell to its negative terminal. It is opposite in direction to the flow of electrons. In circuit diagrams, the arrows shown on symbols for components such as diodes and transistors point in the direction of conventional current flow.

convergence in mathematics, the property of a series of numbers in which the difference between consecutive terms gradually decreases. The sum of a converging series approaches a limit as the number of terms tends to ◊infinity.

> *What can be more curious than that the hand of a man, formed for grasping, that of a mole for digging, [...] and the wing of a bat, should all be constructed on the same pattern, and should include the same bones, in the same relative positions?*
>
> CHARLES DARWIN British naturalist. *On the Origin of Species* 1859

convergent evolution in biology, the independent evolution of similar structures in species (or other taxonomic groups) that are not closely related, as a result of living in a similar way. Thus, birds and bats have wings, not because they are descended from a com-

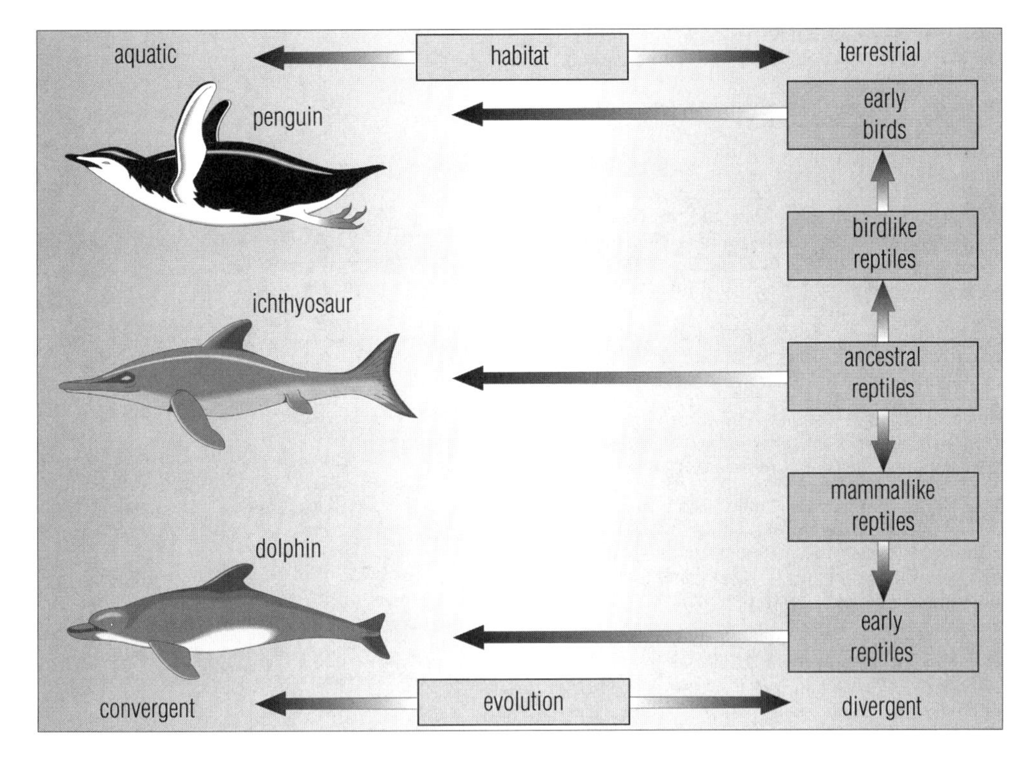

convergent evolution *Convergent evolution produced the superficially similar streamlined bodies of the dolphin and penguin. Despite their very different evolutionary paths – one as a mammal, the other as a bird – both have evolved and adapted to the aquatic environment they now inhabit.*

mon winged ancestor, but because their respective ancestors independently evolved flight.

converging lens lens that converges or brings to a focus those light rays that have been refracted by it. It is a ◊convex lens, with at least one surface that curves outwards, and is thicker towards the centre than at the edge. Converging lenses are used to form real images in many ◊optical instruments, such as cameras and projectors. A converging lens that forms a virtual, magnified image may be used as a ◊magnifying glass or to correct ◊long-sightedness.

converse in mathematics, the reversed order of a conditional statement; the converse of the statement 'if *a*, then *b*' is 'if *b*, then *a*'. The converse does not always hold true; for example, the converse of 'if $x = 3$, then $x^2 = 9$' is 'if $x^2 = 9$, then $x = 3$', which is not true, as x could also be -3.

convertiplane ◊vertical takeoff and landing craft (VTOL) with rotors on its wings that spin horizontally for takeoff, but tilt to spin in a vertical plane for forward flight.

At takeoff it looks like a two-rotor helicopter, with both rotors facing skywards. As forward speed is gained, the rotors tilt slowly forward until they are facing directly ahead. There are several different forms of convertiplane. The LTV-Hillier-Ryan XC-142, designed in the USA, had wings, carrying the four engines and propellers, that rotated. The German VC-400 had two rotors on each of its wingtips. Neither of these designs went into production. A Bell/Boeing design, the V-22 Osprey, uses a pair of tilting engines, with propellers 11.5 m/38 ft across, mounted at the end of the wings. It was originally intended to carry about 50 passengers direct to city centres. Crashes of two prototypes led to design changes 1994, but the project continues.

convex of a surface, curving outwards, or towards the eye. For example, the outer surface of a ball appears convex. In geometry, the term is used to describe any polygon possessing no interior angle greater than 180°. Convex is the opposite of ◊concave.

convex lens lens that possesses at least one surface that curves outwards. It is a ◊converging lens, bringing rays of light to a focus. A convex lens is thicker at its centre than at its edges, and is used to correct long-sightedness.

Common forms include the **biconvex** lens (with both surfaces curved outwards) and the **plano-convex** (with one flat surface and one convex). The whole lens may be further curved overall, making a **concavo-convex** or converging meniscus lens, as in some lenses used in corrective eyewear.

convex mirror curved mirror that reflects light from its outer surface. It diverges reflected light rays to form a reduced, upright, virtual image. Convex mirrors give a wide field of view and are therefore particularly suitable for car wing mirrors and surveillance purposes in shops.

conveyor device used for transporting materials. Widely used throughout industry is the **conveyor belt**, usually a rubber or fabric belt running on rollers. Trough-shaped belts are used, for example in mines, for transporting ores and coal. **Chain conveyors** are also used in coal mines to remove coal from the cutting machines. Overhead endless chain conveyors are used to carry components and bodies in car-assembly works. Other types include **bucket conveyors** and **screw conveyors**, powered versions of the ◊Archimedes screw.

convolvulus or ***bindweed*** any of a group of plants belonging to the morning-glory family. They are characterized by their twining stems and by their petals, which are joined into a funnel-shaped tube. (Genus *Convolvulus,* family Convolvulaceae.)

convulsion series of violent contractions of the muscles over which the patient has no control. It may be associated with loss of consciousness. Convulsions may arise from any one of a number of causes, including brain disease (such as ◊epilepsy), injury, high fever, poisoning, and electrocution.

cookie on the World Wide Web, a short piece of text that a Web site stores in a Cookies folder or a cookie.txt file on the user's computer, either for tracking or configuration purposes, for example, to improve the targeting of banner advertisements. Cookies can also store user preferences and passwords. The cookie is sent back to the server when the browser requests a new page.

Cookies are derived from 'magic cookies', the identification tokens used by some UNIX systems.

Originally, a cookie was an aphorism or short, witty saying obtained by typing 'cookie' at a computer's main system prompt. A cookie was then chosen at random – like a 'fortune cookie' – from a database called a 'cookie file'.

cookie recipe urban legend that circulates around the Net. The story is about a protagonist who ate some delicious cookies for dessert after a meal in a fancy department store or restaurant. When asked for the recipe, the waiter refuses, but finally relents saying it will cost 'two fifty'. When the bill comes, the protagonist discovers the restaurant has charged $250. Feeling stung, he/she posts the recipe to the Net to ensure the maximum distribution (and therefore revenge) possible. A cookie recipe is attached.

coolabah Australian riverside tree *Eucalyptus microtheca* of the myrtle family Myrtaceae. It is common in the inland and usually associated with areas subject to occasional inundation.

coonhound breed of large hound developed from European hounds imported into in Virginia, USA, at the start of the 17th century, and used for hunting various types of small game. It has a short black coat with red markings on muzzle, chest, and feet; long drooping ears; and a long tail. It stands up to 68 cm/27 in tall.

cooperative farming system in which individual farmers pool their resources (excluding land) to buy commodities such as seeds and fertilizers, and services such as marketing. It is a system of farming found throughout the world and is particularly widespread in Denmark and the ex-Soviet republics. In a ◊collective farm, land is also held in common.

coordinate in geometry, a number that defines the position of a point relative to a point or axis (reference line). ◊Cartesian coordinates define a point by its perpendicular distances from two or more axes drawn through a fixed point mutually at right angles to each other. ◊Polar coordinates define a point in a plane by its distance from a fixed point and direction from a fixed line.

cartesian coordinates

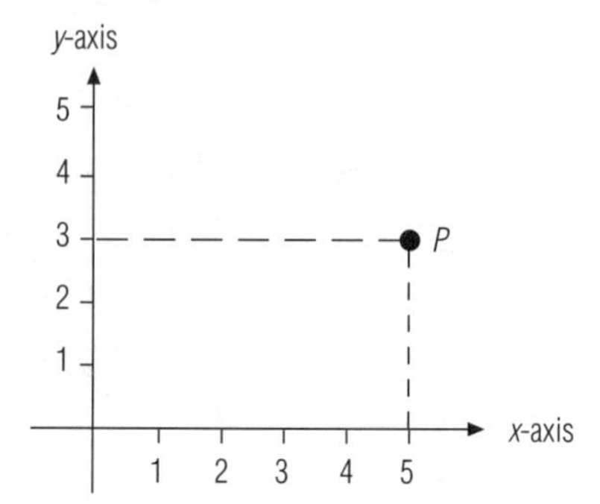

the cartesian coordinates of *P* are (5,3)

polar coordinates

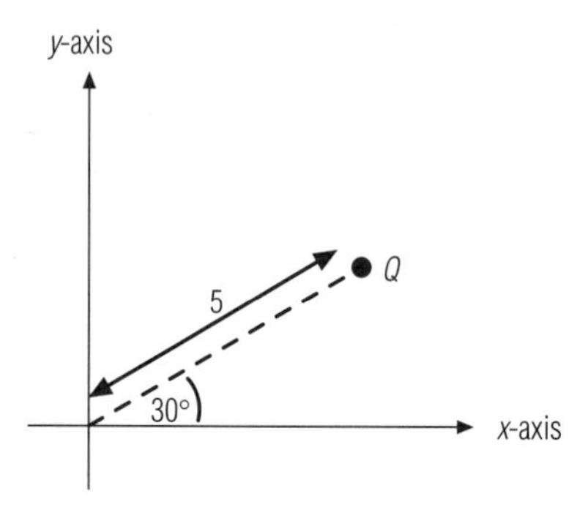

the polar coordinates of *Q* are (5,30°)

***coordinate** Coordinates are numbers that define the position of points in a plane or in space. In the Cartesian coordinate system, a point in a plane is charted based upon its location along intersecting horizontal and vertical axes. In the polar coordinate system, a point in a plane is defined by its distance from a fixed point and direction from a fixed line.*

coordinate geometry or ***analytical geometry*** system of geometry in which points, lines, shapes, and surfaces are represented by algebraic expressions. In plane (two-dimensional) coordinate geometry, the plane is usually defined by two axes at right angles to each other, the horizontal x-axis and the vertical y-axis, meeting at O, the origin. A point on the plane can be represented by a pair of ◊Cartesian coordinates, which define its position in terms of its distance along the x-axis and along the y-axis from O. These distances are respectively the x and y coordinates of the point.

Lines are represented as equations; for example, $y = 2x + 1$ gives a straight line, and $y = 3x^2 + 2x$ gives a ◊parabola (a curve). The graphs of varying equations can be drawn by plotting the coordinates of points that satisfy their equations, and joining up the points. One of the advantages of coordinate geometry is that geometrical solutions can be obtained without drawing but by manipulating algebraic expressions. For example, the coordinates of the point of intersection of two straight lines can be determined by finding the unique values of x and y that satisfy both of the equations for the lines, that is, by solving them as a pair of ◊simultaneous equations. The curves studied in simple coordinate geometry are the ◊conic sections (circle, ellipse, parabola, and hyperbola), each of which has a characteristic equation.

coot freshwater bird of the genus *Fulica* in the rail family, order Gruiformes. Coots are about 38 cm/1.2 ft long, and mainly black. They have a white bill, extending up the forehead in a plate, and big feet with four lobed toes.

copepod ◊crustacean of the subclass Copepoda, mainly microscopic and found in plankton.

coplanar in geometry, describing lines or points that all lie in the same plane.

copper orange-pink, very malleable and ductile, metallic element, symbol Cu (from Latin *cuprum*), atomic number 29, relative atomic mass 63.546. It is used for its durability, pliability, high thermal and electrical conductivity, and resistance to corrosion.

It was the first metal used systematically for tools by humans; when mined and worked into utensils it formed the technological basis for the Copper Age in prehistory. When alloyed with tin it forms bronze, which is stronger than pure copper and may hold a sharp edge; the systematic production and use of this alloy was the basis for the prehistoric Bronze Age. Brass, another hard copper alloy, includes zinc. The element's name comes from the Greek for Cyprus (*Kyprios*), where copper was mined.

COPPER PAGE

http://www.copper.org/

guide to copper and copper alloys, providing a wealth of information ranging from a historical overview of the copper industry to a database of literary resources about copper technology.

copper(II) carbonate $CuCO_3$ green solid that readily decomposes to form black copper(II) oxide on heating.

$$CuCO_3 \rightarrow CuO + CO_2$$

It dissolves in dilute acids to give blue solutions with effervescence caused by the giving off of carbon dioxide.

$$CuCO_3 + H_2SO_4 \rightarrow CuSO_4 + CO_2 + H_2O$$

Copernicus, Nicolaus
(1473–1543)

Latinized form of Mikolaj Kopernik Polish astronomer who believed that the Sun, not the Earth, is at the centre of the Solar System, thus defying the Christian church doctrine of the time. For 30 years, he worked on the hypothesis that the rotation and the orbital motion of the Earth are responsible for the apparent movement of the heavenly bodies. His great work De revolutionibus orbium coelestium/On the Revolutions of the Heavenly Spheres was the important first step to the more accurate picture of the Solar System built up by Tycho Brahe, Kepler, Galileo, and later astronomers.

Copernicus proposed replacing Ptolemy's ideas with a model in which the planets (including the Earth) orbited a centrally situated Sun. He proposed that the Earth described one full orbit of the Sun in a year, whereas the Moon orbited the Earth. The Earth rotated daily about its axis (which was inclined at 23.5° to the plane of orbit), thus accounting for the apparent daily rotation of the sphere of the fixed stars.

Mary Evans Picture Library

This model was a distinct improvement on the Ptolemaic system for a number of reasons. It explained why the planets Mercury and Venus displayed only 'limited motion'; their orbits were inside that of the Earth's. Similarly, it explained that the planets Mars, Jupiter, and Saturn displayed such curious patterns in their movements ('retrograde motion', loops, and kinks) because they travel in outer orbits at a slower pace than the Earth. The movement of the Earth on its axis accounted for the precession of the equinoxes, previously discovered by Hipparchus.

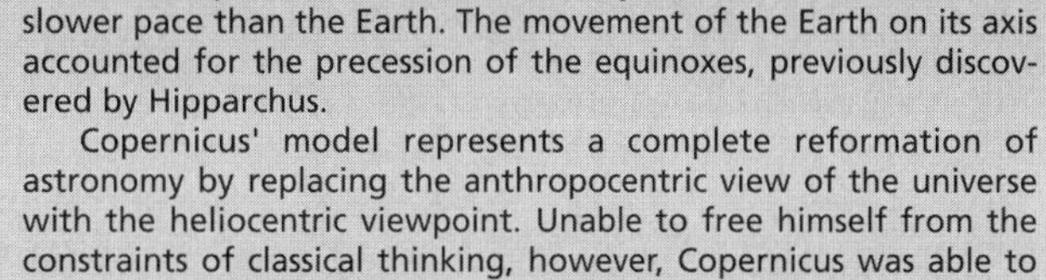

Copernicus' model represents a complete reformation of astronomy by replacing the anthropocentric view of the universe with the heliocentric viewpoint. Unable to free himself from the constraints of classical thinking, however, Copernicus was able to imagine only circular planetary orbits. This forced him to retain the system of epicycles, with the Earth revolving around a centre that revolved around another centre, which in turn orbited the Sun. Kepler rescued the model by introducing the concept of elliptical orbits. Copernicus also held to the notion of spheres, in which the planets were supposed to travel. It was Brahe who finally rid astronomy of that concept.

French mathematician, encyclopedist, and theoretical physicist. In association with Denis Diderot, he helped plan the great Encyclopédie, for which he also wrote the 'Discours préliminaire' 1751. He framed several theorems and principles – notably d'Alembert's principle – in dynamics and celestial mechanics, and devised the theory of partial differential equations.

The principle that now bears his name was first published in his Traité de dynamique 1743, and was an extension of the third of Isaac Newton's laws of motion. D'Alembert maintained that the law was valid not merely for a static body, but also for mobile bodies. Within a year he had found a means of applying the principle to the theory of equilibrium and the motion of fluids. Using also the theory of partial differential equations, he studied the properties of sound, and air compression, and also managed to relate his principle to an investigation of the motion of any body in a given figure.

copper ore any mineral from which copper is extracted, including native copper, Cu; chalcocite, Cu_2S; chalcopyrite, $CuFeS_2$; bornite, Cu_5FeS_4; azurite, $Cu_3(CO_3)_2(OH)_2$; malachite, $Cu_2CO_3(OH)_2$; and chrysocolla, $CuSiO_3.2H_2O$.

Native copper and the copper sulphides are usually found in veins associated with igneous intrusions. Chrysocolla and the carbonates are products of the weathering of copper-bearing rocks. Copper was one of the first metals to be worked, because it occurred in native form and needed little refining. Today the main producers are the USA, Russia, Kazakhstan, Georgia, Uzbekistan, Armenia, Zambia, Chile, Peru, Canada, and Congo (formerly Zaire).

copper(II) oxide CuO black solid that is readily reduced to copper by carbon, carbon monoxide, or hydrogen if heated with any of these.

$$CuO + C \rightarrow Cu + CO \quad CuO + CO \rightarrow Cu + CO_2$$

$$CuO + H_2 \rightarrow Cu + H_2O$$

It is usually made in the laboratory by heating copper(II) carbonate, nitrate, or hydroxide.

$$2Cu(NO_3)_2 \rightarrow 2CuO + 4NO_2 + O_2$$

Copper(II) oxide is a typical basic oxide, dissolving readily in most dilute acids.

copper(II) sulphate $CuSO_4$ substance usually found as a blue, crystalline, hydrated salt $CuSO_4.5H_2O$ (also called blue vitriol). It is made from the action of dilute sulphuric acid on copper(II) oxide, hydroxide, or carbonate.

$$CuO + H_2SO_4 + 4H_2O \rightarrow CuSO_4.5H_2O$$

When the hydrated salt is heated gently it loses its water of crystallization and the blue crystals turn to a white powder. The reverse reaction is used as a chemical test for water.

$$CuSO_4.5H_2O \leftrightarrow CuSO_4 + 5H_2O$$

coppicing woodland management practice of severe pruning where trees are cut down to near ground level at regular intervals, typically every 3–20 years, to promote the growth of numerous shoots from the base.

This form of ◊forestry was once commonly practised in Europe, principally on hazel and chestnut, to produce large quantities of thin branches for firewood, fencing, and so on; alder, eucalyptus, maple, poplar, and willow were also coppiced. The resulting thicket was known as a coppice or copse. See also ◊pollarding.

coprocessor in computing, an additional ◊processor that works with the main ◊central processing unit to carry out a specific function. The two most common coprocessors are the **mathematical coprocessor**, used to speed up calculations, and the **graphic coprocessor**, used to improve the handling of graphics.

copulation act of mating in animals with internal ◊fertilization. Male mammals have a ◊penis or other organ that is used to introduce spermatozoa into the reproductive tract of the female. Most birds transfer sperm by pressing their cloacas (the openings of their reproductive tracts) together.

copy protection techniques used to prevent illegal copying of computer programs. Copy protection is not as common as it used to be because it also prevents legal copying (for backup purposes). Alternative techniques to prevent illegal use include ◊dongles, passwords and the need to uninstall a program before it can be installed on another machine.

coral marine invertebrate of the class Anthozoa in the phylum Cnidaria, which also includes sea anemones and jellyfish. It has a

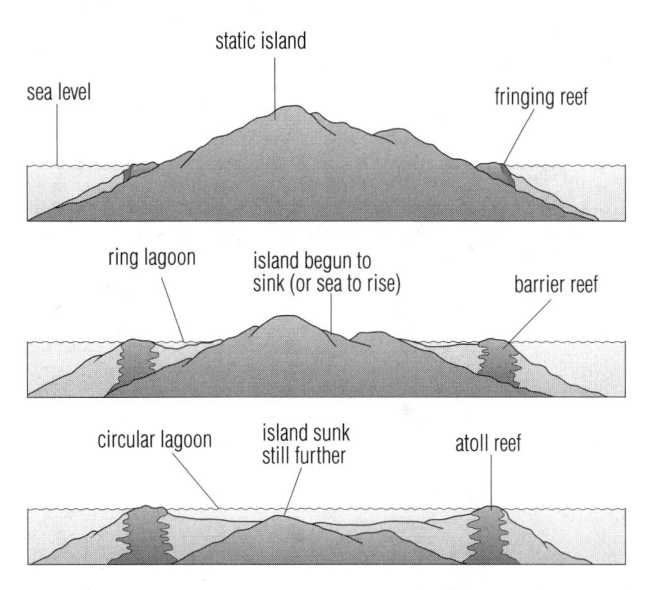

coral atoll *The formation of a coral atoll by the gradual sinking of a volcanic island. The reefs fringing the island build up as the island sinks, eventually producing a ring of coral around the spot where the island sank.*

skeleton of lime (calcium carbonate) extracted from the surrounding water. Corals exist in warm seas, at moderate depths with sufficient light. Some coral is valued for decoration or jewellery, for example, Mediterranean red coral *Corallum rubrum.*

Corals live in a symbiotic relationship with microscopic ◊algae (zooxanthellae), which are incorporated into the soft tissue. The algae obtain carbon dioxide from the coral polyps, and the polyps receive nutrients from the algae. Corals also have a relationship to the fish that rest or take refuge within their branches, and which excrete nutrients that make the corals grow faster. The majority of corals form large colonies although there are species that live singly. Their accumulated skeletons make up large coral reefs and atolls. The Great Barrier Reef, to the NE of Australia, is about 1,600 km/1,000 mi long, has a total area of 20,000 sq km/7,700 sq mi, and adds 50 million tonnes of calcium to the reef each year. The world's reefs cover an estimated 620,000 sq km/240,000 sq mi.

Coral reefs provide a habitat for a diversity of living organisms. In 1997, some 93,000 species were identified. One third of the world's marine fishes live in reefs. The world's first global survey of coral reefs, carried out in 1997, found around 95% of reefs had experienced some damage from overfishing, pollution, dynamiting, poisoning, and the dragging of ships' anchors.

diseases Since the 1990s, coral reefs have been destroyed by previously unknown diseases. The **white plague** attacked 17 species of coral in the Florida Keys, USA, in 1995. The **rapid wasting disease**, discovered in 1997, affects coral reefs from Mexico to Trinidad. In the Caribbean, the fungus *Aspergillus* attacks sea fans, a soft coral. It was estimated in 1997 that around 90% of the coral around the Galapagos islands had been destroyed as a result of 'bleaching', a whitening of coral reefs which occurs when the coloured algae evacuate the coral. This happens either because the corals produce toxins that are harmful to the algae or because they do not produce sufficient nutrients. Without the algae, the coral crumbles and dies away. Bleaching is widespread all over the Caribbean and the Indo-Pacific.

WELCOME TO CORAL FOREST

http://www.blacktop.com/coralforest/

Site dedicated to explaining the importance of coral reefs for the survival of the planet. It is an impassioned plea on behalf of the world's endangered coral reefs and includes a full description of their biodiversity, maps of where coral reefs are to be found (no less than 109 countries), and many photos.

coralroot any leafless orchid of the genus *Corallorhiza*, having branched coral-colored roots and small yellowish or purplish flowers. These orchids are either parasitic on the roots of other plants, or saprophytes, living on decaying organic matter.

coral snake venomous snake. *Elaps corallinus* is a typical specimen; it occurs in the tropical forests of South America, and its small body, less than 1 m/3.3 ft in length, is ringed with coral-red. It is highly poisonous.

classification Coral snakes are in the family Elapidae, suborder Serpentes, order Squamata, class Reptilia.

coral tree any of several tropical trees with bright red or orange flowers and producing a very lightweight wood. (Genus *Erythrina*, family Fabaceae.)

Corba (acronym for ***Common object request broker architecture***) in computing, agreed specification that enables software components or 'objects' from different suppliers running on different computers using different operating systems to interoperate with one another. Corba has been extended via the Internet Inter-Orb Protocol (IIOP) to work over the Internet. Corba is promulgated as a standard by the Object Management Group (OMG).

cord unit for measuring the volume of wood cut for fuel. One cord equals 128 cubic feet (3.456 cubic metres), or a stack 8 feet (2.4 m) long, 4 feet (1.2 m) wide, and 4 feet high.

cordillera group of mountain ranges and their valleys, all running in a specific direction, formed by the continued convergence of two tectonic plates (see ◊plate tectonics) along a line. The term is applied especially to the principal mountain belt of a continent. The Andes of South America are an example.

core in earth science, the innermost part of Earth. It is divided into an outer core, which begins at a depth of 2,898 km/1,800 mi, and an inner core, which begins at a depth of 4,982 km/3,095 mi. Both parts are thought to consist of iron-nickel alloy. The outer core is liquid and the inner core is solid.

The fact that seismic shear waves (see ◊seismic wave) disappear at the mantle–outer core boundary indicates that the outer core is molten, since shear waves cannot travel through fluid. Scientists infer the iron-nickel rich composition of the core from Earth's density and its ◊moment of inertia. The temperature of the core, as estimated from the melting point of iron at high pressure, is thought to be at least 4,000°C/7,232°F, but remains controversial. Earth's magnetic field is believed to be the result of the motions involving the inner and outer cores.

Corel Canadian software company founded in 1983 by British citizen Michael Cowpland. Its drawing program Corel Draw led the market from its first release. Corel bought the desktop publishing package Ventura Publisher 1995 and then the word processor WordPerfect 1996.

corgi breed of dog. See ◊Welsh corgi.

coriander pungent fresh herb belonging to the parsley family, native to Europe and Asia; also a spice made from its dried ripe seeds. The spice is used commercially as a flavouring in meat products, bakery goods, tobacco, gin, liqueurs, chilli, and curry powder. Both are commonly used in cooking in the Middle East, India, Mexico, and China. (*Coriandrum sativum*, family Umbelliferae.)

Coriolis effect the effect of the Earth's rotation on the atmosphere and on all objects on the Earth's surface. In the northern hemisphere it causes moving objects and currents to be deflected to the right; in the southern hemisphere it causes deflection to the left. The effect is named after its discoverer, French mathematician Gaspard de Coriolis (1792–1843).

cork light, waterproof outer layers of the bark covering the branches and roots of almost all trees and shrubs. The cork oak (*Quercus suber*), a native of S Europe and N Africa, is cultivated in Spain and Portugal; the exceptionally thick outer layers of its bark provide the cork that is used commercially.

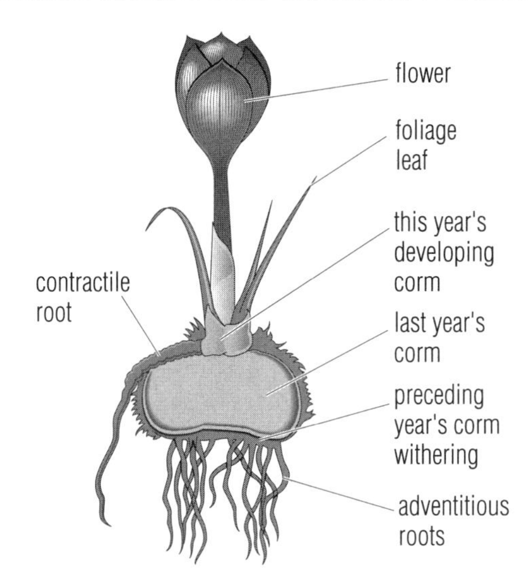

corm Corms, found in plants such as the gladiolus and crocus, are underground storage organs. They provide the food for growth during adverse conditions such as cold or drought.

corm short, swollen, underground plant stem, surrounded by protective scale leaves, as seen in the genus *Crocus*. It stores food, provides a means of ◊vegetative reproduction, and acts as a ◊perennating organ.

During the year, the corm gradually withers as the food reserves are used for the production of leafy, flowering shoots formed from axillary buds. Several new corms are formed at the base of these shoots, above the old corm.

cormorant any of various diving seabirds, mainly of the genus *Phalacrocorax*, order Pelecaniformes, about 90 cm/3 ft long, with webbed feet, a long neck, hooked beak, and glossy black plumage. Cormorants generally feed on fish and shellfish, which they catch by swimming and diving under water, sometimes to a considerable depth. They collect the food in a pouch formed by the dilatable skin at the front of the throat. Some species breed on inland lakes and rivers.

P. carbo has a bright shiny head and neck, with bluish-black feathers, speckled with white. The general colour above is a greenish black, the throat white, and the bill and feet are dark grey. It is found in all parts of the world in coastal regions.

There are about 30 species of cormorant worldwide, including a flightless form *Nannopterum harrisi* in the Galápagos Islands; the **shag**, or **green cormorant**, *P. aristotelis*; and the **small European cormorant**, *Halietor pygmaeus*, which is a freshwater bird. The **guanay cormorant** *P. bougainvillei*, of the Peruvian coast, is the main producer of the ◊guano of those regions.

corn general term for the main ◊cereal crop of a region – for example, wheat in the UK, oats in Scotland and Ireland, maize in the USA. Also, another word for ◊maize.

corncrake or landrail bird *Crex crex* of the rail family Rallidae, order Gruiformes. About 25 cm/10 in long, the bill and tail are short, the legs long and powerful, and the toes have sharp claws. It is drably coloured, shy, and has a persistent rasping call. The corncrake can swim and run easily, but its flight is heavy. It lives in meadows and crops in temperate regions, but has become rare where mechanical methods of cutting corn are used.

cornea transparent front section of the vertebrate ◊eye. The cornea is curved and behaves as a fixed lens, so that light entering the eye is partly focused before it reaches the lens.

There are no blood vessels in the cornea and it relies on the fluid in the front chamber of the eye for nourishment. Further protection for the eye is provided by the ◊conjunctiva. In humans, diseased or opaque parts may be replaced by grafts of corneal tissue from a donor.

cornflower native European and Asian plant belonging to the same genus as the ◊knapweeds but distinguished from them by its deep azure-blue flowers. Formerly a common weed in N European wheat fields, it is now widely grown in gardens as a ◊herbaceous plant with flower colours ranging from blue through shades of pink and purple to white. (*Centaurea cyanus*, family Compositae.)

cornified layer the upper layer of the skin where the cells have died, their cytoplasm being replaced by keratin, a fibrous protein also found in nails and hair. Cornification gives the skin its protective waterproof quality.

corolla collective name for the petals of a flower. In some plants the petal margins are partly or completely fused to form a **corolla tube**, for example in bindweed *Convolvulus arvensis*.

corona faint halo of hot (about 2,000,000°C/3,600,000°F) and tenuous gas around the Sun, which boils from the surface. It is visible at solar ◊eclipses or through a **coronagraph**, an instrument that blocks light from the Sun's brilliant disc. Gas flows away from the corona to form the ◊solar wind.

Corona Australis or ***Southern Crown*** small constellation of the southern hemisphere, located near the constellation ◊Sagittarius. It is similar in size and shape to ◊Corona Borealis but is not as bright.

Corona Borealis or ***Northern Crown*** small but easily recognizable constellation of the northern hemisphere, between ◊Hercules and ◊Boötes, traditionally identified with the jewelled crown of Ariadne that was cast into the sky by Bacchus in Greek mythology. Its brightest star is Alphecca (or Gemma), which is 78 light years from Earth.

It contains several variable stars. R Coronae Borealis is normally fairly constant in brightness but fades at irregular intervals and stays faint for a variable length of time. T Coronae Borealis is normally faint, but very occasionally blazes up and for a few days may be visible to the naked eye. It is a recurrent ◊nova.

coronary artery disease ***Latin** corona **'crown', from the arteries encircling the heart*** condition in which the fatty deposits of ◊atherosclerosis form in the coronary arteries that supply the heart muscle, narrowing them and restricting the blood flow.

These arteries may already be hardened (arteriosclerosis). If the heart's oxygen requirements are increased, as during exercise, the blood supply through the narrowed arteries may be inadequate, and the pain of ◊angina results. A ◊heart attack occurs if the blood supply to an area of the heart is cut off, for example because a blood clot (thrombus) has blocked one of the coronary arteries. The subsequent lack of oxygen damages the heart muscle (infarct), and if a large area of the heart is affected, the attack may be fatal. Coronary artery disease tends to run in families and is linked to smoking, lack of exercise, and a diet high in saturated (mostly animal) fats, which tends to increase the level of blood ◊cholesterol. It is a common cause of death in many industrialized countries; older men are the most vulnerable group. The condition is treated with drugs or bypass surgery.

Coronella genus of snakes inhabiting Europe, Asia, and America. All are harmless.

corpuscular theory hypothesis about the nature of light championed by Isaac Newton, who postulated that it consists of a stream of particles or corpuscles. The theory was superseded at the beginning of the 19th century by English physicist Thomas Young's wave theory. ◊Quantum theory and wave mechanics embody both concepts.

corpus luteum glandular tissue formed in the mammalian ◊ovary after ovulation from the Graafian follicle, a group of cells associated with bringing the egg to maturity. It secretes the hormone progesterone in anticipation of pregnancy.

After the release of an egg the follicle enlarges under the action of luteinizing hormone, released from the pituitary. The corpus luteum secretes the hormone progesterone, which maintains the uterus wall ready for pregnancy. If pregnancy occurs, the corpus luteum continues to secrete progesterone until the fourth month; if pregnancy does not occur the corpus luteum breaks down.

corrasion the grinding away of solid rock surfaces by particles carried by water, ice, and wind. It is generally held to be the most significant form of ◊erosion. As the eroding particles are carried along they become eroded themselves due to the process of attrition.

correlation the degree of relationship between two sets of information. If one set of data increases at the same time as the other, the relationship is said to be positive or direct. If one set of data increases as the other decreases, the relationship is negative or inverse. Correlation can be shown by plotting a best-fit line on a ◊scatter diagram.

In statistics, such relations are measured by the calculation of ◊coefficients of correlation. These generally measure correlation on a scale with 1 indicating perfect positive correlation, 0 no correlation at all, and –1 perfect inverse correlation. Correlation coefficients for assumed linear relations include the Pearson product moment correlation coefficient (known simply as the correlation coefficient), Kendall's tau correlation coefficient, or Spearman's rho correlation coefficient, which is used in nonparametric statistics (where the data are measured on ordinal rather than interval scales). A high correlation does not always indicate dependence between two variables; it may be that there is a third (unstated) variable upon which both depend.

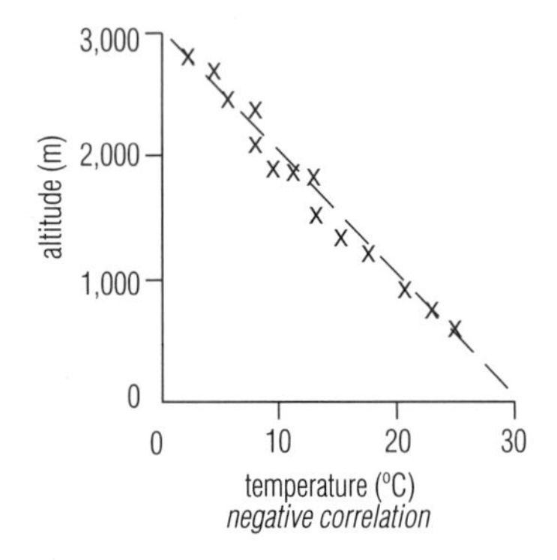

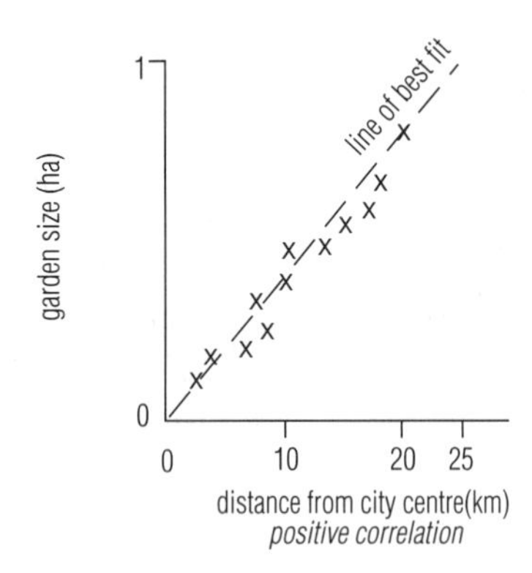

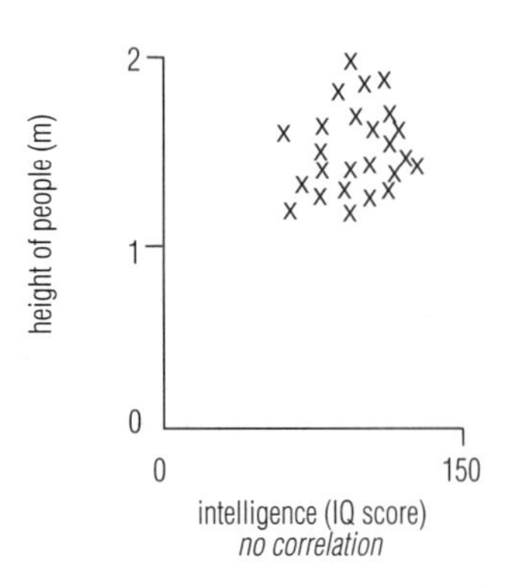

correlation *Scattergraphs showing different kinds of correlation. In this way, a causal relationship between two variables may be proved or disproved, provided there are no hidden factors.*

correspondence in mathematics, the relation between two sets where an operation on the members of one set maps some or all of them onto one or more members of the other. For example, if *A* is the set of members of a family and *B* is the set of months in the year, *A* and *B* are in correspondence if the operation is: '...has a birthday in the month of...'.

corresponding angles a pair of equal angles lying on the same side of a transversal (a line that cuts through two or more lines in the same plane), and making an interior and exterior angle with the intersected lines.

corrie (*Welsh* **cwm**; *French, North American* **cirque**) Scottish term for a steep-sided hollow in the mountainside of a glaciated area. The weight and movement of the ice has ground out the bottom and worn back the sides. A corrie is open at the front, and its sides and back are formed of ◊arêtes. There may be a lake in the bottom, called a tarn.

A corrie is formed as follows: (1) snow accumulates in a hillside hollow (enlarging the hollow by nivation), and turns to ice; (2) the hollow is deepened by abrasion and plucking; (3) the ice in the corrie rotates under the influence of gravity, deepening the hollow still further; (4) since the ice is thinner and moves more slowly at the foot of the hollow, a rock lip forms; (5) when the ice melts, a lake or tarn may be formed in the corrie. The steep back wall may be severely weathered by freeze-thaw, weathering providing material for further abrasion.

corrosion in chemistry, the eating away and eventual destruction of metals and alloys by chemical attack. The rusting of ordinary iron and steel is the most common form of corrosion. Rusting takes place in moist air, when the iron combines with oxygen and water to form a brown-orange deposit of ◊rust (hydrated iron oxide). The rate of corrosion is increased where the atmosphere is polluted with sulphur dioxide. Salty road and air conditions accelerate the rusting of car bodies.

Corrosion is largely an electrochemical process, and acidic and salty conditions favour the establishment of electrolytic cells on the metal, which cause it to be eaten away. Other examples of corrosion include the green deposit that forms on copper and bronze, called verdigris, a basic copper carbonate. The tarnish on silver is a corrosion product, a film of silver sulphide.

corrosion in earth science, an alternative name for ◊solution, the process by which water dissolves rocks such as limestone.

corruption of data introduction or presence of errors in data. Most computers use a range of ◊verification and ◊validation routines to prevent corrupt data from entering the computer system or detect corrupt data that are already present.

cortex in biology, the outer part of a structure such as the brain, kidney, or adrenal gland. In botany the cortex includes nonspecialized cells lying just beneath the surface cells of the root and stem.

corticosteroid any of several steroid hormones secreted by the cortex of the ◊adrenal glands; also synthetic forms with similar properties. Corticosteroids have anti-inflammatory and ◊immunosuppressive effects and may be used to treat a number of conditions, including rheumatoid arthritis, severe allergies, asthma, some skin diseases, and some cancers. Side effects can be serious, and therapy must be withdrawn very gradually.

The two main groups of corticosteroids include **glucocorticoids** (◊cortisone, hydrocortisone, prednisone, and dexamethasone), which are essential to carbohydrate, fat, and protein metabolism, and to the body's response to stress; and **mineralocorticoids** (aldosterone, fluorocortisone), which control the balance of water and salt in the body.

corticotrophin-releasing hormone (CRH) hormone produced by the ◊hypothalamus that stimulates the adrenal glands to produce the steroid cortisol, essential for normal ◊metabolism. CRH is also produced by the ◊placenta and a surge in CRH may trigger the beginning of labour.

cortisol hormone produced by the ◊adrenal glands. It plays a role in helping the body combat stress and is at its highest level in the blood at dawn.

cortisone natural corticosteroid produced by the ◊adrenal gland, now synthesized for its anti-inflammatory qualities and used in the treatment of rheumatoid arthritis.

Cortisone was discovered by Tadeus Reichstein of Basel, Switzerland, and put to practical clinical use for rheumatoid arthritis by Philip Hench (1896–1965) and Edward Kendall (1886–1972) in the USA (all three shared a Nobel prize 1950).

A product of the adrenal gland, it was first synthesized from a constituent of ox bile, and is now produced commercially from a Mexican yam and from a by-product of the sisal plant. It is used for treating allergies and certain cancers, as well as rheumatoid arthritis. The side effects of cortisone steroids include muscle wasting, fat redistribution, diabetes, bone thinning, and high blood pressure.

corundum native aluminium oxide, Al_2O_3, the hardest naturally occurring mineral known apart from diamond (corundum rates 9 on the Mohs' scale of hardness); lack of ◊cleavage also increases its durability. Its crystals are barrel-shaped prisms of the trigonal system. Varieties of gem-quality corundum are **ruby** (red) and **sapphire** (any colour other than red, usually blue). Poorer-quality and synthetic corundum is used in industry, for example as an ◊abrasive.

Corundum forms in silica-poor igneous and metamorphic rocks. It is a constituent of emery, which is metamorphosed bauxite.

cosecant in trigonometry, a ◊function of an angle in a right-angled triangle found by dividing the length of the hypotenuse (the longest side) by the length of the side opposite the angle. Thus the cosecant of an angle *A*, usually shortened to cosec *A*, is always greater than (or equal to) 1. It is the reciprocal of the sine of the angle, that is, cosec $A = 1/\sin A$.

cosine in trigonometry, a ◊function of an angle in a right-angled triangle found by dividing the length of the side adjacent to the angle by the length of the hypotenuse (the longest side). It is usually shortened to **cos**.

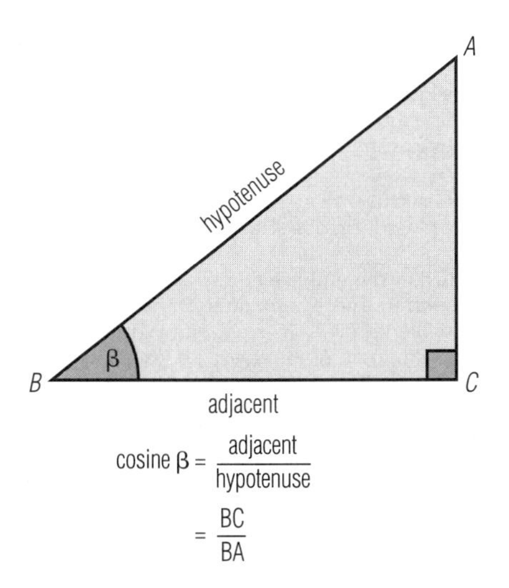

cosine The cosine of angle β is equal to the ratio of the length of the adjacent side to the length of the hypotenuse (the longest side, opposite to the right angle).

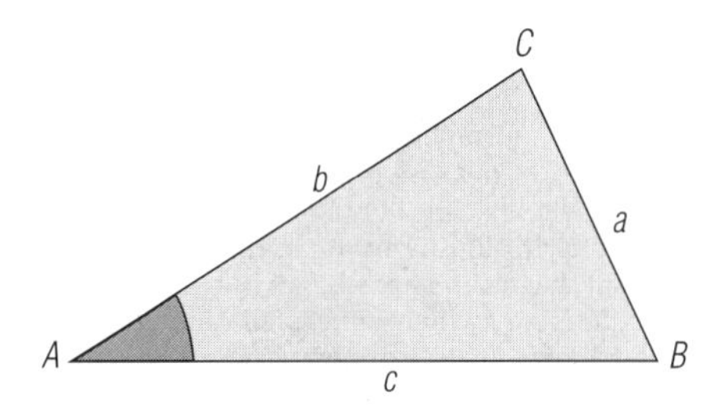

cosine rule The cosine rule is a rule of trigonometry that relates the sides and angles of triangles. It can be used to find a missing length or angle in a triangle.

cosine rule in trigonometry, a rule that relates the sides and angles of triangles. The rule has the formula:

$$a^2 = b^2 + c^2 - 2bc \cos A$$

where *a*, *b*, and *c* are the sides of the triangle, and *A* is the angle opposite *a*.

cosmic background radiation or **3° radiation** electromagnetic radiation left over from the original formation of the universe in the Big Bang around 15 billion years ago. It corresponds to an overall background temperature of 3K (–270°C/–454°F), or 3°C above absolute zero. In 1992 the Cosmic Background Explorer satellite, COBE, detected slight 'ripples' in the strength of the background radiation that are believed to mark the first stage in the formation of galaxies.

Cosmic background radiation was first detected in 1965 by US physicists Arno Penzias (1933–) and Robert Wilson (1936–), who in 1978 shared the Nobel Prize for Physics for their discovery.

cosmic radiation streams of high-energy particles from outer space, consisting of protons, alpha particles, and light nuclei, which collide with atomic nuclei in the Earth's atmosphere, and produce secondary nuclear particles (chiefly ◊mesons, such as pions and muons) that shower the Earth.

Those of low energy seem to be galactic in origin, and those of high energy of extragalactic origin. The galactic particles may come from ◊supernova explosions or ◊pulsars. At higher energies, other sources are necessary, possibly the giants jets of gas which are emitted from some galaxies.

cosmid fragment of ◊DNA from the human genome inserted into a bacterial cell. The bacterium replicates the fragment along with its own DNA. In this way the fragments are copied for a gene library. Cosmids are characteristically 40,000 base pairs in length. The most commonly used bacterium is *Escherichia coli*. A ◊yeast artificial chromosome works in the same way.

cosmogony ***Greek 'universe' and 'creation'*** study of the origin and evolution of cosmic objects, especially the Solar System.

cosmological principle in astronomy, a hypothesis that any observer anywhere in the ◊universe has the same view that we

BRIEF HISTORY OF COSMOLOGY

http://www-history.mcs.st-and.ac.uk/~history/HistTopics/Cosmology.html

Based at St Andrews University, Scotland, a site chronicling the history of cosmology from the time of the Babylonians to the Hubble Space Telescope. It includes links to the biographies of the key historical figures responsible for the advancement of the subject, and in addition has a brief list of references for further reading.

have; that is, that the universe is not expanding from any centre but all galaxies are moving away from one another.

cosmology branch of astronomy that deals with the structure and evolution of the universe as an ordered whole. Its method is to construct 'model universes' mathematically and compare their large-scale properties with those of the observed universe.

Modern cosmology began in the 1920s with the discovery that the universe is expanding, which suggested that it began in an explosion, the ◊Big Bang. An alternative – now discarded – view, the ◊steady-state theory, claimed that the universe has no origin, but is expanding because new matter is being continually created.

cosmonaut term used in the West for any astronaut from the former Soviet Union.

Cosmos name used from the early 1960s for nearly all Soviet artificial satellites. Over 2,300 Cosmos satellites had been launched by mid 1995.

Our loyalties are to the species and the planet. We speak for Earth. Our obligation to survive is owed not just to ourselves but also to that Cosmos, ancient and vast, from which we spring.

CARL SAGAN US astronomer.
Cosmos 1980

CoSy (contraction of ***conferencing system***) ◊command-line interface electronic conferencing software developed at the University of Guelph in the Canadian province of Ontario. It is used on London's ◊Compulink Information eXchange (CIX) service and for the Open University's conferencing, as well as many others worldwide.

cotangent in trigonometry, a ◊function of an angle in a right-angled triangle found by dividing the length of the side adjacent to the angle by the length of the side opposite it. It is usually written as cotan, or cot and it is the reciprocal of the tangent of the angle, so that cot A = 1/tan A, where A is the angle in question.

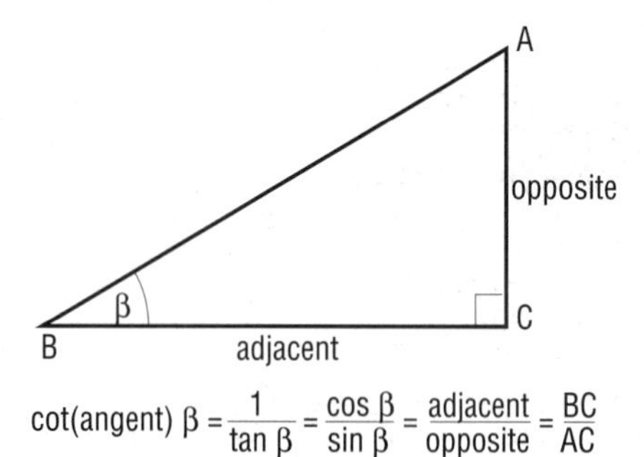

cotangent The cotangent of angle β is equal to the ratio of the length of the adjacent side to the length of the opposite side.

cot death or ***sudden infant death syndrome*** (SIDS) death of an apparently healthy baby, almost always during sleep. It is most common in the winter months, and strikes more boys than girls. The cause is not known but risk factors that have been identified include prematurity, respiratory infection, overheating, and sleeping position.

There was a 60% reduction in the number of cot deaths in the UK in the first nine months of 1993, following a massive advertising campaign advising parents to put their babies to sleep on their backs, ensure they do not overheat, and avoid smoking near them. Earlier, in New Zealand (1991), where the cot death rate had been the highest in the world, the rate was halved by a similar campaign.

cotoneaster any of a group of shrubs or trees found in Europe and Asia, belonging to the rose family and closely related to the hawthorn and medlar. The fruits, though small and unpalatable, are usually bright red and conspicuous, often surviving through the winter. Some of the shrubs are cultivated for their attractive appearance. (Genus *Cotoneaster*, family Rosaceae.)

cotton tropical and subtropical ◊herbaceous plant belonging to the mallow family. Fibres surround the seeds inside the ripened fruits, or bolls, and these are spun into yarn for cloth. (Genus *Gossypium*, family Malvaceae.)

The fibres are separated from the seeds in a machine called a ◊cotton gin. The seeds are used to produce cooking oil and livestock feed, and the pigment gossypol may be useful as a male contraceptive in a modified form. Cotton disease (byssinosis), caused by cotton dust, affects the lungs of those working in the industry.

Cotton production uses 50% of world pesticides and represents 5% of world agricultural output.

cotton gin machine that separates cotton fibres from the seed boll. Production of the gin (then called an engine) by US inventor Eli Whitney in 1793 was a milestone in textile history.

The modern gin consists of a roller carrying a set of circular saws. These project through a metal grill in a hopper containing the seed bolls. As the roller rotates, the saws pick up the cotton fibres, leaving the seeds behind.

cotton grass or ***bog cotton*** any grasslike plant of a group belonging to the sedge family. White tufts cover the fruiting heads in midsummer; these break off and are carried long distances on the wind. Cotton grass is found in wet places throughout the Arctic and temperate regions of the northern hemisphere, most species growing in acid bogs. (Genus *Eriophorum*, family Cyperaceae.)

cotton spinning creating thread or fine yarn from the cotton plant by spinning the raw fibre contained within the seed-pods. The fibre is separated from the pods by a machine called a ◊cotton gin. It is then cleaned and the fibres are separated out (carding). Finally the fibres are drawn out to the desired length and twisted together to form strong thread.

cotton stainer any plant-feeding ◊bug of the family Pyrrhocoridae that pierces and stains cotton bolls; see also ◊Hemiptera.

cottonwood any of several North American ◊poplar trees with seeds topped by a thick tuft of silky hairs. The eastern cottonwood (*Populus deltoides*), growing to 30 m/100 ft, is native to the eastern USA. The name 'cottonwood' is also given to the downy-leaved Australian tree *Bedfordia salaoina*. (True cottonwood genus *Populus*, family Salicaceae.)

cotton-worm caterpillar of the owlet moth, closely allied to the army worm. Cotton-worms are found in both North and South America, where they ravage the cotton crops whilst ignoring other plants.

classification The cotton-worm *Aletia xylinae* is in the family Noctuidae, order Lepidoptera, class Insecta, phylum Arthropoda.

cotyledon structure in the embryo of a seed plant that may form a 'leaf' after germination and is commonly known as a seed leaf. The number of cotyledons present in an embryo is an important character in the classification of flowering plants (◊angiosperms).

Monocotyledons (such as grasses, palms, and lilies) have a single cotyledon, whereas dicotyledons (the majority of plant species) have two. In seeds that also contain ◊endosperm (nutritive tissue), the cotyledons are thin, but where they are the primary food-storing tissue, as in peas and beans, they may be quite large. After germination the cotyledons either remain below ground (hypogeal) or, more commonly, spread out above soil level (epigeal) and become the first green leaves. In gymnosperms there may be up to a dozen cotyledons within each seed.

couch grass European grass that spreads rapidly by underground stems. It is considered a troublesome weed in North America, where it has been introduced. (*Agropyron repens*, family Gramineae.)

cougar another name for the ◊puma, a large North American cat.

coulomb SI unit (symbol C) of electrical charge. One coulomb is the quantity of electricity conveyed by a current of one ◊ampere in one second.

Coulomb, Charles Augustin de (1736–1806) French scientist, inventor of the ◊torsion balance for measuring the force of electric and magnetic attraction. The coulomb was named after him. In the fields of structural engineering and friction, Coulomb greatly influenced and helped to develop engineering in the 19th century.

Coulomb's law of 1787 states that the force between two electric charges is proportional to the product of the charges and inversely proportional to the square of the distance between them.

count rate in physics, the number of particles emitted per unit time by a radioactive source. It is measured by a counter, such as a ◊Geiger counter, or ◊ratemeter.

couple in mechanics, a pair of forces acting on an object that are equal in magnitude and opposite in direction, but do not act along the same straight line. The two forces produce a turning effect or moment that tends to rotate the object; however, no single resultant (unbalanced) force is produced and so the object is not moved from one position to another.

The moment of a couple is the product of the magnitude of either of the two forces and the perpendicular distance between those forces. If the magnitude of the force is F newtons and the distance is d metres then the moment, in newton-metres, is given by:

$$\text{moment} = Fd$$

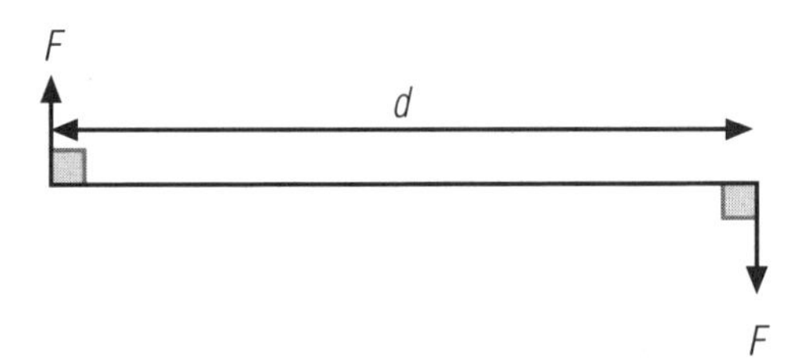

couple Two equal but opposite forces (F) will produce a turning effect on a rigid body, provided that they do not act through the same straight line. The turning effect, or moment, is equal to the magnitude of one of the turning forces multiplied by the perpendicular distance (d) between those two forces.

courgette small variety of ◊marrow, belonging to the gourd family. It is cultivated as a vegetable and harvested before it is fully mature, at 15–20 cm/6–8 in. In the USA and Canada it is known as a zucchini. (*Cucurbita pepo*, family Cucurbitaceae.)

courtship behaviour exhibited by animals as a prelude to mating. The behaviour patterns vary considerably from one species to another, but are often ritualized forms of behaviour not obviously related to courtship or mating (for example, courtship feeding in birds).

Courtship ensures that copulation occurs with a member of the opposite sex of the right species. It also synchronizes the partners' readiness to mate and allows each partner to assess the suitability of the other.

Cousteau, Jacques Yves (1910–1997) French oceanographer who pioneered the invention of the aqualung 1943 and techniques in underwater filming. In 1951 he began the first of many research voyages in the ship *Calypso*. His film and television documentaries and books established him as a household name.

covalent bond chemical ◊bond produced when two atoms share one or more pairs of electrons (usually each atom contributes an electron). The bond is often represented by a single line drawn between the two atoms. Covalently bonded substances include hydrogen (H_2), water (H_2O), and most organic substances.

cowfish type of ◊boxfish.

cow parsley or ***keck*** tall perennial plant belonging to the carrot family, found in Europe, N Asia, and N Africa. It grows up to 1 m/3 ft tall and has pinnate leaves (leaflets growing either side of a stem), hollow furrowed stems, and heads of delicate white flowers. (*Anthriscus sylvestris*, family Umbelliferae.)

cowrie marine snail of the family Cypreidae, in which the interior spiral form is concealed by a double outer lip. The shells are hard, shiny, and often coloured. Most cowries are shallow-water forms, and are found in many parts of the world, particularly the tropical Indo-Pacific. Cowries have been used as ornaments and fertility charms, and also as currency, for example the Pacific money cowrie *Cypraea moneta*.

cow shark species of shark. Cow sharks have six or seven gill openings, are about 5 m/16.5 ft long, and are found in most parts of the oceans.

classification Cow sharks are in family Hexanchidae, order Hexanchiformes, subclass Elasmobranchii, class Chondrichthyes.

cowslip European plant related to the primrose, with several small deep-yellow fragrant flowers growing from a single stem. It is native to temperate regions of the Old World. The oxlip (*Primula elatior*) is also closely related. (*Primula veris*, family Primulaceae.)

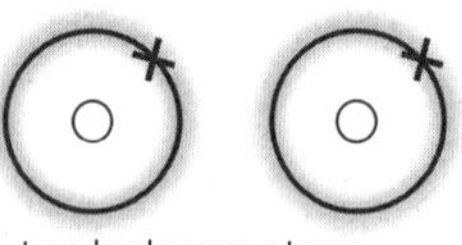

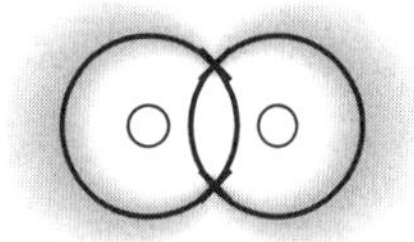

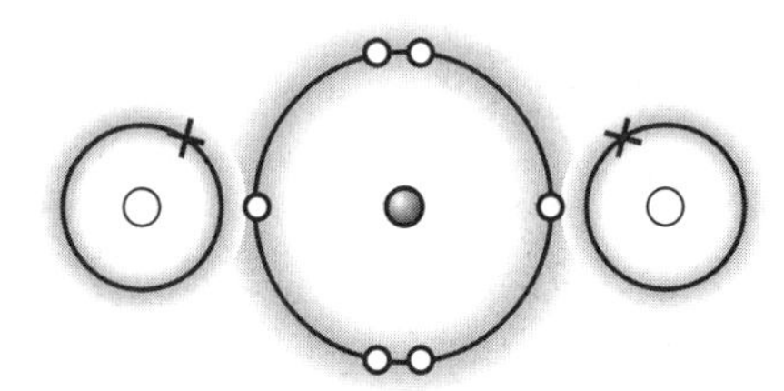

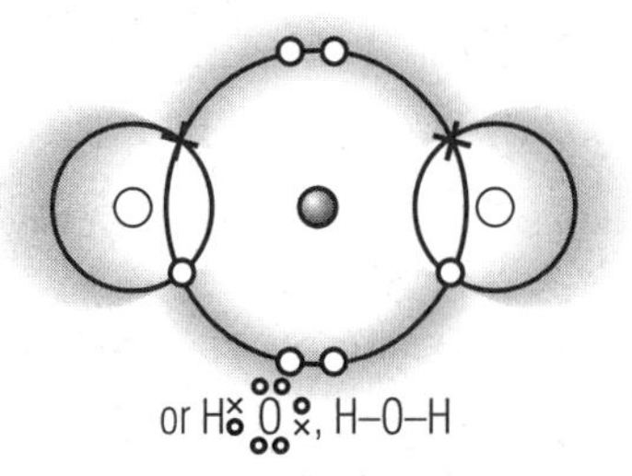

covalent bond *The formation of a covalent bond between two hydrogen atoms to form a hydrogen molecule (H2), and between two hydrogen atoms and an oxygen atom to form a molecule of water (H2O). The sharing means that each atom has a more stable arrangement of electrons (its outer electron shells are full).*

cowrie Although most species of cowrie are tropical, there are several European species. This European cowrie Trivia monacha *is one of two species which can be found along the coast of the British Isles; they are likely to be found only at low water during periods of spring tides. Premaphotos Wildlife*

coyote wild dog *Canis latrans*, in appearance like a small wolf, living in North and Central America. Its head and body are about 90 cm/3 ft long and brown, flecked with grey or black. Coyotes live in open country and can run at 65 kph/40 mph. Their main foods are rabbits and rodents. Although persecuted by humans for over a century, the species is very successful.

coypu South American water rodent *Myocastor coypus*, about 60 cm/2 ft long and weighing up to 9 kg/20 lb. It has a scaly, ratlike tail, webbed hind feet, a blunt-muzzled head, and large orange incisors. The fur ('nutria') is reddish brown. It feeds on vegetation, and lives in burrows in rivers and lake banks.

Taken to Europe and then to North America to be farmed for their fur, many escaped or were released. In Britain, coypus escaped from fur farms and became established on the Norfolk Broads where their adult numbers reached 5,000. They destroyed crops and local vegetation, and undermined banks and dykes. After a 10-year campaign they were eradicated in 1989 at a cost of over £2 million. In 1993 escaped coypu in Louisiana, USA, were causing serious damage to coastal marshland.

CP/M (abbreviation for ***control program/monitor*** or ***control program for microcomputers***) one of the earliest ◊operating systems for microcomputers. It was written by Gary Kildall, who founded Digital Research. In the 1970s it became a standard for microcomputers based on the Intel 8080 and Zilog Z80 8-bit microprocessors. In the 1980s it was superseded by Microsoft's ◊MS-DOS, written for Intel's 16-bit 8086/88 microprocessors.

CPSR abbreviation for ◊Computer Professionals for Social Responsibility.

CPU in computing, abbreviation for ◊central processing unit.

crab any decapod (ten-legged) crustacean of the division Brachyura, with a broad, rather round, upper body shell (carapace) and a small ◊abdomen tucked beneath the body. Crabs are related to lobsters and crayfish. Mainly marine, some crabs live in fresh water or on land. They are alert carnivores and scavengers. They have a typical sideways walk, and strong pincers on the first pair of legs, the other four pairs being used for walking. Periodically, the outer shell is cast to allow for growth. The name 'crab' is sometimes used for similar arthropods, such as the horseshoe crab, which is neither a true crab nor a crustacean.

CRAB

Female crabs can only mate just after moulting, when their shells are still soft. This is a vulnerable time, particularly for female paddle crabs, as some large male paddle crabs are cannibalistic. Most males, however, are keener to mate than to eat. Having found a female that is about to moult, the male paddle crab carries her around until she is ready, then mates with her and protects her until her shell hardens again.

crab apple any of 25 species of wild apple trees, native to temperate regions of the northern hemisphere. Numerous varieties of cultivated apples have been derived from *Malus pumila*, the common native crab apple of SE Europe and central Asia. The fruit of native species is smaller and more bitter than that of cultivated varieties and is used in crab-apple jelly. (Genus *Malus*, family Rosaceae.)

crab louse human pubic ◊louse.
classification The crab louse *Phtirus pubis* is in suborder Anoplura, order Phthiraptera, class Insecta, phylum Arthropoda.

Crab nebula cloud of gas 6,000 light years from Earth, in the constellation ◊Taurus. It is the remains of a star that according to Chinese records, exploded as a ◊supernova observed as a brilliant point of light on 4 July 1054. At its centre is a ◊pulsar that flashes 30 times a second. It was named by Lord Rosse after its crablike shape.

crack street name for a chemical derivative (bicarbonate) of ◊cocaine in hard, crystalline lumps; it is heated and inhaled (smoked) as a stimulant. Crack was first used in San Francisco in the early 1980s, and is highly addictive.

Its use has led to numerous deaths, but it is the fastest-growing sector of the illegal drug trade, since it is less expensive than cocaine.

cracker a hacker (see ◊hacking); the term distinguishes criminal hacking ('cracking') from those who explore to satisfy their intellectual curiosity. The term is used much less than most hackers would like.

crab Crabs have adapted to many habitats. This Sesarma *species crab from Kenya, E Africa, is adapted to the muddy tidal regions dominated by mangroves. Premaphotos Wildlife*

cracking reaction in which a large ◊alkane molecule is broken down by heat into a smaller alkane and a small ◊alkene molecule. The reaction is carried out at a high temperature (600°C or higher) and often in the presence of a catalyst. Cracking is a commonly used process in the ◊petrochemical industry.

It is the main method of preparation of alkenes and is also used to manufacture petrol from the higher-boiling-point ◊fractions that are obtained by fractional ◊distillation (fractionation) of crude oil.

crag in previously glaciated areas, a large lump of rock that a glacier has been unable to wear away. As the glacier passed up and over the crag, weaker rock on the far side was largely protected from erosion and formed a tapering ridge, or **tail**, of debris.

An example of a crag-and-tail feature is found in Edinburgh in Scotland; Edinburgh Castle was built on the crag (Castle Rock), which dominates the city beneath.

crake any of several small birds of the family Rallidae, order Gruiformes, related to the ◊corncrake.

cranberry any of several trailing evergreen plants belonging to the heath family, related to bilberries and blueberries. They grow in marshy places and bear small, acid, crimson berries, high in vitamin C content, used for making sauce and jelly. (Genus *Vaccinium*, family Ericaceae.)

crane in engineering, a machine for raising, lowering, or placing in position heavy loads. The three main types are the jib crane, the overhead travelling crane, and the tower crane. Most cranes have the machinery mounted on a revolving turntable. This may be mounted on trucks or be self-propelled, often being fitted with ◊caterpillar tracks.

The main features of a **jib crane** are a power winch, a rope or cable, and a movable arm or jib. The cable, which carries a pulley block, hangs from the end of the jib and is wound up and down by the winch. The **overhead travelling crane**, chiefly used in workshops, consists of a fixed horizontal arm, along which runs a trolley carrying the pulley block. **Tower cranes**, seen on large building sites, have a long horizontal arm able to revolve on top of a tall tower. The arm carries the trolley.

Cranes can also be mounted on barges or large semisubmersibles for marine construction work.

crane in zoology, a large, wading bird of the family Gruidae, order Gruiformes, with long legs and neck, short powerful wings, a naked or tufted head, and unwebbed feet. The hind toe is greatly elevated, and has a sharp claw. Cranes are marsh- and plains-dwelling birds, feeding on plants as well as insects and small animals. They fly well and are usually migratory. Their courtship includes frenzied, leaping dances. They are found in all parts of the world except South America.

The **common crane** *Grus grus* is still numerous in many parts of Europe, and winters in Africa and India. It stands over 1 m/3ft high. The plumage of the adult bird is grey, varied with black and white, with a red patch of bare skin on the head and neck. All cranes have suffered through hunting and loss of wetlands; the population of the North American **whooping crane** *G. americana* fell to 21 wild birds in 1944. Through careful conservation, numbers have now risen to about 200.

crane fly or ***daddy-longlegs*** any fly of the family Tipulidae, with long, slender, fragile legs. They look like giant mosquitoes, but the adults are quite harmless. The larvae live in soil or water.

cranesbill any of a group of plants containing about 400 species. The plants are named after the long beaklike protrusion attached to the seed vessels. When ripe, this splits into coiling spirals which jerk the seeds out, helping to scatter them. (Genus *Geranium*, family Geraniaceae.)

cranium the dome-shaped area of the vertebrate skull that protects the brain. It consists of eight bony plates fused together by sutures (immovable joints). Fossil remains of the human cranium have aided the development of theories concerning human evolution.

The cranium has been studied as a possible indicator of intelligence or even of personality. The Victorian argument that a large cranium implies a large brain, which in turn implies a more profound intelligence, has been rejected.

crank handle bent at right angles and connected to the shaft of a machine; it is used to transmit motion or convert reciprocating (backwards-and-forwards or up-and-down) movement into rotary movement, or vice versa.

Although similar devices may have been employed in antiquity and as early as the 1st century in China and the 8th century in Europe, the earliest recorded use of a crank in a water-raising machine is by Arab mathematician al-Jazari in the 12th century. Not until the 15th century, however, did the crank become fully assimilated into developing European technology.

crankshaft essential component of piston engines that converts the up-and-down (reciprocating) motion of the pistons into useful rotary motion. The car crankshaft carries a number of cranks. The pistons are connected to the cranks by connecting rods and ◊bearings; when the pistons move up and down, the connecting rods force the offset crank pins to describe a circle, thereby rotating the crankshaft.

crater bowl-shaped depression in the ground, usually round and with steep sides. Craters are formed by explosive events such as the eruption of a volcano, the explosion of bomb, or the impact of a meteorite.

The Moon has more than 300,000 craters over 1 km/6 mi in diameter, formed by meteorite bombardment; similar craters on Earth have mostly been worn away by erosion. Craters are found on many other bodies in the Solar System.

Studies at the Jet Propulsion Laboratory in California, USA, have shown that craters produced by impact or by volcanic activity have distinctive shapes, enabling astronomers to distinguish likely methods of crater formation on planets in the Solar System. Unlike volcanic craters, impact craters have a raised rim and central peak and are almost always circular, irrespective of the meteorite's angle of incidence.

We used to think that if we knew one, we knew two, because one and one are two. We are finding that we must learn a great deal more about 'and'.

ARTHUR STANLEY EDDINGTON British astrophysicist.
Attributed remark

---

craton or ***shield*** core of a continent, a vast tract of highly deformed ◊metamorphic rock around which the continent has been built. Intense mountain-building periods shook these shield areas in Precambrian times before stable conditions set in.

Cratons exist in the hearts of all the continents, a typical example being the Canadian Shield.

crawler on the World Wide Web, automated indexing software that scours the Web for new or updated sites. See also ◊'bot, ◊spider, and ◊agent.

crayfish freshwater decapod (ten-limbed) crustacean belonging to several families structurally similar to, but smaller than, the lobster. Crayfish are brownish-green scavengers and are found in all parts of the world except Africa. They are edible, and some species are farmed.

Accurate reckoning – the entrance into the knowledge of all existing things and all obscure secrets.

AHMES THE SCRIBE Ancient Egyptian Scribe.
Rhind Papyrus

---

creationism theory concerned with the origins of matter and life, claiming, as does the Bible in Genesis, that the world and humanity were created by a supernatural Creator, not more than 6,000 years ago. It was developed in response to Darwin's theory of

crayfish The spectacular blue mountain crayfish Euastacus ontanus *inhabits clear, swiftly-flowing mountain streams in the subtropical forests of S Queensland, Australia. Premaphotos Wildlife*

◊evolution; it is not recognized by most scientists as having a factual basis.

After a trial 1981–82 a US judge ruled unconstitutional an attempt in Arkansas schools to enforce equal treatment of creationism and evolutionary theory. In Alabama from 1996 all biology textbooks must contain a statement that evolution is a controversial theory and not a proven fact.

Creative Labs name of the US and British subsidiaries of the parent computing company Creative Technology, which was founded in Singapore in 1981. Creative Labs manufactures the leading ◊sound card, the SoundBlaster, which it markets alongside the Internet telephony, video, and multimedia products that make up the company's product range.

By the company claimed 20 million users of its products and had 4,400 staff worldwide, 400 of them in research and development.

creep in civil and mechanical engineering, the property of a solid, typically a metal, under continuous stress that causes it to deform below its ◊yield point (the point at which any elastic solid normally stretches without any increase in load or stress). Lead, tin, and zinc, for example, exhibit creep at ordinary temperatures, as seen in the movement of the lead sheeting on the roofs of old buildings.

Copper, iron, nickel, and their alloys also show creep at high temperatures.

creeper any small, short-legged passerine bird of the family Certhidae. They spiral with a mouselike movement up tree trunks, searching for insects and larvae with their thin, down-curved beaks.

The brown creeper *Certhia familiaris* is 12 cm/5 in long, brown above, white below, and is found across North America and Eurasia.

creosote black, oily liquid derived from coal tar, used as a wood preservative. Medicinal creosote, which is transparent and oily, is derived from wood tar.

crescent curved shape of the Moon when it appears less than half illuminated. It also refers to any object or symbol resembling the crescent Moon. Often associated with Islam, it was first used by the Turks on their standards after the capture of Constantinople in 1453, and appears on the flags of many Muslim countries. The **Red Crescent** is the Muslim equivalent of the Red Cross.

cress any of several plants of the cress family, characterized by a pungent taste. The common European garden cress (*Lepidium sativum*) is cultivated worldwide. (Genera include *Lepidium*, *Cardamine*, and *Arabis*; family Cruciferae.)

Cretaceous ***Latin creta 'chalk'*** period of geological time approximately 144.2–65 million years ago. It is the last period of the Mesozoic era, during which angiosperm (seed-bearing) plants evolved, and dinosaurs reached a peak before their extinction at the end of the period. The north European chalk, which forms the white cliffs of Dover, was deposited during the latter half of the Cretaceous.

crevasse deep crack in the surface of a glacier; it can reach several metres in depth. Crevasses often occur where a glacier flows over the break of a slope, because the upper layers of ice are unable to stretch and cracks result. Crevasses may also form at the edges of glaciers owing to friction with the bedrock.

Crick, Francis Harry Compton
(1916–)

English molecular biologist. From 1949 he researched the molecular structure of DNA, and the means whereby characteristics are transmitted from one generation to another. For this work he was awarded a Nobel prize (with Maurice Wilkins and James Watson) in 1962.

Using Wilkins's and others' discoveries, Crick and Watson postulated that DNA consists of a double helix consisting of two parallel chains of alternate sugar and phosphate groups linked by pairs of organic bases. They built molecular models which also explained how genetic information could be coded – in the sequence of organic bases. Crick and Watson published their work on the proposed structure of DNA in 1953. Their model is now generally accepted as correct.

Mary Evans Picture Library

cricket in zoology, an insect belonging to any of various families, especially the Gryllidae, of the order Orthoptera. Crickets are related to grasshoppers. They have somewhat flattened bodies and long antennae. The males make a chirping noise by rubbing together special areas on the forewings. The females have a long needlelike

cricket Most crickets are drably coloured in brown, grey, or black – this Kenyan species Rhicnogryllus lepidus *is an exception. Its bright livery is probably an example of aposematic (warning) coloration. Premaphotos Wildlife* egglaying organ (ovipositor). There are around 900 species known worldwide.

Crinoidea class of echinoderms containing about 600 living species and more than 2,000 fossil forms; the extinct crinoids are commonly called stone lilies, and the existing species ◊sea lilies and ◊feather stars.

Crinoids retain many primitive features, notably in the use of the hydraulic water-vascular system for powering the capture and transportation of food rather than for locomotion. Uniquely amongst living echinoderms, the mouth is on the upper surface. Food particles are collected by the waving arms, and directed to the mouth along food grooves lined with cilia. Most of the living species are non-sessile (free-living), although a stalked stage may occur in development.

classification Crinoidea is in phylum Echinodermata.

crith unit of mass used for weighing gases. One crith is the mass of one litre of hydrogen gas (H_2) at standard temperature and pressure.

critical angle in optics, for a ray of light passing from a denser to a less dense medium (such as from glass to air), the smallest angle of incidence at which the emergent ray grazes the surface of the denser medium – at an angle of refraction of 90°.

When the angle of incidence is less than the critical angle, the ray passes out (is refracted) into the less dense medium; when the angle of incidence is greater than the critical angle, the ray is reflected back into the denser medium (see ◊total internal reflection).

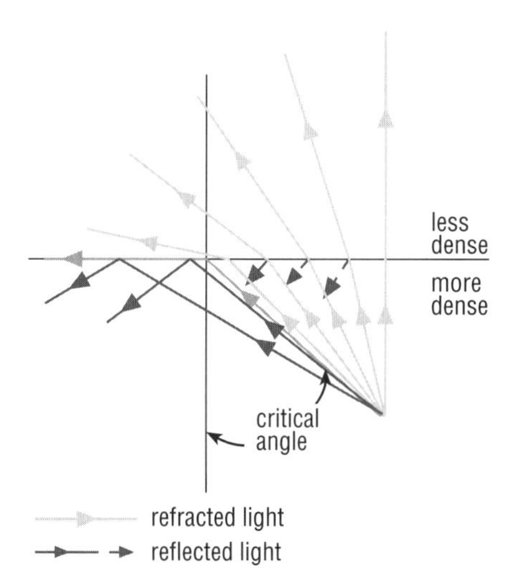

critical angle *The critical angle is the angle at which light from within a transparent medium just grazes the surface of the medium. In the diagram, the red beam is at the critical angle. Blue beams escape from the medium, at least partially. Green beams are totally reflected from the surface.*

critical density in cosmology, the minimum average density that the universe must have for it to stop expanding at some point in the future.

The precise value depends on ◊Hubble's constant and so is not fixed, but it is approximately between 10^{-29} and 2×10^{-29}g/cm^3, equivalent to a few hydrogen atoms per cubic metre. The density parameter (symbol Ω) is the ratio of the actual density to the critical density. If $\Omega < 1$, the universe is open and will expand forever (see heat death). If $\Omega > 1$, the universe is closed and the expansion will eventually halt, to be followed by a contraction (see Big Crunch). Current estimates from visible matter in the universe indicate that Ω is about 0.01, well below critical density, but unseen dark matter may be sufficient to raise Ω to somewhere between 0.1 and 2.

critical mass in nuclear physics, the minimum mass of fissile material that can undergo a continuous ◊chain reaction. Below this mass, too many ◊neutrons escape from the surface for a chain reaction to carry on; above the critical mass, the reaction may accelerate into a nuclear explosion.

critical path analysis procedure used in the management of complex projects to minimize the amount of time taken. The analysis shows which subprojects can run in parallel with each other, and which have to be completed before other subprojects can follow on.

By identifying the time required for each separate subproject and the relationship between the subprojects, it is possible to produce a planning schedule showing when each subproject should be started and finished in order to complete the whole project most efficiently. Complex projects may involve hundreds of subprojects, and computer ◊applications packages for critical path analysis are widely used to help reduce the time and effort involved in their analysis.

critical reaction in a nuclear reactor, a self-sustaining chain reaction in which the number of neutrons being released by the fission of uranium-235 nuclei and the number of neutrons being absorbed by uranium-238 nuclei and by control rods are balanced. If balance is not achieved the reaction will either slow down and cease to generate enough power, or will build up and go out of control, as in a nuclear explosion. Control rods are used to adjust the rate of reaction and maintain balance.

critical temperature temperature above which a particular gas cannot be converted into a liquid by pressure alone. It is also the temperature at which a magnetic material loses its magnetism (the Curie temperature or point).

CRO abbreviation for ◊cathode-ray oscilloscope.

crocodile large scaly-skinned ◊reptile with a long, low cigar-shaped body and short legs. Crocodiles can grow up to 7 m/23 ft in length, and have long, powerful tails that propel them when swimming. They are found near swamps, lakes, and rivers in Asia, Africa, Australia, and Central America, where they are often seen floating in the water like logs, with only their nostrils, eyes, and ears above the surface. They are fierce hunters and active mainly at night. Young crocodiles eat worms and insects, but as they mature they add frogs and small fish to their diet. Adult crocodiles will attack animals the size of antelopes and even, occasionally, people. They can live up to 100 years and are related to the ◊alligator and the smaller cayman.

behaviour In some species, the female lays over 100 hard-shelled eggs in holes or nest mounds of vegetation, which she guards until the eggs hatch, before carrying the hatchlings down to the water in her mouth. When in the sun, crocodiles cool themselves by opening their mouths wide, which also enables scavenging birds to pick their teeth. They can stay underwater for long periods, but must surface to breathe. The nostrils can be closed underwater. They ballast themselves with stones to adjust their buoyancy. Crocodiles have remained virtually unchanged for 200 million years.

types of crocodile There are 15 species of crocodile, all of them endangered, found in tropical parts of Africa, Asia, Australia, and Central America. The largest is the saltwater (indopacific) crocodile *Crocodylus porosus*, which can grow to 7 m/23 ft or more, and is found in E India, Australia, and the W Pacific, in both freshwater and saltwater habitats. The Nile crocodile *C. niloticus* is found in Africa and reaches 6 m/20 ft. The American crocodile *C. acutus*, about 4.6 m/15 ft long, is found from S Florida to Ecuador. The

CROCODILE

Over short distances on land, the crocodile can move at 48 kmph/30 mph, lifting itself up on its legs like a lizard. In the water, it can move at 32 kmph/20 mph.

◊gavial, or gharial, *Gavialis gangeticus* is sometimes placed in a family of its own. It is an Indian species that grows to 6.5 m/21 ft or more, and has a very long narrow snout specialized for capturing and eating fish. The Cuban crocodile *C. rhombifer* has a short snout, grows up to 3.5 m/11.5 ft, and lives in freshwater swamps in Cuba. Morelet's crocodile *C. moreletti* is found in Central America, where it is overhunted, and grows up to 3.5 m/11.5 ft. Johnston's crocodile *C. johnsoni* is an Australian crocodile that feeds mainly on fish and reaches up to 3 m/9.75 ft in length. The Siamese crocodile *C. siamensis* is probably found only in captivity and grows up to 4 m/13 ft in length. The Philippine crocodile *C. mindorensis* is found in the Philippine Islands and grows to just under 3 m/9.75 ft. The mugger *C. palustris* is an Indian crocodile resembling the Nile crocodile but smaller, reaching up to 4 m/13 ft. The Orinoco crocodile *C. intermedius* grows up to 6 m/19.5 ft. False gharial *Tomistoma schlegelli* is found in rivers in India and Indochina and grows up to 4 m/13 ft. African slender-snouted crocodile *C. cataphractus* grows up to 4 m/13 ft and is found in western and central Africa. Dwarf crocodile *Osteolaemus tetraspis* reaches only 2 m/6.6 ft in length and is found in the tropical forests of west and central Africa. New Guinea crocodile *C. novaguineae* reaches 7 m/23 ft in length.

differences between crocodiles and alligators Crocodiles differ from alligators in that they have a narrower, more pointed snout and their fourth tooth on the lower jaw can always be seen, even when their mouth is shut. On average, they are larger.

classification Crocodiles are in the phylum Chordata, subphylum Vertebrata, class Reptilia, subclass Archosauria, order Crocodilia. There are 13 species in the Crocodylidae family, including the Nile crocodile (*Crocodylus niloticus*) and the saltwater crocodile (*C. porosus*).

The crocodile cannot turn its head. Like science, it must always go forward with all-devouring jaws.

PETER KAPITZA Russian physicist.
In A S Eve *Rutherford* 1933

Crocodylia order of the class Reptilia that includes crocodiles, alligators, gavials, and caimans.

There are about 20 living species in three families. The **Gavialidae family** includes only the ◊gavial, a fish-eating reptile found in India. The **Alligatoridae family** includes two living species of ◊alligators and several ◊caimans.The **Crocodylidae family** contains the true ◊crocodiles, comprising rather less than a dozen species ranging over Africa, Asia, northern Australia, and tropical America. The Indian crocodile, known as the mugger and erroneously as the alligator, ranges over India, Sri Lanka, Myanmar, and Malaysia. It is a freshwater variety inhabiting only rivers, lakes, and marshes, and in its characteristics most nearly approaches the caiman and the alligator.

crocus any of a group of plants belonging to the iris family, with single yellow, purple, or white flowers and narrow, pointed leaves. They are native to northern parts of the Old World, especially S Europe and Asia Minor. (Genus *Crocus*, family Iridaceae.)

During the dry season of the year crocuses remain underground in the form of a corm (underground storage organ), and produce fresh shoots and flowers in spring or autumn. At the end of the season of growth new corms are produced. Several species are cultivated as garden plants, the familiar mauve, white, and orange forms being varieties of *Crocus vernus, C. versicolor*, and *C. aureus*. The saffron crocus *C. sativus* belongs to the same genus. The so-called ◊autumn crocus or meadow saffron (*Colchicum autumnale*) is not a true crocus but belongs to the lily family.

Crohn's disease or ***regional ileitis*** chronic inflammatory bowel disease. It tends to flare up for a few days at a time, causing diarrhoea, abdominal cramps, loss of appetite, weight loss, and mild fever. The cause of Crohn's disease is unknown, although stress may be a factor.

Crohn's disease may occur in any part of the digestive system but usually affects the small intestine. It is characterized by ulceration, abscess formation, small perforations, and the development of adhesions binding the loops of the small intestine. Affected segments of intestine may constrict, causing obstruction, or may perforate. It is treated by surgical removal of badly affected segments, and by corticosteroids. Mild cases respond to rest, bland diet, and drug treatment. Crohn's disease first occurs most often in adults aged 20–40.

crop in birds, the thin-walled enlargement of the digestive tract between the oesophagus and stomach. It is an effective storage organ especially in seed-eating birds; a pigeon's crop can hold about 500 cereal grains. Digestion begins in the crop, by the moisturizing of food. A crop also occurs in insects and annelid worms.

crop in computing, to cut away unwanted portions of a picture. The term comes from traditional manual methods of layout and paste-up; cropping is an option made available via photo-finishing and graphics software.

crop circle circular area of flattened grain found in fields especially in SE England, with increasing frequency every summer since 1980. More than 1,000 such formations were reported in the UK in 1991. The cause is unknown.

Most of the research into crop circles has been conducted by dedicated amateur investigators rather than scientists. Physicists who have studied the phenomenon have suggested that an electromagnetic whirlwind, or 'plasma vortex', can explain both the crop circles and some UFO sightings, but this does not account for the increasing geometrical complexity of crop circles, nor for the fact that until 1990 they were unknown outside the UK. Crop circles began to appear in the USA only after a US magazine published an article about them. A few people have confessed publicly to having made crop circles that were accepted as genuine by investigators.

crop rotation system of regularly changing the crops grown on a piece of land. The crops are grown in a particular order to utilize and add to the nutrients in the soil and to prevent the build-up of insect and fungal pests. Including a legume crop, such as peas or beans, in the rotation helps build up nitrate in the soil, because the roots contain bacteria capable of fixing nitrogen from the air.

A simple seven-year rotation, for example, might include a three-year ley followed by two years of wheat and then two years of barley, before returning the land to temporary grass once more. In this way, the cereal crops can take advantage of the build-up of soil fertility that occurs during the period under grass. In the 18th century, a four-year rotation was widely adopted with autumn-sown cereal, followed by a root crop, then spring cereal, and ending with a leguminous crop. Since then, more elaborate rotations have been devised with two, three, or four successive cereal crops, and with the root crop replaced by a cash crop such as sugar beet or potatoes, or by a legume crop such as peas or beans.

crossbill species of ◊finch, genus *Loxia*, family Fringillidae, order Passeriformes, in which the hooked tips of the upper and lower beak cross one another, an adaptation for extracting the seeds from conifer cones. The red or common crossbill *Loxia curvirostra* is found in parts of Eurasia and North America, living chiefly in pine forests.

The parrot crossbill *L. pytopsittacus* of Europe, and the white-winged crossbill *L. leucoptera* of N Asia and North America, feed on pine and larch respectively.

crossing over in biology, a process that occurs during ◊meiosis. While the chromosomes are lying alongside each other in pairs, each partner may twist around the other and exchange corresponding chromosomal segments. It is a form of genetic ◊recombination, which increases variation and thus provides the raw material of evolution.

cross-linking in chemistry, lateral linking between two or more long-chain molecules in a ◊polymer. Cross-linking gives the polymer a higher melting point and makes it harder. Examples of cross-linked polymers include Bakelite and vulcanized rubber.

cross-posting on USENET, the practice of sending a message to more than one ◊newsgroup. A small amount of cross-posting is acceptable if the message is on a topic that is relevant to more than one newsgroup. For example, a message about top tennis player André Agassi's personal life might be posted to both rec.sport.tennis and alt.showbiz.gossip. Large amounts of cross-posting are called spam (see ◊spamming) and may lead to ◊flames and the attention of the ◊CancelMoose.

cross-section the surface formed when a solid is cut through by a plane at right angles to its axis.

croup inflammation of the larynx in small children, with harsh, difficult breathing and hoarse coughing. Croup is most often associated with viral infection of the respiratory tract.

crow any of 35 species of omnivorous birds in the genus *Corvus*, family Corvidae, order Passeriformes, which also includes choughs, jays, and magpies. Crows are usually about 45 cm/1.5 ft long, black, with a strong bill feathered at the base. The tail is long and graduated, and the wings are long and pointed, except in the jays and magpies, where they are shorter. Crows are considered to be very intelligent. The family is distributed throughout the world, though there are very few species in eastern Australia or South America. The common crows are *C. brachyrhynchos* in North America, and *C. corone* in Europe and Asia.

crowfoot any of several white-flowered aquatic plants belonging to the buttercup family, with a touch of yellow at the base of the petals. The divided leaves are said to resemble the feet of a crow. (Genus *Ranunculus*, family Ranunculaceae.)

CRT abbreviation for ◊cathode-ray tube.

crude oil the unrefined form of ◊petroleum.

crumple zone region at the front and rear of a motor vehicle that is designed to crumple gradually during a collision, so reducing the risk of serious injury to passengers. The progressive crumpling absorbs the kinetic energy of the vehicle more gradually than a rigid structure would, thereby diminishing the forces of deceleration acting on the vehicle and on the people inside.

crust the outermost part of the structure of Earth, consisting of two distinct parts, the oceanic crust and the continental crust. The **oceanic** crust is on average about 10 km/6.2 mi thick and consists mostly of basaltic types of rock. By contrast, the **continental** crust is largely made of granite and is more complex in its structure. Because of the movements of ◊plate tectonics, the oceanic crust is in no place older than about 200 million years. However, parts of the continental crust are over 3 billion years old.

Beneath a layer of surface sediment, the oceanic crust is made up of a layer of basalt, followed by a layer of gabbro. The composition of the oceanic crust overall shows a high proportion of **si**licon and **ma**gnesium oxides, hence named **sima** by geologists. The continental crust varies in thickness from about 40 km/25 mi to 70 km/45 mi, being deeper beneath mountain ranges. The surface layer consists of many kinds of sedimentary and igneous rocks. Beneath lies a zone of metamorphic rocks built on a thick layer of granodiorite. **Si**licon and **al**uminium oxides dominate the composition and the name **sial** is given to continental crustal material.

crustacean one of the class of arthropods that includes crabs, lobsters, shrimps, woodlice, and barnacles. The external skeleton is made of protein and chitin hardened with lime. Each segment bears a pair of appendages that may be modified as sensory feelers (antennae), as mouthparts, or as swimming, walking, or grasping structures.

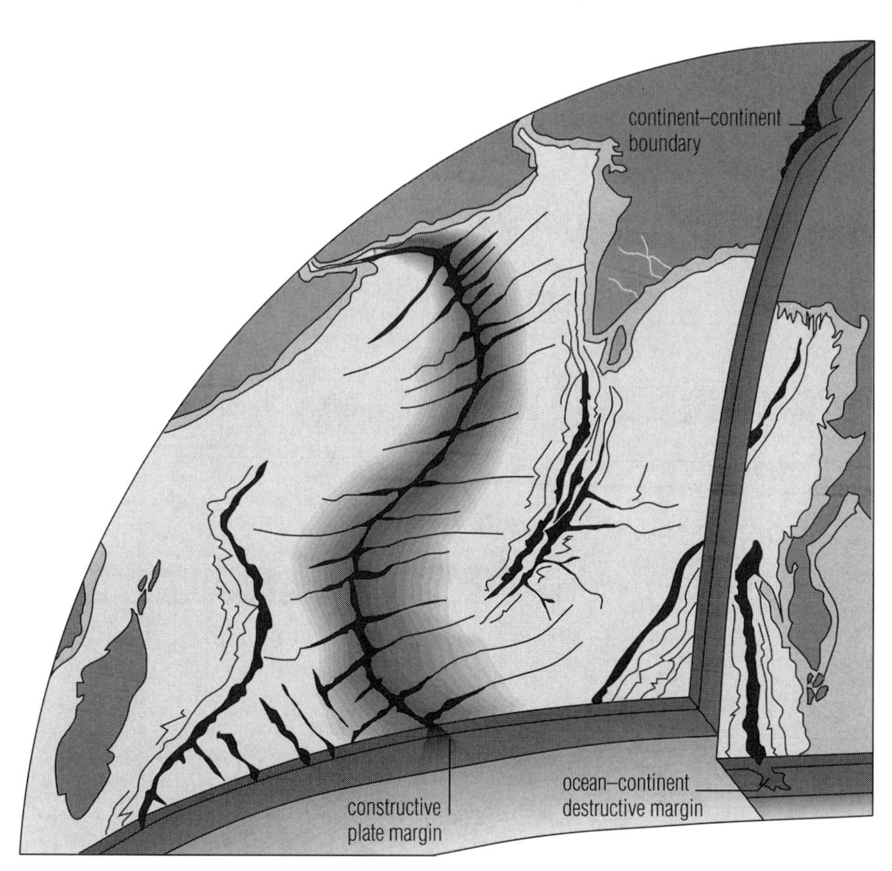

***crust** The crust of the Earth is made up of plates with different kinds of margins. In mid-ocean, there are constructive plate margins, where magma wells up from the Earth's interior, forming new crust. On continent–continent margins, mountain ranges are flung up by the collision of two continents. At an ocean–continent destructive margin, ocean crust is forced under the denser continental crust, forming an area of volcanic instability.*

Crux constellation of the southern hemisphere, popularly known as the Southern Cross, the smallest of the 88 constellations but one of the brightest, and one of the best known as it is represented on the flags of Australia and New Zealand. Its brightest stars are Alpha Crucis (or Acrux), a ◊double star about 400 light years from Earth, and Beta Crucis (or Mimosa).

Near Beta Crucis lies a glittering star cluster known as the Jewel Box, named by John Herschel. The constellation also contains the Coalsack, a dark nebula silhouetted against the bright starry background of the Milky Way.

cryogenics science of very low temperatures (approaching ◊absolute zero), including the production of very low temperatures and the exploitation of special properties associated with them, such as the disappearance of electrical resistance (◊superconductivity).

Low temperatures can be produced by the Joule–Thomson effect (cooling a gas by making it do work as it expands). Gases such as oxygen, hydrogen, and helium may be liquefied in this way, and temperatures of 0.3K can be reached. Further cooling requires magnetic methods; a magnetic material, in contact with the substance to be cooled and with liquid helium, is magnetized by a strong magnetic field. The heat generated by the process is carried away by the helium. When the material is then demagnetized, its temperature falls; temperatures of around 10^{-3}K have been achieved in this way. At temperatures near absolute zero, materials can display unusual properties. Some metals, such as mercury and lead, exhibit superconductivity. Liquid helium loses its viscosity and becomes a 'superfluid' when cooled to below 2K; in this state it flows up the sides of its container. Cryogenics has several practical applications. **Cryotherapy** is a process used in eye surgery, in which a freezing probe is briefly applied to the outside of the eye to repair a break in the retina. Electronic components called ◊Josephson junctions, which could be used in very fast computers, need low temperatures to function. Magnetic levitation (◊maglev) systems must be maintained at low temperatures. Food can be frozen for years, and freezing eggs, sperm and pre-embryos is now routine. In September 1996 South African researchers resuscitated a rat's heart that had been frozen to –196°C.

CRYONICS FREQUENTLY ASKED QUESTION LIST

http://www.cs.cmu.edu/afs/cs/user/tsf/Public-Mail/cryonics/html/overview.html

Answers to all the questions you could possibly ask about the controversial practice of cryonics. If, after reading this, you are tempted to give it a go, be warned. Of the sixty people who have been suspended, forty have been thawed out and buried, as the cryonics companies they had paid went bankrupt.

cryolite rare granular crystalline mineral (sodium aluminium fluoride), Na_3AlF_6, used in the electrolytic reduction of ◊bauxite to aluminium. It is chiefly found in Greenland.

cryptogam obsolete name applied to the lower plants. It included the algae, liverworts, mosses, and ferns (plus the fungi and bacteria in very early schemes of classification). In such classifications seed plants were known as ◊phanerogams.

cryptography science of creating and reading codes; for example, those produced by the Enigma coding machine used by the Germans in World War II and those used in commerce by banks encoding electronic fund-transfer messages, business firms sending computer-conveyed memos between headquarters, and in the growing field of electronic mail. The breaking and decipherment of such codes is known as 'cryptanalysis'. No method of encrypting is completely unbreakable, but decoding can be made extremely complex and time consuming.

CRYPTOGRAPHY

http://rschp2.anu.edu.au:8080/crypt.html

Introduction to the whys and wherefores of encrypting messages, particularly within an Internet framework. The author introduces various common encrypting systems and explains their relative weaknesses. The site also includes a good page on public-key cryptography which is becoming increasingly important as more and more information is transmitted electronically.

cryptosporidium waterborne parasite that causes disease in humans and other animals. It has been found in drinking water in the UK and USA, causing diarrhoea, abdominal cramps, vomiting, and fever, and can be fatal in people with damaged immune systems, such as AIDS sufferers or those with leukaemia. Just 30 cryptosporidia are enough to cause prolonged diarrhoea.

Conventional filtration and chlorine disinfection are ineffective at removing the parasite. Slow sand filtration is the best method of removal, but the existing systems were dismantled in the 1970s because of their slowness.

crystal substance with an orderly three-dimensional arrangement of its atoms or molecules, thereby creating an external surface of clearly defined smooth faces having characteristic angles between them. Examples are table salt and quartz.

Each geometrical form, many of which may be combined in one crystal, consists of two or more faces – for example, dome, prism, and pyramid. A mineral can often be identified by the shape of its crystals and the system of crystallization determined. A single crystal can vary in size from a submicroscopic particle to a mass some 30 m/100 ft in length. Crystals fall into seven crystal systems or groups, classified on the basis of the relationship of three or four imaginary axes that intersect at the centre of any perfect, undistorted crystal.

crystallization formation of crystals from a liquid, gas, or solution.

crystallography the scientific study of crystals. In 1912 it was found that the shape and size of the repeating atomic patterns (unit cells) in a crystal could be determined by passing X-rays through a sample. This method, known as ◊X-ray diffraction, opened up an entirely new way of 'seeing' atoms. It has been found that many substances have a unit cell that exhibits all the symmetry of the whole crystal; in table salt (sodium chloride, NaCl), for instance, the unit cell is an exact cube.

Many materials were not even suspected of being crystals until they were examined by X-ray crystallography. It has been shown that purified biomolecules, such as proteins and DNA, can form crystals, and such compounds may now be studied by this method. Other applications include the study of metals and their alloys, and of rocks and soils.

CRYSTALLOGRAPHY AND MINERALOGY

http://www.iumsc.indiana.edu/docs/crystmin.htm

Understand the shapes and symmetries of crystallography, with these interactive drawings of cubic, tetrahedral, octahedral, and dodecahedral solids (just drag your mouse over the figures to rotate them).

CSCW abbreviation for ◊computer-supported collaborative work.

CSH abbreviation for **corticotrophin-releasing hormone**.

CSLIP (abbreviation for ***Compressed Serial Line Internet Protocol***) in computing, newer version of ◊SLIP allowing slightly faster dial-up connections to the Internet.

CT scanner medical device used to obtain detailed X-ray pictures of the inside of a patient's body. See ◊CAT scan.

CTS/RTS (abbreviation for ***Clear To Send/Ready To Send***) in computing, hardware handshaking (see ◊handshake) used in high-speed modems. In most communications software this is an option that can be ◊toggled on or off. The alternative, software handshaking, is considered less reliable at high speeds.

cu abbreviation for **cubic** (measure).

CUA (abbreviation for ***common user access***) standard designed by ◊Microsoft to ensure that identical actions, such as saving a file or accessing help, can be carried out using the same keystrokes in any piece of software. For example, in programs written to the CUA standard, help is always summoned by pressing the F1 function key. New programs should be easier to use because users will not have to learn new commands to perform standard tasks.

cube in geometry, a regular solid figure whose faces are all squares. It has 6 equal-area faces and 12 equal-length edges.

If the length of one edge is l, the volume V of the cube is given by:

$$V = l^3$$

and its surface area A by:

$$A = 6l^2$$

cube to multiply a number by itself and then by itself again. For example, 5 cubed = $5^3 = 5 \times 5 \times 5 = 125$. The term also refers to a number formed by cubing; for example, 1, 8, 27, 64 are the first four cubes.

cubic centimetre (or metre) the metric measure of volume, corresponding to the volume of a cube whose edges are all 1 cm (or 1 metre) in length.

cubic decimetre metric measure (symbol dm^3) of volume corresponding to the volume of a cube whose edges are all 1 dm (10 cm) long; it is equivalent to a capacity of one litre.

cubic equation any equation in which the largest power of x is x^3. For example, $x^3 + 3x^2y + 4y^2 = 0$ is a cubic equation.

cubic measure measure of volume, indicated either by the word 'cubic' followed by a linear measure, as in 'cubic foot', or the word 'cubed' after a linear measure, as in 'metre cubed'.

cubit earliest known unit of length, which originated between 2800 and 2300 BC. It is approximately 50.5 cm/20.6 in long, which is about the length of the human forearm measured from the tip of the middle finger to the elbow.

cuboid six-sided three-dimensional prism whose faces are all rectangles. A brick is a cuboid.

cuckoo species of bird, any of about 200 members of the family Cuculidae, order Cuculiformes, especially the Eurasian cuckoo *Cuculus canorus,* whose name derives from its characteristic call. Somewhat hawklike, it is about 33 cm/1.1 ft long, bluish-grey and barred beneath (females are sometimes reddish), and typically has a long, rounded tail. Cuckoos feed on insects, including hairy caterpillars that are distasteful to most birds. It is a 'brood parasite', laying its eggs singly, at intervals of about 48 hours, in the nests of small insectivorous birds. As soon as the young cuckoo hatches, it ejects all other young birds or eggs from the nest and is tended by its 'foster parents' until fledging. American species of cuckoo hatch and rear their own young.

The North American **roadrunner** *Geococcyx californianus* is a member of the cuckoo family, and the yellow-billed cuckoo, *Coccysus americanus,* incubates its own eggs.

cuckoo flower or ***lady's smock*** perennial meadow plant of northern temperate regions. From April to June it bears pale lilac flowers, which later turn white. (*Cardamine pratensis,* family Cruciferae.)

cuckoopint or ***lords-and-ladies*** perennial European plant, a wild arum. It has large arrow-shaped leaves that appear in early spring and flower-bearing stalks enclosed by a bract, or spathe (specialized leaf). The bright red berrylike fruits, which are poisonous, appear in late summer. (*Arum maculatum,* family Araceae.)

cuckoo spit the frothy liquid surrounding and exuded by the larvae of the ◊froghopper.

cuckoo-spit insect another name for ◊froghopper.

cucumber trailing annual plant belonging to the gourd family, producing long, green-skinned fruit with crisp, translucent, edible flesh. Small cucumbers, called gherkins, usually the fruit of Cucurbitaceae.)

There are about 735 species belonging to the cucumber family.

cultivar variety of a plant developed by horticultural or agricultural techniques. The term derives from '**culti**vated **var**iety'.

culture in biology, the growing of living cells and tissues in laboratory conditions.

cumin seedlike fruit of the herb cumin, which belongs to the carrot family. It has a bitter flavour and is used as a spice in cooking. (*Cuminum cyminum,* family Umbelliferae.)

cumulative frequency in statistics, the total frequency of a given value up to and including a certain point in a set of data. It is used to draw the cumulative frequency curve, the ogive.

cuprite Cu_2O ore (copper(I) oxide), found in crystalline form or in earthy masses. It is red to black in colour, and is often called ruby copper.

cupronickel copper alloy (75% copper and 25% nickel), used in hardware products and for coinage.

curare black, resinous poison extracted from the bark and juices of various South American trees and plants. Originally used on arrowheads by Amazonian hunters to paralyse prey, it blocks nerve stimulation of the muscles. Alkaloid derivatives (called curarines) are used in medicine as muscle relaxants during surgery.

curie former unit (symbol Ci) of radioactivity, equal to 3.7×10^{10} ◊becquerels. One gram of radium has a radioactivity of about one curie. It was named after French physicist Pierre Curie.

Curie temperature the temperature above which a magnetic material cannot be strongly magnetized. Above the Curie temperature, the energy of the atoms is too great for them to join together to form the small areas of magnetized material, or ◊domains, which combine to produce the strength of the overall magnetization.

curing method of preserving meat by soaking it in salt (sodium chloride) solution, with saltpetre (sodium nitrate) added to give the meat its pink colour and characteristic taste. The nitrates in cured meats are converted to nitrites and nitrosamines by bacteria, and these are potentially carcinogenic to humans.

curium synthesized, radioactive, metallic element of the *actinide* series, symbol Cm, atomic number 96, relative atomic mass 247. It is produced by bombarding plutonium or americium with neutrons. Its longest-lived isotope has a half-life of 1.7×10^7 years.

curlew wading bird of the genus *Numenius* of the sandpiper family, Scolopacidae, order Charadriiformes. The curlew is between 36 cm/14 in and 55 cm/1.8 ft long, and has pale brown plumage with dark bars and mainly white underparts, long legs, and a long, thin, downcurved bill. It feeds on a variety of insects and other invertebrates. Several species live in N Europe, Asia, and North America. The name derives from its haunting flutelike call.

One species, the Eskimo curlew, is almost extinct, never having recovered from relentless hunting in the late 19th century.

currant berry of a small seedless variety of cultivated grape (*Vitis vinifera*). Currants are grown on a large scale in Greece and California and are dried for use in cooking and baking. Because of the similarity of the fruit, the name 'currant' is also given to several species of shrubs (genus *Ribes,* family Grossulariaceae).

Curie, Marie
(1867–1934)

(born *Manya Sklodowska*) Polish scientist who, with husband Pierre Curie, discovered in 1898 two new radioactive elements in pitchblende ores: polonium and radium. They isolated the pure elements in 1902. Both scientists refused to take out a patent on their discovery and were jointly awarded the Nobel Prize for Physics 1903, with Henri Becquerel. Marie Curie was also awarded the Nobel Prize for Chemistry 1911.

From 1896 the Curies worked together on radioactivity, building on the results of Wilhelm Röntgen (who had discovered X-rays) and Becquerel (who had discovered that similar rays are emitted by uranium salts). Marie Curie discovered that thorium emits radiation and found that the mineral pitchblende was even more radioactive than could be accounted for by any uranium and thorium content. In July 1898, the Curies announced the discovery of polonium, followed by the discovery of radium five months later. They eventually prepared 1 g/ 0.04 oz of pure radium chloride – from 8 tonnes of waste pitchblende from Austria.

They also established that beta rays (now known to consist of electrons) are negatively charged particles. In 1910 with André Debierne (1874–1949), who had discovered actinium in pitchblende 1899, Marie Curie isolated pure radium metal in 1911.

Curie, Pierre (1859–1906) French scientist. He shared the Nobel Prize for Physics 1903 with his wife Marie ◊Curie and Henri ◊Becquerel. From 1896 the Curies had worked together on ◊radioactivity, discovering two radioactive elements.

Mary Evans Picture Library

current flow of a body of water or air, or of heat, moving in a definite direction. Ocean currents are fast-flowing currents of seawater generated by the wind or by variations in water density between two areas. They are partly responsible for transferring heat from the Equator to the poles and thereby evening out the global heat imbalance. There are three basic types of ocean current: **drift currents** are broad and slow-moving; **stream currents** are narrow and swift-moving; and **upwelling currents** bring cold, nutrient-rich water from the ocean bottom.

Stream currents include the ◊Gulf Stream and the ◊Japan (or Kuroshio) Current. Upwelling currents, such as the Gulf of Guinea Current and the Peru (Humboldt) current, provide food for plankton, which in turn supports fish and sea birds. At approximate five-to-eight-year intervals, the Peru Current that runs from the Antarctic up the W coast of South America, turns warm, with heavy rain and rough seas, and has disastrous results (as in 1982–83) for Peruvian wildlife and for the anchovy industry. The phenomenon is called **El Niño** (Spanish 'the Child') because it occurs towards Christmas.

current directory in a computer's file system, the ◊directory in which the user is positioned. As users move around a computer system, opening, reading, writing, and storing files, they navigate through that computer's directory structure. Most file commands are assumed to apply to the files in the current directory.

In DOS, adding the command 'prompt pg' to the ◊autoexec.bat file sets the computer to display the name and path of the current directory at the system prompt. On an FTP (file transfer protocol) site, the command 'pwd' will print the name of the current directory on the remote machine.

current, electric see ◊electric current.

cursor on a computer screen, the symbol that indicates the current entry position (where the next character will appear). It usually consists of a solid rectangle or underline character, flashing on and off.

curve in geometry, the ◊locus of a point moving according to specified conditions. The circle is the locus of all points equidistant from a given point (the centre). Other common geometrical curves are the ◊ellipse, ◊parabola, and ◊hyperbola, which are also produced when a cone is cut by a plane at different angles.

Many curves have been invented for the solution of special problems in geometry and mechanics – for example, the cissoid (the inverse of a parabola) and the ◊cycloid.

cuscus tree-dwelling marsupial found in Australia, New Guinea, and Sulawesi. There are five species, all about the size of a cat. They have a prehensile tail and an opposable big toe.

Phalanger spilocuscus is known as the spotted cuscus or tiger cat; *P. ursinus* and *P. celebensis* are natives of the Celebes.
classification Cuscuses are in genus *Phalanger*, in the family Phalangeridae (possums), order Marsupialia, class Mammalia.

CU-SeeMe in computing, software that enables ◊videoconferencing across the Internet. Developed by US computer scientist Richard Cogger, CU-SeeMe was bought in 1996 by US videoconferencing specialist White Pine Software of Nashua, New Hampshire, and is now a commercial product.

Early experiments with CU-SeeMe included broadcasts by the North American Space Agency (NASA) of live and prerecorded video footage of shuttle missions, New Year parties held at cybercafes around the world, and live hook-ups between schools.

cusp point where two branches of a curve meet and the tangents to each branch coincide.

custard apple any of several large edible heart-shaped fruits produced by a group of tropical trees and shrubs which are often cultivated. Bullock's heart (*Annona reticulata*) produces a large dark-brown fruit containing a sweet reddish-yellow pulp; it is a native of the West Indies. (Family Annonaceae.)

cuticle the horny noncellular surface layer of many invertebrates such as insects; in botany, the waxy surface layer on those parts of plants that are exposed to the air, continuous except for ◊stomata and ◊lenticels. All types are secreted by the cells of the ◊epidermis. A cuticle reduces water loss and, in arthropods, acts as an ◊exoskeleton.

cutting technique of vegetative propagation involving taking a section of root, stem, or leaf and treating it so that it develops into a new plant.

He that uses many words for the explaining of any subject, doth, like the cuttle fish, hide himself for the most part in his own ink.

JOHN RAY English naturalist.
On the Creation

---

cuttlefish any of a family, Sepiidae, of squidlike cephalopods with an internal calcareous shell (cuttlebone). The common cuttle *Sepia officinalis* of the Atlantic and Mediterranean is up to 30 cm/1 ft long. It swims actively by means of the fins into which the sides of its oval, flattened body are expanded, and jerks itself backwards by shooting a jet of water from its 'siphon'.

It is capable of rapid changes of colour and pattern. The large head has conspicuous eyes, and the ten arms are provided with suckers. Two arms are very much elongated, and with them the cuttlefish seizes its prey. It has an ink sac from which a dark fluid can be discharged into the water, distracting predators from the cuttle itself. The dark brown pigment sepia is obtained from the ink sacs of cuttlefish.

cutworm common name for the larva of many species of owlet-moth. They cut off the young shoots of cultivated plants. They are related to the army-worm and ◊cotton-worm.
classification Cutworms belong to several genera, including *Agrotis, Prodenia,* and *Euxoa* in the family Noctuidae, class Insecta, phylum Arthropoda.

CWIS in computing, abbreviation for ◊campus-wide information service.

cwt symbol for ◊hundredweight, a unit of weight equal to 112 pounds (50.802 kg); 100 lb (45.36 kg) in the USA.

cyanamide process process used in the manufacture of calcium cyanamide ($CaCN_2$), a colourless crystalline powder used as a fertilizer under the tradename Nitrolime. Calcium carbide is reacted with nitrogen in an electric furnace.

$$CaC_2 + N_2 \rightarrow CaCN_2 + C$$

The calcium cyanamide reacts with water in the soil to form the ammonium ion and calcium carbonate. The ammonium is then oxidized to nitrate, which is taken up by plants. Calcium cyanamide can also be converted commercially into ammonia.

cyanide CN^- ion derived from hydrogen cyanide (HCN), and any salt containing this ion (produced when hydrogen cyanide is neutralized by alkalis), such as potassium cyanide (KCN). The principal cyanides are potassium, sodium, calcium, mercury, gold, and copper. Certain cyanides are poisons.

cyanobacteria (singular ***cyanobacterium***) alternative name for ◊blue-green algae.

cyanocobalamin chemical name for vitamin B_{12}, which is normally produced by microorganisms in the gut. The richest sources are liver, fish, and eggs. It is essential to the replacement of cells, the maintenance of the myelin sheath which insulates nerve fibres, and the efficient use of folic acid, another vitamin in the B complex. Deficiency can result in pernicious anaemia (defective production of red blood cells), and possible degeneration of the nervous system.

cybercafe coffeehouse equipped with public-access Internet terminals. Typically, users pay a small sum to use the terminals for short periods. Cafes usually supply brief tutorials for newcomers. By 1996 many major cities around the world had such cafes. There were more than 1,200 cybercafes in 78 countries by the end of 1997.

CyberCash in computing, one of several schemes for electronic money that can be used to trade on the Internet. Founded in 1994, CyberCash uses the RSA encryption ◊algorithm to protect customer financial information in transit.

The system stores customers' payment information, such as credit card numbers, in an electronic wallet, software which is downloaded from the company's site on the World Wide Web. When a customer wishes to buy something at a commercial Web site, the site generates a payment request, the customer adds a payment method, and the CyberCash server authenticates the transaction. Future plans are to add electronic cheques and cash or debit cards to the choice of payment instruments. See also ◊DigiCash.

cyberlaw in computing, relatively new field of Internet and computer law. Still being defined, the field includes new areas such as the responsibility of ◊Internet Service Providers and ◊bulletin-board system operators for the material that passes through or is stored on their systems and the framework for international electronic commerce, and a new look at traditional areas such as intellectual property rights and copyright and censorship.

cybernetics ***Greek kubernan 'to steer'*** science concerned with how systems organize, regulate, and reproduce themselves, and also how they evolve and learn. In the laboratory, inanimate objects are created that behave like living systems. Applications range from the creation of electronic artificial limbs to the running of the fully automated factory where decisionmaking machines operate up to managerial level.

Cybernetics was founded and named in 1947 by US mathematician Norbert Wiener. Originally, it was the study of control systems using feedback to produce automatic processes.

We should take care not to make the intellect our god; it has, of course, powerful muscles, but no personality.

ALBERT EINSTEIN German-born US physicist.
Out of My Later Life

---

cyberpunk in computing, term coined by US science-fiction writer and editor Gardner Dozois for a particular type of modern science fiction that combines high-technology landscapes with countercultural social and political ideas. Leading writers in this genre include William Gibson, Bruce Sterling, Pat Cadigan, Greg Bear, and Rudy Rucker.

cybersex in computing, online sexual fantasy spun by two or more participants via live, online chat. Futurists hypothesize about a future where 'virtual' sex will take place in ◊virtual reality via body suits and other hardware input devices. In 1996, however, cybersex is limited to text-based systems such as IRC (◊Internet Relay Chat) or the shared worlds created in ◊MUDs (Multiuser dungeons) and ◊MOOs (MUD, object-oriented), both shared role-playing game worlds.

cyberspace the imaginary, interactive 'worlds' created by networked computers; often used interchangeably with 'virtual world'. The invention of the word 'cyberspace' is generally credited to US science-fiction writer William Gibson (1948–) and his first novel *Neuromancer* 1984.

As well as meaning the interactive environment encountered in a virtual reality system, cyberspace is 'where' the global community of computer-linked individuals and groups lives. From the mid-1980s, the development of computer networks and telecommunications, both international (such as the ◊Internet) and local (such as the services known as 'bulletin board' or conferencing systems), made possible the instant exchange of messages using ◊electronic mail and electronic conferencing systems directly from the individual's own home.

cycad any of a group of plants belonging to the ◊gymnosperms, whose seeds develop in cones. Some are superficially similar to palms, others to ferns. Their large cones contain fleshy seeds. There are ten genera and about 80–100 species, native to tropical and subtropical countries. Cycads were widespread during the Mesozoic era (245–65 million years ago). (Order Cycadales.)

The stems of many species yield an edible starchy substance resembling ◊sago. In 1993 cycads were discovered to be pollinated by insects, not by wind as had been previously thought; their cones produce heat that vaporizes a sweet minty odour to attract insects to a supply of nectarlike liquid.

cyclamate derivative of cyclohexysulphamic acid, formerly used as an artificial ◊sweetener, 30 times sweeter than sugar. It was first synthesized 1937.

Its use in foods was banned in the USA and the UK from 1970, when studies showed that massive doses caused cancer in rats.

cyclamen any of a group of perennial plants belonging to the primrose family, with heart-shaped leaves and petals that are twisted at the base and bent back, away from the centre of the downward-facing flower. The flowers are usually white or pink, and several species are cultivated. (Genus *Cyclamen,* family Primulaceae.)

cycle in physics, a sequence of changes that moves a system away from, and then back to, its original state. An example is a vibration that moves a particle first in one direction and then in the opposite direction, with the particle returning to its original position at the end of the vibration.

cyclic compound any of a group of organic chemicals that have rings of atoms in their molecules, giving them a closed-chain structure.

cyclic polygon in geometry, a polygon in which each vertex (corner) lies on the circumference of a circle.

cyclic quadrilateral four-sided figure with all four of its vertices lying on the circumference of a circle. The opposite angles of cyclic quadrilaterals add up to 180°, and are therefore said to be supplementary; each external angle (formed by extending a side of the quadrilateral) is equal to the opposite interior angle.

Cycliophora invertebrate phylum containing only one known species *Symbion pandora*. The phylum was discovered in 1995. *S. pandora* are minute (347 μm) bottle-shaped invertebrates living amongst the mouthparts of lobsters. They have threadlike cilia for gathering food and are attached by an adhesive disc. They have a reproductive cycle consisting of both asexual and sexual forms. The cycliophoran's nearest known relatives are bryozoans.

cyclodextrin ring-shaped ◊glucose molecule chain created in 1993 at Osaka University, Japan. Cyclodextrins are commonly used in food additives, and can also be used as capsules to deliver drugs, as cutters to separate ions and molecules, and as catalysts for chemical reactions.

They generally consist of 6–8 glucose molecules linked together in a ring, leaving a central hole of 0.45–0.8 nanometres, which can hold a small molecule such as benzene. They can be joined together to form tubes even smaller than DNA, the length and width of which can be controlled. They could hypothetically be used in the production of large scale integrated computer systems.

cycloid in geometry, a curve resembling a series of arches traced out by a point on the circumference of a circle that rolls along a straight line. Its applications include the study of the motion of wheeled vehicles along roads and tracks.

cyclone alternative name for a ◊depression, an area of low atmospheric pressure. A severe cyclone that forms in the tropics is called a tropical cyclone or ◊hurricane.

cyclotron circular type of particle ◊accelerator.

Cygnus large prominent constellation of the northern hemisphere, represented as a swan. Its brightest star is first-magnitude Alpha Cygni or ◊Deneb.

Beta Cygni (Albireo) is a yellow and blue ◊double star, visible through small telescopes. The constellation contains the North America nebula (named after its shape), the Veil nebula (the remains of a ◊supernova that exploded about 50,000 years ago), Cygnus A (apparently a double galaxy, a powerful radio source, and the first radio star to be discovered), and the X-ray source Cygnus X-1, thought to mark the position of a black hole. The area is rich in high luminosity objects, nebulae, and clouds of obscuring matter. Deneb marks the tail of the swan, which is depicted as flying along the Milky Way. Some of the brighter stars form the Northern Cross, the upright being defined by Alpha, Gamma, Eta, and Beta, and the crosspiece by Delta, Gamma, and Epsilon Cygni.

cylinder in computing, combination of the tracks on all the platters making up a hard drive or fixed disc that can be accessed without moving the read/write heads.

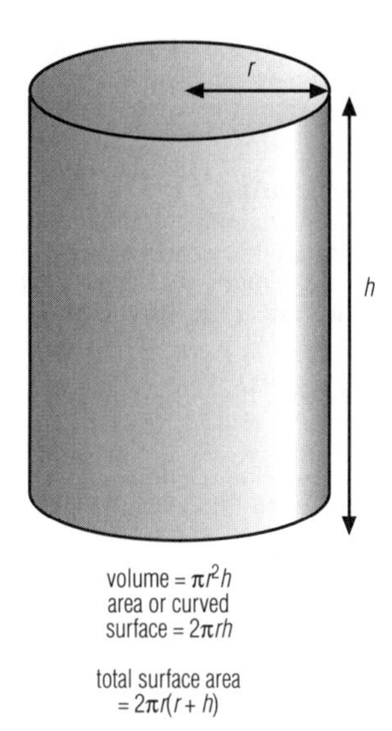

cylinder The volume and area of a cylinder are given by simple formulae relating the dimensions of the cylinder.

cylinder in geometry, a tubular solid figure with a circular base. In everyday use, the term applies to a **right cylinder**, the curved surface of which is at right angles to the base.

The volume V of a cylinder is given by the formula $V = \pi r^2 h$, where r is the radius of the base and h is the height of the cylinder. Its total surface area A has the formula $A = 2\pi r(h + r)$, where $2\pi rh$ is the curved surface area, and $2\pi r^2$ is the area of both circular ends.

cymbidium genus of generally epiphytic orchids found in Africa, Asia, and Australia, producing numerous showy flowers. The three Australian species are popular subjects for hybridization.

cypherpunk (contraction of ***'cipher'*** and ***'cyberpunk'***) in computing, a passionate believer in the importance of free access to strong encryption on the Net, in the interests of guarding privacy and free speech.

cypress any of a group of coniferous trees or shrubs containing about 20 species, originating from temperate regions of the northern hemisphere. They have tiny scalelike leaves and cones made up of woody, wedge-shaped scales containing an aromatic ◊resin. (Genera *Cupressus* and *Chamaecyparis,* family Cupressaceae.)

cystic fibrosis hereditary disease involving defects of various tissues, including the sweat glands, the mucous glands of the bronchi (air passages), and the pancreas. The sufferer experiences repeated chest infections and digestive disorders and generally fails to thrive. In 1989 a gene for cystic fibrosis was identified by teams of researchers in Michigan, USA, and Toronto, Canada. This discovery enabled the development of a screening test for carriers; the disease can also be detected in the unborn child.

inheriting the disease One person in 22 is a carrier of the disease. If two carriers have children, each child has a one-in-four chance of having the disease, so that it occurs in about one in 2,000 pregnancies. Around 10% of newborns with cystic fibrosis develop an intestinal blockage (meconium ileus) which requires surgery.

treatment Cystic fibrosis was once universally fatal at an early age; now, although there is no definitive cure, treatments have

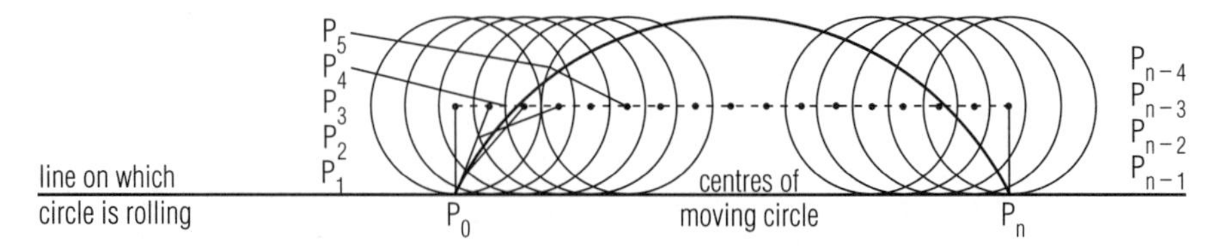

cycloid The cycloid is the curve traced out by a point on a circle as it rolls along a straight line. The teeth of gears are often cut with faces that are arcs of cycloids so that there is rolling contact when the gears are in use.

raised both the quality and expectancy of life. Results in 1995 from a four-year US study showed that the painkiller ibuprofen, available over the counter, slowed lung deterioration in children by almost 90% when taken in large doses.

Management of cystic fibrosis is by diets and drugs, physiotherapy to keep the chest clear, and use of antibiotics to combat infection and minimize damage to the lungs. Some sufferers have benefited from heart-lung transplants.

gene therapy In 1993, UK researchers (at the Imperial Cancer Research Fund, Oxford, and the Wellcome Trust, Cambridge) successfully introduced a corrective version of the gene for cystic fibrosis into the lungs of mice with induced cystic fibrosis, restoring normal function (in 1992, US researchers had introduced such a gene into the lungs of healthy laboratory rats, greatly improving the prospect of a cure). Trials in human subjects began in 1993, and the cystic fibrosis defect in the nasal cavities of three patients in the USA was successfully corrected, though a later trial was halted after a patient became ill following a dose of the genetically altered virus. Patients treated by gene therapy administered in the form of a nasal spray were showing signs of improvement following a preliminary trial in 1996. Cystic fibrosis is seen as a promising test case for ◊gene therapy. It is the commonest fatal hereditary disease amongst white people.

cystitis inflammation of the bladder, usually caused by bacterial infection, and resulting in frequent and painful urination. It is more common in women. Treatment is by antibiotics and copious fluids with vitamin C.

cytochrome protein responsible for part of the process of ◊respiration by which food molecules are broken down in aerobic organisms. Cytochromes are part of the electron transport chain, which uses energized electrons to reduce molecular oxygen (O_2) to oxygen ions (O^{2-}). These combine with hydrogen ions (H^+) to form water (H_2O), the end product of aerobic respiration. As electrons are passed from one cytochrome to another, energy is released and used to make ◊ATP.

cytokine in biology, chemical messenger that carries information from one cell to another, for example the ◊lymphokines.

cytokinin ◊plant hormone that stimulates cell division. Cytokinins affect several different aspects of plant growth and development, but only if ◊auxin is also present. They may delay the process of senescence, or ageing, break the dormancy of certain seeds and buds, and induce flowering.

cytology the study of ◊cells and their functions. Major advances have been made possible in this field by the development of ◊electron microscopes.

cytoplasm the part of the cell outside the ◊nucleus. Strictly speaking, this includes all the ◊organelles (mitochondria, chloroplasts, and so on), but often cytoplasm refers to the jellylike matter in which the organelles are embedded (correctly termed the cytosol). The cytoplasm is the site of protein synthesis.

In many cells, the cytoplasm is made up of two parts: the **ectoplasm** (or plasmagel), a dense gelatinous outer layer concerned with cell movement, and the **endoplasm** (or plasmasol), a more fluid inner part where most of the organelles are found.

cytoskeleton in a living cell, a matrix of protein filaments and tubules that occurs within the cytosol (the liquid part of the cytoplasm). It gives the cell a definite shape, transports vital substances around the cell, and may also be involved in cell movement.

cytotoxic drug any drug used to kill the cells of a malignant tumour; it may also damage healthy cells. Side effects include nausea, vomiting, hair loss, and bone-marrow damage. Some cytotoxic drugs are also used to treat other diseases and to suppress rejection in transplant patients.

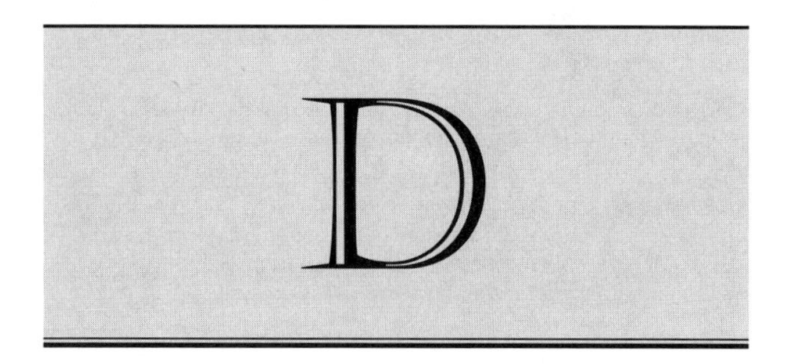

dab small marine flatfish of the flounder family, especially the genus *Limanda*. Dabs live in the North Atlantic and around the coasts of Britain and Scandinavia.

Species include *L. limanda* which grows to about 40 cm/16 in, and the American dab *L. proboscida*, which grows to 30 cm/12 in. Both have both eyes on the right side of their bodies. The left, or blind, side is white, while the rough-scaled right side is light-brown or grey, with dark-brown spots.

DAC abbreviation for ◊digital-to-analogue converter.

dace freshwater fish *Leuciscus leuciscus* of the carp family. Common in England and mainland Europe, it is silvery and grows up to 30 cm/1 ft.

dachshund ***German 'badger-dog'*** small dog of German origin, bred originally for digging out badgers. It has a long body and short legs. Several varieties are bred: standard size (up to 10 kg/22 lb), miniature (5 kg/11 lb or less), long-haired, smooth-haired, and wire-haired.

daddy-longlegs popular name for a ◊crane fly.

Daedalus in space travel, a futuristic project proposed by the British Interplanetary Society to send a ◊robot probe to nearby stars. The probe, 20 times the size of the Saturn V moon rocket, would be propelled by thermonuclear fusion; in effect, a series of small hydrogen-bomb explosions. Interstellar cruise speed would be about 40,000 km/25,000 mi per second.

daffodil any of several Old World species of bulbous plants belonging to the amaryllis family, characterized by their trumpet-shaped yellow flowers which appear in spring. The common daffodil of northern Europe (*Narcissus pseudonarcissus*) has large yellow flowers and grows from a large bulb. There are numerous cultivated forms in which the colours range from white to deep orange. (Genus *Narcissus*, family Amaryllidaceae.)

DAGUERREIAN SOCIETY

http://abell.austinc.edu/dag/home.html

Information about the process involved in this early form of photography, 19th- and 20th-century texts about it, an extensive bibliography of related literature, plenty of daguerreotypes to look at, and information about this society itself can be found here.

daguerreotype in photography, a single-image process using mercury vapour and an iodine-sensitized silvered plate; it was invented by Louis Daguerre 1838.

dahlia any of a group of perennial plants belonging to the daisy family, comprising 20 species and many cultivated forms. Dahlias are stocky plants with tuberous roots and showy flowers that come in a wide range of colours. They are native to Mexico and Central America. (Genus *Dahlia*, family Compositae.)

daisy any of numerous species of perennial plants belonging to the daisy family, especially the field daisy of Europe and North America (*Chrysanthemum leucanthemum*) and the English common daisy (*Bellis perennis*), with a single white or pink flower rising from a rosette of leaves. (Family Compositae.)

daisy bush any of several Australian and New Zealand shrubs with flowers like daisies and felted or hollylike leaves. (Genus *Olearia*, family Compositae.)

daisywheel printing head in a computer printer or typewriter that consists of a small plastic or metal disc made up of many spokes (like the petals of a daisy). At the end of each spoke is a character in relief. The daisywheel is rotated until the spoke bearing the required character is facing an inked ribbon, then a hammer strikes the spoke against the ribbon, leaving the impression of the character on the paper beneath.

The daisywheel can be changed to provide different typefaces; however, daisywheel printers cannot print graphics nor can they print more than one typeface in the same document. For these reasons, they are rapidly becoming obsolete.

DALMATIAN CLUB OF AMERICA

http://www.cet.com/~bholland/dca/

Absolute must for Dalmatian lovers. The site is filled with information about the history of the breed, the breed standard, shows, veterinary advice, and many tips for owners. There is a comprehensive listing of other Dalmatian sites and a useful bibliography. This well-organized site even has a search engine.

Dalmatian breed of dog, about 60 cm/24 in tall, with a distinctive smooth white coat with spots that are black or brown. Dalmatians are born white; the spots appear later. They were formerly used as coach dogs, running beside horse-drawn carriages to fend off highwaymen.

DALMATIAN

Three out of ten dalmatian dogs suffer from some form of hearing disability, and 8% of all dalmations are entirely deaf. Their beautiful spotty coats are caused by intense inbreeding, which has the side-effect of increasing genetic disorders – such as deafness.

Dalton, John
(1766–1844)

English chemist who proposed the theory of atoms, which he considered to be the smallest parts of matter. He produced the first list of relative atomic masses in 'Absorption of Gases' 1805 and put forward the law of partial pressures of gases (Dalton's law).

Mary Evans Picture Library

dam structure built to hold back water in order to prevent flooding, to provide water for irrigation and storage, and to provide

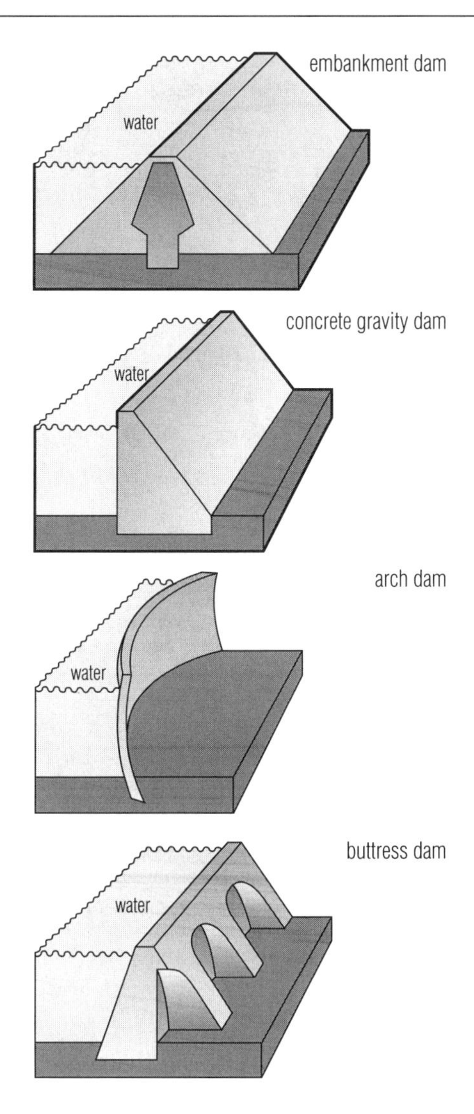

dam *There are two basic types of dam: the gravity dam and the arch dam. The gravity dam relies upon the weight of its material to resist the forces imposed upon it; the arch dam uses an arch shape to take the forces in a horizontal direction into the sides of the river valley. The largest dams are usually embankment dams. Buttress dams are used to hold back very wide rivers or lakes.*

hydroelectric power. The biggest dams are of the earth- and rock-fill type, also called **embankment dams**. Early dams in Britain, built before and about 1800, had a core made from puddled clay (clay which has been mixed with water to make it impermeable). Such dams are generally built on broad valley sites. Deep, narrow gorges dictate a **concrete dam**, where the strength of reinforced concrete can withstand the water pressures involved.

concrete dams A valuable development in arid regions, as in parts of Brazil, is the **underground dam**, where water is stored on a solid rock base, with a wall to ground level, so avoiding rapid evaporation. Many concrete dams are triangular in cross section, with their vertical face pointing upstream. Their sheer weight holds them in position, and they are called **gravity dams**. They are no longer favoured for very large dams, however, because they are expensive and time-consuming to build. Other concrete dams are built in the shape of an arch, which transfers the horizontal force into the sides of the river valley: the **arch dam** derives its strength from the arch shape, just as an arch bridge does, and has been widely used in the 20th century. They require less construction material than other dams and are the strongest type.

buttress dams are used when economy of construction is important or foundation conditions preclude any other type. The upstream portion of a buttress dam may comprise series of cantilevers, slabs, arches or domes supported from the back by a line of buttresses. They are usually made from reinforced and prestressed concrete.

earth dams Earth dams have a watertight core wall, formerly made of puddle clay but nowadays constructed of concrete. Their construction is very economical even for very large structures. **Rock-fill dams** are a variant of the earth dam in which dumped rock takes the place of compacted earth fill.

major dams Rogun (Tajikistan) is the world's highest at 335 m/1,099 ft. New Cornelia Tailings (USA) is the world's biggest in volume, 209 million cu m/7.4 billion cu ft. Owen Falls (Uganda) has the world's largest reservoir capacity, 204.8 billion cu m/7.2 trillion cu ft. Itaipu (Brazil/Paraguay) is the world's most powerful, producing 12,700 megawatts of electricity. The Three Gorges Dam on the Chang Jiang was officially inaugurated in 1994 and is due for completion 2009. A treaty between Nepal and India, ratified by Nepal in 1996, included plans to construct the 315-m/1,035-ft Pancheshwar dam.

In 1997 there were approximately 40,000 large dams (more than 15 m in height) and 800,000 small ones worldwide.

damask textile of woven linen, cotton, wool, or silk, with a reversible figured pattern. It was first made in the city of Damascus, Syria.

damper any device that deadens or lessens vibrations or oscillations; for example, one used to check vibrations in the strings of a piano. The term is also used for the movable plate in the flue of a stove or furnace for controlling the draught.

damselfly long, slender, colourful ◊dragonfly of the suborder Zygoptera, with two pairs of similar wings that are generally held vertically over the body when at rest, unlike those of other dragonflies.

damselfly *As in all members of the suborder Zygoptera, this common blue damselfly* Enallagma cyathigerum *male holds his wings closed above his back. Dragonflies (suborder Anisoptera) hold their wings out to their sides at 90 degrees to the body when at rest, and are generally stouter than damselflies. Premaphotos Wildlife*

damson cultivated variety of plum tree, distinguished by its small oval edible fruits, which are dark purple or blue-black in colour. (*Prunus domestica* var. *institia*.)

dandelion common plant throughout Europe and Asia, belonging to the same family as the daisy. The stalk rises from a rosette of leaves that are deeply indented like a lion's teeth, hence the name (from French *dent de lion*). The flower heads are bright yellow, and the fruit is covered with fine hairs, known as the dandelion 'clock'. (*Taraxacum officinale,* family Compositae.)

The milky juice of the dandelion has laxative properties (causing the bowels to empty), and the young leaves are sometimes eaten in salads. In the Russian species (*Taraxacum koksaghyz*), the juice forms an industrially usable ◊latex, relied upon especially during World War II when Malaysian rubber supplies were blocked by the Japanese (see also ◊rubber).

Dandie Dinmont breed of ◊terrier that originated in the Scottish border country. It is about 25 cm/10 in tall, short-legged and long-bodied, with drooping ears and a long tail. Its hair, about 5 cm/2 in long, can be greyish or yellowish. It is named after the character Dandie Dinmont in Walter Scott's novel *Guy Mannering* 1815.

darcy c.g.s. unit (symbol D) of permeability, used mainly in geology to describe the permeability of rock (for example, to oil, gas, or water).

dark cloud in astronomy, a cloud of cold dust and gas seen in silhouette against background stars or an ◊HII region.

dark matter matter that, according to current theories of ◊cosmology, makes up 90–99% of the mass of the universe but so far remains undetected. Dark matter, if shown to exist, would explain many currently unexplained gravitational effects in the movement of galaxies.

Theories of the composition of dark matter include unknown atomic particles (cold dark matter) or fast-moving neutrinos (hot dark matter) or a combination of both.

In 1993 astronomers identified part of the dark matter in the form of stray planets and ◊brown dwarfs, and possibly, stars that have failed to light up. These objects are known as MACHOs (massive astrophysical compact halo objects) and, according to US astronomers 1996, make up approximately half of the dark matter in the Milky Way's halo.

DARPANET early US computer network. See ◊ARPANET.

dasyure any ◊marsupial of the family Dasyuridae, also known as a 'native cat', found in Australia and New Guinea. Various species have body lengths from 25 cm/10 in to 75 cm/2.5 ft. Dasyures have long, bushy tails and dark coats with white spots. They are agile, nocturnal carnivores, able to move fast and climb.

DARWIN, CHARLES ROBERT

The English naturalist Charles Darwin, who developed the theory of evolution, was once told by his father: 'You care for nothing but shooting, dogs, and rat-catching, and you will be a disgrace to yourself and all your family'.

DAT abbreviation for ◊digital audio tape.

data (singular ***datum***) facts, figures, and symbols, especially as stored in computers. The term is often used to mean raw, unprocessed facts, as distinct from information, to which a meaning or interpretation has been applied.

... mathematics is a natural and a fundamental language. It may well be that it's a property of human beings, that only human beings can think maths. But I think it's probably true that any intelligence in the universe would have this language as well. So maybe it's even greater than – no, not greater than, but more universal than – the human race.

Erik Christopher Zeeman English mathematician.
L Wolpert and A Richards *A Passion for Science* 1988

database in computing, a structured collection of data, which may be manipulated to select and sort desired items of information. For example, an accounting system might be built around a database containing details of customers and suppliers. In larger computers, the database makes data available to the various programs that need it, without the need for those programs to be aware of

Darwin, Charles Robert
(1809–1882)

English naturalist who developed the modern theory of evolution and proposed, with Alfred Russel Wallace, the principle of natural selection. After research in South America and the Galápagos Islands as naturalist on HMS Beagle 1831–36, Darwin published On the Origin of Species *by Means of Natural Selection or the Preservation of Favoured Races in the Struggle for Life 1859.* This book explained the evolutionary process through the principles of natural selection and aroused bitter controversy because it disagreed with the literal interpretation of the Book of Genesis in the Bible.

Darwin's work marked a turning point in many of the sciences, including physical anthropology and palaeontology. But, before the voyage of the Beagle, Darwin, like everyone else at that time, did not believe in the mutability of species. In South America, he saw fossil remains of giant sloths and other animals now extinct, and on the Galápagos Islands he found a colony of finches that he could divide into at least 14 similar species, none of which existed on the mainland. It was obvious to him that one type must have evolved into many others, but how they did so eluded him. Two years after his return he read Malthus's 'An Essay on the Principle of Population' 1798, which proposed that the human population was growing too fast for it to be adequately fed, and that something would have to reduce it, such as war or natural disaster. This work inspired Darwin to see that the same principle could be applied to animal populations.

Darwin's theory of natural selection concerned the variation existing between members of a sexually reproducing population. Those members with variations better fitted to the environment would be more likely to survive and breed, subsequently passing on these favourable characteristics to their offspring. He avoided the issue of human evolution, however, remarking at the end of The Origin of Species that 'much light will be thrown on the origin of man and his history'. It was not until his publication of The *Descent of Man and Selection in Relation to Sex* 1871, that Darwin argued that people evolved just like other organisms. He did not seek the controversy he caused but his ideas soon caught the public imagination. The popular press soon published articles about the 'missing link' between humans and apes.

Darwin also made important discoveries in many other areas, including the fertilization mechanisms of plants, the classification of barnacles, and the formation of coral reefs. He was the first to propose a link between coral reefs and volcanic islands. His ideas remain the primary theory of atoll growth formation.

Mary Evans Picture Library

how the data are stored. The term is also sometimes used for simple record-keeping systems, such as mailing lists, in which there are facilities for searching, sorting, and producing records.

There are three main types (or 'models') of database: hierarchical, network, and relational, of which relational is the most widely used. In a **relational database** data are viewed as a collection of linked tables. A **free-text database** is one that holds the unstructured text of articles or books in a form that permits rapid searching.

A collection of databases is known as a **databank**. A database-management system (DBMS) program ensures that the integrity of the data is maintained by controlling the degree of access of the ◊applications programs using the data. Databases are normally used by large organizations with mainframes or minicomputers.

A telephone directory stored as a database might allow all the people whose names start with the letter B to be selected by one program, and all those living in Chicago by another.

database program Databases are usually created using a database program that enables a user to define the database structure by selecting the number of fields, naming those fields, and allocating the type and amount of data that will valid for each field. Data programs also determine how data can be viewed on screen or extracted into files.

data bus in computing, the electrical pathway, or ◊bus, used to carry data between the components of the computer.

data capture collecting information for computer processing and analysis. Data may be captured automatically – for example, by a ◊sensor that continuously monitors physical conditions such as temperature – or manually; for example, by reading electricity meters.

data communications sending and receiving data via any communications medium, such as a telephone line. The term usually implies that the data are digital (such as computer data) rather than analogue (such as voice messages). However, in the ISDN (◊Integrated Services Digital Network) system, all data – including voices and video images – are transmitted digitally.

See also ◊telecommunications.

data compression in computing, techniques for reducing the amount of storage needed for a given amount of data. They include word tokenization (in which frequently used words are stored as shorter codes), variable bit lengths (in which common characters are represented by fewer ◊bits than less common ones), and run-length encoding (in which a repeated value is stored once along with a count).

In **lossless compression** the original file is retrieved unchanged after decompression. Some types of data (sound and pictures) can be stored by **lossy compression** where some detail is lost during compression, but the loss is not noticeable. Lossy compression allows a greater level of compression. The most popular compression program is ◊PKZIP, widely available as ◊shareware.

data dictionary in computing, a file that holds data about data – for example, lists of files, number of records in each file, and types of fields. Data dictionaries are used by database software to enable access to the data; they are not normally accessible to the user.

Data Encryption Standard (DES) in computing, widely used US government standard for encryption, adopted in 1977 and recertified for five more years 1993. DES was developed by IBM and adopted as a government standard by the National Security Agency. It is a private-key system, so that the sender and recipient encrypt and decrypt the message using the same key.

This means that a secure way has to be found to send the key from one party to the other; any third party who has the key can decrypt the encoded transmissions. Concerns over the long-term security of DES in the face of increasingly available cheap hardware have been somewhat mitigated by new techniques such as triply encrypted DES.

data flow chart diagram illustrating the possible routes that data can take through a system or program; see ◊flow chart.

DataGlove in ◊virtual reality, a glove wired to the computer that allows it to take input from a user's hand gestures. Sensors in the glove detect the wearer's hand movements, and transmit these to the computer in a digital format which the computer can interpret.

DataGlove is a trademark of VPL Research; the general term for such devices is **wired glove**.

data logging in computing, the process, usually automatic, of capturing and recording a sequence of values for later processing and analysis by computer. For example, the level in a water-storage tank might be automatically logged every hour over a seven-day period, so that a computer could produce an analysis of water use. The monitoring is carried out through ◊sensors or similar instruments, connected to the computer via an ◊interface.

The computer logging the data samples the readings at regular time intervals. Data is analysed either continuously (displayed on a changing screen display or as a graph on a ◊plotter) or at the end of the logging period.

data mining analysis of computer data to determine trends. It is used by retailers to find out those items often purchased together. For example, one supermarket chain found that purchases of nappies and beer were linked, and so increased sales by putting beer next to nappies. Store 'loyalty cards' enable retailers to profile customers' week-by-week shopping against their age, sex, and address.

data preparation preparing data for computer input by transferring it to a machine-readable medium. This usually involves typing the data at a keyboard (or at a ◊key-to-disc or key-to-tape station) so that it can be transferred directly to tapes or discs. Various methods of direct data capture, such as ◊bar codes, ◊optical mark recognition (OMR), and ◊optical character recognition (OCR), have been developed to reduce or eliminate lengthy data preparation before computer input.

data processing (DP) or ***electronic data processing*** (EDP) use of computers for performing clerical tasks such as stock control, payroll, and dealing with orders. DP systems are typically ◊batch systems, running on mainframe computers.

data processing cycle For data to be processed the following cycle of operations must be undergone: data are collected, then input into a computer where they are processed to produce the output. As the output may also be the input of a subsequent process the process is deemed cyclical. Whilst being processed data may undergo other operations such as storage and ◊validation.

data protection safeguarding of information about individuals stored on computers, to protect privacy.

data recovery in computing, any of several possible procedures for restoring a computer system and its data after a system crash, burglary, or other damage. The first line of defence in any computer system is ◊backups.

Every system fails at some point, and typically the data on the system is more valuable than the hardware on which it resides. The best course of action depends on the cause of the damage, which may be due to an outside agent, such as a virus, or a simple mistake, such as accidentally deleting important files. Some antivirus software comes with tools to assist users to clean up their systems; for deleted files, utility software such as Norton Utilities may be able to restore ('undelete') the data. In worse cases, specialists may still be able to restore the information by reading the hard disc's platters directly.

data security in computing, precautions taken to prevent the loss or misuse of data, whether accidental or deliberate. These include measures that ensure that only authorized personnel can gain entry to a computer system or file, and regular procedures for storing and 'backing up' data, which enable files to be retrieved or recreated in the event of loss, theft, or damage.

A number of ◊verification and ◊validation techniques may also be used to prevent data from being lost or corrupted by misprocessing.

Encryption involves the translation of data into a form that is meaningless to unauthorized users who do not have the necessary decoding software.

Passwords can be chosen by, or issued to, individual users. These secret words (or combinations of alphanumeric characters) may have to be entered each time a user logs on to a computer system or attempts to access a particular protected file within the system.

Physical access to the computer facilities can be restricted by locking entry doors and storage cabinets.

Master files (files that are updated periodically) can be protected by storing successive versions, or **generations**, of these files and of the transaction files used to update them. The most recent version of the master file may then be recreated, if necessary, from a previous generation. It is common practice to store the three most recent versions of a master file (often called the grandfather, father, and son generations).

Direct-access files are protected by making regular **dumps**, or back-up copies. Because the individual records in direct-access files are constantly being accessed and updated, specific generations of these files cannot be said to exist. The files are therefore dumped at fixed time intervals onto a secure form of backing store. A record, or log, is also kept of all the changes made to a file between security dumps.

Fireproof safes are used to store file generations or sets of security dumps, so that the system can be restarted on a new computer in the event of a fire in the computer department.

Write-protect mechanisms on discs or tapes allow data to be read but not deleted, altered, or overwritten. For example, the protective case of a $3\frac{1}{2}$-inch floppy disc has a write-protect tab that can be slid back with the tip of a pencil or pen to protect the disc's contents.

data terminator or **rogue value** in computing, a special value used to mark the end of a list of input data items. The computer must be able to detect that the data terminator is different from the input data in some way – for instance, a negative number might be used to signal the end of a list of positive numbers, or 'XXX' might be used to terminate the entry of a list of names.

date palm tree, also known as the date palm. The female tree produces the brown oblong fruit, dates, in bunches weighing 9–11 kg/20–25 lb. Dates are an important source of food in the Middle East, being rich in sugar; they are dried for export. The tree also supplies timber and materials for baskets, rope, and animal feed. (Genus *Phoenix*.)

The most important species is *Phoenix dactylifera*; native to northern Africa, soutwest Asia, and parts of India, it grows up to 25 m/80 ft high. A single bunch can contain as many as 1,000 dates. Their juice is made into a kind of wine.

dating science of determining the age of geological structures, rocks, and fossils, and placing them in the context of geological time. The techniques are of two types: relative dating and absolute dating. **Relative dating** can be carried out by identifying fossils of creatures that lived only at certain times (marker fossils), and by looking at the physical relationships of rocks to other rocks of a known age.

Absolute dating is achieved by measuring how much of a rock's radioactive elements have changed since the rock was formed, using the process of ◊radiometric dating.

datura any of a group of plants belonging to the nightshade family, such as the ◊thorn apple, with handsome trumpet-shaped blooms. They have narcotic (pain-killing and sleep-inducing) properties. (Genus *Datura*, family Solanaceae.)

daughterboard in computing, small printed circuit board that plugs into a ◊motherboard to give it new capabilities.

David Dunlap Observatory Canadian observatory at Richmond Hill, Ontario, operated by the University of Toronto, with a 1.88-m/74-in reflector, the largest optical telescope in Canada, opened in 1935.

day time taken for the Earth to rotate once on its axis. The **solar day** is the time that the Earth takes to rotate once relative to the Sun. It is divided into 24 hours, and is the basis of our civil day. The **sidereal day** is the time that the Earth takes to rotate once relative to the stars. It is 3 minutes 56 seconds shorter than the solar day, because the Sun's position against the background of stars as seen from Earth changes as the Earth orbits it.

Thought is only a flash between two long nights, but this flash is everything.

JULES HENRI POINCARÉ French mathematician.

dayflower any plant of the genus *Commelina* of the spiderwort family. All have pointed leaves and creeping stems that form roots. The flowers, usually blue, open in the morning and wither by day's end throughout the summer and fall.

dBASE family of microcomputer programs used for manipulating large quantities of data; also, a related ◊fourth-generation language. The first version, dBASE II, was published by Ashton-Tate in 1981; it has since become the basis for a recognized standard for database applications, known as xBase.

DBS in computing, abbreviation for ◊direct broadcast system.

DC in physics, abbreviation for ***direct current*** (electricity).

DCC abbreviation for ◊digital compact cassette.

DCE (abbreviation for ***data communications equipment***) in computing, another name for a ◊modem.

DDE in computing, abbreviation for ◊dynamic data exchange, a form of communication between processes used in Microsoft Windows.

DDT abbreviation for ***dichloro-diphenyl-trichloroethane***) ($ClC_6H_5)_2CHC(HCl_2)$ insecticide discovered in 1939 by Swiss chemist Paul Müller. It is useful in the control of insects that spread malaria, but resistant strains develop. DDT is highly toxic and persists in the environment and in living tissue. Its use is now banned in most countries, but it continues to be used on food plants in Latin America.

deadly nightshade another name for ◊belladonna, a poisonous plant.

deafness partial or total deficit of hearing in either ear. Of assistance are hearing aids, lip-reading, a cochlear implant in the ear in combination with a special electronic processor, sign language, and 'cued speech' (manual clarification of ambiguous lip movement during speech).

Conductive deafness is due to faulty conduction of sound inwards from the external ear, usually due to infection (see ◊otitis), or a hereditary abnormality of the bones of the inner ear (see ◊otosclerosis).

Perceptive deafness may be inborn or caused by injury or disease of the cochlea, auditory nerve, or the hearing centres in the brain. It becomes more common with age.

deamination removal of the amino group ($-NH_2$) from an unwanted ◊amino acid. This is the nitrogen-containing part, and it is converted into ammonia, uric acid, or urea (depending on the type of animal) to be excreted in the urine.

In vertebrates, deamination occurs in the ◊liver.

death cessation of all life functions, so that the molecules and structures associated with living things become disorganized and indistinguishable from similar molecules found in nonliving things. In medicine, a person is pronounced dead when the brain ceases to control the vital functions, even if breathing and heartbeat are maintained artificially.

medical definition Death used to be pronounced with the permanent cessation of heartbeat, but the advent of life-support equipment has made this point sometimes difficult to determine. For removal of vital organs in transplant surgery, the World Health Organization in 1968 set out that a potential donor should exhibit

no brain–body connection, muscular activity, blood pressure, or ability to breathe spontaneously.
religious belief In religious belief, death may be seen as the prelude to rebirth (as in Hinduism and Buddhism); under Islam and Christianity, there is the concept of a day of judgement and consignment to heaven or hell; Judaism concentrates not on an afterlife but on survival through descendants who honour tradition.

death cap fungus of the ◊amanita group, the most poisonous mushroom known. The fruiting body, or mushroom, has a scaly white cap and a collarlike structure (volva) near the base of the stalk. (*Amanita phalloides,* family Agaricaceae.)

death's-head moth largest British ◊hawk moth with downy wings measuring 13 cm/5 in from tip to tip, and its thorax is marked as though with a skull.

When it is at rest it sometimes gives out a squeaking noise, produced probably by rubbing the palpi (sense organs close to the mouthparts) upon the proboscis.

The caterpillar is about 10 cm/4 in long and is brightly coloured. It feeds on potato plants.
classification Hawk moths *Acherontia atropos* are in the family Sphingidae in order Lepidoptera, class Insecta, phylum Arthropoda.

deathwatch beetle any wood-boring beetle of the family Anobiidae, especially *Xestobium rufovillosum.* The larvae live in oaks and willows, and sometimes cause damage by boring in old furniture or structural timbers. To attract the female, the male beetle produces a ticking sound by striking his head on a wooden surface, and this is taken by the superstitious as a warning of approaching death.

debt-for-nature swap agreement under which a proportion of a country's debts are written off in exchange for a commitment by the debtor country to undertake projects for environmental protection. Debt-for-nature swaps were set up by environment groups in the 1980s in an attempt to reduce the debt problem of poor countries, while simultaneously promoting conservation.

Most debt-for-nature swaps have concentrated on setting aside areas of land, especially tropical rainforest, for protection and have involved private conservation foundations. The first swap took place in 1987, when a US conservation group bought $650,000 of Bolivia's national debt from a bank for $100,000, and persuaded the Bolivian government to set aside a large area of rainforest as a nature reserve in exchange for never having to pay back the money owed. Other countries participating in debt-for-nature swaps are the Philippines, Costa Rica, Ecuador, and Poland. However, the debtor country is expected to ensure that the area of land remains adequately protected, and in practice this does not always happen. The practice has also produced complaints of neocolonialism.

debugging finding and removing errors, or ◊bugs, from a computer program or system.

DEC (acronym for ***Digital Equipment Corporation***) US computer manufacturer. DEC was founded by US computer engineers, Kenneth Olsen and Harlan Anderson, and was the first ◊minicomputer manufacturer. It became the world's second largest computer manufacturer, after ◊IBM, but made huge losses in the early 1990s. DEC's most successful computers were the PDP-11 and the VAX (Virtual Address eXtension). The former was used in the creation of the ◊UNIX operating system.

The original aim was to make the first small computers for engineering and departmental use, and the PDP (Programmed Data Processor) range became known as minicomputers to contrast them with giant mainframes. The success of its 32-bit VAX minis in the 1980s made DEC one of the world's largest computer manufacturers. The company – now called **Digital** – has still to recover, but has developed the world's fastest microprocessor, the Alpha chip, and has a popular search engine called ◊AltaVista on the Internet's World Wide Web.

decagon in geometry, a ten-sided ◊polygon.

decay, radioactive see ◊radioactive decay.

Decca navigation system radio-based aid to marine navigation, available in many parts of the world. The system consists of a master radio transmitter and two or three secondary transmitters situated within 100–200 km/60–120 mi from the master. The signals from the transmitters are detected by the ship's navigation receiver, and slight differences (phase differences) between the master and secondary transmitter signals indicate the position of the receiver. It was first used 1944 for the D-Day landings.

decibel unit (symbol dB) of measure used originally to compare sound intensities and subsequently electrical or electronic power outputs; now also used to compare voltages. An increase of 10 dB is equivalent to a 10-fold increase in intensity or power, and a 20-fold increase in voltage. The decibel scale is used for audibility measurementsi, as one decibel, representing an increase of about 25%, is about the smallest change the human ear can detect. A whisper has an intensity of 20 dB; 140 dB (a jet aircraft taking off nearby) is the threshold of pain.

The difference in decibels between two levels of intensity (or power) L_1 and L_2 is 10 $\log_{10}(L_1/L_2)$; a difference of 1 dB thus corresponds to a change of about 25%. For two voltages V_1 and V_2, the difference in decibels is 20 $\log_{10}(V_1/V_2)$; 1 dB corresponding in this case to a change of about 12%.

Decibel scale

The decibel scale is used primarily to compare sound intensities although it can be used to compare voltages.

Decibels	*Typical sound*
0	threshold of hearing
10	rustle of leaves in gentle breeze
10	quiet whisper
20	average whisper
20–50	quiet conversation
40–45	hotel; theatre (between performances)
50–65	loud conversation
65–70	traffic on busy street
65–90	train
75–80	factory (light/medium work)
90	heavy traffic
90–100	thunder
110–140	jet aircraft at take-off
130	threshold of pain
140–190	space rocket at take-off

deciduous of trees and shrubs, that shed their leaves at the end of the growing season or during a dry season to reduce ◊transpiration (the loss of water by evaporation).

Most deciduous trees belong to the ◊angiosperms, plants in which the seeds are enclosed within an ovary, and the term 'deciduous tree' is sometimes used to mean 'angiosperm tree', despite the fact that many angiosperms are evergreen, especially in the tropics, and a few ◊gymnosperms, plants in which the seeds are exposed, are deciduous (for example, larches). The term **broad-leaved** is now preferred to 'deciduous' for this reason.

decimal fraction a ◊fraction in which the denominator is any higher power of 10. Thus $\frac{3}{10}$, $\frac{51}{100}$, and $\frac{23}{1000}$ are decimal fractions and are normally expressed as 0.3, 0.51, and 0.023. The use of decimals greatly simplifies addition and multiplication of fractions, though not all fractions can be expressed exactly as decimal fractions.

The regular use of the decimal point appears to have been introduced about 1585, but the occasional use of decimal fractions can be traced back as far as the 12th century.

decimal number system or ***denary number system*** the most commonly used number system, to the base ten. Decimal numbers do not necessarily contain a decimal point; 563, 5.63, and –563 are all decimal numbers. Other systems are mainly used in computing and include the ◊binary number system, ◊octal number system, and ◊hexadecimal number system.

Decimal numbers may be thought of as written under column headings based on the number ten. For example, the number 2,567 stands for 2 thousands, 5 hundreds, 6 tens, and 7 ones. Large decimal numbers may also be expressed in ◊floating-point notation.

decimal point the dot dividing a decimal number's whole part from its fractional part (the digits to the left of the point are unit digits). It is usually printed on the line but hand written above the line, for example 3.5. Some European countries use a comma to denote the decimal point, for example 3,56.

decision table in computing, a method of describing a procedure for a program to follow, based on comparing possible decisions and their consequences. It is often used as an aid in systems design.

The top part of the table contains the conditions for making decisions (for example, if a number is negative rather than positive and is less than 1), and the bottom part describes the outcomes when those conditions are met. The program either ends or repeats the operation.

declarative programming computer programming that does not describe how to solve a problem, but rather describes the logical structure of the problem. It is used in the programming language PROLOG. Running such a program is more like proving an assertion than following a ◊procedure.

declination in astronomy, the coordinate on the ◊celestial sphere (imaginary sphere surrounding the Earth) that corresponds to latitude on the Earth's surface. Declination runs from 0° at the celestial equator to 90° at the north and south celestial poles.

decoder in computing, an electronic circuit used to select one of several possible data pathways. Decoders are, for example, used to direct data to individual memory locations within a computer's immediate access memory.

decomposer in biology, any organism that breaks down dead matter. Decomposers play a vital role in the ◊ecosystem by freeing important chemical substances, such as nitrogen compounds, locked up in dead organisms or excrement. They feed on some of the released organic matter, but leave the rest to filter back into the soil as dissolved nutrients, or pass in gas form into the atmosphere, for example as nitrogen and carbon dioxide.

The principal decomposers are bacteria and fungi, but earthworms and many other invertebrates are often included in this group. The ◊nitrogen cycle relies on the actions of decomposers.

decomposition process whereby a chemical compound is reduced to its component substances. In biology, it is the destruction of dead organisms either by chemical reduction or by the action of decomposers, such as bacteria and fungi.

decompression sickness illness brought about by a sudden and substantial change in atmospheric pressure. It is caused by a too rapid release of nitrogen that has been dissolved into the bloodstream under pressure; when the nitrogen forms bubbles it causes the ◊bends. The condition causes breathing difficulties, joint and muscle pain, and cramps, and is experienced mostly by deep-sea divers who surface too quickly.

After a one-hour dive at 30 m/100 ft, 40 minutes of decompression are needed, according to US Navy tables.

decontamination factor in radiological protection, a measure of the effectiveness of a decontamination process. It is the ratio of the original contamination to the remaining radiation after decontamination: 1,000 and above is excellent; 10 and below is poor.

decrepitation in crystallography, unusual features that accompany the thermal decomposition of some crystals, such as lead(II) nitrate. When these are heated, they spit and crackle and may jump out of the test tube before they decompose.

dedicated computer computer built into another device for the purpose of controlling or supplying information to it. Its use has increased dramatically since the advent of the ◊microprocessor: washing machines, digital watches, cars, and video recorders all now have their own processors.

A dedicated system is a general-purpose computer system confined to performing only one function for reasons of efficiency or convenience. A word processor is an example.

Deep Blue name given to the IBM chess-playing computer that first defeated a human grandmaster, the Russian Gary Kasparov, in 1996.

The architect and principal designer of Deep Blue is Feng-Hsiung Hsu, who joined IBM in 1989. Deep Blue's award-winning precursor, Deep Thought, was developed by Hsu and other graduate students at Carnegie-Mellon University in Pittsburgh. In 1988 it was the first computer to achieve a grandmaster rating.

KASPAROV V. DEEP BLUE – THE REMATCH

`http://www.chess.ibm.com/`

Official site of the team that produced the first computer able to beat a world chess champion. This is a complete account of the tussle between Deep Blue and Gary Kasparov. There are some thought provoking articles on the consequences of Deep Blue's victory. There is also some video footage of the games.

deep-sea trench another term for ◊ocean trench.

deer any of various ruminant, even-toed, hoofed mammals belonging to the family Cervidae. The male typically has a pair of antlers, shed and regrown each year. Most species of deer are forest-dwellers and are distributed throughout Eurasia and North America, but are absent from Australia and Africa south of the Sahara.

deerhound breed of large, rough-coated dog, formerly used in Scotland for hunting and killing deer. Slim and long-legged, it grows to 75 cm/30 in or more, usually with a bluish-grey coat.

default in computing, a factory setting for user-configurable options. Default settings appear in all areas of computing, from the on-screen colour scheme in a ◊graphical user interface to the directories where software programs store data.

defibrillation use of electrical stimulation to restore a chaotic heartbeat to a rhythmical pattern. In fibrillation, which may occur in most kinds of heart disease, the heart muscle contracts irregularly; the heart is no longer working as an efficient pump. Paddles are applied to the chest wall, and one or more electric shocks are delivered to normalize the beat.

In patients suffering with ◊arrhythmia, implantable defibrillators are inserted into the chest with leads threading through veins into the right side of the heart. The first was implanted in 1980 and by 1996 around 50,000–80,000 had been implanted worldwide.

In nature there are neither rewards nor punishments – there are consequences.

ROBERT INGERSOLL US lawyer and orator.
Lectures and Essays, 'Some Reasons Why'

deforestation destruction of forest for timber, fuel, charcoal burning, and clearing for agriculture and extractive industries, such as mining, without planting new trees to replace those lost (reafforestation) or working on a cycle that allows the natural forest to regenerate. Deforestation causes fertile soil to be blown away or washed into rivers, leading to ◊soil erosion, drought, flooding, and loss of wildlife. It may also increase the carbon dioxide content of the atmosphere and intensify the ◊greenhouse effect, because there are fewer trees absorbing carbon dioxide from the air for photosynthesis.

Many people are concerned about the rate of deforestation as great damage is being done to the habitats of plants and animals.

Deforestation ultimately leads to famine, and is thought to be partially responsible for the flooding of lowland areas – for example, in Bangladesh – because trees help to slow down water movement.

defragmentation program or ***disc optimizer*** in computing, a program that rearranges data on disc so that files are not scattered in many small sections. See also ◊fragmentation.

degaussing neutralization of the magnetic field around a body by encircling it with a conductor through which a current is maintained. Ships were degaussed in World War II to prevent them from detonating magnetic mines.

degeneration in biology, a change in the structure or chemical composition of a tissue or organ that interferes with its normal functioning. Examples of degeneration include fatty degeneration, fibroid degeneration (cirrhosis), and calcareous degeneration, all of which are part of natural changes that occur in old age.

The causes of degeneration are often unknown. Heredity often has a role in the degeneration of organs; for example, fibroid changes in the kidney can be seen in successive generations. Defective nutrition and continued stress on particular organs can cause degenerative changes. Alcoholism can result in cirrhosis of the liver and tuberculosis causes degeneration of the lungs.

de Gennes, Pierre-Gilles 1932 French physicist who worked on liquid crystals and polymers. He showed how mathematical models, developed for studying simpler systems, are applicable to such complicated systems. He won the Nobel Prize for Physics in 1991.

It had been known for a long time that liquid crystals scatter light in an unusual way but all early explanations failed. De Gennes found the explanation in the special way that the molecules of a liquid crystal are arranged. According to de Gennes, the molecules are arranged in a similar way to the molecules of a magnet, so that they point in the same direction. De Gennes found similar analogies between the behaviour of molecules in magnetic materials and polymers. This led to the formulation of laws from which simple relations between different properties of polymers can be deduced. In this way, predictions can be made about unknown properties – predictions which have been confirmed by experiment.

degree in mathematics, a unit (symbol °) of measurement of an angle or arc. A circle or complete rotation is divided into 360°. A degree may be subdivided into 60 minutes (symbol '), and each minute may be subdivided in turn into 60 seconds (symbol ').

Temperature is also measured in degrees, which are divided on a decimal scale. See also ◊Celsius, and ◊Fahrenheit.

A degree of latitude is the length along a meridian such that the difference between its north and south ends subtend an angle of 1° at the centre of the Earth. A degree of longitude is the length between two meridians making an angle of 1° at the centre of the Earth.

dehydration in chemistry, the removal of water from a substance to give a product with a new chemical formula; it is not the same as drying.

There are two types of dehydration. For substances such as hydrated copper sulphate ($CuSO_4.5H_2O$) that contain ◊water of crystallization, dehydration means removing this water to leave the anhydrous substance. This may be achieved by heating, and is reversible.

Some substances, such as ethanol, contain the elements of water (hydrogen and oxygen) joined in a different form. **Dehydrating agents** such as concentrated sulphuric acid will remove these elements in the ratio 2:1.

dehydration process to preserve food. Moisture content is reduced to 10–20% in fresh produce, and this provides good protection against moulds. Bacteria are not inhibited by drying, so the quality of raw materials is vital.

The process was developed commercially in France about 1795 to preserve sliced vegetables, using a hot-air blast. The earliest large-scale application was to starch products such as pasta, but after 1945 it was extended to milk, potatoes, soups, instant coffee, and prepared baby and pet foods. A major benefit to food manufacturers is reduction of weight and volume of the food products, thus lowering distribution cost.

Deimos one of the two moons of Mars. It is irregularly shaped, 15 × 12 × 11 km/9 × 7.5 × 7 mi, orbits at a height of 24,000 km/15,000 mi every 1.26 days, and is not as heavily cratered as the other moon, Phobos. Deimos was discovered in 1877 by US astronomer Asaph Hall (1829–1907), and is thought to be an asteroid captured by Mars's gravity.

delete remove or erase. In computing, the deletion of a character removes it from the file; the deletion of a file normally means removing its directory entry, rather than actually deleting it from the disc. Many systems now have an ◊undelete facility that allows the restoration of the directory entry. While deleted files may not have been removed from the disc, they can be overwritten.

deliquescence phenomenon of a substance absorbing so much moisture from the air that it ultimately dissolves in it to form a solution.

Deliquescent substances make very good drying agents and are used in the bottom chambers of ◊desiccators. Calcium chloride ($CaCl_2$) is one of the commonest.

delirium in medicine, a state of acute confusion in which the subject is incoherent, frenzied, and out of touch with reality. It is often accompanied by delusions or hallucinations.

Delirium may occur in feverish illness, some forms of mental illness, brain disease, and as a result of drug or alcohol intoxication. In chronic alcoholism, attacks of **delirium tremens** (DTs), marked by hallucinations, sweating, trembling, and anxiety, may persist for several days.

Delphi in computing, text-based UK and US national online information service. In 1992, Delphi was the first national US service to open a ◊gateway to the Internet. Founded 1982 as the world's first online encyclopedia, Delphi was bought by News International 1993, and launched its UK service in 1994. In 1996 the US arm of Delphi was sold back to one of its original owners. The UK service continues in the hands of News International.

delphinium any of a group of plants containing about 250 species, including the butterfly or Chinese delphinium (*Delphinium grandiflorum*), an Asian form and one of the ancestors of the garden delphinium. Most species have blue, purple, or white flowers on a long spike. (Genus *Delphinium,* family Ranunculaceae.)

delta tract of land at a river's mouth, composed of silt deposited as the water slows on entering the sea. Familiar examples of large deltas are those of the Mississippi, Ganges and Brahmaputra, Rhône, Po, Danube, and Nile; the shape of the Nile delta is like the Greek letter *delta* Δ, and thus gave rise to the name.

The **arcuate delta** of the Nile is only one form. Others are **bird-foot deltas**, like that of the Mississippi which is a seaward extension of the river's ◊levee system; and **tidal deltas**, like that of the Mekong, in which most of the material is swept to one side by sea currents.

Delta rocket US rocket used to launch many scientific and communications satellites since 1960, based on the Thor ballistic missile. Several increasingly powerful versions produced as satellites became larger and heavier. Solid-fuel boosters were attached to the first stage to increase lifting power.

delta wing aircraft wing shaped like the Greek letter *delta* Δ. Its design enables an aircraft to pass through the ◊sound barrier with little effect. The supersonic airliner ◊Concorde and the US ◊space shuttle have delta wings.

dementia mental deterioration as a result of physical changes in the brain. It may be due to degenerative change, circulatory disease, infection, injury, or chronic poisoning. **Senile dementia**, a progressive loss of mental faculties such as memory and orientation, is typically a disease process of old age, and can be accompanied by ◊depression.

Dementia is distinguished from amentia, or severe congenital mental insufficiency.

demo or ***demonstration software*** in computing, preview version of software that allows users to try out the main features of a particular program before buying it. Especially common among ◊shareware producers, demo software usually blocks some features of the full version, so that a demo database might be able to handle only a small number of records.

The word 'demo' is also used to refer to fancy graphics and sound routines which are created by young programmers to demonstrate their skills to friends, admirers, and potential employers such as computer game publishers.

demodulation in radio, the technique of separating a transmitted audio frequency signal from its modulated radio carrier wave. At the transmitter, the audio frequency signal (representing speech or music, for example) may be made to modulate the amplitude (AM broadcasting) or frequency (FM broadcasting) of a continuously transmitted radio-frequency carrier wave. At the receiver, the signal from the aerial is demodulated to extract the required speech or sound component. In early radio systems, this process was called detection. See ◊modulation.

Demon Internet in computing, Britain's first and largest mass-market ◊Internet Service Provider. Founded 1992 by English hardware salesman Cliff Stanford with 200 founding subscribers who each paid £120 in advance for a year's service, Demon set the price (£10 a month plus VAT) for Internet access in the UK. By 1996, Demon Internet had 65,000 customers.

demonstration software in computing, see ◊demo.

denaturation irreversible changes occurring in the structure of proteins such as enzymes, usually caused by changes in pH or temperature, by radiation or chemical treatments. An example is the heating of egg albumen resulting in solid egg white.

The enzymes associated with digestion and metabolism become inactive if given abnormal conditions. Heat will damage their complex structure so that the usual interactions between enzyme and substrate can no longer occur.

dendrite part of a ◊nerve cell or neuron. The dendrites are slender filaments projecting from the cell body. They receive incoming messages from many other nerve cells and pass them on to the cell body.

If the combined effect of these messages is strong enough, the cell body will send an electrical impulse along the axon (the thread-like extension of a nerve cell). The tip of the axon passes its message to the dendrites of other nerve cells.

dendrochronology or ***tree-ring dating*** analysis of the ◊annual rings of trees to date past events by determining the age of timber. Since annual rings are formed by variations in the water-conducting cells produced by the plant during different seasons of the year, they also provide a means of establishing past climatic conditions in a given area.

Samples of wood are obtained by driving a narrow metal tube into a tree to remove a core extending from the bark to the centre. Samples taken from timbers at an archaeological site can be compared with a master core on file for that region or by taking cores from old living trees; the year when they were felled can be determined by locating the point where the rings of the two samples correspond and counting back from the present.

Moisture levels will affect growth, the annual rings being thin in dry years, thick in moist ones, although in Europe ring growth is most affected by temperature change and insect defoliation.

In North America, studies are now extremely extensive, covering many wood types, including sequoia, juniper, and sagebrush. Sequences of tree rings extending back over 8,000 years have been obtained for the southwest USA and northern Mexico by using cores from the bristle-cone pine *Pinus aristata* of the White Mountains, California, which can live for over 4,000 years in that region. The dryness of the southwest USA has preserved wood in its archaeological sites. In wet temperate regions, wood is usually absorbed by soil acidity so that this dating technique cannot be used.

Deneb or ***Alpha Cygni*** brightest star in the constellation ◊Cygnus, and the 19th brightest star in the night sky. It is one of the greatest supergiant stars known, with a true luminosity of about 60,000 times that of the Sun. Deneb is about 1,800 light years from Earth.

The name Deneb is derived from the Arabic word for tail.

denier unit used in measuring the fineness of yarns, equal to the mass in grams of 9,000 metres of yarn. Thus 9,000 metres of 15 denier nylon, used in nylon stockings, weighs 15 g/0.5 oz, and in this case the thickness of thread would be 0.00425 mm/0.0017 in. The term is derived from the French silk industry; the *denier* was an old French silver coin.

denitrification process occurring naturally in soil, where bacteria break down ◊nitrates to give nitrogen gas, which returns to the atmosphere.

denominator in mathematics, the bottom number of a fraction, so called because it names the family of the fraction. The top number, or numerator, specifies how many unit fractions are to be taken.

density measure of the compactness of a substance; it is equal to its mass per unit volume and is measured in kg per cubic metre/lb per cubic foot. Density is a ◊scalar quantity. The average density D of a mass m occupying a volume V is given by the formula:

$$D = m/V$$

◊Relative density is the ratio of the density of a substance to that of water at 4°C/32.2°F.

In photography, density refers to the degree of opacity of a negative; in population studies, it is the quantity or number per unit of area; in electricity, current density is the amount of current passing through a cross-sectional area of a conductor (usually given in amperes per sq in or per sq cm).

density wave in astrophysics, a concept proposed to account for the existence of spiral arms in ◊galaxies. In the density wave theory, stars in a spiral galaxy move in elliptical orbits in such a way that they crowd together in waves of temporarily enhanced density that appear as spiral arms. The idea was first proposed by Swedish astronomer Bertil Lindblad in the 1920s and developed by US astronomers C C Lin and Frank Shu in the 1960s.

dental caries in medicine, another name for ◊caries.

dental formula way of showing the number of teeth in an animal's mouth. The dental formula consists of eight numbers separated by a line into two rows. The four above the line represent the teeth on one side of the upper jaw, starting at the front. If this reads 2 1 2 3 (as for humans) it means two incisors, one canine, two premolars, and three molars (see ◊tooth). The numbers below the line represent the lower jaw. The total number of teeth can be calculated by adding up all the numbers and multiplying by two.

dentistry care and treatment of the teeth and gums. **Orthodontics** deals with the straightening of the teeth for aesthetic and clinical reasons, and **periodontics** with care of the supporting tissue (bone and gums).

The bacteria that start the process of dental decay are normal, nonpathogenic members of a large and varied group of microorganisms present in the mouth. They are strains of oral streptococci, and it is only in the presence of sucrose (from refined sugar) in the mouth that they become damaging to the teeth. ◊Fluoride in the water supply has been one attempted solution to prevent decay,

DENTISTRY NOW

http://www.DentistryNow.com/Mainpage.htm#Main Part

Canadian-based site that includes an index of dentists worldwide. There is also an index of university courses where dentistry can be studied. In addition, there is a section on common dental problems and a good site for kids called 'tooth fairy' to introduce them to the importance of cleaning their teeth.

and in 1979 a vaccine was developed from a modified form of the bacterium *Streptococcus mutans.*

dentition type and number of teeth in a species. Different kinds of teeth have different functions; a grass-eating animal will have large molars for grinding its food, whereas a meat-eater will need powerful canines for catching and killing its prey. The teeth that are less useful to an animal's lifestyle may be reduced in size or missing altogether. An animal's dentition is represented diagrammatically by a ◊dental formula.

Young children have **deciduous dentition**, popularly known as 'milk teeth', the first ones erupting at about six months of age. **Mixed dentition** is present from the ages of about six (when the first milk teeth are shed) to about 12. **Permanent dentition** (up to 32 teeth) is usually complete by the mid-teens, although the third molars (wisdom teeth) may not appear until around the age of 21.

herbivore (sheep)

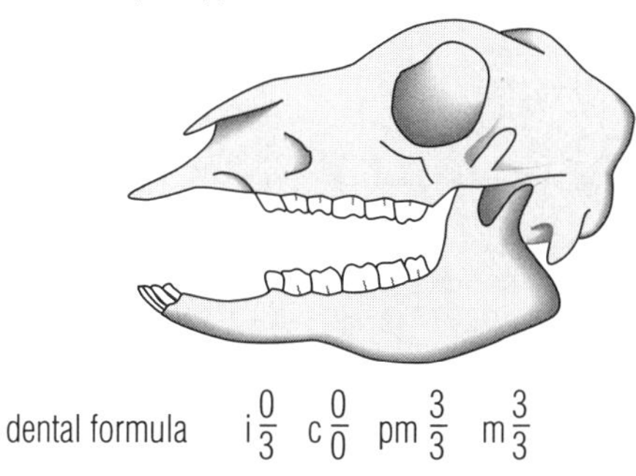

carnivore (dog)

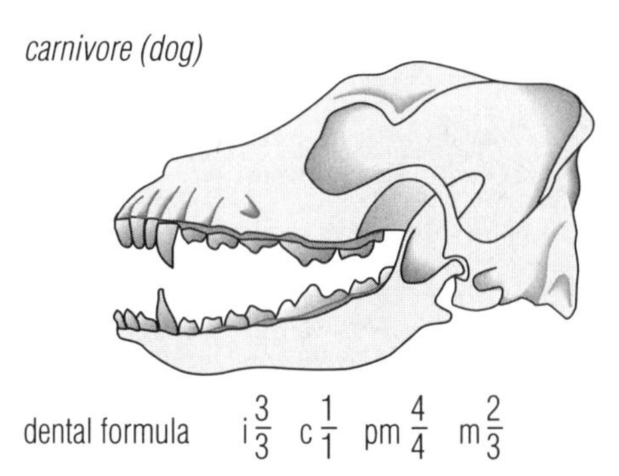

***dentition** The dentition and dental formulae of a typical herbivore (sheep) and carnivore (dog). The dog has long pointed canines for puncturing and gripping its prey and has modified premolars and molars (carnassials) for shearing flesh. In the sheep, by contrast, there is a wide gap, or diastema, between the incisors, developed for cutting through blades of grass, and the grinding premolars and molars; the canines are absent.*

denudation natural loss of soil and rock debris, blown away by wind or washed away by running water, laying bare the rock below. Over millions of years, denudation causes a general lowering of the landscape.

deodar Himalayan ◊cedar tree, often planted as a rapid-growing ornamental. Its fragrant, durable wood is valuable as timber. (*Cedrus deodara,* family Pinaceae.)

deoxyribonucleic acid full name of ◊DNA.

depolarizer oxidizing agent used in dry-cell batteries that converts hydrogen released at the negative electrode into water. This prevents the build-up of gas, which would otherwise impair the efficiency of the battery. ◊Manganese(IV) oxide is used for this purpose.

depression or ***cyclone*** or ***low*** in meteorology, a region of low atmospheric pressure. In mid latitudes a depression forms as warm, moist air from the tropics mixes with cold, dry polar air, producing warm and cold boundaries (◊fronts) and unstable weather – low cloud and drizzle, showers, or fierce storms. The warm air, being less dense, rises above the cold air to produce the area of low pressure on the ground. Air spirals in towards the centre of the depression in an anticlockwise direction in the northern hemisphere, clockwise in the southern hemisphere, generating winds up to gale force. Depressions tend to travel eastwards and can remain active for several days.

depression in medicine, an emotional state characterized by sadness, unhappy thoughts, apathy, and dejection. Sadness is a normal response to major losses such as bereavement or unemployment. After childbirth, ◊postnatal depression is common. However, clinical depression, which is prolonged or unduly severe, often requires treatment, such as antidepressant medication, ◊cognitive therapy, or, in very rare cases, electroconvulsive therapy (ECT), in which an electrical current is passed through the brain.

Periods of depression may alternate with periods of high optimism, over-enthusiasm, and confidence. This is the manic phase in a disorder known as **manic depression** or **bipolar disorder**. A manic depressive state is one in which a person switches repeatedly from one extreme to the other. Each mood can last for weeks or months. Typically, the depressive state lasts longer than the manic phase.

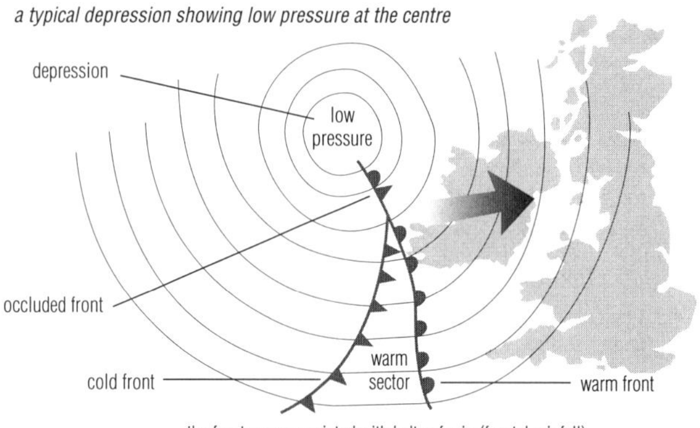

depression

derivative or ***differential coefficient*** in mathematics, the limit of the gradient of a chord linking two points on a curve as the distance between the points tends to zero; for a function of a single variable, $y = f(x)$, it is denoted by $f'(x)$, $Df(x)$, or dy/dx, and is equal to the gradient of the curve.

dermatitis inflammation of the skin (see ◊eczema), usually related to allergy. **Dermatosis** refers to any skin disorder and may be caused by contact or systemic problems.

dermatology medical speciality concerned with the diagnosis and treatment of skin disorders.

derrick simple lifting machine consisting of a pole carrying a block and tackle. Derricks are commonly used on ships that carry freight. In the oil industry the tower used for hoisting the drill pipes is known as a derrick.

derris climbing leguminous plant (see ◊legume) of southeast Asia. Its roots contain rotenone, a strong insecticide. (*Derris elliptica,* family Fabaceae.)

DES in computing, abbreviation for ◊Data Encryption Standard.

desalination removal of salt, usually from sea water, to produce fresh water for irrigation or drinking. Distillation has usually been the method adopted, but in the 1970s a cheaper process, using certain polymer materials that filter the molecules of salt from the water by reverse osmosis, was developed.

Desalination plants occur along the shores of the Middle East where fresh water is in short supply.

DESCARTES, RENÉ

http://www.knight.org/advent/cathen/04744b.htm

Extensive treatment of the life and philosophical, scientific, and mathematical achievements of Renatus Cartesius. The restless travels of the young savant are described before a detailed discussion of his contribution to learning.

Descartes, René (1596–1650)

French philosopher and mathematician. He believed that commonly accepted knowledge was doubtful because of the subjective nature of the senses, and attempted to rebuild human knowledge using as his foundation the dictum cogito ergo sum ('I think, therefore I am'). He also believed that the entire material universe could be explained in terms of mathematical physics, and founded coordinate geometry as a way of defining and manipulating geometrical shapes by means of algebraic expressions. Cartesian coordinates, the means by which points are represented in this system, are named after him. Descartes also established the science of optics, and helped to shape contemporary theories of astronomy and animal behaviour.

Descartes identified the 'thinking thing' (res cogitans), or mind, with the human soul or consciousness; the body, though somehow interacting with the soul, was a physical machine, secondary to, and in principle separable from, the soul. He held that everything has a cause; nothing can result from nothing. He believed that, although all matter is in motion, matter does not move of its own accord; the initial impulse comes from God. He also postulated two quite distinct substances: spatial substance, or matter, and thinking substance, or mind. This is called 'Cartesian dualism', and it preserved him from serious controversy with the church.

Mary Evans Picture Library

Cogito, ergo sum.
I think, therefore I am.

RENÉ DESCARTES French philosopher and mathematician.
Le discours de la méthode

desert arid area with sparse vegetation (or, in rare cases, almost no vegetation). Soils are poor, and many deserts include areas of shifting sands. Deserts can be either hot or cold. Almost 33% of the Earth's land surface is desert, and this proportion is increasing.

The **tropical desert** belts of latitudes from 5° to 30° are caused by the descent of air that is heated over the warm land and therefore has lost its moisture. Other natural desert types are the **continental deserts**, such as the Gobi, that are too far from the sea to receive any moisture; **rain-shadow deserts**, such as California's Death Valley, that lie in the lee of mountain ranges, where the ascending air drops its rain only on the windward slopes; and **coastal deserts**, such as the Namib, where cold ocean currents cause local dry air masses to descend. Desert surfaces are usually rocky or gravelly, with only a small proportion being covered with sand. Deserts can be created by changes in climate, or by the human-aided process of desertification.

Largest deserts in the world

Desert	Location	Area[1] sq km	sq mi
Sahara	northern Africa	9,065,000	3,500,000
Gobi	Mongolia/northeastern China	1,295,000	500,000
Patagonian	Argentina	673,000	260,000
Rub al-Khali	southern Arabian peninsula	647,500	250,000
Chihuahuan	Mexico/southwestern USA	362,600	140,000
Taklimakan	northern China	362,600	140,000
Great Sandy	northwestern Australia	338,500	130,000
Great Victoria	southwestern Australia	338,500	130,000
Kalahari	southwestern Africa	260,000	100,000
Kyzyl Kum	Uzbekistan	259,000	100,000
Thar	India/Pakistan	259,000	100,000
Sonoran	Mexico/southwestern USA	181,300	70,000
Simpson	Australia	103,600	40,000
Mojave	southwestern USA	51,800	20,000

[1] Desert areas are very approximate, because clear physical boundaries may not occur.

Characteristics common to all deserts include irregular rainfall of less than 250 mm/19.75 in per year, very high evaporation rates often 20 times the annual precipitation, and low relative humidity and cloud cover. Temperatures are more variable; tropical deserts have a big diurnal temperature range and very high daytime temperatures (58°C/136.4°F) has been recorded at Azizia in Libya), whereas mid-latitude deserts have a wide annual range and much lower winter temperatures (in the Mongolian desert the mean temperature is below freezing point for half the year).

Desert soils are infertile, lacking in ◊humus and generally grey or red in colour. The few plants capable of surviving such conditions are widely spaced, scrubby and often thorny. Long-rooted plants (phreatophytes) such as the date palm and musquite commonly grow along dry stream channels. Salt-loving plants (◊halophytes) such as saltbushes grow in areas of highly saline soils and near the edges of playas (dry saline lakes). Others, such as the ◊xerophytes are drought-resistant and survive by remaining leafless during the dry season or by reducing water losses with small waxy leaves. They frequently have shallow and widely branching root systems and store water during the wet season (for example, succulents and cacti with pulpy stems).

desertification spread of deserts by changes in climate, or by human-aided processes. Desertification can sometimes be reversed by special planting (marram grass, trees) and by the use of water-absorbent plastic grains, which, added to the soil, enable crops to be grown.

The processes leading to desertification include overgrazing, destruction of forest belts, and exhaustion of the soil by intensive cultivation without restoration of fertility – all of which may be prompted by the pressures of an expanding population or by concentration in land ownership. About 135 million people are directly affected by desertification, mainly in Africa, the Indian subcontinent, and South America.

desiccator airtight vessel, traditionally made of glass, in which materials may be stored either to dry them or to prevent them, once dried, from reabsorbing moisture.

The base of the desiccator is a chamber in which is placed a substance with a strong affinity for water (such as calcium chloride or silica gel), which removes water vapour from the desiccator atmosphere and from substances placed in it.

desktop in computing, a graphical representation of file systems, in which applications and files are represented by pictures (icons), which can be triggered by a single or double click with a ◊mouse button. Such a ◊graphical user interface can be compared with the ◊command line interface, which is character-based.

desktop publishing (DTP) use of microcomputers for small-scale typesetting and page makeup. DTP systems are capable of producing camera-ready pages (pages ready for photographing and printing), made up of text and graphics, with text set in different typefaces and sizes. The page can be previewed on the screen before final printing on a laser printer.

A DTP program is able to import text and graphics from other packages; run text as columns, over pages, and around artwork and other insertions; enable a wide range of ◊fonts; and allow accurate positioning of all elements required to make a page.

desktop video in computing, a ◊videoconferencing system that can be used by an individual from a desktop computer. A desktop conferencing system needs a computer, an attached video camera, microphone, and speakers, and a telephone or network connection.

Early videoconferencing systems required such expensive equipment that participants had to gather in the room where the equipment was kept. Systems introduced in the mid-1990s, however, made videoconferencing as convenient, private, and easy to use as ordinary telephone calls.

The first desktop videoconferencing system on the Internet was ◊CU-SeeMe.

destination page in computing, page designated by a ◊hypertext link.

detergent surface-active cleansing agent. The common detergents are made from ◊fats (hydrocarbons) and sulphuric acid, and their long-chain molecules have a type of structure similar to that of ◊soap molecules: a salt group at one end attached to a long hydrocarbon 'tail'. They have the advantage over soap in that they do not produce scum by forming insoluble salts with the calcium and magnesium ions present in hard water.

To remove dirt, which is generally attached to materials by means of oil or grease, the hydrocarbon 'tails' (soluble in oil or grease) penetrate the oil or grease drops, while the 'heads' (soluble in water but insoluble in grease) remain in the water and, being salts, become ionized. Consequently the oil drops become negatively charged and tend to repel one another; thus they remain in suspension and are washed away with the dirt.

Detergents were first developed from coal tar in Germany during World War I, and synthetic organic detergents were increasingly used after World War II.

Domestic powder detergents for use in hot water have alkyl benzene as their main base, and may also include bleaches and fluorescers as whiteners, perborates to free stain-removing oxygen, and water softeners. Environment-friendly detergents contain no phosphates or bleaches. Liquid detergents for washing dishes are based on epoxyethane (ethylene oxide). Cold-water detergents consist of a mixture of various alcohols, plus an ingredient for breaking down the surface tension of the water, so enabling the liquid to penetrate fibres and remove the dirt. When these surface-active agents (surfactants) escape the normal processing of sewage, they cause troublesome foam in rivers; phosphates in some detergents can also cause the excessive enrichment (◊eutrophication) of rivers and lakes.

determinant in mathematics, an array of elements written as a square, and denoted by two vertical lines enclosing the array. For a 2 × 2 matrix, the determinant is given by the difference between the products of the diagonal terms. Determinants are used to solve sets of ◊simultaneous equations by matrix methods.

When applied to transformational geometry, the determinant of a 2 × 2 matrix signifies the ratio of the area of the transformed shape to the original and its sign (plus or minus) denotes whether the image is direct (the same way round) or indirect (a mirror image).

For example, the determinant of the matrix

$$(a\ b) = |\ a\ b\ | = ad - bc(c\ d)\ |\ c\ d\ |$$

detonator or ***blasting cap*** or ***percussion cap*** small explosive charge used to trigger off a main charge of high explosive. The relatively unstable compounds mercury fulminate and lead azide are often used in detonators, being set off by a lighted fuse or, more commonly, an electric current.

detritus in biology, the organic debris produced during the ◊decomposition of animals and plants.

deuterium naturally occurring heavy isotope of hydrogen, mass number 2 (one proton and one neutron), discovered by Harold Urey in 1932. It is sometimes given the symbol D. In nature, about one in every 6,500 hydrogen atoms is deuterium. Combined with oxygen, it produces 'heavy water' (D_2O), used in the nuclear industry.

deuteron nucleus of an atom of deuterium (heavy hydrogen). It consists of one proton and one neutron, and is used in the bombardment of chemical elements to synthesize other elements.

developer in computing, designer of a computer system, most commonly used to mean a software developer.

developing in photography, the process that produces a visible image on exposed photographic ◊film, involving the treatment of the exposed film with a chemical developer.

The developing liquid consists of a reducing agent that changes the light-altered silver salts in the film into darker metallic silver. The developed image is made permanent with a fixer, which dissolves away any silver salts which were not affected by light. The developed image is a negative, or reverse image: darkest where the strongest light hit the film, lightest where the least light fell. To produce a positive image, the negative is itself photographed, and the development process reverses the shading, producing the final print. Colour and black-and-white film can be obtained as direct reversal, slide, or transparency material. Slides and transparencies are used for projection or printing with a positive-to-positive process such as Cibachrome.

development in biology, the process whereby a living thing transforms itself from a single cell into a vastly complicated multicellular organism, with structures, such as limbs, and functions, such as respiration, all able to work correctly in relation to each other. Most of the details of this process remain unknown, although some of the central features are becoming understood.

Apart from the sex cells (◊gametes), each cell within an organism contains exactly the same genetic code. Whether a cell develops into a liver cell or a brain cell depends therefore not on which ◊genes it contains, but on which genes are allowed to be expressed. The development of forms and patterns within an organism, and the production of different, highly specialized cells, is a problem of control, with genes being turned on and off according to the stage of development reached by the organism.

developmental psychology study of development of cognition and behaviour from birth to adulthood.

device driver in computing, small piece of software required to tell the operating system how to interact with a particular input or output device or peripheral.

Much work has been done to standardize devices and their interfaces to eliminate the need for individual device drivers. Peripherals such as CD-ROM drives, for example, work with a single standard device driver (in Microsoft Windows, MSCDEX.EXE). Other devices, such as modems and printers, still need an individual driver tailored to work with that specific model.

devil ray any of several large rays of the genera *Manta* and *Mobula*, in which two 'horns' project forwards from the sides of the huge mouth. These flaps of skin guide the plankton, on which the fish feed, into the mouth.

The largest of these rays can be 7 m/23 ft across, and weigh 1,000 kg/2,200 lb. They live in warm seas.

devil's coach horse large, black, long-bodied, omnivorous beetle *Ocypus olens*, about 3 cm/1.2 in long. It has powerful jaws and is capable of giving a painful bite. It emits an unpleasant smell when threatened.

devil's coach horse The devil's coach horse beetle (family Staphylinidae) adopts a threatening posture with its tail raised and emits an unpleasant smell when it senses danger. Premaphotos Wildlife

Devonian period of geological time 408–360 million years ago, the fourth period of the Palaeozoic era. Many desert sandstones from North America and Europe date from this time. The first land plants flourished in the Devonian period, corals were abundant in the seas, amphibians evolved from air-breathing fish, and insects developed on land.

The name comes from the county of Devon in southwest England, where Devonian rocks were first studied.

dew precipitation in the form of moisture that collects on the ground. It forms after the temperature of the ground has fallen below the ◊dew point of the air in contact with it. As the temperature falls during the night, the air and its water vapour become chilled, and condensation takes place on the cooled surfaces.

dew point temperature at which the air becomes saturated with water vapour. At temperatures below the dew point, the water vapour condenses out of the air as droplets. If the droplets are large they become deposited on the ground as dew; if small they remain in suspension in the air and form mist or fog.

dewpond drinking pond for farm animals on arid hilltops such as chalk downs. In the UK, dewponds were excavated in the 19th century and lined with mud and clay. It is uncertain where the water comes from but it may be partly rain, partly sea mist, and only a small part dew.

dhole wild dog *Cuon alpinus* found in Asia from Siberia to Java. With head and body up to 1 m/39 in long, variable in colour but often reddish above and lighter below, the dhole lives in groups of from 3 to 30 individuals. The species is becoming rare and is protected in some areas.

Dholes can chase prey for long distances; a pack is capable of pulling down deer and cattle as well as smaller prey. They are even known to have attacked tigers and leopards.

diabase alternative name for ◊dolerite (a form of basalt that contains very little silica), especially dolerite that has metamorphosed.

diabetes disease *diabetes mellitus* in which a disorder of the islets of Langerhans in the ◊pancreas prevents the body producing the hormone ◊insulin, so that sugars cannot be used properly.

Treatment is by strict dietary control and oral or injected insulin, depending on the type of diabetes.

There are two forms of diabetes: Type 1, or insulin-dependent diabetes, which usually begins in childhood (early onset) and is an autoimmune condition; and Type 2, or noninsulin-dependent diabetes, which occurs in later life (late onset).

diagenesis in geology, the physical and chemical changes by which a sediment becomes a ◊sedimentary rock. The main processes involved include compaction of the grains, and the cementing of the grains together by the growth of new minerals deposited by percolating groundwater.

dialler in computing, element of an Internet software package that makes the connection to the ◊online service or ◊Internet Service Provider. In Windows systems, this is usually the WINSOCK.DLL file, with or without a front end (part of the program that interacts with the user) to make configuration easier.

dialog box in ◊graphical user interfaces, a small on-screen window with blanks for user input.

dial-up connection in computing, connection to an ◊online system or ◊Internet Service Provider made by dialling via a ◊modem over a telephone line.

dialysis technique for removing waste products from the blood suffering chronic or acute kidney failure. There are two main methods, haemodialysis and peritoneal dialysis.

In **haemodialysis**, the patient's blood is passed through a pump, where it is separated from sterile dialysis fluid by a semipermeable membrane. This allows any toxic substances which have built up in the bloodstream, and which would normally be filtered out by the kidneys, to diffuse out of the blood into the dialysis fluid. Haemodialysis is very expensive and usually requires the patient to attend a specialized unit.

Peritoneal dialysis uses one of the body's natural semipermeable membranes for the same purpose. About two litres of dialysis fluid is slowly instilled into the peritoneal cavity of the abdomen, and drained out again, over about two hours. During that time toxins from the blood diffuse into the peritoneal cavity across the peritoneal membrane. The advantage of peritoneal dialysis is that the patient can remain active while the dialysis is proceeding. This is known as continuous ambulatory peritoneal dialysis (CAPD).

In the long term, dialysis is expensive and debilitating, and ◊transplants are now the treatment of choice for patients in chronic kidney failure.

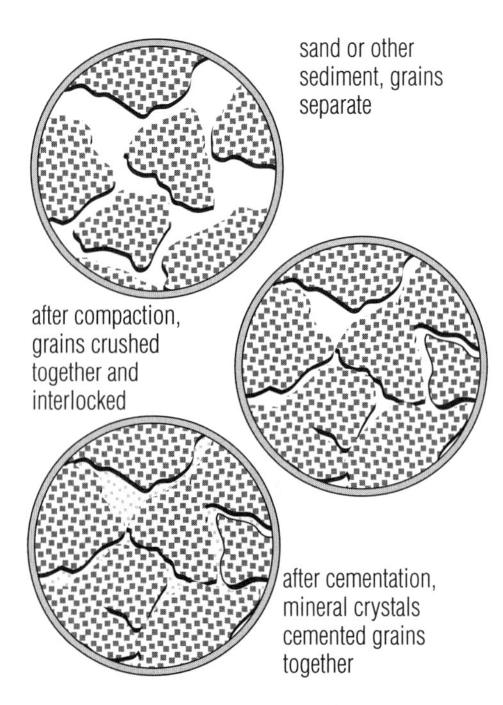

diagenesis The formation of sedimentary rock by diagenesis. Sand and other sediment grains are compacted and cemented together.

diamagnetic material a material weakly repelled by a magnet. All substances are diamagnetic but the behaviour is often masked by stronger forms of magnetism such as ◊paramagnetic or ◊ferromagnetic behaviour.

Diamagnetism is caused by changes in the orbits of electrons in the substance induced by the applied field. Diamagnetic materials have a small negative magnetic ◊susceptibility.

diameter straight line joining two points on the circumference of a circle that passes through the centre of that circle. It divides a circle into two equal halves.

diamond generally colourless, transparent mineral, an ◊allotrope of carbon. It is regarded as a precious gemstone, and is the hardest substance known (10 on the ◊Mohs' scale). Industrial diamonds, which may be natural or synthetic, are used for cutting, grinding, and polishing.

Diamond crystallizes in the cubic system as octahedral crystals, some with curved faces and striations. The high refractive index of 2.42 and the high dispersion of light, or 'fire', account for the spectral displays seen in polished diamonds.

history Diamonds were known before 3000 BC and until their discovery in Brazil in 1725, India was the principal source of supply. Present sources are Australia, Congo (formerly Zaire), Botswana, Russia (Yakut), South Africa, Namibia, and Angola; the first two produce large volumes of industrial diamonds. Today, about 80% of the world's rough gem diamonds are sold through the De Beers Central Selling Organization in London.

sources Diamonds may be found as alluvial diamonds on or close to the Earth's surface in riverbeds or dried watercourses; on the sea bottom (off southwest Africa); or, more commonly, in diamond-bearing volcanic pipes composed of 'blue ground', ◊kimberlite or lamproite, where the original matrix has penetrated the Earth's crust from great depths. They are sorted from the residue of crushed ground by X-ray and other recovery methods.

varieties There are four chief varieties of diamond: well-crystallized transparent stones, colourless or only slightly tinted, valued as gems; **boart**, poorly crystallized or inferior diamonds; **balas**, an industrial variety, extremely hard and tough; and **carbonado**, or industrial diamond, also called black diamond or carbon, which is opaque, black or grey, and very tough. Industrial diamonds are also produced synthetically from graphite. Some synthetic diamonds conduct heat 50% more efficiently than natural diamonds and are five times greater in strength. This is a great advantage in their use to disperse heat in electronic and telecommunication devices and in the production of laser components.

practical uses Because diamonds act as perfectly transparent windows and do not absorb infrared radiation, they were used aboard NASA space probes to Venus in 1978. The tungsten-carbide tools used in steel mills are cut with industrial diamond tools.

cutting Rough diamonds are often dull or greasy before being polished; around 50% are considered 'cuttable' (all or part of the diamond may be set into jewellery). Gem diamonds are valued by weight (◊carat), cut (highlighting the stone's optical properties), colour, and clarity (on a scale from internally flawless to having a large inclusion clearly visible to the naked eye). They are sawn and polished using a mixture of oil and diamond powder. The two most popular cuts are the brilliant, for thicker stones, and the marquise, for shallower ones. India is the world's chief cutting centre.

Noted rough diamonds include the Cullinan, or Star of Africa (3,106 carats, over 500 g/17.5 oz before cutting, South Africa, 1905); Excelsior (995.2 carats, South Africa, 1893); and Star of Sierra Leone (968.9 carats, Yengema, 1972).

experiments a moderate force applied to the small tips of two opposing diamonds can be used to attain extreme pressures of millions of atmospheres or more, allowing scientists to subject small amounts of material to conditions that exist deep within planet interiors.

diamorphine technical term for ◊heroin.

DIANE (acronym for ***direct information access network for Europe***) collection of information suppliers, or 'hosts', for the European computer network.

diapause period of suspended development that occurs in some species of insects, characterized by greatly reduced metabolism. Periods of diapause are often timed to coincide with the winter months, and improve the insect's chances of surviving adverse conditions.

diaphragm in mammals, a thin muscular sheet separating the thorax from the abdomen. It is attached by way of the ribs at either side and the breastbone and backbone, and a central tendon. Arching upwards against the heart and lungs, the diaphragm is important in the mechanics of breathing. It contracts at each inhalation, moving downwards to increase the volume of the chest cavity, and relaxes at exhalation.

diaphragm or ***cap*** or ***Dutch cap*** barrier ◊contraceptive that is passed into the vagina to fit over the cervix (neck of the uterus), preventing sperm from entering the uterus. For a cap to be effective, a ◊spermicide must be used and the diaphragm left in place for 6–8 hours after intercourse. This method is 97% effective if practised correctly.

diarrhoea frequent or excessive action of the bowels so that the faeces are liquid or semiliquid. It is caused by intestinal irritants (including some drugs and poisons), infection with harmful organisms (as in dysentery, salmonella, or cholera), or allergies.

Diarrhoea is the biggest killer of children in the world. In 1996 the World Health Organization reported that 3.1 million deaths had been caused by diarrhoeal disease during 1995. The commonest cause of dehydrating diarrhoea is human rotavirus infection, responsible for about 870,000 infant deaths annually. Dehydration as a result of diarrhoeal disease can be treated by giving a solution of salt and glucose by mouth in large quantities (to restore the electrolyte balance in the blood). Since most diarrhoea is viral in origin, antibiotics are ineffective.

diastole in biology, the relaxation of a hollow organ. In particular, the term is used to indicate the resting period between beats of the heart when blood is flowing into it.

diastolic pressure in medicine, measurement due to the pressure of blood against the arterial wall during diastole (relaxation of the heart). It is the lowest ◊blood pressure during the cardiac cycle. The average diastolic pressure in healthy young adults is about 80 mmHg. The variation of diastolic pressure due to changes in body position and mood is greater than that of systolic pressure. Diastolic pressure is also a more accurate predictor of hypertension (high blood pressure).

diatom microscopic ◊alga found in all parts of the world in either fresh or marine waters. Diatoms consist of single cells that secrete a hard cell wall made of ◊silica. (Division Bacillariophyta.)

The cell wall of a diatom is made up of two overlapping valves known as **frustules**, which are impregnated with silica, and which fit together like the lid and body of a pillbox. Diatomaceous earths (diatomite) are made up of the valves of fossil diatoms, and are used in the manufacture of dynamite and in the rubber and plastics industries.

diatomic molecule molecule composed of two atoms joined together. In the case of an element such as oxygen (O_2), the atoms are identical.

dichloro-diphenyl-trichloroethane full name of the insecticide ◊DDT.

dichotomous key method of identifying an organism. The investigator is presented with pairs of statements, for example 'flower has less than five stamens' and 'flower has five or more stamens'. By successively eliminating statements the field naturalist moves closer to a positive identification. Dichotomous keys assume a good knowledge of the subject under investigation.

dicotyledon major subdivision of the ◊angiosperms, containing the great majority of flowering plants. Dicotyledons are characterized by the presence of two seed leaves, or ◊cotyledons, in the embryo, which is usually surrounded by the ◊endosperm. They generally have broad leaves with netlike veins.

diecasting form of ◊casting in which molten metal is injected into permanent metal moulds or dies.

dielectric an insulator or nonconducter of electricity, such as rubber, glass, and paraffin wax. An electric field in a dielectric material gives rise to no net flow of electricity. However, the applied field causes electrons within the material to be displaced, creating an electric charge on the surface of the material. This reduces the field strength within the material by a factor known as the dielectric constant (or relative permittivity) of the material. Dielectrics are used in capacitors, to reduce dangerously strong electric fields, and have optical applications.

diesel engine ◊internal-combustion engine that burns a lightweight fuel oil. The diesel engine operates by compressing air until it becomes sufficiently hot to ignite the fuel. It is a piston-in-cylinder engine, like the ◊petrol engine, but only air (rather than an air-and-fuel mixture) is taken into the cylinder on the first piston stroke (down). The piston moves up and compresses the air until it is at a very high temperature. The fuel oil is then injected into the hot air, where it burns, driving the piston down on its power stroke. For this reason the engine is called a compression-ignition engine.

Diesel engines have sometimes been marketed as 'cleaner' than petrol engines because they do not need lead additives and produce fewer gaseous pollutants. However, they do produce high levels of the tiny black carbon particles called particulates, which are believed to be carcinogenic and may exacerbate or even cause asthma.

The principle of the diesel engine was first explained in England by Herbert Akroyd (1864–1937) in 1890, and was applied practically by Rudolf Diesel in Germany in 1892.

diesel oil lightweight fuel oil used in diesel engines. Like petrol, it is a petroleum product. When used in vehicle engines, it is also known as **derv** (diesel-engine road vehicle).

diet range of foods eaten by an animal each day; it is also a particular selection of food, or the total amount and choice of food for a specific person or people. Most animals require seven kinds of food in their diet: proteins, carbohydrates, fats, vitamins, minerals, water, and roughage. A diet that contains all of these things in the correct amounts and proportions is termed a balanced diet. The amounts and proportions required varies with different animals, according to their size, age, and lifestyle. The ◊digestive systems of animals have evolved to meet particular needs; they have also adapted to cope with the foods available in the surroundings in which they live. The necessity of finding and processing an appropriate diet is a very basic drive in animal evolution. **Dietetics** is the science of feeding individuals or groups; a dietition is a specialist in this science.

Dietary requirements may vary over the lifespan of an animal, according to whether it is growing, reproducing, highly active, or approaching death. For instance, increased carbohydrate for additional energy, or increased minerals, may be necessary during periods of growth.

An adequate diet for humans is one that supplies the body's daily nutritional needs (see ◊nutrition), and provides sufficient energy to meet individual levels of activity. The average daily requirement for men is 2,500 calories, but this will vary with age, occupation, and weight; in general, women need fewer calories than men. The energy requirements of active children increase steadily with age, reaching a peak in the late teens. At present, about 450 million people in the world – mainly living in famine or poverty stricken areas, especially in Third World countries – subsist on fewer than 1,500 calories per day. The average daily intake in developed countries is 3,300 calories.

The act of putting into your mouth what the earth has grown is perhaps your most direct interaction with the earth.

FRANCES LAPPÉ US ecologist.
Diet for a Small Planet pt 1

dietetics specialized branch of human nutrition, dealing with the promotion of health through the proper kinds and quantities of food.

Therapeutic dietetics has a large part to play in the treatment of certain illnesses, such as allergies, arthritis, and diabetes; it is sometimes used alone, but often in conjunction with drugs. The preventative or curative effects of specific diets, such as the 'grape cure' or raw vegetable diets sometimes prescribed for cancer patients, are disputed by orthodox medicine.

difference in mathematics, the result obtained when subtracting one number from another. Also, those elements of one ◊set that are not elements of another.

difference engine mechanical calculating machine designed (and partly built in 1822) by the British mathematician Charles ◊Babbage to produce reliable tables of life expectancy. A precursor of the ◊analytical engine, it was to calculate mathematical functions by solving the differences between values given to ◊variables within equations. Babbage designed the calculator so that once the initial values for the variables were set it would produce the next few thousand values without error.

differential arrangement of gears in the final drive of a vehicle's transmission system that allows the driving wheels to turn at different speeds when cornering. The differential consists of sets of bevel gears and pinions within a cage attached to the crown wheel. When cornering, the bevel pinions rotate to allow the outer wheel to turn faster than the inner.

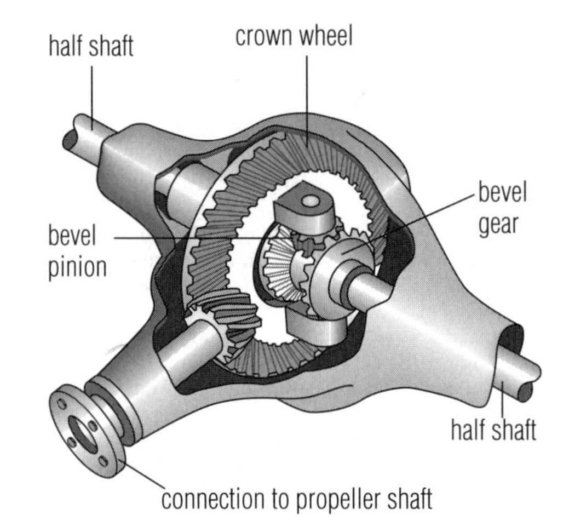

differential *The differential lies midway between the driving wheels of a motorcar. When the car is turning, the bevel pinions spin, allowing the outer wheel to turn faster than the inner wheel.*

differential calculus branch of ◊calculus involving applications such as the determination of maximum and minimum points and rates of change.

differentiation in embryology, the process by which cells become increasingly different and specialized, giving rise to more complex structures that have particular functions in the adult organism. For instance, embryonic cells may develop into nerve, muscle, or bone cells.

differentiation in mathematics, a procedure for determining the ◊derivative or gradient of the tangent to a curve $f(x)$ at any point x.

Diffie-Hellman key exchange system in computing, the basis of ◊public-key cryptography, proposed by researchers Whitfield Diffie and Martin Hellman 1976.

diffraction the slight spreading of a light beam into a pattern of light and dark bands when it passes through a narrow slit or past

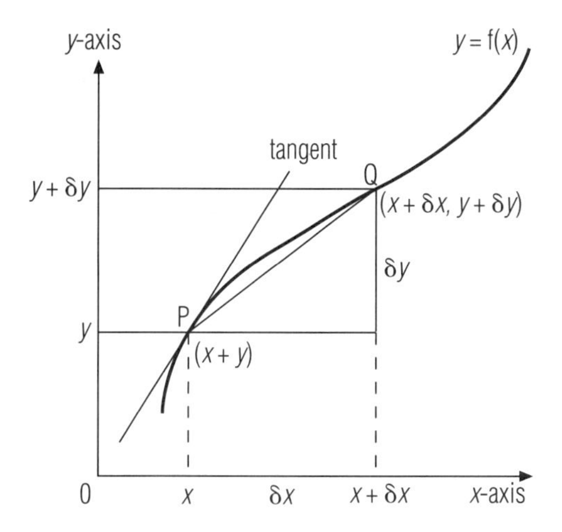

***differentiation** A mathematical procedure for determining the gradient, or slope, of the tangent to any curve f(x) at any point x. Part of a branch of mathematics called differential calculus, it is used to solve problems involving continuously varying quantities (such as the change in velocity or altitude of a rocket), to find the rates at which these variations occur and to obtain maximum and minimum values for the quantities.*

the edge of an obstruction. A **diffraction grating** is a plate of glass or metal ruled with close, equidistant parallel lines used for separating a wave train such as a beam of incident light into its component frequencies (white light results in a spectrum).

The regular spacing of atoms in crystals are used to diffract X-rays, and in this way the structure of many substances has been elucidated, including that of proteins (see ◊X-ray diffraction). Sound waves can also be diffracted by a suitable array of solid objects.

diffusion spontaneous and random movement of molecules or particles in a fluid (gas or liquid) from a region in which they are at a high concentration to a region of lower concentration, until a uniform concentration is achieved throughout. No mechanical mixing or stirring is involved. For instance, if a drop of ink is added to water, its molecules will diffuse until their colour becomes evenly distributed throughout.

In biological systems, diffusion plays an essential role in the transport, over short distances, of molecules such as nutrients, respiratory gases, and neurotransmitters. It provides the means by which small molecules pass into and out of individual cells and microorganisms, such as amoebae, that possess no circulatory system. Diffusion of water across a semi-permeable membrane is termed ◊osmosis.

One application of diffusion is the separation of isotopes, particularly those of uranium. When uranium hexafluoride diffuses through a porous plate, the ratio of the 235 and 238 isotopes is changed slightly. With sufficient number of passages, the separation is nearly complete. There are large plants in the USA and UK for obtaining enriched fuel for fast nuclear reactors and the fissile uranium-235, originally required for the first atom bombs. Another application is the diffusion pump, used extensively in vacuum work, in which the gas to be evacuated diffuses into a chamber from which it is carried away by the vapour of a suitable medium, usually oil or mercury.

Laws of diffusion were formulated by Thomas Graham in 1829 (for gases) and Adolph Fick 1829–1901 (for solutions).

digestion process whereby food eaten by an animal is broken down mechanically, and chemically by ◊enzymes, mostly in the stomach and ◊intestines, to make the nutrients available for absorption and cell metabolism.

In some single-celled organisms, such as amoebae, a food particle is engulfed by the cell and digested in a ◊vacuole within the cell.

digestive system in the body, all the organs and tissues involved in the digestion of food. In animals, these consist of the mouth, stomach, intestines, and their associated glands. The process of digestion breaks down the food by physical and chemical means into the different elements that are needed by the body for energy and tissue building and repair. Digestion begins in the mouth and is completed in the ◊stomach; from there most nutrients are absorbed into the small intestine from where they pass through the intestinal wall into the bloodstream; what remains is stored and concentrated into faeces in the large intestine. Birds have two additional digestive organs – the ◊crop and ◊gizzard. In smaller, simpler animals such as jellyfish, the digestive system is simply a cavity (coelenteron or enteric cavity) with a 'mouth' into which food is taken; the digestible portion is dissolved and absorbed in this cavity, and the remains are ejected back through the mouth.

The digestive system of humans consists primarily of the ◊alimentary canal, a tube which starts at the mouth, continues with the pharynx, oesophagus (or gullet), stomach, large and small intestines, and rectum, and ends at the anus. The food moves through this canal by ◊peristalsis whereby waves of involuntary muscular contraction and relaxation produced by the muscles in the wall of the gut cause the food to be ground and mixed with various digestive juices. Most of these juices contain digestive enzymes, chemicals that speed up reactions involved in the breakdown of food. Other digestive juices empty into the alimentary canal from the salivary glands, gall bladder, and pancreas, which are also part of the digestive system.

The fats, proteins, and carbohydrates (starches and sugars) in foods contain very complex molecules that are broken down (see ◊diet; ◊nutrition) for absorption into the bloodstream: starches and complex sugars are converted to simple sugars; fats are converted to fatty acids and glycerol; and proteins are converted to amino acids and peptides. Foods such as vitamins, minerals, and water do not need to undergo digestion prior to absorption into the bloodstream. The small intestine, which is the main site of digestion and absorption, is subdivided into the duodenum, jejunum, and ileum.

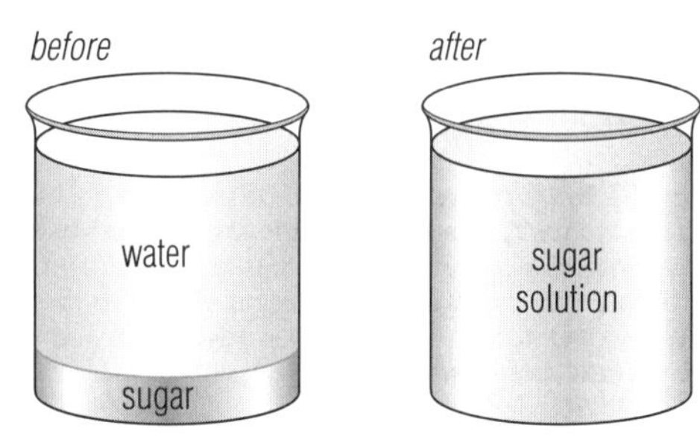

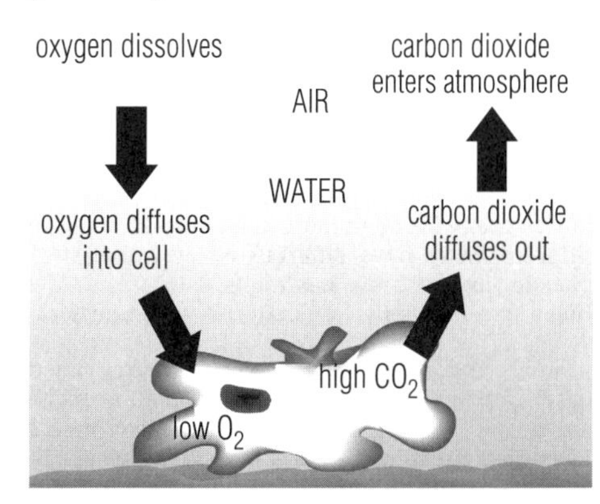

***diffusion** Diffusion is the movement of molecules from a region of high concentration into a region of lower concentration.*

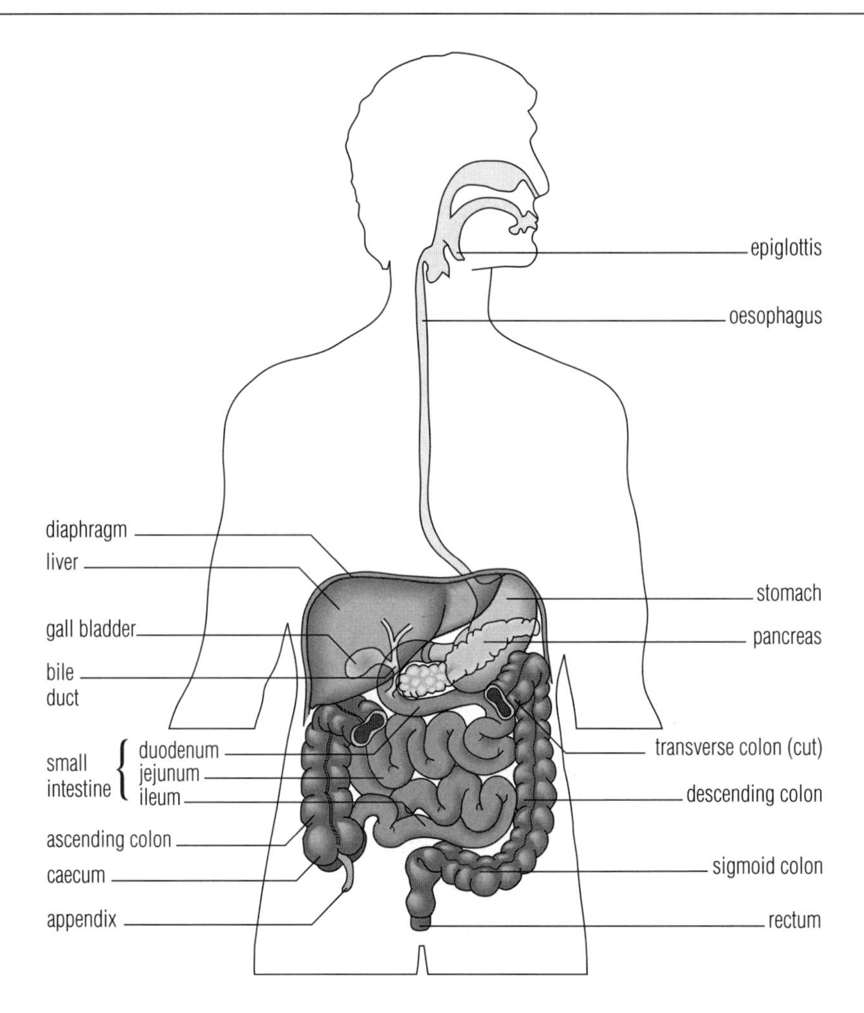

***digestive system** The human digestive system. When food is swallowed, it is moved down the oesophagus by the action of muscles (peristalsis) into the stomach. Digestion starts in the stomach as the food is mixed with enzymes and strong acid. After several hours, the food passes to the small intestine. Here more enzymes are added and digestion is completed. After all nutrients have been absorbed, the indigestible parts pass into the large intestine and thence to the rectum. The liver has many functions, such as storing minerals and vitamins and making bile, which is stored in the gall bladder until needed for the digestion of fats. The pancreas supplies enzymes. The appendix appears to have no function in human beings.*

Covering the surface of its mucous membrane lining are a large number of small prominences called villi which increase the surface for absorption and allow the digested nutrients to diffuse into small blood-vessels lying immediately under the epithelium.

DigiCash in computing, one of several competing systems for electronic money suitable for use on the Internet. Invented by Belgian-based US cryptographer David Chaum, DigiCash uses ◊public-key cryptography techniques to assure anonymity. Trials of the system began in 1994 using software developed for Windows, UNIX, and the Mac.

digit in mathematics, any of the numbers from 0 to 9 in the decimal system. Different bases have different ranges of digits. For example, the ◊hexadecimal system has digits 0 to 9 and A to F, whereas the binary system has two digits (or ◊bits), 0 and 1.

digital in electronics and computing, a term meaning 'coded as numbers'. A digital system uses two-state, either on/off or high/low voltage pulses, to encode, receive, and transmit information. A **digital display** shows discrete values as numbers (as opposed to an analogue signal, such as the continuous sweep of a pointer on a dial).

Digital electronics is the technology that underlies digital techniques. Low-power, miniature, integrated circuits (chips) provide the means for the coding, storage, transmission, processing, and reconstruction of information of all kinds.

digital audio tape (DAT) digitally recorded audio tape produced in cassettes that can carry up to two hours of sound on each side and are about half the size of standard cassettes. DAT players/recorders were developed in 1987. Prerecorded cassettes are copy-protected. The first DAT for computer data was introduced in 1988.

DAT machines are constructed like video cassette recorders (though they use metal audio tape), with a movable playback head, the tape winding in a spiral around a rotating drum. The tape can also carry additional information; for example, a time code for instant location of any point on the track. The music industry delayed releasing prerecorded DAT cassettes because of fears of bootlegging, but a system has now been internationally agreed whereby it is not possible to make more than one copy of any prerecorded compact disc or DAT. DAT is mainly used in recording studios for making master tapes. The system was developed by Sony.

By 1990, DATs for computer data had been developed to a capacity of around 2.5 gigabytes per tape, achieved by using helical scan recording (in which the tape covers about 90% of the total head area of the rotating drum). This enables data from the tape to be read over 200 times faster than it can be written. Any file can be located within 60 seconds.

Digestive System: Pioneering Experiments on the Digestive System

BY JULIAN ROWE

An army marches on its stomach

On 6 June 1822 at Fort Mackinac, Michigan, USA, an 18-year-old French Canadian was accidentally wounded in the abdomen by the discharge of a musket. He was brought to the army surgeon, US physician William Beaumont (1785–1853), who noted several serious wounds and, in particular, a hole in the abdominal wall and stomach. The surgeon observed that through this hole in the patient 'was pouring out the food he had taken for breakfast'.

The patient, Alexis St Martin, a trapper by profession, was serving with the army as a porter and general servant. Not surprisingly, St Martin was at first unable to keep food in his stomach. As the wound gradually healed, firm dressings were needed to retain the stomach contents. Beaumont tended his patient assiduously and tried during the ensuing months to close the hole in his stomach, without success. After 18 months, a small, protruding fleshy fold had grown to fill the aperture (fistula). This 'valve' could be opened simply by pressing it with a finger.

Digestion ... inside and outside

At this point, it occurred to Beaumont that here was an ideal opportunity to study the process of digestion. His patient must have been an extremely tough character to have survived the accident at all. For the next nine years he was the subject of a remarkable series of pioneering experiments, in which Beaumont was able to vary systematically the conditions under which digestion took place and discover the chemical principles involved.

Beaumont attacked the problem of digestion in two ways. He studied how various substances were actually digested in the stomach, and also how they were 'digested' outside the stomach in the digestive juices he extracted from St Martin. He found it was easy enough to drain out the digestive juices from his fortuitously wounded patient 'by placing the subject on his left side, depressing the valve within the aperture, introducing a gum elastic tube and then turning him ... on introducing the tube the fluid soon began to run.'

A typical experiment

Beaumont was basically interested in the rate and temperature of digestion, and also the chemical conditions that favoured different stages of the process of digestion. He describes a typical experiment (he performed hundreds), where (a) digestion in the stomach is contrasted (b) with artificial digestion in glass containers kept at suitable temperatures, like this:

(a) 'At 9 o'clock he breakfasted on bread, sausage, and coffee, and kept exercising. 11 o'clock, 30 minutes, stomach two-thirds empty, aspects of weather similar, thermometer 298°F, temperature of stomach $101\frac{1}{28}$ and $100\frac{3}{48}$. The appearance of contraction and dilation and alternative piston motions were distinctly observed at this examination. 12 o'clock, 20 minutes, stomach empty.'

(b) 'February 7. At 8 o'clock, 30 minutes a.m. I put twenty grains of boiled codfish into three drachms of gastric juice and placed them on the bath.'

'At 1 o'clock, 30 minutes, p.m., the fish in the gastric juice on the bath was almost dissolved, four grains only remaining: fluid opaque, white, nearly the colour of milk. 2 o'clock, the fish in the vial all completely dissolved.'

All a matter of chemistry

Beaumont's research showed clearly for the first time just what happens during digestion and that digestion, as a process, can take place independently outside the body. He wrote that gastric juice: 'so far from being inert as water as some authors assert, is the most general solvent in nature of alimentary matter – even the hardest bone cannot withstand its action. It is capable, even out of the stomach, of effecting perfect digestion, with the aid of due and uniform degree of heat (100°Fahrenheit) and gentle agitation ... I am impelled by the weight of evidence ... to conclude that the change effected by it on the aliment, is purely chemical.'

Our modern understanding of the physiology of digestion as a process whereby foods are gradually broken down into their basic components follows logically from his work. An explanation of how the digestive juices flowed in the first place came in 1889, when Russian physiologist Ivan Pavlov (1849–1936) showed that their secretion in the stomach was controlled by the nervous system. By preventing the food eaten by a dog from actually entering the stomach, he found that the secretions of gastric juices began the moment the dog started eating, and continued as long as it did so. Since no food had entered the stomach, the secretions must be mediated by the nervous system.

Later, it was found that the further digestion that takes place beyond the stomach was hormonally controlled. But it was Beaumont's careful scientific work, which was published in 1833 with the title *Experiments and Observations on the Gastric Juice and Physiology of Digestion*, that triggered subsequent research in this field.

digital camera camera that uses a ◊charge-coupled device (CCD) to take pictures which are stored as digital data rather than on film. The output from digital cameras can be downloaded onto a computer for retouching or storage, and can be readily distributed as computer files. Leading manufacturers of digital cameras include Canon and Kodak.

digital city in computing, area in ◊cyberspace, either text-based or graphical, that uses the model of a city to make it easy for visitors and residents to find specific types of information.

digital compact cassette (DCC) digitally recorded audio cassette that is roughly the same size as a standard cassette. It cannot be played on a normal tape recorder, though standard tapes can be played on a DCC machine; this is known as 'backwards compatibility'. The playing time is 90 minutes.

A DCC player has a stationary playback and recording head similar to that in ordinary tape decks, though the tape used is chrome video tape. The cassettes are copy-protected and can be individually programmed for playing order. Some DCC decks have a liquid-crystal digital-display screen, which can show track titles and other information encoded on the tape.

digital composition or **compositing** in computing, computerized film editing. Some film special effects require shots to be cut together – composited. A sequence showing an actor hanging off the edge of a skyscraper, for example, may be put together out of footage of the actor in a safe location inserted into a shot looking down the side of the skyscraper, which may itself be a model. Traditional techniques for creating such a shot involved photographing the foreground shot with the background shot playing behind it, with an inevitable degradation of quality in the background material. In digital compositing, the same footage is digitized, and the work of merging the two sequences is done by manipulating computer files. The composite image is then transferred back onto film with no loss of quality.

digital computer computing device that operates on a two-state system, using symbols that are internally coded as binary numbers (numbers made up of combinations of the digits 0 and 1); see ◊computer.

digital data transmission in computing, a way of sending data by converting all signals (whether pictures, sounds, or words) into numeric (normally binary) codes before transmission, then reconverting them on receipt. This virtually eliminates any distortion or degradation of the signal during transmission, storage, or processing.

digitalis in botany, any of a group of plants belonging to the figwort family, which includes the ◊foxgloves. The leaves of the common foxglove (*Digitalis purpurea*) are the source of the drug **digitalis** used in the treatment of heart disease. (Genus *Digitalis,* family Scrophulariaceae.)

digitalis in medicine, drug that increases the efficiency of the heart by strengthening its muscle contractions and slowing its rate. It is derived from the leaves of the common European woodland plant *Digitalis purpurea* (foxglove).

It is purified to digoxin, digitoxin, and lanatoside C, which are effective in cardiac regulation but induce the side effects of nausea, vomiting, and pulse irregularities. Pioneered in the late 1700s by William Withering, an English physician and botanist, digitalis was the first cardiac drug.

digital monitor in computing, display ◊monitor using standard cathode-ray tube technology that converts a ◊digital signal from the computer into an ◊analogue signal for display.

Digital monitors are unable to display the continuously variable range of colours offered by analogue monitors.

digital recording technique whereby the pressure of sound waves is sampled more than 30,000 times a second and the values converted by computer into precise numerical values. These are recorded and, during playback, are reconverted to sound waves.

This technique gives very high-quality reproduction. The numerical values converted by computer represent the original sound-wave form exactly and are recorded on compact disc. When this is played back by ◊laser, the exact values are retrieved.

When the signal is fed via an amplifier to a loudspeaker, sound waves exactly like the original ones are reproduced.

digital retouching in computing, technique for touching up digital photographs, similar to airbrushing in the analogue world. It is commonly used in the film industry to remove scratches or to cover up filming mistakes.

The retoucher points out the error to the computer and the computer calculates new colour values for the affected ◊pixels from the colours of neighbouring pixels.

digital sampling electronic process used in ◊telecommunications for transforming a constantly varying (analogue) signal into one composed of discrete units, a digital signal. In the creation of recorded music, sampling enables the composer, producer, or remix engineer to borrow discrete vocal or instrumental parts from other recorded work (it is also possible to sample live sound).

A telephone microphone changes sound waves into an analogue signal that fluctuates up and down like a wave. In the digitizing process the waveform is sampled thousands of times a second and each part of the sampled wave is given a binary code number (made up of combinations of the digits 0 and 1) related to the height of the wave at that point, which is transmitted along the telephone line. Using digital signals, messages can be transmitted quickly, accurately, and economically.

digital signal processor (DSP) in computing, special-purpose integrated circuit that handles voice. DSPs are used in voice modems, which add answering machine facilities to a personal computer, and also in computer dictation systems.

digital signature in computing, method of using encryption to certify the source and integrity of a particular electronic document. Because all ◊ASCII characters look the same no matter who types them, methods have to be found to certify the origins of particular messages if they are to be legally binding for electronic commerce or other transactions. One type of digital signature commonly seen on the Net is generated by the program ◊Pretty Good Privacy (PGP), which adds a digest of the message to the signature.

digital-to-analogue converter electronic circuit that converts a digital signal into an ◊analogue (continuously varying) signal. Such a circuit is used to convert the digital output from a computer into the analogue voltage required to produce sound from a conventional loudspeaker.

digital versatile disc or ***digital video disc*** (DVD) disc format for storing digital information. DVDs can hold 14 times the data stored on current CDs. Pre-recorded CVDs have a storage capacity of 4.7 gigabytes and can hold a full-length feature film. As with CDs, information is etched in the form of microscopic pits onto a plastic disc (though the pits are half the size), which is then coated with aluminium. DVDs may have two pitted surfaces per side whereas CDs have only one. The data is read optically using a laser as the disc rotates. A double layer disc can hold 4 hours of video. The Japanese company TDK produced the rewriteable DVD-RAM, capable of holding 2.6 gigabytes, in 1996.

digital video interactive powerful compression system used for storing video images on computer; see ◊DVI.

digitize in computing, to turn ◊analogue signals into the binary data a computer can read. Any type of analogue signal can be digitized, including pictures, sound, video, or film. The result is files that can be manipulated, stored, or transmitted by computers.

digitizer in computing, a device that converts an analogue video signal into a digital format so that video images can be input, stored, displayed, and manipulated by a computer. The term is sometimes used to refer to a ◊graphics tablet.

dihybrid inheritance in genetics, a pattern of inheritance observed when two characteristics are studied in succeeding generations.

The first experiments of this type, as well as in ◊monohybrid inheritance, were carried out by Austrian biologist Gregor ◊Mendel using pea plants.

dik-dik any of several species of tiny antelope, genus *Madoqua,* found in Africa south of the Sahara in dry areas with scattered brush. Dik-diks are about 60 cm/2 ft long and 35 cm/1.1 ft tall, and are often seen in pairs. Males have short, pointed horns. The dik-dik is so named because of its alarm call.

dilatation and curettage (D and C) common gynaecological procedure in which the cervix (neck of the womb) is widened, or dilated, giving access so that the lining of the womb can be scraped away (curettage). It may be carried out to terminate a pregnancy, treat an incomplete miscarriage, discover the cause of heavy menstrual bleeding, or for biopsy.

dill herb belonging to the carrot family, whose bitter seeds and aromatic leaves are used in cooking and in medicine. (*Anethum graveolens,* family Umbelliferae.)

dilution process of reducing the concentration of a solution by the addition of a solvent.

The extent of a dilution normally indicates the final volume of solution required. A fivefold dilution would mean the addition of sufficient solvent to make the final volume five times the original.

dimension in science, any directly measurable physical quantity such as mass (M), length (L), and time (T), and the derived units obtainable by multiplication or division from such quantities.

For example, acceleration (the rate of change of velocity) has dimensions (LT^{-2}), and is expressed in such units as km s^{-2}. A quantity that is a ratio, such as relative density or humidity, is dimensionless.

In geometry, the dimensions of a figure are the number of measures needed to specify its size. A point is considered to have zero dimension, a line to have one dimension, a plane figure to have two, and a solid body to have three.

dimethyl sulphoxide (DMSO) $(CH_3)_2SO$ colourless liquid used as an industrial solvent and an antifreeze. It is obtained as a by-product of the processing of wood to paper.

DIN (abbreviation for ***Deutsches Institut für Normung***) German national standards body, which has set internationally accepted standards for (among other things) paper sizes and electrical connectors.

dingbat non-alphanumeric character, such as a star, bullet, or arrow. Dingbats have been combined into ◊PostScript and ◊TrueType fonts for use with word processors and graphics programs.

dingo wild dog of Australia. Descended from domestic dogs brought from Asia by Aborigines thousands of years ago, it belongs to the same species *Canis familiaris* as other domestic dogs. It is reddish brown with a bushy tail, and often hunts at night. It cannot bark.

dinitrogen oxide alternative name for ◊nitrous oxide, or 'laughing gas', one of the nitrogen oxides.

Dinorwig location of Europe's largest pumped-storage hydroelectric scheme, completed 1984, in Gwynedd, North Wales. It is used as a backup to meet heavy demands for electricity. Six turbogenerators are installed, with a maximum output of some 1,880 megawatts. The working head of water for the station is 530 m/1,740 ft.

The main machine hall is twice as long as a football field and as high as a 16-storey building.

dinosaur *Greek deinos 'terrible', sauros 'lizard'* any of a group (sometimes considered as two separate orders) of extinct reptiles living between 205 million and 65 million years ago. Their closest living relations are crocodiles and birds. Many species of dinosaur evolved during the millions of years they were the dominant large land animals. Most were large (up to 27 m/90 ft), but some were as small as chickens. They disappeared 65 million years ago for reasons not fully understood, although many theories exist.

classification Dinosaurs are divisible into two unrelated stocks, the orders **Saurischia** ('lizard-hip') and **Ornithischia** ('bird-hip'). Members of the former group possess a reptile-like pelvis and are mostly bipedal and carnivorous, although some are giant amphibious quadrupedal herbivores. Members of the latter group have a bird-like pelvis, are mainly four-legged, and entirely herbivorous.

The Saurischia are divided into: **theropods** ('beast-feet'), including all the bipedal carnivorous forms with long hindlimbs and short forelimbs (◊tyrannosaurus, megalosaurus); and **sauropodomorphs** ('lizard-feet forms'), including sauropods, the large quadrupedal herbivorous and amphibious types with massive limbs, long tails and necks, and tiny skulls (diplodocus, brontosaurus).

The Ornithischia were almost all plant-eaters, and eventually outnumbered the Saurischia. They are divided into four suborders: **ornithopods** ('bird-feet'), Jurassic and Cretaceous bipedal forms (Iguanodon) and Cretaceous hadrosaurs with duckbills; **stegosaurs** ('plated' dinosaurs), Jurassic quadrupedal dinosaurs with a double row of triangular plates along the back and spikes on the tail (stegosaurus); **ankylosaurs** ('armoured' dinosaurs), Cretaceous quadrupedal forms, heavily armoured with bony plates (nodosaurus); and **ceratopsians** ('horned' dinosaurs), Upper Cretaceous quadrupedal horned dinosaurs with very large skulls bearing a neck frill and large horns (triceratops).

These two main dinosaur orders form part of the superorder Archosaurus ('ruling reptiles'), comprising a total of five orders. The other three are **Pterosaurs** ('winged lizards'), including ◊pterodactyls, of which no examples exist today, **crocodilians**, and **birds**. All five orders are thought to have evolved from a 'stem-order', the **Thecondontia**.

species Brachiosaurus, a long-necked plant-eater of the sauropod group, was about 12.6 m/40 ft to the top of its head, and weighed 80 tonnes. Compsognathus, a meat-eater, was only the size of a chicken, and ran on its hind legs. Stegosaurus, an armoured plant-eater 6 m/20 ft long, had a brain only about 3 cm/1.25 in long. Not all dinosaurs had small brains. At the other extreme, the hunting dinosaur stenonychosaurus, 2 m/6 ft long, had a brain size comparable to that of a mammal or bird of today, stereoscopic vision, and grasping hands. Many dinosaurs appear to have been equipped for a high level of activity. ◊Tyrannosaurus was a huge, two-footed, meat-eating theropod dinosaur of the Upper Cretaceous in North America and Asia. The largest carnivorous dinosaur was *Giganotosaurus carolinii.* It lived in Patagonia about 97 million years ago, was 12.5 m/41 ft long, and weighed 6–8 tonnes. Its skeleton was discovered 1995.

theories of extinction A popular theory of dinosaur extinction suggests that the Earth was struck by a giant meteorite or a swarm of comets 65 million years ago and this sent up such a cloud of debris and dust that climates were changed and the dinosaurs could not adapt quickly enough. The evidence for this includes a bed of rock rich in ◊iridium – an element rare on Earth but common in extraterrestrial bodies – dating from the time.

An alternative theory suggests that changes in geography brought about by the movements of continents and variations in sea level led to climate changes and the mixing of populations between previously isolated regions. This resulted in increased competition and the spread of disease.

archaeological findings The term 'dinosaur' was coined 1842 by Richard Owen, although there were findings of dinosaur bones as far back as the 17th century. In 1822 G A Mantell (1790–1852) found teeth of iguanodon in a quarry in Sussex. The first dinosaur to be described in a scientific journal was in 1824, when William Buckland, professor of geology at Oxford University, published his finding of a 'megalosaurus or great fossil lizard' found at Stonesfield, a village northwest of Oxford, although a megalosaurus bone had been found in 1677.

One of the largest dinosaur species found in the UK was a Sauropod, *Cetiosaurus oxoniensis,* discovered in 1870 near Bletchingdon, north of Oxford. It was around 15 m/49 ft long, although specimens have been discovered in North Africa up to 18 m/60 ft long. In 1992 another large dinosaur, *Iguanodon bernissartensis,* was discovered near Ockley in Surrey, England, by amateur fossil hunters.

An almost complete fossil of a dinosaur skeleton was found in 1969 in the Andean foothills, South America; it had been a two-legged carnivore 2 m/6 ft tall and weighed more than 100 kg/220 lb. More than 230 million years old, it is the oldest known dinosaur. In 1982 a number of nests and eggs were found in 'colonies' in Montana, suggesting that some bred together like modern seabirds. In 1987 finds were made in China that may add much to the traditional knowledge of dinosaurs, chiefly gleaned from North American specimens. In 1989 and 1990 an articulated *Tyrannosaurus rex* was unearthed by a palaeontological team in Montana, with a full skull, one of only six known. Short stretches of dinosaur DNA were extracted in 1994 from unfossilized bone retrieved from coal deposits approximately 80 million years old.

recent discoveries The discovery of a small dinosaur was announced in China in 1996. Sinosauropteryx lived about 120 million years old and was 0.5 m/1.6 ft tall. It had short forelegs, a long tail, and short feathers, mainly on its neck and shoulders.

In 1997 US scientists claimed that 65 million-year-old remains discovered in the Atlantic Ocean were proof that a massive asteroid impact on Earth killed the dinosaurs. A sea-drilling expedition discovered three samples that have the signature of an asteroid impact. Previous evidence from sediment suggested that the dinosaurs did not become extinct at exactly the same time as an impact occurred. The new evidence appeared to substantiate the theories of geologists such as Walter Alvarez, who championed the theory that the dinosaurs disappeared from fossil history because of such an impact.

US palaeontologists discovered in 1997 a dinosaur wishbone in place in the skeleton of a velociraptor. This was the first time a wishbone had been found in place and scientists claimed that this constitutes strong evidence for birds having evolved from dinosaurs.

diode combination of a cold anode and a heated cathode, or the semiconductor equivalent, which incorporates a *p–n* junction; see ◊semiconductor diode. Either device allows the passage of direct current in one direction only, and so is commonly used in a ◊rectifier to convert alternating current (AC) to direct current (DC).

dioecious of plants with male and female flowers borne on separate individuals of the same species. Dioecism occurs, for example, in the willows *Salix.* It is a way of avoiding self-fertilization.

dioptre optical unit in which the power of a ◊lens is expressed as the reciprocal of its focal length in metres. The usual convention is that convergent lenses are positive and divergent lenses negative. Short-sighted people need lenses of power about –0.7 dioptre; a typical value for long sight is about +1.5 dioptre.

diorite igneous rock intermediate in composition between mafic (consisting primarily of dark-coloured minerals) and felsic (consisting primarily of light-coloured minerals) – the coarse-grained plutonic equivalent of ◊andesite. Constituent minerals include ◊feldspar and amphibole or pyroxene with only minor amounts of ◊quartz.

dioxin any of a family of over 200 organic chemicals, all of which are heterocyclic hydrocarbons (see ◊cyclic compounds).

The term is commonly applied, however, to only one member of the family, 2,3,7,8-tetrachlorodibenzo-*p*-dioxin (2,3,7,8-TCDD), a highly toxic chemical that occurs, for example, as an impurity in the defoliant Agent Orange, used in the Vietnam War, and sometimes in the weedkiller 2,4,5-T. It has been associated with a disfiguring skin complaint (chloracne), birth defects, miscarriages, and cancer.

Disasters involving accidental release of large amounts of dioxin into the environment have occurred at Seveso, Italy, and Times Beach, Missouri, USA. Small amounts of dioxins are released by the burning of a wide range of chlorinated materials (treated wood, exhaust fumes from fuels treated with chlorinated additives, and plastics) and as a side-effect of some techniques of paper-making. The possibility of food becoming contaminated by dioxins in the environment has led the European Community to decrease significantly the allowed levels of dioxin emissions from incinerators. Dioxin may be produced as a by-product in the manufacture of the bactericide ◊hexachlorophene.

DIP abbreviation for ◊document image processing.

diphtheria acute infectious disease in which a membrane forms in the throat (threatening death by ◊asphyxia), along with the production of a powerful toxin that damages the heart and nerves. The organism responsible is a bacterium (*Corynebacterium diphtheriae*). It is treated with antitoxin and antibiotics.

Although its incidence has been reduced greatly by immunization, an epidemic in the former Soviet Union resulted in 47,802 cases and 1,746 deaths in 1994, and 1,500 deaths in 1995. In 1995 the World Health Organization (WHO) declared the epidemic 'an international public health emergency' after 20 linked cases were identified in other parts of Europe. The epidemic showed signs of abating in 1996, with a 59% decrease in the number of cases for the first three months, compared with the same period in 1995.

diploblastic in biology, having a body wall composed of two layers. The outer layer is the **ectoderm**, the inner layer is the **endoderm**. This pattern of development is shown by ◊coelenterates.

diplodocus plant-eating sauropod dinosaur that lived about 145 million years ago, the fossils of which have been found in the western USA. Up to 27 m/88 ft long, most of which was neck and tail, it weighed about 11 tonnes. It walked on four elephantine legs, had nostrils on top of the skull, and peglike teeth at the front of the mouth.

diploid having paired ◊chromosomes in each cell. In sexually reproducing species, one set is derived from each parent, the ◊gametes, or sex cells, of each parent being ◊haploid (having only one set of chromosomes) due to ◊meiosis (reduction cell division).

diplomonad single-celled organisms with two nuclei and no mitochondria. They produce energy by anaerobic metabolism, such as glycolysis. The human intestinal parasite *Giardia lamblia* is a diplomonad.

dip, magnetic angle at a particular point on the Earth's surface between the direction of the Earth's magnetic field and the horizontal. It is measured using a **dip circle**, which has a magnetized needle suspended so that it can turn freely in the vertical plane of the magnetic field. In the northern hemisphere the needle dips below the horizontal, pointing along the line of the magnetic field towards its north pole. At the magnetic north and south poles, the needle dips vertically and the angle of dip is 90°. See also ◊angle of declination.

dipole the uneven distribution of magnetic or electrical characteristics within a molecule or substance so that it behaves as though it possesses two equal but opposite poles or charges, a finite distance apart.

The uneven distribution of electrons within a molecule composed of atoms of different ◊electronegativities may result in an apparent concentration of electrons towards one end of the molecule and a deficiency towards the other, so that it forms a dipole consisting of apparently separated but equal positive and negative charges. The product of one charge and the distance between them is the **dipole moment**. A bar magnet has a magnetic dipole and behaves as though its magnetism were concentrated in separate north and south magnetic poles.

dipole in radio, a rod aerial, usually one half-wavelength or a whole wavelength long.

dipole, magnetic see ◊magnetic dipole.

dipper or ***water ouzel*** any of various birds of the genus *Cinclus*, family Cinclidae, order Passeriformes, found in hilly and mountainous regions across Eurasia and North America, where there are clear, fast-flowing streams. It can swim, dive, or walk along the bottom, using the pressure of water on its wings and tail to keep it down, while it searches for insect larvae and other small animals. Both wings and tail are short, the beak is fairly short and straight, and the general colour of the bird is brown, the throat and part of the breast being white.

diprotodon extinct giant Australian marsupial. It was about the size of a large rhinoceros, with well developed incisor teeth and huge skull. It is the largest known marsupial and lived during the Pleistocene epoch.

DIP switch (abbreviation of ***dual in-line package***) in computing, tiny switch that controls settings on devices such as printers and modems. The owner's manual will usually specify how DIP switches should be set.

On printers, these switches are typically used to specify which emulation to use; on modems, they set the modem to match the ◊COM port to which it is connected. They should not need to be changed once the device has been installed and is working properly.

Diptera order consisting of the two-winged 'true' ◊flies. The order contains some 75,000 species arranged in approximately 100 families.

Diptera is divided into three suborders.

suborder Nematocera The adults are usually long-legged delicate flies. Their long filamentous antennae have many segments. The larvae usually have a well-defined chitinous head. The pupae are usually free and active. This suborder includes: midges, mosquitoes, gnats, and craneflies

suborder Brachycera These are more robust flies than the nematocerans, with various types of antennae, usually made up of vari-

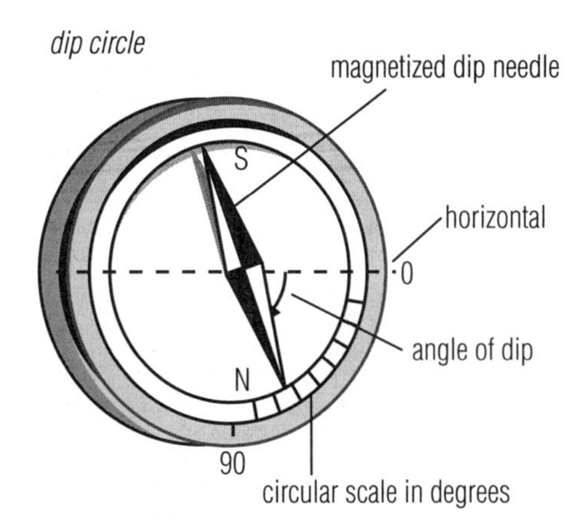

dip *A dip circle is used to measure the angle between the direction of the Earth's magnetic field and the horizontal at any point on the Earth's surface.*

ous, dissimilar segments. The larva has a less heavily chitinized head which is retractile. The pupa is usually free. This suborder includes: assassin flies, bee flies, and horse flies.
suborder Cyclorrhapha The antennae are usually of two small segments with a larger, pendulous third segment; an arista (bristle-like extension) is usually present. The palps usually consist of one segment. The larva is a maggot without a head capsule. The pupa usually has a puparium which is formed from the last-stage larval skin. This suborder includes: house flies, blowflies, botflies, tsetse flies, and fruit flies.
classification Diptera is in the subclass Pterygota, class Insecta, phylum Arthropoda.

direct access or ***random access*** type of ◊file access. A direct-access file contains records that can be accessed by the computer directly because each record has its own address on the storage disc.

direct broadcast system (DBS) in computing, combination of a small satellite dish and receiver which allows consumers to receive television and radio broadcasts from a satellite rather than via terrestrial broadcasting towers and repeaters.

direct connection in computing, connection between two computers via cable to transfer files without the intermediary of a network or online service. Each computer must be running communications software using the same protocols for file transfers. If the computers are in the same room, they can be connected using a special type of serial cable known as a null modem cable; if they are connected via telephone lines each must have a modem so that one can dial the other.

There are several software packages designed for this purpose; the market leader is Laplink.

direct current (DC) electric current that flows in one direction, and does not reverse its flow as ◊alternating current does. The electricity produced by a battery is direct current.

directed number a number with a positive (+) or negative (–) sign attached, for example +5 or –5. On a graph, a positive sign shows a movement to the right or upwards; a negative sign indicates movement downwards or to the left.

direct memory access (DMA) in computing, a technique used for transferring data to and from external devices without going through the ◊central processing unit (CPU) and thus speeding up transfer rates. DMA is used for devices such as ◊scanners.

direct memory access channel in computing, channel used for the fast transfer of data; usually abbreviated as ◊DMA channel.

Director in computing, multimedia software ◊authoring tool published by Macromedia, a company of multimedia software specialists based in San Francisco, USA.

directory in computing, a list of file names, together with information that enables a computer to retrieve those files from ◊backing storage. The computer operating system will usually store and update a directory on the backing storage to which it refers. So, for example, on each ◊disc used by a computer a directory file will be created listing the disc's contents.

directory tree in computing, collective name for a ◊directory and all its subdirectories.

dirigible another name for ◊airship.

disaccharide ◊sugar made up of two monosaccharides or simple sugars. Sucrose, $C_{12}H_{22}O_{11}$, or table sugar, is a disaccharide.

disc in astronomy, the flat, roughly circular region of a spiral or lenticular ◊galaxy containing stars, ◊nebulas, and dust clouds orbiting about the nucleus. Discs contain predominantly young stars and regions of star formation. The disc of our own Galaxy is seen from Earth as the band of the ◊Milky Way.

disc or ***disk*** in computing, a common medium for storing large volumes of data (an alternative is ◊magnetic tape). A **magnetic disc** is rotated at high speed in a disc-drive unit as a read/write (playback or record) head passes over its surfaces to record or read the magnetic variations that encode the data. Recently, **optical discs**,

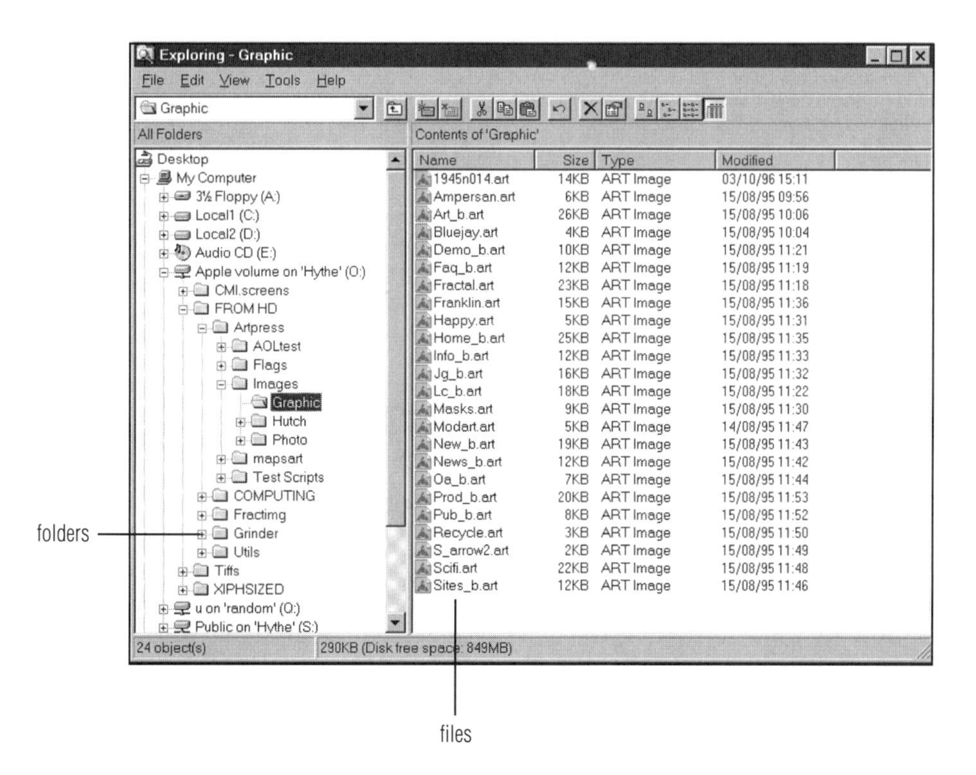

directory *A graphical illustration of the directory filing system on a computer. On the left hand side of the screen are the sub-directories available from the root; on the right are the files contained within the active directory.*

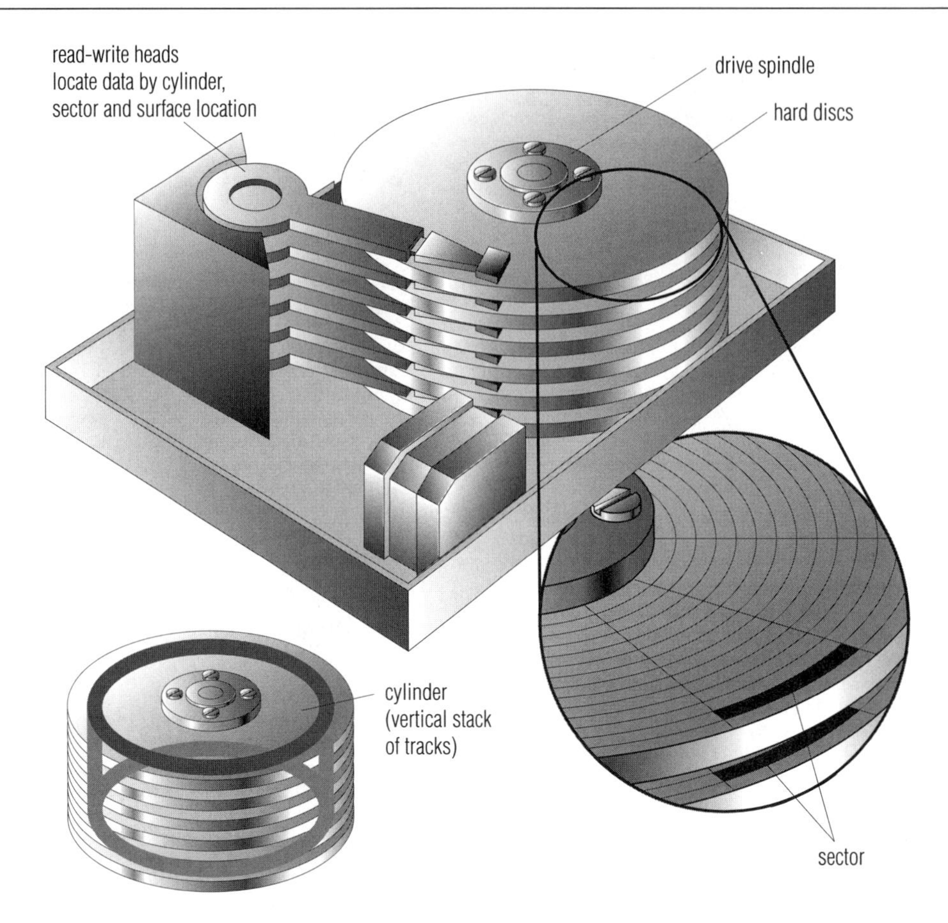

disc A hard disc. Data is stored in sectors within cylinders and is read by a head which passes over the spinning surface of each disc.

such as ◊CD-ROM (compact-disc read-only memory) and ◊WORM (write once, read many times), have been used to store computer data. Data are recorded on the disc surface as etched microscopic pits and are read by a laser-scanning device. Optical discs have an enormous capacity – about 550 megabytes (million ◊bytes) on a compact disc, and thousands of megabytes on a full-size optical disc.

Magnetic discs come in several forms: **fixed hard discs** are built into the disc-drive unit, occasionally stacked on top of one another. A fixed disc cannot be removed: once it is full, data must be deleted in order to free space or a complete new disc drive must be added to the computer system in order to increase storage capacity. In 1997, small hard drives typically stored two thousand megabytes or 2 gigabytes (GB) of data, but could store up to 9 GB or more. Arrays of such discs were also used to store minicomputer and mainframe data in RAID storage systems, replacing large fixed discs and removable hard discs.

Removable hard discs are still found in minicomputer and mainframe systems. The discs are contained, individually or as stacks (disc packs), in a protective plastic case, and can be taken out of the drive unit and kept for later use. By swapping such discs around, a single hard-disc drive can be made to provide a potentially infinite storage capacity. However, access speeds and capacities tend to be lower that those associated with large fixed hard discs. A **floppy disc** (or diskette) is the most common form of backing store for microcomputers. It is much smaller in size and capacity than a hard disc, normally holding 0.5–2 megabytes of data. The floppy disc is so called because it is manufactured from thin flexible plastic coated with a magnetic material. The earliest form of floppy disc was packaged in a card case and was easily damaged; more recent versions are contained in a smaller, rigid plastic case and are much more robust. All floppy discs can be removed from the drive unit.

disc compression technique, based on ◊data compression, that makes hard discs and floppy discs appear to have more storage capacity than is normally available. If the data stored on a disc can be compressed to occupy half the original amount of disc space, it will appear that the disc is twice its original size. The processes of compression (to store data) and decompression (so that data can be used) are hidden from the user by the software.

Several commercial disc compression products are available, for example DoubleSpace in ◊MS-DOS 6.0 and Stacker.

disc drive mechanical device that reads data from, and writes data to, a magnetic ◊disc.

disc formatting in computing, preparing a blank magnetic disc in order that data can be stored on it. Data are recorded on a disc's surface on circular tracks, each of which is divided into a number of sectors. In formatting a disc, the computer's operating system adds control information such as track and sector numbers, which enables the data stored to be accessed correctly by the disc-drive unit.

Some floppy discs, called **hard-sectored discs**, are sold already formatted. However, because different makes of computer use different disc formats, discs are also sold unformatted, or **soft-sectored**, and computers are provided with the necessary ◊utility program to format these discs correctly before they are used.

discharge in a river, the volume of water passing a certain point per unit of time. It is usually expressed in cubic metres per second (cumecs). The discharge of a particular river channel may be calculated by multiplying the channel's cross-sectional area (in square metres) by the velocity of the water (in metres per second).

discharge tube device in which a gas conducting an electric current emits visible light. It is usually a glass tube from which virtually all the air has been removed (so that it 'contains' a near vacu-

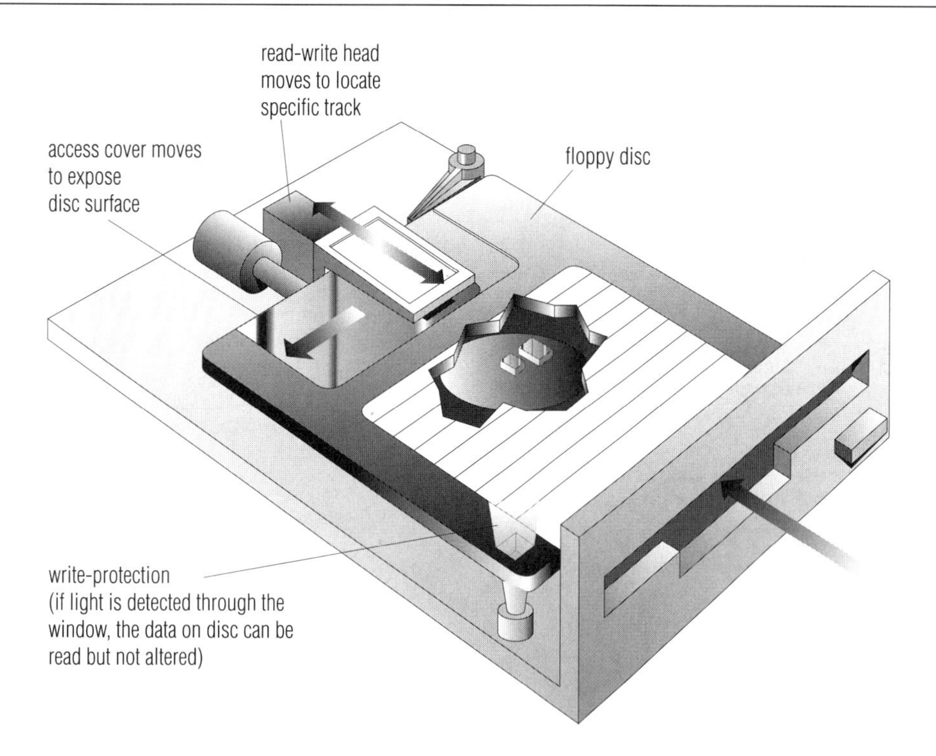

disc drive *A floppy disc drive. As the disc is inserted into the drive, its surface is exposed to the read-write head, which moves over the spinning disc surface to locate a specific track.*

um), with electrodes at each end. When a high-voltage current is passed between the electrodes, the few remaining gas atoms in the tube (or some deliberately introduced ones) ionize and emit coloured light as they conduct the current along the tube. The light originates as electrons change energy levels in the ionized atoms.

By coating the inside of the tube with a phosphor, invisible emitted radiation (such as ultraviolet light) can produce visible light; this is the principle of the fluorescent lamp.

Discman Sony trademark for a portable compact-disc player; the equivalent of a ◊Walkman, it also comes in a model with a liquid-crystal display for data discs.

disc optimizer in computing, another name for a ◊defrag-mentation program, a program that gathers together files that have become fragmented for storage on different areas of a disc. See also ◊fragmentation.

discrete data data that can take only whole-number or fractional values. The opposite is ◊continuous data, which can take all in-between values. Examples of discrete data include frequency and population data. However, measurements of time and other dimensions can give rise to continuous data.

disease condition that disturbs or impairs the normal state of an organism. Diseases can occur in all life forms, and normally affect the functioning of cells, tissues, organs, or systems. Diseases are usually characterized by specific symptoms and signs, and can be mild and short-lasting – such as the common cold – or severe enough to decimate a whole species – such as ◊Dutch elm disease. Diseases can be classified as infectious or noninfectious. Infectious diseases are caused by microorganisms, such as bacteria and viruses, invading the body; they can be spread across a species, or transmitted between one or more species. All other diseases can be grouped together as noninfectious diseases. These can have many causes: they may be inherited (◊congenital diseases); they may be caused by the ingestion or absorption of harmful substances, such as toxins; they can result from poor nutrition or hygiene; or they may arise from injury or ageing. The causes of some diseases are still unknown.

Some diseases occur mainly in certain climates or geographical regions of the world. These are **endemic** diseases. For example, African sleeping sickness, which is carried by the tsetse fly, is found mainly in the very hot, humid regions of Africa. Similarly, malaria, a disease spread by mosquitoes, is usually found in or near the marsh or stagnant water which provide breeding grounds for the insect. Other diseases may be seasonal – such as influenza, which tends to occur mainly in winter, or intestinal illnesses that result from food contamination in summer.

Some age groups may be more prone to certain diseases, such as measles in children, meningitis in young adults, and coronary heart disease in the elderly. Other diseases may tend to occur only in certain racial types and are usually genetic in origin, such as sickle-cell disease which is found mainly among people of black African descent. Other diseases, such as black lung, or coal-workers' pneumoconiosis, result from occupational hazards; some of the 'new' diseases that have appeared in recent years – such as ◊sick building syndrome and legionnaire's disease, result from modern building designs, while the cause of ME (myalgic encephalomyelitis), or chronic fatigue syndrome, is still unknown.

CRIMEAN WAR

In the Crimean War more soldiers died from disease than in the fighting. Disease (mainly typhus, cholera, and dysentery) accounted for many the deaths of 104,494 soldiers compared with 63,261 from wounds; 662,917 were out of action through illness compared with 150,533 wounded.

disinfectant agent that kills, or prevents the growth of, bacteria and other microorganisms. Chemical disinfectants include carbolic acid (phenol, used by Joseph Lister in surgery in the 1870s), ethanol, methanol, chlorine, and iodine.

dispersal phase of reproduction during which gametes, eggs, seeds, or offspring move away from the parents into other areas. The result is that overcrowding is avoided and parents do not find themselves in competition with their own offspring. The mechanisms are various, including a reliance on wind or water currents and, in the case of animals, locomotion. The ability of a species to spread widely through an area and to colonize new habitats has survival value in evolution.

dispersion in chemistry, the distribution of the microscopic particles of a ◊colloid. In colloidal sulphur the dispersion is the tiny particles of sulphur in the aqueous system.

dispersion in physics, the separation of waves of different frequencies by passage through a dispersive medium, in which the speed of the wave depends upon its frequency or wavelength. In optics, the splitting of white light into a spectrum; for example, when it passes through a prism or diffraction grating. It occurs because the prism or grating bends each component wavelength through a slightly different angle. A rainbow is formed when sunlight is dispersed by raindrops.

dispersion in statistics, the extent to which data are spread around a central point (typically the ◊arithmetic mean).

displacement activity in animal behaviour, an action that is performed out of its normal context, while the animal is in a state of stress, frustration, or uncertainty. Birds, for example, often peck at grass when uncertain whether to attack or flee from an opponent; similarly, humans scratch their heads when nervous.

displacement reaction chemical reaction in which a less reactive element is replaced in a compound by a more reactive one.

For example, the addition of powdered zinc to a solution of copper(II) sulphate displaces copper metal, which can be detected by its characteristic colour. See also ◊electrochemical series.

display in computing, an ◊output device that looks like a television set and displays commands to the computer and their results.

display control interface in computing, standard developed by Microsoft and Intel for the ◊device drivers that control ◊graphics cards.

dissection cutting apart of bodies to study their organization, or tissues to gain access to a site in surgery. Postmortem dissection was considered a sin in the Middle Ages. In the UK before 1832, hanged murderers were the only legal source of bodies, supplemented by graverobbing (Burke and Hare were the most notorious grave robbers). The Anatomy Act of 1832 authorized the use of deceased institutionalized paupers unclaimed by next of kin, and by the 1940s bequests of bodies had been introduced.

dissociation in chemistry, the process whereby a single compound splits into two or more smaller products, which may be capable of recombining to form the reactant.

Where dissociation is incomplete (not all the compound's molecules dissociate), a ◊chemical equilibrium exists between the compound and its dissociation products. The extent of incomplete dissociation is defined by a numerical value (dissociation constant).

distance learning form of education using technology to teach pupils who are dispersed geographically. Britain's Open University, founded in 1969, is the oldest and most successful distance-learning institution in the world, using a mixture of postal mail, television, electronic conferencing, and the Internet to offer degree courses to students all over the world. Experiments in the 1990s used ◊videoconferencing and other multimedia techniques to widen the university's range.

distance modulus in astronomy, a method of finding the distance to an object in the universe, such as a star or ◊galaxy, using the difference between the actual and observed brightness of the object. The actual brightness is deduced from the object's type and its size. The apparent brightness is obtained by direct observation.

distance ratio in a machine, the distance moved by the input force, or effort, divided by the distance moved by the output force, or load. The ratio indicates the movement magnification achieved, and is equivalent to the machine's ◊velocity ratio.

distance-time graph graph used to describe the motion of a body by illustrating the relationship between the distance that it travels and the time taken. Plotting distance (on the vertical axis) against time (on the horizontal axis) produces a graph the gradient of which is the body's speed. If the gradient is constant (the graph is a straight line), the body has uniform or constant speed; if the gradient varies (the graph is curved), then so does the speed and the body may be said to be accelerating or decelerating.

distemper any of several infectious diseases of animals characterized by catarrh, cough, and general weakness. Specifically, it refers to a virus disease in young dogs, also found in wild animals, which can now be prevented by vaccination. In 1988 an allied virus killed over 10,000 common seals in the Baltic and North seas.

distillation technique used to purify liquids or to separate mixtures of liquids possessing different boiling points. **Simple distillation** is used in the purification of liquids (or the separation of substances in solution from their solvents) – for example, in the production of pure water from a salt solution.

The solution is boiled and the vapours of the solvent rise into a separate piece of apparatus (the condenser) where they are cooled and condensed. The liquid produced (the distillate) is the pure solvent; the non-volatile solutes (now in solid form) remain in the distillation vessel to be discarded as impurities or recovered as required. Mixtures of liquids (such as ◊petroleum or aqueous ethanol) are separated by **fractional distillation**, or fractionation. When the mixture is boiled, the vapours of its most volatile component rise into a vertical ◊fractionating column where they condense to liquid form. However, as this liquid runs back down the column it is reheated to boiling point by the hot rising vapours of the next-most-volatile component and so its vapours ascend the column once more. This boiling-condensing process occurs repeatedly inside the column, eventually bringing about a temperature gradient along its length. The vapours of the more volatile components therefore reach the top of the column and enter the condenser for collection before those of the less volatile components. In the fractional distillation of petroleum, groups of compounds (fractions) possessing similar relative molecular masses and boiling points are tapped off at different points on the column.

The earliest-known reference to the process is to the distillation of wine in the 12th century by Adelard of Bath. The chemical retort used for distillation was invented by Muslims, and was first seen in the West about 1570.

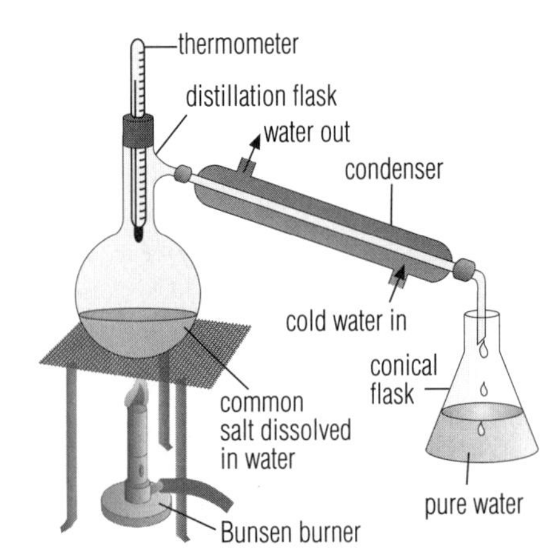

distillation Laboratory apparatus for simple distillation. Other forms of distillation include steam distillation, in which steam is passed into the mixture being distilled, and vacuum distillation, in which air is removed from above the mixture to be distilled.

distributed processing computer processing that uses more than one computer to run an application. ◊Local area networks, ◊client–server architecture, and ◊parallel processing involve distributed processing.

distribution in statistics, the pattern of ◊frequency for a set of data.

distributive operation in mathematics, an operation, such as multiplication, that bears a relationship to another operation, such as addition, such that $a \times (b + c) = (a \times b) + (a \times c)$. For example, $3 \times (2 + 4) = (3 \times 2) + (3 \times 4) = 18$. Multiplication may be said to be distributive over addition. Addition is not, however, distributive over multiplication because $3 + (2 \times 4) \rightarrow (3 + 2) \times (3 + 4)$.

distributor device in the ignition system of a piston engine that distributes pulses of high-voltage electricity to the ◊spark plugs in the cylinders. The electricity is passed to the plug leads by the tip of a rotor arm, driven by the engine camshaft, and current is fed to the rotor arm from the ignition coil. The distributor also houses the contact point or breaker, which opens and closes to interrupt the battery current to the coil, thus triggering the high-voltage pulses. With electronic ignition the distributor is absent.

dithering in computer graphics, a technique for varying the patterns of dots in an image in order to give the impression of shades of grey. Each dot, however, is of the same size and the same intensity, unlike grey scaling (where each dot can have a different shade) and photographically reproduced half-tones (where the dot size varies).

diuretic any drug that increases the output of urine by the kidneys. It may be used in the treatment of high blood pressure and to relieve ◊oedema associated with heart, lung, kidney or liver disease, and some endocrine disorders.

diver or ***loon*** any of four species of marine bird of the order Gaviiformes, specialized for swimming and diving, found in northern regions of the northern hemisphere. The legs are set so far back that walking is almost impossible, but they are powerful swimmers and good flyers, and only come ashore to nest. They have straight bills, short tail-feathers, webbed feet, and long bodies; they feed on fish, crustaceans, and some water plants. During the breeding period they live inland and the female lays two eggs which hatch into down-covered chicks. Of the four species, the largest is the white-billed diver *Gavia adamsii,* an Arctic species 75 cm/2.5 ft long.

diversification in agriculture and business, the development of distinctly new products or markets. A company or farm may diversify in order to spread its risks or because its original area of operation is becoming less profitable. In the UK agricultural diversification has included offering accommodation and services to tourists – for example, bed and breakfast, camping and caravanning sites, and pony trekking.

diverticulitis inflammation of diverticula (pockets of herniation) in the large intestine. It is usually triggered by infection and causes diarrhoea or constipation, and lower abdominal pain. Usually it can be controlled by diet and antibiotics.

diving apparatus any equipment used to enable a person to spend time underwater. Diving bells were in use in the 18th century, the diver breathing air trapped in a bell-shaped chamber. This was followed by cumbersome diving suits in the early 19th century. Complete freedom of movement came with the ◊aqualung, invented by Jacques ◊Cousteau in the early 1940s. For work at greater depths the technique of saturation diving was developed in the 1970s in which divers live for a week or more breathing a mixture of helium and oxygen at the pressure existing on the seabed where they work (as in work on North Sea platforms and tunnel building).

The first diving suit, with a large metal helmet supplied with air through a hose, was invented in the UK by the brothers John and Charles Deane in 1828. Saturation diving was developed for working in offshore oilfields. Working divers are ferried down to the work site by a lock-out ◊submersible. By this technique they avoid the need for lengthy periods of decompression after every dive. Slow decompression is necessary to avoid the dangerous consequences of an attack of the bends, or ◊decompression sickness.

division basic operation of arithmetic, the inverse of ◊multiplication.

dizziness another word for ◊vertigo.

DLL in computing, the abbreviation for ◊dynamic link library.

DMA channel (abbreviation for ***direct memory access channel***) in computing, type of channel used for the fast transfer of data between a computer and peripherals such as CD-ROM drives. Most ISA (industry standard architecture) personal computers (PCs) have eight DMA channels, of which typically six are available for use by add-on peripherals, most of which require dedicated channels.

DNA (abbreviation for ***deoxyribonucleic acid***) complex giant molecule that contains, in chemically coded form, the information needed for a cell to make proteins. DNA is a ladderlike double-stranded ◊nucleic acid which forms the basis of genetic inheritance in all organisms, except for a few viruses that have only ◊RNA. DNA is organized into ◊chromosomes and, in organisms other than bacteria, it is found only in the cell nucleus.

If you want to understand function, study structure.

FRANCIS CRICK English molecular biologist.
What Mad Pursuit 1988

DNA fingerprinting or ***DNA profiling*** another name for ◊genetic fingerprinting.

DNS in computing, abbreviation for domain ◊name server.

Dobermann or ***Dobermann pinscher*** breed of smooth-coated dog with a docked tail, much used as a guard dog. It stands up to 70 cm/27.5 in tall, has a long head with a flat, smooth skull, and is often black with brown markings. It takes its name from the man who bred it in 19th-century Germany.

dock or ***sorrel*** in botany, any of a number of plants belonging to the buckwheat family. They are tall, annual or perennial herbs, often with lance-shaped leaves and small greenish flowers. Native to temperate regions, there are 30 North American and several British species. (Genus *Rumex,* family Polygonaceae.)

dock port accommodation for commercial and naval vessels, usually simple linear quayage (wharfs or piers) adaptable to ships of any size, but with specialized equipment for handling bulk cargoes, refrigerated goods, container traffic, and oil tankers.

document in computing, data associated with a particular application. For example, a **text document** might be produced by a ◊word processor and a graphics document might be produced with a ◊CAD package. An ◊OMR or ◊OCR document is a paper document containing data that can be directly input to the computer using a ◊document reader.

documentation in computing, the written information associated with a computer program or ◊applications package. Documentation is usually divided into two categories: program documentation and user documentation.

Program documentation is the complete technical description of a program, drawn up as the software is written and intended to support any later maintenance or development of that program. It typically includes details of when, where, and by whom the software was written; a general description of the purpose of the software, including recommended input, output, and storage methods; a detailed description of the way the software functions, including full program listings and ◊flow charts; and details of software testing, including sets of ◊test data with expected results. **User documentation** explains how to operate the software. It typically includes a nontechnical explanation of the purpose of the software; instructions for loading, running, and using the software; instructions for preparing any necessary input data; instructions for requesting and interpreting output data; and explanations of any error messages that the program may produce.

document image processing (**DIP**) scanning documents for storage on ◊CD-ROM. The scanned images are indexed electronically,

DNA: Discovery of the Structure of DNA

BY JULIAN ROWE

The first announcement
'We wish to suggest a structure for the salt of deoxyribose nucleic acid (DNA). This structure has novel features which are of considerable biological interest.'

So began a 900-word article that was published in the journal *Nature* in April 1953. Its authors were British molecular biologist Francis Crick (1916–) and US biochemist James Watson (1928–). The article described the correct structure of DNA, a discovery that many scientists have called the most important since Austrian botanist and monk Gregor Mendel (1822–1884) laid the foundations of the science of genetics. DNA is the molecule of heredity, and by knowing its structure, scientists can see exactly how forms of life are transmitted from one generation to the next.

The problem of inheritance
The story of DNA really begins with British naturalist Charles Darwin (1809–1882). When, in November 1859, he published 'On the Origin of Species by Means of Natural Selection' outlining his theory of evolution, he was unable to explain exactly how inheritance came about. For at that time it was believed that offspring inherited an average of the features of their parents. If this were so, as Darwin's critics pointed out, any remarkable features produced in a living organism by evolutionary processes would, in the natural course of events, soon disappear.

The work of Gregor Mendel, only rediscovered 18 years after Darwin's death, provided a clear demonstration that inheritance was not a 'blending' process at all. His description of the mathematical basis to genetics followed years of careful plant-breeding experiments. He concluded that each of the features he studied, such as colour or stem length, was determined by two 'factors' of inheritance, one coming from each parent. Each egg or sperm cell contained only one factor of each pair. In this way a particular factor, say for the colour red, would be preserved through subsequent generations.

Genes
Today, we call Mendel's factors **genes**. Through the work of many scientists, it came to be realized that genes are part of the chromosomes located in the nucleus of living cells and that DNA, rather than protein as was first thought, was a hereditary material.

The double helix
In the early 1950s, scientists realized that X-ray crystallography, a method of using X-rays to obtain an exact picture of the atoms in a molecule, could be successfully applied to the large and complex molecules found in living cells.

It had been known since 1946 that genes consist of DNA. At King's College, London, New Zealand–British biophysicist Maurice Wilkins (1916–) had been using X-ray crystallography to examine the structure of DNA, together with his colleague, British X-ray crystallographer Rosalind Franklin (1920–1958), and had made considerable progress.

While in Copenhagen, US scientist James Watson had realized that one of the major unresolved problems of biology was the precise structure of DNA. In 1952, he came as a young postdoctoral student to join the Medical Research Council Unit at the Cavendish Laboratory, Cambridge, where Francis Crick was already working. Convinced that a gene must be some kind of molecule, the two scientists set to work on DNA.

Helped by the work of Wilkins, they were able to build an accurate model of DNA. They showed that DNA had a double helical structure, rather like a spiral staircase. Because the molecule of DNA was made from two strands, they envisaged that as a cell divides, the strands unravel, and each could serve as a template as new DNA was formed in the resulting daughter cells. Their model also explained how genetic information might be coded in the sequence of the simpler molecules of which DNA is comprised. Here for the first time was a complete insight into the basis of heredity. James Watson commented that this result was 'too pretty not to be true!'

Cracking the code
Later, working with South African–British molecular biologist Sidney Brenner (1927–), Crick went on to work out the genetic code, and so ascribe a precise function to each specific region of the molecule of DNA. These triumphant results created a tremendous flurry of scientific activity around the world. The pioneering work of Crick, Wilkins, and Watson was recognized in the award of the Nobel Prize for Physiology or Medicine in 1962.

The unravelling of the structure of DNA lead to a new scientific discipline, molecular biology, and laid the foundation stones for genetic engineering – a powerful new technique that is revolutionizing biology, medicine, and food production through the purposeful adaptation of living organisms.

which provides much faster access than is possible with either paper or ◊microform. See also ◊optical character recognition.

document reader in computing, an input device that reads marks or characters, usually on preprepared forms and documents. Such devices are used to capture data by ◊optical mark recognition (OMR), ◊optical character recognition (OCR), and ◊mark sensing.

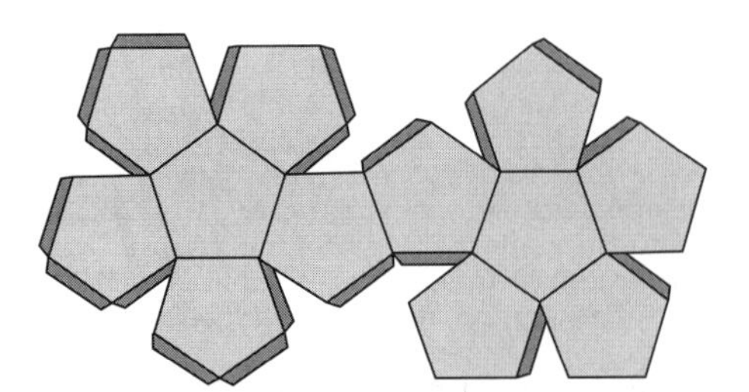

dodecahedron A dodecahedron is a solid figure which has 12 pentagonal faces and 12 vertices. It is one of the five regular polyhedra (with all faces the same size and shape).

dodder parasitic plant belonging to the morning-glory family, without leaves or roots. The thin stem twines around the host plant, and penetrating suckers withdraw nourishment. (Genus *Cuscuta*, family Convolvulaceae.)

dodecahedron regular solid with 12 pentagonal faces and 12 vertices. It is one of the five regular ◊polyhedra, or Platonic solids.

dodo extinct flightless bird *Raphus cucullatus*, order Columbiformes, formerly found on the island of Mauritius, but

DODO

Most pictures of the dodo show a bird with two right legs. This is because the only person to paint a live dodo, the Belgian artist Roelandt Savery, made an elementary error in drawing the parts that he could not see.

exterminated by early settlers around 1681. Although related to the pigeons, it was larger than a turkey, with a bulky body, rudimentary wings, and short curly tail-feathers. The bill was blackish in colour, forming a horny hook at the end.

dog any carnivorous mammal of the family Canidae, including wild dogs, wolves, jackals, coyotes, and foxes. Specifically, the domestic dog *Canis familiaris,* the earliest animal descended from the wolf. Dogs were first domesticated around 14,000 years ago, and migrated with humans to all the continents. They have been selectively bred into many different varieties for use as working animals and pets.

characteristics The dog has slender legs and walks on its toes (**digitigrade**). The forefeet have five toes, the hind feet four, with non-retractile claws. The head is small and the muzzle pointed, but the shape of the head differs greatly in various breeds. The average life of a dog is from 10 to 14 years, though some live to be 20. The dog has a very acute sense of smell and can readily be trained, for it has a good intelligence.

wild dogs Of the wild dogs, some are solitary, such as the long-legged maned wolf *Chrysocyon brachurus* of South America, but others hunt in groups, such as the African hunting dog *Lycaonpictus* (classified as a vulnerable species) and the ◊wolf. ◊Jackals scavenge for food, and the raccoon dog *Nyctereutes procyonoides* of east Asia includes plant food as well as meat in its diet. The Australian wild dog is the ◊dingo.

DOG LOVERS' PAGE

http://www.petnet.com.au/dogs/dogbreedindex.html

Information on breeds, ranging from the world's smallest breed (the Chihuahua) to the tallest (the Irish Wolfhound). This Australian-based site includes photographs and details of over 100 breeds of dog.

dog in computing, reference to a cartoon published in *The New Yorker* that showed a dog poised over a computer keyboard remarking to another dog, 'On the Internet, no one knows you're a dog.' It is used to illustrate the point that the Internet gives users the opportunity to adopt a different persona, for example many women surf using a male identity.

dogbane any sometimes poisonous herbaceous North American plant of the genus *Apocynum,* with opposite leaves, small white or pink flowers, and milky juice.

dogfish any of several small sharks found in the northeast Atlantic, Pacific, and Mediterranean.

dog's mercury plant belonging to the spurge family, common in woods of Europe and southwest Asia. It grows to 30 cm/1 ft, has oval, light-green leaves, and spreads over woodland floors in patches of plants of a single sex. Male flowers are small, greenish yellow, and held on an upright spike above the leaves; female flowers droop below the upper leaves. (*Mercurialis perennis,* family Euphorbiaceae.)

dogwhelk marine mollusc, some of which live on rocky shores whilst others burrow. The shell aperture is grooved to house the inhalant siphon keeping it clear of the mud.

The **common dogwhelk** *Thais lapillus* is a predator of barnacles and mussels and the coloured bands in the shell are believed to be influenced by the food material. The egg capsules are attached singly in crevices and have the appearance of corn grains.

The **netted dogwhelk** *Nassarius reticulatus* is recognizable by the crenulations on the shell. The egg capsules are flat and are deposited in eel-grass. Like the closely related thick-lipped dogwhelk *N. incrassatus,* it is a scavenger.

classifiation Dogwhelks are in order Neogastropoda, subclass Prosobranchia, class Gastropoda, phylum Mollusca.

dogwood any of a group of trees and shrubs belonging to the dogwood family, native to temperate regions of North America, Europe, and Asia. The flowering dogwood (*Cornus florida*) of the eastern USA is often cultivated as an ornamental for its beautiful blooms consisting of clusters of small greenish flowers surrounded by four large white or pink petal-like bracts (specialized leaves). (Genus *Cornus,* family Cornaceae.)

Heads of small white flowers, each with four petals joined as a tube, are produced in midsummer, followed by black berries. The dogwood is characteristic of lime soils in southern England, and is found over much of southern Europe. *C. sanguinea* is native to Britain and common in old hedgerows and woods. It takes its name from the redness of the twigs. The introduced red-osier dogwood (*C. sericea*) has longer twigs of a brighter red, with white berries rather than black. Various other species of dogwood are planted in gardens.

Dolby system electronic circuit that reduces the background high-frequency noise, or hiss, during replay of magnetic tape recordings. The system was developed by US engineer Raymond Dolby (1933–) in 1966.

doldrums area of low atmospheric pressure along the Equator, in the intertropical convergence zone where the northeast and southeast trade winds converge. The doldrums are characterized by calm or very light winds, during which there may be sudden squalls and stormy weather. For this reason the areas are avoided as far as possible by sailing ships.

dolerite igneous rock formed below the Earth's surface, a form of basalt, containing relatively little silica (mafic in composition).

Dolerite is a medium-grained (hypabyssal) basalt and forms in shallow intrusions, such as dykes, which cut across the rock strata, and sills, which push between beds of sedimentary rock. When exposed at the surface, dolerite weathers into spherical lumps.

dolmen prehistoric ◊megalith in the form of a chamber built of three or more large upright stone slabs, capped by a horizontal flat stone. Dolmens are the burial chambers of Neolithic (New Stone Age) chambered tombs and passage graves, revealed by the removal of the covering burial mound. They are found in Europe and Africa, and occasionally in Asia as far east as Japan.

dolomite in mineralogy, white mineral with a rhombohedral structure, calcium magnesium carbonate ($CaMg(CO_3)_2$). Dolomites are common in geological successions of all ages and are often formed when ◊limestone is changed by the replacement of the mineral calcite with the mineral dolomite.

dolphin any of various highly intelligent aquatic mammals of the family Delphinidae, which also includes porpoises. There are about 60 species. Most inhabit tropical and temperate oceans, but there are some freshwater forms in rivers in Asia, Africa, and South America. The name 'dolphin' is generally applied to species having a beaklike snout and slender body, whereas the name 'porpoise' is reserved for the smaller species with a blunt snout and stocky body. Dolphins use sound (◊echolocation) to navigate, to find prey, and for communication. The common dolphin *Delphinus delphis* is found in all temperate and tropical seas. It is up to 2.5 m/8 ft long, and is dark above and white below, with bands of grey, white, and yellow on the sides. It has up to 100 teeth in its jaws, which make the 15 cm/6 in 'beak' protrude forward from the rounded head. The corners of its mouth are permanently upturned, giving the appearance of a smile, though dolphins cannot actually smile. Dolphins feed on fish and squid.

river dolphins There are five species of river dolphin, two South American and three Asian, all of which are endangered. The two South American species are the **Amazon river dolphin** or **boto** *Inia geoffrensis,* the largest river dolphin (length 2.7 m/8.9 ft, weight 180 kg/396 lb) and the **La Plata river dolphin** *Pontoporia blainvillei* (length 1.8 m/5.9 ft, weight 50 kg/110 lb). The **tucuxi** *Sotalia fluviatilis* is not a true river dolphin, but lives in the Amazon and Orinoco rivers, as well as in coastal waters.

The Asian species are the **Ganges river dolphin** *Platanista gangetica,* the **Indus river dolphin** *Platanista minor* (length 2 m/6.6 ft, weight 70 kg/154 lb) (fewer than 500 remaining), and the

Yangtze river dolphin or *baiji Lipotes vexillifer* (length 2 m/6.6 ft, weight 70 kg/154 lb) (fewer than 100 remaining).

As a result of living in muddy water, river dolphins' eyes have become very small. They rely on echolocation to navigate and find food. Some species of dolphin can swim at up to 56 kph/35 mph, helped by special streamlining modifications of the skin.

All dolphins power themselves by beating the tail up and down, and use the flippers to steer and stabilize. The flippers betray dolphins' land-mammal ancestry with their typical five-toed limbbone structure. Dolphins have great learning ability and are popular performers in aquariums. The species most frequently seen is the bottle-nosed dolphin *Tursiops truncatus,* found in all warm seas, mainly grey in colour and growing to a maximum 4.2 m/14 ft. The US Navy began training dolphins for military purposes in 1962, and in 1987 six dolphins were sent to detect mines in the Persian Gulf. Marine dolphins are endangered by fishing nets, speedboats, and pollution. In 1990 the North Sea states agreed to introduce legislation to protect them.

Also known as **dolphin** is the totally unrelated true fish *Coryphaena hippurus,* up to 1.5 m/5 ft long.

dolphinfish carnivorous marine fish. Dolphinfish are large and brilliantly coloured, with hues of metallic yellow, blue, and silver; their bodies are elongated, compressed, and covered with small scales. They are usually about 2 m/6.6 ft in length and feed largely on flying fish.
classification Dolphinfish are in the genus *Coryphaena,* family Corphaenidae, belonging to the order Perciformes, class Osteichthyes.

domain on the Internet, segment of an address that specifies an organization, its type, or its country of origin. Domain names are read backwards, starting at the end. All countries except the USA use a final two-letter code such as **ca** for Canada and **uk** for the UK. US addresses end in one of seven 'top-level' domains, which specify the type of organization: **com** (commercial), **mil** (military), **org** (usually a nonprofit-making organization), and so on.

These names are for humans; to enable mail and other messages to be sorted by machine, computers use IP (Internet protocol) numbers. To route a message, the computer looks up the domain name on a domain name server (DNS), which tells the computer the number.

In 1998 there were 30 million domains on the Internet.

domain small area in a magnetic material that behaves like a tiny magnet. The magnetism of the material is due to the movement of electrons in the atoms of the domain. In an unmagnetized sample of material, the domains point in random directions, or form closed loops, so that there is no overall magnetization of the sample. In a magnetized sample, the domains are aligned so that their magnetic effects combine to produce a strong overall magnetism.

domain name server in computing, see ◊name server.

dominance in genetics, the masking of one allele (an alternative form of a gene) by another allele. For example, if a ◊heterozygous person has one allele for blue eyes and one for brown eyes, his or her eye colour will be brown. The allele for blue eyes is described as ◊recessive and the allele for brown eyes as dominant.

The preservation of favourable variations and the rejection of injurious variations, I call Natural Selection, or Survival of the Fittest. Variations neither useful nor injurious would not be affected by natural selection and would be left a fluctuating element.

CHARLES DARWIN British naturalist.
On the Origin of Species 1859

Dominion Astrophysical Observatory Canadian observatory near Victoria, British Columbia, the site of a 1.85-m/73-in reflector opened in 1918, operated by the National Research Council of Canada. The associated Dominion Radio Astrophysical Observatory at Penticton, British Columbia, operates a 26-m/84-ft radio dish and an aperture synthesis radio telescope.

dongle in computing, a device that ensures the legal use of an application program. It is usually attached to the printer port of the computer (between the port and the printer cable) and the program will not run in its absence.

donkey another name for ◊ass.

Doom popular computer game released in 1994. It is one of a series of games from the Texas-based company id Software, which specializes in 3-D graphics, alien monsters, and complex mazes which players must navigate to find secret treasures and hidden keys along the way. *Doom* can be played competitively over a network as well as on a single computer.

Because the company has encouraged players to create their own additional levels for *Doom* (and its other games) by releasing the necessary source code, a whole culture has grown up around *Doom* and id's other games.

dopamine neurotransmitter, hydroxytyramine $C_8H_{11}NO_2$, an intermediate in the formation of adrenaline. There are special nerve cells (neurons) in the brain that use dopamine for the transmission of nervous impulses. One such area of dopamine neurons lies in the basal ganglia, a region that controls movement. Patients suffering from the tremors of Parkinson's disease show nerve degeneration in this region. Another dopamine area lies in the limbic system, a region closely involved with emotional responses. It has been found that schizophrenic patients respond well to drugs that limit dopamine excess in this area.

Doppler, Christian Johann
(1803–1853)

Austrian physicist who in 1842 described the **Doppler effect** and derived the observed frequency mathematically in Doppler's principle.

Doppler effect change in the observed frequency (or wavelength) of waves due to relative motion between the wave source and the observer. The Doppler effect is responsible for the perceived change in pitch of a siren as it approaches and then recedes, and for the ◊red shift of light from distant galaxies. It is named after the Austrian physicist Christian Doppler.

Dorado constellation of the southern hemisphere, represented as a goldfish. It is easy to locate, since the Large ◊Magellanic Cloud marks its southern border. Its brightest star is Alpha Doradus, just under 200 light years from Earth.

One of the most conspicuous objects in the Large Magellanic Cloud is the 'Great Looped Nebula' that surrounds 30 Doradus.

dor beetle oval-shaped, stout beetle measuring on average 20 mm/0.8 in, and with a metallic sheen. In general, dor beetles feed on dung, mostly of herbivorous animals.

They usually build burrows almost 50 cm/19.5 in deep under an accumulation of dung. The blind end of each burrow is sealed with a portion of dung into which the female deposits a single egg. The developing larva feeds on the dung. Both adults and larvae possess stridulating (sound-producing) organs.
classification Dor beetles are in the family Geotrupidae (superorder Scarabaeoidea), order Coleoptera, class Insecta, phylum Arthropoda.

dormancy in botany, a phase of reduced physiological activity exhibited by certain buds, seeds, and spores. Dormancy can help a plant to survive unfavourable conditions, as in annual plants that pass the cold winter season as dormant seeds, and plants that form dormant buds.

For various reasons many seeds exhibit a period of dormancy even when conditions are favourable for growth. Sometimes this

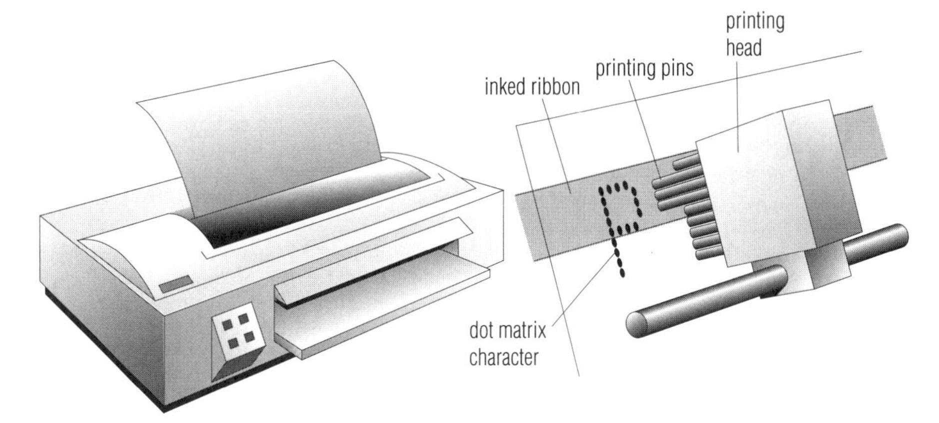

dot matrix printer *Characters and graphics printed by a dot matrix printer are produced by a block of pins which strike the ribbon and make up a pattern using many small dots.*

dormancy can be broken by artificial methods, such as penetrating the seed coat to facilitate the uptake of water (chitting) or exposing the seed to light. Other seeds require a period of ◊after-ripening.

dormouse small rodent, of the family Gliridae, with a hairy tail. There are about ten species, living in Europe, Asia, and Africa. They are arboreal (live in trees) and nocturnal, hibernating during winter in cold regions. They eat berries, nuts, pollen, and insects.

dorsal in vertebrates, the surface of the animal closest to the backbone. For most vertebrates and invertebrates this is the upper surface, or the surface furthest from the ground. For bipedal primates such as humans, where the dorsal surface faces backwards, then the word is 'back'.

Not all animals can be said to have a dorsal surface, just as many animals cannot be said to have a front; for example, jellyfish, anemones, and sponges do not have a dorsal surface.

dory marine fish *Zeus faber* found in the Mediterranean and Atlantic. It grows up to 60 cm/2 ft, and has nine or ten long spines at the front of the dorsal fin, and four at the front of the anal fin. It is considered to be an excellent food fish and is also known as **John Dory**.

The dory is olive brown or grey, with a conspicuous black spot ringed with yellow on each side. A stalking predator, it shoots out its mobile jaws to catch fish.

DOS (acronym for ***disc operating system***) computer ◊operating system specifically designed for use with disc storage; also used as an alternative name for a particular operating system, ◊MS-DOS.

dot in computing, full stop that separates ◊IP addresses, sections of ◊domain names, and the hierarchies in ◊newsgroup names, as well as file names and their extensions.

dot matrix printer computer printer that produces each character individually by printing a pattern, or matrix, of very small dots. The printing head consists of a vertical line or block of either 9 or 24 printing pins. As the printing head is moved from side to side across the paper, the pins are pushed forwards selectively to strike an inked ribbon and build up the dot pattern for each character on the paper beneath.

A dot matrix printer is more flexible than a ◊daisywheel printer because it can print graphics and text in many different typefaces. It is cheaper to buy and maintain than a ◊laser printer or ◊ink-jet printer, and, because its pins physically strike the paper, is capable of producing carbon copies. However, it is noisy in operation and cannot produce the high-quality printing associated with the non-impact printers.

dot pitch in computing, distance between the dots which make up the picture on a computer monitor. The smaller the dot pitch, the better and finer-grained the picture.

dotterel bird *Eudromias morinellus* of the plover family, in order Charadriiformes, nesting on high moors and tundra in Europe and Asia, and migrating south for the winter. About 23 cm/9 in long, its plumage is patterned with black, brown, and white in summer, duller in winter, but always with white eyebrows and breastband. The female is larger than the male, and mates up to five times with different partners, each time laying her eggs and leaving them in the sole care of the male, who incubates and rears the brood. Three pale-green eggs with brown markings are laid in hollows in the ground.

double bond two covalent bonds between adjacent atoms, as in the ◊alkenes (–C=C–) and ◊ketones (–C=O–).

double click in computing, to click (press and release a ◊mouse button) twice in quick succession. Double clicking on an ◊icon shown on a ◊graphical user interface (GUI) is used to start an application. In most GUIs it is possible to set the maximum time interval between the two clicks.

double coconut treelike ◊palm plant, also known as **coco de mer**, of the Seychelles. It produces a two-lobed edible nut, one of the largest known fruits. (*Lodoicea maldivica.*)

double decomposition reaction between two chemical substances (usually ◊salts in solution) that results in the exchange of a constituent from each compound to create two different compounds.

For example, if silver nitrate solution is added to a solution of sodium chloride, there is an exchange of ions yielding sodium nitrate and silver chloride.

double precision in computing, a type of floating-point notation that has higher precision, that is, more significant decimal places. The term 'double' is not strictly correct, deriving from such numbers using twice as many ◊bits as standard floating-point notation.

double star two stars that appear close together. Many stars that appear single to the naked eye appear double when viewed through a telescope. Some double stars attract each other due to gravity, and orbit each other, forming a genuine ◊binary star, but other double stars are at different distances from Earth, and lie in the same line of sight only by chance. Through a telescope both types look the same.

Double stars of the second kind, which are of little astronomical interest, are referred to as 'optical pairs'; those of the first as 'physical pairs' or, more usually, 'visual binaries'. They are the principal source from which is derived our knowledge of stellar masses.

dough mixture consisting primarily of flour, water, and yeast, which is used in the manufacture of bread.

The preparation of dough involves thorough mixing (kneading) and standing in a warm place to 'prove' (increase in volume) so that the ◊enzymes in the dough can break down the starch from the flour into smaller sugar molecules, which are then fermented by the yeast. This releases carbon dioxide, which causes the dough to rise.

Douglas fir any of some six species of coniferous evergreen tree belonging to the pine family. The most common is *Pseudotsuga menziesii*, native to western North America and east Asia. It grows up to 60–90 m/200–300 ft in height, has long, flat, spirally-arranged needles and hanging cones, and produces hard, strong timber. *P. glauca* has shorter, bluish needles and grows to 30 m/100 ft in mountainous areas. (Genus *Pseudotsuga*, family Pinaceae.)

douroucouli or ***night monkey*** or ***owl monkey*** nocturnal South American monkey. Its eyes are very large and its thick coat is a grey brown. Douroucoulis sleep in tree holes or thick vegetation and feed mainly on fruit and leaves, and some insects. They are the only nocturnal monkeys.

Unlike most South American monkeys, their long tails are not prehensile. The incisors in the lower jaw project forwards and they have a total of 36 teeth.

classification Douroucoulis are in the genus *Aotus* in the family Cebidae, order Primates.

dove another name for ◊pigeon.

Down's syndrome condition caused by a chromosomal abnormality (the presence of an extra copy of chromosome 21), which in humans produces mental retardation; a flattened face; coarse, straight hair; and a fold of skin at the inner edge of the eye (hence the former name 'mongolism'). The condition can be detected by prenatal testing.

Those afflicted are usually born to mothers over 40 (one in 100), and in 1995 French researchers discovered a link between Down's syndrome incidence and paternal age, with men over 40 having an increased likelihood of fathering a Down's syndrome baby.

The syndrome is named after J L H Down (1828–1896), an English physician who studied it. All people with Down's syndrome who live long enough eventually develop early-onset ◊Alzheimer's disease, a form of dementia. This fact led to the discovery in 1991 that some forms of early-onset Alzheimer's disease are caused by a gene defect on chromosome 21.

DOWN'S SYNDROME WEB PAGE

http://www.nas.com/downsyn/

Well-organized source of information about the syndrome – with articles, essays, lists of organizations worldwide, toy catalogues, the 'Brag Book' photo gallery, and links to other helpful Web sites.

downtime in computing, time when a computer system is unavailable for use, due to maintenance or a system crash. Some downtime is inevitable on almost all systems.

dowsing ascertaining the presence of water or minerals beneath the ground with a forked twig or a pendulum. Unconscious muscular action by the dowser is thought to move the twig, usually held with one fork in each hand, possibly in response to a local change in the pattern of electrical forces. The ability has been known since at least the 16th century and, though not widely recognized by science, it has been used commercially and in archaeology.

INTRODUCTION TO DOWSING

http://home.interstat.net/~slawcio/dowsing.html

American Association of Dowsers site that includes all the information you might need to make some dowsing rods and how to use them.

dpi abbreviation for ***dots per inch***, measure of the ◊resolution of images produced by computer screens and printers.

Draco in astronomy, a large but faint constellation represented as a dragon coiled around the north celestial pole. Due to ◊precession the star Alpha Draconis (Thuban) was the pole star 4,800 years ago.

This star seems to have faded, for it is no longer the brightest star in the constellation as it was at the beginning of the 17th century. Gamma Draconis is more than a magnitude brighter. It was extensively observed by James Bradley, who from its apparent changes in position discovered the ◊aberration of starlight and ◊nutation.

drag resistance to motion a body experiences when passing through a fluid – gas or liquid. The aerodynamic drag aircraft experience when travelling through the air represents a great waste of power, so they must be carefully shaped, or streamlined, to reduce drag to a minimum. Cars benefit from ◊streamlining, and aerodynamic drag is used to slow down spacecraft returning from space. Boats travelling through water experience hydrodynamic drag on their hulls, and the fastest vessels are ◊hydrofoils, whose hulls lift out of the water while cruising.

drag and drop in computing, in ◊graphical user interfaces, feature that allows users to select a file name or icon using a mouse and move it to the name or icon representing a program so that the computer runs the program using that file as input data.

This method is convenient for computer users, as it eliminates unnecessary typing. Moving the name of a text file, for example, to a copy of a word processor will start up the word processor with that file loaded and ready for editing.

dragon name popularly given to various sorts of lizard. These include the ◊flying dragon *Draco volans* of southeast Asia; the komodo dragon *Varanus komodoensis* of Indonesia, at over 3 m/10 ft the largest living lizard; and some Australian lizards with bizarre spines or frills.

dragonet small, spiny-rayed fish that lives in temperate and tropical seas. The males are larger and brightly coloured with filamentlike rays extending from the first dorsal fin. Females are dull in colour. Dragonets have smooth bodies without scales.

The gill openings are reduced to a single small hole near the nape of the neck and the ventral fins are under the throat. The **sculpin** *Callionymus draco* is about 30 cm/12 in long, and is brown and white in colour; the common dragonet *C. lyra* is yellow, sapphire, and violet in hue.

classification Dragonets are in the genus *Callionymus*, order Perciformes, class Osteichthyes.

dragonfly any of numerous insects of the order Odonata, including the ◊damselfly. They all have long narrow bodies, two pairs of almost equal-sized, glassy wings with a network of veins; short, bristlelike antennae; powerful, 'toothed' mouthparts; and very large compound eyes which may have up to 30,000 facets. They can fly at speeds of up to 64–96 kph/40–60 mph.

Dragonflies hunt other insects by sight, both as adults and as aquatic nymphs. The largest species have a wingspan of 18 cm/7 in, but fossils related to dragonflies have been found with wings of up to 70 cm/2.3 ft across.

DRAM (acronym for ***dynamic random-access memory***) computer memory device in the form of a silicon chip commonly used to provide the ◊immediate-access memory of microcomputers. DRAM loses its contents unless they are read and rewritten every 2 milliseconds or so. This process is called **refreshing** the memory. DRAM is slower but cheaper than ◊SRAM, an alternative form of silicon-chip memory.

drawing program in computing, software that allows a user to draw freehand and create complex graphics. Additional features may include special ◊fonts, ◊clip art, or painting facilities that allow a user to simulate on the computer the drawing characteristics of specific real-world implements such as charcoal, watercolours, or pastels. The market-leading drawing package is Corel Draw.

dream series of events or images perceived through the mind during sleep. Their function is unknown, but Sigmund ◊Freud saw them as wish fulfilment (nightmares being failed dreams prompted by fears of 'repressed' impulses). Dreams occur in periods of rapid eye movement (REM) by the sleeper, when the cortex of the brain is approximately as active as in waking hours. Dreams occupy about a fifth of sleeping time.

If a high level of acetylcholine (chemical responsible for transmission of nerve impulses) is present, dreams occur too early in sleep, causing wakefulness, confusion, and ◊depression, which suggests that a form of memory search is involved. Prevention of dreaming, by taking sleeping pills, for example, has similar unpleasant results. For the purposes of (allegedly) foretelling the future, dreams fell into disrepute in the scientific atmosphere of the 18th century.

drill large Old World monkey *Mandrillus leucophaeus* similar to a baboon and in the same genus as the ◊mandrill. Drills live in the forests of Cameroon and Nigeria. Brownish-coated, black-faced, and stoutly built, with a very short tail, the male can have a head and body up to 75 cm/2.5 ft long, although females are much smaller.

drilling common woodworking and metal machinery process that involves boring holes with a drill bit. The commonest kind of drill bit is the fluted drill, which has spiral grooves around it to allow the cut material to escape. In the oil industry, rotary drilling is used to bore oil wells. The drill bit usually consists of a number of toothed cutting wheels, which grind their way through the rock as the drill pipe is turned, and mud is pumped through the pipe to lubricate the bit and flush the ground-up rock to the surface.

In rotary drilling, a drill bit is fixed to the end of a length of drill pipe and rotated by a turning mechanism, the rotary table. More lengths of pipe are added as the hole deepens. The long drill pipes are handled by lifting gear in a steel tower or ◊derrick.

drinking water water that has been subjected to various treatments, including ◊filtration and ◊sterilization, to make it fit for human consumption; it is not pure water.

DRINKING WATER

Darkling beetles living in the Namib desert in Africa drink by standing on their heads at the crests of sand dunes at dawn. Mists rolling in from the coast condense to form water on their bodies; the water then trickles down into their mouths.

drive bay in computing, slot in a computer designed to hold a disc drive such as a hard drive, floppy drive, or CD-ROM drive. Like most computer components, disc drives have decreased in size. Older drives were $5\frac{1}{4}$ inches in size, but most newer drives are $3\frac{1}{2}$ inches. Kits to fit a $3\frac{1}{2}$ inch drive into a $5\frac{1}{4}$ inch bay are readily available.

driver in computing, a program that controls a peripheral device. Every device connected to the computer needs a driver program.

The driver ensures that communication between the computer and the device is successful.

For example, it is often possible to connect many different types of printer, each with its own special operating codes, to the same type of computer. This is because driver programs are supplied to translate the computer's standard printing commands into the special commands needed for each printer.

dromedary variety of Arabian ◊camel. The dromedary or one-humped camel has been domesticated since 400 BC. During a long period without water, it can lose up to one-quarter of its body weight without ill effects.

DROMEDARY CAMEL

http://www.seaworld.org/animal_bytes/dromedary_camelab.html

Illustrated guide to the dromedary camel including information about genus, size, life span, habitat, gestation, diet, and a series of fun facts.

drop-down list in computing, in a ◊graphical user interface, a list of options that hangs down from a blank space in a ◊dialog box or other on-screen form when a computer awaits user input.

To select one of the choices, highlight it and click. The list will disappear and the selected item will appear in the blank. If the list is longer than the space available, small arrows and a scroll bar will appear on the right-hand side.

drug any of a range of substances, natural or synthetic, administered to humans and animals as therapeutic agents: to diagnose, prevent, or treat disease, or to assist recovery from injury. Traditionally many drugs were obtained from plants or animals; some minerals also had medicinal value. Today, increasing numbers of drugs are synthesized in the laboratory.

Drugs are administered in various ways, including: orally, by injection, as a lotion or ointment, as a ◊pessary, by inhalation, or by transdermal patch.

A miracle drug is any drug that will do what the label says it will do.

Eric Hodgins
Episode

drug, generic any drug produced without a brand name that is identical to a branded product. Usually generic drugs are produced when the patent on a branded drug has expired, and are cheaper than their branded equivalents.

drug misuse illegal use of drugs for nontherapeutic purposes. Under the UK Misuse of Drugs regulations drugs used illegally include: narcotics, such as heroin, morphine, and the synthetic opioids; barbiturates; amphetamines and related substances; ◊benzodiazepine tranquillizers; cocaine, LSD, and cannabis. **Designer drugs**, for example ecstasy, are usually modifications of the amphetamine molecule, altered in order to evade the law as well as for different effects, and may be many times more powerful and dangerous. Crack, a highly toxic derivative of cocaine, became available to drug users in the 1980s. Some athletes misuse drugs such as ephedrine and ◊anabolic steroids.

Sources of traditional drugs include the 'Golden Triangle' (where Myanmar, Laos, and Thailand meet), Mexico, Colombia, China, and the Middle East.

drupe fleshy ◊fruit containing one or more seeds which are surrounded by a hard, protective layer – for example cherry, almond, and plum. The wall of the fruit (◊pericarp) is differentiated into the outer skin (exocarp), the fleshy layer of tissues (mesocarp), and the hard layer surrounding the seed (endocarp).

The coconut is a drupe, but here the pericarp becomes dry and fibrous at maturity. Blackberries are an aggregate fruit composed of a cluster of small drupes.

dry-cleaning method of cleaning textiles based on the use of volatile solvents, such as trichloroethene (trichloroethylene), that dissolve grease. No water is used. Dry-cleaning was first developed in France 1849.

Some solvents are known to damage the ozone layer and one, tetrachloroethene (perchloroethylene), is toxic in water and gives off toxic fumes when heated.

dry ice solid carbon dioxide (CO_2), used as a refrigerant. At temperatures above –79°C/–110.2°F, it sublimes (turns into vapour without passing through a liquid stage) to gaseous carbon dioxide.

dry rot infection of timber in damp conditions by fungi (see ◊fungus), such as *Merulius lacrymans,* that form a threadlike surface. Whitish at first, the fungus later reddens as reproductive spores are formed. Tentacles from the fungus also work their way into the timber, making it dry-looking and brittle. Dry rot spreads rapidly through a building.

Commonly abused drugs

The two main laws about drugs are the Medicines Act and the Misuse of Drugs Act. The Medicines Act controls the way medicines are made and supplied. The Misuse of Drugs Act bans the non-medical use of certain drugs. The Misuse of Drugs Act places drugs in different classes – A, B, and C. The penalties for offences involving a drug depend on the class it is in and will also vary according to individual circumstances. Class A drugs carry the highest penalty, class C the lowest.

Name	Source	Forms and appearance	Legal position	Methods of use°
Amphetamine (also called speed, uppers, whizz, billy, sulphate)	a totally synthetic product	powder form; tablets and capsules	class B, schedule 2 controlled substance	taken orally in drink or licking off the finger; sniffed; smoked; dissolved in water for injecting
Anabolic steroids	synthetic products designed to imitate certain natural hormones within the human body	capsules and tablets; liquid	not controlled	injections; also taken orally
Cannabis (also called dope, grass, hash, ganja, shit, blow, weed)	plants of the genus *Cannabis saliva*	herbal: dried plant material, similar to a coarse cut tobacco (marijuana); resin: blocks of various colours and texture (hashish); oil: extracted from resin, with a distinctive smell	class B, schedule 1 controlled substance	smoked in a variety of ways; can be put into cooking or made into a drink; occasionally eaten on its own
Cocaine (also called coke, charlie, snow, white lady)	leaves of the coca bush, *Erythroxylum coca*	white crystalline powder; very rarely in paste form	class A, schedule 2 controlled substance	sniffed; injected; smoked (paste)
Crack and freebase cocaine (also called rock, wash, cloud nine; base, freebase)	derived from cocaine hydrochloride	crystals (crack cocaine); powder (freebase cocaine)	class A, schedule 2 controlled substances	smoking
Ecstasy (also called disco burgers, Dennis the Menace, diamonds, New Yorkers, E, Adam, XTC, Fantasy, Doves, rhubarb and custard (red and yellow capsules))	a totally synthetic product	tablets and capsules; rarely, powder; ecstasy is not always available in pure form, which increases the risks	class A, schedule 1	orally; occasionally injections
Heroin (also called smack, junk, H, skag, brown, horse, gravy)	from raw opium produced by the opium poppy	powder	class A, schedule 2 controlled substance	smoking; injections; also sniffed or taken orally
Lysergic acid diethylamide (LSD) (also called acid, trips, tab, blotters, dots)	derived from ergot (a fungus of certain cereal grains)	colourless crystals; for street use, mainly impregnated into squares of blotted paper or into squares of clear gelatine; or into tiny pills	class A, schedule 1 controlled substance	orally (paper squares and pills); under the eyelid (gelatine squares)
Magic (hallucinogenic) mushrooms (also called shrooms, mushies)	natural mushrooms (mainly fly agaric and liberty cap)	several varieties; identification is difficult	possession and eating of fresh mushrooms is not an offence; preparation is[1]	eaten; infused to make a drink
Methadone (also called doll, dolly, red rock, tootsie roll; phy-amps, phy (ampoules))	a totally synthetic product	powder; tablets, ampoules, linctus, mixture	class A, schedule 2 controlled substance	orally; injections
Methylamphetamine (also called ice, meth, crystal, glass, ice-cream (crystal form); meth, methedrine (powder or tablets))	a totally synthetic drug, closely related to amphetamine sulphate	crystals or, less commonly, tablets or powder	class B, schedule 2 controlled substance	burning crystals and inhaling the fumes; drinking, sniffing, licking off the finger (powder and tablets)
Nitrites (poppers) (also called nitro, nitrite)	various synthetic volatile chemicals	in small glass bottles under trade names of Liquid Gold, Hi-Tech, Rave, Locker Room, Ram, Rush, etc	controlled by the Medicines Act	inhalation
Solvents (also called glue, gas, can, cog (depending on substance and container))	domestic and commercial products	liquid petroleum gases (LPGs): aerosols, camping gas cylinders, lighter gas refills; liquid solvents: fire extinguisher fluid, corrective fluids, certain paints and removers, nail polish and remover, anti-freeze, petrol; solvent-based glues: impact adhesives used for wood, plastic, laminate surfaces, vinyl floor tiles	not controlled	sprayed into the mouth and inhaled (LPGs); sniffing
Tranquillizers (also called tranx, barbs, barbies, blockers, tueys, traffic lights, golf balls (tranquillizers); jellies, jelly beans, M&Ms, rugby balls (temazepam in jelly capsules))	pharmaceutical drugs aimed at treating patients with problems of anxiety, insomnia, and depression; based on benzodiazepine or barbiturate	tablets or capsules	benzodiazepine based: class C controlled substances; barbiturate based: class B controlled substances	taken orally or injected

[1] Preparation (such as crushing, slicing, drying, etc) is punishable as an offence relating to psilocin and psilocybin – the active ingredients of most hallucinogenic mushrooms, both class A, schedule 1 controlled substances.

Effects of use	Adverse effects	Tolerance potential	Habituation potential	Withdrawal effects	Overdose potential
increased energy, strength, concentration, euphoria and elation; suppression of appetite; wakefulness	increased blood pressure with risk of stroke; diarrhoea or increased urination; disturbance of sleep patterns; weight loss; depression; paranoia; psychosis	tolerance develops rapidly	physical dependence: rare; psychological dependence: common	mental agitation, depression, panic	fatal overdose possible, even at low doses
increase in body bulk and muscle growth; feelings of stamina and strength	bone growth abnormalities; hypertension and heart disease; liver and kidney malfunction; hepatitis; sexual abnormalities and impotence	tolerance may develop	no physical dependence; profound psychological dependence	sudden collapse of muscle strength and stamina; irritability, violent mood swings	overdose can lead to collapse, convulsions, coma, and death
relaxation, feelings of happiness, congeniality, increased concentration, sexual arousal	impaired judgement; loss of short-term memory; dizziness; confusion; anxiety; paranoia; potential for cancer and breathing disorders	tolerance develops rapidly	true physical dependence: rare; psychological dependence: common	disturbed sleep patterns; anxiety, panic	it is not thought possible to overdose fatally
feelings of energy, strength, exhilaration, confidence; talkativeness	agitation, panic, feelings of being threatened; damage to nasal passages, exhaustion, weight loss; collapsed veins, ulceration; delusions, violence	tolerance develops rapidly	strong physical and psychological dependence	severe cravings; feelings of anxiety and panic; depression	it is possible to overdose fatally
elation and euphoria, feelings of power, strength, and well-being	depression, feelings of being threatened; paranoia, psychosis; violence	some tolerance develops	strong physical and psychological dependence	severe depression; aggression, panic; risk of suicide	overdose can lead to coma and death
feelings of euphoria, energy, stamina, sociability, sexual arousal	mood swings, nausea and vomiting, overheating, dehydration, convulsions, sudden death	tolerance develops	physical dependence: none; physiological dependence: low	no physical symptoms; irritability, depression	overdose can lead to coma and death
feelings of euphoria, inner peace, freedom from fear and deprivation	depressed breathing, severe constipation, nausea, and vomiting; effect on general state of health, lower immunity; vein collapse and ulceration; risk of infection from needles	tolerance develops rapidly	profound physical and psychological dependence	sweating, flu-like symptoms 'going cold turkey'; severe cravings; professional assistance necessary	overdose can lead to coma and death
hallucinations	risk of accident while hallucinating; flashbacks; risk of developing a latent psychiatric disorder	tolerance develops and disappears rapidly	no physical dependence; some psychological dependence	no physical effect; few psychological effects	it is not thought possible to overdose
hallucinations; feelings of euphoaria, well-being, gaiety	long-term mental problems; risk of poisoning	tolerance develops rapidly	no dependence	few withdrawal effects	little overdose potential
feelings of relaxation, bodily warmth, freedom from pain and worry	sweating, nausea, itching, tiredness; disruption of menstrual cycle in women	tolerance develops slowly	strong physical and psychological dependence	fever, flu-like symptoms; diarrhoea; aggression	overdose can lead to respiratory depression, collapse, coma, and death
feelings of euphoria, great strength and energy, sustained for long periods without rest or food	increased blood pressure with risk of stroke and heart failure, diarrhoea or increased urination, disturbance of sleep, hallucinations, aggression, psychosis, delusions, paranoia	tolerance develops rapidly	physical dependence: not uncommon; psychological dependence: profound	severe cravings; depression; fear; panic and mental agitation	serious risk of fatal overdose, even at very low levels
feelings of excitement and exhilaration; sexual arousal and increased sensitivity of sexual organs	nausea and vomiting; headaches and dizziness; skin problems; damage to vision if touches the eyes; poisonous if swallowed	tolerance develops quickly	no significant physical or psychological dependence	no significant effects	little risk of overdose
deep intoxication, hallucinations, excitability	over-stimulation of the heart, and death; asphyxiation from swelling of throat tissues or inhalation of vomit; problems with speech, balance, short-term memory, cognitive skills; possible personality changes	tolerance may develop	no physical dependence; strong psychological dependence	no physical symptoms; anxiety and mood swings	overdose can lead to collapse, coma, and death
in higher doses: feelings of euphoria, elimination of fear and feeling of deprivation	violent mood swings, bizarre sexual behaviour, deep depression, disorientation, lethargy	tolerance develops rapidly	profound physical and psychological tolerance	confusion, head-aches, depression; sudden withdrawal may lead to con-vulsions and death	overdose can lead to convulsions, depression of breathing, collapse, coma, and death

DSP in computing, abbreviation for ◊digital signal processor.

DTP abbreviation for ◊desktop publishing.

dubnium synthesized, radioactive, metallic element of the ◊transactinide series, symbol Db, atomic number 105, relative atomic mass 261. Six isotopes have been synthesized, each with very short (fractions of a second) half-lives. Two institutions claim to have been the first to produce it: the Joint Institute for Nuclear Research in Dubna, Russia, 1967; and the University of California at Berkeley, USA, who disputed the Soviet claim, 1970. Its temporary name was unnilpentium.

duck any of about 50 species of short-legged waterbirds with webbed feet and flattened bills, of the family Anatidae, order Anseriformes, which also includes the larger geese and swans. Ducks were domesticated for eggs, meat, and feathers by the ancient Chinese and the ancient Maya (see ◊poultry). Most ducks live in fresh water, feeding on worms and insects as well as vegetable matter. They are generally divided into dabbling ducks and diving ducks.
anatomy The three front toes of a duck's foot are webbed and the hind toe is free; the legs are scaly. The broad rounded bill is skin-covered with a horny tip provided with little plates (lamellae) through which the duck is able to strain its food from water and mud.
species of duck The mallard *Anas platyrhynchos,* 58 cm/1.9 ft, found over most of the northern hemisphere, is the species from which all domesticated ducks originated. The male (drake) has a glossy green head, brown breast, grey body, and yellow bill. The female (duck) is speckled brown, with a duller bill. The male moults and resembles the female for a while just after the breeding season. There are many other species of duck including ◊teal, ◊eider, mandarin duck, ◊merganser, muscovy duck, pintail duck, ◊shelduck, and ◊shoveler. They have different-shaped bills according to their diet and habitat; for example, the shoveler has a wide spade-shaped bill for scooping insects off the surface of water.

duck-billed platypus another name for the ◊platypus.

duckweed any of a family of tiny plants found floating on the surface of still water throughout most of the world, except the polar regions and tropics. Each plant consists of a flat, circular, leaflike structure 0.4 cm/0.15 in or less across, with a single thin root up to 15 cm/6 in long below. (Genus chiefly *Lemna,* family Lemnaceae.)

The plants bud off new individuals and soon cover the surface of the water. Flowers rarely appear, but when they do, they are extremely small and are found in a pocket at the edge of the plant.

ductile material material that can sustain large deformations beyond its elastic limit (see ◊elasticity) without fracture. Metals are very ductile, and may be pulled out into wires, or hammered or rolled into thin sheets without breaking. Compare ◊brittle material.

ductless gland alternative name for an ◊endocrine gland.

dugong marine mammal *Dugong dugong* of the order Sirenia (sea cows), found in the Red Sea, the Indian Ocean, and western Pacific. It can grow to 3.6 m/11 ft long, and has a tapering body with a notched tail and two fore-flippers. It has a very long hind gut (30 m/98 ft in adults) which functions similarly to the rumen in ◊ruminants. It is largely herbivorous, feeding mostly on sea grasses and seaweeds, and is thought to have given rise to the mermaid myth.

Previously thought to be the only truly herbivorous marine mammal, Australian research 1995 showed that some dugongs eat sea squirts, which make up 25.5% of the wet weight of faeces from dugongs in Moreton Bay, eastern Australia.

duiker *Afrikaans diver* any of several antelopes of the family Bovidae, common in Africa. Duikers are shy and nocturnal, and grow to 30–70 cm/12–28 in tall.

duikerbok small African antelope with crested head, large muzzle, and short, conical horns in the males only.
classification Duikerboks are in genus *Cephalophus* in the family Bovidae (cattle and antelopes) of order Artiodactyla.

dulse any of several edible red seaweeds, especially *Rhodymenia palmata,* found on middle and lower shores of the north Atlantic. They may have a single broad blade up to 30 cm/12 in long rising directly from the holdfast which attaches them to the sea floor, or may be palmate (with five lobes) or fan-shaped. The frond is tough and dark red, sometimes with additional small leaflets at the edge.

Dumas, Jean Baptiste André (1800–1884)

French chemist. He made contributions to organic analysis and synthesis, and to the determination of atomic weights (relative atomic masses) through the measurement of vapour densities. In 1833, Dumas worked out an absolute method for the estimation of the amount of nitrogen in an organic compound, which still forms the basis of modern methods of analysis. He went on to correct the atomic masses of 30 elements – half the total number known at that time – referring to the hydrogen value as 1.

In 1826, Dumas began working on atomic theory, and concluded that 'in all elastic fluids observed under the same conditions, the molecules are placed at equal distances' – that is, they are present in equal numbers. His theory of substitution in organic compounds, which he proved by experiments, established that atoms of apparently opposite electrical charge replaced each other. This refuted the dualistic theory of chemistry proposed by Swedish chemist Jöns Berzelius.

Studying blood, Dumas showed that urea is present in the blood of animals from which the kidneys have been removed, proving that one of the functions of the kidneys is to remove urea from the blood, not to produce it.

Mary Evans Picture Library

dumb terminal in computing, a ◊terminal that has no processing capacity of its own. It works purely as a means of access to a main ◊central processing unit. Compare with a ◊personal computer used as an intelligent terminal – for example in ◊client–server architecture.

dump in computing, the process of rapidly transferring data to external memory or to a printer. It is usually done to help with debugging (see ◊bug) or as part of an error-recovery procedure designed to provide ◊data security. A ◊screen dump makes a printed copy of the current screen display.

dune mound or ridge of wind-drifted sand common on coasts and in sandy deserts. Loose sand is blown and bounced along by the wind, up the windward side of a dune. The sand particles then fall to rest on the lee side, while more are blown up from the windward side. In this way a dune moves gradually downwind.

In sandy deserts, the typical crescent-shaped dune is called a **barchan. Seif dunes** are longitudinal and lie parallel to the wind direction, and **star-shaped dunes** are formed by irregular winds.

dung waste matter excreted by living animals. Dung may also serve as a marker through the addition of scents from the anal glands, whether for determining territorial boundaries or as an indication of status within a group.

Some animals, such as rabbits, may reingest dung immediately

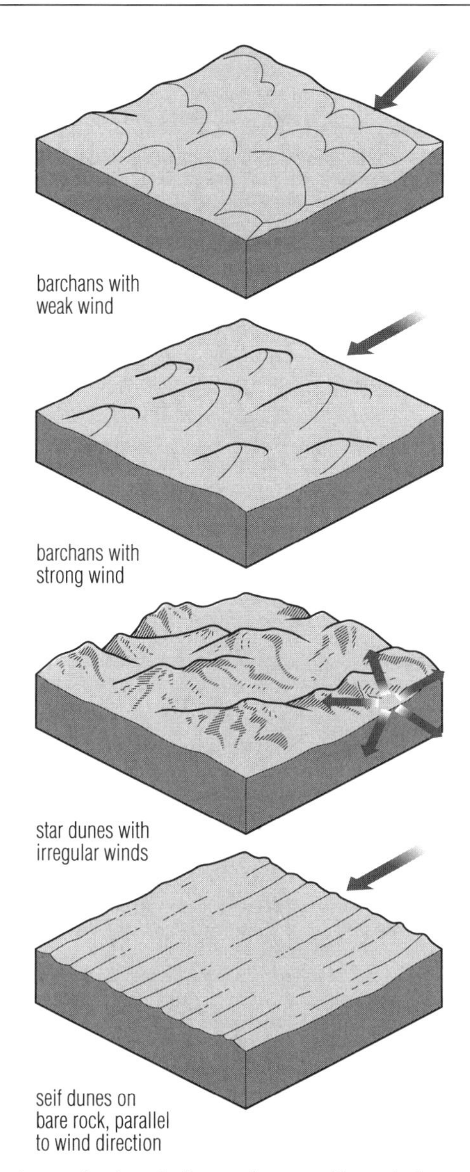

***dune** The shape of a dune indicates the prevailing wind pattern. Crescent-shaped dunes form in sandy desert with winds from a constant direction. Seif dunes form on bare rocks, parallel to the wind direction. Irregular star dunes are formed by variable winds.*

after excretion and continue digesting it, a process known as **refection**. In addition to being broken down by bacteria, animal dung provides food for many invertebrates, especially beetles and flies, and provides a habitat for certain species of fungi and plants such as stinging nettles.

dunlin small gregarious shore bird *Calidris alpina* of the sandpiper family Scolopacidae, order Charadriformes, about 18 cm/7 in long, nesting on moors and marshes in the far northern regions of Eurasia and North America. Chestnut above and black below in summer, it is greyish in winter; the bill and feet are black.

Dunnite US high explosive named after Major Dunn, its developer; also called 'Explosive D'. Made of ammonium picrate powder, it was relatively insensitive, and so was widely used in armour-piercing shells as it withstood the shock of impact against armour without detonating, allowing the shell to pierce the armour before the fuse initiated the explosive.

dunnock or ***hedge sparrow*** European bird *Prunella modularis* family Prunellidae, similar in size and colouring to the sparrow, but with a slate-grey head and breast, and more slender bill. It is characterized in the field by a hopping gait, with continual twitches of the wings whilst feeding. It nests in bushes and hedges.

duodecimal system system of arithmetic notation using 12 as a base, at one time considered superior to the decimal number system in that 12 has more factors (2, 3, 4, 6) than 10 (2, 5).

It is now superseded by the universally accepted decimal system.

duodenum in vertebrates, a short length of ◊alimentary canal found between the stomach and the small intestine. Its role is in digesting carbohydrates, fats, and proteins. The smaller molecules formed are then absorbed, either by the duodenum or the ileum.

Entry of food to the duodenum is controlled by the pyloric sphincter, a muscular ring at the outlet of the stomach. Once food has passed into the duodenum it is mixed with bile released from the gall bladder and with a range of enzymes secreted from the pancreas, a digestive gland near the top of the intestine. The bile neutralizes the acidity of the gastric juices passing out of the stomach and aids fat digestion.

duplex or ***echo*** in printing, the ability to print on both sides of the page; in computer communications, setting that ◊toggles the ability to send and receive signals simultaneously. **Full duplex** means two-way communication is enabled; **half duplex** means it is disabled.

duralumin lightweight aluminium ◊alloy widely used in aircraft construction, containing copper, magnesium, and manganese.

durra or ***doura*** grass, also known as Indian millet, grown as a cereal in parts of Asia and Africa. *Sorghum vulgare* is the chief cereal in many parts of Africa. See also ◊sorghum. (Genus *Sorghum*.)

dust bowl area in the Great Plains region of North America (Texas to Kansas) that suffered extensive wind erosion as the result of drought and poor farming practice in once-fertile soil. Much of the topsoil was blown away in the droughts of the 1930s and the 1980s.

Similar dust bowls are being formed in many areas today, noticeably across Africa, because of overcropping and overgrazing.

Dutch cap common name for a barrier method of contraception; see ◊diaphragm.

Dutch elm disease disease of elm trees *Ulmus*, principally Dutch, English, and American elm, caused by the fungus *Certocystis ulmi*. The fungus is usually spread from tree to tree by the elm-bark beetle, which lays its eggs beneath the bark. The disease has no cure, and control methods involve injecting insecticide into the trees annually to prevent infection, or the destruction of all elms in a broad band around an infected area, to keep the beetles out.

The disease was first described in the Netherlands and by the early 1930s had spread across Britain and continental Europe, as well as North America.

DVD abbreviation for ◊digital versatile disc or **digital video disc**.

DVI (abbreviation for ***digital video interactive***) in computing, a powerful compression and decompression system for digital video and audio. DVI enables 72 minutes of full-screen, full-motion video and its audio track to be stored on a CD-ROM. Originally developed by the US firm RCA, DVI is now owned by Intel and has active support from IBM and Microsoft. It can be used on the hard disc of a PC as well as on a CD-ROM.

Dvorak keyboard alternative keyboard layout to the normal typewriter keyboard layout (◊QWERTY). In the Dvorak layout the most commonly used keys are situated in the centre, so that keying is faster.

DWANGO (acronym for ***Dial-up Wide Area Network Game Operation***) in computing, server that enables computer users with modems to play each other at action games such as DOOM, Duke Nuken 3D, and Monster Truck Madness without the variable delays involved in moving data over the Internet.

dye substance that, applied in solution to fabrics, imparts a colour resistant to washing. **Direct dyes** combine with the material of the fabric, yielding a coloured compound; **indirect dyes** require the presence of another substance (a mordant), with which the fabric must first be treated; **vat dyes** are colourless soluble substances that on exposure to air yield an insoluble coloured compound.

Naturally occurring dyes include indigo, madder (alizarin), logwood, and cochineal, but industrial dyes (introduced in the 19th century) are usually synthetic: acid green was developed 1835 and bright purple in 1856.

Industrial dyes include ◊azo dyestuffs, ◊acridine, ◊anthracene, and ◊aniline.

dyke in earth science, a sheet of ◊igneous rock created by the intrusion of magma (molten rock) across layers of pre-existing rock. (By contrast, a sill is intruded *between* layers of rock.) It may form a ridge when exposed on the surface if it is more resistant than the rock into which it intruded. A dyke is also a human-made embankment built along a coastline (for example, in the Netherlands) to prevent the flooding of lowland coastal regions.

dynamic data exchange (DDE) in computing, a form of interprocess communication used in Microsoft ◊Windows, providing exchange of commands and data between two applications. DDE was used principally to include live data from one application in another – for example, spreadsheet data in a word-processed report. After Windows 3.1 DDE was replaced by ◊object linking and embedding.

DDE links between files rely on the files remaining in the same locations in the computer's directory.

Dynamic HTML in computing, the fourth version of hypertext markup language (◊HTML), the language used to create Web pages. It is called Dynamic HTML because it enables dynamic effects to be incorporated in pages without the delays involved in downloading Java ◊applets and without referring back to the server.

dynamic IP address in computing, a temporary ◊IP address assigned from a pool of available addresses by an ◊Internet Service Provider when a customer logs on to begin an online session.

Companies and other organizations which have their own networks typically have their own permanent IP addresses. Customers of a dial-up service provider, however, only need addresses for the length of time that they are actually on line. This method allows the finite number of available IP addresses to be used most efficiently.

dynamic link library (DLL) in computing, files of executable functions that can be loaded on demand in Microsoft ◊Windows and linked at run time. Windows itself uses DLL files for handling international keyboards, for example, and Windows word-processing programs use DLL files for functions such as spelling and hyphenation checks, and thesaurus.

dynamics or ***kinetics*** in mechanics, the mathematical and physical study of the behaviour of bodies under the action of forces that produce changes of motion in them.

dynamite explosive consisting of a mixture of nitroglycerine and diatomaceous earth (diatomite, an absorbent, chalklike material). It was first devised by Alfred Nobel.

dynamo simple generator or machine for transforming mechanical energy into electrical energy. A dynamo in basic form consists of a powerful field magnet between the poles of which a suitable conductor, usually in the form of a coil (armature), is rotated. The mechanical energy of rotation is thus converted into an electric current in the armature.

Present-day dynamos work on the principles described by English physicist Michael Faraday in 1830, that an ◊electromotive force is developed in a conductor when it is moved in a magnetic field.

dyne c.g.s. unit (symbol dyn) of force. 10^5 dynes make one newton. The dyne is defined as the force that will accelerate a mass of one gram by one centimetre per second per second.

dysentery infection of the large intestine causing abdominal cramps and painful ◊diarrhoea with blood. There are two kinds of dysentery: **amoebic** (caused by a protozoan), common in the tropics, which may lead to liver damage; and **bacterial**, the kind most often seen in the temperate zones.

Both forms are successfully treated with antibacterials and fluids to prevent dehydration.

DYSLEXIA

Hans Christian Andersen, the Danish writer of fairy tales, was a bad speller and was probably dyslexic.

dyslexia ***Greek 'bad', 'pertaining to words'*** malfunction in the brain's synthesis and interpretation of written information, popularly known as 'word blindness'.

Dyslexia may be described as specific or developmental to distinguish it from reading or writing difficulties which are acquired. It results in poor ability in reading and writing, though the person may excel in other areas, for example, in mathematics. A similar disability with figures is called **dyscalculia**. **Acquired dyslexia** may occur as a result of brain injury or disease.

dysprosium ***Greek** dusprositos **'difficult to get near'*** silver-white, metallic element of the ◊lanthanide series, symbol Dy, atomic number 66, relative atomic mass 162.50. It is among the most magnetic of all known substances and has a great capacity to absorb neutrons.

It was discovered in 1886 by French chemist Paul Lecoq de Boisbaudran (1838–1912).

E

2E abbreviation for *east*.

eagle any of several genera of large birds of prey of the family Accipitridae, order Falconiformes, including the golden eagle *Aquila chrysaetos* of Eurasia and North America, which has a 2 m/6 ft wingspan. Eagles occur worldwide, usually building eyries or nests in forests or mountains, and all are fierce and powerful birds of prey. The harpy eagle is the largest eagle.

The white-headed bald eagle *Haliaetus leucocephalus* is the symbol of the USA; rendered infertile through the ingestion of agricultural chemicals, it is now rare, except in Alaska.

Another endangered species is the Philippine eagle, sometimes called the Philippine monkey-eating eagle (although its main prey is the flying lemur). Loss of large tracts of forest, coupled with hunting by humans, have greatly reduced its numbers.

ear organ of hearing in animals. It responds to the vibrations that constitute sound, which are translated into nerve signals and passed to the brain. A mammal's ear consists of three parts: outer ear, middle ear, and inner ear. The **outer ear** is a funnel that collects sound, directing it down a tube to the **ear drum** (tympanic membrane), which separates the outer and **middle ears**. Sounds vibrate this membrane, the mechanical movement of which is transferred to a smaller membrane leading to the **inner ear** by three small bones, the auditory ossicles. Vibrations of the inner ear membrane move fluid contained in the snail-shaped cochlea, which vibrates hair cells that stimulate the auditory nerve connected to the brain. There are approximately 30,000 sensory hair cells (**stereocilia**). Exposure to loud noise and the process of ageing damages the stereocilia, resulting in hearing loss. Three fluid-filled canals of the inner ear detect changes of position; this mechanism, with other sensory inputs, is responsible for the sense of balance.

When a loud noise occurs, muscles behind the eardrum contract automatically, suppressing the noise to enhance perception of sound and prevent injury.

EAR ANATOMY

http://weber.u.washington.edu/~otoweb/ear_anatomy.html

Concise medical information on the anatomy and function of the ear. It includes separate sections on perforated eardrums and their treatment, ear tubes and their use, tinnitus, and a series of different hearing tests.

earth electrical connection between an appliance and the ground. In the event of a fault in an electrical appliance, for example, involving connection between the live part of the circuit and the outer casing, the current flows to earth, causing no harm to the user.

In most domestic installations, earthing is achieved by a connection to a metal water-supply pipe buried in the ground.

Earth third planet from the Sun. It is almost spherical, flattened slightly at the poles, and is composed of three concentric layers: the ◊core, the ◊mantle, and the ◊crust. About 70% of the surface (including the north and south polar icecaps) is covered with water. The Earth is surrounded by a life-supporting atmosphere and is the only planet on which life is known to exist.
mean distance from the Sun 149,500,000 km/92,860,000 mi
equatorial diameter 12,756 km/7,923 mi
circumference 40,070 km/24,900 mi
rotation period 23 hr 56 min 4.1 sec

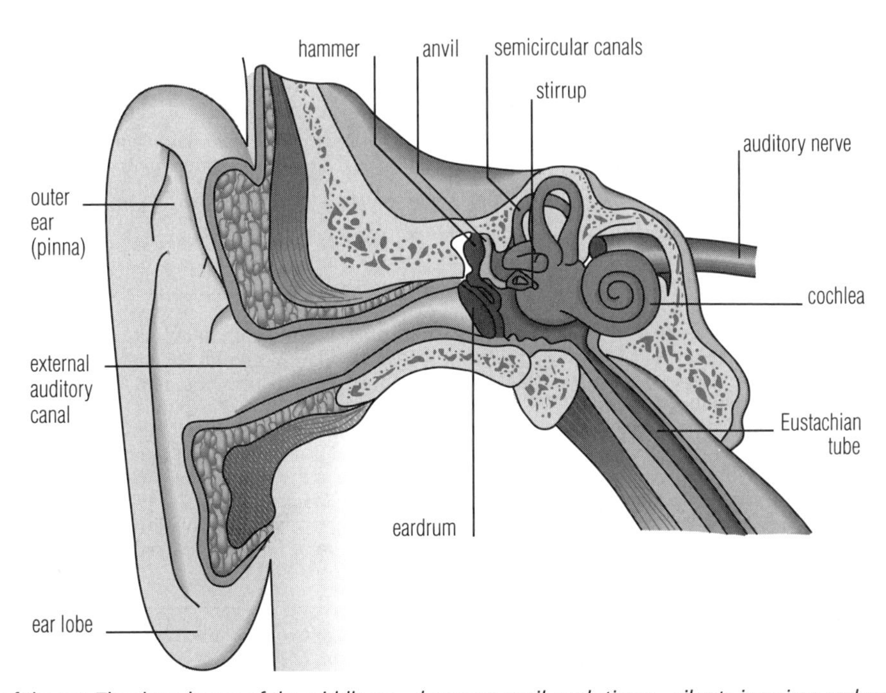

ear *The structure of the ear. The three bones of the middle ear – hammer, anvil, and stirrup – vibrate in unison and magnify sounds about 20 times. The spiral-shaped cochlea is the organ of hearing. As sound waves pass down the spiral tube, they vibrate fine hairs lining the tube, which activate the auditory nerve connected to the brain. The semicircular canals are the organs of balance, detecting movements of the head.*

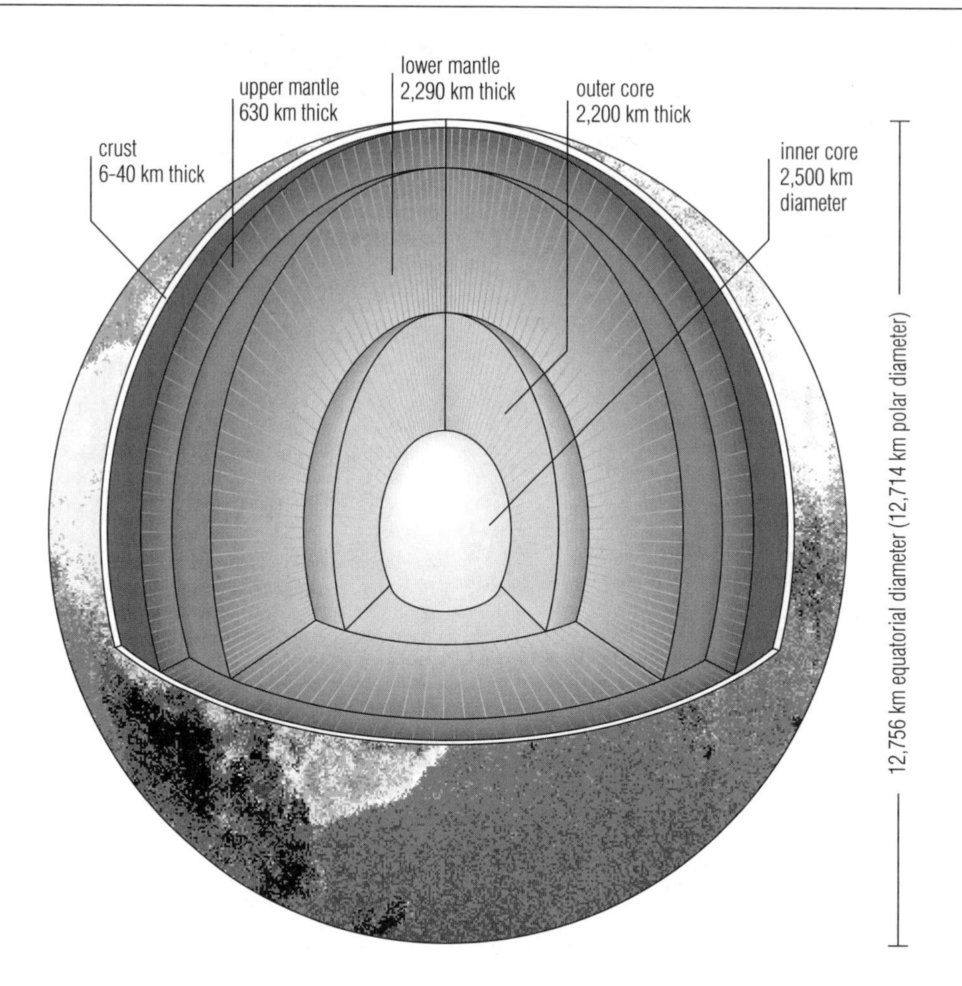

Earth *Inside the Earth. The surface of the Earth is a thin crust about 6 km/4 mi thick under the sea and 40 km/25 mi thick under the continents. Under the crust lies the mantle about 2,900 km/1,800 mi thick and with a temperature of 1,500–3,000°C/2,700–5,400°F. The inner core is probably solid iron and nickel at about 5,000°C/9,000°F.*

year (complete orbit, or sidereal period) 365 days 5 hr 48 min 46 sec. Earth's average speed around the Sun is 30 kps/18.5 mps; the plane of its orbit is inclined to its equatorial plane at an angle of 23.5°, the reason for the changing seasons
atmosphere nitrogen 78.09%; oxygen 20.95%; argon 0.93%; carbon dioxide 0.03%; and less than 0.0001% neon, helium, krypton, hydrogen, xenon, ozone, radon
surface land surface 150,000,000 sq km/57,500,000 sq mi (greatest height above sea level 8,872 m/29,118 ft Mount Everest); water surface 361,000,000 sq km/139,400,000 sq mi (greatest depth 11,034 m/36,201 ft ◊Mariana Trench in the Pacific). The interior is thought to be an inner core about 2,600 km/1,600 mi in diameter, of solid iron and nickel; an outer core about 2,250 km/1,400 mi thick, of molten iron and nickel; and a mantle of mostly solid rock about 2,900 km/1,800 mi thick, separated from the Earth's crust by the ◊Mohorovičić discontinuity. The crust and the topmost layer of the mantle form about twelve major moving plates, some of which carry the continents. The plates are in constant, slow motion, called tectonic drift. US geophysicists announced in 1996 that they had detected a difference in the spinning time of the Earth's core and the rest of the planet; the core is spinning slightly faster.
satellite the Moon
age 4.6 billion years. The Earth was formed with the rest of the ◊Solar System by consolidation of interstellar dust. Life began 3.5–4 billion years ago

To remember the most common elements of the planet's crust, in descending order:

ONLY SILLY ASSES IN COLLEGE STUDY PAST MIDNIGHT

OXYGEN, SILICON, ALUMINIUM, IRON, CALCIUM, SODIUM, POTASSIUM, MAGNESIUM

EARTH AND MOON VIEWER

http://www.fourmilab.ch/earthview/vplanet.html

View a map of the Earth showing the day and night regions at this moment, or view the Earth from the Sun, the Moon, or any number of other locations. Alternatively, take a look at the Moon from the Earth or the Sun, or from above various formations on the lunar surface.

The earth only has so much bounty to offer and inventing ever larger and more notional prices for that bounty does not change its real value.

BEN ELTON English writer and comedian.
Stark, 'Dinner in Los Angeles'

earthquake abrupt motion that propagates through the Earth and along its surfaces. Earthquakes are caused by the sudden release in rocks of strain accumulated over time as a result of ◊tectonics. The study of earthquakes is called ◊seismology. Most earthquakes occur along ◊faults (fractures or breaks) and ◊Benioff zones. Plate tectonic movements generate the major proportion: as two plates move past each other they can become jammed. When sufficient strain has accumulated, the rock breaks, releasing a series of elastic waves (◊seismic waves) as the plates spring free. The force of earthquakes (magnitude) is measured on the ◊Richter scale, and their effect (intensity) on the ◊Mercalli scale. The point at which an earthquake originates is the **seismic focus** or **hypocentre**; the point on the Earth's surface directly above this is the **epicentre**.

The Alaskan (USA) earthquake of 27 March 1964 ranks as one of the greatest ever recorded, measuring 8.3 to 8.8 on the Richter scale. The 1906 San Francisco earthquake is among the most famous in history. Its magnitude was 8.3 on the Richter scale. The deadliest, most destructive earthquake in historical times is thought to have been in China in 1556. In 1987, a California earthquake was successfully predicted by measurement of underground pressure waves; prediction attempts have also involved the study of such phenomena as the change in gases issuing from the ◊crust, the level of water in wells, slight deformation of the rock surface, a sequence of minor tremors, and the behaviour of animals. The possibility of earthquake prevention is remote. However, rock slippage might be slowed at movement points, or promoted at stoppage

Earth science: chronology

10000–8000 BC	Holocene (post-glacial) period of hunters and gatherers. Harvesting and storage of wild grains in southwest Asia. Herding of reindeer in northern Eurasia. Domestic sheep in northern Iraq.
8000	Neolithic revolution with cultivation of domesticated wheats and barleys, sheep, and goats in southwest Asia. Domestication of pigs in New Guinea.
7000–6000	Domestic goats, sheep, and cattle in Anatolia, Greece, Persia, and the Caspian basin. Planting and harvesting techniques transferred from Asia Minor to Europe.
5000	Beginning of Nile valley civilization. Millet cultivated in China.
3400	Flax used for textiles in Egypt. Widespread corn production in the Americas.
3200	Records of ploughing, raking, and manuring by Egyptians.
c. 3100	River Nile dammed during the rule of King Menes.
3000	First record of asses used as beasts of burden in Egypt. Sumerian civilization used barley as main crop with wheat, dates, flax, apples, plums, and grapes.
2900	Domestication of pigs in eastern Asia.
2640	Reputed start of Chinese silk industry.
2500	Domestic elephants in the Indus valley. Potatoes a staple crop in Peru.
2350	Wine-making in Egypt.
2250	First known irrigation dam.
1600	Important advances in the cultivation of vines and olives in Crete.
1500	*Shadoof* (mechanism for raising water) used for irrigation in Egypt.
1400	Iron ploughshares in use in India.
1300	Aqueducts and reservoirs used for irrigation in Egypt.
1200	Domestic camels in Arabia.
1000–500	Evidence of crop rotation, manuring, and irrigation in India.
600	First windmills used for corn grinding in Persia.
350	Rice cultivation well established in parts of western Africa. Hunting and gathering in the east, central, and south parts of the continent.
c. 200	Use of gears to create ox-driven water wheel for irrigation. Archimedes screw used for irrigation.
100	Cattle-drawn iron ploughs in use in China.
AD 65	*De Re Rustica/On Rural Things*, Latin treatise on agriculture and irrigation.
500	'Three fields in two years' rotation used in China.
630	Cotton introduced into Arabia.
800	Origins of the 'open field' system in northern Europe.
900	Wheeled ploughs in use in western Europe. Horse collar, originating in China, allowed horses to be used for ploughing as well as carrying.
1000	Frisians (NW Netherlanders) began to build dykes and reclaim land. Chinese began to introduce Champa rice which cropped much more quickly than other varieties.
11th century	Three-field system replaced the two-field system in western Europe. Concentration on crop growing.
1126	First artesian wells, at Artois, France.
12th century	Increasing use of water mills and windmills. Horses replaced oxen for pulling work in many areas.
12th–14th centuries	Expansion of European population brought more land into cultivation. Crop rotations, manuring, and new crops such as beans and peas helped increase productivity. Feudal system at its height.
13th–14th centuries	Agricultural recession in western Europe with a series of bad harvests, famines, and pestilence.
1347	Black Death killed about a third of the European population.
16th century	Decline of the feudal system in western Europe. More specialist forms of production were now possible with urban markets. Manorial estates and serfdom remained in eastern Europe. Chinese began cultivation of non-indigenous crops such as corn, sweet potatoes, potatoes, and peanuts.
17th century	Potato introduced into Europe. Norfolk crop rotation became widespread in England, involving wheat, turnips, barley and then ryegrass/clover.
1700–1845	Agricultural revolution began in England. Two million hectares of farmland in England enclosed. Removal of open fields in other parts of Europe followed.
c. 1701	Jethro Tull developed the seed drill and the horse-drawn hoe.
1747	First sugar extracted from sugar beet in Prussia.
1762	Veterinary school founded in Lyon, France.
1783	First plough factory in England.
1785	Cast-iron ploughshare patented.
1793	Invention of the cotton gin.
1800	Early threshing machines developed in England.
1820s	First nitrates for fertilizer imported from South America.
1830	Reaping machines developed in Scotland and the US. Steel plough made by John Deere in Illinois, US.
1840s	Extensive potato blight in Europe.
1850s	Use of clay pipes for drainage well established throughout Europe.
1862	First steam plough used in the Netherlands.
1850–1890s	Major developments in transport and refrigeration technology altered the nature of agricultural markets with crops, dairy products, and wheat being shipped internationally.
1890s	Development of stationary engines for ploughing.
1892	First petrol-driven tractor in the USA.
1921	First attempt at crop dusting with pesticides from an aeroplane near Dayton, Ohio, US.
1938	First self-propelled grain combine harvester used in the USA.
1942–62	Huge increase in the use of pesticides, later curbed by disquiet about their effects and increasing resistance of pests to standard controls such as DDT.
1945 onwards	Increasing use of scientific techniques, crop specialization and larger scale of farm enterprises.
1985	First cases of bovine spongiform encephalopathy (BSE) recorded by UK vets.
1992	Number of cases of BSE in cattle was at its peak (700 cases per week).
1995	Increase in the use of genetic engineering with nearly 3,000 transgenic crops being field-tested.
1996	Organic farming was on the increase in EU countries. The rise was 11% per year in Britain, 50% in Germany, and 40% in Italy.

points, by the extraction or injection of large quantities of water underground, since water serves as a lubricant. This would ease overall pressure.

earth science scientific study of the planet Earth as a whole. The mining and extraction of minerals and gems, the prediction of weather and earthquakes, the pollution of the atmosphere, and the forces that shape the physical world all fall within its scope of study. The emergence of the discipline reflects scientists' concern that an understanding of the global aspects of the Earth's structure and its past will hold the key to how humans affect its future, ensuring that its resources are used in a sustainable way. It is a synthesis of several traditional subjects such as ◊geology, ◊meteorology, ◊oceanography, ◊geophysics, ◊geochemistry, and ◊palaeontology.

Earth Summit (official name ***United Nations Conference on Environment and Development***) international meetings aiming at drawing measures towards environmental protection of the world. The first summit took place in Rio de Janeiro, Brazil, in June 1992. Treaties were made to combat global warming and protect wildlife ('biodiversity') (the latter was not signed by the USA). The second Earth Summit was held in New York in June 1997 to review progress on the environment. The meeting agreed to work towards a global forest convention in 2000 with the aim of halting the destruction of tropical and old-growth forests.

The Rio summit, which cost $23 million to stage (of which 60% was spent on security), was attended by 10,000 official delegates, 12,000 representatives of nongovernmental organizations, and 7,000 journalists.

In 1993, the Clinton administration overturned certain decisions made by George Bush at the Earth Summit. The USA, which had failed to ratify the Convention of Biological Diversity pact along with other nations, came under renewed pressure to endorse it in April 1995 after India threatened to prevent US pharmaceutical and cosmetic companies from gaining access to its natural resources.

By 1996 most wealthy nations estimated that they would exceed their emissions targets, including Spain by 24%, Australia by 25%, and the USA by 3%. Britain and Germany were expected to meet their targets.

The second summit (1997) failed to agree a new deal to address the world's escalating environmental crisis. Dramatic falls in aid to the so-called Third World countries, which the 1992 Earth Summit promised to increase, were at the heart of the breakdown. British prime minister Tony Blair condemned the USA, Japan, Canada and Australia for failing to deliver on commitments to stabilise rising emissions of climate-changing greenhouse gases. The European Community as a whole was on target to meet its stabilisation commitment because of cuts in emissions in Germany and the UK.

Deforestation was the main problem tackled at the second summit. The World Bank and the World Wide Fund for Nature signed an agreement aimed at protecting 250 million hectares/617 million acres of forest (10% of the world's forests). The importance of the issue was highlighted by the fact that deforestation progressed rapidly in developing countries since the first summit, with 15,000 sq km/9,300 sq mi a year lost in the Amazon region alone.

earthworm ◊annelid worm of the class Oligochaeta. Earthworms are hermaphroditic and deposit their eggs in cocoons. They live by burrowing in the soil, feeding on the organic matter it contains. They are vital to the formation of humus, aerating the soil and levelling it by transferring earth from the deeper levels to the surface as castings.

EARWIG

In the 18th century, earwigs were prescribed for hearing difficulties. Dr James' *Medicinal Dictionary* of 1743, comments: 'These insects, being dried, pulverized, and mixed with the urine of a hare, are esteemed to be good for deafness, being introduced into the ear.'

earwig nocturnal insect of the order Dermaptera. The forewings are short and leathery and serve to protect the hindwings, which are large and are folded like a fan when at rest. Earwigs seldom fly. They have a pincerlike appendage in the rear. The male is distinguished by curved pincers; those of the female are straight. Earwigs are regarded as pests because they feed on flowers and fruit, but they also eat other insects, dead or alive. Eggs are laid beneath the soil, and the female cares for the young even after they have hatched. The male dies before the eggs have hatched.

Easter Island or ***Rapa Nui, Spanish Isla de Pascua*** Chilean island in the south Pacific Ocean, part of the Polynesian group, about 3,500 km/2,200 mi west of Chile; area about 166 sq km/64 sq mi; population (1994) 2,800. It was first reached by Europeans on Easter Sunday 1722. On it stand over 800 huge carved statues (*moai*) and the remains of boat-shaped stone houses, the work of Neolithic peoples from Polynesia. The chief centre is Hanga-Roa.

In 1996, following seven years of work, a New Zealand linguist, Dr Steven Fischer, deciphered a script discovered on the island. This script showed the inhabitants were the first in Oceania to write. According to Dr Fischer, the script, known as 'rongorongo', was made up of chants in Rapanui, the island's Polynesian tongue, and tell the story of creation.

The carved statues are believed to have been religious icons. However, archaeological evidence suggests that, prior to European contact, the island suffered an environmental or cultural crisis resulting in the inhabitants renouncing their earlier religious values, which caused them to damage or overturn many of the statues.

Eastman, George
(1854–1932)

US entrepreneur and inventor who founded the Eastman Kodak photographic company in 1892. He patented flexible film 1884, invented the Kodak box camera 1888, and introduced daylight-loading film in 1892. By 1900 his company was selling a pocket camera for as little as one dollar.

eau de cologne refreshing toilet water (weaker than perfume), made of alcohol and aromatic oils. Its invention is ascribed to Giovanni Maria Farina (1685–1766), who moved from Italy to Cologne 1709 to manufacture it.

EBCDIC (abbreviation for ***extended binary-coded decimal interchange code***) in computing, a code used for storing and communicating alphabetic and numeric characters. It is an 8-bit code, capable of holding 256 different characters, although only 85 of these

ecdysis *A bush cricket or katydid (family Tettigoniidae) sheds its skin at night in a rainforest in Costa Rica. The insect is very vulnerable to attack at this time, so it is an advantage for ecdysis to take place during the hours of darkness. Premaphotos Wildlife*

Did We Save the World at Rio de Janeiro?

BY NIGEL DUDLEY

The Earth Summit

The United Nations Conference on Environment and Development – usually called UNCED or the Earth Summit – took place in Rio de Janeiro, Brazil, in June 1992. It was the largest gathering of heads of state in history and almost certainly also the largest assembly of professional environmentalists and non-governmental organizations. In a historic meeting, the world's leaders agreed two important conventions to combat environmental destruction – on climate change and biodiversity – along with a comprehensive environmental strategy called 'Agenda 21' and a carefully-worded set of 'Forest Principles'. The meeting raised high expectations. It was hailed as a new start for the environment – indeed as a step towards a safer, more equitable world. Yet did anything really change? Five years on, in June 1997, a follow-up meeting took place in New York. Some of the results make depressing reading.

The Climate Change Convention has made extremely slow progress. After agreeing targets for reducing greenhouse gases that many experts believed were too low, several of the richer countries have failed to meet even these aims, and, instead, pollution continues to increase. The Convention on Biological Diversity has, in turn, become mired in infighting about rights to genetic material and has suffered chronic underfunding: if anything, threats of widespread extinctions are greater now than they were in 1992. Debates about a global forest convention – rejected at the time of the Earth Summit – are still going on today, against a background of continuing deforestation and forest degradation. Worse still, there has been a well-orchestrated backlash against the environment, promoted by some business interests intent on preventing their short-term profits being affected by long-term environmental planning.

Decisions into practice

Yet the bad news has to be balanced by some good. While global negotiations have sometimes seemed like little more than an excuse for a handful of people to fly around the world for talking sessions, changes at the national, regional, and grassroots levels have been more encouraging. A 'Local Agenda 21' initiative has challenged community groups, non-governmental organizations, individuals, and local authorities to put the principles of the Earth Summit into practice. Literally thousands of projects have sprung up around the world, ranging from neighbourhood energy conservation schemes to community forest management and introduction of new techniques to enable resources to be used sustainably. These are important not only because of their practical impact, but because they have enabled the environmental message to be carried into thousands of schools and colleges, women's groups, religious organizations, trade unions, and so on.

Indeed, if there is one fundamental change that can be identified since the Rio meeting, it is that social issues have become much more fundamentally related to environmental concerns. Whereas early environmental projects sometimes virtually ignored people – leading to circumstances such as local people being expelled from their traditional areas to establish nature reserves – today there is increasing effort to integrate human and environmental needs, leading to stronger and more durable solutions. Rather than impose solutions from above, experience has also shown that results are likely to be far better if local people are involved in both the planning and the management of conservation projects, instead of being left to gaze at them from the outside. Such changes make the conservationist's role more complicated, and involve some inevitable give and take, but ultimately result in ways of reducing our impact on the environment that invite general support rather than opposition.

Things are working better at a regional level as well. Although we have no global forest convention, there have been several regional initiatives to define and implement 'sustainable forest management', where timber needs are balanced with other requirements, such as biodiversity conservation, environmental services such as control of soil erosion, production of food, medicinal herbs, and a range of non-timber forest products, recreational uses of forests, and even the spiritual and religious importance of particular forests.

Global partnerships

In the absence of global leadership, local and national partnerships have been developing to fill the vacuum. Some of them are unexpected. For example, sections of the timber trade have been working with non-governmental organizations in several countries to develop 'certification of good forest management'; a forest gains a certificate if it is judged to be well-managed by an independent inspector against an agreed set of standards. Consumers have the chance to buy timber that they know has been produced without damaging the environment or local human communities. Efforts are being co-ordinated by the Forest Stewardship Council, an international organization based in Mexico which had not even been thought of at Rio, and already over 5 million hectares of forest have been certified around the world.

Conditions are still getting worse

These and other similar developments are certainly signs of hope. But down at the sharp end of environmental problems, it is hard to avoid the conclusion that, for many countries, conditions are still getting worse. Almost every developing country is still losing its forests. Desertification – the creation of new deserts as a result of overgrazing, poor irrigation schemes, climate change, and forest loss – is occurring in over a hundred countries. Forest fires are increasing in intensity, particularly in the tropical moist forests that should almost never burn under natural conditions. Commercially important fish species have, in many cases, been over-caught to the extent that stocks have collapsed, and the loss of mangroves in coastal regions has resulted in a similar rapid decline in the coastal species that rely on them to provide cover for breeding. The ozone hole continues to grow, and the evidence for climate change is increasing: extreme weather events appear to be growing in strength and frequency. At the moment we stand at a crossroads. Awareness of the importance of environmental issues has never been stronger. Governments and local communities are becoming increasingly willing to take action. But at the same time, ground continues to be lost – literally – and efforts at conservation and restoration will, in the future, be played out against a backdrop of a planet that will probably be far poorer in species and habitats than it was when the 20th century began.

are defined in the standard version. It is still used in many mainframe computers, but almost all mini-and microcomputers now use ◊ASCII code.

ebony any of a group of hardwood trees belonging to the ebony family, especially some tropical ◊persimmons native to Africa and Asia. (Genus chiefly *Diospyros*, family Ebenaceae.)

e-cash (contraction of ***electronic cash***) generic name for new electronic money systems such as Mondex, ◊CyberCash, and ◊DigiCash.

eccentricity in geometry, a property of a ◊conic section (circle, ellipse, parabola, or hyperbola). It is the distance of any point on the curve from a fixed point (the focus) divided by the distance of that point from a fixed line (the directrix). A circle has an eccentricity of zero; for an ellipse it is less than one; for a parabola it is equal to one; and for a hyperbola it is greater than one.

ecdysis periodic shedding of the ◊exoskeleton by insects and other arthropods to allow growth. Prior to shedding, a new soft and expandable layer is laid down underneath the existing one. The old layer then splits, the animal moves free of it, and the new layer expands and hardens.

ECG abbreviation for ◊electrocardiogram.

echidna or ***spiny anteater*** toothless, egg-laying, spiny mammal of the order Monotremata, found in Australia and New Guinea. There are two species: *Tachyglossus aculeatus*, the short-nosed echidna, and the rarer *Zaglossus bruijni*, the long-nosed echidna. They feed entirely upon ants and termites, which they dig out with their powerful claws and lick up with their prehensile tongues. When attacked, an echidna rolls itself into a ball, or tries to hide by burrowing in the earth.

echinoderm marine invertebrate of the phylum Echinodermata ('spiny-skinned'), characterized by a five-radial symmetry. Echinoderms have a water-vascular system which transports substances around the body. They include starfishes (or sea stars), brittle-stars, sea lilies, sea urchins, and sea cucumbers. The skeleton is external, made of a series of limy plates. Echinoderms generally move by using tube-feet, small water-filled sacs that can be protruded or pulled back to the body.

Echinodermata phylum of invertebrate animals, members of which are called echinoderms.

echo in computing, user input that is printed to the screen so the user can read it.

echo repetition of a sound wave, or of a radar or sonar signal, by reflection from a surface. By accurately measuring the time taken for an echo to return to the transmitter, and by knowing the speed of a radar signal (the speed of light) or a sonar signal (the speed of sound in water), it is possible to calculate the range of the object causing the echo (◊echolocation).

echolocation or ***biosonar*** method used by certain animals, notably bats, whales, and dolphins, to detect the positions of objects by using sound. The animal emits a stream of high-pitched sounds, generally at ultrasonic frequencies (beyond the range of human hearing), and listens for the returning echoes reflected off objects to determine their exact location.

The location of an object can be established by the time difference between the emitted sound and its differential return as an echo to the two ears. Echolocation is of particular value under conditions when normal vision is poor (at night in the case of bats, in murky water for dolphins). A few species of bird can also echolocate, including cave-dwelling birds such as some species of swiftlets and the South American Oil Bird.

The frequency range of bats' echolocation calls is 20–215 kHz. Many species produce a specific and identifiable pattern of sound. Bats vary in the way they use echolocation: some emit pure sounds lasting up to 150 milliseconds, while others use a series of shorter 'chirps'. Sounds may be emitted through the mouth or nostrils depending on species.

Echolocation was first described in the 1930s, though it was postulated by Italian biologist Lazzaro Spallanzani (1729–1799).

echo sounder or ***sonar device*** device that detects objects under water by means of ◊sonar – by using reflected sound waves. Most boats are equipped with echo sounders to measure the water depth beneath them. An echo sounder consists of a transmitter, which emits an ultrasonic pulse, and a receiver, which detects the pulse after reflection from the seabed. The time between transmission and receipt of the reflected signal is a measure of the depth of water. Fishing boats use echo sounders to detect shoals of fish and navies use them to find enemy submarines.

eclampsia convulsions occurring during pregnancy following ◊pre-eclampsia.

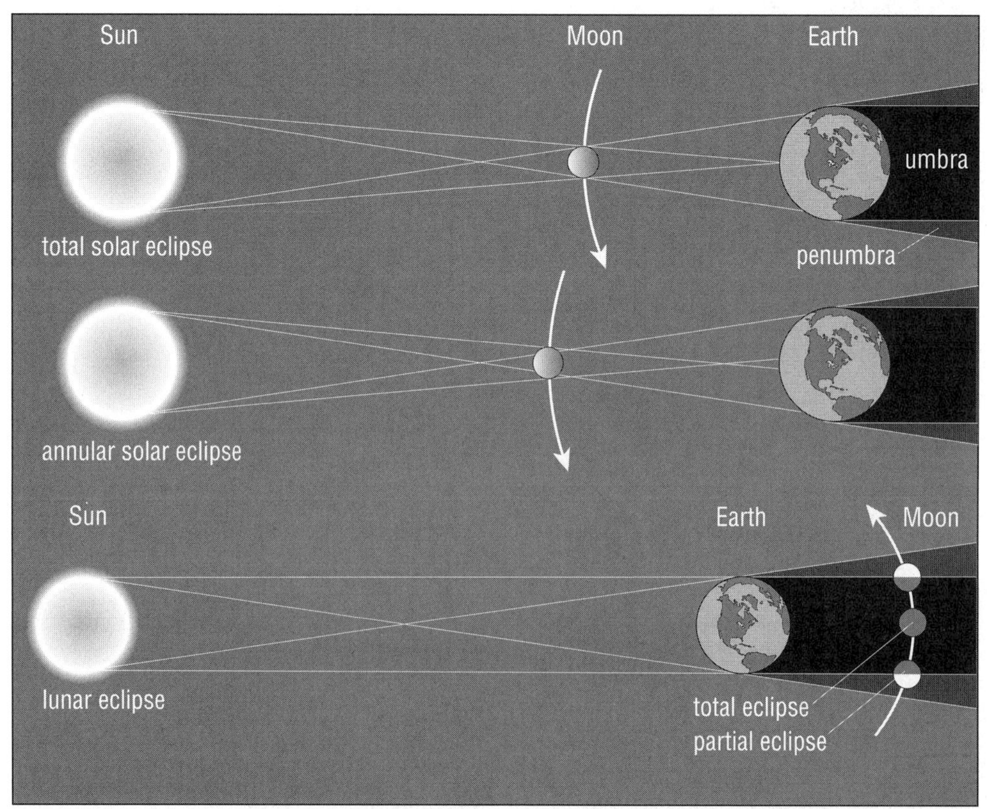

eclipse *The two types of eclipse: lunar and solar. A lunar eclipse occurs when the Moon passes through the shadow of the Earth. A solar eclipse occurs when the Moon passes between the Sun and the Earth, blocking out the Sun's light. During a total solar eclipse, when the Moon completely covers the Sun, the Moon's shadow sweeps across the Earth's surface from west to east at a speed of 3,200 kph/2,000 mph.*

Solar and Lunar Eclipses

Table does not include partial eclipses of the Moon.

Month	Day	Type of eclipse	Duration of maximum eclipse	Region for observation
1998				
February	26	solar total	17 hr 29 min	southern and eastern USA, Central America, northern South America
August	22	solar annular	2 hr 7 min	Southeast Asia, Oceania, Australasia
1999				
February	16	solar annular	6 hr 35 min	southern Indian Ocean, Antarctica, Australia
August	11	solar total	11 hr 4 min	Europe, North Africa, Arabia, western Asia
2000				
January	21	lunar total	4 hr 44 min	the Americas, Europe, Africa, western Asia
February	5	solar partial	12 hr 50 min	Antarctica
July	1	solar partial	19 hr 34 min	southeastern Pacific Ocean
July	16	lunar total	13 hr 56 min	southeastern Asia, Australasia
July	31	solar partial	2 hr 14 min	Arctic regions
December	25	solar partial	17 hr 36 min	USA, eastern Canada, Central America, Caribbean
2001				
January	9	lunar total	20 hr 21 min	Africa, Europe, Asia
June	21	solar total	12 hr 4 min	central and southern Africa
December	14	solar annular	20 hr 52 min	Pacific Ocean

eclipse passage of an astronomical body through the shadow of another.

The term is usually employed for solar and lunar eclipses, which may be either partial or total, but also, for example, for eclipses by Jupiter of its satellites. An eclipse of a star by a body in the Solar System is called an occultation.

A **solar eclipse** occurs when the Moon passes in front of the Sun as seen from Earth, and can happen only at new Moon. During a total eclipse the Sun's ◊corona can be seen. A total solar eclipse can last up to 7.5 minutes. When the Moon is at its farthest from Earth it does not completely cover the face of the Sun, leaving a ring of sunlight visible. This is an **annular eclipse** (from the Latin word *annulus* 'ring'). Between two and five solar eclipses occur each year.

A **lunar eclipse** occurs when the Moon passes into the shadow of the Earth, becoming dim until emerging from the shadow. Lunar eclipses may be partial or total, and they can happen only at full Moon. Total lunar eclipses last for up to 100 minutes; the maximum number each year is three.

eclipsing binary binary (double) star in which the two stars periodically pass in front of each other as seen from Earth.

When one star crosses in front of the other the total light received on Earth from the two stars declines. The first eclipsing binary to be noticed was ◊Algol.

ecliptic path, against the background of stars, that the Sun appears to follow each year as it is orbited by the Earth. It can be thought of as the plane of the Earth's orbit projected on to the ◊celestial sphere (imaginary sphere around the Earth).

The ecliptic is tilted at about 23.5° with respect to the celestial equator, a result of the tilt of the Earth's axis relative to the plane of its orbit around the Sun.

ecliptic coordinates in astronomy, a system for measuring the position of astronomical objects on the ◊celestial sphere with reference to the plane of the Earth's orbit, the ◊ecliptic.

Ecliptic latitude (symbol β) is measured in degrees from the ecliptic ($\beta = 0°$) to the north ($\beta = 90°$) and south ($\beta = -90°$) ecliptic poles.

Ecliptic longitude (symbol λ) is measured in degrees eastward along the ecliptic ($\lambda = 0°$ to $360°$) from a fixed point known as the first point of ◊Aries or the ◊vernal equinox. Ecliptic coordinates are often used to measure the positions of the Sun and planets with respect to the Earth.

Ecliptic latitude and longitude are sometimes known as celestial latitude and longitude or ◊declination and ◊ascension. The ecliptic longitude of the Sun (solar longitude) is a convenient measure of the position of the Earth in its orbit.

WHAT THE HECK IS AN E. COLI?

http://falcon.cc.ukans.edu/~jbrown/ecoli.html

Explains the basics behind this bacterium, including information on the dangerous strain of the bacterium and how it developed. It contains guidelines to reduce the risk of infection. There are a number of links throughout the article to sites expanding on issues raised.

E. coli abbreviation for *Escherichia coli.*

ecology ***Greek oikos 'house'*** study of the relationship among organisms and the environments in which they live, including all living and nonliving components. The chief environmental factors governing the distribution of plants and animals are temperature, humidity, soil, light intensity, daylength, food supply, and interaction with other organisms. The term was coined by the biologist Ernst Haeckel in 1866.

Ecology may be concerned with individual organisms (for example, behavioural ecology, feeding strategies), with populations (for example, population dynamics), or with entire communities (for example, competition between species for access to resources in an ecosystem, or predator–prey relationships). Applied ecology is concerned with the management and conservation of habitats and the consequences and control of pollution.

We are unravelling nature like an old jumper.

PENNY KEMP English ecologist.
A Green Manifesto for the 1990s ch 4

EcoNet in computing, one of several international computer networks dedicated to environmental issues.

Ecology: chronology

c. 325 BC	Greek scholar Theophrastus wrote about the relationship between organisms, and between organisms and their environment – the first ecological study.
1735	Swedish naturalist Carl Linnaeus developed his system for classifying and naming plants and animals.
1798	English cleric Thomas Malthus produced the earliest theoretical study of population dynamics.
1859	English naturalist Charles Darwin published his 'On the Origin of Species'.
1869	German zoologist Ernst Haeckel first defined the term 'ecology'.
1899	US botanist Henry Cowles published his classic paper on succession in sand dunes on Lake Michigan, USA.
1913	British Ecological Society founded.
1915	Ecological Society of America founded.
1916	US ecologist Frederic Clements coined the phrase 'climax communities' for large areas of rather uniform vegetation which he attributed to climactic factors.
1926	Russian botanist N I Vavilov published *Centres of Origin of Cultivated Plants*, concluding that there are relatively few such centres, many of which are located in mountainous areas.
1934	Russian ecologist G F Gause first stated the principles of competitive exclusion, related to a species' niche.
1935	British ecologist Arthur Tansley first coined the term 'ecosystem'.
1938	The coelacanth, a marine fish believed to have become extinct 65 million years ago, was 'rediscovered' in the Indian Ocean.
1940	Population biologist Charles Elton developed the idea of trophic levels in a community of organisms.
1950	The theory that natural selection may favour either individuals with high reproductive rates and rapid development (*r*-selection) or individuals with low reproductive rates and better competitive ability (*k*-selection) was first discussed.
1967	US biologists MacArthur and Wilson proposed their 'Theory of Island Biogeography' which related population and community size to island size. The theory is still widely used in the design of nature reserves today.
1979	English naturalist James Lovelock proposed his Gaia hypothesis, viewing the planet as a single organism.
1993	UN Convention on Biological Diversity came into force.

ecosystem in ecology, an integrated unit consisting of the ◊community of living organisms and the nonliving, or physical, environment in a particular area. The relationships among species in an ecosystem are usually complex and finely balanced, and removal of any one species may be disastrous. The removal of a major predator, for example, can result in the destruction of the ecosystem through overgrazing by herbivores.

Ecosystems can be identified at different scales – for example, the global ecosystem consists of all the organisms living on Earth, the Earth itself (both land and sea), and the atmosphere above; a freshwater-pond ecosystem consists of the plants and animals living in the pond, the pond water and all the substances dissolved or suspended in that water, and the rocks, mud, and decaying matter that make up the pond bottom.

Energy and nutrients pass through organisms in an ecosystem in a particular sequence (see ◊food chain): energy is captured through ◊photosynthesis, and nutrients are taken up from the soil or water by plants; both are passed to herbivores that eat the plants and then to carnivores that feed on herbivores. These nutrients are returned to the soil through the ◊decomposition of excrement and dead organisms, thus completing a cycle that is crucial to the stability and survival of the ecosystem.

ECSTASY.ORG

http://www.ecstasy.org/

Huge clearing house for information on this popular recreational drug. Contents of this much-visited site include instructions for paramedics and hospital staff, how to get an ecstasy sample chemically assessed, how to recognise danger signs, notes on the dance drug scene, and what to do if arrested for possession. On the question of the dangers of the drug, the site is noncommittal, presenting a huge sample of contradictory scientific opinion assessing the toxicity and long-term dangers of MDMA.

ecstasy or *MDMA* (3,4-methylenedioxymethamphetamine) illegal drug in increasing use from the 1980s. It is a modified ◊amphetamine with mild psychedelic effects, and works by depleting serotonin (a neurotransmitter) in the brain. Its long-term effects are unknown, but animal experiments have shown brain damage.

Ecstasy was first synthesized in 1914 by the Merck pharmaceutical company in Germany, and was one of eight psychedelics tested by the US army in 1953, but was otherwise largely forgotten until the mid-1970s. It can be synthesized from nutmeg oil. Since 1996 chemical recipes for making the drug have been circulated on the Internet.

ECT abbreviation for ◊electroconvulsive therapy.

ectoparasite ◊parasite that lives on the outer surface of its host.

ectopic in medicine, term applied to an anatomical feature that is displaced or found in an abnormal position. An ectopic pregnancy is one occurring outside the womb, usually in a Fallopian tube.

ectoplasm outer layer of a cell's ◊cytoplasm.

ectotherm 'cold-blooded' animal (see ◊poikilothermy), such as a lizard, that relies on external warmth (ultimately from the Sun) to raise its body temperature so that it can become active. To cool the body, ectotherms seek out a cooler environment.

eczema inflammatory skin condition, a form of dermatitis, marked by dryness, rashes, itching, the formation of blisters, and the exudation of fluid. It may be allergic in origin and is sometimes complicated by infection.

eddy current electric current induced, in accordance with ◊Faraday's laws, in a conductor located in a changing magnetic field. Eddy currents can cause much wasted energy in the cores of transformers and other electrical machines.

edelweiss perennial alpine plant belonging to the daisy family, with a white, woolly, star-shaped flower, found in the high mountains of Europe and Asia. (*Leontopodium alpinum*, family Compositae.)

edge connector in computing, an electrical connection formed by taking some of the metallic tracks on a ◊printed circuit board to the edge of the board and using them to plug directly into a matching socket.

EDISON, THOMAS

http://hfm.umd.umich.edu/histories/edison/tae.html

Short, illustrated biography of the US inventor, plus a chronology and bibliography. It includes sections on such topics as his childhood and the electric light.

Edison, Thomas Alva
(1847–1931)

US scientist and inventor, whose work in the fields of communications and electrical power greatly influenced the world in which we live. With more than 1,000 patents, Edison produced his most important inventions in Menlo Park, New Jersey 1876–87, including the phonograph and the electric light bulb in 1879. He also constructed a system of electric power distribution for consumers, the telephone transmitter, and the megaphone.

telegraphy and telephony Edison's first success came in the area of telegraphy. Perceiving the need for rapid communications after the Civil War, his first invention was an automatic repeater for telegraphic messages. He then invented a tape machine called a 'ticker', which communicated stock exchange prices across the country.

Turning his attention to the transmission of the human voice over long distances in 1876, he patented an electric transmitter system that proved to be less commercially successful than the telephone of Bell and Gray, patented a few months later. Undeterred, Edison set about improving their system, culminating in his invention of the carbon transmitter, which so increased the volume of the telephone signal that it was used as a microphone in the Bell telephone.

Mary Evans Picture Library

the light bulb While experimenting with the carbon microphone in the 1870s, Edison had toyed briefly with the idea of using a thin carbon filament as a light source in an incandescent electric lamp. He returned to the idea in 1879. His first major success came on 19 October of that year when, using carbonized sewing cotton mounted on an electrode in a vacuum (one millionth of an atmosphere), he obtained a source that remained aglow for 45 hours without overheating – a major problem with all other materials used. He and his assistants tried 6,000 other organic materials before finding a bamboo fibre that gave a bulb life of 1,000 hours.

generators and the first power stations To produce a serious rival to gas illumination, a power source was required as well as a cheap and reliable lamp. The alternatives were either generators or heavy and expensive batteries. At that time, the best generators rarely converted more than 40% of the mechanical energy into electrical energy. Edison's first generator consisted of a drum armature of soft iron wire and a simple bi-polar magnet, and was designed to operate one arc lamp and some incandescent lamps in series.

A few months later he built a much more ambitious generator, the largest built to date; weighing 500 kg/1,103 lb and with an efficiency of 82%. Edison's team were at the forefront of development in generator technology over the next decade, during which efficiency was raised above 90%. To complete his electrical system he designed cables to carry power into the home from small (by modern standards) generating stations, and also invented an electricity meter to record its use.

the phonograph In 1877 he began the era of recorded sound by inventing the phonograph, a device in which the vibrations of the human voice were engraved by a needle on a revolving cylinder coated with tin foil.

Edge connectors are often used to connect the computer's main circuit board, or motherboard, to the expansion boards that provide the computer with extra memory or other facilities.

EDI in computing, abbreviation for ◊electronic data interchange.

EDIFACT (acronym from ***electronic data interchange for administration, commerce, and trade***) in computing, the ISO and ANSI standard system for handling EDI transactions.

editing in computing, act of creating, changing, and formatting word processor documents or pages for distribution on the World Wide Web.

EDO RAM (abbreviation for ***extended data out random-access memory***) in computing, faster type of ◊RAM introduced in the mid-1990s.

EDP in computing, abbreviation for **electronic** ◊data processing.

Educational Resources Information Center in computing, database of resources for education available on the Internet. See ◊ERIC.

edutainment (contraction of ***education and entertainment***) ◊multimedia-related term, used to describe computer software that is both educational and entertaining. Examples include educational software for children that teaches them to spell or count while playing games, and ◊CD-ROMs about machines that contain animations showing how the machines work. Compare ◊infotainment.

Edwards Air Force Base military USAF centre in California, situated on a dry lake bed, often used as a landing site by the Space Shuttle.

EEG abbreviation for ◊electroencephalogram.

eel any fish of the order Anguilliformes. Eels are snakelike, with elongated dorsal and anal fins. They include the freshwater eels of Europe and North America (which breed in the Atlantic), the marine conger eels, and the morays of tropical coral reefs.

A new species of moray eel was discovered 1995 off the coasts of Oman and Somalia. It is up to 60 cm/2 ft in length with a large black blotch around the gill openings.

eelgrass or ***tape grass*** or ***glass wrack*** any of several aquatic plants, especially *Zostera marina*. Eelgrass is found in tidal mud flats and is one of the few flowering plants to adapt to marine conditions, being completely submerged at high tide. (Genus *Zostera*, family Zosteraceae.)

eelpout freshwater fish also called a ◊burbot.

EEPROM (acronym for ***electrically erasable programmable read-only memory***) computer memory that can record data and retain it indefinitely. The data can be erased with an electrical charge and new data recorded.

Some EEPROM must be removed from the computer and erased and reprogrammed using a special device. Other EEPROM, called **flash memory**, can be erased and reprogrammed without removal from the computer.

EFF abbreviation for ◊***Electronic Frontier Foundation***.

Effelsberg site, near Bonn, Germany, of the world's largest fully steerable radio telescope, the 100-m/328-ft radio dish of the Max Planck Institute for Radio Astronomy, opened in 1971.

efficiency in physics, a general term indicating the degree to which a process or device can convert energy from one form to another without loss. It is normally expressed as a fraction or percentage, where 100% indicates conversion with no loss. The efficiency of a machine, for example, is the ratio of the work done by the machine to the energy put into the machine; in practice it is always less than 100% because of frictional heat losses. Certain electrical machines with no moving parts, such as transformers, can approach 100% efficiency.

Since the ◊mechanical advantage, or force ratio, is the ratio of the load (the output force) to the effort (the input force), and the ◊velocity ratio is the distance moved by the effort divided by the distance moved by the load, for certain machines the efficiency can also be defined as the mechanical advantage divided by the velocity ratio.

efflorescence loss of water or crystallization of crystals exposed to air, resulting in a dry powdery surface.

EFTPOS (acronym for ***electronic funds transfer at point of sale***), a form of electronic funds transfer.

EGA (abbreviation for ***enhanced graphics array***) computer colour display system superior to ◊CGA (colour graphics adapter), providing 16 colours on screen and a resolution of 640 x 350, but not as good as ◊VGA.

egestion the removal of undigested food or faeces from the gut. In most animals egestion takes place via the anus, although the invertebrate flatworms must use the mouth because their gut has no exit. Egestion is the last part of a complex feeding process that starts with food capture and continues with digestion and assimilation.

egg in animals, the ovum, or female ◊gamete (reproductive cell).

After fertilization by a sperm cell, it begins to divide to form an embryo. Eggs may be deposited by the female (◊ovipary) or they may develop within her body (◊vivipary and ◊ovovivipary). In the oviparous reptiles and birds, the egg is protected by a shell, and well supplied with nutrients in the form of yolk.

eggar-moth large reddish-brown moth with a highly developed hindwing. The length across the wings is 3.8–11.5 cm/1.5–4.5 in.

The walls of the cocoons sometimes have a smooth, shell-like appearance, hence the name.

classification The eggar-moth in the family Lasiocampidae, order Lepidoptera, class Insecta, phylum Arthropoda.

eggplant another name for ◊aubergine.

ego ***Latin 'I'*** in psychology, the processes concerned with the self and a person's conception of himself or herself, encompassing values and attitudes. In Freudian psychology, the term refers specifically to the element of the human mind that represents the conscious processes concerned with reality, in conflict with the ◊id (the instinctual element) and the ◊superego (the ethically aware element).

No other manmade device since the shields and lances of the ancient knights fulfils a man's ego like an automobile.

ROOTES English car manufacturer.
Quoted in 'Who Said That?', BBC TV, 14 Jan 1958

egret any of several ◊herons with long tufts of feathers on the head or neck. They belong to the order Ciconiiformes.

eider large marine ◊duck of the genus *Somateria*, family Anatidae, order Anseriformes. They are found on the northern coasts of the Atlantic and Pacific Oceans. The **common eider** *S. molissima* is highly valued for its soft down, which is used in quilts and cushions for warmth. The adult male has a black cap and belly and a green nape. The rest of the plumage is white with a pink breast and throat, while the female is a mottled brown. The bill is large and flattened and both bill and feet are olive green.

Other species are the **king eider-duck** *S. spectablis,* **Steller's eider** *Polysticta stelleri,* and the **spectacled eider** *S. fischeri.*

eidophor television projection system that produces pictures up to 10 m/33 ft square, at sports events or rock concerts, for example. The system uses three coloured beams of light, one of each primary colour (red, blue, and green), which scan the screen. The intensity of each beam is controlled by the strength of the corresponding colour in the television picture.

Imagination is more important than knowledge.

ALBERT EINSTEIN German-born US physicist.
On Science

Section through a fertilized egg

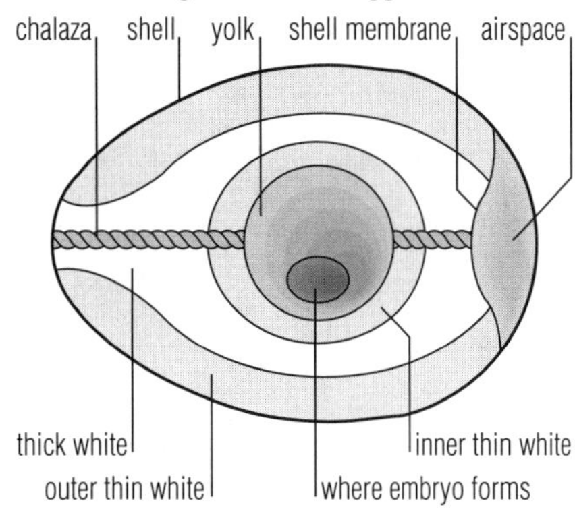

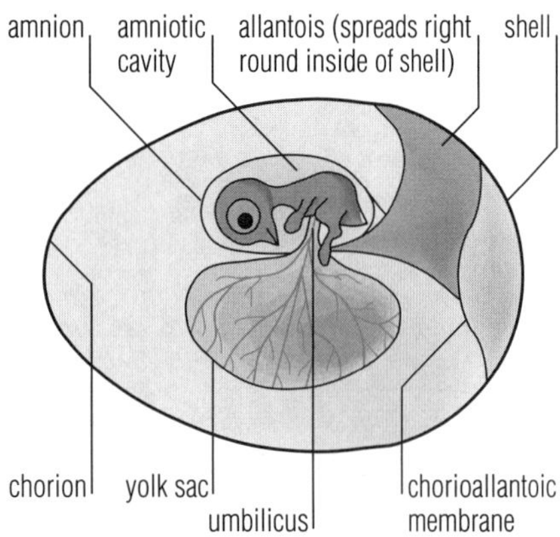

egg *Section through a fertilized bird egg. Inside a bird's egg is a complex structure of liquids and membranes designed to meet the needs of the growing embryo. The yolk, which is rich in fat, is gradually absorbed by the embryo. The white of the egg provides protein and water. The chalaza is a twisted band of protein which holds the yolk in place and acts as a shock absorber. The airspace allows gases to be exchanged through the shell. The allantois contains many blood vessels which carry gases between the embryo and the outside.*

EINSTEIN'S LEGACY

http://www.ncsa.uiuc.edu/Cyberia/NumRel/EinsteinLegacy.html

Illustrated introduction to the man and his greatest legacy – relativity and the concept of space-time. There is a film and audio clip version of the page courtesy of a US scientist and details about how current research is linked to Einstein's revolutionary ideas.

Einstein, Albert
(1879–1955)

German-born US physicist whose theories of relativity revolutionized our understanding of matter, space, and time. Einstein established that light may have a particle nature and deduced the **photoelectric law**, for which he was awarded the Nobel Prize for Physics in 1921. He also investigated Brownian motion, confirming the existence of atoms. His last conception of the basic laws governing the universe was outlined in his unified field theory, made public in 1953.

Brownian motion Einstein's first major achievement concerned Brownian movement, the random movement of fine particles that can be seen through a microscope, which was first observed in 1827 by Robert Brown when studying a suspension of pollen grains in water. The motion of the pollen grains increased when the temperature increased but decreased if larger particles were used. Einstein explained this phenomenon as being the effect of large numbers of molecules (in this case, water molecules) bombarding the particles. He was able to make predictions of the movement and sizes of the particles, which were later verified experimentally by the French physicist Jean Perrin (1870–1942).

Einstein's explanation of Brownian motion and its subsequent experimental confirmation was one of the most important pieces of evidence for the hypothesis that matter is composed of atoms. Experiments based on this work were used to obtain an accurate value of Avogadro's number (the number of atoms in one mole of a substance) and the first accurate values of atomic size.

the photoelectric effect and the Nobel Prize Einstein's work on photoelectricity began with an explanation of the radiation law proposed in 1901 by Max Planck: $E=hv$, where E is the energy of radiation, h is Planck's constant, and v is the frequency of radiation. Einstein suggested that packets of light energy are capable of behaving as particles called 'light quanta' (later called photons). Einstein used this hypothesis to explain the photoelectric effect, proposing that light particles striking the surface of certain metals cause electrons to be emitted.

It had been found experimentally that electrons are not emitted by light of less than a certain frequency ν^0; that when electrons are emitted, their energy increases with an increase in the frequency of the light; and that an increase in light intensity produces more electrons but does not increase their energy. Einstein suggested that the kinetic energy of each electron, $\frac{1}{2}mv^2$, is equal to the difference in the incident light energy $h\nu$ and the light energy needed to overcome the threshold of emission, $h\nu^0$. This can be written mathematically as:

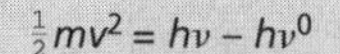

$$\frac{1}{2}mv^2 = h\nu - h\nu^0$$

the speed of light and the special theory of relativity The **special theory of relativity** started with the premises that (1) the laws of nature are the same for all observers in unaccelerated motion, and (2) the speed of light is independent of the motion of its source. Until then, there had been a steady accumulation of knowledge that suggested that light and other electromagnetic radiation do not behave as predicted by classical physics. It proved impossible, for example, to measure the expected changes in the speed of light relative to the motion of the Earth. The **Michelson–Morley experiment** demonstrated that the velocity of light is constant and does not vary with the motion of either the source or the observer. Their experiment also confirmed that no 'ether' exists in the Universe as a medium to carry light waves, as was required by classical physics.

These findings did not worry Einstein, who viewed light as behaving like particles. He recognized that light, being different from waves that travel in a medium, has a measured speed that is independent of the speed of the observer. Thus, contrary to everyday experience with phenomena such as sound waves, the velocity of light is the same for an observer travelling at high speed *towards* a light source as it is for an observer travelling rapidly *away* from the light source. To Einstein it followed that, if the speed of light is the same for both these observers, the time and distance framework they use to measure the speed of light cannot be the same. Time and distance vary, depending on the velocity of each observer.

From the notions of relative motion and the constant velocity of light, Einstein derived that, in a system in motion relative to an observer, length would be observed to decrease, time would slow down, and mass would increase. The magnitude of these effects is negligible at ordinary velocities and Newton's laws still held good. But at velocities approaching that of light, they become substantial. If a system were to move at the velocity of light, to an observer carried with it, its length would be zero, time would be at a stop, and its mass would be infinite. Einstein therefore concluded that no system can move at a velocity equal to or greater than the velocity of light.

Einstein's conclusions regarding time dilation and mass increase were later verified with observations of fast-moving atomic clocks and cosmic rays. Einstein went on to show in 1907 that mass is related to energy by the famous equation $E = mc^2$, which indicates the enormous amount of energy that is stored as mass, some of which is released in radioactivity and nuclear reactions, for example in the Sun.

gravity and the general theory of relativity In the **general theory of relativity**, the properties of space–time were to be conceived as modified locally by the presence of a body with mass; and light rays should bend when they pass by a massive object. A planet's orbit around the Sun arises from its natural trajectory in modified space–time – there is no need to invoke, as Isaac Newton did, a force of gravity. General relativity theory was inspired by the simple idea that it is impossible in a small region to distinguish between acceleration and gravitation effects (as in a lift one feels heavier when it accelerates upwards).

Einstein used the general theory to account for an anomaly in the orbit of the planet Mercury that could not be explained by Newtonian mechanics. Furthermore, the general theory made two predictions concerning light and gravitation. The first was that a red shift is produced if light passes through an intense gravitational field, and this was subsequently detected in astronomical observations in 1925. The second was a prediction that the apparent positions of stars would shift when they are seen near the Sun because the Sun's intense gravity would bend the light rays from the stars as they pass the Sun. Einstein was triumphantly vindicated when observations of a solar eclipse in 1919 showed apparent shifts of exactly the amount he had predicted.

Mary Evans Picture Library

einsteinium synthesized, radioactive, metallic element of the actinide series, symbol Es, atomic number 99, relative atomic mass 254.09.

It was produced by the first thermonuclear explosion, in 1952, and discovered in fallout debris in the form of the isotope Es-253 (half-life 20 days). Its longest-lived isotope, Es-254, with a half-life of 276 days, allowed the element to be studied at length. It is now synthesized by bombarding lower-numbered ◊transuranic elements in particle accelerators. It was first identified by A Ghiorso and his team who named it in 1955 after Albert Einstein, in honour of his theoretical studies of mass and energy.

EIS (abbreviation for ***executive information systems***) software applications that extract information from an organization's computer applications and data files and present the data in a form required by management.

EISA (abbreviation for ***extended industry standard architecture***) in computing, one of several types of ◊data bus created to improve on the original ISA (industry standard architecture) design introduced with the IBM PC AT microcomputer in 1984. The EISA bus adds speed and capacity because it is a 32-bit bus (ISA is a 16-bit bus), although it can still accept ISA-compatible expansion cards.

The EISA bus was developed by a consortium of PC manufacturers to counter IBM's proprietary MCA (micro channel architecture) bus in 1987, but it has since been superseded by the ◊PCI (peripheral component interconnect) bus designed by Intel. See also ◊local bus.

ejecta in astronomy, any material thrown out of a ◊crater due to volcanic eruption or the impact of a ◊meteorite or other object. Ejecta from impact craters on the ◊Moon often form long bright streaks known as rays, which in some cases can be traced for thousands of kilometres across the lunar surface.

ejector seat device for propelling an aircraft pilot out of the plane to parachute to safety in an emergency, invented by the British engineer James Martin (1893–1981). The first seats of 1945 were powered by a compressed spring; later seats used an explosive charge. By the early 1980s, 35,000 seats had been fitted worldwide, and the lives of 5,000 pilots saved by their use.

Seats that can be ejected on takeoff and landing or at low altitude were a major breakthrough of the 1980s. They are as effective as those originally designed for parachuting from high altitudes.

Ekman spiral effect in oceanography, theoretical description of a consequence of the ◊Coriolis effect on ocean currents, whereby currents flow at an angle to the winds that drive them. It derives its name from the Swedish oceanographer Vagn Ekman (1874–1954).

In the northern hemisphere, surface currents are deflected to the right of the wind direction. The surface current then drives the subsurface layer at an angle to its original deflection. Consequent subsurface layers are similarly affected, so that the effect decreases with increasing depth. The result is that most water is transported at about right-angles to the wind direction. Directions are reversed in the southern hemisphere.

eland largest species of ◊antelope, *Taurotragus oryx*. Pale fawn in colour, it is about 2 m/6 ft high, and both sexes have spiral horns about 45 cm/18 in long. It is found in central and southern Africa.

Elasmobranchii subclass of class Chondrichthyes (cartilaginous fish), which includes the ◊sharks and ◊rays. They are characterized by having five to seven pairs of gill clefts that open separately to the exterior and are not covered by a protective fold of skin.

elastic collision in physics, a collision between two or more bodies in which the total ◊kinetic energy of the bodies is conserved (remains constant); none is converted into any other form of energy. The molecules of a gas may be considered to collide elastically, but large objects may not because some of their kinetic energy will be converted on collision to heat and sound, or used to deform the object.

elasticity in physics, the ability of a solid to recover its shape once deforming forces (stresses modifying its dimensions or shape) are removed. An elastic material obeys ◊Hooke's law, which states that its deformation is proportional to the applied stress up to a certain point, called the **elastic limit**, beyond which additional stress will deform it permanently. Elastic materials include metals and rubber; however, all materials have some degree of elasticity.

elastomer any material with rubbery properties that stretches easily and then quickly returns to its original length when released.

Natural and synthetic rubbers and such materials as polychloroprene and butadiene copolymers are elastomers. The convoluted molecular chains making up these materials are uncoiled by a stretching force, but return to their original position when released because there are relatively few crosslinks between the chains.

elater beetle another name for ◊click beetle.

E layer (formerly called the Kennelly–Heaviside layer) the lower regions (90–120 km/56–75 mi) of the ◊ionosphere, which reflect radio waves, allowing their reception around the surface of the Earth. The E layer approaches the Earth by day and recedes from it at night.

elder in botany, any of a group of small trees or shrubs belonging to the honeysuckle family, native to North America, Europe, Asia, and North Africa. Some are grown as ornamentals for their showy yellow or white flower clusters and their colourful black or scarlet berries. (Genus *Sambucus,* family Caprifoliaceae.)

The American elder (*Sambucus canadensis*) reaches tree size and has blue berries. The common elder (*S. nigra*) of Europe, North Africa, and western Asia has pinnate leaves (leaflets growing either side of a stem) and heavy heads of small, sweet-scented, white flowers in early summer. These are followed by clusters of small, black, shiny berries. The scarlet-berried *S. racemosa* is found in parts of Europe, Asia, and North America.

electrical energy form of energy carried by an electric current. It may be converted into other forms of energy such as heat, light, and motion.

The electrical energy W watts converted in a circuit component through which a current I amperes passes and across which there is a potential difference of V volts is given by the formula:

$$W = IV$$

electrical relay an electromagnetic switch; see ◊relay.

electrical safety measures taken to protect human beings from electric shock or from fires caused by electrical faults. They are of paramount importance in the design of electrical equipment. Safety measures include the fitting of earth wires, and fuses or circuit breakers; the insulation of wires; the double insulation of portable equipment; and the use of residual-current devices (RCDs), which will break a circuit and cut off all currents if there is any imbalance between the currents in the live and neutral wires connected to an appliance (caused, for example, if some current is being conducted through a person).

The effects of electric shock vary from a tingling sensation to temporary paralysis and even death, and depend upon the amount of current passing through the body, and upon whether it passes through the central nervous system thereby affecting brain and heart function. Fires are usually caused by overheated cables or loose connections.

electric arc a continuous electric discharge of high current between two electrodes, giving out a brilliant light and heat. The phenomenon is exploited in the carbon-arc lamp, once widely used in film projectors. In the electric-arc furnace an arc struck between very large carbon electrodes and the metal charge provides the heating. In arc ◊welding an electric arc provides the heat to fuse the metal. The discharges in low-pressure gases, as in neon and sodium lights, can also be broadly considered as electric arcs.

electric bell a bell that makes use of electromagnetism. At its heart is a wire-wound coil on an iron core (an electromagnet) which, when a direct current (from a battery) flows through it, attracts an iron ◊armature. The armature acts as a switch, whose movement causes contact with an adjustable contact point to be broken, so breaking the circuit. A spring rapidly returns the armature to the contact point, once again closing the circuit, and so bringing about the oscillation. The armature oscillates back and forth, and the clapper or hammer fixed to the armature strikes the bell.

electric charge property of some bodies that causes them to exert forces on each other. Two bodies both with positive or both with negative charges repel, each other, whereas bodies with opposite or 'unlike' charges attract each other, since each is in the ◊electric field of the other. In atoms, ◊electrons possess a negative charge, and ◊protons an equal positive charge. The ◊SI unit of electric charge is the coulomb (symbol C).

A body can be charged by friction, induction, or chemical change and shows itself as an accumulation of electrons (negative charge) or loss of electrons (positive charge) on an atom or body. Atoms have no charge but can sometimes gain electrons to become negative ions or lose them to become positive ions. So-called ◊static electricity, seen in such phenomena as the charging of nylon shirts when they are pulled on or off, or in brushing hair, is in fact the gain or loss of electrons from the surface atoms. A flow of charge (such as electrons through a copper wire) constitutes an electric current; the flow of current is measured in amperes (symbol A).

electric current the flow of electrically charged particles through a conducting circuit due to the presence of a ◊potential difference. The current at any point in a circuit is the amount of charge flowing per second; its SI unit is the ampere (coulomb per second).

Current carries electrical energy from a power supply, such as a battery of electrical cells, to the components of the circuit, where it is converted into other forms of energy, such as heat, light, or motion. It may be either ◊direct current or ◊alternating current.

heating effect When current flows in a component possessing resistance, electrical energy is converted into heat energy. If the resistance of the component is R ohms and the current through it is I amperes, then the heat energy W (in joules) generated in a time t seconds is given by the formula:

$$W = I^2Rt$$

magnetic effect A magnetic field is created around all conductors that carry a current. When a current-bearing conductor is made into a coil it forms an electromagnet with a ◊◊magnetic field that is similar to that of a bar magnet, but which disappears as soon as the current is switched off. The strength of the magnetic field is directly proportional to the current in the conductor – a property that allows a small ◊electromagnet to be used to produce a pattern of magnetism on recording tape or disc that accurately represents the sound or data to be stored. The direction of the field created around a conducting wire may be predicted by using ◊Maxwell's screw rule.

motor effect A conductor carrying current in a magnetic field experiences a force, and is impelled to move in a direction perpendicular to both the direction of the current and the direction of the magnetic field. The direction of motion may be predicted by Fleming's left-hand rule (see ◊Fleming's rules). The magnitude of the force experienced depends on the length of the conductor and on the strengths of the current and the magnetic field, and is greatest when the conductor is at right angles to the field. A conductor wound into a coil that can rotate between the poles of a magnet forms the basis of an ◊electric motor.

electric eel South American freshwater bony fish. It grows to almost 3 m/10 ft and the electric shock produced, normally for immobilizing prey, is enough to stun an adult human. Electric eels are not true eels.

classification *Electrophorus electricus* is in the order Cypriniformes, class Osteichthyes.

electric field in physics, a region in which a particle possessing electric charge experiences a force owing to the presence of another electric charge. The strength of an electric field, E, is measured in volts per metre (V m^{-1}). It is a type of ◊electromagnetic field.

electric fish any of several unrelated fishes that have electricity-producing powers, including the South American 'electric eel'. These include *Electrophorus electricus,* which is not a true eel, and in which the lateral tail muscles are modified to form electric organs capable of generating 650 volts; the current passing from tail to head is strong enough to stun another animal. Not all electric fishes produce such strong discharges; most use weak electric fields to navigate and to detect nearby objects.

electricity all phenomena caused by ◊electric charge, whether static or in motion. Electric charge is caused by an excess or deficit

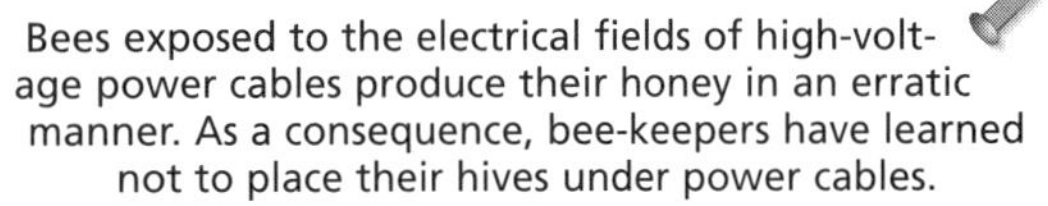

BEE

Bees exposed to the electrical fields of high-voltage power cables produce their honey in an erratic manner. As a consequence, bee-keepers have learned not to place their hives under power cables.

of electrons in the charged substance, and an electric current is the movement of charge through a material. Substances may be electrical conductors, such as metals, that allow the passage of electricity through them readily, or insulators, such as rubber, that are extremely poor conductors. Substances with relatively poor conductivities that can be improved by the addition of heat or light are known as ◊semiconductors.

electrical properties of solids The first artificial electrical phenomenon to be observed was that some naturally occurring materials such as amber, when rubbed with a piece of cloth, would then attract small objects such as dust and pieces of paper. Rubbing the object caused it to become electrically charged so that it had an excess or deficit of electrons. When the amber is rubbed with a piece of cloth electrons are transferred from the cloth to the amber so that the amber has an excess of electrons and is negatively charged, and the cloth has a deficit of electrons and is positively charged. This accumulation of charge is called ◊static electricity.

This charge on the object exerts an electric field in the space around itself that can attract or repel other objects. It was discovered that there are only two types of charge, positive and negative, and that they neutralize one another. Objects with a like charge always repel one another while objects with an unlike charge attract each other. Neutral objects (such as pieces of paper) can be attracted to charged bodies by electrical induction. For example, the charge on a negatively charged body causes a separation of charge across the neutral body. The positive charges tend to move towards the side near the negatively charged body and the negative charges tend to move towards the opposite side so that the neutral body is weakly attracted to the charged body by ◊induction.

The ◊electroscope is a device used to demonstrate the presence of electric charges and to measure its size and whether it is positive or negative. The electroscope was invented by Michael Faraday.

current, charge, and energy An ◊electric current in a material is the passage of charge through it. In metals and other conducting materials, the charge is carried by free electrons that are not bound tightly to the atoms and are thus able to move through the material. For charge to flow in a circuit there must be a ◊potential difference (pd) applied across the circuit. This is often supplied in the form of a battery that has a positive terminal and a negative terminal. Under the influence of the potential difference, the electrons are repelled from the negative terminal side of the circuit and attracted to the positive terminal of the battery. A steady flow of electrons around the circuit is produced.

Current flowing through a circuit can be measured using an ◊ammeter and is measured in ◊amperes (or amps). A ◊coulomb (C) is the unit of charge, defined as the charge passing a point in a wire each second when the current is exactly 1 amp. The unit of charge is named after Charles Augustin de ◊Coulomb. Direct current (DC) flows continuously in one direction; ◊alternating current (AC) flows alternately in each direction.

In a circuit the battery provides energy to make charge flow through the circuit. The amount of energy supplied to each unit of charge is called the electromotive force (emf). The unit of emf is the ◊volt (V). A battery has an emf of 1 volt when it supplies I joule of energy to each coulomb of charge flowing through it. The energy carried by flowing charges can be used to do work, for example to light a bulb, to cause current to flow through a resistor, to emit radiation, or to produce heat. When the energy carried by a current is made to do work in this way, a potential difference can be measured across the circuit component concerned by a voltmeter or a ◊cathode-ray oscilloscope. The potential difference is also measured in volts. Power, measured in ◊watts, is the product of current and voltage.

Although potential difference and current measure different things, they are related to one another. This relationship was discovered by Georg ◊Ohm, and is expressed by ◊Ohm's law: the current through a wire is proportional to the potential difference across its ends. The potential difference divided by the current is a constant for a given piece of wire. This constant for a given material is called the ◊resistance.

conduction in liquids and gases In liquids, current can flow by the movement of charged ions through a solution or molten salt (the electrolyte), resulting in the migration of ions to the electrodes: positive ions (cations) to the negative electrode (cathode) and negative ions (anions) to the positive electrode (anode). This process is called ◊electrolysis and represents bi-directional flow of charge as opposite charges move to oppositely charged electrodes. In metals, charges are only carried by free electrons and therefore move in only one direction.

electromagnetism ◊Magnetic fields are produced either by current-carrying conductors or by permanent magnets. In current-carrying wires, the magnetic field lines are concentric circles around the wire. Their direction depends on the direction of the current and their strength on the size of the current.

If a conducting wire is moved within a magnetic field, the magnetic field acts on the free electrons within the conductor, displacing them and causing a current to flow. The force acting on the electrons and causing them to move is greatest when the wire is perpendicular to the magnetic field lines. The direction of the current is given by the ◊left-hand rule. The generation of a current by the relative movement of a conductor in a magnetic field is called ◊electromagnetic induction. This is the basis of how a ◊dynamo works.

generation of electricity Electricity is the most useful and most convenient form of energy, readily convertible into heat and light and used to power machines. Electricity can be generated in one place and distributed anywhere because it readily flows through wires. It is generated at power stations where a suitable energy source is harnessed to drive ◊turbines that spin electricity generators. Current energy sources are coal, oil, water power (hydroelectricity), natural gas, and ◊nuclear energy. Research is under way to increase the contribution of wind, tidal, solar, and geothermal power. Nuclear fuel has proved a more expensive source of electricity than initially anticipated and worldwide concern over radioactivity may limit its future development.

Electricity is generated at power stations at a voltage of about 25,000 volts, which is not a suitable voltage for long-distance transmission. For minimal power loss, transmission must take place at very high voltage (400,000 volts or more). The generated voltage is therefore increased ('stepped up') by a ◊transformer. The resulting high-voltage electricity is then fed into the main arteries of the grid system, an interconnected network of power stations and distribution centres covering a large area. After transmission to a local substation, the line voltage is reduced by a step-down transformer and distributed to consumers.

Among specialized power units that convert energy directly to electrical energy without the intervention of any moving mechanisms, the most promising are thermionic converters. These use conventional fuels such as propane gas, as in portable military power packs, or, if refuelling is to be avoided, radioactive fuels, as in uncrewed navigational aids and spacecraft.

UK electricity generation was split into four companies in 1990 in preparation for privatization. The nuclear power stations remain in the hands of the state through Nuclear Electric (accounting for 20% of electricity generated); National Power (50%) and PowerGen (30%) generate electricity from fossil-fuel and renewable sources. Transmission lines and substations are owned by the National Grid, which was privatized in 1996.

Electricity generated on a commercial scale was available from the early 1880s and used for electric motors driving all kinds of machinery, and for lighting, first by carbon arc, but later by incandescent filaments (first of carbon and then of tungsten), enclosed in glass bulbs partially filled with inert gas under vacuum. Light is also produced by passing electricity through a gas or metal vapour or a fluorescent lamp. Other practical applications include telephone, radio, television, X-ray machines, and many other applications in ◊electronics.

An important consideration in the design of electrical equipment is ◊electrical safety. This includes measures to minimize the risk of electric shock or fire caused by electrical faults. Safety measures include the fitting of earth wires, and fuses or circuit breakers, and the insulation of wires.

history The fact that amber has the power, after being rubbed, of attracting light objects, such as bits of straw and feathers, is said to have been known to Thales of Miletus and to the Roman naturalist Pliny. William Gilbert, Queen Elizabeth I's physician, found that many substances possessed this power, and he called it 'electric' after the Greek word meaning 'amber'.

In the early 1700s, it was recognized that there are two types of electricity and that unlike kinds attract each other and like kinds repel. The charge on glass rubbed with silk came to be known as positive electricity, and the charge on amber rubbed with wool as negative electricity. These two charges were found to cancel each other when brought together.

In 1800 Alessandro ◊Volta found that a series of cells containing brine, in which were dipped plates of zinc and copper, gave an electric current, which later in the same year was shown to evolve hydrogen and oxygen when passed through water (see ◊electrolysis). Humphry Davy, in 1807, decomposed soda and potash (both thought to be elements) and isolated the metals sodium and potassium, a discovery that led the way to ◊electroplating. Other properties of electric currents discovered were the heating effect, now used in lighting and central heating, and the deflection of a magnetic needle, described by Hans Oersted in 1820 and elaborated by André ◊Ampère in 1825. This work made possible the electric telegraph.

For Michael Faraday, the fact that an electric current passing through a wire caused a magnet to move suggested that moving a wire or coil of wire rapidly between the poles of a magnet would induce an electric current. He demonstrated this in 1831, producing the first dynamo, which became the basis of electrical engineering. The characteristics of currents were crystallized about 1827 by Georg ◊Ohm, who showed that the current passing along a wire was equal to the electromotive force across the wire multiplied by a constant, which was the conductivity of the wire. The unit of resistance (ohm) is named after Ohm, the unit of emf (volt) is named after Volta, and the unit of current (amp) after Ampère.

The work of the late 1800s indicated the wide interconnections of electricity (with magnetism, heat, and light), and about 1855 James Clerk ◊Maxwell formulated a single electromagnetic theory. The universal importance of electricity was decisively proved by the discovery that the atom, until then thought to be the ultimate particle of matter, is composed of a positively charged central core, the nucleus, about which negatively charged electrons rotate in various orbits.

electric motor a machine that converts electrical energy into mechanical energy. There are various types, including direct-current and induction motors, most of which produce rotary motion. A linear induction motor produces linear (in a straight line) rather than rotary motion.

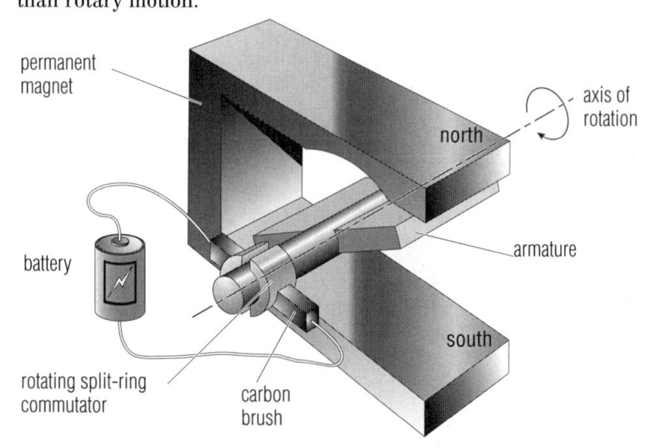

electric motor *In an electric motor, magnetic fields generated by electric currents push against each other, causing a shaft (the armature) to rotate.*

A simple **direct-current motor** consists of a horseshoe-shaped permanent ◊magnet with a wire-wound coil (◊armature) mounted so that it can rotate between the poles of the magnet. A ◊commutator reverses the current (from a battery) fed to the coil on each half-turn, which rotates because of the mechanical force exerted on a conductor carrying a current in a magnetic field.

An **induction motor** employs ◊alternating current. It comprises a stationary current-carrying coil (stator) surrounding another coil (rotor), which rotates because of the current induced in it by the magnetic field created by the stator; it thus requires no commutator.

electric organs specialized organs that discharge electricity and occur only in fish; see ◊electric fish.

electric power the rate at which an electrical machine uses electrical ◊energy or converts it into other forms of energy – for example, light, heat, mechanical energy. Usually measured in watts (equivalent to joules per second), it is equal to the product of the voltage and the current flowing.

An electric lamp that passes a current of 0.4 amperes at 250 volts uses 100 watts of electrical power and converts it into light and heat – in ordinary terms it is a 100-watt lamp. An electric motor that requires 6 amperes at the same voltage consumes 1,500 watts (1.5 kilowatts), equivalent to delivering about 2 horsepower of mechanical power.

electric ray another name for the ◊torpedo.

electrocardiogram (ECG) graphic recording of the electrical activity of the heart, as detected by electrodes placed on the skin. Electrocardiography is used in the diagnosis of heart disease.

electrochemical series or ***electromotive series*** list of chemical elements arranged in descending order of the ease with which they can lose electrons to form cations (positive ions). An element can be displaced (◊displacement reaction) from a compound by any element above it in the series.

electrochemistry the branch of science that studies chemical reactions involving electricity. The use of electricity to produce chemical effects, ◊electrolysis, is employed in many industrial processes, such as the manufacture of chlorine and the extraction of aluminium. The use of chemical reactions to produce electricity is the basis of electrical ◊cells, such as the dry cell and the Leclanché cell.

Since all chemical reactions involve changes to the electronic structure of atoms, all reactions are now recognized as electrochemical in nature. Oxidation, for example, was once defined as a process in which oxygen was combined with a substance, or hydrogen was removed from a compound; it is now defined as a process in which electrons are lost.

Electrochemistry is also the basis of new methods of destroying toxic organic pollutants. For example, the development of electrochemical cells that operate with supercritical water (see ◊fluid, supercritical) to combust organic waste materials.

electroconvulsive therapy (ECT) or ***electroshock therapy*** treatment mainly for severe ◊depression, given under anaesthesia and with a muscle relaxant. An electric current is passed through one or both sides of the brain to induce alterations in its electrical activity. The treatment can cause distress and loss of concentration and memory, and so there is much controversy about its use and effectiveness.

ECT was first used in 1938 but its success in treating depression lead to its excessive use for a wide range of mental illnesses against which it was ineffective. Its side effects included broken bones and severe memory loss.

The procedure in use today is much improved, using the minimum shock necessary to produce a seizure, administered under general anaesthetic with muscle relaxants to prevent spasms and fractures. It is the seizure rather than the shock itself that produces improvement. The smaller the shock administered the less damage there is to memory.

electrode any terminal by which an electric current passes in or out of a conducting substance; for example, the anode or cathode in a battery or the carbons in an arc lamp. The terminals that emit and collect the flow of electrons in thermionic ◊valves (electron tubes) are also called electrodes: for example, cathodes, plates, and grids.

electrodynamics the branch of physics dealing with electric charges, electric currents and associated forces. ◊Quantum electrodynamics (QED) studies the interaction between charged particles and their emission and absorption of electromagnetic radiation. This field combines quantum theory and relativity theory, making accurate predictions about subatomic processes involving charged particles such as electrons and protons.

electroencephalogram (EEG) graphic record of the electrical discharges of the brain, as detected by electrodes placed on the scalp. The pattern of electrical activity revealed by electroencephalography is helpful in the diagnosis of some brain disorders, in particular epilepsy.

electrolysis in chemistry, the production of chemical changes by passing an electric current through a solution or molten salt (the electrolyte), resulting in the migration of ions to the electrodes: positive ions (cations) to the negative electrode (cathode) and negative ions (anions) to the positive electrode (anode).

During electrolysis, the ions react with the electrode, either receiving or giving up electrons. The resultant atoms may be liberated as a gas, or deposited as a solid on the electrode, in amounts that are proportional to the amount of current passed, as discovered by English chemist Michael Faraday. For instance, when acidified water is electrolysed, hydrogen ions (H^+) at the cathode receive electrons to form hydrogen gas; hydroxide ions (OH^-) at the anode give up electrons to form oxygen gas and water.

One application of electrolysis is **electroplating**, in which a solution of a salt, such as silver nitrate ($AgNO_3$), is used and the object to be plated acts as the negative electrode, thus attracting silver ions (Ag^+). Electrolysis is used in many industrial processes, such as coating metals for vehicles and ships, and refining bauxite into aluminium; it also forms the basis of a number of electrochemical analytical techniques, such as polarography.

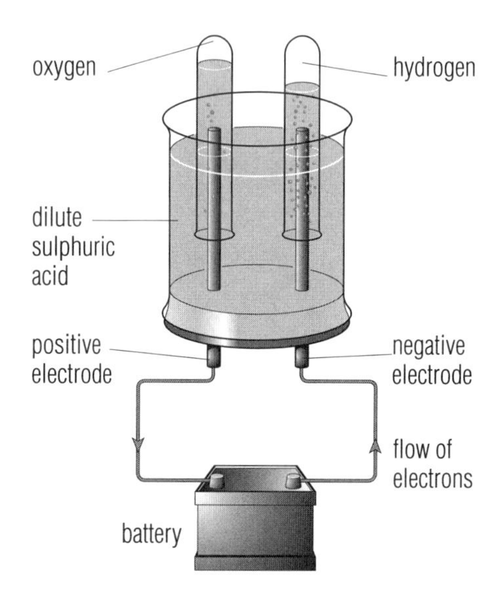

electrolysis Passing an electric current through acidified water (such as diluted sulphuric acid) breaks down the water into its constituent elements – hydrogen and oxygen.

electrolyte solution or molten substance in which an electric current is made to flow by the movement and discharge of ions in accordance with Faraday's laws of ◊electrolysis.

The term 'electrolyte' is frequently used to denote a substance that, when dissolved in a specified solvent, usually water, dissociates into ◊ions to produce an electrically conducting medium.

In medicine the term is often used for the ion itself (sodium or potassium, for example). Electrolyte balance may be severely disrupted in illness or injury.

electromagnet coil of wire wound around a soft iron core that acts as a magnet when an electric current flows through the wire. Electromagnets have many uses: in switches, electric bells, solenoids, and metal-lifting cranes.

electromagnetic field in physics, the region in which a particle with an ◊electric charge experiences a force. If it does so only when moving, it is in a pure **magnetic field**; if it does so when stationary, it is in an **electric field**. Both can be present simultaneously.

electromagnetic force one of the four fundamental ◊forces of nature, the other three being gravity, the strong nuclear force, and the weak nuclear force. The ◊elementary particle that is the carrier for the electromagnetic force is the photon.

electromagnetic induction in electronics, the production of an ◊electromotive force (emf) in a circuit by a change of magnetic flux through the circuit or by relative motion of the circuit and the magnetic flux. In a closed circuit an ◊induced current will be produced. All dynamos and generators make use of this effect. When magnetic tape is driven past the playback head (a small coil) of a tape-recorder, the moving magnetic field induces an emf in the head, which is then amplified to reproduce the recorded sounds.

If the change of magnetic flux is due to a variation in the current flowing in the same circuit, the phenomenon is known as self-induction; if it is due to a change of current flowing in another circuit it is known as mutual induction.

electromagnetic spectrum the complete range, over all wavelengths, of ◊electromagnetic waves.

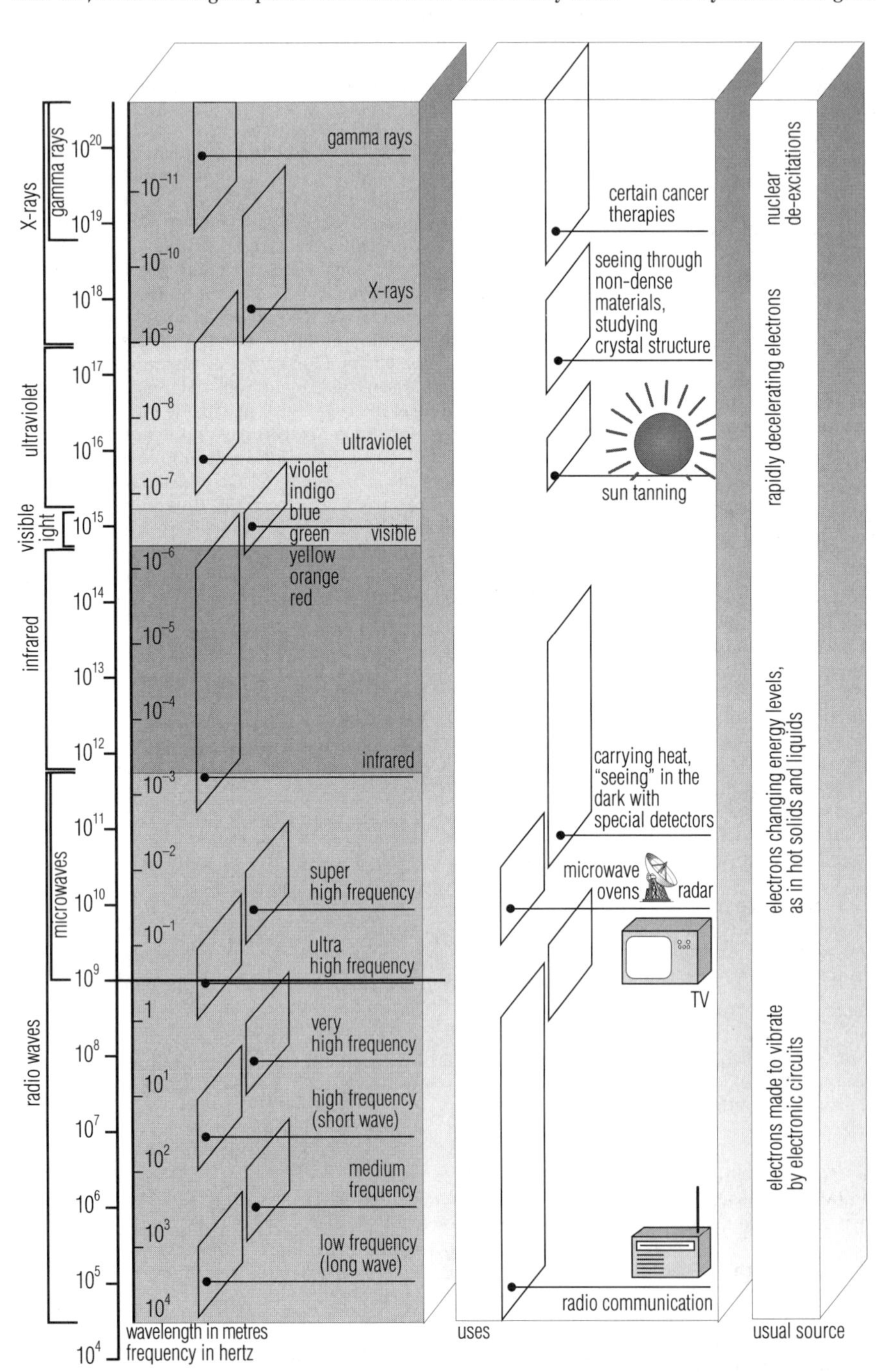

electromagnetic waves Radio waves have the lowest frequency. Infrared radiation, visible light, ultraviolet radiation, X-rays, and gamma rays have progressively higher frequencies.

To remember the different categories of radiation, in order of increasing wavelength:

CARY GRANT EXPECTS UNANIMOUS VOTES IN MOVIE REVIEWS TONIGHT

COSMIC, GAMMA, X-RAYS, ULTRAVIOLET, VISIBLE, INFRARED, MICROWAVE, RADIO, TELEVISION

electromagnetic system of units former system of absolute electromagnetic units (emu) based on the ◊c.g.s. system and having, as its primary electrical unit, the unit magnetic pole. It was replaced by ◊SI units.

electromagnetic waves oscillating electric and magnetic fields travelling together through space at a speed of nearly 300,000 km/186,000 mi per second. The (limitless) range of possible wavelengths or ◊frequencies of electromagnetic waves, which can be thought of as making up the **electromagnetic spectrum**, includes radio waves, infrared radiation, visible light, ultraviolet radiation, X-rays, and gamma rays.

electromotive force (emf) loosely, the voltage produced by an electric battery or generator in an electrical circuit or, more precisely, the energy supplied by a source of electric power in driving a unit charge around the circuit. The unit is the ◊volt.

electron stable, negatively charged ◊elementary particle; it is a constituent of all atoms, and a member of the class of particles known as leptons. The electrons in each atom surround the nucleus in groupings called shells; in a neutral atom the number of electrons is equal to the number of protons in the nucleus. This electron structure is responsible for the chemical properties of the atom (see atomic structure).

Electrons carry a charge of 1.602192×10^{-19} coulomb and have a mass of 9.109×10^{-31} kg, which is $\frac{1}{1836}$ times the mass of a ◊proton. A beam of electrons will undergo ◊diffraction (scattering) and produce interference patterns in the same way as ◊electromagnetic waves such as light; hence they may be regarded as waves as well as particles.

> *The electron is not as simple as it looks.*
>
> WILLIAM HENRY BRAGG British physicist. Recounted by Sir George Paget Thompson at electron diffraction conference 1967

electronegativity the ease with which an atom can attract electrons to itself. Electronegative elements attract electrons, so forming negative ions.

Linus Pauling devised an electronegativity scale to indicate the relative power of attraction of elements for electrons. Fluorine, the most nonmetallic element, has a value of 4.0 on this scale; oxygen, the next most nonmetallic, has a value of 3.5.

In a covalent bond between two atoms of different electronegativities, the bonding electrons will be located close to the more electronegative atom, creating a ◊dipole.

electron gun a part in many electronic devices consisting of a series of ◊electrodes, including a cathode for producing an electron beam. It plays an essential role in ◊cathode-ray tubes (television tubes) and ◊electron microscopes.

electronic banking in computing, system whereby a user can execute banking transactions via a modem, either directly or through an online service or the Internet.

electronic book in computing, software with or without specialized hardware that provides the equivalent of a book's worth of information. The term is used generally to apply even to simple text files created by scanning printed books or manuals such as those created and archived by ◊Project Gutenberg.

'Electronic Book' refers to a specific product released by Sony, a special player for small-sized CD-ROMs containing educational and reference material.

electronic cash in computing, see ◊e-cash.

electronic commerce in computing, business-to-business use of networks such as the Internet to handle legally binding transactions. Traditionally, electronic commerce has required expensive membership of an electronic data interchange (EDI) service. In the mid-1990s, electronic commerce began to shift to the Internet to take advantage of its global reach and inexpensive connections. By 1996 many legal issues remained to be resolved.

Electronic Communications Privacy Act in computing, US law passed in 1986 that protects the privacy of e-mail and other electronic communications.

electronic conferencing in computing, public discussions conducted on an online service or via ◊USENET; any participant may log in at any time and read the collected messages and add new ones.

Because of its time-independent, many-to-many nature, electronic conferencing can be used to provide some of the same functions as real-life meetings, classrooms, and unstructured socializing without requiring the participants to meet face-to-face. While electronic conferencing is no substitute for live interaction, it does allow people who are widely geographically separated or who might otherwise never meet to exchange ideas.

electronic data interchange (EDI) in computing, system for managing business-to-business transactions such as invoicing and ordering to eliminate the wastefulness of paper-based transaction systems.

Traditionally, most EDI systems relied on proprietary protocols and private data networks, with the disadvantages that individual systems were incompatible. The growth of the Intenet is now opening the way for the rapid adoption of global electronic commerce.

electronic flash discharge tube that produces a high-intensity flash of light, used for photography in dim conditions. The tube contains an inert gas such as krypton. The flash lasts only a few thousandths of a second.

Electronic Frontier Foundation (EFF) US organization that lobbies for the extension of civil liberties and constitutional rights into ◊cyberspace. It was founded by former Grateful Dead lyricist John Perry Barlow and Lotus founder Mitch Kapor in 1991 after a series of US raids on suspected computer hackers. Its offices are in San Francisco.

electronic mail or ***E-mail*** messages sent electronically from computer to computer via network connections such as ◊Ethernet or the Internet, or via telephone lines to a host system. Messages once sent are stored on the network or by the host system until the recipient picks them up.

Subscribers to an electronic mail system type messages in ordinary letter form on a word processor, or microcomputer, and 'drop' the letters into a central computer's memory bank by means of a computer/telephone connector (a ◊modem). The recipient 'collects' the letter by calling up the central computer and feeding a unique password into the system.

electronic point of sale (EPOS) system used in retailing in which a bar code on a product is scanned at the cash till and the information relayed to the store computer. The computer will then relay back the price of the item to the cash till. The customer can then be given an itemized receipt while the computer removes the item from stock figures.

EPOS enables efficient computer stock control and reordering as well as giving a wealth of information about turnover, profitability on different lines, stock ratios, and other important financial indicators.

electronic publishing the distribution of information using computer-based media such as ◊multimedia and ◊hypertext in the creation of electronic 'books'. Critical technologies in the development of electronic publishing were ◊CD-ROM, with its massive yet compact storage capabilities, and the advent of computer networking with its ability to deliver information instantaneously anywhere in the world.

electronics branch of science that deals with the emission of◊ electrons from conductors and ◊semiconductors, with the subsequent manipulation of these electrons, and with the construction of electronic devices. The first electronic device was the thermionic ◊valve, or vacuum tube, in which electrons moved in a vacuum, and led to such inventions as ◊radio, television, radar, and the digital ◊computer. Replacement of valves with the comparatively tiny and reliable ◊transistor from 1948 revolutionized electronic development. Modern electronic devices are based on minute ◊integrated circuits (silicon chips), wafer-thin crystal slices holding tens of thousands of electronic components.

By using solid-state devices such as integrated circuits, extremely complex electronic circuits can be constructed, leading to the development of ◊digital watches, pocket ◊calculators, powerful ◊microcomputers, and ◊word processors.

electronic shopping in computing, using an online service or Internet service such as the World Wide Web to select and buy merchandise.

electron microscope instrument that produces a magnified image by using a beam of ◊electrons instead of light rays, as in an optical ◊microscope. An **electron lens** is an arrangement of electromagnetic coils that control and focus the beam. Electrons are not visible to the eye, so instead of an eyepiece there is a fluorescent screen or a photographic plate on which the electrons form an

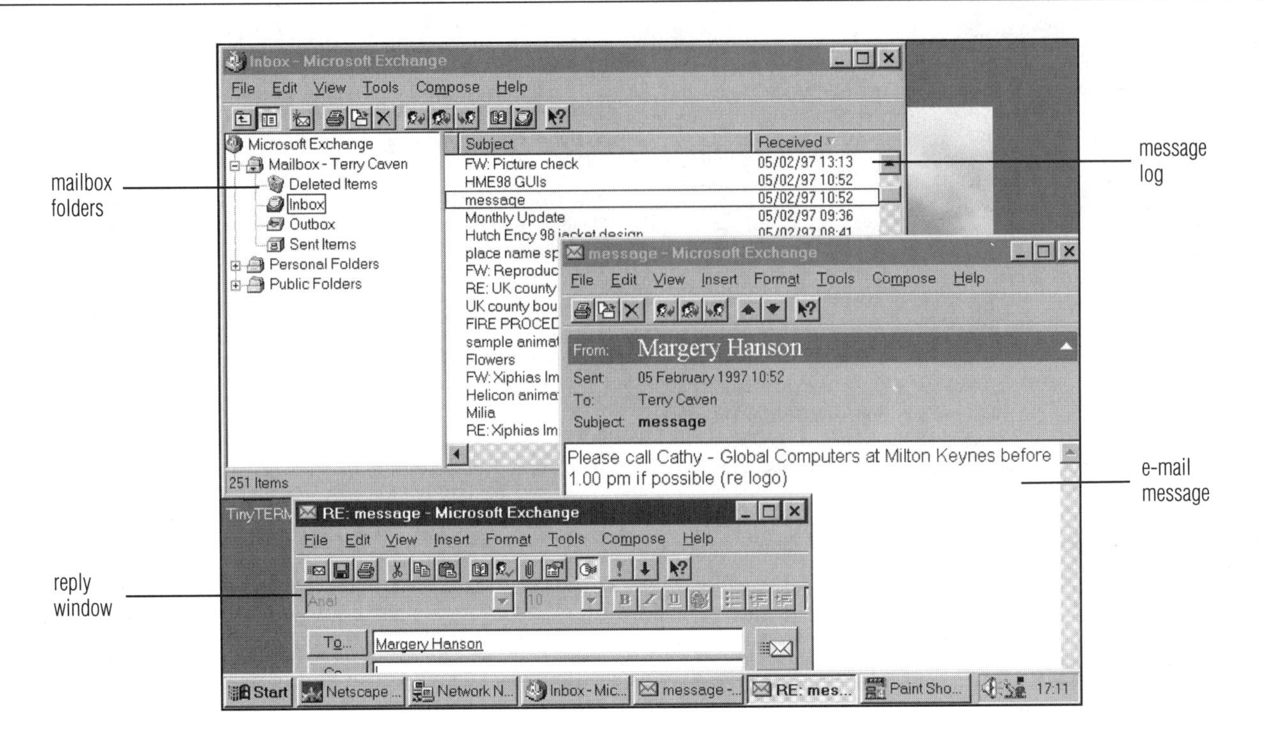

electronic mail *A typical E-mail user interface. Because messages can be created 'off-line' and are sent at high speed, line connection time, and therefore costs, can be kept to a minimum.*

image. The wavelength of the electron beam is much shorter than that of light, so much greater magnification and resolution (ability to distinguish detail) can be achieved. The development of the electron microscope has made possible the observation of very minute organisms, viruses, and even large molecules.

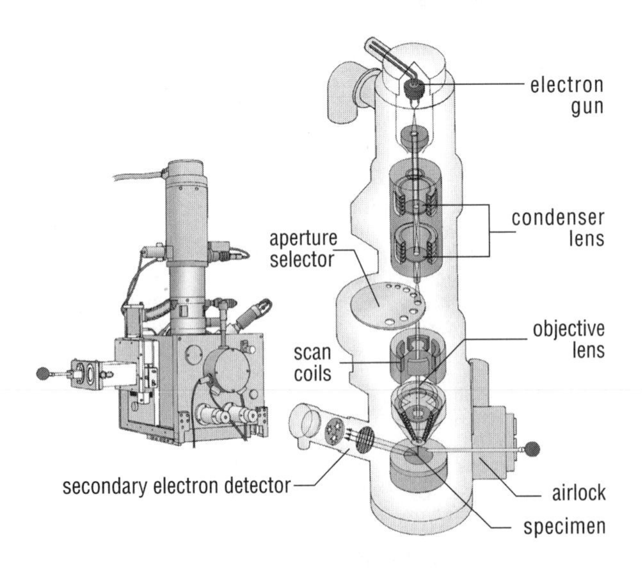

electron microscope *The scanning electron microscope. Electrons from the electron gun are focused to a fine point on the specimen surface by the lens systems. The beam is moved across the specimen by the scan coils. Secondary electrons are emitted by the specimen surface and pass through the detector, which produces an electrical signal. The signal is passed to an electronic console, and produces an image on a screen.*

A transmission electron ◊microscope passes the electron beam through a very thin slice of a specimen. A ◊scanning electron microscope looks at the exterior of a specimen. A ◊scanning transmission electron microscope (STEM) can produce a magnification of 90 million times. See also ◊atomic force microscope.

electron probe microanalyser modified ◊electron microscope in which the target emits X-rays when bombarded by electrons. Varying X-ray intensities indicate the presence of different chemical elements. The composition of a specimen can be mapped without the specimen being destroyed.

electrons, delocalized electrons that are not associated with individual atoms or identifiable chemical bonds, but are shared collectively by all the constituent atoms or ions of some chemical substances (such as metals, graphite, and ◊aromatic compounds).

A metallic solid consists of a three-dimensional arrangement of metal ions through which the delocalized electrons are free to travel. Aromatic compounds are characterized by the sharing of delocalized electrons by several atoms within the molecule.

electrons, localized a pair of electrons in a ◊covalent bond that are located in the vicinity of the nuclei of the two contributing atoms. Such electrons cannot move beyond this area.

electron volt unit (symbol eV) for measuring the energy of a charged particle (◊ion or ◊electron) in terms of the energy of motion an electron would gain from a potential difference of one volt. Because it is so small, more usual units are mega-(million) and giga-(billion) electron volts (MeV and GeV).

electrophoresis the ◊diffusion of charged particles through a fluid under the influence of an electric field. It can be used in the biological sciences to separate ◊molecules of different sizes, which diffuse at different rates. In industry, electrophoresis is used in paint-dipping operations to ensure that paint reaches awkward corners.

electroplating deposition of metals upon metallic surfaces by electrolysis for decorative and/or protective purposes. It is used in the preparation of printers' blocks, 'master' audio discs, and in many other processes.

A current is passed through a bath containing a solution of a salt of the plating metal, the object to be plated being the cathode (negative terminal); the anode (positive terminal) is either an inert substance or the plating metal. Among the metals most commonly used for plating are zinc, nickel, chromium, cadmium, copper, silver, and gold.

In **electropolishing**, the object to be polished is made the anode in an electrolytic solution and by carefully controlling conditions the high spots on the surface are dissolved away, leaving a high-quality stain-free surface. This technique is useful in polishing irregular stainless-steel articles.

electroporation in biotechnology, technique of introducing foreign DNA into pollen with a strong burst of electricity, used in creating genetically engineered plants.

electropositivity in chemistry, a measure of the ability of elements (mainly metals) to donate electrons to form positive ions. The greater the metallic character, the more electropositive the element.

electrorheological fluid another name for ◊smart fluid, a liquid suspension that gels when an electric field is applied across it.

electroscope apparatus for detecting ◊electric charge. The simple gold-leaf electroscope consists of a vertical conducting (metal) rod ending in a pair of rectangular pieces of gold foil, mounted inside and insulated from an earthed metal case or glass jar. An electric charge applied to the end of the metal rod makes the gold leaves diverge, because they each receive a similar charge (positive or negative) and so repel each other.

The polarity of the charge can be found by bringing up another charge of known polarity and applying it to the metal rod. A like charge has no effect on the gold leaves, whereas an opposite charge neutralizes the charge on the leaves and causes them to collapse.

electrostatic precipitator device that removes dust or other particles from air and other gases by electrostatic means. An electric discharge is passed through the gas, giving the impurities a negative electric charge. Positively charged plates are then used to attract the charged particles and remove them from the gas flow. Such devices are attached to the chimneys of coal-burning power stations to remove ash particles.

electrostatics the study of stationary electric charges and their fields (not currents). See ◊static electricity.

electrovalent bond another name for an ◊ionic bond, a chemical bond in which the combining atoms lose or gain electrons to form ions.

electrum naturally occurring alloy of gold and silver used by early civilizations to make the first coins, about the 6th century BC.

element substance that cannot be split chemically into simpler substances. The atoms of a particular element all have the same number of protons in their nuclei (their ◊atomic number). Elements are classified in the ◊periodic table of the elements. Of the known elements, 92 are known to occur in nature (those with atomic numbers 1–92). Those elements with atomic numbers above 92 do not occur in nature and are synthesized only, produced in particle accelerators. Of the elements, 81 are stable; all the others, which include atomic numbers 43, 61, and from 84 up, are radioactive.

Elements are classified as metals, nonmetals, or metalloids (weakly metallic elements) depending on a combination of their physical and chemical properties; about 75% are metallic. Some elements occur abundantly (oxygen, aluminium); others occur moderately or rarely (chromium, neon); some, in particular the radioactive ones, are found in minute (neptunium, plutonium) or very minute (technetium) amounts.

Symbols (devised by Swedish chemist Jöns ◊Berzelius) are used to denote the elements; the symbol is usually the first letter or letters of the English or Latin name (for example, C for carbon, Ca for calcium, Fe for iron, from the Latin *ferrum*). The symbol represents one atom of the element.

According to current theories, hydrogen and helium were produced in the ◊Big Bang at the beginning of the universe. Of the other elements, those up to atomic number 26 (iron) are made by nuclear fusion within the stars. The more massive elements, such as lead and uranium, are produced when an old star explodes; as its centre collapses, the gravitational energy squashes nuclei together to make new elements.

Two or more elements bonded together form a **compound** so that they cannot be separated by physical means. Compounds are held together by ionic or covalent bonds. The number of atoms of an element that combine together to form a molecule is it **atomicity**. A molecule of oxygen (O_2) has atomicity 2; sulphur (S_8) has atomicity 8.

element in mathematics, a member of a ◊set.

elementary particle in physics, a subatomic particle that is not made up of smaller particles, and so can be considered one of the fundamental units of matter. There are three groups of elementary particles: quarks, leptons, and gauge bosons.

Quarks, of which there are 12 types (up, down, charm, strange, top, and bottom, plus the antiparticles of each), combine in groups of three to produce heavy particles called baryons, and in groups of two to produce intermediate-mass particles called mesons. They and their composite particles are influenced by the strong nuclear force.

Leptons are light particles. Again, there are 12 types: the electron, muon, tau; their neutrinos, the electron neutrino, muon neutrino, and tau neutrino; and the antiparticles of each. These particles are influenced by the weak nuclear force.

Gauge bosons carry forces between other particles. There are four types: gluon, photon, weakon, and graviton. The gluon carries the strong nuclear force, the photon the electromagnetic force, the weakons the weak nuclear force, and the graviton the force of gravity (see ◊forces, fundamental).

elements, the four earth, air, fire, and water. The Greek philosopher Empedocles believed that these four elements made up the fundamental components of all matter and that they were destroyed and renewed through the action of love and discord.

This belief was shared by Aristotle who also claimed that the elements were mutable and contained specific qualities: cold and dry for earth, hot and wet for air, hot and dry for fire, and cold and wet for water. The transformation of the elements formed the basis of medieval alchemy, and the belief that base metals could be turned into gold. The theory of the elements prevailed until the 17th century when Robert Boyle redefined an element as a substance 'simple or unmixed, not made of other bodies' and proposed the existence of a greater number than four.

elephant large grazing mammal with thick, grey wrinkled skin, large ears, a long flexible trunk, and huge curving tusks. There are fingerlike projections at the end trunk used for grasping food and carrying it to its mouth. The trunk is also used for carrying water to the mouth. The elephant is herbivorous and, because of its huge size, much of its time must be spent feeding on leaves, shoots, bamboo, reeds, grasses and fruits and, where possible, cultivated crops such as maize and bananas. They are the largest living land animal.

Elephants usually live in herds containing between 20–40 females (cows), led by a mature, experienced cow. Most bull elephants live alone or in small groups; young males remain with the herd until they reach sexual maturity. Elephants have the longest gestation period of any animal (18–23 months between conception and birth) and usually produce one calf , which takes between 10–15 years to reach maturity. Elephants can live up to 60 years in

> TO DISTINGUISH BETWEEN INDIAN AND AFRICAN ELEPHANTS:
>
> INDIAN ELEPHANTS HAVE LITTLE EARS, AFRICAN ELEPHANTS HAVE LARGE EARS

ELEPHANT

http://www.fws.gov/~r9extaff/biologues/bio_elep.html

Presentation on the Asian and African elephant with details about their physical appearance, food, social life, status of protection, and more. It is a useful, quick reference tool for all those concerned about wildlife and its protection.

the wild, but those in captivity have been known to reach over 65. There are two species of elephant, the African and the Indian or Asian elephant.

Elephants have one of the lowest metabolic rates among placental mammals. Their tusks, which are initially tipped with enamel but later consist entirely of ivory, continue growing throughout life. They are preceded by milk tusks, which are shed at an early age.
species differences The African elephant is much the larger of the two specie, growing to heights of 4 m/13 ft and weighing up to 8 tonnes compared with the 2.7 m/9 ft and 4 tonnes of the Indian elephant. The African elephant has larger ears and longer tusks than its Asian relative (many Asian elephants, particularly the females are tuskless). The African elephant has a sloping forehead and a hollow back, whereas the Asian elephant has two domes on its forehead just above its ears, and an arched back. The trunk of the African elephant is ridged with two fingerlike projections; the Asian species only has a smooth trunk with one finger. The African species has four nails on its front foot and three on its hind (back) foot, whereas the Asian elephant has five on its front foot and four on its hind. African elephants live only in Africa, south of the Sahara desert. The Indian or Asian elephant can be found in parts of India and Southeast Asia.

Young Asian elephants are hairy, and in this respect somewhat resemble the extinct mammoth genus; the adults have smooth, nearly naked skin. The African species is of fiercer disposition and can move rapidly over rough ground.
endangered species Elephants are slaughtered for ivory, and this, coupled with the fact that they reproduce slowly and do not breed readily in captivity, is leading to their extinction. In Africa, overhunting caused numbers to collapse during the 1980s and the elephant population of E Africa is threatened with extinction. There were 1.3 million African elephants in 1981; fewer than 700,000 in 1988; 600,000 in 1990; and fewer than 580,000 in 1997. They were placed on the CITES list of most endangered species in 1989, and a world ban on trade in ivory was imposed in 1990, resulting in an apparent drop in poaching. In 1997, at the 10th CITES convention, the elephant was downlisted to CITES Appendix II (vulnerable) and the ban on ivory exportation was lifted.

The Asian elephant was also listed on the CITES endangered list; its wild population in 1996 was only 35,000–54,000. There are about 10,000 working elephants in Asia, most of which are caught from the wild and 'tamed' by starvation and brutality.

It was estimated in 1997 that in Sri Lanka alone elephants might be extinct within ten years. The country's government maintained that there were 4,000 animals left, whereas the Wildlife and Nature Protection Society of Sri Lanka claimed there were only 2,500.
classification Elephants belong to the phylum Chordata, class Mammalia (mammals), order Proboscidea, family Elephantidae. There are two species, the African elephant(*Loxodonta africana*), and the Indian or Asian elephant (*Elephas maximus*).

elephant bird another name for extinct members of the genus ◊Aepyornis.

elephantiasis in the human body, a condition of local enlargement and deformity, most often of a leg, though the scrotum, vulva, or breast may also be affected.

The commonest form of elephantiasis is the tropical variety (filariasis) caused by infestation by parasitic roundworms (filaria); the enlargement is due to damage of the lymphatic system which impairs immunity. This leaves sufferers susceptible to infection from bacteria and fungi, entering through skin splits. The swelling reduces dramatically if the affected area is kept rigorously clean and treated with antibiotic cream, combined with rest, after drug treatment has killed all filarial worms.

elephant's tusk shell burrowing marine mollusc with a tusk-shaped shell open at both ends, from the larger of which the long foot appears and is used in creeping movements. The elephant's tusk shell has tentacles around its mouth, lacks eyes and heart, and lives in muddy sand sometimes at great depths of the sea.
classification Elephant's tusk shells are in genus *Dentalium*, class Scaphopoda, phylum Mollusca.

elephant-trunk fish African freshwater fish *Gnathonemus petersii* in the order Mormyriformes. They grow to about 23 cm/9 in in length, have small eyes and fins and are mainly nocturnal. They generate an electric field which they use to detect obstacles. Elephant-trunk fish live in lakes in West and Central Africa.

Their brains are very large in proportion to their bodies: 3.1% of body mass compared with less than 1% for most fish and 2.3% for humans.

elevation a drawing to scale of one side of an object or building.

elevation of boiling point raising of the boiling point of a liquid above that of the pure solvent, caused by a substance being dissolved in it. The phenomenon is observed when salt is added to boiling water; the water ceases to boil because its boiling point has been elevated.

How much the boiling point is raised depends on the number of molecules of substance dissolved. For a single solvent, such as pure water, all substances in the same molecular concentration produce the same elevation of boiling point. The elevation e produced by the presence of a solute of molar concentration C is given by the equation $e = KC$, where K is a constant (called the ebullioscopic constant) for the solvent concerned.

elevator any mechanical device for raising or lowering goods or materials. Such a device used for lifting people in buildings is known as an elevator in the USA and as a lift in Britain.

elk large deer *Alces alces* inhabiting northern Europe, Asia, Scandinavia, and North America, where it is known as the moose. It is brown in colour, stands about 2 m/6 ft at the shoulders, has very large palmate antlers, a fleshy muzzle, short neck, and long legs. It feeds on leaves and shoots. In North America, the ◊wapiti is called an elk.

elkhound Norwegian dog resembling the ◊husky but much smaller. Its coat is thick, with a full undercoat and the tail is bushy. Elkhounds are grey, with a darker shade on the back, and are about 50 cm/20 in high, weighing approximately 22 kg/48 lb.

elk, Irish extinct Pleistocene species of deer *Cervus megaceros*, the bones of which are found in Irish bogs and also in certain parts of Great Britain and mainland Europe. It stood about 2 m/6.6 ft in height, and is characterized by the enormous size of its antlers, which sometimes had a spread of almost 3.3 m/11 ft.
classification The Irish Elk is in family Cervidae, order Artiodactyla.

It is closely allied to the present day fallow deer and became extinct after the coming of humans to Europe.

ellipse curve joining all points (loci) around two fixed points (foci) such that the sum of the distances from those points is always constant. The diameter passing through the foci is the major axis, and the diameter bisecting this at right angles is the minor axis. An ellipse is one of a series of curves known as ◊conic sections. A slice across a cone that is not made parallel to, and does not pass through, the base will produce an ellipse.

elliptical galaxy in astronomy, one of the main classes of ◊galaxy in the Hubble classification and characterized by a featureless elliptical profile. Unlike spiral galaxies, elliptical galaxies have very little gas or dust and no stars have recently formed within them. They range greatly in size from giant ellipticals, which are often found at the centres of clusters of galaxies and may be strong radio

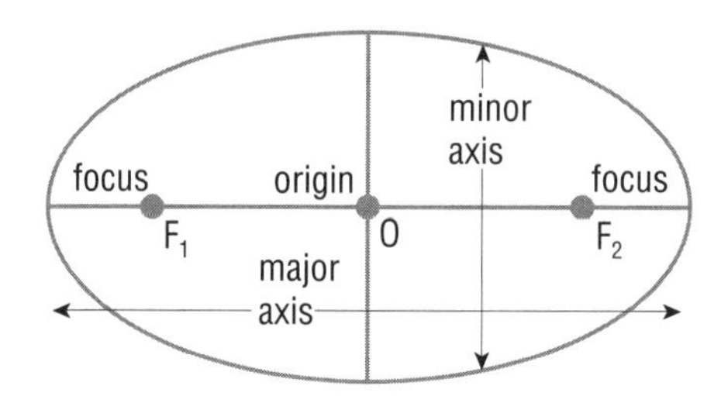

ellipse *Technical terms used to describe the ellipse; for all points on the ellipse, the sum of the distances from the two foci, F1 and F2, is the same.*

sources, to tiny dwarf ellipticals, containing about a million stars, which are the most common galaxies of any type. More than 60% of known galaxies are elliptical.

elm any of a group of trees found in temperate regions of the northern hemisphere and in mountainous parts of the tropics. All have doubly-toothed leaf margins and clusters of small flowers. (Genus *Ulmus,* family Ulmaceae.)

Species include the wych elm (*Ulmus glabra*), native to Britain; the North American white elm (*U. americana*); and the red or slippery elm (*U. fulva*). Most elms (apart from the wych elm) reproduce not by seed but by suckering (new shoots arising from the root system). This nonsexual reproduction results in an enormous variety of forms.

The fungus disease *Ceratocystis ulmi,* known as **Dutch elm disease** because of a severe outbreak in the Netherlands in 1924, has reduced the numbers of elm trees in Europe and North America. It is carried from tree to tree by beetles. Elms were widespread throughout Europe to about 4000 BC, when they suddenly disappeared and were not common again until the 12th century. This may have been due to an earlier epidemic of Dutch elm disease. In 1997 the US National Arboretum developed a Valley Forge elm that is resistant to the disease. It is expected to be available to the public in 2000.

Elm in computing, mail reader commonly used on online systems. It is typically found on older, text-based systems running under UNIX.

El Niño ***Spanish 'the child'*** warm ocean surge of the ◊Peru Current, so called because it tends to occur at Christmas, recurring about every 5–8 years in the eeastern Pacific off South America. It involves a change in the direction of ocean currents, which prevents the upwelling of cold, nutrient-rich waters along the coast of Ecuador and Peru, killing fishes and plants. It is an important factor in global weather.

El Niño is believed to be caused by the failure of trade winds and, consequently, of the ocean currents normally driven by these winds. Warm surface waters then flow in from the east. The phenomenon can disrupt the climate of the area disastrously, and has played a part in causing famine in Indonesia, drought and bush fires in the Galápagos Islands, rainstorms in California and South America, and the destruction of Peru's anchovy harvest and wildlife in 1982–83. El Niño contributed to algal blooms in Australia's drought-stricken rivers and an unprecedented number of typhoons in Japan in 1991. It is also thought to have caused the 1997 drought in Australia and contributed to certain ecological disasters such as bush fires in Indonesia.

El Niño usually lasts for about 18 months, but the 1990 occurrence lasted until June 1995; US climatologists estimated this duration to be the longest in 2,000 years. The last prolonged El Niño of 1939–41 caused extensive drought and famine in Bengal. It is understood that there might be a link between El Niño and ◊global warming.

EL NINO THEME PAGE

http://www.pmel.noaa.gov/toga-tao/el-nino/home.html

Wealth of scientific information about El Nino (a 'disruption of the ocean-atmosphere system in the tropical Pacific') with animated views of the monthly sea changes brought about by it, El Nino-related climate predictions, and forecasts from meteorological centres around the world. It also offers an illuminating FAQ section with basic and more advanced questions as well as an interesting historical overview of the phenomenon starting from 1550.

elongation in astronomy, the angular distance between the Sun and a planet or other solar-system object. This angle is 0° at ◊conjunction, 90° at ◊quadrature, and 180° at ◊opposition.

elution in chemistry, washing of an adsorbed substance from the adsorbing material; it is used, for example, in the separation processes of chromatography and electrophoresis.

elytra horny wing cases characteristic of beetles. The elytra are adapted from the beetles' forewings (only the hindwings are used for flight). They fold over the back, generally meeting in the middle in a straight line, and serve to protect the hindwings and the soft posterior parts of the body.

Elytra are also to be found in ◊earwigs.

In Hemiptera (true ◊bugs) the forewings are hardened over half their length and are known as **hemelytra**. In Orthoptera (crickets, grasshoppers, and locusts) the forewings are greatly thickened and are known as **tegmina**.

EMACS or, more properly, ***GNU EMACS*** in computing, a heavyweight ◊text editor used mainly by UNIX hackers. The name is dervied from Editing Macros, but is humorously, and recursively, said to stand for EMACS Makes A Computer Slow. EMACS was written by Richard Stallman at the MIT AI Lab and is published as ◊public-domain software, Emacs was created by the US Free Software Foundation.

e-mail (abbreviation for ***electronic mail***) in computing, a system that enables the users of a computer network to send messages to other users. The messages (which may contain enclosed text files, artwork, or multimedia clips) are usually placed in a reserved area ('mailbox') on a central computer until they are retrieved by the receiving user. Passwords are frequently used to prevent unauthorized access to stored messages (see ◊data security). The high speed of transmission for e-mail messages means that they cost less than comparable phone calls or faxes.

embolism blockage of a blood vessel by an obstruction called an embolus (usually a blood clot, fat particle, or bubble of air).

VISIBLE EMBRYO

http://visembryo.ucsf.edu/

Learn about the first four weeks of human development.

embryo early developmental stage of an animal or a plant following fertilization of an ovum (egg cell), or activation of an ovum by ◊parthenogenesis. In humans, the term embryo describes the fertilized egg during its first seven weeks of existence; from the eighth week onwards it is referred to as a fetus.

In animals the embryo exists either within an egg (where it is nourished by food contained in the yolk), or in mammals, in the ◊uterus of the mother. In mammals (except marsupials) the embryo is fed through the ◊placenta. The plant embryo is found within the seed in higher plants. It sometimes consists of only a few cells, but usually includes a root, a shoot (or primary bud), and one or two ◊cotyledons, which nourish the growing seedling.

embryology study of the changes undergone by an organism from its conception as a fertilized ovum (egg) to its emergence into the world at hatching or birth. It is mainly concerned with the

El Niño – The Christmas Child

BY NIGEL DUDLEY

In 1997, more tropical forest burned than at any other time in recorded history. Vast fires in Indonesia and the Amazon appeared on television screens all over the world, but fires also blazed throughout Africa, Papua New Guinea, and Australia. Unlike some temperate and boreal forests, most tropical moist forests do not readily burn under natural conditions. Reasons for the increase are complex, and include changes to the forest through over-logging and uncontrolled use of fire as a land management tool. However, the 1997 fires were given a fresh impetus by a catastrophic drought that affected vast areas of Africa, Asia, and South America. This drought was, in turn, associated with a hitherto fairly obscure climatic event known as El Niño (literally 'The Child') that principally affects large parts of the Pacific coast of South America.

Theory

El Niño is a current of warm water in the southeastern Pacific, which usually reaches the Pacific coast of South America around Christmas – hence the name. Although normally benign, it can periodically become extremely damaging when it is associated with another climatic phenomenon known as the Southern Oscillation, a complex series of events including changes in wind temperature, ocean currents, and sea levels. The combination, often known as an El Niño/Southern Oscillation or ENSO event, creates an invasion of warm water into usually cool areas, causing wet and stormy weather in the west Pacific and sometimes as far as North America, and drought conditions in Africa, Brazil, Australia, and parts of Asia and the Pacific.

Immediate impact

The immediate impact on the Latin American coastline can be catastrophic. Warmer water kills plankton, squid, and smaller fish such as anchovy and sardine. Lack of food in turn kills larger fish such as herring and hake. This also sometimes causes extreme short-term declines in the seabird colonies found along the coast and on offshore islands – which may also have their nests inundated by rising waters. Coral reefs sometimes suffer bleaching, perhaps as a result of warmer water. In human terms, fisheries face extreme hardship and agricultural crops are likely to collapse. Along the coasts of Chile and off the southern USA, sardine populations can virtually disappear for a period. Particularly spectacular ENSO events occurred in 1940–41, 1957–58, and 1982–83. Loss of sardines caused a collapse in the industry in California during the 1940s, as recorded in John Steinbeck's *Cannery Row*.

Wider implications

The wider climatic conditions have other impacts. Increased forest fires have been mentioned above, but it is only in recent years that their full significance has been recognized. For example, over several months in the summer of 1997, an area of Southeast Asia from the Philippines to Australia was enveloped in smog caused by forest fires on the Indonesian islands of Java, Borneo, Sulawesi, Irian Jaya, and Sumatra. Over 2 million hectares of forests and other land were destroyed. More than 40,000 Indonesians became ill as a result and over a million suffered eye infections; smog also resulted in plane crashes and shipping accidents. Primary forest and at least 19 protected areas were damaged in Indonesia, and endangered species such as the orang-utan were threatened further. Rains after the fire caused soil erosion and consequent damage to offshore fisheries as coral was smothered with debris. Business, including tourism, suffered badly and initial estimates put costs at a massive US$20 billion. Although most fires were set deliberately – often illegally – by commercial interests such as plantation owners, impacts were exacerbated by the El Niño climatic effect. ENSO events are also associated with particularly bad burning seasons in Australia and the Amazon.

Drought also causes direct human hardship as a result of crop failures. At the end of 1997, hundreds of thousands of people in the Indonesian-controlled area of Irian Jaya and in neighbouring Papua New Guinea faced starvation after the worst drought for 50 years. Associated fires in Papua New Guinea created such a dense pall of smoke that pilots bringing in emergency food supplies were sometimes unable to land.

The future

Currently, El Niño and the associated ENSO events appear to be growing both more frequent and more severe. An increasing number of scientists are linking this to changes caused by pollution-related climatic change, although others think that it may be largely the result of natural fluctuations that will reverse later. Whatever the cause, the impacts are becoming more intense, and off the coast of western South America and the southern USA marine life seems to be undergoing longer-term changes. Since the 1950s, there has been an 80 percent decline in zooplankton along California's coast between San Diego and San Francisco, accompanied by a 1.2–1.6°C temperature rise. This decline is affecting seabird populations, particularly the sooty shearwater, which has declined by 90 %.

Timing of ENSO events is also becoming more erratic. In the past, severe El Niño events tended to last around 18 months, and would be separated by long breaks so that marine and bird life could become re-established. However, the 1990 ENSO event appears to be continuing, with a series of unpredictable peaks and troughs. It apparently reached maturity in early 1992 and started to decline in the expected manner, but untypically increased in strength again the following November. It stayed through 1994 and it appears as if the 1997 events may be simply another continuation. A link between climate change and changes in ENSO events is suspected but is as yet unproven.

Climate change specialists are also interested in El Niño because the main oceanic effects – a rise in sea level, higher water temperatures, and reduced offshore flow – are those most often associated with predictions relating to climate change. Scientists and fisheries experts have studied El Niño to find out what might happen under conditions of global warming.

Given their importance there is a strong incentive to predict ENSO events. For example, scientists are currently using the occurrence of a temperature-sensitive algae, *Emiliana huxleyi*, in layers of sediment to create an accurate profile of changing sea temperatures in the eastern Pacific between 1915 and 1988. However, many aspects of this extremely important phenomenon remain as elusive as when early settlers named the warming after the birth of the Christ child, which they were celebrating when it arrived.

---

changes in cell organization in the embryo and the way in which these lead to the structures and organs of the adult (the process of ◊differentiation).

Applications of embryology include embryo transplants, both commercial (for example, in building up a prize dairy-cow herd quickly at low cost) and in obstetric medicine (as a method for helping couples with fertility problems to have children).

embryo sac large cell within the ovule of flowering plants that represents the female ◊gametophyte when fully developed. It typically contains eight nuclei. Fertilization occurs when one of these nuclei, the egg nucleus, fuses with a male ◊gamete.

emerald a clear, green gemstone variety of the mineral ◊beryl. It occurs naturally in Colombia, the Ural Mountains in Russia,

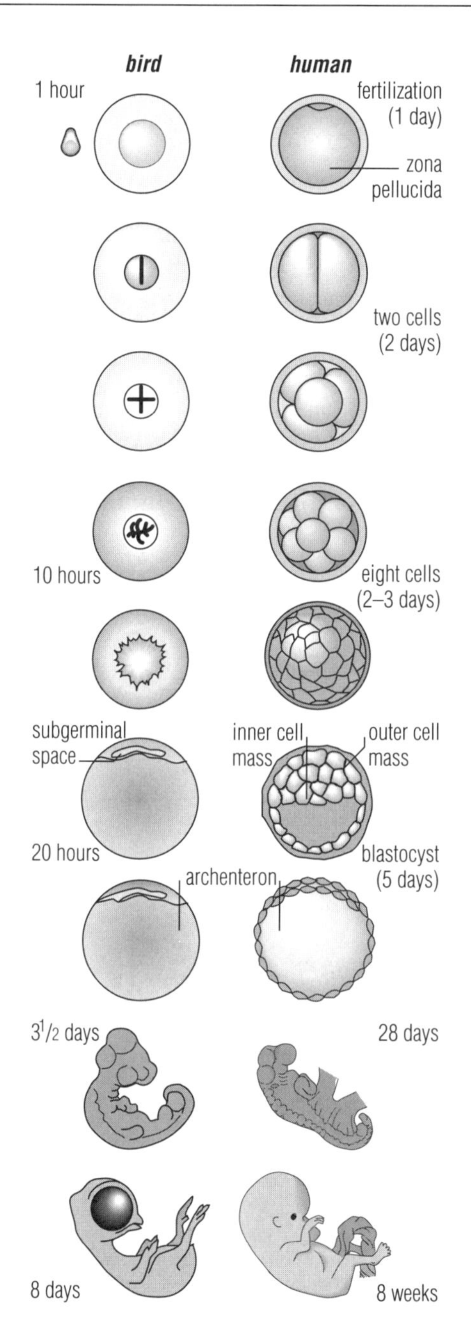

embryo The development of a bird and a human embryo. In the human, division of the fertilized egg, or ovum, begins within hours of conception. Within a week, a hollow, fluid-containing ball – a blastocyte – with a mass of cells at one end has developed. After the third week, the embryo has changed from a mass of cells into a recognizable shape. At four weeks, the embryo is 3 mm/0.1 in long, with a large bulge for the heart and small pits for the ears. At six weeks, the embryo is 1.5 cm/0.6 in long with a pulsating heart and ear flaps. By the eighth week, the embryo (now technically a fetus) is 2.5 cm/1 in long and recognizably human, with eyelids and small fingers and toes.

Zimbabwe, and Australia. The green colour is caused by the presence of the element chromium in the beryl.

emergent properties features of a system that are due to the way in which its components are structured in relation to each other, rather than to the individual properties of those components. Thus the distinctive characteristics of ◊chemical compounds are emergent properties of the way in which the constituent elements are organized, and cannot be explained by the particular properties of those elements taken in isolation. In biology, ◊ecosystem stability is an emergent property of the interaction between the constituent species, and not a property of the species themselves.

emery black to greyish form of impure ◊corundum that also contains the minerals magnetite and hematite. It is used as an ◊abrasive.

emetic any substance administered to induce vomiting. Emetics are used to empty the stomach in many cases of deliberate or accidental drug overdose. The most frequently used is ipecacuanha.

emf in physics, abbreviation for ◊electromotive force.

emission line in astronomy, bright line in the spectrum of a luminous object caused by ◊atoms emitting light at sharply defined ◊wavelengths.

emission spectroscopy in analytical chemistry, a technique for determining the identity or amount present of a chemical substance by measuring the amount of electromagnetic radiation it emits at specific wavelengths; see ◊spectroscopy.

emoticon (contraction of ***emotion and icon***) in computing, symbol composed of punctuation marks designed to express some form of emotion in the form of a human face. Emoticons were invented by ◊e-mail users to overcome the fact that communication using text only cannot convey nonverbal information (body language or vocal intonation) used in ordinary speech.

The following examples should be viewed sideways::-) smiling :-O shouting :-(glum 8-) wearing glasses and smiling.

emotion in psychology, a powerful feeling; a complex state of body and mind involving, in its bodily aspect, changes in the viscera (main internal organs) and in facial expression and posture, and in its mental aspect, heightened perception, excitement and, sometimes, disturbance of thought and judgement. The urge to action is felt and impulsive behaviour may result.

emphysema incurable lung condition characterized by disabling breathlessness. Progressive loss of the thin walls dividing the air spaces (alveoli) in the lungs reduces the area available for the exchange of oxygen and carbon dioxide, causing the lung tissue to expand. The term 'emphysema' can also refer to the presence of air in other body tissues.

Emphysema is most often seen at an advanced stage of chronic ◊bronchitis, although it may develop in other long-standing diseases of the lungs. It destroys lung tissue, leaving behind scar tissue in the form of air blisters called bullae. As the disease progresses, the bullae occupy more and more space in the chest cavity, inflating the lungs and causing severe breathing difficulties. The bullae may be removed surgically, and since early 1994 US trials have achieved measured success using lasers to eliminate them in a procedure called lung-reduction pneumenoplasty (LRP). Lasers are particularly useful where the emphysema is diffuse and bullae are interspersed within healthy tissue. As LRP is a less invasive process, survival rates are improved (90% compared with 75% for conventional surgery) and patients recover quicker.

EMS in computing, abbreviation for ◊***expanded memory specification***.

emu flightless bird *Dromaius novaehollandiae*, family Dromaiidae, order Casuariidae, native to Australia. It stands about 1.8 m/6 ft high and has coarse brown plumage, small rudimentary

EMU

While a male emu is incubating his eggs, he cannot move from the nest to eat or drink. He must survive on his existing fat reserves for eight weeks. When the chicks are hatched they remain in their father's care for at least seven months.

wings, short feathers on the head and neck, and powerful legs, which are well adapted for running and kicking.

The female has a curious bag or pouch in the windpipe that enables her to emit a characteristic loud booming note. Emus are monogamous, and the male wholly or partially incubates the eggs.

In Western Australia emus are farmed for their meat, skins, feathers, and oil.

emulator in computing, an item of software or firmware that allows one device to imitate the functioning of another. Emulator software is commonly used to allow one make of computer to run programs written for a different make of computer. This allows a user to select from a wider range of ◊applications programs, and perhaps to save money by running programs designed for an expensive computer on a cheaper model.

Many printers contain emulator firmware that enables them to imitate Hewlett Packard and Epson printers, because so much software is written to work with these widely used machines.

emulsifier food additive used to keep oils dispersed and in suspension, in products such as mayonnaise and peanut butter. Egg yolk is a naturally occurring emulsifier, but most of the emulsifiers in commercial use today are synthetic chemicals.

emulsion a stable dispersion of a liquid in another liquid – for example, oil and water in some cosmetic lotions.

encapsulate in computing, term used to describe the technique that uses one ◊protocol as an envelope for another for transmission across a network.

encapsulated PostScript (EPS) computer graphics file format used by the ◊PostScript page-description language. It is essentially a PostScript file with a special structure designed for use by other applications.

encephalin a naturally occurring chemical produced by nerve cells in the brain that has the same effect as morphine or other derivatives of opium, acting as a natural painkiller. Unlike morphine, encephalins are quickly degraded by the body, so there is no build-up of tolerance to them, and hence no addiction. Encephalins are a variety of ◊peptides, as are ◊endorphins, which have similar effects.

encephalitis inflammation of the brain, nearly always due to viral infection but it may also occur in bacterial and other infections. It varies widely in severity, from shortlived, relatively slight effects of headache, drowsiness, and fever to paralysis, coma, and death.

Encke's comet comet with the shortest known orbital period, 3.3 years. It is named after German mathematician and astronomer Johann Franz Encke (1791–1865), who calculated its orbit in 1819 from earlier sightings.

It was first seen in 1786 by the French astronomer Pierre Méchain (1744–1804). It is the parent body of the Taurid meteor shower and a fragment of it may have hit the Earth in the ◊Tunguska Event in 1908.

In 1913, it became the first comet to be observed throughout its entire orbit when it was photographed near ◊aphelion (the point in its orbit furthest from the Sun) by astronomers at Mount Wilson Observatory in California, USA.

endangered species plant or animal species whose numbers are so few that it is at risk of becoming extinct. Officially designated endangered species are listed by the ◊International Union for the Conservation of Nature (IUCN).

Endangered species are not a new phenomenon; extinction is an integral part of evolution. The replacement of one species by another usually involves the eradication of the less successful form, and ensures the continuance and diversification of life in all forms. However, extinctions induced by humans are thought to be destructive, causing evolutionary dead-ends that do not allow for succession by a more fit species. The great majority of recent extinctions have been directly or indirectly induced by humans; most often by the loss, modification, or pollution of the organism's habitat, but also by hunting for 'sport' or for commercial purposes.

According to a 1995 report to Congress by the US Fish and Wildlife Service, although seven of the 893 species listed as endangered under the US Endangered Species Act 1968–93 have become extinct, 40% are no longer declining in number. In February 1996, a private conservation group, Nature Conservancy, reported around 20,000 native US plant and animal species to be rare or imperilled.

According to the Red Data List of endangered species, published in 1996 by the IUCN, 25% of all mammal species (including 46% of primates, 36% of insectivores, and 33% of pigs and antelopes), and 11% of all bird species are threatened with extinction.

endive cultivated annual plant, the leaves of which are used in salads and cooking. One variety has narrow, curled leaves; another has wide, smooth leaves. It is related to ◊chicory. (*Cichorium endivia,* family Compositae.)

endocrine gland gland that secretes hormones into the bloodstream to regulate body processes. Endocrine glands are most highly developed in vertebrates, but are also found in other animals, notably insects. In humans the main endocrine glands are the pituitary, thyroid, parathyroid, adrenal, pancreas, ovary, and testis.

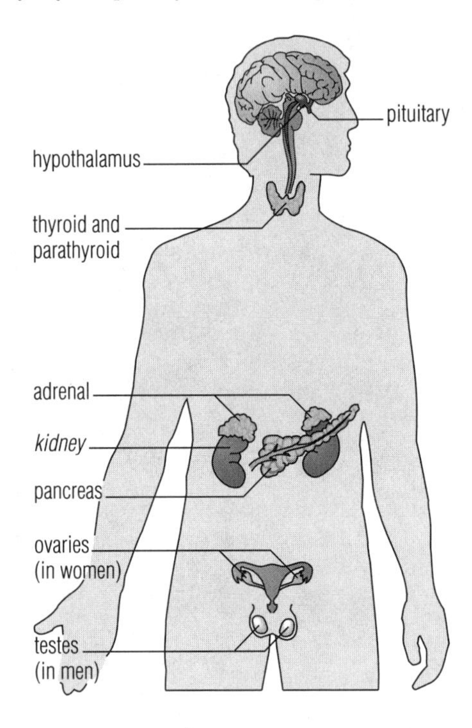

***endocrine gland** The main human endrocrine glands. These glands produce hormones – chemical messengers – which travel in the bloodstream to stimulate certain cells.*

endolymph fluid found in the inner ◊ear, filling the central passage of the cochlea as well as the semicircular canals.

Sound waves travelling into the ear pass eventually through the three small bones of the middle ear and set up vibrations in the endolymph. These vibrations are detected by receptors in the cochlea, which send nerve impulses to the hearing centres of the brain.

endometriosis common gynaecological complaint in which patches of endometrium (the lining of the womb) are found outside the uterus.

This ectopic (abnormally positioned) tissue is present most often in the ovaries, although it may invade any pelvic or abdominal site, as well as the vagina and rectum. Endometriosis may be treated with analgesics, hormone preparations, or surgery. Between 30 and 40% of women treated for infertility are suffering from the condition.

endoparasite ◊parasite that lives inside the body of its host.

endoplasm inner, liquid part of a cell's ◊cytoplasm.

endoplasmic reticulum (ER) a membranous system of tubes, channels, and flattened sacs that form compartments within ◊eukaryotic cells. It stores and transports proteins within cells and also carries various enzymes needed for the synthesis of ◊fats. The ◊ribosomes, or the organelles that carry out protein synthesis, are attached to parts of the ER.

Under the electron microscope, ER looks like a series of channels and vesicles, but it is in fact a large, sealed, baglike structure crumpled and folded into a convoluted mass. The interior of the 'bag', the ER lumen, stores various proteins needed elsewhere in the cell, then organizes them into transport vesicles formed by a small piece of ER membrane budding from the main membrane.

endorphin natural substance (a polypeptide) that modifies the action of nerve cells. Endorphins are produced by the pituitary gland and hypothalamus of vertebrates. They lower the perception of pain by reducing the transmission of signals between nerve cells.

Endorphins not only regulate pain and hunger, but are also involved in the release of sex hormones from the pituitary gland. Opiates act in a similar way to endorphins, but are not rapidly degraded by the body, as natural endorphins are, and thus have a long-lasting effect on pain perception and mood. Endorphin release is stimulated by exercise.

endoscopy examination of internal organs or tissues by an instrument allowing direct vision. An endoscope is equipped with an eyepiece, lenses, and its own light source to illuminate the field of vision. The endoscope used to examine the digestive tract is a flexible fibreoptic instrument swallowed by the patient.

There are various types of endoscope in use – some rigid, some flexible – with names prefixed by their site of application (for example, bronchoscope and laryngoscope). The value of endoscopy is in permitting diagnosis without the need for exploratory surgery. Biopsies (tissue samples) and photographs may be taken by way of the endoscope as an aid to diagnosis, or to monitor the effects of treatment. Some surgical procedures can be performed using fine instruments introduced through the endoscope. Keyhole surgery is increasingly popular as a cheaper, safer option for some conditions than conventional surgery.

endoskeleton the internal supporting structure of vertebrates, made up of cartilage or bone. It provides support, and acts as a system of levers to which muscles are attached to provide movement. Certain parts of the skeleton (the skull and ribs) give protection to vital body organs.

Sponges are supported by a network of rigid, or semirigid, spiky structures called spicules; a bath sponge is the proteinaceous skeleton of a sponge.

endosperm nutritive tissue in the seeds of most flowering plants. It surrounds the embryo and is produced by an unusual process that parallels the ◊fertilization of the ovum by a male gamete. A second male gamete from the pollen grain fuses with two female nuclei within the ◊embryo sac. Thus endosperm cells are triploid (having three sets of chromosomes); they contain food reserves such as starch, fat, and protein that are utilized by the developing seedling.

In 'non-endospermic' seeds, absorption of these food molecules by the embryo is completed early, so that the endosperm has disappeared by the time of germination.

endotherm 'warm-blooded', or ***homeothermic***, animal. Endotherms have internal mechanisms for regulating their body temperatures to levels different from the environmental temperature. See ◊homeothermy.

endothermic reaction chemical reaction that requires an input of energy in the form of heat for it to proceed; the energy is absorbed from the surroundings by the reactants.

The dissolving of sodium chloride in water and the process of photosynthesis are both endothermic changes. See ◊energy of reaction.

endotoxin in biology, heat stable complex of protein and lipopolysaccharide that is produced following the death of certain bacteria. Endotoxins are typically produced by the Gram negative bacteria and can cause fever. They can also cause shock by rendering the walls of the blood vessels permeable so that fluid leaks into the tissues and blood pressure falls sharply.

end user the user of a computer program; in particular, someone who uses a program to perform a task (such as accounting or playing a computer game), rather than someone who writes programs (a programmer).

Energiya most powerful Soviet space rocket, first launched 15 May 1987.

Used to launch the Soviet space shuttle, the Energiya booster is capable, with the use of strap-on boosters, of launching payloads of up to 190 tonnes into Earth orbit.

energy capacity for doing ◊work. Energy can exist in many different forms. For example, potential energy (PE) is energy deriving from position; thus a stretched spring has elastic PE, and an object raised to a height above the Earth's surface, or the water in an elevated reservoir, has gravitational PE. Moving bodies possess kinetic energy (KE). Energy can be converted from one form to another, but the total quantity in a system stays the same (in accordance with the ◊conservation of energy principle). Energy cannot be created or destroyed. For example, as an apple falls it loses gravitational PE but gains KE.

Although energy is never lost, after a number of conversions it tends to finish up as the kinetic energy of random motion of molecules (of the air, for example) at relatively low temperatures. This is 'degraded' energy that is difficult to convert back to other forms.

energy is the capacity to do work A body with no energy can do no ◊work. For example, a flat battery in a torch will not light the torch. If the battery is fully charged, it should contain enough chemical energy to do the work involved in illuminating the torch bulb. When one body A does work on another body B, A transfers energy to B. The energy transferred is equal to the work done by A on B. Energy is therefore measured in ◊joules. The rate of doing work or consuming energy is called power and is measured in ◊watts (joules per second).

energy types and transfer Energy can be converted from any form into another. A ball resting on a slope possesses ◊potential energy that is gradually changed into ◊kinetic energy of rotation and translation as the ball rolls down. As a pendulum swings, energy is constantly being changed from a potential form at the highest points of the swing to kinetic energy at the lowest point. At positions in between these two extremes, the system possesses both kinetic and potential energy in varying proportions.

A weightlifter changes chemical energy from his muscles into potential energy of the weight when the weight is lifted. If the weightlifter releases the weight, the potential energy is converted to kinetic energy as it falls, and this in turn is converted to heat energy and sound energy as it hits the floor. A lump of coal and a tank of petrol, together with the oxygen needed for their combustion, have chemical energy. Other sorts of energy include electrical and nuclear energy, and light and sound.

resources So-called energy resources are stores of convertible energy. Nonrenewable resources include the fossil fuels (coal, oil, and gas) and nuclear-fission 'fuels' – for example, uranium 235. The term 'fuel' is used for any material from which energy can be obtained. We use up fuel reserves such as coal and oil, and convert the energy they contain into other, useful forms. The chemical energy released by burning fuels can be used to do work.

Renewable resources, such as wind, tidal, and geothermal power, have so far been less exploited. Hydroelectric projects are well established, and wind turbines and tidal systems are being developed.

energy conservation and efficiency All forms of energy are interconvertible by appropriate processes. Energy is transferred from one form to another, but the sum total of the energy after the conversion is always the same as the initial energy. This is the principle of conservation of energy. This principle can be illustrated by the use of energy flow diagrams, called Sankey diagrams, which show the energy transformations that take place.

Energy Resources

BY PETER LAFFERTY

Humans are using up the world's energy resources in a way no other animal has ever done. We use them to provide light and heating in our homes, to plough the land, to cook our food, to travel, to run our factories, and in countless other ways. Whether we are rural workers in a developing country or urban workers in a wealthy industrial country, we all need energy, although the sources of the energy and the amounts used vary greatly from one society to another.

Sources of energy

There are different forms or types of energy. Fuels such as coal, oil (petroleum), and wood contain chemical energy. When these fuels are burnt, the chemical energy changes to heat and light energy. Electricity is the most important form of energy in the industrialized world, because it can be transported over long distances via cables and transmission lines. It is also a very convenient form of energy, since it can power a wide variety of household appliances and industrial machinery. It is produced by converting the chemical energy from coal, oil, or natural gas in power stations.

Energy resources fall into two broad groups: renewable and nonrenewable. Renewable resources are those which replenish themselves naturally and will either always be available – hydroelectric power, solar energy, wind and wave power, tidal energy, and geothermal energy – or will continue to be available provided supplies are given sufficient time to replenish themselves – peat and firewood. Nonrenewable resources are those of which there are limited supplies and which once used are gone forever. These include coal, oil, natural gas, and uranium.

Fossil fuels

Coal, oil, and natural gas are called fossil fuels because they are the fossilized remains of plants and animals that lived hundreds of millions of years ago. Burning fossil fuels releases chemicals that cause acid rain, and is gradually increasing the carbon dioxide in the atmosphere, causing global warming.

Fossil fuel resources are not evenly distributed around the world. Over half the world's known oil reserves are in the Middle East; about 40% of the reserves of natural gas are in the Commonwealth of Independent States (CIS), and 25% in the Middle East. About two-thirds of the world's coal is shared between North America, the CIS, and China.

Uranium

Uranium is a radioactive metallic element and a very concentrated source of energy; large reserves are found in Australia, North America, and South Africa. Used to produce electricity in a nuclear power station, a single ton of uranium can produce as much energy as 15,000 tons of coal, or 10,000 tons of oil. Used in the type of nuclear power station now in operation, the world's known uranium supplies have about the same energy content as the known oil reserves. However, these power stations, known as thermal stations or reactors, use only a small part of the energy available in uranium. The next generation of reactors, known as fast or breeder reactors, release virtually all its energy. These reactors would increase the world's uranium energy reserves by sixty times. However, although nuclear power stations do not produce carbon dioxide or cause acid rain, they do produce radioactive waste that is dangerous and difficult to process or store safely.

Solar energy

Many renewable resources take advantage of the energy in sunlight. The Sun's energy can be tapped directly by photovoltaic cells that convert light into electricity. Other solar energy plants use mirrors to direct sunlight onto pipes containing a liquid. The liquid boils and is used to drive an electricity generator. The Sun's energy also drives the wind and waves, so energy produced by wind farms and wave-driven generators is also derived from the Sun.

Gravitational energy

Hydroelectricity and tidal power stations make use of gravitational forces. The Earth's gravity pulls water downward through the turbines in a hydroelectric power station. In a tidal power station, the Moon's gravity lifts water as the tides rise, giving the water potential energy (energy due to position) which is released as the water flows through a turbine. Geothermal energy (the heat energy of hot rocks deep beneath the Earth's surface) is due to gravity compressing and heating the rocks when the Earth formed.

The worldwide energy pie

Globally, the largest contributions to current energy resources come from oil (31%), coal (26%), and natural gas (19%). Renewable energy currently supplies about 20% of the world's energy needs, with hydroelectricity supplying 6% of the world's needs and traditional biofuels (firewood, crop wastes, peat, and dung) supplying 12%. A small contribution is made by new renewables, such as the conversion of crops such as sugar into alcohol fuel and the burning of waste material.

The contribution solar, wave, tidal, and geothermal power can make to the world's energy resources is currently limited. This is because renewable energy depends on the development of means of capturing and concentrating it. In addition, renewable energy is not always available when needed – rivers can dry up, the wind does not always blow.

Future demands

It is clear that, in the future, demand for energy will be higher than at present, due to population growth and increased industrialization. Furthermore, the energy available must be at a reasonable cost or economic growth will be held back. This is especially important for developing countries, where the inability to meet high-energy costs hinders development.

Future solutions

In principle, known resources of nonrenewable energy should be sufficient for several hundred years or more. At the present rate of consumption, oil reserves will last about 40 years; gas reserves will last about 60 years; coal reserves will last about 250 years; and uranium reserves, if used in fast reactors, would last for more than 1,000 years. It is also likely that further fossil-fuel reserves will be discovered as currently known supplies run out. However, in practice, the outlook is uncertain. Increasing concern about pollution might make dirty coal-fired power stations unacceptable in the future.

One alternative is to make greater use of nuclear power, moving to fast reactors and then developing nuclear fusion plants that would mimic the power production process found in the Sun. However, anxiety about safety and waste disposal is already limiting the use of nuclear energy, so it is unlikely to provide an answer in the future.

There is considerable room for development in the use of renewable resources, but with most of the world's energy production based around nonrenewable fuel supplies, the widespread introduction of efficient renewable energy will require a complete restructuring of the ways we produce and use energy.

When a petrol engine is used to power a car, about 75% of the energy from the fuel is wasted. The total energy input equals the total energy output, but a lot of energy is wasted as heat so that the engine is only about 25% efficient. The combustion of the petrol-air mixture produces heat energy as well as kinetic energy. All forms of energy tend to be transformed into heat and can not then readily be converted into other, useful forms of energy.

heat transfer A difference in temperature between two objects in thermal contact leads to the transfer of energy as ◊heat. Heat is energy transferred due to a temperature difference. Heat is transferred by the movement of particles (that possess kinetic energy) by conduction, convection, and radiation. ◊Conduction involves the movement of heat through a solid material by the movement of free electrons. For example, thermal energy is lost from a house by conduction through the walls and windows. ◊Convection involves the transfer of energy by the movement of fluid particles. All objects radiate heat in the form of radiation of electromagnetic waves. Hotter objects emit more energy than cooler objects.

Methods of reducing energy transfer as heat through the use of ◊insulation are important because the world's fuel reserves are limited and heating homes costs a lot of money in fuel bills. Heat transfer from the home can be reduced by a variety of methods, such as loft insulation, cavity wall insulation, and double glazing. The efficiency of insulating materials in the building industry are compared by measuring their heat-conducting properties, represented by a ◊U-value. A low U-value indicates a good insulating material.

$E = mc^2$ It is now recognised that mass can be converted into energy under certain conditions, according to Einstein's theory of relativity. This conversion of mass into energy is the basis of atomic power. ◊Einstein's special theory of ◊relativity 1905 correlates any gain, E, in energy with a gain, m, in mass, by the equation $E = mc^2$, in which c is the speed of light. The conversion of mass into energy in accordance with this equation applies universally, although it is only for nuclear reactions that the percentage change in mass is large enough to detect.

energy, alternative energy from sources that are renewable and ecologically safe, as opposed to sources that are nonrenewable with toxic by-products, such as coal, oil, or gas (fossil fuels), and uranium (for nuclear power). The most important alternative energy source is flowing water, harnessed as ◊hydroelectric power. Other sources include the oceans' tides and waves (see ◊tidal power station and ◊wave power), ◊wind power (harnessed by windmills and wind turbines), the Sun (◊solar energy), and the heat trapped in the Earth's crust (◊geothermal energy) (see also ◊cold fusion).

energy conservation methods of reducing energy use through insulation, increasing energy efficiency, and changes in patterns of use. Profligate energy use by industrialized countries contributes greatly to air pollution and the ◊greenhouse effect when it draws on nonrenewable energy sources.

It has been calculated that increasing energy efficiency alone could reduce carbon dioxide emissions in several high-income countries by 1–2% a year. The average annual decrease in energy consumption in relation to gross national product 1973–87 was 1.2% in France, 2% in the UK, 2.1% in the USA, and 2.8% in Japan.

energy level the permitted energy that an electron can have in any particular atom. Energy levels can be calculated using ◊quantum theory. The permitted energy levels depend mainly on the distance of the electron from the nucleus. See ◊orbital, atomic.

energy of reaction energy released or absorbed during a chemical reaction, also called **enthalpy of reaction** or **heat of reaction**. In a chemical reaction, the energy stored in the reacting molecules is rarely the same as that stored in the product molecules. Depending on which is the greater, energy is either released (an exothermic reaction) or absorbed (an endothermic reaction) from the surroundings (see ◊conservation of energy). The amount of energy released or absorbed by the quantities of substances represented by the chemical equation is the energy of reaction.

Energy Star in computing, US programme requiring all computer equipment to conserve electrical power. Key features of Energy Star-compliant hardware include a built-in function to put the computer and monitor into suspended animation after a specified period of disuse and limits on the amount of power computers and printers can draw.

engine device for converting stored energy into useful work or movement. Most engines use a fuel as their energy store. The fuel is burnt to produce heat energy – hence the name 'heat engine' – which is then converted into movement. Heat engines can be classified according to the fuel they use (◊petrol engine or ◊diesel engine), or according to whether the fuel is burnt inside (◊internal combustion engine) or outside (◊steam engine) the engine, or according to whether they produce a reciprocating or rotary motion (◊turbine or ◊Wankel engine).

engine in computing, core piece of software around which other features and functions are built. A database ◊search engine, for example, accepts user input and handles the processing necessary to find matches between the user input and the database records.

In a computer game, the term 'engine' is also used to refer to the core software that allows users to move around the game's levels and pick up weapons and treasure.

The scientist describes what is: the engineer creates what never was.

Theodore von Kármán Hungarian-born US aerodynamicist.
Biogr. Mem. FRS 1980 26 110

---

engineering the application of science to the design, construction, and maintenance of works, machinery, roads, railways, bridges, harbour installations, engines, ships, aircraft and airports, spacecraft and space stations, and the generation, transmission, and use of electrical power. The main divisions of engineering are aerospace, chemical, civil, computer, electrical, electronic, gas, marine, materials, mechanical, mining, production, radio, and structural.

Did you know that if a beaver two feet long with a tail a foot and a half long can build a dam twelve feet high and six feet wide in two days, all you would need to build the Kariba Dam is a beaver sixty-eight feet long with a fifty-one foot tail?

Norton Juster US writer.
The Phantom Tollbooth ch 14

To remember that to tighten a bolt or nut you turn it clockwise (right), and to take it off you turn it anti-clockwise (left):

Righty tighty, lefty loosie

engineering drawing technical drawing that forms the plans for the design and construction of engineering components and structures. Engineering drawings show different projections, or views of objects, with the relevant dimensions, and show how all the separate parts fit together. They are often produced by computers using computer-aided design (◊CAD).

English toy terrier or ***black-and-tan terrier*** breed of toy dog closely resembling the ◊Manchester terrier but smaller and with erect ears. It weighs no more than 3.5 kg/8 lb and is 25–30 cm/10–12 in high.

enset *Ensete ventricosum* relative of the banana with edible corms and stems. It was domesticated between 5,000 and 10,000 years ago but is now grown only in southern Ethiopia. It is resistant to drought and when mashed and fermented can be kept for months, or even years, before being made into a wide variety of foods. Its fibre can also be used to produce material.

Wild enset grows over much of western and southern Africa. In 1993 a US research team was set up to explore the possibility of cultivating enset in the drought-stricken north of Ethiopia.

enthalpy in chemistry, alternative term for ◊energy of reaction, the heat energy associated with a chemical change.

entomology study of insects.

entropy in ◊thermodynamics, a parameter representing the state of disorder of a system at the atomic, ionic, or molecular level; the greater the disorder, the higher the entropy. Thus the fast-moving disordered molecules of water vapour have higher entropy than those of more ordered liquid water, which in turn have more entropy than the molecules in solid crystalline ice.

In a closed system undergoing change, entropy is a measure of the amount of energy unavailable for useful work. At ◊absolute zero (–273.15°C/–459.67°F/0 K), when all molecular motion ceases and order is assumed to be complete, entropy is zero.

... a living organism ... feeds upon negative entropy ... Thus the device by which an organism maintains itself stationary at a fairly high level of orderliness (fairly low level of entropy) really consists in continually sucking orderliness from its environment.

Erwin Schrödinger Austrian physicist.
What is Life? 1944

E number code number for additives that have been approved for use by the European Commission (EC). The E written before the number stands for European. E numbers do not have to be displayed on lists of ingredients, and the manufacturer may choose to list additives by their name instead. E numbers cover all categories of additives apart from flavourings.

Additives, other than flavourings, that are not approved by the European Commission, but are still used in Britain, are represented by a code number without an E.

envelope in geometry, a curve that touches all the members of a family of lines or curves. For example, a family of three equal circles all touching each other and forming a triangular pattern (like a clover leaf) has two envelopes: a small circle that fits in the space in the middle, and a large circle that encompasses all three circles.

environment in ecology, the sum of conditions affecting a particular organism, including physical surroundings, climate, and influences of other living organisms. See also ◊biosphere and ◊habitat.

In common usage, 'the environment' often means the total global environment, without reference to any particular organism. In genetics, it is the external influences that affect an organism's development, and thus its ◊phenotype.

The sun, the moon and the stars would have disappeared long ago ... had they happened to be within the reach of predatory human hands.

Havelock Ellis British psychologist.
The Dance of Life ch 7

environmental audit another name for ◊green audit, the inspection of a company to assess its environmental impact.

environmentalism theory emphasizing the primary influence of the environment on the development of groups or individuals. It stresses the importance of the physical, biological, psychological, or cultural environment as a factor influencing the structure or behaviour of animals, including humans.

In politics this has given rise in many countries to Green parties, which aim to 'preserve the planet and its people'.

Environmentally Sensitive Area (ESA) scheme introduced by the UK Ministry of Agriculture in 1984, as a result of EC legislation, to protect some of the most beautiful areas of the British countryside from the loss and damage caused by agricultural change. The first areas to be designated ESAs were in the Pennine Dales, the North Peak District, the Norfolk Broads, the Breckland, the Suffolk River Valleys, the Test Valley, the South Downs, the Somerset Levels and Moors, West Penwith, Cornwall, the Shropshire Borders, the Cambrian Mountains, and the Lleyn Peninsula.

The total area designated as ESA's was estimated 1993 at 785,600 hectares. The scheme is voluntary, with farmers being encouraged to adapt their practices so as to enhance or maintain the natural features of the landscape and conserve wildlife habitat. A farmer who joins the scheme agrees to manage the land in this way for at least five years. In return for this agreement, the Ministry of Agriculture pays the farmer a sum that reflects the financial losses incurred as a result of reconciling conservation with commercial farming.

Environmental Protection Agency (EPA) US agency set up 1970 to control water and air quality, industrial and commercial wastes, pesticides, noise, and radiation. In its own words, it aims to protect 'the country from being degraded, and its health threatened, by a multitude of human activities initiated without regard to long-ranging effects upon the life-supporting properties, the economic uses, and the recreational value of air, land, and water'.

environment–heredity controversy see ◊nature–nurture controversy.

enzyme biological ◊catalyst produced in cells, and capable of speeding up the chemical reactions necessary for life. They are large, complex ◊proteins, and are highly specific, each chemical reaction requiring its own particular enzyme. The enzyme's specificity arises from its **active site**, an area with a shape corresponding to part of the molecule with which it reacts (the substrate). The enzyme and the substrate slot together forming an enzyme–substrate complex that allows the reaction to take place, after which the enzyme falls away unaltered.

The activity and efficiency of enzymes are influenced by various factors, including temperature and pH conditions. Temperatures above 60°C/140°F damage (denature) the intricate structure of enzymes, causing reactions to cease. Each enzyme operates best within a specific pH range, and is denatured by excessive acidity or alkalinity.

Digestive enzymes include amylases (which digest starch), lipases (which digest fats), and proteases (which digest protein). Other enzymes play a part in the conversion of food energy into ◊ATP; the manufacture of all the molecular components of the body; the replication of ◊DNA when a cell divides; the production of hormones; and the control of movement of substances into and out of cells.

Enzymes have many medical and industrial uses, from washing powders to drug production, and as research tools in molecular biology. They can be extracted from bacteria and moulds, and ◊genetic engineering now makes it possible to tailor an enzyme for a specific purpose.

ENZYME

The reason that pigs never suffer from gout is because of an enzyme. In humans, this painful disease is caused by a build-up of uric acid. Pigs have an enzyme that breaks uric acid into soluble components. Humans do not have this enzyme, so they suffer from gout and pigs do not.

Eocene second epoch of the Tertiary period of geological time, 56.5–35.5 million years ago. Originally considered the earliest division of the Tertiary, the name means 'early recent', referring to the early forms of mammals evolving at the time, following the extinction of the dinosaurs.

eotvos unit unit (symbol E) for measuring small changes in the intensity of the Earth's ◊gravity with horizontal distance.

ephemeral plant plant with a very short life cycle, sometimes as little as six to eight weeks. It may complete several generations in one growing season.

epicentre the point on the Earth's surface immediately above the seismic focus of an ◊earthquake. Most damage usually takes place at an earthquake's epicentre. The term sometimes refers to a point directly above or below a nuclear explosion ('at ground zero').

epicyclic gear or ***sun-and-planet gear*** gear system that consists of one or more gear wheels moving around another. Epicyclic gears are found in bicycle hub gears and in automatic gearboxes.

epicycloid in geometry, a curve resembling a series of arches traced out by a point on the circumference of a circle that rolls around another circle of a different diameter. If the two circles have the same diameter, the curve is a ◊cardioid.

epicycloid A seven-cusped epicycloid, formed by a point on the circumference of a circle (of diameter d*) that rolls around another circle (of diameter 7*d*/3).*

epidemic outbreak of infectious disease affecting large numbers of people at the same time. A widespread epidemic that sweeps across many countries (such as the ◊Black Death in the late Middle Ages) is known as a **pandemic**.

epidermis outermost layer of ◊cells on an organism's body. In plants and many invertebrates such as insects, it consists of a single layer of cells. In vertebrates, it consists of several layers of cells.

The epidermis of plants and invertebrates often has an outer noncellular ◊cuticle that protects the organism from desiccation.

epididymis in male vertebrates, a long coiled tubule in the ◊testis, in which sperm produced in the seminiferous tube are stored. In men, it is a duct about 6 m/20 ft long, convoluted into a small space.

epigeal seed germination in which the ◊cotyledons (seed leaves) are borne above the soil.

epiglottis small flap located behind the root of the tongue in mammals. It closes off the end of the windpipe during swallowing to prevent food from passing into it and causing choking.

The action of the epiglottis is a highly complex reflex process involving two phases. During the first stage a mouthful of chewed food is lifted by the tongue towards the top and back of the mouth. This is accompanied by the cessation of breathing and by the blocking of the nasal areas from the mouth. The second phase involves the epiglottis moving over the larynx while the food passes down into the oesophagus.

epilepsy medical disorder characterized by a tendency to develop fits, which are convulsions or abnormal feelings caused by abnormal electrical discharges in the cerebral hemispheres of the ◊brain. Epilepsy can be controlled with a number of anticonvulsant drugs.

The term epilepsy covers a range of conditions from mild 'absences', involving momentary loss of awareness, to major convulsions. In some cases the abnormal electrical activity is focal (confined to one area of the brain); in others it is generalized throughout the cerebral cortex. Fits are classified according to their clinical type. They include: the **grand mal** seizure with convulsions and loss of consciousness; the fleeting absence of awareness **petit mal**, almost exclusively a disorder of childhood; **Jacksonian** seizures, originating in the motor cortex; and **temporal-lobe** fits, which may be associated with visual hallucinations and bizarre disturbances of the sense of smell.

Epilepsy affects 1–3% of the world's population. It may arise spontaneously or may be a consequence of brain surgery, organic brain disease, head injury, metabolic disease, alcoholism or withdrawal from some drugs. Almost a third of patients have a family history of the condition.

Most epileptics have infrequent fits that have little impact on their daily lives. Epilepsy does not imply that the sufferer has any impairment of intellect, behaviour, or personality.

UNDERSTANDING EPILEPSY

http://www.epinet.org.au/efvunder.html

Clear and concise guide to epilepsy provided by EpiNet courtesy of the Epilepsy Foundation of Victoria, Australia. It covers the basics of the subject under the headings of 'What is Epilepsy?', 'Diagnosis', and 'Recognizing Seizures'. It also provides a guide to the different types of seizure and suggests first aid measures for each.

epiphyte any plant that grows on another plant or object above the surface of the ground, and has no roots in the soil. An epiphyte does not parasitize the plant it grows on but merely uses it for support. Its nutrients are obtained from rainwater, organic debris such as leaf litter, or from the air.

The greatest diversity of epiphytes is found in tropical areas and includes many orchids.

epithelium in animals, tissue of closely packed cells that forms a surface or lines a cavity or tube. Epithelium may be protective (as in the skin) or secretory (as in the cells lining the wall of the gut).

epoch subdivision of a geological period in the geological time scale. Epochs are sometimes given their own names (such as the

To remember the different epochs, in descending order of age:

To remember the different epochs, in ascending order:

Heavy people put more on every plate

or

Happy plump pregnant mothers only eat pickles

or

Happy people play music, others eat pizza

Put eggs on my plate please honey

or

Please eliminate old men playing poker honestly

Holocene, Pleistocene, Pliocene, Miocene, Oligocene, Eocene, Paleocene

Palaeocene, Eocene, Oligocene, Miocene, Pliocene, Pleistocene, Holocene

Palaeocene, Eocene, Oligocene, Miocene, and Pliocene epochs comprising the Tertiary period), or they are referred to as the late, early, or middle portions of a given period (as the Late Cretaceous or the Middle Triassic epoch).

Geological time is broken up into **geochronological units** of which epoch is just one level of division. The hierarchy of geochronological divisions is eon, era, period, epoch, age, and chron. Epochs are subdivisions of periods and ages are subdivisions of epochs. Rocks representing an epoch of geological time comprise a **series**.

epoxy resin synthetic ◊resin used as an ◊adhesive and as an ingredient in paints. Household epoxy resin adhesives come in component form as two separate tubes of chemical, one tube containing resin, the other a curing agent (hardener). The two chemicals are mixed just before application, and the mix soon sets hard.

EPROM (acronym for ***erasable programmable read-only memory***) computer memory device in the form of an ◊integrated circuit (chip) that can record data and retain it indefinitely. The data can be erased by exposure to ultraviolet light, and new data recorded. Other kinds of computer memory chips are ◊ROM (read-only memory), ◊PROM (programmable read-only memory), and ◊RAM (random-access memory).

EPS in computing, abbreviation for ◊encapsulated PostScript.

Epsilon Aurigae ◊eclipsing binary star in the constellation Auriga. One of the pair is an 'ordinary' star, but the other seems to be an enormous distended object whose exact nature remains unknown. The period (time between eclipses) is 27 years, the longest of its kind. The last eclipse was in 1982–84.

Epsom salts $MgSO_4.7H_2O$ hydrated magnesium sulphate, used as a relaxant and laxative and added to baths to soothe the skin. The name is derived from a bitter saline spring at Epsom, Surrey, England, which contains the salt in solution.

equation in chemistry, representation of a chemical reaction by symbols and numbers; see ◊chemical equation.

equation in mathematics, expression that represents the equality of two expressions involving constants and/or variables, and thus usually includes an equals (=) sign. For example, the equation $A = \pi r^2$ equates the area A of a circle of radius r to the product πr^2.

The algebraic equation $y = mx + c$ is the general one in coordinate geometry for a straight line.

If a mathematical equation is true for all variables in a given domain, it is sometimes called an identity and denoted by ≡.

Thus $(x + y)^2 = x^2 + 2xy + y^2$ for all $x, y \in R$.

An **indeterminate equation** is an equation for which there is an infinite set of solutions – for example, $2x = y$. A **diophantine equation** is an indeterminate equation in which both the solution and the terms must be whole numbers (after Diophantus of Alexandria, c. AD 250).

equation of motion mathematical equation that gives the position and velocity of a moving object at any time. Given the mass of an object, the forces acting on it, and its initial position and velocity, an equation of motion is used to calculate its position and velocity at any later time. The equation must be based on ◊Newton's laws of motion or, if speeds near that of light are involved, on the theory of relativity.

equations of motion or ***kinematic equations*** mathematical equations that give the position or velocity at any time of an object moving with constant acceleration. The five common equations are:

$$v = u + ats = \frac{1}{2}(u + v)ts = ut + \frac{1}{2}at^2v^2 = u^2 + 2$$

as in which a is the object's constant acceleration, u is its initial velocity, v is its velocity after a time t, and s is the distance travelled by it in that time.

Equator or ***terrestrial equator*** the ◊great circle whose plane is perpendicular to the Earth's axis (the line joining the poles). Its length is 40,092 km/24,901.8 mi, divided into 360 degrees of longitude. The Equator encircles the broadest part of the Earth, and represents 0° latitude. It divides the Earth into two halves, called the northern and the southern hemispheres.

The **celestial equator** is the circle in which the plane of the Earth's Equator intersects the ◊celestial sphere.

equatorial coordinates in astronomy, a system for measuring the position of astronomical objects on the ◊celestial sphere with reference to the plane of the Earth's equator.

◊Declination (symbol Δ), analogous to latitude, is measured in degrees from the equator to the north (Δ = 90°) or south (Δ =–90°) celestial poles. Right ◊ascension (symbol α), analogous to longitude, is normally measured in hours of time (α = 0 h to 24 h) eastward along the equator from a fixed point known as the first point of ◊Aries or the ◊vernal equinox.

equatorial mounting in astronomy, a method of mounting a telescope to simplify the tracking of celestial objects. One axis (the polar axis) is mounted parallel to the rotation axis of the Earth so that the ◊telescope can be turned about it to follow objects across the sky. The declination axis moves the telescope in ◊declination and is clamped before tracking begins. Another advantage over the simpler altazimuth mounting is that the orientation of the image is fixed, permitting long-exposure photography.

Equidae horse family in the order Perissodactyla, which includes the odd-toed hoofed animals. Besides the domestic horse, wild asses, wild horses, onagers, and zebras, there are numerous extinct species known from fossils.

All species in the family are inhabitants of flat, open country, except the mountain zebra from the hills of southern Africa. They have long legs, adapted to carry the animals at speed over firm ground; only one toe, terminating in a hoof, is present on each limb. They are all herbivorous and have keen senses to detect their enemies. Their jaws are strong and able to chew the tough grasses and herbs on which they feed in the wild state.

equilateral geometrical figure, having all sides of equal length.

equilibrium in physics, an unchanging condition in which an undisturbed system can remain indefinitely in a state of balance. In a **static equilibrium**, such as an object resting on the floor, there is no motion. In a **dynamic equilibrium**, in contrast, a steady state is maintained by constant, though opposing, changes. For example, in a sealed bottle half-full of water, the constancy of the water level is a result of molecules evaporating from the surface and condensing on to it at the same rate.

equinox the points in spring and autumn at which the Sun's path, the ◊ecliptic, crosses the celestial equator, so that the day and night are of approximately equal length. The **vernal equinox** occurs about 21 March and the **autumnal equinox**, 23 September.

era any of the major divisions of geological time, each including several periods, but smaller than an eon. The currently recognized eras all fall within the Phanerozoic eon – or the vast span of time, starting about 570 million years ago, when fossils are found to become abundant. The eras in ascending order are the Palaeozoic, Mesozoic, and Cenozoic. We are living in the Recent epoch of the Quaternary period of the Cenozoic era.

Geological time is broken up into **geochronological units** of which era is just one level of division. The hierarchy of geochronological divisions is eon, era, period, ◊epoch, age, and chron. Eras are subdivisions of eons and periods are subdivisions of eras. Rocks representing an era of geological time comprise an **erathem**.

erasable optical disc in computing, another name for a ◊floptical disc.

Eratosthenes' sieve a method for finding ◊prime numbers. It involves writing in sequence all numbers from 2. Then, starting with 2, cross out every second number (but not 2 itself), thus eliminating numbers that can be divided by 2. Next, starting with 3, cross out every third number (but not 3 itself), and continue the process for 5, 7, 11, 13, and so on. The numbers that remain are primes.

erbium soft, lustrous, greyish, metallic element of the ◊lanthanide series, symbol Er, atomic number 68, relative atomic mass 167.26. It occurs with the element yttrium or as a minute part of various minerals. It was discovered in 1843 by Carl Mosander (1797–1858), and named after the town of Ytterby, Sweden, near which the lanthanides (rare-earth elements) were first found.

Erbium has been used since 1987 to amplify data pulses in optical fibre, enabling faster transmission. Erbium ions in the fibreglass, charged with infrared light, emit energy by amplifying the data pulse as it moves along the fibre.

erg c.g.s. unit of work, replaced in the SI system by the◊ joule. One erg of work is done by a force of one ◊dyne moving through one centimetre.

ergonomics study of the relationship between people and the furniture, tools, and machinery they use at work. The object is to improve work performance by removing sources of muscular stress and general fatigue: for example, by presenting data and control panels in easy-to-view form, making office furniture comfortable, and creating a generally pleasant environment.

Good ergonomic design makes computer systems easier to use and minimizes the health hazards and physical stresses of working with computers for many hours a day: it helps data entry workers to avoid conditions like ◊strain injury (RSI), eyestrain, and back and muscle aches.

ergot any of a group of parasitic fungi (especially of the genus *Claviceps*), whose brown or black grainlike masses replace the kernels of rye or other cereals. *C. purpurea* attacks the rye plant. Ergot poisoning is caused by eating infected bread, resulting in burning pains, gangrene, and convulsions.

The large grains of the fungus contain the poison ergotamine.

ERIC (abbreviation for ***Educational Resources Information Center***) in computing, database of resources for education available on the Internet. Established 1966, ERIC is a federally funded network of educational information. Sixteen clearing houses index educational materials for the database, which is housed at the University of Saskatchewan, Canada. The database is distributed in a variety of formats including printed books, CD-ROM, and microfiche.

erica any plant of a large group that includes the heathers. There are about 500 species, distributed mainly in South Africa with some in Europe. (Genus *Erica*, family Ericaceae.)

Eridanus in astronomy, the sixth largest constellation, which meanders from the celestial equator (see ◊celestial sphere) deep into the southern hemisphere of the sky. Eridanus is represented as a river. Its brightest star is ◊Achernar, a corruption of the Arabic for 'the end of the river'.

ermine the ◊stoat during winter, when its coat becomes white.

In northern latitudes the coat becomes completely white, except for a black tip on the tail, but in warmer regions the back may remain brownish. The fur is used commercially.

ERNIE acronym for ***electronic random number indicator***, machine designed and produced by the UK Post Office Research Station to select a series of random 9-figure numbers to indicate prizewinners among Premium Bond holders.

Eros in astronomy, an asteroid, discovered in 1898, that can pass 22 million km/14 million mi from the Earth, as observed in 1975. Eros was the first asteroid to be discovered that has an orbit coming within that of Mars. It is elongated, measures about 36 × 12 km/22 × 7 mi, rotates around its shortest axis every 5.3 hours, and orbits the Sun every 1.8 years.

The Near Earth Asteroid Rendezvous (NEAR) launched February 1996 is estimated to take three years to reach Eros. It will spend a year circling the asteroid in an attempt to determine what it is made of.

erosion wearing away of the Earth's surface, caused by the breakdown and transportation of particles of rock or soil (by contrast, ◊weathering does not involve transportation). Agents of erosion include the sea, rivers, glaciers, and wind.

Water, consisting of sea waves and currents, rivers, and rain; ice, in the form of glaciers; and wind, hurling sand fragments against exposed rocks and moving dunes along, are the most potent forces of erosion.

People also contribute to erosion by bad farming practices and the cutting down of forests, which can lead to the formation of dust bowls.

There are several processes of erosion including hydraulic action, ◊corrosion, attrition, and ◊solution.

erratic in geology, a displaced rock that has been transported by a glacier or some other natural force to a site of different geological composition.

error in computing, a fault or mistake, either in the software or on the part of the user, that causes a program to stop running (crash) or produce unexpected results. Program errors, or bugs, are largely eliminated in the course of the programmer's initial testing procedure, but some will remain in most programs. All computer operating systems are designed to produce an **error message** (on the display screen, or in an error file or printout) whenever an error is detected, reporting that an error has taken place and, wherever possible, diagnosing its cause.

Truth comes out of error more easily than out of confusion.

FRANCIS BACON English politician, philosopher, and essayist.
Quoted in R L Weber, *A Random Walk in Science*

---

error detection in computing, the techniques that enable a program to detect incorrect data. A common method is to add a check digit to important codes, such as account numbers and product codes. The digit is chosen so that the code conforms to a rule that the program can verify. Another technique involves calculating the sum (called the ◊hash total) of each instance of a particular item of data, and storing it at the end of the data.

error message message produced by a computer to inform the user that an error has occurred.

erythroblast in biology, a series of nucleated cells that go through various stages of development in the bone marrow until they form red blood cells (erythrocytes). This process is known as erythropoiesis. Erythroblasts can appear in the blood of people with blood cancers.

erythrocyte another name for ◊red blood cell.

Eryx genus of egg-laying snakes, closely allied to the ◊boa constrictor. The species of *Eryx* differ from boas in having a very short, obtuse tail and narrower ventral plates. They occur in Asia and Africa.

ESA abbreviation for ◊European Space Agency.

escalator automatic moving staircase that carries people between floors or levels. It consists of treads linked in an endless belt arranged to form strips (steps), powered by an electric motor that moves both steps and handrails at the same speed. Towards the top and bottom the steps flatten out for ease of passage. The first escalator was exhibited in Paris in 1900.

escape sequence in computing, string of characters sent to a ◊modem to switch it from sending data to a state in which it can accept and act upon commands.

Most modems use the escape sequence patented by Hayes, which consists of three plus signs (+++) with a brief pause on either side to distinguish the characters from data.

escape velocity in physics, minimum velocity with which an object must be projected for it to escape from the gravitational pull of a planetary body. In the case of the Earth, the escape velocity is 11.2 kps/6.9 mps; the Moon, 2.4 kps/1.5 mps; Mars, 5 kps/3.1 mps; and Jupiter, 59.6 kps/37 mps.

escarpment or ***cuesta*** large ridge created by the erosion of dipping sedimentary rocks. It has one steep side (scarp) and one gently sloping side (dip). Escarpments are common features of chalk landscapes, such as the Chiltern Hills and the North Downs in England. Certain features are associated with chalk escarpments, including dry valleys (formed on the dip slope), combes (steep-sided valleys on the scarp slope), and springs.

Escherichia coli rod-shaped Gram-negative ◊bacterium (see bacteria) that lives, usually harmlessly, in the colon of most warm-blooded animals. It is the commonest cause of urinary tract infections in humans. It is sometimes found in water or meat where faecal contamination has occurred and can cause severe gastric problems.

The mapping of the genome of *E. coli*, consisting of 4,403 genes, was completed in 1997.

classification *Escherichia coli* is the only species in the bacterial family Enterobacteriaceae.

ESP abbreviation for ◊extrasensory perception.

esparto species of grass native to southern Spain, southern Portugal, and the Balearics, but now widely grown in dry, sandy locations throughout the world. The plant is just over 1 m/3 ft high, producing greyish-green leaves, which are used for making paper, ropes, baskets, mats, and cables. (*Stipa tenacissima.*)

ESPRIT (abbreviation for ***European Strategic Programme for Research and Development in Information Technology***) European Union programme that funds technology research at an early stage of development. ESPRIT's goals include encouraging the development of international standards and cooperation between European companies, universities, and research centres in order to develop the infrastructure necessary for Europe to be able to compete with Japan and the USA.

essential amino acid water-soluble organic molecule vital to a healthy diet; see ◊amino acid.

essential fatty acid organic compound consisting of a hydrocarbon chain and important in the diet; see ◊fatty acid.

ester organic compound formed by the reaction between an alcohol and an acid, with the elimination of water. Unlike ◊salts, esters are covalent compounds.

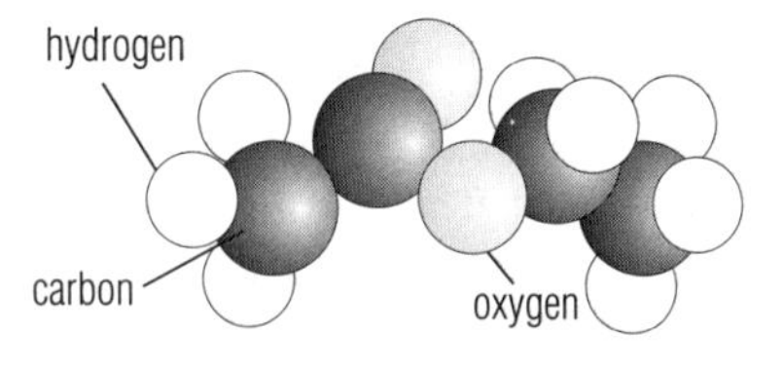

ester Molecular model of the ester ethyl ethanoate (ethyl acetate) CH3CH2COOCH3

estradiol alternative spelling of oestradiol, a type of ◊oestrogen (female sex hormone).

ethanal common name ***acetaldehyde*** CH_3CHO one of the chief members of the group of organic compounds known as ◊aldehydes. It is a colourless inflammable liquid boiling at 20.8°C/69.6°F. Ethanal is formed by the oxidation of ethanol or ethene and is used to make many other organic chemical compounds.

ethanal trimer common name ***paraldehyde*** $(CH_3CHO)_3$ colourless liquid formed from ethanal. It is soluble in water.

ethane CH_3CH_3 colourless, odourless gas, the second member of the alkane series of hydrocarbons (paraffins).

ethane-1,2-diol technical name for ◊glycol.

ethanoate common name ***acetate*** $CH_3CO_2^-$ negative ion derived from ethanoic (acetic) acid; any salt containing this ion. In textiles, acetate rayon is a synthetic fabric made from modified cellulose (wood pulp) treated with ethanoic acid; in photography, acetate film is a non-flammable film made of cellulose ethanoate.

ethanoic acid common name ***acetic acid*** CH_3CO_2H one of the simplest fatty acids (a series of organic acids). In the pure state it is a colourless liquid with an unpleasant pungent odour; it solidifies to an icelike mass of crystals at 16.7°C/62.4°F, and hence is often called glacial ethanoic acid. Vinegar contains 5% or more ethanoic acid, produced by fermentation.

Cellulose (derived from wood or other sources) may be treated with ethanoic acid to produce a cellulose ethanoate (acetate) solution, which can be used to make plastic items by injection moulding or extruded to form synthetic textile fibres.

ethanol common name ***ethyl alcohol*** C_2H_5OH alcohol found in beer, wine, cider, spirits, and other alcoholic drinks. When pure, it is a colourless liquid with a pleasant odour, miscible with water or ether; it burns in air with a pale blue flame. The vapour forms an explosive mixture with air and may be used in high-compression internal combustion engines.

It is produced naturally by the fermentation of carbohydrates by yeast cells. Industrially, it can be made by absorption of ethene and subsequent reaction with water, or by the reduction of ethanal in the presence of a catalyst, and is widely used as a solvent.

Ethanol is used as a raw material in the manufacture of ether, chloral, and iodoform. It can also be added to petrol, where it improves the performance of the engine, or be used as a fuel in its own right (as in Brazil). Crops such as sugar cane may be grown to provide ethanol (by fermentation) for this purpose.

ethene common name ***ethylene*** C_2H_4 colourless, flammable gas, the first member of the ◊alkene series of hydrocarbons. It is the most widely used synthetic organic chemical and is used to produce the plastics polyethene (polyethylene), polychloroethene, and polyvinyl chloride (PVC). It is obtained from natural gas or coal gas, or by the dehydration of ethanol.

Ethene is produced during plant metabolism and is classified as a plant hormone. It is important in the ripening of fruit and in ◊abscission. Small amounts of ethene are often added to the air surrounding fruit to artificially promote ripening. Tomato and marigold plants show distorted growth in concentrations as low as 0.01 parts per million. Plants also release ethene when they are under stress. German physicists invented a device in 1997 that measures stress levels in plants by measuring surrounding ethene levels.

ether in chemistry, any of a series of organic chemical compounds having an oxygen atom linking the carbon atoms of two hydrocarbon radical groups (general formula R-O-R'); also the common name for ethoxyethane $C_2H_5OC_2H_5$ (also called diethyl ether).

This is used as an anaesthetic and as an external cleansing agent before surgical operations. It is also used as a solvent, and in the extraction of oils, fats, waxes, resins, and alkaloids.

Ethoxyethane is a colourless, volatile, inflammable liquid, slightly soluble in water, and miscible with ethanol. It is prepared by treatment of ethanol with excess concentrated sulphuric acid at 140°C/284°F.

ether or ***aether*** in the history of science, a hypothetical medium permeating all of space. The concept originated with the Greeks, and has been revived on several occasions to explain the properties and propagation of light. It was supposed that light and other electromagnetic radiation – even in outer space – needed a medium, the ether, in which to travel. The idea was abandoned with the acceptance of ◊relativity.

Ethernet in computing, the most popular protocol for ◊local area networks. Ethernet was developed principally by the Xerox Corporation, but can now be used on most computers. It normally allows data transfer at rates up to 10 Mbps. but 100 Mbps Fast Ethernet – often called 100Base-T – is already in widespread use while Gigabit Ethernet is being tipped for future success.

Ethernet and Fast Ethernet are IEEE standards, and Gigabit Ethernet should become a standard in 1998.

Ethiopian wolf ***Simien jackal*** or ***Abyssinian wolf*** member of the family Canidae, *Canis simensis,* and genetically close to the grey wolf and coyote. They are found mainly in Ethiopia's Bale and Simien Mountains, but numbers are seriously in decline with only about 400 left in 1997.

ethnography study of living cultures, using anthropological techniques like participant observation (where the anthropologist lives in the society being studied) and a reliance on informants. Ethnography has provided much data of use to archaeologists as analogies.

ethnology study of contemporary peoples, concentrating on their geography and culture, as distinct from their social systems. Ethnologists make a comparative analysis of data from different cultures to understand how cultures work and why they change, with a view to deriving general principles about human society.

ethnopaediatrics in anthropology, the study of child-rearing practices in different cultures. Areas studied include the comparison of feeding regimes, sleeping arrangements, and degree of verbal stimulation.

ethology comparative study of animal behaviour in its natural setting. Ethology is concerned with the causal mechanisms (both the stimuli that elicit behaviour and the physiological mechanisms controlling it), as well as the development of behaviour, its function, and its evolutionary history.

Ethology was pioneered during the 1930s by the Austrians Konrad Lorenz and Karl von Frisch who, with the Dutch zoologist Nikolaas Tinbergen, received the Nobel prize in 1973. Ethologists believe that the significance of an animal's behaviour can be understood only in its natural context, and emphasize the importance of field studies and an evolutionary perspective. A recent development within ethology is ◊sociobiology, the study of the evolutionary function of ◊social behaviour.

ethyl alcohol common name for ethanol.

ethylene common name for ethene.

ethylene glycol alternative name for glycol.

ethyne common name ***acetylene*** CHCH colourless inflammable gas produced by mixing calcium carbide and water. It is the simplest member of the ◊alkyne series of hydrocarbons. It is used in the manufacture of the synthetic rubber neoprene, and in oxyacetylene welding and cutting.

Ethyne was discovered by Edmund Davy in 1836 and was used in early gas lamps, where it was produced by the reaction between water and calcium carbide. Its combustion provides more heat, relatively, than almost any other fuel known (its calorific value is five times that of hydrogen). This means that the gas gives an intensely hot flame; hence its use in oxyacetylene torches.

etiolation in botany, a form of growth seen in plants receiving insufficient light. It is characterized by long, weak stems, small leaves, and a pale yellowish colour (◊chlorosis) owing to a lack of chlorophyll. The rapid increase in height enables a plant that is surrounded by others to quickly reach a source of light, after which a return to normal growth usually occurs.

Etruscan art the art of the inhabitants of Etruria, central Italy, a civilization that flourished 8th–2nd centuries BC. The Etruscans produced sculpture, painting, pottery, metalwork, and jewellery. Etruscan terracotta coffins (sarcophagi), carved with reliefs and topped with portraits of the dead reclining on one elbow, were to influence the later Romans and early Christians.
painting Most examples of Etruscan painting come from excavated tombs, whose frescoes depict scenes of everyday life, mythology, and mortuary rites, typically in bright colours and a vigorous, animated style. Scenes of feasting, dancing, swimming, fishing, and playing evoke a confident people who enjoyed life to the full, and who even in death depicted themselves in a joyous and festive manner. The decline of their civilization, in the shadow of Rome's expansion, is reflected in their later art, which loses its original *joie de vivre* and becomes sombre.
influences Influences from archaic Greece and the Middle East are evident, as are those from the preceding Iron Age Villanovan culture, but the full flowering of Etruscan art represents a unique synthesis of existing traditions and artistic innovation, which was to have a profound influence on the development of Western art.

eucalyptus any tree of a group belonging to the myrtle family, native to Australia, where they are commonly known as gumtrees. About 90% of Australian timber belongs to the eucalyptus genus, which contains about 500 species. The trees have dark hardwood timber which is used for heavy construction work such as railway and bridge building. They are mostly tall, aromatic, evergreen trees with pendant leaves and white, pink, or red flowers. (Genus *Eucalyptus,* family Myrtaceae.)

Compounds isolated from eucalyptus leaves were found in 1996 to be highly effective in killing *Streptococcus mutans,* the bacteria found in the mouth that cause dental decay.

EUCALYPTUS

The tallest tree ever measured was an Australian eucalyptus (*Eucalyptus regnans*), reported in 1872. It was 132 m/435 ft tall.

Eudora in computing, popular program for handling and receiving Internet e-mail. Published by the Californian company Qualcomm, Eudora uses ◊POP3, and by the mid-1990s was one of the most commonly used e-mail programs on the Net.

eugenics ***Greek eugenes 'well-born'*** study of ways in which the physical and mental characteristics of the human race may be improved. The eugenic principle was abused by the Nazi Party in Germany during the 1930s and early 1940s to justify the attempted extermination of entire social and ethnic groups and the establishment of selective breeding programmes. Modern eugenics is concerned mainly with the elimination of genetic disease.

The term was coined by the English scientist Francis Galton in 1883, and the concept was originally developed in the late 19th century with a view to improving human intelligence and behaviour.

In 1986 Singapore became the first democratic country to adopt an openly eugenic policy by guaranteeing pay increases to female university graduates when they give birth to a child, while offering grants towards house purchases for nongraduate married women on condition that they are sterilized after the first or second child. In China in June 1995, a law was passed making it illegal for carriers of certain genetic diseases to marry unless they agree to sterilization or long-term contraception. All couples wishing to marry must undergo genetic screening.

Knowledge is the death of research.

WALTHER NERNST German physical chemist.
On examinations, in C G Gillespie (ed)
The Dictionary of Scientific Biography 1981

Euglena genus of single-celled organisms in the ◊protozoan phylum Sarcomastigophora that live in fresh water. They are usually oval or cigar-shaped, less than 0.5 mm/0.2 in long, and have a nucleus, green pigment in chloroplasts, a contractile vacuole, a light-sensitive eyespot, and one or two ◊flagella, with which they swim. A few species are colourless or red.
classification Euglena are members of the order Euglenida in class Phytomastigophora, subphylum Mastigophora, phylum Sarcomastigophora.

This organism combines animal and plant characteristics. Its plantlike characteristics include the chloroplasts and its consequent ability to photosynthesize, and the rigid cellulose wall of some species. Its main animal characteristic is its motility, but it can also absorb food from its environment, and many species have a flexible body covering.

eukaryote in biology, one of the two major groupings into which all organisms are divided. Included are all organisms, except bacteria and cyanobacteria (◊blue-green algae), which belong to the ◊prokaryote grouping.

The cells of eukaryotes possess a clearly defined nucleus, bounded by a membrane, within which DNA is formed into distinct chromosomes. Eukaryotic cells also contain mitochondria, chloroplasts, and other structures (organelles) that, together with a defined nucleus, are lacking in the cells of prokaryotes.

Euratom acronym for ◊European Atomic Energy Community, forming part of the ***European Union organization***.

eureka in chemistry, alternative name for the copper–nickel alloy ◊constantan, which is used in electrical equipment.

eureka ***Greek 'I've got it!'*** exclamation supposedly made by ◊Archimedes on his discovery of fluid displacement.

Eurocodes series of codes giving design rules for all types of engineering structures, except certain very specialized forms, such as nuclear reactors. The codes will be given the status of ENs (European standards) and will be administered by CEN (European Committee for Standardization). ENs will eventually replace national codes, in Britain currently maintained by the BSI (British Standards Institute), and will include parameters to reflect local requirements.

Europa in astronomy, the fourth-largest moon of the planet Jupiter, diameter 3,140 km/1,950 mi, orbiting 671,000 km/417,000 mi from the planet every 3.55 days. It is covered by ice and criss-crossed by thousands of thin cracks, each some 50,000 km/30,000 mi long.

NASA's robot probe, *Galileo,* began circling Europa in February 1997 and is expected to send back around 800 images from 50 different sites by 1999. NASA plans to send a $250 million robot probe in 2001 or 2002 to circle 100 km/60 mi above the surface, using radar and lasers to establish whether water exists beneath the icy surface. If found, a lander would be launched by 2006 to release a robot that would melt its way through the ice and release a submarine to take pictures. This operation is also billed at $250 million. One of the first discoveries was that what were thought to be cracks covering the surface of the moon are in fact low ridges. Further investigation is needed to determine their origin.

European Atomic Energy Community (Euratom) organization established by the second Treaty of Rome 1957, which seeks the cooperation of member states of the European Union in nuclear research and the rapid and large-scale development of nonmilitary nuclear energy.

European corn borer moth whose larvae are a serious menace in countries, such as the USA, where maize is an important crop.

European Southern Observatory observatory operated jointly by Belgium, Denmark, France, Germany, Italy, the Netherlands, Sweden, and Switzerland with headquarters near Munich. Its telescopes, located at La Silla, Chile, include a 3.6-m/142-in reflector opened 1976 and the 3.58-m/141-in New Technology Telescope opened in 1990. By 1988 work began on the Very Large Telescope, consisting of four 8-m/315-in reflectors mounted independently but capable of working in combination.

European Space Agency (ESA) organization of European countries (Austria, Belgium, Denmark, Finland, France, Germany, Ireland, Italy, the Netherlands, Norway, Spain, Sweden, Switzerland, and the UK) that engages in space research and technology. It was founded in 1975, with headquarters in Paris.

ESA has developed various scientific and communications satellites, the *Giotto* space probe, and the ◊Ariane *rockets. ESA built* ◊Spacelab, *and plans to build its own space station, Columbus,* for attachment to a US space station.

The ESA's earth-sensing satellite ERS-2 was launched successfully in 1995. It will work in tandem with ERS-1 launched in 1991, and should improve measurements of global ozone.

European Strategic Programme for Research and Development in Information Technology full name for ◊ESPRIT.

europium soft, greyish, metallic element of the ◊lanthanide series, symbol Eu, atomic number 63, relative atomic mass 151.96. It is used in lasers and as the red phosphor in colour televisions; its compounds are used to make control rods for nuclear reactors. It was named in 1901 by French chemist Eugène Demarçay (1852–1904) after the continent of Europe, where it was first found.

eusociality form of social life found in insects such as honey bees and termites, in which the colony is made up of special castes (for example, workers, drones, and reproductives) whose membership is biologically determined. The worker castes do not usually reproduce. Only one mammal, the naked mole rat, has a social organization of this type. A eusocial shrimp was discovered in 1996 living in the coral reefs of Belize. *Synalpheus regalis* lives in colonies of up to 300 individuals, all the offspring of a single reproductive female. See also ◊social behaviour.

Eustachian tube small air-filled canal connecting the middle ◊ear with the back of the throat. It is found in all land vertebrates and equalizes the pressure on both sides of the eardrum.

Eutelsat acronym for ***European Telecommunications Satellite Organization***.

euthanasia in medicine, mercy killing of someone with a severe and incurable condition or illness. Euthanasia is an issue that creates much controversy on medical and ethical grounds.

In Australia, a bill legalizing voluntary euthanasia for terminally ill patients was passed by the Northern Territory state legislature in May 1995.

In the Netherlands, where approximately 2,700 patients formally request it each year, euthanasia is technically illegal. However, provided guidelines issued by the Royal Dutch Medical Association are followed, doctors are not prosecuted.

In the USA, the Supreme Court ruled in June 1997 that the terminally ill do not have the fundamental right to have doctors help them to die. The Court upheld state laws in Washington and New York that forbid assisted suicides.

A patient's right to refuse life-prolonging treatment is recognized in several countries, including the UK.

Eutheria former division of class Mammalia, comprising all ◊mammals mammals with a placenta.

eutrophication excessive enrichment of rivers, lakes, and shallow sea areas, primarily by nitrate fertilizers washed from the soil by rain, by phosphates from fertilizers, and from nutrients in municipal sewage, and by sewage itself. These encourage the growth of algae and bacteria which use up the oxygen in the water, thereby making it uninhabitable for fishes and other animal life.

evaporation process in which a liquid turns to a vapour without its temperature reaching boiling point. A liquid left to stand in a saucer eventually evaporates because, at any time, a proportion of its molecules will be fast enough (have enough kinetic energy) to escape through the attractive intermolecular forces at the liquid surface into the atmosphere. The temperature of the liquid tends to fall because the evaporating molecules remove energy from the liquid. The rate of evaporation rises with increased temperature because as the mean kinetic energy of the liquid's molecules rises, so will the number possessing enough energy to escape.

A fall in the temperature of the liquid, known as the **cooling effect**, accompanies evaporation because as the faster molecules escape through the surface the mean energy of the remaining molecules falls. The effect may be noticed when wet clothes are worn, and sweat evaporates from the skin. ◊Refrigeration makes use of the cooling effect to extract heat from foodstuffs, and in the body it plays a part in temperature control.

Euthanasia

BY ROY PORTER

Euthanasia literally means a good or an easy death. Within traditional Christian culture, a good death (as prescribed by the *ars moriendi* – the art of dying well) was a Christian death, departing in a state of grace, denouncing Satan, praying to God, repenting one's sins and (for Roman Catholics) receiving the sacraments.

In the 17th century Francis Bacon argued that relief of suffering was a desideratum in terminal care and that the physician may sometimes hasten death. The idea of dying well was gradually secularized. Dying, it was said, should be like sleep, for a peaceful death betokened a serene conscience, a life well lived. In the new idea of euthanasia emerging in the 19th century, it was the duty of the doctor to ensure a peaceful death, by careful management and judicious application of opiates. At the wishes of family or patient, the family doctor was doubtless the frequent agent of informal (and strictly illegal) euthanasia.

This unofficial trend has been rendered more problematic in recent times. The Nazi 'final solution' perverted euthanasia for supremely evil purposes and created suspicion that any legalization of euthanasia would lead in time to (possibly compulsory) public euthanasia programmes for problem people and the senile. Moreover, death now increasingly occurs in public institutions, notably hospitals and hospices. This may make humane euthanasia more difficult, as physicians and nursing staff involved in such practices may fear exposure and legal prosecution from Christian pressure groups such as 'Life'.

Yet the circumstances of modern death have also increased backing for euthanasia. Thanks to life-support systems, it is now relatively easy to keep many 'dead' people artificially alive, with respirators and support systems. Repugnance is widespread for the 'cruelty' of this meaningless prolongation of life amongst those in a 'permanent vegetative state'. Hence there is a growing desire to devise acceptable procedures for mercy killing, promoted by bodies such as 'Exit'. Euthanasia may be squared with the professional ethics of the physician and with normal morality through the argument that while it is the doctor's duty to save life, that duty does not run so far as to prolong life through artificial means in all circumstances.

Change has been most marked in The Netherlands, where since 1984 the medical colleges have accepted medical euthanasia under strictly controlled circumstances. By 1995 a survey suggested that active euthanasia (a physician humanely intervening to end a terminally ill patient's life at the request of that patient) was taking place in around 1.8% of all deaths. Public acceptance of this practice had been facilitated by the development of 'living wills'. Since 1994 Dutch physicians have been legally obliged to honour such 'living wills'. 'Living wills' have been legally binding in South Australia since the 'Natural Death Act' of 1983.

Such proposals have met with a much more divided reception elsewhere. In Britain, euthanasia remains illegal, living wills have no standing, and 'Exit' has been subject to prosecution – as has the maverick American pathologist, Dr Jack Kevorkian, who has practised doctor-assisted suicide at the patient's request. A law (1996) in Australia's Northern Territories permitting voluntary euthanasia has since been overturned by the Federal Government. In 1997 the British Medical Association came out against voluntary euthanasia.

The advance of modern medicine presents deep ethical dilemmas in the case of death. There is no easy way to balance the sanctity of human life against the right of personal autonomy. Fundamental legal and moral questions are also raised.

evening primrose any of a group of plants that typically have pale yellow flowers which open in the evening. About 50 species are native to North America, several of which now also grow in Europe. Some are cultivated for their oil, which is rich in gamma-linoleic acid (GLA). The body converts GLA into substances which resemble hormones, and **evening primrose oil** is beneficial in relieving the symptoms of ◊premenstrual tension. It is also used in treating eczema and chronic fatigue syndrome. (Genus *Oenothera*, family Onagraceae.)

event-driven in computing, computer system that does not do anything until events are detected, such as mouse-clicks. Microsoft Windows is an event-driven operating environment.

evergreen in botany, a plant such as pine, spruce, or holly, that bears its leaves all year round. Most ◊conifers are evergreen. Plants that shed their leaves in autumn or during a dry season are described as ◊deciduous.

everlasting flower any flower head with coloured bracts that retains its colour when cut and dried.

Nature does not make jumps.

CAROLUS LINNAEUS Swedish naturalist and physician.
Philosophia Botanica

evolution the slow, gradual process of change from one form to another, as in the evolution of the universe from its formation to its present state, or in the evolution of life on Earth. In biology, it is the process by which life has developed by stages from single-celled organisms into the multiplicity of animal and plant life, extinct and existing, that inhabit the Earth. The development of the concept of evolution is usually associated with the English naturalist Charles ◊Darwin who attributed the main role in evolutionary change to ◊natural selection acting on randomly occurring variations. However, these variations in species are now known to be ◊adaptations produced by spontaneous changes or ◊mutations in the genetic material of organisms.

EVOLUTION: THEORY AND HISTORY

http://www.ucmp.berkeley.edu/history/evolution.html

Dedicated to the study of the history and theories associated with evolution, this site explores topics on classification, taxonomy, and dinosaur discoveries, and then looks at the key figures in the field and reviews their contributions.

evolution and creationism Organic evolution traces the development of simple unicellular forms to more complex forms, ultimately to the flowering plants and vertebrate animals, including humans. The Earth contains an immense diversity of living organisms: about a million different species of animals and half a million species of plants have so far been described. Some religions deny the theory of evolution considering it conflicts with their belief that God created all things (see ◊creationism). But most people accept that there is overwhelming evidence the diversity of life arose by a gradual process of evolutionary divergence and not by individual acts of divine creation. There are several lines of evidence: the fossil record, the existence of similarities or homologies between different groups of organisms, embryology, and geographical distribution.

We must, however, acknowledge, as it seems to me, that man with all his noble qualities, still bears in his bodily frame the indelible stamp of his lowly origin.

CHARLES DARWIN British naturalist.
Last words of *Descent of Man* 1871

evolutionary stable strategy (ESS) in ◊sociobiology, an assemblage of behavioural or physical characters (collectively termed a 'strategy') of a population that is resistant to replacement by any forms bearing new traits, because the new traits will not be capable of successful reproduction.

ESS analysis is based on ◊game theory and can be applied both to genetically determined physical characters (such as horn length), and to learned behavioural responses (for example, whether to fight or retreat from an opponent). An ESS may be conditional on the context, as in the rule 'fight if the opponent is smaller, but retreat if the opponent is larger'.

evolutionary toxicology study of the effects of pollution on evolution. A polluted habitat may cause organisms to select for certain traits, as in **industrial melanism** for example, where some insects, such as the peppered moth, are darker in polluted areas, and therefore better camouflaged against predation.

Pollutants may also trigger mutations, for example, voles living around the Chernobyl exploded nuclear reactor have a very high mutation rate despite appearing healthy and reproducing successfully. Fish in polluted rivers also exhibit mutations.

excavation or ***dig*** in archaeology, the systematic recovery of data through the exposure of buried sites and artefacts. Excavation is destructive, and is therefore accompanied by a comprehensive recording of all material found and its three-dimensional locations (its context). As much material and information as possible must be recovered from any dig. A full record of all the techniques employed in the excavation itself must also be made, so that future archaeologists will be able to evaluate the results of the work accurately.

Besides being destructive, excavation is also costly. For both these reasons, it should be used only as a last resort. It can be partial, with only a sample of the site investigated, or total. Samples are chosen either intuitively, in which case excavators investigate those areas they feel will be most productive, or statistically, in which case the sample is drawn using various statistical techniques, so as to ensure that it is representative.

An important goal of excavation is a full understanding of a site's stratigraphy; that is, the vertical layering of a site.

These layers or levels can be defined naturally (for example, soil changes), culturally (for example, different occupation levels), or arbitrarily (for example, 10 cm/4 in levels). Excavation can also be done horizontally, to uncover larger areas of a particular layer and reveal the spatial relationships between artefacts and features in that layer. This is known as open-area excavation and is used especially where single-period deposits lie close to the surface, and the time dimension is represented by lateral movement rather than by the placing of one building on top of the preceding one.

Most excavators employ a flexible combination of vertical and horizontal digging adapting to the nature of their site and the questions they are seeking to answer.

excavator machine designed for digging in the ground, or for earth-moving in general. Diggers with hydraulically powered digging arms are widely used on building sites. They may run on wheels or on ◊caterpillar tracks. The largest excavators are the draglines used in mining to strip away earth covering the coal or mineral deposit.

Excel ◊spreadsheet program produced by ◊Microsoft in 1985. Versions are available for PC-compatibles running ◊Windows and for the Apple Macintosh. Excel pioneered many advanced features in the ease of use of spreadsheets, and has displaced ◊Lotus 1–2–3 as the standard spreadsheet program.

exclusion principle in physics, a principle of atomic structure originated by Austrian–US physicist Wolfgang ◊Pauli. It states that no two electrons in a single atom may have the same set of ◊quantum numbers.

Hence, it is impossible to pack together certain elementary particles, such as electrons, beyond a certain critical density, otherwise they would share the same location and quantum number. A white dwarf star, which consists of electrons and other elementary particles, is thus prevented from contracting further by the exclusion principle and never collapses.

excretion in biology, the removal of waste products from the cells of living organisms. In plants and simple animals, waste products are removed by diffusion, but in higher animals they are removed by specialized organs. In mammals the kidneys are the principle organs of excretion. Water and metabolic wastes are also excreted in the faeces and, in humans, through the sweat glands; carbon dioxide and water are removed via the lungs.

TO REMEMBER THE EXCRETORY ORGANS OF THE BODY:

SKILLSKIN / KIDNEYS / INTESTINES / LIVER / LUNGS

executable file in computing, a file – always a program of some kind – that can be run by the computer directly. The file will have been generated from a ◊source program by an ◊assembler or ◊compiler. It will therefore not be coded in ◊ASCII and will not be readable as text. On ◊MS-DOS systems executable files have an .EXE or .COM extension.

exfoliation in biology, the separation of pieces of dead bone or skin in layers.

exobiology study of life forms that may possibly exist elsewhere in the universe, and of the effects of extraterrestrial environments on Earth organisms. Techniques include space probe experiments designed to detect organic molecules, and the monitoring of radio waves from other star systems.

exocrine gland gland that discharges secretions, usually through a tube or a duct, on to a surface. Examples include sweat glands which release sweat on to the skin, and digestive glands which release digestive juices on to the walls of the intestine. Some animals also have ◊endocrine glands (ductless glands) that release hormones directly into the bloodstream.

exon in genetics, a sequence of bases in ◊DNA that codes for a protein. Exons make up only 2% of the body's total DNA. The remainder is made up of ◊introns. During RNA processing the introns are cut out of the molecule and the exons spliced together.

exoskeleton the hardened external skeleton of insects, spiders, crabs, and other arthropods. It provides attachment for muscles and protection for the internal organs, as well as support. To permit growth it is periodically shed in a process called ◊ecdysis.

exosphere the uppermost layer of the ◊atmosphere. It is an ill-defined zone above the thermosphere, beginning at about 700 km/435 mi and fading off into the vacuum of space. The gases are extremely thin, with hydrogen as the main constituent.

exothermic reaction a chemical reaction during which heat is given out (see ◊energy of reaction).

expanded memory in computing, additional memory in an ◊MS-DOS-based computer, usually installed on an expanded-memory board. Expanded memory requires an expanded-memory manager, which gives access to a limited amount of memory at any one time, and is slower to use than ◊extended memory. Software is available under both MS-DOS and ◊Windows to simulate expanded memory for those applications that require it.

expansion in physics, the increase in size of a constant mass of substance caused by, for example, increasing its temperature

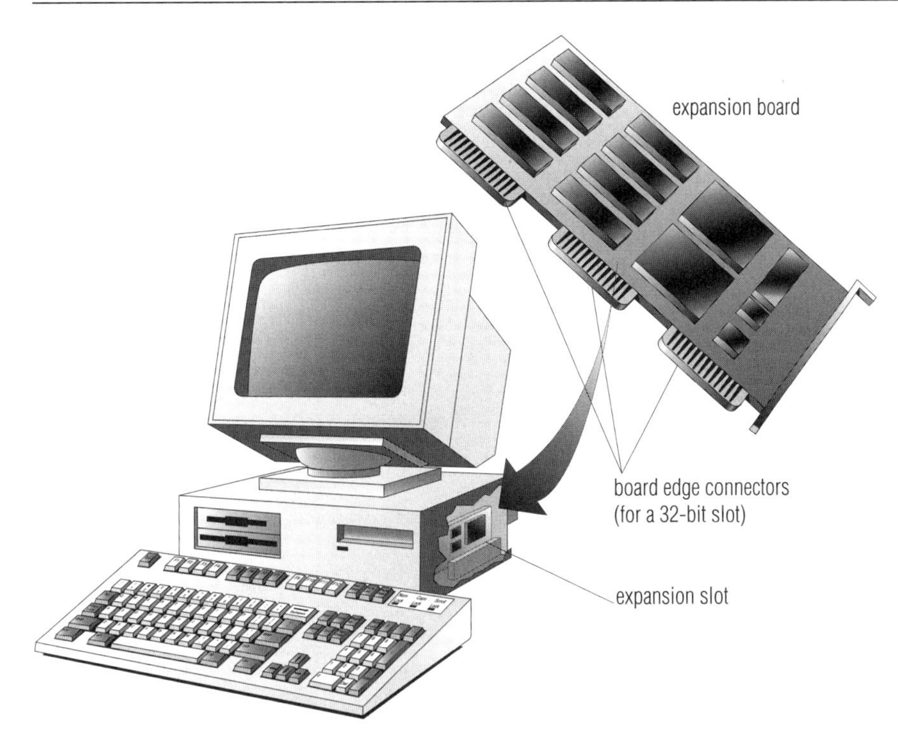

expansion board An expansion board may be fitted into any free expansion slot in a computer to provide additional facilities or functionality.

(◊thermal expansion) or its internal pressure. The **expansivity**, or coefficient of thermal expansion, of a material is its expansion (per unit volume, area, or length) per degree rise in temperature.

expansion board or ***expansion card*** printed circuit board that can be inserted into a computer in order to enhance its capabilities (for example, to increase its memory) or to add facilities (such as graphics).

expectorant any substance, often added to cough mixture, to encourage secretion of mucus in the airways to make it easier to cough up. It is debatable whether expectorants have an effect on lung secretions.

experiment in science, a practical test designed with the intention that its results will be relevant to a particular theory or set of theories. Although some experiments may be used merely for gathering more information about a topic that is already well understood, others may be of crucial importance in confirming a new theory or in undermining long-held beliefs.

The manner in which experiments are performed, and the relation between the design of an experiment and its value, are therefore of central importance. In general, an experiment is of most value when the factors that might affect the results (variables) are carefully controlled; for this reason most experiments take place in a well-managed environment such as a laboratory or clinic.

experimental psychology the application of scientific methods to the study of mental processes and behaviour.

EXPERIMENTAL PSYCHOLOGY LAB

http://www.uni-tuebingen.de/uni/sii/Ulf/Lab/WebExpPsyLab.html

Take part in online psychology experiments carried out by the University of Tübingen's Psychology Institute. There are a variety of experiments here, some requiring Java and ActiveX, and some only in German.

This covers a wide range of fields of study, including: **human and animal learning**, in which learning theories describe how new behaviours are acquired and modified; **cognition**, the study of a number of functions, such as perception, attention, memory, and language; and **physiological psychology**, which relates the study of cognition to different regions of the brain. **Artificial intelligence** refers to the computer simulation of cognitive processes, such as language and problem-solving.

expert system computer program for giving advice (such as diagnosing an illness or interpreting the law) that incorporates knowledge derived from human expertise. It is a kind of ◊knowledge-based system containing rules that can be applied to find the solution to a problem. It is a form of ◊artificial intelligence.

expire in computing, function for removing old USENET articles from an off-line reader program. Sometimes also called 'prune' or 'purge', this function is necessary to make room for new articles.

explanation in science, an attempt to make clear the cause of any natural event by reference to physical laws and to observations.

The extent to which any explanation can be said to be true is one of the chief concerns of philosophy, partly because observations may be wrongly interpreted, partly because explanations should help us predict how nature will behave. Although it may be reasonable to expect that a physical law will hold in the future, that expectation is problematic in that it relies on induction, a much criticized feature of human thought; in fact no explanation, however 'scientific', can be held to be true for all time, and thus the difference between a scientific and a common-sense explanation remains the subject of intense philosophical debate.

Explorer series of US scientific satellites. *Explorer 1,* launched January 1958, was the first US satellite in orbit and discovered the Van Allen radiation belts around the Earth.

explosive any material capable of a sudden release of energy and the rapid formation of a large volume of gas, leading, when compressed, to the development of a high-pressure wave (blast).

types of explosive Combustion and explosion differ essentially only in rate of reaction, and many explosives (called **low explosives**) are capable of undergoing relatively slow combustion under suitable conditions. **High explosives** produce uncontrollable blasts. The first low explosive was ◊gunpowder; the first high explosive was ◊nitroglycerine.

exponent or ***index*** in mathematics, a number that indicates the number of times a term is multiplied by itself; for example $x^2 = x$ x x, $4^3 = 4$ x 4 x 4.

Exponents obey certain rules. Terms that contain them are multiplied together by adding the exponents; for example, x^2 x $x^5 = x^7$. Division of such terms is done by subtracting the exponents; for example, $y^5 \div y^3 = y^2$. Any number with the exponent 0 is equal to 1; for example, $x^0 = 1$ and $99^0 = 1$.

exponential in mathematics, descriptive of a ◊function in which the variable quantity is an exponent (a number indicating the power to which another number or expression is raised).

Exponential functions and series involve the constant e = 2.71828.... . Scottish mathematician John Napier devised natural ◊logarithms in 1614 with e as the base.

Exponential functions are basic mathematical functions, written as e*x* or exp *x*. The expression e*x* has five definitions, two of which are: (i) e*x* is the solution of the differential equation d*x*/d*t* = *x* (*x* =

1 if $t = 0$); (ii) ex is the limiting sum of the infinite series $1 + x + (x2/2!) + (x3/3!) + ... + (xn/n!)$.

export file in computing, a file stored by the computer in a standard format so that it can be accessed by other programs, possibly running on different makes of computer.

For example, a word-processing program running on an Apple ◊Macintosh computer may have a facility to save a file on a floppy disc in a format that can be read by a word-processing program running on an IBM PC-compatible computer. When the file is being read by the second program or computer, it is often referred to as an **import file**.

exposure meter instrument used in photography for indicating the correct exposure – the length of time the camera shutter should be open under given light conditions. Meters use substances such as cadmium sulphide and selenium as light sensors. These materials change electrically when light strikes them, the change being proportional to the intensity of the incident light. Many cameras have a built-in exposure meter that sets the camera controls automatically as the light conditions change.

extended memory in computing, memory in an ◊MS-DOS-based system that exceeds the 1 Mbyte that DOS supports. Extended memory is not accessible to the ◊operating system and requires an extended memory manager.

extensor a muscle that straightens a limb.

exterior angle one of the four external angles formed when a straight line or transveral cuts through a pair of (usually parallel) lines. Also, an angle formed by extending a side of a polygon.

external modem in computing, a ◊modem that is a self-contained unit sitting outside a personal computer (PC) and connected to it by a cable. There are two main types of external modem: mains-powered desktop modems and: credit-card sized modems that fit the PCMCIA slots in notebook and handheld computers.

External modems have the advantage over internal ones in that they are easy to move from computer to computer as needed. However, high-speed modems outstrip the capabilities of the serial ports on older and cheaper PCs by taking in data too fast for the computer to be able to read it.

extinction in biology, the complete disappearance of a species or higher taxon. Extinctions occur when an animal becomes unfit for survival in its natural habitat usually to be replaced by another, better-suited animal. An organism becomes ill-suited for survival because its environment is changed or because its relationship to other organisms is altered. For example, a predator's fitness for survival depends upon the availability of its prey.

past extinctions Mass extinctions are episodes during which large numbers of species have become extinct virtually simultaneously, the best known being that of the dinosaurs, other large reptiles, and various marine invertebrates about 65 million years ago between the end of the ◊Cretaceous period and the beginning of the Tertiary period. The latter, known as the **K–T extinction**, has been attributed to catastrophic environmental changes following a meteor impact or unusually prolonged and voluminous volcanic eruptions.

Another mass extinction occurred about 10,000 years ago when many giant species of mammal died out. This is known as the 'Pleistocene overkill' because their disappearance was probably hastened by the hunting activities of prehistoric humans. The greatest mass extinction occurred about 250 million years ago, marking the Permian–Triassic boundary (see ◊geological time), when up to 96% of all living species became extinct. Mass extinctions apparently occur at periodic intervals of approximately 26 million years.

current extinctions Humans have the capacity to profoundly influence many habitats and today a large number of extinctions are attributable to human activity. Some species, such as the ◊dodo of Mauritius, the ◊moas of New Zealand, and the passenger ◊pigeon of North America, were exterminated by hunting. Others became extinct when their habitat was destroyed. ◊Endangered species are close to extinction. The rate of extinction is difficult to estimate, but appears to have been accelerated by humans. Conservative estimates put the rate of loss due to deforestation alone at 4,000 to 6,000 species a year. Overall, the rate could be as high as one species an hour, with the loss of one species putting those dependent on it at risk. Australia has the worst record for extinction: 18 mammals have disappeared since Europeans settled there, and 40 more are threatened.

> *If we are still here to witness the destruction of our planet [by the Sun] some five billion years or more hence, then we will have achieved something so unprecedented in the history of life that we should be willing to sing our swan song with joy.*
>
> STEPHEN JAY GOULD US palaeontologist and writer. *The Panda's Thumb* 1980

extracellular matrix strong material naturally occurring in animals and plants, made up of protein and long-chain sugars (polysaccharides) in which cells are embedded. It is often called a 'biological glue', and forms part of ◊connective tissues such as bone and skin.

The cell walls of plants and bacteria, and the ◊exoskeletons of insects and other arthropods, are also formed by types of extracellular matrix.

extrasensory perception (ESP) any form of perception beyond and distinct from the known sensory processes. The main forms of ESP are clairvoyance (intuitive perception or vision of events and situations without using the senses); precognition (the ability to foresee events); and telepathy or thought transference (communication between people without using any known visible, tangible, or audible medium). Verification by scientific study has yet to be achieved.

extroversion or ***extraversion*** personality dimension described by the psychologists Carl ◊Jung and, later, Hans Eysenck. The typical extrovert is sociable, impulsive, and carefree. The opposite of extroversion is ◊introversion.

extruded shape in computer graphics, a three-dimensional shape created by extending a two-dimensional shape along a third dimension.

extrusion common method of shaping metals, plastics, and other materials. The materials, usually hot, are forced through the hole in a metal die and take its cross-sectional shape. Rods, tubes, and sheets may be made in this way.

extrusive rock or ***volcanic rock*** ◊igneous rock formed on the surface of the Earth by volcanic activity. The term includes fine-grained crystalline or glassy rocks formed from hot lava quenched at or near Earth's surface and rocks composed of solid debris, called pyroclastics, deposted by explosive eruptions.

Large amounts of extrusive rock called ◊basalt form at the Earth's ◊ocean ridges from lava that fills the void formed when two tectonic plates spread apart. Explosive volcanoes that deposit pyroclastics generally occur where one tectonic plate descends beneath another. ◊Andesite is often formed by explosive volcanoes. Magmas that give rise to pyroclastic extrusive rocks are explosive because they are viscous. The island of Montserrat, West Indies, is an example of an explosive volcano that spews pyroclastics of andesite composition. Magmas that produce crystalline or glassy volcanic rocks upon cooling are less viscous. The low viscosity allows the extruding lava to flow easily. Fluid-like lavas that flow from the volcanoes of the Hawaiian Islands have low viscosity and cool to form basalt.

eye the organ of vision. In the human eye, the light is focused by the combined action of the curved **cornea**, the internal fluids, and the **lens**. The insect eye is compound – made up of many separate facets – known as ommatidia, each of which collects light and directs it separately to a receptor to build up an image. Invertebrates have much simpler eyes, with no lenses. Among molluscs, cephalopods have complex eyes similar to those of vertebrates.

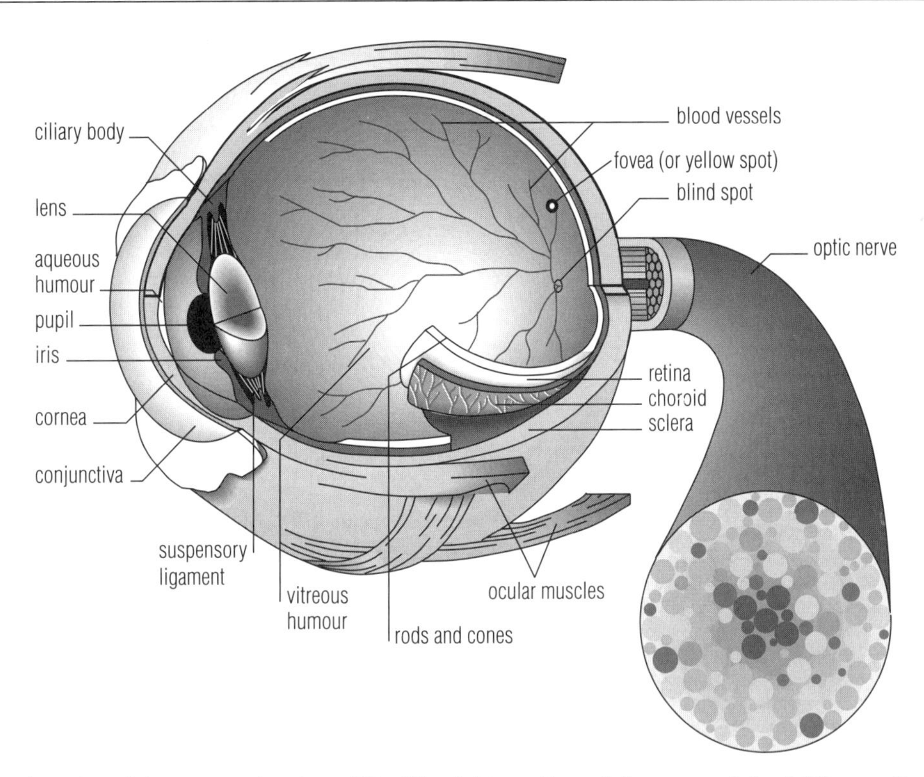

eye *The human eye. The retina of the eye contains about 137 million light-sensitive cells in an area of about 650 sq mm/1 sq in. There are 130 million rod cells for black and white vision and 7 million cone cells for colour vision. The optic nerve contains about 1 million nerve fibres. The focusing muscles of the eye adjust about 100,000 times a day. To exercise the leg muscles to the same extent would need an 80 km/50 mi walk.*

The mantis shrimp's eyes contain ten colour pigments with which to perceive colour; some flies and fishes have five, while the human eye has only three.

human eye This is a roughly spherical structure contained in a bony socket. Light enters it through the cornea, and passes through the circular opening (**pupil**) in the iris (the coloured part of the eye).

The ciliary muscles act on the lens (the rounded transparent structure behind the iris) to change its shape, so that images of objects at different distances can be focused on the retina. This is at the back of the eye, and is packed with light-sensitive cells (rods and cones), connected to the brain by the optic nerve.

Don't believe what your eyes are telling you, all they show is limitation. Look with your understanding, find out what you already know, and you'll see the way to fly.

RICHARD BACH US writer.
Jonathan Livingstone Seagull

eyebright any of a group of annual plants belonging to the figwort family. They are 2–30 cm/1–12 in high and have whitish flowers streaked with purple. The name indicates their traditional use as an eye-medicine. (Genus *Euphrasia*, family Scrophulariaceae.)

eye, defects of the abnormalities of the eye that impair vision. Glass or plastic lenses, in the form of spectacles or contact lenses, are the usual means of correction. Common optical defects are ◊short-sightedness or myopia; farsightedness or hypermetropia; lack of ◊accommodation or presbyopia; and ◊astigmatism. Other eye defects include ◊colour blindness.

e-zine (contraction of ***electronic magazine***) in computing, periodical sent by ◊e-mail. E-zines can be produced very cheaply, as there are no production costs for design and layout, and minimal costs for distribution. Like printed magazines, e-zines typically have multiple contributors and an editor responsible for selecting content.

One of the best-known e-zines is the ◊*Computer Underground Digest*, which tracks battles over freedom of speech online and issues concerning hacking and computer crime.

Many things about our bodies would not seem so filthy and obscene if we did not have the idea of nobility in our heads.

G C Lichtenberg German physicist and philosopher. *Aphorisms*, 'Notebook D' 6

°F symbol for degrees ◊Fahrenheit.

F1 on personal computers (PCs), the key to access ◊online help.

face in geometry, a plane surface of a solid enclosed by edges. A cube has six square faces, a cuboid has six rectangular faces, and a tetrahedron has four triangular faces.

facies in geology, body of rock strata possessing unifying characteristics indicative of the environment in which the rocks were formed. The term is also used to describe the environment of formation itself or unifying features of the rocks that comprise the facies.

Features that define a facies can include collections of fossils, sequences of rock layers, or the occurrence of specific minerals. Sedimentary rocks deposited at the same time, but representing different facies belong to a single **chronostratigraphic unit** (see ◊stratigraphy). But these same rocks may belong to different **lithostratigraphic units**. For example, beach sand is deposited at the same time that mud is deposited further offshore. The beach sand eventually turns to sandstone while the mud turns to shale. The resulting ◊sandstone and ◊shale **strata** comprise two different facies, one representing the beach environment and the other the offshore environment, formed at the same time; the sandstone and shale belong to the same chronostratigraphic unit but distinct lithostratigraphic units.

facsimile transmission full name for ◊fax **or telefax**.

factor a number that divides into another number exactly. For example, the factors of 64 are 1, 2, 4, 8, 16, 32, and 64. In algebra, certain kinds of polynomials (expressions consisting of several or many terms) can be factorized. For example, the factors of $x^2 + 3x + 2$ are $x + 1$ and $x + 2$, since $x^2 + 3x + 2 = (x + 1)(x + 2)$. This is called factorization. See also ◊number.

factorial of a positive number, the product of all the whole numbers (integers) inclusive between 1 and the number itself. A factorial is indicated by the symbol '!'. Thus $6! = 1 \times 2 \times 3 \times 4 \times 5 \times 6 = 720$. Factorial zero, 0!, is defined as 1.

factory farming intensive rearing of poultry or other animals for food, usually on high-protein foodstuffs in confined quarters. Chickens for eggs and meat, and calves for veal are commonly factory farmed. Some countries restrict the use of antibiotics and growth hormones as aids to factory farming because they can persist in the flesh of the animals after they are slaughtered. The emphasis is on productive yield rather than animal welfare so that conditions for the animals are often very poor. For this reason, many people object to factory farming on moral as well as health grounds.

Egg-laying hens are housed in 'batteries' of cages arranged in long rows. If caged singly, they lay fewer eggs, so there are often four to a cage with a floor area of only 2,400 sq cm/372 sq in. In the course of a year, battery hens average 261 eggs each, whereas for free-range chickens the figure is 199.

faeces remains of food and other waste material eliminated from the digestive tract of animals by way of the anus. Faeces consist of quantities of fibrous material, bacteria and other microorganisms, rubbed-off lining of the digestive tract, bile fluids, undigested food, minerals, and water.

Fahrenheit, Gabriel Daniel (1686–1736) Polish-born Dutch physicist who invented the first accurate thermometer in 1724 and devised the Fahrenheit temperature scale. Using his thermometer, Fahrenheit was able to determine the boiling points of liquids and found that they vary with atmospheric pressure.

Fahrenheit scale temperature scale invented in 1714 by Gabriel Fahrenheit which was commonly used in English-speaking countries until the 1970s, after which the ◊Celsius scale was generally adopted, in line with the rest of the world. In the Fahrenheit scale, intervals are measured in degrees (°F); °F = (°C x 9/5) + 32.

Fahrenheit took as the zero point the lowest temperature he could achieve anywhere in the laboratory, and, as the other fixed point, body temperature, which he set at 96°F. On this scale, water freezes at 32°F and boils at 212°F.

fainting sudden, temporary loss of consciousness caused by reduced blood supply to the brain. It may be due to emotional shock or physical factors, such as pooling of blood in the legs from standing still for long periods.

falcon any bird of prey of the genus *Falco,* family Falconidae, order Falconiformes. Falcons are the smallest of the hawks (15–60 cm/6–24 in). They have short curved beaks with one tooth in the upper mandible; the wings are long and pointed, and the toes elongated. They nest in high places and kill their prey on the wing by 'stooping' (swooping down at high speed). They include the peregrine and kestrel.

The peregrine falcon *F. peregrinus,* up to about 50 cm/1.8 ft long, has become re-established in North America and Britain after near extinction (by pesticides, gamekeepers, and egg collectors). When stooping on its intended prey, it is the fastest creature in the world, timed at 240 kph/150 mph. The US government announced preliminary measures to remove the American peregrine falcon from the endangered species list in 1995 following a successful breeding programme.

Fallopian tube or ***oviduct*** in mammals, one of two tubes that carry eggs from the ovary to the uterus. An egg is fertilized by sperm in the Fallopian tubes, which are lined with cells whose ◊cilia move the egg towards the uterus.

fallout harmful radioactive material released into the atmosphere in the debris of a nuclear explosion and descending to the surface of the Earth. Such material can enter the food chain, cause◊radiation sickness, and last for hundreds of thousands of years (see ◊half-life).

fallow land ploughed and tilled, but left unsown for a season to allow it to recuperate. In Europe, it is associated with the medieval three-field system. It is used in some modern ◊crop rotations and in countries that do not have access to fertilizers to maintain soil fertility.

fallow deer one of two species of deer. Fallow deer are characterized by the expansion of the upper part of their antlers in palmate form. Usually they stand about 1 m/3.3 ft high, and have small heads, large ears, and rather long tails. In colour they are fawn, with a number of large white spots, or they may be yellowish-brown or, more rarely, dark brown.

The commonest species *Dama dama* is a native of North Africa and the countries bordering the Mediterranean, but was introduced into Britain at an early period. *D. mesopotamica* is a native of the mountains of Iran.

classification The fallow deer is in genus *Dama* in the Cervidae family of the mammalian order Artiodactyla.

false of a statement, untrue. Falseness is used in proving propositions by considering the negative of the proposition to be true, then making deductions until a contradiction is reached which proves the negative to be false and the proposition to be true.

false-colour imagery graphic technique that displays images in false (not true-to-life) colours so as to enhance certain features. It is widely used in displaying electronic images taken by spacecraft; for example, Earth-survey satellites such as *Landsat*. Any colours can be selected by a computer processing the received data.

falsificationism in philosophy of science, the belief that a scientific theory must be under constant scrutiny and that its merit lies only in how well it stands up to rigorous testing. It was first expounded by philosopher Karl Popper in his *Logic of Scientific Discovery* 1934.

Such thinking also implies that a theory can be held to be scientific only if it makes predictions that are clearly testable. Critics of this belief acknowledge the strict logic of this process, but doubt whether the whole of scientific method can be subsumed into so narrow a programme. Philosophers and historians such as Thomas Kuhn and Paul Feyerabend have attempted to use the history of science to show that scientific progress has resulted from a more complicated methodology than Popper suggests.

family in biological classification, a group of related genera (see ◊genus). Family names are not printed in italic (unlike genus and species names), and by convention they all have the ending -idae (animals) or -aceae (plants and fungi). For example, the genera of hummingbirds are grouped in the hummingbird family, Trochilidae. Related families are grouped together in an ◊order.

family planning spacing or preventing the birth of children. Access to family-planning services (see ◊contraceptive) is a significant factor in women's health as well as in limiting population growth. If all those women who wished to avoid further childbirth were able to do so, the number of births would be reduced by 27% in Africa, 33% in Asia, and 35% in Latin America; and the number of women who die during pregnancy or childbirth would be reduced by about 50%.

The average number of pregnancies per woman is two in the industrialized countries, where 71% use family planning, as compared to six or seven pregnancies per woman in the Third World. According to a World Bank estimate, doubling the annual $2 billion spent on family planning would avert the deaths of 5.6 million infants and 250,000 mothers each year.

fanjet another name for ◊turbofan, the jet engine used by most airliners.

fantail variety of domestic ◊pigeon, often white, with a large, widely fanning tail.

FAQ (abbreviation for ***frequently asked questions***) in computing, file of answers to commonly asked questions on any topic. First used on USENET, where regular posters to ◊newsgroups got tired of answering the same questions over and over and wrote these information files to end the repetition. 'Newbies' are recommended to read the FAQ for any newsgroup they join before posting. Most FAQs are available via FTP from rtfm.mit.edu or via the World-Wide Web from **http://www.cis.ohio-state.edu/htbin/search-usenet-faqs**. By 1996 FAQ was a common term for any information file, on line or off line.

farad SI unit (symbol F) of electrical capacitance (how much electric charge a ◊capacitor can store for a given voltage). One farad is a capacitance of one ◊coulomb per volt. For practical purposes the microfarad (one millionth of a farad, symbol μF) is more commonly used.

faraday unit of electrical charge equal to the charge on one mole of electrons. Its value is 9.648×10^4 coulombs.

Faraday's constant constant (symbol *F*) representing the electric charge carried on one mole of electrons. It is found by multiplying Avogadro's constant by the charge carried on a single electron, and is equal to 9.648×10^4 coulombs per mole.

One **faraday** is this constant used as a unit. The constant is used to calculate the electric charge needed to discharge a particular quantity of ions during ◊electrolysis.

Faraday's laws three laws of electromagnetic induction, and two laws of electrolysis, all proposed originally by English scientist Michael Faraday:

induction (1) a changing magnetic field induces an electromagnetic force in a conductor; (2) the electromagnetic force is proportional to the rate of change of the field; (3) the direction of the induced electromagnetic force depends on the orientation of the field.

electrolysis (1) the amount of chemical change during electrolysis is proportional to the charge passing through the liquid; (2) the amount of chemical change produced in a substance by a given amount of electricity is proportional to the electrochemical equivalent of that substance.

far point the farthest point that a person can see clearly. The eye is unable to focus a sharp image on the retina of an object beyond this point. The far point for a normal eye should be at infinity; any eye that has a far point nearer than this is short-sighted (see ◊short-sightedness).

fast breeder or ***breeder reactor*** alternative names for ◊fast reactor, a type of nuclear reactor.

fasting the practice of voluntarily going without food. It can be undertaken as a religious observance, a sign of mourning, a political protest (hunger strike), or for slimming purposes.

Fasting or abstinence from certain types of food or beverages occurs in most religious traditions. It is seen as an act of self-discipline that increases spiritual awareness by lessening dependence on the material world. In the Roman Catholic church, fasting is seen as a penitential rite, a means to express repentance for sin. The most commonly observed Christian fasting is in Lent, from Ash Wednesday to Easter Sunday, and recalls the 40 days Jesus spent in the wilderness. Roman Catholics and Orthodox Christians usually fast before taking communion and monastic communities observe regular weekly fasts. Devout Muslims go without food or water between sunrise and sunset during the month of Ramadan. Jews fast for *Yom Kippur* and before several other festivals. Many devout Hindus observe a weekly day of partial or total fast.

Total abstinence from food for a limited period is prescribed by some ◊naturopaths to eliminate body toxins or make available for recuperative purposes the energy normally used by the digestive system. Prolonged fasting can be dangerous. The liver breaks up its fat stores, releasing harmful by-products called ketones, which results in a condition called ketosis, which develops within three days. An early symptom is a smell of pear drops on the breath. Other symptoms include nausea, vomiting, fatigue, dizziness, severe depression, and irritability. Eventually the muscles and other body tissues become wasted, and death results.

fast reactor or ***fast breeder reactor*** ◊nuclear reactor that makes use of fast neutrons to bring about fission. Unlike other reactors used by the nuclear-power industry, it has little or no ◊moderator, to slow down neutrons. The reactor core is surrounded by a 'blanket' of uranium carbide. During operation, some of this uranium is converted into plutonium, which can be extracted and later used as fuel.

Fast breeder reactors can extract about 60 times the amount of energy from uranium that thermal reactors do. In the 1950s, when uranium stocks were thought to be dwindling, the fast breeder was considered to be the reactor of the future. Now, however, when new uranium reserves have been found and because of various technical difficulties in their construction, development of the fast breeder has slowed in most parts of the world.

fat in the broadest sense, a mixture of ◊lipids – chiefly triglycerides (lipids containing three ◊fatty acid molecules linked to a molecule of glycerol). More specifically, the term refers to a lipid mixture that is solid at room temperature (20°C); lipid mixtures that are liquid at room temperature are called **oils**. The higher the proportion of saturated fatty acids in a mixture, the harder the fat.

Boiling fats in strong alkali forms soaps (saponification). Fats are essential constituents of food for many animals, with a calorific value twice that of carbohydrates; however, eating too much fat, especially fat of animal origin, has been linked with heart disease

Faraday, Michael
(1791–1867)

English chemist and physicist. In 1821, he began experimenting with electromagnetism, and discovered the induction of electric currents and made the first dynamo, the first electric motor, and the first transformer. Faraday isolated benzene from gas oils and produced the basic laws of electrolysis in 1834. He also pointed out that the energy of a magnet is in the field around it and not in the magnet itself, extending this basic conception of field theory to electrical and gravitational systems.

chemistry and the discovery of benzene Faraday was mainly interested in chemistry during his early years at the Royal Institution. He investigated the effects of including precious metals in steel in 1818, producing high-quality alloys that later stimulated the production of special high-grade steels. In 1823, Faraday produced liquid chlorine by heating crystals of chlorine hydrate in an inverted U-tube, one limb of which was heated and the other placed in a freezing mixture. After the production of liquid carbon dioxide in 1835, he used this coolant to liquefy other gases. In the same year, Faraday isolated benzene from gas oils and demonstrated the use of platinum as a catalyst. He also demonstrated the importance in chemical reactions of surfaces and inhibitors, foreshadowing a huge area of the modern chemical industry.

laws of electrolysis Faraday's laws of electrolysis established the link between electricity and chemical affinity, one of the most fundamental concepts in science. Electrolysis is the production of chemical changes by passing an electric current through a solution. It was Faraday who coined the terms anode, cathode, cation, anion, electrode, and electrolyte. He postulated that, during the electrolysis of an aqueous electrolyte, positively-charged cations move towards the negatively-charged cathode and negatively-charged anions migrate to the positively-charged anode. Faraday demonstrated that the ions are discharged at each electrode according to the following rules:

(a) the quantity of a substance produced is proportional to the amount of electricity passed;

(b) the relative quantities of different substances produced by the same amount of electricity are proportional to their equivalent weights (that is, the relative atomic mass divided by the oxidation state or valency).

electromagnetism and the electric motor In 1821, only one year after Hans Oersted had discovered with a compass needle that a current of electricity flowing through a wire produces a magnetic field, Faraday was asked to investigate the phenomenon of electromagnetism by the editor of the *Philosophical Magazine*. Faraday conceived that circular lines of magnetic force are produced around the wire to explain the orientation of Oersted's compass needle.

Faraday's conviction that an electric current gives rise to lines of magnetic force arose from his idea that electricity was a form of vibration and not a moving fluid. He believed that electricity was a state of varying strain in the molecules of the wire conductor, and that this gave rise to a similar strain in the medium surrounding the conductor. It was reasonable to consider therefore that the transmitted strain might set up a similar strain in the molecules of another nearby conductor.

private collection

Faraday set about devising an apparatus that would demonstrate the conversion of electrical energy into motive force. His device consisted of two vessels of mercury connected to a battery. Above the vessels and connected to each other were suspended a magnet and a wire, which were free to move and dipped just below the surface of the mercury. In the mercury were fixed a wire and a magnet respectively. When the current was switched on, it flowed through both the fixed and free wires, generating a magnetic field in them. This caused the free magnet to revolve around the fixed wire, and the free wire to revolve around the fixed magnet.

The experiment demonstrated the basic principles governing the electric motor. Although the practical motors that subsequently developed had a very different form to Faraday's apparatus, he is usually credited with the invention of the electric motor.

electromagnetic induction and the transformer Faraday hunted for the effect of electromagnetic induction from 1824 onwards, expecting to find that a magnetic field would induce a steady electric current in a conductor. Faraday eventually succeeded in producing induction in 1831. He wound two coils around an iron bar and connected one to a battery and the other to a galvanometer (an instrument for detecting small electric currents by their magnetic effect). Nothing happened when the current flowed through the first coil, but Faraday noticed that the galvanometer responded whenever the current was switched on or off. Faraday found an immediate explanation with his lines of force. If the lines of force were cut – that is, if the magnetic field changed – then an electric current would be induced in a conductor placed within the magnetic field. The iron bar helped to concentrate the magnetic field, as Faraday later came to understand, and a current was induced in the second coil by the magnetic field momentarily set up as current entered or left the first coil. With this device, Faraday had discovered the transformer, a modern transformer being no different in essence even though the alternating current required had not then been discovered.

Faraday is thus also credited with the simultaneous discovery of electromagnetic induction, although the same discovery had been made in the same way by Joseph Henry in 1830. However, busy teaching, Henry had not been able to publish his findings before Faraday did, although both men are now credited with the independent discovery of induction.

Arago's wheel and the electric generator In 1824, Francois Arago found that a rotating non-magnetic disc, specifically of copper, caused the deflection of a magnetic needle placed above it. This was in fact a demonstration of electromagnetic induction, but nobody at that time could explain 'Arago's wheel'. Faraday realized that the motion of the copper wheel relative to the magnet in Arago's experiment caused an electric current to flow in the disc, which in turn set up a magnetic field and deflected the magnet. He set about constructing a similar device in which the current produced could be led off, and built the first electric generator in 1831. It consisted of a copper disc that was rotated between the poles of a magnet; Faraday touched wires to the edge and centre of the disc and connected them to a galvanometer, which registered a steady current.

electrostatic charge In 1832 Faraday showed that an electrostatic charge gives rise to the same effects as current electricity. He demonstrated in 1837 that electrostatic force consists of a field of curved lines of force, and that different substances have specific inductive capacities – that is, they take up different amounts of electric charge when subjected to an electric field.

In 1838, he proposed a theory of electricity elaborating his idea of varying strain in molecules. In a good conductor, a rapid build-up and breakdown of strain took place, transferring energy quickly from one molecule to the next. This also accounted for the decomposition of compounds in electrolysis. At the same time, Faraday wrongly rejected the notion that electricity involved the movement of any kind of electrical fluid (the motion of electrons is involved). However, in that this motion causes a rapid transfer of electrical

energy through a conductor, Faraday's ideas were valid.
polarization of light Finally, Faraday considered the nature of light and in 1846 arrived at a form of the electromagnetic theory of light that was later developed by Scottish physicist James Clerk Maxwell. In 1845, Lord Kelvin suggested that Faraday investigate the action of electricity on polarized light. Faraday had in fact already carried out such experiments with no success, but this could have been because electrical forces were not strong. Faraday now used an electromagnet to give a strong magnetic field instead and found that it causes the plane of polarization to rotate, the angle of rotation being proportional to the strength of the magnetic field.
paramagnetism and diamagnetism Several further discoveries resulted from this experiment. Faraday realized that the glass block used to transmit the beam of light must also transmit the magnetic field, and he noticed that the glass tended to set itself at right-angles to the poles of the magnet rather than lining up with it as an iron bar would. He showed that the differing responses of substances to a magnetic field depended on the distribution of the lines of force through them. He called materials that are attracted to a magnetic field paramagnetic, and those that are repulsed diamagnetic. Faraday then went on to point out that the energy of a magnet is in the field around it and not in the magnet itself, and he extended this basic conception of field theory to electrical and gravitational systems.

in humans. In many animals and plants, excess carbohydrates and proteins are converted into fats for storage. Mammals and other vertebrates store fats in specialized connective tissues (◊adipose tissues), which not only act as energy reserves but also insulate the body and cushion its organs.

As a nutrient, fat serves five purposes: it is a source of energy (9 kcal/g); makes the diet palatable; provides basic building blocks for cell structure; provides essential fatty acids (linoleic and linolenic); and acts as a carrier for fat-soluble vitamins (A, D, E, and K). Foods rich in fat are butter, lard, margarine, and cooking oils. Products high in monounsaturated or polyunsaturated fats are thought to be less likely to contribute to cardiovascular disease.

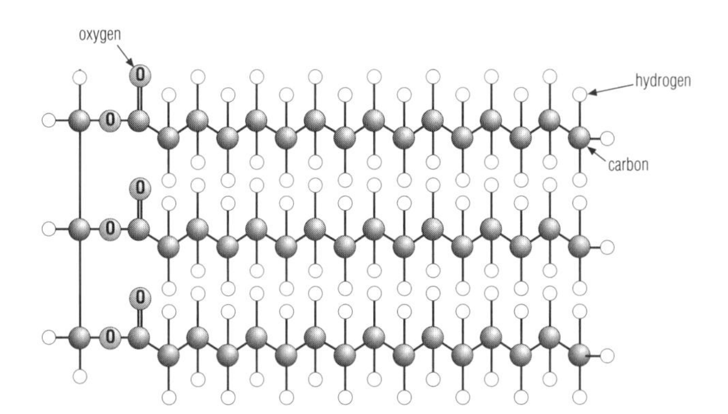

fat The molecular structure of typical fat. The molecule consists of three fatty acid molecules linked to a molecule of glycerol.

FAT in computing, abbreviation for ◊file allocation table.

fat hen plant belonging to the goosefoot family, widespread in temperate regions. It grows up to 1 m/3 ft tall and has lance- or diamond-shaped leaves and compact heads of small inconspicuous flowers. Now considered a weed, fat hen was once valued for its fatty seeds and edible leaves. (*Chenopodium album*, family Chenopodiaceae.)

fathom *Anglo-Saxon faethm 'to embrace'* in mining, seafaring, and handling timber, a unit of depth measurement (1.83 m/6 ft) used prior to metrication; it approximates to the distance between an adult man's hands when the arms are outstretched.

fatigue in muscle, reduced response brought about by the accumulation of lactic acid in muscle tissue due to excessive cellular activity.

fatty acid or ***carboxylic acid*** organic compound consisting of a hydrocarbon chain, up to 24 carbon atoms long, with a carboxyl group (–COOH) at one end. The covalent bonds between the carbon atoms may be single or double; where a double bond occurs the carbon atoms concerned carry one instead of two hydrogen atoms. Chains with only single bonds have all the hydrogen they can carry, so they are said to be **saturated** with hydrogen. Chains with one or more double bonds are said to be **unsaturated** (see ◊polyunsaturate). Fatty acids are produced in the small intestine when fat is digested.

Saturated fatty acids include palmitic and stearic acids; unsaturated fatty acids include oleic (one double bond), linoleic (two double bonds), and linolenic (three double bonds). Linoleic acid accounts for more than one third of some margarines. Supermarket brands that say they are high in polyunsaturates may contain as much as 39%. Fatty acids are generally found combined with glycerol in ◊lipids such as tryglycerides.

fault in geology, a fracture in the Earth either side of which rocks have moved past one another. Faults involve displacements, or offsets, ranging from the microscopic scale to hundreds of kilometres. Large offsets along a fault are the result of the accumulation of smaller movements (metres or less) over long periods of time. Large motions cause detectable ◊earthquakes.

Faults are planar features. Fault orientation is described by the inclination of the fault plane with respect to horizontal and its direction in the horizontal plane (see ◊strike). Faults at high angle with respect to horizontal (in which the fault plane is steep) are classified as either **normal faults**, where one block has apparently moved downhill along the inclined fault plane, or **reverse faults**, where one block appears to have moved uphill along the fault

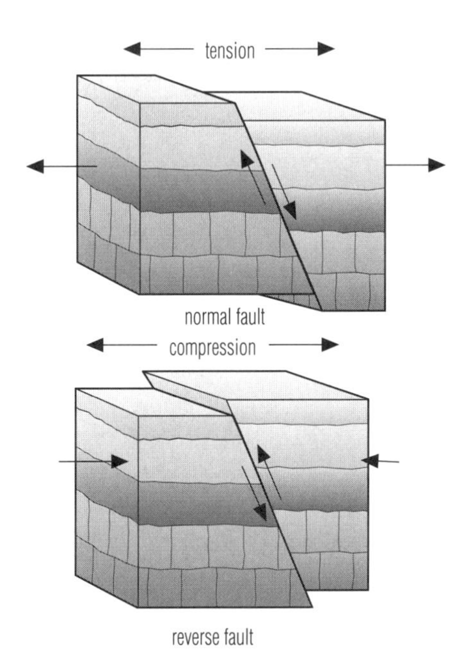

fault Faults are caused by the movement of rock layers, producing such features as block mountains and rift valleys. A normal fault is caused by a tension or stretching force acting in the rock layers. A reverse fault is caused by compression forces. Faults can continue to move for thousands or millions of years.

plane. Normal faults occur where rocks on either side have moved apart. Reverse faults occur where rocks on either side have been forced together. A reverse fault that forms a low angle with the horizontal plane is called a **thrust fault**.

A **lateral fault**, or **tear fault**, occurs where the relative movement along the fault plane is sideways. A particular kind of fault found only in ocean ridges is the **transform fault** (a term coined by Canadian geophysicist John Tuzo Wilson in 1965). On a map, an ocean ridge has a stepped appearance. The ridge crest is broken into sections, each section offset from the next. Between each section of the ridge crest the newly generated plates are moving past one another, forming a transform fault.

Faults produce lines of weakness on the Earth's surface (along their strike) that are often exploited by processes of ◊weathering and ◊erosion. Coastal caves and geos (narrow inlets) often form along faults and, on a larger scale, rivers may follow the line of a fault.

favourites menu option on Microsoft's Internet Explorer Web browser that allows users to go quickly to sites that have been ◊bookmarked, as with the bookmarks feature in Netscape Navigator. The access software for the AOL and CompuServe online services uses Favourite Places to provide the same feature.

fax (common name for ***facsimile transmission*** or ***telefax***) the transmission of images over a ◊telecommunications link, usually the telephone network. When placed on a fax machine, the original image is scanned by a transmitting device and converted into coded signals, which travel via the telephone lines to the receiving fax machine, where an image is created that is a copy of the original. Photographs as well as printed text and drawings can be sent. The standard transmission takes place at 4,800 or 9,600 bits of information per second.

The world's first fax machine, the *pantélégraphe,* was invented by Italian physicist Giovanni Caselli in 1866, over a century before the first electronic model came on the market. Standing over 2 m/6.5 ft high, it transmitted by telegraph nearly 5,000 handwritten documents and drawings between Paris and Lyon in its first year.

fax modem ◊modem capable of transmitting and receiving data in the form of a fax.

A normal fax machine sends data in binary form down a telephone line, in a similar way to a modem. A modem can therefore act as a fax machine, given suitable software. This means a document does not need to be printed before faxing and an incoming fax can be viewed before printing out on a plain-paper printer. However, the computer must be permanently turned on in order to receive faxes.

FDDI (abbreviation for ***fibre-optic digital device interface***) in computing, a series of network protocols, developed by the ◊American National Standards Institute, concerned with high-speed networks using ◊fibre optic cable.

FDDI supports data transmission rates of up to 100 Mb per second and is being introduced in many sites as a replacement for ◊Ethernet. FDDI not only makes possible transmission of large amounts of data, for example colour pictures, but also allows the transmission of voice and video data. See also ◊optical fibres.

feather rigid outgrowth of the outer layer of the skin of birds, made of the protein keratin. Feathers provide insulation and facilitate flight. There are several types, including long quill feathers on

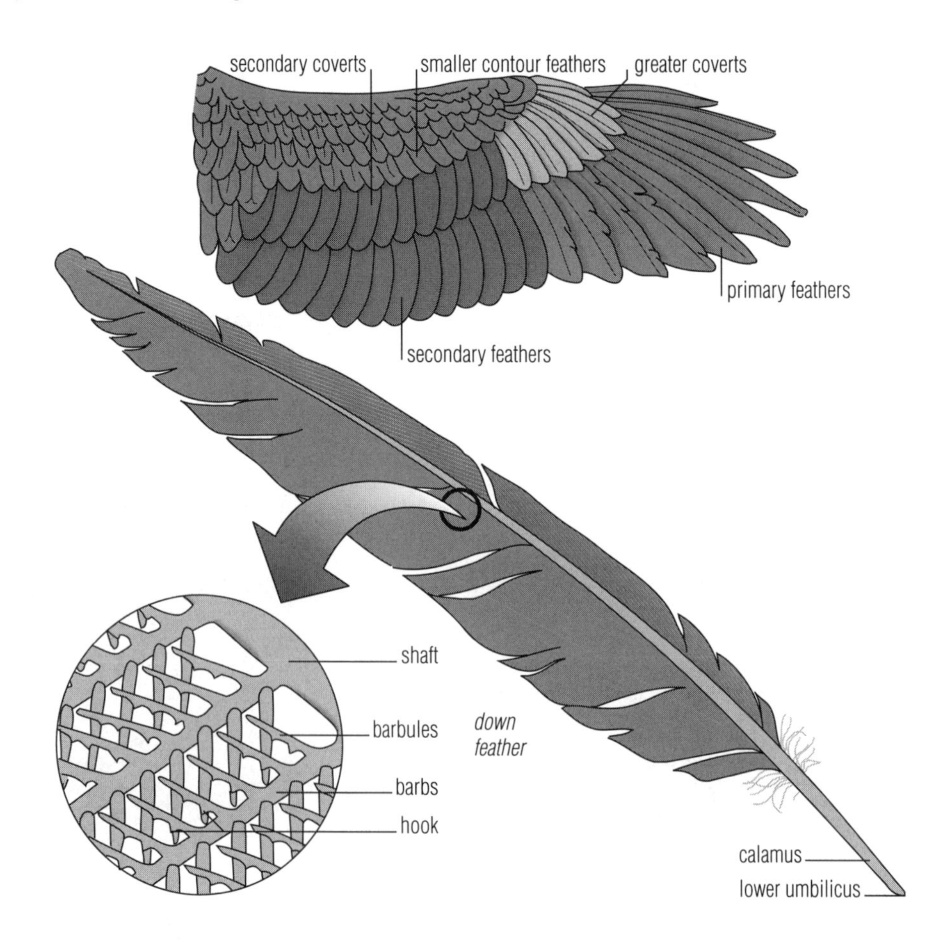

feather *Types of feather. A bird's wing is made up of two types of feather: contour feathers and flight feathers. The primary and secondary feathers are flight feathers. The coverts are contour feathers, used to streamline the bird. Semi-plume and down feathers insulate the bird's body and provide colour.*

the wings and tail, fluffy down feathers for retaining body heat, and contour feathers covering the body. The colouring of feathers is often important in camouflage or in courtship and other displays. Feathers are normally replaced at least once a year.

There is an enormous variation between species in the number of feathers, for example a whistling swan has over 25,000 contour feathers, whereas a ruby-throated hummingbird has less than 950.

Feathers generally consist of two main parts, axis and barbs, the former of which is divided into the quill, which is bare and hollow, and the shaft, which bears the barbs. The quill is embedded in the skin, and has at its base a small hole through which the nourishment passes during the growth of the feather. The barbs which constitute the vane are lath-shaped and taper to a point, and each one supports a series of outgrowths known as barbules, so that each barb is like a tiny feather. Adjacent barbs are linked to each other by hooks on the barbules.

feather star any of an unattached, free-swimming group of sea lilies, order Comatulida. The arms are branched into numerous projections (hence 'feather' star), and grow from a small cup-shaped body. Below the body are appendages that can hold on to a surface, but the feather star is not permanently attached.

fecundity the rate at which an organism reproduces, as distinct from its ability to reproduce (◊fertility). In vertebrates, it is usually measured as the number of offspring produced by a female each year.

Federal Aviation Administration (FAA) agency of the US Department of Transportation that controls air traffic. Its responsibilities include regulating air transportation, aviation safety, developing and operating a system of air-traffic control, requiring airports and airlines to provide antihijacking security, and conducting aviation research. The agency is also responsible for investigating aeroplane accidents. It was established in 1958 as the Federal Aviation Agency and was renamed upon its assignment to Transportation in 1967. It is directed by an administrator (assistant secretary) appointed directly by the president.

feedback in biology, another term for biofeedback.

feedback general principle whereby the results produced in an ongoing reaction become factors in modifying or changing the reaction; it is the principle used in self-regulating control systems, from a simple ◊thermostat and steam-engine ◊governor to automatic computer-controlled machine tools. A fully computerized control system, in which there is no operator intervention, is called a **closed-loop feedback** system. A system that also responds to control signals from an operator is called an **open-loop feedback** system.

In self-regulating systems, information about what *is* happening in a system (such as level of temperature, engine speed, or size of workpiece) is fed back to a controlling device, which compares it with what *should* be happening. If the two are different, the device takes suitable action (such as switching on a heater, allowing more steam to the engine, or resetting the tools). The idea that the Earth is a self-regulating system, with feedback operating to keep nature in balance, is a central feature of the ◊Gaia hypothesis.

Fehling's test chemical test to determine whether an organic substance is a reducing agent (substance that donates electrons to other substances in a chemical reaction). It is usually used to detect reducing sugars (monosaccharides, such as glucose, and the disaccharides maltose and lactose) and aldehydes.

If the test substance is heated with a freshly prepared solution containing copper(II) sulphate, sodium hydroxide and sodium potassium tartrate, the production of a brick-red precipitate indicates the presence of a reducing agent.

feldspar a group of ◊silicate minerals. Feldspars are the most abundant mineral type in the Earth's crust. They are the chief constituents of ◊igneous rock and are present in most metamorphic and sedimentary rocks. All feldspars contain silicon, aluminium, and oxygen, linked together to form a framework. Spaces within this framework structure are occupied by sodium, potassium, calcium, or occasionally barium, in various proportions. Feldspars form white, grey, or pink crystals and rank 6 on the ◊Mohs' scale of hardness.

The four extreme compositions of feldspar are represented by the minerals **orthoclase**, $KAlSi_3O_8$; **albite**, $NaAlSi_3O_8$; **anorthite**, $CaAl_2Si_2O_8$; and **celsian**, $BaAl_2Si_2O_8$. **Plagioclase feldspars** contain variable amounts of sodium (as in albite) and calcium (as in anorthite) with a negligible potassium content. **Alkali feldspars** (including orthoclase) have a high potassium content, less sodium, and little calcium.

The type known as moonstone has a pearl-like effect and is used in jewellery. Approximately 4,000 tonnes of feldspar are used in the ceramics industry annually.

feldspathoid any of a group of silicate minerals resembling feldspars but containing less silica. Examples are nepheline ($NaAlSiO_4$ with a little potassium) and leucite ($KAlSi_2O_6$). Feldspathoids occur in igneous rocks that have relatively high proportions of sodium and potassium. Such rocks may also contain alkali feldspar, but they do not generally contain quartz because any free silica would have combined with the feldspathoid to produce more feldspar instead.

felsic rock a ◊plutonic rock composed chiefly of light-coloured minerals, such as quartz, feldspar and mica. It is derived from **feldspar**, **lenad** (meaning feldspathoid), and **silica**. The term **felsic** also applies to light-coloured minerals as a group, especially quartz, feldspar, and feldspathoids.

femtosecond ◊SI unit of time. It is 10^{-15} seconds (one millionth of a billionth).

femur the ***thigh-bone***; also the upper bone in the hind limb of a four-limbed vertebrate.

fence lizard or ***swift*** name given to several species of lizard found in North and Central America. They are viviparous (the young develop in the mother before birth).

Fermat, Pierre de (1601–1665)

French mathematician who, with Blaise Pascal, founded the theory of probability and the modern theory of numbers. Fermat also made contributions to analytical geometry. In 1657, Fermat published a series of problems as challenges to other mathematicians, in the form of theorems to be proved.

Fermat's last theorem states that equations of the form $xn + yn = zn$ *where* x, y, z, and n are all integers have no solutions if $n > 2$. Fermat scribbled the theorem in the margin of a mathematics textbook and noted that he could have shown it to be true had he enough space in which to write the proof. The theorem remained unproven for 300 years (and therefore, strictly speaking, constituted a conjecture rather than a theorem). In 1993, Andrew Wiles, the English mathematician of Princeton University, USA, announced a proof; this turned out to be premature, but he put forward a revised proof in 1994. Fermat's last theorem was finally laid to rest in June 1997 when Wiles collected the Wolfskehl prize (the legacy bequeathed in the 19th century for the problem's solution).

Mary Evans Picture Library

classification Fence lizards are all in genus *Sceloporus*, family Iguanidae, suborder Sauria, order Squamata, class Reptilia.

fennec small nocturnal desert ◊fox *Fennecus zerda* found in North Africa and Arabia. It has a head and body only 40 cm/1.3 ft long, and its enormous ears act as radiators to lose excess heat. It eats insects and small animals.

fennel any of several varieties of a perennial plant with feathery green leaves, belonging to the carrot family. Fennels have an aniseed (liquorice) flavour, and the leaves and seeds are used in seasoning. The thickened leafstalks of sweet fennel (*F. vulgare dulce*) are eaten as a vegetable. (*Foeniculum vulgare*, family Umbelliferae.)

Fermat's principle in physics, the principle that a ray of light, or other radiation, moves between two points along the path that takes the minimum time.

The principle is named after French mathematician Pierre de Fermat, who used it to deduce the laws of ◊reflection and ◊refraction.

FERMAT'S LAST THEOREM

http://www-groups.dcs.st-and.ac.uk/~history/HistTopics/Fermat's_last_theorem.html

Account of Fermat's last theorem and of the many attempts made to prove it. It is extensively hyperlinked to related mathematicians and also includes a list of 17 references for further reading.

fermentation the breakdown of sugars by bacteria and yeasts using a method of respiration without oxygen (◊anaerobic). Fermentation processes have long been utilized in baking bread, making beer and wine, and producing cheese, yoghurt, soy sauce, and many other foodstuffs.

In baking and brewing, yeasts ferment sugars to produce ◊ethanol and carbon dioxide; the latter makes bread rise and puts bubbles into beers and champagne. Many antibiotics are produced by fermentation; it is one of the processes that can cause food spoilage.

Wine is the most healthful and most hygienic of beverages.

LOUIS PASTEUR French chemist and microbiologist.
Etudes sur la Vin Pt 1 Ch 2

fermi unit of length equal to 10^{-15}m, used in atomic and nuclear physics. The unit is named after Enrico Fermi.

Fermilab (shortened form of ***Fermi National Accelerator Laboratory***)

US centre for ◊particle physics at Batavia, Illinois, near Chicago. It is named after Italian–US physicist Enrico Fermi. Fermilab was opened in 1972, and is the home of the Tevatron, the world's most powerful particle ◊accelerator. It is capable of boosting protons and antiprotons to speeds near that of light (to energies of 20 TeV).

fermion in physics, a subatomic particle whose spin can only take values that are half-integers, such as $\frac{1}{2}$ or $\frac{3}{2}$. Fermions may be classified as leptons, such as the electron, and baryons, such as the proton and neutron. All elementary particles are either fermions or ◊bosons.

The exclusion principle, formulated by Austrian–US physicist Wolfgang Pauli in 1925, asserts that no two fermions in the same system (such as an atom) can possess the same position, energy state, spin, or other quantized property.

fermium synthesized, radioactive, metallic element of the ◊actinide series, symbol Fm, atomic number 100, relative atomic mass 257.10. Ten isotopes are known, the longest-lived of which, Fm-257, has a half-life of 80 days. Fermium has been produced only in minute quantities in particle accelerators.

Fermi, Enrico (1901–1954)

Italian-born US physicist who proved the existence of new radioactive elements produced by bombardment with neutrons, and discovered nuclear reactions produced by low-energy neutrons. This research won him the Nobel Prize for Physics in 1938 and was the basis for studies leading to the atomic bomb and nuclear energy. Fermi built the first nuclear reactor in 1942 at Chicago University and later took part in the Manhattan Project to construct an atom bomb. His theoretical work included the study of the weak nuclear force, one of the fundamental forces of nature, and beta decay.

neutron bombardment and the Nobel Prize Following the work of the Joliot-Curies, who discovered artificial radioactivity in 1934 using alpha particle bombardment, Fermi began producing new radioactive isotopes by neutron bombardment. Unlike the alpha particle, which is positively charged, the neutron is charge neutral. Fermi realized that less energy would be wasted when a bombarding neutron encounters a positively charged target nucleus.

He also found that a block of paraffin wax or a jacket of water around the neutron source produced slow, or 'thermal', neutrons. Slow-neutrons are more effective at producing artificial radioactive elements because they remain longer near the target nucleus and have a greater chance of being absorbed.

He did, however, misinterpret the results of experiments involving neutron bombardment of uranium, failing to recognize that nuclear fission had occurred. Instead, he maintained that the bombardment produced two new transuranic elements. It was left to Lise Meitner and Otto Frisch to explain nuclear fission in 1938.

nuclear reactors and the atomic bomb In the USA, Fermi continued the work on the fission of uranium (initiated by neutrons) by building the first nuclear reactor, then called an **atomic pile,** because it had a moderator consisting of a pile of purified graphite blocks (to slow the neutrons) with holes drilled in them to take rods of enriched uranium. Other neutron-absorbing rods of cadmium, called control rods, could be lowered into or withdrawn from the pile to limit the number of slow-neutrons available to initiate the fission of uranium. The reactor was built on the squash court of Chicago University. On the afternoon of 2 December 1942, the control rods were withdrawn for the first time and the reactor began to work, using a self-sustaining nuclear chain reaction. Two years later, the USA, through a team led by Arthur Compton and Fermi, had constructed an atomic bomb, which used the same reaction but without control, resulting in a nuclear explosion.

beta decay and the neutrino Fermi's experimental work on beta decay in radioactive materials provided further evidence for the existence of the neutrino, predicted by Austrian physicist Wolfgang Pauli. The decay, which takes place spontaneously in the unstable nuclei of radioactive elements, results from the conversion of a neutron into a proton, an electron (beta particle) and an antineutrino:

$$n \rightarrow p + e^{-} + \nu$$

Mary Evans Picture Library

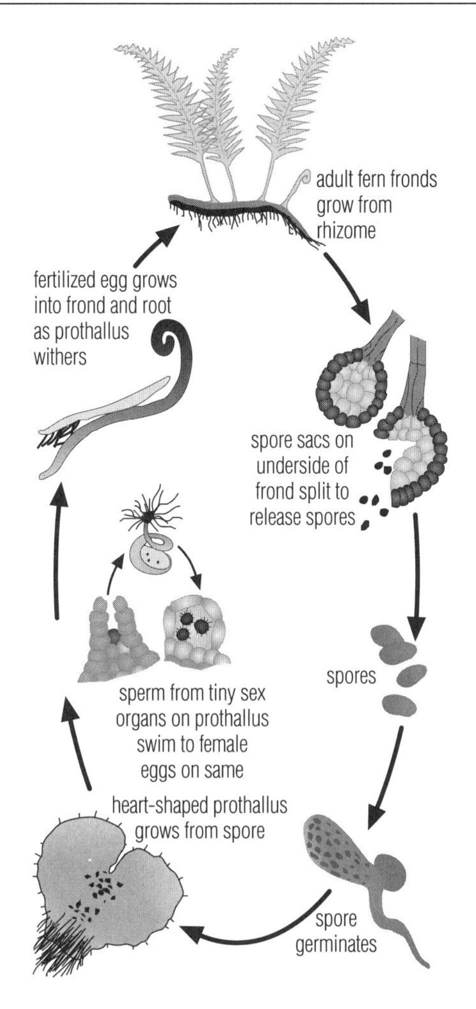

fern *The life cycle of a fern. Ferns have two distinct forms that alternate during their life cycle. For the main part of its life, a fern consists of a short stem (or rhizome) from which roots and leaves grow. The other part of its life is spent as a small heart-shaped plant called a prothallus.*

It was discovered in 1952 in the debris of the first thermonuclear explosion. The element was named in 1955 in honour of US physicist Enrico ◊Fermi.

fern any of a group of plants related to horsetails and clubmosses. Ferns are spore-bearing, not flowering, plants and most are perennial, spreading by slow-growing roots. The leaves, known as fronds, vary widely in size and shape. Some taller types, such as tree ferns, grow in the tropics. There are over 7,000 species. (Order Filicales.)

ferret domesticated variety of the Old World ◊polecat.

About 35 cm/1.2 ft long, it usually has yellowish-white fur and pink eyes, but may be the dark brown colour of a wild polecat.

Ferrets may breed with wild polecats. They have been used since ancient times to hunt rabbits and rats.

ferric ion traditional name for the trivalent condition of iron, Fe^{3+}; the modern name is iron(III). Ferric salts are usually reddish or yellow in colour and form reddish-yellow solutions. $Fe_2(SO_4)_3$ is iron(III) sulphate (ferric sulphate).

ferrimagnetism form of ◊magnetism in which adjacent molecular magnets are aligned anti-parallel, but have unequal strength, producing a strong overall magnetization. Ferrimagnetism is found in certain inorganic substances, such as ◊ferrites.

ferrite ceramic ferrimagnetic material. Ferrites are iron oxides to which small quantities of ◊transition metal oxides (such as cobalt and nickel oxides) have been added. They are used in transformer cores, radio antennae, and, formerly, in computer memories.

ferro-alloy alloy of iron with a high proportion of elements such as manganese, silicon, chromium, and molybdenum. Ferro-alloys are used in the manufacture of alloy steels. Each alloy is generally named after the added metal – for example, ferrochromium.

ferroelectric material ceramic dielectric material that, like ferromagnetic materials, has a ◊domain structure that makes it exhibit magnetism and usually the ◊piezoelectric effect. An example is Rochelle salt (potassium sodium tartrate tetrahydrate, $KNaC_4H_4O_6.4H_2O$).

ferromagnetism form of ◊magnetism in which magnetism can be acquired in an external magnetic field and usually retained in its absence, so that ferromagnetic materials are used to make permanent magnets. A ferromagnetic material may therefore be said to have a high magnetic ◊permeability and ◊susceptibility (which depends upon temperature). Examples are iron, cobalt, nickel, and their alloys.

Ultimately, ferromagnetism is caused by spinning electrons in the atoms of the material, which act as tiny weak magnets. They align parallel to each other within small regions of the material to form ◊domains, or areas of stronger magnetism. In an unmagnetized material, the domains are aligned at random so there is no overall magnetic effect. If a magnetic field is applied to that material, the domains align to point in the same direction, producing a strong overall magnetic effect. Permanent magnetism arises if the domains remain aligned after the external field is removed. Ferromagnetic materials exhibit ◊hysteresis.

ferrous ion traditional name for the divalent condition of iron, Fe^{2+}; the modern name is iron(II). Ferrous salts are usually green, and form yellow-green solutions. $FeSO_4$ is iron(II) sulphate (ferrous sulphate).

fertility an organism's ability to reproduce, as distinct from the rate at which it reproduces (◊fecundity). Individuals become infertile (unable to reproduce) when they cannot generate gametes (eggs or sperm) or when their gametes cannot yield a viable ◊embryo after fertilization.

fertility drug any of a range of drugs taken to increase a female's fertility, developed in Sweden in the mid-1950s. They increase the chances of a multiple birth.

The most familiar is gonadotrophin, which is made from hormone extracts taken from the human pituitary gland: follicle-stimulating hormone and luteinizing hormone. It stimulates ovulation in women. As a result of a fertility drug, in 1974 the first sextuplets to survive were born to Susan Rosenkowitz of South Africa.

fertilization in ◊sexual reproduction, the union of two ◊gametes (sex cells, often called egg and sperm) to produce a ◊zygote, which combines the genetic material contributed by each parent. In self-fertilization the male and female gametes come from the same plant; in cross-fertilization they come from different plants. Self-fertilization rarely occurs in animals; usually even ◊hermaphrodite animals cross-fertilize each other.

In terrestrial insects, mammals, reptiles, and birds, fertilization occurs within the female's body. In humans it usually takes place in the ◊Fallopian tube. In the majority of fishes and amphibians, and most aquatic invertebrates, fertilization occurs externally, when both sexes release their gametes into the water. In most fungi, gametes are not released, but the hyphae of the two parents grow towards each other and fuse to achieve fertilization. In higher plants, ◊pollination precedes fertilization.

fertilizer substance containing some or all of a range of about 20 chemical elements necessary for healthy plant growth, used to compensate for the deficiencies of poor or depleted soil. Fertilizers may be **organic**, for example farmyard manure, composts, bonemeal, blood, and fishmeal; or **inorganic**, in the form of compounds, mainly of nitrogen, phosphate, and potash, which have been used on a very much increased scale since 1945.

Because externally applied fertilizers tend to be in excess of

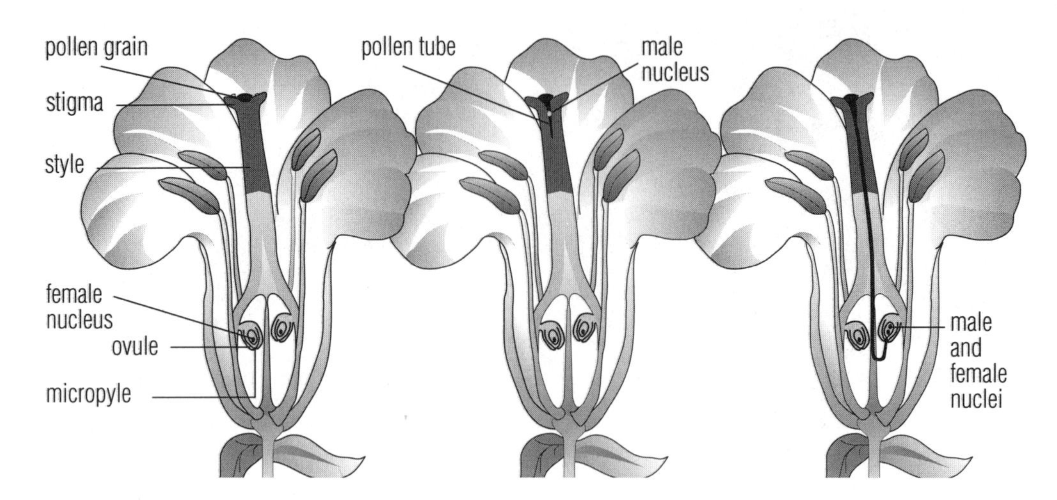

fertilization *In a flowering plant pollen grains land on the surface of the stigma, and if conditions are acceptable the pollen grain germinates, forming a pollen tube, through which the male gametes pass, entering the ovule via the micropyle in order to reach the female egg.*

plant requirements and drain away to affect lakes and rivers (see ⟩eutrophication), attention has turned to the modification of crop plants themselves. Plants of the legume family, including the bean, clover, and lupin, live in symbiosis with bacteria located in root nodules, which fix nitrogen from the atmosphere. Research is now directed to producing a similar relationship between such bacteria and crops such as wheat.

fescue any grass of a widely distributed group. Many are used in temperate regions for lawns and pasture. Many upland species are viviparous, producing young plantlets instead of flowers. (Genus *Festuca*, family Gramineae.)

fetal therapy diagnosis and treatment of conditions arising in the unborn child. While some anomalies can be diagnosed antenatally, fetal treatments are only appropriate in a few cases – mostly where the development of an organ is affected.

Fetal therapy was first used 1963 with exchange transfusion for haemolytic disease of the newborn, once a serious problem (see also ⟩rhesus factor). Today the use of fetal therapy remains limited. Most treatments involve 'needling': introducing fine instruments through the mother's abdominal and uterine walls under ultrasound guidance. Open-womb surgery (hysterotomy) remains controversial because of the risks involved. It is available only in some centres in the United States.

fetch-execute cycle or ***processing cycle*** in computing, the two-phase cycle used by the computer's central processing unit to process the instructions in a program. During the **fetch phase**, the next program instruction is transferred from the computer's immediate-access memory to the instruction register (memory location used to hold the instruction while it is being executed). During the **execute phase**, the instruction is decoded and obeyed. The process is repeated in a continuous loop.

fetishism in psychology, the transfer of erotic interest to an object, such as an item of clothing, whose real or fantasized presence is necessary for sexual gratification. The fetish may also be a part of the body not normally considered erogenous, such as the feet.

fetus or ***foetus*** stage in mammalian ⟩embryo development. The human embryo is usually termed a fetus after the eighth week of development, when the limbs and external features of the head are recognizable.

fever condition of raised body temperature, usually due to infection.

fibre, dietary or ***roughage*** plant material that cannot be digested by human digestive enzymes; it consists largely of cellulose, a carbohydrate found in plant cell walls. Fibre adds bulk to the gut contents, assisting the muscular contractions that force food along the intestine. A diet low in fibre causes constipation and is believed to increase the risk of developing diverticulitis, diabetes, gall-bladder disease, and cancer of the large bowel – conditions that are rare in nonindustrialized countries, where the diet contains a high proportion of unrefined cereals.

Soluble fibre consists of indigestible plant carbohydrates (such as pectins, hemicelluloses, and gums) that dissolve in water. A high proportion of the fibre in such foods as oat bran, pulses, and

Feynman, Richard P(hillips) (1918–1988)

US physicist whose work laid the foundations of quantum electrodynamics. For his work on the theory of radiation he shared the Nobel Prize for Physics 1965 with Julian Schwinger and Sin-Itiro Tomonaga (1906–1979). He also contributed to many aspects of particle physics, including quark theory and the nature of the weak nuclear force.

For his work on quantum electrodynamics, he developed a simple and elegant system of **Feynman diagrams** to represent interactions between particles and how they moved from one space-time point to another. He had rules for calculating the probability associated with each diagram.

His other major discoveries are the theory of superfluidity (frictionless flow) in liquid helium, developed in the early 1950s; his work on the weak interaction (with US physicist Murray Gell-Mann) and the strong force; and his prediction that the proton and neutron are not elementary particles. Both particles are now known to be composed of quarks.

Californian Institute of Technology

vegetables is of this sort. Its presence in the diet has been found to reduce the amount of cholesterol in blood over the short term, although the mechanism for its effect is disputed.

fibreglass glass that has been formed into fine fibres, either as long continuous filaments or as a fluffy, short-fibred glass wool. Fibreglass is heat- and fire-resistant and a good electrical insulator. It has applications in the field of fibre optics and as a strengthener for plastics in ◊GRP (glass-reinforced plastics).

The long filament form is made by forcing molten glass through the holes of a spinneret, and is woven into textiles. Glass wool is made by blowing streams of molten glass in a jet of high-pressure steam, and is used for electrical, sound, and thermal insulation, especially for the roof space in houses.

fibre optics branch of physics dealing with the transmission of light and images through glass or plastic fibres known as ◊optical fibres.

fibrin insoluble protein involved in blood clotting. When an injury occurs fibrin is deposited around the wound in the form of a mesh, which dries and hardens, so that bleeding stops. Fibrin is developed in the blood from a soluble protein, fibrinogen.

The conversion of fibrinogen to fibrin is the final stage in blood clotting. Platelets, a type of cell found in blood, release the enzyme thrombin when they come into contact with damaged tissue, and the formation of fibrin then occurs. Calcium, vitamin K, and a variety of enzymes called factors are also necessary for efficient blood clotting.

fibula the rear lower bone in the hind leg of a vertebrate. It is paired and often fused with a smaller front bone, the tibia.

Fidonet in computing, early network of ◊bulletin board systems (BBSs) which sends mail and news around the world via an arrangement whereby the individual systems call each other to exchange data every night. Such systems are called 'store-and-forward'.

field in computing, a specific item of data. A field is usually part of a **record**, which in turn is part of a ◊file.

field in physics, a region of space in which an object exerts a force on another separate object because of certain properties they both possess. For example, there is a force of attraction between any two objects that have mass when one is in the gravitational field of the other.

Other fields of force include ◊electric fields (caused by electric charges) and ◊magnetic fields (caused by circulating electric currents), either of which can involve attractive or repulsive forces.

field enclosed area of land used for farming. Traditionally fields were measured in ◊acres; the current unit of measurement is the hectare (2.47 acres).

In Britain, regular field systems were functioning before the Romans' arrival. The open-field system was in use at the time of the Norman Conquest. Enclosure began in the 14th century and continued into the 19th century.

In the Middle Ages, the farmland of an English rural community was often divided into three large fields (the **open-field system**). These were worked on a simple rotation basis of one year wheat, one year barley, and one year ◊fallow. The fields were divided into individually owned strips of the width that one plough team with oxen could plough (about 20 m/66 ft). At the end of each strip would be a turning space, either a road or a **headland**. Through repeated ploughing a **ridge-and-furrow** pattern became evident. A farmer worked a number of strips, not necessarily adjacent to each other, in one field.

The open-field communities were subsequently reorganized, the land enclosed, and the farmers' holdings redistributed into individual blocks which were then divided into separate fields. This enclosure process reached its peak during the 18th century. 20th-century developments in agricultural science and technology have encouraged farmers to amalgamate and enlarge their fields, often to as much as 40 hectares/100 acres.

The open field system was also found in France, Germany, Greece, and Slavonic lands.

fieldfare gregarious thrush *Turdus pilaris* of the family Muscicapidae, order Passeriformes; it has chestnut upperparts with a pale-grey lower back and neck, and a dark tail. The bird's underparts are a rich ochre colour, spotted with black. Its nest is of long fine grass with an intervening layer of mud; it may be built in birch or fir trees at a height of 5 m/16 ft or less. It feeds on berries, insects, and other invertebrates.

field-length check ◊validation check in which the characters in an input field are counted to ensure that the correct number of characters have been entered. For example, a six-figure date field may be checked to ensure that it does contain exactly six digits.

field of view angle over which an image may be seen in a mirror or an optical instrument such as a telescope. A wide field of view allows a greater area to be surveyed without moving the instrument, but has the disadvantage that each of the objects seen is smaller. A ◊convex mirror gives a larger field of view than a plane or flat mirror. The field of view of an eye is called its **field of vision** or visual field.

A telephoto lens has a small field of view, around 14°, and produces a highly magnified image. A 'fish-eye' lens has a very wide angle of view, around 180°. Our eyes have a field of vision of around 45°.

field studies study of ecology, geography, geology, history, archaeology, and allied subjects, in the natural environment as opposed to the laboratory.

The most exciting phrase to hear in science, the one that heralds new discoveries, is not 'Eureka!' (I've found it!) but 'That's funny...'

ISAAC ASIMOV Russian-born US writer.

fifth-generation computer anticipated new type of computer based on emerging microelectronic technologies with high computing speeds and ◊parallel processing. The development of very large-scale integration (◊VLSI) technology, which can put many more circuits on to an integrated circuit (chip) than is currently possible, and developments in computer hardware and software design may produce computers far more powerful than those in current use.

It has been predicted that such a computer will be able to communicate in natural spoken language with its user; store vast knowledge databases; search rapidly through these databases, making intelligent inferences and drawing logical conclusions; and process images and 'see' objects in the way that humans do.

In 1981 Japan's Ministry of International Trade and Industry launched a ten-year project to build the first fifth-generation computer, the 'parallel inference machine', consisting of over a thousand microprocessors operating in parallel with each other. By 1992, however, the project was behind schedule and had only produced 256-processor modules. It has since been suggested that research into other technologies, such as ◊neural networks, may present more promising approaches to artificial intelligence. Compare earlier ◊computer generations.

fig any of a group of trees belonging to the mulberry family, including the many cultivated varieties of *F. carica,* originally from W Asia. They produce two or three crops of fruit a year. Eaten fresh or dried, figs have a high sugar content and laxative properties. (Genus *Ficus,* family Moraceae.)

In the wild, *F. carica* is dependent on the fig wasp for pollination, and the wasp in turn is parasitic on the flowers. The tropical **banyan** (*F. benghalensis*) has less attractive edible fruit, and roots that grow down from its branches. The **bo tree** under which Buddha became enlightened is the Indian peepul or wild fig (*F. religiosa*).

fighting fish any of a southeast Asian genus *Betta* of fishes of the gourami family, especially *B. splendens,* about 6 cm/2 in long and a popular aquarium fish. It can breathe air, using an accessory breathing organ above the gill, and can live in poorly oxygenated water. The male has large fins and various colours, including shining greens, reds, and blues. The female is yellowish brown with short fins.

The male builds a nest of bubbles at the water's surface and displays to a female to induce her to lay. Rival males are attacked, and in a confined space, fights may occur. In Thailand, public contests are held.

figwort any of a group of Old World plants belonging to the figwort family, which also includes foxgloves and snapdragons. Members of the genus have square stems, opposite leaves, and open two-lipped flowers in a cluster at the top of the stem. (Genus *Scrophularia,* family Scrophulariaceae.)

filament in astronomy, a dark, winding feature occasionally seen on images of the Sun in hydrogen light. Filaments are clouds of relatively cool gas suspended above the Sun by magnetic fields and seen in silhouette against the hotter ◊photosphere below. During total ◊eclipses they can be seen as bright features against the sky at the edge of the Sun where they are known as ◊prominences.

file in computing, a collection of data or a program stored in a computer's external memory (for example, on ◊disc). It might include anything from information on a company's employees to a program for an adventure game. **Serial files** hold information as a sequence of characters, so that, to read any particular item of data, the program must read all those that precede it. **Random-access files** allow the required data to be reached directly. Files are usually located via a ◊directory.

file access in computing, the way in which the records in a file are stored, retrieved, or updated by computer. There are four main types of file organization, each of which allows a different form of access to the records.

Records in a **serial file** are not stored in any particular order, so a specific record can be accessed only by reading through all the previous records.

Records in a **sequential file** are sorted by reference to a key field (see ◊sorting) and the computer can use a searching technique, such as a binary search, to access a specific record.

An **indexed sequential file** possesses an index, which records the position of each block of records and is created and updated with that file. By consulting the index, the computer can obtain the address of the block containing the required record, and search just that block rather than the whole file.

A **direct-access** or **random-access file** contains records that can be accessed directly by the computer.

file allocation table (FAT) in computing, a table used by the operating system to record the physical arrangement of files on disc. As a result of ◊fragmentation, files can be split into many parts sited at different places on the disc.

file extension in computing, the last three letters of a file name in DOS or Windows, which indicate the type of data the file contains. Extensions in common use include .TXT for 'text', .GIF for 'graphics interchange format', and .EXE for 'executable'.

In Windows, the operating system may be configured to associate specific file extensions with specific programs, so that double-clicking on a file name starts the right program and opens the file for editing.

file format in computing, specific way data is stored in a file. Most computer programs use proprietary file formats which cannot be read by other programs. As this is inconvenient for users, in recent years software publishers have developed filters which convert older file formats into the ones the program in use can read.

Often ◊file extensions are used to indicate which program was used to create a particular file. Some formats, such as GIF (graphics interchange format), have become so popular and widely used that they are supported by many programs.

Before transmitting data over a public network to another user, it is important to check that the receiving user can read the format the data is in. For this purpose, the most commonly readable format is plain ◊ASCII for text and either GIF or ◊JPEG (Joint Photographic Experts Group) for graphics.

file generation in computing, a specific version of a file. When ◊file updating takes place, a new generation of the file is created, containing accurate, up-to-date information. The old generation of the file will often be stored to provide ◊data security in the event that the new generation of the file is lost or damaged.

file librarian or ***media librarian*** job classification for ◊computer personnel. A file librarian stores and issues the data files used by the computer department.

file merging in computing, combining two or more sequentially ordered files into a single sequentially ordered file.

file searching ◊searching a computer memory (usually ◊backing storage) for a file.

file server computer on a ◊network that handles (and usually stores) the data used by other computers on the network. See also ◊client–server architecture.

file sorting arranging files in sequence; see ◊sorting.

file transfer in computing, the transmission of a file (data stored on disc, for example) from one machine to another. Both machines must be physically linked (for example, by a telephone line via a ◊modem or ◊acoustic coupler) and both must be running appropriate communications software.

file updating in computing, reviewing and altering the records in a file to ensure that the information they contain is accurate and up-to-date. Three basic processes are involved: adding new records, deleting existing records, and amending existing records.

The updating of a **direct-access file** is a continuous process because records can be accessed individually and changed at any time. This type of updating is typical of large interactive database systems, such as airline ticket-booking systems. Each time a ticket is booked, files are immediately updated so that double booking is impossible.

In large commercial applications, however, millions of customer records may be held in a large sequentially ordered file, called the **master file**. Each time the records in the master file are to be updated (for example, when quarterly bills are being drawn up), a **transaction file** must be prepared. This will contain all the additions, deletions, and amendments required to update the master file. The transaction file is sorted into the same order as the master file, and then the computer reads both files and produces a new updated **generation** of the master file, which will be stored until the next file updating takes place.

film, photographic strip of transparent material (usually cellulose acetate) coated with a light-sensitive emulsion, used in cameras to take pictures. The emulsion contains a mixture of light-sensitive silver halide salts (for example, bromide or iodide) in gelatin. When the emulsion is exposed to light, the silver salts are invisibly altered, giving a latent image, which is then made visible by the process of ◊developing. Films differ in their sensitivities to light, this being indicated by their speeds. Colour film consists of several layers of emulsion, each of which records a different colour in the light falling on it.

In **colour film** the front emulsion records blue light, then comes a yellow filter, followed by layers that record green and red light respectively. In the developing process the various images in the layers are dyed yellow, magenta (red), and cyan (blue), respectively.

When they are viewed, either as a transparency or as a colour print, the colours merge to produce the true colour of the original scene photographed.

filter in chemistry, a porous substance, such as blotting paper, through which a mixture can be passed to separate out its solid constituents.

filter in computing, a program that transforms data. Filters are often used when data output from one ◊application program is input into a different program, which requires a different data format. For example files transferred between two different word-processing programs are run through either an output filter supplied with the first program or an input filter supplied with the second program.

Filters are also used to expand coding structures, which have been simplified for keyboard input, into the often more verbose form required by such standards as SGML (◊Standard Generalized Markup Language).

filter in electronics, a circuit that transmits a signal of some frequencies better than others. A low-pass filter transmits signals of low frequency and direct current; a high-pass filter transmits high-frequency signals; a band-pass filter transmits signals in a band of frequencies.

filter in optics, a device that absorbs some parts of the visible ◊spectrum and transmits others. For example, a green filter will absorb or block all colours of the spectrum except green, which it allows to pass through. A yellow filter absorbs only light at the blue and violet end of the spectrum, transmitting red, orange, green, and yellow light.

filtrate liquid or solution that has passed through a filter.

filtration technique by which suspended solid particles in a fluid are removed by passing the mixture through a filter, usually porous paper, plastic, or cloth. The particles are retained by the filter to form a residue and the fluid passes through to make up the filtrate. For example, soot may be filtered from air, and suspended solids from water.

fin in aquatic animals, flattened extension from the body that aids balance and propulsion through the water.

In fish they may be paired, such as the pectoral and ventral fins, or singular, such as the caudal and dorsal fins, all being supported by a series of cartilaginous or bony rays.

The fins in cetaceans (whales and dolphins) are simple extensions of the soft tissue and have no bony rays. The flippers of seals are modified five-fingered limbs and contain the same bones as the limbs of other vertebrates.

finch any of various songbirds of the family Fringillidae, in the order Passeriformes (perching birds). They are seed-eaters with stout conical beaks. The name may also be applied to members of the Emberizidae (buntings), and Estrildidae (weaver-finches).

FINCHWORLD

http://www.finchworld.com/

Huge source of information on finches. Contents include practical advice for those buying a finch for the first time, tips on diet, housing, recreation activities, and avian illnesses. There are links to finch sites around the world and news of ornithological research on these songbirds.

finite having a countable number of elements, the opposite of infinite.

finsen unit unit (symbol FU) for measuring the intensity of ultraviolet (UV) light; for instance, UV light of 2 FUs causes sunburn in 15 minutes.

fiord alternative spelling of ◊fjord.

fir any of a group of ◊conifer trees belonging to the pine family. The true firs include the balsam fir (*A. balsamea*) of northern North America and the silver fir (*A. alba*) of Europe and Asia. Douglas firs of the genus *Pseudotsuga* are native to western North America and the Far East. (True fir genus *Abies*, family Pinaceae.)

fireball in astronomy, a very bright ◊meteor, often bright enough to be seen in daylight and occasionally leading to the fall of a ◊meteorite. Some fireballs are caused by ◊satellites or other space debris burning up in the Earth's atmosphere.

firebrat any insect of the order Thysanura (◊bristletail).

fire clay a ◊clay with refractory characteristics (resistant to high temperatures), and hence suitable for lining furnaces (firebrick). Its chemical composition consists of a high percentage of silicon and aluminium oxides, and a low percentage of the oxides of sodium, potassium, iron, and calcium.

firedamp gas that occurs in coal mines and is explosive when mixed with air in certain proportions. It consists chiefly of methane (CH_4, natural gas or marsh gas) but always contains small quantities of other gases, such as nitrogen, carbon dioxide, and hydrogen, and sometimes ethane and carbon monoxide.

fire-danger rating unit index used by the UK Forestry Commission to indicate the probability of a forest fire. 0 means a fire is improbable, 100 shows a serious fire hazard.

fire extinguisher device for putting out a fire. Fire extinguishers work by removing one of the three conditions necessary for fire to continue (heat, oxygen, and fuel), either by cooling the fire or by excluding oxygen.

The simplest fire extinguishers contain water, which when propelled onto the fire cools it down. Water extinguishers cannot be used on electrical fires, as there is a danger of electrocution, or on burning oil, as the oil will float on the water and spread the blaze.

Many domestic extinguishers contain liquid carbon dioxide under pressure. When the handle is pressed, carbon dioxide is released as a gas that blankets the burning material and prevents oxygen from reaching it. Dry extinguishers spray powder, which then releases carbon dioxide gas. Wet extinguishers are often of the soda-acid type; when activated, sulphuric acid mixes with sodium bicarbonate, producing carbon dioxide. The gas pressure forces the solution out of a nozzle, and a foaming agent may be added.

Some extinguishers contain halons (hydrocarbons with one or more hydrogens substituted by a halogen such as chlorine, bromine, or fluorine). These are very effective at smothering fires, but cause damage to the ozone layer, and their use is now restricted.

firefly any winged nocturnal beetle of the family Lampyridae. They all emit light through the process of ◊bioluminescence.

fire protection methods available for fighting fires. Industrial and commercial buildings are often protected by an automatic sprinkler system: heat or smoke opens the sprinkler heads on a network of water pipes which spray the source of the fire. In circumstances where water is ineffective and may be dangerous, for example, for oil and petrol storage-tank fires, foam systems are used; for industrial plants containing flammable vapours, carbon dioxide is used; where electricity is involved, vaporizing liquids create a nonflammable barrier; for some chemicals only various dry powders can be used.

firewall in computing, security system built to block access to a particular computer or network while still allowing some types of data to flow in and out onto the Internet.

A firewall allows a company's employees to access sites on the World Wide Web or exchange e-mail while at the same time preventing hackers from gaining access to the company's data.

firewheel Australian tree *Stenocarpus sinuatus* native to rainforests of New South Wales and Queensland, which bears wheel-like whorls of bright red flowers.

FireWire in computing, Apple's implementation of the IEEE ◊1394 serial connection system.

firewood the principal fuel for some 2 billion people, mainly in the Third World. In principle a renewable energy source, firewood is being cut far faster than the trees can regenerate in many areas of Africa and Asia, leading to ◊deforestation.

In Mali, for example, wood provides 97% of total energy consumption, and deforestation is running at an estimated 9,000 hectares a year. The heat efficiency of firewood can be increased by use of well-designed stoves, but for many people they are either unaffordable or unavailable. With wood for fuel becoming scarcer the UN Food and Agricultural Organization has estimated that by the year 2000 3 billion people worldwide will face chronic problems in getting food cooked.

firmware computer program held permanently in a computer's ◊ROM (read-only memory) chips, as opposed to a program that is read in from external memory as it is needed.

First Amendment amendment to the US Constitution that guarantees freedom of religion, of speech, of assembly, and of the press. Adopted in 1791, the First Amendment is often quoted on the Internet, even by non-US citizens, in arguments over international attempts at censorship.

First Virtual Bank joint project with the bank FirstUSA which allows shoppers on the World Wide Web to open a central account using credit cards. Shoppers use their account numbers to make purchases at any of a number of participating merchants.

FISH FAQ

http://www.wh.whoi.edu/homepage/faq.html

Answers to hundreds of questions about fish and shellfish. Unfortunately the questions are not grouped, or searchable, in any way, but they do include answers to such things as 'Do fish sleep?' and 'Are Hawaiian monk seals coming back?'.

FISH

Spurdog, alewife, twaite shad, jollytail, tadpole madtom, bummalow, walleye pollock, wrestling halfbeak, mummichog, jolthead porgy, sweetlip emperor, and slippery dick are all common names for species of fish.

fish aquatic vertebrate that uses gills to obtain oxygen from fresh or sea water. There are three main groups: the bony fishes or Osteichthyes (goldfish, cod, tuna); the cartilaginous fishes or Chondrichthyes (sharks, rays); and the jawless fishes or Agnatha (hagfishes, lampreys).

Fishes of some form are found in virtually every body of water in the world except for the very salty water of the Dead Sea and some of the hot larval springs. Of the 30,000 fish species, approximately 2,500 are freshwater.

bony fishes These constitute the majority of living fishes (about 20,000 species). The skeleton is bone, movement is controlled by mobile fins, and the body is usually covered with scales. The gills are covered by a single flap. Many have a ◊swim bladder with which the fish adjusts its buoyancy. Most lay eggs, sometimes in vast numbers; some ◊cod can produce as many as 28 million. These are laid in the open sea, and probably no more than 28 of them will survive to become adults. Those species that produce small numbers of eggs very often protect them in nests, or brood them in their mouths. Some fishes are internally fertilized and retain eggs until hatched inside the body, then giving birth to live young. Most bony fishes are ray-finned fishes, but a few, including lungfishes and coelacanths, are fleshy-finned.

cartilaginous fishes These are efficient hunters. There are fewer than 600 known species of sharks and rays. The skeleton is cartilage, the mouth is generally beneath the head, the nose is large and sensitive, and there is a series of open gill slits along the neck region. They have no swimbladder and, in order to remain buoyant, must keep swimming. They may lay eggs ('mermaid's purses') or bear live young. Some types of cartilaginous fishes, such as sharks, retain the shape they had millions of years ago.

jawless fishes Jawless fish have a body plan like that of some of the earliest vertebrates that existed before true fishes with jaws evolved. There is no true backbone but a ◊notochord. The lamprey attaches itself to the fishes on which it feeds by a suckerlike rasping mouth. Hagfishes are entirely marine, very slimy, and feed on carrion and injured fishes.

The world's largest fish is the whale shark *Rhineodon typus,* more than 20 m/66 ft long; the smallest is the dwarf pygmy goby *Pandaka pygmaea*), 7.5–9.9 mm long. The study of fishes is called ichthyology.

fish as food The nutrient composition of fish is similar to that of meat, except that there are no obvious deposits of fat. Examples of fish comparatively high in fat are salmon, mackerel, and herring. White fish such as cod, haddock, and whiting contain only 0.4–4% fat. Fish are good sources of B vitamins and iodine, and the fatty fish livers are good sources of A and D vitamins. Calcium can be obtained from fish with soft skeletons, such as sardines. Roe and caviar have a high protein content (20–25%).

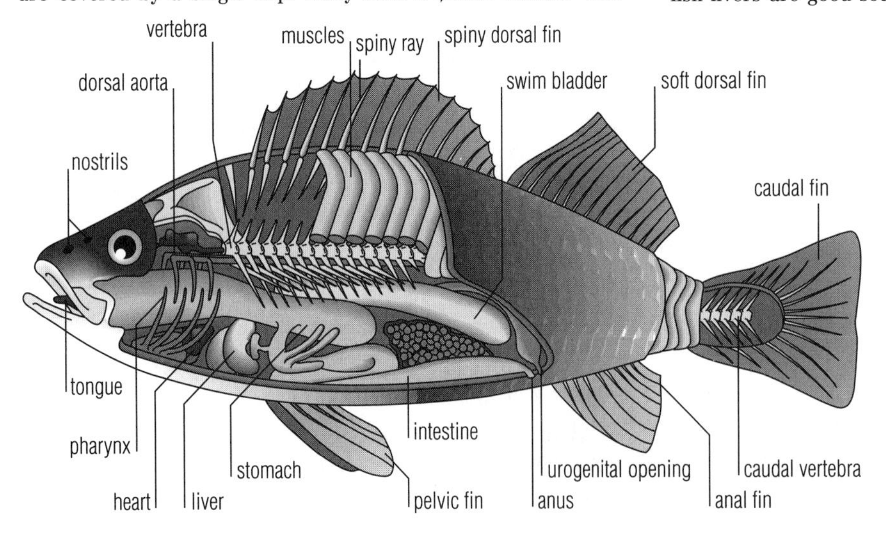

fish The anatomy of a fish. All fishes move through water using their fins for propulsion. The bony fishes, like the specimen shown here, constitute the largest group of fishes with about 20,000 species.

fisher or ***pekan*** North American ◊marten *Martes pennanti,* dark brown with greyish foreparts and blackish rump and tail. It is less arboreal than the smaller American marten *M. americana.*

fish farming or ***aquaculture*** raising fish (including molluscs and crustaceans) under controlled conditions in tanks and ponds, sometimes in offshore pens. It has been practised for centuries in the Far East, where Japan today produces some 100,000 tonnes of fish a year; the US, Norway, and Canada are also big producers. In the 1980s 10% of the world's consumption of fish was farmed, notably carp, catfish, trout, Atlantic salmon, turbot, eel, mussels, clams, oysters, and shrimp.

A total 600,000 tonnes of salmon was produced by world fish in 1995, accounting for 37% of salmon consumed. Fish farms are environmentally controversial because of the risk of escapees that could spread disease and alter the genetic balance of wild populations. In 1995 1,500,000 fish escaped from Norwegian fish farms.

By 1998 shrimp farms worldwide were producing 3 million tonnes of shrimps per year.

fishing and fisheries fisheries can be classified by (1) type of water: freshwater (lake, river, pond); marine (inshore, midwater, deep sea); (2) catch: for example, salmon fishing; (3) fishing method: diving, stunning or poisoning, harpooning, trawling, drifting. The world's total fish catch is about 100 million tonnes a year (1995).

marine fishing Most of the world's catch comes from the oceans, and marine fishing accounts for around 20% of the world's animal-based protein. A wide range of species is included in the landings of the world's marine fishing nations, but the majority belong to the herring and cod groups. The majority of the crustaceans landed are shrimps, and squid and bivalves, such as oysters, are dominant among the molluscs.

Almost all marine fishing takes place on or above the continental shelf, in the photic zone, the relatively thin surface layer (50 m/165 ft) of water that can be penetrated by light, allowing photosynthesis by plant ◊plankton to take place. **Pelagic fishing** exploits not only large fish such as tuna, which live near the surface in the open sea and are caught in purse-seine nets, with an annual catch of over 30 million tonnes, but also small, shoaling and plankton-feeding fish that live in the main body of the water.

Examples are herring, sardines, anchovies, and mackerel, which are caught with drift nets, purse seines, and pelagic trawls. The fish are often used for fish meal rather than for direct human consumption. **Demersal fishes**, such as haddock, halibut, plaice, and cod, live primarily on or near the ocean floor, and feed on various invertebrate marine animals. Over 20 million tonnes of them are caught each year by trawling.

freshwater fishing Such species as salmon and eels, which migrate to and from fresh and salt water for spawning, may be fished in either system. About a third of the total freshwater catch comes from ◊fish farming methods, which are better developed in freshwater than in marine systems. There is large demand for a , trout, carp, eel, bass, pike, perch, and catfish. These are caught in ponds, lakes, rivers, or swamps. In Africa, although marine fishing is generally more important, certain areas have significant freshwater fisheries; Lake Victoria annually yields a catch of 100,000 tonnes, which is four times the total catch from the whole eastern African seaboard. In western Europe there is very little food production from fresh water; instead the fish are usually exploited for recreational purposes or sport.

methods The gear and methods used to catch fish are very varied and show much geographical and historical variation. The method chosen for a particular situation will depend on the species being hunted and the nature of the habitat (for example, the speed of the current, the depth of water, and the roughness of the sea bed). It is often useful to divide gear types into active (for example trawls, seines, harpoons, dredges) and passive (drift nets, traps, hooks and lines). Passive gear relies on the fish's own movements to bring them into contact with it, and may involve some method of artificial attraction such as baits or lights. Most fishing gear is operated from boats, ranging from one-person canoes to trawlers about 100 m/330 ft long.

trawling Much of the world's fish catch is caught by trawls. These may be used on the sea bed (demersal) or in midwater (pelagic), but in all cases the equipment consists essentially of a tapered bag of netting which is towed through the water. The mouth of the net is kept open in the vertical plane by having floats on the headline and weights on the footrope. On bottom trawls these weights are usually hollow iron spheres that roll over the sea bed. In addition there may be tickler chains which help to dislodge or disturb fish from the sea bed in advance of the trawl so that they are more likely to be caught. There are three methods of keeping the net open in the horizontal plane: (1) pair trawling, in which the two trawl warps are towed by separate vessels; (2) beam trawling, in which the net is supported on a rigid frame consisting of a horizontal wooden or metal beam with a shoe at either end; and (3) otter trawling, in which an otter board (a weighted board with lines and baited hooks attached) is incorporated into each warp and acts as a hydroplane to push the warp out sideways. Most modern trawlers are otter trawlers, and many haul the net up over the stern rather than the side. The main problem with pelagic trawls is to control the depth of fishing and to relate this to the concentrations of fish. The most effective means of tracking fish for this purpose is by using an ◊echo sounder.

seine nets Seine nets operate by trapping fish within encircling gear. The **Danish seine** resembles a light trawl with very long side pieces, or wings, but is operated differently. The method consists of dropping a large buoy and then paying out up to 4 km/2.5 mi of warp in dogleg shape, then the net, followed by a further length of rope warp in reverse dogleg, bringing the boat back to the buoy, which is then picked up before hauling in the rope warps. As the ropes straighten on the sea bed, they channel the fish into a narrow path between them. The fish are then swept up by the net as it is hauled towards the boat. This method requires smooth sandy ground. **Beach seines** are similar to Danish seines but may consist simply of a wall of netting. They are set in a line or semicircle parallel to a beach and can then be hauled onto the beach, trapping the fish between the net and the shore. Salmon are often caught this way in estuaries. **Purse seines**, nets that close like a purse, are used to catch pelagic fish such as herring, mackerel, and tuna, which form dense shoals near the surface. Once a shoal has been located, usually by echo sounder, the net is shot around it by one vessel and later hauled in towards another. The nets are large, often as long as 30 nautical miles, and are not usually hauled aboard like trawls. Instead the fish are scooped or pumped out of the net into the ship's hold. They have caused a crisis in the South Pacific where Japan, Taiwan, and South Korea fish illegally in other countries' fishing zones.

gill nets Gill nets passively depend on the fish entangling themselves in the meshes of the net, usually being held fast by their gill covers. An example is the drift net used for pelagic fish, but in many areas it is now superseded by purse seines and pelagic trawls. Drift nets are walls of netting suspended from floats on the surface. Those used in the East Anglian herring fishery were only 70 m/230 ft long but were set in fleets up to 4 km/2.5 mi long. Herring were caught in them as they came up to the surface at night to feed. Other types of gill net can be used near the sea bed, and one, the trammel, is still used quite commonly in inshore fisheries along the south coast of England. This typically consists of a curtain of large- and small-mesh netting into which the fish swim, forcing the small-mesh net into the large and becoming trapped in a net bag.

traps Netting panels can be arranged to form traps into which fish are guided or attracted; those used on the northeast coast of England to catch salmon are good examples. Many crustaceans, such as lobsters and crabs, are normally taken in baited baskets set in strings of several hundred laid over suitable ground. Earthenware jars are used as octopus traps in the Mediterranean.

lines and hooks Although a distinct method of catching fish, lines and hooks can be regarded as a special type of trap. Natural or artificial baits are used and the gear may be fished anywhere from the sea bed to the surface. Hooks and lines fished off the sea bed may be towed from moving boats, which is called trolling. The largest lines, called long lines, are those used by the Japanese to catch tuna in ocean areas. These are up to 80 km/50 mi long and the baited hooks hang well below the surface from the buoyed lines.

dredges Dredges act like small trawls to collect molluscs and other sluggish or sessile organisms; some are hydraulic and use jets of water to dislodge the molluscs from the bottom and wash them into the dredge bag or directly onto the boat via a conveyor belt.

other methods Molluscs may also be gathered by hand, either on foot at low water or by divers below the shoreline. Rakes may be used to dig out cockles from within the sand. Other methods include dip, lift and cast nets, harpoons, and spears.

fission in physics, the splitting of a heavy atomic nucleus into two or more major fragments. It is accompanied by the emission of two or three neutrons and the release of large amounts of ◊nuclear energy.

Fission occurs spontaneously in nuclei of uranium-235, the main fuel used in nuclear reactors. However, the process can also be induced by bombarding nuclei with neutrons because a nucleus that has absorbed a neutron becomes unstable and soon splits. The neutrons released spontaneously by the fission of uranium nuclei may therefore be used in turn to induce further fissions, setting up a ◊chain reaction that must be controlled if it is not to result in a nuclear explosion.

> *Anyone who expects a source of power from the transformation of these atoms is talking moonshine.*
>
> Ernest Rutherford New Zealand physicist. *Physics Today* Oct 1970

fistula in medicine, an abnormal pathway developing between adjoining organs or tissues, or leading to the exterior of the body. A fistula developing between the bowel and the bladder, for instance, may give rise to urinary-tract infection by intestinal organisms.

fit in medicine, popular term for ◊convulsion.

fitness in genetic theory, a measure of the success with which a genetically determined character can spread in future generations. By convention, the normal character is assigned a fitness of one, and variants (determined by other ◊alleles) are then assigned fitness values relative to this. Those with fitness greater than one will spread more rapidly and will ultimately replace the normal allele; those with fitness less than one will gradually die out.

fixed font in computing, a ◊font that uses fixed, rather than proportional, spacing. It is a necessary option in off-line reader software and e-mail programs, since some ASCII art and tables do not display correctly without it.

fixed point temperature that can be accurately reproduced and used as the basis of a temperature scale. In the Celsius scale, the fixed points are the temperature of melting ice, defined to be 0°C (32°F), and the temperature of boiling water (at standard atmospheric pressure), defined to be 100°C (212°F).

fixed-point notation system in which numbers are represented using a set of digits with the decimal point always in its correct position. For very large and very small numbers this requires a lot of digits. In computing, the size of the numbers that can be handled in this way is limited by the capacity of the computer, and so the slower ◊floating-point notation is often preferred.

fjord or *fiord* narrow sea inlet enclosed by high cliffs. Fjords are found in Norway, New Zealand, and western parts of Scotland. They are formed when an overdeepened U-shaped glacial valley is drowned by a rise in sea-level. At the mouth of the fjord there is a characteristic lip causing a shallowing of the water. This is due to reduced glacial erosion and the deposition of moraine at this point.

Fiordland is the deeply indented southwest coast of South Island, New Zealand; one of the most beautiful inlets is Milford Sound.

flaccidity in botany, the loss of rigidity (turgor) in plant cells, caused by loss of water from the central vacuole so that the cytoplasm no longer pushes against the cellulose cell wall. If this condition occurs throughout the plant then wilting is seen.

Flaccidity can be induced in the laboratory by immersing the plant cell in a strong saline solution. Water leaves the cell by ◊osmosis causing the vacuole to shrink. In extreme cases the actual cytoplasm pulls away from the cell wall, a phenomenon known as plasmolysis.

flag in botany, another name for ◊iris, especially yellow flag (*Iris pseudacorus*), which grows wild in damp places throughout Europe; it is a true water plant but adapts to garden borders. It has a thick rhizome (underground stem), stiff bladelike leaves, and stems up to 150 cm/5 ft high. The flowers are large and yellow.

flag in computing, an indicator that can be set or unset in order to signal whether a particular condition is true – for example, whether the end of a file has been reached, or whether an overflow error has occurred. The indicator usually takes the form of a single binary digit, or bit (either 0 or 1).

flagellum small hairlike organ on the surface of certain cells. Flagella are the motile organs of certain protozoa and single-celled algae, and of the sperm cells of higher animals. Unlike ◊cilia, flagella usually occur singly or in pairs; they are also longer and have a more complex whiplike action.

Each flagellum consists of contractile filaments producing snakelike movements that propel cells through fluids, or fluids past cells. Water movement inside sponges is also produced by flagella.

flame angry public or private ◊electronic mail message. Users of the ◊Internet use flames to express disapproval of breaches of ◊netiquette or the voicing of an unpopular opinion. An offensive message posted to, for example, a USENET ◊newsgroup, will cause those offended to flame the culprit. Such flames maintain a level of discipline among the Internet's users.

flame test in chemistry, the use of a flame to identify metal ◊cations present in a solid.

A nichrome or platinum wire is moistened with acid, dipped in a compound of the element, either powdered or in solution, and then held in a hot flame. The colour produced in the flame is characteristic of metals present; for example, sodium burns with an orange-yellow flame, and potassium with a lilac one.

Flame test

element	*colour of flame*
sodium	orange-yellow
potassium	lilac
calcium	red or yellow-red
strontium, lithium	crimson
barium, manganese (manganese chloride)	pale green
copper, thallium, boron (boric acid)	bright green
lead, arsenic, antimony	livid blue
copper (copper (II) chloride)	bright blue

flame tree any of various trees with brilliant red flowers, including the smooth-stemmed semi-deciduous *Brachychiton acerifolium* with scarlet bell-shaped flowers, native to Australia, but spread throughout the tropics.

flame war in computing, heated electronic argument where few good points are made and most of the participants ◊flame each other repeatedly.

Several flame wars have passed into USENET legend, including the deliberate 1994 invasion of the **rec.pets.cats** newsgroup by the **alt.tasteless** newsgroup.

flamingo long-legged and long-necked wading bird, family Phoenicopteridae, of the stork order Ciconiiformes. Largest of the family is the greater or roseate flamingo *Phoenicopterus ruber*, found in Africa, the Caribbean, and South America, with delicate pink plumage and 1.25 m/4 ft tall. They sift the mud for food with their downbent bills, and build colonies of high, conelike mud nests, with a little hollow for the eggs at the top.

flannel flower Australian flower *Actinotus helianthi* found in New South Wales and Queensland, having white flannel-like bracts below the flower.

flare, solar brilliant eruption on the Sun above a ◊sunspot, thought to be caused by release of magnetic energy. Flares reach maximum brightness within a few minutes, then fade away over about an hour. They eject a burst of atomic particles into space at up to 1,000 kps/600 mps. When these particles reach Earth they can cause radio blackouts, disruptions of the Earth's magnetic field, and ◊aurorae.

flash flood flood of water in a normally arid area brought on by a sudden downpour of rain. Flash floods are rare and usually occur in mountainous areas. They may travel many kilometres from the site of the rainfall.

Because of the suddenness of flash floods, little warning can be given of their occurrence. In 1972 a flash flood at Rapid City, South Dakota, USA, killed 238 people along Rapid Creek.

flash memory type of ◊EEPROM memory that can be erased and reprogrammed without removal from the computer.

flashover small fire that erupts suddenly into a much larger one. They can occur if there is sufficient build up of heat from a small fire to ignite the mixture of gas and smoke produced. This increases temperatures further igniting surroundings. Flashovers can occur extremely rapidly with temperatures of 1,100°C being reached in seconds.

FlashPix in computing, a ◊file format for digital imaging intended as a universal standard for both individual multimedia applications and external communications over online services. It was developed collaboratively by Kodak, Hewlett-Packard, Live Picture, and Microsoft in 1996.

flash point in physics, the lowest temperature at which a liquid or volatile solid heated under standard conditions gives off sufficient vapour to ignite on the application of a small flame.

The **fire point** of a material is the temperature at which full combustion occurs. For safe storage of materials such as fuel or oil, conditions must be well below the flash and fire points to reduce fire risks to a minimum.

flash upgrade in computing, technique for upgrading firmware by updating the software embedded in it. It is used particularly for modems and ◊EPROMs.

flatfish bony fishes of the order Pleuronectiformes, having a characteristically flat, asymmetrical body with both eyes (in adults) on the upper side. Species include flounders, turbots, halibuts, plaice, and the European soles.

flat screen type of display suitable for portable computers such as LCD (◊liquid crystal display) or gas plasma screens (see ◊plasma display). Flat-screen, or flat-panel, displays are compact and lightweight compared to traditional cathode-ray tube monitors and TV sets.

It is predicted that eventually all TV screens will be made using this type of technology.

flatworm invertebrate of the phylum Platyhelminthes. Some are free-living, but many are parasitic (for example, tapeworms and flukes). The body is simple and bilaterally symmetrical, with one opening to the intestine. Many are hermaphroditic (with both male and female sex organs) and practise self-fertilization.

flax any of a group of plants including the cultivated *L. usitatissimum;* **linen** is produced from the fibre in its stems. The seeds yield **linseed oil**, used in paints and varnishes. The plant, of almost worldwide distribution, has a stem up to 60 cm/24 in high, small leaves, and bright blue flowers. (Genus *Linum,* family Linaceae.)

After extracting the oil, what is left of the seeds is fed to cattle. The stems are retted (soaked) in water after harvesting, and then dried, rolled, and scutched (pounded), separating the fibre from the central core of woody tissue. The long fibres are spun into linen thread, twice as strong as cotton, yet more delicate, and suitable for lace; shorter fibres are used to make string or paper.

Annual world production of flax fibre amounts to approximately 60,000 tonnes, with Russia, Ukraine, Belarus, and Latvia accounting for half of the total. Other producers are Belgium, the Netherlands, and Northern Ireland.

> *Colour possessed me. I didn't have to pursue it. It will possess me always ... This is the meaning of this happy hour: colour and I are one. I am a painter.*
>
> PAUL KLEE Swiss artist. *The Diaries of Paul Klee*, April 1914

FLEA

As a flea jumps, its rate of acceleration is 20 times that of the space shuttle during launching. It reaches a speed of 100 m per sec within the first 500th of a second.

flea wingless insect of the order Siphonaptera, with blood-sucking mouthparts. Fleas are parasitic on warm-blooded animals. Some fleas can jump 130 times their own height.

Species include the human flea *Pulex irritans;* the rat flea *Xenopsylla cheopsis,* the transmitter of plague and typhus; and (fostered by central heating) the cat and dog fleas *Ctenocephalides felis* and *C. canis.*

fleabane any of several plants of two related groups, belonging to the daisy family. Common fleabane (*P. dysenterica*) has golden-yellow flower heads and grows in wet and marshy places throughout Europe. (Genera *Pulicaria* and *Erigeron,* family Compositae.)

Fleming, Alexander (1881–1955) Scottish bacteriologist who discovered the first antibiotic drug, ◊penicillin, in 1928. In 1922 he had discovered lysozyme, an antibacterial enzyme present in saliva, nasal secretions, and tears. While studying this, he found an unusual mould growing on a neglected culture dish, which he isolated and grew into a pure culture; this led to his discovery of penicillin. It came into use in 1941. In 1945 he won the Nobel Prize for Physiology or Medicine with Howard W Florey and Ernst B Chain, whose research had brought widespread realization of the value of penicillin.

Fleming, (John) Ambrose (1849–1945)

English electrical physicist and engineer who invented the thermionic valve in 1904 and devised Fleming's rules. Knighted 1929.

Mary Evans Picture Library

Fleming's rules memory aids used to recall the relative directions of the magnetic field, current, and motion in an electric generator or motor, using one's fingers. The three directions are represented by the thu*m*b (for *m*otion), *f*orefinger (for *f*ield), and se*c*ond finger (for ◊*c*onventional ◊*c*urrent), all held at right angles to each other. The right hand is used for generators and the left for motors.

The rules were devised by the English physicist John Fleming.

flesh fly medium-sized fly varying from golden-brown to dark grey. The larvae feed on carrion and animal waste, though the larvae of *Wohlfahrtia* often invade the skin of children and young animals and other larvae will cause myiasis (invasion of the tissues) when the skin is already broken by a cut or abrasion.

classification Flesh flies are members of the genera *Sarcophaga* and *Wohlfahrtia* in the family Sarcophagidae of the insect order Diptera, class Insecta, phylum Arthropoda.

flexor any muscle that bends a limb. Flexors usually work in opposition to other muscles, the extensors, an arrangement known as antagonistic.

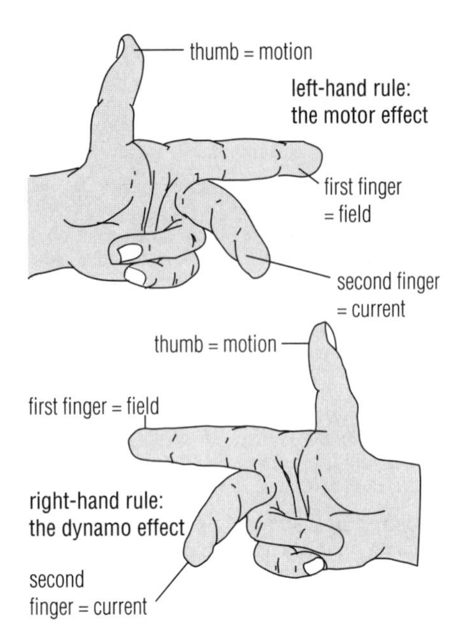

Fleming's rules Fleming's rules give the direction of the magnetic field, motion, and current in electrical machines. The left hand is used for motors, and the right hand for generators and dynamos.

flight or ***aviation*** method of transport in which aircraft carry people and goods through the air. People first took to the air in ◊balloons and began powered flight in 1852 in ◊airships, but the history of flying, both for civilian and military use, is dominated by the ◊aeroplane. The earliest planes were designed for gliding; the advent of the petrol engine saw the first powered flight by the ◊Wright brothers in 1903 in the USA. This inspired the development of aircraft throughout Europe. Biplanes were succeeded by monoplanes in the 1930s. The first jet plane (see ◊jet propulsion) was produced in 1939, and after the end of World War II the development of jetliners brought about a continuous expansion in passenger air travel. In 1969 came the supersonic aircraft ◊Concorde.

history In the 14th century the English philosopher Roger Bacon spoke of constructing an aircraft by means of a hollow globe and liquid fire. He was followed in the 15th century by Albert of Saxony, who also spoke of balloon flight by means of fire in a light sphere. During the 16th and 17th centuries a number of fantastic ideas were put forward; one was that swans' eggs be filled with sulphur or mercury and thereby drawn up to the Sun.

early ideas Francisco de Lana in 1670 proposed that four hollow balls made of very thin brass should be emptied of air. To them should be attached a small boat and sail, and in that way a balloon would be contrived which could carry a person. The idea was not feasible, since the globes, made of brass only 0.1 mm thick, would have collapsed by reason of their own weight. But although de Lana saw this difficulty, he argued that their shape would prevent that.

balloons It was not until the next century that the real ◊balloon was invented. The beginning of the development of the balloon was the work of two brothers, Joseph and Etienne Montgolfier, who came to the conclusion that a paper bag filled with a 'substance of a cloud-like nature' would float in the atmosphere. They made a number of experiments which attracted attention and further efforts from others. Progress was made gradually, and the first person-carrying ascent took place in October 1783, when Pilatre de Rozier went up in a Montgolfier captive balloon. The first woman to ascend was Madame Thible, who went up from Lyons in 1784. In 1859 a flight of over 1,600 km/994 mi was made in the USA.

adding power It had long been recognized that the difficulty with balloons was navigating through the air. Oars were tried, but were not successful. The first attempt to navigate the balloon by means of a small, light engine came in 1852, the experiment being made by Henri Giffard. From 1897 the development of the airship was the special work of Ferdinand ◊Zeppelin. In 1900 he made his first flight with a dirigible balloon carrying five men. It was made of aluminium, supported by gas-bags, and driven by two motors, each of about 12 kW. His first experiment met with some success, a second, more powerful version was wrecked, and a third met with great success. This airship carried 11 passengers and attained a speed of about 55 kph/34 mph, travelling about 400 km/248 mi in 11 hours, but was wrecked by a storm in 1908, caught fire, and was completely destroyed.

powered flight In the late 19th century experiments were being made with soaring machines and hang gliders, chiefly by Otto Lilienthal, who, with an arrangement formed on the plan of birds' wings, attempted to imitate their 'soaring flight'. Following up Lilienthal's ideas, the Wright brothers produced their first powered aeroplane in 1903. Their first successful machine was simply an aeroplane that flew in a straight line, but this received many modifications; and in 1908 they went to France to carry on experiments, during which Wilbur Wright created a record by remaining in the air for over an hour while carrying a passenger. He also attained a speed of 60 kph/37 mph.

In Europe, at the beginning of the 20th century, France led in aeroplane design and Louis Blériot brought aviation much publicity by crossing the English Channel in 1909, as did the Reims air races of that year. The first powered flight in the UK was made by Samuel Franklin Cody 1908. In 1912 Sopwith and Bristol both built small biplanes. The first big twin-engined aeroplane was the Handley Page bomber 1917. The stimulus of World War I (1914–18) and rapid development of the petrol engine led to increased power, and speeds rose to 320 kph/200 mph. Streamlining the body of planes became imperative: the body, wings, and exposed parts were reshaped to reduce drag. Eventually the biplane was superseded by the internally braced monoplane structure, for example, the Hawker Hurricane and Supermarine Spitfire fighters and Avro Lancaster and Boeing Flying Fortress bombers of World War II (1939–45).

jet aircraft The German Heinkel 178, built in 1939, was the first jet plane; it was driven, not by a ◊propeller as all planes before it, but by a jet of hot gases. The first British jet aircraft, the Gloster E.28/39, flew from Cranwell, Lincolnshire, on 15 May 1941, powered by a jet engine invented by British engineer Frank Whittle. Twin-jet Meteor fighters were in use by the end of WWII. The rapid development of the jet plane led to enormous increases in power and speed until air-compressibility effects were felt near the speed of sound, which at first seemed to be a flight speed limit (the sound barrier). The sound barrier was first broken in the USA in 1947 by a rocket-powered aircraft piloted by Chuck Yeager. To attain ◊supersonic speed, streamlining the aircraft body became insufficient: wings were swept back, engines buried in wings and tail units, and bodies were even eliminated in all-wing delta designs. In the 1950s the first jet airliners, such as the Comet (first introduced in 1949), were introduced into service. Today jet planes dominate both military and civilian aviation, although many light planes still use piston engines and propellers. The late 1960s saw the introduction of the ◊jumbo jet, and in 1976 the Anglo-French Concorde, which makes a transatlantic crossing in under three hours, came into commercial service.

other developments During the 1950s and 1960s research was done on V/STOL (vertical and/or short takeoff and landing) aircraft. The British Harrier jet fighter has been the only VTOL aircraft to achieve commercial success, but STOL technology has fed into subsequent generations of aircraft. The 1960s and 1970s also saw the development of variable geometry ('swing-wing') aircraft, the wings of

HOW DO PLANES FLY?

http://observe.ivv.nasa.gov/nasa/exhibits/planes/planes_1a.html

Lucid exposition of the whys and hows connected with the flight of an aeroplane. Kids are invited to perform a number of related experiments and the answers are given with a series of fun interactive steps.

which can be swept back in flight to achieve higher speeds. In the 1980s much progress was made in 'fly-by-wire' aircraft with computer-aided controls. International partnerships have developed both civilian and military aircraft. The airbus is a wide-bodied airliner built jointly by companies from France, Germany, the UK, the Netherlands, and Spain. The Eurofighter 2000 is a joint project between the UK, Italy, Germany, and Spain. The B-2 bomber, (a stealth bomber) developed by the US Air Force in 1989, is invisible to radar. The altitude record for a solar-powered plane was set in 1997 by Pathfinder, a 30-m/98-ft wingspan aircraft, which reached 20,528 m/67,349 ft above sea level over Hawaii.

flight simulator computer-controlled pilot-training device, consisting of an artificial cockpit mounted on hydraulic legs, that simulates the experience of flying a real aircraft. Inside the cockpit, the trainee pilot views a screen showing a computer-controlled projection of the view from a real aircraft, and makes appropriate adjustments to the controls. The computer monitors these adjustments, changes both the alignment of the cockpit on its hydraulic legs, and the projected view seen by the pilot. In this way a trainee pilot can progress to quite an advanced stage of training without leaving the ground.

flint compact, hard, brittle mineral (a variety of chert), brown, black, or grey in colour, found as nodules in limestone or shale deposits. It consists of cryptocrystalline (grains too small to be visible even under a light microscope) ◊silica, SiO_2, principally in the crystalline form of quartz. Implements fashioned from flint were widely used in prehistory.

The best flint, used for Neolithic tools, is **floorstone**, a shiny black flint that occurs deep within chalk.

Because of their hardness (7 on the ◊Mohs' scale), flint splinters are used for abrasive purposes and, when ground into powder, added to clay during pottery manufacture. Flints have been used for making fire by striking the flint against steel, which produces a spark, and for discharging guns. Flints in cigarette lighters are made from cerium alloy.

flip-flop in computing, another name for a ◊bistable circuit.

floating state of equilibrium in which a body rests on or is suspended in the surface of a fluid (liquid or gas). According to ◊Archimedes' principle, a body wholly or partly immersed in a fluid will be subjected to an upward force, or upthrust, equal in magnitude to the weight of the fluid it has displaced.

If the ◊density of the body is greater than that of the fluid, then its weight will be greater than the upthrust and it will sink. However, if the body's density is less than that of the fluid, the upthrust will be the greater and the body will be pushed upwards towards the surface. As the body rises above the surface the amount of fluid that it displaces (and therefore the magnitude of the upthrust) decreases. Eventually the upthrust acting on the submerged part of the body will equal the body's weight, equilibrium will be reached, and the body will float.

floating-point notation system in which numbers are represented by means of a decimal fraction and an exponent. For example, in floating-point notation, 123,000,000,000 would be represented as 0.123×10^{12}, where 0.123 is the fraction, or mantissa, and 12 the exponent. The exponent is the power of 10 by which the fraction must be multiplied in order to obtain the true value of the number.

In computing, floating-point notation enables programs to work with very large and very small numbers using only a few digits; however, it is slower than ◊fixed-point notation and suffers from small rounding errors.

flocculation in soils, the artificially induced coupling together of particles to improve aeration and drainage. Clay soils, which have very tiny particles and are difficult to work, are often treated in this way. The method involves adding more lime to the soil.

floppy disc in computing, a storage device consisting of a light, flexible disc enclosed in a cardboard or plastic jacket. The disc is placed in a disc drive, where it rotates at high speed. Data are recorded magnetically on one or both surfaces.

Floppy discs were invented by IBM in 1971 as a means of loading programs into the computer. They were originally 20 cm/8 in in diameter and typically held about 240 ◊kilobytes of data. Present-day floppy discs, widely used on ◊microcomputers, are usually either 13.13 cm/5.25 in or 8.8 cm/3.5 in in diameter, and generally hold 0.5–2 ◊megabytes, depending on the disc size, recording method, and whether one or both sides are used.

Floppy discs are inexpensive, and light enough to send through the post, but have slower access speeds and are more fragile than hard discs. (See also ◊disc).

FLOPS (abbreviation for ***floating point operations per second***) measure of the speed at which a computer program can be run.

floptical disc or ***erasable optical disc*** in computing, a type of optical disc that can be erased and loaded with new data, just like a magnetic disc. By contrast, most optical discs are read-only. A single optical disc can hold as much as 1,000 megabytes of data, about 800 times more than a typical floppy disc. Floptical discs need a special disc drive, but some such drives are also capable of accepting standard 3.5 inch floppy discs.

floral diagram diagram showing the arrangement and number of parts in a flower, drawn in cross section. An ovary is drawn in the centre, surrounded by representations of the other floral parts, indicating the position of each at its base. If any parts such as the petals or sepals are fused, this is also indicated. Floral diagrams allow the structure of different flowers to be compared, and are usually shown with the floral formula.

floral formula symbolic representation of the structure of a flower. Each kind of floral part is represented by a letter (K for calyx, C for corolla, P for perianth, A for androecium, G for gynoecium) and a number to indicate the quantity of the part present, for example, C5 for a flower with five petals. The number is in brackets if the parts are fused. If the parts are arranged in distinct whorls within the flower, this is shown by two separate figures, such as A5 + 5, indicating two whorls of five stamens each.

A flower with radial symmetry is known as **actinomorphic**; a flower with bilateral symmetry as **zygomorphic**.

floret small flower, usually making up part of a larger, composite flower head. There are often two different types present on one flower head: disc florets in the central area, and ray florets around the edge which usually have a single petal known as the ligule. In the common daisy, for example, the disc florets are yellow, while the ligules are white.

flotation, law of law stating that a floating object displaces its own weight of the fluid in which it floats. See ◊Archimedes principle.

flotation process common method of preparing mineral ores for subsequent processing by making use of the different wetting properties of various components. The ore is finely ground and then mixed with water and a specially selected wetting agent. Air is bubbled through the mixture, forming a froth; the desired ore particles attach themselves to the bubbles and are skimmed off, while unwanted dirt or other ores remain behind.

flounder small flatfish *Platychthys flesus* of the NE Atlantic and Mediterranean, although it sometimes lives in estuaries. It is dull in colour and grows to 50 cm/1.6 ft.

FLOUNDER

Flounders attempt to merge into any background. They will even assume an approximate chequered pattern if the floor of their tank resembles a chess-board.

flour beetle beetle that is a major pest of stored agricultural products, such as flour. They are found worldwide in granaries and stores where both the adult beetles and the larvae feed on damaged grain or flour. Neither adults nor larvae can eat intact grains.

classification Flour beetles are in the genus *Tribolium,* family Tenebrionidae, class Insecta, phylum Arthropoda.

flow chart diagram, often used in computing, to show the possible paths that data can take through a system or program.

A **system flow chart**, or **data flow chart**, is used to describe the flow of data through a complete data-processing system. Different graphic symbols represent the clerical operations involved and the different input, storage, and output equipment required. Although the flow chart may indicate the specific programs used, no details are given of how the programs process the data.

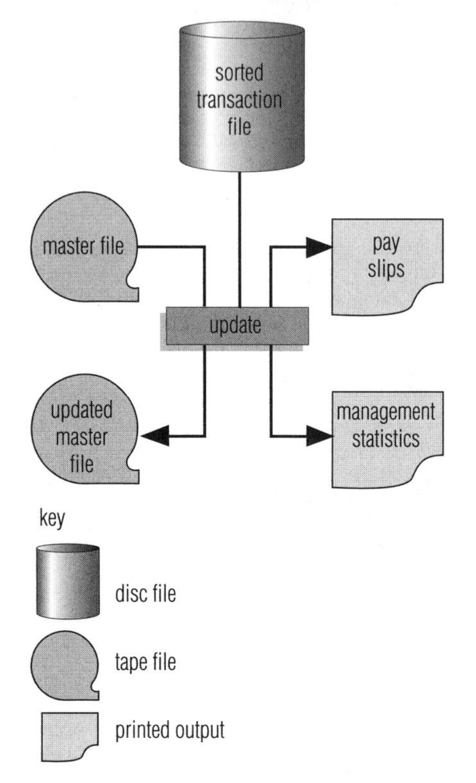

flow chart A system flow chart describes the flow of data through a data-processing system. This chart shows the data flow in a basic accounting system.

A **program flow chart** is used to describe the flow of data through a particular computer program, showing the exact sequence of operations performed by that program in order to process the data. Different graphic symbols are used to represent data input and output, decisions, branches, and ◊subroutines.

flow control in data communications, hardware or software signals that control the flow of data to ensure that it is not transmitted too quickly for the receiving computer to handle.

flower the reproductive unit of an angiosperm or flowering plant, typically consisting of four whorls of modified leaves: ◊sepals, ◊petals, ◊stamens, and ◊carpels. These are borne on a central axis or ◊receptacle. The many variations in size, colour, number, and arrangement of parts are closely related to the method of pollination. Flowers adapted for wind pollination typically have reduced or absent petals and sepals and long, feathery ◊stigmas that hang outside the flower to trap airborne pollen. In contrast, the petals of insect-pollinated flowers are usually conspicuous and brightly coloured.

structure The sepals and petals form the **calyx** and **corolla** respectively and together comprise the **perianth** with the function of protecting the reproductive organs and attracting pollinators.

The stamens lie within the corolla, each having a slender stalk, or filament, bearing the pollen-containing anther at the top. Collectively they are known as the **androecium** (male organs). The

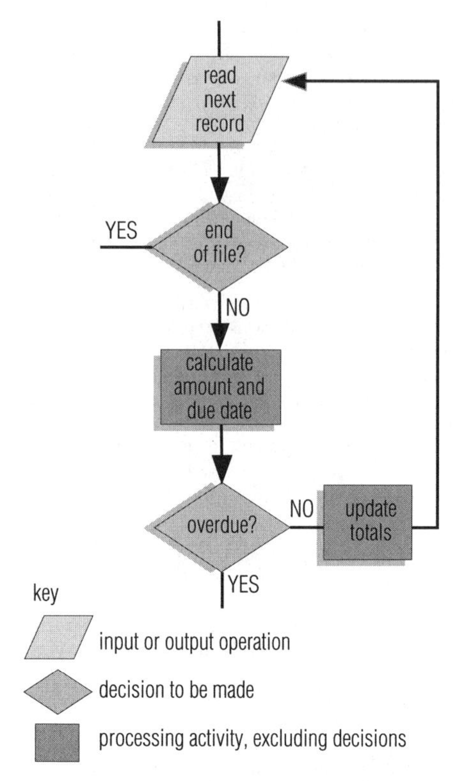

flow chart A program flow chart shows the sequence of operations needed to achieve a task, in this case reading customer accounts and calculating the amount due for each customer. After an account has been processed, the program loops back to process the next one.

inner whorl of the flower comprises the carpels, each usually consisting of an ovary in which are borne the ◊ovules, and a stigma borne at the top of a slender stalk, or style. Collectively the carpels are known as the **gynoecium** (female organs).

types of flower In size, flowers range from the tiny blooms of duckweeds scarcely visible to the naked eye to the gigantic flowers of the Malaysian *Rafflesia,* which can reach over 1 m/3 ft across. Flowers may either be borne singly or grouped together in inflorescences. The stalk of the whole ◊inflorescence is termed a **peduncle**, and the stalk of an individual flower is termed a **pedicel**. A

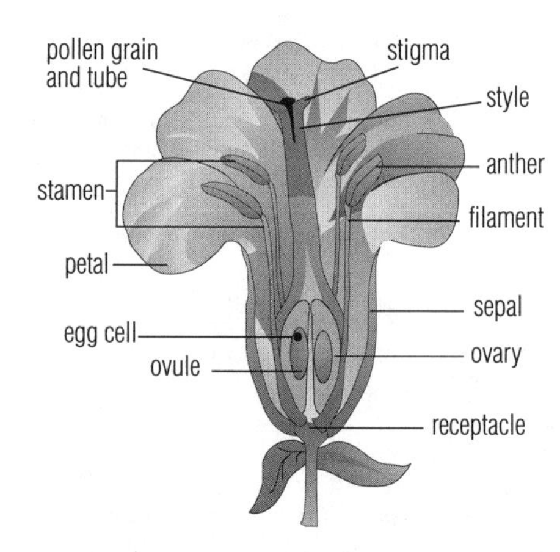

flower Cross section of a typical flower showing its basic components: sepals, petals, stamens (anthers and filaments), and carpel (ovary and stigma). Flowers vary greatly in the size, shape, colour, and arrangement of these components.

flower is termed hermaphrodite when it contains both male and female reproductive organs. When male and female organs are carried in separate flowers, they are termed **monoecious**; when male and female flowers are on separate plants, the term **dioecious** is used.

flowering plant term generally used for ◊angiosperms, which bear flowers with various parts, including sepals, petals, stamens, and carpels.

Sometimes the term is used more broadly, to include both angiosperms and ◊gymnosperms, in which case the ◊cones of conifers and cycads are referred to as 'flowers'. Usually, however, the angiosperms and gymnosperms are referred to collectively as ◊seed plants, or spermatophytes.

In 1996 UK palaeontologists found fossils in southern England of what may be the world's oldest flowering plant. *Bevhalstia pebja,* a wetland herb about 25 cm/10 in high, has been dated as early Cretaceous, about 130 million years old.

flue-gas desulphurization process of removing harmful sulphur pollution from gases emerging from a boiler. Sulphur compounds such as sulphur dioxide are commonly produced by burning ◊fossil fuels, especially coal in power stations, and are the main cause of ◊acid rain.

The process is environmentally beneficial but expensive, adding about 10% to the cost of electricity generation.

fluid any substance, either liquid or gas, in which the molecules are relatively mobile and can 'flow'.

fluid mechanics the study of the behaviour of fluids (liquids and gases) at rest and in motion. Fluid mechanics is important in the study of the weather, the design of aircraft and road vehicles, and in industries, such as the chemical industry, which deal with flowing liquids or gases.

fluid, supercritical fluid brought by a combination of heat and pressure to the point at which, as a near vapour, it combines the properties of a gas and a liquid. Supercritical fluids are used as solvents in chemical processes, such as the extraction of lubricating oil from refinery residues or the decaffeination of coffee, because they avoid the energy-expensive need for phase changes (from liquid to gas and back again) required in conventional distillation processes.

fluke any of various parasitic flatworms of the classes Monogenea and Digenea, that as adults live in and destroy the livers of sheep, cattle, horses, dogs, and humans. Monogenetic flukes can complete their life cycle in one host; digenetic flukes require two or more hosts, for example a snail and a human being, to complete their life cycle.

An estimated 40 million people worldwide are infected by foodborne flukes, mostly from undercooked or raw fish or shellfish, according to a 1994 WHO report.

fluorescence in scientific usage, very short-lived ◊luminescence (a glow not caused by high temperature). Generally, the term is used for any luminescence regardless of the persistence. ◊Phosphorescence lasts a little longer.

Fluorescence is used in strip and other lighting, and was developed rapidly during World War II because it was a more efficient means of illumination than the incandescent lamp. Recently, small bulb-size fluorescence lamps have reached the market. It is claimed that, if widely used, their greater efficiency could reduce demand for electricity. Other important applications are in fluorescent screens for television and cathode-ray tubes.

fluorescence microscopy technique for examining samples under a ◊microscope without slicing them into thin sections. Instead, fluorescent dyes are introduced into the tissue and used as a light source for imaging purposes. Fluorescent dyes can also be bonded to monoclonal antibodies and used to highlight areas where particular cell proteins occur.

fluoridation addition of small amounts of fluoride salts to drinking water by certain water authorities to help prevent tooth decay. Experiments in Britain, the USA, and elsewhere have indicated that a concentration of fluoride of 1 part per million in tap water retards the decay of children's teeth by more than 50%.

Much concern has been expressed about the risks of medicating the population at large by the addition of fluoride to the water supply, but the medical evidence demonstrates conclusively that there is no risk to the general health from additions of 1 part per million of fluoride to drinking water.

fluoride negative ion (F^-) formed when hydrogen fluoride dissolves in water; compound formed between fluorine and another element in which the fluorine is the more electronegative element (see ◊electronegativity, halide).

In parts of India, the natural level of fluoride in water is 10 parts per million. This causes fluorosis, or chronic fluoride poisoning, mottling teeth and deforming bones.

fluorine pale yellow, gaseous, nonmetallic element, symbol F, atomic number 9, relative atomic mass 19. It is the first member of the halogen group of elements, and is pungent, poisonous, and highly reactive, uniting directly with nearly all the elements. It occurs naturally as the minerals fluorite (CaF_2) and cryolite (Na_3AlF_6). Hydrogen fluoride is used in etching glass, and the freons, which all contain fluorine, are widely used as refrigerants.

Fluorine was discovered by the Swedish chemist Karl Scheele in 1771 and isolated by the French chemist Henri Moissan in 1886. Combined with uranium as UF_6, it is used in the separation of uranium isotopes.

The Infrared Space Observatory detected hydrogen fluoride molecules in an interstellar gas cloud in the constellation Sagittarius in 1997. It was the first time fluorine had been detected in space.

fluorite or ***fluorspar*** a glassy, brittle halide mineral, calcium fluoride CaF_2, forming cubes and octahedra; colourless when pure, otherwise violet, blue, yellow, brown, or green.

Fluorite is used as a flux in iron and steel making; colourless fluorite is used in the manufacture of microscope lenses. It is also used for the glaze on pottery, and as a source of fluorine in the manufacture of hydrofluoric acid.

fluorocarbon compound formed by replacing the hydrogen atoms of a hydrocarbon with fluorine. Fluorocarbons are used as inert coatings, refrigerants, synthetic resins, and as propellants in aerosols.

There is concern that the release of fluorocarbons – particularly those containing chlorine (chlorofluorocarbons, CFCs) – depletes the ◊ozone layer, allowing more ultraviolet light from the Sun to penetrate the Earth's atmosphere, and increasing the incidence of skin cancer in humans.

FLY

Fruit flies (*Drosophila*) reach sexual maturity in less than two weeks. This means that there are 25 generations a year. Every female can lay 100 eggs or more.

fly any insect of the order Diptera. A fly has a single pair of wings, antennae, and compound eyes; the hind wings have become modified into knoblike projections (halteres) used to maintain equilibrium in flight. There are over 90,000 species.

The mouthparts project from the head as a proboscis used for sucking fluids, modified in some species, such as mosquitoes, to pierce a victim's skin and suck blood. Discs at the ends of hairs on their feet secrete a fluid enabling them to walk up walls and across ceilings. Flies undergo complete metamorphosis; their larvae (maggots) are without true legs, and the pupae are rarely enclosed in a cocoon. The sexes are similar and coloration is rarely vivid, though some are metallic green or blue. The fruitfly, genus *Drosophila,* is much used in genetic experiments as it is easy to keep, fast-breeding, and has easily visible chromosomes.

flying dragon lizard *Draco volans* of the family Agamidae. It lives in sotheast Asia, and can glide on flaps of skin spread and

fly A tachinid fly Blepharella snyderi *in Kakamega Forest, Kenya. The females of the family Tachinidae lay their eggs on or near the bodies of other arthropods (future hosts), or on food which is likely to be eaten by a host at some point. The fly's larvae eventually live as internal parasites in other insects.*

supported by its ribs. This small (7.5 cm/3 in head and body) arboreal lizard can glide between trees for 6m/20 ft or more.

flying fish any marine bony fishes of the family Exocoetidae, order Beloniformes, best represented in tropical waters. They have winglike pectoral fins that can be spread to glide over the water.

flying fox another name for the fruit bat, a fruit-eating ◊bat of the suborder Megachiroptera.

flying gurnard any of various marine fishes of the order Dactylopteriformes (especially the genus *Dactylopterus*), having wing-like pectoral fins and capable of gliding for short distances.

They are not related to flying fishes.

flying lemur commonly used, but incorrect, name for ◊colugo.

It cannot fly, and it is not a lemur.

flying lizard another name for ◊flying dragon.

flying squirrel any of 43 known species of squirrel, not closely related to the true squirrels. They are characterized by a membrane along the side of the body from forelimb to hindlimb (in some species running to neck and tail) which allows them to glide through the air. Several genera of flying squirrel are found in the Old World; the New World has the genus *Glaucomys*. Most species are eastern Asian.

The giant flying squirrel *Petaurista* grows up to 1.1 m/3.5 ft including tail.

flystrike or ***blowfly strike*** or ***sheep strike*** infestation of the flesh of living sheep by blowfly maggots, especially those of the blue blowfly. It is one of the most costly sheep diseases in Australia, affecting all the grazing areas of New South Wales. Control has mainly been by insecticide, but non-chemical means, such as docking of tails and mulesing, are increasingly being encouraged. Mulesing involves an operation to remove the wrinkles of skin which trap moisture and lay the sheep open to infestation.

flythrough in ◊virtual reality, animation allowing users to view a model of a proposed or actual site as if they were inside it and moving through it.

For the 1996 Olympics in Atlanta, USA, flythroughs assisted site planners to identify areas in the main stadium where camera positions would be blocked by the audience, allowing solutions to be found in advance of construction.

flywheel heavy wheel in an engine that helps keep it running and smooths its motion. The ◊crankshaft in a petrol engine has a flywheel at one end, which keeps the crankshaft turning in between the intermittent power strokes of the pistons. It also comes into contact with the ◊clutch, serving as the connection between the engine and the car's transmission system.

FM in physics, abbreviation for ◊frequency modulation.

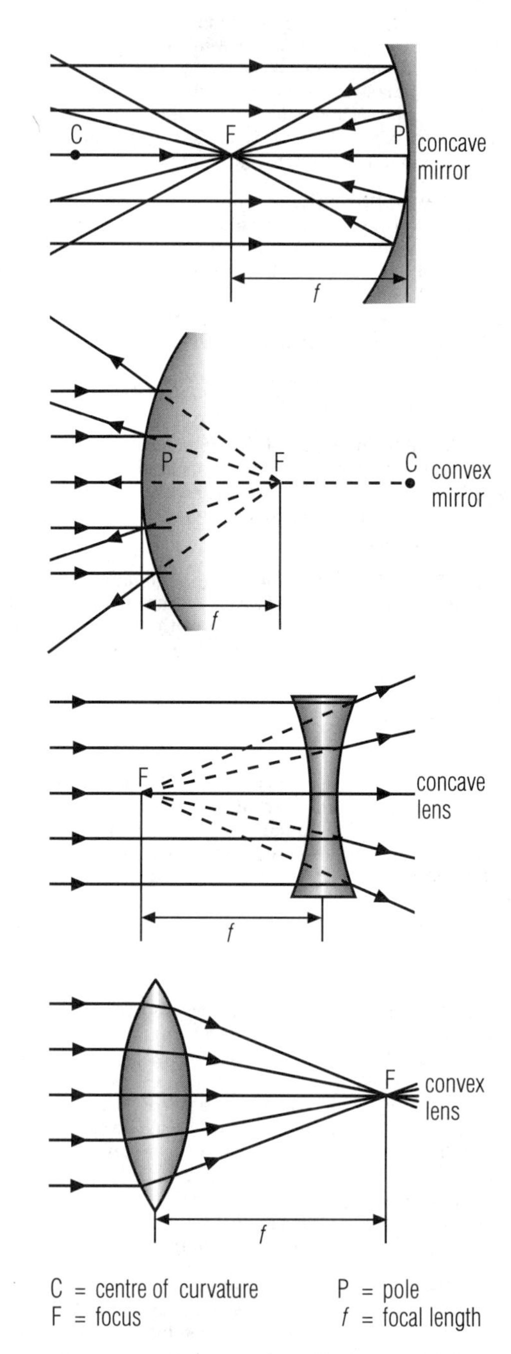

focal length *The distance from the pole (P), or optical centre, of a lens or spherical mirror to its principal focus (F). The focal length of a spherical mirror is equal to half the radius of curvature* ($f = \frac{CP}{2}$)*. The focal length of a lens is inversely proportional to the power of that lens (the greater the power the shorter the focal length).*

FM synthesizer (abbreviation for ***frequency modulation synthesizer***) in computing, method for generating synthetic sounds based on techniques used to transmit FM radio signals.

FMV abbreviation for ◊full-motion video.

f-number or ***f-stop*** measure of the relative aperture of a telescope or camera lens; it indicates the light-gathering power of the lens. In photography, each successive f-number represents a halving of exposure speed.

focal length or ***focal distance*** the distance from the centre of a lens or curved mirror to the focal point. For a concave mirror or convex lens, it is the distance at which rays of light parallel to the principal axis of the mirror or lens are brought to a focus (for a mirror, this is half the radius of curvature). For a convex mirror or concave lens, it is the distance from the centre to the point from which rays of light parallel to the principal axis of the mirror or lens diverge.

With lenses, the greater the power (measured in dioptres) of the lens, the shorter its focal length. The human eye has a lens of adjustable focal length to allow the light from objects of varying distance to be focused on the retina.

focus in astronomy, either of two points lying on the major axis of an elliptical ◊orbit on either side of the centre. One focus marks the centre of mass of the system and the other is empty. In a circular orbit the two foci coincide at the centre of the circle and in a parabolic orbit the second focus lies at infinity. See ◊Kepler's Laws.

focus or ***focal point*** in optics, the point at which light rays converge, or from which they appear to diverge. Other electromagnetic rays, such as microwaves, and sound waves may also be brought together at a focus. Rays parallel to the principal axis of a lens or mirror are converged at, or appear to diverge from, the ◊principal focus.

focus in photography, the distance that a lens must be moved in order to focus a sharp image on the light-sensitive film at the back of the camera. The lens is moved away from the film to focus the image of closer objects. The focusing distance is often marked on a scale around the lens; however, some cameras now have an automatic focusing (autofocus) mechanism that uses an electric motor to move the lens.

fog cloud that collects at the surface of the Earth, composed of water vapour that has condensed on particles of dust in the atmosphere. Cloud and fog are both caused by the air temperature falling below ◊dew point. The thickness of fog depends on the number of water particles it contains. Officially, fog refers to a condition when visibility is reduced to 1 km/0.6 mi or less, and mist or haze to that giving a visibility of 1–2 km or about 1 mi.

There are two types of fog. An **advection fog** is formed by the meeting of two currents of air, one cooler than the other, or by warm air flowing over a cold surface. Sea fogs commonly occur where warm and cold currents meet and the air above them mixes. A **radiation fog** forms on clear, calm nights when the land surface loses heat rapidly (by radiation); the air above is cooled to below its dew point and condensation takes place. A **mist** is produced by condensed water particles, and a haze by smoke or dust.

In drought areas, for example, Baja California, Canary Islands, Cape Verde Islands, Namib Desert, Peru, and Chile, coastal fogs enable plant and animal life to survive without rain and are a potential source of water for human use (by means of water collectors exploiting the effect of condensation).

Industrial areas uncontrolled by pollution laws have a continual haze of smoke over them, and if the temperature falls suddenly, a dense yellow smog forms. At some airports since 1975 it has been possible for certain aircraft to land and take off blind in fog, using radar navigation.

fold in geology, a bend in ◊beds or layers of rock. If the bend is arched up in the middle it is called an **anticline**; if it sags downwards in the middle it is called a **syncline**. The line along which a bed of rock folds is called its axis. The axial plane is the plane joining the axes of successive beds.

folder in computing, name for a computer directory in Windows 95 and on the Macintosh operating system.

folic acid a ◊vitamin of the B complex. It is found in liver and green leafy vegetables, and is also synthesized by the intestinal bacteria. It is essential for growth, and plays many other roles in the body. Lack of folic acid causes anaemia because it is necessary for the synthesis of nucleic acids and the formation of red blood cells.

follicle in botany, a dry, usually many-seeded fruit that splits along one side only to release the seeds within. It is derived from a single ◊carpel. Examples include the fruits of the larkspurs *Delphinium* and columbine *Aquilegia*. It differs from a pod, which always splits open (dehisces) along both sides.

follicle in zoology, a small group of cells that surround and nourish a structure such as a hair (hair follicle) or a cell such as an egg (Graafian follicle; see ◊menstrual cycle).

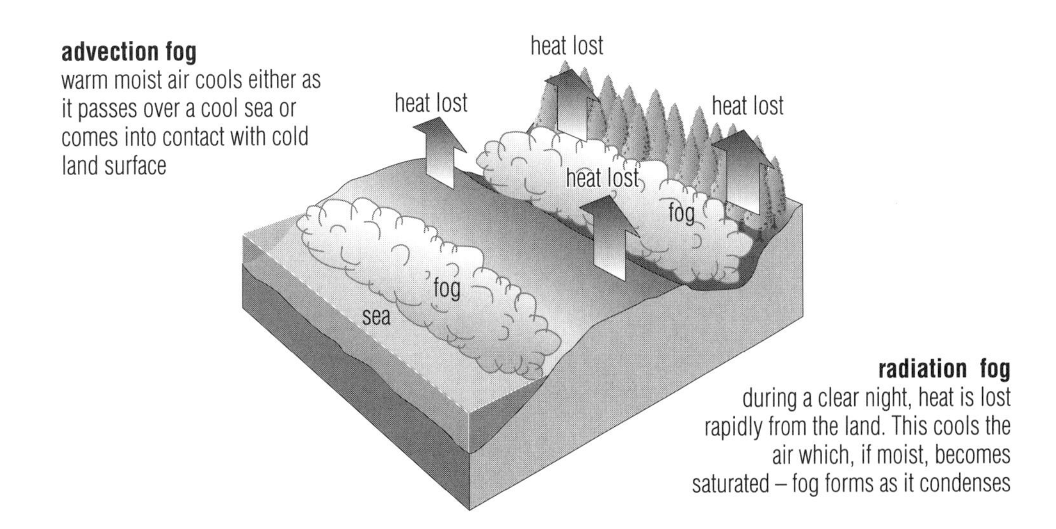

fog *Advection fog occurs when two currents of air, one cooler than the other meet, or by warm air flowing over a cold surface. Radiation fog forms through rapid heat loss from the land, causing condensation to take place and a mist to appear.*

follicle-stimulating hormone (FSH) a ◊hormone produced by the pituitary gland. It affects the ovaries in women, stimulating the production of an egg cell.

Luteinizing hormone is needed to complete the process. In men, FSH stimulates the testes to produce sperm. It is used to treat some forms of infertility.

follow-up post in computing, publicly posted reply to a USENET message; unlike a personal e-mail reply, follow-up post can be read by anyone.

Full-featured ◊newsreaders include a facility for setting the names of the newsgroups to which follow-ups should be posted. If, for example, an original message was posted to a number of groups, several of which were inappropriate, the person posting the follow-up might want to restrict further replies to only those groups where the message actually belongs.

Fomalhaut or ***Alpha Piscis Austrini*** the brightest star in the southern constellation ◊Piscis Austrinus and the 18th brightest star in the night sky. It is 22 light years from Earth, with a true luminosity 13 times that of the Sun.

Fomalhaut is one of a number of stars around which ◊IRAS (the Infra-Red Astronomy Satellite) detected excess infrared radiation, presumably from a region of solid particles around the star. This material may be a planetary system in the process of formation.

font or *fount* complete set of printed or display characters of the same typeface, size, and style (bold, italic, underlined, and so on).

Fonts used in computer setting are of two main types: bit-mapped and outline. **Bit-mapped fonts** are stored in the computer memory as the exact arrangement of ◊pixels or printed dots required to produce the characters in a particular size on a screen or printer. **Outline fonts** are stored in the computer memory as a set of instructions for drawing the circles, straight lines, and curves that make up the outline of each character. They require a powerful computer because each character is separately generated from a set of instructions and this requires considerable computation. Bit-mapped fonts become very ragged in appearance if they are enlarged and so a separate set of bit maps is required for each font size. In contrast, outline fonts can be scaled to any size and maintain exactly the same appearance.

food anything eaten by human beings and other animals and plants to sustain life and health. The building blocks of food are nutrients, and humans can utilize the following nutrients: **carbohydrates**, as starches found in bread, potatoes, and pasta; as simple sugars in sucrose and honey; as fibres in cereals, fruit, and vegetables; **proteins** as from nuts, fish, meat, eggs, milk, and some vegetables; **fats** as found in most animal products (meat, lard, dairy products, fish), also in margarine, nuts and seeds, olives, and edible oils; **vitamins**, found in a wide variety of foods, except for vitamin B_{12}, which is found mainly in foods of animal origin; **minerals**, found in a wide variety of foods (for example, calcium from milk and broccoli, iodine from seafood, and iron from liver and green vegetables); **water** ubiquitous in nature; **alcohol**, found in fermented distilled beverages, from 40% in spirits to 0.01% in low-alcohol lagers and beers.

Food is needed both for energy, measured in ◊calories or kilojoules, and nutrients, which are converted to body tissues. Some nutrients, such as fat, carbohydrate, and alcohol, provide mainly energy; other nutrients are important in other ways; for example, fibre is an aid to metabolism. Proteins provide energy and are necessary for building cell and tissue structure.

> *Food probably has a very great influence on the condition of men … . Who knows if a well-prepared soup was not responsible for the pneumatic pump or a poor one for a war?*
>
> G C Lichtenberg German physicist and philosopher. *Aphorisms*, 'Notebook A' 14

Food and Agriculture Organization (FAO) United Nations specialized agency that coordinates activities to improve food and timber production and levels of nutrition throughout the world. It is also concerned with investment in agriculture and dispersal of emergency food supplies. It has headquarters in Rome and was founded in 1945.

The USA cut its FAO funding in 1990 from $61.4 million to $18 million because of its alleged politicization.

WORLD FOOD PROGRAMME

http://www.unicc.org/wfp/

Information on the work of the UN agency charged with addressing the needs of the one out of seven people on Earth who are starving. This is a good source of frequently updated information on international efforts to assist the victims of artificially created and natural disasters and of disaster mitigation initiatives.

food chain in ecology, a sequence showing the feeding relationships between organisms in a particular ◊ecosystem. Each organism depends on the next lowest member of the chain for its food.

Energy in the form of food is shown to be transferred from ◊autotrophs, or producers, which are principally plants and photosynthetic microorganisms, to a series of ◊heterotrophs, or consumers. The heterotrophs comprise the ◊herbivores, which feed on the producers; ◊carnivores, which feed on the herbivores; and ◊decomposers, which break down the dead bodies and waste products of all four groups (including their own), ready for recycling.

In reality, however, organisms have varied diets, relying on different kinds of foods, so that the food chain is an oversimplification. The more complex **food web** shows a greater variety of relationships, but again emphasizes that energy passes from plants to herbivores to carnivores.

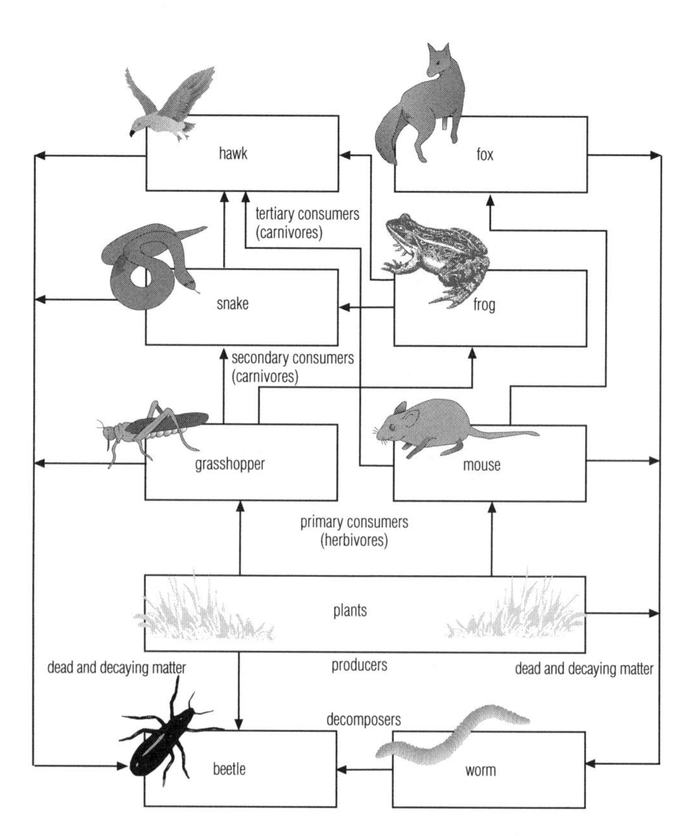

food chain *The complex interrelationships between animals and plants in a food chain. Food chains are normally only three or four links long. This is because most of the energy at each link is lost in respiration, and so cannot be passed on to the next link.*

Environmentalists have used the concept of the food chain to show how poisons and other forms of pollution can pass from one animal to another, threatening rare species. For example, the pesticide DDT has been found in lethal concentrations in the bodies of animals at the top of the food chain, such as the golden eagle *Aquila chrysaetos.*

food irradiation the exposure of food to low-level ◊irradiation to kill microorganisms; a technique used in ◊food technology. Irradiation is highly effective, and does not make the food any more radioactive than it is naturally. Irradiated food is used for astronauts and immunocompromised patients in hospitals. Some vitamins are partially destroyed, such as vitamin C, and it would be unwise to eat only irradiated fruit and vegetables.

The main cause for concern is that it may be used by unscrupulous traders to 'clean up' consignments of food, particularly shellfish, with high bacterial counts. Bacterial toxins would remain in the food, so that it could still cause illness, although irradiation would have removed signs of live bacteria. Stringent regulations would be needed to prevent this happening. Other damaging changes may take place in the food, such as the creation of ◊free radicals, but research so far suggests that the process is relatively safe.

food poisoning any acute illness characterized by vomiting and diarrhoea and caused by eating food contaminated with harmful bacteria (for example, ◊listeriosis), poisonous food (for example, certain mushrooms, puffer fish), or poisoned food (such as lead or arsenic introduced accidentally during processing). A frequent cause of food poisoning is ◊Salmonella bacteria. Salmonella comes in many forms, and strains are found in cattle, pigs, poultry, and eggs.

Deep freezing of poultry before the birds are properly cooked is a common cause of food poisoning. Attacks of salmonella also come from contaminated eggs that have been eaten raw or cooked only lightly. Pork may carry the roundworm *Trichinella,* and rye the parasitic fungus ergot. The most dangerous food poison is the bacillus that causes ◊botulism. This is rare but leads to muscle paralysis and, often, death. ◊Food irradiation is intended to prevent food poisoning.

food technology the application of science to the commercial processing of foodstuffs. Food is processed to make it more palatable or digestible, for which the traditional methods include boiling, frying, flour-milling, bread-, yoghurt-, and cheese-making, and brewing; or to prevent the growth of bacteria, moulds, yeasts, and other microorganisms; or to preserve it from spoilage caused by the action of ◊enzymes within the food that change its chemical composition, resulting in changes in flavour, odour, colour, and texture. These changes are not always harmful or undesirable; examples of desirable changes are the ripening of cream in butter manufacture, flavour development of cheese, and the hanging of meat to tenderize the muscle fibres. Fatty or oily foods suffer oxidation of the fats, which makes them rancid.

Preservation enables foods that are seasonally produced to be available all the year. Traditional forms of **food preservation** include salting, smoking, pickling, drying, bottling, and preserving in sugar. Modern food technology also uses many novel processes and additives, which allow a wider range of foodstuffs to be preserved. All foods undergo some changes in quality and nutritional value when subjected to preservation processes. No preserved food is identical in quality to its fresh counterpart, hence only food of the highest quality should be preserved.

In order to grow, bacteria, yeasts, and moulds need moisture, oxygen, a suitable temperature, and food. The various methods of food preservation aim to destroy the microorganisms within the food, to remove one or more of the conditions essential for their growth, or to make the foods unsuitable for their growth. Adding large amounts of salt or sugar reduces the amount of water available to microorganisms, because the water tied up by these solutes cannot be used for microbial growth. This is the principle in salting meat and fish, and in the manufacture of jams and jellies. These conditions also inhibit the enzyme activity in food. Preservatives may also be developed in the food by the controlled growth of microorganisms to produce fermentation that may make alcohol, or acetic or lactic acid. Examples of food preserved in this way are vinegar, sour milk, yoghurt, sauerkraut, and alcoholic beverages.

Refrigeration below 5°C/41°F (or below 3°C/37°F for cooked foods) slows the processes of spoilage, but is less effective for foods with a high water content. This process cannot kill microorganisms, nor stop their growth completely, and a failure to realize its limitations causes many cases of food poisoning. Refrigerator temperatures should be checked as the efficiency of the machinery (see ◊refrigeration) can decline with age, and higher temperatures are dangerous.

Deep freezing (–18°C/–1°F or below) stops almost all spoilage processes, except residual enzyme activity in uncooked vegetables and most fruits, which are blanched (dipped in hot water to destroy the enzymes) before freezing. Preservation by freezing works by rendering the water in foodstuffs unavailable to microorganisms by converting it to ice. Microorganisms cannot grow or divide while frozen, but most remain alive and can resume activity once defrosted. Some foods are damaged by freezing, notably soft fruits and salad vegetables, the cells of which are punctured by ice crystals, leading to loss of crispness. Fatty foods such as cow's milk and cream tend to separate. Freezing has little effect on the nutritive value of foods, though a little vitamin C may be lost in the blanching process for fruit and vegetables. Various processes are used for deep freezing foods commercially.

Pasteurization is used mainly for milk. By holding the milk at 72°C/161.6°F for 15 seconds, all disease-causing bacteria can be destroyed. Less harmful bacteria survive, so the milk will still go sour within a few days.

Ultra-heat treatment is used to produce UHT milk. This process uses higher temperatures than pasteurization, and kills all bacteria present, giving the milk a long shelf life but altering the flavour.

Drying is effective because both microorganisms and enzymes need water to be active. This is one of the oldest, simplest, and most effective way of preserving foods. In addition, drying concentrates the soluble ingredients in foods, and this high concentration prevents the growth of bacteria, yeasts, and moulds. Dried food will deteriorate rapidly if allowed to become moist, but provided they are suitably packaged, products will have a long shelf life. Traditionally, foods were dried in the sun and wind, but commercially today, products such as dried milk and instant coffee are made by spraying the liquid into a rising column of dry, heated air; solid foods, such as fruit, are spread in layers on a heated surface.

Freeze-drying is carried out under vacuum. It is less damaging to food than straight dehydration in the sense that foods reconstitute better, and is used for quality instant coffee and dried vegetables. The foods are fast frozen, then dried by converting the ice to vapour under very low pressure. The foods lose much of their weight, but retain the original size and shape. They have a spongelike texture, and rapidly reabsorb liquid when reconstituted. Refrigeration is unnecessary during storage; the shelf life is similar to dried foods, provided the product is not allowed to become moist. The success of the method is dependent on a fast rate of freezing, and rapid conversion of the ice to vapour. Hence, the most acceptable results are obtained with thin pieces of food, and the method is not recommended for pieces thicker than 3 cm/1 in. Fruit, vegetables, meat, and fish have proved satisfactory. This method of preservation is commercially used but the products are most often used as constituents of composite dishes, such as packet meals.

Canning relies on high temperatures to destroy microorganisms and enzymes. The food is sealed in a can to prevent recontamination. The effect of heat processing on the nutritive value of food is variable. For instance, the vitamin-C content of green vegetables is much reduced, but, owing to greater acidity, in fruit juices vitamin C is quite well retained. There is also a loss of 25–50% of water-soluble vitamins if the liquor is not used. Vitamin B (thiamine) is easily destroyed by heat treatment, particularly in alkaline conditions. Acid products retain thiamine well, because they require only minimum heat during sterilization. The sterilization process seems to have little effect on retention of vitamins A and B_2. During storage of canned foods, the proportion of vitamins B and C decreases gradually. Drinks may be canned to preserve the carbon dioxide that makes them fizzy.

Pickling utilizes the effect of acetic (ethanoic) acid, found in vinegar, in stopping the growth of moulds. In sauerkraut, lactic acid, produced by bacteria, has the same effect. Similar types of nonharmful, acid-generating bacteria are used to make yoghurt and cheese.

Curing of meat involves soaking in salt (sodium chloride) solution, with saltpetre (sodium nitrate) added to give the meat its pink colour and characteristic taste. Bacteria convert the nitrates in cured meats to nitrites and nitrosamines, which are potentially carcinogenic to humans.

Irradiation is a method of preserving food by subjecting it to low-level radiation (see ◊food irradiation).

Puffing is a method of processing cereal grains. They are subjected to high pressures, then suddenly ejected into a normal atmospheric pressure, causing the grain to expand sharply. This is used to make puffed wheat cereals and puffed rice cakes.

Chemical treatments are widely used, for example in margarine manufacture, in which hydrogen is bubbled through vegetable oils in the presence of a ◊catalyst to produce a more solid, spreadable fat. The catalyst is later removed. Chemicals introduced in processing that remain in the food are known as **food additives** and include flavourings, preservatives, anti-oxidants, emulsifiers, and colourings.

food test any of several types of simple test, easily performed in the laboratory, used to identify the main classes of food.
starch–iodine test Food is ground up in distilled water and iodine is added. A dense black colour indicates that starch is present.
sugar–Benedict's test Food is ground up in distilled water and placed in a test tube with Benedict's reagent. The tube is then heated in a boiling water bath. If glucose is present the colour changes from blue to brick-red.
protein–Biuret test Food is ground up in distilled water and a mixture of copper(II) sulphate and sodium hydroxide is added. If protein is present a mauve colour is seen.

foot in geometry, point where a line meets a second line to which it is perpendicular.

foot imperial unit of length (symbol ft), equivalent to 0.3048 m, in use in Britain since Anglo-Saxon times. It originally represented the length of a human foot. One foot contains 12 inches and is one-third of a yard.

foot-and-mouth disease contagious eruptive viral disease of cloven-hoofed mammals, characterized by blisters in the mouth and around the hooves. In cattle it causes deterioration of milk yield and abortions. It is an airborne virus, which makes its eradication extremely difficult.

foot-candle unit of illuminance, now replaced by the ◊lux. One foot-candle is the illumination received at a distance of one foot from an international candle. It is equal to 10.764 lux.

foot-pound imperial unit of energy (ft-lb), defined as the work done when a force of one pound moves through a distance of one foot. It has been superseded for scientific work by the joule: one foot-pound equals 1.356 joule.

footprint in computing, the area on the desk or floor required by a computer or other peripheral device.

footprint of a satellite, the area of the Earth over which its signals can be received.

footrot contagious disease of sheep caused by a bacterium and spreading easily in warm, wet conditions. It is characterized by inflammation and lameness and is controlled by vaccination, foot-bathing, and segregation.

forage crop plant that is grown to feed livestock; for example, grass, clover, and kale (a form of cabbage). Forage crops cover a greater area of the world than food crops, and grass, which dominates this group, is the world's most abundant crop, though much of it is still in an unimproved state.

foraminifera any marine protozoan of the order Foraminiferida, with shells of calcium carbonate. Their shells have pores through which filaments project. Some form part of the ◊plankton, others live on the sea bottom.

The many-chambered *Globigerina* is part of the plankton. Its shells eventually form much of the chalky ooze of the ocean floor.

forbidden line in astronomy, emission line seen in the spectra of certain astronomical objects that are not seen under the conditions prevailing in laboratory experiments. They indicate that the hot gas emitting them is at extremely low density. Forbidden lines are seen, for example, in the tenuous gas of the Sun's ◊corona, in ◊HII regions, and in the nucleuses of certain active galaxies.

force any influence that tends to change the state of rest or the uniform motion in a straight line of a body. The action of an unbalanced or resultant force results in the acceleration of a body in the direction of action of the force, or it may, if the body is unable to move freely, result in its deformation (see ◊Hooke's law). Force is a vector quantity, possessing both magnitude and direction; its SI unit is the newton.

According to Newton's second law of motion the magnitude of a resultant force is equal to the rate of change of ◊momentum of the body on which it acts; the force F producing an acceleration a m s^{-2} on a body of mass m kilograms is therefore given by: $F = ma$ See also ◊Newton's laws of motion.

force feedback in ◊virtual reality, realistic simulation of the physical sense of touch. This is an area of active research, as many applications of virtual reality are useless or impossible without it.

For example, force feedback is essential in medical training systems to teach the students how hard to press with a scalpel in delicate areas of the human body. Even simulated games need force feedback in order to allow objects to respond realistically to falling or being hit.

force multiplier machine designed to multiply a small effort in order to move a larger load. The number of times a machine multiplies the effort is called its ◊mechanical advantage. Examples of a force multiplier include crowbar, wheelbarrow, nutcrackers, and bottle opener.

force ratio the magnification of a force by a machine; see ◊mechanical advantage.

forces, fundamental in physics, the four fundamental interactions believed to be at work in the physical universe. There are two long-range forces: **gravity**, which keeps the planets in orbit around the Sun, and acts between all particles that have mass; and the **electromagnetic force**, which stops solids from falling apart, and acts between all particles with ◊electric charge. There are two very short-range forces which operate only inside the atomic nucleus: the **weak nuclear force**, responsible for the reactions that fuel the Sun and for the emission of ◊beta particles from certain nuclei; and the **strong nuclear force**, which binds together the protons and neutrons in the nuclei of atoms. The relative strengths of the four forces are: strong, 1; electromagnetic, 10^{-2}; weak, 10^{-6}; gravitational, 10^{-40}. By 1971, US physicists Steven Weinberg and Sheldon Glashow, Pakistani physicist Abdus Salam, and others had developed a theory that suggested that the weak and electromagnetic forces were aspects of a single force called the **electroweak force**; experimental support came from observation at ◊CERN in the 1980s. Physicists are now working on theories to unify all four forces.

People can have the Model T in any colour – so long as it's black.

HENRY FORD US automobile manufacturer.
A Nevins *Ford*

Fordism mass production characterized by a high degree of job specialization, as typified by the Ford Motor Company's early use of assembly lines. Mass-production techniques were influenced by US management consultant F W Taylor's book *Principles of Scientific Management* 1911.

Post-Fordism management theory and practice emphasize flexibility and autonomy of decisionmaking for nonmanagerial staff. It is concerned more with facilitating and coordinating tasks than with control.

forensic entomology branch of ◊forensic science, involving the study of insects on and around the corpse. Insects rapidly infest a corpse, and do so in an accepted sequence beginning with flies laying eggs. Further insects follow to feed on the decomposing flesh and fly maggots. Forensic entomologists are able to determine time of death by analysing insect colonization. They can also tell whether or not a corpse has been moved by examining the faunal community in the 'seepage area' beneath the body.

FORENSIC ENTOMOLOGY

Cocaine speeds up the growth of certain insects. This enables forensic entomologists to determine if a corpse is that of a cocaine user.

forensic medicine in medicine, branch of medicine concerned with the resolution of crimes. Examples of forensic medicine include the determination of the cause of death in suspicious circumstances or the identification of a criminal by examining tissue found at the scene of a crime. Forensic psychology involves the establishment of a psychological profile of a criminal that can assist in identification.

forensic science the use of scientific techniques to solve criminal cases. A multidisciplinary field embracing chemistry, physics, botany, zoology, and medicine, forensic science includes the identification of human bodies or traces. Ballistics (the study of projectiles, such as bullets), another traditional forensic field, makes use of such tools as the comparison microscope and the electron microscope.

Traditional methods such as fingerprinting are still used, assisted by computers; in addition, blood analysis, forensic dentistry, voice and speech spectrograms, and ◊genetic fingerprinting are increasingly applied. Chemicals, such as poisons and drugs, are analysed by ◊chromatography. ESDA (electrostatic document analysis) is a technique used for revealing indentations on paper, which helps determine if documents have been tampered with. ◊Forensic entomology is also a branch of forensic science.

forest area where trees have grown naturally for centuries, instead of being logged at maturity (about 150–200 years). A natural, or old-growth, forest has a multistorey canopy and includes young and very old trees (this gives the canopy its range of heights). There are also fallen trees contributing to the very complex ecosystem, which may support more than 150 species of mammals and many thousands of species of insects.

The Pacific forest of the west coast of North America is one of the few remaining old-growth forests in the temperate zone.

It consists mainly of conifers and is threatened by logging – less than 10% of the original forest remains.

forest fly external blood-sucking parasite, chiefly of horses. It is about 8 mm/0.3 in long and bears brown and yellow flecks.
classification The forest fly *Hippobosca equina* is a member of the insect family Hippoboscidae in order Diptera, class Insecta, phylum Arthropoda.

forestry the science of forest management. Recommended forestry practice aims at multipurpose crops, allowing the preservation of varied plant and animal species as well as human uses (lumbering, recreation). Forestry has often been confined to the planting of a single species, such as a rapid-growing conifer providing softwood for paper pulp and construction timber, for which world demand is greatest. In tropical countries, logging contributes to the destruction of ◊rainforests, causing global environmental problems. Small unplanned forests are ◊woodland.

The earliest planned forest dates from 1368 at Nürnberg, Germany; in Britain, planning of forests began in the 16th century. In the UK, Japan, and other countries, forestry practices have been criticized for concentration on softwood conifers to the neglect of native hardwoods.

forgery in computing, the art of falsifying either the contents or the origins of a message. On the Internet, where a person's

Forensic Science: Recent Advances

BY JOHN BROAD

The scientific support for Inspector Morse

The forensic scientist has to provide evidence that will stand scrutiny in a court of law. This is why crime laboratories need to keep up with the latest research and maintain the highest standards.

Every contact leaves a trace. The ordinary microscope is still a vital instrument for examining trace evidence – hairs, fibres, fragments of glass or paint – but the scanning electron microscope is also used. It provides high magnification with good resolution, and can also incorporate a **microprobe** that identifies the actual elements, particularly metallic ones, in the surface being examined.

Surface elements absorb electrons and emit X-rays, which the microprobe converts into an X-ray emission spectrum with a characteristic pattern that reveals the elements present. In this way, it is possible to detect and identify particles invisible to the optical microscope, such as those scattered from a firearm when discharged. These particles can indicate the type and make of ammunition used.

Anti-crime antibodies

Advances have also been made in analytical techniques called **immunoassays,** which use antibodies to detect and measure drugs, poisons, proteins, and even explosives such as TNT and Semtex. When a foreign chemical, such as a disease organism, enters the human body, antibodies are produced that recognize and react with the foreign substance.

The same process occurs in animals, which can be used to produce antibodies against a wide variety of chemicals. An animal is injected with a target substance, such as cocaine, and the resulting antibodies can be separated and used to recognize the substance against a background of body fluids, or in a body swab.

When trying to detect a target compound, such as the presence of explosive residue on a person's hands, it is vital to take account of possible contamination, since the substance might have been picked up casually. Also, it is important that the method used detects the target compound and no other. The value of antibodies is that they are specific to the compound that triggered their production. With other analytical methods, such as **thin-layer chromatography,** care has to be taken to eliminate other compounds that could show the same experimental result.

DNA fingerprinting

One impressive recent scientific advance is DNA fingerprinting, or DNA profiling. This has been used effectively in assault, rape, and murder cases and in paternity disputes. It figured prominently in the O J Simpson trial. For forensic purposes two DNA fingerprints are necessary – one from the suspect and another taken from the crime scene. There must be a convincing match to establish a link.

DNA profiling involves using special enzymes to cut up precisely a sample of DNA extracted from body cells. Only a very small amount of DNA – from just a few cells – is needed; the polymerase chain reaction (PCR) can amplify it into sufficient material to profile. The resulting 'bits' are separated by gel electrophoresis and can be blotted onto a special membrane, marked radioactively with specialized probes and then visualized as the familiar sequence of bars (known as an autoradiograph). Alternatively, the DNA bands can be labelled or highlighted in the actual gel then scanned with a laser. The result is a graphical print out showing peaks where the bands occur. These peaks can be converted into digital codes for storage on a computer. This is the method employed in building up national DNA databases, which store, for rapid recall, the DNA profiles of known offenders, suspects, and tissue samples taken from crime scenes. Rapid comparison can establish that a suspect could be linked with a number of scenes.

Controversies
Should blood for DNA profiling purposes be taken not only from suspects but from suspicious persons, unwanted persons, even undesirable persons? If a person refuses to submit to supplying a sample for DNA analysis (as is their entitlement) what inference should be drawn from this refusal? Should everyone be DNA profiled shortly after birth so that the ordinary citizen can more easily be monitored? Despite refinements and increased effectiveness, DNA profiling must be treated with caution. If two profiles convincingly match, they may not come from the same person. The chances of this are very small – but not as small as once thought. Identical twins have the same DNA profile, but DNA profiles of people from small communities with significant in-breeding can be deceptively similar. If two bands on adjacent profiles correspond, but not exactly, statistical analysis may be needed to decide if there is a match. In the end, despite the statistical back-up, the matching of two DNA profiles is a matter of expert opinion.

Electrostatic document analysis
Electrostatic document analysis (ESDA) is a recent technique used for revealing indentations on paper. Left on an underlying sheet, these indicate what has been written on the paper above.

The method uses a high electrostatic voltage to transfer the indentations onto imaging film where they are visualized by photographic toner. If the resulting impressions are markedly uneven – perhaps one half is more heavily indented than the other – then the writing under examination (on the top sheet) may have been written at two different times, showing that the document has been tampered with. The release of the 'Guildford Four' and the 'Birmingham Six' in the UK was clinched when the ESDA machine revealed 'doctored' written evidence at the original trial.

Computers on the beat
The increasing power of computers has been responsible for great advances in forensic science, as in other sciences. Computers have revolutionized the storage and retrieval of information. For example, all car registration numbers and owners' names are stored on computer for almost instant access. Information can be rapidly communicated to police officers in the field. This has increased the power of those who hold the information and, in addition to speeding the response to crime, has helped in the monitoring of 'undesirables' and in maintaining order on the streets.

Identification of a suspect fingerprint, by comparison with thousands stored on file, was once a time-consuming process. Nowadays, with a computer, the process takes minutes, even seconds. Even so, the final decision on fingerprint identification is still made visually by a trained expert. No matter how sophisticated the hardware, the human senses are still vital in forensic work.

The carnage on the roads
A vast amount of police time is consumed by road traffic problems. It would be correct to say that two scientific instruments – the Breathalyser and the Radar Speed Device (or gun) have had more impact on the general public than all other forensic scientific hardware put together. Scientific evidence presented in a court of law must be accurate, reliable, and based on thorough research in order to convince a Magistrate or jury. Confirmatory or back-up tests must be available when necessary. Scientific efforts designed to reduce the number of deaths and injuries caused by the drunk driver have resulted in the development of a roadside breath screening device; an evidential breath alcohol testing instrument; and a confirmatory blood test. The Breathalyser relies on a fuel cell to estimate the level of alcohol in a driver's breath. The intoximeter, used in the police station, is based on the absorption of infra red light (at 3.4 microns) by alcohol molecules and accurately establishes the breath alcohol level. The confirmatory blood test employs headspace gas chromatography to find the blood alcohol concentration or BAC. All these procedures use a different scientific principle for their method of operation.

The radar speed device has aimed to bring about some reduction in the devastation caused by speeding, particularly on modern motorways and highways. Its operation depends upon the Doppler effect and its output – often a photograph and accompanying data when it is linked to a camera – is valid as evidence in a court of law. All radar speed devices are checked rigorously in order to maintain the required accuracy and reliability. More efficient, but much more expensive, is the laser speed device that relies on the reflection of impulses of light from a moving source to calculate speed. Both these devices are easy to operate and to carry around.

If public pressure results in the legislation of some, or maybe all, recreational drugs then scientists will be obliged to design roadside screening tests for particular, or all, drugs with back-up evidential and confirmatory procedures. Apparently it is much more dangerous to drive a vehicle when 'stoned' with drugs than when drunk. Work has already begun on these problems. Trials are due to start in the UK on the use of absorbent pads impregnated with special chemicals. When pressed against the skin, sweat diffuses into the pad and a colour reaction denotes the presence of a particular drug. This is a noninvasive technique, but for evidential testing, blood or urine would be required and subjected to an immune assay and perhaps chromatography or even mass spectrometry. In the USA, efforts have so far centred on urine testing and the opinion has been expressed that eventually each police car will be followed closely by a 'urine bus' or 'slash van'.

Surveillance techniques
The recent developments and use of sophisticated surveillance techniques seen in car parks, town centres, business premises and around the residences of the rich and powerful, bear testimony to the trend. Most notable is CCTV (closed circuit television). Whether this actually reduces crime or merely drives the criminals elsewhere is open to debate. However the concern raised about such techniques, which extend to telephone tapping, is about the invasion of privacy. This could represent one or many steps down the slippery slope of monitoring the innocent and the guilty going about their daily business. Apart from the implications of 'Big Brother', there is the perennial vexed question of who monitors those who do the monitoring?

A dream or a nightmare?
The aim of the forensic scientist is to link a suspect with a crime scene, victim or incident, using the very latest in scientific technology. If there is no suspect then data can be collected and stored only as far as time and money allow. Nothing further can be done until the breakthrough occurs as a result of a stroke of luck or a tip-off. In the end, all that is required for a watertight case might be available – a suspect whose name and address is known and an impressive array of evidence and statements. But if the suspect has gone to ground and all enquiries are met with a wall of silence, no amount of scientific evidence will find someone determined to 'melt away'. Maybe it would be wise to tag everybody electronically at birth and monitor their every move on a central computer. This would solve the problem of the vanishing suspect and the missing person, but is it a dream or a ghastly nightmare?

identity is shaped by his/her words as sent out via e-mail or public conferencing systems such as USENET, sending out a forged message in another person's name can seriously damage them. Forged messages are, however, used by the ◊CancelMoose to manage ◊spamming and keep it from spreading.

forget-me-not any of a group of plants belonging to the borage family, including *M. sylvatica* and *M. scorpioides*, with small bright blue flowers. (Genus *Myosotis*, family Boraginaceae.)

forging one of the main methods of shaping metals, which involves hammering or a more gradual application of pressure. A blacksmith hammers red-hot metal into shape on an anvil, and the traditional place of work is called a forge. The blacksmith's mechanical equivalent is the drop forge. The metal is shaped by the blows from a falling hammer or ram, which is usually accelerated by steam or air pressure. Hydraulic presses forge by applying pressure gradually in a squeezing action.

formaldehyde common name for ◊methanal.

formalin aqueous solution of formaldehyde (methanal) used to preserve animal specimens.

formatting in computing, short for ◊disc formatting.

Formica trademark of the Formica Corporation for a heat-proof plastic laminate, widely used as a veneer on wipe-down kitchen surfaces and children's furniture. It is made from formaldehyde resins similar to ◊Bakelite. It was first put on the market in 1913.

formic acid common name for ◊methanoic acid.

forms on the World Wide Web, facility for accepting structured user input and inserting it into a program such as a database. Most newer graphical Web browsers can handle forms, as can the older, text-based browser Lynx. Forms are needed to manage database queries at sites such as ◊AltaVista, and to fill out registration forms for those sites that require them.

Web page designers implement forms by using a special set of hypertext markup language (HTML) tags and attaching a script, which parses the data and feeds it to the program specified in a form the program can use. The results, such as a user name and password or a list of matches, are sent back to the user.

formula in chemistry, a representation of a molecule, radical, or ion, in which the component chemical elements are represented by their symbols. An **empirical formula** indicates the simplest ratio of the elements in a compound, without indicating how many of them there are or how they are combined. A **molecular formula** gives the number of each type of element present in one molecule. A **structural formula** shows the relative positions of the atoms and the bonds between them. For example, for ethanoic acid, the empirical formula is CH_2O, the molecular formula is $C_2H_4O_2$, and the structural formula is CH_3COOH.

formula in mathematics, a set of symbols and numbers that expresses a fact or rule. $A = \pi r^2$ is the formula for calculating the area of a circle. Einstein's famous formula relating energy and mass is $E = mc^2$.

forsythia any of a group of temperate eastern Asian shrubs, which bear yellow bell-shaped flowers in early spring before the leaves appear. (Genus *Forsythia*, family Oleaceae.)

FORTRAN (or *fortran*, acronym for ***formula translation***) high-level computer-programming language suited to mathematical and scientific computations. Developed by John Backus at IBM in 1956, it is one of the earliest computer languages still in use. A recent version, Fortran 90, is now being used on advanced parallel computers. ◊BASIC was strongly influenced by FORTRAN and is similar in many ways.

fossil ***Latin fossilis 'dug up'*** a cast, impression, or the actual remains of an animal or plant preserved in rock. Fossils were created during periods of rock formation, caused by the gradual accumulation of sediment over millions of years at the bottom of the sea bed or an inland lake. Fossils may include footprints, an internal cast, or external impression. A few fossils are preserved intact, as with ◊mammoths fossilized in Siberian ice, or insects trapped in tree resin that is today amber. The study of fossils is called ◊palaeontology. Palaeontologists are able to deduce much of the geological history of a region from fossil remains.

About 250,000 fossil species have been discovered – a figure that is believed to represent less than 1 in 20,000 of the species that ever lived. **Microfossils** are so small they can only be seen with a microscope. They include the fossils of pollen, bone fragments, bacteria, and the remains of microscopic marine animals and plants, such as foraminifera and diatoms.

FOSSIL HOMINIDS FAQ

http://earth.ics.uci.edu:8080/faqs/fossil-hominids.html

Basic information about hominid species, the most important hominid fossils, and creationist arguments, plus links to related sites.

If A is a success in life, then A equals x plus y plus z. Work is x; y is play; and z is keeping your mouth shut.

ALBERT EINSTEIN German-born US physicist. *Observer* 15 Jan 1950

fossil fuel fuel, such as coal, oil, and natural gas, formed from the fossilized remains of plants that lived hundreds of millions of years ago. Fossil fuels are a ◊nonrenewable resource and will eventually run out. Extraction of coal and oil causes considerable environmental pollution, and burning coal contributes to problems of ◊acid rain and the ◊greenhouse effect.

four-colour process colour ◊printing using four printing plates, based on the principle that any colour is made up of differing proportions of the primary colours blue, red, and green. The first stage in preparing a colour picture for printing is to produce separate films, one each for the blue, red, and green respectively in the picture (colour separations). From these separations three printing plates are made, with a fourth plate for black (for shading or outlines and type). Ink colours complementary to those represented on the plates are used for printing – yellow for the blue plate, cyan for the red, and magenta for the green.

Fourdrinier machine papermaking machine patented by the Fourdrinier brothers Henry and Sealy in England in 1803. On the machine, liquid pulp flows onto a moving wire-mesh belt, and water drains and is sucked away, leaving a damp paper web. This is passed

Fourier, Jean Baptiste Joseph (1768–1830)

French applied mathematician whose formulation of heat flow in 1807 contains the proposal that, with certain constraints, any mathematical function can be represented by trigonometrical series. This principle forms the basis of **Fourier analysis**, used today in many different fields of physics. His idea, not immediately well received, gained currency and is embodied in his *Théorie analytique de la chaleur/The Analytical Theory of Heat* 1822.

Light, sound, and other wavelike forms of energy can be studied using Fourier's method, a developed version of which is now called harmonic analysis.

first through a series of steam-heated rollers, which dry it, and then between heavy calendar rollers, which give it a smooth finish.

Such machines can measure up to 90 m/300 ft in length, and are still in use.

four-stroke cycle the engine-operating cycle of most petrol and ◊diesel engines. The 'stroke' is an upward or downward movement of a piston in a cylinder. In a petrol engine the cycle begins with the induction of a fuel mixture as the piston goes down on its first stroke. On the second stroke (up) the piston compresses the mixture in the top of the cylinder. An electric spark then ignites the mixture, and the gases produced force the piston down on its third, power, stroke. On the fourth stroke (up) the piston expels the burned gases from the cylinder into the exhaust.

fourth-generation language in computing, a type of programming language designed for the rapid programming of ◊applications but often lacking the ability to control the individual parts of the computer. Such a language typically provides easy ways of designing screens and reports, and of using databases. Other 'generations' (the term implies a class of language rather than a chronological sequence) are ◊machine code (first generation); ◊assembly languages, or low-level languages (second); and conventional high-level languages such as ◊BASIC and ◊PASCAL (third).

fowl chicken or chickenlike bird. Sometimes the term is also used for ducks and geese. The red jungle fowl *Gallus gallus* is the ancestor of all domestic chickens. It is a forest bird of Asia, without the size or egg-laying ability of many domestic strains. ◊Guinea fowl are of African origin.

FOX

The average British fox obtains only about a third of its food from hunting. The remainder is acquired by scavenging.

fox one of the smaller species of wild dog of the family Canidae, which live in Africa, Asia, Europe, North America, and South America. Foxes feed on a wide range of animals from worms to rabbits, scavenge for food, and also eat berries. They are very adaptable, maintaining high populations close to urban areas.

Most foxes are nocturnal, and make an underground den, or 'earth'. The common or red fox *Vulpes vulpes* is about 60 cm/2 ft long plus a tail ('brush') 40 cm/1.3 ft long. The fur is reddish with black patches behind the ears and a light tip to the tail. Other foxes include the Arctic fox *Alopex lagopus*, the ◊fennec, the grey foxes genus *Urocyon* of North and Central America, and the South American genus *Dusicyon*, to which the extinct Falkland Islands dog belonged.

ADAM'S FOX BOX

http://tavi.acomp.usf.edu/foxbox

Beautifully designed site providing an impressive amount of information about foxes: articles, songs, stories, poems, images, and a video clip. The site also provides many pointers to other fox-related sites, including one explaining how to dance the foxtrot!

foxglove any of a group of flowering plants found in Europe and the Mediterranean region. They have showy spikes of bell-like flowers, and grow up to 1.5 m/5 ft high. (Genus *Digitalis*, family Scrophulariaceae.)

The wild species (*D. purpurea*), native to Britain, produces purple to reddish flowers. Its leaves were the original source of digitalis, a drug used for some heart problems.

foxhound small, keen-nosed hound, up to 60 cm/2 ft tall and black, tan, and white in colour. There are two recognized breeds: the English foxhound, bred for some 300 years to hunt foxes, and the American foxhound, not quite as stocky, used for foxes and other game.

fox terrier breed of ◊terrier evolved for use in foxhunts to attack foxes in their earths (dens) where the larger foxhound was unable to reach them. Two types are distinguished – the smooth- and wire-haired; but both have a mainly white coat with large brown and/or black markings. They weigh about 8 kg/17.5 lb, stand up to 39 cm/15 in tall, and carry their short, usually docked, tails upright.

f.p.s. system system of units based on the foot, pound, and second as units of length, mass, and time, respectively. It has now been replaced for scientific work by the ◊SI system.

fractal ***from Latin fractus 'broken'*** irregular shape or surface produced by a procedure of repeated subdivision. Generated on a computer screen, fractals are used in creating models of geographical or biological processes (for example, the creation of a coastline by erosion or accretion, or the growth of plants).

Sets of curves with such discordant properties were developed in the 19th century in Germany by Georg Cantor and Karl Weierstrass. The name was coined by the French mathematician Benoit Mandelbrod. Fractals are also used for computer art.

TO REMEMBER THE RULE OF DIVISION OF FRACTIONS:

THE NUMBER YOU ARE DIVIDING BY
TURN UPSIDE DOWN AND MULTIPLY.

fraction in chemistry, a group of similar compounds, the boiling points of which fall within a particular range and which are separated during fractional ◊distillation (fractionation).

fraction ***from Latin fractus 'broken'*** in mathematics, a number that indicates one or more equal parts of a whole. Usually, the number of equal parts into which the unit is divided (denominator) is written below a horizontal line, and the number of parts comprising the fraction (numerator) is written above; thus $\frac{2}{3}$ or $\frac{3}{4}$. Such fractions are called **vulgar** or **simple** fractions. The denominator can never be zero.

A **proper fraction** is one in which the numerator is less than the denominator. An **improper fraction** has a numerator that is larger than the denominator, for example $\frac{3}{2}$. It can therefore be expressed as a mixed number, for example, $1\frac{1}{2}$. A combination such as $\frac{5}{0}$ is not regarded as a fraction (an object cannot be divided into zero equal parts), and mathematically any number divided by 0 is equal to infinity. A **decimal fraction** has as its denominator a power of 10, and these are omitted by use of the decimal point and notation, for example 0.04, which is $\frac{4}{100}$. The digits to the right of the decimal point indicate the numerators of vulgar fractions whose denominators are 10, 100, 1,000, and so on. Most fractions can be expressed exactly as decimal fractions ($\frac{1}{3}$ = 0.333...). Fractions are also known as the **rational numbers**; that is, numbers formed by a ratio. **Integers** may be expressed as fractions with a denominator of 1.

fractionating column device in which many separate ◊distillations can occur so that a liquid mixture can be separated into its components.

Various designs exist but the primary aim is to allow maximum contact between the hot rising vapours and the cooling descending liquid. As the mixture of vapours ascends the column it becomes progressively enriched in the lower-boiling-point components, so these separate out first.

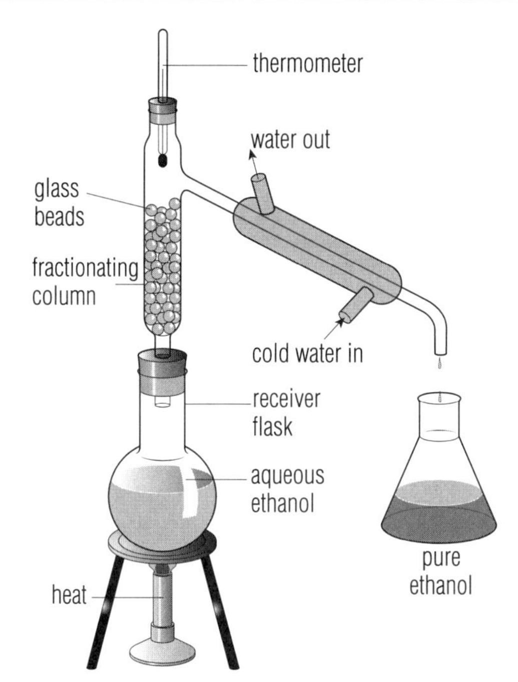

fractionating column Laboratory apparatus for fractional distillation. Fractional distillation is the main means of separating the components of crude oil.

This tomb holds Diophantus. Ah, how great a marvel! the tomb tells scientifically the measure of his life. God granted him to be a boy for the sixth part of his life, and adding a twelfth part to this, He clothed his cheeks with down; He lit him the light of wedlock after a seventh part, and five years after his marriage He granted him a son. Alas! late-born wretched child; after attaining the measure of half his father's life, chill Fate took him. After consoling his grief by this science of numbers for four years he ended his life.

DIOPHANTUS Greek mathematician.
Arithmetical riddle supposedly inscribed on Diophantus' tombstone, quoted in *The Greek Anthology V*

fractionation or *fractional distillation* process used to split complex mixtures (such as crude oil) into their components, usually by repeated heating, boiling, and condensation; see ◊distillation.

fragmentation in computing, the breaking up of files into many smaller sections stored on different parts of a disc. The computer ◊operating system stores files in this way so that maximum use can be made of disc space. Each section contains a pointer to where the next section is stored. The ◊file allocation table keeps a record of this.

Fragmentation slows down access to files. It is possible to defragment a disc by copying files. In addition, ◊defragmentation programs, or disc optimizers, allow discs to be defragmented without the need for files to be copied to a second storage device.

frame a single photograph in a sequence representing motion, or movement, on film; in a ◊network, a unit of data; in word processing or desktop publishing, a marked-out area on a page that can contain text or graphics.

What appears to be motion on a cinema or TV screen is actually a rapid sequence of single shots. Because of limitations in the human eye – known as the Phi phenomenon – those individual shots, if played in sequence at a rate of 24 to 30 frames per second, make the motion thus captured appear continuous.

frame buffer in computing, a ◊buffer used to store a screen image.

frame relay in ◊wide-area networks, a standard for the transmission of data that is optimized for high speeds up to about 1.5 Mbits/second.

francium radioactive metallic element, symbol Fr, atomic number 87, relative atomic mass 223. It is one of the alkali metals and occurs in nature in small amounts as a decay product of actinium. Its longest-lived isotope has a half-life of only 21 minutes. Francium was discovered and named in 1939 by Marguérite Perey to honour her country.

frangipani any of a group of tropical American trees, especially the species *P. rubra,* belonging to the dogbane family. Perfume is made from the strongly scented waxy flowers. (Genus *Plumeria,* family Apocynaceae.)

frankincense resin of various African and Asian trees, burned as incense. Costly in ancient times, it is traditionally believed to be one of the three gifts brought by the Magi to the infant Jesus. (Genus *Boswellia,* family Burseraceae.)

Frasch process process used to extract underground deposits of sulphur. Superheated steam is piped into the sulphur deposit and melts it. Compressed air is then pumped down to force the molten sulphur to the surface. The process was developed in the USA in 1891 by German-born Herman Frasch (1851–1914).

free fall the state in which a body is falling freely under the influence of ◊gravity, as in freefall parachuting (skydiving). In a vacuum, a freely falling body accelerates at a rate of 9.806 m sec^{-2}/32.174 ft sec^{-2}; the value varies slightly at different latitudes and altitudes. A body falling through air, accelerates until it reaches a maximum speed called the ◊terminal velocity; thereafter, there is no further acceleration.

In orbit, astronauts and spacecraft are still held by gravity and are in fact falling freely toward the Earth. Because of their speed (orbital velocity), the amount they fall towards the Earth just equals the amount the Earth's surface curves away; in effect they remain at the same height, apparently weightless.

Free-Net free community-based online system such as a network of public ◊bulletin boards and/or municipally owned systems. Free-Net is a registered service mark of the National Public Telecomputing Network. The first Free-Net was set up in 1986 in Cleveland, Ohio, USA.

free radical in chemistry, an atom or molecule that has an unpaired electron and is therefore highly reactive. Most free radicals are very short-lived. They are by-products of normal cell chemistry and rapidly oxidize other molecules they encounter. Free radicals are thought to do considerable damage. They are neutralized by protective enzymes.

Free radicals are often produced by high temperatures and are found in flames and explosions.

freesia any of a South African group of plants belonging to the iris family, commercially grown for their scented, funnel-shaped flowers. (Genus *Freesia,* family Iridaceae.)

Free Software Foundation (FSF) US organization, based in Boston, which creates and distributes good-quality free software and utilities. FSF is the publisher of the ◊GNU software, which includes compilers, operating systems, utilities, editors, databases, and PostScript viewers. All the software is free of licensing fees and restrictions.

The FSF was founded in 1983 by US artificial intelligence specialist Richard Stallman as a way of bringing back the cooperative spirit of the computing community's early days that had vanished by the early 1980s with the advent of widely sold proprietary software. The project's ultimate goal is to make commercial software obsolete by providing free software to do everything computer users want to do.

freeware in computing, free software which may or may not be in the public domain (see ◊public-domain software). One of the best-known examples of freeware is the encryption program ◊Pretty Good Privacy (PGP).

freeze-drying method of preserving food; see ◊food technology. The product to be dried is frozen and then put in a vacuum chamber that forces out the ice as water vapour, a process known as sublimation.

Many of the substances that give products such as coffee their typical flavour are volatile, and would be lost in a normal drying process because they would evaporate along with the water. In the freeze-drying process these volatile compounds do not pass into the ice that is to be sublimed, and are therefore largely retained.

freeze-thaw form of physical ◊weathering, common in mountains and glacial environments, caused by the expansion of water as it freezes. Water in a crack freezes and expands in volume by 9% as it turns to ice. This expansion exerts great pressure on the rock causing the crack to enlarge. After many cycles of freeze-thaw, rock fragments may break off to form ◊scree slopes.

For freeze-thaw to operate effectively the temperature must fluctuate regularly above and below 0°C/32°F. It is therefore uncommon in areas of extreme and perpetual cold, such as the polar regions.

freezing change from liquid to solid state, as when water becomes ice. For a given substance, freezing occurs at a definite temperature, known as the **freezing point**, that is invariable under similar conditions of pressure, and the temperature remains at this point until all the liquid is frozen. The amount of heat per unit mass that has to be removed to freeze a substance is a constant for any given substance, and is known as the latent heat of fusion.

freezing point for any given liquid, the temperature at which any further removal of heat will convert the liquid into the solid state. The temperature remains at this point until all the liquid has solidified. It is invariable under similar conditions of pressure – for example, the freezing point of water under standard atmospheric pressure is 0°C/32°F.

freezing point, depression of lowering of a solution's freezing point below that of the pure solvent; it depends on the number of molecules of solute dissolved in it. For a single solvent, such as pure water, all solute substances in the same molar concentration produce the same lowering of freezing point. The depression d produced by the presence of a solute of molar concentration C is given by the equation $d = KC$, where K is a constant (called the cryoscopic constant) for the solvent concerned.

Antifreeze mixtures for car radiators and the use of salt to melt ice on roads are common applications of this principle. Animals in arctic conditions, for example insects or fish, cope with the extreme cold either by manufacturing natural 'antifreeze' and staying active, or by allowing themselves to freeze in a controlled fashion, that is, they manufacture proteins to act as nuclei for the formation of ice crystals in areas that will not produce cellular damage, and so enable themselves to thaw back to life again.

Measurement of freezing-point depression is a useful method of determining the molecular weights of solutes. It is also used to detect the illicit addition of water to milk.

frequency in physics, the number of periodic oscillations, vibrations, or waves occurring per unit of time. The SI unit of frequency is the hertz (Hz), one hertz being equivalent to one cycle per second.

Human beings can hear sounds from objects vibrating in the range 20–15,000 Hz. Ultrasonic frequencies well above 15,000 Hz can be detected by such mammals as bats. Infrasound (low frequency sound) can be detected by some animals and birds. Pigeons can detect sounds as low as 0.1 Hz; elephants communicate using sounds as low as 1 Hz.

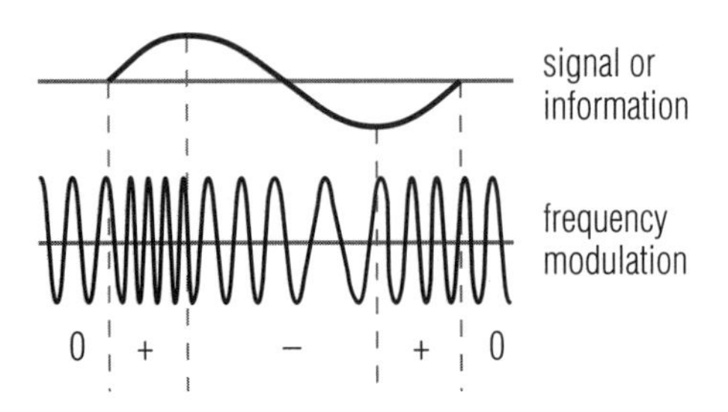

frequency modulation In FM radio transmission, the frequency of the carrier wave is modulated, rather than its amplitude (as in AM broadcasts). The FM system is not affected by the many types of interference which change the amplitude of the carrier wave, and so provides better quality reception than AM broadcasts.

Freud, Sigmund
(1856–1939)

Austrian physician who pioneered the study of the unconscious mind. He developed the methods of free association and interpretation of dreams that are basic techniques of psychoanalysis. The influence of unconscious forces on people's thoughts and actions was Freud's discovery, as was his controversial theory of the repression of infantile sexuality as the root of neuroses in the adult. His books include *Die Traumdeutung/The Interpretation of Dreams* 1900, *Jenseits des Lustprinzips/Beyond the Pleasure Principle* 1920, *Das Ich und das Es/The Ego and the Id* 1923, and *Das Unbehagen in der Kultur/Civilization and its Discontents* 1930. His influence has permeated the world to such an extent that it may be discerned today in almost every branch of thought.

From 1886 to 1938 Freud had a private practice in Vienna, and his theories and writings drew largely on case studies of his own patients, who were mainly upper-middle-class, middle-aged women. Much of the terminology of psychoanalysis was coined by Freud, and many terms have passed into popular usage, not without distortion. His theories have changed the way people think about human nature and brought about a more open approach to sexual matters. Antisocial behaviour is now understood to result in many cases from unconscious forces, and these new concepts have led to wider expression of the human condition in art and literature. Nevertheless, Freud's theories have caused disagreement among psychologists and psychiatrists, and his methods of psychoanalysis cannot be applied in every case.

Mary Evans Picture Library

frequency in statistics, the number of times an event occurs. For example, when two dice are thrown repeatedly and the two scores added together, each of the numbers 2 to 12 may have a frequency of occurrence. The set of data including the frequencies is called a **frequency distribution**, usually presented in a frequency table or shown diagramatically, by a frequency polygon.

frequency modulation (FM) method by which radio waves are altered for the transmission of broadcasting signals. FM varies the frequency of the carrier wave in accordance with the signal being transmitted. Its advantage over AM (◊amplitude modulation) is its better signal-to-noise ratio. It was invented by the US engineer Edwin Armstrong.

frequently asked questions in computing, expansion of the abbreviation ◊FAQ.

The ideas of Freud were popularized by people who only imperfectly understood them, who were incapable of the great effort required to grasp them in their relationship to larger truths, and who therefore assigned to them a prominence out of all proportion to their true importance.

Alfred North Whitehead English philosopher and mathematician.
Dialogues Dialogue XXVIII 3 June 1943

friction in physics, the force that opposes the relative motion of two bodies in contact. The **coefficient of friction** is the ratio of the force required to achieve this relative motion to the force pressing the two bodies together.

Friction is greatly reduced by the use of lubricants such as oil, grease, and graphite. Air bearings are now used to minimize friction in high-speed rotational machinery. In other instances friction is deliberately increased by making the surfaces rough – for example, brake linings, driving belts, soles of shoes, and tyres.

FRIENDS OF THE EARTH HOME PAGE

http://www.foe.co.uk/

Appeal for raised awareness of environmental issues with masses of information and tips for action from Friends of the Earth. The site hosts lengthy accounts of several campaigns undertaken by FoE on climate, industry, transport, and sustainable development. It also maintains an archive of press releases from FoE on some of the most controversial environmental problems encountered in the course of last year around the world.

Friends of the Earth (FoE or FOE) environmental pressure group, established in the UK 1971, that aims to protect the environment and to promote rational and sustainable use of the Earth's resources. It campaigns on such issues as acid rain; air, sea, river, and land pollution; recycling; disposal of toxic wastes; nuclear power and renewable energy; the destruction of rainforests; pesticides; and agriculture. FoE has branches in 30 countries.

Green consumerism is a target for exploitation. There's a lot of green froth on top, but murkiness lurks underneath.

JONATHON PORRITT English environmental campaigner. Speech at a Friends of the Earth Conference 1989

frilled lizard yellowish-brown Australian lizard with a large frill of skin to the sides of the neck and throat. The lizard is about 90 cm/35 in long.

When the lizard is angry or alarmed it erects its frill, which may be as much as 25cm/10 in in diameter, thus giving itself the appearance of being larger than it really is. Frilled lizards are generally tree-living but may spend some time on the ground, where they run with their forelimbs in the air.

classification The frilled lizard *Chlamydosaurus kingi* belongs to the family Agamidae, suborder Sauria, order Squamata, class Reptilia.

Frisch–Peierls memorandum document revealing, for the first time, how small the critical mass (the minimum quantity of substance required for a nuclear chain reaction to begin) of uranium needed to be if the isotope uranium-235 was separated from naturally occurring uranium; the memo thus implied the feasibility of using this isotope to make an atom bomb. It was written by Otto Frisch and Rudolf Peierls at the University of Birmingham in 1940.

fritillary in botany, any of a group of plants belonging to the lily family. The snake's head fritillary (*F. meleagris*) has bell-shaped flowers with purple-chequered markings. (Genus *Fritillaria*, family Liliaceae.)

fritillary in zoology, any of a large grouping of butterflies of the family Nymphalidae. Mostly medium-sized, fritillaries are usually orange and reddish with a black criss-cross pattern or spots above and with silvery spots on the underside of the hindwings.

They take their name from the Latin word *fritillus* ('dice box') because of their spotted markings.

frog any amphibian of the order Anura (Greek 'tailless'). There are about 24 different families of frog, containing more than 3,800 species. There are no clear rules for distinguishing between frogs and ◊toads.

FROG

Certain frogs do not spend time as tadpoles. The female of a species of tiny Brazilian frog lays a single egg which hatches as a single froglet, rather than going through a tadpole phase first.

Frogs usually have squat bodies, with hind legs specialized for jumping, and webbed feet for swimming. Most live in or near water, though as adults they are air-breathing. A few live on land or even in trees. Their colour is usually greenish in the genus *Rana*, but other Ranidae are brightly coloured, for instance black and orange or yellow and white. Many use their long, extensible tongues to capture insects. The eyes are large and bulging. Frogs vary in size from the North American little grass frog *Limnaoedus ocularis*, 12 mm/0.5 in long, to the giant aquatic frog *Telmatobius culeus*, 50 cm/20 in long, of Lake Titicaca, South America. Frogs are widespread, inhabiting all continents except Antarctica, and they have adapted to a range of environments including deserts, forests, grasslands, and even high altitudes, with some species in the Andes and Himalayas existing above 5,000 m/19,600 ft.

courtship and reproduction In many species the males attract the females in great gatherings, usually by croaking. In some tropical species, the male's inflated vocal sac may exceed the rest of his body in size. Other courtship 'lures' include thumping on the ground and 'dances'.

Some lay eggs in large masses (spawn) in water. The jelly surrounding the eggs provides support and protection and retains warmth. Some South American frogs build mud-pool 'nests', and African tree frogs make foam nests from secreted mucus. In other species, the eggs may be carried in pockets on the mother's back, brooded by the male in his vocal sac or, as with the Eurasian midwife toad *Alytes obstetricans*, wrapped round the male's hind legs until hatching.

life cycle The tadpoles hatch from the eggs in about a fortnight. At first they are fishlike animals with external gills and a long swimming tail, but no limbs. The first change to take place is the disappearance of the external gills and the development of internal gills, which are still later supplanted by lungs. The hind legs appear before the front legs, and the last change to occur is the diminution and final disappearance of the tail. The tadpole stage lasts about three or four months. At the end of this time the animal leaves the water. Some species, such as the edible frog, are always aquatic. By autumn the frog grows big and sluggish. It stores fat in a special gland in the abdomen; it is this fat that it lives on during hibernation.

species Certain species of frog have powerful skin poisons (alkaloids) to deter predators. 'True frogs' are placed in the worldwide family Ranidae, with 800 species, of which the genus *Rana* is the best known. The North American bullfrog *Rana catesbeiana*, with a croak that carries for miles, is able to jump nine times its own length. The flying frogs, genus *Rhacophorus*, of Malaysia, using webbed fore and hind feet, can achieve a 12 m/40 ft glide. The hairy frog *Astylosternus robustus* is found in West Africa; it has long outgrowths on its flanks, which seem to aid respiration. A four-year rainforest study in E Madagascar revealed 106 new frog species in 1995. Indian zoologists discovered the first known leaf-eating frog in 1996, in Tamil Nadu, southern India. *R. hexadactyla* feeds mainly on leaves, flowers, and algae. New species are constantly being discovered. In 1997 a species *Eleutherodactylus pluvicanorus* was discovered in Bolivia; it is 4 cm long and ground-dwelling.

SOMEWHAT AMUSING WORLD OF FROGS

http://www.csu.edu.au/faculty/commerce/account/frogs/frog.htm

Fascinating facts about frogs – did you know, for example, that most frogs will drown eventually if denied access to land?

frogbit small water plant *Hydrocharis morsus-ranae* with submerged roots, floating leaves, and small green and white flowers.

froghopper or ***spittlebug*** leaping plant-bug, of the family Cercopidae, in the same order (Homoptera) as leafhoppers and aphids. Froghoppers live by sucking the juice from plants. The pale green larvae protect themselves (from drying out and from predators) by secreting froth ('cuckoo spit') from their anuses.

frogmouth nocturnal bird, related to the nightjar, of which the commonest species, the tawny frogmouth *Podargus strigoides,* is found throughout Australia, including Tasmania. Well camouflaged, it sits and awaits its prey.

frond large leaf or leaflike structure; in ferns it is often pinnately divided. The term is also applied to the leaves of palms and less commonly to the plant bodies of certain seaweeds, liverworts, and lichens.

front in meteorology, the boundary between two air masses of different temperature or humidity. A **cold front** marks the line of advance of a cold air mass from below, as it displaces a warm air mass; a **warm front** marks the advance of a warm air mass as it rises up over a cold one. Frontal systems define the weather of the mid-latitudes, where warm tropical air is constantly meeting cold air from the poles.

Warm air, being lighter, tends to rise above the cold; its moisture is carried upwards and usually falls as rain or snow, hence the changeable weather conditions at fronts. Fronts are rarely stable and move with the air mass. An **occluded front** is a composite form, where a cold front catches up with a warm front and merges with it.

front-end processor small computer used to coordinate and control the communications between a large mainframe computer and its input and output devices.

frost condition of the weather that occurs when the air temperature is below freezing, 0°C/32°F. Water in the atmosphere is deposited as ice crystals on the ground or exposed objects. As cold air is heavier than warm, ground frost is more common than hoar frost, which is formed by the condensation of water particles in the same way that dew collects.

frostbite the freezing of skin or flesh, with formation of ice crystals leading to tissue damage. The treatment is slow warming of the affected area; for example, by skin-to-skin contact or with lukewarm water. Frostbitten parts are extremely vulnerable to infection, with the risk of gangrene.

FRS abbreviation for ***Fellow of the ◊Royal Society***.

fructose $C_6H_{12}O_6$ a sugar that occurs naturally in honey, the nectar of flowers, and many sweet fruits; it is commercially prepared from glucose.

It is a monosaccharide, whereas the more familiar cane or beet sugar is a disaccharide, made up of two monosaccharide units: fructose and glucose. It is sweeter than cane sugar and can be used to sweeten foods for people with diabetes.

fruit ***from Latin frui 'to enjoy'*** in botany, the ripened ovary in flowering plants that develops from one or more seeds or carpels and encloses one or more seeds. Its function is to protect the seeds during their development and to aid in their dispersal. Fruits are often edible, sweet, juicy, and colourful. When eaten they provide vitamins, minerals, and enzymes, but little protein. Most fruits are borne by perennial plants.

Fruits are divided into three agricultural categories on the basis of the climate in which they grow. **Temperate fruits** require a cold season for satisfactory growth; the principal temperate fruits are apples, pears, plums, peaches, apricots, cherries, and soft fruits, such as strawberries. **Subtropical fruits** require warm conditions but can survive light frosts; they include oranges and other citrus fruits, dates, pomegranates, and avocados. **Tropical fruits** cannot tolerate temperatures that drop close to freezing point; they include bananas, mangoes, pineapples, papayas, and litchis. Fruits can also be divided botanically into **dry** (such as the ◊capsule, ◊follicle, ◊schizocarp, ◊nut, ◊caryopsis, pod or legume, ◊lomentum, and ◊achene) and those that become **fleshy** (such as the ◊drupe and the ◊berry). The fruit structure consists of the pericarp or fruit wall, which is usually divided into a number of distinct layers. Sometimes parts other than the ovary are incorporated into the fruit structure, resulting in a false fruit or ◊pseudocarp, such as the apple and strawberry. True fruits include the tomato, orange, melon, and banana. Fruits may be dehiscent, which open to shed their seeds, or indehiscent, which remain unopened and are dispersed as a single unit. Simple fruits (for example, peaches) are derived from a single ovary, whereas compositae or multiple fruits (for example, blackberries) are formed from the ovaries of a number of flowers. In ordinary usage, 'fruit' includes only sweet, fleshy items; it excludes many botanical fruits such as acorns, bean pods, thistledown, and cucumbers.

methods of seed dispersal Efficient seed dispersal is essential to avoid overcrowding and enable plants to colonize new areas; the natural function of a fruit is to aid in the dissemination of the seeds which it contains. A great variety of dispersal mechanisms exist: winged fruits are commonly formed by trees, such as ash and elm, where they are in an ideal position to be carried away by the wind; some wind-dispersed fruits, such as clematis and cotton, have plumes of hairs; others are extremely light, like the poppy, in which the capsule acts like a pepperpot and shakes out the seeds as it is blown about by the wind. Some fruits float on water; the coconut can be dispersed across oceans by means of its buoyant fruit. Geraniums, gorse, and squirting cucumbers have explosive mechanisms, by which seeds are forcibly shot out at dehiscence. Animals often act as dispersal agents either by carrying hooked or sticky fruits (burs) attached to their bodies, or by eating succulent fruits, the seeds passing through the alimentary canal unharmed.

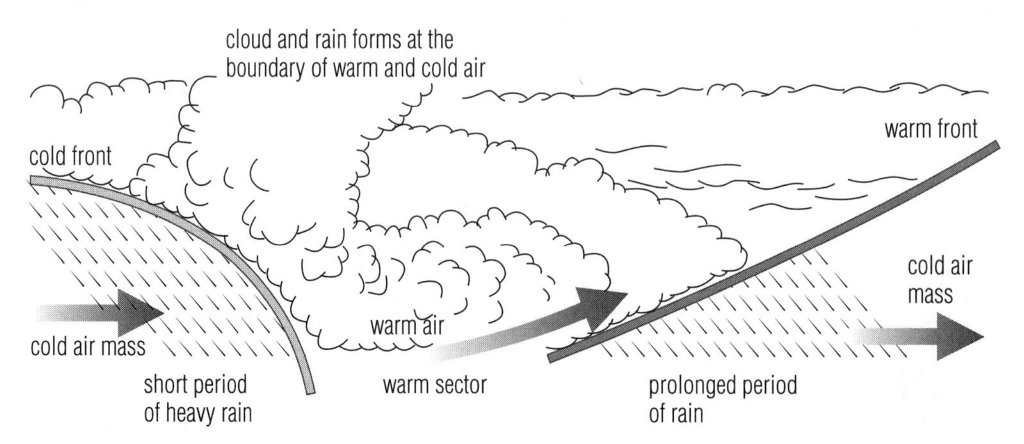

front *The boundaries between two air masses of different temperature and humidity. A warm front is when warm air displaces cold air; if cold air replaces warm air, it is a cold front.*

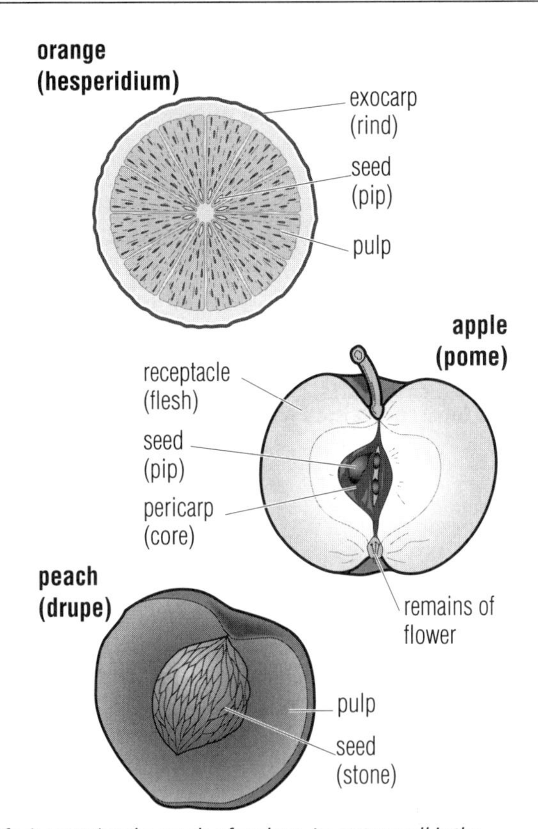

***fruit** A fruit contains the seeds of a plant. Its outer wall is the exocarp, or epicarp; its inner layers are the mesocarp and endocarp. The orange is a hesperidium, a berry having a leathery rind and containing many seeds. The peach is a drupe, a fleshy fruit with a hard seed, or 'stone', at the centre. The apple is a pome, a fruit with a fleshy outer layer and a core containing the seeds.*

Recorded world fruit production in the mid-1980s was approximately 300 million tonnes per year. Technical advances in storage and transport have made tropical fruits available to consumers in temperate areas, and fresh temperate fruits available all year in major markets.

frustule the cell wall of a ◊diatom (microscopic alga). Frustules are intricately patterned on the surface with spots, ridges, and furrows, each pattern being characteristic of a particular species.

frustum ***from Latin for 'a piece cut off'*** in geometry, a 'slice' taken out of a solid figure by a pair of parallel planes. A conical frustum, for example, resembles a cone with the top cut off. The volume and area of a frustum are calculated by subtracting the volume or area of the 'missing' piece from those of the whole figure.

FSF in computing, abbreviation for the ◊Free Software Foundation.

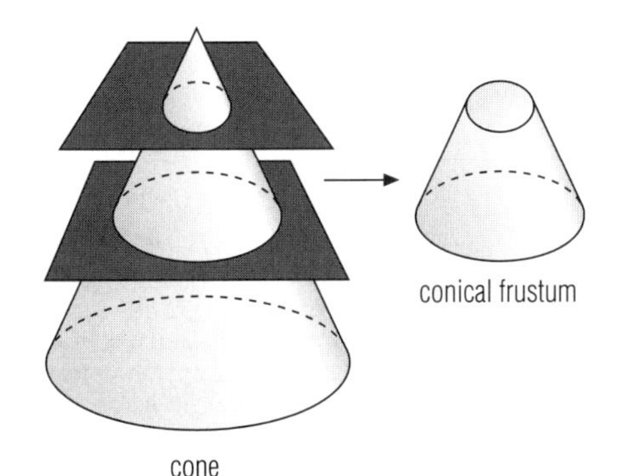

***frustum** The frustum, a slice taken out of a cone.*

FSH abbreviation for ◊follicle-stimulating hormone.

f-stop in photography, another name for ◊f-number.

ft symbol for ◊foot, **a measure of distance**.

FTP (abbreviation for ***File Transfer Protocol***) in computing, rules for transferring files between computers on the ◊Internet. The use of FTP avoids incompatibility between individual computers. To use FTP over the Internet, a user must have an Internet connection, an FTP client or World Wide Web◊ browser, and an account on the system holding the files. Many commercial and noncommercial systems allow anonymous FTP either to distribute new versions of software products or as a public service.

FTPmail in computing, an ◊FTP server that can be operated by e-mail. This service is useful for people with only limited access to the Internet.

fuchsia any shrub or ◊herbaceous plant of a group belonging to the evening-primrose family. Species are native to South and Central America and New Zealand, and bear red, purple, or pink bell-shaped flowers that hang downwards. (Genus *Fuchsia*, family Onagraceae.)

The genus was named in 1703 after German botanist Leonhard Fuchs (1501–1566).

fuel any source of heat or energy, embracing the entire range of materials that burn in air (combustibles). A **nuclear fuel** is any material that produces energy by nuclear fission in a nuclear reactor.

fuel cell cell converting chemical energy directly to electrical energy.

It works on the same principle as a battery but is continually fed with fuel, usually hydrogen. Fuel cells are silent and reliable (no moving parts) but expensive to produce.

Hydrogen is passed over an ◊electrode (usually nickel or platinum) containing a ◊catalyst, which strips electrons off the atoms. These pass through an external circuit while hydrogen ions (charged atoms) pass through an ◊electrolyte to another electrode, over which oxygen is passed. Water is formed at this electrode (as a by-product) in a chemical reaction involving electrons, hydrogen ions, and oxygen atoms. If the spare heat also produced is used for hot water and space heating, 80% efficiency in fuel is achieved.

fuel injection injecting fuel directly into the cylinders of an internal combustion engine, instead of by way of a carburettor. It is the standard method used in ◊diesel engines, and is now becoming standard for petrol engines. In the diesel engine, oil is injected into the hot compressed air at the top of the second piston stroke and explodes to drive the piston down on its power stroke. In the petrol engine, fuel is injected into the cylinder at the start of the first induction stroke of the ◊four-stroke cycle.

full duplex in computing, modem setting which means that two-way communication is enabled, so that everything you type is echoed back to the screen. See ◊duplex.

fullerene form of carbon, discovered in 1985, based on closed cages of carbon atoms. The molecules of the most symmetrical of the fullerenes are called ◊buckminsterfullerenes (or buckyballs). They are perfect spheres made up of 60 carbon atoms linked together in 12 pentagons and 20 hexagons fitted together like those of a spherical football. Other fullerenes, with 28, 32, 50, 70, and 76 carbon atoms, have also been identified.

Fullerenes can be made by arcing electricity between carbon rods. They may also occur in candle flames and in clouds of interstellar gas. Fullerene chemistry may turn out to be as important as organic chemistry based on the benzene ring. Already, new molecules based on the buckyball enclosing a metal atom, and 'buckytubes' (cylinders of carbon atoms arranged in hexagons), have been made. Applications envisaged include using the new molecules as lubricants, semiconductors, and superconductors, and as the starting point for making new drugs.

fuller's earth soft, greenish-grey rock resembling clay, but without clay's plasticity. It is formed largely of clay minerals, rich in montmorillonite, but a great deal of silica is also present. Its absorbent properties make it suitable for removing oil and grease, and it was formerly used for cleaning fleeces ('fulling'). It is still used in the textile industry, but its chief application is in the purification of oils. Beds of fuller's earth are found in the southern USA, Germany, Japan, and the UK.

full-motion video (FMV) in computing, video system that can display continuous motion. Some slow-speed CD-ROM drives and low-bandwidth networks are unable to handle the mass of data required for full-motion video, so video playback tends to jerk unevenly.

fulmar any of several species of petrels of the family Procellariidae, which are similar in size and colour to herring gulls. The northern fulmar *Fulmarus glacialis* is found in the North Atlantic and visits land only to nest, laying a single egg.

fulminate any salt of fulminic (cyanic) acid (HOCN), the chief ones being silver and mercury. The fulminates detonate (are exploded by a blow); see ◊detonator.

FULMAR

The world's official oldest wild bird is over 50 years old. It is a female fulmar, which nests each year on an uninhabited Orkney island. She has been monitored

fumitory any of a group of plants native to Europe and Asia. The common fumitory (*F. officinalis*) grows to 50 cm/20 in and produces pink flowers tipped with blackish red; it has been used in medicine for stomach and liver complaints. (Genus *Fumeria*, family Fumariaceae.)

function in computing, a small part of a program that supplies a specific value – for example, the square root of a specified number, or the current date. Most programming languages incorporate a number of built-in functions; some allow programmers to write their own. A function may have one or more arguments (the values on which the function operates). A **function key** on a keyboard is one that, when pressed, performs a designated task, such as ending a program.

function in mathematics, a function f is a non-empty set of ordered pairs $(x, f(x))$ of which no two can have the same first element. Hence, if $f(x) = x^2$ two ordered pairs are (–2,4) and (2,4). The set of all first elements in a function's ordered pairs is called the **domain**; the set of all second elements is the **range**. In the algebraic expression $y = 4x^3 + 2$, the dependent variable y is a function of the independent variable x, generally written as $f(x)$.

Functions are used in all branches of mathematics, physics, and science generally; for example, the formula $t = 2\pi\sqrt{(l/g)}$ shows that for a simple pendulum the time of swing t is a function of its length l and of no other variable quantity (π and g, the acceleration due to gravity, are ◊constants).

Form and function are a unity, two sides of one coin. In order to enhance function, appropriate form must exist or be created.

IDA ROLF US biochemist and physical therapist. *Rolfing*, Preface

functional group in chemistry, a small number of atoms in an arrangement that determines the chemical properties of the group and of the molecule to which it is attached (for example, the carboxyl group COOH, or the amine group NH_2). Organic compounds can be considered as structural skeletons, with a high carbon content, with functional groups attached.

functional programming computer programming based largely on the definition of ◊functions. There are very few functional programming languages, HOPE and ML being the most widely used, though many more conventional languages (for example, C) make extensive use of functions.

function key key on a keyboard that, when pressed, performs a designated task, such as ending a computer program.

fundamental constant physical quantity that is constant in all circumstances throughout the whole universe. Examples are the electric charge of an electron, the speed of light, Planck's constant, and the gravitational constant.

Physical constants, or fundamental constants, are standardized values whose parameters do not change.

Constant	*Symbol*	*Value in SI units*
acceleration of free fall	g	9.80665 m s^{-2}
Avogadro's constant	N_A	6.0221367×10^{23} mol^{-1}
Boltzmann's constant	k	1.380658×10^{-23} J K^{-1}
elementary charge	e	$1.60217733 \times 10^{-19}$ C
electronic rest mass	m_e	$9.1093897 \times 10^{-31}$ kg
Faraday's constant	F	9.6485309×10^{4} C mol^{-1}
gas constant	R	8.314510 J K^{-1} mol^{-1}
gravitational constant	G	6.672×10^{-11} N m^2 kg^{-2}
Loschmidt's number	N_L	2.686763×10^{25} m^{-3}
neutron rest mass	m_n	$1.6749286 \times 10^{-27}$ kg
Planck's constant	h	$6.6260755 \times 10^{-34}$ J s
proton rest mass	m_p	$1.6726231 \times 10^{-27}$ kg
speed of light in a vacuum	c	2.99792458×10^{8} m s^{-1}
standard atmosphere	atm	1.01325×10^{5} Pa
Stefan–Boltzmann constant	θ	5.67051×10^{-8} W m^{-2} K^{-4}

fundamental forces see ◊forces, fundamental.

fundamental particle another term for ◊elementary particle.

fundamental vibration standing wave of the longest wavelength that can be established on a vibrating object such as a stretched string or air column. The sound produced by the fundamental vibration is the lowest-pitched (usually dominant) note heard.

The fundamental vibration of a string has a stationary ◊node at each end and a single ◊antinode at the centre where the amplitude of vibration is greatest.

fungicide any chemical ◊pesticide used to prevent fungus diseases in plants and animals. Inorganic and organic compounds containing sulphur are widely used.

fungus (plural ***fungi***) any of a unique group of organisms that includes moulds, yeasts, rusts, smuts, mildews, mushrooms, and toadstools. About 50,000 species have been identified. They are not considered to be plants for three main reasons: they have no leaves or roots; they contain no chlorophyll (green colouring) and are therefore unable to make their own food by ◊photosynthesis; and they reproduce by ◊spores. Some fungi are edible but many are highly poisonous; they often cause damage and sometimes disease

FUNGUS

Leaf-cutter ants feed their larvae on a fungus that they cultivate themselves in underground gardens. The ants depend on the fungus for food, and the fungus cannot reproduce without the ants. When a new queen takes flight to establish another colony, she takes some of the fungus with her in a special mouth pouch.

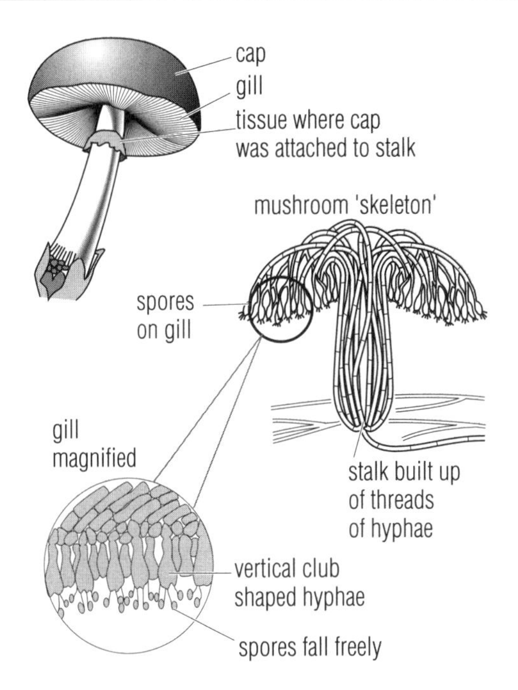

fungus Fungi grow from spores as fine threads, or hyphae. These have no distinct cellular structure. Mushrooms and toadstools are the fruiting bodies formed by the hyphae. Gills beneath the caps of these aerial structures produce masses of spores.

to the organic matter they live and feed on, but some fungi are exploited in the production of food and drink (for example, yeasts in baking and brewing) and in medicine (for example, penicillin). (Kingdom Fungi.)

Fungi are either ◊parasites, existing on living plants or animals, or ◊saprotrophs, living on dead matter. Many of the most serious plant diseases are caused by fungi, and several fungi attack humans and animals. Athlete's foot, ◊thrush, and ◊ringworm are fungal diseases.

Before the classification Fungi came into use, they were included within the division Thallophyta, along with ◊algae and ◊bacteria. Two familiar fungi are bread mould, which illustrates the typical many-branched body (mycelium) of the organism, made up of threadlike chains of cells called hyphae; and mushrooms, which are the sexually reproductive fruiting bodies of an underground mycelium.

The mycelium of a true fungus is made up of many intertwined hyphae. When the fungus is ready to reproduce, the hyphae become closely packed into a solid mass called the fruiting body, which is usually small and inconspicuous but can be very large; mushrooms, toadstools, and bracket fungi are all examples of large fruiting bodies. These carry and distribute the spores. Most species of fungi reproduce both asexually (on their own) and sexually (involving male and female parents).

FUNGI

http://www.herb.lsa.umich.edu/kidpage/factindx.htm

University-run network of hyperlinked pages on fungi from their earliest fossil records to their current ecology and life cycles, from how they are classified systematically to how they are studied. Although this page is not shy of technical terms, there are clear explanations and pictures to help the uninitiated.

fur the ◊hair of certain animals. Fur is an excellent insulating material and so has been used as clothing. This is, however, vociferously criticized by many groups on humane grounds, as the methods of breeding or trapping animals are often cruel. Mink, chinchilla, and sable are among the most valuable, the wild furs being finer than the farmed.

Fur such as mink is made up of a soft, thick, insulating layer called underfur and a top layer of longer, lustrous guard hairs.

Furs have been worn since prehistoric times and have long been associated with status and luxury (ermine traditionally worn by royalty, for example), except by certain ethnic groups like the Inuit. The fur trade had its origin in North America, where in the late 17th century the Hudson's Bay Company was established. The chief centres of the fur trade are New York, London, St Petersburg, and Kastoria in Greece. It is illegal to import furs or skins of endangered species listed by ◊CITES (such as the leopard). Many synthetic fibres are widely used as substitutes.

furlong unit of measurement, originating in Anglo-Saxon England, equivalent to 220 yd (201.168 m).

A furlong consists of 40 rods, poles, or perches; 8 furlongs equal one statute ◊mile. Its literal meaning is 'furrow-long', and refers to the length of a furrow in the common field characteristic of medieval farming.

furnace structure in which fuel such as coal, coke, gas, or oil is burned to produce heat for various purposes. Furnaces are used in conjunction with ◊boilers for heating, to produce hot water, or steam for driving turbines – in ships for propulsion and in power stations for generating electricity. The largest furnaces are those used for smelting and refining metals, such as the ◊blast furnace, electric furnace, and ◊open-hearth furnace.

furniture beetle wood-boring beetle. See ◊woodworm

FurryMUCK in computing, popular ◊MUD site where the players take on the imaginary shapes and characters of furry, anthropomorphic animals.

furze another name for ◊gorse, a shrub.

fuse in electricity, a wire or strip of metal designed to melt when excessive current passes through. It is a safety device that halts surges of current which would otherwise damage equipment and cause fires. In explosives, a fuse is a cord impregnated with chemicals so that it burns slowly at a predetermined rate. It is used to set off a main explosive charge, sufficient length of fuse being left to allow the person lighting it to get away to safety.

fusel oil liquid with a characteristic unpleasant smell, obtained as a by-product of the distillation of the product of any alcoholic fermentation, and used in paints, varnishes, essential oils, and plastics. It is a mixture of fatty acids, alcohols, and esters.

fusion in physics, the fusing of the nuclei of light elements, such as hydrogen, into those of a heavier element, such as helium. The resultant loss in their combined mass is converted into energy. Stars and thermonuclear weapons are powered by nuclear fusion.

FUSION

http://www.pppl.gov/~rfheeter/

All-text site, but packed with information about fusion research and its applications. It is quite well-organized and it includes a glossary of commonly used terms to aid the uninitiated.

fuzzy logic in mathematics and computing, a form of knowledge representation suitable for notions (such as 'hot' or 'loud') that cannot be defined precisely but depend on their context. For example, a jug of water may be described as too hot or too cold, depending on whether it is to be used to wash one's face or to make tea.

The central idea of fuzzy logic is **probability of set membership**. For instance, referring to someone 175 cm/5 ft 9 in tall, the statement 'this person is tall' (or 'this person is a member of the set of tall people') might be about 70% true if that person is a man, and about 85% true if that person is a woman.

G

g symbol for ◊gram.

gabbro mafic (consisting primarily of dark-coloured crystals) igneous rock formed deep in the Earth's crust. It contains pyroxene and calcium-rich feldspar, and may contain small amounts of olivine and amphibole. Its coarse crystals of dull minerals give it a speckled appearance.

Gabbro is the plutonic version of basalt (that is, derived from magma that has solidified below the Earth's surface), and forms in large, slow-cooling intrusions.

Gabcikovo Dam hydroelectric dam on the river Danube, at the point where it crosses the frontier between Hungary and the Slovak Republic. A treaty agreeing to its construction was signed by Hungary and Czechoslovakia in 1977, but work was suspended in 1989 after Hungary withdrew its support for a scheme to divert water from the river. Czechoslovakia resumed work 1991, despite warnings from scientists and environmentalists that the scheme would destroy valuable wetlands in the Danube valley.

A dramatic reduction in river flow resulted from Czechoslovakia's first attempts to divert water November 1992, prompting the setting-up of an investigative committee, under the auspices of the European Community, and involving both parties concerned, to reassess the project.

gadfly fly that bites cattle, such as a ◊botfly or ◊horsefly.

gadolinium silvery-white metallic element of the lanthanide series, symbol Gd, atomic number 64, relative atomic mass 157.25. It is found in the products of nuclear fission and used in electronic components, alloys, and products needing to withstand high temperatures.

Gaia hypothesis theory that the Earth's living and nonliving systems form an inseparable whole that is regulated and kept adapted for life by living organisms themselves. The planet therefore functions as a single organism, or a giant cell. The hypothesis was elaborated by British scientist James Lovelock and first published in 1968.

The only laws of matter are those which our minds must fabricate, and the only laws of mind are fabricated by matter.

James Clerk Maxwell Scottish physicist. Attributed remark

gain in audio, the volume control.

gain in electronics, the ratio of the amplitude of the output signal produced by an amplifier to that of the input signal.

In a ◊voltage amplifier the voltage gain is the ratio of the output voltage to the input voltage; in an inverting ◊operational amplifier (op-amp) it is equal to the ratio of the resistance of the feedback resistor to that of the input resistor.

gal symbol for ◊gallon, ◊galileo.

galactic coordinates in astronomy, a system for measuring the position of astronomical objects on the ◊celestial sphere with reference to the galactic equator (or ◊great circle).

Galactic latitude (symbol b) is measured in degrees from the galactic equator (b = 0°) to the north (b = 90°) and south (b = – 90°) galactic poles.

Galactic longitude (symbol l) is measured in degrees eastward (l = 0° to 360°) from a fixed point in the constellation of ◊Sagittarius that approximates to the centre of the Galaxy. Galactic coordinates are often used when astronomers are studying the distribution of material in the ◊Galaxy.

galactic halo in astronomy, the outer, sparsely populated region of a galaxy, roughly spheroid in shape and extending far beyond the bulk of the visible stars. In our own Galaxy, the halo contains the globular clusters, and may harbour large quantities of ◊dark matter.

galactic plane in astronomy, a plane passing through the ◊Sun and the centre of the ◊Galaxy defining the mid-plane of the galactic disc. Viewed from the Earth, the galactic plane is a ◊great circle (galactic equator) marking the approximate centre line of the ◊Milky Way.

galago small African prosimian also known as ◊bushbaby.

galaxy congregation of millions or billions of stars, held together by gravity. **Spiral galaxies**, such as the ◊Milky Way, are flattened in shape, with a central bulge of old stars surrounded by a disc of younger stars, arranged in spiral arms like a Catherine wheel.

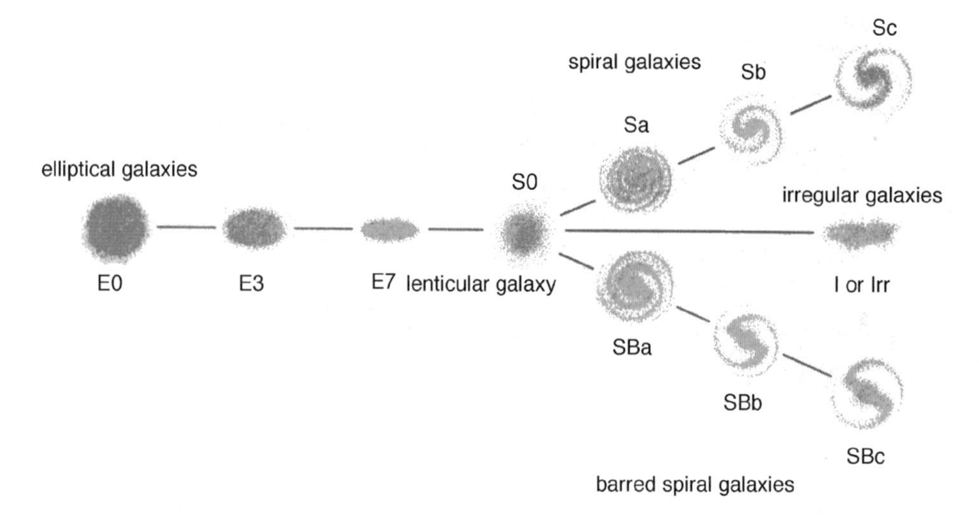

galaxy Galaxies were classified by US astronomer Edwin Hubble in 1925. He placed the galaxies in a 'tuning-fork' pattern, in which the two prongs correspond to the barred and non-barred spiral galaxies.

Barred spirals are spiral galaxies that have a straight bar of stars across their centre, from the ends of which the spiral arms emerge. The arms of spiral galaxies contain gas and dust from which new stars are still forming.

Elliptical galaxies contain old stars and very little gas. They include the most massive galaxies known, containing a trillion stars. At least some elliptical galaxies are thought to be formed by mergers between spiral galaxies. There are also irregular galaxies. Most galaxies occur in clusters, containing anything from a few to thousands of members.

Our own Galaxy, the Milky Way, is about 100,000 light years in diameter, and contains at least 100 billion stars. It is a member of a small cluster, the ◊Local Group. The Sun lies in one of its spiral arms, about 25,000 light years from the centre.

By the end of a five-year study in 1995, US astronomers had identified 600 previously uncatalogued galaxies, mostly 200–400 million light years away, leading to the conclusion that there may be 30–100% more galaxies than previously estimated. Two galaxies were discovered obscured by galactic dust at the edge of the Milky Way. One, named MB1, is a spiral galaxy 17,000 light years across; the other, MB2, is an irregular-shaped dwarf galaxy about 4,000 light years across. In 1996 US astronomers discovered a further new galaxy 17 million light years away. The galaxy NGC2915 is a blue compact dwarf galaxy and 95% of its mass is in the form of dark matter. In 1997 an international team of astronomers detected the furthest known object in the universe, which is a galaxy lying 13 billion light years away.

galena mineral consisting of lead sulphide, PbS, the chief ore of lead. It is lead-grey in colour, has a high metallic lustre and breaks into cubes because of its perfect cubic cleavage. It may contain up to 1% silver, and so the ore is sometimes mined for both metals. Galena occurs mainly among limestone deposits in Australia, Mexico, Russia, Kazakhstan, the UK, and the USA.

galileo unit (symbol gal) of acceleration, used in geological surveying. One galileo is 10^{-2} metres per second per second. The

Galileo
(1564–1642)

properly Galileo Galilei Italian mathematician, astronomer, and physicist. He developed the astronomical telescope and was the first to see sunspots, the four main satellites of Jupiter, and the appearance of Venus going through phases, thus proving it was orbiting the Sun. Galileo discovered that freely falling bodies, heavy or light, have the same, constant acceleration and that this acceleration is due to gravity. He also determined that a body moving on a perfectly smooth horizontal surface would neither speed up nor slow down. He invented a thermometer, a hydrostatic balance, and a compass, and discovered that the path of a projectile is a parabola.

Galileo's work founded the modern scientific method of deducing laws to explain the results of observation and experiment, although the story of his dropping cannonballs from the Leaning Tower of Pisa is questionable. His observations were an unwelcome refutation of the ideas of the Greek philosopher Aristotle taught at the church-run universities, largely because they made plausible for the first time the Sun-centred theory of Polish astronomer Nicolaus Copernicus. Galileo's persuasive *Dialogo sopra i due massimi sistemi del mondo/Dialogues on the Two Chief Systems of the World* 1632 was banned by the church authorities in Rome and he was made to recant by the Inquisition. Evangelista Torricelli is the most famous of his pupils.

astronomy and the invention of the telescope In July 1609, hearing that a Dutch scientist had made a telescope, Galileo worked out the principles involved and made a number of telescopes. He compiled fairly accurate tables of the orbits of four of Jupiter's satellites and proposed that their frequent eclipses could serve as a means of determining longitude on land and at sea. His observations on sunspots and Venus going through phases supported Copernicus's theory that the Earth rotated and orbited the Sun. Galileo's results published in *Sidereus Nuncius/The Starry Messenger* 1610 were revolutionary.

He believed, however – following both Greek and medieval tradition – that orbits must be circular, not elliptical, in order to maintain the fabric of the cosmos in a state of perfection. This preconception prevented him from deriving a full formulation of the law of inertia, which was later to be attributed to the contemporary French mathematician René Descartes.

Mary Evans Picture Library

the pendulum Galileo made several fundamental contributions to mechanics. He rejected the impetus theory that a force or push is required to sustain motion. While watching swinging lamps in Pisa cathedral, Galileo determined that each oscillation of a pendulum takes the same amount of time despite the difference in amplitude, and recognized the potential importance of this observation to timekeeping. In a later publication, he presented his derivation that the square of the period of a pendulum varies with its length (and is independent of the mass of the pendulum bob).

mechanics and the law of falling bodies Galileo discovered before Newton that two objects of different weights – an apple and a melon, for instance – falling from the same height would hit the ground at the same time. He realized that gravity not only causes a body to fall, but also determines the motion of rising bodies and, furthermore, that gravity extends to the centre of the Earth. Galileo then showed that the motion of a projectile is made up of two components: one component consists of uniform motion in a horizontal direction, and the other component is vertical motion under acceleration or deceleration due to gravity.

Galileo used this explanation to refute objections to Copernicus. The Church argued that a turning Earth would not carry along birds and clouds. Galileo explained that motion of a bird, like a projectile, has a horizontal component that is provided by the motion of the Earth and that this horizontal component of motion always exists to keep such objects in position even though they are not attached to the ground.

Galileo came to an understanding of uniform velocity and uniform acceleration by measuring the time it takes for bodies to move various distances. He had the brilliant idea of slowing vertical motion by measuring the movement of balls rolling down inclined planes, realizing that the vertical component of this motion is a uniform acceleration due to gravity. It took Galileo many years to arrive at the correct expression of the law of falling bodies, which he presented in *Discorsi e dimostrazioni matematiche intorno a due nove scienze/Discourses and Mathematical Discoveries Concerning Two New Sciences* 1638 as:

$$s=1/2at^2$$

where s is speed, a is the acceleration due to gravity, and t is time. He found that the distance travelled by a falling body is proportional to the square of the time of descent.

A summation of his life's work, *Discourses* also included the facts that the trajectory of a projectile is a parabola, and that the law of falling bodies is perfectly obeyed only in a vacuum, and that air resistance always causes a uniform terminal velocity to be reached.

Earth's gravitational field often differs by several milligals (thousandths of gals) in different places, because of the varying densities of the rocks beneath the surface.

GALILEO PROJECT INFORMATION

http://nssdc.gsfc.nasa.gov/planetary/galileo.html

Site dedicated to the Galileo Project and the opportunities it has offered scientists and astronomers to enhance our understanding of the universe. The site includes information about the mission's objectives, scientific results, images, and links to other relevant sites on the Web.

Galileo spacecraft launched from the space shuttle *Atlantis* October 1989, on a six-year journey to Jupiter. *Galileo's* probe entered the atmosphere of Jupiter December 1995. It radioed information back to the orbiter for 57 minutes before it was destroyed by atmospheric pressure. The orbiter will continue circling Jupiter until 1997. Despite technical problems data is still being relayed to Earth, but very slowly. The first pictures of Jupiter are due July 1996.

It flew past Venus February 1990 and passed within 970 km/600 mi of Earth December 1990, using the gravitational fields of these two planets to increase its velocity. It flew past the asteroids Gaspra in 1991 and Ida in 1993, taking close-up photographs.

At the end of July 1995, and 55 million km/34 million mi from Jupiter, *Galileo* entered a dust storm and began detecting up to 20,000 particles a day (previously the maximum detected was 200). The dust is associated with Jupiter, and may come from the planet itself, its rings, or its satellites.

***Galileo** The* Galileo *spacecraft about to be detached from the Earth-orbiting space shuttle* Atlantis *at the beginning of its six-year journey to Jupiter. National Aeronautical Space Agency*

gall abnormal outgrowth on a plant that develops as a result of attack by insects or, less commonly, by bacteria, fungi, mites, or nematodes. The attack causes an increase in the number of cells or an enlargement of existing cells in the plant. Gall-forming insects generally pass the early stages of their life inside the gall.

Gall wasps are responsible for the conspicuous bud galls forming on oak trees, 2.5–4 cm/1–1.5 in across, known as 'oak apples'. The organisms that cause galls are host-specific. Thus, for example, gall wasps tend to parasitize oaks, and ◊sawflies willows.

gall bladder small muscular sac, part of the digestive system of most, but not all, vertebrates. In humans, it is situated on the underside of the liver and connected to the small intestine by the bile duct. It stores bile from the liver.

galley ship powered by oars, and usually also equipped with sails. Galleys typically had a crew of hundreds of rowers arranged in banks. They were used in warfare in the Mediterranean from antiquity until the 18th century.

France maintained a fleet of some 40 galleys, crewed by over 10,000 convicts, until 1748. The maximum speed of a galley is estimated to have been only four knots (7.5 kph/4.5 mph), because only 20% of the rower's effort was effective, and galleys could not be used in stormy weather because of their very low waterline.

gallium grey metallic element, symbol Ga, atomic number 31, relative atomic mass 69.72. It is liquid at room temperature. Gallium arsenide (GaAs) crystals are used in microelectronics, since electrons travel a thousand times faster through them than through silicon. The element was discovered in 1875 by Lecoq de Boisbaudran (1838–1912).

gallium arsenide compound of gallium and arsenic, formula GaAs, used in lasers, photocells, and microwave generators. Its semiconducting properties make it a possible rival to ◊silicon for use in microprocessors. Chips made from gallium arsenide require less electric power and process data faster than those made from silicon.

gall midge minute and fragile long-legged flies, with longish hairy antennae. The larvae are small maggots, ranging in colour from white or yellow, to orange and bright red, that feed on developing fruits which become deformed and decay, and frequently produce ◊galls on plants.

Some forms live within galls formed by other insects, such as beetles, or other species of Cecidomyiidae. The ◊hessian fly of North America and New Zealand, is a pest of wheat. The **pearl midge** *Contarinia pyrivota* is a serious fruit pest in Europe.

classification Gall midges are in the family Cecidomyiidae (suborder Nematocera) of the insect order Diptera, class Insecta, phylum Arthropoda.

gallon imperial liquid or dry measure, equal to 4.546 litres, and subdivided into four quarts or eight pints. The US gallon is equivalent to 3.785 litres.

gallstone pebblelike, insoluble accretion formed in the human gall bladder or bile ducts from cholesterol or calcium salts present in bile. Gallstones may be symptomless or they may cause pain, indigestion, or jaundice. They can be dissolved with medication or removed, either by means of an endoscope or, along with the gall bladder, in an operation known as cholecystectomy.

gall wasp small (only a few millimetres long), dark-coloured insect with a compressed abdomen. Most gall wasps form ◊galls, though a few live within the galls formed by other species; these are called **inquilines**. Others feed on gall-formers and inquilines.

classification Gall wasps are in the family Cynipidae, order Hymenoptera, class Insecta, phylum Arthropoda.

The exact reactions which lead to gall formation in the host plant are little understood. Basically it is a reaction of the cells of the plant to the presence of the larva.

complex life history The oak-apple gall is caused by species *Biorhiza pallida* and both winged males and wingless or vestigial-winged females emerge from these galls. After mating the females lay their eggs in the root of the same host plant, thus producing root galls. In the following spring only wingless females are produced. These females climb up the oak tree and produce the characteristic oak-apples, thus repeating the life cycle.

Rose galls are often produced by *Diplolepis rosae*. These gall wasps usually reproduce asexually; the females are about 4 mm/0.2 in long; parts of their abdomens and legs are yellow-red, while the rest of the body is black. Males of this species have been observed only rarely. The galls are a mass of reddish filaments within which are found a number of sealed chambers enclosing larvae. The larvae feed on the gall tissue.

galvanizing process for rendering iron rust-proof, by plunging it into molten zinc (the dipping method), or by electroplating it with zinc.

galvanometer instrument for detecting small electric currents by their magnetic effect.

games console computer capable only of playing games, which are supplied as cartridges or CD-ROM discs that slot directly into the console.

gamete cell that functions in sexual reproduction by merging with another gamete to form a ◊zygote. Examples of gametes include sperm and egg cells. In most organisms, the gametes are haploid (they contain half the number of chromosomes of the parent), owing to reduction division or ◊meiosis.

In higher organisms, gametes are of two distinct types: large immobile ones known as eggs or egg cells (see ◊ovum) and small ones known as ◊sperm. They come together at ◊fertilization. In some lower organisms the gametes are all the same, or they may belong to different mating strains but have no obvious differences in size or appearance.

game theory group of mathematical theories, developed in 1944 by Oscar Morgenstern (1902–1977) and John Von Neumann, that seeks to abstract from invented game-playing scenarios and their outcome the essence of situations of conflict and/or cooperation in the real political, business, and social world.

A feature of such games is that the rationality of a decision by one player will depend on what the others do; hence game theory has particular application to the study of oligopoly (a market largely controlled by a few producers).

A theory is a good theory if it satisfies two requirements: it must accurately describe a large class of observations on the basis of a model that contains only a few arbitrary elements, and it must make definite predictions about the results of future predictions.

STEPHEN HAWKING English physicist. *A Brief History of Time* 1988

gametophyte the ◊haploid generation in the life cycle of a plant that produces gametes; see ◊alternation of generations.

gamma radiation very high-frequency electromagnetic radiation, similar in nature to X-rays but of shorter wavelength, emitted by the nuclei of radioactive substances during decay or by the interactions of high-energy electrons with matter. Cosmic gamma rays have been identified as coming from pulsars, radio galaxies, and quasars, although they cannot penetrate the Earth's atmosphere.

Gamma rays are stopped only by direct collision with an atom and are therefore very penetrating; they can, however, be stopped by about 4 cm/1.5 in of lead or by a very thick concrete shield. They are less ionizing in their effect than alpha and beta particles, but are dangerous nevertheless because they can penetrate deeply into body tissues such as bone marrow. They are not deflected by either magnetic or electric fields.

Gamma radiation is used to kill bacteria and other microorganisms, sterilize medical devices, and change the molecular structure of plastics to modify their properties (for example, to improve their resistance to heat and abrasion).

gamma-ray astronomy the study of gamma rays from space. Much of the radiation detected comes from collisions between hydrogen gas and cosmic rays in our Galaxy. Some sources have been identified, including the Crab nebula and the Vela pulsar (the most powerful gamma-ray source detected).

Gamma rays are difficult to detect and are generally studied by use of balloon-borne detectors and artificial satellites. The first gamma-ray satellites were *SAS II* (1972) and *COS B* (1975), although gamma-ray detectors were carried on the *Apollo 15* and *16* missions. *SAS II* failed after only a few months, but *COS B* continued working until 1982, carrying out a complete survey of the galactic disc.

ganglion (plural ***ganglia***) solid cluster of nervous tissue containing many cell bodies and ◊synapses, usually enclosed in a tissue sheath; found in invertebrates and vertebrates.

In many invertebrates, the central nervous system consists mainly of ganglia connected by nerve cords. The ganglia in the head (cerebral ganglia) are usually well developed and are analogous to the brain in vertebrates. In vertebrates, most ganglia occur outside the central nervous system.

gangrene death and decay of body tissue (often of a limb) due to bacterial action; the affected part gradually turns black and causes blood poisoning.

Gangrene sets in as a result of loss of blood supply to the area. This may be due to disease (diabetes, atherosclerosis), an obstruction of a major blood vessel (as in ◊thrombosis), injury, or frostbite. Bacteria colonize the site unopposed, and a strong risk of blood poisoning often leads to surgical removal of the tissue or the affected part (amputation).

gannet any of three species of North Atlantic seabirds; the largest is *Sula bassana*. When fully grown, it is white with buff colouring on the head and neck; the beak is long and thick and compressed at the point; the wings are black-tipped with a span of 1.7 m/5.6 ft. It breeds on cliffs in nests made of grass and seaweed, laying a single white egg. Gannets feed on fish that swim near the surface, such as herrings and pilchards. (Family Sulidae, order Pelecaniformaes.)

Diving swiftly and sometimes from a considerable height upon their prey, they enter the water with closed wings and neck outstretched. They belong to the same family as the ◊booby.

The gannets are the largest seabirds of the North Atlantic; they are found also in the southeast Pacific and in temperate waters off Africa.

gannet *The Cape gannet* Morus capensis. *It forms dense breeding colonies around the coasts of S Africa, incubating its single egg with its feet. Gannets are accomplished 'plunge divers' and make a spectacular sight when fishing. Premaphotos Wildlife*

Ganymede in astronomy, the largest moon of the planet Jupiter, and the largest moon in the Solar System, 5,260 km/3,270 mi in diameter (larger than the planet Mercury). It orbits Jupiter every 7.2 days at a distance of 1.1 million km/700,000 mi. Its surface is a mixture of cratered and grooved terrain. Molecular oxygen was identified on Ganymede's surface in 1994.

The space probe *Galileo* detected a magnetic field around Ganymede in 1996; this suggests it may have a molten core. *Galileo* photographed Ganymede at a distance of 7,448 km/4,628 mi. The resulting images were 17 times clearer than those taken by *Voyager 2* in 1979, and show the surface to be extensively cratered and ridged, probably as a result of forces similar to those that create mountains on Earth. *Galileo* also detected molecules containing both carbon and nitrogen on the surface March 1997. Their presence may indicate that Ganymede harboured life at some time.

gar any of a group of primitive bony fishes, which also includes ◊sturgeons. Gar have long, beaklike snouts and elongated bodies covered in heavy, bony scales. All four species of gar live in freshwater rivers and lakes of the Mississippi drainage. See also ◊needlefish. (Order Semionotiformes.)

gardenia any of a group of subtropical and tropical trees and shrubs found in Africa and Asia, belonging to the madder family, with evergreen foliage and flattened rosettes of fragrant waxen-looking flowers, often white in colour. (Genus *Gardenia,* family Rubiaceae.)

garfish European marine fish with a long spearlike snout. The common garfish (*Belone belone*) has an elongated body measuring 75 cm/2.5 ft in length. (Family Belonidae, order Beloniformes.)

garlic perennial Asian plant belonging to the lily family, whose strong-smelling and sharp-tasting bulb, made up of several small segments, or cloves, is used in cooking. The plant has white flowers. It is widely cultivated and has been used successfully as a fungicide in the cereal grass ◊sorghum. It also has antibacterial properties. (*Allium sativum,* family Liliaceae.)

In tests carried out in 1994, US doctors found freshly pressed garlic extract killed a number of bacteria, including drug-resistant strains, even when diluted to one part in 250. Its effectiveness is probably due to allicin, a simple organic disulphide.

garnet group of ◊silicate minerals with the formula $X_3Y_2(SiO_4)_3$, where X is calcium, magnesium, iron, or manganese, and Y is usually aluminium or sometimes iron or chromium. Garnets are used as semiprecious gems (usually pink to deep red) and as abrasives. They occur in metamorphic rocks such as gneiss and schist.

Garnets consisting of neodymium, yttrium, and aluminium (referred to as Nd-YAG) produce infrared laser light when sufficiently excited. Nd-YAG lasers are inexpensive and used widely in industry and scientific research.

garpike freshwater bony fish. It has an elongated snout and its body is covered with thick scales. Garpikes are predators found in North America.
classification Garpikes are in the family Lepisosteidae, order Lepisosteiformes, class Osteichthyes.

gas in physics, a form of matter, such as air, in which the molecules move randomly in otherwise empty space, filling any size or shape of container into which the gas is put.

A sugar-lump sized cube of air at room temperature contains 30 trillion molecules moving at an average speed of 500 metres per second (1,800 kph/1,200 mph). Gases can be liquefied by cooling, which lowers the speed of the molecules and enables attractive forces between them to bind them together.

GAS UTILITIES HISTORY

http://www.geocities.com/Athens/Acropolis/4007/gsframe.htm

Despite the irritating frames layout of this web site it does contain plenty of information on the origins of the natural gas industry. The information database held here is searchable by country, time, or company, and the information pops up in another frame.

gas collection method used to collect a gas in a laboratory preparation. The properties of the gas, and whether it is required dry, dictate the method used. Dry ammonia is collected by downward displacement of air.

gas constant in physics, the constant R that appears in the equation $PV = nRT$, which describes how the pressure P, volume V, and temperature T of an ideal gas are related (n is the amount of gas in moles). This equation combines ◊Boyle's law and ◊Charles's law.

R has a value of 8.3145 joules per kelvin per mole.

gas-cooled reactor type of nuclear reactor; see ◊advanced gas-cooled reactor.

gas engine internal-combustion engine in which a gas (coal gas, producer gas, natural gas, or gas from a blast furnace) is used as the fuel.

The first practical gas engine was built in 1860 by Jean Etienne Lenoir, and the type was subsequently developed by Nikolaus August Otto, who introduced the ◊four-stroke cycle.

gas exchange movement of gases between an organism and the atmosphere, principally oxygen and carbon dioxide. All aerobic organisms (most animals and plants) take in oxygen in order to burn food and manufacture ◊ATP. The resultant oxidation reactions release carbon dioxide as a waste product to be passed out into the environment. Green plants also absorb carbon dioxide during ◊photosynthesis, and release oxygen as a waste product.

Specialized respiratory surfaces have evolved during evolution to make gas exchange more efficient. In humans and other tetrapods (four-limbed vertebrates), gas exchange occurs in the ◊lungs, aided by the breathing movements of the ribs. Many adult amphibia and terrestrial invertebrates can absorb oxygen directly through the skin. The bodies of insects and some spiders contain a system of air-filled tubes known as ◊tracheae. Fish have ◊gills as their main respiratory surface. In plants, gas exchange generally takes place via the ◊stomata and the air-filled spaces between the cells in the interior of the leaf.

gas giant in astronomy, any of the four large outer ◊planets of the Solar System, ◊Jupiter, ◊Saturn, ◊Uranus, and ◊Neptune, which consist largely of gas and have no solid surface.

gas laws physical laws concerning the behaviour of gases. They include ◊Boyle's law and ◊Charles's law, which are concerned with the relationships between the pressure, temperature, and volume of an ideal (hypothetical) gas. These two laws can be combined to give the **general** or **universal gas law**, which may be expressed as:

$$(\text{pressure} \times \text{volume})/\text{temperature} = \text{constant}$$

Van der Waals' law includes corrections for the nonideal behaviour of real gases.

gasohol motor fuel that is 90% petrol and 10% ethanol (alcohol). The ethanol is usually obtained by fermentation, followed by distillation, using maize, wheat, potatoes, or sugar cane. It was used in early cars before petrol became economical, and its use was revived during the 1940s war shortage and the energy shortage of the 1970s, for example in Brazil.

gasoline the US term for ◊petrol.

gas syringe graduated piece of glass apparatus used to measure accurately the volumes of gases.

gastroenteritis inflammation of the stomach and intestines, giving rise to abdominal pain, vomiting, and diarrhoea. It may be caused by food or other poisoning, allergy, or infection. Dehydration may be severe and it is a particular risk in infants.

gastrolith stone that was once part of the digestive system of a dinosaur or other extinct animal. Rock fragments were swallowed to assist in the grinding process in the dinosaur digestive tract, much as some birds now swallow grit and pebbles to grind food in their crop. Once the animal has decayed, smooth round stones remain – often the only clue to their past use is the fact that they are geologically different from their surrounding strata.

gastropod any member of a very large group of ◊molluscs (soft-bodied invertebrate animals). Gastropods have a single shell (in a spiral or modified spiral form) and eyes on stalks, and they move on a flattened, muscular foot. They have well-developed heads and rough, scraping tongues called radulae. Some are marine, some freshwater, and others land creatures, but they all tend to live in damp places. (Class Gastropoda.)

gas turbine engine in which burning fuel supplies hot gas to spin a ◊turbine. The most widespread application of gas turbines has been in aviation. All jet engines (see under ◊jet propulsion) are modified gas turbines, and some locomotives and ships also use gas turbines as a power source.

They are also used in industry for generating and pumping purposes.

In a typical gas turbine a multivaned compressor draws in and compresses air. The compressed air enters a combustion chamber at high pressure, and fuel is sprayed in and ignited. The hot gases produced escape through the blades of (typically) two turbines and spin them around. One of the turbines drives the compressor; the other provides the external power that can be harnessed.

gate, logic in electronics, see ◊logic gate.

Gates, Bill (William) Henry, III (1955–) US businessman and computer programmer. He co-founded ◊Microsoft Corporation in 1975 and was responsible for supplying MS-DOS, the operating system and the Basic language that ◊IBM used in the IBM PC.

In 1997 Gates controlled a $39.8 billion shareholding in Microsoft, making him the world's richest individual.

When the IBM deal was struck in 1980, Microsoft did not actually have an operating system, but Gates bought one from another company, renamed it MS-DOS, and modified it to suit IBM's new computer. Microsoft also retained the right to sell MS-DOS to other computer manufacturers, and because the IBM PC was not only successful but easily copied by other manufacturers, MS-DOS found its way onto the vast majority of PCs. The revenue from MS-DOS helped Microsoft to expand into other areas of software, guided by Gates.

gateway in computing, the point of contact between two ◊wide-area networks.

gauge any scientific measuring instrument – for example, a wire gauge or a pressure gauge. The term is also applied to the width of a railway or tramway track.

gauge boson or *field particle* any of the particles that carry the four fundamental forces of nature (see ◊forces, fundamental).

Gauge bosons are ◊elementary particles that cannot be subdivided, and include the photon, the graviton, the gluons, and the weakons.

gaur Asiatic wild ox, dark grey-brown in colour with white 'socks' and standing 2 m/6 ft tall at the shoulders. It originally roamed across a vast area of land stretching from India to SE Asia and Malaysia, but population numbers and the area where it can be found are now much smaller. (Species *Bos gaurus*.)

gauss c.g.s. unit (symbol Gs) of magnetic induction or magnetic flux density, replaced by the SI unit, the ◊tesla, but still commonly used. It is equal to one line of magnetic flux per square centimetre. The Earth's magnetic field is about 0.5 Gs, and changes to it over time are measured in gammas (one gamma equals 10^{-5} gauss).

gavial large reptile related to the crocodile. It grows to about 7 m/23 ft long, and has a very long snout with about 100 teeth in its jaws. Gavials live in rivers in northern India, where they feed on fish and frogs. They have been extensively hunted for their skins, and are now extremely rare. (Species *Gavialis gangeticus*.)

gayal species of ox found in the highland regions of east India and Burma. The animal is often found wild, but just as frequently in a semi-domesticated condition. It is smaller than the ◊gaur, and its horns are much straighter.

The gayal and the gaur frequently interbreed.

classification The gayal *Bos frontalis* is in family Bovidae (cattle and antelopes) of order Artiodactyla.

GAVIAL

http://www.sdcs.k12.ca.us/roosevelt/gavialhome.html

Well-presented information about the gavial. Intended to educate children about this harmless reptile, the site has information on habitat, diet, interaction with humans, and reproduction.

Gay-Lussac, Joseph Louis (1778–1850)

French physicist and chemist who investigated the physical properties of gases, and discovered new methods of producing sulphuric and oxalic acids. In 1802 he discovered the approximate rule for the expansion of gases now known as Charles's law; see also gas laws.

Mary Evans Picture Library

gazelle any of a number of lightly built, fast-running antelopes found on the open plains of Africa and southern Asia. (Especially species of the genus *Gazella*.)

gear toothed wheel that transmits the turning movement of one shaft to another shaft. Gear wheels may be used in pairs, or in threes if both shafts are to turn in the same direction. The gear ratio – the ratio of the number of teeth on the two wheels – determines the torque ratio, the turning force on the output shaft compared with the turning force on the input shaft. The ratio of the angular velocities of the shafts is the inverse of the gear ratio.

The common type of gear for parallel shafts is the **spur gear**, with straight teeth parallel to the shaft axis. The **helical gear** has teeth cut along sections of a helix or corkscrew shape; the double form of the helix gear is the most efficient for energy transfer. **Bevel gears**, with tapering teeth set on the base of a cone, are used to connect intersecting shafts.

gecko any of a group of lizards. Geckos are common worldwide in warm climates, and have large heads and short, stout bodies. Many have no eyelids. Their sticky toe pads enable them to climb vertically and walk upside down on smooth surfaces in their search for flies, spiders, and other prey. (Family Gekkonidae.)

There are about 850 known species of gecko. There are 102 Australian species, 17 new species having been discovered there

GECKO

One species of Hawaiian gecko reproduces without sex to create all-female populations. These females have fewer parasites (and thus better health) than their mixed-gender neighbours.

1986–96. A new species of gecko *Tarentola mindiae* was identified in Egypt's Western Desert in 1997.

geebung Australian shrub or tree of the genus *Persoonia* of the Proteaceae family with small bell-shaped cream or yellow flowers.

geek in computing, stereotypical exceptionally bright, obsessive computer user or programmer. See also ◊anorak and ◊nerd.

Geiger, Hans (Wilhelm) (1882–1945)

German physicist who produced the Geiger counter. He spent the period 1906–12 in Manchester, England, working with Ernest Rutherford on radioactivity. In 1908 they designed an instrument to detect and count alpha particles, positively charged ionizing particles produced by radioactive decay.

In 1928 Geiger and Walther Müller produced a more sensitive version of the counter, which could detect all kinds of ionizing radiation.

Geiger counter any of a number of devices used for detecting nuclear radiation and/or measuring its intensity by counting the number of ionizing particles produced (see ◊radioactivity). It detects the momentary current that passes between ◊electrodes in a suitable gas when a nuclear particle or a radiation pulse causes the ionization of that gas. The electrodes are connected to electronic devices that enable the number of particles passing to be measured. The increased frequency of measured particles indicates the intensity of radiation. The device is named after the German physicist Hans ◊Geiger.

The Geiger–Müller, Geiger–Klemperer, and Rutherford–Geiger counters are all devices often referred to loosely as Geiger counters.

Geissler tube high-voltage ◊discharge tube in which traces of gas ionize and conduct electricity. Since the electrified gas takes on a luminous colour characteristic of the gas, the instrument is also used in ◊spectroscopy. It was developed in 1858 by the German physicist Heinrich Geissler.

gel solid produced by the formation of a three-dimensional cage structure, commonly of linked large-molecular-mass polymers, in which a liquid is trapped. It is a form of ◊colloid. A gel may be a jellylike mass (pectin, gelatin) or have a more rigid structure (silica gel).

gelignite type of ◊dynamite.

gem mineral valuable by virtue of its durability (hardness), rarity, and beauty, cut and polished for ornamental use, or engraved. Of 120 minerals known to have been used as gemstones, only about 25 are in common use in jewellery today; of these, the diamond, emerald, ruby, and sapphire are classified as precious, and all the others semiprecious; for example, the topaz, amethyst, opal, and aquamarine.

Among the synthetic precious stones to have been successfully produced are rubies, sapphires, emeralds, and diamonds (first produced by General Electric in the USA in 1955). Pearls are not technically gems.

Gemini prominent zodiacal constellation in the northern hemisphere represented as the twins Castor and Pollux. Its brightest star is ◊Pollux; Castor is a system of six stars. The Sun passes through Gemini from late June to late July. Each December, the Geminid meteors radiate from Gemini. In astrology, the dates for Gemini are between about 21 May and 21 June (see ◊precession).

Gemini project US space programme (1965–66) in which astronauts practised rendezvous and docking of spacecraft, and working outside their spacecraft, in preparation for the Apollo Moon landings. Gemini spacecraft carried two astronauts and were launched by Titan rockets.

gemma (plural ***gemmae***) unit of ◊vegetative reproduction, consisting of a small group of undifferentiated green cells. Gemmae are found in certain mosses and liverworts, forming on the surface of the plant, often in cup-shaped structures, or gemmae cups. Gemmae are dispersed by splashes of rain and can then develop into new plants. In many species, gemmation is more common than reproduction by ◊spores.

gemsbok species of antelope that inhabits the desert regions of southern Africa. It stands about 1.2 m/4 ft in height, and its general colour is greyish. Its horns are just over 1 m/3.3 ft long.
classification The gemsbok *Oryx gazella* is in family Bovidae, order Artiodactyla.

gene unit of inherited material, encoded by a strand of ◊DNA and transcribed by ◊RNA. In higher organisms, genes are located on the ◊chromosomes. A gene consistently affects a particular character in an individual – for example, the gene for eye colour. Also termed a Mendelian gene, after Austrian biologist Gregor ◊Mendel, it occurs at a particular point, or locus, on a particular chromosome and may have several variants, or ◊alleles, each specifying a particular form of that character – for example, the alleles for blue or brown eyes. Some alleles show ◊dominance. These mask the effect of other alleles, known as ◊recessive.

In the 1940s, it was established that a gene could be identified with a particular length of DNA, which coded for a complete protein molecule, leading to the 'one gene, one enzyme' principle. Later it was realized that proteins can be made up of several ◊polypeptide chains, each with a separate gene, so this principle was modified to 'one gene, one polypeptide'. However, the fundamental idea remains the same, that genes produce their visible effects simply by coding for proteins; they control the structure of those proteins via the genetic code, as well as the amounts produced and the timing of production.

In modern genetics, the gene is identified either with the ◊cistron (a set of ◊codons that determines a complete polypeptide) or with the unit of selection (a Mendelian gene that determines a particular character in the organism on which ◊natural selection can act). Genes undergo ◊mutation and ◊recombination to produce the variation on which natural selection operates.

We are survival machines – robot vehicles blindly programmed to preserve the selfish molecules known as genes. This is a truth which still fills me with astonishment.

RICHARD DAWKINS English zoologist.
The Selfish Gene Preface

gene amplification technique by which selected DNA from a single cell can be duplicated indefinitely until there is a sufficient amount to analyse by conventional genetic techniques.

Gene amplification uses a procedure called the polymerase chain reaction. The sample of DNA is mixed with a solution of enzymes called polymerases, which enable it to replicate, and with a plentiful supply of nucleotides, the building blocks of DNA. The mixture is repeatedly heated and cooled. At each warming, the double-stranded DNA present separates into two single strands, and with each cooling the polymerase assembles a new paired strand for each single strand. Each cycle takes approximately 30 minutes to complete, so that after 10 hours there is one million times more DNA present than at the start.

The technique has been used to analyse DNA from a man who died in 1959, showing the presence of sequences from the HIV virus in his cells. It can also be used to test for genetic defects in a single cell taken from an embryo, before the embryo is reimplanted in ◊in vitro fertilization.

gene bank collection of seeds or other forms of genetic material, such as tubers, spores, bacterial or yeast cultures, live animals and plants, frozen sperm and eggs, or frozen embryos. These are stored for possible future use in agriculture, plant and animal breeding, or in medicine, genetic engineering, or the restocking of wild habitats where species have become extinct. Gene banks will be increasingly used as the rate of extinction increases, depleting the Earth's genetic variety (biodiversity).

gene imprinting genetic phenomenon whereby a small number of genes function differently depending on whether they were inherited from the father or the mother. If two copies of an imprinted gene are inherited from one parent and none from the other, a genetic abnormality results, whereas no abnormality occurs if, as is normal, a copy is inherited from both parents. Gene imprinting is known to play a part in a number of genetic disorders and childhood diseases, for example, the Prader–Willi syndrome (characterized by mild mental retardation and compulsive eating).

gene pool total sum of ◊alleles (variants of ◊genes) possessed by all the members of a given population or species alive at a particular time.

General Electric US electrical and electronics company founded in 1878 in New Jersey as the ◊Edison Electric Light Company to back the experiments of the inventor Thomas Edison. In 1892 the company merged with a competitor to form General Electric. Its headquarters are in Fairfield, Connecticut.

general MIDI (***musical instrument digital interface***) or GM standard set of 96 instrument and percussion 'voices' that can be used to encode musical tracks which can be reproduced on any GM-compatible synthesizer, or ◊MIDI.

general protection fault computing error message; see ◊GPF.

generate in mathematics, to produce a sequence of numbers from either the relationship between one number and the next or the relationship between a member of the sequence and its position. For example, $un+1 = 2un$ generates the sequence 1, 2, 4, 8, ... ; $an = n(n+1)$ generates the sequence of numbers 2, 6, 12, 20, ...

generation in computing, stage of development in computer electronics (see ◊computer generation) or a class of programming language (see ◊fourth-generation language).

generator machine that produces electrical energy from mechanical energy, as opposed to an ◊electric motor, which does the opposite. A simple generator (dynamo) consists of a wire-wound coil (◊armature) that is rotated between the poles of a permanent magnet. The movement of the wire in the magnetic field induces a current in the coil by ◊electromagnetic induction, which can be fed by means of a ◊commutator as a continuous direct current into an external circuit. Slip rings instead of a commutator produce an alternating current, when the generator is called an alternator.

genet any of several small, nocturnal, carnivorous mammals belonging to the mongoose and civet family. Most species live in Africa, but *G. genetta* is also found in Europe and the Middle East. It is about 50 cm/1.6 ft long with a 45 cm/1.5 ft tail, and greyish yellow in colour with rows of black spots. It is a good climber. (Genus *Genetta,* family Viverridae.)

gene therapy medical technique for curing or alleviating inherited diseases or defects; certain infections, and several kinds of cancer in which affected cells from a sufferer would be removed from the body, the ◊DNA repaired in the laboratory (◊genetic engineering), and the functioning cells reintroduced. In 1990 a genetically engineered gene was used for the first time to treat a patient.

The first human being to undergo gene therapy, in 1990, was one of the so-called 'bubble babies' – a four-year-old American girl suffering from a rare enzyme (ADA) deficiency that cripples the immune system. Unable to fight off infection, such children are nursed in a germ-free bubble; they usually die in early childhood.

Cystic fibrosis is the commonest inherited disorder and the one most keenly targeted by genetic engineers; it has been pioneered in patients in the USA and UK. Gene therapy is not the final answer to inherited disease; it may cure the patient but it cannot prevent him or her from passing on the genetic defect to any children. However, it does hold out the promise of a cure for various other conditions, including heart disease and some cancers; US researchers have successfully used a gene gun to target specific tumour cells. In 1995 tumour growth was halted in mice when DNA-coated gold bullets were fired into tumour cells.

By the end of 1995, although 600 people had been treated with gene therapy, nobody had actually been cured. Even in the ADA trials, the most successful to date, the children were still receiving injections of synthetic ADA, possibly the major factor in their improvement.

genetic code the way in which instructions for building proteins, the basic structural molecules of living matter, are 'written' in the genetic material ◊DNA. This relationship between the sequence of bases (the subunits in a DNA molecule) and the sequence of ◊amino acids (the subunits of a protein molecule) is the basis of heredity. The code employs ◊codons of three bases each; it is the same in almost all organisms, except for a few minor differences recently discovered in some protozoa.

Only 2% of DNA is made up of base sequences, called **exons**, that code for proteins. The remaining DNA is known as 'junk' DNA or **introns**.

genetic disease any disorder caused at least partly by defective genes or chromosomes. In humans there are some 3,000 genetic diseases, including cystic fibrosis, Down's syndrome, haemophilia, Huntington's chorea, some forms of anaemia, spina bifida, and Tay-Sachs disease.

genetic engineering deliberate manipulation of genetic material by biochemical techniques. It is often achieved by the introduction of new ◊DNA, usually by means of a virus or ◊plasmid. This can be for pure research, ◊gene therapy, or to breed functionally specific plants, animals, or bacteria. These organisms with a foreign gene added are said to be transgenic (see ◊transgenic organism). At the beginning of 1995 more than 60 plant species had been genetically engineered, and nearly 3,000 transgenic crops had been field-tested.

practical uses In genetic engineering, the splicing and reconciliation of genes is used to increase knowledge of cell function and reproduction, but it can also achieve practical ends. For example, plants grown for food could be given the ability to fix nitrogen, found in some bacteria, and so reduce the need for expensive fertilizers, or simple bacteria may be modified to produce rare drugs. A foreign gene can be inserted into laboratory cultures of bacteria to generate commercial biological products, such as synthetic insulin, hepatitis-B vaccine, and interferon. Gene splicing was invented in 1973 by the US scientists Stanley Cohen and Herbert Boyer, and patented in the USA in 1984.

new developments Developments in genetic engineering have led to the production of growth hormone, and a number of other bone-marrow stimulating hormones. New strains of animals have also been produced; a new strain of mouse was patented in the USA in 1989 (the application was rejected in the European Patent Office). A ◊vaccine against a sheep parasite (a larval tapeworm) has been developed by genetic engineering; most existing vaccines protect against bacteria and viruses.

The first genetically engineered food went on sale in 1994; the 'Flavr Savr' tomato, produced by the US biotechnology company Calgene, was available in California and Chicago.

safety measures There is a risk that when transplanting genes between different types of bacteria (*Escherichia coli,* which lives in the human intestine, is often used) new and harmful strains might be produced. For this reason strict safety precautions are observed, and the altered bacteria are disabled in some way so they are unable to exist outside the laboratory.

genetic fingerprinting or ***genetic profiling*** technique used for determining the pattern of certain parts of the genetic material ◊DNA that is unique to each individual. Like conventional fingerprinting, it can accurately distinguish humans from one another, with the exception of identical siblings from multiple births. It can be applied to as little material as a single cell.

Genetic fingerprinting involves isolating DNA from cells, then comparing and contrasting the sequences of component chemicals between individuals. The DNA pattern can be ascertained from a sample of skin, hair, or semen. Although differences are minimal (only 0.1% between unrelated people), certain regions of DNA, known as **hypervariable regions**, are unique to individuals.

genetics branch of biology concerned with the study of ◊heredity and variation; it attempts to explain how characteristics of living organisms are passed on from one generation to the next. The science of genetics is based on the work of Austrian biologist Gregor ◊Mendel whose experiments with the cross-breeding (hybridization) of peas showed that the inheritance of characteristics and traits takes place by means of discrete 'particles' (◊genes). These are present in the cells of all organisms, and are now recognized as being the basic units of heredity. All organisms possess ◊genotypes (sets of variable genes) and ◊phenotypes (characteristics produced by certain genes). Modern geneticists investigate the structure, function, and transmission of genes.

Before the publication of Mendel's work in 1865, it had been assumed that the characteristics of both parents were blended during inheritance, but Mendel showed that the genes remain intact, although their combinations change. As a result of his experiments with the cultivation of the common garden pea, Mendel introduced the concept of hybridization (see ◊monohybrid inheritance). Since Mendel, the study of genetics has advanced greatly, first through ◊breeding experiments and light-microscope observations (classical genetics), later by means of biochemical and electron microscope studies (molecular genetics).

In 1944, Canadian-born bacteriologist Oswald Avery, together with his colleagues at the Rockefeller Institute, Colin McLeod and Maclyn McCarty, showed that the genetic material was deoxyribonucleic acid ((◊DNA), and not protein as was previously thought. A further breakthrough was made in 1953 when James ◊Watson and Francis ◊Crick published their molecular model for the structure of DNA, the double helix, based on X-ray diffraction photographs. The following decade saw the cracking of the genetic code. The ◊genetic code is said to be universal since the same code applies to all organisms from bacteria and viruses to higher plants and animals, including humans. Today the deliberate manipulation of genes by biochemical techniques, or ◊genetic engineering, is commonplace.

genetic screening in medicine, the determination of the genetic make-up of an individual to determine if he or she is at risk of developing a hereditary disease later in life. Genetic screening can also be used to determine if an individual is a carrier for a particular genetic disease and, hence, can pass the disease on to any children. Genetic counselling should be undertaken at the same time as genetic screening of affected individuals. Diseases that can be screened for include cystic fibrosis, Huntington's chorea, and certain forms of cancer.

genitalia reproductive organs of sexually reproducing animals, particularly the external/visible organs of mammals: in males, the penis and the scrotum, which contains the testes, and in females, the clitoris and vulva.

genome the full complement of ◊genes carried by a single (haploid) set of ◊chromosomes. The term may be applied to the genetic information carried by an individual or to the range of genes found in a given species. The human genome is made up of 75,000 genes.

The first fully decoded genome in a life domain to which humans belong was announced in 1997. Scientists completed a genetic blueprint for *Saccharomyces cerevisiae*, common brewer's yeast, which shares a high number with genetic sequences with humans. The scientists produced a map of all the yeast's genes and the full genetic coding. Genomes have been identified for seven organisms, including the bacteria *Haemofilus influenzae* and *Escherichia coli*, and the mycoplasmas *Mycoplasma genitalium* and *Mycoplasma pneumoniae*.

GENE

There are more than 80,000 genes in human DNA.

We are built to make mistakes, coded for error.

LEWIS THOMAS US physician and educator.
The Medusa and the Snail, 'To Err is Human'

genotype the particular set of ◊alleles (variants of genes) possessed by a given organism. The term is usually used in conjunction with ◊phenotype, which is the product of the genotype and all environmental effects. See also ◊nature–nurture controversy.

genus (plural *genera*) group of ◊species with many characteristics in common.

Thus all doglike species (including dogs, wolves, and jackals) belong to the genus *Canis* (Latin 'dog').

Species of the same genus are thought to be descended from a common ancestor species. Related genera are grouped into ◊families.

geochemistry science of chemistry as it applies to geology. It deals with the relative and absolute abundances of the chemical elements and their ◊isotopes in the Earth, and also with the chemical changes that accompany geologic processes.

geochronology the branch of geology that deals with the dating of the Earth by studying its rocks and contained fossils. The ◊geological time chart is a result of these studies. Absolute dating methods involve the measurement of radioactive decay over time in certain chemical elements found in rocks, whereas relative dating methods establish the sequence of deposition of various rock layers by identifying and comparing their contained fossils.

geode in geology, a subspherical cavity into which crystals have grown from the outer wall into the centre. Geodes often contain very well-formed crystals of quartz (including amethyst), calcite, or other minerals.

geodesic dome hemispherical dome, a type of space-frame, whose surface is formed out of short rods arranged in triangles. The rods lie on geodesics (the shortest lines joining two points on a curved surface). This type of dome allows large spaces to be enclosed using the minimum of materials, and was patented by US engineer Buckminster Fuller in 1954.

geodesy methods of surveying the Earth for making maps and correlating geological, gravitational, and magnetic measurements. Geodesic surveys, formerly carried out by means of various measuring techniques on the surface, are now commonly made by using radio signals and laser beams from orbiting satellites (see ◊global positioning system).

geographical information system (GIS) computer software that makes possible the visualization and manipulation of spatial data, and links such data with other information such as customer records.

geography the study of the Earth's surface; its topography, climate, and physical conditions, and how these factors affect people and society. It is usually divided into **physical geography**, dealing with landforms and climates, and **human geography**, dealing with the distribution and activities of peoples on Earth.

history Early preclassical geographers concentrated on map-making, surveying, and exploring. In classical Greece theoretical ideas first became a characteristic of geography. Aristotle and ◊Pythagoras believed the Earth to be a sphere, Eratosthenes was the first to calculate the circumference of the world, and Herodotus investigated the origin of the Nile floods and the relationship between climate and human behaviour.

During the medieval period the study of geography progressed little in Europe, but the Muslim world retained much of the Greek tradition, embellishing the 2nd-century maps of ◊Ptolemy. During the early Renaissance the role of the geographer as an explorer and surveyor became important once again.

The foundation of modern geography as an academic subject stems from the writings of Friedrich Humboldt and Johann Ritter, in the late 18th and early 19th centuries, who for the first time defined geography as a major branch of scientific inquiry.

To remember the geological periods in descending order of age:

Camels often sit down carefully. Perhaps their joints creak? Early oiling might prevent permanent rheumatism.

or

China owls seldom deceive clay pigeons. They just chase each other making preposterous puns.

Cambrian, Ordovician, Silurian, Devonian, Carboniferous, Permian, Triassic, Jurassic, Cretaceous, Eocene, Oligocene, Miocene, Pliocene, Pleistocene (Recent)

geological time time scale embracing the history of the Earth from its physical origin to the present day. Geological time is traditionally divided into eons (Archaean or Archaeozoic, Proterozoic, and Phanerozoic in ascending chronological order), which in turn are subdivided into eras, periods, epochs, ages, and finally chrons.

The terms eon, era, period, epoch, age and chron are **geochronological units** representing intervals of geological time. Rocks representing an interval of geological time comprise a **chronostratigraphic unit**. Each of the hierarchical geochronological terms has a chronostratigraphic equivalent. Thus, rocks formed during an eon (a geochronological unit) are members of an eonothem (the chronostratigraphic unit equivalent of eon). Rocks of an era belong to an erathem. The chronostratigraphic equivalents of period, epoch, age, and chron are system, series, stage, and chronozone, repectively.

Having in the natural history of this earth, seen a succession of worlds, we may conclude that there is a system in nature. ... The result, therefore of our present enquiry is, that we find no vestige of a beginning – no prospect of an end.

James Hutton Scottish geologist.
Transactions of the Royal Society of Edinburgh 1788

geology science of the Earth, its origin, composition, structure, and history. It is divided into several branches: **mineralogy** (the minerals of Earth), **petrology** (rocks), **stratigraphy** (the deposition of successive beds of sedimentary rocks), **palaeontology** (fossils), and **tectonics** (the deformation and movement of the Earth's crust).

Geology is regarded as part of earth science, a more widely embracing subject that brings in meteorology, oceanography, geophysics, and geochemistry.

GEOLOGYLINK

http://www.geologylink.com/

Comprehensive information on geology that features a daily update on current geologic events, virtual classroom tours, and virtual field trips to locations around the world. You will also find an in-depth look at a featured event, geologic news and reports, an image gallery, glossary, maps, and an area for asking geology professors your most perplexing questions, plus a list of references and links.

geomagnetic reversal another term for ◊polar reversal.

geometric mean in mathematics, the *n*th root of the product of *n* positive numbers. The geometric mean *m* of two numbers *p* and *q* is such that $m = \sqrt{(p \times q)}$. For example, the mean of 2 and 8 is $\sqrt{(2 \times 8)} = \sqrt{16} = 4$.

geometric progression or ***geometric sequence*** in mathematics, a sequence of terms (progression) in which each term is a constant multiple (called the **common ratio**) of the one preceding it. For example, 3, 12, 48, 192, 768, ... is a geometric progression with a common ratio 4, since each term is equal to the previous term multiplied by 4. Compare ◊arithmetic progression.

The sum of *n* terms of a **geometric series**

$$1 + r + r^2 + r^3 + \ldots + rn^{-1}$$

is given by the formula

$$Sn = (_1 - rn)/(^1 - r)$$

for all $r \neq 1$. For $r = 1$, the geometric series can be summed to infinity:

$$S\infty = 1/(1 - r).$$

In nature, many single-celled organisms reproduce by splitting in two so that one cell gives rise to 2, then 4, then 8 cells, and so on, forming a geometric sequence 1, 2, 4, 8, 16, 32, ..., in which the common ratio is 2.

geometric tortoise South African tortoise *Psammobates geometricus* which grows to only 10–12 cm/4–5 in in length. It is acutely threatened by the loss of its habitat, which has shrunk to just 4% of the reptile's original range. Remaining areas are mainly fragmented and have been invaded by alien plant species that the tortoise is unable to eat.

geometry branch of mathematics concerned with the properties of space, usually in terms of plane (two-dimensional) and solid (three-dimensional) figures. The subject is usually divided into **pure geometry**, which embraces roughly the plane and solid geometry dealt with in Greek mathematician Euclid's *Stoicheia/Elements*, and **analytical** or ◊coordinate geometry, in which problems are solved using algebraic methods. A third, quite distinct, type includes the non-Euclidean geometries.

pure geometry This is chiefly concerned with properties of figures that can be measured, such as lengths, areas, and angles and is therefore of great practical use. An important idea in Euclidean geometry is the idea of **congruence**. Two figures are said to be congruent if they have the same shape and size (and area). If one figure is imagined as a rigid object that can be picked up, moved and placed on top of the other so that they exactly coincide, then the two figures are congruent. Some simple rules about congruence may be stated: two line segments are congruent if they are of equal length; two triangles are congruent if their corresponding sides are equal in length or if two sides and an angle in one is equal to those in the other; two circles are congruent if they have the same radius; two polygons are congruent if they can be divided into congruent triangles assembled in the same order.

The idea of picking up a rigid object to test congruence can be expressed more precisely in terms of elementary 'movements' of figures: a translation (or glide) in which all points move the same distance in the same direction (that is, along parallel lines); a rotation through a defined angle about a fixed point; a reflection (equivalent to turning the figure over).

Two figures are congruent to each other if one can be transformed into the other by a sequence of these elementary movements. In Euclidean geometry a fourth kind of movement is also studied; this is the enlargement in which a figure grows or shrinks in all directions by a uniform scale factor. If one figure can be transformed into another by a combination of translation, rotation, reflection, and enlargement then the two are said to be similar. All circles are similar. All squares are similar. Triangles are similar if corresponding angles are equal.

coordinate geometry A system of geometry in which points, lines, shapes, and surfaces are represented by algebraic expressions. In plane (two-dimensional) coordinate geometry, the plane is usually defined by two axes at right angles to each other, the horizontal x-axis and the vertical y-axis, meeting at 0, the origin. A point on the plane can be represented by a pair of ◊Cartesian coordinates, which define its position in terms of its distance along the x-axis and along the y-axis from 0. These distances are respectively the x and y coordinates of the point.

Lines are represented as equations; for example, $y = 2x + 1$ gives a straight line, and $y = 3x^2 + 2x$ gives a ◊parabola (a curve). The graphs of varying equations can be drawn by plotting the coordinates of points that satisfy their equations, and joining up the points. One of the advantages of coordinate geometry is that geometrical solutions can be obtained without drawing but by manipulating algebraic expressions. For example, the coordinates of the point of intersection of two straight lines can be determined by finding the unique values of x and y that satisfy both of the equations for the lines, that is, by solving them as a pair of ◊simultaneous equations. The curves studied in simple coordinate geometry are the ◊conic sections (circle, ellipse, parabola, and hyperbola), each of which has a characteristic equation.

Geometry probably originated in ancient Egypt, in land measurements necessitated by the periodic inundations of the river Nile, and was soon extended into surveying and navigation. Early geometers were the Greek mathematicians Thales, Pythagoras, and Euclid. Analytical methods were introduced and developed by the French philosopher René ◊Descartes in the 17th century. From the 19th century, various non-Euclidean geometries were devised by Carl Friedrich Gauss, János Bolyai, and Nikolai Lobachevsky. These were later generalized by Bernhard Riemann and found to have applications in the theory of relativity.

Ubi materia, ibi geometria.
Where there is matter, there is geometry.

JOHANNES KEPLER German astronomer and mathematician. Attributed remark

geomorphology branch of geology that deals with the nature and origin of surface landforms such as mountains, valleys, plains, and plateaus.

geophagy eating soil. It is a practice found in many animal species but the reasons for geophagy are poorly understood. Canadian and Japanese researchers established 1996, that chimpanzees eat earth containing clay minerals to combat diarrhoea and other digestive ailments. Other primates are also thought to eat earth for health reasons.

geophysics branch of earth science using physics to study the Earth's surface, interior, and atmosphere. Studies also include winds, weather, tides, earthquakes, volcanoes, and their effects.

geostationary orbit circular path 35,900 km/22,300 mi above the Earth's Equator on which a ◊satellite takes 24 hours, moving from west to east, to complete an orbit, thus appearing to hang stationary over one place on the Earth's surface. Geostationary orbits are used particularly for communications satellites and weather satellites. They were first thought of by the author Arthur C Clarke. A **geosynchronous orbit** lies at the same distance from Earth but is inclined to the Equator.

geothermal energy energy extracted for heating and electricity generation from natural steam, hot water, or hot dry rocks in the Earth's crust. Water is pumped down through an injection well where it passes through joints in the hot rocks. It rises to the surface through a recovery well and may be converted to steam or run through a heat exchanger. Dry steam may be directed through turbines to produce electricity. It is an important source of energy in volcanically active areas such as Iceland and New Zealand.

geraldton wax shrub *Chamelaucium uncinatum* with delicate needle leaves and waxy pink myrtle flowers. It is endemic to Western Australia and widely cultivated in other states

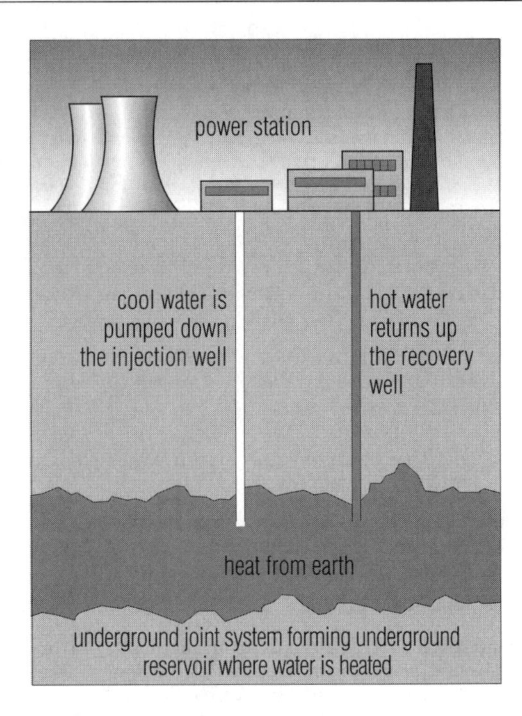

***geothermal energy** Geothermal energy is derived from the natural heat present below the surface of the Earth. Cool water is pumped down where it is heated up in large underground reservoirs before being pumped back to the surface.*

geranium any of a group of plants either having divided leaves and white, pink, or purple flowers (geraniums), or having a hairy stem, and white, pink, red, or black-purple flowers (◊pelarg-oniums). Some geraniums are also called ◊cranesbill. (Genera *Geranium* and *Pelargonium*, family Geraniaceae.)

gerbil any of numerous rodents with elongated back legs, good at hopping or jumping. Gerbils range from mouse- to rat-size, and have hairy tails. Many of the 13 genera live in dry, sandy, or sparsely vegetated areas of Africa and Asia. (Family Cricetidae.)

The Mongolian jird or gerbil (*Meriones unguiculatus*) is a popular pet.

***gerbil** A hairy-footed gerbil* Gerbillurus paeba *foraging across the sands of the Kalahari Desert in Southern Africa. Found in Africa and Asia, gerbils live in burrows and feed at night, mostly on seeds and roots. Premaphotos Wildlife*

gerenuk antelope about 1 m/3 ft high at the shoulder, with a very long neck. It browses on leaves, often balancing on its hind legs to do so. Sandy brown in colour, it is well camouflaged in its East African habitat of dry scrub. (Species *Litocranius walleri.*)

geriatrics medical speciality concerned with diseases and problems of the elderly.

germ colloquial term for a microorganism that causes disease, such as certain ◊bacteria and ◊viruses. Formerly, it was also used to mean something capable of developing into a complete organism (such as a fertilized egg, or the ◊embryo of a seed).

germanium brittle, grey-white, weakly metallic (◊metalloid) element, symbol Ge, atomic number 32, relative atomic mass 72.6. It belongs to the silicon group, and has chemical and physical properties between those of silicon and tin. Germanium is a semiconductor material and is used in the manufacture of transistors and integrated circuits. The oxide is transparent to infrared radiation, and is used in military applications. It was discovered in 1886 by German chemist Clemens Winkler (1838–1904).

In parts of Asia, germanium and plants containing it are used to treat a variety of diseases, and it is sold in the West as a food supplement despite fears that it may cause kidney damage.

German measles or ***rubella*** mild, communicable virus disease, usually caught by children. It is marked by a sore throat, pinkish rash, and slight fever, and has an incubation period of two to three weeks. If a woman contracts it in the first three months of pregnancy, it may cause serious damage to the unborn child.

German shepherd or ***Alsatian*** breed of dog. It is about 63 cm/25 in tall and has a wolflike appearance, a thick coat with many varieties of colouring, and a distinctive way of moving. German shepherds are used as police dogs because of their courage and intelligence.

They were introduced from Germany into Britain and the USA after World War I.

German silver or ***nickel silver*** silvery alloy of nickel, copper, and zinc. It is widely used for cheap jewellery and the base metal for silver plating. The letters EPNS on silverware stand for **e**lectro**p**lated **n**ickel **s**ilver.

germination in botany, the initial stages of growth in a seed, spore, or pollen grain. Seeds germinate when they are exposed to favourable external conditions of moisture, light, and temperature, and when any factors causing dormancy have been removed.

The process begins with the uptake of water by the seed. The embryonic root, or radicle, is normally the first organ to emerge, followed by the embryonic shoot, or plumule. Food reserves, either within the ◊endosperm or from the ◊cotyledons, are broken down to nourish the rapidly growing seedling. Germination is considered to have ended with the production of the first true leaves.

germ layer in ◊embryology, a layer of cells that can be distinguished during the development of a fertilized egg. Most animals have three such layers: the inner, middle, and outer. These differentiate to form the various body tissues.

The inner layer **(endoderm)** gives rise to the gut, the middle one **(mesoderm)** develops into most of the other organs, while the outer one **(ectoderm)** gives rise to the skin and nervous system. Simple animals, such as sponges, lack a mesoderm.

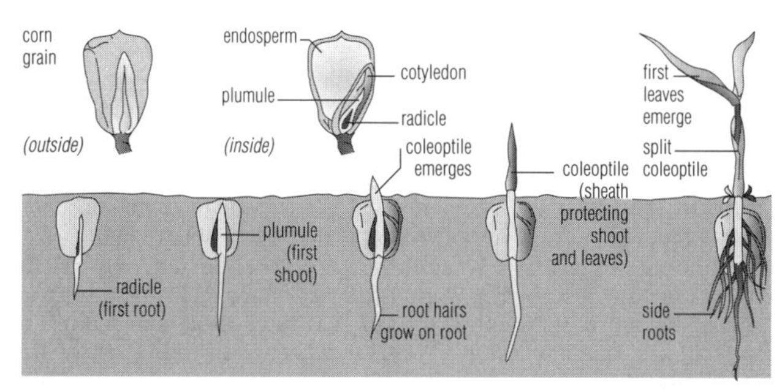

germination *The germination of a corn grain. The plumule and radicle emerge from the seed coat and begin to grow into a new plant. The coleoptile protects the emerging bud and the first leaves.*

germ-line therapy hypothetical application of ◊gene therapy to sperm and egg cells to remove the risk of an inherited disease being passed to offspring. It is controversial because of the fear it will be used to produce 'designer babies', and may result in unforseen side effects.

Gestalt *German 'form'* concept of a unified whole that is greater than, or different from, the sum of its parts; that is, a complete structure whose nature is not explained simply by analysing its constituent elements. A chair, for example, will generally be recognized as a chair despite great variations between individual chairs in such attributes as size, shape, and colour.

Gestalt psychology regards all mental phenomena as being arranged in organized, structured wholes, as opposed to being composed of simple sensations. For example, learning is seen as a reorganizing of a whole situation (often involving insight), as opposed to the behaviourists' view that it consists of associations between stimuli and responses. Gestalt psychologists' experiments show that the brain is not a passive receiver of information, but that it structures all its input in order to make sense of it, a belief that is now generally accepted; however, other principles of Gestalt psychology have received considerable criticism.

The term 'Gestalt' was first used in psychology by the Austrian philosopher and psychologist Christian von Ehrenfels in 1890. Max Wertheimer, Wolfgang Köhler, and Kurt Koffka (1886–1941) were cofounders of Gestalt psychology.

gestation in all mammals except the ◊monotremes (platypus and spiny anteaters), the period from the time of implantation of the embryo in the uterus to birth. This period varies among species; in humans it is about 266 days, in elephants 18–22 months, in cats about 60 days, and in some species of marsupial (such as opossum) as short as 12 days.

gesture recognition in computing, technique whereby a computer accepts human gestures transmitted via hardware such as a ◊DataGlove as meaningful input to which it can respond. Gesture recognition is a key technology needed in the development of ◊virtual reality systems if they are to allow humans to interact fully and naturally with objects in computerized worlds.

geyser natural spring that intermittently discharges an explosive column of steam and hot water into the air due to the build-up of steam in underground chambers. One of the most remarkable geysers is Old Faithful, in Yellowstone National Park, Wyoming, USA. Geysers also occur in New Zealand and Iceland.

g-force force that pilots and astronauts experience when their craft accelerate or decelerate rapidly. One g is the ordinary pull of gravity.

Early astronauts were subjected to launch and reentry forces of up to six g or more; in the space shuttle, more than three g is experienced on liftoff. Pilots and astronauts wear g-suits that prevent their blood pooling too much under severe g-forces, which can lead to unconsciousness.

gherkin young or small green ◊cucumber, used for pickling.

ghost gum inland Australian species of gum *Eucalyptus papuana* with a smooth white trunk; made familiar by the paintings of Albert Namatjira.

giant molecular structure or ***macromolecular structure*** solid structure made up of many similar molecules; examples include diamond, graphite, silica, and polymers.

giant star in astronomy, a class of stars to the top right of the ◊Hertzsprung– Russell diagram characterized by great size and ◊luminosity. Giants have exhausted their supply of hydrogen fuel and derive their energy from the fusion of helium and heavier elements. They are roughly 10–300 times bigger than the Sun with 30–1,000 times the luminosity. The cooler giants are known as red giants.

gibberellin plant growth substance (see also ◊auxin) that promotes stem growth and may also affect the breaking of dormancy in certain buds and seeds, and the induction of flowering. Application of gibberellin can stimulate the stems of dwarf plants to additional growth, delay the ageing process in leaves, and promote the production of seedless fruit (◊parthenocarpy).

gibbon any of a group of several small southern Asian apes. The **common** or **lar gibbon** (*H. lar*) is about 60 cm/2 ft tall, with a body that is hairy except for the buttocks, which distinguishes it from other types of apes. Gibbons have long arms and no tail. They spend most of their time in trees and are very agile when swinging from branch to branch. On the ground they walk upright, and are more easily caught by predators. (Genus *Hylobates,* including the subgenus *Symphalangus.*)

The **siamang** (*S. syndactylus*) is the largest of the gibbons, growing to 90 cm/36 in tall; it is entirely black. Gibbons are found from Assam through the Malay peninsula to Borneo, but are becoming rare, with certain species classified as endangered.

Gibbs' function in ◊thermodynamics, an expression representing part of the energy content of a system that is available to do external work, also known as the free energy G. In an equilibrium system at constant temperature and pressure, $G = H-TS$, where H is the enthalpy (heat content), T the temperature, and S the ◊entropy (decrease in energy availability). The function was named after US physicist Josiah Willard Gibbs.

gidgee small Australian tree *Acacia cambagei* which gives off an unpleasant odour at the approach of rain; also known as stinking wattle.

GIF (acronym for ***Graphics Interchange Format***) in computing, popular and economical picture file format developed by CompuServe. GIF (pronounced with a hard 'g') is one of the two most commonly used file formats for pictures on the World Wide Web (the other is ◊JPEG) because pictures saved in this format take up a relatively small amount of space. The term is often used simply to mean 'pictures'.

giga- prefix signifying multiplication by 10^9 (1,000,000,000 or 1 billion), as in **gigahertz**, a unit of frequency equivalent to 1 billion hertz.

gigabyte in computing, a measure of ◊memory capacity, equal to 1,024 ◊megabytes. It is also used, less precisely, to mean 1,000 billion ◊bytes.

Giganotosaurus carolinii carnivorous dinosaur of the ◊Cretaceous period. They weighed 6–8 tonnes and were about 12.5 m/40 ft in length, making them the largest predators ever to walk the Earth. *Giganotosaurus* lived in Patagonia about 97 million years ago. Argentinian palaeontologists discovered about 70% of its skeleton in 1995.

GIGO (acronym for ***garbage in, garbage out***) expression used in computing to emphasize that inaccurate input data will result in inaccurate output data.

GII abbreviation for the ◊Global Information Infrastructure.

gila monster lizard native to the southwestern USA and Mexico. It is one of the only two existing venomous lizards, the other being the Mexican beaded lizard of the same genus. It has poison glands in its lower jaw, but its bite is not usually fatal to humans. (Species *Heloderma suspectum.*)

gill in biology, the main respiratory organ of most fishes and immature amphibians, and of many aquatic invertebrates. In all types, water passes over the gills, and oxygen diffuses across the gill membranes into the circulatory system, while carbon dioxide passes from the system out into the water.

In aquatic insects, these gases diffuse into and out of air-filled canals called tracheae.

gill imperial unit of volume for liquid measure, equal to one-quarter of a pint or 5 fluid ounces (0.142 litre). It is used in selling alcoholic drinks.

In S England it is also called a noggin, but in N England the large noggin is used, which is two gills.

gillyflower old name for the ◊carnation and related plants, used in the works of Chaucer, Shakespeare, and Spenser.

ginger southeast Asian reedlike perennial plant; the hot-tasting spicy underground root is used as a food flavouring and in preserves. (*Zingiber officinale,* family Zingiberaceae.)

ginkgo or ***maidenhair tree*** tree belonging to the ◊gymnosperm (or naked-seed-bearing) division of plants. It may reach a height of 30 m/100 ft by the time it is 200 years old. (*Ginkgo biloba.*)

The only living member of its group (Ginkgophyta), widespread in Mesozoic times (245–65 million years ago), it has been cultivated in China and Japan since ancient times, and is planted in many parts of the world. Its leaves are fan-shaped, and it bears fleshy, yellow, foul-smelling fruit enclosing edible kernels.

ginseng plant with a thick forked aromatic root used in alternative medicine as a tonic. (*Panax ginseng,* family Araliaceae.)

Giotto space probe built by the European Space Agency to study ◊Halley's comet. Launched by an Ariane rocket in July 1985, *Giotto* passed within 600 km/375 mi of the comet's nucleus on 13 March 1986. On 2 July 1990, it flew 23,000 km/14,000 mi from Earth, which diverted its path to encounter another comet, Grigg-Skjellerup, on 10 July 1992.

giraffe world's tallest mammal. It stands over 5.5 m/18 ft tall, the neck accounting for nearly half this amount. The giraffe has two to four small, skin-covered, hornlike structures on its head and a long, tufted tail. The fur has a mottled appearance and is reddish brown and cream. Giraffes are found only in Africa, south of the Sahara Desert. They eat leaves and vegetation that is out of reach of smaller mammals, and are ruminants; that is, they chew the cud. (Species *Giraffa camelopardalis,* family Giraffidae.)

GIRAFFE

http://www.seaworld.org/animal_bytes/giraffeab.html

Illustrated guide to the giraffe including information about genus, size, life span, habitat, gestation, diet, and a series of fun facts.

GIS in computing, abbreviation for ◊geographical information system.

gizzard muscular grinding organ of the digestive tract, below the ◊crop of birds, earthworms, and some insects, and forming part of the ◊stomach. The gizzard of birds is lined with a hardened horny layer of the protein keratin, preventing damage to the muscle layer during the grinding process. Most birds swallow sharp grit which aids maceration of food in the gizzard.

glacial trough or ***U-shaped valley*** steep-sided, flat-bottomed valley formed by a glacier. The erosive action of the glacier and of the debris carried by it results in the formation not only of the trough itself but also of a number of associated features, such as truncated spurs (projections of rock that have been sheared off by the ice) and hanging valleys (smaller glacial valleys that enter the trough at a higher level than the trough floor). Features characteristic of glacial deposition, such as drumlins and eskers, are commonly found on the floor of the trough, together with linear lakes called ribbon lakes.

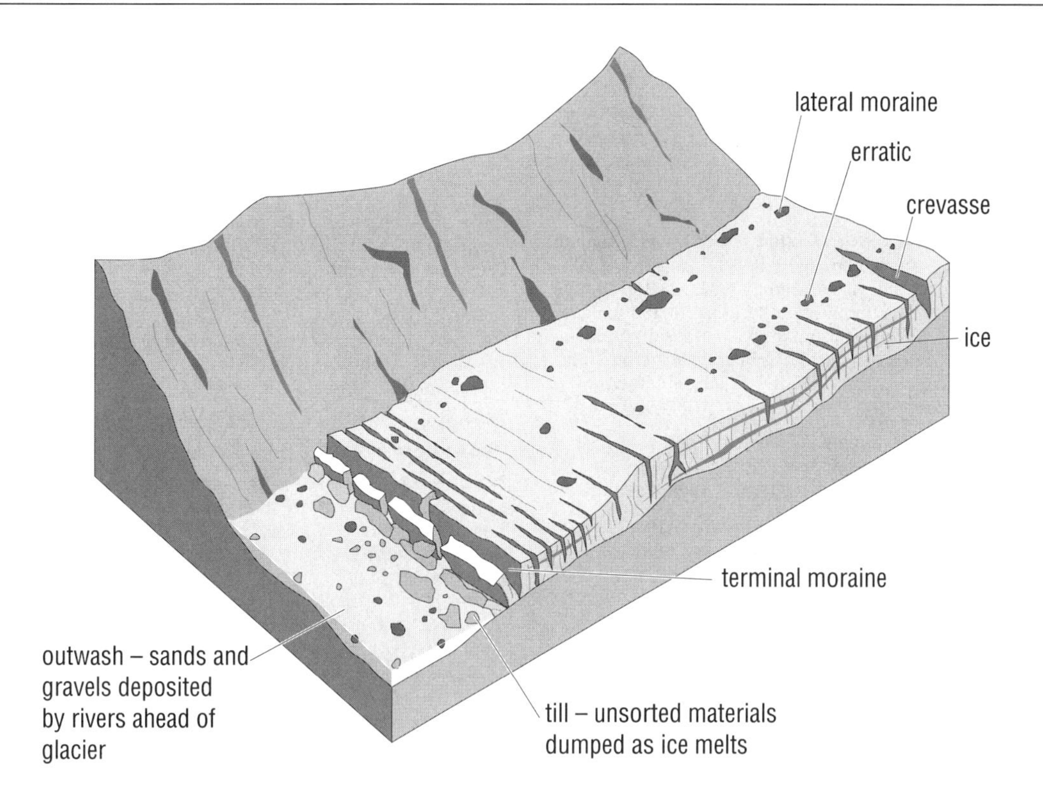

glacial deposition *A glacier picks up large boulders and rock debris from the valley and deposits them at the snout of the glacier when the ice melts. Some deposited material is carried great distances by the ice to form erratics.*

glacier tongue of ice, originating in mountains in snowfields above the snowline, which moves slowly downhill and is constantly replenished from its source. The geographic features produced by the erosive action of glaciers are characteristic and include ◊glacial troughs (U-shaped valleys), ◊corries, and ◊arêtes. In lowlands, the laying down of ◊moraine (rocky debris once carried by glaciers) produces a variety of landscape features.

Glaciers form where annual snowfall exceeds annual melting and drainage. The snow compacts to ice under the weight of the layers above.

Under pressure the ice moves plastically (changing its shape permanently). When a glacier moves over an uneven surface, deep crevasses are formed in rigid upper layers of the ice mass; if it reaches the sea or a lake, it breaks up to form icebergs. A glacier that is formed by one or several valley glaciers at the base of a mountain is called a **piedmont** glacier. A body of ice that covers a large land surface or continent, for example Greenland or Antarctica, and flows outward in all directions is called an **ice sheet**.

In Oct 1996 a volcano erupted under Europe's largest glacier, Vatnajökull in Iceland, causing flooding.

GLACIERS

http://www-nsidc.colorado.edu/NSIDC/EDUCATION/GLACIERS/

Comprehensive information about glaciers from the US National Snow and Ice Data Centre. There are explanations of why glaciers form, different kinds of glaciers, and what they may tell us about climate change. There are a number of interesting facts and a bibliography about the compacted tongues of ice which cover 10% of the land surface of our planet.

gladiolus any plant of a group of southern European and African cultivated perennials belonging to the iris family, with brightly coloured funnel-shaped flowers borne on a spike; the swordlike leaves spring from a corm (swollen underground stem). (Genus *Gladiolus,* family Iridaceae.)

gland specialized organ of the body that manufactures and secretes enzymes, hormones, or other chemicals. In animals, glands vary in size from small (for example, tear glands) to large (for example, the pancreas), but in plants they are always small, and may consist of a single cell. Some glands discharge their products internally, ◊endocrine glands, and others, ◊exocrine glands, externally. Lymph nodes are sometimes wrongly called glands.

glandular fever or ***infectious mononucleosis*** viral disease characterized at onset by fever and painfully swollen lymph nodes; there may also be digestive upset, sore throat, and skin rashes. Lassitude persists for months and even years, and recovery can be slow. It is caused by the Epstein–Barr virus.

glass transparent or translucent substance that is physically neither a solid nor a liquid. Although glass is easily shattered, it is one of the strongest substances known. It is made by fusing certain types of sand (silica); this fusion occurs naturally in volcanic glass (see ◊obsidian).

In the industrial production of common types of glass, the type of sand used, the particular chemicals added to it (for example, lead, potassium, barium), and refinements of technique determine the type of glass produced. Types of glass include: soda glass; flint glass, used in cut-crystal ware; optical glass; stained glass; heat-resistant glass; and glasses that exclude certain ranges of the light spectrum. Blown glass is either blown individually from molten glass (using a tube up to 1.5 m/4.5 ft long), as in the making of expensive crafted glass, or blown automatically into a mould – for example, in the manufacture of light bulbs and bottles; pressed glass is simply pressed into moulds, for jam jars, cheap vases, and light fittings; while sheet glass, for windows, is made by putting the molten glass through rollers to form a 'ribbon', or by floating molten glass on molten tin in the 'float glass' process; ◊fibreglass is made from fine glass fibres. Metallic glass is produced by treating alloys so that they take on the properties of glass while retaining the malleability and conductivity characteristic of metals.

glass lizard another name for ◊glass snake.

glass-reinforced plastic (GRP) a plastic material strengthened by glass fibres, sometimes erroneously called ◊fibreglass. Glass-reinforced plastic is a favoured material for boat hulls and for the bodies and some structural components of high-performance cars and aircraft; it is also used in the manufacture of passenger cars.

Products are usually moulded, mats of glass fibre being sandwiched between layers of a polyester plastic, which sets hard when mixed with a curing agent.

glass snake or ***glass lizard*** any of a worldwide group of legless lizards. Their tails are up to three times the head–body length and are easily broken off. (Genus *Ophisaurus,* family Anguidae.)

Glauber's salt crystalline sodium sulphate decahydrate $Na_2SO_4.10H_2O$, produced by the action of sulphuric acid on common salt. It melts at 87.8°F/31°C; the latent heat stored as it solidifies makes it a convenient thermal energy store. It is used in medicine as a laxative.

glaucoma condition in which pressure inside the eye (intraocular pressure) is raised abnormally as excess fluid accumulates. It occurs when the normal outflow of fluid within the chamber of the eye (aqueous humour) is interrupted. As pressure rises, the optic nerve suffers irreversible damage, leading to a reduction in the field of vision and, ultimately, loss of eyesight.

The most common type, **chronic glaucoma**, usually affects people over the age of 40, when the trabecular meshwork (the filtering tissue at the margins of the eye) gradually becomes blocked and drainage slows down. The condition cannot be cured, but, in many cases, it is controlled by drug therapy. Laser treatment to the trabecular meshwork often improves drainage for a time; surgery to create an artificial channel for fluid to leave the eye offers more long-term relief. A tiny window may be cut in the iris during the same operation.

Acute glaucoma is a medical emergency. A precipitous rise in pressure occurs when the trabecular meshwork suddenly becomes occluded (blocked). This is treated surgically to remove the cause of the obstruction. Acute glaucoma is extremely painful. Treatment is required urgently since damage to the optic nerve begins within hours of onset.

GLAUCOMA RESEARCH FOUNDATION

http://www.glaucoma.org/

Excellent source of well-presented information on glaucoma. There is an informative list of frequently asked questions about the disease and the latest research findings. There are links to a number of optometric and opthalmological organizations and a search engine.

Global Information Infrastructure (GII) in computing, planned worldwide high-bandwidth network. US vice president Al Gore proposed the GII in a 1994 speech to the International Telecommunications Union, saying that it would promote the functioning of democracy, help nations to cooperate with each other, and be the key to economic growth for national and international economies.

Global Network Navigator (GNN) in computing, subscription-based online service for the World Wide Web, pioneered by US book publishers O'Reilly & Associates and bought by ◊America Online in 1995. Its Virtual Places software, released in 1996, allows users to interact with each other using avatars and live messages at any Virtual Places-enabled site on the Web. GNN also offers news and resource listings.

global positioning system (GPS) US satellite-based navigation system, a network of 24 satellites in six orbits, each circling the Earth once every 24 hours. Each satellite sends out a continuous time signal, plus an identifying signal. To fix position, a user needs to be within range of four satellites, one to provide a reference signal and three to provide directional bearings. The user's receiver can then calculate the position from the difference in time between receiving the signals from each satellite.

The position of the receiver can be calculated to better than 0.5 m/1.6 ft, although only the US military can tap the full potential of the system. Other users can obtain a position to within 100 m/330 ft. This is accurate enough to be of use to boats, walkers, and motorists, and suitable receivers are on the market.

global variable in computing, a ◊variable that can be accessed by any program instruction. See also ◊local variable.

global warming an increase in average global temperature of approximately 1°F/0.5°C over the past century. Global temperature has been highly variable in Earth history and many fluctuations in global temperature have occurred in historical times, but this most recent episode of warming coincides with the spread of industralization, prompting the hypothesis that it is the result of an accelerated ◊greenhouse effect caused by atmospheric pollutants, especially carbon dioxide gas. Recent melting and collapse of the Larsen Ice Shelf, Antarctica, is a consequence of global warming. Melting of ice is expected to raise sea level in the coming decades.

Natural, perhaps chaotic, climatic variations have not been ruled out as the cause of the current global rise in temperature, but scientists are still assessing the likely influence of anthropogenic (human-made) pollutants. Assessing the impact of humankind on global climate is complicated by the natural variability on both geological and human time scales. The present episode of global warming has thus far still left England approximately 1°C cooler than during the peak of the so-called Medieval Warm Period (1000 to 1400 AD). The latter was part of a purely natural climatic fluctuation on a global scale. With respect to historical times, the interval between the Medieval Warm Period and the rise in temperatures we see today was unusually cold throughout the world.

In addition to a rise in average global temperature, global warming as caused seasonal variations to be more pronounced in recent decades. Examples are the most severe winter on record in the eastern US in 1976–77 and the record heat waves in the Netherlands and Denmark the following year.

A 1995 United Nations summit in Berlin agreed to take action to reduce gas emissions harmful to the environment. Delegates at the summit, from more than 120 countries, approved a two-year negotiating process aimed at setting specific targets and timetables for reducing nations' emissions of carbon dioxide and other greenhouse gases after the year 2000. The Kyoto Protocol of 1997 commits the world's industrialized countries to cut their annual emissions of harmful gases by 5.2% by 2012.

GLOBAL WARMING

http://pooh.chem.wm.edu/chemWWW/courses/chem105/projects/group1/page1.html

Interesting step-by-step explanation of the chemistry behind global warming. There is information on the causes of global warming, the environmental effects, and the social and economic consequences. The views of those who challenge the assertion that the world is warming up are also presented. The graphics accompanying the site are attractive and easy to follow.

globefish another name for the ◊puffer fish.

Global Positioning System

BY EDWARD YOUNG

Satellites

The Global Positioning System, or GPS, has become indispensable as a tool for navigation and scientific discovery. It relies on 24 ◊satellites that can be accessed for accurate positioning by users all over the world. GPS is funded and controlled by the United States Department of Defense. Essential to the satellites are on-board ◊atomic clocks accurate to 3 billionths of one second.

A Master Control site at Falcon Air Force Base in Colorado Springs, USA, sends data to each satellite. Ephemeris data consists of satellite positions. Almanac data consists of the projected orbits of each satellite and information about their overall status. Clock corrections are also transmitted as necessary. In turn the satellites routinely send signals to five monitor stations so that their positions can be tracked. The monitor stations also 'downlink' the Almanac information sent to the satellites from the Master Control site.

GPS satellites transmit two ◊microwave carrier signals. The first high-frequency signal carries position and time data for civilian users worldwide. The US Department of Defense to limit horizontal and vertical accuracy to 100 and 156 meters respectively degrades this signal. A second lower-frequency signal can be used to correct for the effects of the ◊ionosphere for more precise positioning by authorized users, including allied military, government agencies and approved civilian users.

Receivers

GPS receivers search for the coded high-frequency radio signals transmitted by the satellites. Once located, complete transmission from a satellite takes 12.5 minutes. The signal includes the position of the satellite and time at which the signal was sent. Transit time of the satellite radio signal gives the distance, or range, of the satellite from the receiver because the speed of electromagnetic radiation (◊radio waves) is known. Time passes more quickly for the satellites than on Earth, as predicted by Einstein's theory of general ◊relativity, and so 'relativistic' corrections to the transit time must be made. Signals from four satellites are required for precise positioning and time measurement.

Applications

GPS receivers are used or navigation of ships, motor vehicles, aircraft, space vehicles, and persons on the ground. In addition, the precise time information and radio signals that carry the information are useful for scientific experiments.

◊Geodetics is one example of a scientific discipline benefiting from GPS technology. The relative motions of Earth's ◊tectonic plates cause ◊mountains and ◊earthquakes. Plate tectonic motions are in the order of millimetres per year. The small displacements over long time periods makes measuring tectonic motions by traditional geodetic methods laborious and time consuming. Scientists are now using signals by GPS satellites to track Earth's tectonic plates and to establish how they deform to form mountains, depressions, and earthquakes as they move. Measurements that might have taken months or years to perform by traditional geodetic methods can be obtained in just a few days with greater accuracy using GPS. In this application, referred to as space geodesy, the characteristics of the satellite radio waves themselves are used rather than the data they transmit. Slight differences in the phase, or wave-crest positions, of GPS radio waves detected by two receiver stations can be converted into the distance between the stations. In this way distances between points can be measured with millimeter accuracy if receivers are within approximately 30 kilometers of one another.

As part of one space geodesy study, changing ◊topography and relative motions measured by GPS reveal where strain is building in the tectonically active Aegean region. This information is helping scientists to identify potential earthquake hazards. In a similar study, GPS is being used to establish the ever-changing elevation and position of the worlds highest peak, Mount Everest.

globular cluster spherical or near-spherical ◊star cluster containing from approximately 10,000 to millions of stars. More than a hundred globular clusters are distributed in a spherical halo around our Galaxy. They consist of old stars, formed early in the Galaxy's history. Globular clusters are also found around other galaxies.

glomerulus in the kidney, the cluster of blood capillaries at the threshold of the renal tubule, or nephron, responsible for filtering out the fluid that passes down the tubules and ultimately becomes urine. In the human kidney there are approximately one million tubules, each possessing its own glomerulus.

The structure of the glomerulus allows a wide range of substances including amino acids and sugar, as well as a large volume of water, to pass out of the blood. As the fluid moves through the tubules, most of the water and all of the sugars are reabsorbed, so that only waste remains, dissolved in a relatively small amount of water. This fluid collects in the bladder as urine.

glottis in medicine, narrow opening at the upper end of the larynx that contains the vocal cords.

glove box in high technology, a protective device used when handling toxic, radioactive, or sterile materials within an enclosure containing a window for viewing. Gloves fixed to ports in the walls of a box allow manipulation of objects within the box. The risk that the operator might inhale fine airborne particles of poisonous materials is removed by maintaining a vacuum inside the box, so that any airflow is inwards.

glow-worm wingless female of any of a large number of luminous beetles (fireflies). The luminous organs, situated under the abdomen, at the end of the body, give off a greenish glow at night and attract winged males for mating. There are about 2,000 species of glow-worms, distributed worldwide. (Family Lampyridae.)

glucagon in biology, a hormone secreted by the alpha cells of the islets of Langerhans in the ◊pancreas, which increases the concentration of glucose in the blood by promoting the breakdown of glycogen in the liver. Secretion occurs in response to a lowering of blood glucose concentrations.

Glucagon injections can be issued to close relatives of patients with ◊diabetes who are being treated with insulin. Hypoglycaemia may develop in such patients in the event of inadequate control of diabetes. An injection of glucagon can be used to reverse hypoglycaemia before serious symptoms, such as unconsciousness, develop.

glucose or ***dextrose*** or ***grape sugar*** $C_6H_{12}0_6$ sugar present in the blood and manufactured by green plants during ◊photosynthesis. The ◊respiration reactions inside cells involves the oxidation of glucose to produce ◊ATP, the 'energy molecule' used to drive many of the body's biochemical reactions.

In humans and other vertebrates optimum blood glucose levels are maintained by the hormone ◊insulin.

GLOW-WORM

The simplest way to tell the sex of a glow-worm is by measuring the rate at which its emits light pulses. A male flashes once every 5.8 seconds, a female every 2.1 seconds.

Glucose is prepared in syrup form by the hydrolysis of cane sugar or starch, and may be purified to a white crystalline powder. Glucose is a monosaccharide sugar (made up of a single sugar unit), unlike the more familiar sucrose (cane or beet sugar), which is a disaccharide (made up of two sugar units: glucose and fructose).

glue type of ◊adhesive.

glue ear or ***secretory otitis media*** condition commonly affecting small children, in which the Eustachian tube, which normally drains and ventilates the middle ear, becomes blocked with mucus. The resulting accumulation of mucus in the middle ◊ear muffles hearing. It is the leading cause of deafness (usually transient) in children.

Glue ear resolves spontaneously after some months, but because the loss of hearing can interfere with a child's schooling the condition is often treated by a drainage procedure (myringotomy) and the surgical insertion of a small ventilating tube, or **grommet**, into the eardrum (tympanic membrane). This allows air to enter the middle ear, thereby enabling the mucus to drain freely once more along the Eustachian tube and into the back of the throat. The grommet is gradually extruded from the eardrum over several months, and the eardrum then heals naturally.

glue-sniffing or ***solvent misuse*** inhalation of the fumes from organic solvents of the type found in paints, lighter fuel, and glue, for their hallucinatory effects. As well as being addictive, solvents are dangerous for their effects on the user's liver, heart, and lungs. It is believed that solvents produce hallucinations by dissolving the cell membrane of brain cells, thus altering the way the cells conduct electrical impulses.

I am now convinced that theoretical physics is actual philosophy.

MAX BORN German-born British physicist.
Autobiography

---

gluon in physics, a ◊gauge boson that carries the ◊strong nuclear force, responsible for binding quarks together to form the strongly interacting subatomic particles known as ◊hadrons. There are eight kinds of gluon.

Gluons cannot exist in isolation; they are believed to exist in balls ('glueballs') that behave as single particles.

Glueballs may have been detected at CERN in 1995 but further research is required to confirm their existence.

gluten protein found in cereal grains, especially wheat and rye. Gluten enables dough to expand during rising. Sensitivity to gliadin, a type of gluten, gives rise to ◊coeliac disease.

glyceride ◊ester formed between one or more acids and glycerol (propan-1,2,3-triol). A glyceride is termed a mono-, di-, or triglyceride, depending on the number of hydroxyl groups from the glycerol that have reacted with the acids.

Glycerides, chiefly triglycerides, occur naturally as esters of ◊fatty acids in plant oils and animal fats.

glycerine another name for ◊glycerol.

glycerol or ***glycerine*** or ***propan-1,2,3-triol*** $HOCH_2CH(OH)CH_2OH$ thick, colourless, odourless, sweetish liquid. It is obtained from vegetable and animal oils and fats (by treatment with acid, alkali, superheated steam, or an enzyme), or by fermentation of glucose, and is used in the manufacture of high explosives, in antifreeze solutions, to maintain moist conditions in fruits and tobacco, and in cosmetics.

glycine $CH_2(NH_2)COOH$ the simplest amino acid, and one of the main components of proteins. When purified, it is a sweet, colourless crystalline compound.

Glycine was found in 1994 in the star-forming region Sagittarius B2. The discovery is important because of its bearing on the origins of life on Earth.

glycogen polymer (a polysaccharide) of the sugar ◊glucose made and retained in the liver as a carbohydrate store, for which reason it is sometimes called animal starch. It is a source of energy when needed by muscles, where it is converted back into glucose by the hormone ◊insulin and metabolized.

glycol or **ethylene glycol** or **ethane-1,2-diol** $HOCH_2CH_2OH$ thick, colourless, odourless, sweetish liquid. It is used in antifreeze solutions, in the preparation of ethers and esters (used for explosives), as a solvent, and as a substitute for glycerol.

glycoside in biology, compound containing a sugar and a non-sugar unit. Many glycosides occur naturally, for example, ◊digitalis is a preparation of dried and powdered foxglove leaves that contains a mixture of cardiac glycosides. One of its constituents, digoxin, is used in the treatment of congestive heart failure and cardiac arrhythmias.

GM synthesizer in computing, synthesizer standard; see ◊general MIDI.

GMT abbreviation for ◊Greenwich Mean Time.

gnat any of a group of small two-winged biting insects belonging to the mosquito family. The eggs are laid in water, where they hatch into wormlike larvae, which pass through a pupal stage (see ◊pupa) to emerge as adults. (Family Culicidae.)

Species include *Culex pipiens,* abundant in England; the carrier of malaria *Anopheles maculipennis;* and the banded mosquito *Aedes aegypti,* which transmits yellow fever.

Only the female is capable of drawing blood; the male does not have piercing jaws.

gneiss coarse-grained ◊metamorphic rock, formed under conditions of high temperature and pressure, and often occurring in association with schists and granites. It has a foliated, or layered, structure consisting of thin bands of micas and/or amphiboles dark in colour alternating with bands of granular quartz and feldspar that are light in colour. Gneisses are formed during regional ◊metamorphism; **paragneisses** are derived from metamorphism of sedimentary rocks and **orthogneisses** from metamorphism of granite or similar igneous rocks.

GNN in computing, abbreviation for ◊Global Network Navigator.

gnu another name for ◊wildebeest.

GNU in computing, suite of free UNIX-like software distributed by the ◊Free Software Foundation. The software includes operating systems, compilers, text editors (such as EMACS), and other useful utilities.

goat ruminant mammal (it chews the cud), closely related to sheep. Both male and female goats have horns and beards. They are sure-footed animals, and feed on shoots and leaves more than on grass. (Genus *Capra,* family Bovidae.)

Domestic varieties are descended from the **scimitar-horned wild goat** (*C. aegagrus*) and have been kept for over 9,000 years in southern Europe and Asia. They are kept for milk or for mohair (angora and cashmere goats). Wild species include the **ibex** (*C. ibex*) of the Alps and **markhor** (*C. falconeri*) of the Himalayas, 1 m/3 ft high and with long twisted horns. The **Rocky Mountain goat** (*Oreamnos americanus*) is a 'goat antelope' and is not closely related to true goats.

goat moth large yellowish-grey or brown moth with irregular markings of white and black on the upper wings and a wingspan of about 7–8 cm/2.8–3 in. It is common in Europe and the Middle East.

When the moth is frightened it emits a disagreeable odour like that of a male goat hence its name.

classification The goat moth *Cossus ligniperda* is in order Lepidoptera, class Insecta, phylum Arthropoda.

goblet cell in biology, cup-shaped cell present in the epithelium of the respiratory and gastrointestinal tracts. Goblet cells secrete mucin, the main constituent of mucous, which lubricates the mucous membranes of these tracts.

goby small marine bony fish. Nearly all gobies are found in the shallow coastal waters of the temperate and tropical oceans.

The first dorsal fin consists of a few flexible spines and the second dorsal fin is opposed to the anal fin. The caudal fin is generally rounded, with the pelvic fins united to form a cup-shaped sucker.

classification Gobies are in the family Gobiidae, order Perciformes, class Osteichthyes.

ABOUT GODDARD SPACE FLIGHT CENTRE

http://pao.gsfc.nasa.gov/gsfc/welcome/history/history.htm

Primarily composed of a biography and achievement history of the American rocket pioneer, the web site also describes the Goddard Space Flight Centre and details the type of mission they are responsible for undertaking. As part of NASA's huge Internet resources there are of course many links to related topics.

Goddard Space Flight Center NASA installation at Greenbelt, Maryland, USA, responsible for the operation of NASA's unmanned scientific satellites, including the ◊Hubble Space Telescope. It is also home of the National Space Science Data centre, a repository of data collected by satellites.

goitre enlargement of the thyroid gland seen as a swelling on the neck. It is most pronounced in simple goitre, which is caused by iodine deficiency. More common is toxic goitre or ◊hyperthyroidism, caused by overactivity of the thyroid gland.

According to a World Health Organization's survey of 1997 up to 60% of people in some Indian states were suffering with goitre.

gold heavy, precious, yellow, metallic element; symbol Au, atomic number 79, relative atomic mass 197.0. It is unaffected by temperature changes and is highly resistant to acids. For manufacture, gold is alloyed with another strengthening metal (such as copper or silver), its purity being measured in ◊carats on a scale of 24.

In 1990 the three leading gold-producing countries were South Africa, 605.4 tonnes; USA, 295 tonnes; and Russia, 260 tonnes. In 1989 gold deposits were found in Greenland with an estimated yield of 12 tonnes per year.

Gold occurs naturally in veins, but following erosion it can be transported and redeposited. It has long been valued for its durability, malleability, and ductility, and its uses include dentistry and jewellery. As it will not corrode, it is also used in the manufacture of electric contacts for computers and other electrical devices.

goldcrest smallest European bird, about 9 cm/3.5 in long and weighing 5 g/0.011 lb; a ◊warbler. It is olive green, with a bright orange-yellow streak running from the beak to the back of the head and a black border above the eye. The tail is brown, marked with black and white, and the cheeks, throat, and breast are a greyish white. (Species *Regulus regulus*, family Muscicapidae, order Passeriformes.)

The goldcrest builds its nest in conifers. It is found all over Europe, particularly frequenting fir woods. In winter it can be found feeding with tit flocks, moving through deciduous woodlands.

golden-eye fly alternative name for green ◊lacewing.

golden retriever breed of dog. See ◊retriever.

goldenrod one of several tall and leafy North American perennial plants, belonging to the daisy family. Flower heads are mostly composed of many small yellow flowers, or florets. (Genus *Solidago*, family Compositae.)

goldenseal North American plant *Hydrastis canadensis* of the buttercup family whose thick yellow root is used medicinally as an astringent and a tonic. The root contains the alkaloid hydrastine, employed by herbalists to stop uterine bleeding.

golden section visually satisfying ratio, first constructed by the Greek mathematician Euclid and used in art and architecture. It is found by dividing a line AB at a point O such that the rectangle produced by the whole line and one of the segments is equal to the square drawn on the other segment. The ratio of the two segments is about 8:13 or 1:1.618, and a rectangle whose sides are in this ratio is called a **golden rectangle**. The ratio of consecutive Fibonacci numbers tends to the golden ratio.

In van Gogh's picture *Mother and Child*, for example, the Madonna's face fits perfectly into a golden rectangle.

goldfinch songbird found in Eurasia, North Africa, and North America. (Species *Carduelis carduelis*, family Fringillidae, order Passeriformes.)

goldfish fish belonging to the ◊carp family, found in East Asia. It is greenish-brown in its natural state, but has been bred by the Chinese for centuries, taking on highly coloured and sometimes freakishly shaped forms. Goldfish can occur in a greater range of colours than any other animal tested. (Species *Carassius auratus*, family Cyprinidae.)

Golgi apparatus or ***Golgi body*** stack of flattened membranous sacs found in the cells of ◊eukaryotes. Many molecules travel through the Golgi apparatus on their way to other organelles or to the endoplasmic reticulum. Some are modified or assembled inside the sacs. The Golgi apparatus is named after the Italian physician Camillo Golgi.

Goliath beetle large beetle found only in tropical countries. The biggest Goliath beetle *Goliathus giganteus*, found in equatorial Africa, may be more than 150 mm/6 in long and is one of the largest insects.

They lay their eggs in the rotting wood of trees, and most adults feed on the tender floral parts or suck the sap exuded from trees. Its 'brushed velvet' wingcases are maroon and the pronotum (shield) is black with roughly longitudinal white stripes.

classification Goliath beetles belong to the subfamily Cetoniinae of the family Scarabaeidae in order Coleoptera, class Insecta, phylum Arthropoda.

gonad the part of an animal's body that produces the sperm or egg cells (ova) required for sexual reproduction. The sperm-producing gonad is called a ◊testis, and the egg-producing gonad is called an ◊ovary.

gonadotrophin any hormone that supports and stimulates the function of the gonads (sex glands); some gonadotrophins are used as fertility drugs.

Gondwanaland or ***Gondwana*** southern landmass formed 200 million years ago by the splitting of the single world continent ◊Pangaea. (The northern landmass was ◊Laurasia.) It later fragmented into the continents of South America, Africa, Australia, and Antarctica, which then drifted slowly to their present positions. The baobab tree found in both Africa and Australia is a relic of this ancient land mass.

A database of the entire geology of Gondwanaland has been constructed by geologists in South Africa. The database, known as

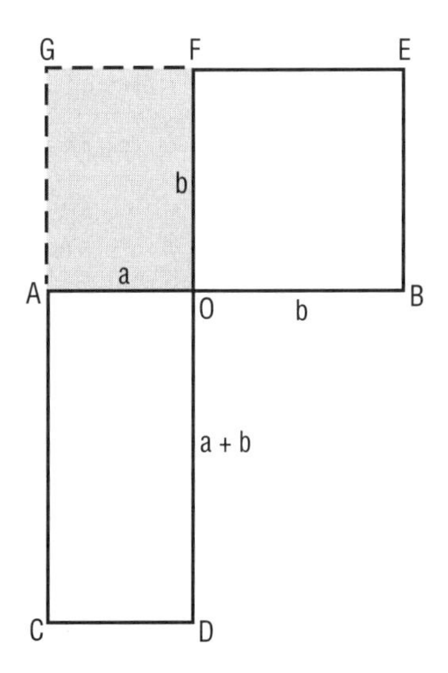

golden section *The golden section is the ratio a:b, equal to 8:13. A golden rectangle is one, like that shaded in the picture, that has its length and breadth in this ratio. These rectangles are said to be pleasant to look at and have been used instinctively by artists in their pictures.*

Golgi, Camillo (1843–1926)

Mary Evans Picture Library

Italian cell biologist who produced the first detailed knowledge of the fine structure of the nervous system. He shared the 1906 Nobel Prize for Physiology or Medicine with Santiago Ramón y Cajal, who followed up Golgi's work.

Golgi's use of silver salts in staining cells proved so effective in showing up the components and fine processes of nerve cells that even the synapses – tiny gaps between the cells – were visible. The Golgi apparatus, a series of flattened membranous cavities found in the cytoplasm of cells, was first described by him in 1898.

Gondwana Geoscientific Indexing Database (GO-GEOID), displays information as a map of Gondwana 155 million years ago, before the continents drifted apart.

gonorrhoea common sexually transmitted disease arising from infection with the bacterium *Neisseria gonorrhoeae,* which causes inflammation of the genito-urinary tract. After an incubation period of two to ten days, infected men experience pain while urinating and a discharge from the penis; infected women often have no external symptoms.

Untreated gonorrhoea carries the threat of sterility to both sexes; there is also the risk of blindness in a baby born to an infected mother. The condition is treated with antibiotics, though ever-increasing doses are becoming necessary to combat resistant strains.

Good King Henry perennial plant belonging to the goosefoot family, growing to 50 cm/1.6 ft, with triangular leaves which are mealy when young. Spikes of tiny greenish-yellow flowers appear above the leaves in midsummer. (*Chenopodium bonus-henricus,* family Chenopodiaceae.)

Goonhilly British Telecom satellite-tracking station in Cornwall, England. It is equipped with a communications-satellite transmitter–receiver in permanent contact with most parts of the world.

goose any of several large aquatic birds belonging to the same family as ducks and swans. There are about 12 species, found in North America, Greenland, Europe, North Africa, and Asia north of the Himalayas. Both sexes are similar in appearance: they have short, webbed feet, placed nearer the front of the body than in other members of the family, and a slightly hooked beak. Geese feed entirely on grass and plants, build nests of grass and twigs on the ground, and lay 5–9 eggs, white or cream-coloured, according to the species. (Genera mainly *Anser* and *Branta,* family Anatidae, order Anseriformes.)

The **barnacle goose** (*B. leucopsis*) is about 60 cm/2 ft long and weighs about 2 kg/4.5 lb. It is black and white, marbled with blue and grey, and the beak is black. The **bean goose** (*A. fabalis*) is a grey species of European wild goose with an orange or yellow and black beak. It breeds in northern Europe and Siberia. The **Brent goose** (*B. bernicla*) is a small goose, black or brown, white, and grey in colour. It is almost completely herbivorous, feeding on eel grass and algae. The world population of Brent geese was 25,000 in 1996. The **greylag goose** (*A. anser*) is the ancestor of domesticated geese.

Other species include the **Canada goose** (*B. canadensis*) (common to North America and introduced into Europe in the 18th century), the **pink-footed goose** (*A. brachyrhynchus*), the **white-fronted goose** (*A. albifrons*), and the **ne-ne** or **Hawaiian goose** (*B. sandvicensis*).

gooseberry edible fruit of a low-growing bush (*Ribes uva-crispa*) found in Europe and Asia, related to the ◊currant. It is straggling in its growth, and has straight sharp spines in groups of three and rounded, lobed leaves. The flowers are green and hang on short stalks. The sharp-tasting fruits are round, hairy, and generally green, but there are reddish and white varieties.

goosefoot any of a group of plants belonging to the goosefoot family, closely related to spinach and beets. The seeds of white goosefoot (*C. album*) were used as food in Europe from Neolithic times, and also from early times in the Americas. White goosefoot grows to 1 m/3 ft tall and has lance- or diamond-shaped leaves and packed heads of small inconspicuous flowers. The green part is eaten as a spinach substitute. (Genus *Chenopodium,* family Chenopodiaceae.)

gopher any of a group of burrowing rodents. Gophers are a kind of ground squirrel represented by some 20 species distributed across western North America, Europe, and Asia. Length ranges from 15 cm/6 in to 90 cm/16 in, excluding the furry tail; colouring ranges from plain yellowish to striped and spotted species. (Genus *Citellus,* family Sciuridae.)

The name **pocket gopher** is applied to the eight genera of the North American family Geomyidae.

Gopher (derived from ***go for***; alternatively, named for the mascot of the University of Minnesota, where it was invented) menu-based server on the ◊Internet that indexes resources and retrieves them according to user choice via any one of several built-in methods such as ◊FTP or ◊Telnet. Gopher servers can also be accessed via the World Wide Web and searched via special servers called ◊Veronica.

Gopherspace in computing, name for the knowledge base composed of all the documents indexed on all the ◊Gophers in the world.

gopher tortoise land tortoise occurring in the southern USA. It has a domed shell and scaly legs. Its forelegs are flattened for digging the burrow where it lives. Gopher tortoises reach lengths of up to 37 cm/14.5 in.

classification The gopher tortoise *Gopherus polyphemus* is in family Cheloniidae, order Testudinae, class Reptilia.

Gordon setter breed of dog. See ◊setter.

gorge narrow steep-sided valley (or canyon) that may or may not have a river at the bottom. A gorge may be formed as a ◊waterfall retreats upstream, eroding away the rock at the base of a river valley; or it may be caused by rejuvenation, when a river begins to cut downwards into its channel once again (for example, in response to a fall in sea level). Gorges are common in limestone country, where they may be formed by the collapse of the roofs of underground caverns.

gorilla largest of the apes, found in the dense forests of West Africa and mountains of central Africa. The male stands about 1.8 m/6 ft high and weighs about 200 kg/450 lbs. Females are about half this size. The body is covered with blackish hair, silvered on the back in older males. Gorillas live in family groups; they are vegetarian, highly intelligent, and will attack only in self-defence. They are dwindling in numbers, being shot for food by some local people, or by poachers taking young for zoos, but protective measures

GORILLA

http://www.seaworld.org/animal_bytes/gorillaab.html

Illustrated guide to the gorilla including information about genus, size, life span, habitat, gestation, diet, and a series of fun facts.

are having some effect. (Species *Gorilla gorilla*.)

Gorillas construct stoutly built nests in trees for overnight use. The breast-beating movement, once thought to indicate rage, actually signifies only nervous excitement. There are three races – western lowland, eastern lowland, and mountain gorillas – and US scientists suggested in 1994 that there may be two separate species of gorilla.

gorse or ***furze*** or ***whin*** any of a group of plants native to Europe and Asia, consisting of thorny shrubs with spine-shaped leaves growing thickly along the stems and bright-yellow coconut-scented flowers. (Genus *Ulex*, family Leguminosae.)

GOSHAWK

The male goshawk mates up to 600 times with his partner for every clutch of eggs. This ensures that the sperm of any rival are completely swamped.

goshawk or ***northern goshawk*** woodland hawk similar in appearance to the peregrine falcon, but with shorter wings and legs. It is native to most of Europe, Asia, and North America, and is used in falconry. The male is much smaller than the female. It is ash grey on the upper part of the body and whitish underneath with brown horizontal stripes; it has a dark head and cheeks with a white stripe above the eye. The tail has dark bands across it. (Species *Accipiter gentilis*, order Falconiformes.)

Gossamer Albatross the first human-powered aircraft to fly across the English Channel, in June 1979. It was designed by Paul MacCready and piloted and pedalled by Bryan Allen. The Channel crossing took 2 hours 49 minutes. The same team was behind the first successful human-powered aircraft (*Gossamer Condor*) two years earlier.

Gouraud shading in computer animation, technique for calculating the correct colours and intensity of lighting playing on an on-screen three-dimensional object.

Gouraud shading works by measuring the colour and brightness at the vertices of the polygons that make up the object and mixing these to get values for the areas inside the polygons. Specialized hardware makes this process relatively fast. The technique is named after its inventor, Henri Gouraud and was developed 1973.

gourd any of a group of plants that includes melons and pumpkins. In a narrower sense, the name applies only to the genus *Lagenaria*, of which the bottle gourd or ◊calabash (*L. siceraria*) is best known. (Family Cucurbitaceae.)

gout hereditary form of ◊arthritis, marked by an excess of uric acid crystals in the tissues, causing pain and inflammation in one or more joints (usually of the feet or hands). Acute attacks are treated with anti-inflammatories.

The disease, ten times more common in men, poses a long-term threat to the blood vessels and the kidneys, so ongoing treatment may be needed to minimize the levels of uric acid in the bloodstream. It is aggravated by heavy drinking.

governor in engineering, any device that controls the speed of a machine or engine, usually by regulating the intake of fuel or steam.

Scottish inventor James ◊Watt invented the steam-engine governor in 1788. It works by means of heavy balls, which rotate on the end of linkages and move in or out because of ◊centrifugal force according to the speed of rotation. The movement of the balls closes or opens the steam valve to the engine. When the engine speed increases too much, the balls fly out, and cause the steam valve to close, so the engine slows down. The opposite happens when the engine speed drops too much.

GP in medicine, abbreviation for ***general practitioner***.

GPF (abbreviation for ***general protection fault***) in Windows 3.1, error message returned by a computer when it crashes. A GPF is the same as a UAE (unexpected application error) in Windows 3.0. It often indicates that one application has tried to use memory reserved for another.

Graafian follicle fluid-filled capsule that surrounds and protects the developing egg cell inside the ovary during the ◊menstrual cycle. After the egg cell has been released, the follicle remains and is known as a corpus luteum.

gradient on a graph, the slope of a straight or curved line. The slope of a curve at any given point is represented by the slope of the ◊tangent at that point.

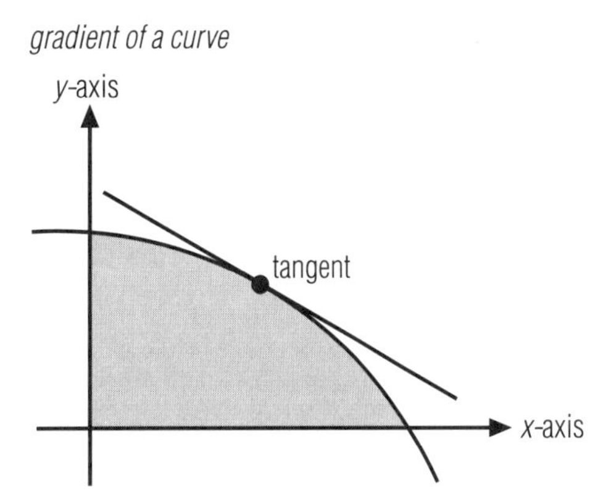

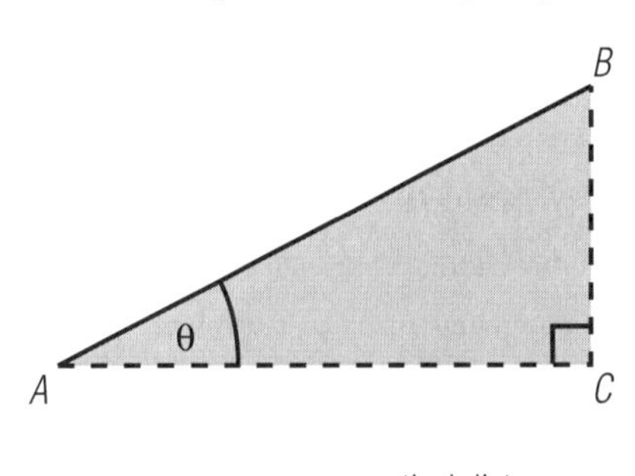

$$\text{gradient of } AB = \frac{\text{vertical distance}}{\text{horizontal distance}}$$

$$\frac{BC}{AC} = \tan\theta$$

***gradient** The gradient of a curve keeps changing, so in order to calculate the gradient you have to draw a straight line that touches a point on the curve (the tangent). The gradient for that point on the curve will then be the same as the gradient of the straight line.*

grafting in medicine, the operation by which an organ or other living tissue is removed from one organism and transplanted into the same or a different organism.

In horticulture, it is a technique widely used for propagating plants, especially woody species. A bud or shoot on one plant, termed the **scion**, is inserted into another, the **stock**, so that they continue growing together, the tissues combining at the point of union. In this way some of the advantages of both plants are obtained.

Grafting is usually only successful between species that are closely related and is most commonly practised on roses and fruit trees. The grafting of nonwoody species is more difficult but it is sometimes used to propagate tomatoes and cacti. See also ◊transplant.

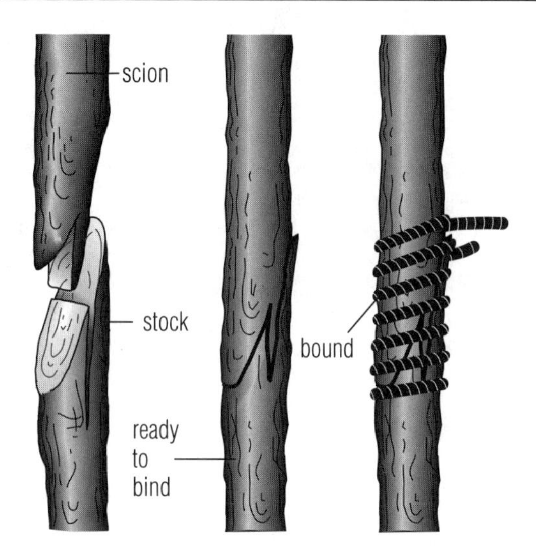

grafting *Grafting, a method of artificial propagation in plants, is commonly used in the propagation of roses and fruit trees. A relatively small part, the scion, of one plant is attached to another plant so that growth continues. The plant receiving the transplanted material is called the stock.*

grain the smallest unit of mass in the three English systems (avoirdupois, troy, and apothecaries' weights) used in the UK and USA, equal to 0.0648 g. It was reputedly the weight of a grain of wheat. One pound avoirdupois equals 7,000 grains; one pound troy or apothecaries' weight equals 5,760 grains.

gram metric unit of mass; one-thousandth of a kilogram.

gramophone old-fashioned name for a record player or stereo. It was developed from US inventor Thomas Edison's **phonograph**, which remains the traditional US name.

grampus common name for Risso's dolphin, a slate-grey dolphin found in tropical and temperate seas. These dolphins live in large schools and can reach 4 m/13 ft in length. They have blunt snouts with only a few teeth, and feed on squid and small fish. The name grampus is sometimes also used for the killer ◊whale. (Species *Grampus griseus.*)

Grande Dixence dam the world's highest dam, located in Switzerland, which measures 285 m/935 ft from base to crest. Completed in 1961, it contains 6 million cu m/8 million cu yd of concrete.

grand unified theory in physics, a sought-for theory that would combine the theory of the strong nuclear force (called ◊quantum chromodynamics) with the theory of the weak nuclear and electromagnetic forces. The search for the grand unified theory is part of a larger programme seeking a unified field theory, which would combine all the forces of nature (including gravity) within one framework.

granite coarse-grained intrusive ◊igneous rock, typically consisting of the minerals quartz, feldspar, and biotite mica. It may be pink or grey, depending on the composition of the feldspar. Granites are chiefly used as building materials.

Granites often form large intrusions in the core of mountain ranges, and they are usually surrounded by zones of ◊metamorphic rock (rock that has been altered by heat or pressure). Granite areas have characteristic moorland scenery. In exposed areas the bedrock may be weathered along joints and cracks to produce a tor, consisting of rounded blocks that appear to have been stacked upon one another.

grape fruit of any grape ◊vine, especially *V. vinifera.* (Genus *Vitis,* family Vitaceae.)

grapefruit round, yellow, juicy, sharp-tasting fruit of the evergreen grapefruit tree. The tree grows up to 10 m/more than 30 ft and has dark shiny leaves and large white flowers. The large fruits grow in grapelike clusters (hence the name). Grapefruits were first established in the West Indies and subsequently cultivated in Florida by the 1880s; they are now also grown in Israel and South Africa. Some varieties have pink flesh. (*Citrus paradisi,* family Rutaceae.)

graph pictorial representation of numerical data, such as statistical data, or a method of showing the mathematical relationship between two or more variables by drawing a diagram.

There are often two axes, or reference lines, at right angles intersecting at the origin – the zero point, from which values of the variables (for example, distance and time for a moving object) are assigned along the axes. Pairs of simultaneous values (the distance moved after a particular time) are plotted as points in the area between the axes, and the points then joined by a smooth curve to produce a graph. The horizontal axis is usually referred to as the x-axis, and the vertical axis as the y-axis.

Cartesian coordinates On a line graph values are plotted using coordinates, components used to define the position of a point by its perpendicular distance from a set of two or more axes, or reference lines. For a two-dimensional area defined by two axes at right angles, the coordinates of a point are given by its perpendicular distances from the y-axis and x-axis, written in the form (x,y). For example, a point P that lies three units from the y-axis and four units from the x-axis has Cartesian coordinates (3,4).

straight-line graph This type of graph is produced by plotting the variables of a ◊linear equation with the general form

$$y = mx + c$$

where m is the slope of the line represented by the equation and c is the y-intercept, or the value of y where the line crosses the y-axis in the ◊Cartesian coordinate system.

histogram These are graphs used in statistics, showing frequency of data, in which the horizontal axis details discrete units or class boundaries, and the vertical axis represents the frequency. Blocks are drawn such that their areas (rather than their height as in a ◊bar chart) are proportional to the frequencies within a class or across several class boundaries. There are no spaces between blocks.

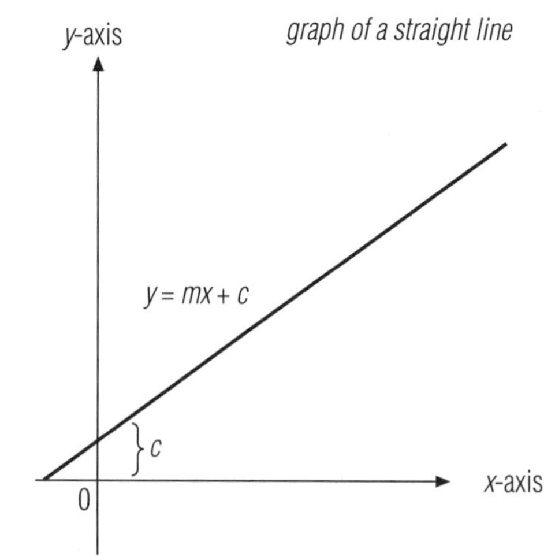

the equation of the straight-line graph takes the form $y = mx + c$, where m is the gradient (slope) of the line, and c is the y-intercept (the value of y where the line cuts the y-axis)for example, a graph of the equation $y = x 4$ will have a gradient of –1 and will cut the y-axis at $y = 4$

graph *A graph is pictorial illustration of numerical data. It is a useful tool for interpreting data and is often used to spot trends or approximate a solution.*

applications Graphs have many practical applications in all disciplines, for example **distance–time graphs** are used to describe the motion of a body by illustrating the relationship between the distance that it travels and the time taken. Plotting distance (on the vertical axis) against time (on the horizontal axis) produces a graph the gradient of which is the body's speed. If the gradient is constant (the graph is a straight line), the body has uniform or constant speed; if the gradient varies (the graph is curved), then so does the speed and the body may be said to be accelerating or decelerating.

Speed–time graphs are used to describe the motion of a body by illustrating how its speed or velocity changes with time. The gradient of the graph gives the object's acceleration: if the gradient is zero (the graph is horizontal) then the body is moving with constant speed or uniform velocity; if the gradient is constant, the body is moving with uniform acceleration. The area under the graph gives the total distance travelled by the body.

Conversion graphs are used for changing values from one unit to another, for example from Celsius to Fahrenheit, with the two axes representing the different units.

graphical user interface (GUI) or ***WIMP*** in computing, a type of ◊user interface in which programs and files appear as icons (small pictures), user options are selected from pull-down menus, and data are displayed in windows (rectangular areas), which the operator can manipulate in various ways. The operator uses a pointing device, typically a ◊mouse, to make selections and initiate actions.

The concept of the graphical user interface was developed by the Xerox Corporation in the 1970s, was popularized with the Apple Macintosh computers in the 1980s, and is now available on many types of computer – most notably as Windows, an operating system for IBM PC-compatible microcomputers developed by the software company Microsoft.

graphic equalizer control used in hi-fi systems that allows the distortions introduced in the sound output by unequal amplification of different frequencies to be corrected.

The frequency range of the signal is divided into separate bands, usually third-octave bands. The amplification applied to each band is adjusted by a sliding contact; the position of the contact indicates the strength of the amplification applied to each frequency range.

graphic file format format in which computer graphics are stored and transmitted. There are two main types: ◊raster graphics in which the image is stored as a ◊bit map (arrangement of dots), and ◊vector graphics, in which the image is stored using geometric formulas. There are many different file formats, some of which are used by specific computers, operating systems or applications. Some formats use file compression, particularly those that are able to handle more than one colour.

graphics used with computers, see ◊computer graphics.

graphics board in computing, another name for ◊graphics card.

graphics card in computing, a peripheral device that processes and displays graphics.

Graphics Interchange Format in computing, picture file format usually abbreviated to ◊GIF.

graphics tablet or ***bit pad*** in computing, an input device in which a stylus or cursor is moved, by hand, over a flat surface. The computer can keep track of the position of the stylus, so enabling the operator to input drawings or diagrams into the computer.

A graphics tablet is often used with a form overlaid for users to mark boxes in positions that relate to specific registers in the computer, although recent developments in handwriting recognition may increase its future versatility.

graphite blackish-grey, laminar, crystalline form of ◊carbon. It is used as a lubricant and as the active component of pencil lead.

The carbon atoms are strongly bonded together in sheets, but the bonds between the sheets are weak, allowing other atoms to enter regions between the layers causing them to slide over one another. Graphite has a very high melting point (3,500°C/6,332°F), and is a good conductor of heat and electricity. It absorbs neutrons and is therefore used to moderate the chain reaction in nuclear reactors.

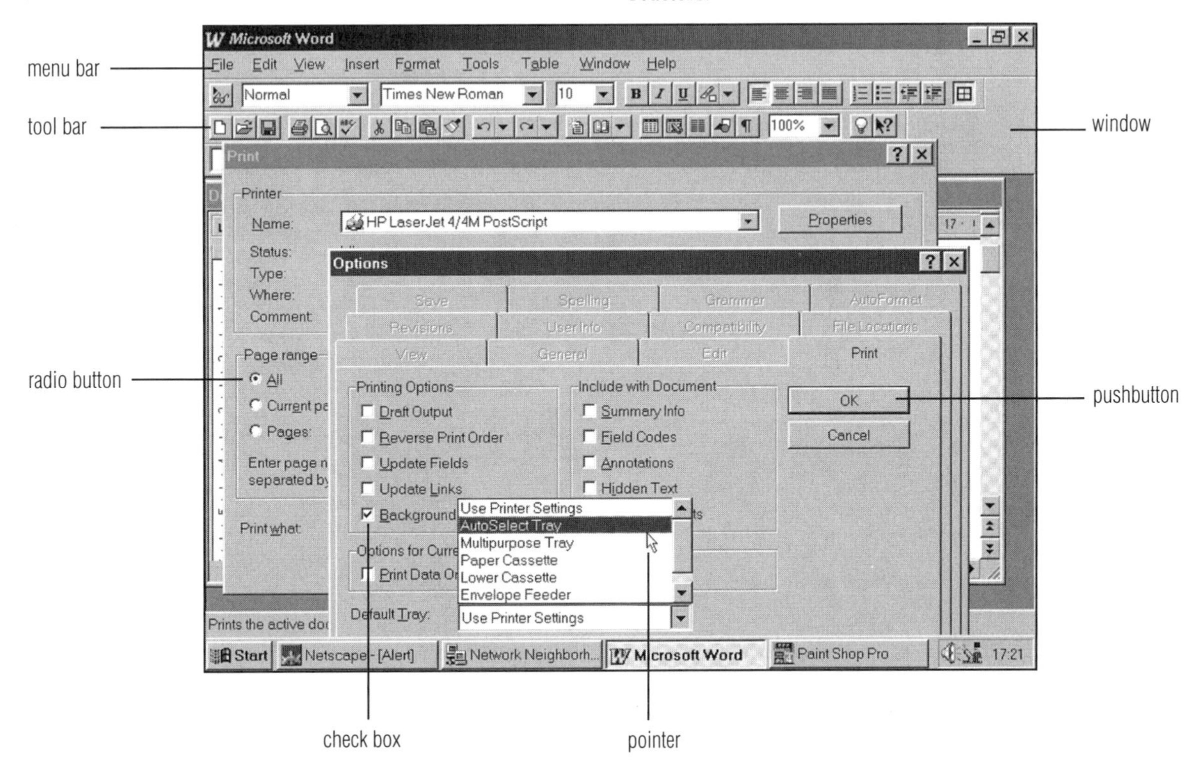

graphical user interface *A typical GUI, where the user is taken around the system by simply clicking on representative buttons or icons using the mouse.*

graphics tablet A graphics tablet enables images drawn freehand to be translated directly to the computer screen.

graph plotter alternative name for a ◊plotter.

grass any of a very large family of plants, many of which are economically important because they provide grazing for animals and food for humans in the form of cereals. There are about 9,000 species distributed worldwide except in the Arctic regions. Most are perennial, with long, narrow leaves and jointed, hollow stems; flowers with both male and female reproductive organs are borne on spikelets; the fruits are grainlike. Included in the family are bluegrass, wheat, rye, maize, sugarcane, and bamboo. (Family Gramineae.)

grasshopper any of several insects with strongly developed hind legs, enabling them to leap into the air. The hind leg in the male usually has a row of protruding joints that produce the characteristic chirping sound when rubbed against the hard wing veins. ◊Locusts, ◊crickets, and katydids are related to grasshoppers. (Families Acrididae and Tettigoniidae, order Orthoptera.)

The **short-horned grasshoppers** constitute the family Acrididae, and include locusts. All members of the family feed voraciously on vegetation. Eggs are laid in a small hole in the ground, and the unwinged larvae become adult after about six moults.

grass of Parnassus plant, unrelated to grasses, found growing in marshes and on wet moors in Europe and Asia. It is low-growing, with a rosette of heart-shaped stalked leaves, and has five-petalled white flowers with conspicuous veins growing singly on stem tips in late summer. (*Parnassia palustris,* family Parnassiaceae.)

grass snake olive-green, grey or brownish non-venomous snake *Natrix natrix* found near water in lowland areas with woodland. They are about 80 cm/32 in long and feed mainly on frogs, toads, and newts, which they hunt in the water. They are the largest British reptiles. There is also a grass snake in the USA.

The female lays 10–40 eggs within a pile of rotting vegetation, where they are incubated by the heat generated by the rotting process. Eggs are elongated, white, leathery, and about 3 cm/1.2 in long.

grass tree Australian plant belonging to the lily family. The tall, thick stems have a grasslike tuft at the top above which rises a flower spike resembling a spear; this often appears after bushfires and in some species can grow to a height of 3 m/10 ft. (Genus *Xanthorrhoea,* family Liliaceae.)

gravel coarse ◊sediment consisting of pebbles or small fragments of rock, originating in the beds of lakes and streams or on beaches. Gravel is quarried for use in road building, railway ballast, and for an aggregate in concrete. It is obtained from quarries known as gravel pits, where it is often found mixed with sand or clay.

Some gravel deposits also contain placer deposits of metal ores (chiefly tin) or free metals (such as gold and silver).

gravimetric analysis in chemistry, a technique for determining, by weighing, the amount of a particular substance present in a sample. It usually involves the conversion of the test substance into a compound of known molecular weight that can be easily isolated and purified.

gravimetry study of the Earth's gravitational field. Small variations in the gravitational field (gravimetric anomalies) can be caused by varying densities of rocks and structure beneath the surface. Such variations are measured by a device called a gravimeter, which consists of a weighted spring that is pulled further downwards where the gravity is stronger (at a Bouguer anomaly). Gravimetry is used by geologists to map the subsurface features of the Earth's crust, such as underground masses of heavy rock such as granite, or light rock such as salt.

gravitational field the region around a body in which other bodies experience a force due to its gravitational attraction. The gravitational field of a massive object such as the Earth is very strong and easily recognized as the force of gravity, whereas that of an object of much smaller mass is very weak and difficult to detect. Gravitational fields produce only attractive forces.

gravitational field strength (symbol g) the strength of the Earth's gravitational field at a particular point. It is defined as the gravitational force in newtons that acts on a mass of one kilogram. The value of g on the Earth's surface is taken to be 9.806 N kg^{-1}.

The symbol g is also used to represent the acceleration of a freely falling object in the Earth's gravitational field.

Near the Earth's surface and in the absence of friction due to the air, all objects fall with an acceleration of 9.806 m s^{-2}.

gravitational lensing bending of light by a gravitational field, predicted by Einstein's general theory of relativity. The effect was first detected in 1917 when the light from stars was found to be bent as it passed the totally eclipsed Sun. More remarkable is the splitting of light from distant quasars into two or more images by intervening galaxies. In 1979 the first double image of a quasar produced by gravitational lensing was discovered and a quadruple image of another quasar was later found.

gravitational potential energy energy possessed by an object when it is placed in a position from which, if it were free to do so, it would fall under the influence of gravity. The gravitational potential energy E_p of an object of mass m kg placed at a height h m above the ground is given by the formula:

$$E_p = mgh$$

where g is the gravitational field strength in N kg^{-1} of the Earth at the place.

In a ◊hydroelectric power station, gravitational potential energy of water held in a high-level reservoir is used to drive turbines to produce electricity.

graviton in physics, the ◊gauge boson that is the postulated carrier of the gravitational force.

gravity force of attraction that arises between objects by virtue of their masses. On Earth, gravity is the force of attraction between

GRAVITY

The maximum speed with which a falling raindrop can hit you is about 29 kmph/18 mph. In a vacuum, the further an object falls, the more speed it gains, but in the real world, air resistance eventually balances out the accelerating effect of gravity.

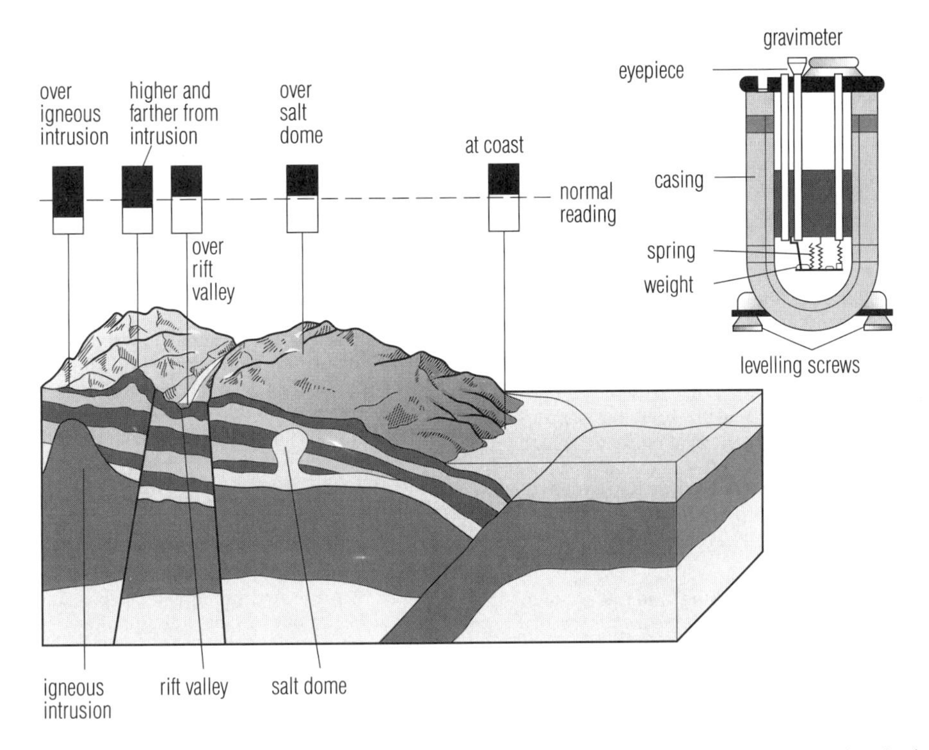

***gravimetry** The gravimeter is an instrument for measuring the force of gravity at a particular location. Variations in the force of gravity acting on a weight suspended by a spring cause the spring to stretch. The gravimeter is used in aerial surveys. Geological features such as intrusions and salt domes are revealed by the stretching of the spring.*

any object in the Earth's gravitational field and the Earth itself. It is regarded as one of the four fundamental ◊forces of nature, the other three being the ◊electromagnetic force, the ◊strong nuclear force, and the ◊weak nuclear force. The gravitational force is the weakest of the four forces, but it acts over great distances. The particle that is postulated as the carrier of the gravitational force is the ◊graviton.

One of the earliest gravitational experiments was undertaken by Nevil Maskelyne in 1774 and involved the measurement of the attraction of Mount Schiehallion (Scotland) on a plumb bob.

measuring forces of attraction An experiment for determining the force of attraction between two masses was first planned in the mid-18th century by the Reverend J Mitchell, who did not live to work on the apparatus he had designed and completed. After Mitchell's death the apparatus came into the hands of Henry Cavendish, who largely reconstructed it but kept to Mitchell's original plan. The attracted masses consisted of two small balls, connected by a stiff wooden beam suspended at its middle point by a long, fine wire. The whole of this part of the apparatus was enclosed in a case, carefully coated with tinfoil to secure, as far as possible, a uniform temperature within the case. Irregular distribution of temperature would have resulted in convection currents of air which would have had a serious disturbing effect on the suspended system. To the beam was attached a small mirror with its plane vertical. A small glazed window in the case allowed any motion of the mirror to be observed by the consequent deviations of a ray of light reflected from it. The attracting masses consisted of two equal, massive lead spheres. Using this apparatus, Cavendish, in 1797, obtained for the gravitational constant G the value 6.6×10^{-11} N m^2 kg^{-2}. The apparatus was refined by Charles Vernon Boys and he obtained the improved value 6.6576×10^{-11} N m^2 kg^{-2}. The value generally used today is 6.6720×10^{-11} N m^2 kg^{-2}.

gravure one of the three main ◊printing methods, in which printing is done from a plate etched with a pattern of recessed cells in which the ink is held. The greater the depth of a cell, the greater the strength of the printed ink. Gravure plates are expensive to make, but the process is economical for high-volume printing and reproduces illustrations well.

gray SI unit (symbol Gy) of absorbed radiation dose. It replaces the rad (1 Gy equals 100 rad), and is defined as the dose absorbed when one kilogram of matter absorbs one joule of ionizing radiation. Different types of radiation cause different amounts of damage for the same absorbed dose; the SI unit of **dose equivalent** is the ◊sievert.

grayling freshwater fish with a long multirayed dorsal (back) fin and silver to purple body colouring. It is found in northern parts of Europe, Asia, and North America, where it was once common in the Great Lakes. (Species *Thymallus thymallus,* family Salmonidae.)

grayling butterfly butterfly widely distributed over the British Isles. It has dark brown wings with two black eye-spots on each of the forewings and one black eyespot centred with white on the hindwings. It is found on heaths and in dry stony places, especially on chalk and in clearings in woods.

classification The grayling butterfly *Hipparchia semele* is in order Lepidoptera, class Insecta, phylum Arthropoda.

Great Artesian Basin the largest area of artesian water in the world. It underlies much of Queensland, New South Wales, and South Australia, and in prehistoric times formed a sea. It has an area of 1,750,000 sq km/676,250 sq mi.

Great Bear popular name for the constellation ◊Ursa Major.

great circle circle drawn on a sphere such that the diameter of the circle is a diameter of the sphere. On the Earth, all meridians of longitude are half great circles; among the parallels of latitude, only the Equator is a great circle.

The shortest route between two points on the Earth's surface is along the arc of a great circle. These are used extensively as air routes although on maps, owing to the distortion brought about by ◊projection, they do not appear as straight lines.

Great Dane breed of large, short-haired dog, often fawn or brindle in colour, standing up to 76 cm/30 in tall, and weighing up to 70 kg/154 lb. It has a large head and muzzle, and small, erect ears. It was formerly used in Europe for hunting boar and stags.

Great Red Spot prominent oval feature, 14,000 km/8,500 mi wide and some 30,000 km/20,000 mi long, in the atmosphere of the planet ◊Jupiter, south of the Equator. It was first observed in the 19th century. Space probes show it to be an anticlockwise vortex of cold clouds, coloured possibly by phosphorus.

Great Wall array of galaxies arranged almost in a perfect plane, consisting of some 2,000 galaxies (about 500 million × 200 million light years across). It was discovered by US astronomers in Cambridge, Massachusetts, in 1989.

grebe any of a group of 19 species of water birds. The **great crested grebe** (*Podiceps cristatus*) is the largest of the Old World grebes. It feeds on fish, and lives on ponds and marshes in Europe, Asia, Africa, and Australia. It grows to 50 cm/20 in long and has a white breast, with chestnut and black feathers on its back and head. Dark ear tufts and a prominent collar or crest of feathers around the base of the head appear during the breeding season; these are lost in winter. (Family Podicipedidae, order Podicipediformes.)

Grebes have broad, flat feet, and the toes are partially webbed, the legs being set extremely far back on the body. The wings are short and rounded, there is practically no tail, and flight is low. Both sexes are similar in appearance.

greeking method used in ◊desktop publishing and other page make-up systems for showing type below a certain size on screen. Rather than the actual characters being displayed, either a grey bar or graphics symbols are used. Greeking is usually employed when a general impression of the page lay-out is required.

green audit inspection of a company to assess the total environmental impact of its activities or of a particular product or process.

For example, a green audit of a manufactured product looks at the impact of production (including energy use and the extraction of raw materials used in manufacture), use (which may cause pollution and other hazards), and disposal (potential for recycling, and whether waste causes pollution).

Such 'cradle-to-grave' surveys allow a widening of the traditional scope of economics by ascribing costs to variables that are usually ignored, such as despoliation of the countryside or air pollution.

Green Bank site in West Virginia, USA, of the National Radio Astronomy Observatory. Its main instruments are a 43-m/140-ft fully steerable dish, opened in 1965, and three 26-m/85-ft dishes. A 90-m/300-ft partially steerable dish, opened in 1962, collapsed 1988 because of metal fatigue; a replacement dish, 100 m/330 ft across, was under construction in 1995.

greenbottle type of ◊blowfly.

greenbrier or ***catbrier*** any of several climbing woody vines of the genus *Smilax* of the lily family, having smooth, shiny green oval leaves and usually black berries. The prickly stems of these plants often form impenetrable thickets.

green computing the gradual movement by computer companies toward incorporating energy-saving measures in the design of systems and hardware. The increasing use of energy-saving devices, so that a computer partially shuts down during periods of inactivity, but can reactivate at the touch of a key, could play a significant role in ◊energy conservation.

It is estimated that worldwide electricity consumption by computers amounts to 240 billion kilowatt hours per year, equivalent to the entire annual consumption of Brazil. In the USA, carbon dioxide emissions could be reduced by 20 million tonnes per year – equivalent to the carbon dioxide output of 5 million cars – if all computers incorporated the latest 'sleep technology' (which shuts down most of the power-consuming features of a computer if it is unused for any length of time).

Although it was initially predicted that computers would mean 'paperless offices', in practice the amount of paper consumed continues to rise. Other environmentally-costly features of computers include their rapid obsolescence, health problems associated with monitors and keyboards, and the unfavourable economics of component recycling.

greenfinch olive-green songbird common in Europe and North Africa. It has bright-yellow markings on the outer tail feathers and wings; males are much brighter in colour than females. (Species *Carduelis chloris*, family Fringillidae, order Passeriformes.)

greenfly plant-sucking insect, a type of ◊aphid.

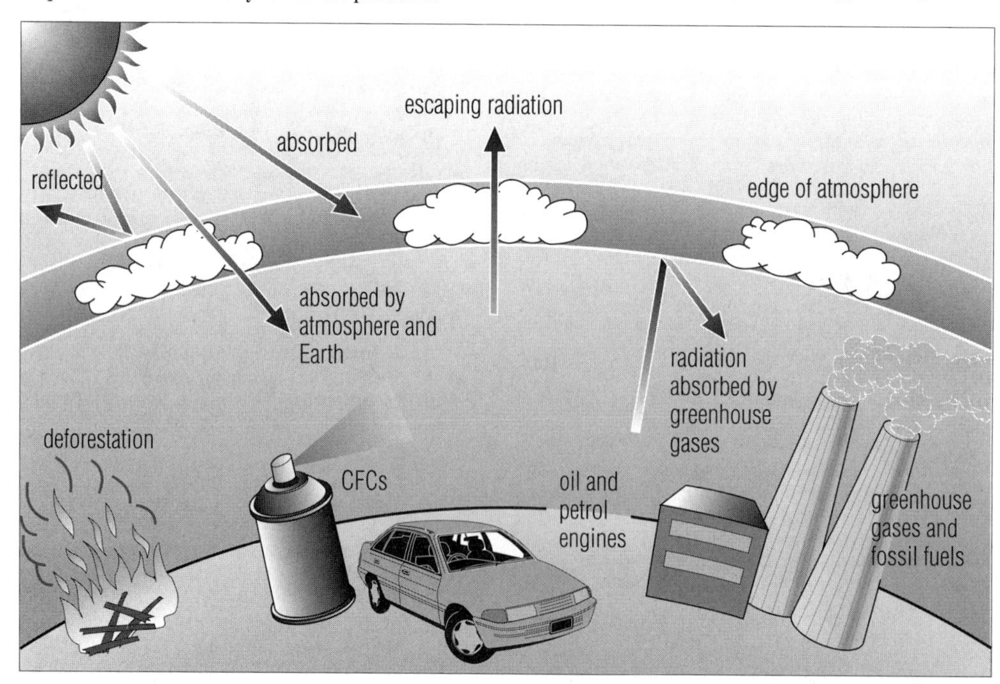

greenhouse effect *The warming effect of the Earth's atmosphere is called the greenhouse effect. Radiation from the Sun enters the atmosphere but is prevented from escaping back into space by gases such as carbon dioxide (produced for example, by the burning of fossil fuels), nitrogen oxides (from car exhausts), and CFCs (from aerosols and refrigerators). As these gases build up in the atmosphere, the Earth's average temperature is expected to rise.*

greenhouse effect phenomenon of the Earth's atmosphere by which solar radiation, trapped by the Earth and re-emitted from the surface as infrared radiation, is prevented from escaping by various gases in the air. Greenhouse gases trap heat because they readily absorb infrared radiation. The result is a rise in the Earth's temperature (◊global warming). The main greenhouse gases are carbon dioxide, methane, and ◊chlorofluorocarbons (CFCs) as well as water vapour. Fossil-fuel consumption and forest fires are the principal causes of carbon dioxide build-up; methane is a byproduct of agriculture (rice, cattle, sheep).

The United Nations Environment Programme estimates that by 2025, average world temperatures will have risen by 1.5°C/2.7°F with a consequent rise of 20 cm/7.9 in in sea level. Low-lying areas and entire countries would be threatened by flooding and crops would be affected by the change in climate. However, predictions about global warming and its possible climatic effects are tentative and often conflict with each other.

At the 1992 Earth Summit it was agreed that by 2000 countries would stabilize carbon dioxide emissions at 1990 levels, but to halt the acceleration of global warming, emissions would probably need to be cut by 60%. Any increases in carbon dioxide emissions are expected to come from transport. The Berlin Mandate, agreed unanimously at the climate conference in Berlin in 1995, committed industrial nations to the continuing reduction of greenhouse gas emissions after 2000, when the existing pact to stabilize emissions runs out. The stabilization of carbon dioxide emissions at 1990 levels by 2000 will not be achieved by a number of developed countries, including Spain, Australia, and the USA, according to 1997 estimates. Australia is in favour of different targets for different nations, and refused to sign a communiqué at the South Pacific Forum meeting in the Cook Islands in 1997 which insisted on legally binding reductions in greenhouse gas emissions.

Dubbed the 'greenhouse effect' by Swedish scientist Svante Arrhenius, it was first predicted in 1827 by French mathematician Joseph Fourier.

Our planet is not fragile at its own time scale, and we, pitiful latecomers in the last microsecond of our planetary year, are stewards of nothing in the long run. Yet no political movement is more vital and timely than modern environmentalism – because we must save ourselves (and our neighbor species) from our own immediate folly.

STEPHEN JAY GOULD US palaeontologist and writer. *Bully for Brontosaurus* 1991

green movement collective term for the individuals and organizations involved in efforts to protect the environment. The movement encompasses political parties such as the Green Party and organizations like ◊Friends of the Earth and ◊Greenpeace.

Despite a rapid growth of public support, and membership of environmental organizations running into many millions worldwide, political green groups have failed to win significant levels of the vote in democratic societies.

GreenNet in computing, international computer network used by environmental activists to exchange information and news.

Greenpeace international environmental pressure group, founded in 1971, with a policy of nonviolent direct action backed by scientific research. During a protest against French atmospheric nuclear testing in the South Pacific in 1985, its ship *Rainbow Warrior* was sunk by French intelligence agents, killing a crew member. In 1995 it played a prominent role in opposing the disposal of waste from an oil rig in the North Sea, and again attempted to disrupt French nuclear tests in the Pacific. In 1997 Greenpeace had a membership in 43 'chapters' worldwide.

green revolution in agriculture, the change in methods of arable farming instigated in the 1940s and 1950s in Third World countries. The intent was to provide more and better food for their populations, albeit with a heavy reliance on chemicals and machinery. It was abandoned by some countries in the 1980s. Much of the food produced was exported as ◊cash crops, so that local diet did not always improve.

The green revolution tended to benefit primarily those land-owners who could afford the investment necessary for such intensive agriculture. Without a dosage of 70–90 kg/154–198 lb of expensive nitrogen fertilizers per hectare, the high-yield varieties will not grow properly. Hence, rich farmers tended to obtain bigger yields while smallholders were unable to benefit from the new methods.

In terms of production, the green revolution was initially successful in southeast Asia; India doubled its wheat yield in 15 years, and the rice yield in the Philippines rose by 75%. However, yields have levelled off in many areas; some countries that cannot afford the dams, fertilizers, and machinery required, have adopted ◊intermediate technologies.

greenshank greyish shorebird of the sandpiper group. It has long olive-green legs and a long, slightly upturned bill, with white underparts and rump and dark grey wings. It breeds in northern Europe and regularly migrates through the Aleutian Islands, southwest of Alaska. (Species *Tringa nebularia,* family Scolopacidae, order Charadriiformes.)

Greenwich Mean Time (GMT) local time on the zero line of longitude (the ***Greenwich meridian***), which passes through the Old Royal Observatory at Greenwich, London. It was replaced in 1986 by coordinated universal time (UTC), but continued to be used to measure longitudes and the world's standard time zones; see ◊time.

grenadier another name for ◊rat-tail, a deep-sea fish.

grep in computing, UNIX command that allows full-text searching within files. On the Net, grep is sometimes used as an all-purpose synonym for 'search'.

grevillea genus of almost exclusively Australian trees and shrubs of the family Proteaceae bearing attractive spider-flowers. There are some 250 species widely distributed throughout the continent.

greyhound ancient breed of dog, with a long narrow head, slight build, and long legs. It stands up to 75 cm/30 in tall. It is renowned for its swiftness, and can exceed 60 kph/40 mph. Greyhounds were bred to hunt by sight, their main quarry being hares. Hunting hares with greyhounds is the basis of the ancient sport of coursing. Track-based greyhound racing is a popular spectator sport.

The Italian greyhound is similar in build to the ordinary greyhound, but very much smaller, weighing only about 3.6 kg/8 lb.

grey matter in biology, those parts of the brain and spinal cord that are made up of interconnected and tightly packed nerve cell nucleuses. The outer layers of the cerebellum contains most of the grey matter in the brain. It is the region of the brain that is responsible for advanced mental functions. Grey matter also constitutes the inner core of the spinal cord. This is in contrast to white matter, which is made of the axons of nerve cells.

grey scales method of representing continuous tone images on a screen or printer. Each dot in the ◊bit map is represented by a number of bits and can have a different shade of grey. Compare with ◊dithering when shades are simulated by altering the density and the pattern of black dots on a white background.

grid network of crossing parallel lines. **Rectangular grids** are used for drawing graphs. **Isometric grids** are used for drawing representations of solids in two dimensions in which lengths in the drawing match the lengths of the object.

grid network by which electricity is generated and distributed over a region or country. It contains many power stations and switching centres and allows, for example, high demand in one area to be met by surplus power generated in another.

The term is also used for any grating system, as in a cattle grid for controlling the movement of livestock across roads, and a conductor in a storage battery or electron gun.

grid reference a cadastral numbering system to specify location on a map. The numbers representing grid lines at the bottom of the map (eastings) are given before those at the side (northings). Successive decimal digits refine the location within the grid system.

griffon small breed of dog originating in Belgium. Red, black, or black and tan in colour and weighing up to 5 kg/11 lb, griffons are square-bodied and round-headed. There are rough- and smooth-coated varieties.

The name is also applied to several larger breeds of hunting dogs with rough coats, including two bred in northern France to pursue wild boar.

griffon Bruxelloise breed of terrierlike toy dog originally bred in Belgium. It weighs up to 4.5 kg/10 lb and has a harsh and wiry coat that is red or black in colour. The smooth-haired form of the breed is called the **petit Brabançon**.

The griffon Bruxelloise has a large rounded head with semi-erect ears; large black eyes with black eye-rims; and a short nose surrounded with black hair that converges upwards to meet the hair round the eyes. Its chest is wide and deep; legs are straight and of medium length; the tail is traditionally docked and carried upwards.

griffon vulture Old World vulture found in southern Europe, west and central Asia, and parts of Africa. It has a bald head with a neck ruff, and is 1.1 m/3.5 ft long with a wingspan of up to 2.7 m/9 ft. (Species *Gyps fulvus*, family Accipitridae.)

grooming in biology, the use by an animal of teeth, tongue, feet, or beak to clean fur or feathers. Grooming also helps to spread essential oils for waterproofing. In many social species, notably monkeys and apes, grooming of other individuals is used to reinforce social relationships.

Groom Lake or ***Area 51*** dry lake-bed site in Nevada, USA, of US Air Force base used for the development of secret projects. In the 1980s it was used for testing of Stealth aircraft and Star Wars (Strategic Defense Initiative) projects and later for studying ex-Soviet aircraft purchased from Russia.

The base was established in the 1950s as a testing ground for the U-2 spy plane. In 1984 it was designated so secret that it no longer appeared on maps, and its existence was officially denied until 1992. Because of the strangely shaped aircraft and lights to be seen, the nearest public vantage point to Groom Lake has since 1989 attracted UFO watchers.

grosbeak any of various thick-billed ◊finches. The **pine grosbeak** (*Pinicola enucleator*) breeds in Arctic forests. Its plumage is similar to that of the pine ◊crossbill. (Family Fringillidae, order Passeriformes.)

ground beetle large, adorned, brilliantly metallic beetle. Ground beetles are mainly terrestrial with few species being capable of flight. About 20,000 species are known to exist; nearly all are carnivorous as adults and larvae.

The larvae are of particular economic importance, destroying large numbers of soil insects and worms.

classification Ground beetles are in the family Carabidae, order Coleoptera, class Insecta, phylum Arthropoda.

ground controlled interception GCI British term for the ground command of fighter aircraft during and after the Battle of Britain 1940.

Using advanced ◊radar, the British could see German air formations at considerable ranges which enabled ground controllers to direct fighter aircraft into the path of the enemy, doing away with the need to fly standing patrols across likely approaches.

groundnut another name for ◊peanut.

ground water water collected underground in porous rock strata and soils; it emerges at the surface as springs and streams. The groundwater's upper level is called the **water table**. Sandy or other kinds of beds that are filled with groundwater are called **aquifers**. Recent estimates are that usable ground water amounts to more than 90% of all the fresh water on Earth; however, keeping such supplies free of pollutants entering the recharge areas is a critical environmental concern.

Most groundwater near the surface moves slowly through the ground while the water table stays in the same place. The depth of the water table reflects the balance between the rate of infiltration, called recharge, and the rate of discharge at springs or rivers or pumped water wells. The force of gravity makes underground water run 'downhill' underground just as it does above the surface. The greater the slope and the permeability, the greater the speed. Velocities vary from 100 cm/40 in per day to 0.5 cm/0.2 in.

group in chemistry, a vertical column of elements in the ◊periodic table. Elements in a group have similar physical and chemical properties; for example, the group I elements (the alkali metals: lithium, sodium, potassium, rubidium, caesium, and francium) are all highly reactive metals that form univalent ions. There is a gradation of properties down any group: in group I, melting and boiling points decrease, and density and reactivity increase.

group in mathematics, a finite or infinite set of elements that can be combined by an operation; formally, a group must satisfy certain conditions. For example, the set of all integers (positive or negative whole numbers) forms a group with regard to addition because: (1) addition is associative, that is, the sum of two or more integers is the same regardless of the order in which the integers are added; (2) adding two integers gives another integer; (3) the set includes an identity element 0, which has no effect on any integer to which it is added (for example, $0 + 3 = 3$); and (4) each integer has an inverse (for instance, 7 has the inverse -7), such that the sum of an integer and its inverse is 0. **Group theory** is the study of the properties of groups.

grouper any of several species of large sea perch (spiny-finned fish), found in warm waters. Some species grow to 2 m/6.5 ft long, and can weigh 300 kg/660 lbs. (Family Serranidae.)

The spotted **giant grouper** (*Promicrops itaiara*) is 2–2.5 m/6–8 ft long, may weigh over 300 kg/700 lb and is sluggish in movement. Formerly game fish, groupers are now commercially exploited as food.

groupware in computing, software designed to be used collaboratively by a small group of users, each with his/her own computer and a copy of the software. Examples of groupware are Lotus Notes and Novell GroupWise, both of which provide facilities for sending e-mail and sharing documents.

Standard business applications such as word processors are spoken of as 'groupware-enabled' if they provide facilities for a number of users to make revisions and incorporate them all into a final version. See also ◊computer-supported collaborative work.

grouse plump fowl-like game bird belonging to a subfamily of the pheasant family, which also includes the ptarmigan, capercaillie, and prairie chicken. Grouse are native to North America and northern Europe. They spend most of their time on the ground. During the mating season the males undertake elaborate courtship displays in small individual territories (◊leks). (Subfamily Tetraonidae, family Phasianidae, order Galliformes.)

growth in biology, the increase in size and weight during the development of an organism. Growth is an increase in biomass (mass of organic material, excluding water) and is associated with cell division.

All organisms grow, although the rate of growth varies over a lifetime. Typically, an organism shows an S-shaped curve, in which growth is at first slow, then fast, then, towards the end of life, nonexistent. Growth may even be negative during the period before death, with decay occurring faster than cellular replacement.

The concept of an average, the equation to a curve, the description of a froth or cellular tissue, all come within the scope of mathematics for no other reason than that they are summations of more elementary principles or phenomena. Growth and Form are throughout of this composite nature; therefore the laws of mathematics are bound to underlie them, and her methods to be peculiarly fitted to interpret them.

D'Arcy Wentworth Thompson British zoologist.
On Growth and Form

growth and decay curve graph showing exponential change (growth where the increment itself grows at the same rate) as occurs with compound interest and populations.

growth ring another name for ◊annual ring.

GRP abbreviation for ◊glass-reinforced plastic.

grub legless larval stages of Coleoptera (beetles) and Hymenoptera (bees, ants and wasps). See ◊larvae.

g-scale scale for measuring force by comparing it with the force due to ◊gravity (*g*), often called ◊g-force.

guan any of several large, pheasantlike birds native to the forests of South and Central America. They are sociable birds, almost the size of a turkey, with long, strong legs. Their colour is olive-green or brown. The family also includes the curassows. (Family Cracidae.)

guanaco hoofed ruminant (cud-chewing) mammal belonging to the camel family, found in South America on the pampas and mountain plateaux. It grows up to 1.2 m/4 ft at the shoulder, with the head and body measuring about 1.5 m/5 ft in length. It is sandy brown in colour, with a blackish face, and has fine wool. It lives in small herds and is the ancestor of the domestic ◊llama and ◊alpaca. It is also related to the other wild member of the camel family, the ◊vicuna. (Species *Lama guanacoe*, family Camelidae.)

guano dried excrement of fish-eating birds that builds up under nesting sites. It is a rich source of nitrogen and phosphorous, and is widely collected for use as fertilizer. Some 80% comes from the sea cliffs of Peru.

guarana Brazilian woody climbing plant. A drink with a high caffeine content is made from its roasted seeds, and it is the source of the drug known as zoom in the USA. Starch, gum, and several oils are extracted from it for commercial use. (*Paullinia cupana*, family Sapindaceae.)

guard cell in plants, a specialized cell on the undersurface of leaves for controlling gas exchange and water loss. Guard cells occur in pairs and are shaped so that a pore, or stomata, exists between them. They can change shape with the result that the pore disappears. During warm weather, when a plant is in danger of losing excessive water, the guard cells close, cutting down evaporation from the interior of the leaf.

guava tropical American tree belonging to the myrtle family; the astringent yellow pear-shaped fruit is used to make guava jelly, or it can be stewed or canned. It has a high vitamin C content. (*Psidium guajava*, family Myrtaceae.)

gudgeon any of an Old World group of freshwater fishes of the carp family, especially the species *G. gobio* found in Europe and northern Asia on the gravel bottoms of streams. It is olive-brown, spotted with black, and up to 20 cm/8 in long, with a distinctive barbel (sensory bristle, or 'whisker') at each side of the mouth. (Genus *Gobio*, family Cyprinidae.)

guelder rose or ***snowball tree*** cultivated shrub or small tree, native to Europe and North Africa, with round clusters of white flowers which are followed by shiny red berries. (*Viburnum opulus*, family Caprifoliaceae.)

guenon African monkey with characteristically greenish, yellow, or brown coat with brilliant markings. Guenons are slim and graceful in movement with very long tails that are not prehensile.

classification Guenons are in the genus *Cercopithecus*, family Cercopithecidae, order Primates.

GUI in computing, abbreviation for ◊graphical user interface.

guillemot any of several diving seabirds belonging to the auk family that breed on rocky North Atlantic and Pacific coasts. The **common guillemot** (*U. aalge*) has a long straight beak and short tail and wings; the feet are three-toed and webbed, the feathers are sooty brown and white. It breeds in large colonies on sea cliffs. The **black guillemot** (*C. grylle*) of northern coasts is much smaller and mostly black in summer, with orange legs when breeding. Guillemots build no nest, but lay one large, almost conical egg. (Genera *Uria* and *Cepphus*, family Alcidae, order Charadriiformes.)

guiltware or ***nagware*** in computing, variety of ◊shareware software that attempts to make the user register (and pay for) the software by exploiting the user's sense of guilt.

On-screen messages are displayed, usually when the program is started, reminding users that they have an unregistered version of the program that they should pay for if they intend to continue using it. Some programs will also display the message at random intervals while the program is in use.

guinea fowl any of a group of chickenlike African birds, including the **helmet guinea fowl** (*Numida meleagris*), which has a horny growth on the head, white-spotted feathers, and fleshy cheek wattles (loose folds of skin). It is the ancestor of the domestic guinea fowl. Guinea fowl are mostly gregarious ground-feeders, eating insects, leaves, and snails; at night they roost in trees. (Family Numididae, order Galliformes.)

guinea pig species of ◊cavy, a type of rodent.

Guinea worm parasitic, microscopic ◊nematode worm found in India and Africa, affecting some 650,000 people in Nigeria alone. It enters the body via drinking water and migrates to break out through the skin. (Species *Dracunculus medinensis*.)

GUINEA PIG CARE

http://www.bcyellowpages.com/advert/b/BCHES/guinea.htm

From housing to health, this site contains lots of useful information for guinea pig lovers, including a health chart to help you diagnose illnesses.

Gulf Stream warm ocean ◊current that flows north from the warm waters of the Gulf of Mexico. Part of the current is diverted east across the Atlantic, where it is known as the **North Atlantic Drift**, and warms what would otherwise be a colder climate in the British Isles and northwestern Europe.

gull any of a group of seabirds that are usually 25–75 cm/10–30 in long, white with grey or black on the back and wings, and have large beaks. Immature birds are normally a mottled brown colour. Gulls are sociable, noisy birds and they breed in colonies. (Genus principally *Larus*, subfamily Larinae, family Laridae, order Charadriiformes.)

gull *The sharp, heavy beak typical of gulls can be clearly seen on this lesser black-backed gull.* *Premaphotos Wildlife*

The **common black-headed gull** (*L. ridibundus*), common on both sides of the Atlantic Ocean, is grey and white with (in summer) a dark-brown head and a red beak; it breeds in large colonies on wetlands, making a nest of dead rushes and laying, on average, three eggs. The **great black-headed gull** (*L. ichthyaetus*) is native to Asia. The **herring gull** (*L. argentatus*), common in the northern hemisphere, has white and pearl-grey plumage and a yellow beak. The **oceanic great black-backed gull** (*L. marinus*), found in the Atlantic, is over 75 cm/2.5 ft long.

The **kelp gull** or **Southern black-backed gull** (*L. dominicanus*) is common throughout the southern hemisphere. It feeds mainly on limpets, which are swallowed whole, with the shell later spat out and left in a heap around the nest area.

gum in botany, complex polysaccharides (carbohydrates) formed by many plants and trees, particularly by those from dry regions. They form four main groups: plant exudates (gum arabic); marine plant extracts (agar); seed extracts; and fruit and vegetable extracts. Some are made synthetically.

Gums are tasteless and odourless, insoluble in alcohol and ether but generally soluble in water. They are used for adhesives, fabric sizing, in confectionery, medicine, and calico printing.

gum in mammals, the soft tissues surrounding the base of the teeth. Gums are liable to inflammation (gingivitis) or to infection by microbes from food deposits (periodontal disease).

gum arabic substance obtained from certain species of ◊acacia trees, especially *A. senegal*, with uses in medicine, confectionery, and adhesive manufacture.

gumtree common name for the ◊eucalyptus tree.

gun any kind of firearm or any instrument consisting of a metal tube from which a projectile is discharged; see also ◊pistol and ◊small arms.

gun metal type of ◊bronze, an alloy high in copper (88%), also containing tin and zinc, so-called because it was once used to cast cannons. It is tough, hard-wearing, and resists corrosion.

gunpowder or ***black powder*** the oldest known ◊explosive, a mixture of 75% potassium nitrate (saltpetre), 15% charcoal, and 10% sulphur. Sulphur ignites at a low temperature, charcoal burns readily, and the potassium nitrate provides oxygen for the explosion. As gunpowder produces lots of smoke and burns quite slowly, it has progressively been replaced since the late 19th century by high explosives, although it is still widely used for quarry blasting, fuses, and fireworks. Gunpowder has high ◊activation energy; a gun based on gunpowder alone requires igniting by a flint or a match.

Gunpowder is believed to have been invented in China in the 10th century, but may also have been independently discovered by the Arabs. Certainly the Arabs produced the first known working gun, in 1304. Gunpowder was used in warfare from the 14th century but it was not generally adapted to civil purposes until the 17th century, when it began to be used in mining.

guppy brightly coloured fish *Poecilia reticulata*, with an elongated body and round tail fin, measuring about 6 cm/2.25 in in length. It occurs naturally in fresh and brackish water in parts of South America and the West Indies. Guppies have also been introduced into many other tropical areas as a natural pest control because they feed on mosquito and other insect larvae. The guppy gives birth to live young, and is a popular aquarium species.

gurnard any of a group of coastal fish that creep along the sea bottom with the help of three fingerlike appendages detached from the pectoral fins. Gurnards are both tropical and temperate zone fish. (Genus *Trigla*, family Trigilidae.)

gut or ***alimentary canal*** in the ◊digestive system, the part of an animal responsible for processing food and preparing it for entry into the blood.

The gut consists of a tube divided into segments specialized to perform different functions. The front end (the mouth) is adapted for food intake and for the first stages of digestion. The stomach is a storage area, although digestion of protein by the enzyme pepsin starts here; in many herbivorous mammals this is also the site of cellulose digestion. The small intestine follows the stomach and is specialized for digestion and for absorption. The large intestine, consisting of the colon, caecum, and rectum, has a variety of functions, including cellulose digestion, water absorption, and storage of faeces. From the gut nutrients are carried to the liver via the hepatic portal vein, ready for assimilation by the cells.

Gutenberg, Johannes (Gensfleisch) (c. 1398–1468)

German printer, the inventor of printing from movable metal type (although Laurens Janszoon Coster has a rival claim), based on the Chinese wood-block-type method.

Gutenberg began work on the process in the 1440s and in 1450 set up a printing business in Mainz. By 1456 he had produced the first printed Bible (known as the Gutenberg Bible). It is not known what other books he printed.

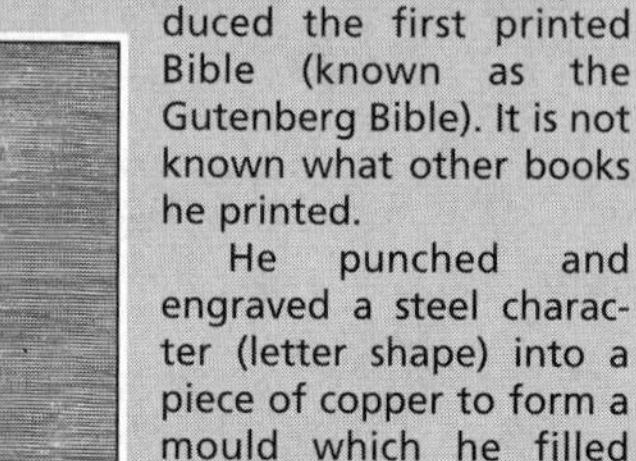

He punched and engraved a steel character (letter shape) into a piece of copper to form a mould which he filled with molten metal. The letters were in the Gothic style and of equal height. By 1500, more than 180 European towns had working presses of this kind.

Mary Evans Picture Library

gutta-percha juice of various tropical trees of the sapodilla family (such as the Malaysian *Palaquium gutta*), which can be hardened to form a flexible, rubbery substance used for electrical insulation, dentistry, and golf balls; it has now been largely replaced by synthetic materials.

guttation secretion of water on to the surface of leaves through specialized pores, or ◊hydathodes. The process occurs most frequently during conditions of high humidity when the rate of transpiration is low. Drops of water found on grass in early morning are often the result of guttation, rather than dew. Sometimes the water contains minerals in solution, such as calcium, which leaves a white crust on the leaf surface as it dries.

gymnosperm Greek ***'naked seed'*** in botany, any plant whose seeds are exposed, as opposed to the structurally more advanced ◊angiosperms, where they are inside an ovary. The group includes conifers and related plants such as cycads and ginkgos, whose seeds develop in ◊cones. Fossil gymnosperms have been found in rocks about 350 million years old.

gynaecology medical speciality concerned with disorders of the female reproductive system.

gynoecium or **gynaecium** collective term for the female reproductive organs of a flower, consisting of one or more ◊carpels, either free or fused together.

gyre circular surface rotation of ocean water in each major sea (a type of ◊current). Gyres are large and permanent, and occupy the northern and southern halves of the three major oceans. Their movements are dictated by the prevailing winds and the ◊Coriolis effect. Gyres move clockwise in the northern hemisphere and anticlockwise in the southern hemisphere.

gyroscope mechanical instrument, used as a stabilizing device and consisting, in its simplest form, of a heavy wheel mounted on an axis fixed in a ring that can be rotated about another axis, which is also fixed in a ring capable of rotation about a third axis.

Applications of the gyroscope principle include the gyrocompass, the gyropilot for automatic steering, and gyro-directed torpedoes.

The components of the gyroscope are arranged so that the three axes of rotation in any position pass through the wheel's centre of gravity. The wheel is thus capable of rotation about three mutually perpendicular axes, and its axis may take up any direction. If the axis of the spinning wheel is displaced, a restoring movement develops, returning it to its initial direction.

GZip in computing, compression software, properly called ◊GNU Zip, commonly used on the Internet. Files compressed using GZip can be recognized by the file extension '.GZ'. The software is published by the ◊Free Software Foundation and was originally developed for UNIX, although a DOS version is readily available.

Haber, Fritz
(1868–1934)

German chemist whose conversion of atmospheric nitrogen to ammonia opened the way for the synthetic fertilizer industry. His study of the combustion of hydrocarbons led to the commercial 'cracking' or fractional distillation of natural oil (petroleum) into its components (for example, diesel, petrol, and paraffin). In electrochemistry, he was the first to demonstrate that oxidation and reduction take place at the electrodes; from this he developed a general electrochemical theory.

At the outbreak of World War I in 1914, Haber was asked to devise a method of producing nitric acid for making high explosives. Later he became one of the principals in the German chemical-warfare effort, devising weapons and gas masks, which led to protests against his Nobel prize in 1918.

Mary Evans Picture Library

ha symbol for ◊hectare.

Haber process or ***Haber–Bosch process*** industrial process by which ammonia is manufactured by direct combination of its elements, nitrogen and hydrogen. The reaction is carried out at 400–500°C/752–932°F and at 200 atmospheres pressure. The two gases, in the proportions of 1:3 by volume, are passed over a ◊catalyst of finely divided iron.

Around 10% of the reactants combine, and the unused gases are recycled. The ammonia is separated either by being dissolved in water or by being cooled to liquid form.

$$N_2 + 3H_2 \leftrightarrow 2NH_3$$

habitat localized ◊environment in which an organism lives, and which provides for all (or almost all) of its needs. The diversity of habitats found within the Earth's ecosystem is enormous, and they are changing all the time. Many can be considered inorganic or physical; for example, the Arctic ice cap, a cave, or a cliff face. Others are more complex; for instance, a woodland or a forest floor. Some habitats are so precise that they are called **microhabitats**, such as the area under a stone where a particular type of insect lives. Most habitats provide a home for many species.

hacking unauthorized access to a computer, either for fun or for malicious or fraudulent purposes. Hackers generally use microcomputers and telephone lines to obtain access. In computing, the term is used in a wider sense to mean using software for enjoyment or self-education, not necessarily involving unauthorized access. The most destructive form of hacking is the introduction of a computer ◊virus.

Hacking can be divided into four main areas: ◊viruses, phreaking, software piracy (stripping away the protective coding that should prevent the software being copied), and accessing operating systems.

A 1996 US survey co-sponsored by the FBI showed 41% of academic, corporate, and government organizations interviewed had had their computer systems hacked into during 1995.

haddock marine fish belonging to the cod family and found off the N Atlantic coastline. It is brown with silvery underparts and black markings above the pectoral fins. It can grow up to 1 m/3 ft in length. Haddock are important food fish; about 45 million kg/100 million lb are taken annually off the New England fishing banks alone. (Species *Melanogrammus aeglefinus,* family Gadidae.)

Hadrian's Wall Roman frontier system built AD 122–26 to mark England's northern boundary and abandoned about 383; its ruins run 185 km/115 mi from Wallsend on the river Tyne to Maryport, W Cumbria. In some parts, the wall was covered with a glistening, white coat of mortar. The fort at South Shields, Arbeia, built to defend the eastern end, has been under reconstruction.

hadron in physics, a subatomic particle that experiences the strong nuclear force. Each is made up of two or three indivisible particles called ◊quarks. The hadrons are grouped into the ◊baryons (protons, neutrons, and hyperons) and the ◊mesons (particles with masses between those of electrons and protons).

Hadron–Electron Ring Accelerator (HERA) particle ◊accelerator built under the streets of Hamburg, Germany, occupying a tunnel 6.3 km/3.9 mi in length. It is the world's most powerful collider of protons and electrons, designed to accelerate protons to energies of 820 GeV (billion electron volts), and electrons to 30 GeV. HERA began operating 1992.

HERA can propel electrons into the proton interior, where they interact with the proton's constituent particles: three ◊quarks and a number of ◊gluons.

haematology medical speciality concerned with disorders of the blood.

haemoglobin protein used by all vertebrates and some invertebrates for oxygen transport because the two substances combine reversibly. In vertebrates it occurs in red blood cells (erythrocytes), giving them their colour.

In the lungs or gills where the concentration of oxygen is high, oxygen attaches to haemoglobin to form **oxyhaemoglobin**. This process effectively increases the amount of oxygen that can be carried in the bloodstream. The oxygen is later released in the body tissues where it is at a low concentration, and the deoxygenated blood returned to the lungs or gills. Haemoglobin will combine also with carbon monoxide to form carboxyhaemoglobin, but in this case the reaction is irreversible.

haemolymph circulatory fluid of those molluscs and insects that have an 'open' circulatory system. Haemolymph contains water, amino acids, sugars, salts, and white cells like those of blood. Circulated by a pulsating heart, its main functions are to transport digestive and excretory products around the body. In molluscs, it also transports oxygen and carbon dioxide.

haemolysis destruction of red blood cells. Aged cells are constantly being lysed (broken down), but increased wastage of red cells is seen in some infections and blood disorders. It may result in ◊jaundice (through the release of too much haemoglobin) and in ◊anaemia.

haemophilia any of several inherited diseases in which normal blood clotting is impaired. The sufferer experiences prolonged bleeding from the slightest wound, as well as painful internal bleeding without apparent cause.

Haemophilias are nearly always sex-linked, transmitted through the female line only to male infants; they have afflicted a number of European royal households. Males affected by the most common form are unable to synthesize Factor VIII, a protein involved in the clotting of blood. Treatment is primarily with Factor VIII (now mass-produced by recombinant techniques), but the haemophiliac remains at risk from the slightest incident of bleeding. The disease is a painful one that causes deformities of joints.

haemorrhage loss of blood from the circulatory system. It is 'manifest' when the blood can be seen, as when it flows from a wound, and 'occult' when the bleeding is internal, as from an ulcer or internal injury.

Rapid, profuse haemorrhage causes ◊shock and may prove fatal if the circulating volume cannot be replaced in time. Slow, sustained bleeding may lead to ◊anaemia. Arterial bleeding is potentially more serious than blood lost from a vein. It may be stemmed by applying pressure directly to the wound.

haemorrhoids distended blood vessels (◊varicose veins) in the area of the anus, popularly called **piles**.

haemostasis natural or surgical stoppage of bleeding. In the natural mechanism, the damaged vessel contracts, restricting the flow, and blood ◊platelets plug the opening, releasing chemicals essential to clotting.

hafnium Latin *Hafnia* 'Copenhagen' silvery, metallic element, symbol Hf, atomic number 72, relative atomic mass 178.49. It occurs in nature in ores of zirconium, the properties of which it resembles. Hafnium absorbs neutrons better than most metals, so it is used in the control rods of nuclear reactors; it is also used for light-bulb filaments.

It was named in 1923 by Dutch physicist Dirk Coster (1889–1950) and Hungarian chemist Georg von Hevesy after the city of Copenhagen, where the element was discovered.

hail precipitation in the form of pellets of ice (hailstones). It is caused by the circulation of moisture in strong convection currents, usually within cumulonimbus ◊clouds.

Water droplets freeze as they are carried upwards. As the circulation continues, layers of ice are deposited around the droplets until they become too heavy to be supported by the currents and they fall as a hailstorm.

HAIL

Hailstones can kill. In the Gopalganji region of Bangladesh in 1988, 92 people died after being hit by huge hailstones weighing up to 1 kg/2.2 lb.

hair fine filament growing from mammalian skin. Each hair grows from a pit-shaped follicle embedded in the second layer of the skin, the dermis. It consists of dead cells impregnated with the protein keratin.

The average number of hairs on a human head varies from 98,000 (red-heads) to 120,000 (blondes). Each grows at the rate of 5–10 mm/0.2–0.4 in per month, lengthening for about three years before being replaced by a new one. A coat of hair helps to insulate land mammals by trapping air next to the body. The thickness of this layer can be varied at will by raising or flattening the coat. In some mammals a really heavy coat may be so effective that it must be shed in summer and a thinner one grown. Hair also aids camouflage, as in the zebra and the white winter coats of Arctic animals; and protection, as in the porcupine and hedgehog; bluffing enemies by apparently increasing the size, as in the cat; sexual display, as in humans and the male lion; and its colouring or erection may be used for communication. In 1990 scientists succeeded for the first time in growing human hair in vitro.

hairstreak any of a group of small butterflies, related to blues and coppers. Hairstreaks live in both temperate and tropical regions. Most of them are brownish or greyish-blue with hairlike tips streaked with white at the end of their hind wings. (Genera *Callophrys* and other related genera, family Lycaenidae.)

hake any of various marine fishes belonging to the cod family, found in N European, African, and American waters. They have silvery elongated bodies and grow up to 1 m/3 ft in length. They have two dorsal fins and one long anal fin. The silver hake (*M. bilinearis*) is an important food fish. (Genera *Merluccius* and *Urophycis*, family Gadidae.)

hakea shrub or tree of the Australian genus *Hakea*, family Proteaceae, characterized by hard woody fruit with winged seeds.

Hale-Bopp, Comet see ◊Comet Hale-Bopp.

half duplex in computing, a ◊modem setting which controls whether or not characters echo to (appear on) the screen. See ◊full duplex.

half-life during ◊radioactive decay, the time in which the strength of a radioactive source decays to half its original value. In theory, the decay process is never complete and there is always some residual radioactivity. For this reason, the half-life of a radioactive isotope is measured, rather than the total decay time. It may vary from millionths of a second to billions of years.

Radioactive substances decay exponentially; thus the time taken for the first 50% of the isotope to decay will be the same as the time taken by the next 25%, and by the 12.5% after that, and so on.

For example, carbon-14 takes about 5,730 years for half the material to decay; another 5,730 for half of the remaining half to decay; then 5,730 years for half of that remaining half to decay, and so on. Plutonium-239, one of the most toxic of all radioactive substances, has a half-life of about 24,000 years.

halftone in computing, term used in the publishing industry for a black-and-white photograph, indicating the many shades of grey that must be reproduced.

halftone process technique used in printing to reproduce the full range of tones in a photograph or other illustration. The intensity of the printed colour is varied from full strength to the lightest shades, even if one colour of ink is used. The picture to be reproduced is photographed through a screen ruled with a rectangular mesh of fine lines, which breaks up the tones of the original into areas of dots that vary in frequency according to the intensity of the tone. In the darker areas the dots run together; in the lighter areas they have more space between them.

halibut any of a group of large flatfishes found in the Atlantic and Pacific oceans. The largest of the flatfishes, they may grow up to 2 m/6 ft in length and weigh 90–135 kg/200–300 lb. They are a very dark mottled brown or green above and pure white on the underside. The Atlantic halibut (*H. hippoglossus*) is caught offshore at depths from 180 m/600 ft to 730 m/2,400 ft. (Genus *Hippoglossus*, family Pleuronectidae.)

halide any compound produced by the combination of a ◊halogen, such as chlorine or iodine, with a less electronegative element (see ◊electronegativity). Halides may be formed by ◊ionic bonds or by ◊covalent bonds.

halite mineral form of sodium chloride, NaCl. Common ◊salt is the mineral halite. When pure it is colourless and transparent, but it is often pink, red, or yellow. It is soft and has a low density.

Halite occurs naturally in evaporite deposits that have precipitated on evaporation of bodies of salt water. As rock salt, it forms beds within a sedimentary sequence; it can also migrate upwards through surrounding rocks to form salt domes. It crystallizes in the cubic system.

Hall effect production of a voltage across a conductor or semiconductor carrying a current at a right angle to a surrounding magnetic field. It was discovered in 1897 by the US physicist Edwin Hall (1855–1938). It is used in the **Hall probe** for measuring the strengths of magnetic fields and in magnetic switches.

Halley's comet comet that orbits the Sun about every 76 years, named after Edmond Halley who calculated its orbit. It is the brightest and most conspicuous of the periodic comets. Recorded

Halley, Edmond (1656–1742)

English astronomer. He not only identified the comet that was later to be known by his name, but also compiled a star catalogue, detected the proper motion of stars, using historical records, and began a line of research that, after his death, resulted in a reasonably accurate calculation of the astronomical unit.

Halley calculated that the comet sightings reported in 1456, 1531, 1607, and 1682 all represented reappearances of the same comet. He reasoned that the comet would follow a parabolic path and announced in 1705 in his Synopsis Astronomia Cometicae that it would reappear in 1758. When it did, public acclaim for the astronomer was such that his name was irrevocably attached to it.

He made many other notable contributions to astronomy, including the discovery of the proper motions of Aldebaran, Arcturus, and Sirius, and working out a method of obtaining the solar parallax by observations made during a transit of Venus. He was Astronomer Royal from 1720.

Mary Evans Picture Library

sightings go back over 2,000 years. It travels around the Sun in the opposite direction to the planets. Its orbit is inclined at almost 20° to the main plane of the Solar System and ranges between the orbits of Venus and Neptune. It will next reappear 2061.

The comet was studied by space probes at its last appearance in 1986. The European probe *Giotto* showed that the nucleus of Halley's comet is a tiny and irregularly shaped chunk of ice, measuring some 15 km/10 m long by 8 km/5 m wide, coated by a layer of very dark material, thought to be composed of carbon-rich compounds. This surface coating has a very low ◊albedo, reflecting just 4% of the light it receives from the Sun. Although the comet is one of the darkest objects known, it has a glowing head and tail produced by jets of gas from fissures in the outer dust layer. These vents cover 10% of the total surface area and become active only when exposed to the Sun. The force of these jets affects the speed of the comet's travel in its orbit.

HALLEY'S COMET

http://www.fis.uc.pt/astronomy/solar/halley.htm

Attractive site devoted to the comet – with facts and statistics, images, and information about the spacecraft that have visited it.

hallmark official mark stamped on British gold, silver, and (from 1913) platinum, instituted in 1327 (royal charter of London Goldsmiths) in order to prevent fraud. After 1363, personal marks of identification were added. Now tests of metal content are carried out at authorized assay offices in London, Birmingham, Sheffield, and Edinburgh; each assay office has its distinguishing mark, to which is added a maker's mark, date letter, and mark guaranteeing standard.

hallucinogen any substance that acts on the ◊central nervous system to produce changes in perception and mood and often hallucinations. Hallucinogens include ◊LSD, peyote, and ◊mescaline. Their effects are unpredictable and they are illegal in most countries.

In some circumstances hallucinogens may produce panic or even suicidal feelings, which can recur without warning several days or months after taking the drug. In rare cases they produce an irreversible psychotic state mimicking schizophrenia. Spiritual or religious experiences are common, hence the ritual use of hallucinogens in some cultures. They work by chemical interference with the normal action of neurotransmitters in the brain.

Reality is not only more fantastic than we think, but also much more fantastic than we imagine.

JBS Haldane British physiologist. Attributed remark

halogen any of a group of five nonmetallic elements with similar chemical bonding properties: fluorine, chlorine, bromine, iodine, and astatine. They form a linked group in the ◊periodic table of the elements, descending from fluorine, the most reactive, to astatine, the least reactive. They combine directly with most metals to form salts, such as common salt (NaCl). Each halogen has seven electrons in its valence shell, which accounts for the chemical similarities displayed by the group.

halon organic chemical compound containing one or two carbon atoms, together with ◊bromine and other ◊halogens. The most commonly used are halon 1211 (bromochlorodifluoromethane) and halon 1301 (bromotrifluoromethane). The halons are gases and are widely used in fire extinguishers. As destroyers of the ◊ozone layer, they are up to ten times more effective than ◊chlorofluorocarbons (CFCs), to which they are chemically related.

Levels in the atmosphere are rising by about 25% each year, mainly through the testing of fire-fighting equipment. The use of halons in fire extinguishers was banned in 1994.

halophyte plant adapted to live where there is a high concentration of salt in the soil, for example, in salt marshes and mud flats.

hamadryad or ***king cobra*** or ***giant cobra*** large and poisonous cobra found from India to China and the Philippines, sometimes reaching a length of 5 m/16 ft. It is one of the longest and most venomous of snakes, and is yellow with black crossbands.
classification The hamadryad *Ophiophagus hannah* is in family Elapidae, suborder Serpentes, order Squamata, class Reptilia.

hammerhead any of several species of shark found in tropical seas, characterized by having eyes at the ends of flattened hammerlike extensions of the skull. Hammerheads can grow to 4 m/13 ft in length. (Genus *Sphyrna*, family Sphyrnidae.)

hamster any of a group of burrowing rodents with a thickset body, short tail, and cheek pouches to carry food. Several genera are found across Asia and in SE Europe. Hamsters are often kept as pets. (Genera include *Cricetus* and *Mesocricetus*, family Cricetidae.)

Species include the European and Asian **black-bellied** or **common hamster** (*C. cricetus*), about 25 cm/10 in long, which can be a crop pest and stores up to 90 kg/200 lb of seeds in its burrow. The **golden hamster** (*M. auratus*) lives in W Asia and SE Europe. All golden hamsters now kept as pets originated from one female and 12 young captured in Syria 1930.

hand unit used in measuring the height of a horse from front hoof to shoulder (withers). One hand equals 10.2 cm/4 in.

handfish very rare Tasmanian fish *Brachionichthys hirsutus* that moves along the seafloor on handlike fins. It is about 10 cm/4 in long and is found only in the coastal waters of S Tasmania.

handle in computing, term used on ◊Internet Relay Chat and other live chat services for a nickname.

A given user's handle may or may not be the same as his/her ◊user-ID; on many systems users are allowed to pick any name they like to use on chat systems as long as it is not already taken by another user.

Handles are also used on CB and ham radio, and hackers use handles, for cultural reasons as much as to disguise their real identities.

handshake in computing, an exchange of signals between two devices that establishes the communications channels and protocols necessary for the devices to send and receive data.

handwriting recognition in computing, ability of a computer to accept handwritten input and turn it into ◊digital data that can be processed and displayed or stored as ◊ASCII characters on the computer screen.

Handwriting recognition would free computer users from having to use the keyboard, but it is difficult to implement. A few machines, such as the Apple Newton, have the ability built in, but the technology is still at an early stage of development and such machines typically require users to train the machine by entering a sample alphabet. Technical limitations mean written input has to be printed in small boxes.

Hannover German fighter aircraft made by the *Hannoverische Wagenfabrik* from 1917. A two-seater biplane, the Hannover CLII, was also used for ground attack and as an escort for bombers. Improved as the CLIII A, it became a major part of the German Air Force, over 550 being built in 1917–18.

Hansa-Brandenburg German seaplanes, the most popular floatplanes with the German Navy in World War I. They continued in use after the war, and the chief designer, Ernst Heinkel, become famous as a manufacturer of warplanes during World War II.

haploid having a single set of ◊chromosomes in each cell. Most higher organisms are ◊diploid – that is, they have two sets – but their gametes (sex cells) are haploid. Some plants, such as mosses, liverworts, and many seaweeds, are haploid, and male honey bees are haploid because they develop from eggs that have not been fertilized. See also ◊meiosis.

hard copy computer output printed on paper.

hard disc in computing, a storage device usually consisting of a rigid metal ◊disc coated with a magnetic material. Data are read from and written to the disc by means of a disc drive. The hard disc may be permanently fixed into the drive or in the form of a disc pack that can be removed and exchanged with a different pack. Hard discs vary from large units with capacities of more than 3,000 megabytes, intended for use with mainframe computers, to small units with capacities as low as 20 megabytes, intended for use with microcomputers.

hardening of oils transformation of liquid oils to solid products by ◊hydrogenation.

Vegetable oils contain double covalent carbon-to-carbon bonds and are therefore examples of ◊unsaturated compounds. When hydrogen is added to these double bonds, the oils become saturated. The more saturated oils are waxlike solids.

hardness physical property of materials that governs their use. Methods of heat treatment can increase the hardness of metals. A scale of hardness was devised by German–Austrian mineralogist Friedrich Mohs in the 1800s, based upon the hardness of certain minerals from soft talc (Mohs' hardness 1) to diamond (10), the hardest of all materials.

hard-sectored disc floppy disc that is sold already formatted, so that ◊disc formatting is not necessary. Usually sectors are marked by holes near the hub of the disc. This system is now obsolete.

hardware the mechanical, electrical, and electronic components of a computer system, as opposed to the various programs, which constitute ◊software.

hard water water that does not lather easily with soap, and produces a deposit or 'scale' in kettles. It is caused by the presence of certain salts of calcium and magnesium.

Temporary hardness is caused by the presence of dissolved hydrogencarbonates (bicarbonates); when the water is boiled, they are converted to insoluble carbonates that precipitate as 'scale'. **Permanent hardness** is caused by sulphates and silicates, which are not affected by boiling. Water can be softened by ◊distillation, ◊ion exchange (the principle underlying commercial water softeners), addition of sodium carbonate or of large amounts of soap, or boiling (to remove temporary hardness).

Hardy–Weinberg equilibrium in population genetics, the theoretical relative frequency of different ◊alleles within a given population of a species, when the stable endpoint of evolution in an undisturbed environment is reached.

hare mammal closely related to the rabbit, similar in appearance but larger. Hares have very long black-tipped ears, long hind legs, and short upturned tails. (Genus *Lepus,* family Leporidae, order Lagomorpha.)

Throughout the long breeding season (June–August) there are chases and 'boxing matches' among males and females; the expression 'mad as a March hare' arises from this behaviour.

harebell perennial plant of the ◊bellflower family, with bell-shaped blue flowers, found on dry grassland and heaths. It is known in Scotland as the bluebell. (*Campanula rotundifolia,* family Campanulaceae.)

Hare's apparatus in physics, a specific kind of ◊hydrometer used to compare the relative densities of two liquids, or to find the density of one if the other is known. It was invented by US chemist Robert Hare (1781–1858).

It consists of a vertical E-shaped glass tube, with the long limbs dipping into the two liquids and a tap on the short limb. With the tap open, air is removed from the tops of the tubes and the liquids are pushed up the tubes by atmospheric pressure. When the tap is closed, the heights of the liquids are inversely proportional to their relative densities.

Harrier the only truly successful vertical takeoff and landing fixed-wing aircraft, often called the **jump jet**. It was built in Britain and made its first flight 1966. It has a single jet engine and a set of swivelling nozzles. These deflect the jet exhaust vertically downwards for takeoff and landing, and to the rear for normal flight. Designed to fly from confined spaces with minimal ground support, it refuels in midair.

harrier any of a group of birds of prey. Harriers have long wings and legs, a small head with a short beak, an owl-like frill of thickset feathers around the face, and soft plumage. They eat frogs, birds, snakes, and small mammals, and are found mainly in marshy areas throughout the world. (Genus *Circus,* family Accipitridae, order Falconiformes.)

harrier breed of hound, similar to a ◊foxhound but smaller, used in packs for hare-hunting.

harrow agricultural implement used to break up the furrows left by the ◊plough and reduce the soil to a fine consistency or tilth, and to cover the seeds after sowing. The traditional harrow consists of spikes set in a frame; modern harrows use sets of discs.

hartebeest large African antelope with lyre-shaped horns set close on top of the head in both sexes. It can grow to 1.5 m/5 ft tall at the rather humped shoulders and up to 2 m/6 ft long. Although they are clumsy-looking runners, hartebeest can reach speeds of 65 kph/40 mph. (Species *Alcelaphus buselaphus,* family Bovidae.)

hart's-tongue fern with straplike undivided fronds, up to 60 cm/24 in long, which have clearly visible brown spore-bearing organs on the undersides. The plant is native to Europe, Asia, and E North America, and is found on walls, in shady rocky places, and in woods. (*Phyllitis scolopendrium,* family Polypodiaceae.)

harvestman small animal (an ◊arachnid) related to spiders with very long, thin legs and a small body. Harvestmen are different from true spiders in that they do not have a waist or narrow part to the oval body. They feed on small insects and spiders, and lay their eggs in autumn, to hatch the following spring or early summer. They are found from the Arctic to the tropics. (Order Opiliones.)

harvest mite another name for the ◊chigger, a parasitic mite.

Harwell main research establishment of the United Kingdom Atomic Energy Authority, situated near the village of Harwell in Oxfordshire.

hash function in computing, an ◊algorithm that calculates a value from the content of a message which can then be used to detect alterations to the original message.

Similar to a ◊checksum but with greater security, hash functions play an important role in secure cryptographic systems (see ◊cryptography), where authentication is as important as hiding the data from third parties.

hashing in computing, the process used to convert a record, usually in a database, into a number that can be used to retrieve the record, or check its validity. The 'hashing algorithm', which may be based on manipulating the ASCII values of letters, will be devised so that different records give a useful range of results. Hashing is faster than storing things alphabetically, for example, where some areas may have lots of very similar records (for example, under c, s, or t) while others are little used (q, x, z).

hashish drug made from the resin contained in the female flowering tops of hemp (◊cannabis).

hash total in computing, a ◊validation check in which an otherwise meaningless control total is calculated by adding together numbers (such as payroll or account numbers) associated with a set of records. The hash total is checked each time data are input, in order to ensure that no entry errors have been made.

hassium synthesized, radioactive element of the ◊transactinide series, symbol Hs, atomic number 108, relative atomic mass 265. It was first synthesized in 1984 by the Laboratory for Heavy Ion Research in Darmstadt, Germany. Its temporary name was unniloctium.

haustorium (plural ***haustoria***) specialized organ produced by a parasitic plant or fungus that penetrates the cells of its host to absorb nutrients. It may be either an outgrowth of hyphae (see ◊hypha), as in the case of parasitic fungi, or of the stems of flowering parasitic plants, as in dodders (*Cuscuta*). The suckerlike haustoria of a dodder penetrate the vascular tissue of the host plant without killing the cells.

Havana cat breed of domestic shorthaired cat bred in Britain during the 1950s from a cross between two varieties of Siamese. It has a deep, rich-brown fur, wedge-shaped head, green eyes, long legs, and svelte build. In America, the Havana Brown is less Oriental in look than its British counterparts. It is sturdier with a medium-length body, rounder face, oval eyes, and longer fur. The name is derived from the colouring which resembles the Cuban cigar tobacco.

hawfinch European ◊finch, about 18 cm/7 in long. It feeds on berries and seeds, and can crack cherry stones with its large, powerful beak. The male bird has brown plumage, a black throat and black wings with a bold white shoulder stripe, a short white-tipped tail, and a broad band of grey at the back of the neck. (Species *Coccothraustes coccothraustes,* family Fringillidae, order Passeriformes.)

Hawfinches spend most of their time in the treetops, where they eat the fruits of pine, hornbeam, plum, cherry, hawthorn, laurel, and holly trees. They build their nests of twigs and mosses in lichen-covered trees, 2–10 m/6.5–33 ft above the ground. They are abundant in southern Europe and are also found in the temperate parts of Asia.

hawk any of a group of small to medium-sized birds of prey, belonging to the same family as eagles, kites, ospreys, and vultures.

HAWK

Some hawks have evolved a unique way of hanging on to their prey. The African harrier hawk and the Central and South American crane hawk can bend their legs both backwards and forwards from the middle joint. Their prey (mostly frogs, lizards, and baby birds) thus has great difficulty eluding them in even the trickiest crevices.

Hawks have short, rounded wings and a long tail compared with ◊falcons, and keen eyesight; the ◊sparrow hawk and ◊goshawk are examples. (Especially genera *Accipiter* and *Buteo,* family Accipitridae.)

Hawking, Stephen (William) (1942–)

English physicist whose work in general relativity – particularly gravitational field theory – led to a search for a quantum theory of gravity to explain black holes and the Big Bang, singularities that classical relativity theory does not adequately explain. His book *A Brief History of Time* 1988 gives a popular account of cosmology and became an international bestseller. His latest book is *The Nature of Space and Time*, written with Roger Penrose.

Hawking's objective of producing an overall synthesis of quantum mechanics and relativity theory began around the time of the publication in 1973 of his seminal book *The Large Scale Structure of Space-Time*, written with G F R Ellis. His most remarkable result, published in 1974, was that black holes could in fact emit particles in the form of thermal radiation – the so-called **Hawking radiation**.

If we find why it is that we and the universe exist, it would be the ultimate triumph of human reason – for then we would know the mind of God.

STEPHEN HAWKING English physicist.

hawk moth any member of a family of ◊moths with more than 1,000 species distributed throughout the world, but found mainly in tropical regions. Some South American hawk moths closely resemble hummingbirds. (Family Sphingidae.)

HAWK MOTH

The tongue of the Malagasy hawk moth is more than twice the length of its body. This enables it to feed from flowers without having to land on them, thus avoiding predators waiting in ambush.

hawthorn any of a group of shrubs or trees belonging to the rose family, growing abundantly in E North America, and also in Europe and Asia. All have alternate, toothed leaves and bear clusters of showy white, pink, or red flowers. Their small applelike fruits can be red, orange, blue, or black. Hawthorns are popular as ornamentals. (Genus *Crataegus,* family Rosaceae.)

hay preserved grass used for winter livestock feed. The grass is cut and allowed to dry in the field before being removed for storage in a barn.

The optimum period for cutting is when the grass has just come into flower and contains most feed value. During the natural drying process, the moisture content is reduced from 70–80% down to a safe level of 20%. In normal weather conditions, this takes from two to five days, during which time the hay is turned by machine to ensure even drying. Hay is normally baled before removal from the field.

Hayashi track in astronomy, a path on the ◊Hertzsprung–Russell diagram taken by protostars as they emerge from the clouds of dust and gas out of which they were born. A protostar appears on the right (cool) side of the Hertzsprung- Russell diagram and follows a Hayashi track until it arrives on the main sequence where hydrogen burning can start. It is named after the Japanese astrophysicist Chushiro Hayashi, who studied the theory of protostars in the 1960s.

hay fever allergic reaction to pollen, causing sneezing, with inflammation of the nasal membranes and conjunctiva of the eyes. Symptoms are due to the release of ◊histamine. Treatment is by antihistamine drugs.

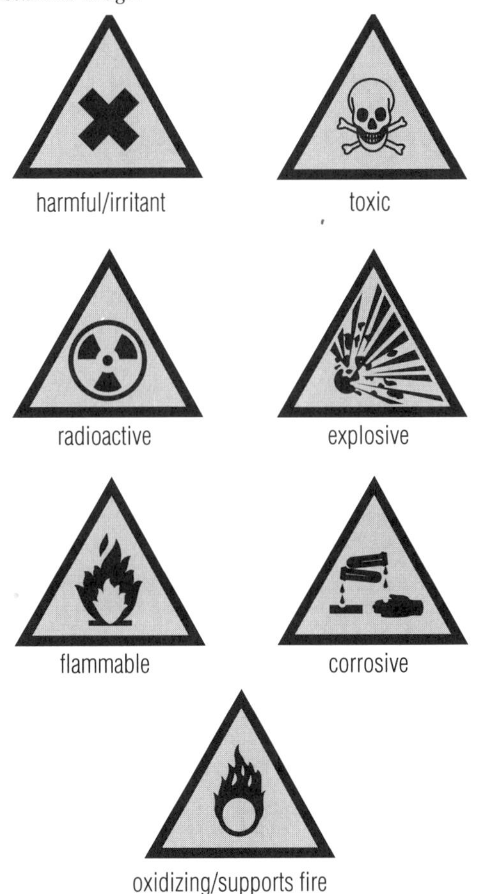

***hazard label** The internationally recognized symbols, warning of the potential dangers of handling certain substances.*

hazardous waste waste substance, usually generated by industry, that represents a hazard to the environment or to people living or working nearby. Examples include radioactive wastes, acidic resins, arsenic residues, residual hardening salts, lead from car exhausts, mercury, nonferrous sludges, organic solvents, asbestos, chlorinated solvents, and pesticides. The cumulative effects of toxic waste can take some time to become apparent (anything from a few hours to many years), and pose a serious threat to the ecological stability of the planet; its economic disposal or recycling is the subject of research.

haze factor unit of visibility in mist or fog. It is the ratio of the brightness of the mist compared with that of the object.

hazel any of a group of shrubs or trees that includes the European common hazel or cob (*C. avellana*), of which the filbert is the cultivated variety. North American species include the American hazel (*C. americana*). (Genus *Corylus,* family Corylaceae.)

HCI abbreviation for ◊human–computer interaction.

HDTV abbreviation for ◊high-definition television.

headache pain felt within the skull. Most headaches are caused by stress or tension, but some may be symptoms of brain or ◊systemic disease, including ◊fever.

Chronic daily headache may be caused by painkiller misuse, according to the European Headache Foundation in 1996. People who take daily analgesics to treat chronic headaches may actually be causing the headaches by doing so. See also ◊migraine.

header in computing, line or lines of text that appear at the beginning of each e-mail or USENET message sent across the Internet. The header includes important routing and identifying information, such as the sender's name, recipient's name (either a person or a newsgroup), date, time, and machine used when the message was composed, and the path by which the message arrived at its destination.

In the case of USENET postings, it also indicates if the message is intended for more than one group. The exact material is determined by ◊RFC (requests for comments) and discussion.

head louse parasitic insect that lives in human hair. See ◊louse.

health, world the health of people worldwide is monitored by the World Health Organization (WHO). Outside the industrialized world in particular, poverty and degraded environmental conditions mean that easily preventable diseases are widespread: WHO estimated in 1990 that 1 billion people, or 20% of the world's population, were diseased, in poor health, or malnourished. In North Africa and the Middle East, 25% of the population were ill.

vaccine-preventable diseases Every year, 46 million infants are not fully immunized; 2.8 million children die and 3 million are disabled due to vaccine-preventable diseases (polio, tetanus, diphtheria, whooping cough, tuberculosis, and measles).

diarrhoea Every year, there are 750 million cases in children, causing 4 million deaths. Oral rehydration therapy can correct dehydration and prevent 65% of deaths due to diarrhoeal disease. The basis of therapy is prepackaged sugar and salt. Treatment to cure the disease costs less than 20 cents, but fewer than one-third of children are treated in this way.

tuberculosis 1.6 billion people carry the bacteria, and there are 3 million deaths every year. Some 95% of all patients could be cured within six months using a specific antibiotic therapy which costs less than $30 per person.

prevention and cure Increasing health spending in industrialized countries by only $2 per head would enable immunization of all children to be performed, polio to be eradicated, and drugs provided to cure all cases of diarrhoeal disease, acute respiratory infection, tuberculosis, malaria, schistosomiasis, and most sexually transmitted diseases.

That physician will hardly be thought very careful of the health of others who neglects his own.

GALEN Greek physician and anatomist.
Of Protecting the Health bk V

---

hearing aid any device to improve the hearing of partially deaf people. Hearing aids usually consist of a battery-powered transistorized microphone/amplifier unit and earpiece. Some miniaturized aids are compact enough to fit in the ear or be concealed in the frame of eyeglasses.

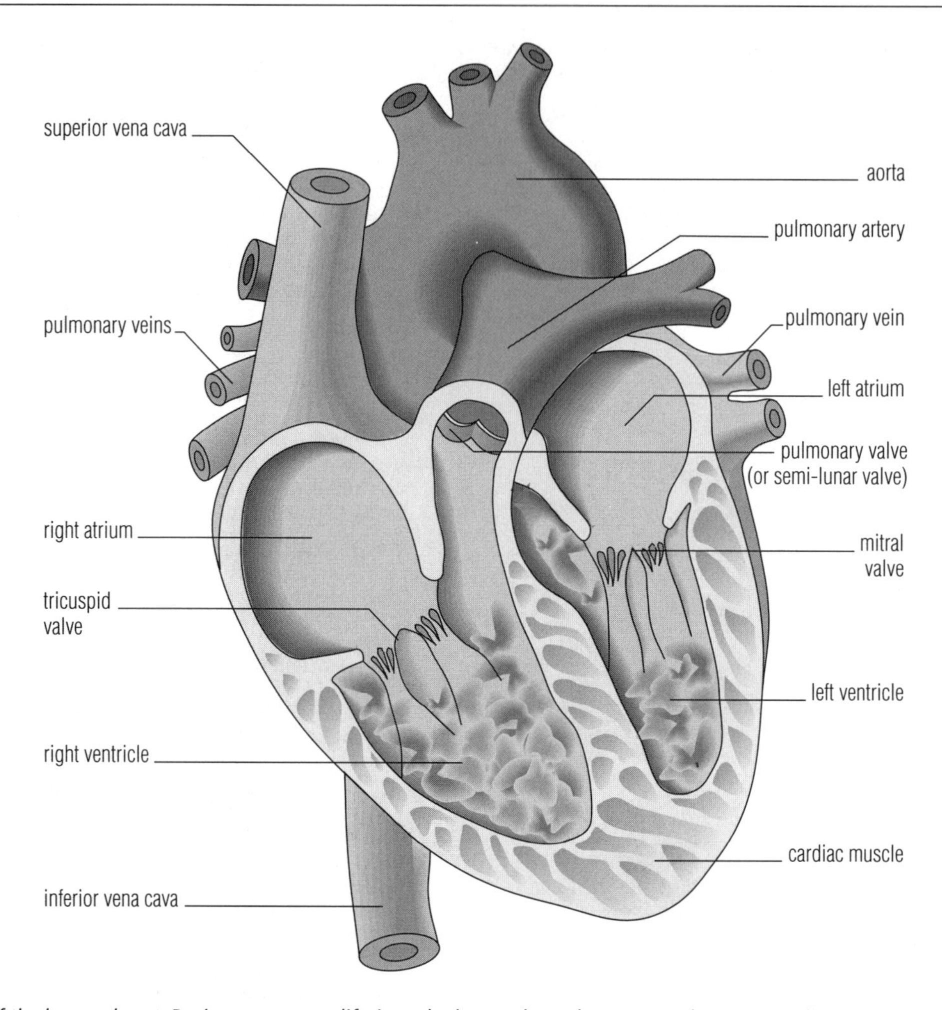

heart *The structure of the human heart. During an average lifetime, the human heart beats more than 2,000 million times and pumps 500 million l/110 million gal of blood. The average pulse rate is 70–72 beats per minute at rest for adult males, and 78–82 beats per minute for adult females.*

heart muscular organ that rhythmically contracts to force blood around the body of an animal with a circulatory system. Annelid worms and some other invertebrates have simple hearts consisting of thickened sections of main blood vessels that pulse regularly. An earthworm has ten such hearts. Vertebrates have one heart. A fish heart has two chambers – the thin-walled **atrium** (once called the auricle) that expands to receive blood, and the thick-walled **ventricle** that pumps it out. Amphibians and most reptiles have two atria and one ventricle; birds and mammals have two atria and two ventricles. The beating of the heart is controlled by the autonomic nervous system and an internal control centre or pacemaker, the **sinoatrial node**.

the cardiac cycle The cardiac cycle is the sequence of events during one complete cycle of a heart beat. This consists of the simultaneous contraction of the two atria, a short pause, then the simultaneous contraction of the two ventricles, followed by a longer pause while the entire heart relaxes. The contraction phase is called 'systole' and the relaxation phase which follows is called 'diastole'. The whole cycle is repeated 70–80 times a minute under resting conditions.

When the atria contract, the blood in them enters the two relaxing ventricles, completely filling them. The mitral and tricuspid valves, which were open, now begin to shut and as they do so, they create vibrations in the heart walls and tendons, causing the first heart sound. The ventricles on contraction push open the pulmonary and aortic valves and eject blood into the respective vessels. The closed mitral and tricuspid valves prevent return of blood into the atria during this phase. As the ventricles start to relax, the aortic and pulmonary valves close to prevent backward flow of blood, and their closure causes the second heart sound. By now, the atria have filled once again and are ready to start contracting to begin the next cardiac cycle.

heart attack or ***myocardial infarction*** sudden onset of gripping central chest pain, often accompanied by sweating and vomiting, caused by death of a portion of the heart muscle following obstruction of a coronary artery by thrombosis (formation of a blood clot). Half of all heart attacks result in death within the first two hours, but in the remainder survival has improved following the widespread use of thrombolytic (clot-buster) drugs.

After a heart attack, most people remain in hospital for seven to ten days, and may make a gradual return to normal activity over the following months. How soon a patient is able to return to work depends on the physical and mental demands of their job. Despite widespread fears to the contrary, it is safe to return to normal sexual activity within about a month of the attack.

AMERICAN HEART ASSOCIATION

http://www.amhrt.org/

Home page of the American Heart Association offers a risk assessment test, information about healthy living, including the effects of diet, and access to resources for both patients and carers.

heartbeat the regular contraction and relaxation of the heart, and the accompanying sounds. As blood passes through the heart a double beat is heard. The first is produced by the sudden closure of the valves between the atria and the ventricles. The second, slightly delayed sound, is caused by the closure of the valves found at the entrance to the major arteries leaving the heart. Diseased valves may make unusual sounds, known as heart murmurs.

heartburn burning sensation behind the breastbone (sternum). It results from irritation of the lower oesophagus (gullet) by excessively acid stomach contents, as sometimes happens during pregnancy and in cases of duodenal ulcer or obesity. It is often due to a weak valve at the entrance to the stomach that allows its contents to well up into the oesophagus.

heart–lung machine apparatus used during heart surgery to take over the functions of the heart and the lungs temporarily. It has a pump to circulate the blood around the body and is able to add oxygen to the blood and remove carbon dioxide from it. A heart–lung machine was first used for open-heart surgery in the USA in 1953.

heat form of energy possessed by a substance by virtue of the vibrating movement (kinetic energy) of its molecules or atoms. Heat energy is transferred by conduction, convection, and radiation. It always flows from a region of higher ◊temperature (heat intensity) to one of lower temperature. Its effect on a substance may be simply to raise its temperature, or to cause it to expand, melt (if a solid), vaporize (if a liquid), or increase its pressure (if a confined gas).
measurement Quantities of heat are usually measured in units of energy, such as joules (J) or calories (cal). The **specific heat** of a substance is the ratio of the quantity of heat required to raise the temperature of a given mass of the substance through a given range of temperature to the heat required to raise the temperature of an equal mass of water through the same range. It is measured by a ◊calorimeter.
conduction, convection, and radiation Conduction is the passing of heat along a medium to neighbouring parts with no visible motion accompanying the transfer of heat – for example, when the whole length of a metal rod is heated when one end is held in a fire. Convection is the transmission of heat through a fluid (liquid or gas) in currents – for example, when the air in a room is warmed by a fire or radiator. Radiation is heat transfer by infrared rays. It can pass through a vacuum, travels at the same speed as light, can be reflected and refracted, and does not affect the medium through which it passes. For example, heat reaches the Earth from the Sun by radiation.

For the transformation of heat, see ◊thermodynamics.

heat capacity in physics, the quantity of heat required to raise the temperature of an object by one degree. The **specific heat capacity** of a substance is the heat capacity per unit of mass, measured in joules per kilogram per kelvin ($J\ kg^{-1}\ K^{-1}$).

heat death in cosmology, a possible fate of the universe in which it continues expanding indefinitely while all the stars burn out and no new ones are formed. See ◊critical density.

heath in botany, any of a group of woody, mostly evergreen shrubs, including ◊heather, many of which have bell-shaped pendant flowers. They are native to Europe, Africa, and North America. (Common Old World genera *Erica* and *Calluna*, family Ericaceae.)

heather low-growing evergreen shrub of the ◊heath family, common on sandy or acid soil. The common heather (*Calluna vulgaris*) is a carpet-forming shrub, growing up to 60 cm/24 in high and bearing pale pink-purple flowers. It is found over much of Europe and has been introduced to North America.

heat of reaction alternative term for ◊energy of reaction.

heat pump machine, run by electricity or another power source, that cools the interior of a building by removing heat from interior air and pumping it out or, conversely, heats the inside by extracting energy from the atmosphere or from a hot-water source and pumping it in.

heat shield any heat-protecting coating or system, especially the coating (for example, tiles) used in spacecraft to protect the astronauts and equipment inside from the heat of re-entry when returning to Earth. Air friction can generate temperatures of up to 1,500°C/2,700°F on re-entry into the atmosphere.

heat storage any means of storing heat for release later. It is usually achieved by using materials that undergo phase changes, for example, Glauber's salt and sodium pyrophosphate, which melts at 70°C/158°F. The latter is used to store off-peak heat in the home: the salt is liquefied by cheap heat during the night and then freezes to give off heat during the day.

Other developments include the use of plastic crystals, which change their structure rather than melting when they are heated. They could be incorporated in curtains or clothing.

heatstroke or ***sunstroke*** rise in body temperature caused by excessive exposure to heat.

Mild heatstroke is experienced as feverish lassitude, sometimes with simple fainting; recovery is prompt following rest and replenishment of salt lost in sweat. Severe heatstroke causes collapse akin to that seen in acute ◊shock, and is potentially lethal without prompt treatment, including cooling the body carefully and giving fluids to relieve dehydration.

heat treatment in industry, the subjection of metals and alloys to controlled heating and cooling after fabrication to relieve internal stresses and improve their physical properties. Methods include ◊annealing, ◊quenching, and ◊tempering.

Heaviside's dolphin or **benguela dolphin** *Cephalorhynchus heavisidii* one of the least-known dolphins, confined to the coastal waters of Namibia. It is thought that about 100 a year are killed in purse seine nets from fishing boats, which could endanger this apparently rare species.

heavy horse powerful ◊horse specially bred for hauling wagons and heavy agricultural implements. After a decline following the introduction of the tractor, they are again being bred in increasing numbers and used for specialized work.

heavy industry industry that processes large amounts of bulky raw materials. Examples are the iron and steel industry, shipbuilding, and aluminium smelting. Heavy industries are often tied to locations close to their supplies of raw materials.

heavy water or ***deuterium oxide*** D_2O water containing the isotope deuterium instead of hydrogen (relative molecular mass 20 as opposed to 18 for ordinary water).

Its chemical properties are identical with those of ordinary water, but its physical properties differ slightly. It occurs in ordinary water in the ratio of about one part by mass of deuterium to 5,000 parts by mass of hydrogen, and can be concentrated by electrolysis, the ordinary water being more readily decomposed by this means than the heavy water. It has been used in the nuclear industry because it can slow down fast neutrons, thereby controlling the chain reaction.

hectare metric unit of area equal to 100 ares or 10,000 square metres (2.47 acres), symbol ha.

Trafalgar Square, London's only metric square, was laid out as one hectare.

hedge or ***hedgerow*** row of closely planted shrubs or low trees, generally acting as a land division and windbreak. Hedges also serve as a source of food and as a refuge for wildlife, and provide a ◊habitat not unlike the understorey of a natural forest.

hedgehog insectivorous mammal native to Europe, Asia, and Africa. The body, including the tail, is 30 cm/1 ft long. It is greyish brown in colour, has a piglike snout, and its back and sides are covered with sharp spines. When threatened it rolls itself into a ball bristling with spines. Hedgehogs feed on insects, slugs, mice, frogs, young birds, and carrion. Long-eared hedgehogs and desert hedgehogs are placed in different genera. (Genus *Erinaceus,* order Insectivora, family Erinaceidae.)

Hedgehogs normally shelter by day and go out at night. They find food more by smell and sound than by sight. The young are born in the late spring or early summer, and are blind, helpless, and covered with soft spines. For about a month they feed on their mother's milk, after which she teaches them to find their own food. In the autumn, hedgehogs make a nest of leaves and moss in the roots of a tree or in a hole in the ground and hibernate until spring.

hedge sparrow another name for the ◊dunnock, **a small European bird**.

height of a plane figure or solid, the perpendicular distance from the vertex to the base; see ◊altitude.

Heisenberg, Werner (Karl) (1901–1976)

German physicist who developed quantum theory and formulated the uncertainty principle, which concerns matter, radiation, and their reactions, and places absolute limits on the achievable accuracy of measurement. He was awarded a Nobel prize in 1932 for work he carried out when only 24.

Heisenberg was concerned not to try to picture what happens inside the atom but to find a mathematical system that explained it. His starting point was the spectral lines given by hydrogen, the simplest atom. Assisted by Max Born, Heisenberg presented his ideas in 1925 as a system called **matrix mechanics**. He obtained the frequencies of the lines in the hydrogen spectrum by mathematical treatment of values within matrices or arrays. His work was the first precise mathematical description of the workings of the atom and with it Heisenberg is regarded as founding quantum mechanics, which seeks to explain atomic structure in mathematical terms.

Heisenberg also was able to predict from studies of the hydrogen spectrum that hydrogen exists in two allotropes – ortho-hydrogen and para-hydrogen – in which the two nuclei of the atoms in a hydrogen molecule spin in the same or opposite directions respectively. The allotropes were discovered in 1929 (see allotropy).

In 1927, Heisenberg made the discovery of the **uncertainty principle**, for which he is best known. The uncertainty principle states that there is a theoretical limit to the precision with which a particle's position and momentum can be measured. In other words, it is impossible to specify precisely both the position and the simultaneous momentum (mass multiplied by velocity) of a particle. There is always a degree of uncertainty in either, and as one is determined with greater precision, the other can only be found less exactly. Heisenberg also formulated that multiplying the degrees of uncertainty of the position and momentum yields a value approximately equal to Planck's constant. The idea that the result of an action can be expressed only in terms of the probability of a certain effect was revolutionary, and it discomforted even Albert Einstein (1879–1955), but has remained valid.

In 1927, Heisenberg used the Pauli exclusion principle, which states that no two electrons can have all four quantum numbers the same, to show that ferromagnetism (the ability of some materials to acquire magnetism in the presence of an external magnetic field) is caused by electrostatic interaction between the electrons.

Mary Evans Picture Library

Helicobacter pylori spiral-shaped swimming bacterium that causes gastritis and stomach ◊ulcers when it colonizes the stomach lining. Without antibiotic treatment, infection can be permanent. *H. pylori* may also contribute towards stomach cancer.

Approximately 60% of 60-year-olds in the USA and western Europe are infected with *H. pylori;* the infection is rarely found in children. In developing countries 60–70% of children under 10 exhibit infection.

In 1997 scientists completed a full genetic blueprint (or genome) of *H. pylori.*

HELISPOT

http://www.helispot.com/

Helicopters of all shapes, sizes, and purposes are shown in photographs on this Web site. Police, military, news, and rescue helicopters are all featured in the hundreds of photographs listed. In addition there is an e-mail forum and many links to other helicopter resources

helicopter powered aircraft that achieves both lift and propulsion by means of a rotary wing, or rotor, on top of the fuselage. It can take off and land vertically, move in any direction, or remain stationary in the air. It can be powered by piston or jet engine. The ◊autogiro was a precursor.

The rotor of a helicopter has two or more blades of aerofoil cross-section like an aeroplane's wings. Lift and propulsion are achieved by angling the blades as they rotate. Experiments using the concept of helicopter flight date from the early 1900s, with the first successful liftoff and short flight in 1907. Ukrainian–US engineer Igor Sikorsky built the first practical single-rotor craft in the USA in 1939.

A single-rotor helicopter must also have a small tail rotor to counter the torque, or tendency of the body to spin in the opposite direction to the main rotor. Twin-rotor helicopters, like the Boeing Chinook, have their rotors turning in opposite directions to prevent the body from spinning. Helicopters are now widely used in passenger service, rescue missions on land and sea, police pursuits and traffic control, fire-fighting, and agriculture. In war they carry troops and equipment into difficult terrain, make aerial reconnaissance and attacks, and carry the wounded to aid stations. A fire-fighting helicopter was tested in Japan 1996, designed to reach skyscrapers beyond the reach of fire-engine ladders.

heliography old method of signalling, used by armies in the late 19th century, which employed sunlight reflected from a mirror to pass messages in ◊Morse code. On a clear day, a heliograph could send over distances in excess of 50 km/30 mi.

Also, an early photographic process by which a permanent image was formed on a glass plate.

helioseismology study of the Sun's structure by analysing vibrations and monitoring effects on the Sun's surface. Compare with ◊seismology.

heliosphere region of space through which the ◊solar wind flows outwards from the Sun. The **heliopause** is the boundary of this region, believed to lie about 100 astronomical units from the Sun, where the flow of the solar wind merges with the interstellar gas.

heliotrope decorative plant belonging to the borage family, with distinctive spikes of blue, lilac, or white flowers, including the Peruvian or cherry pie heliotrope (*H. peruvianum*). (Genus *Heliotropium,* family Boraginaceae.)

helium Greek *helios* 'Sun' colourless, odourless, gaseous, nonmetallic element, symbol He, atomic number 2, relative atomic mass 4.0026. It is grouped with the ◊inert gases, is nonreactive, and forms no compounds. It is the second-most abundant element (after hydrogen) in the universe, and has the lowest boiling (–268.9°C/–452°F) and melting points (–272.2°C/–458°F) of all the elements. It is present in small quantities in the Earth's atmosphere from gases issuing from radioactive elements (from ◊alpha decay) in the Earth's crust; after hydrogen it is the second lightest element.

Helium is a component of most stars, including the Sun, where the nuclear-fusion process converts hydrogen into helium with the

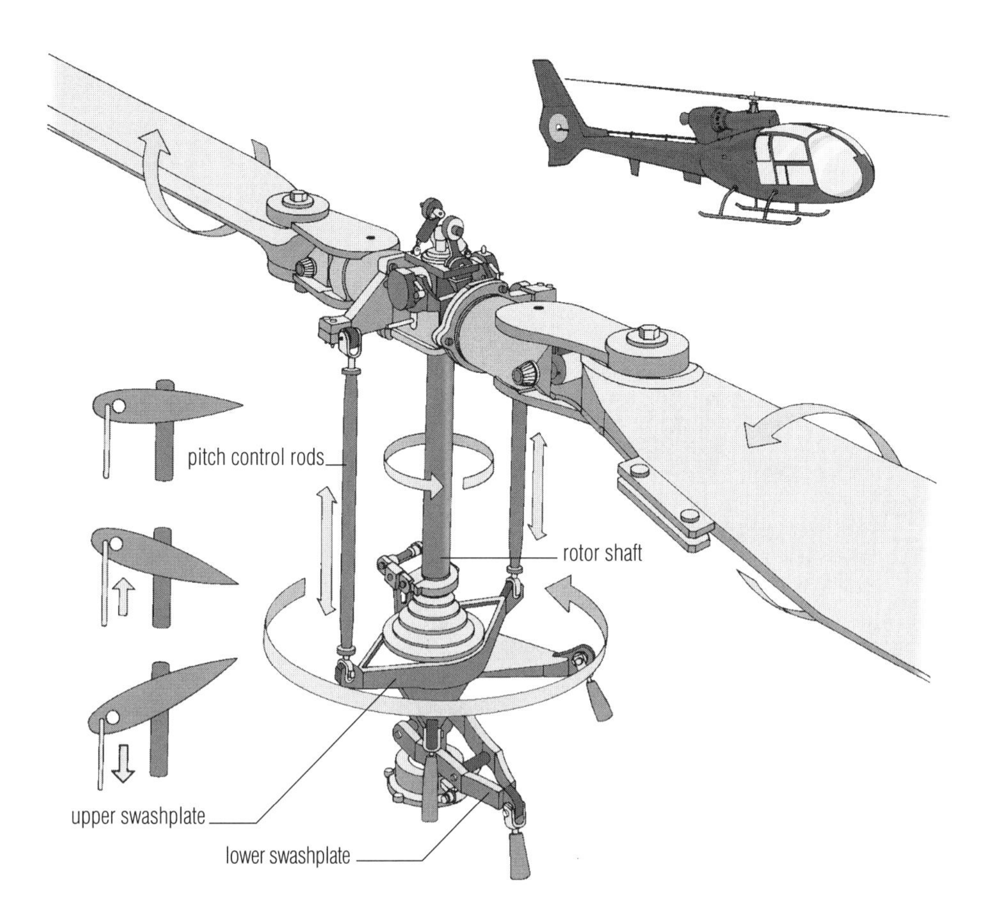

helicopter *The helicopter is controlled by varying the rotor pitch (the angle of the rotor blade as it moves through the air). For backwards flight, the blades in front of the machine have greater pitch than those behind the craft. This means that the front blades produce more lift and a backwards thrust. For forwards flight, the situation is reversed. In level flight, the blades have unchanging pitch.*

production of heat and light. It is obtained by compression and fractionation of naturally occurring gases. It is used for inflating balloons and as a dilutant for oxygen in deep-sea breathing systems. Liquid helium is used extensively in low-temperature physics (cryogenics).

helix in mathematics, a three-dimensional curve resembling a spring, corkscrew, or screw thread. It is generated by a line that encircles a cylinder or cone at a constant angle.

hellebore poisonous European ◊herbaceous plant belonging to the buttercup family. The stinking hellebore (*H. foetidus*) has greenish flowers early in the spring. (Genus *Helleborus,* family Ranunculaceae.)

helleborine one of several temperate Old World orchids, including the marsh helleborine (*E. palustris*) and the hellebore orchid (*E. helleborine*) introduced to North America. (Genera *Epipactis* and *Cephalanthera,* family Orchidaceae.)

helminth in medicine, collective name used to describe parasitic worms. There are several classes of helminth that can cause infections in humans, including ascarids (ascariasis), ◊tapeworms, and ◊threadworms.

helper application in computing, in Web ◊browsers, an external application that adds the ability to display certain types of files. Common helper applications include ◊RealAudio, which allows browsers to play live sound tracks such as radio broadcasts or recorded lectures; ◊Acrobat; and mIRC, which allows access to ◊Internet Relay Chat via the World Wide Web.

hematite principal ore of iron, consisting mainly of iron(III) oxide, Fe_2O_3. It occurs as **specular hematite** (dark, metallic lustre), **kidney ore** (reddish radiating fibres terminating in smooth, rounded surfaces), and a red earthy deposit.

Hemiptera large insect order consisting of the ◊bugs and containing about 55,000 species.

classification Hemiptera is in class Insecta, phylum Arthropoda.

The order is divided into two suborders.

suborder Homoptera These are the plant bugs. They are small to moderate-sized (from 3 mm/0.1 in to several centimetres/nearly an inch) insects with two pairs of wings usually present, held sloping over the body at rest. The forewings are usually evenly thickened with chitin; wingless forms are frequent. This suborder contains a large number of agricultural pests, some of which are vectors of plant viruses.

The important groups include the families Aleyrodidae, whiteflies; the Aphididae, greenflies; Cicadidae, cicadas; the Cicadellidae, leafhoppers; and the Psyllidae, plant lice. The superfamily Coccoidea, mealy bugs and scale insects, contains a further 16 families, many of which are important, including the lac insects from which shellac is obtained.

The plant bugs attack and destroy a wide range of plant life including grains, cereals, vegetables, fruit trees, and other trees. The aphids, one of the most important groups, transmit over 50 plant diseases.

suborder Heteroptera This suborder, characterized by having forewings unevenly thickened with chitin, is separated into two divisions: **Gymnocerata** have antennae that are usually longer than the head. They are terrestrial or water-skating forms. There are over 40 families, a few containing pests of agricultural importance. The Pyrrhocorridae, the cotton-stainers cause the most damage; the Coreidae, Pentastomidae, and Tingidae are of lesser importance. Most of the other families are predacious, feeding on other insects. Three families: Cimicidae, bedbugs; Polyctenidae; and Reduviidae, assassin bugs, are active bloodsuckers of mammals and birds. (Polyctenidae attacks only bats.)

suborder Cryptocerata Members of this suborder have antennae that are shorter than the head and are usually concealed. These bugs are usually predacious, and are truly aquatic, being adapted for swimming. This division includes the families Belostomatidae, giant water bugs; Corixidae, water boatmen; Nepidae, water-scorpions; and Notonectidae, backswimmers.

hemlock plant belonging to the carrot family, native to Europe, W Asia, and N Africa. It grows up to 2 m/6 ft high and produces delicate clusters of small white flowers. The whole plant, especially the root and fruit, is poisonous, causing paralysis of the nervous system. The name 'hemlock' is also given to some North American and Asiatic conifers (genus *Tsuga*) belonging to the pine family. (*Conium maculatum,* family Umbelliferae.)

hemp annual plant originally from Asia, now cultivated in most temperate countries for the fibres produced in the outer layer of the stem, which are used in ropes, twines, and, occasionally, in a type of linen or lace. The drug ◊cannabis is obtained from certain varieties of hemp. (*Cannabis sativa,* family Cannabaceae.)

The name 'hemp' is also given to other similar types of fibre: **sisal hemp** and **henequen** obtained from the leaves of *Agave* species native to Yucatán and cultivated in many tropical countries, and **manila hemp** obtained from *Musa textilis,* a plant native to the Philippines and the Maluku Islands, Indonesia.

henbane poisonous plant belonging to the nightshade family, found on waste ground throughout most of Europe and W Asia. It is a branching plant, up to 80 cm/31 in high, with hairy leaves and a sickening smell. The yellow flowers are bell-shaped. Henbane is used in medicine as a source of the drugs hyoscyamine and scopolamine. (*Hyoscyamus niger,* family Solanaceae.)

henna small shrub belonging to the loosestrife family, found in Iran, India, Egypt, and N Africa. The leaves and young twigs are ground to a powder, mixed to a paste with hot water, and applied to the fingernails and hair to give an orange-red hue. The colour may then be changed to black by applying a preparation of indigo. (*Lawsonia inermis,* family Lythraceae.)

Henna can also be used to dye both natural and synthetic textiles.

henry SI unit (symbol H) of ◊inductance (the reaction of an electric current against the magnetic field that surrounds it). One henry is the inductance of a circuit that produces an opposing voltage of one volt when the current changes at one ampere per second.

It is named after the US physicist Joseph Henry.

hepatitis any inflammatory disease of the liver, usually caused by a virus. Other causes include alcohol, drugs, gallstones, ◊lupus erythematous, and amoebic ◊dysentery. Symptoms include weakness, nausea, and jaundice.

Five different hepatitis viruses have been identified; A, B, C, D, and E. The hepatitis A virus (HAV) is the commonest cause of viral hepatitis, responsible for up to 40% of cases worldwide. It is spread by contaminated food. Hepatitis B, or serum hepatitis, is a highly contagious disease spread by blood products or in body fluids. It often culminates in liver failure, and is also associated with liver cancer, although only 5% of those infected suffer chronic liver damage. During 1995, 1.1 million people died of hepatitis B. Around 300 million people are ◊carriers. Vaccines are available against hepatitis A and B.

Hepatitis C is mostly seen in people needing frequent transfusions. Hepatitis D, which only occurs in association with hepatitis B, is common in the Mediterranean region. Hepatitis E is endemic in India and South America.

herb any plant (usually a flowering plant) tasting sweet, bitter, aromatic, or pungent, used in cooking, medicine, or perfumery; technically, a herb is any plant in which the aerial parts do not remain above ground at the end of the growing season.

HERB

The herb Catnip (*Nepeta cataria*) has a remarkable effect on cats. Members of the cat family roll over, extend their claws and twist in excitement when they smell the pungent odour of catnip. This reaction is believed to be caused by a chemical called trans-neptalacone, which is similar to a substance found in the female cat's urine.

herbaceous plant plant with very little or no wood, dying back at the end of every summer. The herbaceous perennials survive winters as underground storage organs such as bulbs and tubers.

herbalism in alternative medicine, the prescription and use of plants and their derivatives for medication. Herbal products are favoured by alternative practitioners as 'natural medicine', as opposed to modern synthesized medicines and drugs, which are regarded with suspicion because of the dangers of side effects and dependence.

Many herbal remedies are of proven efficacy both in preventing and curing illness. Medical herbalists claim to be able to prescribe for virtually any condition, except those so advanced that surgery is the only option.

herbarium collection of dried, pressed plants used as an aid to identification of unknown plants and by taxonomists in the ◊classification of plants. The plant specimens are accompanied by information, such as the date and place of collection, by whom collected, details of habitat, flower colour, and local names.

Herbaria range from small collections containing plants of a limited region, to the large university and national herbaria (some at ◊botanical gardens) containing millions of specimens from all parts of the world.

herbicide any chemical used to destroy plants or check their growth; see ◊weedkiller.

herbivore animal that feeds on green plants (or photosynthetic single-celled organisms) or their products, including seeds, fruit, and nectar. The most numerous type of herbivore is thought to be the zooplankton, tiny invertebrates in the surface waters of the oceans that feed on small photosynthetic algae. Herbivores are more numerous than other animals because their food is the most abundant. They form a vital link in the food chain between plants and carnivores.

herb Robert wild ◊geranium found throughout Europe and central Asia and naturalized in North America. About 30 cm/12 in high, it has hairy leaves and small pinkish to purplish flowers. (*Geranium robertianum,* family Geraniaceae.)

Herculaneum ancient city of Italy between Naples and Pompeii. Along with Pompeii, it was buried when Vesuvius erupted AD 79. It was excavated from the 18th century onwards.

Hercules in astronomy, the fifth-largest constellation, lying in the northern hemisphere. Despite its size it contains no prominent stars. Its most important feature is the best example in the northern hemisphere of a ◊globular cluster of stars 22,500 light years from Earth, which lies between Eta and Zeta Herculis.

Hercules beetle largest beetle in the world: males measure up to 17 cm/6.6 in in length; females are smaller. Hercules beetles are found mainly in the tropical and subtropical regions.

They are black, nocturnal, and exhibit extreme forms of sexual dimorphism (marked visual differences between males and females). Males have a dark-coloured, slender, curved horn of up to 10 mm/3.9 in long, on their heads.

classification Hercules beetles *Dynastes hercules* belong to the subfamily Dynastinae, family Scarabaeidae, order Coleoptera, class Insecta, phylum Arthropoda.

heredity in biology, the transmission of traits from parent to offspring. See also ◊genetics.

hermaphrodite organism that has both male and female sex organs. Hermaphroditism is the norm in such species as earthworms and snails, and is common in flowering plants. Cross-fertilization is the rule among hermaphrodites, with the parents functioning as male and female simultaneously, or as one or the other sex at different stages in their development. Human hermaphrodites are extremely rare.

hermit crab type of ◊crab.

hernia or ***rupture*** protrusion of part of an internal organ through a weakness in the surrounding muscular wall, usually in the groin. The appearance is that of a rounded soft lump or swelling.

heroin or ***diamorphine*** powerful ◊opiate analgesic, an acetyl derivative of ◊morphine. It is more addictive than morphine but causes less nausea. It has an important place in the control of severe pain in terminal illness, severe injuries, and heart attacks, but is widely used illegally.

Heroin was discovered in Germany in 1898.The major regions of opium production, for conversion to heroin, are the 'Golden Crescent' of Afghanistan, Iran, and Pakistan, and the 'Golden Triangle' across parts of Myanmar (Burma), Laos, and Thailand.

In 1971 there were 3,000 registered heroin addicts in the UK; in 1989 there were over 100,000.

heron large to medium-sized wading bird belonging to the same family as bitterns, egrets, night herons, and boatbills. Herons have sharp bills, broad wings, long legs, slender bodies, and soft plumage. They are found mostly in tropical and subtropical regions, but also in temperate zones, on lakes, fens, and mudflats, where they wade searching for prey. (Genera include *Ardea, Butorides,* and *Nycticorax;* family Ardeidae, order Ciconiiformes.)

They capture small animals, such as fish, molluscs, and worms, by spearing them with their long bills. Herons nest in trees or bushes, on ivy-covered rocks, or in reedbeds, making a loose fabric of sticks lined with grass or leaves; they lay greenish or drab-coloured eggs, varying in number from two to seven according to the different species.

herpes any of several infectious diseases caused by viruses of the herpes group. **Herpes simplex I** is the causative agent of a common inflammation, the cold sore. **Herpes simplex II** is responsible for genital herpes, a highly contagious, sexually transmitted disease characterized by painful blisters in the genital area. It can be transmitted in the birth canal from mother to newborn. **Herpes zoster** causes ◊shingles; another herpes virus causes chickenpox.

A number of antivirals treat these infections, which are particularly troublesome in patients whose immune systems have been suppressed medically; for example, after a transplant operation. The drug acyclovir, originally introduced for the treatment of genital herpes, has now been shown to modify the course of chickenpox and the related condition shingles, by reducing the duration of the illness.

herpetology the scientific study of ◊reptiles, including their classification, anatomy, physiology, behaviour, and ecology.

herring any of various marine fishes belonging to the herring family, but especially the important food fish *Clupea harengus.* A silvered greenish blue, it swims close to the surface, and may be 25–40 cm/10–16 in long. Herring travel in schools several kilometres long and wide. They are found in large quantities off the east coast of North America, and the shores of NE Europe. Overfishing and pollution have reduced their numbers. (Family Clupeidae.)

HERRING

The wolf herring *Chirocentrus dorab* is the largest fish in the herring order. It grows up to 3.7 m/12 ft long.

Herschel, (Frederick) William (1738–1822)

German-born English astronomer. He was a skilled telescope-maker, and pioneered the study of binary stars and nebulae. He discovered the planet Uranus in 1781 and infrared solar rays in 1801. He catalogued over 800 double stars, and found over 2,500 nebulae, catalogued by his sister Caroline Herschel; this work was continued by his son John Herschel. By studying the distribution of stars, William established the basic form of our Galaxy, the Milky Way.

Herschel discovered the motion of binary stars around one another, and recorded it in his *Motion of the Solar System in Space* 1783. In 1789 he built, in Slough, Berkshire, a 1.2-m/4-ft telescope of 12 m/40 ft focal length (the largest in the world at the time), but he made most use of a more satisfactory 46-cm/18-in instrument. He discovered two satellites of Uranus and two of Saturn.

Mary Evans Picture Library

hertz SI unit (symbol Hz) of frequency (the number of repetitions of a regular occurrence in one second). Radio waves are often measured in megahertz (MHz), millions of hertz, and the ◊clock rate of a computer is usually measured in megahertz. The unit is named after Heinrich Hertz.

Hertz, Heinrich Rudolf (1857–1894)

German physicist who studied electromagnetic waves, showing that their behaviour resembles that of light and heat waves.

Hertz confirmed James Clerk Maxwell's theory of electromagnetic waves. In 1888, he realized that electric waves could be produced and would travel through air, and he confirmed this experimentally. He went on to determine the velocity of these waves (which were later called radio waves) and, on showing that it was the same as that of light, devised experiments to show that the waves could be reflected, refracted, and diffracted.

Mary Evans Picture Library

Hertzsprung–Russell diagram in astronomy, a graph on which the surface temperatures of stars are plotted against their luminosities. Most stars, including the Sun, fall into a narrow band called the ◊main sequence. When a star grows old it moves from the main sequence to the upper right part of the graph, into the

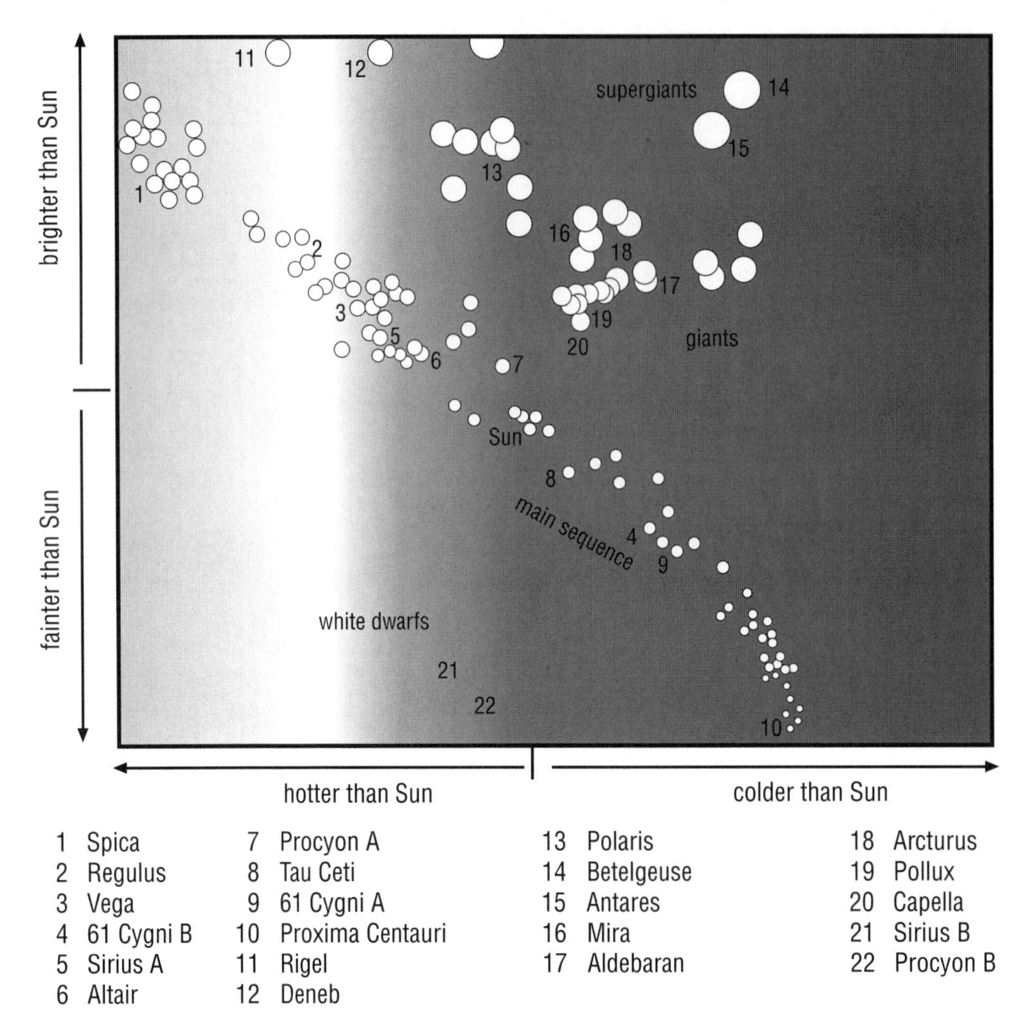

Hertzsprung–Russell diagram *The Hertzsprung–Russell diagram relates the brightness (or luminosity) of a star to its temperature. Most stars fall within a narrow diagonal band called the main sequence. A star moves off the main sequence when it grows old. The Hertzsprung–Russell diagram is one of the most important diagrams in astrophysics.*

Hertzsprung, Ejnar (1873–1967)

Danish astronomer and physicist. He introduced the concept of the absolute magnitude (brightness) of a star, and described the\ relationship between the absolute magnitude and the temperature of a star, formulating his results in the form of a diagram, known as the Hertzsprung–Russell diagram, that has become a standard reference.

His astronomical interests were very wide, but his observations were mainly of variable stars, double stars, and clusters.

area of the giants and supergiants. At the end of its life, as the star shrinks to become a white dwarf, it moves again, to the bottom left area. It is named after the Dane Ejnar Hertzsprung (1873–1967) and the American Henry Norris Russell (1877–1957), who independently devised it in the years 1911–13.

hessian fly or ***barley midge*** tiny black fly, a species of ◊gall midge that feeds on wheat and is considered a serious pest of winter wheat in the USA and New Zealand.

classification The hessian fly *Mayetiola destructor* belongs to the family Cecidomyiidae (suborder Nematocera) of order Diptera, class Insecta, phylum Arthropoda.

Hessian flies lay their eggs on the top surface of leaves of young wheat. The larvae feed on the leaf sheaths and also suck the plant sap. Wheat infested during the autumn appears stunted and the leaves are dark bluish-green. Intensely parasitized plants die during the winter.

The use of resistant wheat varieties in combination with delayed planting in the autumn (observing 'fly-safe' dates) has greatly reduced the damage caused by hessian flies in most wheat-producing areas of the USA.

Other allied pests include the sorghum midge *Contarinia sorghicola*, and the clover seed midge, *Dasyneura leguminicola*. The sorghum midge is a serious pest in the southern USA and the more humid areas of the Gulf States.

heterogeneous reaction in chemistry, a reaction where there is an interface between the different components or reactants. Examples of heterogeneous reactions are those between a gas and a solid, a gas and a liquid, two immiscible liquids, or two different solids.

heterosis or ***hybrid vigour*** improvement in physical capacities that sometimes occurs in the ◊hybrid produced by mating two genetically different parents.

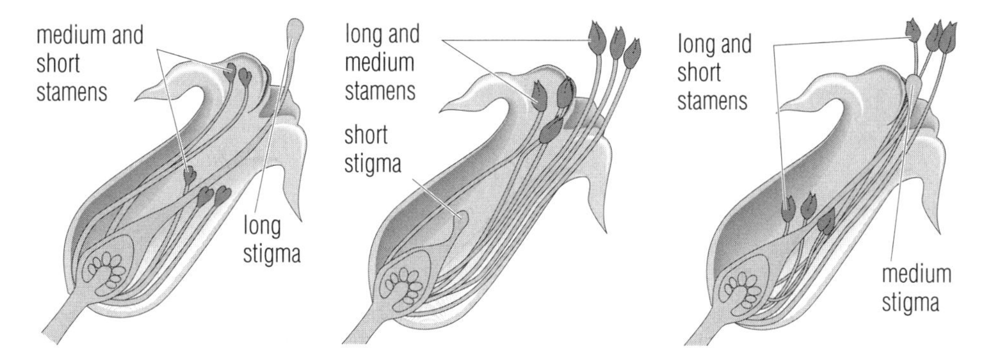

heterostyly *Heterostyly, in which lengths of the stamens and stigma differ in flowers of different plants of the same species. This is a device to ensure cross-pollination by visiting insects.*

The parents may be of different strains or varieties within a species, or of different species, as in the mule, which is stronger and has a longer life span than either of its parents (donkey and horse). Heterosis is also exploited in hybrid varieties of maize, tomatoes, and other crops.

heterostyly in botany, having ◊styles of different lengths.

Certain flowers, such as primroses (*Primula vulgaris*), have different-sized ◊anthers and styles to ensure cross-fertilization (through ◊pollination) by visiting insects.

heterotroph any living organism that obtains its energy from organic substances produced by other organisms. All animals and fungi are heterotrophs, and they include herbivores, carnivores, and saprotrophs (those that feed on dead animal and plant material).

heterozygous in a living organism, having two different ◊alleles for a given trait. In ◊homozygous organisms, by contrast, both chromosomes carry the same allele. In an outbreeding population an individual organism will generally be heterozygous for some genes but homozygous for others.

For example, in humans, alleles for both blue-and brown-pigmented eyes exist, but the 'blue' allele is ◊recessive to the dominant 'brown' allele.

Only individuals with blue eyes are predictably homozygous for this trait; brown-eyed people can be either homozygous or heterozygous.

heuristics in computing, a process by which a program attempts to improve its performance by learning from its own experience.

The separate branches of natural knowledge have a real and intimate connection.

ALEXANDER VON HUMBOLDT German botanist and geologist.
Cosmos 1845

Hewlett-Packard (often abbreviated to ***HP***) major manufacturer of computer and telecommunications hardware, founded 1939 by William Hewlett and David Packard and based in Palo Alto, California, USA. In 1996 the company was manufacturing more than 24,000 products, including medical equipment, analytical instruments, calculators, PCs, printers, workstations, and palmtops.

HP's sales grew dramatically from $13.2 billion in 1990 to $38.4 billion in 1996.

hexachlorophene $(C_6HCl_3OH)_2CH_2$ white, odourless bactericide, used in minute quantities in soaps and surgical disinfectants.

Trichlorophenol is used in its preparation, and, without precise temperature control, the highly toxic TCDD (tetrachlorodibenzodioxin; see ◊dioxin) may form as a by-product.

hexadecimal number system or ***hex*** number system to the base 16, used in computing. In hex the decimal numbers 0–15 are represented by the characters 0, 1, 2, 3, 4, 5, 6, 7, 8, 9, A, B, C, D, E, F.

Hexadecimal numbers are easy to convert to the computer's internal ◊binary code and are more compact than binary numbers.

Each place in a number increases in value by a power of 16 going from right to left; for instance, 8F is equal to 15 + (8 × 16) = 143 in decimal. Hexadecimal numbers are often preferred by programmers writing in low-level languages because they are more easily converted to the computer's internal binary (base-two) code than are decimal numbers, and because they are more compact than binary numbers and therefore more easily keyed, checked, and memorized.

hexagon six-sided ◊polygon.

HF in physics, abbreviation for **high** ◊frequency. HF radio waves have frequencies in the range 3–30 MHz.

HGV abbreviation for ***heavy goods vehicle***.

hibernation state of dormancy in which certain animals spend the winter. It is associated with a dramatic reduction in all metabolic processes, including body temperature, breathing, and heart rate. It is a fallacy that animals sleep throughout the winter.

The body temperature of the Arctic ground squirrel falls to below 0°C/32°F during hibernation. Hibernating bats may breathe only once every 45 minutes, and can go for up to 2 hours without taking a breath.

hibiscus any of a group of plants belonging to the mallow family. Hibiscuses range from large ◊herbaceous plants to trees. Popular as ornamental plants because of their brilliantly coloured, red to white, bell-shaped flowers, they include *H. syriacus* and *H. rosa-sinensis* of Asia and the rose mallow (*H. palustris*) of North America. (Genus *Hibiscus,* family Malvaceae.)

hickory tree belonging to the walnut family, native to North America and Asia. It provides a valuable timber, and all species produce nuts, though some are inedible. The pecan (*C. illinoensis*) is widely cultivated in the southern USA, and the shagbark (*C. ovata*) in the northern USA. (Genus *Carya,* family Juglandaceae.)

hidden file computer file in an ◊MS-DOS system that is not normally displayed when the directory listing command is given. Hidden files include certain system files, principally so that there is less chance of modifying or deleting them by accident, but any file can be made hidden if required.

hide or ***hyde*** Anglo-Saxon unit of measurement used to measure the extent of arable land; it varied from about 296 ha/120 acres in the east of England to as little as 99 ha/40 acres in Wessex. One hide was regarded as sufficient to support a peasant and his household; it was the area that could be ploughed in a season by one plough and one team of oxen.

The hide was the basic unit of assessment for taxation and military service; under Norman rule it became the basis for the feudal tax of hidage.

hierarchy in computing, on USENET, the structure for naming ◊newsgroups. All newsgroups on USENET are assigned to a major group. The ◊Big Seven hierarchies were the first to be set up, and setting up a new newsgroup in these involved following more or less formal procedures. The ◊alt hierarchy was set up to allow more flexibility. The biz hierarchy was set up 1994, after the first incidence of ◊spamming, to give advertising its own place.

A number of other hierarchies are available, set up for specific countries (**de** is Germany, **dk** is Denmark, **uk** is Britain); for Internet Service Providers (Demon, CompuServe, and AOL all have their own local groups); or for individual companies.

hi-fi (abbreviation for ***high-fidelity***) faithful reproduction of sound from a machine that plays recorded music or speech. A typical hi-fi system includes a turntable for playing vinyl records, a cassette tape deck to play magnetic tape recordings, a tuner to pick up radio broadcasts, an amplifier to serve all the equipment, possibly a compact-disc player, and two or more loudspeakers.

Advances in mechanical equipment and electronics, such as digital recording techniques and compact discs, have made it possible to eliminate many distortions in sound-reproduction processes.

Higgs boson or ***Higgs particle*** postulated ◊elementary particle whose existence would explain why particles have mass. The current theory of elementary particles, called the ◊standard model, cannot explain how mass arises. To overcome this difficulty, Peter Higgs (1929–) of the University of Edinburgh and Thomas Kibble (1932–) of Imperial College, London proposed in 1964 a new particle that binds to other particles and gives them their mass. The Higgs boson has not yet been detected experimentally.

high-definition television (HDTV) ◊television system offering a significantly greater number of scanning lines, and therefore a clearer picture, than that provided by conventional systems. Typically, HDTV has about twice the horizontal and vertical resolution of current 525-line (such as the American standard, NTSC) or 625-line standands (such as the British standard, PAL); a frame rate of at least 24 Hz; and a picture aspect ratio of 9:16 instead of the current 3:4. HDTV systems have been in development since the mid-1970s.

The Japanese HDTV system, or HiVision as it is trade-named in Japan, uses 1,125 scanning lines and an aspect ratio of 16:9 instead of the squarish 4:3 that conventional television uses. A European HDTV system, called HD-MAC, using 1,250 lines, is under development. In the USA, a standard incorporating digital techniques is being discussed.

highest common factor (HCF) in a set of numbers, the highest number that will divide every member of the set without leaving a remainder. For example, 6 is the highest common factor of 36, 48 and 72.

high-level language in computing, a programming language designed to suit the requirements of the programmer; it is independent of the internal machine code of any particular computer. High-level languages are used to solve problems and are often described as **problem-oriented languages** – for example, ◊BASIC was designed to be easily learnt by first-time programmers; ◊COBOL is used to write programs solving business problems; and ◊FORTRAN is used for programs solving scientific and mathematical problems. In contrast, low-level languages, such as ◊assembly languages, closely reflect the machine codes of specific computers, and are therefore described as **machine-oriented languages**.

Unlike low-level languages, high-level languages are relatively easy to learn because the instructions bear a close resemblance to everyday language, and because the programmer does not require a detailed knowledge of the internal workings of the computer. Each instruction in a high-level language is equivalent to several machine-code instructions. High-level programs are therefore more compact than equivalent low-level programs. However, each high-level instruction must be translated into machine code – by either a ◊compiler or an ◊interpreter program – before it can be executed by a computer. High-level languages are designed to be **portable** – programs written in a high-level language can be run on any computer that has a compiler or interpreter for that particular language.

high memory in computing, the first 64 kilobytes in the ◊extended memory of an ◊MS-DOS system. The operating system itself is usually installed in this area to allow more conventional memory (below 640 kilobytes) for applications.

High-Sierra format in computing, standard format for writing CD-ROM discs; see ◊ISO 9660.

high-tech industry any industry that makes use of advanced technology. The largest high-tech group is the fast-growing electronics industry and especially the manufacture of computers, microchips, and telecommunications equipment.

The products of these industries have low bulk but high value, as do their components. Silicon Valley in the USA and Silicon Glen in Scotland are two areas with high concentrations of such firms.

highway in Britain, any road over which there is a right of way. In the USA, any public road, especially a main road.

high-yield variety crop that has been specially bred or selected to produce more than the natural varieties of the same species. During the 1950s and 1960s, new strains of wheat and maize were developed to reduce the food shortages in poor countries (the ◊Green Revolution). Later, IR8, a new variety of rice that increased yields by up to six times, was developed in the Philippines. Strains of crops resistant to drought and disease were also developed. High-yield varieties require large amounts of expensive artificial fertilizers and sometimes pesticides for best results.

HII region in astronomy, a region of extremely hot ionized hydrogen, surrounding one or more hot stars, visible as a bright patch or emission ◊nebula in the sky. The gas is ionized by the intense ultraviolet radiation from the stars within it. HII regions are often associated with interstellar clouds in which new stars are being born. An example is the ◊Orion Nebula. It takes its name from a spectroscopic notation in which HI represents neutral hydrogen (H) and HII represents ionized hydrogen (H^+).

hill figure in Britain, any of a number of figures, usually of animals, cut from the turf to reveal the underlying chalk. Their origins are variously attributed to Celts, Romans, Saxons, Druids, or Benedictine monks, although most are of modern rather than ancient construction.

hillfort European Iron Age site with massive banks and ditches for defence, used as both a military camp and a permanent settlement. Examples found across Europe, in particular France, central Germany, and the British Isles, include Heuneberg near Sigmaringen, Germany, Spinans Hill in County Wicklow, Ireland, and Maiden Castle, Dorset, England.

Iron Age Germanic peoples spread the tradition of forts with massive defences, timberwork reinforcements, and sometimes elaborately defended gateways with guardrooms, the whole being overlooked from a rampart walk. The ramparts usually follow the natural line of a hilltop and are laid out to avoid areas of dead ground.

hinge joint in vertebrates, a joint where movement occurs in one plane only. Examples are the elbow and knee, which are controlled by pairs of muscles, the ◊flexors and ◊extensors.

hinterland area that is served by a port or settlement (the central place) and included in its sphere of influence. The city of Rotterdam, the Netherlands, is the hinterland of a port.

hinting in computing, a method of reducing the effects of ◊aliasing in the appearance of ◊outline fonts. Hinting makes use of a series of priorities so that noticeable distortions, such as uneven stem weight, are corrected. ◊PostScript Type 1 and ◊TrueType fonts are hinted.

Hipparcos (acronym for ***high precision parallax collecting satellite***) satellite launched by the European Space Agency in 1989. Named after the Greek astronomer Hipparchus, it is the world's first ◊astrometry satellite and is providing precise positions, distances, colours, brightnesses, and apparent motions for over 100,000 stars.

hippopotamus ***Greek 'river horse'*** large herbivorous, short-legged, even-toed hoofed mammal. The **common hippopotamus** (*Hippopotamus amphibius*) is found in Africa. It weighs up to 3,200 kg/7,040 lb, stands about 1.6 m/5.25 ft tall, and has a

Hippocrates
(c. 460–c. 377 BC)

Greek physician, often called the founder of medicine. Important Hippocratic ideas include cleanliness (for patients and physicians), moderation in eating and drinking, letting nature take its course, and living where the air is good. He believed that health was the result of the 'humours' of the body being in balance; imbalance caused disease. These ideas were later adopted by Galen.

He was born and practised on the island of Kos, where he founded a medical school. He travelled throughout Greece and Asia Minor, and died in Larisa, Thessaly. He is known to have discovered aspirin in willow bark. The *Corpus Hippocraticum/Hippocratic Collection*, a group of some 70 works, is attributed to him but was probably not written by him, although the works outline his approach to medicine. They include *Aphorisms* and the **Hippocratic Oath**, which embodies the essence of medical ethics.

Mary Evans Picture Library

HIPPOPOTAMUS

A hippopotamus can open its jaws to an angle of 150° – almost a straight line (180°).

brown or slate-grey skin. It is an endangered species. (Family Hippopotamidae.)

Hippos are social animals and live in groups. Because they dehydrate rapidly (at least twice as quickly as humans), they must stay close to water. When underwater, adults need to breath every 2–5 minutes and calves every 30 seconds. When out of water, their skin exudes an oily red fluid that protects them against the Sun's ultraviolet rays. The hippopotamus spends the day wallowing in rivers or waterholes, only emerging at night to graze. It can eat up to 25–40 kg/55–88 lb of grass each night. The **pygmy hippopotamus** (*Choeropsis liberiensis*) lives in W Africa.

There are an estimated 157,000 hippos in Africa (1993 figure), but they are under threat from hunters because of the value of their meat, hides, and large canine teeth (up to 0.5 m/1.6 ft long), which are used as a substitute for ivory.

HIPPOPOTAMUS

http://www.seaworld.org/animal_bytes/hippopotamusab.html

Illustrated guide to the hippopotamus including information about genus, size, life span, habitat, gestation, diet, and a series of fun facts.

Hispano-Suiza car designed by a Swiss engineer Marc Birkigt (1878–1947) who emigrated to Barcelona where he founded a factory which produced cars during the period 1900–38, legendary for their handling, elegance, and speed.

During World War I the Hispano-Suiza company produced a light-alloy aero-engine for the French air force.

histamine inflammatory substance normally released in damaged tissues, which also accounts for many of the symptoms of ◊allergy. It is an amine, $C_5H_9N_3$. Substances that neutralize its activity are known as ◊antihistamines. Histamine was first described in 1911 by British physiologist Henry Dale (1875–1968).

histogram in statistics, a graph showing frequency of data, in which the horizontal axis details discrete units or class boundaries, and the vertical axis represents the frequency. Blocks are drawn such that their areas (rather than their height as in a ◊bar chart) are proportional to the frequencies within a class or across several class boundaries. There are no spaces between blocks.

histology study of plant and animal tissue by visual examination, usually with a ◊microscope.

◊Stains are often used to highlight structural characteristics such as the presence of starch or distribution of fats.

histology in medicine, the laboratory study of cells and tissues.

history in computing, a list of sites visited by a Web ◊browser during the current session. The history is usually stored as a list of page titles and is accessed via the browser's menu system. The purpose is to make it easy for users to go back to a recently visited site.

hit in computing, request sent to a ◊file server.

Sites on the World Wide Web often measure their popularity in numbers of hits. However, this is misleading, as a single Web page may be made up of many files, each of which counts as a hit when a user downloads the whole page. Counting individual visits is a better indication of a site's success.

HIV (abbreviation for ***human immunodeficiency virus***) the infectious agent that is believed to cause ◊AIDS. It was first discovered in 1983 by Luc Montagnier of the Pasteur Institute in Paris, who called it lymphocyte-associated virus (LAV). Independently, US scientist Robert Gallo of the National Cancer Institute in Bethesda, Maryland, claimed its discovery in 1984 and named it human T-lymphocytotrophic virus 3 (HTLV-III).

transmission Worldwide, heterosexual activity accounts for three-quarters of all HIV infections. In addition to heterosexual men and women, high-risk groups are homosexual and bisexual men, prostitutes, intravenous drug-users sharing needles, and haemophiliacs and other patients treated with contaminated blood products. The virus has a short life outside the body, which makes transmission of the infection by methods other than sexual contact, blood transfusion, and shared syringes extremely unlikely.

US researchers in 1995 developed an explanation of why HIV is transmitted mainly by heterosexual sex in Africa and Asia, and by homosexual sex and intravenous drug use in Europe and the USA. They found that the HIV variant subtype B – responsible for 90% of European and US cases – did not grow well in reproductive tract cells, whereas subtype E – common in developing countries – did grow well. If subtype E becomes more prevalent in Europe and the USA, infection patterns will probably change. The first case of subtype E in Britain was documented in May 1996.

the development of HIV Many people who have HIV in their blood are not ill; in fact, it was initially thought that during the delay between infection with HIV and the development of AIDS the virus lay dormant. However, US researchers estimated in 1995 that HIV reproduces at a rate of a billion viruses a day, even in individuals with no symptoms, but is held at bay by the immune system producing enough white blood cells (CD4 cells) to destroy them. Gradually, the virus mutates so much that the immune system is unable to continue to counteract; people with advanced AIDS have virtually no CD4 cells remaining. These results indicate the importance of treating HIV-positive individuals before symptoms develop, rather than delaying treatment until the onset of AIDS.

About 15% of babies born to HIV-positive mothers are themselves HIV-positive. A very small number of these babies (less than 3%) test negative for the virus some months later, a phenomenon yet to be explained.

HIV INFOWEB

www.infoweb.org/

Up-to-date information about HIV and AIDS treatments and other important resources for patients and their families, such as housing.

HIV statistics In 1997 there were an estimated just under 30 million (1% of the world's adult population) HIV infections in the world. In Sub-Saharan Africa there were an estimated 20 million HIV sufferers, in S and SE Asia 6 million, in South America 1.3 million, in North America 860,000, and Western Europe 150,000 (figures released by the United Nations AIDS programme, UNAIDS).

hoatzin tropical bird found only in the Amazon, resembling a small pheasant in size and appearance. The beak is thick and the facial skin blue. Adults are olive-coloured with white markings above and red-brown below. The hoatzin is the only bird in its family. (Species *Opisthocomus hoatzin,* family Opisthocomidae, order Galliformes.)

The young are hatched naked, with claws on their wings, which they use to crawl reptile-fashion about the tree; these claws later fall off. They fly only reluctantly and prefer to climb among branches using their wings – they cannot grip with their feet. Hoatzin are chiefly arboreal, nesting on low trees or shrubs, and feeding on leaves and fruit.

hobby small ◊falcon found across Europe and N Asia. It is about 30 cm/1 ft long, with a grey-blue back, streaked front, and chestnut thighs. It is found in open woods and heaths, and feeds on insects and small birds. (Species *Falco subbuteo.*)

Hodgkin's disease or ***lymphadenoma*** rare form of cancer mainly affecting the lymph nodes and spleen. It undermines the immune system, leaving the sufferer susceptible to infection.

However, it responds well to radiotherapy and ◊cytotoxic drugs, and long-term survival is usual.

HODGKIN'S DISEASE INFORMATION

http://www.cancer.org/cidSpecificCancers/hodgkins/index.html

Comprehensive information on lymphadenoma from the American Cancer Society. Written in easily understandable language, this guide explains the normal functions of the lymphatic system before describing the disease. There is information on risk factors, causes, diagnosis, and treatment of Hodgkin's disease, and latest research news. Further sources of information are also indicated.

Hoffman, Albert

Swiss-born, US physician who, in 1943, accidentally discovered lysergic acid diethylamide (LSD), the most potent psychoactive drug ever known. LSD causes hallucinatory effects, paranoia, and depression, and is illegal except for research purposes.

Hoffman, together with Swiss chemist Arthur Stoll, produced lysergic acid diethylamide (LSD) in 1938 while trying to synthesize a new drug for the treatment of headaches. Since the new drug appeared to have no analgesic (pain-relieving) effect on laboratory animals, it remained untouched on a shelf for five years. Hoffman decided to perform further tests on LSD in 1943, during which he accidentally ingested an unknown amount of the drug. He described his first experience of LSD intoxication as 'a kind of drunkenness which was not unpleasant and which was characterized by extreme activity of the imagination'. LSD was later shown to block or inhibit the action of the neurotransmitter seratonin in the brain.

Hoffman's voltameter in chemistry, an apparatus for collecting gases produced by the ◊electrolysis of a liquid.

It consists of a vertical E-shaped glass tube with taps at the upper ends of the outer limbs and a reservoir at the top of the central limb. Platinum electrodes fused into the lower ends of the outer limbs are connected to a source of direct current. At the beginning of an experiment, the outer limbs are completely filled with electrolyte by opening the taps. The taps are then closed and the current switched on. Gases evolved at the electrodes bubble up the outer limbs and collect at the top, where they can be measured.

hog any member of the ◊pig family. The **river hog** (*Potamochoerus porcus*) lives in Africa, south of the Sahara. Reddish or black, up to 1.3 m/4.2 ft long plus tail, and 90 cm/3 ft at the shoulder, this gregarious animal roots for food in many types of habitat. The **giant forest hog** (*Hylochoerus meinerzthageni*) lives in thick forests of central Africa and grows up to 1.9 m/6 ft long. The ◊wart hog **is another African wild pig. The pygmy hog** *Sus salvanus,* the smallest of the pig family, is about 65 cm long (25 cm at the shoulder) and weighs 8–9 kg.

hognose North American colubrine, nonvenomous snake with a flattened head and a projecting snout for burrowing.

classification The hognose is in genus *Heterodon,* family Elapidae, suborder Serpentes, order Squamata, class Reptilia.

hogweed any of a group of plants belonging to the carrot family. The giant hogweed (*H. mantegazzianum*) grows over 3 m/9 ft high. (Genus *Heracleum,* family Umbelliferae.)

holdfast organ found at the base of many seaweeds, attaching them to the sea bed. It may be a flattened, suckerlike structure, or dissected and fingerlike, growing into rock crevices and firmly anchoring the plant.

holism in philosophy, the concept that the whole is greater than the sum of its parts.

holistic medicine umbrella term for an approach that virtually all alternative therapies profess, which considers the overall health and lifestyle profile of a patient, and treats specific ailments not primarily as conditions to be alleviated but rather as symptoms of more fundamental disease.

A physician is obligated to consider more than a diseased organ, more even than the whole man – he must view the man in his world.

HARVEY CUSHING US surgeon.
Quoted in René Dubos *Man Adapting*

holly any of a group of trees or shrubs that includes the English Christmas holly (*I. aquifolium*), an evergreen with spiny, glossy leaves, small white flowers, and poisonous scarlet berries on the female tree. Leaves of the Brazilian holly (*I. paraguayensis*) are used to make the tea **yerba maté**. (Genus *Ilex,* family Aquifoliaceae.)

hollyhock tall flowering plant belonging to the mallow family. *A. rosea,* originally a native of Asia, produces spikes of large white, yellow, pink, or red flowers, 3 m/10 ft high when cultivated as a biennial; it is a popular cottage garden plant. (Genus *Althaea,* family Malvaceae.)

holmium ***Latin*** *Holmia* ***'Stockholm'*** silvery, metallic element of the ◊lanthanide series, symbol Ho, atomic number 67, relative atomic mass 164.93. It occurs in combination with other rare-earth metals and in various minerals such as gadolinite. Its compounds are highly magnetic.

The element was discovered in 1878, spectroscopically, by the Swiss chemists J L Soret and Delafontaine, and independently in 1879 by Swedish chemist Per Cleve (1840–1905), who named it after Stockholm, near which it was found.

Holocene epoch of geological time that began 10,000 years ago, the second and current epoch of the Quaternary period. During this epoch the glaciers retreated, the climate became warmer, and humans developed significantly.

hologram three-dimensional image produced by holography. Small, inexpensive ◊holograms appear on credit cards and software licences to guarantee their authenticity.

holography method of producing three-dimensional (3-D) images, called ◊holograms, by means of ◊laser light. Holography uses a photographic technique (involving the splitting of a laser beam into two beams) to produce a picture, or hologram, that contains 3-D information about the object photographed. Some holograms show meaningless patterns in ordinary light and produce a 3-D image only when laser light is projected through them, but reflection holograms produce images when ordinary light is reflected from them (as found on credit cards).

Although the possibility of holography was suggested as early as 1947 (by Hungarian-born British physicist Dennis Gabor), it could not be demonstrated until a pure coherent light source, the laser, became available in 1963. The first laser-recorded holograms were created by Emmett Leith and Juris Upatnieks at the University of Michigan, USA, and Yuri Denisyuk in the Soviet Union.

The technique of holography is also applicable to sound, and bats may navigate by ultrasonic holography. Holographic techniques also have applications in storing dental records, detecting stresses and strains in construction and in retail goods, detecting forged paintings and documents, and producing three-dimensional body scans. The technique of detecting strains is of widespread application. It involves making two different holograms of an object on one plate, the object being stressed between exposures. If the object has distorted during stressing, the hologram will be greatly changed, and the distortion readily apparent.

Using holography, digital data can be recorded page by page in a crystal. In 1993 10,000 pages (100 megabytes) of digital data were stored in an iron-doped lithium nobate crystal measuring 1 cm^3.

homeopathy alternative spelling of ◊homoeopathy.

homeostasis maintenance of a constant internal state in an organism, particularly with regard to pH, salt concentration, temperature, and blood sugar levels. Stable conditions are important for the efficient functioning of the ◊enzyme reactions within the cells, which affect the performance of the entire organism.

homeothermy maintenance of a constant body temperature in endothermic (warm-blooded) animals, by the use of chemical processes to compensate for heat loss or gain when external temperatures change. Such processes include generation of heat by the breakdown of food and the contraction of muscles, and loss of heat by sweating, panting, and other means.

Mammals and birds are homeotherms, whereas invertebrates, fish, amphibians, and reptiles are cold-blooded or poikilotherms. Homeotherms generally have a layer of insulating material to retain heat, such as fur, feathers, or fat (see ◊blubber). Their metabolism functions more efficiently due to homeothermy, enabling them to remain active under most climatic conditions.

home page in computing, opening page on a particular site on the World Wide Web. The term is also used for the page which loads automatically when a user opens a Web ◊browser, and for a user's own personal Web pages.

Many Internet Service Providers provide free space to allow all their users to create and maintain their own home pages.

homoeopathy or ***homeopathy*** system of alternative medicine based on the principle that symptoms of disease are part of the body's self-healing processes, and on the practice of administering extremely diluted doses of natural substances found to produce in a healthy person the symptoms manifest in the illness being treated. Developed by German physician Samuel Hahnemann (1755–1843), the system is widely practised today as an alternative to allopathic (orthodox) medicine, and many controlled tests and

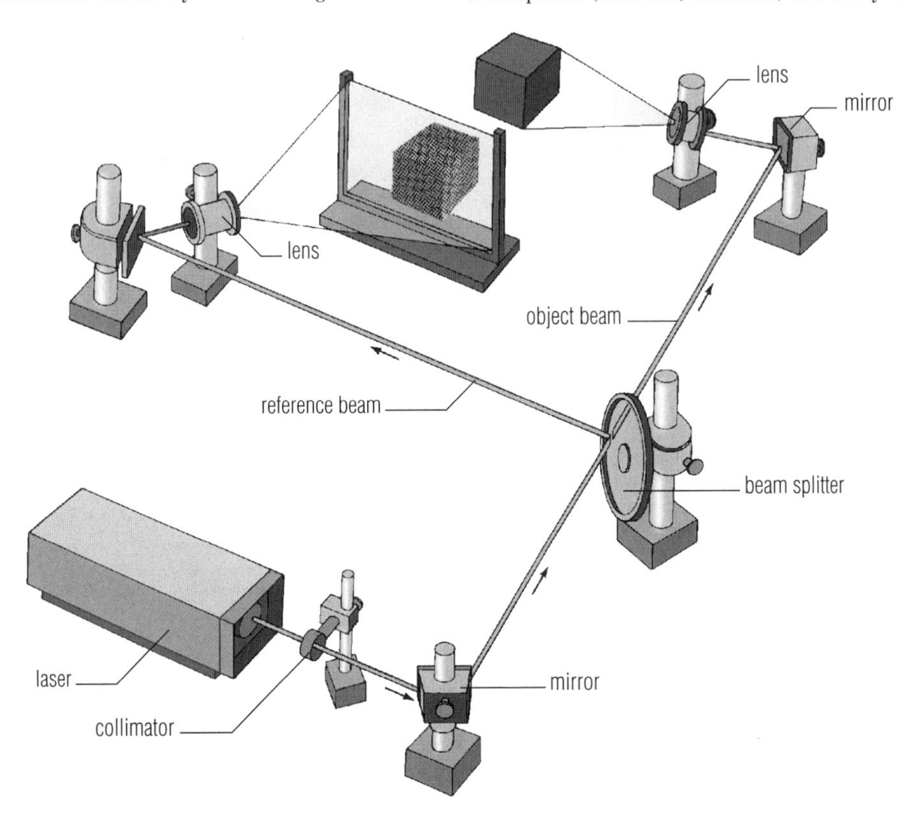

holography *Recording a transmission hologram. Light from a laser is divided into two beams. One beam goes directly to the photographic plate. The other beam reflects off the object before hitting the photographic plate. The two beams combine to produce a pattern on the plate which contains information about the 3-D shape of the object. If the exposed and developed plate is illuminated by laser light, the pattern can be seen as a 3-D picture of the object.*

achieved cures testify its efficacy.

In 1992, the German health authority, the *Bundesgesundheitsamt,* banned 50 herbal and homeopathic remedies containing ◊alkaloids because they are toxic, and set dose limits on 550 other natural remedies.

homogeneous reaction in chemistry, a reaction where there is no interface between the components. The term applies to all reactions where only gases are involved or where all the components are in solution.

Homo habilis tool-using hominid living about 2.5 million years ago in Africa; see ◊human species, origins of.

The Farmer will never be happy again; / He carries his heart in his boots; / For either the rain is destroying his grain / Or the drought is destroying his roots.

ALAN PATRICK HERBERT English writer and politician. 'The Farmer'

homologous in biology, a term describing an organ or structure possessed by members of different taxonomic groups (for example, species, genera, families, orders) that originally derived from the same structure in a common ancestor. The wing of a bat, the arm of a monkey, and the flipper of a seal are homologous because they all derive from the forelimb of an ancestral mammal.

homologous series any of a number of series of organic chemicals with similar chemical properties in which members differ by a constant relative molecular mass.

Alkanes (paraffins), alkenes (olefins), and alkynes (acetylenes) form such series in which members differ in mass by 14, 12, and 10 atomic mass units respectively. For example, the alkane homologous series begins with methane (CH_4), ethane (C_2H_6), propane (C_3H_8), butane (C_4H_{10}), and pentane (C_5H_{12}), each member differing from the previous one by a CH_2 group (or 14 atomic mass units).

homozygous in a living organism, having two identical ◊alleles for a given trait. Individuals homozygous for a trait always breed true; that is, they produce offspring that resemble them in appearance when bred with a genetically similar individual; inbred varieties or species are homozygous for almost all traits.

◊Recessive alleles are only expressed in the homozygous condition. See also ◊heterozygous.

honey sweet syrup produced by honey ◊bees from the nectar of flowers. It is stored in honeycombs and made in excess of their needs as food for the winter. Honey comprises various sugars, mainly laevulose and dextrose, with enzymes, colouring matter, acids, and pollen grains. It has antibacterial properties and was widely used in ancient Egypt, Greece, and Rome as a wound salve. It is still popular for sore throats, in hot drinks or in lozenges.

honeycomb moth another name for the ◊wax moth.

honeycomb worm colonial marine worm *Sabellaria alveolata* named after the hexagonal tubes made of cemented sand and shell fragments in which it lives low down on rocky sandy beaches. It feeds on organic material suspended in the water. It spends its larval stage as plankton.

honeyeater or ***honey-sucker*** any of a group of small, brightly coloured birds with long, curved beaks and long tails, native to Australia. They have a long tongue divided into four at the end to form a brush for collecting nectar from flowers. (Family Meliphagidae.)

Larger honeyeaters, such as the **blue-faced honeyeater** (*Entomyza cyanotis*) of NE Australia, which is 30 cm/12 in long, also eat insects and fruit. The blood-bird is a honeyeater.

Honeyeaters from Australasia colonized Hawaii, where four distinct species evolved of which only one, the *Kauaioo,* survives; it too was thought to be extinct but was rediscovered in 1960.

honey guide in botany, line or spot on the petals of a flower that indicate to pollinating insects the position of the nectaries (see ◊nectar) within the flower. The orange dot on the lower lip of the toadflax flower (*Linaria vulgaris*) is an example. Sometimes the markings reflect only ultraviolet light, which can be seen by many insects although it is not visible to the human eye.

honey possum or ***honey mouse*** or ***noolbenger*** tiny marsupial that is native of western Australia. It lives in trees, and feeds on insects and honey, which it extracts from flowers with its long extensile tongue.

classification The honey possum *Tarsipes spenserae* is the only member of the family Tarsipedidae, order Marsupialia, class Mammalia.

honeysuckle vine or shrub found in temperate regions of the world. The common honeysuckle or woodbine (*L. periclymenum*) of Europe is a climbing plant with sweet-scented flowers, reddish and yellow-tinted outside and creamy white inside; it now grows in the northeastern USA. (Genus *Lonicera,* family Caprifoliaceae.)

The North American trumpet honeysuckle (*L. sempervirens*) has unusual vaselike flowers and includes scarlet and yellow varieties.

hoof horny covering that protects the sensitive parts of the foot of an animal. The possession of hooves is characteristic of the orders Artiodactyla (even-toed ungulates such as deer and cattle), and Perissodactyla (horses, tapirs, and rhinoceroses).

alkane	alcohol	aldehyde	ketone	carboxylic acid	alkene
CH_4 methane	CH_3OH methanol	HCHO methanal	—	HCOOH methanoic acid	—
CH_3CH_3 ethane	CH_3CH_2OH ethanol	CH_3CHO ethanal	—	CH_3COOH ethanoic acid	CH_2CH_2 ethene
$CH_3CH_2CH_3$ propane	$CH_3CH_2CH_2OH$ propanol	CH_3CH_2CHO propanal	CH_3COCH_3 propanone	CH_3CH_2COOH propanoic acid	CH_2CHCH_3 propene
methane	methanol	methanal	propanone	methanoic acid	ethene

homologous series

The flexibility of the hoof is promoted by a fluid secreted by the keratogenous (horn-producing) membrane. The cloven hoof of the Artiodactyla has been evolved for walking and climbing on irregular surfaces by the formation of a separate hoof on each digit of the foot. Horses walk on the third digit of the foot, while cattle walk on the third and fourth.

Hooke, Robert (1635–1703)

English scientist and inventor, originator of Hooke's law, and considered the foremost mechanic of his time. His inventions included a telegraph system, the spirit level, marine barometer, and sea gauge. He coined the term 'cell' in biology.

He studied elasticity, furthered the sciences of mechanics and microscopy, invented the hairspring regulator in timepieces, perfected the air pump, and helped improve such scientific instruments as microscopes, telescopes, and barometers. His work on gravitation and in optics contributed to the achievements of his contemporary Isaac Newton.

Hooke's law law stating that the deformation of a body is proportional to the magnitude of the deforming force, provided that the body's elastic limit (see ◊elasticity) is not exceeded. If the elastic limit is not reached, the body will return to its original size once the force is removed. The law was discovered by Robert Hooke 1676.

For example, if a spring is stretched by 2 cm by a weight of 1 N, it will be stretched by 4 cm by a weight of 2 N, and so on; however, once the load exceeds the elastic limit for the spring, Hooke's law will no longer be obeyed and each successive increase in weight will result in a greater extension until finally the spring breaks.

hookworm parasitic roundworm (see ◊worm) with hooks around its mouth. It lives mainly in tropical and subtropical regions, but also in humid areas in temperate climates. The eggs are hatched in damp soil, and the larvae bore into the host's skin, usually through the soles of the feet. They make their way to the small intestine, where they live by sucking blood. The eggs are expelled with faeces, and the cycle starts again. The human hookworm causes anaemia, weakness, and abdominal pain. It is common in areas where defecation occurs outdoors. (Genus *Necator.*)

hoopoe bird slightly larger than a thrush, with a long, thin, slightly downward-curving bill and a bright pinkish-buff crest tipped with black that expands into a fan shape on top of the head. The wings and tail are banded with black and white, and the rest of the plumage is buff-coloured. The hoopoe is found throughout southern Europe and Asia down to southern Africa, India, Malaya. (Species *Upupa epops,* family Upupidae, order Coraciiformes.)

hoop pine softwood timber tree *Araucaria cunninghamii* of NE New South Wales and Queensland in Australia, and Papua New Guinea.

Hoover Dam highest concrete dam in the USA, 221 m/726 ft, on the Colorado River at the Arizona–Nevada border. It was built 1931–36. Known as **Boulder Dam** 1933–47, its name was restored by President Truman as the reputation of the former president, Herbert Hoover, was revived. It impounds Lake Mead, and has a hydroelectric power capacity of 1,300 megawatts.

hop in computing, on the Internet, an intermediate stage of the journey taken by a message travelling from one site to another.

Internet messages must travel through many machines to get to their destinations. The exact route is recorded in the ◊bang path.

Hope's apparatus in physics, an apparatus used to demonstrate the temperature at which water has its maximum density. It is named after Thomas Charles Hope (1766–1844).

It consists of a vertical cylindrical vessel fitted with horizontal thermometers through its sides near the top and bottom, and surrounded at the centre by a ledge that holds a freezing mixture (ice and salt). When the cylinder is filled with water, this gradually cools, the denser water sinking to the bottom; eventually the upper thermometer records 0°C/32°F (the freezing point of water) and the lower one has a constant reading of 4°C/39°F (the temperature at which water is most dense).

hops A female hop plant showing the fruits which, in cultivated varieties, are used in brewing. The genus Humulus *is a small one, with* H. lupulus *coming from Europe and W Asia and one species each from North America and E Asia. Premaphotos Wildlife*

hops female fruit heads of the hop plant *Humulus lupulus,* family Cannabiaceae; these are dried and used as a tonic and in flavouring beer. In designated areas in Europe, no male hops may be grown, since seedless hops produced by the unpollinated female plant contain a greater proportion of the alpha acid that gives beer its bitter taste.

horehound any of a group of plants belonging to the mint family. The white horehound (*M. vulgare*), found in Europe, N Africa, and W Asia and naturalized in North America, has a thick hairy stem and clusters of dull white flowers; it has medicinal uses. (Genus *Marrubium,* family Labiatae.)

horizon in astronomy, the ◊great circle dividing the visible part of the sky from the part hidden by the Earth.

horizon the limit to which one can see across the surface of the sea or a level plain, that is, about 5 km/3 mi at 1.5 m/5 ft above sea level, and about 65 km/40 mi at 300 m/1,000 ft.

hormone ***Greek 'arousing'*** secretion of the ◊endocrine glands, concerned with control of body functions. The major glands are the thyroid, parathyroid, pituitary, adrenal, pancreas, ovary, and testis. Hormones bring about changes in the functions of various organs according to the body's requirements. The ◊hypothalamus, which adjoins the pituitary gland, at the base of the brain, is a control centre for overall coordination of hormone secretion; the thyroid hormones determine the rate of general body chemistry; the adrenal hormones prepare the organism during stress for 'fight or flight'; and the sexual hormones such as oestrogen govern reproductive functions.

There are also hormone-secreting cells in the kidney, liver, gastrointestinal tract, thymus (in the neck), pineal (in the brain), and placenta. Many diseases due to hormone deficiency can be relieved with hormone preparations.

hormone-replacement therapy (HRT) use of ◊oestrogen and progesterone to help limit the unpleasant effects of the menopause in women. The treatment was first used in the 1970s.

At the menopause, the ovaries cease to secrete natural oestrogen. This results in a number of symptoms, including hot flushes, anxiety, and a change in the pattern of menstrual bleeding. It is also associated with osteoporosis, or a thinning of bones, leading to an increased incidence of fractures, frequently of the hip, in older women. Oestrogen preparations, taken to replace the decline in natural hormone levels, combined with regular exercise can help to maintain bone strength in women. In order to improve bone density, however, HRT must be taken for five years, during which time the woman will continue to menstruate. Many women do not find this acceptable.

horn broad term for a hardened processes on the heads of some members of order Artiodactyla: deer, antelopes, cattle, goats, and sheep; and the rhinoceroses in order Perissodactyla. They are used usually for sparring rather than serious fighting, often between members of the same species rather than against predators.

The structure of horn shows immense variation, some being primarily made of bone, as in the antlers of deer; others of the substance called horn, as in cattle or antelopes; and others of compressed hair, as in rhinoceroses. Antlers are usually shed and regrown every year, while true horns are grown for life.

In most horned species they are possessed by both sexes, but in some species horns are limited to males.

hornbeam any of a group of trees belonging to the birch family. They have oval leaves with toothed edges and hanging clusters of flowers, each with a nutlike seed attached to the base. The trunk is usually twisted, with smooth grey bark. (Genus *Carpinus*, family Betulaceae.)

hornbill any of a group of omnivorous birds found in Africa, India, and Malaysia. They are about 1 m/3 ft long, and have powerful downcurved beaks, usually surmounted by a bony growth or casque. During the breeding season, the female walls herself into a hole in a tree and does not emerge until the young are hatched. There are about 45 species. (Family Bucerotidae, order Coraciiformes.)

Hornbills feed chiefly on the ground, their food consisting of insects, small mammals, and reptiles. The **great hornbill** (*Buceros bicornis*) of SE Asia can reach up to 1.3 m/4.3 ft in length.

The **southern ground hornbill** lives in groups of about three to five birds (though sometimes as many as ten) with only one breeding pair, and the rest acting as helpers. On average, only one chick is reared successfully every nine years. Lifespan can be 40 years or more.

hornblende green or black rock-forming mineral, one of the amphiboles. It is a hydrous ◊silicate composed mainly of calcium, iron, magnesium, and aluminium in addition to the silicon and oxygen that are common to all silicates. Hornblende is found in both igneous and metamorphic rocks and can be recognized by its colour and prismatic shape.

horned toad or ***horned lizard*** common name for several species of lizard. Horned toads have large spines or horns on their heads and pointed scales, vary from 7–12 cm/3–5 in long, and are found in arid areas of North and Central America. An example is the desert-dwelling spiny moloch.

When attacked, horned toads inflate the body with air, gape, hiss, and bite.

classification Horned toads are in the genus *Phrynosoma*, family Iguanidae, suborder Sauria, order Squamata, class Reptilia.

horned viper northeast African snake. It is remarkable for the possession of a large spiky scale above each eye

classification The horned viper *Cerastes cornutus* belongs to the family Viperidae, suborder Serpentes, order Squamata, class Reptilia.

hornet type of ◊wasp.

hornfels ◊metamorphic rock formed by rocks heated by contact with a hot igneous body. It is fine-grained, brittle, and lacks foliation (a planar structure).

Hornfels may contain minerals only formed under conditions of great heat, such as andalusite, Al_2SiO_5, and cordierite, $(Mg,Fe)_2Al_4Si_5O_{18}$. This rock, originating from sedimentary rock strata, is found in contact with large igneous ◊intrusions where it represents the heat-altered equivalent of the surrounding clays. Its hardness makes it suitable for road building and railway ballast.

horn fly small fly that is a pest of cattle and other animals.

classification Horn flies are in the family Muscidae, class Insecta, phylum Arthropoda.

hornwort nonvascular plant (with no 'veins' to carry water and food), related to the ◊liverworts and ◊mosses. Hornworts are found in warm climates, growing on moist shaded soil. (Class Anthocerotae, order Bryophyta.)

The name is also given to a group of aquatic flowering plants which are found in slow-moving water. They have whorls of finely divided leaves and may grow up to 2 m/7 ft long. (Genus *Ceratophyllum*, family Ceratophyllaceae.)

Like liverworts and mosses, the bryophyte hornworts exist in two different reproductive forms, sexual and asexual, which appear alternately (see ◊alternation of generations). A leafy plant body, or gametophyte, produces gametes, or sex cells, and a small horned form, or sporophyte, which grows upwards from the gametophyte, produces spores. Unlike the sporophytes of mosses and liverworts, the hornwort sporophyte survives after the gametophyte has died.

horse hoofed, odd-toed, grazing mammal belonging to the same family as zebras and asses. The many breeds of domestic horse of Euro-Asian origin range in colour from white to grey, brown, and black. The yellow-brown **Mongolian wild horse**, or **Przewalski's horse** (*Equus przewalskii*), named after its Polish 'discoverer' about 1880, is the only surviving species of wild horse. (Species *Equus caballus*, family Equidae.)

Przewalski's horse became extinct in the wild because of hunting and competition with domestic animals for food; about 800 survive in captivity. However, in the late 1990s 55 Przewalski's horses were successfully reintroduced to the wild in Mongolia.

horse chestnut any of a group of trees, especially *A. hippocastanum*, originally from SE Europe but widely planted elsewhere. Horse chestnuts have large palmate (five-lobed) leaves, showy upright spikes of white, pink, or red flowers, and large, shiny, inedible seeds (**conkers**) in prickly green capsules. The horse chestnut is not related to the true chestnut. In North America it is called buckeye. (Genus *Aesculus*, family Hippocastanaceae.)

horsefly any of over 2,500 species of fly. The females suck blood from horses, cattle, and humans; the males live on plants and suck nectar. The larvae are carnivorous. (Family Tabanidae.)

horsefly *A female horsefly, or cleg,* Haematopota pluvialis *feeding on human blood. The bite of these insects is painful and in humans often produces a large swelling that can take a long time to heal. Premaphotos Wildlife*

HORSEFLY

There is a fly larva that eats toads. The horsefly larva *Tabanus punctifer* lives in the soft mud at the edges of ponds. It kills tiny, newly metamorphosed spadefoot toads by injecting them with venom, and then sucks out their body fluids.

horsepower imperial unit (abbreviation hp) of power, now replaced by the ◊watt. It was first used by the engineer James ◊Watt, who employed it to compare the power of steam engines with that of horses.

horseradish hardy perennial plant, native to SE Europe but naturalized elsewhere. The thick cream-coloured root is strong-tasting and is often made into a savoury sauce to accompany food. (*Armoracia rusticana,* family Cruciferae.)

horsetail plant related to ferns and club mosses; some species are also called **scouring rush**. There are about 35 living species, bearing their spores on cones at the stem tip. The upright stems are ribbed and often have spaced whorls of branches. Today they are of modest size, but hundreds of millions of years ago giant treelike forms existed. (Genus *Equisetum,* order Equisetales.)

horticulture art and science of growing flowers, fruit, and vegetables. Horticulture is practised in gardens and orchards, along with millions of acres of land devoted to vegetable farming. Some areas, like California, have specialized in horticulture because they have the mild climate and light fertile soil most suited to these crops.

host in biology, an organism that is parasitized by another. In ◊commensalism, the partner that does not benefit may also be called the host.

hot key in computing, a key stroke (or sequence of key strokes) that triggers a memory-resident program. Such programs are called ◊terminate and stay resident. Hot keys should be chosen so that they do not conflict with key sequences in commonly used applications.

hotlist in computing, stored list of favourite sites which allows users to move quickly to frequently used resources. See also ◊bookmark.

HOTOL (acronym for ***horizontal takeoff and landing***) reusable hypersonic spaceplane invented by British engineer Alan Bond in 1983 but never put into production.

HOTOL was to be a single-stage vehicle that could take off and land on a runway. It featured a revolutionary dual-purpose engine that enabled it to carry far less oxygen than a conventional spaceplane: it functioned as a jet engine during the initial stage of flight, taking in oxygen from the surrounding air; when the air became too thin, it was converted into a rocket, burning oxygen from an onboard supply. The project was developed by British Aerospace and Rolls-Royce but foundered for lack of capital in 1988.

hot spot in geology, isolated rising plume of molten mantle material that may rise to the surface of the Earth's crust creating features such as volcanoes, chains of ocean islands, seamounts, and rifts in continents. Hot spots occur beneath the interiors of tectonic plates and so differ from areas of volcanic activity at plate margins (see ◊plate tectonics). Examples of features made by hot spots are Iceland in the Atlantic Ocean, and in the Pacific Ocean the Hawaiian Islands and Emperor Seamount chain, and the Galápagos Islands.

Hot spots are responsible for large amounts of volcanic activity within tectonic plates rather than at plate margins. Volcanism from a hot spot formed the unique features of Yellowstone National Park, Wyoming, USA. The same hot spot that built Iceland atop the mid-Atlantic ridge in the North Atlantic Ocean also produced the voluminous volcanic rocks of the Isle of Skye, Scotland, at a time before these regions were rifted apart by the opening of the Atlantic Ocean.

Chains of volcanic seamounts trace the movements of tectonic plates as they pass over hot spots. Immediately above a hot spot on oceanic crust a volcano will form. This volcano is then carried away by plate tectonic movement, and becomes extinct. A new volcano forms beside it, again above the hot spot. The result is an active volcano and a chain of increasingly old and eroded extinct volcanoes stretching away along the line traced by the plate movement. The chain of volcanoes comprising the Hawaiian Islands and Emperor Seamounts formed in this way.

hot-swapping in computing, a technique that allows a user to exchange components without having to shut down the entire system.

The most common example of hot-swapping is ◊PCMCIA (personal computer memory card interface adapter) components: a user with only one PCMCIA slot can exchange a modem for a network card or hard disc while the machine is running. Special software recognizes the components and allows their immediate use.

hour period of time comprising 60 minutes; 24 hours make one calendar day.

housefly fly found in and around human dwellings, especially *M. domestica,* a common worldwide species. Houseflies are grey and have mouthparts adapted for drinking liquids and sucking moisture from food and manure. (Genus *Musca.*)

HOUSEFLY

In nine months, a housefly could lay enough eggs to produce a layer of flies that would cover all of Germany to a depth of 14 m/47 ft.

hovercraft vehicle that rides on a cushion of high-pressure air, free from all contact with the surface beneath, invented by British engineer Christopher Cockerell in 1959. Hovercraft need a smooth terrain when operating overland and are best adapted to use on waterways. They are useful in places where harbours have not been established.

Large hovercraft (SR-N4) operate a swift car-ferry service across the English Channel, taking only about 35 minutes between Dover and Calais. They are fitted with a flexible 'skirt' that helps maintain the air cushion.

A military version made of fibreglass, the M-10, is tough manoeuvrable, and less noisy. *See illustration on page 377.*

hoverfly brightly coloured winged insect. Hoverflies usually have spots, stripes, or bands of yellow or brown against a dark-coloured background, sometimes with dense hair covering the body surface. Many resemble bees, bumble bees, and wasps (displaying Batesian ◊mimicry) and most adults feed on nectar and pollen.

classification Hoverflies are members of the large family Syrphidae (numbering over 2,500 species), suborder Cyclorrhapha, order Diptera, class Insecta, phylum Arthropoda.

One of the most characteristic features of hoverflies is the presence of a longitudinal false vein in the wing.

larva The larvae show remarkable variations in appearance and feeding habits. They may feed externally on plants or they may be internal feeders, attacking the bulbs; for example the **narcissus fly**, *Merodon equestris.* Many are carnivorous, feeding on ◊scale insects, greenfly (◊aphids), and other insects that harm commercial crops.

The larvae may also feed on rotting wood or the decaying organic matter in stagnant pools; for example the **rat-tailed maggot**, larva of the drone fly *Eristalis tenax* is found in polluted pools. They breathe by extending their tail breathing tubes to reach the surface of the water.

howler monkey widely distributed large Central and South American monkey. Howler monkeys are tree-dwelling and feed on fruit and leaves. They have a prominent, hairless face and deep jaw, and the tail is long and prehensile.

Howler monkeys howl at dawn to demarcate territory. The howling is produced by the unusually developed egg-shaped hyoid bone at the upper end of the wind-pipe, in a swelling beneath the chin; the whole forms a hollow, resonant soundbox.
classification Howler monkeys are in genus *Alouatta,* family Cebidae, order Primates.

HP in computing, abbreviation for ◊Hewlett-Packard.

hp abbreviation for ◊horsepower.

HPGL (abbreviation for ***Hewlett Packard Graphics Language***) file format used in ◊vector graphics. HPGL is often generated by ◊CAD systems.

href in computing, a tag in HTML (hypertext markup language) that indicates that the following text is a link either to another portion of the same document or to an external document on the same or a remote site.

ht abbreviation for height.

HTML (abbreviation for ***Hypertext Markup Language***) standard for structuring and describing a document on the ◊World Wide Web. The HTML standard provides labels for constituent parts of a document (for example headings and paragraphs) and permits the inclusion of images, sounds, and 'hyperlinks' to other documents. A ◊browser program is then used to convert this information into a graphical document on-screen. The specifications for HTML version 4, called Dynamic HTML, were adopted at the end of 1997.

HTML is a specific example of ◊SGML (the international standard for text encoding). As such it is not a rigid standard but is constantly being improved to incorporate new features and allow greater freedom of design.

HTML extension in computing, any proprietary addition to the standard specification of HTML (hypertext markup language).

Both Microsoft and Netscape, publishers of the two leading Web ◊browsers, have built in such extensions, which are controversial as they clash with the basic ideal that the Net should operate on open standards which allow interoperability. In general, any browser should be able to log on to any site and be able to access most of its information, but the features implemented with proprietary extensions will only display correctly with a browser that supports those extensions.

HTTP (abbreviation for ***Hypertext Transfer Protocol***) in computing, the ◊protocol used for communications between client (the Web ◊browser) and ◊server on the World Wide Web.

hub in computing, central distribution point in a computer ◊network.

Hubble classification in astronomy, a scheme for classifying ◊galaxies according to their shapes, originally devised by the US astronomer Edwin Hubble in the 1920s.

Elliptical galaxies are classed from type E0 to E7, where the figure denotes the degree of ellipticity. An E0 galaxy appears circular to an observer, while an E7 is highly elliptical (this is based on the apparent shape; the true shape, distorted by foreshortening, may be quite different.) **Spiral galaxies** are classed as type Sa, Sb, or Sc, where Sa is a tightly wound spiral with a large central bulge and Sc is loosely wound with a small bulge. Intermediate types are denoted by Sab or Sbc. **Barred spiral galaxies**, which have a prominent bar across their centres, are similarly classed as type SBa, SBb, or SBc with intermediates SBab or SBbc. **Lenticular galaxies**, which have no spiral arms, are classed as type S0. **Irregular galaxies**, type Irr, can be subdivided into Irr I, which resemble poorly formed spirals, and Irr II which are otherwise

The Hubble classification was once believed to reveal an evolutionary sequence (from ellipticals to spirals) but this is now known not to be the case. Our own ◊Milky Way Galaxy is classified as type Sb or Sc, but may have a bar.

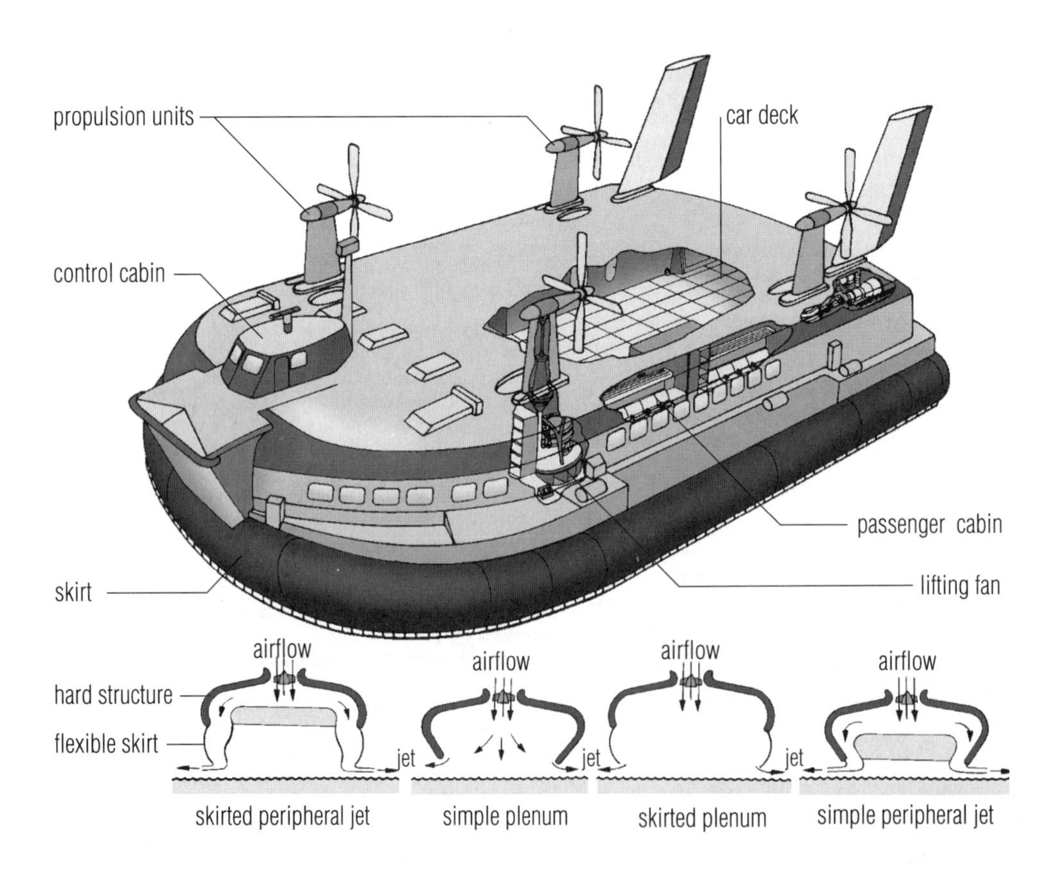

hovercraft *There are several alternative ways of containing the cushion of air beneath the hull of a hovercraft. The passenger-carrying hovercraft that sails across the English Channel has a flexible skirt; other systems are the open plenum and the peripheral jet.*

Hubble's constant in astronomy, a measure of the rate at which the universe is expanding, named after Edwin Hubble. Observations suggest that galaxies are moving apart at a rate of 50–100 kps/30–60 mps for every million ◊parsecs of distance. This means that the universe, which began at one point according to the ◊Big Bang theory, is between 10 billion and 20 billion years old (probably closer to 20). Observations by the Hubble Space Telescope in 1996 produced a revised constant of 73 kps/45 mps.

Hubble's law the law that relates a galaxy's distance from us to its speed of recession as the universe expands, announced in 1929 by Edwin Hubble. He found that galaxies are moving apart at speeds that increase in direct proportion to their distance apart. The rate of expansion is known as Hubble's constant.

Hubble Space Telescope (HST) space-based astronomical observing facility, orbiting the Earth at an altitude of 610 km/380 mi. It consists of a 2.4 m/94 in telescope and four complimentary scientific instruments, is roughly cylindrical, 13 m/43 ft long, and 4 m/13 ft in diameter, with two large solar panels. HST produces a wealth of scientific data, and allows astronomers to observe the birth of stars, find planets around neighbouring stars, follow the expanding remnants of exploding stars, and search for black holes in the centre of galaxies. HST is a cooperative programme between the European Space Agency (ESA) and the US agency NASA, and is the first spacecraft specifically designed to be serviced in orbit as a permanent space-based observatory. It was launched in 1990.

By having a large telescope above Earth's atmosphere, astronomers are able to look at the universe with unprecedented clarity. Celestial observations by HST are unhampered by clouds and other atmospheric phenomena that distort and attenuate starlight. In particular, the apparent twinkling of starlight caused by density fluctuations in the atmosphere limits the clarity of ground-based telescopes. HST performs at least ten times better than such telescopes and can see almost back to the edge of the universe and to the beginning of time (see ◊Big Bang).

Before HST could reach its full potential, a flaw in the shape of its main mirror, discovered two months after the launch, had to be corrected. In 1993, as part of a planned servicing and instrument upgrade mission, NASA astronauts aboard the space shuttle *Endeavor* installed a set of corrective lenses to compensate for the error in the mirror figure. COSTAR (corrective optics space telescope axial replacement), a device containing ten coin-sized mirrors, now feeds a corrected image from the main mirror to three of the HST's four scientific instruments. HST is also being used to detail the distribution of dust and stars in nearby galaxies, watch the collisions of galaxies in detail, infer the evolution of galaxies, and measure the age of the universe.

In December 1995 HST was trained on an 'empty' area of sky near the Plough, now termed the **Hubble Deep Field**. Around 1,500 galaxies, mostly new discoveries, were photographed.

Two new instruments were added in February 1997. The Near Infared Camera and Multi-Object Spectrometer (NICMOS) will enable Hubble to see things even further away (and therefore older) than ever before. The Space Telescope Imaging Spectograph will work 30 times faster than its predecessor as it can gather information about different stars at the same time. Three new cameras had to be fitted shortly afterwards as one of the original ones was found to be faulty.

In May 1997, three months after astronauts installed new equipment, US scientists reported that Hubble had made an extraordinary finding. Within 20 minutes of searching, it discovered evidence of a black hole 300 million times the mass of the Sun. It is located in the middle of galaxy M84 about 50 million light-years from Earth. Further findings in December 1997 concerned different shapes of dying stars. Previously, astronomers had thought that most stars die with a round shell of burning gas expanding into space. The photographs taken by the HST show shapes such as pinwheels and jet exhaust. This may be indicative of how the Sun will die.

huckleberry berry-bearing bush closely related to the ◊blueberry in the USA and bilberry in Britain. Huckleberry bushes have edible dark-blue berries. (Genus *Gaylussacia,* family Ericaceae.)

A human being: an ingenious assembly of portable plumbing.

CHRISTOPHER MORLEY US editor, poet, and essayist. *Human Being*

human body the physical structure of the human being. It develops from the single cell of the fertilized ovum, is born at 40 weeks, and usually reaches sexual maturity between 11 and 18 years of age. The bony framework (skeleton) consists of more than 200 bones, over half of which are in the hands and feet. Bones are held together by joints, some of which allow movement. The circulatory system supplies muscles and organs with blood, which provides oxygen and food and removes carbon dioxide and other waste products. Body functions are controlled by the nervous system and hormones. In the upper part of the trunk is the thorax, which contains the lungs and heart. Below this is the abdomen, containing the digestive system (stomach and intestines); the liver, spleen, and pancreas; the urinary system (kidneys, ureters, and bladder); and, in women, the reproductive organs (ovaries, uterus, and vagina). In men, the prostate gland and seminal vesicles only of the reproductive system are situated in the abdomen, the testes being in the scrotum, which, with the penis, is suspended in front of and below the abdomen. The bladder empties through a small channel (urethra); in the female this opens in the upper end of the vulval cleft, which also contains the opening of the vagina, or birth canal; in the male, the urethra is continued into the penis. In both sexes, the lower bowel terminates in the anus, a ring of strong muscle situated between the buttocks.

skeleton The skull is mounted on the spinal column, or spine, a chain of 24 vertebrae. The ribs, 12 on each side, are articulated to the spinal column behind, and the upper seven meet the breastbone (sternum) in front. The lower end of the spine rests on the pelvic girdle, composed of the triangular sacrum, to which are attached the hipbones (ilia), which are fused in front. Below the sacrum is the tailbone (coccyx). The shoulder blades (scapulae) are held in place behind the upper ribs by muscles, and connected in front to the breastbone by the two collarbones (clavicles).

Each shoulder blade carries a cup (glenoid cavity) into which fits the upper end of the armbone (humerus). This articulates below with the two forearm bones (radius and ulna). These are articulated at the wrist (carpals) to the bones of the hand (metacarpals and phalanges). The upper end of each thighbone (femur) fits into a depression (acetabulum) in the hipbone; its lower end is articulated at the knee to the shinbone (tibia) and calf bone (fibula), which are articulated at the ankle (tarsals) to the bones of the foot (metatarsals and phalanges). At a moving joint, the end of each bone is formed of tough, smooth cartilage, lubricated by ◊synovial fluid. Points of special stress are reinforced by bands of fibrous tissue (ligaments).

Muscles are bundles of fibres wrapped in thin, tough layers of connective tissue (fascia); these are usually prolonged at the ends into strong, white cords (tendons, sinews) or sheets (aponeuroses), which connect the muscles to bones and organs, and by way of which the muscles do their work. Membranes of connective tissue also enfold the organs and line the interior cavities of the body. The thorax has a stout muscular floor, the diaphragm, which expands and contracts the lungs in the act of breathing.

The blood vessels of the **circulatory system**, branching into multitudes of very fine tubes (capillaries), supply all parts of the muscles and organs with blood, which carries oxygen and food necessary for life. The food passes out of the blood to the cells in a clear fluid (lymph); this is returned with waste matter through a system of lymphatic vessels that converge into collecting ducts that drain into large veins in the region of the lower neck. Capillaries join together to form veins which return blood, depleted of oxygen, to the heart.

A finely branching **nervous system** regulates the function of the muscles and organs, and makes their needs known to the controlling centres in the central nervous system, which consists of the brain and spinal cord. The inner spaces of the brain and the cord contain cerebrospinal fluid. The body processes are regulated both by the nervous system and by hormones secreted by the endocrine

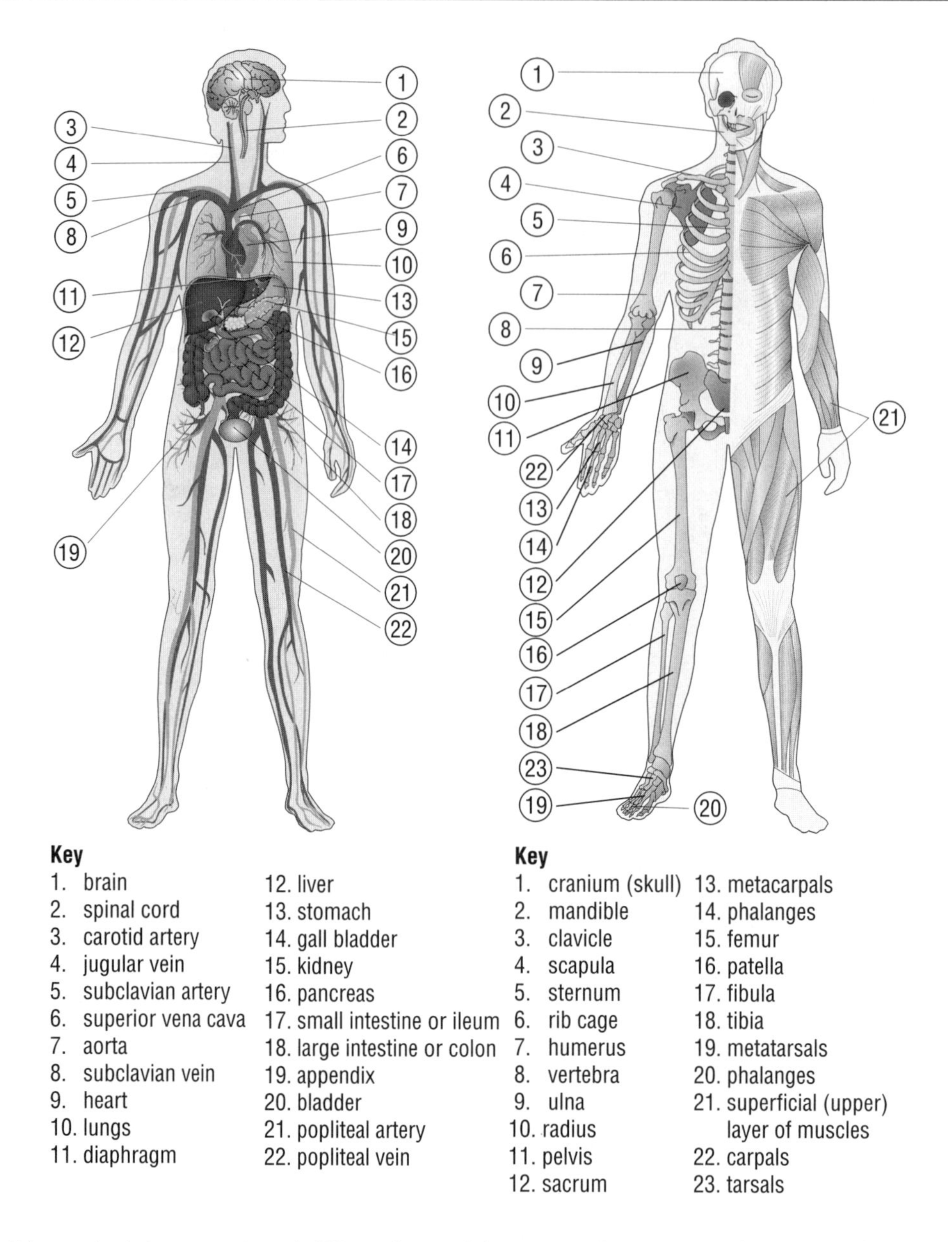

human body *The adult human body has approximately 650 muscles, 100 joints, 100,000 km/60,000 mi of blood vessels and 13,000 nerve cells. There are 206 bones in the adult body, nearly half of them in the hands and feet.*

glands. Cavities of the body that open onto the surface are coated with mucous membranes, which secrete a lubricating fluid (mucus).

The exterior surface of the body is covered with **skin**. Within the skin are the sebaceous glands, which secrete sebum, an oily fluid that makes the skin soft and pliable, and the sweat glands, which secrete water and various salts. From the skin grow hairs, chiefly on the head, in the armpits, and around the sexual organs; and nails shielding the tips of the fingers and toes; both hair and nails are modifications of skin tissue. The skin also contains nerve receptors for sensations of touch, pain, heat, and cold.

The human **digestive system** is nonspecialized and can break down a wide variety of foodstuffs. Food is mixed with saliva in the mouth by chewing and is swallowed. It enters the stomach, where it is gently churned for some time and mixed with acidic gastric juice. It then passes into the small intestine. In the first part of this, the duodenum, it is broken down further by the juice of the pancreas and duodenal glands, and mixed with bile from the liver, which splits up the fat. The jejunum and ileum continue the work of digestion and absorb most of the nutritive substances from the food. The large intestine completes the process, reabsorbing water into the body, and ejecting the useless residue as faeces.

The body, to be healthy, must maintain water and various salts in the right proportions; the process is called **osmoregulation**. The blood is filtered in the two kidneys, which remove excess water, salts, and metabolic wastes. Together these form urine, which has

HUMAN ANATOMY ONLINE

`http://www.innerbody.com/indexbody.html`

Fun, interactive, and educational site on the human body. The site is divided into many informative sections, including hundreds of images, and animations for Java compatible browsers.

a yellow pigment derived from bile, and passes down through two fine tubes (ureters) into the bladder, a reservoir from which the urine is emptied at intervals (micturition) through the urethra. Heat is constantly generated by the combustion of food in the muscles and glands, and by the activity of nerve cells and fibres. It is dissipated through the skin by conduction and evaporation of sweat, through the lungs in the expired air, and in other excreted substances. Average body temperature is about 38°C/100°F (37°C/98.4°F in the mouth).

human–computer interaction exchange of information between a person and a computer, through the medium of a ◊user interface, studied as a branch of ergonomics.

Human Genome Project research scheme, begun in 1988, to map the complete nucleotide (see ◊nucleic acid) sequence of human ◊DNA. There are approximately 80,000 different ◊genes in the human genome, and one gene may contain more than 2 million nucleotides. The programme aims to collect 10–15,000 genetic specimens from 722 ethnic groups whose genetic make-up is to be preserved for future use and study. The knowledge gained is expected to help prevent or treat many crippling and lethal diseases, but there are potential ethical problems associated with knowledge of an individual's genetic make-up, and fears that it will lead to genetic discrimination. Many indigenous people have condemned the project as 'bio-prospecting' – taking genetic material and exploiting it for economic gain – after attempts were made to patent Human T-Lymphotropic Virus Type 2 taken from a Guayami woman with leukaemia, in 1993.

The Human Genome Organization (HUGO) coordinating the project expects to spend $1 billion over the first five years, making this the largest research project ever undertaken in the life sciences. Work is being carried out in more than 20 centres around the world. By the beginning of 1991, about 2,000 genes had been mapped. By late 1994 a genetic map of the complete genome had been completed.

Concern that, for example, knowledge of an individual's genes may make that person an unacceptable insurance risk has led to planned legislation on genome privacy in the USA, and 3% of HUGO's funds have been set aside for researching and reporting on the ethical implications of the project.

Each strand of DNA carries a sequence of chemical building blocks, the nucleotides. There are only four different types, but the number of possible combinations is immense. The different combinations of nucleotides produce different proteins in the cell, and thus determine the structure of the body and its individual variations. To establish the nucleotide sequence, DNA strands are broken into fragments, which are duplicated (by being introduced into cells of yeast or the bacterium *Escherichia coli*) and distributed to the research centres.

Genes account for only a small amount of the DNA sequence. Over 90% of DNA appears not to have any function, although it is perfectly replicated each time the cell divides, and handed on to the next generation. Many higher organisms have large amounts of redundant DNA and it may be that this is an advantage, in that there is a pool of DNA available to form new genes if an old one is lost by mutation.

HUMAN GENETIC DISEASE: A LAYMANS APPROACH

http://mcrcr2.med.nyu.edu/murphp01/lysosome/hgd.htm

Comprehensive manual of cell biology for the family. It includes discussions of cell structure, DNA, chromosomes, and the detection of genetic defects. It also outlines the main goals of state-of-the-art genetic research.

human reproduction an example of ◊sexual reproduction, where the male produces sperm and the female eggs. These gametes contain only half the normal number of chromosomes, 23 instead of 46, so that on fertilization the resulting cell has the correct genetic complement. Fertilization is internal, which increases the chances of conception; unusually for mammals, copulation and pregnancy can occur at any time of the year. Human beings are also remarkable for the length of childhood and for the highly complex systems of parental care found in society. The use of contraception and the development of laboratory methods of insemination and fertilization are issues that make human reproduction more than a merely biological phenomenon.

human species, origins of evolution of humans from ancestral ◊primates. The African apes (gorilla and chimpanzee) are shown by anatomical and molecular comparisons to be the closest living relatives of humans. The oldest known **hominids** (of the human group), the australopithecines, found in Africa, date from 3.5–4.4 million years ago. The first to use tools came 2 million years later, and the first humanoids to use fire and move out of Africa appeared 1.7 million years ago. Neanderthals were not direct ancestors of the human species. Modern humans are all believed to descend from one African female of 200,000 years ago, although there is a rival theory that humans evolved in different parts of the world simultaneously.

Miocene apes Genetic studies indicate that the last common ancestor between chimpanzees and humans lived 5 to 10 million years ago. There are only fragmentary remains of ape and hominid fossils from this period. Dispute continues over the hominid status of *Ramapithecus*, the jaws and teeth of which have been found in India and Kenya in late Miocene deposits, dated between 14 and 10 million years. The lower jaw of a fossil ape found in the Otavi Mountains, Namibia, comes from deposits dated between 10 and 15 million years ago, and is similar to finds from E Africa and Turkey. It is thought to be close to the initial divergence of the great apes and humans.

Australopithecines Bones of the earliest known human ancestor, a hominid named *Australopithecus ramidus* 1994, were found in Ethiopia and dated as 4.4million years old. *A. afarensis*, found in Ethiopia and Kenya, date from 3.9 to 4.4 million years ago. These hominids walked upright and they were either direct ancestors or an offshoot of the line that led to modern humans. They may have been the ancestors of *Homo habilis* (considered by some to be a species of *Australopithecus*), who appeared about 2 million years later, had slightly larger bodies and brains, and were probably the first to use stone tools. Also living in Africa at the same time was *A. africanus*, a gracile hominid thought to be a meat-eater, and *A.robustus*, a hominid with robust bones, large teeth, heavy jaws, and thought to be a vegetarian. They are not generally considered to be our ancestors.

Homo erectus ***Over 1.7 million years ago,*** *Homo erectus*, believed by some to be descended from *H. habilis*, appeared in Africa. *H. erectus* had prominent brow ridges, a flattened cranium, with the widest part of the skull low down, and jaws with a rounded tooth row, but the chin, characteristic of modern humans, is lacking. They also had much larger brains (900–1,200 cu cm), and were probably the first to use fire and the first to move out of Africa. Their remains are found as far afield as China, W Asia, Spain, and S Britain. Modern human *H. sapiens sapiens* and the Neanderthals *H. sapiens neanderthalensis* are probably descended from *H. erectus*.

Neanderthals Neanderthals were large-brained and heavily built, probably adapted to the cold conditions of the ice ages. They lived in Europe and the Middle East, and disappeared about 40,000 years ago, leaving *H. sapiens sapiens* as the only remaining species of the hominid group. Possible intermediate forms between Neanderthals and *H.sapiens sapiens* have been found at Mount Carmel in Israel and at Broken Hill in Zambia, but it seems that *H.sapiens sapiens* appeared in Europe quite rapidly and either wiped out the Neanderthals or interbred with them.

modern humans There are currently two major views of human evolution: the **'out of Africa' model**, according to which *H. sapiens* emerged from *H.erectus*, or a descendant species, in Africa and then spread throughout the world; and the **multiregional model**, according to which selection pressures led to the emergence of similar advanced types of *H. sapiens* from *H. erectus* in different parts of the world at around the same time. Analysis of DNA in

Evolution: Out of Africa and the Eve Hypothesis

BY CHRIS STRINGER

Introduction

Most palaeoanthropologists recognize the existence of two human species during the last million years – *Homo erectus*, now extinct, and *Homo sapiens*, the species which includes recent or 'modern' humans. In general, they believe that *Homo erectus* was the ancestor of *Homo sapiens*. How did the transition occur?

The multiregional model

There are two opposing views. The multiregional model says that *Homo erectus* gave rise to *Homo sapiens* across its whole range, which, about 700,000 years ago, included Africa, China, Java (Indonesia), and, probably, Europe. *Homo erectus*, following an African origin about 1.7 million years ago, dispersed around the Old World, developing the regional variation that lies at the roots of modern 'racial' variation. Particular features in a given region persisted in the local descendant populations of today.

For example, Chinese *Homo erectus* specimens had the same flat faces, with prominent cheekbones, as modern Oriental populations. Javanese *Homo erectus* had robustly built cheekbones and faces that jutted out from the braincase, characteristics found in modern Australian Aborigines. No definite representatives of *Homo erectus* have yet been discovered in Europe. Here, the fossil record does not extend back as far as those of Africa and eastern Asia, although a possible *Homo erectus* jawbone more than a million years old was recently excavated in Georgia.

Nevertheless, the multiregional model claims that European *Homo erectus* did exist, and evolved into a primitive form of *Homo sapiens*. Evolution in turn produced the Neanderthals: the ancestors of modern Europeans. Features of continuity in this European lineage include prominent noses and midfaces.

Genetic continuity

The multiregional model was first described in detail by Franz Weidenreich, a German palaeoanthropologist. It was developed further by the American Carleton Coon, who tended to regard the regional lineages as genetically separate. Most recently, the model has become associated with such researchers as Milford Wolpoff (USA) and Alan Thorne (Australia), who have re-emphasized the importance of gene flow between the regional lines. In fact, they regard the continuity in time and space between the various forms of *Homo erectus* and their regional descendants to be so complete that they should be regarded as representing only one species – *Homo sapiens*.

The opposing view

The opposing view is that *Homo sapiens* had a restricted origin in time and space. This is an old idea. Early in the 20th century, workers such as Marcellin Boule (France) and Arthur Keith (UK) believed that the lineage of *Homo sapiens* was very ancient, having developed in parallel with that of *Homo erectus* and the Neanderthals. However, much of the fossil evidence used to support their ideas has been re-evaluated, and few workers now accept the idea of a very ancient and separate origin for modern *Homo sapiens*.

The Garden of Eden

Modern proponents of this approach focus on a recent and restricted origin for modern *Homo sapiens*. This was dubbed the 'Garden of Eden' or 'Noah's Ark' model by the US anthropologist William Howells in 1976 because of the idea that all modern human variation had a localized origin from one centre. Howells did not specify the centre of origin, but research since 1976 points to Africa as especially important in modern human origins.

The consequent 'Out of Africa' model claims that *Homo erectus* evolved into modern *Homo sapiens* in Africa about 100,000–150,000 years ago. Part of the African stock of early modern humans spread from the continent into adjoining regions and eventually reached Australia, Europe, and the Americas (probably by 45,000, 40,000, and 15,000 years ago respectively). Regional ('racial') variation only developed during and after the dispersal, so that there is no continuity of regional features between *Homo erectus* and present counterparts in the same regions.

Like the multiregional model, this view accepts that *Homo erectus* evolved into new forms of human in inhabited regions outside Africa, but argues that these non-African lineages became extinct without evolving into modern humans. Some, such as the Neanderthals, were displaced and then replaced by the spread of modern humans into their regions.

... and an African Eve?

In 1987, research on the genetic material called mitochondrial DNA (mtDNA) in living humans led to the reconstruction of a hypothetical female ancestor for all present-day humanity. This 'Eve' was believed to have lived in Africa about 200,000 years ago. Recent re-examination of the 'Eve' research has cast doubt on this hypothesis, but further support for an 'Out of Africa' model has come from genetic studies of nuclear DNA, which also point to a relatively recent African origin for present-day *Homo sapiens*.

Studies of fossil material of the last 50,000 years also seem to indicate that many 'racial' features in the human skeleton have developed only over the last 30,000 years, in line with the 'Out of Africa' model, and at odds with the million-year timespan one would expect from the multiregional model.

---

recent human populations suggests that *H. sapiens* originated about 200,000 years ago in Africa from a single female ancestor, 'Eve'. The oldest known fossils of *H.sapiens* also come from Africa, dating from 150,000–100,000 years ago. Separation of human populations would have occurred later, with separation of Asian, European, and Australian populations taking place between 100,000 and 50,000 years ago.

Humber Bridge suspension bridge with twin towers 163 m/535 ft high, which spans the estuary of the river Humber in NE England. When completed in 1980, it was the world's longest bridge with a span of 1,410 m/4,628 ft.

Built at a cost of £150 million, toll revenues over the following 15 years proved inadequate to pay even the interest on the debt.

humerus the upper bone of the forelimb of tetrapods. In humans, the humerus is the bone above the elbow.

humidity the quantity of water vapour in a given volume of the atmosphere (absolute humidity), or the ratio of the amount of water vapour in the atmosphere to the saturation value at the same temperature (relative humidity). At ◊dew point the relative humidity is 100% and the air is said to be saturated. Condensation (the conversion of vapour to liquid) may then occur. Relative humidity is measured by various types of ◊hygrometer.

hummingbird any of various small, brilliantly coloured birds found in the Americas. The name comes from the sound produced by the rapid vibration of their wings when hovering near flowers to feed. Hummingbirds have long, needlelike bills and tongues to obtain nectar from flowers and capture insects. They are the only birds able to fly backwards. The Cuban **bee hummingbird** (*Mellisuga helenae*), the world's smallest bird, is 5.5 cm/2 in long and weighs less than 2.5 g/0.1 oz. There are over 300 species. (Family Trochilidae, order Apodiformes.)

The long cleft tongue of a hummingbird is in the form of a double tube, which can be extended a considerable distance beyond the bill and withdrawn again very rapidly; the sternum (breastbone) is greatly developed, forming a suitable base for the wing muscles; the plumage has a metallic lustre.

hummingbird moth type of ◊hawk moth.

humours, theory of theory prevalent in the West in classical and medieval times that the human body was composed of four kinds of fluid: phlegm, blood, choler or yellow bile, and melancholy or black bile. Physical and mental characteristics were explained by different proportions of humours in individuals.

An excess of phlegm produced a 'phlegmatic', or calm, temperament; of blood a 'sanguine', or passionate, one; of yellow bile a 'choleric', or irascible, temperament; and of black bile a 'melancholy', or depressive, one. The Greek physician Galen connected the theory to that of the four elements (see ◊elements, the four): the phlegmatic was associated with water, the sanguine with air, the choleric with fire, and the melancholic with earth. An imbalance of the humours could supposedly be treated by diet.

humus component of ◊soil consisting of decomposed or partly decomposed organic matter, dark in colour and usually richer towards the surface. It has a higher carbon content than the original material and a lower nitrogen content, and is an important source of minerals in soil fertility.

hundredweight imperial unit (abbreviation cwt) of mass, equal to 112 lb (50.8 kg). It is sometimes called the long hundredweight, to distinguish it from the short hundredweight or **cental**, equal to 100 lb (45.4 kg).

hunting dog or ***painted dog*** wild dog that once roamed over virtually the whole of sub-Saharan Africa. A pack might have a range of almost 4,000 km/2,500 mi, hunting zebra, antelope, and other game. Individuals can run at 50 kph/30 mph for up to 5 km/3 mi, with short bursts of even higher speeds. The number of hunting dogs that survive has been reduced to a fraction of the original population. According to a 1997 International Union for the Conservation of Nature (IUCN) report, there were fewer than 3,000 hunting dogs remaining in the wild, with many existing populations too small to be viable. (Species *Lycaon pictus*, family Canidae.)

The maximum pack size found today is usually eight to ten, whereas in the past several hundred might have hunted together. Habitat destruction and the decline of large game herds have played a part in its decline, but the hunting dog has also suffered badly from the effects of distemper, a disease which was introduced into E Africa early in the 20th century.

Huntington's chorea rare hereditary disease of the nervous system that mostly begins in middle age. It is characterized by involuntary movements (◊chorea), emotional disturbances, and rapid mental degeneration progressing to ◊dementia. There is no known cure but the genetic mutation giving rise to the disease was located 1993, making it easier to test individuals for the disease and increasing the chances of developing a cure.

FACING HUNTINGTON'S DISEASE

http://neuro-chief-e.mgh.harvard.edu/MCMENEMY/facinghd.html

Excellent source of basic information about HD. Sympathetically presented by the British Huntington's Disease association, the guide covers how HD is passed on, how it is diagnosed, the risks to children conceived by couples at risk, and the course of this degenerative illness.

hurricane revolving storm in tropical regions, called **typhoon** in the N Pacific. It originates at latitudes between 5° and 20° N or S of the Equator, when the surface temperature of the ocean is above 27°C/80°F. A central calm area, called the eye, is surrounded by inwardly spiralling winds (anticlockwise in the northern hemisphere) of up to 320 kph/200 mph. A hurricane is accompanied by lightning and torrential rain, and can cause extensive damage. In meteorology, a hurricane is a wind of force 12 or more on the ◊Beaufort scale.

During 1995 the Atlantic Ocean region suffered 19 tropical storms, 11 of them hurricanes. This was the third-worst season since 1871, causing 137 deaths. The most intense hurricane recorded in the Caribbean/Atlantic sector was Hurricane Gilbert in 1988, with sustained winds of 280 kph/175 mph and gusts of over 320 kph/200 mph.

husky any of several breeds of sledge dog used in Arctic regions, growing to 70 cm/27.5 in high, and weighing about 50 kg/110 lbs, with pricked ears, thick fur, and a bushy tail. The Siberian husky is the best known.

Hutton, James (1726–1797)

Scottish geologist, known as the 'founder of geology', who formulated the concept of uniformitarianism. In 1785 he developed a theory of the igneous origin of many rocks.

His *Theory of the Earth* 1788 proposed that the Earth was incalculably old. Uniformitarianism suggests that past events could be explained in terms of processes that work today. For example, the kind of river current that produces a certain settling pattern in a bed of sand today must have been operating many millions of years ago, if that same pattern is visible in ancient sandstones.

Mary Evans Picture Library

hyacinth any of a group of bulb-producing plants belonging to the lily family, native to the E Mediterranean and Africa. The cultivated hyacinth (*H. orientalis*) has large, scented, cylindrical heads of pink, white, or blue flowers. (Genus *Hyacinthus*, family Liliaceae.)

The ◊water hyacinth is unrelated, a floating plant from South America.

Hyades V-shaped cluster of stars that forms the face of the bull in the constellation ◊Taurus. It is 150 light years away and contains

HUYGENS, CHRISTIAAN

http://www-history.mcs.st-and.ac.uk/~history/Mathematicians/Huygens.html

Extensive biography of the great Dutch astronomer, physicist, and mathematician. The site contains a description of his contributions to astronomy, physics, and mathematics. Also included are the title page of his book *Horologium Oscillatorium* (1673) and the first page of his book *De Ratiociniis in Ludo Aleae* (1657). Several references for further reading are also listed, and the Web site also features a portrait of Huygens.

Huygens, Christiaan (1629–1695)

or Huyghens Dutch mathematical physicist and astronomer. He proposed the wave theory of light, developed the pendulum clock in 1657, discovered polarization, and observed Saturn's rings. He made important advances in pure mathematics, applied mathematics, and mechanics, which he virtually founded. His work in astronomy was an impressive defence of the Copernican view of the Solar System.

mechanics Huygens' first studies in applied mathematics dealt with mechanics, the branch of physics pertaining to motions and forces. Working on impact and collision, Huygens used the idea of relative frames of reference, considering the motion of one body relative to the other. He anticipated the law of conservation of momentum stating that in a system of bodies under impact the centre of gravity is conserved. In *De Motu Corporum* 1656, he was also able to show that the quantity $\frac{1}{2}mv^2$ is conserved in an elastic collision.

Huygens also studied centrifugal force and showed, in 1659, its similarity to gravitational force, although he lacked the Newtonian concept of acceleration. He considered projectiles and gravity, developing the mathematically primitive ideas of Galileo. He found an accurate experimental value for the distance covered by a falling body in one second. In fact, his gravitational theories successfully deal with several difficult points that Newton carefully avoided. In the 1670s, Huygens studied motion in resisting media, becoming convinced by experiment that the resistance in such media as air is proportional to the square of the velocity.

the pendulum clock In 1657, Huygens developed a clock regulated by a pendulum, an idea that he published and patented. By 1658, major towns in Holland had pendulum tower clocks.

Mary Evans Picture Library

Huygens worked at the theory first of the simple pendulum and then of harmonically oscillating systems throughout the rest of his life, publishing the *Horologium Oscillatorium* 1673. He derived the relationship between the period of a simple pendulum and its length.

the theory of light The *Traité de la Lumière/Treatise on Light* 1678 contained Huygens' famous wave or pulse theory of light. Two years earlier, Huygens had been able to use his principle of secondary wave fronts to explain reflection and refraction, showing that refraction is related to differing velocities of light in media. He theorized that light is transmitted as a pulse moving through a medium, or ether, by setting up a whole train of vibrations in the ether in a serial displacement. His publication was partly a counter to Newton's particle theory of light. The thoroughness of Huygens' analysis of this model is impressive, but although he observed the effects due to polarization, he could not yet use his ideas to explain this phenomenon.

astronomy and the telescope Huygens' comprehensive study of geometric optics led to the invention of a telescope eyepiece that reduced chromatic aberration. It consisted of two thin plano-convex lenses, rather than one fat lens, with the field lens having a focal length three times greater than that of the eyepiece lens. Its main disadvantage was that cross-wires could not be fitted to measure the size of an image. Huygens then developed a micrometer to measure the angular diameter of celestial objects.

With a home-made telescope, he discovered Titan, one of Saturn's moons, in 1655. Later that year he observed that Titan's period of revolution was about 16 days and that it moved in the same plane as the so-called 'arms' of Saturn. This phenomenon had been somewhat of an enigma to many earlier astronomers, but because of Huygens' superior 7-m telescope, he partially unravelled the detail of Saturn's rings. In 1659, he published a Latin anagram that, when interpreted, read 'It (Saturn) is surrounded by a thin flat ring, nowhere touching and inclined to the ecliptic'. The theory behind Huygens' hypothesis followed later in *Systema Saturnium* 1659, which included observations on the planets, their satellites, the Orion nebula and the determination of the period of Mars, and provided further evidence for the Copernican view of the Solar System.

over 200 stars, although only about 12 are visible to the naked eye.

The Hyades is a much older cluster than the Pleiades, for not only have some of the brighter stars evolved into ◊red giants, some have gone even further and are now ◊white dwarfs. ◊Aldebaran, which marks the eye of the bull and which appears to be in the middle of the cluster, is not actually a member of the cluster. It is only 68 light years away, while the cluster is 130 light years away.

hybrid offspring from a cross between individuals of two different species, or two inbred lines within a species. In most cases, hybrids between species are infertile and unable to reproduce sexually. In plants, however, doubling of the chromosomes (see ◊polyploid) can restore the fertility of such hybrids.

hybridization the production of a ◊hybrid.

hydathode specialized pore, or less commonly, a hair, through which water is secreted by hydrostatic pressure from the interior of a plant leaf onto the surface. Hydathodes are found on many different plants and are usually situated around the leaf margin at vein endings. Each pore is surrounded by two crescent-shaped cells and resembles an open ◊stoma, but the size of the opening cannot be varied as in a stoma. The process of water secretion through hydathodes is known as ◊guttation.

Hydra in astronomy, the largest constellation, winding across more than a quarter of the sky between ◊Cancer and ◊Libra in the southern hemisphere. Hydra is named after the multiheaded monster slain by Hercules. Despite its size, it is not prominent; its brightest star is second-magnitude Alphard.

hydra in zoology, any of a group of freshwater polyps, belonging among the ◊coelenterates. The body is a double-layered tube (with six to ten hollow tentacles around the mouth), 1.25 cm/0.5 in long when extended, but capable of contracting to a small knob. Usually fixed to waterweed, hydras feed on minute animals that are caught and paralysed by stinging cells on the tentacles. (Genus *Hydra*, family Hydridae, phylum Coelenterata, subphylum Cnidaria.)

Hydras reproduce asexually in the summer and sexually in the winter. They have no specialized organs except those of reproduction.

hydrangea any of a group of flowering shrubs belonging to the saxifrage family, native to Japan. Cultivated varieties of *H. macrophylla* normally produce round heads of pink flowers, but these may be blue if there are certain chemicals in the soil, such as alum or iron. The name comes from the Greek for 'water vessel', after the cuplike seed capsules. (Genus *Hydrangea*, family Hydrangeaceae.)

hydrate chemical compound that has discrete water molecules combined with it. The water is known as **water of crystallization** and the number of water molecules associated with one molecule of the compound is denoted in both its name and chemical formula: for example, $CuSO_4.5H_2O$ is copper(II) sulphate pentahydrate.

hydration in chemistry, the combination of water and another substance to produce a single product. It is the opposite of ◊dehydration.

hydraulic radius measure of a river's ◊channel efficiency (its ability to discharge water), used by water engineers to assess the likelihood of flooding. The hydraulic radius of a channel is defined as the ratio of its cross-sectional area to its wetted perimeter (the part of the cross-section that is in contact with the water).

The greater the hydraulic radius, the greater the efficiency of the channel and the less likely the river is to flood. The highest values occur when channels are deep, narrow, and semi-circular in shape.

hydraulics field of study concerned with utilizing the properties of water and other liquids, in particular the way they flow and transmit pressure, and with the application of these properties in engineering. It applies the principles of ◊hydrostatics and hydrodynamics. The oldest type of hydraulic machine is the **hydraulic press**, invented by Joseph Bramah in England in 1795. The hydraulic principle of pressurized liquid increasing a force is commonly used on vehicle braking systems, the forging press, and the hydraulic systems of aircraft and excavators.

A hydraulic press consists of two liquid-connected pistons in cylinders, one of narrow bore, one of large bore. A force applied to the narrow piston applies a certain pressure (force per unit area) to the liquid, which is transmitted to the larger piston. Because the area of this piston is larger, the force exerted on it is larger. Thus the original force has been magnified, although the smaller piston must move a great distance to move the larger piston only a little.

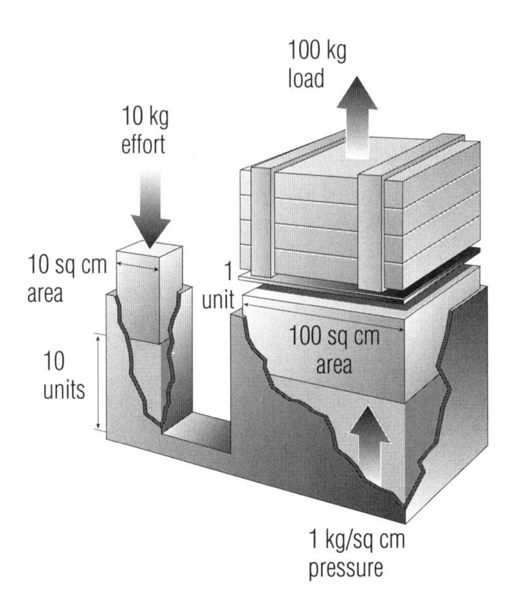

***hydraulics** The hydraulic jack transmits the pressure on a small piston to a larger one. A larger total force is developed by the larger piston but it moves a smaller distance than the small piston.*

hydride chemical compound containing hydrogen and one other element, and in which the hydrogen is the more electronegative element (see ◊electronegativity).

Hydrides of the more reactive metals may be ionic compounds containing a hydride anion (H^-).

hydrocarbon any of a class of chemical compounds containing only hydrogen and carbon (for example, the alkanes and alkenes). Hydrocarbons are obtained industrially principally from petroleum and coal tar.

hydrocephalus potentially serious increase in the volume of cerebrospinal fluid (CSF) within the ventricles of the brain. In infants, since their skull plates have not fused, it causes enlargement of the head, and there is a risk of brain damage from CSF pressure on the developing brain.

Hydrocephalus may be due to mechanical obstruction of the outflow of CSF from the ventricles or to faulty reabsorption. Treatment usually involves surgical placement of a shunt system to drain the fluid into the abdominal cavity. In infants, the condition is often seen in association with ◊spina bifida. Hydrocephalus may occur as a consequence of brain injury or disease.

hydrochloric acid HCl solution of hydrogen chloride (a colourless, acidic gas) in water. The concentrated acid is about 35% hydrogen chloride and is corrosive. The acid is a typical strong, monobasic acid forming only one series of salts, the chlorides. It has many industrial uses, including recovery of zinc from galvanized scrap iron and the production of chlorine. It is also produced in the stomachs of animals for the purposes of digestion.

hydrocyanic acid or ***prussic acid*** solution of hydrogen cyanide gas (HCN) in water. It is a colourless, highly poisonous, volatile liquid, smelling of bitter almonds.

hydrodynamics branch of physics dealing with fluids (liquids and gases) in motion.

hydroelectric power electricity generated by moving water. In a typical scheme, water stored in a reservoir, often created by damming a river, is piped into water ◊turbines, coupled to electricity generators. In ◊pumped storage plants, water flowing through the turbines is recycled. A ◊tidal power station exploits the rise and fall of the tides. About one-fifth of the world's electricity comes from hydroelectric power.

Hydroelectric plants have prodigious generating capacities. The Grand Coulee plant in Washington State, USA, has a power output of around 10,000 megawatts. The Itaipu power station on the Paraná River (Brazil/Paraguay) has a potential capacity of 12,000 megawatts.

Work on the world's largest hydroelectric project, the Three Gorges Dam on the Chang Jiang, was officially inaugurated in 1994. By 1996, around 600,000 sq km/231,660 sq mi of land had been flooded worldwide for hydroelectric reservoirs.

hydrofoil wing that develops lift in the water in much the same way that an aeroplane wing develops lift in the air. A hydrofoil boat is one whose hull rises out of the water owing to the lift, and the boat skims along on the hydrofoils. The first hydrofoil was fitted to a boat 1906. The first commercial hydrofoil went into operation 1956. One of the most advanced hydrofoil boats is the Boeing ◊jetfoil. Hydrofoils are now widely used for fast island ferries in calm seas.

hydrogen ***Greek** hydro + gen **'water generator'*** colourless, odourless, gaseous, nonmetallic element, symbol H, atomic number 1, relative atomic mass 1.00797. It is the lightest of all the elements and occurs on Earth chiefly in combination with oxygen as water. Hydrogen is the most abundant element in the universe, where it accounts for 93% of the total number of atoms and 76% of the total mass. It is a component of most stars, including the Sun, whose heat and light are produced through the nuclear-fusion process that converts hydrogen into helium. When subjected to a pressure 500,000 times greater than that of the Earth's atmosphere, hydrogen becomes a solid with metallic properties, as in one of the inner zones of Jupiter. Hydrogen's common and industrial uses include the hardening of oils and fats by hydrogenation, the creation of high-temperature flames for welding, and as rocket fuel. It has been proposed as a fuel for road vehicles.

> TO REMEMBER THE FOUR ELEMENTS THAT MAKE LIFE'S BUILDING BLOCKS:
>
> HONC IF YOU LIKE LIFE!
>
> HYDROGEN, OXYGEN, NITROGEN, CARBON

Its isotopes ◊deuterium and ◊tritium (half-life 12.5 years) are used in nuclear weapons, and deuterons (deuterium nuclei) are used in synthesizing elements. The element's name refers to the generation of water by the combustion of hydrogen, and was coined in 1787 by French chemist Louis Guyton de Morveau (1737–1816).

hydrogenation addition of hydrogen to an unsaturated organic molecule (one that contains ◊double bonds or ◊triple bonds). It is widely used in the manufacture of margarine and low-fat spreads by the addition of hydrogen to vegetable oils.

Vegetable oils contain double carbon-to-carbon bonds and are therefore examples of unsaturated compounds. When hydrogen is added to these double bonds, the oils become saturated and more solid in consistency.

hydrogen bomb bomb that works on the principle of nuclear ◊fusion. Large-scale explosion results from the thermonuclear release of energy when hydrogen nuclei are fused to form helium nuclei. The first hydrogen bomb was exploded at Enewetak Atoll in the Pacific Ocean by the USA in 1952.

In some sort of crude sense ... the physicists have known sin; and this is a knowledge which they cannot lose.

J Robert Oppenheimer US physicist.
On the hydrogen bomb, lecture at MIT
25 Nov 1947 *Physics in the Contemporary World*

hydrogen burning in astronomy, any of several processes by which hydrogen is converted to ◊helium by ◊nuclear fusion in the core of a star. In the Sun, the main process is the proton–proton chain, while in heavier stars the carbon cycle is more important. In both processes, four protons are converted to a helium nucleus with the emission of ◊positrons, ◊neutrinos, and gamma ◊rays. The temperature must exceed several million K for hydrogen burning to start and the least massive stars (◊brown dwarfs) never become hot enough.

hydrogen carbonate or ***bicarbonate*** compound containing the ion HCO_3^-, an acid salt of carbonic acid (solution of carbon dioxide in water). When heated or treated with dilute acids, it gives off carbon dioxide. The most important compounds are ◊sodium hydrogen carbonate (bicarbonate of soda), and ◊calcium hydrogen carbonate.

hydrogen cyanide HCN poisonous gas formed by the reaction of sodium cyanide with dilute sulphuric acid; it is used for fumigation.

The salts formed from it are cyanides – for example sodium cyanide, used in hardening steel and extracting gold and silver from their ores. If dissolved in water, hydrogen cyanide gives hydrocyanic acid.

hydrogen peroxide H_2O_2 in medicine, a liquid used, in diluted form, as an antiseptic. Oxygen is released when hydrogen peroxide is added to water and the froth helps to discharge dead tissue from wounds and ulcers. It is also used as a mouthwash and as a bleach.

hydrogen sulphate HSO_4^- compound containing the hydrogen sulphate ion. Hydrogen sulphates are ◊acid salts.

hydrogen sulphide H_2S poisonous gas with the smell of rotten eggs. It is found in certain types of crude oil where it is formed by decomposition of sulphur compounds. It is removed from the oil at the refinery and converted to elemental sulphur.

hydrogen trioxide H_2O_3 relatively stable compound of hydrogen and oxygen present in the atmosphere and possibly also in living tissue. It was first synthesized in 1994; previously it had been assumed to be too unstable.

It is produced in a reaction similar to that used for the commercial production of hydrogen peroxide (H_2O_2) but ozone (O_3) is used instead of oxygen. Hydrogen trioxide is stable at low temperatures but begins to decompose slowly at –40°C forming the high energy form of oxygen, singlet oxygen.

hydrograph graph showing how the discharge of a river varies with time. By studying hydrographs, water engineers can predict when flooding is likely and take action to prevent its taking place.

A hydrograph shows the time lag, or delay, between peak rainfall and the resultant peak in discharge, and the length of time taken for that discharge to peak. The shorter the time lag and the higher the peak, the more likely it is that flooding will occur. Factors likely to give short time lags and high peaks include heavy rainstorms, steep slopes, deforestation, poor soil quality, and the covering of surfaces with impermeable substances such as tarmac and concrete. Actions taken by water engineers to increase time lags and lower peaks include planting trees in the drainage basin of a river.

hydrography study and charting of Earth's surface waters in seas, lakes, and rivers.

hydrological cycle alternative name for the ◊water cycle, by which water is circulated between the Earth's surface and its atmosphere.

hydrology study of the location and movement of inland water, both frozen and liquid, above and below ground. It is applied to major civil engineering projects such as irrigation schemes, dams, and hydroelectric power, and in planning water supply.

hydrolysis chemical reaction in which the action of water or its ions breaks down a substance into smaller molecules. Hydrolysis occurs in certain inorganic salts in solution, in nearly all nonmetallic chlorides, in esters, and in other organic substances. It is one of the mechanisms for the breakdown of food by the body, as in the conversion of starch to glucose.

hydrometer in physics, an instrument used to measure the relative density of liquids (the density compared with that of water). A hydrometer consists of a thin glass tube ending in a sphere that leads into a smaller sphere, the latter being weighted so that the hydrometer floats upright, sinking deeper into less dense liquids than into denser liquids. Hydrometers are used in brewing and to test the strength of acid in car batteries.

The hydrometer is based on ◊Archimedes' principle.

hydrophilic ***Greek 'water-loving'*** in chemistry, a term describing ◊functional groups with a strong affinity for water, such as the carboxyl group (–COOH).

If a molecule contains both a hydrophilic and a ◊hydrophobic group (a group that repels water), it may have an affinity for both aqueous and nonaqueous molecules. Such compounds are used to stabilize ◊emulsions or as ◊detergents.

hydrophily type of ◊pollination where the pollen is transported by water. Water-pollinated plants occur in 31 genera in 11 different families. They are found in habitats as diverse as rainforests and seasonal desert pools. Pollen is either dispersed underwater or on the water's surface.

Pollen may be released directly onto the water's surface, as in the sea grass *Halodule pinifolia*, forming pollen rafts, or as in the freshwater plant *Vallisneria*, the pollen may be released within floating male flowers. In Caribbean turtle grass, *Thalassia testudinum*, pollen is released underwater embedded in strands of mucilage. Denser than water, it is carried by the current.

hydrophobia another name for the disease ◊rabies.

hydrophobic ***Greek 'water-hating'*** in chemistry, a term describing ◊functional groups that repel water (the opposite of hydrophilic).

hydrophone underwater ◊microphone and ancillary equipment capable of picking up waterborne sounds. It was originally developed to detect enemy submarines but is now also used, for example, for listening to the sounds made by whales.

hydrophyte plant adapted to live in water, or in waterlogged soil.

Hydrophytes may have leaves with a very reduced or absent ◊cuticle and no ◊stomata (since there is no need to conserve water), a reduced root and water-conducting system, and less supporting tissue since water buoys plants up. There are often numerous spaces between the cells in their stems and roots to make ◊gas exchange with all parts of the plant body possible. Many have highly divided leaves, which lessens resistance to flowing water; an example is spiked water milfoil *Myriophyllum spicatum*.

hydroplane on a submarine, a movable horizontal fin angled downwards or upwards when the vessel is descending or ascending. It is also a highly manoeuvrable motorboat with its bottom rising in steps to the stern, or a ◊hydrofoil boat that skims over the surface of the water when driven at high speed.

hydroponics cultivation of plants without soil, using specially prepared solutions of mineral salts. Beginning in the 1930s, large crops were grown by hydroponic methods, at first in California but since then in many other parts of the world.

Julius von Sachs (1832–1897), in 1860, and W Knop, in 1865, developed a system of plant culture in water whereby the relation of mineral salts to plant growth could be determined, but it was not until about 1930 that large crops could be grown. The term was first coined by US scientist W F Gericke.

hydrosphere the water component of the Earth, usually encompassing the oceans, seas, rivers, streams, swamps, lakes, groundwater, and atmospheric water vapour.

hydrostatics in physics, the branch of ◊statics dealing with fluids in equilibrium – that is, in a static condition. Practical applications include shipbuilding and dam design.

hydroxide any inorganic chemical compound containing one or more hydroxyl (OH) groups and generally combined with a metal. Hydroxides include sodium hydroxide (caustic soda, NaOH), potassium hydroxide (caustic potash, KOH), and calcium hydroxide (slaked lime, $Ca(OH)_2$).

hydroxyl group an atom of hydrogen and an atom of oxygen bonded together and covalently bonded to an organic molecule. Common compounds containing hydroxyl groups are alcohols and phenols.

In chemical reactions, the hydroxyl group (–OH) frequently behaves as a single entity.

hydroxypropanoic acid technical name for ◊lactic acid.

HYENA

The female spotted hyena has a higher level of the male sex hormone testosterone than the male. She is also up to 12% heavier than the male.

hyena any of three species of carnivorous doglike mammals living in Africa and Asia. Hyenas have extremely powerful jaws. They are scavengers, feeding on the remains of animals killed by predators such as lions, although they will also attack and kill live prey. (Genera *Hyaena* and *Crocuta*, family Hyaenidae, order Carnivora.)

The species are the **striped hyena** (*H. hyaena*) found from Asia Minor to India; the **brown hyena** (*H. brunnea*), found in S Africa; and the **spotted** or **laughing hyena** (*C. crocuta*), common south of the Sahara. The ◊aardwolf also belongs to the hyena family.

SPOTTED HYENAS

http://www.csulb.edu/~persepha/hyena.html

Facts about hyenas illustrated by a number of photos. There is also a highly informative 'Frequently Asked Questions' section dealing with questions such as 'Do they really laugh?' and 'Will they eat people?', as well as a folklore section dealing with the depiction of hyenas through history.

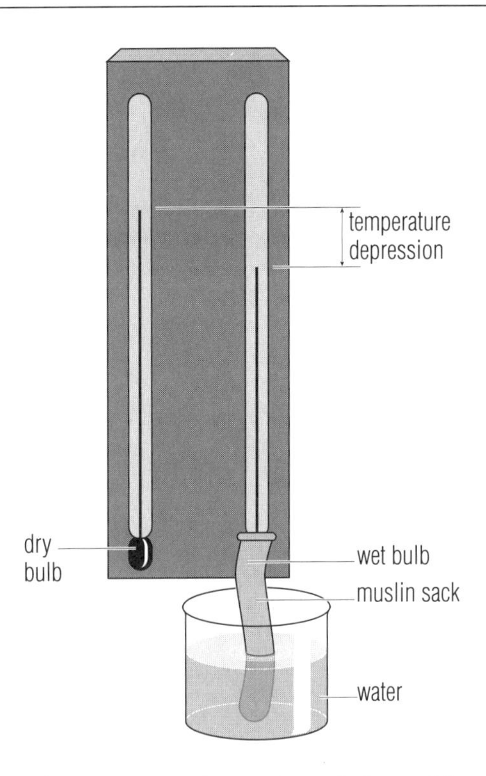

***hygrometer** The most common hygrometer, or instrument for measuring the humidity of a gas, is the wet and dry bulb hygrometer. The wet bulb records a lower temperature because water evaporates from the muslin, taking heat from the wet bulb. The degree of evaporation and hence cooling depends upon the humidity of the surrounding air or other gas.*

hygrometer in physics, any instrument for measuring the humidity, or water vapour content, of a gas (usually air). A wet and dry bulb hygrometer consists of two vertical thermometers, with one of the bulbs covered in absorbent cloth dipped into water. As the water evaporates, the bulb cools, producing a temperature difference between the two thermometers. The amount of evaporation, and hence cooling of the wet bulb, depends on the relative humidity of the air.

Other hygrometers work on the basis of a length of natural fibre, such as hair or a fine strand of gut, changing with variations in humidity. In a ◊dew-point hygrometer, a polished metal mirror gradually cools until a fine mist of water (dew) forms on it. This gives a measure of the dew point, from which the air's relative humidity can be calculated.

hyoscine or ***scopolamine*** drug that acts on the autonomic nervous system and prevents muscle spasm. It is frequently included in premedication to dry up lung secretions and as a postoperative sedative. It is also used to treat ulcers, to relax the womb in labour, for travel sickness, and to dilate the pupils before an eye examination. It is an alkaloid, $C_{17}H_{21}NO_2$, obtained from various plants of the nightshade family (such as ◊belladonna).

hyperactivity condition of excessive activity in young children, combined with restlessness, inability to concentrate, and difficulty in learning. There are various causes, ranging from temperamental predisposition to brain disease. In some cases food additives have come under suspicion; in such instances modification of the diet may help. Mostly there is improvement at puberty, but symptoms may persist in the small proportion diagnosed as having ◊attention-deficit hyperactivity disorder.

hyperbola in geometry, a curve formed by cutting a right circular cone with a plane so that the angle between the plane and the base is greater than the angle between the base and the side of the cone. All hyperbolae are bounded by two asymptotes (straight

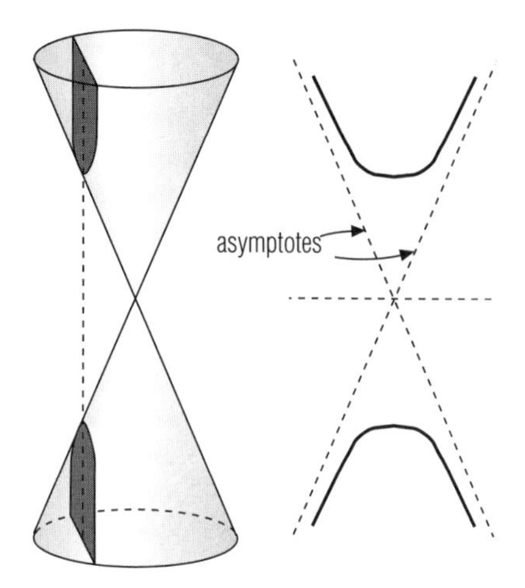

hyperbola The hyperbola is produced when a cone is cut by a plane. It is one of a family of curves called conic sections: the circle, ellipse, and parabola. These curves are produced when the plane cuts the cone at different angles and positions.

lines which the hyperbola moves closer and closer to but never reaches).

A hyperbola is a member of the family of curves known as ◊conic sections.

A hyperbola can also be defined as a path traced by a point that moves such that the ratio of its distance from a fixed point (focus) and a fixed straight line (directrix) is a constant and greater than 1; that is, it has an ◊eccentricity greater than 1.

Hypercard computer application developed for the Apple ◊Macintosh, in which data are stored as if on cards in a card-index system. A group of cards forms a stack. Additional features include the ability to link cards in different ways and, by the use of software buttons (icons that can be clicked or double clicked with a mouse), to access other data. Hypercard is very similar to ◊hypertext, although it does not conform to the rigorous definition of hypertext.

hypercharge in physics, a property of certain ◊elementary particles, analogous to electric charge, that accounts for the absence of some expected behaviour (such as decay).

hyperlink link from one document to another or, within the same document, from one place to another. It can be activated by clicking on the link with a ◊mouse. The link is usually highlighted in some way, for example by the inclusion of a small graphic. Documents linked in this way are described as ◊hypertext. Examples of programs that use hypertext and hyperlinks are ◊Windows help files, ◊Acrobat, and ◊Mosaic.

hypermedia in computing, system that uses links to lead users to related graphics, audio, animation, or video files in the same way that ◊hypertext systems link related pieces of text. The World Wide Web is an example of a hypermedia system, as is ◊HyperCard.

hyperon in physics, any of a group of highly unstable ◊elementary particles that includes all the baryons with the exception of protons and neutrons. They are all composed of three quarks. The lambda, xi, sigma, and omega particles are hyperons.

hypertension abnormally high ◊blood pressure due to a variety of causes, leading to excessive contraction of the smooth muscle cells of the walls of the arteries. It increases the risk of kidney disease, stroke, and heart attack.

Hypertension is one of the major public health problems of the developed world, affecting 15–20% of adults in industrialized countries (1996). It may be of unknown cause (**essential hypertension**), or it may occur in association with some other condition, such as kidney disease (**secondary or symptomatic hypertension**). It is controlled with a low-salt diet and drugs.

hypertext system for viewing information (both text and pictures) on a computer screen in such a way that related items of information can easily be reached. For example, the program might display a map of a country; if the user clicks (with a ◊mouse) on a particular city, the program will display information about that city.

hyperthyroidism or ***thyrotoxicosis*** overactivity of the thyroid gland due to enlargement or tumour. Symptoms include accelerated heart rate, sweating, anxiety, tremor, and weight loss. Treatment is by drugs or surgery.

hypha (plural ***hyphae***) delicate, usually branching filament, many of which collectively form the mycelium and fruiting bodies of a ◊fungus. Food molecules and other substances are transported along hyphae by the movement of the cytoplasm, known as 'cytoplasmic streaming'.

Typically hyphae grow by increasing in length from the tips and by the formation of side branches. Hyphae of the higher fungi (the ascomycetes and basidiomycetes) are divided by cross walls or septa at intervals, whereas those of lower fungi (for example, bread mould) are undivided. However, even the higher fungi are not truly cellular, as each septum is pierced by a central pore, through which cytoplasm, and even nuclei, can flow. The hyphal walls contain ◊chitin, a polysaccharide.

hypnosis artificially induced state of relaxation or altered attention characterized by heightened suggestibility. There is evidence that, with susceptible persons, the sense of pain may be diminished, memory of past events enhanced, and illusions or hallucinations experienced. Posthypnotic amnesia (forgetting what happened during hypnosis) and posthypnotic suggestion (performing an action after hypnosis that had been suggested during it) have also been demonstrated.

Hypnosis has a number of uses in medicine. Hypnotically induced sleep, for example, may assist the healing process, and hypnotic suggestion (hypnotherapy) may help in dealing with the symptoms of emotional and psychosomatic disorders. The Austrian physician Friedrich Anton Mesmer is said to be the discoverer of hypnosis, but he called it 'animal magnetism', believing it to be a physical force or fluid. The term 'hypnosis' was coined by James Braid (1795–1860), a British physician and surgeon who was the first to regard it as a psychological phenomenon. The Scottish surgeon James Esdaile (1805–1859), working in India, performed hundreds of operations in which he used hypnosis to induce analgesia (insensitivity to pain) or general anaesthesia (total insensitivity).

hypnotic any substance (such as ◊barbiturate, benzodiazepine, alcohol) that depresses brain function, inducing sleep. Prolonged use may lead to physical or psychological addiction.

hypo in photography, a term for sodium thiosulphate, discovered 1819 by John Herschel, and used as a fixative for photographic images since 1837.

hypocycloid in geometry, a cusped curve traced by a point on the circumference of a circle that rolls around the inside of another larger circle. (Compare ◊epicycloid.)

hypodermic syringe instrument used for injecting fluids beneath the skin into either muscles or blood vessels. It consists of a small graduated tube with a close-fitting piston and a nozzle onto which a hollow needle can be fitted.

hypogeal term used to describe seed germination in which the ◊cotyledons remain below ground. It can refer to fruits that develop underground, such as peanuts *Arachis hypogea.*

hypoglycaemia condition of abnormally low level of sugar (glucose) in the blood (below 60 g/100 ml), which starves the brain. It causes weakness, sweating, and mental confusion, sometimes fainting.

Hypoglycaemia is most often seen in ◊diabetes. Low blood sugar occurs when the diabetic has taken too much insulin. It is treated by administering glucose.

hypotenuse the longest side of a right-angled triangle, opposite the right angle. It is of particular application in Pythagoras' theorem (the square of the hypotenuse equals the sum of the squares of the other two sides), and in trigonometry where the ratios ◊sine and ◊cosine are defined as the ratios opposite/hypotenuse and adjacent/hypotenuse respectively.

hypothalamus region of the brain below the ◊cerebrum which regulates rhythmic activity and physiological stability within the body, including water balance and temperature. It regulates the production of the pituitary gland's hormones and controls that part of the ◊nervous system governing the involuntary muscles.

hypothermia condition in which the deep (core) temperature of the body falls below 35°C. If it is not discovered, coma and death ensue. Most at risk are the aged and babies (particularly if premature).

hypothesis in science, an idea concerning an event and its possible explanation. The term is one favoured by the followers of the philosopher Karl Popper, who argue that the merit of a scientific hypothesis lies in its ability to make testable predictions.

Historians will have to face the fact that natural selection determined the evolution of cultures in the same manner as it did that of species.

KONRAD LORENZ Austrian zoologist.
On Aggression 1966

hypothyroidism or ***myxoedema*** deficient functioning of the thyroid gland, causing slowed mental and physical performance, weight gain, sensitivity to cold, and susceptibility to infection.

This may be due to lack of iodine in the diet or a defect of the thyroid gland, both being productive of ◊goitre; or to the pituitary gland providing insufficient stimulus to the thyroid gland. Treatment of thyroid deficiency is by the hormone thyroxine. When present from birth, hypothyroidism can lead to cretinism if untreated.

hypsometer ***Greek hypsos 'height'*** instrument for testing the accuracy of a thermometer at the boiling point of water. It was originally used for determining altitude by comparing changes in the boiling point with changes in atmospheric pressure.

Hyracotherium extinct mammal belonging to the order Perissodactyla and considered to be an ancestor of the horse. It occurs in Eocene strata in Europe and was a small animal about 1 m/3.3 ft in length, with complete dentition, four digits on the forelimbs, and three on the hind-limbs, and orbits not enclosed by bone.

hyrax any of a group of small, rodentlike, herbivorous mammals that live among rocks in desert areas, and in forests in Africa, Arabia, and Syria. They are about the size of a rabbit, with a plump body, short legs, short ears, brownish fur, and long, curved front teeth. (Family Procaviidae, order Hyracoidea.)

They have four toes on the front limbs, and three on the hind limbs, each of which has a tiny hoof. There are nine species of hyrax. They are related to elephants.

hyssop aromatic herb belonging to the mint family, found in Asia, S Europe, and around the Mediterranean. It has blue flowers, oblong leaves, and stems that are woody near the ground but herbaceous (fleshy) above. (*Hyssopus officinalis,* family Labiatae.)

hysterectomy surgical removal of all or part of the uterus (womb). The operation is performed to treat fibroids (benign tumours growing in the uterus) or cancer; also to relieve heavy menstrual bleeding. A woman who has had a hysterectomy will no longer menstruate and cannot bear children.

hysteresis phenomenon seen in the elastic and electromagnetic behaviour of materials, in which a lag occurs between the application or removal of a force or field and its effect.

If the magnetic field applied to a magnetic material is increased and then decreased back to its original value, the magnetic field inside the material does not return to its original value. The internal field 'lags' behind the external field. This behaviour results in a loss of energy, called the **hysteresis loss**, when a sample is repeatedly magnetized and demagnetized. Hence the materials used in transformer cores and electromagnets should have a low hysteresis loss. Similar behaviour is seen in some materials when varying electric fields are applied (**electric hysteresis**). **Elastic hysteresis** occurs when a varying force repeatedly deforms an elastic material. The deformation produced does not completely disappear when the force is removed, and this results in energy loss on repeated deformations.

Hytelnet (contraction of ***hypertext browser for Telnet-accessible sites on the Internet***) in computing, program developed in 1990 which indexes Telnet-accessible sites on the Internet so that users can quickly look up the necessary access information.

Versions of Hytelnet exist for PCs, DEC VAXes, and UNIX machines. Hytelnet is distributed as ◊shareware; it is updated via the HYTEL-L electronic mailing list.

Hz in physics, the symbol for ◊hertz.

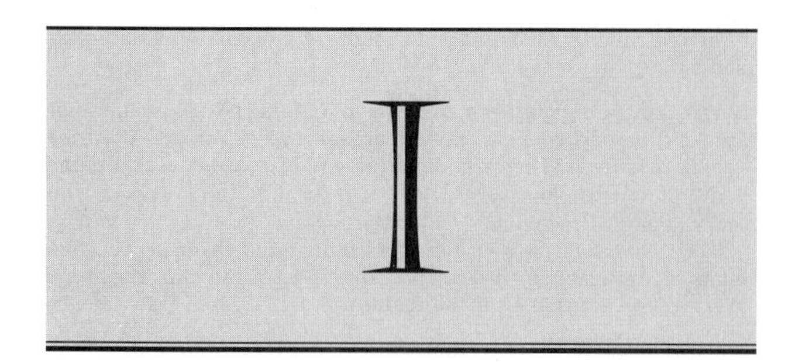

IAB in computing, abbreviation for ◊*Internet Architecture Board*.

IAEA abbreviation for ◊*International Atomic Energy Agency*.

ibex any of various wild goats found in mountainous areas of Europe, NE Africa, and Central Asia. They grow to 100 cm/3.5 ft, and have brown or grey coats and heavy horns. They are herbivorous and live in small groups.

ibis any of various wading birds, about 60 cm/2 ft tall, belonging to the same family as spoonbills. Ibises have long legs and necks, and long, downward-curved beaks, rather blunt at the end; the upper part is grooved. Their plumage is generally black and white. Various species occur in the warmer regions of the world. (Family Threskiornidae, order Ciconiiformes.)

The **scarlet ibis** (*Guara ruber*), a South American species, is brilliant scarlet with a few black patches. The scarlet colour is caused by an accumulation of pigment from the aquatic invertebrates that it feeds on.

IBM (abbreviation for ***International Business Machines***) multinational company, the largest manufacturer of computers in the world. The company is a descendant of the Tabulating Machine Company, formed 1896 by US inventor Herman Hollerith to exploit his punched-card machines. It adopted its present name in 1924. By 1991 it had an annual turnover of $64.8 billion and employed about 345,000 people, but in 1992 and 1993 it made considerable losses. The company acquired Lotus Development Corporation in 1995. By 1997 IBM had, under new management, recovered financially, with an annual turnover of $76 billion, which means it is still a dominant industry player.

Its aquisition of the Lotus Development Corporation gave IBM access to its wide range of innovative software, including the 1–2–3 spreadsheet and Notes, a market leader in groupware.

IBM-compatible in computing, a ◊clone of an IBM PC; synonymous with PC-compatible.

Although there were successful personal computers before the PC, IBM set the most common standard for these machines when it launched the PC in 1981. It created a clone industry by using readily available parts in the IBM PC, instead of developing proprietary parts itself. The success of the PC established Intel processors and Microsoft software as industry standards.

IC abbreviation for ◊integrated circuit.

Icarus in astronomy, an ◊Apollo asteroid 1.5 km/1 mi in diameter, discovered 1949. It orbits the Sun every 409 days at a distance of 28–300 million km/18–186 million mi (0.19–2.0 astronomical units). It was the first asteroid known to approach the Sun closer than does the planet Mercury. In 1968 it passed 6 million km/4 million mi from the Earth.

ice solid formed by water when it freezes. It is colourless and its crystals are hexagonal. The water molecules are held together by hydrogen bonds.

The freezing point of ice, used as a standard for measuring temperature, is 0° for the Celsius and Réaumur scales and 32° for the Fahrenheit. Ice expands in the act of freezing (hence burst pipes), becoming less dense than water (0.9175 at 5°C/41°F).

ice form of methamphetamine that is smoked for its stimulating effect; its use has been illegal in the USA since 1989. Its use may be followed by a period of depression and psychosis.

ANTARCTICA

The ice that covers Antarctica is 4,776 m/15,669 ft

ice age any period of glaciation occurring in the Earth's history, but particularly that in the Pleistocene epoch, immediately preceding historic times. On the North American continent, ◊glaciers reached as far south as the Great Lakes, and an ice sheet spread over N Europe, leaving its remains as far south as Switzerland.

There were several glacial advances separated by interglacial stages during which the ice melted and temperatures were higher than today.

Formerly there were thought to have been only three or four glacial advances, but recent research has shown about 20 major incidences. For example, ocean-bed cores record the absence or presence in their various layers of such cold-loving small marine animals as radiolaria, which indicate a fall in ocean temperature at regular intervals. Other ice ages have occurred throughout geological time: there were four in the Precambrian era, one in the Ordovician, and one at the end of the Carboniferous and beginning of the Permian. The occurrence of an ice age is governed by a combination of factors (the **Milankovitch hypothesis**): (1) the Earth's change of attitude in relation to the Sun, that is, the way it tilts in a 41,000-year cycle and at the same time wobbles on its axis in a 22,000-year cycle, making the time of its closest approach to the Sun come at different seasons; and (2) the 92,000-year cycle of eccentricity in its orbit round the Sun, changing it from an elliptical to a near circular orbit, the severest period of an ice age coinciding with the approach to circularity. There is a possibility that the Pleistocene ice age is not yet over. It may reach another maximum in another 60,000 years.

iceberg floating mass of ice, about 80% of which is submerged, rising sometimes to 100 m/300 ft above sea level. Glaciers that reach the coast become extended into a broad foot; as this enters the sea, masses break off and drift towards temperate latitudes, becoming a danger to shipping.

iceman nickname given to the preserved body of a prehistoric man discovered in a glacier on the Austrian–Italian border 1991. On the basis of the clothing and associated artefacts, the body was at first believed to be 4,000 years old, from the Bronze Age. Carbon dating established its age at about 5,300 years. The discovery led to a reappraisal of the boundary between the Bronze and the Stone Age.

ichneumon fly any of a large group of parasitic wasps. There are several thousand species in Europe, North America, and other regions. They have slender bodies, and the females have unusually long, curved ovipositors (egg-laying instruments) that can pierce several inches of wood. The eggs are laid in the eggs, larvae, or pupae of other insects, usually butterflies or moths. (Family Ichneumonidae.)

ichthyology the scientific study of ◊fish, including their classification, general biology, behaviour, ecology, and research into commercial fisheries.

ICHTHYOLOGY RESOURCES

http://muse.bio.cornell.edu/cgi-bin/hl?fish

Everything anyone should need for studying fish, including historical information about development, as well as up-to-date listings of currently endangered species. There is also a gallery of fish images and paintings for the less scientifically inclined.

icon in computing, a small picture on the computer screen, or ◊VDU, representing an object or function that the user may manipulate or otherwise use. It is a feature of ◊graphical user interface (GUI) systems. Icons make computers easier to use by allowing the user to point to and click with a ◊mouse on pictures, rather than type commands.

icosahedron (plural *icosahedra*) regular solid with 20 equilateral (equal-sided) triangular faces. It is one of the five regular ◊polyhedra, or Platonic solids.

id in Freudian psychology, the mass of motivational and instinctual elements of the human mind, whose activity is largely governed by the arousal of specific needs. It is regarded as the ◊unconscious element of the human psyche, and is said to be in conflict with the ◊ego and the ◊superego.

IDEA (acronym for ***International Data Encryption Algorithm***) in computing, an encryption ◊algorithm, developed in 1990 in Zürich, Switzerland. For reasons of speed, it is used in the encryption program ◊Pretty Good Privacy (PGP) along with ◊RSA.

identity in mathematics, a number or operation that leaves others unchanged when combined with them. Zero is the identity for addition; one is the identity for multiplication. For example:

$$7 + 0 = 7$$
$$7 \times 1 = 7$$

id Software computer software company that publishes popular games such as ◊*Doom* and ◊*Quake*, based in Texas, USA. An entire subculture has built up around id's games because of its habit of releasing ◊source code to enable fans to write their own additional game levels using settings of their own choice.

The company's first major product was the 1992 game *Wolfenstein 3-D*, in which players move around a series of complicated mazes retrieving treasure and shooting Nazi troops and guard dogs. *Quake*, released in 1996, uses complex, carefully styled 3-D graphics, adds vertical movement and underwater caves, and includes a gruesome collection of fierce aliens. Both *Doom* and *Quake* can be played competitively over networks, including the Internet.

IEEE abbreviation for ◊***Institute of Electrical and Electronic Engineers***, US institute which sets technical standards for electrical equipment and computer data exchange.

IETF in computing, abbreviation for Internet Engineering Task Force.

igneous rock rock formed from cooling magma or lava, and solidifying from a molten state. Igneous rocks are largely composed of silica (SiO_2) and they are classified according to their crystal size, texture, method of formation, or chemical composition, for example by the proportions of light and dark minerals.

ignis fatuus another name for ◊will-o'-the-wisp.

ignition coil ◊transformer that is an essential part of a petrol engine's ignition system. It consists of two wire coils wound around an iron core. The primary coil, which is connected to the car battery, has only a few turns. The secondary coil, connected via the ◊distributor to the ◊spark plugs, has many turns. The coil takes in a low voltage (usually 12 volts) from the battery and transforms it to a high voltage (about 15,000–20,000 volts) to ignite the engine.

When the engine is running, the battery current is periodically interrupted by means of the contact breaker in the distributor. The collapsing current in the primary coil induces a current in the secondary coil, a phenomenon known as ◊electromagnetic induction. The induced current in the secondary coil is at very high voltage, typically about 15,000–20,000 volts. This passes to the spark plugs to create sparks.

ignition temperature or ***fire point*** minimum temperature to which a substance must be heated before it will spontaneously burn independently of the source of heat; for example, ethanol has an ignition temperature of 425°C/798°F and a ◊flash point of 12°C/54°F.

iguana any of about 700 species of lizard, chiefly found in the Americas. The **common iguana** (*I. iguana*) of Central and South America is a vegetarian and may reach 2 m/6 ft in length. (Especially genus *Iguana*, family Iguanidae.)

iguanodon plant-eating ◊dinosaur whose remains are found in deposits of the Lower ◊Cretaceous age, together with the remains of other dinosaurs of the same order (ornithiscians) such as stegosaurus and◊ triceratops. It was 5–10 m/16–32 ft long and, when standing upright, 4 m/13 ft tall. It walked on its hind legs, using its long tail to balance its body. (Order *Ornithiscia*.)

ileum part of the small intestine of the ◊digestive system, between the duodenum and the colon, that absorbs digested food.

Its wall is muscular so that waves of contraction (peristalsis) can mix the food and push it forward. Numerous fingerlike projections, or villi, point inwards from the wall, increasing the surface area available for absorption. The ileum has an excellent blood supply, which receives the food molecules passing through the wall and transports them to the liver via the hepatic portal vein.

i.Link in computing, Sony Corporation's branded name for the IEEE ◊1394 serial port and bus. Sony has licensed the name and logo to other companies including Hitachi, Matsushita, Sharp, and Victor of Japan (JVC).

illumination or ***illuminance*** the brightness or intensity of light falling on a surface. It depends upon the brightness, distance, and angle of any nearby light sources. The SI unit is the ◊lux.

ILS abbreviation for ◊instrument landing system, **an automatic system for assisting aircraft landing at airports**.

IMA in computing, abbreviation for ◊***Interactive Multimedia Association***.

image in mathematics, a point or number that is produced as the result of a ◊transformation or mapping.

image picture or appearance of a real object, formed by light that passes through a lens or is reflected from a mirror. If rays of light actually pass through an image, it is called a **real image**. Real images, such as those produced by a camera or projector lens, can be projected onto a screen. An image that cannot be projected onto a screen, such as that seen in a flat mirror, is known as a **virtual image**.

image compression in computing, one of a number of methods used to reduce the amount of information required to represent an image, so that it takes up less computer memory and can be transmitted more rapidly and economically via telecommunications systems. It plays a major role in fax transmission and in videophone and multimedia systems.

image intensifier electronic device that brightens a dark image. Image intensifiers are used for seeing at night; for example, in military situations.

The intensifier first forms an image on a photocathode, which emits electrons in proportion to the intensity of the light falling on it. The electron flow is increased by one or more amplifiers. Finally, a fluorescent screen converts the electrons back into visible light, now bright enough to see.

image map in computing, on the World Wide Web, a large image with multiple hot spots on which users click to navigate around the site.

image processing technique for cleaning up and digitally retouching photographs.

A lot of the fundamental work involved in developing image processing techniques was done at the Jet Propulsion Laboratory in Pasadena, California, USA, which manages unmanned space flight for NASA. Pictures taken in-flight of planets have drop-out areas where data is missing due to static or other interference. These pictures are also often taken using parts of the spectrum which the human eye cannot see. Accordingly, computer ◊algorithms had to be developed to fill in the missing data and compute the correct colours. The images produced in this way are made available publicly and often appear in the media.

imaginary number term often used to describe the non-real element of a ◊complex number. For the complex number $(a + ib)$, ib is the imaginary number where $i = \sqrt{-1}$, and b any real number.

The imaginary number is a fine and wonderful recourse of the divine spirit, almost an amphibian between being and not being.

GOTTFRIED WILHELM LEIBNIZ German mathematician and philosopher. Attributed remark

imago sexually mature stage of an ◊insect.

immediate access memory in computing, ◊memory provided in the ◊central processing unit to store the programs and data in current use.

immersive in ◊virtual reality, term describing the sense that the user is completely surrounded by and immersed in the virtual world.

immiscible describing liquids that will not mix with each other, such as oil and water. When two immiscible liquids are shaken together, a turbid mixture is produced. This normally forms separate layers on being left to stand.

immunity the protection that organisms have against foreign microorganisms, such as bacteria and viruses, and against cancerous cells (see ◊cancer). The cells that provide this protection are called white blood cells, or leucocytes, and make up the immune system. They include neutrophils and ◊macrophages, which can engulf invading organisms and other unwanted material, and natural killer cells that destroy cells infected by viruses and cancerous cells. Some of the most important immune cells are the ◊B cells and ◊T cells. Immune cells coordinate their activities by means of chemical messengers or ◊lymphokines, including the antiviral messenger ◊interferon. The lymph nodes play a major role in organizing the immune response.

Immunity is also provided by a range of physical barriers such as the skin, tear fluid, acid in the stomach, and mucus in the airways. ◊AIDS is one of many viral diseases in which the immune system is affected.

immunization conferring immunity to infectious disease by artificial methods. The most widely used technique is ◊vaccination.

Immunization is an important public health measure. If most of the population has been immunized against a particular disease, it is impossible for an epidemic to take hold.

Vaccination against smallpox was developed by Edward ◊Jenner in 1796. In the late 19th century Louis ◊Pasteur developed vaccines against cholera, typhoid, typhus, plague, and yellow fever. In 1991, the WHO and UNICEF announced that four out of five children around the world are now immunized against six killer diseases: measles, tetanus, polio, diphtheria, whooping cough, and tuberculosis. Ten years ago this figure was only one in five children.

When meditating over a disease, I never think of finding a remedy for it, but, instead, a means of preventing it.

LOUIS PASTEUR French chemist and microbiologist. Address to the Fraternal Association of Former Students of the Ecole Centrale des Arts et Manufactures, Paris, 15 May 1884

immunocompromised lacking a fully effective immune system. The term is most often used in connection with infections such as ◊AIDS where the virus interferes with the immune response (see ◊immunity).

Other factors that can impair the immune response are pregnancy, diabetes, old age, malnutrition and extreme stress, making someone susceptible to infections by microorganisms (such as listeria) that do not affect normal, healthy people. Some people are immunodeficient; others could be on ◊immunosuppressive drugs.

immunodeficient lacking one or more elements of a working immune system. Immune deficiency is the term generally used for patients who are born with such a defect, while those who acquire such a deficiency later in life are referred to as ◊immunocompromised **or immunosuppressed**.

A serious impairment of the immune system is sometimes known as SCID, or Severe Combined Immune Deficiency. At one time children born with this condition would have died in infancy. They can now be kept alive in a germ-free environment, then treated with a bone-marrow transplant from a relative, to replace the missing immune cells. At present, the success rate for this type of treatment is still fairly low. See also ◊gene therapy.

immunoglobulin human globulin ◊protein that can be separated from blood and administered to confer immediate immunity on the recipient. It participates in the immune reaction as the antibody for a specific ◊antigen (disease-causing agent).

Normal immunoglobulin (gamma globulin) is the fraction of the blood serum that, in general, contains the most antibodies, and is obtained from plasma pooled from about a thousand donors. It is given for short-term (two to three months) protection when a person is at risk, mainly from hepatitis A (infectious hepatitis), or when a pregnant woman, not immunized against ◊German measles, is exposed to the rubella virus.

Specific immunoglobulins are injected when a susceptible (non-immunized) person is at risk of infection from a potentially fatal disease, such as hepatitis B (serum hepatitis), rabies, or tetanus. These immunoglobulins are prepared from blood pooled from donors convalescing from the disease.

immunosuppressive any drug that suppresses the body's normal immune responses to infection or foreign tissue. It is used in the treatment of autoimmune disease (see ◊autoimmunity); as part of chemotherapy for leukaemias, lymphomas, and other cancers; and to help prevent rejection following organ transplantation.

Immunosuppressed patients are at greatly increased risk of infection.

impact printer computer printer that creates characters by striking an inked ribbon against the paper beneath. Examples of impact printers are dot-matrix printers, daisywheel printers, and most types of line printer.

Impact printers are noisier and slower than nonimpact printers, such as ink-jet and laser printers, but can be used to produce carbon copies.

impala African ◊antelope found from Kenya to South Africa in savannas and open woodland. The body is sandy brown. Males have lyre-shaped horns up to 75 cm/2.5 ft long. Impalas grow up to 1.5 m/5 ft long and 90 cm/3 ft tall. They live in herds and spring high in the air when alarmed. (Species *Aepyceros melampus*, family Bovidae.)

impedance the total opposition of a circuit to the passage of alternating electric current. It has the symbol Z. For an ◊alternating current (AC) it includes the resistance R and the reactance X (caused by ◊capacitance or ◊inductance); the impedance can then be found using the equation $Z^2 = R^2 + X^2$.

Imperial College of Science, Technology, and Medicine (formerly ***Imperial College of Science and Technology***) institution established in South Kensington, London, 1907, for advanced scientific training and research, applied especially to industry. Part of the University of London, it comprises three separate colleges, the City and Guilds College (engineering faculty), the Royal College of Science (pure science), and the Royal School of Mines (mining). St Mary's Hospital Medical School was added in 1988, resulting in the change of name.

imperial system traditional system of units developed in the UK, based largely on the foot, pound, and second (f.p.s.) system.

implantation in mammals, the process by which the developing ◊embryo attaches itself to the wall of the mother's uterus and stimulates the development of the ◊placenta. In humans it occurs 6–8 days after ovulation.

Immunization: Vaccines in Foods

BY J M DUNWELL

Vaccines

◊Vaccines are ◊proteins that stimulate the immune response of animals, and help protect against infectious diseases caused by ◊viruses and ◊bacteria. More than 75 such diseases can now be prevented in this way and millions of deaths are prevented annually by the use of these products.

Oral immunization

Many infectious diseases are spread by contaminated water or food, and enter the body through the membranes of the ◊digestive system. Stimulating the immune system of the cells lining the digestive system best combats such diseases. Most people are familiar with the ◊polio vaccine provided on sugar cubes. This is probably the best example of vaccination provided by mouth (oral immunization) rather than by injection. The most obvious advantage of the oral route is that it does not require the use of sterile needles.

Need for new production methods

Like many pharmaceutical products, vaccines are relatively expensive to produce. Recently, it has been suggested that plants may be a useful new production system because large amounts of ◊antigen (the active protein component of the vaccine) could be produced at low cost, using agricultural methods rather than ◊cell culture-based systems that are expensive and complex. This would be particularly valuable for developing countries which often lack capital-intensive infrastructure, and where, for example, the problem of ◊diarrhoeal diseases of children is especially acute.

Transgenic plants as a production system

There are two potential methods for producing vaccines in plants. In the first, the ◊DNA fragment encoding the antigen could be introduced directly into the nuclear ◊chromosome of the plant, using any of the methods developed for the production of ◊transgenic plants. The transgenic plants containing the antigen would then be used directly as an oral vaccine, or the antigen could be purified and then consumed. Alternatively, the DNA could be incorporated into a specific virus that only infects plants; this virus could be multiplied within the plant chosen for production. The plant virus particles containing the antigen would then be purified prior to their being used as a vaccine.

Both methods have already been shown to be effective in primary studies, including some conducted on animals. Crops used for animal feed, such as ◊alfalfa, ◊cereals, and ◊legumes are obvious choices for animal vaccines. For humans, plants chosen for production include ◊potato and ◊banana; the latter species may be preferred for tropical countries since it would permit vaccine production in the regions of greatest need. It would also allow inexpensive storage of vaccine material without the need for costly refrigeration facilities.

Future prospects

In the next stage of development, the promising results must be confirmed in a wider range of animal and human studies, before such production systems can be approved for widespread use. Remaining challenges include maximizing the level of production of the antigenic protein, stabilizing this protein during post-harvest storage of plant material, and enhancing its immunogenic capacity. In addition, it will be necessary to determine whether 'edible vaccines' should be provided as a specific medical (or veterinary) product or as a routine food source.

In some species, such as seals and bats, implantation is delayed for several months, during which time the embryo does not grow; thus the interval between mating and birth may be longer than the ◊gestation period.

import file in computing, a file that can be read by a program even though it was produced as an ◊export file by a different program or make of computer.

imprinting in ◊ethology, the process whereby a young animal learns to recognize both specific individuals (for example, its mother) and its own species.

Imprinting is characteristically an automatic response to specific stimuli at a time when the animal is especially sensitive to those stimuli (the **sensitive period**). Thus, goslings learn to recognize their mother by following the first moving object they see after hatching; as a result, they can easily become imprinted on other species, or even inanimate objects, if these happen to move near them at this time. In chicks, imprinting occurs only between 10 and 20 hours after hatching. In mammals, the mother's attachment to her infant may be a form of imprinting made possible by a sensitive period; this period may be as short as the first hour after giving birth.

improper fraction ◊fraction whose numerator is larger than its denominator.

impulse in mechanics, the product of a force and the time over which it acts. An impulse applied to a body causes its ◊momentum to change and is equal to that change in momentum. It is measured in newton seconds (N s).

For example, the impulse J given to a football when it is kicked is given by:

$$J = Ft$$

where F is the kick force in newtons and t is the time in seconds for which the boot is in contact with the ball.

in abbreviation for ◊inch, **a measure of distance**.

inbreeding in ◊genetics, the mating of closely related individuals. It is considered undesirable because it increases the risk that offspring will inherit copies of rare deleterious ◊recessive alleles (genes) from both parents and so suffer from disabilities.

incandescence emission of light from a substance in consequence of its high temperature. The colour of the emitted light from liquids or solids depends on their temperature, and for solids generally the higher the temperature the whiter the light. Gases may become incandescent through ◊ionizing radiation, as in the glowing vacuum ◊discharge tube.

The oxides of cerium and thorium are highly incandescent and for this reason are used in gas mantles. The light from an electric filament lamp is due to the incandescence of the filament, rendered white-hot when a current passes through it.

inch imperial unit of linear measure, a twelfth of a foot, equal to 2.54 centimetres.

It was defined in statute by Edward II of England as the length of three barley grains laid end to end.

incisor sharp tooth at the front of the mammalian mouth. Incisors are used for biting or nibbling, as when a rabbit or a sheep eats grass. Rodents, such as rats and squirrels, have large continually-growing incisors, adapted for gnawing. The elephant tusk is a greatly enlarged incisor. In humans, the incisors are the four teeth at the front centre of each jaw.

inclination angle between the ◊ecliptic and the plane of the orbit of a planet, asteroid, or comet. In the case of satellites orbiting a planet, it is the angle between the plane of orbit of the satellite and the equator of the planet.

inclusive fitness in ◊genetics, the success with which a given variant (or allele) of a ◊gene is passed on to future generations by a particular individual, after additional copies of the allele in the individual's relatives and their offspring have been taken into account.

The concept was formulated by W D Hamilton as a way of explaining the evolution of ◊altruism in terms of ◊natural selection. See also ◊fitness and ◊kin selection.

incontinence failure or inability to control evacuation of the bladder or bowel (or both in the case of double incontinence). It may arise as a result of injury, childbirth, disease, or senility.

incremental backup in computing, a ◊backup copy of only those files that have been modified or created since the last incremental or full backup.

Independent Television Commission (ITC) (formerly the Independent Broadcasting Authority) public body responsible for licensing and regulating commercial television services in the UK, including ITV, Channel Four, and Channel Five, as well as Teletext and a number of cable and satellite services. It is not responsible for licensing S4C, the fourth Welsh channel. Its duties include implementing a code of practice, ensuring adequate quality of services, reporting on complaints, and ensuring competition. It is funded by payments from its licensees; it was paid £315 million in 1995. Members of the Commission are appointed by the Government.

indeterminacy principle alternative name for ◊uncertainty principle.

In effect, we have redefined the task of science to be the discovery of laws that will enable us to predict events up to the limits set by the uncertainty principle.

STEPHEN HAWKING English physicist.

index (plural *indices*) ***Latin 'sign, indicator'*** in mathematics, another term for ◊exponent, the number that indicates the power to which a term should be raised.

indexed sequential file in computing, a type of file ◊access in which an index is used to obtain the address of the ◊block containing the required record.

indexing in computing, computerized service on the Internet that automatically scans ◊servers and compiles lists of the information they hold to make it easier for users to find what they are looking for.

Indexing servers for ◊FTP (File Transfer Protocol) are called ◊Archie servers. On the World Wide Web, the best-known indexing service is ◊Yahoo, which organizes sites by categories and subcategories, and also allows free-form searching.

Indian corn another name for ◊maize.

indicator in chemistry, a compound that changes its structure and colour in response to its environment. The commonest chemical indicators detect changes in ◊pH (for example, ◊litmus), or in the oxidation state of a system (redox indicators).

indicator species plant or animal whose presence or absence in an area indicates certain environmental conditions, such as soil type, high levels of pollution, or, in rivers, low levels of dissolved oxygen. Many plants show a preference for either alkaline or acid soil conditions, while certain trees require aluminium, and are found only in soils where it is present. Some lichens are sensitive to sulphur dioxide in the air, and absence of these species indicates atmospheric pollution.

indigo violet-blue vegetable dye obtained from various tropical plants such as the anil, but now replaced by a synthetic product. It was once a major export crop of India. (Plant genus *Indigofera*, family Leguminosae.)

indium ***Latin indicum 'indigo'*** soft, ductile, silver-white, metallic element, symbol In, atomic number 49, relative atomic mass 114.82. It occurs in nature in some zinc ores, is resistant to abrasion, and is used as a coating on metal parts. It was discovered in 1863 by German metallurgists Ferdinand Reich (1799–1882) and Hieronymus Richter (1824–1898), who named it after the two indigo lines of its spectrum.

indri largest living ◊lemur of Madagascar. It is black and white, almost tailless, with long arms and legs, and grows to 70 cm/2.3 ft long. It lives in the trees and is active by day. Its howl is doglike or human in tone. Like all lemurs, its survival is threatened by the widespread deforestation of Madagascar. (Species *Indri indri*, family Indriidae.)

induced current electric current that appears in a closed circuit when there is relative movement of its conductor in a magnetic field. The effect is known as the **dynamo effect**, and is used in all ◊dynamos and generators to produce electricity. See ◊electromagnetic induction.

There is no battery or other source of power in a circuit in which an induced current appears: the energy supply is provided by the relative motion of the conductor and the magnetic field. The magnitude of the induced current depends upon the rate at which the magnetic flux is cut by the conductor, and its direction is given by Fleming's right-hand rule (see ◊Fleming's rules).

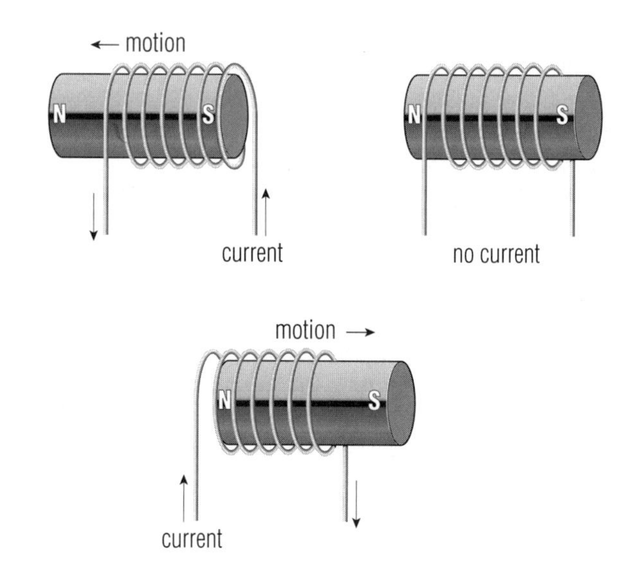

induced current

inductance in physics, the phenomenon where a changing current in a circuit builds up a magnetic field which induces an ◊electromotive force either in the same circuit and opposing the current (self-inductance) or in another circuit (mutual inductance). The SI unit of inductance is the henry (symbol H).

A component designed to introduce inductance into a circuit is called an ◊inductor (sometimes inductance) and is usually in the form of a coil of wire. The energy stored in the magnetic field of the coil is proportional to its inductance and the current flowing through it. See ◊electromagnetic induction.

induction in obstetrics, deliberate intervention to initiate labour before it starts naturally; then it usually proceeds normally.

Induction involves rupture of the fetal membranes (amniotomy) and the use of the hormone oxytocin to stimulate contractions of the womb. In biology, induction is a term used for various processes, including the production of an ◊enzyme in response to a particular chemical in the cell, and the ◊differentiation of cells in an ◊embryo in response to the presence of neighbouring tissues.

In obstetrics, induction is recommended as a medical necessity where there is risk to the mother or baby in waiting for labour to begin of its own accord.

induction in physics, an alteration in the physical properties of a body that is brought about by the influence of a field. See ◊electromagnetic induction and ◊magnetic induction.

induction coil type of electrical transformer, similar to an ◊ignition coil, that produces an intermittent high-voltage alternating current from a low-voltage direct current supply.

It has a primary coil consisting of a few turns of thick wire wound around an iron core and passing a low voltage (usually from a battery). Wound on top of this is a secondary coil made up of many turns of thin wire. An iron armature and make-and-break mechanism (similar to that in an ◊electric bell) repeatedly interrupts the current to the primary coil, producing a high, rapidly alternating current in the secondary circuit.

inductor device included in an electrical circuit because of its inductance.

Industrial Light & Magic (ILM) company that creates special effects for films and which has broken new ground in computer animation techniques (see ◊animation, computer). ILM was set up in 1975 by US director George Lucas to create special effects for his *Star Wars* films, and is based in San Rafael, California, USA.

The company's best-known computer-generated effects include the sea creature in *The Abyss* 1990, the liquid-metal man in *Terminator 2* 1991, and the dinosaurs in *Jurassic Park* 1993.

industry the extraction and conversion of raw materials, the manufacture of goods, and the provision of services. Industry can be either low technology, unspecialized, and labour-intensive, as in countries with a large unskilled labour force, or highly automated, mechanized, and specialized, using advanced technology, as in the industrialized countries. Major recent trends in industrial activity have been the growth of electronic, robotic, and microelectronic technologies, the expansion of the offshore oil industry, and the prominence of Japan and other Pacific-region countries in manufacturing and distributing electronics, computers, and motor vehicles.

Indus Valley civilization one of the four earliest ancient civilizations of the Old World (the other three being the Sumerian civilization 3500 BC; Egypt 3000 BC; and China 2200 BC), developing in the NW of the Indian subcontinent about 2500 BC.

Mohenjo Daro and Harappa were the two main city complexes, but many more existed along the Indus Valley, now in Pakistan. Remains include grid-planned streets with municipal drainage, public and private buildings, baths, temples, and a standardized system of weights and measures – all of which testify to centralized political control. Evidence exists for trade with Sumer and Akkad. The Aryan invasion of about 1500 BC probably led to its downfall.

inequality in mathematics, a statement that one quantity is larger or smaller than another, employing the symbols $<$ and $>$. Inequalities may be solved by finding sets of numbers that satisfy them. For example, the solution set to the inequality $2x + 5 < 19$ consists of all values of x less than 7. Inequality relationships involving variables are sometimes called **inequations**.

inert gas or ***noble gas*** any of a group of six elements (helium, neon, argon, krypton, xenon, and radon), so named because they were originally thought not to enter into any chemical reactions. This is now known to be incorrect: in 1962, xenon was made to combine with fluorine, and since then, compounds of argon, krypton, and radon with fluorine and/or oxygen have been described.

The extreme unreactivity of the inert gases is due to the stability of their electronic structure. All the electron shells (◊energy levels) of inert gas atoms are full and, except for helium, they all have eight electrons in their outermost (◊valency) shell. The apparent stability of this electronic arrangement led to the formulation of the ◊octet rule to explain the different types of chemical bond found in simple compounds.

inertia in physics, the tendency of an object to remain in a state of rest or uniform motion until an external force is applied, as described by Isaac Newton's first law of motion (see ◊Newton's laws of motion).

inertial navigation navigation system that makes use of gyroscopes and accelerometers to monitor and measure a vehicle's movements. A computer calculates the vehicle's position relative to its starting position using the information supplied by the sensors. Inertial navigation is used in aircraft, submarines, spacecraft, and guided missiles.

infant mortality rate measure of the number of infants dying under one year of age, usually expressed as the number of deaths per 1,000 live births. Improved sanitation, nutrition, and medical care have considerably lowered figures throughout much of the world; for example in the 18th century in the USA and UK infant mortality was about 500 per thousand, compared with under 10 per thousand in 1989. The lowest infant mortality rate is in Japan, at 4.5 per 1,000 live births. In much of the Third World, however, the infant mortality rate remains high.

infection invasion of the body by disease-causing organisms (pathogens, or germs) that become established, multiply, and produce symptoms. Bacteria and viruses cause most diseases, but diseases are also caused by other microorganisms, protozoans, and other parasites.

Most pathogens enter and leave the body through the digestive or respiratory tracts. Polio, dysentery, and typhoid are examples of diseases contracted by ingestion of contaminated foods or fluids. Organisms present in the saliva or nasal mucus are spread by airborne or droplet infection; fine droplets or dried particles are inhaled by others when the affected individual talks, coughs, or sneezes. Diseases such as measles, mumps, and tuberculosis are passed on in this way.

A less common route of entry is through the skin, either by contamination of an open wound (as in tetanus) or by penetration of the intact skin surface, as in a bite from a malaria-carrying mosquito. Relatively few diseases are transmissible by skin-to-skin contact. Glandular fever and herpes simplex (cold sore) may be passed on by kissing, and the group now officially bracketed as sexually transmitted diseases (◊STDs) are mostly spread by intimate contact.

inferiority complex in psychology, a ◊complex or cluster of repressed fears, described by Alfred Adler, based on physical inferiority. The term is popularly used to describe general feelings of inferiority and the overcompensation that often ensues.

inferior planet planet (Mercury or Venus) whose orbit lies within that of the Earth, best observed when at its greatest elongation from the Sun, either at eastern elongation in the evening (setting after the Sun) or at western elongation in the morning (rising before the Sun).

inferno in astrophysics, a unit for describing the temperature inside a star. One inferno is 1 billion K, or approximately 1 billion °C.

infertility in medicine, inability to reproduce. In women, this may be due to blockage in the Fallopian tubes, failure of ovulation, a deficiency in sex hormones, or general ill health. In men, impotence, an insufficient number of sperm or abnormal sperm may be the cause of infertility. Clinical investigation will reveal the cause of the infertility in about 75% of couples and assisted conception may then be appropriate.

infinite series in mathematics, a series of numbers consisting of a denumerably infinite sequence of terms. The sequence $n, n^2, n^3, \ldots$ gives the series $n + n^2 + n^3 + \ldots$. For example, $1 + 2 + 3 + \ldots$ is a divergent infinite arithmetic series, and $8 + 4 + 2 + 1 + \frac{1}{2} + \ldots$ is a convergent infinite geometric series that has a sum to infinity of 16.

infinity mathematical quantity that is larger than any fixed assignable quantity; symbol ∞. By convention, the result of dividing any number by zero is regarded as infinity.

inflammation defensive reaction of the body tissues to disease or damage, including redness, swelling, and heat. Denoted by the suffix *-itis* (as in appendicitis), it may be acute or chronic, and may be accompanied by the formation of pus. This is an essential part of the healing process.

Inflammation occurs when damaged cells release a substance (◊histamine) that causes blood vessels to widen and leak into the surrounding tissues. This phenomenon accounts for the redness, swelling, and heat. Pain is due partly to the pressure of swelling and also to irritation of nerve endings. Defensive white blood cells congregate within an area of inflammation to engulf and remove foreign matter and dead tissue.

inflation in cosmology, a phase of extremely fast expansion thought to have occurred within 10–32 seconds of the ◊Big Bang and in which almost all the matter and energy in the universe was created. The inflationary model based on this concept accounts for the density of the universe being very close to the ◊critical density, the smoothness of the ◊cosmic background radiation, and the homogeneous distribution of matter in the universe. Inflation was proposed by US astronomer Alan Guth in the early 1980s.

inflorescence in plants, a branch, or system of branches, bearing two or more individual flowers. Inflorescences can be divided into two main types: cymose (or definite) and racemose (or indefinite). In a **cymose inflorescence**, the tip of the main axis produces a single flower and subsequent flowers arise on lower side branches, as in forget-me-not *Myosotis* and chickweed *Stellaria;* the oldest flowers are, therefore, found at the tip. A **racemose inflorescence** has an active growing region at the tip of its main axis, and bears flowers along its length, as in hyacinth *Hyacinthus;* the oldest flowers are found near the base or, in cases where the inflorescence is flattened, towards the outside.

The stalk of the inflorescence is called a peduncle; the stalk of each individual flower is called a pedicel.

Types of racemose inflorescence include the **raceme**, a spike of similar, stalked flowers, as seen in lupin *Lupinus.* A **corymb**, seen in candytuft *Iberis amara,* is rounded or flat-topped because the pedicels of the flowers vary in length, the outer pedicels being longer than the inner ones. A **panicle** is a branched inflorescence made up of a number of racemes; such inflorescences are seen in many grasses, for example, the oat *Avena.* The pedicels of an **umbel**, seen in members of the carrot family (Umbelliferae), all arise from the same point on the main axis, like the spokes of an umbrella. Other types of racemose inflorescence include the ◊catkin, a pendulous inflorescence, made up of many small stalkless flowers; the ◊spadix, **in which tiny flowers are borne on a fleshy axis; and the** ◊capitulum, in which the axis is flattened or rounded, bears many small flowers, and is surrounded by large petal-like bracts.

influenza any of various viral infections primarily affecting the air passages, accompanied by ◊systemic effects such as fever, chills, headache, joint and muscle pains, and lassitude. Treatment is with bed rest and analgesic drugs such as aspirin or paracetamol.

Depending on the virus strain, influenza varies in virulence and duration, and there is always the risk of secondary (bacterial) infection of the lungs (pneumonia). Vaccines are effective against known strains but will not give protection against newly evolving viruses. The 1918–19 influenza pandemic (see ◊epidemic) killed about 20 million people worldwide.

infobahn (from German autobahn 'motorway') in computing, short name for ◊information superhighway.

information service in computing, commercial online service which offers access to (usually high-priced) periodical databases and other information sources. The two major services are ◊America Online (AOL) and ◊CompuServe.

information superhighway popular collective name for the ◊Internet and other related large-scale computer networks. The term was first used in 1993 by US vice president Al Gore in a speech outlining plans to build a high-speed national data communications network.

information technology (IT) collective term for the various technologies involved in processing and transmitting information. They include computing, telecommunications, and microelectronics.

Word processing, databases, and spreadsheets are just some of the computing ◊software packages that have revolutionized work in the office environment. Not only can work be done more quickly than before, but IT has given decisionmakers the opportunity to consider far more data when making decisions.

infotainment (contraction of ***information and entertainment***) term applied to software that seeks to inform and entertain simultaneously. Many non-fiction ◊CD-ROM titles are classified as infotainment, such as multimedia encyclopedias or reference discs. Compare ◊edutainment.

infrared absorption spectrometry technique used to determine the mineral or chemical composition of artefacts and organic substances, particularly amber. A sample is bombarded by infrared radiation, which causes the atoms in it to vibrate at frequencies characteristic of the substance present, and absorb energy at those frequencies from the infrared spectrum, thus forming the basis for identification.

infrared astronomy study of infrared radiation produced by relatively cool gas and dust in space, as in the areas around forming stars. In 1983, the Infra-Red Astronomy Satellite (IRAS) surveyed the entire sky at infrared wavelengths. It found five new comets, thousands of galaxies undergoing bursts of star formation, and the possibility of planetary systems forming around several dozen stars.

Planets and gas clouds emit their light in the far and mid-infrared region of the spectrum. The Infrared Space Observatory (ISO), launched in 1995, observes a broad wavelength (3–200 micrometres) in this region. It is 10,000 times more sensitive than IRAS, and will search for ◊brown dwarfs (cool masses of gas smaller than the Sun).

infrared radiation invisible electromagnetic radiation of wavelength between about 0.75 micrometres and 1 millimetre – that is, between the limit of the red end of the visible spectrum and the shortest microwaves. All bodies above the ◊absolute zero of temperature absorb and radiate infrared radiation. Infrared radiation is used in medical photography and treatment, and in industry, astronomy, and criminology.

Infrared absorption spectra are used in chemical analysis, particularly for organic compounds. Objects that radiate infrared radiation can be photographed or made visible in the dark on specially sensitized emulsions. This is important for military purposes and in detecting people buried under rubble. The strong absorption by many substances of infrared radiation is a useful method of applying heat.

Infrared Space Observatory (ISO) orbiting telescope with a 60-cm/24-in diameter mirror. It was launched in November 1995 by the European Space Agency and will spend 18 months in an elongated orbit giving it a range from the Earth of 1,000–70,500 km/620–43,800 mi and keeping it as much as possible outside the radiation belts that would swamp its detectors.

Since its launch, ISO has made the first-ever discovery of water vapour from a source beyond the Solar System (in planetary nebula NGC 2027); has traced the spiral arms of the Whirlpool Galaxy and detected sites of star formation there; and obtained the first comprehensive spectrum of Saturn's atmosphere.

infrared telescope in astronomy, a ◊telescope designed to receive ◊electromagnetic waves in the infrared part of the spectrum. Infrared telescopes are always reflectors (glass lenses are opaque to infrared waves) and are normally of the ◊Cassegrain telescope type.

Since all objects at normal temperatures emit strongly in the infrared, careful design is required to ensure that the weak signals from the sky are not swamped by radiation from the telescope itself. Infrared telescopes are sited at high mountain observatories above the obscuring effects of water vapour in the atmosphere. Modern large telecopes are often designed to work equally well in both visible and infrared light.

infrastructure on the Internet, the underlying structure of telephone links, leased lines, and computer programs that makes communication possible.

ingestion process of taking food into the mouth. The method of food capture varies but may involve biting, sucking, or filtering. Many single-celled organisms have a region of their cell wall that acts as a mouth. In these cases surrounding tiny hairs (cilia) sweep food particles together, ready for ingestion.

inhibition, neural in biology, the process in which activity in one ◊nerve cell suppresses activity in another. Neural inhibition in networks of nerve cells leading from sensory organs, or to muscles, plays an important role in allowing an animal to make fine sensory discriminations and to exercise fine control over movements.

ink-jet printer computer printer that creates characters and graphics by spraying very fine jets of quick-drying ink onto paper. Ink-jet printers range in size from small machines designed to work with microcomputers to very large machines designed for high-volume commercial printing.

Because they produce very high-quality printing and are virtually silent, small ink-jet printers (along with ◊laser printers) are replacing impact printers, such as dot-matrix and daisywheel printers, for use with microcomputers.

in-line graphics on the ◊World Wide Web, images included in Web pages which can be downloaded and viewed on the fly. Web ◊browsers display these graphics automatically without any action required by the user. Those with slow connections, however, may choose to turn these off in the interests of speed and just view the text.

in-line video in computing, on the ◊World Wide Web, video files included in Web pages which can be played back on the fly. Web ◊browsers typically require a ◊helper application or ◊plug-in to be installed to play these files. Most sites which include in-line video have links to the necessary software for users who are not already equipped.

inoculation injection into the body of dead or weakened disease-carrying organisms or their toxins (◊vaccine) to produce immunity by inducing a mild form of a disease.

inorganic chemistry branch of chemistry dealing with the chemical properties of the elements and their compounds, excluding the more complex covalent compounds of carbon, which are considered in ◊organic chemistry.

The origins of inorganic chemistry lay in observing the characteristics and experimenting with the uses of the substances (compounds and elements) that could be extracted from mineral ores. These could be classified according to their chemical properties: elements could be classified as metals or nonmetals; compounds as acids or bases, oxidizing or reducing agents, ionic compounds (such as salts), or covalent compounds (such as gases). The arrangement of elements into groups possessing similar properties led to Mendeleyev's ◊periodic table of the elements, which prompted chemists to predict the properties of undiscovered elements that might occupy gaps in the table. This, in turn, led to the discovery of new elements, including a number of highly radioactive elements that do not occur naturally.

inorganic compound compound found in organisms that are not typically biological.

Water, sodium chloride, and potassium are inorganic compounds because they are widely found outside living cells. The term is also applied to those compounds that do not contain carbon and that are not manufactured by organisms. However, carbon dioxide is considered inorganic, contains carbon, and is manufactured by organisms during respiration. See ◊organic compound.

input device device for entering information into a computer. Input devices include keyboards, joysticks, mice, light pens, touch-sensitive screens, scanners, graphics tablets, speech-recognition devices, and vision systems. Compare ◊output device.

types of input device **Keyboards**, the most frequently used input devices, are used to enter instructions and data via keys. There are many variations on the layout and labelling of keys. Extra numeric keys may be added, as may special-purpose function keys, whose effects can be defined by programs in the computer.

The **graphics tablet** is an input device in which a stylus or cursor is moved, by hand, over a flat surface. The computer can keep track of the position of the stylus, enabling the operator to input drawings or diagrams into the computer. The **joystick** signals to a computer the direction and extent of displacement of a hand-held lever.

Light pens resemble ordinary pens and are used to indicate locations on a computer screen. With certain computer-aided design (◊CAD) programs, the light pen can be used to instruct the computer to change the shape, size, position, and colours of sections of a screen image.

Scanners produce a digital image of a document for input and storage in a computer, using technology similar to that of a photocopier. Small scanners can be passed over the document surface by hand; larger versions have a flat bed, like that of a photocopier, on which the input document is placed and scanned.

Input devices that are used commercially – for example, by banks, postal services, and supermarkets – must be able to read and capture large volumes of data very rapidly. Such devices include **document readers** for magnetic-ink character recognition (MICR), ◊optical character recognition (OCR), and optical mark recognition (OMR); mark-sense readers; bar-code scanners; magnetic-strip readers; and point-of-sale (POS) terminals. Punched-card and paper-tape readers were used in earlier commercial applications but are now obsolete.

insanity in medicine and law, any mental disorder in which the patient cannot be held responsible for their actions. The term is no longer used to refer to psychosis.

INSECT

The world's smallest winged insect is smaller than the eye of a house fly. It is the Tanzanian parasitic wasp, which has a wingspan of 0.2 mm/0.008 in.

insect any of a vast group of small invertebrate animals with hard, segmented bodies, three pairs of jointed legs, and, usually, two pairs of wings; they belong among the ◊arthropods and are distributed throughout the world. An insect's body is divided into three segments: head, thorax, and abdomen. On the head is a pair of feelers, or antennae. The legs and wings are attached to the thorax, or middle segment of the body. The abdomen, or end segment of the body, is where food is digested and excreted and where the reproductive organs are located.

Insects vary in size from 0.02 cm/0.007 in to 35 cm/13.5 in in length. The world's smallest insect is believed to be a 'fairy fly' wasp in the family Mymaridae, with a wingspan of 0.2 mm/0.008 in. (Class Insecta.)

Many insects hatch out of their eggs as ◊larvae (an immature stage, usually in the form of a caterpillar, grub, or maggot) and have to pass through further major physical changes (◊metamorphosis) before reaching adulthood. An insect about to go through metamorphosis hides itself or makes a cocoon in which to hide, then rests while the changes take place; at this stage the insect is called a ◊pupa, or a chrysalis if it is a butterfly or moth. When the changes are complete, the adult insect emerges.

The **classification** of insects is largely based upon characteristics of the mouthparts, wings, and metamorphosis. Insects are divided into two subclasses (one with two divisions) and 29 orders. More than 1 million species are known, and several thousand new ones are discovered each year.

The study of insects is called **entomology**.

TO REMEMBER THE PARTS OF AN INSECT'S LEG:

COCKROACHES TRAVEL FAST TOWARDS THEIR CHILDREN.

COXA / TROCHANTER / FEMUR / TIBIA / TARSUS / CLAW

INSECT CONTROL FAQ

http://res.agr.ca/lond/pmrc/faq/insect.html

Table of frequently asked questions about all kinds of insect control, including the Colorado potato beetle and aphids. The questions answered on this site include 'what are the most common insect pests?' and 'what alternatives are there to insecticides to control insects?'.

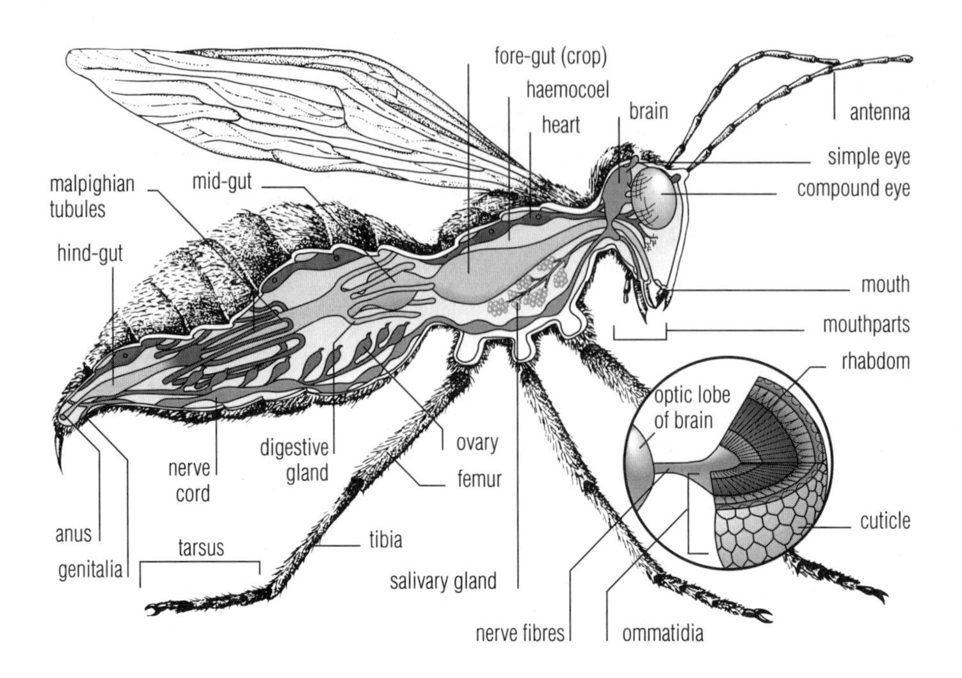

insect *Body plan of an insect. The general features of the insect body include a segmented body divided into head, thorax, and abdomen, jointed legs, feelers or antennae, and usually two pairs of wings. Insects often have compound eyes with a large field of vision.*

insecticide any chemical pesticide used to kill insects. Among the most effective insecticides are synthetic organic chemicals such as ◊DDT and dieldrin, which are chlorinated hydrocarbons. These chemicals, however, have proved persistent in the environment and are also poisonous to all animal life, including humans, and are consequently banned in many countries. Other synthetic insecticides include organic phosphorus compounds such as malathion. Insecticides prepared from plants, such as derris and pyrethrum, are safer to use but need to be applied frequently and carefully.

insectivore any animal whose diet is made up largely or exclusively of ◊insects. In particular, the name is applied to mammals of the order Insectivora, which includes the shrews, hedgehogs, moles, and tenrecs.

According to the Red List of endangered species published by the World Conservation Union (IUCN) for 1996, 36% of insectivore species are threatened with extinction.

insectivorous plant plant that can capture and digest live prey (normally insects), to obtain nitrogen compounds that are lacking in its usual marshy habitat. Some are passive traps, for example, the pitcher plants *Nepenthes* and *Sarracenia.* One pitcher-plant species has container-traps holding 1.6 l/3.5 pt of the liquid that 'digests' its food, mostly insects but occasionally even rodents. Others, for example, sundews *Drosera,* butterworts *Pinguicula,* and Venus flytraps *Dionaea muscipula,* have an active trapping mechanism. Insectivorous plants have adapted to grow in poor soil conditions where the number of microorganisms recycling nitrogen compounds is very much reduced. In these circumstances other plants cannot gain enough nitrates to grow. See also ◊leaf.

Near-carnivorous plants are unable to digest insects, but still trap them on their sticky coated leaves. The insects die and decay naturally, with the nutrients eventually becoming washed into the soil where they finally benefit the plant.

inselberg or ***kopje German 'island mountain'*** prominent steep-sided hill of resistant solid rock, such as granite, rising out of a plain, usually in a tropical area. Its rounded appearance is caused by so-called onion-skin ◊weathering, in which the surface is eroded in successive layers.

The Sugar Loaf in Rio de Janeiro harbour in Brazil, and Ayers Rock in Northern Territory, Australia, are famous examples.

insemination, artificial see ◊artificial insemination.

instinct in ◊ethology, behaviour found in all equivalent members of a given species (for example, all the males, or all the females with young) that is presumed to be genetically determined.

Examples include a male robin's tendency to attack other male robins intruding on its territory and the tendency of many female mammals to care for their offspring. Instincts differ from ◊reflexes in that they involve very much more complex actions, and learning often plays an important part in their development.

instruction register in computing, a special memory location used to hold the instruction that the computer is currently processing. It is located in the control unit of the ◊central processing unit, and receives instructions individually from the immediate-access memory during the fetch phase of the ◊fetch-execute cycle.

instruction set in computing, the complete set of machine-code instructions that a computer's ◊central processing unit can obey.

instrument landing system (ILS) landing aid for aircraft that uses ◊radio beacons on the ground and instruments on the flight deck. One beacon (localizer) sends out a vertical radio beam along the centre line of the runway. Another beacon (glide slope) transmits a beam in the plane at right angles to the localizer beam at the ideal approach-path angle. The pilot can tell from the instruments how to manoeuvre to attain the correct approach path.

insulation process or material that prevents or reduces the flow of electricity, heat or sound from one place to another.

Electrical insulation makes use of materials such as rubber, PVC, and porcelain, which do not conduct electricity, to prevent a current from leaking from one conductor to another or down to the ground. Insulation is a vital safety measure that prevents electric currents from being conducted through people and causing electric shock.

Double insulation is a method of constructing electrical appliances that provides extra protection from electric shock, and renders the use of an earth wire unnecessary. In addition to the usual cable insulation, an appliance that meets the double insulation standard is totally enclosed in an insulating plastic body or structure so that there is no direct connection between any external metal parts and the internal electrical components.

Thermal or **heat insulation** makes use of insulating materials such as fibreglass to reduce the loss of heat through the roof and walls of buildings. The U-value of an insulating layer is a measure of its ability to conduct heat – a material chosen as an insulator should therefore have a low ◊U-value. Air trapped between the fibres of clothes acts as a thermal insulator, preventing loss of body warmth.

insulator any poor ◊conductor of heat, sound, or electricity. Most substances lacking free (mobile) ◊electrons, such as non-metals, are electrical or thermal insulators. Usually, devices of glass or porcelain, called insulators, are used for insulating and supporting overhead wires.

insulin protein ◊hormone, produced by specialized cells in the islets of Langerhans in the pancreas, that regulates the metabolism (rate of activity) of glucose, fats, and proteins. Insulin was discovered by Canadian physician Frederick Banting and Canadian physiologist Charles Best, who pioneered its use in treating ◊diabetes.

Normally, insulin is secreted in response to rising blood sugar levels (after a meal, for example), stimulating the body's cells to store the excess. Failure of this regulatory mechanism in diabetes mellitus requires treatment with insulin injections or capsules taken by mouth. Types vary from pig and beef insulins to synthetic and bioengineered ones. They may be combined with other substances to make them longer-or shorter-acting. Implanted, battery-powered insulin pumps deliver the hormone at a preset rate, to eliminate the unnatural rises and falls that result from conventional, subcutaneous (under the skin) delivery. Human insulin has now been produced from bacteria by ◊genetic engineering techniques, but may increase the chance of sudden, unpredictable ◊hypoglycaemia, or low blood sugar. In 1990 the Medical College of Ohio developed gelatin capsules and an aspirinlike drug which helps the insulin pass into the bloodstream.

TO REMEMBER THE ROLE OF INSULIN:

REMEMBER THAT INSULIN GETS SUGAR INTO CELLS. WITHOUT INSULIN, A PERSON CAN HAVE EXCESS SUGAR IN HIS BLOOD YET DIE OF LACK OF SUGAR.

integer any whole number. Integers may be positive or negative; 0 is an integer, and is often considered positive. Formally, integers are members of the set $Z = \{... -3, -2, -1, 0, 1, 2, 3,... \}$. Fractions, such as $\frac{1}{2}$ and 0.35, are known as non-integral numbers ('not integers').

God made the integers, man made the rest.

LEOPOLD KRONECKER German mathematician. *Jahresberichte der deutschen Mathematiker Vereinigung* bk 2. In F Cajori *A History of Mathematics* 1919

integral calculus branch of mathematics using the process of ◊integration. It is concerned with finding volumes and areas and summing infinitesimally small quantities.

integrated circuit (IC), popularly called ***silicon chip***, a miniaturized electronic circuit produced on a single crystal, or chip, of a semiconducting material – usually silicon. It may contain many millions of components and yet measure only 5 mm/0.2 in square and 1 mm/0.04 in thick. The IC is encapsulated within a plastic or ceramic case, and linked via gold wires to metal pins with which it is connected to a ◊printed circuit board and the other components that make up such electronic devices as computers and calculators.

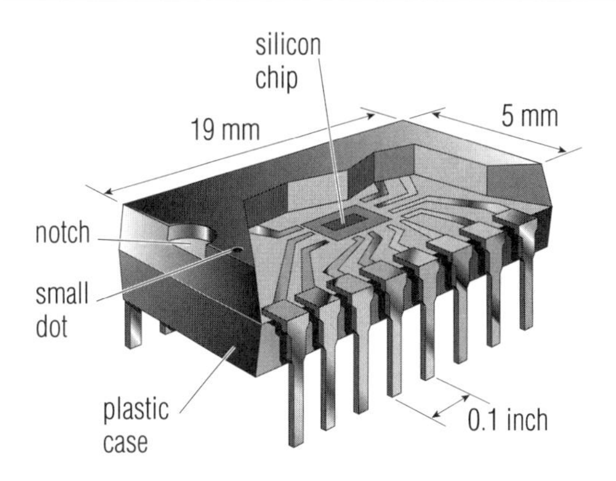

integrated circuit *An integrated circuit (IC), or silicon chip. The IC is a piece of silicon, about the size of a child's fingernail, on which the components of an electrical circuit are etched. The IC is packed in a plastic container with metal legs that connect it to the circuit board.*

Integrated Services Digital Network (ISDN) internationally developed telecommunications system for sending signals in ◊digital format. It involves converting the 'local loop' – the link between the user's telephone (or private automatic branch exchange) and the digital telephone exchange – from an ◊analogue system into a digital system, thereby greatly increasing the amount of information that can be carried. The first large-scale use of ISDN began in Japan in 1988.

ISDN has advantages in higher voice quality, better quality faxes, and the possibility of data transfer between computers faster than current modems. With ISDN's **Basic Rate Access**, a multiplexer divides one voice telephone line into three channels: two B bands and a D band. Each B band offers 64 kilobits per second and can carry one voice conversation or 50 simultaneous data calls at 1,200 bits per second. The D band is a data-signalling channel operating at 16 kilobits per second. With **Primary Rate Access**, ISDN provides 30 B channels.

integrated steelworks modern industrial complex where all the steelmaking processes – such as iron smelting and steel shaping – take place on the same site.

integration in mathematics, a method in ◊calculus of determining the solutions of definite or indefinite integrals.

An example of a definite integral can be thought of as finding the area under a curve (as represented by an algebraic expression or function) between particular values of the function's variable. In practice, integral calculus provides scientists with a powerful tool for doing calculations that involve a continually varying quantity (such as determining the position at any given instant of a space rocket that is accelerating away from Earth). Its basic principles were discovered in the late 1660s independently by the German philosopher Leibniz and the British scientist ◊Newton.

integument in seed-producing plants, the protective coat surrounding the ovule. In flowering plants there are two, in gymnosperms only one. A small hole at one end, the micropyle, allows a pollen tube to penetrate through to the egg during fertilization.

Intel manufacturer of the microprocessors that form the basis of the IBM PC range and its clones. Intel developed the first ◊microprocessor, the 4004, in 1971, and has largely retained compatibility throughout the x86 range from the 8086 to the 80486 and the ◊Pentium or 586 released in 1993. Intel's current strategy is to promote the use of the Pentium II processor, introduced in 1997, while it is developing its next-generation chip, code-named Merced, in conjunction with Hewlett-Packard.

History of Insulin

BY PAULETTE PRATT

The term 'diabetes' (from the Ionian Greek 'to pass through'), together with a succinct clinical description of this once-fatal disease, was bequeathed to us by Aretaeus of Cappadocia. In the 2nd century AD he wrote: 'Diabetes is a dreadful affliction, not very frequent among men, being a melting down of the flesh and limbs into urine. The patients never stop making water and the flow is incessant, like the opening of aqueducts. Life is short, unpleasant and painful, thirst unquenchable, drinking excessive...'

'One cannot stop them either from drinking or making water. If for a while they abstain from drinking, their mouths become parched and their bodies dry; the viscera seem scorched up, the patients are affected by nausea, restlessness and a burning thirst, and within a short time they expire.'

Although Eastern physicians had noted the sweet taste of the urine in diabetes a thousand years earlier, in Europe the connection between sugar in the urine and diabetes was only made in 1670 by an English physician, Thomas Willis. In 1776, Liverpool physician Matthew Dobson showed that the blood serum of diabetic patients also contains a sweet-tasting substance. He proved that this is sugar and deduced that it is formed in the serum rather than in the kidneys. This was the first indication that diabetes is a systemic disease (pervading the whole body) rather than a specific kidney problem as some people had thought.

Soon after, another English physician, John Rollo, was among the first to attach the adjective 'mellitus' (from the Greek and Latin roots for 'honey') to distinguish diabetes from other conditions where there is copious urine output. The next major observation, published in the *London Medical Journal* in 1788, was that diabetes may ensue from damage to the pancreas. In Strasbourg a century later, Oskar Minkowski and Josef von Mering were able to produce experimental diabetes in a dog by removing its pancreas.

This firm association of diabetes with pancreatic deficiency prompted researchers to begin looking for the actual substance involved in order to develop a treatment. In 1893, a Frenchman, Edouard Laguesse, suggested that the pancreatic 'islets of Langerhans 'might be implicated. He named these after the distinguished German pathologist, Paul Langerhans, who, at the age of 22, had been the first person to describe them (1869).

A Belgian physician, Jean de Meyer, gave the name *insuline* (from the Latin insula, meaning island) to the as yet hypothetical blood sugar-lowering substance in 1909. The British physiologist Edward Sharpey Schafer, who argued that the islets must secrete a substance which governs carbohydrate metabolism, used the word 'insulin' for this notional substance in 1916, but the term did not immediately become current: Banting and Best and their collaborators, who discovered insulin in 192l, at first called it 'isletin'.

By the early years of this century then, it was known that diabetes mellitus arises from a lack of this unknown substance. Meanwhile, people developing the disease lived brief, wretched lives. Those who sought treatment were put on starvation diets. The effect of this was to lower blood glucose and produce emaciation; the extension it won to the patient's life was at best a few months.

At this time a number of scientists were on the track of insulin and an unknown Canadian doctor, 29-year-old Frederick Grant Banting, also developed an interest in the hormone. He approached J J R Macleod, then Professor of Physiology at the University of Toronto (and an expert on sugar metabolism), for facilities to conduct experiments in a bid to isolate insulin. Macleod made available laboratory space and experimental dogs for Banting's use; he also offered one of his students as an assistant, Charles H Best (who had won the opportunity to work alongside Banting on the toss of a coin). But it was the addition of biochemist James B. Collip which would ensure their success.

The first trial of the team's pancreatic extract was on 14-year-old Leonard Thompson, who was dying of diabetes in Toronto General Hospital, on January 1, 1922. This failed to relieve his symptoms and, moreover, caused an abscess to form at the injection site. However, a further extract, injected on January 23, brought the boy's blood sugar down to normal. In May, when the group reported the outcome of this modest trial to a meeting of the Association of American Physicians in Washington DC, they received a standing ovation.

From October 1923, once a commercially viable extraction process had been developed, insulin became widely available throughout North America and Europe. Now, for diabetic patients who would otherwise have been doomed, the issue was no longer one of survival but of quality of life on insulin treatment.

Among the people who witnessed the introduction of the life-saving extract into clinical use was a Portuguese doctor, Ernesto Roma, who was on a visit to Boston. Returning to Lisbon in 1926, he set up the World's first diabetic organisation, the Portuguese Association for the Protection of Poor Diabetics, which provided insulin free of charge. The British Diabetic Association was founded in 1934 by a doctor whose life had been saved by insulin, Robin Lawrence, and the writer H G Wells, who also had the disease.

Controversially, only Banting and Macleod received the Nobel prize for the discovery of insulin, in 1923 (in what was one of the Nobel committee's quickest ever recognitions). Despite the acrimony which had arisen within the group, the two men shared the award money with their collaborators. Frederick Banting, whose vision had driven the project, became a World hero and received a knighthood and a generous annuity from the Canadian government. He died in an air crash in Newfoundland in 194l.

Leonard Thompson, the original insulin 'guinea pig', survived until 1935 and at least two of the children who had been treated in Toronto in the trial period outlived all four of the men who had worked on the discovery of insulin. One of these, Ted Ryder, lived until 1993, reaching the age of 76.

Intel is thought to supply the processors for almost 90% of the world's personal computers.

intellectual property material such as computer software, magazine articles, songs, novels, or recordings which can be described as the expression of ideas fixed in a tangible form.

Generally, intellectual property is protected by copyright law, and distribution, sale, and copying of such material is restricted so that the creators can be paid for their work. On the Internet, intellectual property may include the words, graphics, audio files, and other material which comprise pages on the World Wide Web, as well as the words written by individuals in e-mail or on USENET.

intellectual property rights the right of control over the copying, distribution, and sale of ◊intellectual property which is codified in the copyright laws.

The future of intellectual property rights is unclear, as the Internet makes mass distribution and copying quick and easy. In the mid-1990s, many schemes were being considered for using encryption to mark computer files or prevent copying in an effort to safeguard these rights.

intelligence in psychology, a general concept that summarizes the abilities of an individual in reasoning and problem solving, particularly in novel situations. These consist of a wide range of verbal and nonverbal skills and therefore some psychologists dispute a unitary concept of intelligence.

Intelligence is quickness to apprehend as distinct from ability, which is capacity to act wisely on the thing apprehended.

ALFRED NORTH WHITEHEAD English philosopher and mathematician. *Dialogues* 15 Dec 1939

intelligence test test that attempts to measure innate intellectual ability, rather than acquired ability.

It is now generally believed that a child's ability in an intelligence test can be affected by his or her environment, cultural background, and teaching. There is scepticism about the accuracy of intelligence tests, but they are still widely used as a diagnostic tool when children display learning difficulties. 'Sight and sound' intelligence tests, developed by Christopher Brand in 1981, avoid cultural bias and the pitfalls of improvement by practice. Subjects are shown a series of lines being flashed on a screen at increasing speed, and are asked to identify in each case the shorter of a pair; and when two notes are relayed over headphones, they are asked to identify which is the higher. There is a close correlation between these results and other intelligence test scores.

intelligent agent in computing, another name for ◊agent.

intelligent terminal in computing, a ◊terminal with its own processor which can take some of the processing load away from the main computer.

Intelsat (acronym for ***International Telecommunications Satellite Organization***) organization established in 1964 to operate a worldwide system of communications satellites. In 1994 it had 134 member nations and 22 satellites in orbit. Its headquarters are in Washington DC. Intelsat satellites are stationed in geostationary orbit (maintaining their positions relative to the Earth) over the Atlantic, Pacific, and Indian Oceans. The first Intelsat satellite was *Early Bird,* launched in 1965.

intensity in physics, the power (or energy per second) per unit area carried by a form of radiation or wave motion. It is an indication of the concentration of energy present and, if measured at varying distances from the source, of the effect of distance on this. For example, the intensity of light is a measure of its brightness, and may be shown to diminish with distance from its source in accordance with the ◊inverse square law (its intensity is inversely proportional to the square of the distance).

interactive describing a computer system that will respond directly to data or commands entered by the user. For example, most popular programs, such as word processors and spreadsheet applications, are interactive. Multimedia programs are usually highly interactive, allowing users to decide what type of information to display (text, graphics, video, or audio) and enabling them (by means of ◊hypertext) to choose an individual route through the information.

interactive computing in computing, a system for processing data in which the operator is in direct communication with the computer, receiving immediate responses to input data. In ◊batch processing, by contrast, the necessary data and instructions are prepared in advance and processed by the computer with little or no intervention from the operator.

interactive digital video service alternative term for ◊video-on-demand.

interactive media in computing, new technology such as ◊CD-ROM and online systems which allow users to interact with other users or to choose their own path through the material.

The newest attempts to create interactive media are books published on the World Wide Web which allow readers to use ◊hyperlinks to move around the material at will in the order they choose. Other interactive media include plans for films and other projects which allow viewers to choose how to follow the story, which characters to focus on, or which plot threads to follow.

Interactive Multimedia Association (IMA) organization founded in 1987 to promote the growth of the multimedia industry. Based in Anapolis, Maryland, USA, the IMA runs special interest groups, summit meetings, conferences, and trade shows for its member companies.

interactive video (IV) computer-mediated system that enables the user to interact with and control information (including text, recorded speech, or moving images) stored on video disc. IV is most commonly used for training purposes, using analogue video discs, but has wider applications with digital video systems such as CD-I (Compact Disc Interactive, from Philips and Sony) which are based on the CD-ROM format derived from audio compact discs.

Intercast in computing, ◊Intel device that adds TV reception capability to a PC and uses blank lines to deliver data.

A number of leading PC manufacturers expect to bundle Intercast TV tuner boards with new computer systems.

intercostal in biology, the nerves, blood vessels, and muscles that lie between the ribs.

interface in computing, the point of contact between two programs or pieces of equipment. The term is most often used for the physical connection between the computer and a peripheral device, which is used to compensate for differences in such operating characteristics as speed, data coding, voltage, and power consumption. For example, a **printer interface** is the cabling and circuitry used to transfer data from a computer to a printer, and to compensate for differences in speed and coding.

Common standard interfaces include the **Centronics interface**, used to connect parallel devices, and the **RS232 interface**, used to connect serial devices. For example, in many microcomputer systems, an RS232 interface is used to connect the microcomputer to a modem, and a Centronics device is used to connect it to a printer.

interference in physics, the phenomenon of two or more wave motions interacting and combining to produce a resultant wave of larger or smaller amplitude (depending on whether the combining waves are in or out of ◊phase with each other).

Interference of white light (multiwavelength) results in spectral coloured fringes; for example, the iridescent colours of oil films seen on water or soap bubbles (demonstrated by ◊Newton's rings). Interference of sound waves of similar frequency produces the phenomenon of beats, often used by musicians when tuning an instrument. With monochromatic light (of a single wavelength), interference produces patterns of light and dark bands. This is the basis of ◊holography, for example. Interferometry can also be applied to radio waves, and is a powerful tool in modern astronomy.

interferometer in physics, a device that splits a beam of light into two parts, the parts being recombined after travelling different paths to form an interference pattern of light and dark bands.

Interferometers are used in many branches of science and industry where accurate measurements of distances and angles are needed.

In the Michelson interferometer, a light beam is split into two by a semisilvered mirror. The two beams are then reflected off fully silvered mirrors and recombined. The pattern of dark and light bands is sensitive to small alterations in the placing of the mirrors, so the interferometer can detect changes in their position to within one ten-millionth of a metre. Using lasers, compact devices of this kind can be built to measure distances, for example to check the accuracy of machine tools.

In radio astronomy, interferometers consist of separate radio telescopes, each observing the same distant object, such as a galaxy, in the sky. The signal received by each telescope is fed into a computer. Because the telescopes are in different places, the distance travelled by the signal to reach each differs and the overall signal is akin to the interference pattern in the Michelson interferometer. Computer analysis of the overall signal can build up a detailed picture of the source of the radio waves.

In space technology, interferometers are used in radio and radar systems. These include space-vehicle guidance systems, in which the position of the spacecraft is determined by combining the signals received by two precisely spaced antennae mounted on it.

interferometry in astronomy, any of several techniques used in astronomy to obtain high-resolution images of astronomical objects. See ◊speckle interferometry and ◊VLBI (very long baseline interferometry).

interferon naturally occurring cellular protein that makes up part of the body's defences against viral disease. Three types (alpha, beta, and gamma) are produced by infected cells and enter the bloodstream and uninfected cells, making them immune to virus attack.

Interferon was discovered in 1957 by Scottish virologist Alick Isaacs. Interferons are cytokines, small molecules that carry signals from one cell to another. They can be divided into two main types: **type I** (alpha, beta, tau, and omega) interferons are more effective at bolstering cells' ability to resist infection; **type II** (gamma) interferon is more important to the normal functioning of the immune system. Alpha interferon may be used to treat some cancers; interferon beta 1b has been found useful in the treatment of ◊multiple sclerosis.

interior angle one of the four internal angles formed when a transversal cuts two or more (usually parallel) lines. Also, one of the angles inside a ◊polygon.

interlacing technique for increasing resolution on computer graphic displays. The electron beam traces alternate lines on each pass, providing twice the number of lines of a non-interlaced screen. However, screen refresh is slower and screen flicker may be increased over that seen on an equivalent non-interlaced screen.

INTERMEDIATE TECHNOLOGY DEVELOPMENT GROUP

http://www.oneworld.org/itdg/

Full account of the work of the British aid organization working with rural poor to develop relevant and sustainable technologies. Reports of the low-tech projects supported by IT make fascinating reading. Of equal interest is work with other British aid agencies to develop approaches to project monitoring and evaluation which involve local people.

intermediate technology application of mechanics, electrical engineering, and other technologies, based on inventions and designs developed in scientifically sophisticated cultures, but utilizing materials, assembly, and maintenance methods found in technologically less advanced regions (known as the Third World).

Intermediate technologies aim to allow developing countries to benefit from new techniques and inventions of the 'First World', without the burdens of costly maintenance and supply of fuels and spare parts that in the Third World would represent an enormous and probably uneconomic overhead. See also ◊appropriate technology.

Science clears the fields on which technology can build.

Werner Carl Heisenberg German physicist.

intermediate vector boson alternative name for **weakon**, the elementary particle responsible for carrying the ◊weak nuclear force.

intermolecular force or *van der Waals' force* force of attraction between molecules. Intermolecular forces are relatively weak; hence simple molecular compounds are gases, liquids, or low-melting-point solids.

internal-combustion engine heat engine in which fuel is burned inside the engine, contrasting with an external-combustion engine (such as the steam engine) in which fuel is burned in a separate unit. The ◊diesel engine and ◊petrol engine are both internal-combustion engines. Gas ◊turbines and ◊jet and ◊rocket engines are also considered to be internal-combustion engines because they burn their fuel inside their combustion chambers.

internal modem in computing, a ◊modem that fits into a slot inside a personal computer. On older PCs, an internal modem may prove a better choice for high-speed data communications than an external modem, as it may have built-in features which make up for features missing in older computers. Internal modems are generally also cheaper, except for the small-sized ◊PCMCIA types.

The disadvantages are that an internal modems cannot easily be swapped from one computer to another, require greater skill to install (again except for PCMCIA), and have no external displays of lights to give users feedback.

internal resistance or ***source resistance*** the resistance inside a power supply, such as a battery of cells, that limits the current that it can supply to a circuit.

International Atomic Energy Agency (IAEA) agency of the United Nations established in 1957 to advise and assist member countries in the development and peaceful application of nuclear power, and to guard against its misuse. It has its headquarters in Vienna, and is responsible for research centres in Austria and Monaco, and the International Centre for Theoretical Physics, Trieste, Italy, established in 1964. It conducts inspections of nuclear installations in countries suspected of developing nuclear weapons, for example Iraq and North Korea.

international biological standards drugs (such as penicillin and insulin) of which the activity for a specific mass (called the international unit, or IU), prepared and stored under specific conditions, serves as a standard for measuring doses. For penicillin, one IU is the activity of 0.0006 mg of the sodium salt of penicillin, so a dose of a million units would be 0.6 g.

International Civil Aviation Organization agency of the United Nations, established in 1947 to regulate safety and efficiency and air law; headquarters Montréal, Canada.

INTERNATIONAL CIVIL AVIATION ORGANIZATION

http://www.icao.int

Site of the specialized UN agency regulating civil aviation. The role and history of the ICAO are well presented. There is information on rules of the air, international conventions, and standardization of safety standards. The ICAO tries to reassure nervous flyers that air travel is getting safer. There are links to all the online airlines, airports, and pilot training centres in the world.

International Date Line (IDL) imaginary line that approximately follows the 180° line of longitude. The date is put forward a day when crossing the line going west, and back a day when going east. The IDL was chosen at the International Meridian Conference 1884.

International Organization for Standardization (ISO) international organization founded in 1947 to standardize technical terms, specifications, units, and so on. Its headquarters are in Geneva.

International Telecommunication Union body belonging to the Economic and Social Council of the United Nations. It aims to extend international cooperation by improving telecommunications of all kinds.

International Telecommunications Union (ITU) international organization, based in Geneva, Switzerland, which manages telecommunications standards such as ◊modem speeds and ◊protocols. ITU activities include the coordination, development, regulation, and standardization of telecommunications.

The ITU has two permanent standards-making committes, the International Telegraph and Telephone Consultative Committee (CCITT) and the International Radio Consultative Committee (CCIR).

The Internal Combustion Engine

BY PETER LAFFERTY

The steam engine was the first reliable source of power. By the early 1800s, the steam engine had developed to the point where it could propel carriages along the road at reasonable speeds, and undertake long journeys. But the quest was on for a lighter, more powerful engine.

The idea of an internal combustion engine seems to have occurred around 1800 to French inventor Philippe Lebon. He argued that an engine that burnt fuel inside a cylinder would waste less heat and be more efficient than an engine, like the steam engine, which had a separate furnace in which the fuel was burnt. He proposed an engine in which a compressed mixture of gas and air was burnt in a cylinder, the heated gas driving a piston. Unfortunately, Lebon met an untimely end before he could build the engine. He was attacked and stabbed on the Champs Elysées in Paris and died of his wounds.

After Lebon, several inventors tried to build internal combustion engines but failed, mainly because of difficulties with ignition of the fuel–air mixture inside the cylinder.

In 1807 Swiss inventor Isaac de Rivas built an engine which he used to move a trolley. However, the engine was too inefficient to be of any use. In 1829, the Reverend William Cecil read a paper at Cambridge University about his experiments, and William Barnet in 1839 also laid claims to the invention. In 1853 Italians Eugenio Barsanti and Felice Mattecci patented a design, but they never built a working engine. It was not until 1860 that an efficient engine was built by Belgian engineer Etienne Lenoir.

Lenoir's engine consisted of a single cylinder with a storage battery for electrical ignition of the fuel–air mixture. It was a two-stroke engine, producing its power in two strokes, or movements, of the piston. It was fuelled by coal gas, as used then for domestic purposes and street lighting. The engine developed a feeble two horsepower. This was just enough power to drive a road vehicle and in 1863 Lenoir took a 10 km/6 mi journey in the first car powered by an internal combustion engine. The trip took 3 h to complete. The real value of the Lenoir engine was for powering small machines, and by 1865 more than 400 were in use in Paris, driving printing presses, lathes and water pumps. However, technical weaknesses, especially low compression, limited the potential of the Lenoir engine.

The next step forward was taken by German engineer Nikolaus Otto. In 1876, Otto patented the four- stroke engine, which produced its power using four movements of the piston. It used only half the fuel consumed by the Lenoir engine and was more powerful. The Otto engine was an immediate success; more than 35,000 were made in a few years. They were large engines and developed over 450 kW/600 hp.

Unfortunately for him, Otto's patent was invalidated in 1886 when his competitors dug up an earlier patent taken out by French inventor Alphonse Beau de Rochas in 1862. Rochas had described the four-stroke cycle in his patent. A court battle ensued, lasting two years, after which Otto's patent was declared invalid. This gave the go-ahead to other inventors to use the basic ideas without hindrance. German Karl Benz refined the Lenoir engine and introduced the use of liquid fuel, such as alcohol or petrol. He fitted his engine into a three-wheeled vehicle and in July 1886 at Mannheim the first petrol-driven motor car took to the roads. The trip covered about 1.6 km/1 mi mile at a speed of 14.5 kph/9 mph. Meanwhile, just 100 km/60 mi away in Cannstadt, Gottlieb Daimler was building his four-wheeled petrol-driven motor car. He completed a successful test run in August 1886.

A different type of internal combustion engine was invented by German engineer Karl Diesel in 1892. In the diesel engine, the compression used in the piston is much higher than in the petrol engine. This compression raises the temperature of air in the cylinder so high that electric ignition is not needed – the fuel ignites spontaneously when injected into the cylinder. This arrangement avoided the problems associated with electrical ignition and produced a simpler engine.

Working with such high pressures, Diesel was lucky to escape being killed when the cylinder head blew off one of his prototype engines. Undeterred he carried on to perfect the engine, and in 1899 he set up a factory to manufacture the engines. The factory flourished despite Diesel having little business sense. In 1913, at the height of his success, he vanished from the decks of a steamer crossing from Antwerp to England. His body was never found.

International Traffic in Arms Regulations US laws which prohibit the export of strong encryption by classifying it as a munition. Non-US users of common products such as ◊Netscape and ◊Lotus Notes are affected by these laws, as outside the USA American software suppliers must weaken the encryption built in to protect sensitive data.

In the mid-1990s several bills were introduced into the US Congress attempting to change these laws.

International Union for the Conservation of Nature (IUCN) organization established by the United Nations to promote the conservation of wildlife and habitats as part of the national policies of member states.

It has formulated guidelines and established research programmes (for example, International Biological Programme, IBP) and set up advisory bodies (such as Survival Commissions, SSC). In 1980, it launched the **World Conservation Strategy** to highlight particular problems, designating a small number of areas as **World Heritage Sites** to ensure their survival as unspoiled habitats (for example, Yosemite National Park in the USA, and the Simen Mountains in Ethiopia). It also compiles the **Red Data List of Threatened Animals**, classifying species according to their vulnerability to extinction.

According to its list of endangered species published in 1996, 25% of all mammal species (including 36% of insectivores and 33% of pigs and antelopes) and 11% of all bird species are threatened with extinction.

Internet global computer network connecting governments, companies, universities, and many other networks and users. ◊Electronic mail, conferencing, and chat services are all supported across the network, as is the ability to access remote computers and send and retrieve files. In 1997 around 55 million adults had access to the Internet in the USA alone.

The technical underpinnings of the Internet were developed as a project funded by the Advanced Research Project Agency (ARPA) to research how to build a network that would withstand bomb damage. The Internet itself began in 1984 with funding from the US National Science Foundation as a means to allow US universities to share the resources of five regional supercomputing centres. The number of users grew quickly, and in the early 1990s access became cheap enough for domestic users to have their own links on home personal computers. As the amount of information available via the Internet grew, indexing and search services such as Gopher, Archie, Veronica, and WAIS were created by Internet users to help both themselves and others. The newer World Wide Web allows seamless browsing across the Internet via ◊hypertext.

Internet Architecture Board (IAB) in computing, committee that coordinates the development of Internet ◊standards. Set up 1983, the IAB is a technical advisory group of the ◊Internet Society. Its responsibilities include architectural oversight for the ◊protocols and procedures used by the Internet, standards process oversight and appeal, editorial management and publication of ◊RFC (request for comments) documents, and advising the Internet

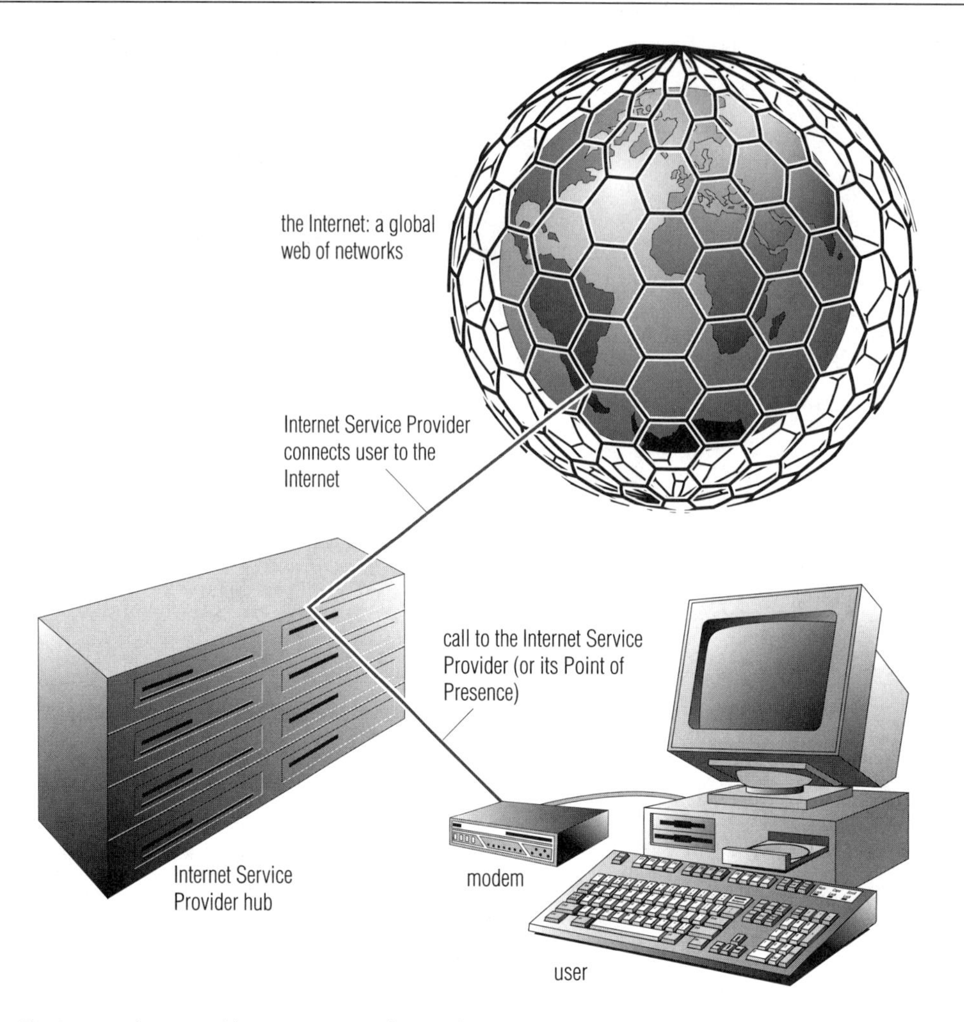

Internet *The Internet is accessed by users via a modem to the service provider's hub, which handles all connection requests. Once connected, the user can access a whole range of information from many different sources, including the World Wide Web.*

Society concerning technical, architectural, procedural, and some policy matters.

Internet-enabled in computing, facility that allows desktop applications to exchange information directly across the Internet. The most common Internet facility to build in is e-mail. Also popular is integrated Web access, so that a user can click on a ◊URL (uniform resource locator) from inside an application such as a word processor or personal information manager and be taken directly to that page on the World Wide Web.

Internet Engineering Task Force (IETF) in computing, international group which supervises the development of RFC (requests for comments), ◊protocols, and other engineering design for the Internet, reporting to the ◊Internet Architecture Board. It was formed 1986 and is based in Reston, Virginia, USA.

Internet Explorer in computing, Web ◊browser created by Microsoft 1995 to compete with ◊Netscape Navigator.Internet Explorer is given away free by Microsoft and bundled with its Windows 95 program. The US Justice Department sued Microsoft 1997 to prevent it from denying firms access to Windows 95 if they chose not to include Internet Explorer.

Internet Hunt in computing, monthly game played to test contestants' skills at finding information on the Internet.

Internet mail in computing, e-mail sent across the Internet. The distinction is primarily made on closed or commercial systems, where Internet mail comes from outside via a ◊gateway. Systems such as CompuServe used to charge extra for receiving or sending Internet mail, but such charges have been phased out.

Internet phone in computing, technology allowing users of the World Wide Web to talk to each other in more or less real time, via microphones and headsets. Network delays mean such connections are not as good quality as traditional telephone connections, but they are much cheaper for long-distance calls since users pay only for their local telephone connection to the Internet.

The earliest products were limited in that they only allowed users to talk to each other if both were logged on to the Vocaltec Web site at the same time. More recent products make it possible for a person using the Internet to dial any telephone in the world.

Internet Relay Chat (IRC) in computing, service that allows users connected to the Internet to chat with each other over many channels. There are probably hundreds of IRC channels active at any one time, covering a variety of topics. Many abbreviations are used to cut down on typing.

Internet Service Provider (ISP) in computing, any company that sells dial-up access to the Internet. Several types of company provide Internet access, including online information services such as ◊CompuServe and ◊America Online (AOL), electronic conferencing systems such as the ◊WELL and ◊Compulink Information eXchange, and local bulletin board systems (BBSs). Most recently founded ISPs, such as ◊Demon Internet and ◊PIPEX, offer only direct access to the Internet without the burden of running services of their own just for their members.

Such companies typically work out cheaper for their users, as they charge a low, flat rate for unlimited usage. By contrast, commercial online services typically charge by the hour or minute.

Internet Society (ISOC) global volunteer group that works to coordinate and develop the Internet and its underlying technology. It was founded 1992 and is based in Reston, Virginia, USA; the president (1996) is Vinton Cerf. Members include individuals, companies, nonprofit-making organizations, and government agencies.

Internet Talk Radio (also known as the ***Internet multicasting service***) service that broadcasts radio programmes of interest to the technical community, such as *Geek of the Week*. Based in Washington DC, USA, the service broadcasts via ◊MBONE.

Internet worm in computing, a virus; see ◊worm.

InterNIC in computing, service that administers ◊domain names and maintains a number of Internet user directories. Users interested in registering a particular domain name can use the InterNIC's resources to check if the domain name or a similar one is already in use.

interplanetary matter gas and dust thinly spread through the Solar System. The gas flows outwards from the Sun as the ◊solar wind.

Fine dust lies in the plane of the Solar System, scattering sunlight to cause the ◊zodiacal light. Swarms of dust shed by comets enter the Earth's atmosphere to cause ◊meteor showers.

interpolation mathematical technique for using two values to calculate intermediate values. It is used in ◊computer graphics to create smooth shadings.

interpreter computer program that translates and executes a program written in a high-level language. Unlike a ◊compiler, which produces a complete machine-code translation of the high-level program in one operation, an interpreter translates the source program, instruction by instruction, each time that program is run.

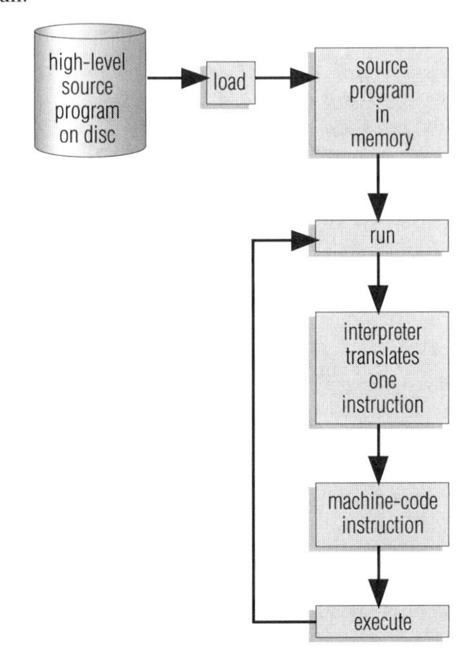

interpreter *The sequence of events when running an interpreter on a high-level language program. Instructions are translated one at a time, making the process a slow one; however, interpreted programs do not need to be compiled and may be executed immediately.*

> *Man is the interpreter of nature, science the right interpretation.*
>
> WILLIAM WHEWELL English physicist and philosopher. *The Philosophy of the Inductive Sciences* 1837

interquartile range in statistics, a measure of ◊dispersion in a frequency distribution, equalling the difference in value between the upper and lower ◊quartiles.

interrupt in computing, a signal received by the computer's central processing unit that causes a temporary halt in the execution of a program while some other task is performed. Interrupts may be generated by the computer's internal electronic clock (clock interrupt), by an input or output device, or by a software routine. After the computer has completed the task to which it was diverted, control returns to the original program.

For example, many computers, while printing a long document, allow the user to carry on with other work. When the printer is ready for more data, it sends an interrupt signal that causes the computer to halt work on the user's program and transmit more data to the printer.

intersection on a graph, the point where two lines or curves meet. The intersections of graphs provide the graphical solutions of equations.

intersection in set theory, the set of elements that belong to both set A and set B.

intersex individual that is intermediate between a normal male and a normal female in its appearance (for example, a genetic male that lacks external genitalia and so resembles a female).

interstellar cirrus in astronomy, wispy cloud-like structures discovered in the mid-1980s by the Infrared Astronomy Satellite (◊IRAS) and believed to be the remains of dust shells blown into space from cool giant or supergiant stars.

interstitial in biology, undifferentiated tissue that is interspersed with the characteristic tissue of an organ. It is often formed of fibrous tissue and supports the organ. Interstitial fluid refers to the fluid present in small amounts in the tissues of an organ.

intestine in vertebrates, the digestive tract from the stomach outlet to the anus. The human **small intestine** is 6 m/20 ft long, 4 cm/1.5 in in diameter, and consists of the duodenum, jejunum, and ileum; the **large intestine** is 1.5 m/5 ft long, 6 cm/2.5 in in diameter, and includes the caecum, colon, and rectum. Both are muscular tubes comprising an inner lining that secretes alkaline digestive juice, a submucous coat containing fine blood vessels and nerves, a muscular coat, and a serous coat covering all, supported by a strong peritoneum, which carries the blood and lymph vessels, and the nerves. The contents are passed along slowly by ◊peristalsis (waves of involuntary muscular action). The term intestine is also applied to the lower digestive tract of invertebrates.

intranet in computing, the use of software and other technology developed for the Internet on internal company ◊networks.

Many company networks (and those of other organizations) use the same ◊protocols as the Internet, namely ◊TCP/IP. Therefore the same technology that enables the World Wide Web can be used on an internal network to build an organization-wide web of internal documents that is familiar, easy to use, and comparatively inexpensive.

intrauterine device (IUD) or **coil**, a contraceptive device that is inserted into the womb (uterus). It is a tiny plastic object, sometimes containing copper. By causing a mild inflammation of the lining of the uterus it prevents fertilized eggs from becoming implanted.

IUDs are not usually given to women who have not had children. They are generally very reliable, as long as they stay in place, with a success rate of about 98%. Some women experience heavier and more painful periods, and there is a very slight risk of a pelvic infection leading to infertility.

intron or ***junk DNA*** in genetics, a sequence of bases in ◊DNA that carries no genetic information. Introns, discovered in 1977, make up 98% of DNA (the rest is made up of ◊exons). Their function is unknown.

10% of the human genome is made up of one base sequence, *Alu,* that occurs in about 1 million separate locations. It is made up of 283 nucleotides, has no determinable function (though some do have an effect on nearby genes), and is a ◊transposon ('jumping gene').

introversion in psychology, preoccupation with the self, generally coupled with a lack of sociability. The opposite of introversion is ◊extroversion.

The term was introduced by the Swiss psychiatrist Carl Jung 1924 in his description of ◊schizophrenia, where he noted that 'interest does not move towards the object but recedes towards the subject'. The term is also used within psychoanalysis to refer to the turning of the instinctual drives towards objects of fantasy rather than the pursuit of real objects. Another term for this sense is fantasy cathexis.

intrusion mass of ◊igneous rock that has formed by 'injection' of molten rock, or magma, into existing cracks beneath the surface of the Earth, as distinct from a volcanic rock mass which has erupted from the surface. Intrusion features include vertical cylindrical structures such as stocks, pipes, and necks; sheet structures such as dykes that cut across the strata and sills that push between them; laccoliths, which are blisters that push up the overlying rock; and batholiths, which represent chambers of solidified magma and contain vast volumes of rock.

intrusive rock ◊igneous rock formed beneath the Earth's surface. Magma, or molten rock, cools slowly at these depths to form coarse-grained rocks, such as granite, with large crystals. (◊Extrusive rocks, which are formed on the surface, are usually fine-grained.) A mass of intrusive rock is called an intrusion.

intuitionism in mathematics, the theory that propositions can be built up only from intuitive concepts that we all recognize easily, such as unity or plurality. The concept of ◊infinity, of which we have no intuitive experience, is thus not allowed.

There are children playing in the street who could solve some of my top problems in physics, because they have modes of sensory perception that I lost long ago.

J ROBERT OPPENHEIMER US physicist.
Attributed remark

Invar trademark for an alloy of iron containing 36% nickel, which expands or contracts very little when the temperature changes.

It is used to make precision instruments (such as pendulums and tuning forks) whose dimensions must not alter.

inverse function ◊function that exactly reverses the transformation produced by a function f; it is usually written as f^{-1}. For example $3x + 2$ and $(x - 2)/3$ are mutually inverse functions. Multiplication and division are inverse operations (see ◊reciprocals).

An inverse function is clearly demonstrated on a calculator by entering any number, pressing x^2, then pressing $\sqrt{x}$ to get the inverse. The functions on a scientific calculator can be inversed in a similar way.

inverse multiplexing in computing, technique for combining individual low-bandwidth channels into a single high-bandwidth channel. It is used to create high-speed telephone links for applications such as ◊videoconferencing which require the transmission of huge quantities of data.

inverse square law in physics, the statement that the magnitude of an effect (usually a force) at a point is inversely proportional to the square of the distance between that point and the object exerting the force.

Light, sound, electrostatic force (Coulomb's law), gravitational force (Newton's law) all obey the inverse square law.

inverse video or ***reverse video*** in computing, a display mode in which images on a display screen are presented as a negative of their normal appearance.

For example, if the computer screen normally displays dark images on a light background, inverse video will change all or part of the screen to a light image on a dark background.

Inverse video is commonly used to highlight parts of a display or to mark out text and pictures that the user wishes the computer to change in some way. For example, the user of a word-processing program might use a pointing device such as a ◊mouse to mark in inverse video a paragraph of text that is to be deleted from the document.

invertebrate animal without a backbone. The invertebrates comprise over 95% of the million or so existing animal species and include sponges, coelenterates, flatworms, nematodes, annelid worms, arthropods, molluscs, echinoderms, and primitive aquatic chordates, such as sea squirts and lancelets.

INVERTEBRATE

Every year, at least two species of British invertebrates become extinct.

inverted file in computing, a file that reorganizes the structure of an existing data file to enable a rapid search to be made for all records having one field falling within set limits.

For example, a file used by an estate agent might store records on each house for sale, using a reference number as the key field for ◊sorting. One field in each record would be the asking price of the house. To speed up the process of drawing up lists of houses falling within certain price ranges, an inverted file might be created in which the records are rearranged according to price. Each record would consist of an asking price, followed by the reference numbers of all the houses offered for sale at this approximate price.

in vitro fertilization (IVF; ***'fertilization in glass'***) allowing eggs and sperm to unite in a laboratory to form embryos. The embryos (properly called pre-embryos in their two- to eight-celled state) are stored by cooling to the temperature of liquid air (cryopreservation) until they are implanted into the womb of the otherwise infertile mother (an extension of ◊artificial insemination). The first baby to be produced by this method was born in 1978 in the UK. In cases where the Fallopian tubes are blocked, fertilization may be carried out by **intra-vaginal culture**, in which egg and sperm are incubated (in a plastic tube) in the mother's vagina, then transferred surgically into the uterus.

in vitro process biological experiment or technique carried out in a laboratory, outside the body of a living organism (literally 'in glass', for example in a test tube). By contrast, an in vivo process takes place within the body of an organism.

in vivo process biological experiment or technique carried out within a living organism; by contrast, an in vitro process takes place outside the organism, in an artificial environment such as a laboratory.

involuntary action behaviour not under conscious control, for example the contractions of the gut during peristalsis or the secretion of adrenaline by the adrenal glands. Breathing and urination reflexes are involuntary, although both can be controlled voluntarily to some extent. These processes are regulated by the ◊autonomic nervous system.

involute ***Latin 'rolled in'*** in geometry, ◊spiral that can be thought of as being traced by a point at the end of a taut nonelastic thread being wound onto or unwound from a spool.

I/O (abbreviation for ***input/output***) see ◊input devices and ◊output devices. The term is also used to describe transfer to and from disc – that is, disc I/O.

Io in astronomy, the third-largest moon of the planet Jupiter, 3,630 km/2,260 mi in diameter, orbiting in 1.77 days at a distance of 422,000 km/262,000 mi. It is the most volcanically active body in the Solar System, covered by hundreds of vents that erupt not lava but sulphur, giving Io an orange-coloured surface.

In July 1995 the Hubble Space Telescope revealed the appearance of a 320-km/200-mi yellow spot on the surface of Io, located on the volcano Ra Patera. Though clearly volcanic in origin, astronomers are unclear as to the exact cause of the new spot.

Using data gathered by the spacecraft *Galileo,* US astronomers concluded in 1996 that Io has a large metallic core. The *Galileo* space probe also detected a 10-megawatt beam of electrons flowing between Jupiter and Io.

In 1997 instruments aboard the spacecraft *Galileo* measured the temperature of Io's volcanoes and detected a minimum tmperature of 1,800 K (in comparison, Earth's hottest volcanoes only reach about 1,600 K).

iodide compound formed between iodine and another element in which the iodine is the more electronegative element (see ◊electronegativity, halide).

iodine ***Greek iodes 'violet'*** greyish-black nonmetallic element, symbol I, atomic number 53, relative atomic mass 126.9044. It is a member of the ◊halogen group. Its crystals give off, when heated, a violet vapour with an irritating odour resembling that of chlorine. It only occurs in combination with other elements. Its salts are known as iodides, which are found in sea water. As a mineral nutrient it is vital to the proper functioning of the thyroid gland, where it occurs in trace amounts as part of the hormone thyroxine. Absence of iodine from the diet leads to ◊goitre. Iodine is used in photography, in medicine as an antiseptic, and in making dyes.

Its radioactive isotope ^{131}I (half-life of eight days) is a dangerous fission product from nuclear explosions and from the nuclear reactors in power plants, since, if ingested, it can be taken up by the thyroid and damage it. It was discovered in 1811 by French chemist B Courtois (1777–1838).

iodoform (chemical name ***triiodomethane***) CHI_3, an antiseptic that crystallizes into yellow hexagonal plates. It is soluble in ether, alcohol, and chloroform, but not in water.

Iomega in computing, leading manufacturer of removable storage and back-up devices, in direct competition with ◊Syquest. Based in Roy, Utah, in the USA, Iomega's two most popular products are the Zip drive, which uses inexpensive 100Mb discs, and the Jaz drive, which uses 1Gb discs.

ion atom, or group of atoms, that is either positively charged (◊cation) or negatively charged (◊anion), as a result of the loss or gain of electrons during chemical reactions or exposure to certain forms of radiation.

ion engine rocket engine that uses ◊ions (charged particles) rather than hot gas for propulsion. Ion engines have been successfully tested in space, where they will eventually be used for gradual rather than sudden velocity changes. In an ion engine, atoms of mercury, for example, are ionized (given an electric charge by an electric field) and then accelerated at high speed by a more powerful electric field.

ion exchange process whereby an ion in one compound is replaced by a different ion, of the same charge, from another compound. It is the basis of a type of ◊chromatography in which the components of a mixture of ions in solution are separated according to the ease with which they will replace the ions on the polymer matrix through which they flow. The exchange of positively charged ions is called cation exchange; that of negatively charged ions is called anion exchange.

Ion-exchange is used in commercial water softeners to exchange the dissolved responsible for the water's hardness with others that do not have this effect. For example, when hard water is passed over an ion-exchange resin, the dissolved calcium and magnesium ions are replaced by either sodium or hydrogen ions, so the hardness is removed.

ion half equation equation that describes the reactions occurring at the electrodes of a chemical cell or in electrolysis. It indicates which ion is losing electrons (oxidation) or gaining electrons (reduction).

Examples are given from the electrolysis of dilute hydrochloric acid (HCl).

$$2Cl^- - 2e^- \rightarrow Cl_2 \text{ (positive electrode)} 2H^+ + 2e^- \rightarrow H_2 \text{ (negative electrode)}$$

ionic bond or **electrovalent bond** bond produced when atoms of one element donate electrons to atoms of another element, forming positively and negatively charged ions respectively. The attraction between the oppositely charged ions constitutes the bond. Sodium chloride (Na^+Cl^-) is a typical ionic compound.

Each ion has the electronic structure of an inert gas (see ◊noble gas structure). The maximum number of electrons that can be gained is usually two.

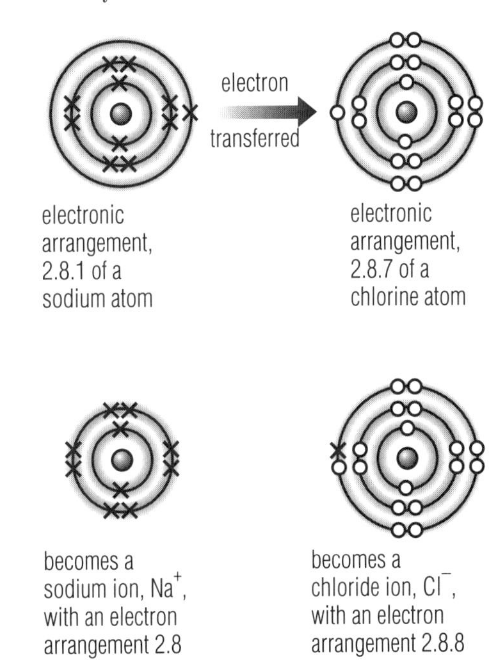

ionic bond *The formation of an ionic bond between a sodium atom and a chlorine atom to form a molecule of sodium chloride. The sodium atom transfers an electron from its outer electron shell (becoming the positive ion Na+) to the chlorine atom (which becomes the negative chloride ion Cl–). The opposite charges mean that the ions are strongly attracted to each other. The formation of the bond means that each atom becomes more stable, having a full quota of electrons in its outer shell.*

ionic compound substance composed of oppositely charged ions. All salts, most bases, and some acids are examples of ionic compounds. They possess the following general properties: they are crystalline solids with a high melting point; are soluble in water and insoluble in organic solvents; and always conduct electricity when molten or in aqueous solution. A typical ionic compound is sodium chloride (Na^+Cl^-).

ionic equation equation showing only those ions in a chemical reaction that actually undergo a change, either by combining together to form an insoluble salt or by combining together to form one or more molecular compounds. Examples are the precipitation of insoluble barium sulphate when barium and sulphate ions are combined in solution, and the production of ammonia and water from ammonium hydroxide.

$Ba^{2+}_{(aq)} + SO_4^{2-}{}_{(aq)} \rightarrow BaSO_{4(s)} NH_4^+{}_{(aq)} + OH^-{}_{(aq)} \rightarrow NH_{3(g)} + H_2O_{(l)}$

The other ions in the mixtures do not take part and are called ◊spectator ions.

ionization process of ion formation. It can be achieved in two ways. The first way is by the loss or gain of electrons by atoms to form positive or negative ions.

$$Na - e^- \rightarrow Na^+ 1/2Cl_2 + e^- \rightarrow Cl^-$$

In the second mechanism, ions are formed when a covalent bond breaks, as when hydrogen chloride gas is dissolved in water. One portion of the molecule retains both electrons, forming a negative ion, and the other portion becomes positively charged. This bond-fission process is sometimes called dissociation.

$$HCl_{(g)} + aq \leftrightarrow H^+{}_{(aq)} + Cl^-{}_{(aq)}$$

ionization chamber device for measuring ◊ionizing radiation. The radiation ionizes the gas in the chamber and the ions formed are collected and measured as an electric charge. Ionization chambers are used for determining the intensity of X-rays or the disintegration rate of radioactive materials.

ionization potential measure of the energy required to remove an ◊electron from an ◊atom. Elements with a low ionization potential readily lose electrons to form ◊cations.

ionization therapy enhancement of the atmosphere of an environment by instrumentally boosting the negative ion content of the air.

Fumes, dust, cigarette smoke, and central heating cause negative ion deficiency, which particularly affects sufferers from respiratory disorders such as bronchitis, asthma, and sinusitis. Symptoms are alleviated by the use of ionizers in the home or workplace. In severe cases, ionization therapy is used as an adjunct to conventional treatment.

ionizing radiation radiation that knocks electrons from atoms during its passage, thereby leaving ions in its path. Alpha and beta particles are far more ionizing in their effect than are neutrons or gamma radiation.

ionosphere ionized layer of Earth's outer ◊atmosphere (60–1,000 km/38–620 mi) that contains sufficient free electrons to modify the way in which radio waves are propagated, for instance by reflecting them back to Earth. The ionosphere is thought to be produced by absorption of the Sun's ultraviolet radiation.

ion plating method of applying corrosion-resistant metal coatings. The article is placed in argon gas, together with some coating metal, which vaporizes on heating and becomes ionized (acquires charged atoms) as it diffuses through the gas to form the coating. It has important applications in the aerospace industry.

IP address (abbreviation for ***Internet protocol address***) in computing, numbered ◊address assigned to an Internet ◊host. Traditionally, IP addresses are ◊32-bit, which means that numbered addresses have four sections separated by dots, each a decimal number between 0 and 255.

ipecacuanha or *ipecac* South American plant belonging to the madder family, the dried roots of which are used in medicine as an emetic (to cause vomiting) and to treat amoebic dysentery (infection of the intestine with amoebae). (*Psychotria ipecacuanha,* family Rubiaceae.)

IR or ***ir*** in physics, abbreviation for ***infrared***.

IRAS acronym for ***Infrared Astronomy Satellite*** joint US–UK–Dutch satellite launched in 1983 to survey the sky at infrared wavelengths, studying areas of star formation, distant galaxies, possible embryo planetary systems around other stars, and discovering five new comets in our own Solar System. It operated for 10 months.

IRC in computing, abbreviation for ◊***Internet Relay Chat***.

iridium ***Latin iridis 'rainbow'*** hard, brittle, silver-white, metallic element, symbol Ir, atomic number 77, relative atomic mass 192.2. It is resistant to tarnish and corrosion. Iridium is one of the so-called platinum group of metals; it occurs in platinum ores and as a free metal (◊native metal) with osmium in osmiridium, a natural alloy that includes platinum, ruthenium, and rhodium.

It is alloyed with platinum for jewellery and used for watch bearings and in scientific instruments. It was named in 1804 by English chemist Smithson Tennant (1761–1815) for its iridescence in solution.

iridium anomaly unusually high concentrations of the element iridium found world-wide in sediments which were deposited at the Cretaceous-Tertiary boundary (◊K-T boundary) 65 million years ago. Since iridium is more abundant in extraterrestrial material, its presence is thought to be evidence for a large meteor impact which may have caused the extinction of the dinosaurs.

iris in anatomy, the coloured muscular diaphragm that controls the size of the pupil in the vertebrate eye. It contains radial muscle that increases the pupil diameter and circular muscle that constricts the pupil diameter. Both types of muscle respond involuntarily to light intensity.

iris in botany, any of a group of perennial northern temperate flowering plants belonging to the iris family. The leaves are usually sword-shaped; the purple, white, or yellow flowers have three upright inner petals and three outward- and downward-curving ◊sepals. The wild yellow iris is called a flag. (Genus *Iris,* family Iridaceae.)

Irish terrier breed of large ◊terrier from the region of Cork and Ballymena. It has a thick, rough coat in reddish and wheaten hues and an athletic build. It stands about 45 cm/18 in at the shoulder.

Irish water spaniel breed of medium-sized gundog with a dark brown, thickly curling coat, used especially for hunting and retrieving waterfowl. It grows to 58.5 cm/23 in and is an ancestor of the smaller (up to 46 cm/18 in), but very similar, American water spaniel.

Irish wolfhound breed of hound used in Ireland for many centuries to hunt wolves and other large game. Of massive size (80 cm/32 in upwards) and powerful build, the wolfhound has a shaggy coat in a range of colours – wheaten, grey, brindled, black, red, or white.

In the 19th century the breed was revived when nearly extinct by crossing it with the ◊deerhound.

iron hard, malleable and ductile, silver-grey, metallic element, symbol Fe (from Latin *ferrum*), atomic number 26, relative atomic mass 55.847. It is the fourth most abundant element (the second most abundant metal, after aluminium) in the Earth's crust. Iron occurs in concentrated deposits as the ores hematite (Fe_2O_3), spathic ore ($FeCO_3$), and magnetite (Fe_3O_4). It sometimes occurs as a free metal, occasionally as fragments of iron or iron–nickel meteorites.

Iron is the most common and most useful of all metals; it is strongly magnetic and is the basis for ◊steel, an alloy with carbon and other elements (see also ◊cast iron). In electrical equipment it is used in all permanent magnets and electromagnets, and forms the cores of transformers and magnetic amplifiers. In the human body, iron is an essential component of haemoglobin, the molecule in red blood cells that transports oxygen to all parts of the body. A deficiency in the diet causes a form of anaemia.

Iron Age developmental stage of human technology when weapons and tools were made from iron. Preceded by the Stone and Bronze ages, it is the last technological stage in the Three Age System framework for prehistory. Iron was produced in Thailand about 1600 BC, but was considered inferior in strength to bronze until about 1000 BC, when metallurgical techniques improved, and the alloy steel was produced by adding carbon during the smelting process.

Ironworking was introduced into different regions over a wide time span, appearing in Thailand about 1600 BC, Asia Minor about 1200 BC, central Europe about 900 BC, China about 600 BC, and in remoter areas during exploration and colonization by the Old

World. It reached the Fiji Islands with an expedition in the late 19th century.

Iron Age cultures include Hallstatt (named after a site in Austria) and ◊La Tène (named after a site in Switzerland).

ironbark any species of ◊eucalyptus tree with hard, tough bark.

iron ore any mineral from which iron is extracted. The chief iron ores are ◊magnetite, **a black oxide**; ◊hematite, **or kidney ore, a reddish oxide**; **limonite**, brown, impure oxyhydroxides of iron; and **siderite**, a brownish carbonate.

Iron ores are found in a number of different forms, including distinct layers in igneous intrusions, as components of contact metamorphic rocks, and as sedimentary beds. Much of the world's iron is extracted in Russia, Kazakhstan, and the Ukraine. Other important producers are the USA, Australia, France, Brazil, and Canada; over 40 countries produce significant quantities of ore.

iron pyrites or ***pyrite*** FeS_2 common iron ore. Brassy yellow, and occurring in cubic crystals, it is often called 'fool's gold', since only those who have never seen gold would mistake it.

irradiation subjecting anything to radiation, including cancer tumours. See also ◊food irradiation.

irrational number a number that cannot be expressed as an exact ◊fraction. Irrational numbers include some square roots (for example, $\sqrt{2}$, $\sqrt{3}$, and $\sqrt{5}$ are irrational) and numbers such as π (the ratio of the circumference of a circle to its diameter, which is approximately equal to 3.14159) and e (the base of ◊natural logarithms, approximately 2.71828).

irregular galaxy in astronomy, a class of ◊galaxy with little structure, which does not conform to any of the standard shapes in the Hubble classification. The two satellite galaxies of the ◊Milky Way, the ◊Magellanic Clouds, are both irregulars. Some galaxies previously classified as irregulars are now known to be normal galaxies distorted by tidal effects or undergoing bursts of star formation (see ◊starburst galaxy).

irrigation artificial water supply for dry agricultural areas by means of dams and channels. Drawbacks are that it tends to concentrate salts at the surface, ultimately causing soil infertility, and that rich river silt is retained at dams, to the impoverishment of the land and fisheries below them.

Irrigation has been practised for thousands of years, in Eurasia as well as the Americas. An example is the channelling of the annual Nile flood in Egypt, which has been done from earliest times to its present control by the Aswan High Dam.

ISA bus (abbreviation for ***industry standard architecture bus***) in computing, 16-bit data ◊bus introduced in 1984 with the IBM PC AT and still in common use in PCs, alongside the superior ◊PCI bus. PC hardware and software manufacturers would like to get rid of the ISA bus as it is not compatible with ◊Plug and Play.

ISBN (abbreviation for ***International Standard Book Number***) code number used for ordering or classifying book titles. Every book printed now has a number on its back cover or jacket, preceded by the letters ISBN. It is a code to the country of origin and the publisher. The number is unique to the book, and will identify it anywhere in the world.

The final digit in each ISBN number is a check digit, which can be used by a computer program to validate the number each time it is input (see ◊validation).

ischaemic heart disease (IHD) disorder caused by reduced perfusion of the coronary arteries due to ◊atherosclerosis. It is the commonest cause of death in the Western world, leading to more than a million deaths each year in the USA and about 160,000 in the UK. See also ◊coronary artery disease.

Early symptoms of IHD include angina or palpitations, but sometimes a heart attack is the first indication that a person is affected.

ISDN abbreviation for ◊***Integrated Services Digital Network***, a telecommunications system.

island area of land surrounded entirely by water. Australia is classed as a continent rather than an island, because of its size.

Islands can be formed in many ways. **Continental islands** were once part of the mainland, but became isolated (by tectonic movement, erosion, or a rise in sea level, for example). **Volcanic islands**, such as Japan, were formed by the explosion of underwater volcanoes. **Coral islands** consist mainly of coral, built up over many years. An **atoll** is a circular coral reef surrounding a lagoon; atolls were formed when a coral reef grew up around a volcanic island that subsequently sank or was submerged by a rise in sea level. **Barrier islands** are found by the shore in shallow water, and are formed by the deposition of sediment eroded from the shoreline.

island arc curved chain of islands produced by volcanic activity at a destructive margin (where one tectonic plate slides beneath another). Island arcs are common in the Pacific where they ring the ocean on both sides; the Aleutian Islands off Alaska are an example.

Such island arcs are often later incorporated into continental margins during mountain-building episodes.

islets of Langerhans groups of cells within the pancreas responsible for the secretion of the hormone insulin. They are sensitive to the blood sugar, producing more hormone when glucose levels rise.

ISO abbreviation for ◊***International Organization for Standardization***.

ISO in photography, a numbering system for rating the speed of films, devised by the International Standards Organization.

ISO 9660 in computing, standard file format for ◊CD-ROM discs, synonymous with High Sierra format. This format is compatible with most systems, so the same disc can contain both Apple Macintosh and PC versions.

isobar line drawn on maps and weather charts linking all places with the same atmospheric pressure (usually measured in millibars).

When used in weather forecasting, the distance between the isobars is an indication of the barometric gradient (the rate of change in pressure).

Where the isobars are close together, cyclonic weather is indicated, bringing strong winds and a depression, and where far apart anticyclonic, bringing calmer, settled conditions.

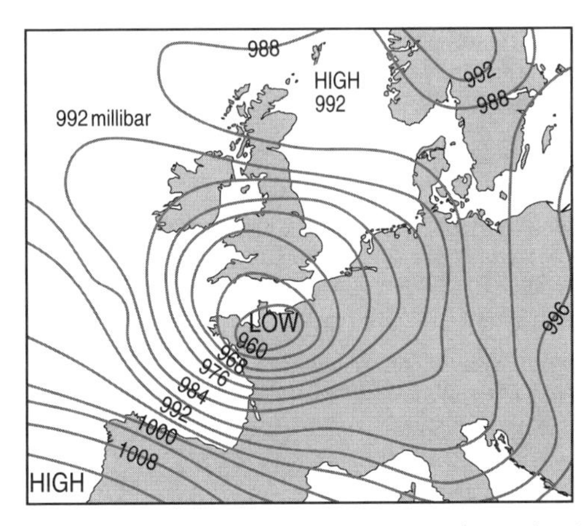

isobar The isobars around a low-pressure area or depression. In the northern hemisphere, winds blow anticlockwise around lows, approximately parallel to the isobars, and clockwise around highs. In the southern hemisphere, the winds blow in the opposite directions.

ISOC in computing, abbreviation for ◊***Internet Society***.

isomer chemical compound having the same molecular composition and mass as another, but with different physical or chemical properties owing to the different structural arrangement of its constituent atoms. For example, the organic compounds butane ($CH_3(CH_2)_2CH_3$) and methyl propane ($CH_3CH(CH_3)CH_3$) are isomers, each possessing four carbon atoms and ten hydrogen atoms but differing in the way that these are arranged with respect to each other.

butane $CH_3(CH_2)_2CH_3$

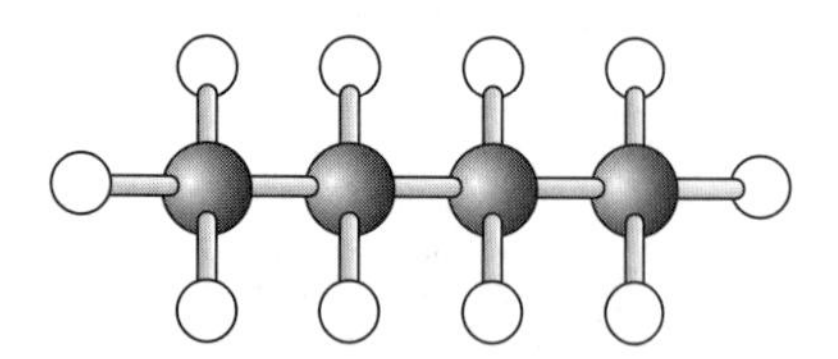

methyl propane $CH_3CH(CH_3)CH_3$

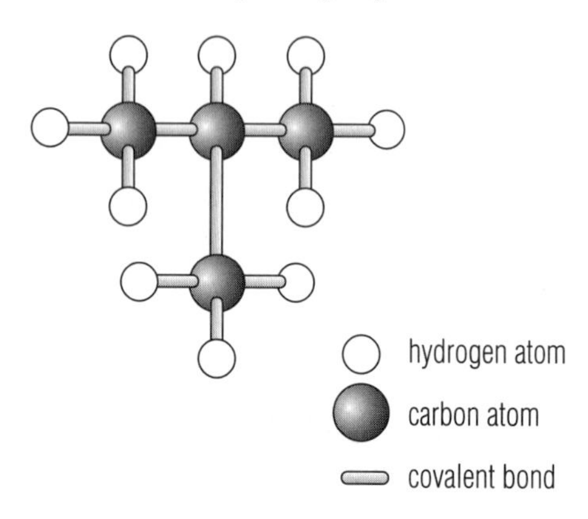

isomer The chemicals butane and methyl propane are isomers. Each has the molecular formula CH3CH(CH3)CH3, but with different spatial arrangements of atoms in their molecules.

Structural isomers have obviously different constructions, but **geometrical** and **optical isomers** must be drawn or modelled in order to appreciate the difference in their three-dimensional arrangement. Geometrical isomers have a plane of symmetry and arise because of the restricted rotation of atoms around a bond; optical isomers are mirror images of each other. For instance, 1,1-dichloroethene (CH_2=CCl_2) and 1,2-dichloroethene (CHCl=CHCl) are structural isomers, but there are two possible geometric isomers of the latter (depending on whether the chlorine atoms are on the same side or on opposite sides of the plane of the carbon–carbon double bond).

isometric transformation a ◊transformation in which length is preserved.

isomorphism the existence of substances of different chemical composition but with similar crystalline form.

isoprene $CH_2CHC(CH_3)CH_2$ (technical name ***methylbutadiene***) colourless, volatile fluid obtained from petroleum and coal, used to make synthetic rubber.

isosceles triangle a ◊triangle with two sides equal, hence its base angles are also equal. The triangle has an axis of symmetry which is an ◊altitude of the triangle.

isostasy the theoretical balance in buoyancy of all parts of the Earth's ◊crust, as though they were floating on a denser layer beneath. There are two theories of the mechanism of isostasy, the Airy hypothesis and the Pratt hypothesis, both of which have validity. In the **Airy hypothesis** crustal blocks have the same density but different depths: like ice cubes floating in water, higher mountains have deeper roots. In the **Pratt hypothesis**, crustal blocks have different densities allowing the depth of crustal material to be the same.

There appears to be more geological evidence to support the Airy hypothesis of isostasy. During an ◊ice age the weight of the ice sheet pushes that continent into the Earth's mantle; once the ice has melted, the continent rises again. This accounts for shoreline features being found some way inland in regions that were heavily glaciated during the Pleistocene period.

isotope one of two or more atoms that have the same atomic number (same number of protons), but which contain a different number of neutrons, thus differing in their atomic masses. They may be stable or radioactive, naturally occurring or synthesized. The term was coined by English chemist Frederick Soddy, pioneer researcher in atomic disintegration.

ISP in computing, abbreviation for ◊Internet Service Provider.

iteration in computing, a method of solving a problem by performing the same steps repeatedly until a certain condition is satisfied. For example, in one method of ◊sorting, adjacent items are repeatedly exchanged until the data are in the required sequence.

iteration in mathematics, method of solving ◊equations by a series of approximations which approach the exact solution more and more closely. For example, to find the square root of N, start with a guess n_1; calculate $N/n_1 = x_1$; calculate $(n_1 + x_1/2 = n_2$; calculate $N/n_1 = x_2$; calculate $(n_2 + x_2)/2 = n_3$. The sequence n_1, n_2, n_3 approaches the exact square root of

N. Iterative methods are particularly suitable for work with computers and programmable calculators.

iteroparity in biology, the repeated production of offspring at intervals throughout the life cycle. It is usually contrasted with ◊semelparity, where each individual reproduces only once during its life.

ITU abbreviation for ◊***International Telecommunications Union***, the standards-setting body for the communications industry.

IUE (acronym for ***International Ultraviolet Explorer***) joint NASA-ESA (US and European Space Agency) orbiting ultraviolet telescope with a 45-cm/18-in mirror, launched in 1978. It was switched off in September 1996.

IUPAC abbreviation for ***International Union of Pure and Applied Chemistry***, organization that recommends the nomenclature to be used for naming substances, the units to be used, and which conventions are to be adopted when describing particular changes.

IVF abbreviation for ◊***in vitro fertilization***.

ivory hard white substance of which the teeth and tusks of certain mammals are made. Among the most valuable are elephants' tusks, which are of unusual hardness and density. Ivory is used in carving and other decorative work, and is so valuable that poachers continue to illegally destroy the remaining wild elephant herds in Africa to obtain it.

Poaching for ivory has led to the decline of the African elephant population from 2 million to approximately 600,000, with the species virtually extinct in some countries. Trade in ivory was halted by Kenya in 1989, but Zimbabwe continued its policy of controlled culling to enable the elephant population to thrive and to release ivory for export. China and Hong Kong have refused to obey an international ban on ivory trading. In 1997, the 138 member nations of the Convention on International Trade in Endangered Species (◊CITES) voted in Harare, amidst much controversy, to remove the ban on trade in ivory in Botswana, Namibia, and Zimbabwe. These three countries would be allowed to sell a limited amount of ivory to Japan, and the money must be channelled into elephant conservation projects. Trade is scheduled to resume in 1999 and would only be allowed from the existing stockpiles of ivory.

Vegetable ivory is used for buttons, toys, and cheap ivory goods. It consists of the hard albumen of the seeds of a tropical palm (*Phytelephas macrocarpa*), and is imported from Colombia.

ivy any of an Old World group of woody climbing, trailing, or creeping evergreen plants. English or European ivy (*H. helix*) has shiny five-lobed leaves and clusters of small, yellowish-green flowers followed by black berries. It climbs by means of rootlike suckers put out from its stem, and causes damage to trees. (Genus *Hedera*, family Araliaceae.)

Ground ivy (*Glechoma hederacea*) is a small, originally European creeping plant belonging to the mint family; the North American poison ivy (*Rhus radicans*) belongs to the cashew family.

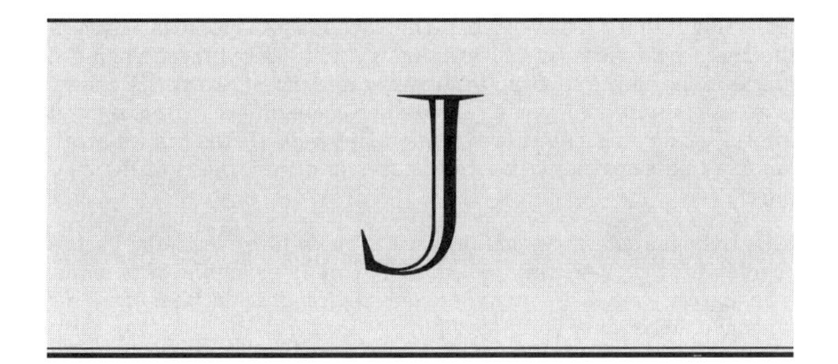

J in physics, the symbol for **joule**, the SI unit of energy.

jabiru stork found in Central and South America. It is 1.5 m/5 ft tall with white plumage. The head is black and red with a massive, slightly upturned bill. The neck is bare of feathers and can be puffed out, a manoeuvre which is probably used in social rituals. (Species *Jabiru mycteria,* family Ciconiidae, order Ciconiiformes.)

jaborandi plant belonging to the rue family, native to South America. It is the source of **pilocarpine**, used in medicine to contract the pupil of the eye. (*Pilocarpus microphyllus,* family Rutaceae.)

jacamar insect-eating bird related to the woodpeckers, found in dense tropical forest in Central and South America. It has a long, straight, sharply-pointed bill, a long tail, and paired toes. The plumage is golden bronze with a steely lustre. Jacamars are usually seen sitting motionless on trees from which they fly out to catch insects on the wing, then return to crack them on a branch before eating them. The largest species is *Jacamerops aurea,* which is nearly 30 cm/12 in long. (Family Galbulidae, order Piciformes.)

jacana or ***lily-trotter*** wading bird with very long toes and claws enabling it to walk on the floating leaves of water plants. There are seven species. Jacanas are found in Mexico, Central America, South America, Africa, S Asia, and Australia, usually in marshy areas. (Family Jacanidae, order Charadriiformes.)

The **Australian jacana** (*Irediparra gallinacea*) is so well adapted to life on water that the eggs are laid on floating vegetation and can themselves float.

jacaranda any of a group of tropical American trees belonging to the bignonia family, with fragrant wood and showy blue or violet flowers, commonly cultivated in the southern USA. (Genus *Jacaranda,* family Bignoniaceae.)

jack in computing, small plug allowing users to connect peripherals to CPUs (◊central processing units).

jack tool or machine for lifting, hoisting, or moving heavy weights, such as motor vehicles. A **screw jack** uses the principle of the screw to magnify an applied effort; in a car jack, for example, turning the handle many times causes the lifting screw to rise slightly, and the effort is magnified to lift heavy weights. A **hydraulic jack** uses a succession of piston strokes to increase pressure in a liquid and force up a lifting ram.

jackal any of several wild dogs found in S Asia, S Europe, and N Africa. Jackals can grow to 80 cm/2.7 ft long, and have greyish-brown fur and a bushy tail. (Genus *Canis.*)

The **golden jackal** (*C. aureus*) of S Asia, S Europe, and N Africa is 45 cm/1.5 ft high and 60 cm/2 ft long. It is greyish-yellow, and darker on the back. A nocturnal animal, it preys on smaller mammals and poultry, although packs will attack larger animals; it will also scavenge. The **side-striped jackal** (*C. adustus*) is found over much of Africa; the **black-backed jackal** (*C. mesomelas*) occurs only in the south of Africa.

jackdaw bird belonging to the crow family, native to Europe and Asia. It is mainly black, but greyish on the sides and back of the head, and about 33 cm/1.1 ft long. It nests in tree holes or on buildings. Usually it lays five bluish-white eggs, mottled with tiny dark brown spots. Jackdaws feed on a wide range of insects, molluscs, spiders, worms, birds' eggs, fruit, and berries. (Species *Corvus monedula,* family Corvidae, order Passeriformes.)

Jack Russell terrier or ***Parson Jack Russell terrier*** breed of small, short-legged, smooth-haired ◊terrier, which takes its name from its originator, the English clergyman and fox-hunting enthusiast John Russell (1795–1883). It was recognized by the UK Kennel Club 1990 as a distinct variant of the fox terrier, which it resembles in having a mainly white coat with brown patches.

jade semiprecious stone consisting of either jadeite, $NaAlSi_2O_6$ (a pyroxene), or nephrite, $Ca_2(Mg,Fe)_5Si_8O_{22}(OH,F)_2$ (an amphibole), ranging from colourless through shades of green to black according to the iron content. Jade ranks 5.5–6.5 on the Mohs' scale of hardness.

The early Chinese civilization discovered jade, bringing it from E Turkestan, and carried the art of jade-carving to its peak. The Olmecs, Aztecs, Maya, and the Maori have also used jade for ornaments, ceremony, and utensils.

jaggies in computing, 'stepped' appearance of curved or diagonal lines in computer graphics caused by ◊aliasing.

jaguar largest species of cat in the Americas, formerly ranging from the southwestern USA to southern South America, but now extinct in most of North America. It can grow up to 2.5 m/8 ft long including the tail. The background colour of the fur varies from creamy white to brown or black, and is covered with black spots. The jaguar is usually solitary. (Species *Panthera onca,* family Felidae.)

jaguarundi wild cat found in forests in Central and South America. Up to 1.1 m/3.5 ft long, it is very slim with rather short legs and short rounded ears. It is uniformly coloured dark brown or chestnut. A good climber, it feeds on birds and small mammals and, unusually for a cat, has been reported to eat fruit. (Species *Felis yaguoaroundi,* family Felidae.)

jansky unit of radiation received from outer space, used in radio astronomy. It is equal to 10^{-26} watts per square metre per hertz, and is named after US engineer Karl Jansky.

Jansky, Karl Guthe
(1905–1950)

US radio engineer who in 1932 discovered that the Milky Way galaxy emanates radio waves; he did not follow up his discovery, but it marked the birth of radioastronomy.

jackal *The black-backed jackal* Canis mesomelas *is a common sight on the savannas of S Africa. It hunts singly for small animals and in packs for larger prey, often scavenging on carcasses left by larger predators such as lions. Premaphotos Wildlife*

Japan Current or ***Kuroshio*** warm ocean ◊current flowing from Japan to North America.

Japanese spaniel or **Japanese chin** ancient breed of toy dog introduced into Japan from China. Daintily built, it stands about 30 cm/12 in tall, and, like the pekingese, has a rounded head, drooping ears, and tail curved over its back. Its long, soft coat is always white, with either black or red markings.

jarrah type of ◊eucalyptus tree of W Australia, with durable timber.

Jarvik 7 the first successful artificial heart intended for permanent implantation in a human being. Made from polyurethane plastic and aluminium, it is powered by compressed air. Barney Clark became the first person to receive a Jarvik 7, in Salt Lake City, Utah, USA, 1982; it kept him alive for 112 days.

jasmine any of a group of subtropical plants with white or yellow flowers. The common jasmine (*J. officinale*) has fragrant pure white flowers that yield jasmine oil, used in perfumes; the Chinese winter jasmine (*J. nudiflorum*) has bright yellow flowers that appear before the leaves. (Genus *Jasminum*, family Oleaceae.)

jaundice yellow discoloration of the skin and whites of the eyes caused by an excess of bile pigment in the bloodstream. Approximately 60% of newborn babies exhibit some degree of jaundice, which is treated by bathing in white, blue, or green light that converts the bile pigment bilirubin into a water-soluble compound that can be excreted in urine. A serious form of jaundice occurs in rhesus disease (see ◊rhesus factor).

Bile pigment is normally produced by the liver from the breakdown of red blood cells, then excreted into the intestines. A build-up in the blood is due to abnormal destruction of red cells (as in some cases of ◊anaemia), impaired liver function (as in ◊hepatitis), or blockage in the excretory channels (as in gallstones or ◊cirrhosis). The jaundice gradually recedes following treatment of the underlying cause.

Java in computing, programming language much like C developed by James Gosling at ◊Sun Microsystems 1995. Java has been adopted as a multipurpose, cross-platform lingua franca for network computing, including the ◊World Wide Web. When users connect to a server that uses Java, they download a small program called an applet onto their computers. The ◊applet then runs on the computer's own processor via a ◊Java Virtual Machine program or JVM.

Java Virtual Machine (JVM) in computing, a program that sits on top of a computer's usual operating system and runs Java ◊applets.

Different computers require different JVMs but they should run the same Java code. This means that servers only need to provide one version of each applet, instead of different 'native code' versions for PCs, Apple Macintoshes, and UNIX workstations, as is the case with other plug-ins.

A JVM is commonly supplied as part of a Web browser but may be included as part of the operating system.

jaw one of two bony structures that form the framework of the mouth in all vertebrates except lampreys and hagfishes (the agnathous or jawless vertebrates). They consist of the upper jawbone (maxilla), which is fused to the skull, and the lower jawbone (mandible), which is hinged at each side to the bones of the temple by ◊ligaments.

jay any of several birds belonging to the crow family, generally brightly coloured and native to Europe, Asia, and the Americas. In the Eurasian **common jay** (*Garrulus glandarius*), the body is fawn with patches of white, blue, and black on the wings and tail. (Family Corvidae, order Passeriformes.)

Jays are shy and retiring in their habits, but have a screeching cry with the power to vary it by mimicking other birds. They feed chiefly on snails, insects, worms, and nuts, particularly acorns. They hide their nests in trees with thick foliage and lay about five or six eggs at a time.

The **blue jay** (*Cyanocitta cristata*), of the eastern and central USA, has a crest and is very noisy and bold.

Jeans mass in astronomy, the mass that a cloud (or part of a cloud) of interstellar gas must have before it can contract under its own weight to form a protostar. The Jeans mass is an expression of the **Jeans criterion**, which says that a cloud will contract when the gravitational force tending to drawing material towards its centre is greater than the opposing force due to gas pressure. It is named after English mathematician James Hopwood Jeans whose work focussed on the kinetic theory of gases and the origins of the cosmos.

jellyfish marine invertebrate, belonging among the ◊coelenterates, with an umbrella-shaped body made of a semi-transparent jellylike substance, often tinted with blue, red, or orange colours, and stinging tentacles that trail in the water. Most adult jellyfish move freely, but during parts of their life cycle many are polyplike and attached to rocks, the seabed, or another underwater surface. They feed on small animals that are paralysed by stinging cells in the jellyfish tentacles. (Phylum Coelenterata, subphylum Cnidaria.)

Most jellyfish cause no more discomfort to humans than a nettle sting, but contact with the tentacles of the subtropical Portuguese man-of-war (*Physalia physalis*) or the Australian box jellyfish (*Chironex fleckeri*) can be life-threatening.

Jenner, Edward (1749–1823)

English physician who pioneered vaccination. In Jenner's day, smallpox was a major killer. His discovery in 1796 that inoculation with cowpox gives immunity to smallpox was a great medical breakthrough.

Mary Evans Picture Library

Jenner observed that people who worked with cattle and contracted cowpox from them never subsequently caught smallpox. In 1798 he published his findings that a child inoculated with cowpox, then two months later with smallpox, did not get smallpox. He coined the word 'vaccination' from the Latin word for cowpox, *vaccinia*.

Jentink's duiker small, shy ◊antelope that plunges into bushes when startled. It is acutely threatened by deforestation in its remaining habitat in W Africa, where it is also hunted. One captive breeding colony exists in Texas, USA, and there are hopes of establishing others as the immediate future for this species in the wild appears to be bleak. (Species *Cephalophus jentinki*, family Bovidae.)

jerboa any of a group of small nocturnal rodents with long and powerful hind legs developed for leaping. There are about 25 species of jerboa, native to desert areas of N Africa and SW Asia. (Family Dipodidae.)

The common N African jerboa (*Jaculus orientalis*) is a typical species. Its body is about 15 cm/6 in long and the tail is 25 cm/10 in long with a tuft at the tip. At speed it moves in a series of long jumps with its forefeet held close to its body.

Jerusalem artichoke a variety of ◊artichoke.

JET (abbreviation for ***Joint European Torus***) research facility at Culham, near Abingdon, Oxfordshire, UK, that conducts experiments on nuclear fusion. It is the focus of the European effort to produce a safe and environmentally sound fusion-power reactor. On 9 November 1991 the JET ◊tokamak, operating with a mixture of deuterium and iritium, produced a 1.7 megawatt pulse of power

JET WORLD WIDE WEB PAGE

http://www.jet.uk/

Web site of the Joint European Torus (JET) at Culham, UK, which includes an introduction to some aspects of nuclear fusion and an online tour of the project.

in an experiment that lasted two seconds. In 1997 isotopes of deuterium and tritium were fused to produce a record 21 megajoule of nuclear fusion power. JET has tested the first large-scale plant of the type needed to process and supply tritium in a future fusion power station.

jet in astronomy, a narrow luminous feature seen protruding from a star or galaxy, and representing a rapid outflow of material. See ◊active galaxy.

jet hard, black variety of lignite, a type of coal. It is cut and polished for use in jewellery and ornaments. Articles made of jet have been found in Bronze Age tombs.

jetfoil advanced type of ◊hydrofoil boat built by Boeing, propelled by water jets. It features horizontal, fully submerged hydrofoils fore and aft and has a sophisticated computerized control system to maintain its stability in all waters

Jetfoils have been in service worldwide since 1975. A jetfoil service operates across the English Channel between Dover and Ostend, Belgium, with a passage time of about 1.5 hours. Cruising speed of the jetfoil is about 80 kph/50 mph.

jet lag the effect of a sudden switch of time zones in air travel, resulting in tiredness and feeling 'out of step' with day and night. In 1989 it was suggested that use of the hormone melatonin helped to lessen the effect of jet lag by resetting the body clock. See also ◊circadian rhythm.

jet propulsion method of propulsion in which an object is propelled in one direction by a jet, or stream of gases, moving in the other. This follows from Isaac ◊Newton's third law of motion: 'To every action, there is an equal and opposite reaction.' The most widespread application of the jet principle is in the jet engine, the most common kind of aircraft engine

Jet Propulsion Laboratory NASA installation at Pasadena, California, operated by the California Institute of Technology. It is the command centre for NASA's deep-space probes such as the ◊Voyager, Magellan, and Galileo missions, with which it communicates via the Deep Space Network of radio telescopes at Goldstone, California; Madrid, Spain; and Canberra, Australia.

jet stream narrow band of very fast wind (velocities of over 150 kph/95 mph) found at altitudes of 10–16 km/6–10 mi in the upper troposphere or lower stratosphere. Jet streams usually occur about the latitudes of the Westerlies (35°–60°).

The jet stream may be used by high flying aircraft to speed their journeys. Their discovery of the existence of the jet stream allowed the Japanese to send gas-filled balloons carrying bombs to the northwestern US during World War II.

jigger or ***sandflea*** flea found in tropical and subtropical countries. The males of the species are free-living and measure about 1 mm/0.03 in in length. The females, which are slightly bigger, are parasites of humans and other animals.

Jiggers burrow into the skin, particularly the soft skin between the toes, and soon become enveloped within a skin fold. As they feed they become enlarged and distended and their legs degenerate; eventually a stage is reached when only the last two segments at the abdomen protrude from the skin fold. The eggs are then voided and fall on the soil or sand.

Apart from their presence within the host skin, which causes acute pain and the formation of ulcers, the fleas may be responsible for secondary infections. Preventive measures include ridding the host of the parasite, and destroying breeding places.

classification The jigger *Tunga penetrans* is in order Siphonaptera, class Insecta, phylum Arthropoda.

Jobs, Steven Paul (1955–)

US computer entrepreneur. He cofounded Apple Computer Inc with Steve Wozniak 1976, and founded NeXT Technology Inc in 1985. In 1986 he bought Pixar Animation Studios, the computer animation studio spun off from George Lucas's LucasFilm.

Jobs has been involved with the creation of three different types of computer: the Apple II personal computer 1977, the Apple Macintosh 1984 – marketed as 'the computer for the rest of us' – and the NeXT workstation 1988.

The NeXT was technically the most sophisticated and powerful design, but it was a commercial disaster, and in 1993 NeXT abandoned hardware manufacturing to concentrate on its highly-regarded UNIX-based object-oriented operating system, NextStep. Apple Computer bought NeXT at the end of 1996 to obtain NextStep, and Jobs returned to Apple in an advisory capacity. However, he soon took over as acting chief executive officer of the struggling firm 1997.

Jodrell Bank site in Cheshire, England, of the Nuffield Radio Astronomy Laboratories of the University of Manchester. Its largest instrument is the 76 m/250 ft radio dish (the Lovell Telescope), completed 1957 and modified 1970. A 38 x 25 m/125 x 82 ft elliptical radio dish was introduced 1964, capable of working at shorter wave lengths

These radio telescopes are used in conjunction with six smaller dishes up to 230 km/143 mi apart in an array called MERLIN (**m**ulti-**e**lement **r**adio-**l**inked **in**terferometer **n**etwork) to produce detailed maps of radio sources.

JODRELL BANK HOME PAGE

http://www.jb.man.ac.uk/index.html

Comprehensive Web site describing the Nuffield Radio Astronomy Laboratories at Jodrell Bank, Cheshire, UK. The radar telescope has been used for astronomy for many years, this Web site describes some of the important discoveries it has led to and also hints at future developments.

John Dory marine bony fish also called a ◊dory.

Johnson Space Center NASA installation at Houston, Texas, home of mission control for crewed space missions. It is the main centre for the selection and training of astronauts.

joint in any animal with a skeleton, a point of movement or articulation. In vertebrates, it is the point where two bones meet. Some joints allow no motion (the sutures of the skull), others allow a very small motion (the sacroiliac joints in the lower back), but most allow a relatively free motion. Of these, some allow a gliding motion (one vertebra of the spine on another), some have a hinge action (elbow and knee), and others allow motion in all directions (hip and shoulder joints) by means of a ball-and-socket arrangement. The ends of the bones at a moving joint are covered with cartilage for greater elasticity and smoothness, and enclosed in an envelope (capsule) of tough white fibrous tissue lined with a membrane which secretes a lubricating and cushioning ◊synovial fluid. The joint is further strengthened by ligaments. In invertebrates with an ◊exoskeleton, the joints are places where the exoskeleton is replaced by a more flexible outer covering, the arthrodial membrane, which allows the limb (or other body part) to bend at that point.

Joint European Torus experimental nuclear-fusion machine, known as ◊JET.

jojoba desert shrub *Simmondsia chinensis* of southwest North America and Mexico, that grows to a height of around 6 m/20 ft. Its seeds are used as a source of jojoba oil, used as a substitute for sperm whale oil, and as a fragrance. It is increasingly grown as a commercial crop.

jonquil species of small ◊daffodil, with yellow flowers. It is native to Spain and Portugal, and is cultivated in other countries. (*Narcissus jonquilla,* family Amaryllidaceae.)

Josephson junction device used in 'superchips' (large and complex integrated circuits) to speed the passage of signals by a phenomenon called 'electron tunnelling'. Although these superchips respond a thousand times faster than the ◊silicon chip, they have the disadvantage that the components of the Josephson junctions operate only at temperatures close to ◊absolute zero. They are named after English theoretical physicist Brian Josephson.

joule SI unit (symbol J) of work and energy, replacing the ◊calorie (one joule equals 4.2 calories)

Joule, James Prescott
(1818–1889)

English physicist. His work on the relations between electrical, mechanical, and chemical effects led to the discovery of the first law of thermodynamics.

He determined the mechanical equivalent of heat (Joule's equivalent) in 1843, and the SI unit of energy, the joule, is named after him. He also discovered Joule's law, which defines the relation between heat and electricity; and with Irish physicist Lord Kelvin in 1852 the Joule–Kelvin (or Joule–Thomson) effect.

Mary Evans Picture Library

Joule–Kelvin effect or *Joule–Thomson effect* in physics, the fall in temperature of a gas as it expands adiabatically (without loss or gain of heat to the system) through a narrow jet. It can be felt when, for example, compressed air escapes through the valve of an inflated bicycle tyre. It is the basic principle of most refrigerators.

joystick in computing, an input device that signals to a computer the direction and extent of displacement of a hand-held lever. It is similar to the joystick used to control the flight of an aircraft.

Joysticks are sometimes used to control the movement of a cursor (marker) across a display screen, but are much more frequently used to provide fast and direct input for moving the characters and symbols that feature in computer games. Unlike a ◊mouse, which can move a pointer in any direction, simple games joysticks are often capable only of moving an object in one of eight different directions.

JPEG (abbreviation for ***Joint Photographic Experts Group***) used to describe a compression standard set up by that group and now widely accepted for the storage and transmission of colour images. The JPEG compression standard reduces the size of image files considerably.

Jughead (acronymn for ***Jonzy's Universal Gopher Hierarchy Excavation and Display***) in computing, a ◊search engine enabling users of the Internet server ◊Gopher to find keywords in ◊Gopherspace directories.

jugular vein one of two veins in the necks of vertebrates; they return blood from the head to the superior (or anterior) ◊vena cava and thence to the heart.

jujube any of a group of trees belonging to the buckthorn family, with berrylike fruits. The common jujube (*Z. jujuba*) of Asia, Africa, and Australia, cultivated in S Europe and California, has fruit the size of small plums, known as Chinese dates when preserved in syrup. See also ◊lotus. (Genus *Zizyphus,* family Thamnaceae.)

Julian date in astronomy, a measure of time used in astronomy in which days are numbered consecutively from noon ◊GMT on 1 January 4713 BC. It is useful where astronomers wish to compare observations made over long time intervals. The Julian date (JD) at noon on 1 January 2000 will be 2451545.0. The modified Julian date (MJD), defined as MJD = JD - 2400000.5, is more commonly used since the date starts at midnight GMT and the smaller numbers are more convenient.

jumbo jet popular name for a generation of huge wide-bodied airliners including the **Boeing 747**, which is 71 m/232 ft long, has a wingspan of 60 m/196 ft, a maximum takeoff weight of nearly 400 tonnes, and can carry more than 400 passengers.

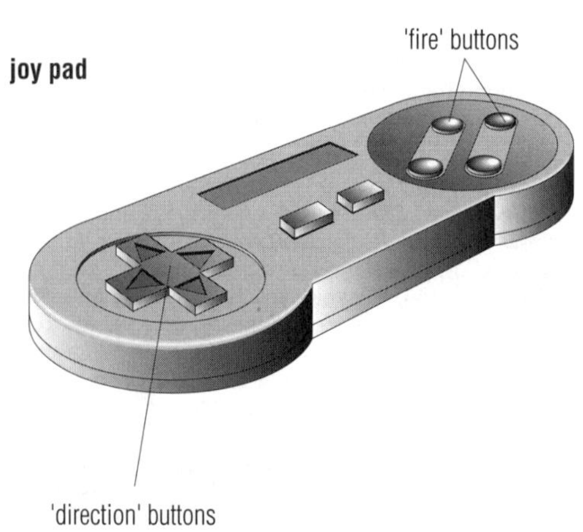

joystick *The directional and other controls on a conventional joystick may be translated to a joy pad, which enables all controls to be activated by buttons.*

jump in computing, a programming instruction that causes the computer to branch to a different part of a program, rather than execute the next instruction in the program sequence. Unconditional jumps are always executed; conditional jumps are only executed if a particular condition is satisfied.

jumper in computing, rectangular plug used to make connections on a circuit board. By pushing a jumper onto a particular set of pins on the board, or removing another, users can adjust the configuration of their computer's circuitry. Most home users, however, prefer to leave the insides of their machines with all the factory settings intact.

jumping hare or ***springhare*** either of two African species of long-eared rodents. The springhare (*P. capensis*) is about 40 cm/16 in long and resembles a small kangaroo with a bushy tail. It inhabits dry sandy country in E central Africa. (Genus *Pedetes*, family Pedetidae.)

Jung, Carl Gustav (1875–1961)

Swiss psychiatrist. He collaborated with Sigmund Freud from 1907 until their disagreement 1914 over the importance of sexuality in causing psychological problems. Jung studied myth, religion, and dream symbolism, saw the unconscious as a source of spiritual insight, and distinguished between introversion and extroversion.

Jung devised the word-association test in the early 1900s as a technique for penetrating a subject's unconscious mind. He also developed his theory concerning emotional, partly repressed ideas which he termed 'complexes'. In place of Freud's emphasis on infantile sexuality, Jung introduced the idea of a 'collective unconscious' which is made up of many archetypes or 'congenital conditions of intuition'.

Mary Evans Picture Library

jungle popular name for rainforest.

juniper any of a group of aromatic evergreen trees or shrubs of the cypress family, found throughout temperate regions. Its berries are used to flavour gin. Some junipers are mistakenly called cedars. (Genus *Juniperus*, family Cupressaceae.)

junk DNA another name for intron, a region of DNA that contains no genetic information.

Jupiter the fifth planet from the Sun, and the largest in the Solar System, with a mass equal to 70% of all the other planets combined, 318 times that of Earth's. It is largely composed of hydrogen and helium, liquefied by pressure in its interior, and probably with a rocky core larger than Earth. Its main feature is the Great Red Spot, a cloud of rising gases, 14,000 km/8,500 mi wide and 30,000 km/20,000 mi long, revolving anticlockwise.
mean distance from the Sun 778 million km/484 million mi
equatorial diameter 142,800 km/88,700 mi
rotation period 9 hr 51 min
year (complete orbit) 11.86 Earth years

JUPITER

http://www.hawastsoc.org/solar/eng/jupiter.htm

Full details of the planet and its moons including a chronology of exploration, various views of the planet and its moons, and links to other planets.

atmosphere consists of clouds of white ammonia crystals, drawn out into belts by the planet's high speed of rotation (the fastest of any planet). Darker orange and brown clouds at lower levels may contain sulphur, as well as simple organic compounds. Further down still, temperatures are warm, a result of heat left over from Jupiter's formation, and it is this heat that drives the turbulent weather patterns of the planet.
surface although largely composed of hydrogen and helium, Jupiter probably has a rocky core larger than Earth.

In 1995, the *Galileo* probe revealed Jupiter's atmosphere to consist of 0.2% water, less than previously estimated.
satellites Jupiter has 16 moons. The four largest moons, Io, Europa (which is the size of our Moon), Ganymede, and Callisto, are the **Galilean satellites**, discovered 1610 by Galileo (Ganymede, which is about the size of Mercury, is the largest moon in the Solar System). Three small moons were discovered 1979 by the Voyager space probes, as was a faint ring of dust around Jupiter's equator 55,000 km/34,000 mi above the cloud tops.

The Great Red Spot was first observed in 1664. Its top is higher than the surrounding clouds; its colour is thought to be due to red phosphorus. Jupiter's strong magnetic field gives rise to a large surrounding magnetic 'shell', or magnetosphere, from which bursts of radio waves are detected. The Southern Equatorial Belt in which the Great Red Spot occurs is subject to unexplained fluctuation. In 1989 it sustained a dramatic and sudden fading.

Comet Shoemaker-Levy 9 crashed into Jupiter July 1994. Impact zones were visible but are not likely to remain.

Jurassic period of geological time 208–146 million years ago; the middle period of the Mesozoic era. Climates worldwide were equable, creating forests of conifers and ferns; dinosaurs were abundant, birds evolved, and limestones and iron ores were deposited.

The name comes from the Jura Mountains in France and Switzerland, where the rocks formed during this period were first studied.

Till now man has been up against Nature, from now on he will be up against his own nature.

DENNIS GABOR Hungarian-born British physicist.
Inventing the Future

justification in printing and word processing, the arrangement of text so that it is aligned with either the left or right margin, or both.

jute fibre obtained from two plants of the linden family: *C. capsularis* and *C. olitorius*. Jute is used for sacks and sacking, upholstery, webbing (woven strips used to support upholstery), string, and stage canvas. (Genus *Corchorus*, family Tiliaceae.)

In the production of bulk packaging and tufted carpet backing, jute is now often replaced by synthetic polypropylene. The world's largest producer of jute is Bangladesh.

k symbol for **kilo-**, as in kg (kilogram) and km (kilometre).

K abbreviation for thousand, as in a salary of £30K.

K symbol for **kelvin**, a scale of temperature.

KA9Q in computing, ◊TCP/IP protocol named after the call sign of Philip Karn, the radio ham who wrote it for ◊packet radio. The system proved also to be useable on telephone connections, and so was adapted to several other computer platforms. It formed the basis of connections to ◊Demon Internet for many years.

Kagoshima Space Centre headquarters of Japan's Institute of Space and Astronautical Science (ISAS), situated in S Kyushu Island.

ISAS is responsible for the development of satellites for scientific research; other aspects of the space programme fall under the National Space Development Agency which runs the ◊Tanegashima Space Centre. Japan's first satellite was launched from Kagoshima 1970. By 1988 ISAS had launched 17 satellites and space probes.

kagu crested bird found in New Caledonia in the S Pacific. About 50 cm/1.6 ft long, it is virtually flightless and nests on the ground. The introduction of cats and dogs has endangered its survival. (Species *Rhynochetos jubatus*, order Gruiformes.)

kakapo nocturnal flightless parrot that lives in burrows in New Zealand. It is green, yellow, and brown with a disc of brown feathers round its eyes, like an owl. It weighs up to 3.5 kg/7.5 lb. When in danger, its main defence is to remain perfectly still. Because of the introduction of predators such as dogs, cats, rats, and ferrets, it is in danger of extinction; in 1997 there were only about 40 birds left. (Species *Strigops habroptilus*, order Psittaciformes.)

kale type of ◊cabbage.

kaleidoscope optical toy invented by the British physicist David Brewster in 1816. It usually consists of a pair of long mirrors at an angle to each other, and arranged inside a triangular tube containing pieces of coloured glass, paper, or plastic. An axially symmetrical (hexagonal) pattern is seen by looking along the tube, which can be varied infinitely by rotating or shaking the tube.

kangaroo any of a group of marsupials (mammals that carry their young in pouches) found in Australia, Tasmania, and New Guinea. Kangaroos are plant-eaters and most live in groups. They are adapted to hopping, the vast majority of species having very large, powerful back legs and feet compared with the small forelimbs. The larger types can jump 9 m/30 ft in a single bound. Most are nocturnal. Species vary from small rat kangaroos, only 30 cm/1 ft long, through the medium-sized wallabies, to the large red and great grey kangaroos, which are the largest living marsupials. These may be 1.8 m/5.9 ft long with 1.1 m/3.5 ft tails. (Family Macropodidae.)

In New Guinea and N Queensland, tree kangaroos (genus *Dendrolagus*) occur. These have comparatively short hind limbs. The great grey kangaroo (*Macropus giganteus*) produces a single young ('joey') about 2 cm/1 in long after a very short gestation, usually in early summer. At birth the young kangaroo is too young even to suckle. It remains in its mother's pouch, attached to a nipple which squirts milk into its mouth at intervals. It stays in the pouch, with excursions as it matures, for about 280 days.

A new species of kangaroo was discovered 1994 in New Guinea. Local people know it as 'bondegezou'. It weighs 15 kg/33 lb and is 1.2 m/3.9 ft in height. As it shows traits of both arboreal and ground-dwelling species, it may be a 'missing link'.

KANGAROO

A fully grown kangaroo can jump 13 m/14 yd. Mathematical calculations have shown that at speeds of over 29 kmph/18 mph, bouncing is a more energy-efficient way of moving for a kangaroo than running would be.

kangaroo paw bulbous plant *Anigozanthos manglesii*, family Hameodoraceae, with a row of small white flowers emerging from velvety green tubes with red bases. It is the floral emblem of Western Australia.

kaolin group of clay minerals, such as ◊kaolinite, $Al_2Si_2O_5(OH)_4$, derived from the alteration of aluminium silicate minerals, such as ◊feldspars and ◊mica. It is used in medicine to treat digestive upsets, and in poultices.

Kaolinite is economically important in the ceramic and paper industries. It is mined in the UK, the USA, France, and the Czech Republic.

kaolinite white or greyish ◊clay mineral, hydrated aluminium silicate, $Al_2Si_2O_5(OH)_4$, formed mainly by the decomposition of feldspar in granite. China clay (kaolin) is derived from it. It is mined in France, the UK, Germany, China, and the USA.

kapok silky hairs that surround the seeds of certain trees, particularly the **kapok tree** (*Bombax ceiba*) of India and Malaysia and the **silk-cotton tree** (*Ceiba pentandra*) of tropical America. Kapok is used for stuffing cushions and mattresses and for sound insulation; oil obtained from the seeds is used in food and soap.

karabash breed of large guard dog developed by the shepherds of Anatolia to protect their flocks. It is strongly built, stands up to 75 cm/29 in, and has a short grey, beige, or brindled coat with black on its muzzle and ears.

Karelian bear dog breed of medium-sized dog, used to protect Russian settlements from bears. About 60 cm/2 in high, the dog has a black or reddish-brown coat with white markings.

It was not exported until 1989, when some were sent to Yellowstone Park, USA, to keep bears away from tourists.

karri giant ◊eucalyptus tree *Eucalyptus diversifolia*, found in the extreme SW of Australia. It may reach over 120 m/400 ft. Its exceptionally strong timber is used for girders.

karst landscape characterized by remarkable surface and underground forms, created as a result of the action of water on permeable limestone. The feature takes its name from the Karst region on the Adriatic coast in Slovenia and Croatia, but the name is applied to landscapes throughout the world, the most dramatic of which is found near the city of Guilin in the Guangxi province of China.

Limestone is soluble in the weak acid of rainwater. Erosion takes place most swiftly along cracks and joints in the limestone and these open up into gullies called grikes. The rounded blocks left upstanding between them are called clints.

karyotype in biology, the set of ◊chromosomes characteristic of a given species. It is described as the number, shape, and size of the chromosomes in a single cell of an organism. In humans for example, the karyotype consists of 46 chromosomes, in mice 40, crayfish 200, and in fruit flies 8.

The diagrammatic representation of a complete chromosome set is called a **karyogram**.

katydid or ***bush cricket*** or ***longhorn grasshopper*** one of over 4,000 insect species, most of which are tropical, related to grasshoppers.

Members of this family have very long antennae and they tend to be wingless. The tympanal organs ('ears') are on the forelegs. They may be either plant-eating or carnivorous.

Some species are winged and the left forewing generally overlaps the right one; stridulation is produced by rubbing the forewings together. The winged species are usually green and are found among herbage, bushes, and trees where they often simulate the colour and shape of a leaf. Wingless forms inhabit places at ground-level, such as in the soil or under stones.

classification Katydids are in the family Tettigoniidae in order Orthoptera, class Insecta, phylum Arthropoda.

kauri pine New Zealand coniferous tree (see ◊conifer). Its fossilized gum deposits are valued in varnishes; the wood is used for carving and handicrafts. (*Agathis australis*, family Araucariaceae.)

kava narcotic, intoxicating beverage prepared from the roots or leaves of a variety of pepper plant, *Piper methysticum*, found in the South Pacific islands.

kayser unit of wave number (number of waves in a unit length), used in spectroscopy. It is expressed as waves per centimetre, and is the reciprocal of the wavelength. A wavelength of 0.1 cm has a wave number of 10 kaysers.

kcal symbol for **kilocalorie** (see ◊calorie).

kea hawklike greenish parrot found in New Zealand. It eats insects, fruits, and discarded sheep offal. The Maori name imitates its cry. (Species *Nestor notabilis*, family Psittacidae, order Psittaciformes.)

Keck Telescope world's largest optical telescope, situated on Mauna Kea, Hawaii. It has a primary mirror 10 m/33 ft in diameter, unique in that it consists of 36 hexagonal sections, each controlled and adjusted by a computer to generate single images of the objects observed. It received its first images in 1990.

An identical telescope next to it, named Keck II, became operational in 1996. It weighs 300 tonnes and has a 10-m/33-ft mirror comprised of 36 hexagons. Both telescopes are jointly owned by the California Institute of Technology and the University of California.

keeshond or ***Dutch barge dog*** sturdily built dog with erect ears and curled tail. It has a long grey top-coat, forming a mane around the neck, and a short, very thick undercoat, with darker 'spectacles' around the eyes. The ideal height is 46 cm/18 in for dogs; 43 cm/17 in for bitches.

keloid in medicine, overgrowth of fibrous tissue, usually produced at the site of a scar. Surgical removal is often unsuccessful, because the keloid returns.

kelp collective name for a group of large brown seaweeds. Kelp is also a term for the powdery ash of burned seaweeds, a source of iodine. (Typical families Fucaceae and Laminariaceae.)

The brown kelp (*Macrocystis pyrifera*), abundant in Antarctic and sub-Antarctic waters, is one of the fastest-growing living things, reaching lengths of up to 100 m/320 ft. It is farmed for the ◊alginate industry, its rapid surface growth allowing cropping several times a year, but it is an unwanted pest in N Atlantic waters.

kelpie breed of small herding dog also known as the Australian sheepdog. Bred from imported collie stock crossed with ◊dingo, it is much valued as a working dog. It has a coarse black or dark-coloured coat and is about 50 cm/20 in tall.

Do not imagine that mathematics is hard and crabbed, and repulsive to common sense. It is merely the etherealization of common sense.

WILLIAM KELVIN British physicist.
In S P Thomson *Life of Lord Kelvin* 1910

kelvin scale temperature scale used by scientists. It begins at ◊absolute zero (–273.15°C) and increases by the same degree intervals as the Celsius scale; that is, 0°C is the same as 273.15 K and 100°C is 373.15 K.

Kelvin, William Thomson, 1st Baron Kelvin (1824–1907)

Irish physicist who introduced the **Kelvin scale**, the absolute scale of temperature. His work on the conservation of energy in 1851 led to the second law of thermodynamics. Knighted 1866, Baron 1892.

Kelvin's knowledge of electrical theory was largely responsible for the first successful transatlantic telegraph cable. In 1847 he concluded that electrical and magnetic fields are distributed in a manner analogous to the transfer of energy through an elastic solid. From 1849 to 1859, Kelvin also developed the work of English scientist Michael Faraday into a full theory of magnetism, arriving at an expression for the total energy of a system of magnets.

Mary Evans Picture Library

Kennedy Space Center ◊NASA launch site on Merritt Island, near Cape Canaveral, Florida, used for *Apollo* and space-shuttle launches. The first flight to land on the Moon (1969) and *Skylab*, the first orbiting laboratory (1973), were launched here.

Kennelly–Heaviside layer former term for the ◊E layer of the ionosphere.

KENNEDY SPACE CENTRE

http://www.ksc.nasa.gov/

NASA's well-presented guide to the history and current operations of the USA's gateway to the universe. There is an enormous quantity of textual and multimedia information of interest for the general reader and for those who are technically minded.

Kepler's laws in astronomy, three laws of planetary motion formulated in 1609 and 1619 by the German mathematician and astronomer Johannes Kepler: (1) the orbit of each planet is an ellipse with the Sun at one of the foci; (2) the radius vector of each planet sweeps out equal areas in equal times; (3) the squares of the periods of the planets are proportional to the cubes of their mean distances from the Sun.

Kepler derived the laws after exhaustive analysis of numerous observations of the planets, especially Mars, made by Tycho ◊Brahe without telescopic aid. Isaac ◊Newton later showed that Kepler's Laws were a consequence of the theory of universal gravitation.

keratin fibrous protein found in the ◊skin of vertebrates and also in hair, nails, claws, hooves, feathers, and the outer coating of horns.

If pressure is put on some parts of the skin, more keratin is produced, forming thick calluses that protect the layers of skin beneath.

Kerberos in computing, system of symmetric ◊key cryptography developed at the Massachussetts Institute of Technology.

kermit in computing, a ◊file-transfer protocol, originally developed at Columbia University and made available without charge. Kermit is available as part of most communications packages and

Kepler, Johannes
(1571–1630)

German mathematician and astronomer. He formulated what are now called **Kepler's laws** of planetary motion: (1) the orbit of each planet is an ellipse with the Sun at one of the foci; (2) the radius vector of each planet sweeps out equal areas in equal times; (3) the squares of the periods of the planets are proportional to the cubes of their mean distances from the Sun. Kepler's laws are the basis of our understanding of the Solar System, and such scientists as Isaac Newton built on his ideas.

Kepler was one of the first advocates of Sun-centred cosmology, as put forward by Copernicus. Unlike Copernicus and Galileo, Kepler rejected the Greek and medieval belief that orbits must be circular in order to maintain the fabric of the cosmos in a state of perfection.

early work Kepler also produced a calendar of predictions for the year 1595 which proved uncanny in its accuracy. In 1596, he published his *Prodromus dissertationum cosmographicarum seu mysterium cosmographicum* in which he demonstrated that the five Platonic solids (the only five regular polyhedrons) could be fitted alternately inside a series of spheres to form a 'nest'. The nest described quite accurately (within 5%) the distances of the planets from the Sun. Kepler regarded this discovery as a divine inspiration that revealed the secret of the Universe. Written in accordance with Copernican theories, it brought Kepler to the attention of all European astronomers.

In 1601, Kepler was bequeathed all of Tycho Brahe's data on planetary motion. He had already made a bet that, given Tycho's unfinished tables, he could find an accurate planetary orbit within a week. It was five years before Kepler obtained his first planetary orbit, that of Mars. His analysis of these data led to the discovery of his three laws. In 1604, his attention was diverted from the planets by his observation of the appearance of a new star, 'Kepler's nova'. Kepler had observed the first supernova visible since the one discovered by Brahe in 1572.

Mary Evans Picture Library

Kepler's laws Kepler's first two laws of planetary motion were published in *Astronomia Nova* 1609. The first law stated that planets travel in elliptical rather than circular, or epicyclic, orbits and that the Sun occupies one of the two foci of the ellipses. The second law established the Sun as the main force governing the orbits of the planets. It stated that the line joining the Sun and a planet traverses equal areas of space in equal periods of time, so that the planets move more quickly when they are nearer the Sun. He also suggested that the Sun itself rotates, a theory that was confirmed using Galileo's observations of sunspots, and he postulated that this established some sort of 'magnetic' interaction between the planets and the Sun, driving them in orbit. This idea, although incorrect, was an important precursor of Newton's gravitational theory.

Kepler's third law was published in *De Harmonices Mundi*. It described in precise mathematical language the link between the distances of the planets from the Sun and their velocities – specifically, that the orbital velocity of a planet is inversely proportional to its distance from the Sun.

Rudolphine Tables and other work Kepler finally completed and published the *Rudolphine Tables* 1627 based on Brahe's observations. These were the first modern astronomical tables, enabling astronomers to calculate the positions of the planets at any time in the past, present or future. The publication also included other vital information, such as a map of the world, a catalogue of stars, and the latest aid to computation, logarithms.

available on most operating systems, but it is now rarely used on the Internet.

kernel the inner, softer part of a ◊nut, or of a seed within a hard shell.

kerosene thin oil obtained from the distillation of petroleum; a highly refined form is used in jet aircraft fuel. Kerosene is a mixture of hydrocarbons of the ◊paraffin series.

Kerry blue terrier compact and sturdy dog, with a soft, full coat of bluish tone. It is 46 cm/18 in high and weighs 15–17 kg/33–37.4 lb. Its ears lie close to the head and it has a thin tail that is held erect.

kestrel or **windhover** small hawk that breeds in Europe, Asia, and Africa. About 30 cm/1 ft long, the male has a bluish grey head and tail and is light chestnut brown back with black spots on the back and pale with black spots underneath. The female is slightly larger and reddish brown above, with bars; she does not have the bluish grey head. The kestrel hunts mainly by hovering in midair while searching for prey. It feeds on small mammals, insects, frogs, and worms. (Species *Falco tinnunculus*, family Falconidae, order Falconiformes.)

It rarely builds its own nest, but uses those of other birds, such as crows and magpies, or scrapes a hole on a cliff-ledge. It is found all over Europe and Asia and most parts of Africa, and most birds migrate southwards in winter.

The **lesser kestrel** (*Falco naumanni*) is an inhabitant of southern Europe. The **American kestrel** or **sparrowhawk** (*F. sparverius*) is somewhat smaller, and occurs in most of North America. It is russet, grey, and tan in colour, with the male having more grey on its wings.

ketone member of the group of organic compounds containing the carbonyl group (C=O) bonded to two atoms of carbon (instead of one carbon and one hydrogen as in ◊aldehydes). Ketones are liquids or low-melting-point solids, slightly soluble in water.

An example is propanone (acetone, CH_3COCH_3), used as a solvent.

Kew Gardens popular name for the Royal Botanic Gardens, Kew, Surrey, England. They were founded 1759 by Augusta of Saxe-Coburg (1719–1772), the mother of King George III, as a small garden and passed to the nation by Queen Victoria 1840. By then they had expanded to almost their present size of 149 hectares/368 acres and since 1841 have been open daily to the public. They contain a collection of over 25,000 living plant species and many fine buildings. The gardens are also a centre for botanical research.

key in cryptography, the password needed to both encode and decipher a file. The key performs a sequence of operations on the original data. The recipient of the encoded file will need to apply another key in order to reverse all the operations in the correct order. Current encryption techniques such as ◊Pretty Good Privacy (PGP) make use of a ◊public key and a secret one.

keyboard in computing, an input device resembling a typewriter keyboard, used to enter instructions and data. There are many variations on the layout and labelling of keys. Extra numeric keys may be added, as may special-purpose function keys, whose effects can be defined by programs in the computer.

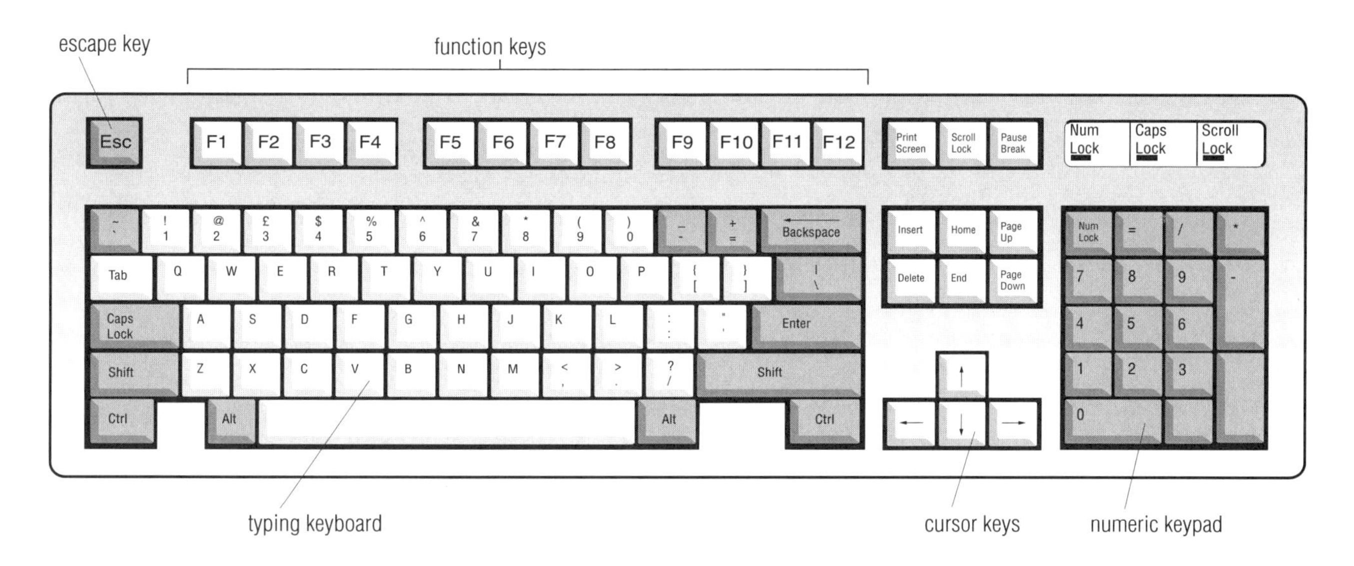

keyboard *A standard 102-key keyboard. As well as providing a QWERTY typing keyboard, the function keys (labelled F1–F12) may be assigned tasks specific to a particular system.*

key escrow in ◊public key cryptography, requirement that users store copies of their private keys with the government or other authorities for release to law enforcement officials upon production of the necessary legal documents.

Key escrow was first proposed in the USA, where it was built into the controversial ◊Clipper chip. In 1996, both the USA and the European Union were considering legislation requiring users of strong encryption to escrow their keys to protect law enforcement interests.

key field in computing, a selected field, or portion, of a record that is used to identify that record uniquely; in a file of records it is the field used as the basis for ◊sorting the file. For example, in a file containing details of a bank's customers, the customer account number would probably be used as the key field.

key frame in animation, a frame which was drawn by the user rather than generated by the computer. Animators feed a sequence of key frames into the computer, allowing the program to draw the intervening stages in a process known as **tweening**.

keyhole limpet primitive mollusc with a conical shell resembling that of the common limpet.

The shell is either cleft, as in *Fissurella,* or has a hole at the tip of the shell, as in *Diodora* (the rough keyhole limpet). A siphon is developed from the mantle and projects through this opening as the exhalant exit. Water is drawn over the gills from the forward end.

classification Keyhole limpets are of the superfamily Fissurellacea in subclass Prosobranchia, class Gastropoda, phylum Mollusca.

key-to-disc system or ***key-to-tape system*** in computing, a system that enables large amounts of data to be entered at a keyboard and transferred directly onto computer-readable discs or tapes.

Such systems are used in ◊batch processing, in which batches of data, prepared in advance, are processed by computer with little or no intervention from the user. The preparation of the data may be controlled by a minicomputer, freeing a larger, mainframe computer for the task of processing.

kg symbol for ◊kilogram.

kiang type of Asiatic wild ◊ass that lives in cold (below –40°C/–40°F in winter), high altitude (above 4,000 m/23,280 ft) deserts. Kiangs may be subspecies of the Asiatic wild ass *Equus hemionus,* or different species *E. kiang* There are three varieties: southern, western, and eastern. All are endangered and until recently the southern kiang was believed to be extinct.

The southern kiang, found between Sikkim in N India and the Tsangpo River in Tibet, has a shoulder height of 104–114 cm/41–45 in. Only about 100 remain. The western population, which ranges across Pakistan, India, and China, numbers around 1,500 in India; the eastern population is more numerous, with 10,000–30,000 in the protected Arjin Mountain Nature Reserves in China.

kidney in vertebrates, one of a pair of organs responsible for fluid regulation, excretion of waste products, and maintaining the ionic composition of the blood. The kidneys are situated on the rear wall of the abdomen. Each one consists of a number of long tubules; the outer parts filter the aqueous components of blood, and the inner parts selectively reabsorb vital salts, leaving waste products in the remaining fluid (urine), which is passed through the ureter to the bladder.

The action of the kidneys is vital, although if one is removed, the other enlarges to take over its function. A patient with two defective kidneys may continue near-normal life with the aid of a kidney machine or continuous ambulatory peritoneal ◊dialysis (CAPD); or a kidney transplant may be recommended.

kidney machine medical equipment used in ◊dialysis.

killer application a program so good or so compelling to certain potential users that they buy the computer that the program runs on for no other reason than to be able to use that program.

Killer applications are very rare. The most successful was VisiCalc, the first spreadsheet to run on a personal computer (the original Apple II microcomputer). VisiCalc succeeded as a killer application because it provided a unique tool for accountants to manipulate numbers easily without the need for programming skills. Another clear example is PageMaker, the first desktop publishing program, which was responsible for selling the Apple ◊Macintosh to the design and publishing community.

The ◊World Wide Web is the killer application for the Internet: by bringing visual excitement and ease of use to the Internet, it inspired people to buy new computers capable of supporting Web ◊browsers.

TO REMEMBER THE SEVEN LAYERS OF THE OSI REFERENCE MODEL FOR COMPUTER NETWORKING:

ACTIVE PENGUINS SEEK THE NEAREST DEEP POOL

APPLICATION, PRESENTATION, SESSION, TRANSPORT, NETWORK, DATA-LINK, PHYSICAL

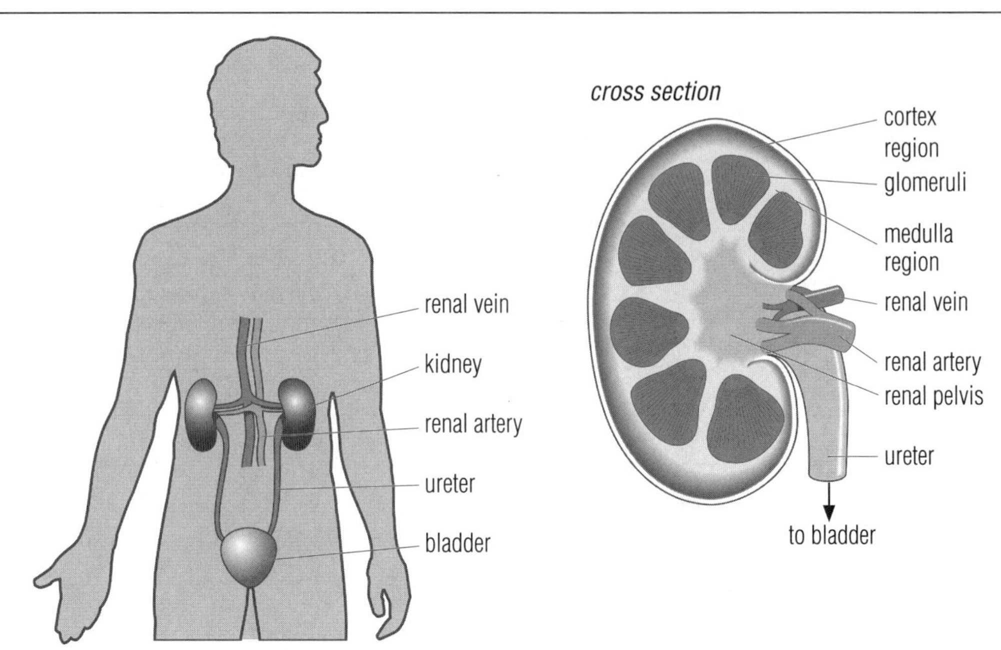

kidney Blood enters the kidney through the renal artery. The blood is filtered through the glomeruli to extract the nitrogenous waste products and excess water that make up urine. The urine flows through the ureter to the bladder; the cleaned blood then leaves the kidney via the renal vein.

KILLER WHALES

http://www.seaworld.org/killer_whale/killerwhales.html

Pictures and videos of this sea creature. There is also a lot of detailed biological information concerning, for example, the habitat, behaviour, and communication of killer whales. There are also plenty of activities and ideas to keep younger children entertained.

killer whale or ***orca*** toothed whale belonging to the dolphin family, found in all seas of the world. It is black on top, white below, and grows up to 9 m/30 ft long. It is the only whale that has been observed to prey on other whales, as well as on seals and seabirds. (Species *Orcinus orca,* family Delphinidae.)

killfile in computing, file specifying material that you do not wish to see when accessing a ◊newsgroup. By entering names, subjects or phrases into a killfile, users can make ◊USENET a more pleasant experience, filtering out tedious threads, offensive subject headings, ◊spamming or contributions from other irritating subscribers.

kiln high-temperature furnace used commercially for drying timber, roasting metal ores, or for making cement, bricks, and pottery. Oil- or gas-fired kilns are used to bake ceramics at up to 1,760°C/3,200°F; electric kilns do not generally reach such high temperatures.

kilo- prefix denoting multiplication by 1,000, as in kilohertz, a unit of frequency equal to 1,000 hertz.

kilobyte (K or KB) in computing, a unit of memory equal to 1,024 ◊bytes. It is sometimes used, less precisely, to mean 1,000 bytes.

In the metric system, the prefix 'kilo-' denotes multiplication by 1,000 (as in kilometre, a unit equal to 1,000 metres). However, computer memory size is based on the ◊binary number system, and the most convenient binary equivalent of 1,000 is 2^{10}, or 1,024.

kilogram SI unit (symbol kg) of mass equal to 1,000 grams (2.24 lb). It is defined as a mass equal to that of the international prototype, a platinum-iridium cylinder held at the International Bureau of Weights and Measures in Sèvres, France.

kilometre unit of length (symbol km) equal to 1,000 metres, equivalent to 3,280.89 ft or 0.6214 (about $\frac{5}{8}$) of a mile.

kilowatt unit (symbol kW) of power equal to 1,000 watts or about 1.34 horsepower.

kilowatt-hour commercial unit of electrical energy (symbol kWh), defined as the work done by a power of 1,000 watts in one hour and equal to 3.6 megajoules. It is used to calculate the cost of electrical energy taken from the domestic supply.

Kimball tag stock-control device commonly used in clothes shops, consisting of a small ◊punched card attached to each item offered for sale. The tag carries information about the item (such as its serial number, price, colour, and size), both in the form of printed details (which can be read by the customer) and as a pattern of small holes. When the item is sold, the tag (or a part of the tag) is removed and kept as a computer-readable record of sales.

kimberlite an igneous rock that is ultramafic (containing very little silica); a type of alkaline ◊peridotite with a porphyritic texture (larger crystals in a fine-grained matrix), containing mica in addition to olivine and other minerals. Kimberlite represents the world's principal source of diamonds.

Kimberlite is found in carrot-shaped pipelike ◊intrusions called **diatremes**, where mobile material from very deep in the Earth's crust has forced itself upwards, expanding in its ascent. The material, brought upwards from near the boundary between crust and mantle, often altered and fragmented, includes diamonds. Diatremes are found principally near Kimberley, South Africa, from which the name of the rock is derived, and in the Yakut area of Siberia, Russia.

kinesis (plural ***kineses***) in biology, a nondirectional movement in response to a stimulus; for example, woodlice move faster in drier surroundings. **Taxis** is a similar pattern of behaviour, but there the response is directional.

kinetic energy the energy of a body resulting from motion. It is contrasted with ◊potential energy.

kinetics the branch of chemistry that investigates the rates of chemical reactions.

kinetics branch of ◊dynamics dealing with the action of forces producing or changing the motion of a body; **kinematics** deals with motion without reference to force or mass.

kinetic theory theory describing the physical properties of matter in terms of the behaviour – principally movement – of its component atoms or molecules. The temperature of a substance is dependent on the velocity of movement of its constituent particles, increased temperature being accompanied by increased movement. A gas consists of rapidly moving atoms or molecules and, according to kinetic theory, it is their continual impact on the walls of the containing vessel that accounts for the pressure of the gas. The slowing of molecular motion as temperature falls, according to kinetic theory, accounts for the physical properties of liquids and solids, culminating in the concept of no molecular motion at ◊absolute zero (0K/–273°C).

By making various assumptions about the nature of gas molecules, it is possible to derive from the kinetic theory the various gas laws (such as ◊Avogadro's hypothesis, ◊Boyle's law, and ◊Charles's law).

King Charles spaniel breed of toy dog favoured by King Charles I of England. It has a characteristic domed head and upturned muzzle. It normally weighs less than 6 kg/13 lb. Different varieties are distinguished by the colours of the coat: King Charles (black with red markings); Prince Charles (white and black with red markings); Ruby (chestnut); Blenheim (white with chestnut markings).

king crab or ***horseshoe crab*** marine ◊arthropod found on the Atlantic coast of North America, and the coasts of Asia. The upper side of the body is entirely covered with a dark, rounded shell, and it has a long spinelike tail. It is up to 60 cm/2 ft long. It is unable to swim, and lays its eggs in the sand at the high-water mark. (Class Arachnida, subclass Xiphosura.)

There were approximately 500,000 *Paralithodes camtschatica* in the Barents Sea at the end of 1996. They were introduced by Russia as a potential food source. Norway claims they are unbalancing the marine ecosystem.

kingcup another name for ◊marsh marigold.

kingdom the primary division in biological ◊classification. At one time, only two kingdoms were recognized: animals and plants. Today most biologists prefer a five-kingdom system, even though it still involves grouping together organisms that are probably unrelated. One widely accepted scheme is as follows: **Kingdom Animalia** (all multicellular animals); **Kingdom Plantae** (all plants, including seaweeds and other algae, except blue-green); **Kingdom Fungi** (all fungi, including the unicellular yeasts, but not slime moulds); **Kingdom Protista** or **Protoctista** (protozoa, diatoms, dinoflagellates, slime moulds, and various other lower organisms with eukaryotic cells); and **Kingdom Monera** (all prokaryotes – the bacteria and cyanobacteria, or ◊blue-green algae). The first four of these kingdoms make up the eukaryotes.

When only two kingdoms were recognized, any organism with a rigid cell wall was a plant, and so bacteria and fungi were considered plants, despite their many differences. Other organisms, such as the photosynthetic flagellates (euglenoids), were claimed by both kingdoms. The unsatisfactory nature of the two-kingdom system became evident during the 19th century, and the biologist Ernst Haeckel was among the first to try to reform it. High-power microscopes have revealed more about the structure of cells; it has become clear that there is a fundamental difference between cells without a nucleus (◊prokaryotes) and those with a nucleus (◊eukaryotes). However, these differences are larger than those between animals and higher plants, and are unsuitable for use as kingdoms. At present there is no agreement on how many kingdoms there are in the natural world.

Although the five-kingdom system is widely favoured, some schemes have as many as 20.

The idea of man as a dominant animal of the earth whose whole behaviour tends to be dominated by his own desire for dominance gripped me. It seemed to explain almost everything.

MACFARLANE BURNET Australian physician.
Dominant Manual 1970

kingfisher any of a group of heavy-billed birds found near streams, ponds, and coastal areas around the world. The head is exceptionally large, and the long, angular bill is keeled; the tail and wings are relatively short, and the legs very short, with short toes. Kingfishers plunge-dive for fish and aquatic insects. The nest is usually a burrow in a riverbank. (Family Alcedinidae, order Coraciiformes.)

There are 88 species of kingfisher, the largest being the Australian ◊kookaburra. The Alcedinidae are sometimes divided into the subfamilies Daceloninae, Alcedininae, and Cerylinae.

kinkajou Central and South American carnivorous mammal belonging to the raccoon family. Yellowish-brown, with a rounded face and slim body, the kinkajou grows to 55 cm/1.8 ft with a 50 cm/1.6 ft tail, and has short legs with sharp claws. It spends its time in trees and has a prehensile tail, which it uses as an extra limb when moving from branch to branch. It feeds largely on fruit. (Species *Potos flavus,* family Procyonidae.)

kin selection in biology, the idea that ◊altruism shown to genetic relatives can be worthwhile, because those relatives share some genes with the individual that is behaving altruistically, and may continue to reproduce. See◊ inclusive fitness.

Alarm-calling in response to predators is an example of a behaviour that may have evolved through kin selection: relatives that are warned of danger can escape and continue to breed, even if the alarm caller is caught.

kiosk in computing, any computer that has been set up to act as an information centre in a public place. Users navigate the display using keyboards or ◊touch screens, but are never allowed to access the computer's operating system. A kiosk in a museum might show an interactive multimedia display, or one in a library might give readers access to catalogues.

Kirchhoff, Gustav Robert
(1824–1887)

German physicist who with R W von Bunsen developed spectroscopic analysis in the 1850s and showed that all elements, heated to incandescence, have their individual spectra. In 1845 he derived the laws now known as Kirchhoff's laws that determine the value of the electric current and potential at any point in a network.

Mary Evans Picture Library

Kirchhoff's laws two laws governing electric circuits devised by the German physicist Gustav Kirchhoff. **Kirchhoff's first law** states that the total current entering any junction in a circuit is the same as the total current leaving it. This is an expression of the conservation of electric charge. **Kirchhoff's second law** states that the sum of the potential drops across each resistance in any closed loop in a circuit is equal to the total electromotive force acting in that loop. The laws are equally applicable to DC and AC circuits.

Kirkwood gaps in astronomy, regions of the ◊asteroid belt, between ◊Mars, and ◊Jupiter, where there are relatively few asteroids.

The orbital periods of particles in the gaps correspond to simple fractions, especially $\frac{1}{3}$, $\frac{2}{3}$, $\frac{3}{7}$, and $\frac{1}{2}$, of the orbital period of Jupiter, indicating that they are caused by the gravitational influence of the larger planet. The gaps are named after Daniel Kirkwood, the 19th century US astronomer who first drew attention to them.

kiss of life (***artificial ventilation***) in first aid, another name for ◊artificial respiration.

kite any of a group of birds of prey found in all parts of the world. Kites have long, pointed wings and, usually, a forked tail. There are about 20 species. (Family Accipitridae, order Falconiformes.)

kite quadrilateral with two pairs of adjacent equal sides. The geometry of this figure follows from the fact that it has one axis of symmetry.

Kitt Peak National Observatory observatory in the Quinlan Mountains near Tucson, Arizona, USA, operated by AURA (Association of Universities for Research into Astronomy). Its main telescopes are the 4-m/158-in Mayall reflector, opened in 1973, and the McMath Solar Telescope, opened in 1962, the world's largest of its type.

Among numerous other telescopes on the site is a 2.3-m/90-in reflector owned by the Steward Observatory of the University of Arizona.

KITT PEAK NATIONAL OBSERVATORY

http://www.noao.edu/kpno/kpno.html

Comprehensive information on the range of research carried out at Kitt Peak. In addition to scientific information of interest mainly to professional astronomers, there are details of current weather conditions at Kitts Peak and information for visitors. Visible and infrared images from satellites in geostationary orbits can be accessed.

kiwi flightless bird found only in New Zealand. It has long hairlike brown plumage, minute wings and tail, and a very long beak with nostrils at the tip. It is nocturnal and insectivorous. It lays two white eggs, each weighing up to 450 g/15.75 oz. (Species *Apteryx australis*, family Apterygidae, order Apterygiformes.)

All kiwi species have declined since European settlement of New Zealand, and the little spotted kiwi is most at risk. It survives only on one small island reservation, which was stocked with birds from the mainland.

KIWI

The kiwi produces the largest egg, proportionally to size, of any bird. The female lays one or two eggs, each weighing a quarter of her body weight.

kiwi fruit or ***Chinese gooseberry*** fruit of a vinelike plant grown commercially on a large scale in New Zealand. Kiwi fruits are egg-sized, oval, and similar in flavour to gooseberries, though much sweeter, with a fuzzy brown skin. (*Actinidithia chinensis*, family Actinidiaceae.)

kleptomania *Greek kleptēs 'thief'* behavioural disorder characterized by an overpowering desire to possess articles for which one has no need. In kleptomania, as opposed to ordinary theft, there is no obvious need or use for what is stolen and sometimes the sufferer has no memory of the theft.

kleptoparasitism habitual stealing of food from another organism. Skuas kleptoparasitize other seabirds, forcing them to relinquish their catches midflight. The Spanish slug *Deroceras hilbrandi* takes prey from the insect-eating plant *Pinguicula vallisneriifolia* whilst leaving the edible plant unharmed. Many small spiders are kleptoparasites living on the webs of bigger spiders.

km symbol for ◊kilometre.

knapweed any of several weedy plants belonging to the daisy family. In the common knapweed (*C. nigra*), also known as a **hardhead**, the hard, dark buds break open at the top into pale purple composite flowers. It is native to Europe and has been introduced to North America. (Genus *Centaurea*, family Compositae.)

knifefish any of a group of fishes in which the body is deep at the front and drawn to a narrow or pointed tail at the rear, the main fin being the well-developed long ventral (stomach) fin that completes the knifelike shape. The ventral fin is rippled for forward or backward movement. Knifefishes produce electrical fields, which they use for navigation. (Genus *Gymnotus* and other allied genera, family Gymnotidae.)

knocking in a spark-ignition petrol engine, a phenomenon that occurs when unburned fuel-air mixture explodes in the combustion chamber before being ignited by the spark. The resulting shock waves produce a metallic knocking sound. Loss of power occurs, which can be prevented by reducing the compression ratio, redesigning the geometry of the combustion chamber, or increasing the octane number of the petrol (usually by the use of tetraethyl lead anti-knock additives, or increasingly by MTBE – methyl tertiary butyl ether in unleaded petrol).

Knossos chief city of ◊Minoan Crete, near present-day Irákleion, 6 km/4 mi SE of Candia. The archaeological site, excavated by Arthur Evans 1899–1935, dates from about 2000–1400 BC, and includes the palace throne room, the remains of frescoes, and construction on more than one level.

Excavation of the palace of the legendary King Minos showed that the story of Theseus' encounter with the Minotaur in a labyrinth was possibly derived from the ritual 'bull-leaping' by young people depicted in the palace frescoes and from the maze-like layout of the palace.

knot wading bird belonging to the sandpiper family. It is about 25 cm/10 in long, with a short bill, neck, and legs. In the winter, it is grey above and white below, but in the breeding season, it is brick-red on the head and chest and black on the wings and back. It feeds on insects and molluscs. (Species *Calidris canutus*, family Scolopacidae, order Charadriiformes.)

Breeding in North American, European, and Asian arctic regions, knots travel widely in winter, to be found as far south as South Africa, Australasia, and southern parts of South America.

knot in navigation, unit by which a ship's speed is measured, equivalent to one ◊nautical mile per hour (one knot equals about 1.15 miles per hour). It is also sometimes used in aviation.

knotgrass annual plant belonging to the dock family. The bases of the small lance-shaped leaves enclose the slender stems, giving a superficial resemblance to grass. Small pinkish flowers are followed by seeds that are eaten by birds. Knotgrass grows worldwide except in the polar regions. (*Polygonum aviculare*, family Polygonaceae.)

knowbot in computing, a program that will search a system or a network, such as the Internet, seeking and retrieving information on behalf of a user and reporting back when it has found it. An example is the Knowbot Information Service, which can process users' queries by e-mail.

knowledge-based system (KBS) computer program that uses an encoding of human knowledge to help solve problems. It was discovered during research into ◊artificial intelligence that adding heuristics (rules of thumb) enabled programs to tackle problems that were otherwise difficult to solve by the usual techniques of computer science.

Chess-playing programs have been strengthened by including knowledge of what makes a good position, or of overall strategies, rather than relying solely on the computer's ability to calculate variations.

koala marsupial (mammal that carries its young in a pouch) found only in E Australia. It feeds almost entirely on eucalyptus shoots. It is about 60 cm/2 ft long, and resembles a bear (it is often incorrectly described as a 'koala bear'). The popularity of its greyish fur led to its almost complete extermination by hunters. Under protection since 1936, it rapidly increased in numbers, but recently numbers have fallen from 400,000 in 1985 to 40,000–80,000 in 1995. (Species *Phascolarctos cinereus*, family Phalangeridae.)

A three-year trial began in November 1996 in southern Australia, overpopulated by koalas, to bring their numbers under control by giving males vasectomies and females hormone implants preventing ovulation. In 1997, the programme of sterilization continued, and a relocation plan, following earlier trials, began.

KOALA

The fingerprints of koala bears are very similar to those of humans, with similar whorls, loop and arches. To avoid any confusion, Australian forensic scientists were warned, in 1996, to watch out for koalas at the scene of the crime.

Koch, (Heinrich Hermann) Robert (1843–1910)

German bacteriologist. Koch and his assistants devised the techniques for culturing bacteria outside the body, and formulated the rules for showing whether or not a bacterium is the cause of a disease. Nobel Prize for Physiology or Medicine 1905.

His techniques enabled him to identify the bacteria responsible for tuberculosis (1882), cholera (1883), and other diseases. He investigated anthrax bacteria in the 1870s and showed that they form spores which spread the infection.

Koch was a great teacher, and many of his pupils, such as Shibasaburo¯ Kitasato, Paul Ehrlich, and Emil von Behring, became outstanding scientists.

Mary Evans Picture Library

kohlrabi variety of kale, which is itself a variety of ◊cabbage; it is used for food and resembles a turnip. The leaves of kohlrabi shoot from a round swelling on the main stem. (*Brassica oleracea caulorapa* or *B. oleracea gongylodes,* family Cruciferae.)

kola alternative spelling of ◊cola, any of a group of tropical trees.

komondor breed of large herding dog originating in Hungary. It has a long, heavy, all-white coat with a rough 'corded' texture. Larger than the rather similar ◊puli, it stands about 80 cm/31.5 in tall.

Königsberg bridge problem long-standing puzzle that was solved by topology (the geometry of those properties of a figure which remain the same under distortion). In the city of Königsberg (now Kaliningrad in Russia), seven bridges connect the banks of the river Pregol'a and the islands in the river. For many years, people were challenged to cross each of the bridges in a single tour and return to their starting point. In 1736 Swiss mathematician Leonhard Euler converted the puzzle into a topological network, in which the islands and river banks were represented as nodes (junctions), and the connecting bridges as lines. By analysing this network he was able to show that it is not traversable – that is, it is impossible to cross each of the bridges once only and return to the point at which one started.

kookaburra or ***laughing jackass*** largest of the world's ◊kingfishers, found in Australia, with an extraordinary laughing call. It feeds on insects and other small creatures. The body and tail measure 45 cm/18 in, the head is greyish with a dark eye stripe, and the back and wings are flecked brown with grey underparts. It nests in shady forest regions, but will also frequent the vicinity of houses, and its cry is one of the most familiar sounds of the bush in E Australia. (Species *Dacelo novaeguineae,* family Alcedinidae, order Coraciiformes.)

Koonalda Cave cave in SW South Australia below the Nullarbor Plain. Anthropologists in the 1950s and 1960s discovered evidence of flint-quarrying and human markings that have been dated as 20,000 years old.

Korat breed of domestic shorthaired cat. It derives its name from the NE province of Korat in Thailand where it originated. It has a blue-grey coat with a silvery sheen with very thick fur.

It has a heart-shaped face with green eyes, and a lithe body.

Imported into the USA 1959, the breed was recognized there 1966, and in the UK in 1975.

kouprey wild ox native to the forests of N Cambodia. Only known to science since 1937, it is in great danger of extinction. Koupreys have cylindrical, widely separated horns and grow to 1.9 m/6 ft in height. (Species *Bos sauveli.*)

Kourou second-largest town of French Guiana, NW of Cayenne, site of the Guiana Space Centre of the European Space Agency; population (1996) 20,000 (20% of the total population of French Guiana).

Kow Swamp area in N Victoria, Australia, W of the town of Echuca, where the remains of about 40 humans have been found, buried in shallow graves. These bones have mostly been dated as 9,000–14,000 years old and were accompanied by human artefacts and objects such as shells. They are of a larger and more robust group of humans than those found from an earlier period at Lake Mungo and are similar to other finds from widely scattered sites in Australia such as Talgai in Queensland.

kph or ***km/h*** symbol for **kilometres per hour**.

krait highly venomous Indian snake related to the *cobra.*

Krebs, Hans Adolf (1900–1981)

German-born British biochemist. He discovered the citric acid cycle, also known as the ◊Krebs cycle, the final pathway by which food molecules are converted into energy in living tissues. For this work he shared the 1953 Nobel Prize for Physiology or Medicine. Knighted 1958.

Krebs first became interested in the process by which the body degrades amino acids. He discovered that nitrogen atoms are the first to be removed (deamination) and are then excreted as urea in the urine. He then investigated the processes involved in the production of urea from the removed nitrogen atoms, and by 1932 he had worked out the basic steps in the urea cycle.

Mary Evans Picture Library

TO REMEMBER THE KREBS CYCLE, ALSO KNOWN AS THE TCA (TRICARBOXYLIC ACID) CYCLE:

ACTORS IN KANSAS SHOULD SEE FOREIGN MOVIES, OF COURSE

ACONITATE, ISOCITRATE, α-KETOGLUTARATE, SUCCINYL-COA, SUCCINATE, FUMARATE, MALATE, OXALOACETATE, CITRATE

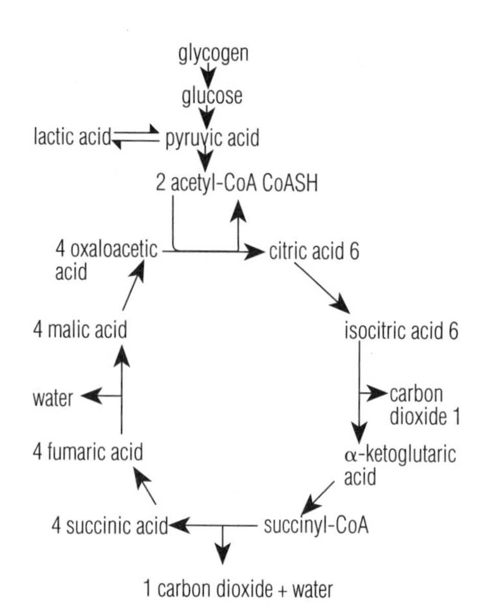

***Krebs cycle** The purpose of the Krebs (or citric acid) cycle is to complete the biochemical breakdown of food to produce energy-rich molecules, which the organism can use to fuel work. Acetyl coenzyme A (acetyl CoA) – produced by the breakdown of sugars, fatty acids, and some amino acids – reacts with oxaloacetic acid to produce citric acid, which is then converted in a series of enzyme-catalysed steps back to oxaloacetic acid. In the process, molecules of carbon dioxide and water are given off, and the precursors of the energy-rich molecules ATP are formed. (The numbers in the diagram indicate the number of carbon atoms in the principal compounds.)*

Krebs cycle or ***citric acid cycle*** or ***tricarboxylic acid cycle*** final part of the chain of biochemical reactions by which organisms break down food using oxygen to release energy (respiration). It takes place within structures called ◊mitochondria in the body's cells, and breaks down food molecules in a series of small steps, producing energy-rich molecules of ◊ATP.

krill any of several Antarctic ◊crustaceans, the most common species being *Euphausia superba*. Similar to a shrimp, it is up to 5 cm/2 in long, with two antennae, five pairs of legs, seven pairs of light organs along the body, and is coloured orange above and green beneath. It is the most abundant animal, numbering perhaps 600 trillion (million million). (Order Euphausiacea.)

Moving in enormous swarms, krill constitute the chief food of the baleen whales, and have been used to produce a protein concentrate for human consumption, and meal for animal feed.

KRILL

Approximately 600 trillion krill thrive in the Southern Ocean. Together they weigh more than the world's entire human population. The schools of shrimplike krill (moving at a rapid 20 cm/7.9 in per sec) form the base of many food chains; whales alone consume 150 million tonnes a year.

krypton ***Greek kryptos 'hidden'*** colourless, odourless, gaseous, nonmetallic element, symbol Kr, atomic number 36, relative atomic mass 83.80. It is grouped with the inert gases and was long believed not to enter into reactions, but it is now known to combine with fluorine under certain conditions; it remains inert to all other reagents. It is present in very small quantities in the air (about 114 parts per million). It is used chiefly in fluorescent lamps, lasers, and gas-filled electronic valves.

Krypton was discovered in 1898 in the residue from liquid air by British chemists William Ramsay and Morris Travers; the name refers to their difficulty in isolating it.

K-T boundary geologists' shorthand for the boundary between the rocks of the ◊Cretaceous and the ◊Tertiary periods 65 million years ago. It coincides with the end of the extinction of the dinosaurs and in many places is marked by a layer of clay or rock enriched in the element iridium. Extinction of the dinosaurs at the K-T boundary and deposition of the iridium layer are thought to be the result of either impact of a meteorite (or comet) that crashed into the Yucatán Peninsula (forming the **Chicxulub crater**) or the result of intense volcanism on the continent of India.

kudu either of two species of African antelope. The **greater kudu** (*T. strepsiceros*) is fawn-coloured with thin white vertical stripes, and stands 1.3 m/4.2 ft at the shoulder, with head and body 2.4 m/8 ft long. Males have long spiral horns. The greater kudu is found in bush country from Angola to Ethiopia. The similar **lesser kudu** (*T. imberbis*) lives in E Africa and is 1 m/3 ft at the shoulder. (Genus *Tragelaphus*, family Bovidae.)

kudzu Japanese creeper belonging to the ◊legume family, which helps fix nitrogen (see ◊nitrogen cycle) and can be used as a feed crop for animals, but became a pest in the southern USA when introduced to check soil erosion. (*Pueraria lobata*, family Leguminosae.)

Kuiper belt ring of small, icy bodies orbiting the Sun beyond the outermost planet. The Kuiper belt, named after US astronomer Gerard Kuiper who proposed its existence in 1951, is thought to be the source of comets that orbit the Sun with periods of less than 200 years. The first member of the Kuiper belt was seen in 1992. In 1995 the first comet-sized objects were discovered; previously the only objects found had diameters of at least 100 km/63 mi (comets generally have diameters of less than 10 km/6.3 mi).

Two new objects were discovered in the Kuiper belt in 1996. The first, 1996 TL66, is 500 km/300 mi in diameter and has an irregular orbit that takes it four–six times further from the Sun than Neptune. The second, 1996 RQ20, is slightly smaller, with an orbit that takes it about three times further from the Sun than Neptune. The orbits of both are at an angle of 20° to the plane of the Solar System.

kumquat small orange-yellow fruit of any of several evergreen trees native to E Asia and cultivated throughout the tropics. The trees grow 2.4–3.6 m/8–12 ft high and have dark green shiny leaves and white scented flowers. The fruit is eaten fresh (the skin is edible), preserved, or candied. The oval or Nagami kumquat is the most common variety. (Genus *Fortunella*, family Rutaceae.)

kurrajong tree widespread in E Australia *Brachychiton populneus* valued as fodder for stock particularly in drought conditions.

kuvasz breed of large guard dog native to Hungary, where it was formerly used to chase wolves and wild boar. Its medium-length, wavy coat is always white. It is strongly built and reaches 75 cm/29 in at the shoulder.

kW symbol for ◊kilowatt.

kwashiorkor severe protein deficiency in children under five years, resulting in retarded growth, lethargy, ◊oedema, diarrhoea, and a swollen abdomen. It is common in Third World countries with a high incidence of malnutrition.

kyanite aluminium silicate, Al_2SiO_5, a pale-blue mineral occurring as blade-shaped crystals. It is an indicator of high-pressure conditions in metamorphic rocks formed from clay sediments. Andalusite, kyanite, and sillimanite are all polymorphs (see ◊polymorphism).

What Killed the Dinosaurs?

BY EDWARD YOUNG

Sixty-five million years ago dinosaurs disappeared. With them went 70% of all species of the time. This 'mass extinction', one of many such events throughout Earth's history, marks the end of the ◊Cretaceous period and the beginning of the ◊Tertiary period, an interval of geological time known as the ◊K-T boundary (K for Kreide, German for Cretaceous). What killed the ◊dinosaurs? Two rival hypotheses dominate. One is that a huge asteroid or comet collided with Earth during K-T time, the other, is that voluminous volcanism in what is now western India was responsible. Distinguishing between these two hypotheses has proven difficult. In either case, the possible link between these events and the demise of the dinosaurs is forcing scientists to reexamine the role of catastrophe in Earth's history.

From Catastrophism to Uniformitarianism

Geology arose as a modern scientific discipline in the early 19th century with the fall of catastrophism, the notion that the history of our planet was shaped by successive catastrophic events and the rise of uniformitarianism, which holds that every-day processes, operating however slowly, are sufficient to explain the principle features of Earth's history.

Naturalists of the late 18th century understood that fossil marine animals and the sedimentary rock in which they were found represented ancient sea floors. But the cast accumulations of fossil-bearing rock revealed in the sides of mountains posed a time problem. The Earth is just 6,000 years old, it was believed, and any casual observer of the oceans could see that sediment did not accumulate fast enough to yield kilometre thicknesses of marine sediment (later turned to rock) unless normal processes were somehow accelerated by cataclysmic events. The great flood described in the Old Testament of the Bible was one catastrophe commonly invoked. Catastrophic upheavals were also called upon to explain mountains.

Near the end of the 18th century a Scottish scientist named James Hutton reasoned that all of the Earth's geological features could be explained by the slow and unchanging forces operating all around us *if* the Earth was very much older than had been previously imagined. This new view, ◊uniformitarianism, proved immensely successful in explaining geological observations, though not at first. Several decades later another Scot, Charles Lyell, refined and popularized Hutton's uniformitarianism. Lyell emphasized that human kind had existed for sufficient time as to bear witness to all kinds of processes that affect Earth history. Hutton and Lyell imagined Earth where the past looked much like the present, with no beginning nor an end. But it was their principle of uniformitarianism that allowed Charles Darwin to concieve of the evolution of species by natural selection and show that Earth's inhabitants at least, had changes irrevocably over geological time.

Without uniformitarianism there could have been no modern geology and no Darwinian theory of evolution and geologists have been understandably reluctant to dismiss the premise that 'the present is the key to the past'.

The new catastrophism

Was Lyell right? Have human beings actually witnessed all of the important agents of Earth change? Since the early years of uniformitarianism we have learned that Earth is 4.5 billion years old and that our ◊Solar System formed from the collapse of a fragment of molecular interstellar cloud approximately 4.6 billion years ago. Human history spans a mere one tenth of one thousandth of this interval and we would have been fortunate indeed if our existence had coincided with all of the forces that periodically effect our planet over millions and even billions of years. Therefore the most likely answer to the question: was Lyell right? must be *no*. Meteorite impact or volcanism on a massive scale may have killed off the dinosaurs. But both of these events, as it turns out, are normal in the course of the evolution of our planet. Should they be considered catastrophes in the context of Earth history?

In July of 1994 ◊comet Shoemaker-Levy 9 collided with ◊Jupiter. The impact was a catastrophe (for Jupiter) the like of which we have not seen before. Despite the fact that this event was unique in terms of human experiences, such colossal collisions are business-as-usual in astronomical terms. Numerous impact craters scar the rocky surfaces of most bodies in our solar system. These craters reveal that prior to 3.8 billion years ago, planets were routinely bombarded by meter to kilmetre-sized objects. The collisions constituted the final stages of rocky ◊accretion as the planets swept up the debris from which they were made. After 3.8 billion years accretion was essentially complete and impacts upon the planets became less frequent, but as Shoemaker-Levy 9 reminded us, they have not ceased entirely.

Impact structures dating back to the time of frequent bombardment have been destroyed on Earth by erosion and ◊tectonic processes that constantly deform and make over the crust. Nonetheless, scientists have thus far managed to identify more than 100 younger impact craters on Earth. It is estimated that, on average, ◊asteroids or ◊comets measuring 5 to 10 kilometres in diameter impact our planet every 50 to 100 million years. Kilometre-sized bodies are thought to hit roughly every million years. Objects in the order of 50 meters in diameter strike about once every 1000 years. For comparison, Shoemaker-Levy 9 was composed of several objects ranging from one to several kilometres in size, their collision with Jupiter released more energy than all of the nuclear weapons on Earth combined.

At present 150 asteroids are known to pass within one Earth orbit of the ◊Sun. They range in size from a few meters to 8 kilometres. A working group under the auspices of the ◊United States National Aeronautics and Space Administration (NASA) suggests that there may be 2100 such bodies in all. The impact of a single asteroid of 2 or more kilometres would have enough energy to profoundly affect Earth's ◊biosphere, hydrosphere, and ◊atmosphere. Fortunately, the likelihood of a collision of severe consequence in the next few hundred years is remote.

Not all catastrophes arrive from space. In 1783, just as catastrophism was about to give way to uniformitarianism, the Laki volcano erupted on Iceland. Unusually harsh winter conditions ensued in North America and Europe. Sulphuric acid rain destroyed Iceland's crops and livestock. Several events of comparable magnitude have confirmed that volcanic eruptions can bring about changes in climate. Have human's experienced the full extent of the influence of volcanism on Earth history? The geological record suggests probably not.

The impact hypothesis

In 1980, physicist and Nobel Prize laureate Luis W. Alvarez and his colleagues Walter Alvarez, Frank Asaro and Helen V. Michel suggested that the K-T mass extinction resulted from the impact of an asteroid or comet (the term bolide describes an impactor of any sort). A bolide greater than 10 kilometres in diameter and travelling 10 kilometres per second, they submitted, would liberate enough energy to trigger environmental disaster across the globe. The Earth would be plunged into months of darkness as the dust propelled into the air by the collision blocked the Sun's rays.

◊Photosynthesis would cease and plant-eating animals would be deprived of food, touching off a breakdown in the food chain. Temperatures would fall and as the shroud of dust prevented the Sun from warming Earth's surface, a phenomenon referred to as 'impact winter'. While land-dwelling animals were fighting starvation and the cold, marine creatures would have to battle acidification of the oceans; aerosols of debris and chemical reactions triggered by an atmospheric shock wave would result in nitric and sulphuric acid rain on a grand scale.

Alvarez and coworkers made their proposal on the basis of an unusual enrichment of the rare element ◊iridium in a layer of clay marking the K-T boundary in Italy. Iridium, chemically similar to ◊platinum and ◊gold, is rare on Earth but much more abundant in primitive rocky meteorites. Since their original report, the 'iridium anomaly', as it has come to be known, has since been found in rocks and sediments deposited at the K-T boundary around the world. There is other evidence supportive of an impact. Ratios of the isotope of the element ◊osmium, a platinum-group element like iridium, from some K-T samples are similar to primitive meteorites. Large amounts of soot in K-T boundary clay is thought to have come from large-scale burning of vegetation ignited by debris ejected from the atmosphere that fell back to earth like a hail of red-hot meteors. ◊Quartz grains shocked by pressures exceeding one hundred thousand atmospheres have been found in K-T deposits. Previously, shocked quartz had only been found at impact craters (Arizona's Meteor Crater for example) or sites where underground nuclear weapons were tested.

In 1991 a circular impact structure of K-T age was found buried beneath one kilometre of Tertiary carbonate rock in Mexico's Yucatan peninsula. The structure is referred to as the Chicxulub crater after the nearby town of Chicxulub Puerto. Experts agree that Chicxulub is the most spectacular crater on Earth and is the best candidate for the K-T impact site envisioned by Alvarez and co-workers. There are only two other impact structures of comparably large size and they are about 2 billion years old and poorly preserved. The precise size of Chicxulub has been debated. Estimates range from 180 to 300 kilometres in diameter. Most recent studies indicate that the hole upon impact was approximately 120 kilometres wide.

Despite its obvious attraction, the impact hypothesis is not without problems. Paleontologists find evidence that the K-T extinction was not instantaneous and may have begun up to a million years before K-T time and continued for tens of thousands of years after. Shocked quartz and even the iridium anomaly are found to occur just above and below the K-T sediment. Disastrous environmental conditions that would have besieged the Earth after impact of a large bolide should have lasted about a decade and so none of these observations is easily explained by the crash of a single asteroid or comet. To redress this apparent short-coming of the hypothesis, some proponents have argued that more than one bolide struck Earth at about K-T time.

The volcanism hypothesis

Another competing catastrophe theory has been put forward to explain not only the chemical and physical features of the K-T boundary but are also the protracted nature of the mass extinction. Beginning in the 1970s it was observed that the largest episode of ◊volcanism in the past 200 million years coincided with the K-T mass extinction. Remains of this volcanism are exposed today in the Deccan plateau region of western India. Here, vast quantities of ◊basalt (a type of dark volcanic rock) known as the Deccan Traps are exposed. Combined, the ancient flows are more than two kilometres thick and comprise roughly 3% of the entire Indian continent. Decan volcanism lasted several hundreds of thousands of years and was most active at the time of the K-T mass extinction. American geologist Dewey M. McLean and French geophysicist Vincent Courtillot, among others, argue that it was the voluminous Deccan volcanism that was responsible for the K-T mass extinction. Recall the climatic effects of the Laki eruption? Imagine the effects of continuous large-scale volcanism for hundreds of thousands of years.

The chemical composition of Earth's ◊mantle is more like that of primitive meteorites than the crust. Deccan lavas came from the mantle. Thus instead of being derived from dust thrown up by an asteroid, the iridium anomaly and other chemical signatures of the K-T boundary could be the result of a deluge of volcanic dust. Similarly, shocked quartz, cited as critical evidence for bolide impact, can apparently be formed by eruption of some types of explosive volcanoes.

The environmental effects of large-scale volcanism and bolide impact may be similar. Volcanism, like impact, would loft large amounts of dust high into the atmosphere where it would be transported around the world. As with the impact hypothesis, the Earth might cool as the dust blocked the Sun's rays. Sulphur released by the volcanoes would cause rain to be acidic. Alternatively, McLean has argued that it was ◊carbon dioxide, a ◊greenhouse gas released during volcanism, that was the killer. Large amounts of carbon dioxide released in to the atmosphere would have changed the chemistry of the oceans and caused ◊global warming, both of which would be harmful to life.

An unlucky coincidence

New theoretical studies suggest that without a vulnerability to extinction, catastrophe may have little influence on the diversity of living organisms. Mass extinctions occur with a crude 30 million year periodicity. The largest, in which 90% of species disappeared, was at the end of the ◊Palaeozoic era approximately 250 million years ago. Smaller extinctions are more common. Theoreticians have shown recently that a pattern of frequent smaller extinctions and less frequent larger extinctions is the inevitable consequence of the dependence of living organisms on one another. The disappearance of an animal's prey, for example, may make the animal less fit for survival. The predator species might then vacate its ecological niche and in turn affect fitness of another species and so it goes until eventually an 'avalanche' of linked extinctions occurs. During an extinction 'avalanche' large numbers of species enter new habitats for which they are not well adapted. Mathematical simulations of such processes show that mass extinction requires the coincidence of *both* large numbers of vulnerable species due to an extinction avalanche, the result of normal evolutionary change, and an unusual amount of stress caused by an environmental catastrophe. Thus it seems the dinosaurs that lived during the Cretaceous period must have been ripe for expiration. Unfortunately for them, their vulnerability coincided with a particularly unlucky time in Earth's history when volcanism was rampant and a collision of the sort that occurs just once every 50 million years actually happened.